Modern Thermodynamics for Chemists and Biochemists

Modern Thermodynamics for Chemists and Biochemists

Dennis Sherwood and Paul Dalby

UNIVERSITY PRESS

Great Clarendon Street, Oxford, OX2 6DP,
United Kingdom

Oxford University Press is a department of the University of Oxford.
It furthers the University's objective of excellence in research, scholarship,
and education by publishing worldwide. Oxford is a registered trade mark of
Oxford University Press in the UK and in certain other countries

© Dennis Sherwood and Paul Dalby 2018

The moral rights of the authors have been asserted

First Edition published in 2018

Impression: 1

All rights reserved. No part of this publication may be reproduced, stored in
a retrieval system, or transmitted, in any form or by any means, without the
prior permission in writing of Oxford University Press, or as expressly permitted
by law, by licence or under terms agreed with the appropriate reprographics
rights organization. Enquiries concerning reproduction outside the scope of the
above should be sent to the Rights Department, Oxford University Press, at the
address above

You must not circulate this work in any other form
and you must impose this same condition on any acquirer

Published in the United States of America by Oxford University Press
198 Madison Avenue, New York, NY 10016, United States of America

British Library Cataloguing in Publication Data

Data available

Library of Congress Control Number: 2017932407

ISBN 978–0–19–878295–7 (hbk.)
ISBN 978–0–19–878470–8 (pbk.)

Printed and bound by
CPI Group (UK) Ltd, Croydon, CR0 4YY

Links to third party websites are provided by Oxford in good faith and
for information only. Oxford disclaims any responsibility for the materials
contained in any third party website referenced in this work.

Foreword

Thermodynamics has evolved dramatically since the precursor of this book, Dennis Sherwood's *Introductory Chemical Thermodynamics*, was published in 1971. This development is completely reflected in the new text, which is really an entirely new book. The title has also very aptly been changed in order to emphasise that one of the most important new areas where thermodynamics can make a major impact is within the bio world: biochemistry and molecular biology. This is emphasised by chapters on the bioenergetics of living cells, macromolecular conformations and interactions, and even an outlook toward where thermodynamics seems to be headed in the future, such as the self-assembly of large complexes.

The sequence of chapters cleverly escalates from everyday experiences to precise definitions, to ideal modelling and then real adjustments. Spontaneity, time, order and information follow naturally, and from these the more complicated chemical and electrochemical reactions, ending up with reactions and structure formation in the living environment – a very long staircase, but with comfortable small steps.

While the content has been brought fully up-to-date and the focus adjusted to fertile modern areas, the old friendly writing style has been preserved. In particular in the beginning where the basic thermodynamic concepts are introduced, we find essentially no equations, only simple verbal explanations based on common observations so that the reader will build a clear intuitive understanding of the topic without the all too frequent mathematical barrier. This approach is especially important for readers in the bio field who often do not have the same strong background in mathematical thinking and modelling as those in the hard sciences and engineering. This is not to say that the book has left out all maths, it just comes later when the concepts have been understood. This is a unique pedagogical approach among thermodynamics textbooks, which undoubtedly will facilitate the reader's entry into thermodynamic thinking.

Every chapter starts with a summary of the concepts presented in that chapter, useful both before reading the chapter for giving direction and after reading it for wrapping up the new items into a whole. The exercises at the end of all chapters further emphasise understanding and relations. They are unconventional by not asking the student to calculate a certain quantity, but to explain an observed behaviour, relating different effects, predict a behaviour and find an error in an argument. In other words, they encourage thinking, rather than mechanical calculational skills. The concluding glossary of thermodynamics terms, and the introductory index of symbols, are very useful for the novice when the many new words and symbols become confusing.

I strongly recommend this introductory thermodynamics textbook for its inviting approach, focus on concepts and relationships, comprehensive coverage, and openness toward the biological sciences.

Bjarne Andresen
Niels Bohr Institute, University of Copenhagen

> Oh, you can't pass heat from the cooler to the hotter
> You can try it if you like, but you far better notter
> 'Cause the cold in the cooler will get cooler as a ruler
> That's the physical law!

From *First and Second Law*, by Michael Flanders and Donald Swann, performed in their musical revue *At the Drop of Another Hat*, 1963

Preface

This book originated as a proposed second edition to *Introductory Chemical Thermodynamics*, published in 1971, with the specific intention of adding material relating to current-day applications of thermodynamics to biology, including topics such as bioenergetics, protein-folding, protein-ligand interactions, and protein aggregation. This has, indeed, been done, but we also took the opportunity to enrich and enhance the discussion of the fundamentals of thermodynamics, the Three Laws, and chemical applications. Accordingly, this book is structured as:

- **Part 1: Fundamentals**: introducing the concepts of work, temperature, heat and energy, state functions and path functions, and some of the mathematical principles that will be used throughout the book.
- **Part 2: The Three Laws**: the core of the book, in which we explore the First Law, internal energy and enthalpy; the Second Law and entropy; and the Third Law and the approach to absolute zero.
- **Part 3: Free energy, spontaneity, and equilibrium**: where we explain the central role of the Gibbs free energy as regards both the spontaneity of change, and also the nature of chemical equilibrium.
- **Part 4: Chemical applications**: covering how the principles discussed so far can be applied to phenomena such as phase equilibria; reactions in solution; acids, bases, and buffer solutions; boiling points and melting points; mixing and osmosis; and electrochemistry.
- **Part 5: Biochemical applications**: where we describe how biological systems capture the free energy within molecules such as glucose, or within light, store it temporarily within molecules such as ATP, and then use that free energy to drive, for example, the synthesis of complex biomolecules; we also explore how proteins fold, and interact with ligands, as well as how proteins self-assemble to form larger-scale structures.

Thermodynamics is notoriously difficult to understand, learn, and use, and so we have taken great care to explain as clearly as possible all the fundamental concepts. As a quantitative branch of science, thermodynamics necessarily uses mathematics to describe how physically measureable phenomena, such as the pressure exerted by a gas, or the concentration of a component within a solution, are related, and how they change as conditions such as the system temperature vary. Much of the required mathematics is explained, and developed, within the text. The only pre-requisites are some knowledge of basic algebra, and of differential and integral calculus (for example, if $y = 3x^2$, then $dy/dx = 6x$, and $\int (1/x)\, dx = \ln x$).

This book has not been written to support a specific curriculum; rather, it has been written to provide "everything a student needs to know about chemical and biochemical thermodynamics" in the context of passing undergraduate examinations, and providing a solid platform for more advanced studies. The content of the book is therefore likely to be broader, and in some respects deeper, than the precise requirements for any specific course. We trust, however that it includes all the required content for very many courses. As a consequence, the book will be of value to undergraduate students of chemistry and biochemistry, and related fields, as well as to students of higher-level programmes who seek a source of reference. Also, the exercises associated with each chapter have been designed to stimulate thinking, rather than as practice problems for a specific examination.

Many people have, of course, contributed to our thinking and to the knowledge we are sharing in this book, and we gratefully acknowledge all our own teachers and mentors. In particular, we wish to thank Professor Alan Cooper, of the University of Glasgow, and Professor Bjarne Andresen, of the Niels Bohr Institute at the University of Copenhagen, for their most helpful suggestions and insights. We also thank Harriet Konishi, Shereen Karmali, Megan Betts and Sonke Adlung at OUP, and also Marie Felina Francois, Indumadhi Srinivasan and everyone in the production team, with whom it has been a pleasure to work—and, of course, our wives and children, who have been remarkably patient, supportive, and understanding as we have been (from their totally legitimate standpoint) both distracted and obsessed by the intricacies of reversible changes, electrode potentials, and entropy.

We trust you will enjoy reading this book and will benefit accordingly. If you notice any errors, think any particular topic is poorly explained, or if you have any ideas for making the book clearer or more useful, please do let us know—our email addresses are dennis@silverbulletmachine.com and p.dalby@ucl.ac.uk. Thank you!

Dennis Sherwood
Exton, Rutland

Paul Dalby
University College, London

Contents

Index of symbols — xxi
Index of units-of-measure — xxvii
List of Tables — xxviii

PART 1 FUNDAMENTALS

1 Systems and states — 3

SUMMARY — 3
1.1 Some very familiar concepts... — 3
1.2 The macroscopic viewpoint — 4
1.3 The system, the surroundings, and the system boundary — 5
1.4 State functions — 5
1.5 Extensive and intensive state functions — 6
1.6 The mole number n — 7
1.7 The 'ideal' concept — 8
1.8 Equilibrium — 8
1.9 Changes in state — 10
1.10 The surroundings have state functions too — 13
1.11 Pressure — 13
1.12 The ideal gas — 16
1.13 Pressure – a molecular interpretation — 17
EXERCISES — 18

2 Work and energy — 20

SUMMARY — 20
2.1 Work – an initial definition — 20
2.2 The work done by an expanding gas — 21
2.3 Path functions — 23
2.4 An important sign convention — 26
2.5 Useful work... — 27
2.6 ...and wasted work — 28
2.7 Quasistatic paths — 28
2.8 Work and Boyle's Law — 31
2.9 P, V diagrams — 31
2.10 Changes at constant pressure — 34

2.11	Thermodynamic cycles	35
2.12	Energy	40
EXERCISES		40

3 Temperature and heat — 43

SUMMARY		43
3.1	Temperature	43
3.2	The ideal gas law	46
3.3	A very important principle	48
3.4	Dalton's law of partial pressures	48
3.5	Some other equations-of-state	50
3.6	Heat	50
3.7	Some more definitions	56
3.8	How to get work from heat	58
3.9	Temperature – a deeper look	67
EXERCISES		74

4 Thermodynamics and mathematics — 76

SUMMARY		76
4.1	What this chapter is about	76
4.2	Functions of more than one variable	77
4.3	Partial derivatives	78
4.4	Systems of constant mass	80
4.5	Partial derivatives and state functions	82
4.6	A look ahead …	84
EXERCISES		88

PART 2 THE THREE LAWS

5 The First Law of Thermodynamics — 93

SUMMARY		93
5.1	The First Law	94
5.2	A molecular interpretation of internal energy	95
5.3	A reminder of some important sign conventions	96
5.4	Internal energy and temperature	96
5.5	The First Law, state functions, and path functions	99
5.6	The First Law and cycles	101
5.7	The mathematics of U	102
5.8	The First Law in open, closed and isolated systems	104
5.9	The First Law in an isolated system	105
5.10	The First Law in a closed system – the adiabatic change, $dq = 0$	106
5.11	The First Law in a closed system – the isochoric change, $dV = 0$	106
5.12	The heat capacity at constant volume, C_V	107

5.13	The First Law in a closed system – the isothermal change, $dT = 0$, for an ideal gas	114
5.14	The First Law in a closed system – the isobaric change, $dP = 0$	114
5.15	Reversible and irreversible paths	117
5.16	Mixing	126
5.17	Friction	128
5.18	Friction and irreversible paths	134
5.19	Real paths	140
	EXERCISES	144

6 Enthalpy and thermochemistry — 146

	SUMMARY	146
6.1	Man's most important technology	147
6.2	Enthalpy	148
6.3	The mathematics of H	151
6.4	Endothermic and exothermic reactions	153
6.5	Enthalpy, directionality and spontaneity	156
6.6	The difference $\Delta H - \Delta U$	158
6.7	Phase changes	161
6.8	Measuring enthalpy changes – calorimetry	163
6.9	Calculating enthalpy changes – Hess's law	165
6.10	Chemical standards	168
6.11	Standard enthalpies of formation	178
6.12	Ionic enthalpies	184
6.13	Bond energies	185
6.14	The variation of ΔH with temperature	190
6.15	The variation of $\Delta_r H^\circ$ with temperature	198
6.16	Flames and explosions	200
	EXERCISES	206

7 Ideal gas processes – and two ideal gas case studies too — 208

	SUMMARY	208
7.1	What this chapter is about	208
7.2	Ideal gases	209
7.3	Some important assumptions	210
7.4	C_V and C_P for ideal gases	210
7.5	Three formulae that apply to all ideal gas changes	218
7.6	The isochoric path, $dV = 0$	218
7.7	The isobaric path, $dP = 0$	219
7.8	The isothermal path, $dT = 0$	220
7.9	The adiabatic path, $dq = 0$	221
7.10	The key ideal gas equations	229
7.11	Case Study: The Carnot cycle	230
7.12	Case Study: The thermodynamic pendulum	236
	EXERCISES	254

8 Spontaneous changes — 258

SUMMARY — 258
- 8.1 A familiar picture — 258
- 8.2 Spontaneity, unidirectionality and irreversibility — 260
- 8.3 Some more examples of spontaneous, unidirectional, and irreversible changes — 261
- 8.4 Spontaneity and causality — 263
- 8.5 The significance of the isolated system — 266

EXERCISE — 266

9 The Second Law of Thermodynamics — 267

SUMMARY — 267
- 9.1 The Second Law — 268
- 9.2 Entropy – a new state function — 269
- 9.3 The Clausius inequality — 272
- 9.4 Real changes — 273
- 9.5 Two examples — 275
- 9.6 The First and Second Laws combined — 280
- 9.7 The mathematics of entropy — 282
- 9.8 Entropy changes for an ideal gas — 285
- 9.9 Entropy changes at constant pressure — 290
- 9.10 Phase changes — 291
- 9.11 The Third Law of Thermodynamics — 296
- 9.12 T, S diagrams — 297

EXERCISES — 299

10 Clausius, Kelvin, Planck, Carathéodory and Carnot — 303

SUMMARY — 303
- 10.1 The Clausius statement of the Second Law — 304
- 10.2 The Kelvin–Planck statement of the Second Law — 309
- 10.3 Heat engines and heat pumps — 313
- 10.4 Irreversibility — 315
- 10.5 A graphical interpretation — 317
- 10.6 The Carathéodory statement — 321
- 10.7 Carnot engines and Carnot pumps — 323
- 10.8 Real engines — 328

EXERCISE — 331

11 Order, information and time — 332

SUMMARY — 332
- 11.1 What this chapter is about — 332
- 11.2 Why do things get muddled? — 333
- 11.3 Order and disorder — 333
- 11.4 Macrostates and microstates — 334

11.5	Three important principles	337
11.6	Measuring disorder	337
11.7	What happens when a gas expands into a vacuum	340
11.8	The Boltzmann equation	341
11.9	Maxwell's demon	344
11.10	Entropy and time	347
11.11	Thermoeconomics	347
11.12	Organodynamics	348
EXERCISES		350

12 The Third Law of Thermodynamics — 351

SUMMARY		351
12.1	The Third Law	351
12.2	Absolute entropies	352
12.3	Approaching absolute zero	357
EXERCISES		362

PART 3 FREE ENERGY, SPONTANEITY, AND EQUILIBRIUM

13 Free energy — 365

SUMMARY		365
13.1	Changes in closed systems	367
13.2	Spontaneous changes in closed systems	369
13.3	Gibbs free energy	374
13.4	The significance of the non-conservative function	375
13.5	Enthalpy- and entropy-driven reactions	377
13.6	The mathematics of G	378
13.7	Helmholtz free energy	385
13.8	The mathematics of A	386
13.9	Maximum available work	386
13.10	How to make non-spontaneous changes happen	390
13.11	Coupled systems …	391
13.12	… an explanation of frictional heat, and tidying a room …	393
13.13	…and the maintenance of life	395
13.14	Climate change and global warming – the 'big picture'	395
13.15	Standard Gibbs free energies	397
13.16	Gibbs free energies at non-standard pressures	399
13.17	The Gibbs free energy of mixtures	401
13.18	The 'Fourth Law' of Thermodynamics	408
EXERCISES		411

14 Chemical equilibrium and chemical kinetics — 413

SUMMARY		413
14.1	Chemical reactions	415
14.2	How $G_{sys}(\tau)$ varies over time	417

14.3	Chemical equilibrium	423
14.4	The pressure thermodynamic equilibrium constant K_p ...	428
14.5	... and the meaning of $\Delta_r G_p^{\circ}$	430
14.6	Non-equilibrium systems	432
14.7	Another equilibrium constant, K_x ...	436
14.8	... and a third equilibrium constant, K_c	437
14.9	Two worked examples – methane and ammonia	439
14.10	How the equilibrium constant varies with temperature	444
14.11	Le Chatelier's principle	448
14.12	Thermodynamics meets kinetics	449
14.13	The Arrhenius equation	455
14.14	The overall effect of temperature on chemical reactions	459
14.15	A final thought	460
EXERCISES		461

PART 4 CHEMICAL APPLICATIONS

15 Phase equilibria — 467

SUMMARY		467
15.1	Vapour pressure	467
15.2	Vapour pressure and external pressure	470
15.3	The Gibbs free energy of phase changes	471
15.4	Melting and boiling	472
15.5	Changing the external pressure P_{ex}	478
15.6	The Clapeyron and Clausius–Clapeyron equations	481
15.7	Phase changes, ideal gases and real gases	485
15.8	The mathematics of G_s, G_l and G_g	497
EXERCISES		500

16 Reactions in solution — 502

SUMMARY		502
16.1	The ideal solution	503
16.2	The Gibbs free energy of an ideal solution	511
16.3	Equilibria of reactions in solution	517
16.4	Dilute solutions, molalities and molar concentrations	520
16.5	Multi-state equilibria and chemical activity	522
16.6	Coupled reactions in solution	526
EXERCISES		530

17 Acids, bases and buffer solutions — 534

SUMMARY		534
17.1	Dissociation	535
17.2	The ionisation of pure water, pH and pOH	536
17.3	Acids	538

17.4	Bases	545
17.5	The Henderson-Hasselbalch approximation	548
17.6	Buffer solutions	549
17.7	Why buffer solutions maintain constant pH	552
17.8	How approximate is the Henderson-Hasselbalch approximation?	553
17.9	Buffer capacity	554
17.10	Other reactions involving hydrogen ions	561
17.11	The mass action effect	566
17.12	Water as a reagent	567
EXERCISES		569

18 Boiling points and melting points — 571

SUMMARY		571
18.1	Non-volatile solutes	571
18.2	Elevation of the boiling point	572
18.3	Depression of the freezing point	575
EXERCISE		579

19 Mixing and osmosis — 580

SUMMARY		580
19.1	Mixing	580
19.2	Osmosis	582
19.3	Osmotic pressure	583
19.4	Reverse osmosis	588
19.5	A note on hydrostatic pressure	588
EXERCISES		590

20 Electrochemistry — 591

SUMMARY		591
20.1	Work and electricity	592
20.2	Electrical work and Gibbs free energy	594
20.3	Half-cells	596
20.4	The electrochemical cell	599
20.5	Anodes and cathodes ...	600
20.6	...and the flow of ions and electrons	601
20.7	The chemistry of the Daniell Cell	602
20.8	Currents, voltages and electrode potentials	603
20.9	A different type of Daniell cell	604
20.10	Different types of electrode	605
20.11	Different types of electrochemical cell	607
20.12	The electromotive force	612
20.13	Oxidising agents and reducing agents	619
20.14	The thermodynamics of electrochemical cells	621

20.15	The Nernst equation	624
20.16	Redox reactions	628
EXERCISES		631

21 Mathematical round up — 633

SUMMARY		633
21.1	The fundamental functions	634
21.2	Pure substances	635
21.3	Heat capacities	636
21.4	The Maxwell relations	637
21.5	The chain rule	640
21.6	The thermodynamic equations-of-state	641
21.7	The difference $C_P - C_V$	645
21.8	The Joule–Thomson coefficient	648
21.9	The compressibility factor	652
EXERCISES		654

22 From ideal to real — 655

SUMMARY		655
22.1	Real gases and fugacity	656
22.2	Real solutions and activity	659
22.3	A final thought	668
EXERCISES		668

PART 5 BIOCHEMICAL APPLICATIONS

23 The biochemical standard state — 673

SUMMARY		673
23.1	Thermodynamics and biochemistry	674
23.2	A note on standards	675
23.3	The biochemical standard state	676
23.4	Why this is important	678
23.5	The implications of the biochemical standard state	679
23.6	Transformed equations	681
23.7	$G(H^+)$ and $G'(H^+)$, and $\Delta_f G^{\circ}(H^+)$ and $\Delta_f G^{\circ\prime}(H^+)$	683
23.8	$\mathbb{E}^{\circ}(H^+, H_2)$ and $\mathbb{E}^{\circ\prime}(H^+, H_2)$	686
23.9	$\Delta_f G^{\circ\prime}(H_2)$	689
23.10	Transformed standard molar Gibbs free energies of formation $\Delta_f G^{\circ\prime}$	691
23.11	$\Delta_r G^{\circ}$ and $\Delta_r G^{\circ\prime}$	694
23.12	K_r and K_r'	696
23.13	G_{sys} and G_{sys}' for an unbuffered reaction	697
23.14	G_{sys} and G_{sys}' for a buffered reaction	704

	23.15	Water as a biochemical reagent	707
	23.16	The rules for converting into the biochemical standard	708
	EXERCISES		708

24 The bioenergetics of living cells — 711

SUMMARY — 711
- 24.1 Creating order without breaking the Second Law: open systems — 711
- 24.2 Metabolic pathways, mass action and pseudo-equilibria — 712
- 24.3 Life's primary 'energy currency' – ATP — 712
- 24.4 NADH is an energy currency too — 715
- 24.5 Glycolysis – substrate-level phosphorylation of ADP — 716
- 24.6 The metabolism of pyruvate — 719
- 24.7 The TCA cycle — 720
- 24.8 Oxidative phosphorylation, and the chemi-osmotic potential — 723
- 24.9 The efficiency of glucose metabolism — 735
- 24.10 Photosynthesis — 737
- EXERCISES — 745

25 Macromolecular conformations and interactions — 747

SUMMARY — 747
- 25.1 Protein structure — 747
- 25.2 The thermodynamics of protein folding — 748
- 25.3 Protein-ligand interactions — 763
- 25.4 Protein folding kinetics — 769
- EXERCISES — 779

26 Thermodynamics today – and tomorrow — 780

SUMMARY — 780
- 26.1 Self-assembly of large complexes — 780
- 26.2 Non-ideal gases and the formation of liquids — 781
- 26.3 Kinetics of nucleated molecular polymerisation — 782
- 26.4 Molecular mechanisms in protein aggregation — 786
- 26.5 The end-point of protein aggregation — 787
- 26.6 The thermodynamics of self-assembly for systems with defined final structures — 787
- 26.7 Towards the design of self-assembling systems — 791
- EXERCISES — 793

Glossary — 795
Bibliography — 844
Index — 847

Index of symbols

Roman upper case

A	Area	14
$\mathcal{A}$	Constant in the Boltzmann equation	68
A	Helmholtz free energy	385
$\boldsymbol{A}$	Molar Helmholtz free energy = A/n	385
A	Pre-exponential factor in the Arrhenius equation	456
C	Celsius, unit-of-measure of temperature, °C	44
C_P	Heat capacity at constant pressure	151
$\boldsymbol{C_P}$	Molar heat capacity at constant pressure = C_P/n	151
C_V	Heat capacity at constant volume	107
$\boldsymbol{C_V}$	Molar heat capacity at constant volume = C_V/n	108
D°	Bond dissociation energy	187
E	Electrical or electrode potential	603
$\mathbb{E}$	Reversible electrode potential	613
$\mathcal{E}$	Molecular energy	68
E_a	Activation energy	456
E_b	Ebullioscopic constant	575
E_f	Cryoscopic constant	578
F	Faraday constant, 9.6485×10^4 C mol^{-1}	595
F	Force	13
G	Gibbs free energy	374
$\boldsymbol{G}$	Molar Gibbs free energy = G/n	375
H	Enthalpy	149
$\boldsymbol{H}$	Molar enthalpy = H/n	149
$\mathbb{H}$	Hartley entropy	346
I	Ionic strength	667
I	Electric current	592
I	Intensity of a measured signal	754
J	Joule, unit-of-measure of work, heat and energy	21
K	Kelvin, unit-of-measure of absolute temperature	44
K_a	Acid dissociation constant	539
K_b	Base dissociation constant	545
K_b	Molality equilibrium constant	521
K_c	Concentration equilibrium constant	438

INDEX OF SYMBOLS

K_d	General dissociation constant	535
K_p	Pressure equilibrium constant	427
K_r	Activity, or generalised, equilibrium constant	523
K_{unf}	Equilibrium constant of protein unfolding	751
K_w	Ionic product of water $= 10^{-14}$	536
K_x	Mole fraction equilibrium constant	437
M	Mass	6
$\mathbf{M}$	Molar mass $= M/n$, the molecular weight in kg mol^{-1}	7
M	Molarity	171
$\mathbb{M}$	Message transmission multiplicity	346
N	Newton, unit-of-measure of force	14
N_A	Avogadro constant, 6.022×10^{23} mol^{-1}	7
$\mathbb{P}$	Mathematical probability	339
P	Pressure	14
Pa	Pascal $= 1$ N m^{-2}, unit-of-measure of pressure	14
$\mathbb{Q}$	Quantity of electric charge	595
R	Ideal gas constant, 8.314 J K^{-1} mol^{-1}	47
R	Electrical resistance	592
S	Entropy	269
$\mathbf{S}$	Molar entropy $= S/n$	269
T	Thermodynamic temperature, as measured in K	44
$\mathcal{T}$	Time	251
$\mathbb{T}$	Thermodynamic temperature, as measured in K	354
U	Internal energy	94
$\mathbf{U}$	Molar internal energy $= U/n$	94
V	Volume	6
$\mathbf{V}$	Molar volume $= V/n$	48
V	Volt, unit-of-measure of electrical potential	593
W	Macrostate multiplicity, or thermodynamic probability	337
X	Arbitrary state function	7
Z	Compressibility factor	652

Roman lower case

a	Chemical activity	523
a	Van der Waals parameter	486
$å$	Average effective ionic diameter	668
aq	In aqueous solution	184
b	Molality	171
b	Van der Waals parameter	487
c	velocity of light	738
d	Exact differential operator, as in dV	11
đ	Inexact differential operator, as in đw	26

INDEX OF SYMBOLS

dm	Decimetre = 10^{-1} m, unit-of-measure of length	170
e	Base of natural logarithms, 2.71828...	68
e^-	Electron	596
f	Arbitrary path function	230
f	Frequency of simple harmonic motion	237
f	Fugacity	657
g	Acceleration due to gravity = 9.81 m sec^{-2}	21
g	Gas phase	154
h	Planck's constant, 6.6261×10^{-34} J sec^{-1}	738
h	Height	21
k	Arbitrary constant	238
k	Reaction rate constant	452
k_B	Boltzmann's constant, 1.381×10^{-23} J K^{-1}	70
$k_{H,i}$	Henry's law constant for component i	510
kg	Kilogram, unit-of-measure of mass	6
kJ	Kilojoule = 10^3 J, unit-of-measure of work, heat and energy	154
l	Liquid phase	155
l	Litre = 1 dm^3 = 10^{-3} m^3, unit-of-measure of volume	170
m	Mass	17
m	Metre, unit-of-measure of distance	21
mol	Mole, unit-of-measure of quantity	7
n	Mole number	7
n_e	Number of electrons	595
n^*	Critical number of monomers for nucleation	782
nm	Nanometre = 10^{-9} m, unit-of-measure of length	724
ox, red	Referring to a redox reaction	619
p	Partial pressure	49
p	Proton motive force, as in Δp	732
pX	$= -\log_{10} X$ (as, for example, p$K_a = -\log_{10} K_a$), or	539
	$= -\log_{10}[X]$ (as, for example, pH $= -\log_{10}[H^+]$)	537
q	Heat	56
s	Solid phase	154
sec	Second, unit-of-measure of time	21
t	Temperature as measured in °C	6
v	Rate of reaction	450
w	Work	26
x	Arbitrary variable	77
x	Linear distance	133
x	Mole fraction	170
y	Arbitrary variable	77
z	Arbitrary variable	88
z	Charge number	667

INDEX OF SYMBOLS

Greek upper case

Symbol	Description	Page
Γ	Mass action ratio	432
Δ	Difference operator	11
$\Delta_r G$	Reaction Gibbs free energy, $= dG_{sys}(\xi)/d\xi$	432
Λ	$= K_c / [H^+]_{eq}^h$	563
Π	Osmotic pressure	587
Σ	Summation operator	11
Φ	Phi-value, alternative to ϕ_F	777
Ψ	Membrane potential, as in $\Delta\Psi$	731
Ω	Macrostate multiplicity, or thermodynamic probability	337
Ω	Ohm, unit-of-measure of electrical resistance	592

Greek lower case

Symbol	Description	Page
α	Parameter in phi-value analysis	776
α_V	Volumetric expansion coefficient	89
β	$= 1/k_B T$	70
β	Buffer capacity	555
β	Parameter in phi-value analysis	776
γ	Activity coefficient	661
γ	Ratio C_P/C_V	224
γ	Parameter in phi-value analysis	776
δ	Infinitesimal increment	11
∂	Partial derivative operator	79
ϵ	Mean bond energy	187
η	Efficiency	234
κ	Transmission coefficient	771
κ_T	Isothermal compressibility	89
λ	Wavelength of electromagnetic radiation	738
μ	Chemical potential	404
μ_{JT}	Joule-Thomson coefficient	650
ν	Frequency of electromagnetic radiation	738
ξ	Extent-of-reaction	417
π	Circumference/diameter ratio of a circle = 3.1416...	667
ρ	Density	7
τ	Time	133
φ	Frictional force parameter	133
ϕ	Fugacity coefficient	657
ϕ_F	Phi-value	771
ω	Frequency of simple harmonic motion $= 2\pi f$	237

INDEX OF SYMBOLS

Subscripts

adi	Referring to an adiabatic change	106
an	Referring to the anode of an electrochemical cell	603
b	Referring to the molality b	515
c	Referring to the molar concentration [I]	438
c	In $\Delta_c H$, referring to a combustion reaction	177
C	Referring to a cold reservoir	230
cat	Referring to the cathode of an electrochemical cell	603
cell	Referring to an electrochemical cell	613
crit	Referring to the critical state of a real gas	480
D	Referring to a dissolved solute	503
eq	Referring to an equilibrium state	422
ex	External to the system	21
f	In $\Delta_f H^\ominus$ or $\Delta_f G^\ominus$, referring to a formation reaction	177
f	Referring to a forward reaction	450
fric	Referring to friction	130
fus	Referring to fusion, solid $\rightleftharpoons$ liquid	161
H	Referring to a hot reservoir	230
H	At constant enthalpy	649
i	Referring to a general component i	7
irrev	Referring to an irreversible change	142
isol	Referring to an isolated system	105
max	Maximum value of	236
p	Referring to the partial pressure p	427
P	At constant pressure	114
ph	Referring to a phase change	294
r	In $\Delta_r H$ or $\Delta_r G$, referring to a chemical reaction in general	177
r	Referring to a reverse reaction	450
rev	Referring to a reversible change	142
S	Referring to a solvent	503
sub	Referring to sublimation, solid $\rightleftharpoons$ vapour	161
sur	Referring to the surroundings	117
sys	Referring to the system	117
T	At constant temperature	114
V	At constant volume	106
vap	Referring to vaporisation, liquid $\rightleftharpoons$ vapour	161
x	Referring to the mole fraction x	437
±	Relating to a cation-anion pair	666

Superscripts

E	Referring to a heat engine	325
P	Referring to a heat pump	325
+	Relating to cations	504
−	Relating to anions	504
⦵	Specifies standard value, as in $\Delta H^{⦵}$ or $\Delta G^{⦵}$	177
⦵′	Specifies biochemical standard state value, as in $\Delta G^{⦵\prime}$	677
′	Referring to the biochemical standard	677
*	Referring to the pure material	505
°	Degree Celsius, as °C, measurement of temperature	44

Other symbols

[I]	Molar concentration of component i	171
[X]	State X	6
{...}	A finite change in a path function, as $\{_1w_2\}_X$	26
⟨...⟩	Average, as in $\langle \mathcal{E} \rangle$	68
|...|	Absolute, and therefore necessarily positive, magnitude	307
⋯	Intermolecular interaction or force	507
‡	Referring to the activated state or complex	457
$\int_{X_1}^{X_2} \ldots dX$	Integral, usually evaluated between two limits, for example, from an initial value X_1 to a final value X_2	12
$\oint df$	Cyclic integral, necessarily evaluated around a closed path	37

Index of units-of-measure

amp	unit-of-measure of electric current	592
bar	Bar, unit-of-measure of pressure = 10^5 Pa	170
C	Coulomb, unit-of-measure of electric charge	595
°C	Centigrade degree, unit-of-measure of temperature	44
dm	Decimetre, unit-of-measure of length = 10^{-1} m	170
J	Joule, unit-of-measure of energy and work = 1 N m	170
K	Kelvin, unit-of-measure of absolute temperature	44
kg	Kilogram, unit-of-measure of mass	170
kJ	Kilojoule, unit-of-measure of energy and work = 10^3 J	154
l	Litre, unit-of-measure of volume = 10^{-6} m^3 = 10^{-3} dm^3	170
m	Metre, unit-of-measure of length	6
mol	Mole, unit-of-measure of amount of substance = 6.022×10^{23} particles	7
N	Newton, unit-of-measure of force = 1 kg m sec^{-2}	14
nm	Nanometre, unit-of-measure of length = 10^{-9} m	724
Pa	Pascal, unit-of-measure of pressure = 1 N m^{-2}	170
sec	Second, unit-of-measure of time	21
V	Volt, unit-of-measure of electrical potential	593
Ω	Ohm, unit of measure of electrical resistance	592

List of Tables

5.1	Values of the molar heat capacity at constant volume, C_V, for selected gases	109
5.2	Thermodynamic data corresponding to the changes depicted in Figure 5.4	119
5.3	Thermodynamic data corresponding to the changes depicted in Figure 5.6	124
5.4	Key thermodynamic variables for the system and the surroundings corresponding to the quasistatic, adiabatic, cyclic change depicted in Figure 5.8 in which friction is present	137
6.1	Values of C_P for selected materials, at 1 bar and 298.15 K	152
6.2	Values of C_p for water, at 1 bar and a variety of temperatures	152
6.3	Some values of $\Delta_{fus}H$ and $\Delta_{vap}H$ for phase changes for some selected materials at the stated temperatures	162
6.4	IUPAC standard states	172
6.5	Common reaction categories	177
6.6	Some standard enthalpies of formation $\Delta_f H^\ominus$	180
6.7	Standard enthalpies of formation $\Delta_f H^\ominus$ for some selected aqueous ions at a temperature of 298.15 K = 25 °C	185
6.8	Bond dissociation energies D^0 for some selected bonds, as shown by the – symbol, at 1 bar and 298.15 K	188
6.9	Mean bond energies ϵ (X–Y) for some selected bonds	189
6.10	The temperature variation of $C_P = a + bT + cT^2 + dT^3 + eT^4$ for some selected gases	195
6.11	The Shomate temperature variation of $C_P = A + BT + CT^2 + DT^3 + E/T^2$ for some selected gases	196
7.1	Molar heat capacities of selected elements and compounds	217
7.2	Summary of ideal gas processes	229
9.1	Some thermodynamic data for phase changes	293
11.1	The number of microstates corresponding to a given macrostate for a system of 10 molecules, as illustrated in Figure 11.1, and the associated probabilities	339
12.1	Some standard molar entropies $S^\ominus$	356
13.1	Some standard molar Gibbs free energies of formation $\Delta_f G^\ominus$	398
14.1	Parameters used for the computation of $G_{sys}(\xi)$	424

15.1	Van der Waals parameters a and b for selected gases	488
17.1	Values of pK_a for some selected acids in aqueous solution at 298 K	540
17.2	Values of pK_b for some selected bases in aqueous solution at 298 K	546
17.3	Some materials suitable for buffer solutions at 298.15 K and 1 bar pressure	554
18.1	Some values of the ebullioscopic constant E_b at 1 atm	575
18.2	Some values of the cryoscopic constant E_f at 1 atm	578
20.1	(Near) standard reversible electrode potentials	616
20.2	(Near) standard reversible redox potentials $E^\ominus$ of aqueous ions	617
23.1	Some standard transformed reversible redox potentials $E^{\ominus\prime}$ for selected biochemical redox reactions in aqueous solution, in accordance with the biochemical standard	688
23.2	Some values of standard molar Gibbs free energy of formation as measured in accordance with conventional standards, $\Delta_f \mathbf{G}^\ominus$, and the biochemical standard, $\Delta_f \mathbf{G}^{\ominus\prime}$, for selected molecules at 298.15 K, and an ionic strength of zero	693
24.1	Overall free energy changes for the complete metabolism of glucose in mitochondria, including the capture of free energy by ATP	737

PART 1

Fundamentals

1 Systems and states

 Summary

Thermodynamics is the macroscopic study of **heat**, **work** and **energy**.

The domain of the universe selected for study comprises the **system**, and the rest of the universe constitutes the **surroundings**. The system and the surroundings are separated by the system **boundary**.

At any time, any system has a number of properties, known as **state functions**, which can be measured, and serve to define the state of the system at any time. **Extensive state functions**, such as mass, depend on the extent of the system; **intensive state functions**, such as temperature, are independent of the extent of the system. All extensive state functions per unit mass are intensive state functions.

Thermodynamic equilibrium is a state in which all state functions are constant over time, and for which all intensive state functions have the same values at all locations within the system.

If measurements are taken on an equilibrium system at different times, and if the value of at least one state function X has changed from an initial value X_1 to a value X_2, then the system has undergone a **change in state**. The corresponding change ΔX in the state function X is defined as

$$\Delta X = X_2 - X_1 \tag{1.2a}$$

in which the initial value X_1 is subtracted from the final value X_2. Mathematically, all state functions are defined by an **exact differential** dX.

A consequence of equation (1.2a) is that the change ΔX in any state function X depends only on the values X_1 and X_2 of X in the initial and final states, and is independent of the path followed during the change in state. The value of ΔX therefore contains no information of how a particular change in state took place.

An **ideal system** - of which an **ideal gas** is one example - is a system in which, fundamentally, there are no intermolecular interactions. Any macroscopic properties, such as the thermodynamic state functions, are linear additions of the state functions of smaller sub-systems, and, ultimately, of the microscopic properties of the molecules themselves. In real systems, molecules do interact, and so ideal systems are a theoretical abstraction. They are, however, much simpler to describe and analyse, and so the study of ideal systems provides a very useful model, which can then be used as a basis of the study of more complex, real, systems.

1.1 Some very familiar concepts...

We all know that iced water feels cold, that freshly made tea or coffee feels hot, and that many of the meals we eat are warm – not as cold as the iced water, not as hot as the tea, but somewhere in-between. From a very early age, we learn that the degree of 'coldness' or 'hotness' we experience is associated with a concept we call 'temperature' – things that feel hot have a high temperature, things that feel cold have a low temperature.

We also know that flames are very hot indeed, far too hot for us to feel directly with our hands. And when we put a saucepan containing cold water in contact with a hot flame – as we do when we're cooking – we know that the water in the saucepan gets steadily warmer: the proximity of the hot flame to the cold water heats the water up.

Putting something hot next to something cold is not the only way things can get warmer: another way is by working. Once again, we all know that when we work hard – for example, by vigorous physical exercise such as running hard, digging a hole, or carrying heavy weights – we quickly become very warm, just as warm as we would by sitting quietly by a log fire. And after we've worked hard for a while, we become tired, and we feel we've lost energy, as if the energy that was in our body earlier in the day has been used up because of the work we have done. So we rest, perhaps have something to eat, and after a while, we feel we have more energy, and can then do some more work.

This is very familiar to all of us – words such as cold, hot, temperature, heat, work and energy are part of our natural every-day language. They are also the fundamental concepts underpinning the science of **thermodynamics**, and to explore that science – as we will do in this book – we need to enrich our understanding of what words such as 'temperature', 'heat', 'work' and 'energy' actually mean, moving beyond subjective feelings such as 'hotness' and 'coldness' to well-formulated scientific definitions. So, our purpose in the first three chapters is to do just that, and to offer some deeper insights into these familiar every-day phenomena.

1.2 The macroscopic viewpoint

Thermodynamics is a very practical branch of science. It's development, during the nineteenth century, was closely associated with the need to gain a better understanding of steam engines, addressing questions such as:

- How much work can a steam engine actually do?
- How might we design better engines – engines that can perform more work for the same amount of coal or wood used as fuel?
- Is there a maximum amount of work a steam engine might do for a given amount of coal or wood? In which case, what might this optimal design be?

Given the importance of steam engines at that time – engines that provided mechanical power to factories, motive power to railways, as well as releasing ships from their reliance on the wind – this is practical stuff indeed.

As a consequence, thermodynamics is concerned with quantities that are readily measurable in real circumstances – quantities such as the mass of an engine, the volume of a boiler, the temperature of the steam in a turbine. These quantities all at a 'human scale', they are all **macroscopic**. Macroscopic quantities may be contrasted with **microscopic** quantities, where in this context, the term 'microscopic' does not relate to what you might observe in the optical instrument known as a microscope; rather, it refers to phenomena associated with the atomic and molecular structures of, for example, the engine, the boiler or the steam. We now know, without any doubt, that atoms and molecules exist, and we now have a deep understanding of their behaviour. But when thermodynamics was developed, the concepts of atoms and molecules were theoretical, and very much under exploration – there was at that time no direct evidence that these invisible particles actually existed, and there were no measurements of their properties.

One of the strengths of thermodynamics is that the intellectual framework, and very many of its practical applications, are rooted firmly in the macroscopic, directly observable, world. As a consequence, thermodynamics does not rely on any assumptions or knowledge of microscopic entities such as atoms and molecules. That said, now that we have some very powerful theories of atomic and molecular behaviour, it is often both possible, and helpful, to interpret the macroscopically observed behaviour of real systems, as expressed and understood by thermodynamics, in terms of the aggregate microscopic behaviour of large numbers of atoms and molecules – that's the realm of the branch of science known as **statistical mechanics**, which forms a bridge between the microscopic world of the atom and molecule, and the macroscopic world of the readily observable.

Accordingly, much of this book will deal with the macroscopic, observable world – but on occasion, especially when the interpretation of macroscopic behaviour is made more insightful by reference to what is happening at an atomic or molecular level, we'll take a microscopic view too.

1.3 The system, the surroundings, and the system boundary

Our universe is huge and complex, and however much we may wish to understand the universe as a whole, we often choose to examine only a small portion of it, and seek to understand that. The areas of study that different people might select can be very diverse in scope, and of very different scales: so, for example, a sociologist might seek to understand the social interactions in a city; an astrophysicist, a star; a biochemist, the structure of a protein. We use the term **system** to define the domain of interest in any specific circumstance, so, for the sociologist, the relevant system will be a chosen city; for the astrophysicist, a particular star; for the biochemist, a specific protein. Everything outside the defined system constitutes the **surroundings**, and the system and the surroundings collectively make up the **universe**. Given the distinction between the system of interest and the surroundings, we use the term **system boundary** to refer to the system's outer perimeter, defining precisely where the system meets the surroundings: everything within the system boundary comprises the system, everything beyond it, the surroundings. The system boundary may be rigid if the system is of fixed size and shape, but this is not a necessary condition – many systems of interest can change their size or shape, changing the boundary accordingly.

1.4 State functions

That said, our study of thermodynamics will start with a system that does have a rigid boundary – a system comprised of a homogeneous **gas**, within a sealed container, the walls of which are assumed to be rigid (for example, steel), rather than flexible (for example, a rubber, inflatable, balloon). The interior surface of the container wall forms the system boundary, as shown in Figure 1.1, with the container itself being in the surroundings.

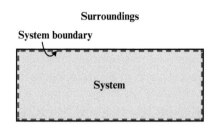

Figure 1.1 A system. This system is a gas within a sealed, rigid, container, with the system boundary being the interior wall (as shown by the somewhat exaggerated dashed line). The gas within the container may be associated with a number of properties, such as its mass M kg, its volume V m³ and its temperature t °C.

At any time, any system will be associated with a number of relevant properties. So, for example, the system of a homogeneous gas within a container will have a mass M kg (that's the mass of just the gas, not including the mass of the container that holds the gas), the gas will occupy a volume V m³, and the gas will have a temperature t °C. Properties of a system that can be measured at any single point in time – of which mass, volume and temperature are three examples – are known as **state functions**. The simultaneous values of all the state functions relevant to any particular system collectively define the **state** of the system at the time of measurement, and a state may be represented by specifying the appropriate state function values within square brackets as $[M, V, t, \ldots]$.

1.5 Extensive and intensive state functions

All state functions may be classified as either **extensive** or **intensive**, according to whether or not a measurement of that state function depends on the size and scale of the system.

So, for example, a system's volume clearly depends on how big the system is, and if an imaginary partition is drawn half-way across a system of volume V, this would result in two sub-systems, each of volume $V/2$. Volume is therefore classified as an extensive function, as is mass M, and to determine the value of any extensive state function, we need to make a measurement on the system as a whole.

In contrast, an intensive state function does not require a measurement to be taken on the system as a whole: rather, a meaningful measurement can be taken at any location within a system. One example of an intensive state function is temperature; another is density = mass/volume, where we see that the intensive state function, density, is the ratio of two extensive functions, mass and volume.

In general, extensive state functions are additive, whereas many intensive state functions are not. To illustrate this, consider two systems: the first a solid of a given material of mass M_1 kg, volume V_1 m³, density $\rho_1 = M_1/V_1$ kg/m³ and temperature t °C; and the second, a solid of a different material of mass M_2 kg, volume V_2 m³, density $\rho_2 = M_2/V_2$ kg/m³ and at the same temperature t °C. If the two systems are combined, then, according to the Law of the Conservation of Mass, the mass of the resulting system is $M_1 + M_2$ kg, and we would expect the volume to be $V_1 + V_2$ m^3. The density of the combined system, however, is $(M_1 + M_2)/(V_1 + V_2)$ kg/m^3, which is not in general equal to the sum $\rho_1 + \rho_2 = M_1/V_1 + M_2/V_2$; furthermore, given that both systems were at the same temperature t °C, the temperature of the combined system is also t °C, and not the sum of the temperatures $2t$ °C. Extensive functions are therefore additive, but many intensive functions are not.

1.6 The mole number n

An extensive state function that will feature strongly throughout this book is the **mole number** n, which specifies the number of **moles** of material within any given system. By definition, 1 mol of material comprises a fixed number of particles, which may be atoms, molecules or ions, depending on the nature of the system in question. The "fixed number" is defined by the **Avogadro constant** $N_A = 6.022141$ particles mol^{-1}. The mole number n defines how much material is within any given system, so for example, the total mass M_i of a system of n_i mol of any pure substance i is given by $M_i = n_i m_i$, where m_i is the mass of a single particle, this being an atom, molecule or ion as appropriate.

As we have just seen, the value of any extensive function for any system depends on the extent of that system, where 'extent' is determined by how much material is contained within the system. For a system comprised of a single pure substance i, all extensive functions therefore depend linearly on the mole number n_i. Accordingly, the mass M of any system is related to the mole number n as

$$M = n\boldsymbol{M}$$

in which $\boldsymbol{M}$, the **molar mass**, is the mass M of a system comprising precisely 1 mol of material, where, as before, the 'material' refers to the particles from which the system is composed, these being atoms, molecules or ions as appropriate.

Our example so far has referred only to the mass M; in fact, for any system of n mol, any extensive state function X is related to its molar equivalent by an equation of the form

$$X = n\boldsymbol{X} \tag{1.1a}$$

from which

$$\boldsymbol{X} = \frac{X}{n} \tag{1.1b}$$

Equations (1.1a) and (1.1b) have a particularly important implication. Since any molar state extensive function $\boldsymbol{X}$ is defined for a specific, fixed, quantity of material, 1 mol, then the value of any molar extensive function $\boldsymbol{X}$ cannot depend on the extent of the corresponding system – that extent is totally defined as 1 mol. Any molar extensive state function $\boldsymbol{X}$ is therefore itself an *intensive* state function. It is therefore always possible to convert any extensive state function X into its intensive counterpart $\boldsymbol{X}$ by dividing X by the appropriate mole number n.

1.7 The 'ideal' concept

In the previous paragraphs, we used our words carefully: so, for example, we said "in general, state functions are directly additive ...", "according to the Law of the Conservation of Mass ..." and "we would expect the volume to be $V_1 + V_2$ m^3". These words might appear to be superfluous: of course adding a mass M_1 kg to a mass M_2 kg results in a combined mass of $(M_1 + M_2)$ kg; of course adding a volume V_1 m^3 to a volume V_2 m^3 results in a system of volume $(V_1 + V_2)$ m^3. Both of these statements are often true, but not always. So, for example, at room temperature, if 1 m^3 of pure ethanol C_2H_5OH is added to 1 m^3 of pure water, the resulting volume is not 2 m^3 – rather, it is about 1.92 m^3. And if two masses of 0.75 kg of uranium-235 are added, the result is not a mass of 1.50 kg – it is a nuclear explosion.

Being able to add the values of extensive state functions is very useful, and so two substances are said to be **ideal** if the value of any extensive state function – such as the mass or the volume – of any mixture of those two substances is the sum of the appropriate values of the corresponding state functions of each substance in its pure state. This concept also applies to a pure substance too, for a system comprising any given mass M kg of a pure substance is, in principle, a mixture of two half-systems, each of mass $M/2$ kg. All extensive state functions of ideal substances are therefore linear with the quantity of matter, usually measured in terms of the mole number, the number of moles of material present, as represented by the symbol n.

As will be seen throughout this book, ideal behaviour is much easier to analyse, and to represent mathematically, than real behaviour. And although ideal behaviour is fundamentally a theoretical abstraction, the behaviour of many real systems approximates to the ideal closely enough for ideal analysis to have real practical value. Also, the theoretical foundations of ideal behaviour act as a very sound basis for adding the additional complexities required for a better understanding of real behaviour. We will identify some further properties of ideal systems elsewhere (see, for example, page 17); in general, throughout this book, unless explicitly stated otherwise, all systems will be assumed to be ideal, and associated with linearly additive extensive state functions.

1.8 Equilibrium

Suppose we observe a system over a time interval, and measure all the system's state functions continuously. If all the state functions maintain the same values throughout that time, then the system is stable and unchanging – it is in **equilibrium**. Then, as time continues, if the value of even just one state function changes, the system is said to have undergone a **change in state**. Once again, that's all obvious – but there is a subtlety: we haven't specified how long that 'time interval' is. If the time interval is long – say, hours, days or years – and the values of all the state functions maintain the same values, then words such as 'stable', 'unchanging' and 'equilibrium' all make sense. But if the time interval is very short – say, nanoseconds – then we would expect many systems to be 'stable' over this very short timescale, but not over a somewhat longer timescale, say, a few milliseconds or seconds. This implies that, if the time interval over which measurements are made is short enough, *all* systems will be identified as stable, unchanging, in equilibrium – at which point, these concepts become unhelpful.

To avoid this problem, this book will make the assumption that the time interval over which any system is being observed is 'long' – that means seconds at the very minimum, and often in principle hours and days – rather than 'short' (picoseconds, nanoseconds, milliseconds).

A special, and limited, case of equilibrium is **thermal equilibrium**, as happens when two systems, or different component parts within a single system, are at the same temperature. **Thermodynamic equilibrium** is a broader concept, requiring all thermodynamic state functions to be in equilibrium. It is therefore possible for a given system to be in thermal equilibrium, but not in thermodynamic equilibrium – as, for example, happens when a gas expands, so changing its volume, but keeping its temperature constant.

A further feature of an equilibrium state is that, at any time, the values of all intensive state functions are the same at all locations within the system, whereas in a non-equilibrium system, it's likely that at least one intensive state function will have different values at different locations. As an example, consider a system composed of a block of metal at a higher temperature, placed in direct physical contact with a block of an equal mass of the same metal at a lower temperature, as shown in Figure 1.2.

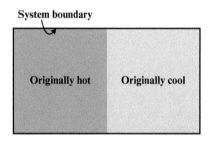

Figure 1.2 A system which is not in equilibrium. This system comprises a hotter block of metal (on the left) in contact with a cooler block of the same mass of the same metal (on the right). Over time, although the mass and volume of this system both remain constant, the temperature at any specific location in the system will change as the originally hotter block becomes cooler, and the originally cooler block becomes hotter. Furthermore, at any one time, the temperature will be different at different locations. Ultimately, both blocks will assume the same temperature, and that temperature will be uniform throughout the system: the system will then be in equilibrium.

An observer of this system would notice that, as time passes, the hotter block becomes cooler, and the cooler one hotter. Although the mass of the system remains constant, as does the volume (assuming that any thermal expansion or compression is negligible), the temperature at any single location within the system changes over time; furthermore, at any one time, the temperature will be different at different locations within the system. These observations verify that the system is not in equilibrium. Ultimately, the system arrives at a state in which, at any location, the temperature no longer changes over time; furthermore, throughout the system, the temperature has the same value. Thermal, and thermodynamic, equilibrium have now been achieved.

Equilibrium is an important concept since it underpins measurement: if a system is not in equilibrium, then the values of at least one state function will be changing over time; furthermore, at any one time, it is also likely that at least one intensive state function will have different values in different locations within the system. Under these conditions, it is impossible to make statements of the form "the value of [this] state function is [this number]", and so the state of the system, as expressed by the set of values of that system's state functions, cannot be defined. The assumption made throughout this book is therefore that, unless specifically otherwise stated, measurements of any system refer to equilibrium states of that system, and the corresponding value of any state function, extensive or intensive, is an equilibrium value.

The study of the thermodynamics of equilibrium states, and of changes from one equilibrium state to another, is, unsurprisingly, known as **equilibrium thermodynamics**. During the mid-twentieth century, the concepts of equilibrium thermodynamics were enhanced and enriched to encompass the behaviour of **non-equilibrium states**, and for his pioneering contributions to the development **non-equilibrium thermodynamics**, the Belgian scientist Ilya Prigogine was awarded the 1977 Nobel Prize in Chemistry. Non-equilibrium thermodynamics is a fascinating branch of science, and still very much an active area of research, with applications to a diversity of fields such as biochemistry and even economics, but beyond the scope of this book – for further information, please refer to the titles suggested in the references.

1.9 Changes in state

1.9.1 Identifying changes in state

Consider now the system illustrated in Figure 1.3, which comprises a homogeneous gas within a cylinder fitted with a piston, rather like the cylinder and piston in an internal combustion engine, or a conventional hand-operated pump for inflating the tyres on a bicycle. Let's assume that there is no friction between the piston and the cylinder; let's also assume that the walls of the cylinder are impermeable and that the piston is very close-fitting, so that nothing can get into, or leave, the interior of the cylinder, either through the walls or by leaking past the piston. But unlike the system shown in Figure 1.1, which has a rigid boundary and a fixed volume, the system in Figure 1.3 has a flexible boundary and can change its volume as the piston moves inwards or outwards.

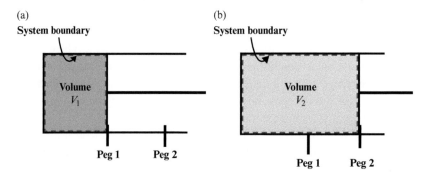

Figure 1.3 A change in state. This system is a gas enclosed in a cylinder fitted with a moveable, frictionless, piston. In (a), the piston is held in place by peg 1, and the volume is V_1; in (b), the piston has moved to the right, and is now held in place by peg 2. The volume of the system has increased to V_2, and there has been a change in state.

Suppose that the piston is held in place by peg 1, as shown in Figure 1.3(a), and that the gas inside the cylinder is in equilibrium. Suppose further that we measure the value of the state function mass to be M_1 and value of the state function volume to be V_1.

If peg 1 is then taken away, let us suppose, with reference to Figure 1.3(b), that the piston moves a small distance to the right, until it stops at peg 2. Instantaneously after peg 1 is taken away and the piston moves, the gas is turbulent and not in equilibrium, but very soon after the piston has stopped at peg 2, the gas returns to equilibrium. Given our assumptions that the walls of the cylinder are impermeable, and that the piston doesn't allow leaks, the mass of the gas is still M_1; but since the gas has expanded, the volume will now be measured as V_2, which will be larger than the original measurement V_1. The value of the state function volume V has changed, from which we infer that the system has undergone a change in state.

1.9.2 Measuring changes in state

The measurement of changes in state is central to thermodynamics, which adopts a convention as to how the corresponding change in any state function is represented. So, for example, a change in volume, represented by ΔV, is defined as

$$\Delta V = V_2 - V_1 \tag{1.2a}$$

in which the initial value V_1 is subtracted from the final value V_2: importantly, the subtraction is *always* that way around

change in state function = value of state function in *final* state
− value of state function in *initial* state

This convention implies that all changes in state functions are signed, algebraic, quantities that convey particular meaning. So, for example, if a gas expands, so that V_2 is greater than V_1, then $\Delta V = V_2 - V_1$ is a positive number; conversely, if the value of ΔV as associated with any particular change is known to be positive, then we may infer that the volume of the system has expanded. Similarly, if a gas is compressed, so that V_2 is less than V_1, then $\Delta V = V_2 - V_1$ is a negative number; conversely, if the value of ΔV as associated with any particular change is known to be negative, then we may infer that the volume of the system has contracted.

The symbol Δ is used for macroscopic changes, such as those that are readily measureable. Sometimes, especially in the development of the theory of thermodynamics, it is useful to consider very small, or even infinitesimally small, changes: accordingly, small changes in state functions are conventionally represented using the symbol δ, for example δV, and infinitesimally small changes are represented by the symbol d, for example dV. As with the definition of the macroscopic change ΔV, as expressed by equation (1.2a), both small changes δV and infinitesimal changes dV are defined as (value of state function in final state) − (value of state function in initial state), and are signed algebraic quantities.

Macroscopic changes ΔX for the change in any state function X from state [1] to state [2] can also be represented as the summation of small changes as

$$\Delta X = \sum_{\text{state [1]}}^{\text{state [2]}} \delta X = X_2 - X_1 \tag{1.2b}$$

or as an integral of infinitesimal changes as

$$\Delta X = \int_{\text{state [1]}}^{\text{state [2]}} dX = X_2 - X_1 \tag{1.2c}$$

The fact that expression defined by equation (1.2c) can be integrated directly to give a result of the form $X_2 - X_1$ implies that dX is what is known mathematically as an **exact differential.**

1.9.3 Changes in state can follow many different paths

Suppose that we observe a system in equilibrium, and measure the volume as V_1. Sometime later, we observe the system again, and measure the volume V_2, from which we can infer, as we have seen, that the system has undergone a change in state, the change in volume being $\Delta V = V_2 - V_1$. We can verify that a change in state has indeed taken place, but *how* did that change happen? How might we gain some insight as to what took place between the initial state, V_1, and the final state V_2? The answer to these questions is that the measurement of *the change in volume ΔV gives no information whatsoever about the change itself*: the observation of the initial state, V_1, and of the final state V_2, associated with the calculation of $\Delta V = V_2 - V_1$, tells us *only* about the overall change in state, but nothing about *how* that change took place, nothing about the **path** taken.

This is illustrated in Figure 1.4, which shows a change in state from an initial volume V_1 to a final volume V_2. This could happen as a single step, as indicated by path 1; alternatively, the

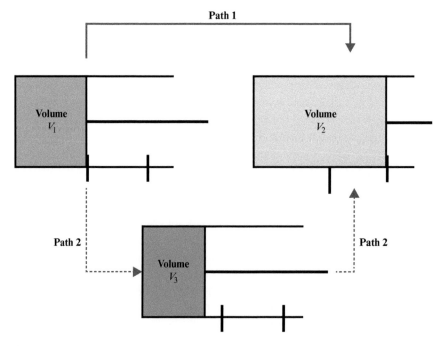

Figure 1.4 A change in a state function is independent of the path of the change. For a change in state from V_1 to V_2, the change in volume $\Delta V = V_2 - V_1$ is the same, no matter what path is followed.

change could have taken place firstly by compressing the gas to an intermediate equilibrium state with a smaller volume V_3, followed by an expansion to the final state with volume V_2, as indicated by path 2. Both paths lead from the same initial state of volume V_1 to the same final state of volume V_2; both paths have the same change in volume $\Delta V = V_1 - V_2$. The actual paths taken, however, were different – but measurements of the volume V in the initial and final states can give no information as to whether path 1 was followed, or path 2. Paths 1 and 2, as shown in Figure 1.4, are of course just two possibilities – anything might have happened. The change in volume ΔV, however, is determined solely by values of the volume V in the initial and final states, regardless of what has happened as the change in state took place, and this applies to all state functions: state functions have no 'memory' of how the system has evolved – they simply describe the state of the system as it is at the time of measurement.

1.10 The surroundings have state functions too

Most of our attention, quite understandably, is on the system of interest, and on the state functions that describe that system at any time. It's worth noting, however, that the surroundings are characterised by state functions too – and although the measurement of an extensive function such as volume might be problematic (the surroundings, in principle, extend to the edge of the universe!), no such problems arise with the measurement of intensive functions, such as temperature, since their values are independent of size.

The measurement of the state functions of the surroundings is of especial relevance in connection with a particular type of path. We have already seen that a change in state for a system can take place along any number of paths, and it so happens that one particular type of path is especially important, a path known as a **reversible** path. We'll explore the properties of reversible paths in more detail later (see pages 117 to 126) – but one feature of a reversible path is relevant here: a reversible path is a path which, when reversed, returns *both* the system *and* the surroundings to their original states, implying that all the state functions, for both the system and the surroundings, are restored to their original values.

1.11 Pressure

1.11.1 Pressure is an intensive state function

With reference to Figure 1.5(a), the piston of the cylinder containing the gas is held in place by peg 1. Suppose that the surroundings of the cylinder are a vacuum. What would we observe if peg 1 is removed?

What would happen is that, as soon as the peg is removed, the piston would move to the right, and probably quite quickly too (the piston is assumed to be frictionless), until the piston is stopped by peg 2, as shown in Figure 1.5(b). Why does the piston move?

The fact that the piston, which was originally at rest (and so has a velocity of zero), starts to move (and so has a non-zero velocity) implies that it accelerates, from which, according to Newton's Second Law of Motion, we infer that the piston was subject to a **force**. This force must come from the gas within the cylinder, and be a force which is exerted over the internal

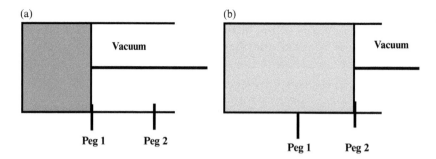

Figure 1.5 Expansion against a vacuum. The piston is initially held in place by peg 1 as in Figure 1.5(a). When peg 1 is removed, the piston moves to the right until stopped by peg 2, as in Figure 1.5(b).

surface of piston. If this force is F N, and if the area of the inner surface of the piston is A m², then we can define the **pressure** P exerted by the gas on the piston as

$$P = \frac{F}{A} \quad \text{Pa}$$

and the force F is given by

$$F = PA \quad \text{N}$$

The pressure P is a property of the system, it can be measured at any point in time, and its value does not depend on the size of the system – all of which imply that pressure P is an intensive state function. The unit of pressure measurement is the Pascal, named after the French mathematician and physicist, Blaise Pascal, with 1 Pa = 1 N m⁻².

In Figure 1.5 (a), when peg 1 is in place securing the piston, the piston does not move because the force exerted by the gas on the internal surface of the piston is counterbalanced by the force exerted by the peg on the external surface of the piston, so holding the piston in position. When peg 1 is removed, this restraining force is no longer present; furthermore, because the surroundings are a vacuum, and the piston is assumed to be frictionless, there is no longer any opposing force at all, and so the piston moves to the right in Figure 1.5(b), until it is stopped by peg 2.

If, however, the surroundings are not a vacuum, but also gaseous – say the atmosphere – then the surroundings will exert a pressure P_{ex} on the external surface of the piston. If the external surface area of the piston is the same as the internal surface area, A m², then the external force, acting on the piston from right to left in Figure 1.6, is $P_{ex} A$ N. When peg 1 is removed, and in the absence of any friction, the net force $F_{L \to R}$ acting on the piston, from left to right in Figure 1.6, is therefore given by

$$F_{L \to R} = PA - P_{ex}A = (P - P_{ex})A$$

If the internal pressure P and the external pressure P_{ex} are equal, then $(P - P_{ex}) = 0$, implying that $F_{L \to R} = 0$. There is no net force on the piston, and so the piston does not move.

If, however, the internal pressure P is greater than the external pressure P_{ex}

$$P > P_{ex}$$

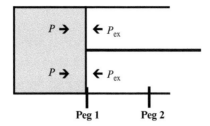

Figure 1.6 Pressure difference. If the pressure P exerted by the gas on the inner surface of the piston is greater than the pressure P_{ex} exerted by the surroundings on the outer surface of the piston, when peg 1 is removed, the piston will move to the right.

then the net force $F_{L \to R}$ is positive, and so the piston will move from left to right in Figure 1.6; but if

$$P < P_{ex}$$

the net force $F_{L \to R}$ is negative, and so the piston will move from right to left in Figure 1.6. The movement of the piston, and the direction of that movement, therefore depends on the *pressure difference* $(P - P_{ex})$ between the gas inside the piston and the pressure exerted by the surroundings.

1.11.2 Atmospheric pressure

As just noted, the gases in the earth's atmosphere exert pressure on the earth's surface – and on us too, but we usually don't notice it because the interior of our bodies are at the same pressure. The actual pressure exerted by the atmosphere depends on a number of circumstances, for example, the local weather and the altitude (which is when we do notice things – in an aeroplane, or sometimes in the lift in a high building, we experience our ears 'popping' as the pressure inside our bodies equilibrates with the external pressure), so the term **standard atmospheric pressure**, sometimes (and rather inaccurately) abbreviated to **atmospheric pressure**, or expressed as the unit-of-measure one **atmosphere**, is a reference pressure defined as 101,325 Pa, this being a representative value of the average atmospheric pressure at sea level. Since 101,325 is rather a clumsy number, a numerically simpler, but nearby, unit-of-measure, the **bar**, defined as 10^5 Pa, is commonly used in thermodynamics, especially in applications associated with engineering.

1.11.3 Boyle's Law – an equation-of-state

The study of the atmosphere plays an important part in the early development of thermodynamics, and indeed of modern science, for in the 1660s, a number of scientists were studying how the pressure of a specific quantity of air changed as its volume changed. Some careful experiments were carried out in which a specific mass of air was allowed to come to equilibrium, and measurements made of the air's pressure, say, P_1, and volume, V_1. The pressure was then changed to a new value P_2, and the air once again allowed to come to equilibrium, taking care to ensure that the temperature of the air remained constant throughout. Once the air had regained equilibrium (which it did quite quickly), the new volume V_2 was recorded. This procedure was repeated several times, so producing a set of pairs of $[P_1, V_2]$, $[P_2, V_2]$, …, $[P_n, V_n]$, …, each pair representing the simultaneous values of the two state functions P and V at successive, different, equilibrium states of the system, all of which were at the same temperature. The results were striking: for a given mass of air, the values P_n and V_n were all different, but the product $P_n V_n$ for any pair of simultaneous values turned out to be the same number:

$$PV = \text{a constant, for a fixed mass of air at constant temperature} \tag{1.3}$$

This relationship, now known as **Boyle's Law** (after Robert Boyle, one of the scientists involved), shows that, for a system comprising a fixed mass of air at constant temperature, the simultaneous values of the two state functions, pressure P and volume V, are not independent, but are related to one another by a simple equation. This implies that if the value of either of the state functions P or V is known, the corresponding value of the other can be calculated, so enabling us to predict the behaviour of the system – and the ability to predict with confidence is at the very heart of science. Equations that show how different state functions of a single system are related to one another are known as **equations-of-state**, of which Boyle's Law is our first example.

1.12 The ideal gas

Careful experimentation, using different gases at wide ranges of pressure, demonstrates that Boyle's Law is not universally true – for example, at a temperature of 150 °C and a pressure of 1 atmosphere, water exists as a gas, and for small variations of pressure, Boyle's Law will hold. But as the pressure is increased, and the volume decreases, keeping the temperature constant at 150 °C as required for Boyle's Law, a point is reached at which the behaviour of the gas deviates significantly from that predicted by Boyle's Law, and ultimately – even at 150 °C – a great enough pressure (in fact, not so great – rather less than 5 times normal atmospheric pressure) causes the gas to condense as liquid water, at which point Boyle's Law has no validity at all. In reality, Boyle's Law is an approximation, and an approximation that usually works better for lower pressures (and greater volumes) than higher pressures (and smaller volumes). Various modifications and extensions have been made to Boyle's Law, resulting in more complex equations-of-state for gases that have broader applicability: one, relatively simple, example is the **van der Waals equation**, which can be written as

$$\left(P + \frac{a}{V^2}\right)(V - b) = c \tag{1.4}$$

where, for a given mass of gas at a constant temperature, a, b and c are constants; some rather more complex equations-of-state for real gases are given on page 50, and the considerably more complex equations-of-state for **solids** and **liquids** are best found in texts on condensed matter physics.

To avoid mathematical complexity, it is very helpful to invoke the concept of the **ideal (or perfect) gas**. Although an ideal gas does not exist in reality, it does provide a very useful 'model' for the development of theory, and so whenever we use the term 'gas' in this book, we are referring to an ideal gas.

Formally, an ideal gas is one whose molecules

- occupy zero volume
- are spherical
- undergo elastic collisions, and
- have no mutual interactions, however far apart or close together.

Most importantly, for an ideal gas, Boyle's Law, equation (1.3), is valid unconditionally, and, as we shall see in this book, this opens the door to the identification of many other important properties of an ideal gas too – properties that give great insight into the behaviour of real systems.

1.13 Pressure – a molecular interpretation

To understand more richly *why* gases exert pressure, we need to consider what is happening at a microscopic, molecular, level. Imagine that the cylinder containing an ideal gas is made of glass, and that the gas has a slight colour. With reference to Figure 1.5, when the piston is held in position towards the left by peg 1, we will see the colour uniformly spread throughout the available space within the cylinder, showing that the gas fills the entire volume. Now think what we would see if peg 1 is removed, and the piston moves to the right to the position defined by peg 2, so allowing the gas to occupy a larger volume. After some initial turbulence, we will see a uniform colour filling the entire, larger, space. From this observation, we can infer that the individual gas molecules are able to move – and move very quickly – over macroscopic, easily observable, distances: if they were not able to move, or if they were able to move only very slowly, the colour would stay 'bunched up' within the volume as defined by the original position of the piston, even after the piston has moved.

The fact that the molecules can move implies that each molecule has a velocity of magnitude v, and, since each molecule also has mass m, each molecule has a corresponding momentum of magnitude mv. So before peg 1 is removed, in accordance with Newton's First Law of Motion, any molecule will be moving with its appropriate velocity in a straight line, until something happens – for example, an elastic collision (by assumption, the molecules of an ideal gas undergo elastic collisions) with another molecule, as a result of which the two colliding molecules move off in different directions, according to the law of conservation of momentum. But instead of colliding with another molecule, a given molecule might undergo an elastic collision with a wall of the cylinder, or the inside surface of the piston. Since the walls of the cylinder are rigid, and the piston is held in position by peg 1, the molecule will bounce back, once again according to the law of conservation of momentum.

Figure 1.7 shows the collision of a molecule of mass m with the inner surface of the piston. The molecule approaches the piston at an angle θ, implying that the component of the velocity parallel to the surface of the piston is $v \sin \theta$ (downwards in Figure 1.7), and that the component perpendicular to the surface of the piston is $v \cos \theta$ (from left to right in Figure 1.7). At the collision, which takes place over a very brief time period $\delta \tau$, the molecule bounces back, but the piston, being held in place by peg 1, and vastly more massive than the molecule, does not

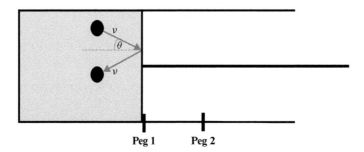

Figure 1.7 Pressure is attributable to molecular collisions. When a molecule of mass m collides with the interior surface of the piston, it undergoes a change in momentum of $-2\,mv\sin\theta$. This results in a tiny force on the piston. Pressure is the macroscopic effect of all these molecular collisions.

move, nor does the container. If the collision is elastic, so that there is no loss of kinetic energy, then, after the collision, the component of the molecule's velocity parallel to the surface of the piston is unchanged, but the component perpendicular to the surface of the piston becomes $-v\sin\theta$, with the – sign indicating that the molecule is now moving, in Figure 1.7, from right to left. During the collision, the molecule therefore undergoes a change in momentum given by

Change in momentum = New momentum − Old momentum = $-2\,mv\sin\theta$

Now, Newton's Second Law of Motion can be stated as "force equals the rate of change in momentum", implying that, during the collision, the molecule experiences, very briefly, a force of magnitude $-2\,mv\sin\theta/\delta\tau$, in which the – sign indicates that this force is from right to left in Figure 1.7, so explaining why the molecule bounces back. We now invoke Newton's Third Law of Motion, which tells us that "to every action, there is an equal an opposite reaction". So, if the piston exerts a force of $-2\,mv\sin\theta/\delta\tau$ on the molecule so causing it to bounce back, then the molecule exerts an equal and opposite force of $+2\,mv\sin\theta/\delta\tau$ on the piston, the + sign indicating that this force is from left to right in Figure 1.7. Each molecular collision therefore exerts a tiny force on the piston – just like the impact of a tennis ball on a tennis racket. Macroscopically, the huge number of molecular collisions each second results in the total force exerted not only on the piston, but on all the walls of the cylinder too – hence the measurable pressure P.

EXERCISES

1. Write down clear, complete, and precise definitions of:
   - system
   - surroundings
   - boundary
   - state
   - state function
   - extensive state function
   - intensive state function
   - change in state

- path
- pressure
- equation-of-state.

2. Classify the following state functions as either extensive or intensive:
   - mass
   - molar volume
   - temperature
   - volume
   - pressure
   - surface area
   - density
   - mole number
   - molar concentration (the number of moles of a pure substance, say, sucrose, dissolved in a given volume, say, 1 m³, of a solute such as pure water).

3. You are observing a system. How would you determine whether the system is, or is not, in equilibrium?

4. You are observing a system of a given mass of a gas, and make a series of measurements of the system pressure P for a number of equilibrium states:

State	Pressure P, Pa
[1]	1.01325×10^5
[2]	1.00000×10^5
[3]	0.81300×10^5
[4]	2.40000×10^5
[5]	1.25000×10^5
[6]	0.65000×10^5

   What is the change ΔP in the system pressure for the following changes in state:
   - From state [1] to state [3]?
   - From state [3] to state [1]?
   - From state [4] to state [5]?
   - From state [4] to state [6]?
   - From state [6] to state [5]?
   - From state [6], via state [5], to state [4]?
   - From state [4], via state [2], then state [5], then state [1], to state [3]?
   - From state [3], via state [1], then state [5], then state [2], then state [6], then state [1], then state [4], to state [3]?

5. What are the key characteristics of an ideal gas?

6. A given mass of an ideal gas, within a given volume, exerts a pressure P. If the average molecular velocity doubles, what pressure does the gas now exert? Why? What do you think might cause the average molecular velocity to increase?

2 Work and energy

 Summary

Work is a phenomenon observed at the boundary of a system when that system changes state, if and only if something occurs at that boundary (such as the motion of the boundary) which may be interpreted as motion against an external force. An example of work is P, V **work of expansion** – the work a system does when it expands against an external force, for example, the force attributable to atmospheric pressure.

Work takes place only as a system changes state. Work is therefore not a property of the state of a system, and so is not a state function. Rather, the amount of work performed by a system on the surroundings, or by the surroundings on the system, is determined by the **path** taken as the change in state takes place, and so work is an example of a **path function**. This implies that, even if the initial and final states of a system are the same, different amounts of work will be performed according to the path followed between those two states. Mathematically, this path-dependence implies that work cannot be represented by an exact differential, but rather by an **inexact differential**, $đw$.

A special form of path is a **quasistatic** path, which takes place

- through a sequence of equilibrium states
- in an infinite number of infinitesimal steps
- infinitely slowly

implying that, throughout the change in state, the state functions of the system undergoing the change are well-defined. Quasistatic changes can therefore be represented as a curve on a graph, such as a P, V **diagram**.

A **thermodynamic cycle** is a sequence of steps which return a system back to its original state. For a complete cycle, the change ΔX in any state function X must be zero, hence

$$\oint dX = 0 \tag{2.6}$$

In general, however, for a path function, for example, work w

$$\oint đw \neq 0 \tag{2.8}$$

Energy is the capacity to perform work.

2.1 Work – an initial definition

The simplest, and most obvious, example of work is the lifting of a mass from the ground. If a mass M kg is on the ground, then gravity is holding it on the surface with a force given by

Modern Thermodynamics for Chemists and Biochemists. Dennis Sherwood and Paul Dalby.
© Oxford University Press 2018. Published 2018 by Oxford University Press.

Mg N, where g m sec^{-2} is the acceleration attributable to gravity. To raise the mass, an upwards force, just greater than Mg, is required to overcome the force of gravity, and if the mass is lifted vertically through a distance h m, the work done is defined as

Work done = Force overcome x Distance moved in direction of the force = Mgh J

where the unit-of-measure is the Joule, J, such that 1 J = 1 N m = 1 kg m² sec^{-2}.

2.2 The work done by an expanding gas

2.2.1 An important equation, $work = P_{ex}dV$

We now return to the system of the gas in the cylinder, with the piston held in place by a peg. The pressure of the gas is P, and let's assume that the surroundings are the atmosphere, which exerts a pressure P_{ex}. If the atmospheric pressure P_{ex} on the external surface of the piston equals the pressure P of the gas on the internal surface of the piston, and if the internal and external areas A of the piston are equal, then, as we saw on page 14, the forces acting on the piston are also equal. In this case, if the peg holding the piston is removed, the piston stays still.

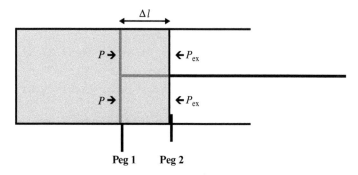

Figure 2.1 Work. If the external pressure P_{ex} on the piston is slightly less than the internal pressure P, when peg 1 is removed, and in the absence of all friction, the piston will move a distance Δl to the right, sweeping out a volume $\Delta V = A\Delta l$, where A is the area of both the interior and exterior faces of the piston. If the external pressure P_{ex} remains constant throughout the change, then the work done by the gas against the surroundings is $P_{ex}A \Delta l = P_{ex} \Delta V$.

But suppose that the external pressure P_{ex} of the atmosphere is a little lower than the internal pressure P of the gas inside the cylinder. When peg 1 is removed, the force exerted by the gas on the piston is greater than the force exerted by the atmosphere, and so, in the absence of all friction, the piston will move to the right, until stopped by peg 2, as shown in Figure 2.1. In so doing, the piston is moving against a force, and through a distance, implying that work is being done by the gas within the cylinder, against the atmosphere outside. If the distance the piston moves is, say, Δl m, then – once again assuming that there is no friction between the piston and the cylinder - the work done by the gas against the surroundings can be quantified as

Work done by the system on the surroundings

= Force overcome × Distance moved in direction of the force

If the external pressure P_{ex} of the atmosphere is constant throughout the change, as is quite reasonable if the change takes place over a short time, then

Work done by the system on the surroundings = $P_{ex} A \times \Delta l = P_{ex} \times A \Delta l$

But $A \Delta l$ is the volume ΔV swept out by the movement of the piston, this being the increase in the volume V of the gas inside the piston, and so

Work done by the system on the surroundings = $P_{ex} \Delta V$ \hfill (2.1)

The use of the symbol ΔV for the change of volume suggests that the change in state is finite; exactly the same reasoning can be used for an infinitesimal change, allowing us to express the infinitesimal work done by the system on the surroundings as $P_{ex} dV$.

These two expressions $P_{ex} \Delta V$ and $P_{ex} dV$ for the work done by a system of an expanding gas against the system's surroundings are important, and will be used many times in this book. And there are four aspects of this result which merit particular attention.

2.2.2 Work is done only against an external force

The first is that the work done by the gas inside the cylinder is calculated by reference to the *external* pressure P_{ex}, a state function of the surroundings, not the internal pressure P, a state function of the system, the gas in the cylinder. This is often something of a surprise – but it makes sense in that work is done in *overcoming* a force, and the force that is being overcome in this instance is that attributable to the external pressure P_{ex}, as exerted by the surroundings. Indeed, if the surroundings are a vacuum (or very close to it), then $P_{ex} = 0$, and so no work is done. As we shall shortly examine in more detail (see page 29), there is a very important circumstance in which P_{ex} is only a very little less than the gas pressure P, in which case $P_{ex} \approx P$ and so $P_{ex} \Delta V \approx P \Delta V$, and, for an infinitesimal change, $P_{ex} dV \approx P dV$. The work done by the system is now expressed in terms only of state functions P and V of the system itself. This is significant, for if we have information on the appropriate equation-of-state (see page 31) specifying how the gas's pressure P, and volume V, are related, we can then compute the actual value of $P \Delta V$, or $P dV$, so enabling us to quantify the amount of work done.

2.2.3 Work done 'by', and work done 'on'

The second important aspect of equation (2.1) is to note that work always involves two 'parties' – 'someone' who *does the work*, and, at the same time, 'someone' *on whom that work is done*. So if I lift a mass from the floor, I do the work, and work is done on the mass. In the case of the gas in the cylinder, if the pressure P of the gas is higher than the pressure P_{ex} of the surroundings, so that the piston moves to the right in Figure 2.1, then, as the gas expands, work is done by the gas on the surroundings; if the pressure P of the gas is less than the pressure P_{ex} of the surroundings, so that the piston moves from right to left in Figure 2.1 so compressing the gas, then work is done on the gas by the surroundings.

2.2.4 Work happens at the system boundary…

The third important aspect of equation (2.1) is rather less obvious, but a moment's thought will verify that work is a phenomenon that can be observed *only at the boundary* of a system, for it is at the boundary that the 'two parties' just referred to – the system and the surroundings – interact. This is helpful, for it tells us where to look if we wish to determine whether work is being done during a change in state, or not: we don't look within the system, we look at the system's boundary.

2.2.5 …as the system changes state

Furthermore, *work only happens as a system changes state*, as is evident from the presence of the symbol Δ in equation (2.1). This therefore leads to a more sophisticated definition of work as

> **Work is a phenomenon, identified at the boundary of a system, as that system changes state, if and only if something occurs at that boundary which may be interpreted as motion against an external force.**

This definition highlights the two facts that work takes place at the system boundary, and can be identified only during a change in state. In addition, this definition allows for work to be associated with a variety of different contexts. One such context, now very familiar, is when a gas expands, driving the piston of the cylinder against an external pressure at the system boundary – which can indeed "be interpreted as motion against an external force". This is therefore an instance of work, which, since it concerns changes in the system's pressure P and volume V, is known as P, V **work of expansion**, or, more simply, P, V **work**. But there are other contexts too: the most familiar example is gravitational work done in lifting a mass against the force of a gravitational field; another relates to the motion of an electric charge against an electrical potential difference, this being 'electrical work'; a third is the work required to stretch a surface, against the force attributable to surface tension (as takes place when a soap bubble expands), known as 'surface work'; a fourth is the work required to change the orientation of a magnetic dipole against the force attributable to a local magnetic field – 'magnetic work'; and yet another is the work done to overcome the force of friction, known as 'frictional work'.

2.3 Path functions

2.3.1 Work is an example of a path function

With reference to Figure 2.2(a), suppose that the piston is originally held in position by peg 1, that the pressure of the gas in the cylinder is P_1, and that the volume is V_1. We may therefore represent the initial state of the gas as $[P_1, V_1]$. If the pressure P_{ex} exerted by the surroundings is a little less than the pressure P_1, then, when peg 1 is removed, and in the absence of all friction, the piston moves to the right, until it is stopped by peg 2, as shown in Figure 2.2(b). The gas inside the cylinder now has a volume V_2, greater than the original volume V_1; the pressure is also likely to have changed too, so let's represent the new pressure as P_2. The system

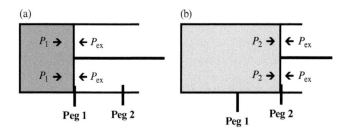

Figure 2.2 A change in state from $[P_1, V_1]$ to $[P_2, V_2]$.

has therefore undergone a change in state from $[P_1, V_1]$ to a different state, which we can represent as $[P_2, V_2]$.

As this change in state took place, the piston moved to the right, from the position defined by peg 1 to that defined by peg 2 in Figure 2.2, and so the system did work against the external pressure P_{ex} of the surroundings, which we can quantify using equation (2.1) as

$$\text{Work done by the system on the surroundings} = P_{ex} \Delta V \tag{2.1}$$

The important words here are "as this change in state took place", for this emphasises a fundamentally important attribute of work, which we have already noticed: *work happens only as a system changes state*. When the piston stops at peg 2, and the system achieves its final state $[P_2, V_2]$, any work that might have been done has been finished – it no longer exists. Unlike variables such as pressure and volume, which are *properties of states*, work is not a property *of* a state, but a phenomenon which can occur only when a system *changes* state.

Phenomena which occur as a system changes state are known as **path functions**, of which work is an example. Path functions are fundamentally different from state functions: as we have seen, a *state function describes a specific property of a system at a point in time*, whereas a *path function describes a phenomenon that occurs as a system changes state over time*. The distinction between a state function and a path function is further clarified by imagining what happens when time 'freezes'. If time is 'frozen', the question "What is the volume V of the gas in the cylinder?" has a sensible answer: we can measure that volume as, say, 0.76 m³, for the gas has a property called 'volume', which we can measure at a specific point in time. In contrast, the question "What is the work of the gas?" is meaningless: the gas does not possess a property called 'work', and we can't measure work at a single point in time. Rather, 'work' is a description of a phenomenon which takes place over a period of time, as a system changes from one state to another. So, the sensible question relating to work is not "What is the work of the gas in state 2?", but rather "What amount of work was done as the system changed from state 1 to state 2?".

2.3.2 The value of a path function depends on the path taken between two states

An important feature of all path functions is that their value *depends upon the path taken* between any two states. This is different from the behaviour of state functions: as we have seen, the value of any state function, such as volume V or pressure P, is a property of a state itself, and the values of the changes ΔV and ΔP of the state functions V and P depend only on the values V_1 and P_1 of the initial state in which the system started, and V_2 and P_2 of the final state in which the system ended up. The changes ΔV and ΔP are therefore determined by reference only to the initial state and the final state, and so are independent of how the change in state took place.

Path functions are significantly different: their values are totally dependent on the path, as we shall now demonstrate by reference to the system illustrated in Figure 2.3, the left-hand side of which shows our now-familiar gas in a cylinder, fitted with a frictionless piston, held in place by peg 1.

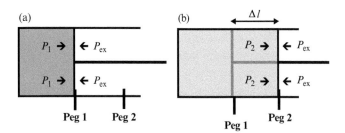

Figure 2.3 Work is a path function. The system is initially in state (a), $[P_1, V_1]$, and, after removal of peg 1, changes to state (b), $[P_2, V_2]$. When the piston moves frictionlessly to the right, from peg 1 to peg 2, the change ΔV in the state function volume depends only on the difference between the final volume V_2 and the initial volume V_1. The work done by the gas on the surroundings, $P_{ex} \Delta V$, depends on P_{ex}, which can take a wide range of different values, even though the change ΔV in volume is the same in each case. The amount of work done by the gas against the surroundings therefore depends on the way in which the change take place: work is therefore a path function.

Suppose that, in the state illustrated by Figure 2.3(a), the volume of the gas inside the cylinder is V_1, and the pressure P_1. Suppose further that the surroundings are a vacuum, so that the external pressure P_{ex} is zero, or very close to it. When peg 1 is removed, the piston moves frictionlessly to the right in Figure 2.3, and stops at peg 2. In this second state, the volume of the gas inside the cylinder is now V_2, greater than the original volume V_1. Since the value V_2 of the state function volume is different as compared to the value V_1, a change in state has occurred, and we can calculate the change of volume, $\Delta V = V_2 - V_1$.

The work done by the gas during this change in state can be computed using equation (2.1):

Work done by gas on the surroundings = $P_{ex} \Delta V$ (2.1)

But, in this case, P_{ex} is zero, so the work done is zero.

Let's do this experiment again, starting with the piston at peg 1, and the volume of gas V_1 and the pressure P_1, but with an external pressure P_{ex} at some (constant) non-zero value, rather lower than the pressure P_1 of the gas inside the cylinder. When the peg is removed, the higher pressure within the cylinder drives the piston frictionlessly against the external pressure P_{ex} to the right in Figure 2.3, until the piston stops once more at peg 2, with the volume of the gas once again V_2. The change $\Delta V = V_2 - V_1$ in the volume of the gas is exactly the same as before, but this time, because P_{ex} now has some constant, non-zero, value, the work done by the gas

Work done by gas on the surroundings = $P_{ex} \Delta V$ (2.1)

also has a non-zero value, depending on precisely what the values of P_{ex} and ΔV actually are: if P_{ex} is small, then the amount of work done is correspondingly small; if P_{ex} is large (but still less than the pressure inside the cylinder), then the amount of work is larger. We therefore see that the change ΔV in the state function volume is the same in all cases, and depends solely on the values of the state function volume in the initial and final states; the amount of work, however, depends on how the change in state took place, the actual path followed during the change. This demonstrates that work is indeed a path function, and that the amount of work done during any change in state depends on the path taken.

We have already seen that small changes in state functions are represented using the symbols δ or d, as exemplified by δV and dV. In recognition of the different nature of path functions as compared to state functions, small quantities of a path function are represented by the single symbol đ – so a small quantity of work is written as đw. The total amount of work done during a change in state from state [1] to state [2] is represented by $\{_1w_2\}_X$, where the notation $\{\ldots\}_X$ indicates that the actual value of total work done during any particular change in state depends, as we have seen, on the path X taken. If path X takes place in small steps, we can write

$$\{_1w_2\}_X = \sum_{state\,[1]}^{state\,[2]} đw \tag{2.2a}$$

and if the path can be considered to be taken in infinitesimal steps, then

$$\{_1w_2\}_X = \int_{state\,[1]}^{state\,[2]} đw \tag{2.2b}$$

It's important to note that although equation (2.2b), representing the total work done during the change from state [1] to state [2], is written as an integral, this does not imply that the expression can be integrated directly to give a result of the form $w_2 - w_1$, for this would require that w_2, representing the 'work in state 2', and w_1, the 'work in state 1', actually exist, and can be subtracted from one another. As we have seen, there is no property w_2, nor w_1, so the summation, equation (2.2a), or the integral, equation (2.2b), needs to be determined for each particular circumstance, according to the actual path followed. The fact that the expression (2.2b) cannot be integrated directly implies that đw, an infinitesimal change of a path function, is what mathematicians call an **inexact differential** – in contrast to the exact differential dX, representing an infinitesimal change of the state function X that we met in equation (1.2c).

2.4 An important sign convention

As we have seen (see page 11), changes in state functions, whether large, as symbolised by ΔV, small (δV), or infinitesimal (dV), are signed quantities, so that a positive value implies (in this case) that the volume of the system has increased, and a negative value, that the volume has decreased. Work too is a signed quantity, and the sign also has significance. Reference to equation (2.1) will show that if the gas expands, ΔV is positive. Since the force against which work is done is always positive, then the product $P_{ex}\Delta V$ is positive. The work done *by* the system on the surroundings is therefore a positive number. In contrast, if work is done by the surroundings *on* the system, the gas is compressed and so ΔV is negative, implying that the work done on the gas is a negative number. This is an important sign convention, which we can summarise as

- Work done *by* a system is represented by a *positive* number.
- Work done *on* a system is represented by a *negative* number.

2.5 Useful work …

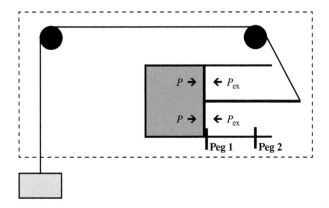

Figure 2.4 Two types of work. If $P > P_{ex}$, then as the piston moves to the right, the system does two types of work: P, V work of expansion $P_{ex}\Delta V$ against the atmosphere, and also gravitational work $Mg\Delta h$ as required to lift the mass.

In Figure 2.4, the system of the now-familiar gas within a cylinder fitted with a frictionless piston is connected to a mass M. Let us now suppose that peg 1 is removed, and the piston moves to the right. In so doing, the system of the gas in the cylinder necessarily does P, V work of expansion against the external pressure P_{ex}, and, in addition, causes the mass M to be raised by a distance Δh. The total work done by the system of the gas inside the cylinder is therefore the sum of the P, V work of expansion $P_{ex}\Delta V$, plus the work $Mg\Delta h$ done in lifting the mass

Total work done by system = $P_{ex}\Delta V + Mg\Delta h$

If the purpose of the system is to lift the mass, then we see that the total work done by the system is not all available – some work, $P_{ex}\Delta V$, has to be done to 'push the atmosphere out of the way' as the piston moves to the right. The **useful work**, or **available work**, that a system can perform is therefore defined as the difference between the total work done by the system, and any P, V work of expansion against the atmosphere that cannot otherwise be harnessed:

Useful work = Total work – P, V work of expansion that cannot otherwise be used

Many chemical reactions, especially those which involve gases, are necessarily accompanied by P, V work of expansion against the atmosphere, and so the distinction between total work and useful work has considerable significance, as will be discussed further on pages 386 to 390.

2.6 ...and wasted work

All the examples we have studied so far have related to the expansion of a gas within a frictionless cylinder. Let's now examine what happens when friction is present between the rim of the piston and the internal surface of the cylinder.

Friction is of two types – **static friction**, and **dynamic friction**. Static friction is what makes surfaces 'sticky', so that they do not easily slide over one another, and it is thanks to static friction that we can (usually) lean a ladder against a wall, and climb the ladder without fearing that it might slip backwards (but, for safety, it's always advisable for someone to be holding the ladder to make sure!). Dynamic friction (sometimes known as kinetic friction) arises when two surfaces are moving against one another, as happens, for example, when a saw moves against wood, or a drill bites into metal, or – as is relevant here – when the rim of the piston scrapes along the inside of the cylinder as it moves.

Both types of friction apply to our system of a gas at pressure P_{ex} inside a cylinder fitted with a piston. If the external pressure P_{ex} equals the internal pressure P, then the forces on each side of the piston are equal, and the piston does not move. If the external pressure P_{ex} is reduced, in the absence of friction, the pressure difference $P - P_{ex}$ will cause the piston to move outwards, but in the presence of static friction, a small pressure difference will not cause the piston to move, for the force exerted by the static friction acts like a peg, resisting the movement of the piston. If the external pressure P_{ex} is progressively reduced, the pressure difference $P - P_{ex}$ increases, until a point is reached such that the force exerted by the pressure difference is large enough to overcome the static friction, and the piston jerks outwards.

Thereafter, the piston slides against the internal surface of the cylinder, against the force exerted by dynamic friction, which acts to oppose the motion. And in so doing, the gas inside the cylinder is doing **frictional work** against the dynamic friction, as well as P, V work of expansion against the external pressure, and perhaps gravitational work in lifting a weight too. Eventually the piston will come to rest, and the total work done by the system can be expressed as

Total work done by system = P, V work of expansion

+ work done against dynamic friction

+ work done in lifting a weight

Once again, if the purpose of our system is to lift a weight, or to do useful work of any sort (such as power a machine, or drive an engine), then not only is the P, V work of expansion not available, but the work done against dynamic friction is wasted too. And wasted in a very particular way: as we shall see in the next chapter, frictional work is inevitably lost as heat – which is why saws get hot as they cut through wood, and drills get hot as they bore through metal. This also explains why so many of the examples used in this book refer to frictionless systems, so ensuring that this waste is not present.

2.7 Quasistatic paths

Let's do the experiment illustrated in Figure 2.3 once more, with exactly the same initial state: the volume of the gas is V_1, the pressure P_1, and the piston is held in place by a peg. The difference in this experiment is that the initial external pressure – let's represent this as $P_{ex,1}$ – is now exactly equal to the internal gas pressure P_1. Since the pressures on both sides of the piston are the same, the force acting on the inside of the piston from the gas within the cylinder is exactly balanced by the force acting on the outside of the piston by the gas in the surroundings. If the peg is removed, the piston will stay absolutely still.

Suppose now that the external pressure is reduced by an infinitesimally small amount, dP_{ex}, to a new, slightly lower, external pressure $P_{ex,2} = P_{ex,1} - dP_{ex}$. The internal pressure on the piston, P_1 is now infinitesimally greater than the external pressure, and so, in the absence of all friction, the piston will move against the external pressure $P_{ex,2}$. The volume of the gas will therefore change from its initial volume V_1 to an infinitesimally larger volume $V_2 = V_1 + dV_1$; simultaneously, since in general the pressure of a gas drops as the volume increases, the pressure will change from its initial value P_1 to a new, infinitesimally lower, value $P_2 = P_1 - dP_1$. When the new gas pressure P_2 equals the external pressure $P_{ex,2}$, the piston will stop moving, and the system will return to equilibrium. Overall, the gas has changed from its original state $[P_1, V_1]$ to a final state $[P_2, V_2]$, and the surroundings have changed from pressure $P_{ex,1}$ to $P_{ex,2}$.

As the gas inside the cylinder expands, the gas does P, V work of expansion against the surroundings. Assuming as usual that the piston, and the system as a whole, are perfectly frictionless, then that the only work done by the gas within the system is P, V work of expansion against the surroundings, given by

Work done by system during change from $[P_1, V_1]$ to $[P_2, V_2] = P_{ex,2}\, dV_1$
$$= (P_{ex,1} - dP_{ex})\, dV_1 = P_{ex,1}\, dV_1 - dP_{ex}\, dV_1$$

Now, the changes dP_{ex} and dV_1 are both very small, in which case the product $dP_{ex}\, dV_1$ is very small indeed, and certainly much smaller than the product $P_{ex,1}\, dV_1$. Hence we may write

Work done by system during change from $[P_1, V_1]$ to $[P_2, V_2] = P_{ex,1}\, dV_1$

But we also know that the initial external pressure $P_{ex,1}$ was equal to the initial internal pressure P_1, hence

Work done by system during change from $[P_1, V_1]$ to $[P_2, V_2] = P_1 dV_1$

The significance of this equation for the work done by the system is that it is expressed in terms *only of variables relating to the system* – P_1, the initial pressure of the system, and dV_1, the change of volume of the system. This is in contrast to the equation (2.1), which expressed the work done by the system in terms of one variable relating to the system, dV, and a second relating not to the system, but to the surroundings, P_{ex}.

Let's consider now what happens when the external pressure is reduced by another very small amount dP_{ex} from $P_{ex,2}$ to $P_{ex,3} = P_{ex,2} - dP_{ex}$. The pressure P_2 of the gas in the cylinder is once again infinitesimally higher than the pressure in the surroundings, and so the piston will once more move frictionlessly and infinitesimally to the right. The system of the gas in the cylinder changes state from $[P_2, V_2]$ to $[P_3, V_3]$, such that $V_3 = V_2 + dV_2$ and $P_3 = P_2 - dP_2$. As before, during this change, the gas in the cylinder does work against the surroundings equal to $P_{ex,2}\, dV_2$, and we can use exactly the same reasoning as before to show that

Work done by system during change from $[P_2, V_2]$ to $[P_3, V_3] = P_2 dV_2$

This implies that the total work done by the gas in changing state from $[P_1, V_1]$ to $[P_3, V_3]$ via $[P_2, V_2]$ is given by

Work done by system during change from $[P_1, V_1]$ to $[P_3, V_3]$ = $P_1\,dV_1 + P_2\,dV_2$

You can see the pattern: if we were to reduce the pressure in the surroundings in a sequence of infinitesimally small steps, each of dP_{ex}, then

Work done by system during change from $[P_1, V_1]$ to $[P_n, V_n]$ = $\sum_{1}^{n-1} P_n\,dV_n$

Since the volume changes dV are so small as to be infinitesimal, then the total work done by the gas as it expands from its initial volume V_1 to its final volume V_n can be expressed as an integral:

$$\text{Work done by system during change from } [P_1, V_1] \text{ to } [P_n, V_n] = \int_{V_1}^{V_n} P\,dV \qquad (2.3)$$

Equation (2.3) allows us to express the total work done by the system for a finite change between any two states in terms of state functions of the system alone. The derivation of equation (2.3), however, rests on three key assumptions about the nature of the change: specifically, that the change took place along a particular path

- through a sequence of equilibrium states
- in an infinite number of infinitesimal steps
- infinitely slowly.

The significance of the first condition – the sequence of equilibrium states – ensures that the state functions P and V are well-defined throughout the change, and the second condition – the infinitesimal steps – allowed us to replace the external pressure P_{ex} by the internal pressure P. The third condition is a corollary of the first two; in principle, this third assumption can be relaxed to a condition which is that the change takes place sufficiently slowly to allow the first two conditions to apply. Any path which obeys these three conditions is known as a **quasistatic path**, and a change in state which takes place along a quasistatic path is known as a **quasistatic change**.

The quasistatic path discussed here has been in the context of an infinitesimal pressure difference across the system boundary, resulting in the performance of P, V work, either by the system on the surroundings, or by the surroundings on the system. In fact, the concept of the quasistatic path applies in other contexts too, of which two are of particular relevance. The first is the flow of heat: a quasistatic flow of heat is one which takes place infinitely slowly through an infinite number of steps such that

- the temperature difference driving the flow of heat is infinitely small, and
- the two systems between which the heat is flowing are in equilibrium throughout.

This implies that any heat flow down a finite temperature gradient is necessarily non-quasistatic – which applies to all real heat flows.

Similarly, a quasistatic chemical reaction takes place infinitely slowly through a sequence of states of chemical equilibrium – which is, once again, a theoretical concept rather than a reality.

2.8 Work and Boyle's Law

Equation (2.3) for the work done by an expanding gas, and the concept of the quasistatic change on which equation (2.3) rests, are important, in that it expresses the P, V work of expansion of a frictionless system in terms of *state functions of the system alone*. And if we know the relevant equation-of-state, such as Boyle's Law, equation (1.3)

$$PV = c \tag{1.3}$$

where c is a constant for a given mass of gas at a given temperature, then we can compute what the work done during any change in state actually is. So, combining equations (2.3) and (1.3), we have

Work done by system during change from $[P_1, V_1]$ to $[P_2, V_2]$

$$= \int_{V_1}^{V_2} P \, dV = \int_{V_1}^{V_2} \frac{c}{V} dV = c \ln\left(\frac{V_2}{V_1}\right) \tag{2.4}$$

This result, equation (2.4), is a special case, valid only if two conditions are simultaneously fulfilled: firstly, that the change in state from $[P_1, V_1]$ to $[P_2, V_2]$ is quasistatic, and, secondly, that Boyle's Law is valid throughout. It does, however, illustrate an important general point: because equation (2.3) is expressed only in terms of variables relating to the system, then, if an appropriate equation-of-state for that system is known (as it often is), then equation (2.3) may be used to compute, and therefore predict, how much work a system can do. It's important, though, to re-emphasise that equation (2.3) is based on the assumption that the change is quasistatic, taking place through a very large (mathematically, infinite) number of very small (infinitesimal) frictionless steps. In reality, such a change is impossible, but theoretically, it provides a very useful way to compute quantities mathematically.

2.9 P, V diagrams

2.9.1 A graphical representation of a frictionless quasistatic change in state ...

As we have now seen, many of the changes in state of a gas in a cylinder are associated with changes in the system's pressure P and volume V. A very useful way of representing these changes is by plotting the simultaneous values of these state functions on a graph, the vertical axis of which represents the system pressure P, and the horizontal axis, the system volume V. Reasonably enough, such a representation is known as a P, V *diagram*. As an example, Figure 2.5 shows a P, V diagram representing a change in state, in which the gas shown in

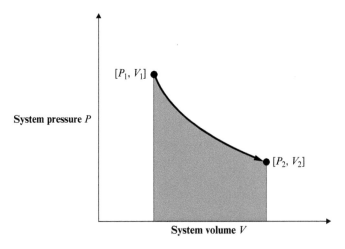

Figure 2.5 A P, V diagram for a frictionless quasistatic change in state. The system undergoes a frictionless quasistatic change in state from $[P_1, V_1]$ to $[P_2, V_2]$, as indicated by the direction of the arrow, under conditions in which Boyle's Law is valid. The work done by the system on the surroundings is represented by the shaded area. Note that, in P, V diagrams, it is conventional to define states as, for example, $[P_1, V_1]$, even though pressure is the vertical (y) axis, and volume the horizontal (x) axis.

Figure 2.3 expands quasistatically, and in the absence of all friction, from an initial equilibrium state $[P_1, V_1]$, to a final, different, equilibrium state $[P_2, V_2]$, under conditions of constant temperature so that Boyle's Law, $PV = c$, equation (1.2), holds.

Since the change in state is assumed to be quasistatic, the change takes place through an infinite succession of intermediate equilibrium states. Each of these has its own, well-defined, values of pressure and volume, and so each can be represented by a specific point on the P, V diagram. A quasistatic change, starting at state $[P_1, V_1]$ and ending in state $[P_2, V_2]$, can therefore be represented on a P, V diagram by a solid line, as shown in Figure 2.5, such that each point on the line represents the corresponding intermediate equilibrium state, and with the direction of the change being shown by the arrow.

In the total absence of friction, all the work done by the system is P, V work of expansion against the surroundings, as given by equation (2.3)

$$\text{Work done by system during change from } [P_1, V_1] \text{ to } [P_2, V_2] = \int_{V_1}^{V_2} P \, dV \qquad (2.3)$$

Since Boyle's Law is assumed to be valid for this particular change, equation (2.3) becomes

$$\text{Work done by system during change from } [P_1, V_1] \text{ to } [P_2, V_2] = c \int_{V_1}^{V_2} \frac{dV}{V} = c \ln\left(\frac{V_2}{V_1}\right) \qquad (2.4)$$

The work done is represented by the shaded area between the graph of the change in state and the horizontal axis, as shown in Figure 2.5.

2.9.2 ... and of a non-quasistatic change

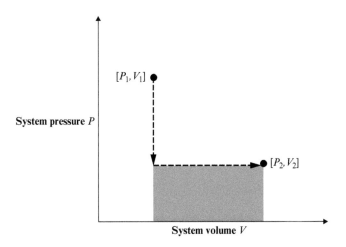

Figure 2.6 A P, V diagram for a frictionless non-quasistatic change in state. The external pressure P_{ex} on the cylinder is abruptly reduced to P_2, and the gas in the cylinder expands non-quasistatically at constant pressure P_2 to its final equilibrium state $[P_2, V_2]$. During the change in state, the system is not in equilibrium, and so it is not possible to define 'the' pressure and 'the' volume of the system at any intermediate state. The path of the change in state is therefore represented by a dashed line, which indicates a plausible path, consistent with the known fact that the work done by the gas during the change is certainly $P_2 \Delta V$, as shown by the shaded area.

Figure 2.6 shows another change in state from $[P_1, V_1]$ to $[P_2, V_2]$, but by a different path: in this case, the gas expands against a constant external pressure $P_{ex} = P_2$, as would happen, for example, if P_{ex} in Figure 2.3(a) and 2.3(b) were equal to P_2, and peg 1 is removed.

In this case, there is a substantial difference between the initial pressure P_1 and the external pressure P_2, so that, when the peg is removed, and in the absence of friction, the piston moves to the right very quickly. This creates initial turbulence within the system, and although the system started in an equilibrium state $[P_1, V_1]$, and ended at another equilibrium state $[P_2, V_2]$, during the change, the system is not in equilibrium. The change in state therefore does not take place through an infinitesimal sequence of equilibrium states, and so is a non-quasistatic change.

During the change, the pressure within the system will have different values at different points within the system; likewise, the volume of the system is uncertain since the molecules are not distributed evenly. As a consequence, it is not possible to specify 'the' pressure or 'the' volume at any intermediate state, implying that we cannot plot the path between $[P_1, V_1]$ and $[P_2, V_2]$ with any certainty. But we do know, from equation (2.1), that in the absence of friction, the work done by the gas against the surroundings is given by $P_{ex} \Delta V = P_2 \Delta V$ since P_{ex} is equal to P_2. Now $P_2 \Delta V$ is the shaded area in Figure 2.6, and so we can represent the frictionless non-quasistatic change from $[P_1, V_1]$ to $[P_2, V_2]$ by the dashed line as shown. The use of a dashed line, rather than a solid line, indicates that the values of the pressure and volume of the gas whilst the change is taking place are indeterminate, but the general shape is meaningful. Certainly, the area under the horizontal dashed line gives the right answer for the work done by the system, and the piston does move against a constant external pressure $P_{ex} = P_2$, so the horizontal dashed line is plausible. But whereas the solid line in Figure 2.5, indicating a quasistatic path, is 'the truth', the dashed line in Figure 2.6, indicating a non-quasistatic path, is just a representation.

2.10 Changes at constant pressure

Figure 2.6 describes a non-quasistatic expansion of a gas against a constant external pressure P_{ex}. As a moment's thought will verify, there are very many real events that take place under conditions of constant external pressure, especially when that pressure is exerted by the atmosphere. So, for example, many chemical reactions take place in vessels open to the atmosphere; most industrial processes – and many engines – operate in the atmosphere; and the same applies to many living processes too, whether under the pressure of the atmosphere, or the local pressure at a given depth in sea. And although the local atmospheric pressure can vary, for example, with the weather and the altitude, and the pressure in the oceans varies with depth, over the time taken for many of these processes to take place – typically fractions of seconds, seconds and minutes – the local pressure is constant. Changes at constant pressure are therefore very common, and important, and will feature strongly in this book.

But there is a subtlety here, a subtlety best explored by re-examining the interaction between a system and its surroundings when a gas expands from an initial state $[P_1, V_1]$ to a final state $[P_2, V_2]$. So, consider Figure 2.7, which represents both the system of interest and the surroundings, which together constitute the universe.

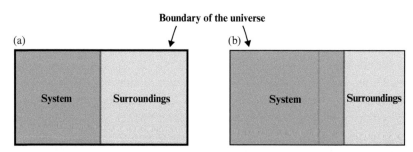

Figure 2.7 Can a system expand at constant pressure? As the gas in the system expands, from the state shown in Figure 2.7(a) to that shown in Figure 2.7(b), the volume of the gas in the system increases, and the pressure of the gas decreases. Simultaneously, the gas in the surroundings is compressed, and its pressure must correspondingly increase. Does this imply that it is impossible for a system to expand at constant pressure?

As the gas in the system expands against the surroundings, the volume of the system increases, and the pressure of the system falls. But, as implied by Figure 2.7, an *increase* in the volume of the system results in a corresponding *decrease* in the volume of the surroundings. The gas in the surroundings is therefore being compressed, and so we would expect the pressure of the gas in the surroundings to increase. That presents a puzzle: how is it possible for the gas in the cylinder to expand against a *constant* pressure? Surely, as the gas in the surroundings – the atmosphere – becomes compressed, the pressure in the atmosphere must increase, so it seems to be impossible for the gas in the cylinder to expand against a constant pressure. Is the concept of a change at constant pressure just theoretical (like the concept of the quasistatic change), or are changes at constant pressure real?

If the situation were indeed as illustrated in Figure 2.7, this would all be true: the pressure of the surroundings would increase as the system expands. Figure 2.7, however, is misleading, for two reasons: firstly, for ease of drawing, the volumes of the system and the surroundings have been depicted as about the same; secondly, the heavy line representing the 'boundary of the universe' is shown as a solid line, suggesting that it is rigid.

When we think more deeply about Figure 2.7, we realise that, in reality, the volume V_{ex} of the surroundings – the rest of the universe – is vastly, vastly, greater than the volume V of the system: V_{ex} is an unimaginably large number. When the piston moves to the right in Figure 2.7, the change of the volume of the system is ΔV, implying that the change of the volume of the surroundings is $-\Delta V$ (with the – sign indicating a decrease in volume as expected). The volume of the surroundings after the change is therefore $V_{ex} - \Delta V$, which, as expected is less than the original volume V_{ex}. But not by much: because the volume V_{ex} of the surroundings is such a huge number, $V_{ex} - \Delta V \approx V_{ex}$. In practice, the volume of the surroundings remains unchanged, despite the expansion of the gas in the system, and so the pressure of the surroundings stays constant.

Furthermore, if the boundary of the universe is not rigid – and we have no reason to believe that it is – then the volume of the atmosphere is not constrained, implying that the atmosphere can increase its volume as much as it 'likes', without any change of pressure.

Both explanations are valid, and both have the same conclusion: events that happen within systems open to the atmosphere, or within the vastness of the ocean, take place at constant pressure. The puzzle is resolved: changes at constant pressure are not theoretical concepts; on the contrary, changes at constant pressure are both possible and real.

2.11 Thermodynamic cycles

2.11.1 From here to there – and back again

Let's now take matters further, and consider what happens when the gas in the cylinder starts in the equilibrium state $[P_1, V_1]$, and follows the frictionless non-quasistatic path illustrated in the P, V diagram shown in Figure 2.6, expanding at constant pressure P_2, to the equilibrium state $[P_2, V_2]$. During this change, as is now familiar, work is done by the gas on the surroundings, equal to $P_2 \Delta V$. Having established an equilibrium state $[P_2, V_2]$, suppose now that the gas in the cylinder undergoes a frictionless *quasistatic* compression, reducing the volume and increasing the pressure. If the conditions are such that the temperature of the system is constant and Boyle's Law is valid, then the system will retrace the quasistatic path illustrated in the P, V diagram shown in Figure 2.5, and return to state $[P_1, V_1]$. During this frictionless quasistatic change, the work done by the gas is given by equation (2.4), but with the limits of the integration switched, since the change in state in this case is from $[P_2, V_2]$ to $[P_1, V_1]$:

$$\text{Work done by system during change from } [P_2, V_2] \text{ to } [P_1, V_1] = c \int_{V_2}^{V_1} \frac{dV}{V} = c \ln\left(\frac{V_1}{V_2}\right)$$

(2.4)

Since V_1 is less than V_2, the ratio V_1/V_2 is less that one, and so we may write

$$\text{Work done by system during change from } [P_2, V_2] \text{ to } [P_1, V_1] = -c \ln\left(\frac{V_2}{V_1}\right) \quad (2.5)$$

In this expression, V_2/V_1 is greater than one, implying that $\ln(V_2/V_1)$ is positive. Since the constant c is also positive, then the product $c \ln(V_2/V_1)$ is positive, and so $-c \ln(V_2/V_1)$ is necessarily negative. This therefore indicates that, during the change in state from $[P_2, V_2]$ to $[P_1, V_1]$, work is done *on* the gas by the surroundings (see page 27), as makes sense since this is a compression.

These changes are all illustrated in Figure 2.8.

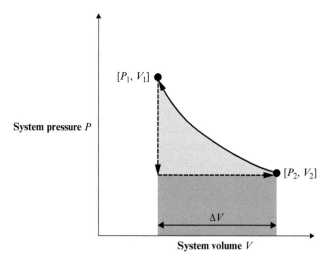

Figure 2.8 **A P, V diagram for a thermodynamic cycle**. The system is initially in the equilibrium state $[P_1, V_1]$, and then follows the frictionless non-quasistatic path (represented by the dashed line), at constant pressure P_2, to the equilibrium state $[P_2, V_2]$. During this change, the system does work $P_2 \Delta V$ on the surroundings, as shown by the dark-blue rectangular area. The system is then compressed along the frictionless quasistatic path (shown by the solid line), returning to the equilibrium state $[P_1, V_1]$, so completing the thermodynamic cycle. During this change, the surroundings do work $c \ln(V_2/V_1)$ on the system, represented by the total area under the curve, comprising both the light-blue and dark-blue areas. For the entire cycle, there is no change to either the state function P, or the state function V, and so $\oint dP = \oint dV = 0$. But as can be seen from the diagram, more work is done on the system in the change from $[P_2, V_2]$ to $[P_1, V_1]$ than was done by the system in the change from $[P_1, V_1]$ to $[P_2, V_2]$, and so $\oint dw \neq 0$, as represented by the near-triangular area in the middle of the diagram.

Any process in which a thermodynamic system starts in a particular equilibrium state, then undergoes one or more changes in state, and finally returns to the original equilibrium state – of which Figure 2.8 is an example – is known as a **thermodynamic cycle**. As we shall see during this book, thermodynamic cycles are of considerable significance: all engines – for example, steam engines, internal combustion engines, and jet engines - operate in cycles, and cycles feature in many important biological processes, such as the tricarboxylic acid cycle (sometimes known as the Krebs cycle) associated with the metabolism of sugars, and the light-dependent Calvin cycle (or Calvin–Benson–Bassham cycle, to give credit to Melvin Calvin's co-researchers at the University of California, Berkeley), central to plant photosynthesis.

Of particular relevance here are two insights: the first concerning the behaviour of state functions throughout a complete thermodynamic cycle; the second, concerning the behaviour of path functions.

2.11.2 Thermodynamic cycles and state functions

Let's firstly consider what happens to the state function V throughout the cycle. The system starts in an equilibrium state of volume V_1, and changes to a different equilibrium state of volume V_2. Accordingly, the volume change $\Delta V_{1 \to 2}$ is given by

$$\Delta V_{1 \to 2} = V_2 - V_1$$

If the change takes place through a sequence of infinitesimal steps, this can also be written as

$$\Delta V_{1 \to 2} = V_2 - V_1 = \int_{V_1}^{V_2} dV$$

Similarly, for the change from $[P_2, V_2]$ back to $[P_1, V_1]$, we may write

$$\Delta V_{2 \to 1} = V_1 - V_2 = \int_{V_2}^{V_1} dV$$

The total change of volume $\Delta V_{1 \to 2 \to 1}$, from $[P_1, V_1]$ to $[P_2, V_2]$ and back to $[P_1, V_1]$ is therefore

$$\Delta V_{1 \to 2 \to 1} = \Delta V_{1 \to 2} + \Delta V_{2 \to 1} = V_2 - V_1 + V_1 - V_2 = 0 = \int_{V_1}^{V_2} dV + \int_{V_2}^{V_1} dV$$

The total change of volume for the entire cycle is therefore zero – which is entirely as expected: volume is a state function, and we know that the value of the change in any state function is the value of that state function in the final state minus the value in the initial state. Since, for a cycle, the volume of the final state necessarily equals that of the initial state, then the change of volume for the entire cycle must be zero.

Let's, however, look more closely at the integral expression of this:

$$\Delta V_{1 \to 2 \to 1} = 0 = \int_{V_1}^{V_2} dV + \int_{V_2}^{V_1} dV$$

Now the sum of those two integrals, the first from V_1 to V_2, and the second from V_2 to V_1, represents an integral over the entire cycle from V_1 to V_2 and back to V_1 again, and can be represented as a cyclic integral $\oint$

$$\int_{V_1}^{V_2} dV + \int_{V_2}^{V_1} dV = \oint dV = 0$$

This shows that, for the state function volume, the cyclic integral, around any thermodynamic cycle, will be zero. This is an important general result which holds for all state functions: if a function X is a state function, then, for any thermodynamic cycle

$$\oint dX = 0 \tag{2.6}$$

Conversely, if it can be demonstrated for a function X that $\oint dX = 0$, then this proves that the function X is a state function, and dX is then said to be an **exact differential**.

2.11.3 Thermodynamic cycles and path functions

Let's now consider the work done during the cycle. During the frictionless non-quasistatic change from $[P_1, V_1]$ to $[P_2, V_2]$, all the work done by the system is P, V work of expansion against the surroundings, given by

Work done by system on surroundings in change from state 1 to state 2 = $P_2 \Delta V$

as represented by the dark-blue shaded area in Figure 2.8.

We have also just shown that the work done by the gas on the surroundings during the quasistatic change from $[P_2, V_2]$ back to $[P_1, V_1]$ is given by

Work done by system on surroundings in change from state 2 to state 1 = $-c \ln\left(\dfrac{V_2}{V_1}\right)$
$$\tag{2.5}$$

where the negative sign indicates that work is in fact done on the system by the surroundings during this compression. This is represented in Figure 2.8 by the total area under the solid line between $[P_2, V_2]$ and $[P_1, V_1]$, as represented by the combination of the dark-blue, and light-blue, areas.

The total work done by the system on the surroundings for the entire thermodynamic cycle can be determined by adding these together:

$$\text{Total work done by gas during the cycle} = P_2 \Delta V - c \ln\left(\dfrac{V_2}{V_1}\right)$$

$$= P_2(V_2 - V_1) - c \ln\left(\dfrac{V_2}{V_1}\right) \tag{2.7}$$

The number represented by the value of equation (2.7) is determined by the volumes V_1 and V_2, the final pressure P_2, and the appropriate Boyle's Law constant c. If it so happens that $V_1 = V_2$, then this number is zero - but if $V_1 = V_2$, then Boyle's Law tells us that $P_1 = P_2$ as well, implying that no change has taken place. If, however, a change *has* taken place, V_1 and V_2 will be different, and the numerical result of equation (2.7) will depend on the specific circumstances. But whatever the specific circumstances, it is invariably true is that this number will *not* be zero. Indeed, as reference to the particular thermodynamic cycle represented by the P, V diagram shown in Figure 2.8 makes quite clear, in this case, the result of equation (2.7) is a negative number, for the work $P_2 \Delta V$ done by the system on the surroundings in the expansion from $[P_1, V_1]$ to $[P_2, V_2]$, as represented by the dark blue shaded area in Figure 2.8, is clearly less than the work $c \ln(V_2/V_1)$ done on the system by the surroundings during the compression, as represented by the total area between the curved line and the horizontal axis in Figure 2.8. This diagram therefore vividly illustrates that more work is required to compress the gas quasistatically from $[P_2, V_2]$ back to $[P_1, V_1]$ than was done by the gas in expanding non-quasistatically from $[P_1, V_1]$ to $[P_2, V_2]$ – the net difference being represented by the nearly triangular figure in the centre of Figure 2.8.

If the amount of work done by the gas on the surroundings during any infinitesimal change is represented as đw, then the total amount of work done by the gas on the surroundings during any thermodynamic cycle may be expressed as a cyclic integral, the value of which must, in general, be some non-zero number:

$$\oint đw \neq 0 \qquad (2.8)$$

There is only one, very special, case in which the value of this cyclic integral is zero, and that is – with reference to Figure 2.8 – when the path from state $[P_2, V_2]$ back to state $[P_1, V_1]$ traces exactly the same frictionless quasistatic path as the original frictionless quasistatic path from state $[P_1, V_1]$ to state $[P_1, V_1]$, but in the reverse direction. This implies that the two paths, as shown on the diagram, would be identical, and so the area of the near-triangular area in the centre would shrink to zero.

Other than this particular case, the inequality of expression (2.8) is in general true: the total work done by a system through a generalised thermodynamic cycle will be some, non-zero, number. We can't predict, in general, what this number might be, for its value depends on the specific circumstances. Furthermore, the value might be a negative number, as in the particular case we have examined as illustrated in Figure 2.8; it could, however, also be a positive number, as indeed happens if the thermodynamic cycle shown on Figure 2.8 were to be carried out in the other direction, with a quasistatic *expansion* from $[P_1, V_1]$ to $[P_2, V_2]$ happening first, and then followed by a non-quasistatic compression at constant pressure P_2. What's important about equation (2.8) is not the specific value in any particular circumstances, but the fact that the result is, in general, some number other than zero.

We know that work is a path function, and Figure 2.8 demonstrates diagrammatically that the amount of work done by a system as it changes state does indeed depend on the path taken. Mathematically, this is represented by equation (2.8), which applies to all path functions, and, as we noted earlier (see page 26), đw is known as an **inexact differential** – hence the symbol đ, to distinguish between an inexact differential đ and an exact differential d. If we are dealing with any path function, then we can predict that the cyclic integral of that function around any thermodynamic cycle will be some number other than zero; likewise, if we can verify that the cyclic integral of any function around any thermodynamic cycle is non-zero, then we may infer that the function in question is a path function.

Equations (2.6) and (2.8) are therefore the fundamental mathematical definitions of state functions and path functions respectively.

2.11.4 An important link between path functions and state functions

Finally in this section, we note an important link between path functions and state functions, as exemplified by the now-familiar equation for the work đw done by a gas at pressure P, expanding quasistatically by a volume dV:

$$đw = P\,dV \qquad (2.9)$$

This equation links a path function, đw, to the product of two state functions, expressed in the form an [intensive state function, pressure] × change of an [extensive state function, volume]. As we shall see in due course, this is an example of a generic relationship of the from

đ(path function) = (intensive state function) × d(extensive state function)

2.12 Energy

As we have examined now in some detail, when a gas in a cylinder expands against the pressure P_{ex} of the surroundings, the gas inside the cylinder does work, quantified as $P_{ex} \Delta V$, on the surroundings. If the piston were linked to other appropriate mechanisms, this work could be harnessed, for example, to lift a weight, or to drive an engine – for all steam, petrol and diesel engines operate by the expansion of gases in cylinders against external pressure.

Before, however, the system does any work – before the piston actually moves – the gas in the cylinder must have possessed the *potential to do work*: it's as if the gas in the cylinder has some sort of hidden property, a property which, once released, can do useful things like work. So let's give this hidden property, this potential to do work, a name – **energy**. Given that work is a manifestation of energy, and that energy is 'work-in-waiting', energy and work have the same unit of measurement, the Joule, J.

Energy is the central concept in this book, and indeed is the central concept in much of science too. We'll gain a much richer understanding of energy as the book evolves, but this is a good place to start – the realisation that if a system is capable of performing work, then, before that work is performed, the potential to perform that work is a property of the original system. And if energy is a property of a system, then maybe there is a state function associated with it. Indeed there is – as we shall see in Chapter 5.

EXERCISES

1. In thermodynamics, what, precisely, is meant by the term 'work'? What is meant by 'work done on a system'? And by 'work done by a system'? What is 'P, V work'? Identify at least three examples of work, other than P, V work.
2. What is a 'path function'? How does a path function differ from a state function?
3. What are the characteristics of a 'quasistatic' path? Why are quasistatic paths important?
4. A system comprises a fixed mass of gas is contained within a cylinder, fitted with a frictionless piston, such that the system is in equilibrium with a volume of 0.5 m³, and at a pressure of 1 bar = 1×10^5 Pa. Calculate the P, V work of expansion done by, or P, V work of compression done on, the system, under the following circumstances, all of which take place at constant temperature:
   - The volume of the system increases to 0.75 m³ quasistatically.
   - The volume of the system increases to 0.75 m³ against a constant external pressure of 0.8×10^5 Pa.
   - The volume of the system increases to 0.75 m³ against a vacuum.
   - The volume of the system decreases to 0.3 m³ quasistatically.

- The volume of the system firstly increases to 0.75 m³ against a vacuum, and then decreases to 0.3 m³ quasistatically.
- The volume of the system firstly increases to 0.75 m³ quasistatically, and then decreases to 0.3 m³ quasistatically.

Calculate the final pressure of the system in each case, and illustrate each of your answers using a P, V diagram.

5. For the system described in question 4, suppose that the external pressure is initially 1×10^5 Pa. If the external pressure is increased quasistatically to 2×10^5 Pa, what is the final equilibrium pressure of the system if the temperature of the system is maintained constant? What is the corresponding final volume? How much work is done on the system by the surroundings for the change from the system's initial state to the system's final state? How would you represent this change on a P, V diagram?

6. In a P, V diagram, why must all real changes be represented by a dashed line, rather than a solid line?

7. For a change from any state [1] to any state [2], the work performed may be expressed as the integral of the infinitesimal amounts of work đw

$$\text{Total work done in change from state [1] to state [2]} = \int_{\text{state [1]}}^{\text{state [2]}} đw$$

How would you explain that, although this integral is valid and meaningful, the statement

$$\int_{\text{state [1]}}^{\text{state [2]}} đw = w\,(\text{state [2]}) - w\,(\text{state [1]})$$

is not only wrong, but meaningless? And if this is indeed wrong and meaningless, how, if at all, can the integral $\int_{\text{state [1]}}^{\text{state [2]}} đw$ be evaluated?

8. Rather harder: As in question 5, the external pressure is initially 1×10^5 Pa, but now this in increased instantaneously to 2×10^5 Pa. Suppose that the system's final equilibrium state is that same as the final state of question 5. Assuming that the change of state takes place at constant temperature, is the change in state of the system quasistatic? How might you represent this change in state on a P, V diagram? How might you estimate the work done on the system by the surroundings? How does this answer compare to the answer to question 5?

9. What are 'static' friction and 'dynamic' friction? In questions 4, 5 and 6, one of the assumptions we made was that the piston was frictionless. Suppose now that the surfaces between the piston and the interior of the cylinder are subject to both these types of friction. The initial state of the system is a volume of 0.5 m³, and a pressure of 1 bar = 1×10^5 Pa, and the external pressure is also 1 bar = 1×10^5 Pa. The external pressure now reduces by an infinitesimal amount – which, in the absence of friction, would cause the piston to move against the (now infinitesimally smaller) external pressure, as

described on page 29. But what happens in the presence of both static and dynamic friction? Describe what you would observe as the external pressure is progressively reduced until the piston has moved such that the final volume of the system is 0.75 m³. What is the final pressure of the system? Represent the path actually followed, in the presence of friction, on a P, V diagram. What is the P, V work of expansion done by the system on the surroundings? How does this quantity of work compare to the quantity of work that would have been performed in the absence of friction?

3 Temperature and heat

 Summary

Temperature is an intensive state function, and a measure of the average energy of the molecules with the system.

Heat is the transfer of energy between two systems at different temperatures. Like work, heat q is a path function, and an infinitesimal flow of heat is represented as an inexact differential dq.

The **Zeroth Law of Thermodynamics** states that

If two systems are each in thermal equilibrium with a third system, then they will be in thermal equilibrium with each other. Each of the three systems may be associated with an intensive state function, temperature, which, in this case, will take the same value for each system.

An **ideal gas** is one that obeys the **equation-of-state**

$$PV = nRT \qquad (3.3)$$

An **isolated** system is surrounded by a boundary which prevents both the flow of matter, and also the flow of energy in the form of heat and work.

A **closed** system is surrounded by a boundary which prevents the flow of matter, but allows the exchange of energy, in the form of both heat and work, between the system and the surroundings.

An **open** system is surrounded by a boundary which allows the exchange of matter, and also the exchange of energy, in the form of both heat and work, between the system and the surroundings.

The average molecular energy $\langle \mathcal{E} \rangle$ of an ideal monatomic gas is proportional to the temperature T as

$$\langle \mathcal{E} \rangle = \frac{3}{2} k_B T \qquad (3.19)$$

where k_B is Boltzmann's constant. Although equation (3.19) is valid only for an ideal monatomic gas, the interpretation of the temperature T as a measure of a system's average molecular energy is of more general applicability.

3.1 Temperature

3.1.1 Temperature is an intensive state function

Another state function of any system is its temperature, which determines how hot, or cold, the system is. Since the temperature of a system does not depend on the system's size or extent, temperature is an intensive state function. And to measure temperature, we need a suitable scale of measurement, and an appropriate instrument.

3.1.2 Temperature scales

Over the years, several scales of measurement have been devised, of which four in particular have been widely accepted. The **Fahrenheit** scale is named after Daniel Fahrenheit, a German physicist, who, in 1724, took the freezing point of pure water as '32' on his scale, and the boiling point as '212', dividing the interval between these two points into 180 equally-spaced 'degrees', designated °F. Twenty years later, the Swedish astronomer Anders Celsius defined the zero of the **Celsius** (alternatively, and now less commonly, **Centigrade**) scale as the freezing point of pure water, and the boiling point as 100, with 100 intermediate degrees, °C. Temperatures measured in °C are conventionally represented by the symbol t.

A third scale, the **thermodynamic scale**, is of especial significance to scientists in general, and this book in particular. The unit of measurement of the thermodynamic scale is the kelvin, K, (without the ° symbol), named in honour of William Thomson, Lord Kelvin, one of the greatest scientists of the nineteenth century, and whose name will appear many times in this book. As we have just seen, the Fahrenheit and Celsius scales both took two reference points – freezing water and boiling water – and then determined the size of the unit of measurement by dividing the interval between these two reference points into 180 or 100 equally-spaced degrees. The thermodynamic scale is derived rather differently: the size of the unit of measurement, one K, is defined to be the same as one °C, and so only one reference point is needed, this being a somewhat less familiar property of water known as the **triple point** – the state in which ice, (liquid) water and water vapour are mutually in thermodynamic equilibrium. The triple point of water is a very special state, for it can exist only at a single, specific combination of temperature and pressure – if the pressure is maintained, then at higher temperatures, the ice is unstable and melts, with liquid water being in thermodynamic equilibrium with water vapour until the water boils; and at lower temperatures, liquid water freezes. The thermodynamic scale of temperature uses the triple point of water as its reference point, defining – for reasons that will be explained later in this book (see page 480) – the corresponding temperature as 273.16 K, with the freezing point of water being very close by, 273.15 K. Temperatures measured in kelvin are conventionally represented by the symbol T, where the use of upper case distinguishes the use of the Kelvin scale from the use of the Celsius scale, which, as we have seen, is represented by the lower case symbol t. Zero on the thermodynamic scale, equivalent to −273.15°C, is known as **absolute zero**: as we shall see in Chapter 12, this extremely low temperature can be approached, but never actually reached.

The fourth scale, known as the **ideal gas scale** and, like the thermodynamic scale, defines its unit-of-measure as the kelvin K, equal to 1°C. Whereas the thermodynamic scale uses the triple point of water as its reference point, the ideal gas scale takes its reference, and zero, point to be absolute zero, −273.15°C.

3.1.3 Temperature measurement

To measure the temperature of a system of interest, an appropriate instrument – a thermometer – is usually placed in contact with the system, and, a very short time later, we take a reading from the thermometer's scale. A moment's thought shows that the temperature we actually measure is the *temperature of the thermometer*, rather than the temperature of the system, and we then assume that the thermometer's temperature is also the temperature of the system. We never think about, or even recognise, this assumption, so let's take a moment here to examine it, and to understand the underlying science.

All temperature measurement is based on a very familiar phenomenon: when a colder system is in contact with a hotter one, the colder one becomes hotter, and the hotter one colder, until both systems achieve the same, stable, temperature, indicating that the two systems are now in thermal equilibrium. The final, equilibrium, temperature is somewhere between the two original temperatures, depending on the masses, and materials, of the two systems: if the materials are the same, and the two masses equal, then the final equilibrium temperature is half-way between the two original temperatures; if the materials are the same, and one mass is rather greater than the other, then the final temperature is towards the original temperature of the greater mass; and if the mass of one of the two systems is negligible compared to the mass of the other, then the final equilibrium temperature is sensibly equal to the original temperature of the greater mass. The implication for this as regards the measurement of temperature is that any instrument used for that measurement – a thermometer – must be very small and light compared to the system whose temperature is to be measured, so that, when the thermometer is in contact with that system, the final equilibrium temperature of the thermometer is as closely as possible equal to the original temperature of the system.

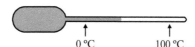

Figure 3.1 A representation of a mercury-in-glass thermometer. As the temperature of the mercury changes, the volume the mercury occupies also changes. Since the volume of the bulb on the left is fixed, this change in volume can be seen as a change in the length of the column of mercury within the sealed capillary tube. The void space in the capillary tube is a vacuum, so ensuring that the mercury can expand or contract freely, without being subject to an external pressure.

To enable the temperature of a thermometer to be measured easily, the thermometer needs to make use of a property that changes with temperature. A good example of this is the (nowadays somewhat old-fashioned!) mercury-in-glass thermometer, depicted in Figure 3.1, which comprises a reservoir of mercury, attached to a very narrow, sealed, capillary tube, the void space of which is a vacuum. Over a wide range of every-day temperatures, mercury is a liquid – a liquid whose volume changes with temperature, such that the higher the temperature, the greater the volume. When the bulb of the thermometer is placed in contact with the system whose temperature is to be measured, the mercury will contract or expand. Since the volume of the bulb is fixed, this contraction or expansion takes place within the capillary tube. Because the diameter of the capillary tube is very small, any changes in the volume of the mercury correspond to easily observable changes in the length of the mercury within the capillary. When the thermometer is placed in a mixture of ice and water, under normal atmospheric pressure, the position of the mercury in the capillary can be marked as 0°C, and when the bulb is placed into gently boiling water, also under normal atmospheric pressure, the (now different) position of the mercury can be marked as 100°C. The interval between these two marks can then be divided into 100 equal divisions, allowing the temperature of any other body (within these ranges), as measured on the Celsius scale, to be determined with ease.

All other types of thermometer work on the same principle, but using other properties which are sensitive to temperature, such as the resistance of an electrical circuit or the colour of a liquid crystal, and many modern thermometers have a digital readout.

3.2 The ideal gas law

As we saw on page 16, Boyle's Law

$$PV = \text{constant} \tag{1.3}$$

describes the relationship between the pressure P and volume V of a fixed mass of an ideal gas, undergoing changes in state at constant temperature. What, though, happens when the temperature changes? Around 1700, the French scientist Guillaume Amontons demonstrated that, at constant volume, the pressure of a given mass of air increases linearly with temperature; rather later – in the 1780s – another French scientist, Jacques Charles, also showed that, at constant pressure, the volume of a given mass of air increases linearly with temperature. These results are valid at temperatures easily accessible in a laboratory, but – as we saw with Boyle's Law – for real gases, these simple relationships break down as the conditions become more extreme. If, however, we consider the hypothetical ideal gas, then we can assert that these linear relationships apply at all temperatures, illustrated graphically in Figure 3.2, in which the horizontal temperature axes are defined using the Celsius scale.

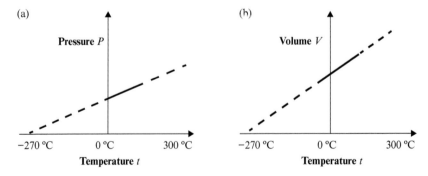

Figure 3.2 The ideal gas law. Graph (a) shows the linear relationship between the pressure P and temperature t of a given mass of an ideal gas at constant volume, and graph (b) shows the linear relationship between the volume V and temperature t of a given mass of ideal gas at constant pressure. If the temperature is measured on the Celsius scale, although the two lines have different slopes, they both intercept the temperature axis at the same point, close to −270 °C.

Experimentally, measurements are usually made at temperatures which are readily attained in a laboratory, say from a few degrees below 0 °C, to temperatures around 100 °C, as indicated by the solid line segments in Figure 3.2. Although the slopes of the volume-temperature and pressure-temperature graphs are different, when extrapolated, they both intercept the temperature axis at the same point, close to −270 °C. One interpretation of this is that, as the temperature approaches −270 °C, the volume of the gas approaches zero, as does its pressure. Since it is difficult to imagine a gas with a volume, or pressure, less than zero, the temperature close to −270 °C – now recognised as −273.15 °C – was designated as **absolute zero**, the lowest possible temperature.

Mathematically, we can represent these two linear relationships as

$$P = c_1(t + 273.15) \quad \text{for a given mass of gas at constant volume}$$

and

$$V = c_2(t + 273.15) \quad \text{for a given mass of gas at constant pressure}$$

In these equations, c_1 and c_2 are appropriate constants, and, in accordance with the extrapolation of experimental data, when $t = -273.15$ °C, both $P = 0$ and $V = 0$.

If, however, we define a new variable T such that

$$T = t + 273.15$$

then we can write two rather simpler relationships

$$P = c_1 T \quad \text{for a given mass of gas at constant volume} \quad (3.1)$$
$$V = c_2 T \quad \text{for a given mass of gas at constant pressure} \quad (3.2)$$

so that when $T = 0$, both $P = 0$ and $V = 0$. The variable T defines a new temperature scale, the ideal gas scale, whose zero is at absolute zero, −273.15 °C, and whose unit of measurement is of the same size as one °C. The ideal gas scale is identical to the thermodynamic scale, subject to the slight (but important) difference in the definition of each scale's reference point – the reference point of the ideal gas scale is (the in practice unattainable) absolute zero, whereas the reference point of the thermodynamic scale is the well-defined, and directly observable, triple point of water. As we noted earlier (see page 44), given the equivalence between the ideal gas scale and the thermodynamic scale, both use the same unit of measurement, the kelvin K.

Equations (3.1) and (3.2) can be combined with Boyle's Law

$$PV = c_3 \quad \text{for a given mass of gas at constant temperature} \quad (1.3)$$

into a single equation, first identified by the French scientist Émile Clapeyron in 1834,

$$PV = nRT \quad (3.3)$$

In which n represents the number of moles of gas (and hence the amount, and – given the appropriate molecular weight – the mass, of gas), and R is a constant, known as the **ideal gas constant**, **universal gas constant**, or **gas constant**, and whose value is 8.314462 J K^{-1} mol^{-1}.

Equation (3.3) is known as the **ideal gas law**, and represents a more complete equation-of-state for an ideal gas than Boyle's Law, and the two associated laws, equations (3.1) and (3.2).

3.3 A very important principle

If we define the molar volume $\mathbf{V} = V/n$, then the ideal gas law becomes

$$P\mathbf{V} = RT \tag{3.4}$$

Whereas V and n are both extensive state functions, the ratio $V/n = \mathbf{V}$ is an intensive state function, implying that – since R is a constant – the equation-of-state (3.4) is expressed in terms only of intensive state functions. Furthermore, if any two of the variables P, $\mathbf{V}$ or T are known, then equation (3.4) allows us to compute the third, unknown, variable. This implies that an ideal gas, as described by the equation-of-state (3.4), has two, and only two, independent state functions.

This is in fact a special case of a very important general principle:

> In the absence of gravitation, motion, electric, magnetic and surface effects, a fixed mass of a pure substance has two, and only two, independent state functions.

From the standpoint of most of chemical and biochemical thermodynamics, the exclusion of gravitation and the rest is not a problem – and if any of these effects (such as an electric effect, as is relevant to electrochemistry) do apply in a particular instance, the principle can be modified accordingly.

This principle is therefore of very broad applicability, and it is important because it tells us that the state of a given, constant, mass of any pure substance can be defined by reference to two, and only two, state functions. If any two state functions are known, all other state functions can be calculated, once we know the appropriate mathematical relationships, of which the ideal gas law, equation (3.4) is one example. And if we need to calculate the value of any of the extensive state functions, then we can do this once we know n, the number of moles present in the system of interest.

We shall refer to this principle many times in this book, for it is the starting point for gaining an understanding of many thermodynamic systems.

3.4 Dalton's law of partial pressures

The ideal gas law, as written in the form

$$PV = nRT \tag{3.3}$$

applies to n mol of a pure substance. Furthermore, two of the assumptions underpinning the definition of an ideal gas are that

- the molecules of an ideal gas occupy no volume, and
- there are no intermolecular interactions.

A consequence of the first assumption is that the total volume V of the system is available to every individual molecule – no space is denied as a result of being 'full of something else'. The second assumption implies that no individual molecule is influenced in any way by any other molecules that are present – the behaviour of each molecule is independent of all the others.

Suppose, then, that a vessel of interior volume V contains molecules of two different ideal gases: say, n_A mol of an ideal gas A and n_B mol of an ideal gas B, giving a total $n = n_A + n_B$ mols for the whole system. Each molecule of ideal gas A has access to the entire volume V, and, at a given temperature T, moves around that volume at random, occasionally impacting the walls of the vessel, and exerting pressure, as described on page 17. If we represent the total pressure exerted by the n_A molecules of ideal gas A as p_A, then equation (3.3) must apply, and we may write

$$p_A V = n_A RT$$

Similarly, the molecules of ideal gas B are also exerting a pressure p_B such that

$$p_B V = n_B RT$$

The total pressure P exerted on the inside of the vessel is attributable to all molecular impacts, and so

$$P = p_A + p_B$$

Multiplying by V, we have

$$PV = p_A V + p_B V = n_A RT + n_B RT = (n_A + n_B) RT$$

Since the total number n of molecules, of both types, is $(n_A + n_B)$, we therefore return to the familiar ideal gas law

$$PV = nRT \tag{3.3}$$

where, for a mixture of n_A mol of an ideal gas A and n_B mol of an ideal gas B,

$$n = n_A + n_B$$

and

$$P = p_A + p_B$$

The two pressures, p_A and p_B, are each known as the **partial pressure** of the appropriate component. Physically, each represents the pressure that each gas would exert, independently, if it were to occupy the whole volume V alone, as is consistent with the assumption of the ideal gas. For a mixture of more than two ideal gases, the total pressure P of the system is the sum of the individual partial pressures p_i for each component

$$P = \sum_i p_i \tag{3.5}$$

This result is known as **Dalton's law of partial pressures**, and, as we shall see (particularly in Chapter 14), it plays a central role in understanding gas phase chemical reactions, which are inherently gaseous mixtures in which the quantities of the various components – and hence their partial pressures – change as the reaction progresses.

Dalton's law represents a special case of the linear additivity of state functions – applied, in this case, to the intensive state functions, the partial pressures of the components A and B. As we saw on page 8, linear additivity is a characteristic of ideal substances, as is consistent with the assumption that both A and B are ideal gases, and that the ideal gas law applies.

3.5 Some other equations-of-state

The ideal gas law, equation (3.3)

$$PV = nRT \tag{3.3}$$

is just one example of an equation-of-state. For solids and liquids, the determination of valid equations-of-state is a challenging problem in condensed matter physics; for real gases, the ideal gas equation can be modified to relax the ideal assumptions noted on page 17 that the molecules occupy no volume, and that there are no intermolecular interactions, as, for example, expressed by the van der Waals equation

$$\left(P + \frac{an^2}{V^2}\right)(V - nb) = nRT \tag{3.6}$$

Berthelot's equation

$$\left(P + \frac{an^2}{TV^2}\right)(V - nb) = nRT \tag{3.7}$$

and

Dieterici's equation

$$P(V - nb) = nRT \exp\left(-\frac{na}{RTV}\right) \tag{3.8}$$

In all of these equations, a is a constant related to the strength of intermolecular forces (which tend to bind neighbouring molecules together), and b represents the actual volume occupied by 1 mole of molecules (which acts to reduce the volume available for the gas to occupy).

From a pragmatic standpoint, the virial equation-of-state

$$PV = nRT\left(1 + B\frac{n}{V} + C\frac{n^2}{V^2} + D\frac{n^3}{V^3} + \ldots\right) \tag{3.9}$$

in which $B, C, D \ldots$ are known as the second, third, fourth $\ldots$ virial coefficients, is very useful: the parameters $B, C, D \ldots$ are not directly related to physical quantities such as the volume actually occupied by the molecules – rather, they are determined empirically from experimental data for any given real gas. The virial equation-of-state therefore makes no fundamental physical assumptions that may, or may not, apply in any specific circumstances, but is rooted in experimental reality.

3.6 Heat

3.6.1 What happens when hot things cool down, and cold things warm up?

When a hot block of metal at, say, 80 °C is in contact with a colder block of the same metal at, say, 20 °C, we observe that the originally hot block cools down, and the originally colder block heats up, until both blocks achieve the same temperature somewhere between 80 °C and 20 °C. That's a very common-place observation - but what is actually happening?

It's very tempting to explain this observation in terms of the transfer of some 'substance X' from one system to another. If we imagine that the presence of this 'substance X' confers 'hotness', with hotter bodies having more of 'substance X' than colder ones, then when the two blocks of metal are in contact, it seems as if the originally hotter block loses some 'substance X', so becoming cooler, whilst, at the same time, the originally colder block gains some 'substance X', so becoming hotter.

And it's even more tempting to imagine that the amount of 'substance X' lost by the originally hotter system is equal to the amount gained by the originally colder system, so that the total quantity of 'substance X' remains the same. This would imply some sort of 'law of conservation', in which the total quantity of 'substance X' is constant, but with this total quantity being distributed around different systems in different ways. Let's, for the moment, call 'substance X' heat.

So far, so good, and the 'law of the conservation of heat' satisfactorily explains what happens in many situations in which hotter systems are in contact with colder ones. But there was another, very familiar, phenomenon where it was a lot harder to understand how a 'law of the conservation of heat' might actually work.

3.6.2 The experiments of Rumford and Joule

That other phenomenon is the way in which a colder system can become hotter, not by contact with something hotter, but as a result of work. We all experience this personally, as we become progressively warmer as we work progressively harder; and we've all noticed a very similar effect whenever our work involves friction – for example, when cutting wood with a saw, our experience is that both the saw *and* the wood become hot. And we become hot too, as a result of all that hard work! This is a real puzzle: how is it possible for three originally cold systems – the saw, the wood and me – all to become hot together, without any of these three systems being in direct contact with something hotter to start with? And the more work we do, the hotter everything becomes, as if the work is somehow creating heat from absolutely nothing. This poses a fundamental problem for a 'law of the conservation of heat': by definition, a 'conservation' law rules out the possibility that heat can be 'created', apparently out of nothing – but this is precisely what appears to be happening with work.

The fact that work can make things hotter has been known since ancient times, but was first explored methodically by one of the more colourful figures in the history of science – Benjamin Thompson, Count Rumford (no relation to William Thomson, Lord Kelvin, despite the similarity in names and their noble titles!). Born in 1753 in a small village in the then-English colony of Massachusetts, Rumford came to London in 1781, spent many years in Bavaria, and died in 1814 in Paris. Whilst involved with the reorganisation of the Bavarian army, he became interested in the manufacture of cannon. The barrel of a cannon is a cylinder of metal, with a deep 'hole' in it – the 'hole' through which the cannon ball will be fired. In Rumford's day, the barrel was typically made from solid bronze, and the 'hole' was then drilled out. Rumford noticed that as the drilling took place, the bronze became progressively hotter, without limit: the more drilling that was done, the hotter the metal became - observations which he published in 1798. Rumford realised that the rise in temperature was attributable to the friction between the drill and the bronze, friction directly associated with the work being done to drive the drill through the metal.

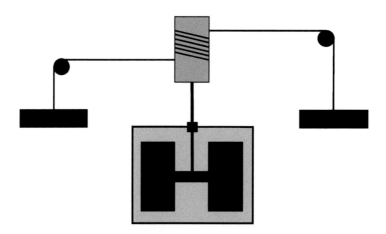

Figure 3.3 A schematic representation of Joule's apparatus. As the masses to the left and right fall, the paddle rotates within some water. The temperature of the water is observed to rise.

Rumford's findings were refined by James Joule, a somewhat less exotic figure, but a great scientist nonetheless. In his laboratories in Manchester, England, Joule carried out a series of meticulous experiments during the early 1840s to investigate the relationship between work and heat. His most ingenious apparatus, illustrated in Figure 3.3, was a mechanism in which two falling masses, each of mass M kg, caused a paddle, immersed in water, to rotate. As the masses fall a distance h m, they do work equal to $2Mgh$ J; this work is harnessed to drive the paddle; and, as the paddle rotates against the water, the internal friction causes the temperature of the water to rise – just like the result of the friction between Rumford's drill and the bronze. Joule's studies, reported in 1845, determined that 4410 J of work (the unit of work is named in his honour) consistently raise the temperature of 1 kg of water by 1 °C, so quantifying what became known as the **mechanical equivalent of heat**. In 1850, Joule published a refined estimate of 4159 J – very close to the currently accepted value of 4186 J.

3.6.3 A modification of Joule's experiment

Let's now make an imaginary modification to Joule's experiment – instead of using falling masses to rotate the paddle, let's connect the shaft of the paddle to the piston of a cylinder, so that as the piston moves to the right in Figure 3.4, the paddle rotates.

The piston is held in place by the now-familiar peg 1. When this peg is removed, and if the gas pressure P inside the cylinder is greater than the external pressure P_{ex}, the piston moves to the right, so causing the paddle to rotate. And in expanding, the gas is doing work, some of which (as we saw on page 27) is required to 'push back the atmosphere', and some used to rotate the paddle. Now the paddle, and the surrounding water, do not 'know' that the paddle is being rotated by the expansion of a gas rather than by the falling of a mass – as far as the water is concerned, the paddle is rotating and the water is getting hotter. And if there happens to be 1 kg of water, then it will take 4186 J of work to raise the temperature by 1 °C, regardless of whether the work is performed by a falling mass, or by the expansion of a gas in a cylinder.

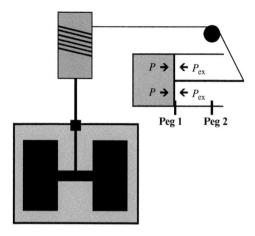

Figure 3.4 An alternative version of Joule's experiment. In this experiment, the pressure P inside the cylinder is greater than the external pressure P_{ex}. If peg 1 is removed, the piston moves to the right, the paddle rotates, and the temperature of the water rises – just as happens when the rotation of the paddle is caused by the falling of a mass.

The heating of the water is attributable to the work done by the expansion of the gas, but, as we saw on page 40, the work performed by the expansion of the gas is itself attributable to the energy of the gas in the cylinder before the peg was removed. In this experiment, the energy originally in the gas has been transformed firstly into work, and then into heat – indeed, if much of the mechanics of the apparatus were contained within a 'black box', we would not observe the intermediate stages, and we would infer that the energy of the gas is directly causing the temperature of the water to rise. This therefore implies that *energy is not only the capacity to perform work: it's also the capacity to transfer heat*.

3.6.4 The conservation of energy

Joule's experiment confirmed the 'mechanical equivalent of heat', the fact that it is possible to transform a given amount of work into a corresponding amount of heat. But our extension of Joule's experiment illustrates that the 'mechanical equivalent of heat' is, more fundamentally, an 'energetic equivalent of heat': since energy is the capacity to perform work as well as the capacity to transfer heat, then energy, work and heat are all, fundamentally, the same.

This insight resolves the paradox of the 'law of the conservation of heat': it isn't heat that is being conserved – it's energy. So, when a hotter system is placed in contact with the colder system, energy is transferred from the hotter system to the colder system, so that the hotter system cools down, and the colder system warms up, until both systems achieve the same temperature. The *process of energy transfer* from the hotter system to the colder one is what we call **heat**, and whilst this process is taking place, the energy lost by the hotter system equals that gained by the colder system, and so energy is conserved.

In Count Rumford's experiments on drilling bronze cannon, the heat produced during the process of drilling isn't 'created from nothing' – it's the transfer of energy resulting from the work needed to operate the drill. So, once again, energy is conserved.

We'll take a more extensive look at the relationships between energy, work and heat, and also at the corresponding conservation law, in the next chapter: for the moment, we'll look at some further aspects of heat and temperature.

3.6.5 Units of measurement for heat

Since energy, as well as being 'work-in-waiting', is also 'heat-in-waiting', then the unit of measurement for heat is most sensibly the same as the unit of measurement used for both energy and work, namely the Joule, J. Historically, however, heat was being studied for a long time before it was realised that heat is in fact a manifestation of energy. These studies required a unit of measurement, and so the 'calorie' was defined as 'the amount of heat required to raise the temperature of 1 gram of pure water by 1 °C (specifically, from 14.5 °C to 15.5 °C) at a pressure of 1 atmosphere'. An associated unit, the 'Calorie' (with an upper case C) or 'kilocalorie', has a similar definition, but as applied to 1 kilogram of pure water. When Joule's experiment established that 4159 J = 1 Calorie, the use of the calorie, and Calorie, as units of measurement for heat became redundant, for any measurement in calories (or Calories) can be converted into a measurement in J using the conversion factor 1 Calorie = 4159 J. As it happened, however, the calorie, and Calorie, remained in use for many years as the units of measurement for heat, long after Joule's experiment, and continue to be used to this day in connection with, for example, the specification of the energy content of many foods, as shown on the labels on food packaging. Scientifically, however, the SI unit for work, energy and heat is the Joule, J.

3.6.6 Thermal contact

If heat can flow from one system to another, then those systems are said to be in **thermal contact**. Thermal contact can be the result of physical proximity, as when something hot is in direct contact with something cold, in which case we speak of the **conduction** of heat (for heat flow within, or between, solids), or the **convection** of heat (for heat flow within, or between, fluids). Heat flow can also take place across distances – for example, the heat flow from the sun to the earth – in which case we speak of **radiation**.

3.6.7 Heat isn't a substance, it's a description of a process

Very importantly, we now appreciate that heat isn't some form of *substance*; rather, it's the name we give to the *process* in which energy is transferred either

- between two systems in thermal contact, one hotter and one colder, so that the hotter system becomes cooler, and the colder system hotter, or
- as the result of work being done on a system, causing that system to become hotter.

Taking the case of the transfer of energy from a hotter system to a colder system, we also know that hotter systems are associated with higher temperatures, and colder systems with lower temperatures. A temperature difference between two systems is known as a **temperature gradient**, so another way of defining heat is the process of energy flow down a temperature gradient.

3.6.8 The 'Zeroth' Law of Thermodynamics

The definition of heat as the process of energy flow down a temperature gradient has two, rather hidden, implications. The first is that if there is no temperature gradient between two systems – that is, if the two systems have the same temperature – then there is no transfer of energy, and no flow of heat, this being the situation which we have already defined (see page 9) as thermal equilibrium. Looked at the other way around, if two systems are in thermal contact and there is no flow heat, then we can infer that they are at the same temperature. This leads to what is sometimes called the 'Zeroth' Law of Thermodynamics:

> If two systems are each in thermal equilibrium with a third system, then they will be in thermal equilibrium with each other. Each of the three systems may be associated with an intensive state function, temperature, which, in this case, will take the same value for each system.

This statement highlights an empirical observation – thermal equilibrium, and the corresponding absence of heat flow – and introduces an associated thermodynamic state function, in this case, temperature. We all experience situations in which we detect heat flows between systems, and we also experience systems which are not associated with the flow of heat: these are every-day empirical observations. The Zeroth Law then tells us that if each system can be associated with an intensive state function, temperature, then if the temperatures of two systems are the same, there will be no heat flow, but if the temperatures of the two systems are different, then heat flow between them. Conversely, if we detect no heat flow between two systems in thermal contact, then they are at the same temperature; if we do detect a flow of heat, then they are at different temperatures. As we shall see for the First and Second Laws of Thermodynamics (see pages 94 and 268, respectively), these too can be expressed in terms of an empirical observation, and the introduction of an associated state function.

3.6.9 Heat flow is unidirectional

The second implication of the fact that a temperature gradient between two systems in thermal contact causes a flow of energy so that the originally hotter system becomes colder, and the originally colder system becomes hotter, is that this statement implicitly contains a statement of directionality – the energy flow is *down* the temperature gradient, from the higher temperature system to the lower, not *up* it. This 'unidirectionality' has very considerable significance, as we shall see in Chapter 8.

3.6.10 Heat is a path function

We now appreciate that a system does not 'have' heat; rather, heat is the name we give to the process in which heat is transferred between two systems, in thermal contact and originally at different temperatures, until such time as the temperature across both systems becomes uniform, and the flow of heat stops. Heat is therefore not a property of a state, but rather a phenomenon that can be measured only as a system changes state – just as we saw on pages

TEMPERATURE AND HEAT

24 in relation to work. Heat, like work, is therefore a **path function**, and the amount of heat transferred during any change of state will depend on the path taken. Accordingly, a small quantity of heat is represented as đq and, like đw, đq is signed, according to whether the system gains or loses heat during the corresponding change of state:

- Heat *gained by* a system is represented by a *positive* number.
- Heat *lost from* a system is represented by a *negative* number.

For any finite change of state from state [1] to state [2], the total amount of heat $\{_1q_2\}_X$ gained by, or lost from, the system can be determined by summation or integration as

$$\{_1q_2\}_X = \sum_{\text{state [1]}}^{\text{state [2]}} đq = \int_{\text{state [1]}}^{\text{state [2]}} đq \qquad (3.10)$$

where the notation $\{\ldots\}_X$ indicates that the specific value of $\{_1q_2\}_X$ for a particular change in state depends on the path X taken.

Also, since heat is a path function, this implies that, for any thermodynamic cycle

$$\oint đq \neq 0 \qquad (3.11)$$

As with work, the net amount of heat gained by, or lost from, a system during a thermodynamic cycle can be evaluated only by reference to the specific path followed, some elements of which will be associated with the gain of heat, some with the loss. Overall, however, for a complete thermodynamic cycle, the result will be either a positive (net heat gained) or a negative (net heat lost) number; it will not be zero.

One further thought in relation to heat as a path function: as we saw earlier (see page 39), the path function work, đw, is linked to the two state functions pressure P and volume V by the equation

$$đw = P\,dV$$

which we expressed in a more generic form

$$d(\text{path function}) = (\text{intensive state function}) \times d(\text{extensive state function})$$

We now know that heat, đq, is a path function, and very much associated with the intensive state function temperature T. So that triggers a hunch that

$$đq = T\,d(\text{an extensive state function})$$

What that extensive state function might be is not so obvious, and, as we shall see in Chapter 9, one of the great triumphs of thermodynamics was the discovery of what that state function actually is.

3.7 Some more definitions

Now we have a deeper understanding of heat, there are a few more technical terms that we need to define, which we do here.

3.7.1 Isolated, closed and open systems

We have already seen that a system is that part of the universe selected for study, and that any system is separated from the surroundings by the system boundary. Within this broad definition of 'system', it is useful to distinguish three categories:

- An **isolated system** is one whose boundary prevents the exchange of both matter and energy with the surroundings.
- A **closed system** is one whose boundary prevents the exchange of matter with the surroundings, but allows the transfer of energy, for example, as work or heat.
- An **open system** is one whose boundary allows the transfer of both matter and energy with the surroundings.

Living things are open systems in that all living things exchange matter with their surroundings (for example, by eating and breathing), as well as energy (photosynthesis in green plants, for example, uses light energy to transform carbon dioxide and water into glucose and oxygen).

The now-familiar system of a gas in a cylinder with a moveable piston is an example of a closed system – the walls of the cylinder and the seal around the piston stop any of the internal gas from leaking out, as well as stopping any matter in the surroundings from getting in, so preventing the exchange of matter, whilst the moveable piston allows the system to perform work on the surroundings (if the piston moves outwards), and also the surroundings to perform work on the system (if the piston moves inwards). Furthermore, if the walls of the cylinder are made of a material which can conduct heat – for example, metal – then heat too can be exchanged with the surroundings.

An example of an isolated system is a gas within a sealed container, the walls of which are fixed, and made of a material that provides perfect thermal insulation. The fact that the container is sealed stops any matter from getting in or out; the rigid walls can't move, and so no work can be done either by the system or on it; and the thermal insulation prevents any exchange of heat. In practice, however, there are no truly perfect thermal insulators, so any real system (with one exception, to be noted very shortly) is only an approximation to a theoretical isolated system: the exception is the universe as a whole, for we are free to position the boundary as far into deep space as we like …

3.7.2 Isothermal and adiabatic changes in state

Systems can change their state in many different ways, but two types of change of state are particularly important:

- An **isothermal** change of state is one in which the temperature of the system is maintained constant throughout the change, and an **isotherm** is a plot on a diagram, such as a P, V diagram, that defines an isothermal path.
- An **adiabatic** change of state is one in which there is no exchange of heat between the system and the surroundings, and an **adiabat** is a plot on a diagram, such as a P, V diagram, that defines an adiabatic path.

An isothermal change usually requires the flow of heat into, or out of, the system, and so can take place within an open or a closed system, but not within an isolated one; changes within an isolated system are necessarily adiabatic. A closed system can undergo an adiabatic change if its boundaries allow work to be done by, or on, the system, but block the transfer of heat; an open system can in principle undergo an adiabatic change under specific circumstances in which heat is not exchanged with the surroundings, even though heat exchange is in principle possible – but this is very rare.

3.7.3 Isobaric and isochoric changes in state

The terms isothermal and adiabatic are used frequently; here are two other terms which, though used rather less frequently, are also worth noting:

- An *isobaric* change of state is one in which the pressure of the system is maintained constant throughout the change, and an *isobar* is a plot on a diagram, such as a P, V diagram, that defines an isobaric path. On a P, V diagram, an isobar is necessarily horizontal.
- An *isochoric* change of state is one in which the volume of the system is maintained constant throughout the change, and an *isochore* is a plot on a diagram, such as a P, V diagram, that defines an isochoric path. On a P, V diagram, an isochore is necessarily vertical.

3.8 How to get work from heat

3.8.1 The steam engine

Around the time that Count Rumford was in Bavaria studying how work is converted into heat during the drilling of bronze cannon, some 2,000 km away to the north-west, the brilliant Scottish engineer James Watt, working firstly in Scotland and then in England, was doing the very opposite – perfecting his machine that converted heat into work, a machine known as a steam engine. Steam engines played a very important role in the development of thermodynamics, for they are devices that transform the heat produced from burning a fuel such as wood or coal, and later oil, into useful work that can be harnessed to drive a locomotive or industrial machinery, as achieved by exploiting the P, V work done by the expansion of a gas, steam. The invention, and development, of a reliable steam engine was of enormous significance: up to that time, the only sources of work were animals and human beings (which need to be fed, and get tired), windmills (which rely on the wind), water-mills (which require a steady flow of water, and need to be sited accordingly), and – to a very much more limited extent - falling weights (as used to power clocks). Steam engines are much more reliable, and can be sited anywhere – all they need is a supply of fuel to burn, and water to boil.

Watt's steam engines were sophisticated machines, and – as with all great inventions – built on a succession of earlier endeavours, such as the engines designed and constructed by Thomas Newcomen (who, with John Cawley, was awarded a patent in 1705), Thomas Savery (around 1698), Denis Papin (1688), the Marquis of Worcester (1663), and Giovanni Branca (1629) – indeed the earliest documented record of heat producing work, of the very first steam engine, is the 'aeolipile', described by Heron (sometimes known as Hero) of Alexandria some 2,000 years ago, as illustrated schematically in Figure 3.5.

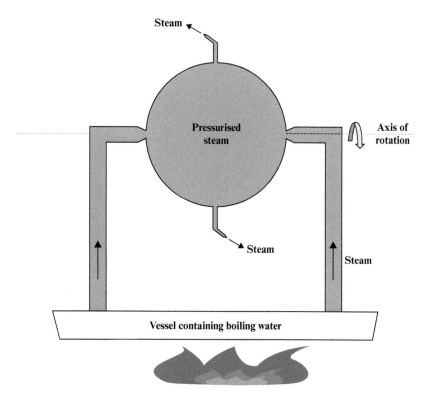

Figure 3.5 Heron's aeolipile. Steam form boiling water passes into a hollow sphere, which can rotate about a horizontal axis. The steam can escape only through two narrow L-shaped pipes, each ending in a narrow outlet: in this diagram, the lower outlet is projecting perpendicularly out of the paper towards the reader, and the upper outlet is projecting out of the paper away from the reader. As the pressurised steam escapes through the outlets, the sphere rotates in the opposite direction. The rotation of the sphere, driven by the P, V work done by the expanding steam, could, in principle, be harnessed to lift a weight.

3.8.2 How to harness the P, V work of an expanding gas

One feature of Heron's aeolipile is that the steam emitted from the outlets expands freely against the atmosphere. In Watt's steam engine, in contrast, the steam is kept enclosed within a cylinder, and expands against a moveable piston. In this section, we will build on our discussion in Chapter 2 to explore in more detail how a heated, expanding, gas can be used to perform useful work.

As we saw in Chapter 2, in principle, the conversion of heat into work, using a gas expanding against a piston, is very simple. If the piston in Figure 3.6 is initially held fixed, the volume V of the system is held constant. On heating the gas – say, air – its temperature T increases, and, assuming that the ideal gas law $PV = nRT$ is reasonably valid, since the volume V is constant, the pressure P will increase accordingly. When the pressure P has reached a given level above atmospheric pressure, peg 1 can be removed, so that the piston moves against the atmosphere. The expansion of the air therefore does P, V work as we discussed on page 27, 'pushing the atmosphere out of the way', and doing other 'useful work' too – if the piston is

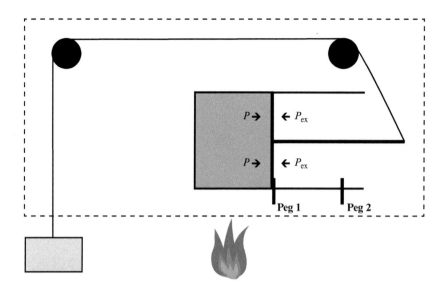

Figure 3.6 Producing work from heat. When the air inside the piston is heated, its pressure P increases. On removing peg 1, the piston then expands against the external atmospheric pressure P_{ex} until it stops at peg 2. As the piston moves to the right, the mass is raised. The device inside the dashed line acts as a machine that converts the heat from the flame into the work required to lift the mass.

suitably connected mechanically to another device, then the motion of the piston during this 'power stroke' can be used to lift a weight, or rotate, say, a mill-wheel. Overall, this device converts heat into work.

There is, however, a very practical problem: the work that this device performs takes place only as the piston moves to the right, and stops when the piston has reached the position defined by peg 2. This could happen quite quickly, and so any work that is performed will be finished within a very short time. It may be that this time is long enough to raise the weight to the required position, but this might just be a fortunate co-incidence – what we seek is a device that can perform work as long as we wish it to, rather than one that can only perform work in a short burst.

One way of increasing the time over which work is performed is to use a longer cylinder, so that the distance between peg 1 and peg 2 is significantly greater. But in practice, how long can a cylinder be? And even if it were possible to use a longer cylinder, another problem will soon arise. As the piston moves against the atmosphere, the volume within the cylinder increases, and so the pressure of the air inside the cylinder drops. Eventually, the pressure of the air inside the cylinder will fall to that of the atmosphere, and so the piston will stop, even if it hasn't yet reached peg 2. This problem, too, can be avoided, in two different ways. The first is to heat the air to a very high pressure, before releasing peg 1, so that the piston will move through a longer distance before the internal and external pressures equilibrate. And the second is to keep the temperature of the air inside the cylinder high, so that the pressure drop attributable to the volume increase is compensated by the higher temperature - which is all about manipulating $PV = nRT$ so that the internal pressure P remains higher than the external atmospheric pressure P_{ex} even as the volume V increases. Both these 'solutions' require a lot of heat (and therefore fuel, which is expensive), and if the pressure inside the cylinder is high, then the cylinder must be strong enough to withstand that high pressure without exploding. In practice, the problems of the piston hitting peg 2, and of having a cylinder of reasonable length and made of a strong enough material, and of working in a way that does not consume vast amounts of fuel, are very real.

But perhaps there is a different type of solution: if we can *get the piston back to where it started*, then we can heat the air a second time, and drive the piston out again. And a third time, and a fourth... What we are seeking is therefore a machine which uses a reasonable quantity of fuel, formed from a suitably strong cylinder of a sensible length, in which the piston operates in a continuous in-out, in-out, in-out cycle, keeping the work going for as long as we wish, without exploding, or having everything stop when the pressure inside the cylinder equilibrates with the external atmosphere.

How, then, do we get the piston back to its original position? The obvious answer is to push it back. But to do this, the piston needs to be pushed back *against* the pressure of the air inside the cylinder – and as the volume of this air decreases, the pressure becomes progressively higher, and so this becomes increasingly more difficult. Pushing the piston back on this 'return stroke' therefore *requires work to be done on the system*, and so we may define the **net work** done by the system over a complete cycle as

> Net work done by the system over a complete cycle
>
> = work done by the system on the surroundings during the 'power stroke'
>
> − work done by the surroundings on the system during the 'return stroke' (3.12)

Suppose that, in any particular case, the work done on the system in returning the piston to its original position is very close to, or even greater than, the work done by the system during the power stroke. For a complete cycle, the net work done by the system is very small, or even negative, with more work having to be done *on* the system than the system does for us. This is clearly a 'bad deal' – we'd be better off grinding our corn using the power of a horse.

The problem we need to solve therefore becomes more focused: how can we return the piston to its original position, but in a way in which the work needing to be done on the system to achieve this is substantially *less* than the work done by the system during the expansion? If we can do this, then, although some work is 'wasted' in returning the piston to its original position, overall, for a complete cycle, more work is done during the power stroke than is required for the return, so that, for a complete cycle of the system, there is a meaningful net amount of work actually done – work that can be harnessed to lift a weight, drive a mill-wheel, or propel a locomotive.

3.8.3 How a steam engine works

One way of reducing the amount of work required to return the piston to its original position is to lower the pressure inside the cylinder for the return stroke – as can be achieved by reducing the temperature of the air inside the cylinder: as is evident from the ideal gas law $PV = nRT$, when the piston is at its maximum extent, a reduction in the temperature T of the gas within the cylinder, whilst the volume V remains unchanged, causes the pressure P to drop accordingly.

Here, then, is a way of solving our problem, as a four-step cycle:

- Step 1: With the piston in its starting position, heat the cylinder, so causing the pressure of the air inside the cylinder to increase.
- Step 2: When the pressure within the cylinder is suitably high, release the piston, so allowing the air to expand and perform P, V work on the surroundings. During this power stroke, the work done by the piston (after, as we saw on page 27, taking into account the work required to 'push the atmosphere out of the way') can be harnessed, say, to lift a weight.
- Step 3: When the piston reaches the end of the cylinder, spray water over the outside of the (now hot) cylinder, so that the cylinder, and its contents, are cooled down.
- Step 4: Push the piston back to its original position.
- Start the cycle again.

This cycle is represented in the P, V diagram shown in Figure 3.7, where, as discussed on page 33, the dashed lines indicate that each step in the four-step cycle is real.

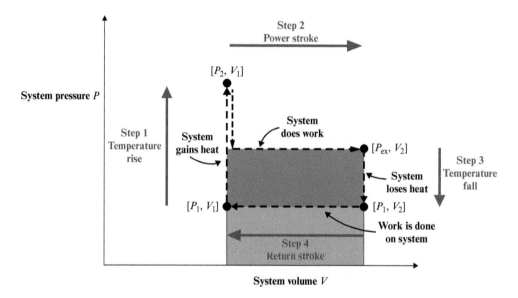

Figure 3.7 How to convert heat, continuously, into useful work. Air in a cylinder, fitted with a piston, is initially in state $[P_1, V_1]$, with a pressure P_1, a volume V_1, and at a given temperature. The piston is held fixed, and the air inside the cylinder is then heated at constant volume V_1, increasing the air's temperature. This causes the air's pressure to increase to P_2, so as to achieve state $[P_2, V_1]$. The piston is then released, and expands at a constant external pressure P_{ex} to achieve state $[P_{ex}, V_2]$. This is the 'power stroke' in which the system does work on the surroundings, work that can be harnessed to lift a weight or drive a wheel. This work is represented by the total area under the upper horizontal dotted line – the sum of the light and dark areas. The piston is then held fixed at its greatest extent, and the hot cylinder, and the air within it, cooled down, by, for example, pouring cold water over the cylinder. This causes the air inside the cylinder to lose heat, lowering its temperature and so reducing its pressure to P_1, to achieve state $[P_1, V_2]$. The piston can now be pushed back (the 'return stroke') to its original volume V_1, so that the system returns to its original state $[P_1, V_1]$. During the return stroke, the surroundings do work on the system, as represented by the light area under the lower horizontal dotted line. The cycle can then start again. As a result of the cycle, the system does net work on the surroundings, as represented by the dark rectangular area in the centre of the diagram – but for this to happen, there must be an alternating sequence of heating and cooling.

The alternating heating and cooling of the gas over each cycle allows more work to be done on the power stroke than is required by the return stroke. As a result, the net work performed each cycle, as defined by equation (3.12), is positive, as indicated by the area of the 'box' in the centre of the P, V diagram shown in Figure 3.7. A further inference from this diagram is that, for any given external pressure P_{ex} – which in practice is usually the pressure of the atmosphere – then the lower the pressure P_1 for the return stroke, the larger the area of this 'box', and so the greater the net work done by the engine. A low pressure P_1 requires a low temperature, implying that the greater the temperature difference between the states represented by $[P_{ex}, V_2]$ and $[P_1, V_2]$, the greater the net work.

This, very rudimentary, engine does perform net work continuously, but at a cost – the alternating heating and cooling of the cylinder and its contents consumes a considerable amount of heat, and therefore (expensive) fuel. How might the performance of this engine be improved, so that it performs the same amount (or indeed more) work, but requiring less heat and so consuming less fuel? This is a question of *efficiency* – how to get the maximum amount of work from the engine for the minimum input of heat, and hence the minimum consumption, and cost, of fuel. The quest for ever-more efficient engines was a (quite literal) driving force behind the development of thermodynamics in the nineteenth century, and will be discussed further in Chapter 10. For the moment, though, we mention two ways in which the efficiency of our current engine can be improved.

The first improvement is not to use air within the cylinder, but to use steam – for steam has a very valuable property: if steam is cooled below 100 °C, it condenses, and gaseous steam becomes liquid water. When this happens, the volume occupied by any given mass of steam suddenly reduces – for example, 1 kg of steam at a temperature just above 100 °C occupies approximately 1.7 m^3 at a pressure of one atmosphere, whereas the same mass of liquid water at a temperature just below 100 °C, also at a pressure of one atmosphere, occupies only about 1 x 10^{-3} m^3, a volume some three orders of magnitude smaller. So, consider a cylinder containing 1 kg of steam at, say, 150 °C. The piston expands against the pressure of the atmosphere, and, as we have seen, does work. Suppose further - at the end of the power stroke, with the piston at its farthest extension – that the volume of the steam is around 2 m^3. If the temperature of the steam is quickly reduced to, say, 20 °C, the steam condenses to a volume more than one thousand times smaller, with the result that the space within the cylinder becomes very close to a vacuum – the pressure is almost zero. Since this pressure is very significantly less than the external atmosphere, *atmospheric pressure will push the piston back*. There is therefore no requirement for the 'surroundings' to do any active work at all to return the piston – the piston will return 'naturally' to its starting position, just like a weight 'naturally' falls to the ground under gravity. This important principle was first described by the French mathematician Denis Papin around 1690, and then used in a landmark steam engine designed and built by the English engineer Thomas Savery in 1698; a design that was subsequently improved by Thomas Newcomen in the 1710s.

A second improvement is to devise a way in which the cylinder itself can be kept hot throughout the whole cycle, so that there is no requirement to re-heat the cylinder from its cold condition prior to each power stroke, thereby using much less fuel. The solution here is a machine comprising two chambers. One chamber is the cylinder with which we are now familiar, a cylinder with a piston, and which can be maintained hot throughout the cycle. This cylinder is connected by a pipe to a second chamber, which is kept cold, so that, once the piston has been pushed in the power stroke to its furthest limit, the hot steam inside the cylinder passes in to the cold chamber, where it condenses back into water. This creates the internal vacuum, which causes the piston to be returned to its original position. This second, cold, chamber, is called the *condenser*, and its invention in 1765 by James Watt was a significant 'breakthrough' in steam engine design.

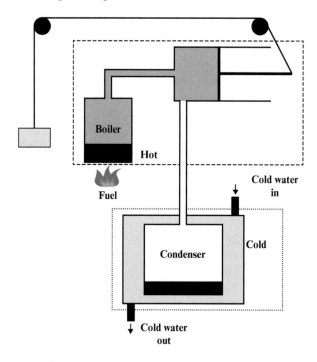

Figure 3.8 A representation of a steam engine with a condenser. This engine has two different, temperature-controlled, areas. The first, hot, area comprises a boiler, in which water is heated to produce pressurised steam, and a cylinder fitted with a piston. The second, cold, area comprises a water-cooled condenser, into which steam is drawn after the piston has reached its full extent. As the hot steam is drawn into the cold condenser, the steam condenses into water, creating a vacuum behind the piston, so enabling atmospheric pressure to return the piston to its original position. The diagram is a representation only, and does not show important components of a real engine, for example, the return of the water from the condenser to the boiler and the valves controlling the flows through the system.

Figure 3.8 shows a representation of a steam engine with a condenser. Water is boiled to produce steam, at a pressure substantially above that of the atmosphere. To produce this high-temperature, pressurised, steam, heat must be supplied by a source such as burning coal or oil. This heat source acts as a **hot reservoir** and it the primary source of energy. The pressurised stream expands, and during this power stroke, the motion of the piston is connected mechanically to some other device, so as to perform useful work, such as the lifting of a weight. At the end of the power stroke, the steam, now at a lower pressure, is removed from the cylinder, and passes into a condenser, this being a chamber which is kept cool so that the hot steam condenses to liquid water. As it does this, two things happen: firstly, in cooling and condensing, the steam loses heat to the condenser's surroundings, which act as a **cold reservoir**, and secondly, a near-vacuum is created within the engine, so that the piston is returned to its original position as a result of atmospheric pressure. The cycle can then start again, with the consumption of more fuel keeping the temperature of the hot reservoir hot, and with an appropriate means – such as exposure to the atmosphere, or a constant flow of cold water – to keep the temperature of the cold reservoir cold.

For each cycle:

- burning a suitable fuel creates a permanently hot reservoir ...
- ...which is used as a source of heat to boil water and to produce hot, pressurised, steam ...
- ...which in turn drives a piston, so producing useful work.
- The steam is then condensed, during which heat is extracted from the steam into the cold reservoir ...
- ...whilst also creating a vacuum, allowing the piston to return to its original position ...
- ...so allowing the cycle to start again.

Overall, heat flows from the hot reservoir to the cold reservoir, and, in so doing, work is done by the engine, which operates in a cycle for as long as the hot reservoir remains hot, and the cold reservoir cold, as illustrated in the schematic shown in Figure 3.9.

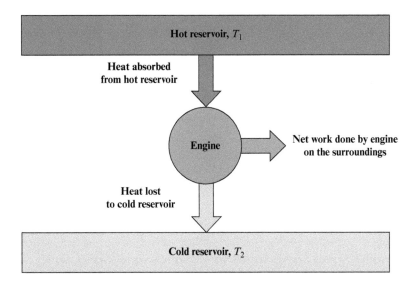

Figure 3.9 A schematic of a generalised 'heat engine'. The natural flow of heat from a hot reservoir to a cold reservoir is harnessed to perform useful work – just as the natural flow of water from a higher level to a lower level may be harnessed by a water-wheel. The two reservoirs are in principle infinitely large so that they maintain the same temperature despite (in the case of the hot reservoir) losing heat to the engine, and (in the case of the cold reservoir) gaining heat from the engine.

Figure 3.9 represents a generic schematic of any device which uses heat to perform work – a **heat engine**. In essence, a heat engine utilises the natural flow of heat from a hotter to a colder body, 'capturing' the heat as it flows, and converting that heat into useful work. This is, in principle, very similar to the way in which a water-wheel operates: as water flows naturally, under gravity, from a higher level to a lower level, the flow can be 'captured' by the paddles of a water-wheel, which, as it turns, can be used to perform useful work, such as the milling of flour.

All heat engines work on this principle, using different fuels to provide the heat, and using different mechanisms: so a steam engine uses wood, coal or oil to boil water to create high pressure steam which can be used to drive a piston, or to rotate a turbine; a petrol or diesel engine uses an ignited hydrocarbon to produce a high-temperature, high pressure mixture of carbon dioxide, steam and air to drive a piston or a turbine; a nuclear power plant uses the energy from nuclear fission to boil water. And all heat engines require a cold reservoir to act as the heat 'sink' – the condenser of a steam engine; for a motor car, the cooling system provided by water circulating past the engine and then exposed to the atmosphere at the car's radiator; for an electricity power station, by means of huge cooling towers. Despite the profound differences between a jet engine and the steam engines designed by pioneers such as Savery, Newcomen and Watt, fundamentally, all heat engines work on the same principle, as represented by Figure 3.9.

One last thought, for the moment, on heat engines. If you refer back to page 61, you'll see that the design of the steam engine, and its successors, resulted from the need to solve the problem of how to get the piston back to its original position. But there is a more fundamental issue – the fact that work is a path function. As we saw in Chapter 2, work is a path function, not a state function, and so work can only be performed as a system changes from one equilibrium state to another. Once the final equilibrium state has been achieved – as represented by the piston being in the position represented by peg 2 in Figure 3.6 – any work done in arriving at that equilibrium state is necessarily over.

That implies that, if we wish to have a machine that performs work continuously over a meaningful period of time – for example, if we want the petrol engine of our car to keep going for a few hours so we can travel from 'here' to 'there' – *then we have to find a way of maintaining a continuous, non-equilibrium, state in a controlled manner.* And the most effective way of doing that is to devise a way in which a series of equilibrium states can be connected together in a cycle, from one to another and back again, so that the path function between these states acts as a continuum. As an example, a typical internal combustion engine in a motor car will go through its cycle 100 times per second.

The requirement of the engine of our car to keep going for the entire journey we wish to carry out is just one example of the need for continuous work. So is the requirement to power industrial machinery, or to keep electricity flowing. And there is another need for doing work too – a need which is perhaps less obvious, but nonetheless very important – the need to maintain order, and to prevent complex structures from falling apart. This has particular significance for life, for all living things are complex, well-ordered structures – structures that can be maintained only by the continuous performance of work, as achieved by the controlled operation of a number of interlinked, cyclic, non-equilibrium processes. The theme of order is of especial significance – and although the steam engine might appear to be rather a distraction, it turns out that the study of the efficiency of steam engines unlocks one of the most important principles in science. But that is a story for a later chapter …

3.9 Temperature – a deeper look

3.9.1 Molecules in motion

Let's now take a deeper look at temperature. So far, we've defined temperature as an intensive state function specifying how hot or cold a system might be, and we have seen that heat flows down a temperature gradient. To obtain a deeper understanding of temperature, we need to take a molecular view, and although we can't see atoms and molecules directly, there is one, familiar, macroscopic phenomenon that provides an important clue – so let's consider what we see when we heat a saucepan of water. Let's suppose that the water is initially at room temperature. There isn't a lot to see: the water is there, in the saucepan, 'doing nothing'. Then, as the water becomes hotter, we begin to see movement and turbulence. The originally still water begins to tumble, bubbles form and rise to the surface, and above the surface we see steam rising. All these macroscopic observations are attributable to microscopic molecular activity, from which we can infer that as the temperature of the water rises, the molecules move progressively faster, more energetically, and more randomly. Suppose further that we hold the temperature of the water constant, at, say 85 °C, for a number of minutes. Even though the temperature of the water is stable, we will still see the water churning, with different parts of the water behaving differently: some parts are relatively still, some are moving faster in one direction, some in another. This suggests that different molecules have different energies, even if the temperature of the water as a whole is uniform.

To delve more deeply, we'll now return to our familiar system of a gas within a cylinder, and imagine what the molecules are actually doing. And for simplicity, let's assume that the molecule of the gas is a single atom, such as helium (He), rather than a multi-atom molecule such as oxygen (O_2), carbon dioxide (CO_2), ammonia (NH_3) or methane (CH_4). We also assume that the gas is ideal, so that there are no intermolecular interactions. This enables us to adopt the principle of 'linear additivity' (see page 8), so allowing us to express properties of the system as a whole in terms of linear sums of the corresponding properties of individual molecules.

We have already seen (page 17), that, at any temperature T, the molecules in the gas are moving – that's why a gas expands when the volume of the cylinder is increased. Each molecule, of mass m, therefore has a velocity of magnitude v, a momentum of magnitude mv, and a kinetic energy given by $½mv^2$. Since the gas is chemically homogeneous – all the molecules are chemically identical – each molecule has the same mass m. But do they all have the same velocity, and as a consequence, the same kinetic energy? They might. But in the light of our observation of the boiling water, it's more likely that some molecules are moving more slowly, and some more quickly. And if different molecules have different velocities – and therefore energies – how many molecules might there be with any particular velocity, or energy? This question is about the *distribution* of velocity and energy across the different molecules – out of the total number of molecules present, how many have [this] velocity and energy, and how many have [that]?

This question was asked, and answered, by the Austrian physicist, Ludwig Boltzmann, whose reasoning we summarise – and very much simplify! – here.

3.9.2 The Boltzmann distribution

Consider a macroscopic system of an ideal monatomic gas, in thermodynamic equilibrium at pressure P, temperature T, and of volume V. Microscopically, this system is formed from a total of N identical monatomic gaseous molecules, and let's suppose that one particular molecule i has an energy $\mathcal{E}_i$, and another molecule j has an energy $\mathcal{E}_j$. It's quite possible that several different molecules will all have the same energy, and so Boltzmann expressed the number $n(\mathcal{E}_i)$ of molecules that have energy $\mathcal{E}_i$ as

$$n(\mathcal{E}_i) = A e^{-\beta \mathcal{E}_i} \tag{3.13}$$

where e is 2.718 …, the base of natural logarithms, and, for a system of a fixed total number of molecules at a given temperature, A and β are constants, representing properties of the macroscopic system as a whole. The assumption Boltzmann is making here is that the distribution of energy is inversely exponential, with progressively fewer molecules having progressively higher energies – hence the name given to equation (3.13), the **Boltzmann distribution.**

3.9.3 A system's total energy $\mathcal{E}_T$, and the average molecular energy $\langle \mathcal{E} \rangle$

Equation (3.13) enables us to express the total energy $\mathcal{E}_n$ associated with the $n(\mathcal{E}_i)$ molecules, each of which has energy $\mathcal{E}_i$, as

$$\mathcal{E}_n = \mathcal{E}_i n(\mathcal{E}_i) = \mathcal{E}_i A e^{-\beta \mathcal{E}_i} = A \mathcal{E}_i e^{-\beta \mathcal{E}_i}$$

If we now assume that the lowest possible energy is $\mathcal{E}_{\text{low}}$, with a population $n(\mathcal{E}_{\text{low}})$, and that the highest possible energy is $\mathcal{E}_{\text{hi}}$, with a population $n(\mathcal{E}_{\text{hi}})$, we can then define the total energy $\mathcal{E}_T$ of the whole system of N molecules as

$$\mathcal{E}_T = \sum_{\mathcal{E}_{\text{low}}}^{\mathcal{E}_{\text{hi}}} \mathcal{E}_n = A \sum_{\mathcal{E}_{\text{low}}}^{\mathcal{E}_{\text{hi}}} \mathcal{E}_i e^{-\beta \mathcal{E}_i}$$

Furthermore, the total number N of molecules is given by

$$N = \sum_{\mathcal{E}_{\text{low}}}^{\mathcal{E}_{\text{hi}}} n(\mathcal{E}_i) = A \sum_{\mathcal{E}_{\text{low}}}^{\mathcal{E}_{\text{hi}}} e^{-\beta \mathcal{E}_i}$$

implying that the average molecular energy $\langle \mathcal{E} \rangle$ is

$$\langle \mathcal{E} \rangle = \frac{\mathcal{E}_T}{N} = \frac{A \sum_{\mathcal{E}_{\text{low}}}^{\mathcal{E}_{\text{hi}}} \mathcal{E}_i e^{-\beta \mathcal{E}_i}}{A \sum_{\mathcal{E}_{\text{low}}}^{\mathcal{E}_{\text{hi}}} e^{-\beta \mathcal{E}_i}} = \frac{\sum_{\mathcal{E}_{\text{low}}}^{\mathcal{E}_{\text{hi}}} \mathcal{E}_i e^{-\beta \mathcal{E}_i}}{\sum_{\mathcal{E}_{\text{low}}}^{\mathcal{E}_{\text{hi}}} e^{-\beta \mathcal{E}_i}} \tag{3.14}$$

3.9.4 Two simplifying assumptions

Quantum theory tells us that each of the energies $\mathcal{E}_i$ are discrete, collectively defining a series of discontinuous energy levels. This implies that if we wish to compute the average molecular energy $\langle \mathcal{E} \rangle$, we need to carry out the summations defined by equation (3.14), over each of the discrete energy levels from the lowest allowed energy level $\mathcal{E}_{\text{low}}$ to the highest $\mathcal{E}_{\text{hi}}$. This is true – but it is also beyond the scope of this book. So, for our purposes, we now make two simplifying assumptions:

- that the energy distribution is continuous, not discrete, implying that we may replace the summations by integrals
- that the lowest possible energy $\mathcal{E}_{\text{low}}$ is zero, and that the highest possible energy $\mathcal{E}_{\text{hi}}$ is infinite, implying that we may use zero and infinity as the limits for the integrals.

Under these assumptions, we can approximate the true value of the average molecular energy $\langle \mathcal{E} \rangle$ by an estimate, represented by the symbol $\langle \tilde{\mathcal{E}} \rangle$, which we can express as

$$\langle \tilde{\mathcal{E}} \rangle = \frac{\int_0^\infty \mathcal{E}\, e^{-\beta \mathcal{E}}\, d\mathcal{E}}{\int_0^\infty e^{-\beta \mathcal{E}}\, d\mathcal{E}}$$

Now

$$\int_0^\infty \mathcal{E}\, e^{-\beta \mathcal{E}}\, d\mathcal{E} = -\frac{1}{\beta} \int_0^\infty \mathcal{E}\, d(e^{-\beta \mathcal{E}}) = -\frac{1}{\beta} \left[\mathcal{E}\, e^{-\beta \mathcal{E}} \right]_0^\infty + \frac{1}{\beta} \int_0^\infty e^{-\beta \mathcal{E}}\, d\mathcal{E} = 0 + \frac{1}{\beta} \int_0^\infty e^{-\beta \mathcal{E}}\, d\mathcal{E}$$

Hence

$$\langle \tilde{\mathcal{E}} \rangle = \frac{1}{\beta} \frac{\int_0^\infty e^{-\beta \mathcal{E}}\, d\mathcal{E}}{\int_0^\infty e^{-\beta \mathcal{E}}\, d\mathcal{E}}$$

and so

$$\langle \tilde{\mathcal{E}} \rangle = \frac{1}{\beta}$$

3.9.5 The right answer, and Boltzmann's great insight

It turns out that, if this analysis is done properly, without making the simplifying assumptions made here, then the correct average molecular energy $\langle \mathcal{E} \rangle$ for a monatomic gas is given by

$$\langle \mathcal{E} \rangle = \frac{3}{2\beta} \tag{3.15}$$

What's important is not the numerical factor of 3/2, but the fact that both the correct value $\langle \mathcal{E} \rangle$ of the average molecular energy of the system, and the estimated value $\langle \tilde{\mathcal{E}} \rangle$, have a reciprocal relationship to the parameter β, which defines the energy distribution of the system according to the Boltzmann distribution, equation (3.13):

$$n(\mathcal{E}_i) = \mathcal{A}\, e^{-\beta \mathcal{E}_i} \tag{3.13}$$

TEMPERATURE AND HEAT

Boltzmann's great insight was to relate the parameter β to the system's temperature T as

$$\frac{1}{\beta} = k_B T \tag{3.16}$$

where k_B is a constant, now known as **Boltzmann's constant**, defined as the ratio of the ideal gas constant R to the Avogadro constant N_A

$$k_B = \frac{R}{N_A} \tag{3.17}$$

Since $R = 8.314$ J K^{-1}mol^{-1}, and $N_A = 6.022 \times 10^{23}$ mol^{-1}, $k_B = 1.381 \times 10^{-23}$ J K^{-1}.

Equation (3.13) can now be expressed as

$$n(\mathcal{E}_i) = \mathcal{A} \, e^{-\mathcal{E}_i/k_B T} \tag{3.18}$$

and equation (3.15) as

$$\langle \mathcal{E} \rangle = \frac{3}{2} k_B T \tag{3.19}$$

Furthermore, if the total number of molecules in the system is N, then, at any temperature T

$$N = \sum_i n(\mathcal{E}_i) = \int_0^\infty \mathcal{A} e^{-\mathcal{E}_i/k_B T} d\mathcal{E}_i = \mathcal{A} \int_0^\infty e^{-\mathcal{E}_i/k_B T} d\mathcal{E}_i = \mathcal{A} \left[-k_B T \, e^{-\mathcal{E}_i/k_B T} \right]_0^\infty = \mathcal{A} k_B T$$

implying that

$$\mathcal{A} = \frac{N}{k_B T}$$

Equations (13.13) and (13.18) for the Boltzmann distribution now become

$$n(\mathcal{E}_i) = \frac{N}{k_B T} e^{-\mathcal{E}_i/k_B T} \tag{3.20}$$

in which the factors $\mathcal{A}$ and β are replaced by the functions of temperature $N/k_B T$ and $1/k_B T$, respectively.

3.9.6 A deeper understanding of temperature

Equations (3.18), (3.19) and (3.20) give us a deeper understanding of temperature, taking us beyond "temperature is a measure of how hot something is". Equation (3.19) tells us that temperature is not just a measure of how hot something is, but, more fundamentally, a measure of the *average molecular energy of a system*. Furthermore, equation (3.19) tells us that the average molecular energy is a *linear* function of temperature. An important consequence of this is that a change in temperature is associated with a corresponding, proportionate, change in the average molecular energy, and, *vice versa*, that a change in the average molecular energy corresponds to a proportionate change in temperature. Furthermore, higher (or lower) temperatures imply, and are implied by, higher (or lower) average molecular energies. So, when a hotter system is placed thermal contact with a colder system it is this difference in average molecular energy – which we measure as a temperature difference, a temperature gradient – which drives the transfer of energy from the hotter system to the colder system, a process of energy transfer that we call heat. During this process, the hotter system loses energy, and the average molecular energy falls, causing the hotter body to cool and its temperature to fall; simultaneously, the colder system absorbs energy, causing the average molecular energy to rise, so becoming hotter as its temperature rises. After some time, the average molecular energies of both systems become the same, both systems have the same temperature, and the flow of energy from the hotter system to the colder one stops.

3.9.7 Visualising the Boltzmann distribution

The last paragraph focused on the average molecular energy $\langle \mathcal{E} \rangle$ of the system as defined by equation (3.19); if we now examine the Boltzmann distribution itself, equation (3.20)

$$n(\mathcal{E}_i) = \frac{N}{k_B T} e^{-\mathcal{E}_i / k_B T} \qquad (3.20)$$

we can gain a greater understanding of the distribution of energy across the molecules within the system. As we have seen, the molecules within a system each have energy, but not necessarily the same energy: and it is equation (3.20) that tells us how the energy is distributed.

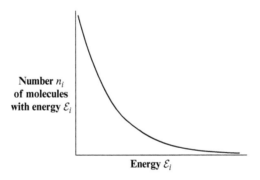

Figure 3.10 The Boltzmann distribution at a single temperature T. This graph shows the Boltzmann distribution for a system at a given temperature T, assuming that the allowed energy levels $\mathcal{E}_i$ are continuous, rather than discrete. The number of molecules n_i with a given energy $\mathcal{E}_i$ decreases exponentially, with many more molecules having lower energy than higher energy.

Figure 3.10 shows what that distribution looks like for a given temperature T: as can be seen, the number n_i of molecules with a particular energy $\mathcal{E}_i$ decreases exponentially as the energy $\mathcal{E}_i$ increases – there are more molecules with relatively lower energies than higher energies. But what happens at a different temperature? How does the distribution of energy across the molecules in the system change as the temperature becomes higher or lower?

These questions are answered in Figure 3.11 – as the temperature increases, progressively more molecules have higher energies. But since the total number of molecules remains constant, as the populations of the higher energy levels increases, the population of the lower

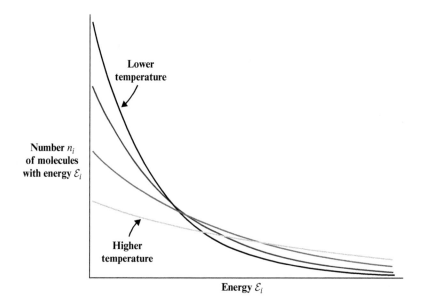

Figure 3.11 The Boltzmann distribution at different temperatures. As the temperature T increases for a system containing a fixed number of molecules N, the number of molecules n_i with higher energies $\mathcal{E}_i$ increases whilst the number of molecules n_i with lower energies $\mathcal{E}_i$ decreases.

energy states must fall, implying that as the temperature increases, the energy distribution becomes progressively 'flatter'. Ultimately, as the temperature becomes infinite, all the allowed energy levels will become equally populated; conversely, as the temperature tends towards absolute zero, all the molecules tend towards the lowest allowed energy level.

3.9.8 A paradox resolved

The distinction between the *average molecular energy* and the *total energy of the system* also explains what is often perceived to be a paradox – the fact that a small system can have a high temperature and low energy, and that a big system can have a huge amount of energy, but a low temperature. It's easy to think that "hot systems have a lot of energy" and "cold systems have low energy" for both these statements seem to make intuitive sense. Indeed they might, but both these statements are misleading, if not downright false. The fallacy lies within the mapping of high (or low) temperature to high (or low) energy, and the fallacy is readily resolved by recognising the (important) distinction between on the one hand, the average molecular energy, and, on the other, the total energy of the system.

The temperature of a system – how hot or cold it is – is a measure of the *average molecular energy*, and so is independent of how big, or small, the system is. Temperature, remember, is an *intensive* state function. The total energy of a system, in contrast, does depend on the size and scale of the system, and the bigger a system is, the more energy it has – energy is an *extensive* state function. So a big system will have a high total energy, simply because it is big – yet at the same time it might have a low average molecular energy, and hence a low temperature. Conversely, a very small system may not have that much energy, but if the average molecular energy is high, it can be very hot indeed.

3.9.9 Can temperatures be negative?

We conclude this discussion of the molecular interpretation of temperature with a brief discussion of a very strange question: is it possible for a temperature to be negative? Given that we have already seen (page 47) that absolute zero is a temperature that can be approached, but never reached, the idea of a negative temperature – a temperature even lower than absolute zero, seems very odd. As indeed it is when temperature is interpreted in terms of 'hotness' and 'coldness' – how can something be even colder than 'absolute' zero? But when temperature is interpreted in terms of the Boltzmann distribution, a rather different insight emerges.

To explain this, consider a system comprising a total of N molecules at a temperature T around 300 K, the temperature of a very pleasant summer's day. According to the Boltzmann distribution, energy is distributed within this system such that the number $n(\mathcal{E}_i)$ of molecules with energy $\mathcal{E}_i$ is given by

$$n(\mathcal{E}_i) = \frac{N}{k_B T} e^{-\mathcal{E}_i/k_B T} \tag{3.20}$$

Let us now assume, for simplicity, that only two energy states are available: a low energy $\mathcal{E}_{low}$ state associated with $n(\mathcal{E}_{low})$ molecules, and a high energy $\mathcal{E}_{hi}$ state associated with $n(\mathcal{E}_{hi})$ molecules. Using equation (3.20), at any temperature T

$$\frac{n(\mathcal{E}_{hi})}{n(\mathcal{E}_{low})} = \frac{e^{-\mathcal{E}_{hi}/k_B T}}{e^{-\mathcal{E}_{low}/k_B T}} = e^{-(\mathcal{E}_{hi}-\mathcal{E}_{low})/k_B T} \tag{3.21}$$

Since $\mathcal{E}_{hi} > \mathcal{E}_{low}$, then $-(\mathcal{E}_{hi} - \mathcal{E}_{low})/k_B T$ in equation (3.21) is a negative number, implying that $n(\mathcal{E}_{hi}) < n(\mathcal{E}_{low})$. There are therefore fewer molecules with higher energy than lower energy, as we would expect.

But suppose now that, mathematically, the parameter T in equations (3.19) and (3.20) happens to be a *negative* number. $\mathcal{E}_{hi}$ is still greater than $\mathcal{E}_{low}$ implying that $(\mathcal{E}_{hi} - \mathcal{E}_{low})$ is still positive. But because we are exploring what might happen if T is negative, the exponential term $-(\mathcal{E}_{hi} - \mathcal{E}_{low})/k_B T$ in equation (3.21) is now positive. As a consequence $n(\mathcal{E}_{hi}) > n(\mathcal{E}_{low})$, implying that there are now more molecules with higher energy than lower energy.

Under 'normal' conditions, and for almost every scientific system on the planet, the Boltzmann distribution applies in the 'usual' way, such that the population within the system associated with any higher energy is smaller than the population within the system associated with any lower energy – as in the simple two energy level case in which we demonstrated that $n(\mathcal{E}_{hi}) < n(\mathcal{E}_{low})$. It so happens, however, that it is now possible to create transient, non-equilibrium, systems in which a population associated with a higher energy is *greater* than the population associated with a lower energy, $n(\mathcal{E}_{hi}) > n(\mathcal{E}_{low})$. This is known as 'population inversion', and we all benefit from it, for the operation of all lasers – lasers such as those in laser printers, DVD players, and bar-code readers (to name only just a few practical applications of laser technology) - relies on the prior creation of a population inversion of the electrons across the various allowed energy levels within suitably excitable atoms: lasers work by exploiting a system in which there are a greater number of electrons at higher energy levels than lower ones. If we wish to describe a system associated with population inversion using Boltzmann's equation, then we need to invoke a negative temperature to do so. In this sense, negative temperatures do indeed 'exist', but as a mathematical abstraction, rather than as associated with 'hotness' or 'coldness'.

EXERCISES

1. The approximate composition of the atmosphere is such that 80% of the volume is nitrogen, 19% is oxygen, and 1% represents other gases. If the atmospheric pressure is 1.01325×10^5 Pa, calculate the partial pressures attributable to nitrogen, oxygen and other gases, assuming that all the component gases, and the atmosphere as a whole, behave as ideal gases.
2. Find a friend whom you trust – and who is patient! Sit down with your friend, and, over a cup of tea or coffee, explain the concepts of 'temperature', 'heat' and 'energy', so that your friend really understands them, and the difference between them.
3. Write down clear, complete, and precise definitions of:
    - open system
    - closed system
    - isolated system
    - isothermal change
    - adiabatic change
    - isobaric change
    - isochoric change.
4. During much of the late eighteenth, and early nineteenth, centuries, most scientists believed that heat was attributable to the presence of a weightless substance they called 'caloric' – the more caloric a body has, the hotter it is, and as a hot body cools down and a cold body warms up, caloric is transferred from the hotter to the colder body. Indeed, if you had done an examination in science around, say, 1810, you could well have had an exam question such as "explain the caloric theory". So, pretend for a moment that it is indeed 1810, and think about these questions:
    - Why do you think scientists believed caloric was weightless? What experiments would you do to measure the weight of caloric to verify this? And how are you going to explain that, quite often, a cold body weighs more than that same body at a hotter temperature?
    - What experimental evidence is there that can successfully be interpreted using the caloric theory?
    - With the benefit of the knowledge we now have in the 21st Century, what evidence would you use to convince an early 19th Century scientist that the caloric theory really isn't a good explanation of heat?
5. "Bodies that are big contain a lot of energy and so they must be hot." Is this statement always true, always false, sometimes true, or sometimes false? Why?
6. If you have access to a spreadsheet, here is an exercise on the Boltzmann distribution, equation (3.20)

$$n(\mathcal{E}_i) = \frac{N}{k_B T} e^{-\mathcal{E}_i/k_B T} \tag{3.20}$$

resulting in graphs of the form shown in Figure 3.11. Using the true numerical value of k_B (1.381×10^{-23} J K^{-1}), and a realistic value of N ($\sim 10^{23}$ or 10^{24}), makes the spreadsheet rather messy, so this exercise explores the functional behaviour of equation (3.20) rather than realistic numerical values. Accordingly, define a cell setting $k_B = 1$ J K^{-1}, and another cell $N = 10^6$. Then set up a column for values of $\mathcal{E}_i$ from 5 J to 500 J in increments of 5 J, corresponding to 599 rows. Set up successive columns for six values of $T = 150$ K, 200 K, 250 K, 300 K, 350 K, and 400 K. In each cell, write a formula for $n(\mathcal{E}_i)$, equation (3.20). Create a line graph to show how $n(\mathcal{E}_i)$ varies with $\mathcal{E}_i$ for a given value of T.

4 Thermodynamics and mathematics

 Summary

Thermodynamics enables us to answer questions such as "if the pressure P of a given mass of an ideal gas changes, by how much does the temperature T change?". To do this, we need to have an appropriate mathematical expression which connects P and T, as well as a method for exploring changes. For an ideal gas, the mathematical expression is the ideal gas law

$$PV = nRT \tag{3.3}$$

and the method for exploring changes is differential calculus.

In applying differential calculus to equation (3.3), however, we need to recognise that changing the temperature T of the system is not the only way in which the pressure P can change: the pressure P can also change if the volume V changes. Any observed change in the pressure P can therefore be attributable solely to a change in the temperature T provided that the volume V is held constant; or to a change in the volume V if the temperature T is held constant; or to a simultaneous change in both the temperature T and the volume V.

Mathematically, the fact that the pressure P of a fixed mass of an ideal gas depends on both the temperature T and the volume V is represented by expressing the pressure as a function of two variables, $P(T, V)$. For changes, an infinitesimal change dP is expressed in terms of infinitesimal changes in temperature dP and volume dV as

$$dP = \left(\frac{\partial P}{\partial T}\right)_V dT + \left(\frac{\partial P}{\partial V}\right)_T dV \tag{4.10}$$

where $(\partial P/\partial T)_V$ and $(\partial P/\partial V)_T$ are known as **partial derivatives**. So, for example, $(\partial P/\partial T)_V$ is the result of using the usual rules of differential calculus to differentiate $P(T, V)$ with respect to T assuming that V is constant.

Very many of the important variables X in thermodynamics are functions of more than one variable $X(Y, Z, \ldots)$, and so equations of the form of equation (4.10) appear frequently.

4.1 What this chapter is about

As with most branches of science and engineering, confidence in the appropriate mathematical tools is very valuable in enriching understanding. A important feature of thermodynamics is that all the state functions – pressure, volume, temperature and the like – are interrelated such that the determination of the value of any one state function (say, pressure) for a system of a given mass requires knowledge of the simultaneous values of two other variables (say, both volume and temperature), as is evident from the ideal gas law

$$PV = nRT \tag{3.3}$$

which can also be written as

$$P = \frac{nRT}{V} \tag{3.3a}$$

Mathematically, pressure (for example) is known as a 'function of more than one variable', and so the purpose of this section is therefore to present those aspects of mathematics which will underpin much of the theory presented in this book. If the behaviour of functions of more than one variable, and the concept of partial derivatives, are familiar, this chapter may be skipped, or kept as reference; if they are less familiar, it is important that they become so. Mathematics is sometimes intimidating: we trust that this chapter will be intelligible, and quite straight-forward – the only pre-requisite is confidence in basic algebra, and some knowledge of differentiation.

4.2 Functions of more than one variable

Given an equation such as

$$y = 5x^3 \tag{4.1}$$

if we know a particular value of the variable x – say, 4 – then we can compute the corresponding variable y as $5 \times 4^3 = 320$. Since the value of y depends on the value of x (and only the value of x), y is said to be a **function** of x, and to make that explicit, y is often written as $y(x)$ or $y = f(x)$.

Knowledge of how $y(x)$ depends on x, for example, equation (4.1)

$$y = 5x^3 \tag{4.1}$$

enables us to express the derivative (more formally, the **total derivative**) dy/dx as

$$\frac{dy}{dx} = 15x^2 \tag{4.2}$$

As can be seen from equation (4.2), dy/dx is itself a function of x, but a function, $15x^2$, different from the original function $y(x) = 5x^3$; graphically, dy/dx represents the slope of the tangent to the graph of $y(x)$ at any point x; physically, dy/dx tells us by how much $y(x)$ changes – that's dy – for a small change dx in x, as illustrated in Figure 4.1.

The quantity dy, known as a **total differential**, may be expressed in two different ways:

$$dy = 15x^2 \, dx \tag{4.3}$$

is one alternative, which is specific to equation (4.1), and

$$dy = \left(\frac{dy}{dx}\right) dx \tag{4.4}$$

is another, more general, representation.

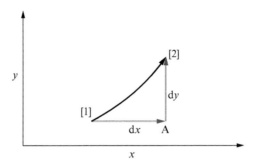

Figure 4.1 A function of one variable. y is a function y(x) of a single variable x. The system is initially in state [1] ≡ [x, y], and undergoes a change in state to state [2] ≡ [x + dx, y + dy], where dx is the distance as measured along the x axis from state [1] to point A, and dy is the distance as measured along the y axis from point A to state [2]. The smaller the distance dx, the smaller the distance dy, and the more closely the area defined by state [1], point A and state 2 becomes a right-angled triangle. As this area approaches a right-angled triangle, the more closely does the distance from point A to state [2], dy, become equal to the distance form state [1] to point A, dx, multiplied by the tangent of the angle at point [1]. Mathematically, the tangent of this angle is dy/dx, as evaluated for the value of x corresponding to state [1], which we may write as [dy/dx]₁. Hence dy = [dy/dx]₁ dx.

As a contrast to equation (4.1)

$$y = 5x^3 \tag{4.1}$$

consider the ideal gas law, equation (3.3)

$$PV = nRT \tag{3.3}$$

written as

$$P = \frac{nRT}{V} \tag{3.3a}$$

This very familiar equation expresses the pressure P of n mol of an ideal gas in terms of the corresponding temperature T, volume V, and the ideal gas constant R. R, the ideal gas constant, is indeed a constant, of value 8.314 J K^{-1} mol^{-1}, but n, T and V can all vary: for any particular ideal gas, the actual pressure P depends on the simultaneous values of n, T and V. The pressure P is therefore not a function of one variable, like equation (4.1), but a function of three variables, n, T and V. To make that explicit, we can express P as $P(n, T, V)$.

4.3 Partial derivatives

The fact that $P = P(n, T, V)$ is a function of three variables implies that if any one of n, T or V changes – or any pair, or all three – then the value of P will change too. For a function of one variable, such as equation (4.1)

$$y = 5x^3 \tag{4.1}$$

any change dy in y is totally attributable to a change dx in x; furthermore, as we saw on page 77, given knowledge of the appropriate function – in this case, equation (4.1) – we can compute dy/dx, and hence determine dy for any particular dx using equation (4.4)

PARTIAL DERIVATIVES

$$dy = \left(\frac{dy}{dx}\right) dx \tag{4.4}$$

For a function of three variables, such as $P(n, T, V)$, a change dP in P can result from a change dn in n, or a change dT in T, or a change dV in V, either individually or in any combination. So, for the ideal gas law, expressed in the form of equation (3.3a)

$$P = \frac{nRT}{V} \tag{3.3a}$$

the change dP is expressed in terms of the changes dn, dT and dV as

$$dP = \left(\frac{\partial P}{\partial n}\right)_{V,T} dn + \left(\frac{\partial P}{\partial T}\right)_{n,V} dT + \left(\frac{\partial P}{\partial V}\right)_{n,T} dV \tag{4.5}$$

Equations (4.4) and (4.5) are doing the same job, and are equivalent. The difference in form is attributable to the fact that, for equation (4.4), the function $y(x)$ is a function of only one variable x, and so any change dy is totally attributable to a change dx in a single variable x; equation to the (4.5) recognises that $P(n, T, V)$ is a function of three variables, n, T and V, and that any change dP is attributable to changes dn, dT and dV in each of these variables.

Equation (4.5) is important, and equations of this form will appear frequently in this book. Two features of equation (4.5) are of particular significance. Firstly, equation (4.5) expresses the change dP in terms of a linear sum of three elements, each element representing the effect of a change in each of dn, dT and dV. Secondly, the coefficients associated with each of dn, dT and dV – such as $(\partial P/\partial T)_{n,V}$, as associated with the term dT – are known as **partial derivatives**, and each represents the derivative of $P(n, T, V)$ with respect to any one variable n, T or V with the other two variables held constant.

To determine these partial derivatives in any particular circumstances, we need to know the equation-of-state that is appropriate to the system of interest. So, for an ideal gas, we may use the ideal gas law, equation (3.3a)

$$P = \frac{nRT}{V} \tag{3.3a}$$

If we differentiate P with respect to n, holding V and T constant, since R is a constant always, we obtain

$$\left(\frac{\partial P}{\partial n}\right)_{V,T} = \frac{RT}{V} \tag{4.6}$$

This quantity refers to conditions in which the volume V and temperature T are held constant, and represents the incremental change $\partial P_{V,T}$ in the pressure of the gas, resulting from an incremental change $\partial n_{V,T}$ in the mole number n.

Likewise, differentiating with respect to T, holding n and V constant, gives

$$\left(\frac{\partial P}{\partial T}\right)_{n,V} = \frac{nR}{V} \tag{4.7}$$

which represents the incremental change $\partial P_{n,V}$ in the pressure of the gas, resulting from an incremental change $\partial T_{n,V}$ in the temperature T of the gas, under conditions of constant mole number n and constant volume V.

Finally, differentiating with respect to V, holding n and T constant, we have

$$\left(\frac{\partial P}{\partial V}\right)_{n,T} = -\frac{nRT}{V^2} \tag{4.8}$$

which represents the incremental change $\partial P_{n,T}$ in the pressure of the gas, resulting from an incremental change $\partial V_{n,T}$ in the volume V of the gas, under conditions of constant mole number n and constant temperature T.

Substituting equations (4.6), (4.7) and (4.8) in equation (4.5) gives

$$dP = \left(\frac{RT}{V}\right) dn + \left(\frac{nR}{V}\right) dT - \left(\frac{nRT}{V^2}\right) dV \tag{4.9}$$

which defines the change dP in pressure resulting from simultaneous changes dn, dT and dV, for an ideal gas obeying the equation-of-state

$$P = \frac{nRT}{V} \tag{3.3a}$$

If the system obeys a different equation-of-state, for example, Berthelot's equation, equation (3.7)

$$\left(P + \frac{an^2}{TV^2}\right)(V - nb) = nRT \tag{3.7}$$

then equation (4.5) still holds, but the partial derivatives will be different from those shown in equations (4.6), (4.7) and (4.8).

An important, and reassuring, aspect of equation (4.9), concerns what happens if we apply equation (4.9) to conditions of, say, constant mole number n and constant temperature T – implying that dn and dT are both zero. Under these circumstances, equation (4.9) becomes

$$dP = -\left(\frac{nRT}{V^2}\right) dV \qquad \text{under conditions of constant } n \text{ and } T$$

which can be expressed differently, and more succinctly, as

$$\left(\frac{\partial P}{\partial V}\right)_{n,T} = -\frac{nRT}{V^2}$$

Reassuringly, this is identical to equation (4.8).

4.4 Systems of constant mass

By definition, as we saw on page 57, a closed thermodynamic system is one in which no matter may cross the system boundary. Closed systems are therefore necessarily of constant mass, implying that dn is always zero. So, in the case of an ideal gas, for a closed system of n mol, the ideal gas equation (3.3a)

$$P = \frac{nRT}{V} \tag{3.3a}$$

is still true, but n is now a constant. Given the fixed mass, the pressure P becomes a function not of three variables, but only of two, the temperature T and the volume V, enabling us to express P as $P(T, V)$. As a consequence, equation (4.5) becomes

$$dP = \left(\frac{\partial P}{\partial T}\right)_V dT + \left(\frac{\partial P}{\partial V}\right)_T dV \tag{4.10}$$

Another way of thinking about a system of constant mass is to express equation (3.3a) in terms of the molar volume $\mathbf{V} = V/n$

$$P = \frac{RT}{\mathbf{V}} \tag{3.4a}$$

Equation (3.4a) is necessarily of constant mass – the mass of 1 mol of the ideal gas under study – and so equation (4.10) takes the form

$$dP = \left(\frac{\partial P}{\partial T}\right)_{\mathbf{V}} dT + \left(\frac{\partial P}{\partial \mathbf{V}}\right)_T d\mathbf{V} \tag{4.11}$$

In equation (3.4a), all the three variables – P, T and $\mathbf{V}$ – are state functions, and any one can be expressed in terms of the other two. As we saw on page 48, this is one instance of the general principle that

> **In the absence of gravitation, motion, electric, magnetic and surface effects, a fixed mass of a pure substance has two, and only two, independent state functions.**

This principle is general, and, as already mentioned, is of great importance. For a given mass of any pure substance – not just an ideal gas – this principle states that, if we know the values of any two state functions, for example the pressure P and temperature T, then *all other state functions are determined*, and can be calculated if we know the appropriate equations. The principle also enables us, for any system of constant mass, to express *any state function* – say the state function represented by the symbol X – *as a function of two, and only two, other state functions*, say, the temperature T and the pressure P, $X(T, P)$. As a consequence, we may express an infinitesimal change dX in X as

$$dX = \left(\frac{\partial X}{\partial T}\right)_P dT + \left(\frac{\partial X}{\partial P}\right)_T dP \tag{4.12}$$

In the next few chapters, we will introduce a number of state functions – in particular, the internal energy U, the enthalpy H, the entropy S, the Gibbs free energy G, and the Helmholtz free energy A, each of which can be represented, for a closed system of constant mass, as a function of only two other state functions, such as the temperature T and the pressure P. As an example, the enthalpy H can be represented as a function $H(T, P)$, and so

$$dH = \left(\frac{\partial H}{\partial T}\right)_P dT + \left(\frac{\partial H}{\partial P}\right)_T dP \tag{4.13}$$

We shall meet this equation on page 151, and many others of a similar form throughout this book.

4.5 Partial derivatives and state functions

As we have seen, for a system of constant mass, we may express the pressure P as function $P(T, V)$ of the temperature T and volume V, so that an infinitesimal change dP can be written as

$$dP = \left(\frac{\partial P}{\partial T}\right)_V dT + \left(\frac{\partial P}{\partial V}\right)_T dV \tag{4.10}$$

Another way of interpreting this equation, and of interpreting the partial derivatives, is in the context that the pressure P, volume V and temperature T are all state functions. And an important feature of all state functions, as we saw on page 12, is that the change in any state function depends only on the initial and final states of the system, and is independent of the path – a feature which is true for finite changes, and also for infinitesimal ones too.

So consider a system of a fixed mass of a gas, which does not have to be ideal, initially in state $[1] \equiv [P, V, T]$, changing to a final state $[2] \equiv [P + dP, V + dV, T + dT]$. As a result of this change of state, the pressure has changed by dP, whilst, simultaneously, the volume has changed by dV, and the temperature has changed by dT. These changes have all taken place together, but the state function concept tells us that the overall change from state $[1] \equiv [P, V, T]$ to state $[2] \equiv [P + dP, V + dV, T + dT]$, and the corresponding changes dP, dV, and dT, are the same for all paths beginning at state $[1]$ and ending at state $[2]$. One such path is the direct path, in which the pressure, volume and temperature change place simultaneously; another is a path that takes place in two steps

- Firstly, a change from state $[1] \equiv [P, V, T]$ such that the system's temperature is increased by dT at constant volume V. As a consequence, the system's pressure increases by dP_1 so that the system achieves an intermediate state $[3] \equiv [P + dP_1, V, T + dT]$.
- Secondly, a change from state $[3] \equiv [P + dP_1, V, T + dT]$ such that the system's volume is allowed to increase by dV at the now higher temperature $T + dT$. The change in volume dV at constant temperature $T + dT$ causes a second, different, change in pressure dP_2, and so the system achieves a final state $= [P + dP_1 + dP_2, V + dV, T + dT]$. If this final state is in fact state $[2] \equiv [P + dP, V + dV, T + dT]$, then $dP_1 + dP_2 = dP$. For any gas, an increase in volume results in a decrease in pressure, and so, for the change from state $[3]$ to state $[2]$, the quantity dP_2 is negative.

These changes are all shown in Figure 4.2, which is a three-dimensional version of Figure 4.1.

The direct change from state $[1] \equiv [P, V, T]$, to the final state $[2] \equiv [P+dP, V+dV, T+dT]$ is associated with changes dP, dV, and dT in each of the state functions. We now examine in detail the indirect change through path $[3]$.

The first step, from state $[1] \equiv [P, V, T]$ to state $[3] \equiv [P + dP_1, V, T + dT]$ is at constant volume; furthermore, since the second step from state $[3] \equiv [P + dP_1, V, T + dT]$ to state $[2] \equiv [P + dP, V + dV, T + dT]$ is at constant temperature, the change in temperature dT for the first step from state $[1]$ to state $[3]$ is determined by the overall change from state $[1]$ to state $[2]$.

PARTIAL DERIVATIVES AND STATE FUNCTIONS

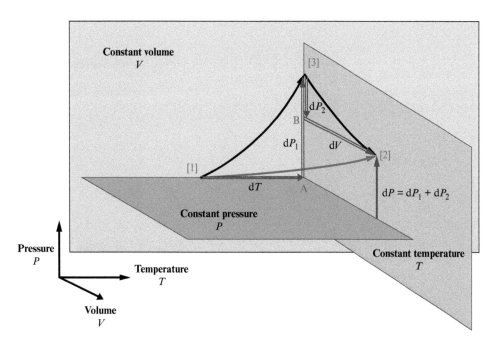

Figure 4.2 A function of two variables. The pressure P of a closed system of a gas is a function $P(T, V)$ of the temperature T and volume V. The system changes from state [1] to state [2] along two different paths: by the direct path, and alternatively by a path through state [3], comprising a change at constant volume from state [1] to state [3], followed by a change at constant temperature from state [3] to state [2]. The overall changes dP, dV, and dT in each of the state functions is the same for both paths.

At constant volume, the change dP_1 in pressure attributable to a change dT in the system's temperature is given mathematically by

$$dP_1 = \left(\frac{\partial P}{\partial T}\right)_V dT$$

In terms of Figure 4.2, the change dT at constant volume is represented as the arrow from state [1] to the point shown as A; simultaneously, if we assume that the path is quasistatic, and that all the state functions are well-defined throughout the change (see page 32), the system follows the path shown by the black arrow from state [1] to state [3], which lies totally within the plane of constant volume. The change dP_1 is equal to the distance from point A to state [3], which, for small changes is given by the tangent of the angle of the heavy black line, multiplied by dT – which exactly what $(\partial P/\partial T)_V \, dT$ represents.

The change from state [3] to state [2] takes place at constant temperature – the now higher temperature of $T + dT$. The system is allowed to expand by dV, as a result of which the system pressure drops by an amount dP_2. The system follows the path shown by the heavy arrow, within the plane of constant temperature. Mathematically

$$dP_2 = \left(\frac{\partial P}{\partial V}\right)_T dV$$

whilst geometrically, dP_1 is equal to the distance from state [3] to point B, which, for small changes is given by the tangent of the angle of the heavy black line, multiplied by dV – which exactly what $(\partial P/\partial V)_T \, dP$ represents.

For the overall change from state [1] to state [2]

$$dP = dP_1 + dP_2 = \left(\frac{\partial P}{\partial T}\right)_V dT + \left(\frac{\partial P}{\partial V}\right)_T dV \tag{4.14}$$

so verifying equation (4.10), and also showing how the state function concept, and the partial derivatives, are linked.

4.6 A look ahead...

4.6.1 Introducing some new state functions

Our understanding of state functions and path functions, and of heat, work and temperature, as discussed in these first three chapters, sets the scene for the rest of this book, so it is appropriate to pause for a moment and provide an introduction to the chapters that follow.

Thermodynamics is about the behaviour of real systems. That behaviour is described primarily by the system's state functions, some of which are already familiar – notably the mole number n, mass M, volume V, pressure P and temperature T. Future chapters will introduce new state functions, and explore their behaviours

- internal energy, U – Chapter 5
- enthalpy, H – Chapter 6
- entropy S – chapters 8, 9, 10, 11 and 12
- Gibbs free energy G – Chapters 13 and 14
- Helmholtz free energy A – Chapter 13.

Embedded within these chapters is much 'real' chemistry, for example

- thermochemistry, and the study of the heat associated with chemical reactions, is a key feature of Chapter 6
- Chapter 7 deals with what happens in systems of ideal gases as a result of changes in, for example, volume, pressure and temperature
- Chapter 13 explains why some chemical reactions are spontaneous, and others are not...
- ...leading to an analysis of chemical equilibrium on Chapter 14...
- ...and a thermodynamic description of a range of phenomena, such as osmosis and electrochemical cells, as described in Chapters 15 to 20.
- Furthermore, Chapters 23, 24, 25 and 26 show how the principles of thermodynamics can be applied to biochemical systems and processes, such as bioenergetics and the folding of proteins.

4.6.2 Pure substances and mixtures

Almost all of the analysis in the following chapters concerns a closed system comprising a fixed mass of a single pure substance. We shall explain why there is a focus on closed systems in the next sub-section; here we explain the significance of the system composed of a single pure substance.

Much of chemistry is not about pure substances, but about mixtures: all chemical reactions necessarily involve a variety of reactants that transform into a variety products, all of which are usually mixed together as the reaction takes place. A 'system' can of course refer to a mixture, and any such system will have values for all the state functions, mass M, volume V, enthalpy H, and all the rest, so why is much of our analysis restricted to systems of a single pure substance?

The answer to this question is based on the concept of the ideal mixture, which, as we saw on page 8, is a mixture in which all the extensive state functions are linearly additive, so that the value of any extensive state function X of a system comprising n_A mol of pure component A and n_B mol of pure component B is given by

$$X = n_A \mathbf{X}_A + n_B \mathbf{X}_B$$

in which $\mathbf{X}_A$ and $\mathbf{X}_B$ are the molar values of the state function X for the pure substances A and B, respectively. If we therefore understand how the state functions of pure substances behave, we can then apply this understanding easily to ideal mixtures, hence the focus throughout much of this book on pure substances.

In reality, many mixtures do not conform to ideal behaviour, and so the way in which the ideal is modified to describe, as far as possible, the real, is discussed in Chapter 22 (see pages 655 to 670).

4.6.3 Closed systems of fixed mass

As we saw on page 57, a closed system, by definition, is one in which the system boundary allows the exchange of energy, in the form of heat and work, between the system and the surroundings, but prevents the exchange of matter. Accordingly, the mass M of a closed system is fixed, as is the mole number n, and as we saw on page 81, if the mole number n of a system is fixed, then all extensive functions behave mathematically as intensive state functions. Given the principle that

> In the absence of gravitation, motion, electric, magnetic and surface effects, a fixed mass of a pure substance has two, and only two, independent state functions.

then, for a system of fixed mass, *any* state function X of a system of fixed mass can be expressed as a function $X(Y, Z)$ of *any* other two state functions Y and Z; furthermore, we can have confidence that the function $X(Y, Z)$ is a *complete* description of the system, from which all other state functions can be derived, provided that we know the appropriate equations.

As an example, consider the state function, the enthalpy H. It is likely that this state function is unfamiliar, and it will be introduced in Chapter 5. For the moment, however, what, precisely, the enthalpy H actually is does not matter; what does matter is that H is indeed a state function (as will be proven on page 149), and so, for a closed system of constant mass, we may express H as a function of any two other state functions (remembering that, for a system of constant mass, all state functions behave mathematically as intensive state functions). There are several other state functions from which we may choose, so let us select the temperature T and pressure P.

THERMODYNAMICS AND MATHEMATICS

Accordingly, we may express H as $H(T, P)$, and a change dH may be written as equation (4.13)

$$dH = \left(\frac{\partial H}{\partial T}\right)_P dT + \left(\frac{\partial H}{\partial P}\right)_T dP \qquad (4.13)$$

Alternatively, we could select the temperature T and volume V, in which case $H = H(T, V)$, and so dH can be written as

$$dH = \left(\frac{\partial H}{\partial T}\right)_V dT + \left(\frac{\partial H}{\partial V}\right)_T dV \qquad (4.15)$$

A third choice is the combination pressure P and volume V, for which $H = H(P, V)$, and so dH becomes

$$dH = \left(\frac{\partial H}{\partial P}\right)_V dP + \left(\frac{\partial H}{\partial V}\right)_P dV \qquad (4.16)$$

Equations (4.13), (4.15) and (4.16) are all different, and the one which is chosen in any particular situation depends on which pair of the three variables temperature T, pressure P and volume V is of relevance. In general, we can easily control, and change, the temperature at which a chemical process takes place; also, it is usually easier to change the pressure of a system than its volume. Temperature and pressure are therefore the most frequently used variables, and so expressions of the form of equation (4.13) are more widely-used than those of the forms of equations (4.15) and (4.16); that said, as we shall see on various occasions in this book, there are circumstances in which alternative choices are more appropriate.

4.6.4 The importance of partial derivatives

To gain a complete understanding of any system, we need to know the values of all the state functions under any given conditions. Even better is to know, in addition, how the state functions change when the conditions change, for this enables us to predict: if we know the value of, say, the enthalpy H_1 of a system at any particular temperature T_1, and also how the enthalpy changes as the temperature changes, then we can then predict what the enthalpy H_2 of the system will be if the temperature changes to T_2. If the enthalpy H were a function of only the single variable, the temperature T, then this requires us to know the total derivative dH/dT, for this tells us, mathematically, the value of the change dH in the enthalpy H that results from a given change dT in the temperature, for any given system.

We know, however, that, as a general principle

> In the absence of gravitation, motion, electric, magnetic and surface effects, a fixed mass of a pure substance has two, and only two, independent state functions.

and so, for a system of constant mass, the enthalpy H will be a function not of a single variable, the temperature T, but of two variables – say the temperature T and the pressure P – in which case we can express H as $H = H(T, P)$.

The question "what is the change dH in the enthalpy H of a system of constant mass resulting from a change dT in the system's temperature T?" therefore has a more complex answer than might appear at first sight – for the answer depends on whether the change dT in temperature takes place whilst the pressure P is constant, or whether both the pressure *and* the temperature change simultaneously.

For a system of constant mass, since $H = H(T, P)$, then, as we saw on page 86, we can express a change dH as

$$dH = \left(\frac{\partial H}{\partial T}\right)_P dT + \left(\frac{\partial H}{\partial P}\right)_T dP \tag{4.13}$$

If the pressure is constant, dP = 0, and so the change dH in the enthalpy H, attributable to a change dT in the system's temperature T, is given in its entirety by the partial derivative $(\partial H/\partial T)_P$. If, however, the change in the system is such that the temperature T and the pressure P both change simultaneously, then the change dH in the enthalpy H, attributable to a change dT in the system's temperature T, combined with a simultaneous change dP in the system's pressure P is given by equation (4.13), implying that we need to know the two partial derivatives, both $(\partial H/\partial T)_P$ and also $(\partial H/\partial P)_T$.

That's why partial derivatives are important. If we wish to understand how systems behave as conditions change, and to predict their behaviour, then, given that all functions of thermodynamic interest are functions of at least two variables, then we need to know the corresponding partial derivatives.

Much of this book is therefore about determining what the appropriate partial derivatives are under various circumstances, and then applying them to situations in which the variables change. Equations of the form of equation (4.13) are therefore to be seen throughout this book.

Furthermore, since is it usually easier to grasp what is happening when only one variable changes at any one time, this book will repeatedly use phrases such as 'system of constant mass' (implying that the mole number n is constant, and so dn = 0), 'at constant pressure' (dP = 0), 'at constant volume' (dV = 0), and 'at constant temperature' (dT = 0). Unfortunately, these phrases add 'clutter' and interrupt the 'main story' – they are, however, necessary, for they specify the precise conditions that apply in any given circumstance, and help us distinguish between the various partial derivatives we shall encounter, for example, the rather similar-looking, but very different, $(\partial P/\partial T)_V$, which is the change ∂P_V in pressure attributable to a change ∂T_V in temperature at constant volume, and $(\partial V/\partial P)_T$, which is the change ∂V_T in volume attributable to a change ∂P_T in pressure at constant temperature.

Although partial derivatives have a somewhat scary appearance, you will soon gain confidence in using them and interpreting them – indeed, they all have meaning. So, for example, $(\partial V/\partial P)_T$, as we have just seen, represents the change ∂V_T in the volume of a system attributable to a change ∂P_T in pressure exerted on that system, whilst the temperature of the system is held constant. Increasing the pressure on any system results in that system being somewhat compressed, and so an increase in the pressure – represented by a positive value for ∂P_T – will result in a decrease in volume, implying that the corresponding value of ∂V_T will be negative. The ratio represented by the partial derivative $(\partial V/\partial P)_T$ will therefore be negative too.

Also, we know that condensed forms of matter – solids and liquids – are virtually incompressible, and so, for any given increase ∂P_T in the pressure exerted on a system, the corresponding change ∂V_T in the volume of the system will be small. For a solid and a liquid, we would therefore expect the partial derivative $(\partial V/\partial P)_T$ will be both negative, and relatively small. A gas, however, is easily compressed, and so a given increase ∂P_T in the system pressure could result in quite a large change ∂V_T in the system volume. So, for a gas, we would expect the partial derivative $(\partial V/\partial P)_T$ to be negative, and relatively large.

So, whenever a partial derivative is encountered – as will happen on many occasions – it is often worth taking a moment to consider the question "in real, physical, terms, what does this particular partial derivative represent?".

4.6.5 The mathematics of the state function X

Throughout this book, soon after a new state function X is introduced, there is a section headed "The mathematics of X", in which we will examine an equation of the form

$$dX = \left(\frac{\partial X}{\partial T}\right)_P dT + \left(\frac{\partial X}{\partial P}\right)_T dP \tag{4.13}$$

or the similar equations (4.15) and (4.16), and the corresponding partial derivatives. This will then set the scene for exploring how knowledge of the behaviour of each state function enriches our understanding of thermodynamics, and of the associated chemistry and biochemistry.

? EXERCISES

1. Suppose that a function $r(x, y, z)$ is defined in terms of the three variables x, y, and z as

    $$r = x^2 y^3 + z$$

    Show that

    $$\left(\frac{\partial r}{\partial x}\right)_{y,z} = 2xy^3$$

    $$\left(\frac{\partial r}{\partial y}\right)_{x,z} = 3x^2 y^2$$

    $$\left(\frac{\partial r}{\partial z}\right)_{x,y} = 1$$

2. Using the results from question 1, show further that

    $$\frac{\partial}{\partial y}\left(\frac{\partial r}{\partial x}\right)_{y,z} = \left(\frac{\partial^2 r}{\partial y \partial x}\right)_z = 6xy^2$$

    $$\frac{\partial}{\partial x}\left(\frac{\partial r}{\partial y}\right)_{x,z} = \left(\frac{\partial^2 r}{\partial x \partial y}\right)_z = 6xy^2$$

    Note that this is a particular case of the general result for second partial derivatives that, for any function $r(x, y, z)$

    $$\frac{\partial}{\partial y}\left(\frac{\partial r}{\partial x}\right)_{y,z} = \left(\frac{\partial^2 r}{\partial y \partial x}\right)_z = \left(\frac{\partial^2 r}{\partial x \partial y}\right)_z = \frac{\partial}{\partial x}\left(\frac{\partial r}{\partial y}\right)_{x,z}$$

3. Physically, what does the partial derivative $(\partial P/\partial V)_T$ represent? How is this partial derivative shown on a P, V diagram? For an ideal gas, is $(\partial P/\partial V)_T$ always positive, always negative, or sometimes positive and sometimes negative? Why?
4. Physically, what does the partial derivative $(\partial V/\partial T)_P$ represent? How is this partial derivative shown on a P, V diagram? For an ideal gas, is $(\partial V/\partial T)_P$ always positive, always negative, or sometimes positive and sometimes negative? Why?
5. Physically, what does the partial derivative $(\partial T/\partial P)_V$ represent? How is this partial derivative shown on a P, V diagram? For an ideal gas, is $(\partial T/\partial P)_V$ always positive, always negative, or sometimes positive and sometimes negative? Why?
6. Starting in each case with the ideal gas equation $PV = nRT$, derive each of the entries in the following table:

	P	V	T	n
$\frac{\partial}{\partial P}$		$\left(\frac{\partial V}{\partial P}\right)_{T,n} = -\frac{V}{P}$	$\left(\frac{\partial T}{\partial P}\right)_{V,n} = \frac{T}{P}$	$\left(\frac{\partial n}{\partial P}\right)_{V,T} = \frac{n}{P}$
$\frac{\partial}{\partial V}$	$\left(\frac{\partial P}{\partial V}\right)_{T,n} = -\frac{P}{V}$		$\left(\frac{\partial T}{\partial V}\right)_{P,n} = \frac{T}{V}$	$\left(\frac{\partial n}{\partial V}\right)_{P,T} = \frac{n}{V}$
$\frac{\partial}{\partial T}$	$\left(\frac{\partial P}{\partial T}\right)_{V,n} = \frac{P}{T}$	$\left(\frac{\partial V}{\partial T}\right)_{P,n} = \frac{V}{T}$		$\left(\frac{\partial n}{\partial T}\right)_{P,V} = -\frac{n}{T}$
$\frac{\partial}{\partial n}$	$\left(\frac{\partial P}{\partial n}\right)_{V,T} = \frac{P}{n}$	$\left(\frac{\partial V}{\partial n}\right)_{P,T} = \frac{V}{n}$	$\left(\frac{\partial T}{\partial n}\right)_{P,V} = -\frac{T}{n}$	

Note the high degree of symmetry in this table, and the various reciprocal relationships, for example, $(\partial V/\partial P)_{T,n} = 1/(\partial P/\partial V)_{T,n}$.

7. Show that the ideal gas law can be written as

$$\ln P + \ln V = \ln n + \ln R + \ln T$$

As a result, show that

$$\frac{dP}{P} + \frac{dV}{V} = \frac{dn}{n} + \frac{dT}{T}$$

and then derive all the entries in the table in question 6.

8. The **volumetric expansion coefficient** α_V of a system of fixed mass is defined as

$$\alpha_V = \frac{1}{V}\left(\frac{\partial V}{\partial T}\right)_P \tag{4.17}$$

Physically, what does the volumetric expansion coefficient represent? Show that, for an ideal gas, $\alpha_V = 1/T$.

9. The **isothermal compressibility** κ_T of a system of fixed mass is defined as

$$\kappa_T = -\frac{1}{V}\left(\frac{\partial V}{\partial P}\right)_T \tag{4.18}$$

Physically, what does the isothermal compressibility represent? Show that, for an ideal gas, $\kappa_T = 1/P$.

10. One way of simplifying the van der Waals equation-of-state

$$\left(P + \frac{an^2}{V^2}\right)(V - nb) = nRT \tag{3.6}$$

is to assume that $a = 0$, so that equation (3.6) becomes

$$P(V - nb) = nRT$$

Physically, what does the assumption $a = 0$ imply? Assuming that this assumption is valid, show that

$$\alpha_V = \frac{nR}{PV} = \frac{nR}{nRT + nbP} = \frac{1}{T + bP/R}$$

What happens, both mathematically and physically, when $b = 0$?

11. A second, different, way of simplifying the van der Waals equation is to assume that $b = 0$, giving

$$\left(P + \frac{n^2 a}{V^2}\right)V = nRT$$

Physically, what does the assumption $b = 0$ imply? Assuming that this assumption is valid, show that

$$\kappa_T = \frac{1}{P - n^2 a/V^2}$$

What happens, both mathematically and physically, when $a = 0$? (Hint: rather than deriving $(\partial V/\partial P)_T$ directly from the modified van der Waals equation, derive $(\partial P/\partial V)_T$ first.)

PART 2

The Three Laws

5 The First Law of Thermodynamics

 Summary

The **First Law of Thermodynamics** states that

Energy can be neither created nor destroyed. For any closed system, we may define an extensive state function U, **the internal energy of the system, such that the change** dU **in the system's internal energy for any change in state is given by**

$$dU = đq - đw \tag{5.1a}$$

where $đq$ **represents the total heat exchanged between the system and its surroundings, and** $đw$ **represents the total work done by the system on the surroundings, or by the surroundings on the system, during that change in state.**

The statement of the First Law as defined by equation (5.1a) assumes the sign convention that

- If heat is *gained by* a system, then $đq$ is *positive*.
- Heat *lost from* a system, then $đq$ is *negative*.
- If work is done *by* a system on the surroundings, then $đw$ is *positive*.
- If work done *on* a system by the surroundings, then $đw$ is *negative*.

A **reversible** change is a change that can be reversed exactly, returning both the system and the surroundings to their original states. A reversible change is necessarily quasistatic, taking place

- through a sequence of equilibrium states
- in an infinite number of infinitesimal steps
- infinitely slowly

and in addition has a further feature: during the change, all **dissipative effects** must be absent, examples of dissipative effects being **dynamic friction** (and other processes that create **dissipative heat**, such as the flow of an electric current through an electrical resistance), and **molecular mixing**. All real changes are necessarily **irreversible**; reversible changes, however, are very useful theoretical ideals.

The extensive state function the **heat capacity at constant volume** C_V for any system is defined as

$$C_V = \left(\frac{\partial U}{\partial T}\right)_V \tag{5.14b}$$

For any change in a system at constant volume, the change ΔU_V is equal to the heat $\{_1q_2\}_V$ reversibly exchanged between the system and the surroundings.

If we compare the work $đw_{rev}$ performed by a system on the surroundings during a reversible change, with the work $đw_{irrev}$ done by the system for a corresponding irreversible change, then

Modern Thermodynamics for Chemists and Biochemists. Dennis Sherwood and Paul Dalby.
© Oxford University Press 2018. Published 2018 by Oxford University Press.

$$đw_{rev} > đw_{irrev} \tag{5.37}$$

The maximum work a system can perform is therefore the reversible work, and all real systems must inevitably fall below this ideal.

Similarly, as a consequence of the First Law,

$$đq_{rev} > đq_{irrev} \tag{5.38}$$

implying that the maximum heat that a system can release to the surroundings is the heat associated with a reversible change; for any real change, the corresponding heat is inevitably less.

5.1 The First Law

Energy can be neither created nor destroyed. For any closed system, we may define an extensive state function U**, the internal energy of the system, such that the change** dU **in the system's internal energy for any change in state is given by**

$$dU = đq - đw \tag{5.1a}$$

where $đq$ **represents the total heat exchanged between the system and its surroundings, and** $đw$ **represents the total work done by the system on the surroundings, or by the surroundings on the system, during that change in state.**

This formal statement of the First Law comprises three elements:

- Firstly, a statement based on empirical observation – that energy cannot be created or destroyed.
- Secondly, the identification of an extensive state function – in this case, a system's internal energy U.
- Thirdly, a mathematical definition of dU for any change in state – in this case, as the difference between $đq$ and $đw$, representing respectively the total heat exchanged, and total work done, as that change in state took place.

As we shall see in Chapter 9, the statement of the Second Law also follows this three-element format.

In essence, the First Law is a statement of the conservation of energy, resolving the puzzle as regards the nature of heat, as discussed on page 53. The First Law takes matters further, however, by introducing a new extensive state function, the internal energy U, and quantifying the value of the change dU for any change in state as the difference between the corresponding values of $đq$ and $đw$.

As an extensive state function, the value of the internal energy U depends on the size and extent of the system under study. If a given system comprises n mols of material, then we may define the **molar internal energy** $\mathbf{U}$ as

$$\mathbf{U} = U/n$$

and so, for 1 mol of material, the First Law can be written as

$$dU = đq - đw \qquad (5.1b)$$

The **molar internal energy** U is necessarily an intensive state function. Also, note that the form of the First Law as defined by equations (5.1b) and (5.1a) relates to the internal energy of a closed system of constant mass, a system that does not exchange matter with the surroundings. This therefore excludes open 'flow through' systems in which matter flows either into, through, or out of the system of interest (as happens, for example, in many industrial chemical engineering processes). As we shall see over the next several chapters, the restriction to closed systems is not unduly limiting, and many real systems come within this scope. This restriction will be removed in Chapter 13 (see page 408), where we will present a modified version of equation (5.1a) which applies to open systems.

From a practical point of view, the First Law is of great importance, for, as will be shown in this chapter and the next, it enables us to answer real questions such as:

- How much heat is released (or absorbed) when a given chemical reaction takes place?
- How much energy is required to break, or to form, a particular chemical bond?
- What is the likely temperature of a flame – for example, the flame of an oxyacetylene torch?
- What is the maximum amount of work that results from the combustion of a given fuel?
- How much energy is released during respiration, in which living organisms oxidise glucose, $C_6H_{12}O_6$, to form carbon dioxide, CO_2 and water, H_2O?

5.2 A molecular interpretation of internal energy

Like all the other thermodynamic state functions, the internal energy U of a system is a macroscopic property, but let's take a moment to interpret the macroscopically observable internal energy U in terms of the microscopic behaviour of the system's molecules.

Within any system, any single molecule will be in motion. For a monatomic gas, the motion is translational: as the molecule moves across the volume available, the molecule will possess translational kinetic energy, attributable to motion, as well as potential energy, attributable to its position in space; in a liquid or a solid, the motion of any single molecule is more constrained, but it still associated with kinetic energy (attributable to, say, oscillations about a given location within a solid), potential energy (attributable to location), and also energies associated with interactions with neighbouring molecules (such as those attributable to mutual attraction or repulsion). Polyatomic molecules have additional modes of movement: they may be tumbling, and possess rotational kinetic energy, and there may be internal vibrations within the molecule itself, each associated with vibrational energy; furthermore, for all molecules, there are energies within each atom, associated with the electrons and the nucleus. If we add together all the individual molecular and atomic energies, the resulting sum is the macroscopic thermodynamic state function, the internal energy U.

5.3 A reminder of some important sign conventions

All the variables in equation (5.1a)

$$dU = đq - đw \tag{5.1a}$$

are algebraic, signed quantities, and all obey a consistent sign convention.

Firstly, as we saw on page 11, dU is defined as the difference

$$dU = U(\text{final state}) - U(\text{initial state}) \tag{1.2a}$$

in which the value of U in the initial state is always subtracted from the value of U in the final state.

Secondly, the sign of $đq$ is determined, as stated on page 56, by whether the system of interest gains heat from the surroundings, or loses heat to the surroundings:

- If heat is *gained by* a system, then $đq$ is *positive*.
- If heat *lost from* a system, then $đq$ is *negative*.

Thirdly, the sign of $đw$ is determined (see page 27) by whether the system of interest does work on the surroundings, or the surroundings do work on the system:

- If work is done *by* a system on the surroundings, then $đw$ is *positive*.
- If work done *on* a system by the surroundings, then $đw$ is *negative*.

During any particular change, the system will interact with its surroundings, and exchange heat and work according to the circumstances: if we can determine what the corresponding values of $đq$ and $đw$ are, perhaps by calculation, then we can use equation (5.1a) to compute dU. As a result, we can then draw inferences concerning the behaviour of the system, as will be demonstrated throughout this book.

5.4 Internal energy and temperature

5.4.1 A microscopic view

One important inference is whether the result of a change dU in the internal energy U of a system makes the system hotter or cooler. To explore this, we draw on the result we presented on pages 69 to 71 that, for a system comprising an ideal gas of monatomic molecules, the average molecular energy $\langle \mathcal{E} \rangle$ is related to the temperature T as

$$\langle \mathcal{E} \rangle = \frac{3}{2} k_B T \tag{3.19}$$

For a gas of 1 mol = 6.022×10^{23} = N_A monatomic molecules, the total energy of the system is therefore

$$N_A \langle \mathcal{E} \rangle = \frac{3}{2} N_A k_B T$$

But

$$k_B = \frac{R}{N_A} \tag{3.17}$$

and so

$$N_A \langle \mathcal{E} \rangle = \frac{3}{2}RT$$

This expression represents the total energy of an ideal gas of 1 mol of monatomic molecules as derived microscopically; macroscopically, this quantity is the molar internal energy U, and so

$$U = \frac{3}{2}RT \qquad (5.2)$$

Equation (5.2) has two important implications.

Firstly, equation (5.2) tells us that the molar internal energy U of a monatomic ideal gas is a function of the temperature T only, and not of any other variables such as pressure and volume. This is a special, and restricted, case of the general principle we stated on page 48 that,

> In the absence of gravitation, motion, electric, magnetic and surface effects, a fixed mass of a pure substance has two, and only two, independent state functions.

According to this principle, we would expect the molar internal energy U of a system to be a function of two variables, and for systems in general, this is true; for an ideal gas, however, the molar internal energy U is a function of temperature only, and so $U = U(T)$.

The second implication of equation (5.2) is that the molar internal energy $U = U(T)$ of an ideal gas is a linear function of temperature, so that $U(T)$ increases proportionally with T. Mathematically, since R is a constant, equation (5.2) can be differentiated with respect to T to give

$$\frac{dU}{dT} = \frac{3}{2}R \qquad (5.3)$$

from which we see that a consequence of the linear form of equation (5.2) is that the derivative dU/dT is a constant, $3R/2$, the significance of which we shall explore shortly (see page 111). The term $3R/2$ that appears in equations (5.2) and (5.3) is valid for a monatomic gas only; for a polyatomic gas, the linear relationship still holds, and the derivative dU/dT is also a constant, but, as we shall see on page 112, the coefficient is different.

For systems other than ideal gases, the molar internal energy U is in general a function of two variables, say, temperature T and the molar volume V, so that $U = U(T, V)$. The actual form of the function depends on the system of interest, and – for solids in particular – can be quite complicated. For systems in general, $U(T, V)$ is not linear with temperature T, but it is generally true that as T increases, U increases too, so that the partial derivative $(\partial U/\partial T)_V$ is always positive, at all temperatures: for all systems, the higher the temperature T, the higher the molar internal energy U.

5.4.2 A macroscopic view

Macroscopically, a change dU in the internal energy U of any system of 1 mol of material is related to the heat $đq$ exchanged between the system and the surroundings, and the work $đw$ done by the system on the surroundings according to the First Law

$$dU = đq - đw \qquad (5.1b)$$

We can apply equation (5.1b) to a variety of different macroscopic circumstances, which, as we shall see, confirm the inference we have just drawn from a microscopic perspective that an increase in temperature is associated with an increase in internal energy. In so doing, we will also build confidence in using the sign conventions concerning dU, đq and đw. Furthermore, we note that all of the following examples refer to any system, not just an ideal gas.

So as a first example, consider a system that

- exchanges no heat with the surroundings, so that đq = 0
- does no work on the surroundings, nor do the surroundings do any work on the system, so that đw = 0
- which together imply that dU = đq − đw = 0.

This system experiences no change at all, and so we would expect the system temperature to remain unchanged.

Secondly, consider a system that

- loses heat to the surroundings, so that đq is negative, and
- does no work on the surroundings, nor do the surroundings do any work on the system, implying that đw is zero
- in which case dU = đq − đw is negative.

The system's internal energy U therefore decreases, and the loss of heat implies that the system's temperature T decreases.

A third example is a system that

- absorbs heat, implying that đq is positive, but
- does no work on the surroundings, nor is any work done on the system by the surroundings, so that đw is zero,
- in which case dU = đq − đw is positive, and the internal energy U increases.

But by absorbing heat, the system gets hotter, and so the system's temperature T necessarily increases too: an increase in the system's internal energy U is associated with an increase in the system's temperature T.

As a fourth example, suppose that

- no heat is exchanged between the system and the surroundings, so that đq is zero, but
- the surroundings do work on the system, so that đw is negative
- implying that dU = đq − đw is positive.

In the preceding, third, example, dU was also positive, and, because heat was absorbed, it is self-evident that the system becomes hotter, and that the temperature T increases. In this fourth example, dU is also positive, this time as a result of work being done on the system, rather than a transfer of heat. However, as we saw on page 13, a change in any state function depends only on the initial and final states, and contains no 'memory' of how the change took place. Since the internal energy U and the temperature T are both state functions, it does not matter whether the change took place along a path which involved the exchange of heat, or along a path involving the performance of work. In the third example, in which the system gains heat from the surroundings, causing the change dU to be positive, it is intuitively obvious that the system's temperature increases; in this fourth case, however, the positive change dU is a result of work being done on the system by the surroundings, and not a result of the flow of heat, but the system's temperature increases nonetheless. This explains why a gas gets hot when it is compressed – as happens when a bicycle tyre is inflated using a conventional hand pump.

And as our final example, consider a system that

- exchanges no heat with the surroundings, so that đq is zero, but
- does work on its surroundings, so that đw is positive
- implying dU = đq – đw is negative.

The internal energy U of the system decreases, and we would expect the system's temperature to fall – a phenomenon known as **adiabatic cooling**, which has practical importance in the design, for example, of air conditioning and climate control equipment.

Finally in this section, we note that although these examples were expressed in terms of the molar internal energy U, they all apply more generally to any system of n mol of any pure material, and with internal energy $U = n\mathbf{U}$.

5.5 The First Law, state functions, and path functions

In our explanation of the fourth example (đq is zero, and đw is negative, implying that dU is positive), we referred to the fact that any change in any state function depends only on the initial and final states of the system, and is independent of the path followed as the change in state took place – a theme we shall explore more deeply in this section.

An important aspect of equations (5.1a) is that it relates a change dU in a state function, U, to the difference đq – đw between two path functions. This is a second instance of the interplay between state functions and path functions – the first being the definition of P, V work as

$$đw = P_{ex}\, dV \qquad (2.1)$$

For a finite change in state, from an initial state [1], with internal energy U_1, to a final state [2], with internal energy U_2, we may write

$$\Delta U = \int_{\text{state [1]}}^{\text{state [2]}} dU = U_2 - U_1$$

According to the First Law, equation (5.1a), ΔU is also given by

$$\Delta U = \int_{\text{state [1]}}^{\text{state [2]}} (đq - đw) = \int_{\text{state [1]}}^{\text{state [2]}} đq - \int_{\text{state [1]}}^{\text{state [2]}} đw$$

and so

$$\Delta U = U_2 - U_1 = \{_1 q_2\}_X - \{_1 w_2\}_X \qquad (5.4)$$

where the notation $\{\ldots\}_X$ indicates that the specific values of $\{_1q_2\}_X$ and $\{_1w_2\}_X$ are each determined according to the actual path X followed during the change in state from state [1] to state [2].

Since the internal energy U is a state function, then, for any finite change in state from an initial state [1] to a final state [2], the value of ΔU depends only on the values U_1 and U_2 of the internal energy in the initial and final equilibrium states, and is independent of the path followed. In contrast, the values of $\{_1q_2\}_X$ and $\{_1w_2\}_X$ *do* depend on the nature of the path X followed.

As an example, consider what happens when a system changes state from an initial equilibrium state [1] to a final equilibrium state [2], but along two different paths, as illustrated in Figure 5.1.

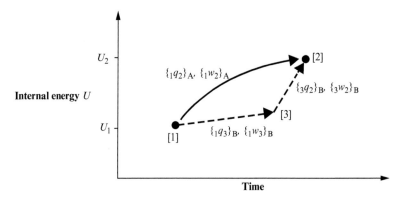

Figure 5.1 State functions and path functions. Here are two different paths, A (solid line) and B (dashed line), from the equilibrium state [1], with internal energy U_1, to the equilibrium state [2], with internal energy U_2. The change $\Delta U = U_2 - U_1$ is the same for both paths (and indeed for all paths), but the total heat exchange, and work done by or on the system, depends on the path.

Consider firstly the change that follows path A, directly from state [1] to state [2]. If during this change the system gains heat $\{_1q_2\}_A$, and does work $\{_1w_2\}_A$ on the surroundings, then according to the First Law, for path A, the change in internal energy $\Delta U_{1\to2}$ is given by

$$\Delta U_{1\to2} = U_2 - U_1 = \{_1q_2\}_A - \{_1w_2\}_A \tag{5.5}$$

Path B is shown by the dashed line, from the equilibrium state [1] to an intermediate equilibrium state [3], and then on to the final equilibrium state [2]. For the first part of path B from state [1] to state [3], the First Law holds, and can be written as

$$\Delta U_{1\to3} = \{_1q_3\}_B - \{_1w_3\}_B$$

Similarly, for the second part, from state [3] to state [2]

$$\Delta U_{3\to2} = \{_3q_2\}_B - \{_3w_2\}_B$$

Hence, for the entire path B, from state [1] to state [2] through the intermediate state [3]

$$\begin{aligned}\Delta U_{1\to3} + \Delta U_{3\to2} &= [\{_1q_3\} - \{_1w_3\}]_B + [\{_3q_2\} - \{_3w_2\}]_B \\ &= [\{_1q_3\} + \{_3q_2\}]_B - [\{_1w_3\} + \{_3w_2\}]_B\end{aligned} \tag{5.6}$$

in which $[\{_1q_3\} + \{_3q_2\}]_B$ represents the total amount of heat gained by, or lost from, the system during the change along path B, and $[\{_1w_3\} + \{_3w_2\}]_B$ represents the total work done by, or on, the system along path B.

Since the internal energy U is a state function, the value of ΔU for any change in state is the same for all paths, and so we may write

$$\Delta U_{1\to 2} = U_2 - U_1 = \Delta U_{1\to 3} + \Delta U_{3\to 2}$$

Combining equations (5.5) and (5.6), we therefore have

$$\Delta U_{1\to 2} = \{_1q_2\}_A - \{_1w_2\}_A = [\{_1q_3\} + \{_3q_2\}]_B - [\{_1w_3\} + \{_3w_2\}]_B$$

Although the individual values of $\{_1q_2\}_A$, $\{_1w_2\}_A$, $\{_1q_3\}_B$, $\{_3q_2\}_B$, $\{_1w_3\}_B$ and $\{_3w_2\}_B$ are all different, the values of the two differences $\{_1q_2\}_A - \{_1w_2\}_A$ and $[\{_1q_3\} + \{_3q_2\}]_B - [\{_1w_3\} + \{_3w_2\}]_B$ are equal, and equal to the value of $\{_1q_2\}_X - \{_1w_2\}_X$ for *any* path between state [1] and state [2].

5.6 The First Law and cycles

For an infinitesimal change, the First Law tells us that

$$dU = đq - đw \qquad (5.1a)$$

Consider now a sequence of infinitesimal steps which collectively form a thermodynamic cycle, so that the final state of the system is identical to the system's initial state. For any step in the cycle, equation (5.1a) will be true, and as a result of that step, the internal energy U of the system will have changed by dU. Furthermore, during that step, there will have been a corresponding exchange $đq$ of heat between the system and the surroundings (if the system gains heat during that step, $đq$ is positive, if the system loses heat during that step $đq$ is negative); and the system will either have done work on the surroundings ($đw$ is positive), or the surroundings will have done work on the system ($đw$ is negative).

The overall result of a sequence of infinitesimal steps may be obtained by integration, which for a complete cycle, may be expressed as a cyclic integral

$$\oint dU = \oint đq - \oint đw$$

where $\oint dU$ represents the total change in the internal energy U for the complete thermodynamic cycle; $\oint đq$ represents the net amount of heat exchanged between the system and the surroundings for a complete cycle (and is positive if, overall, the system gains heat, or negative if, overall, the system loses heat); and $\oint đw$ represents the net amount of work done by the system on the surroundings (if the integral is positive), or on the system by the surroundings (if the integral is negative).

The internal energy U, however, is a state function, and so we know that

$$\oint dU = 0$$

hence

$$0 = \oint đq - \oint đw$$

and so, for any thermodynamic cycle

$$\oint dq = \oint dw \qquad (5.7)$$

Equation (5.7) tells us that, for *any* thermodynamic cycle – indeed for *all* thermodynamic cycles – the net total heat absorbed by the system from the surroundings (or released from the system to the surroundings) is equal to the net total work done by the system on the surroundings (or by the surroundings on the system).

5.7 The mathematics of U

5.7.1 U as a function $U(T, V)$ of T and V

The discussion so far has been qualitative, helping us gain a 'feel' for how the variables dU, đq and đw interact. To gain a quantitative understanding, we present here an initial mathematical analysis of the internal energy U, and the associated variable, the molar internal energy ***U***, of a closed system composed of a pure substance, under given conditions of temperature, pressure and volume. Both U and ***U*** are state functions, the difference between being that U is an extensive state function, and depends on the quantity of matter in the system, as defined by the mole number n, whereas ***U*** = U/n is an intensive state function, and represents the internal energy of a system comprising 1 mol of the same pure substance, under the same conditions of temperature, pressure and volume.

We now invoke the principle stated on page 48,

> In the absence of gravitation, motion, electric, magnetic and surface effects, a fixed mass of a pure substance has two, and only two, independent state functions.

This implies that we may express ***U*** as a function of any two other state functions, for example, the temperature T and the molar volume ***V***. If ***U*** is therefore expressed as ***U***$(T, $***V***$)$, then, as we saw on page 88, the infinitesimal change d***U*** may be written as

$$d\boldsymbol{U} = \left(\frac{\partial \boldsymbol{U}}{\partial T}\right)_V dT + \left(\frac{\partial \boldsymbol{U}}{\partial \boldsymbol{V}}\right)_T d\boldsymbol{V} \qquad (5.8a)$$

For a closed system of a constant mass of n mol, we may multiply equation (5.8a) throughout by n, giving

$$n d\boldsymbol{U} = n\left(\frac{\partial \boldsymbol{U}}{\partial T}\right)_V dT + n\left(\frac{\partial \boldsymbol{U}}{\partial \boldsymbol{V}}\right)_T d\boldsymbol{V}$$

Since n is a constant, this equation may be written as

$$d(n\boldsymbol{U}) = \left(\frac{\partial (n\boldsymbol{U})}{\partial T}\right)_V dT + \left(\frac{\partial (n\boldsymbol{U})}{\partial (n\boldsymbol{V})}\right)_T d(n\boldsymbol{V})$$

Recognising that $U = n\boldsymbol{U}$ and $V = n\boldsymbol{V}$, we can therefore express the total internal energy U of the system of n mol as a function $U(T, V)$ of the temperature T and the total system volume V as

$$dU = \left(\frac{\partial U}{\partial T}\right)_V dT + \left(\frac{\partial U}{\partial V}\right)_T dV \qquad (5.8b)$$

Equations (5.8a) and (5.8b) are equivalent, bearing in mind that equation (5.8a) in expressed in terms of the molar quantities $\boldsymbol{U}$ and $\boldsymbol{V}$ and so intrinsically refers to a constant mass of material; in equation (5.8b), however, the assumption of constant mass is assumed, but not obviously stated.

Equations (5.8a) and (5.8b) apply to any closed system of a pure substance of constant mass, where the actual form of the partial derivatives $(\partial U/\partial T)_V$, $(\partial \boldsymbol{U}/\partial \boldsymbol{V})_V$, $(\partial U/\partial T)_V$ and $(\partial U/\partial V)_T$ can be determined given knowledge of the equation-of-state $U = U(T, V)$ for the system of interest.

Regardless of the equation-of-state that might be used, it's important to note that all partial derivatives, of which $(\partial U/\partial T)_V$ and $(\partial U/\partial V)_T$ are examples, specify the path along which a change takes place, as defined by 'at constant volume' or 'at constant temperature'. This does **not** imply that partial derivatives are path functions, like work and heat. Work and heat are path functions because the quantity of work done or heat exchanged *depends* on the path taken, which can be *any* path between any two equilibrium states. In contrast, the value of a partial derivative such as $(\partial U/\partial T)_V$ is *defined only for one specific path* – in this particular case, a path at constant volume.

5.7.2 U as a function $U(T, P)$ of T and P

In deriving equation (5.8b), we chose to express U as a function $U(T, V)$ of the temperature T and the system volume V. As we saw on page 86, U can be expressed as a function of any other two intensive state functions, and so an alternative is to express U as a function $U(T, P)$ of the system temperature T and system pressure P. In this case dU becomes

$$dU = \left(\frac{\partial U}{\partial T}\right)_P dT + \left(\frac{\partial U}{\partial P}\right)_T dP \qquad (5.9)$$

Equations (5.8b) and (5.9) are structurally identical, but the details are different: the coefficient for dT in equation (5.8b) is $(\partial U/\partial T)_V$, a partial derivative at constant volume, whereas the coefficient for dT in equation (5.9) is $(\partial U/\partial T)_P$, a partial derivative at constant pressure; furthermore, the second term in equation (5.8b) relates to a change dV in volume, but the corresponding term in equation (5.9) relates to a change dP in pressure.

5.7.3 U, partial derivatives, and paths

Although equations (5.8b) and (5.9) are different, they are describing the same change dU – if a closed system of constant mass undergoes a change, then, with the single exception of the mole number n, all the state functions in the initial state $[1] \equiv [P_1, V_1, T_1, U_1]$ are likely to assume different values in the final state $[2] \equiv [P_2, V_2, T_2, U_2]$. Since the internal energy U is a state function, the change $\Delta U = U_2 - U_1$ is independent of the path taken between the initial and final states. Equations (5.8b) and (5.9) describe two particular paths (out of many possible paths), equation (5.8b)

$$dU = \left(\frac{\partial U}{\partial T}\right)_V dT + \left(\frac{\partial U}{\partial V}\right)_T dV \qquad (5.8b)$$

referring to a path comprised of a step at constant volume, followed by a step at constant temperature, whilst equation (5.9)

$$dU = \left(\frac{\partial U}{\partial T}\right)_P dT + \left(\frac{\partial U}{\partial P}\right)_T dP \tag{5.9}$$

refers to a path comprised of a step at constant pressure, followed by a step at constant temperature. Physically, equations (5.8b) and (5.9) describe the same overall change and are equivalent; mathematically, as we shall see, we may choose between using equation (5.8b) and equation (5.9) according to whether U is more naturally expressed as $U(T, V)$, a function of the temperature T and the total system volume V, or $U(T, P)$, a function of the temperature T and the system pressure P, or according to our knowledge of how a particular change actually takes place.

5.8 The First Law in open, closed and isolated systems

As we saw on page 57, an open system is one which can exchange heat, work and also matter with the surroundings. The application of the First Law to an open system therefore needs to take into account all energy transfers associated with any flows of matter, for example, changes in kinetic energy attributable to the movement of mass (such as the flow of a fluid horizontally through a pipe), and changes in potential energy (associated with for example, matter falling in a gravitational field). Flows are of importance in many fields of engineering, notably in the design of components such as pumps, turbines and compressors, as well as large-scale installations, as found in power stations, oil refineries and chemical plant.

Open systems, however, are beyond our scope: this volume will focus on applications within closed systems (those that permit the transfer of work and heat across the system boundary, but not matter), and isolated systems (those that do not permit any transfer of work, heat or matter across the system boundary). This exclusion of open systems may at first sight appear to be unduly restrictive, but in fact, very many systems of interest can be regarded as closed, especially when we remember that the boundary of a system is not necessarily rigid.

So, for example, a system comprising a liquid (say, water) can be regarded as a closed system, even when the water boils, and water vapour 'escapes'. If the boundary of the system is defined as a given mass of liquid water, at, say, room temperature, then the system boundary corresponds to the surface of the liquid within the container. If the water is heated by a heat source outside the system boundary, there is then a flow of heat across the boundary, causing the temperature of the water to rise. As the water heats up, eventually to boiling point, water molecules, in gaseous rather than liquid form, 'escape' from the surface of the liquid water, but if the boundary of the system 'moves' with the water vapour, then all the water molecules that formed the original mass of liquid water will remain within the boundary of the system, which still conforms to the definition of 'closed'. Furthermore, as boiling takes place, the water vapour does P, V work of expansion against the atmosphere, which can be thought of as work done by the system on the surroundings. So, even though the water is open to the atmosphere, and in reality the water vapour molecules mix with the molecules of nitrogen, oxygen and carbon dioxide in the atmosphere, the system of the water can be regarded as a closed system, and the First Law, as expressed by equation (5.1a) can be used.

Almost all chemical reactions carried out in the laboratory, and very many biochemical reactions in living organisms, can be regarded as taking place within closed systems, and so the principles discussed in this volume have very broad applicability. Moreover, these processes are almost always open to the atmosphere, and so take place under conditions of constant (atmospheric) pressure – and although the atmospheric pressure does change (for example, with the weather), it does so over times of the order of hours and days: many chemical reactions, both within the laboratory and within living organisms, take place over much shorter time periods (seconds, or fractions of seconds), during which the atmospheric pressure remains unchanged. Changes at constant pressure – as we shall shortly see – are therefore of particular importance.

To explore how the First Law can be used in real situations, the following sections will examine some initial applications, firstly in isolated systems, and then in some particular cases of changes within closed systems. In so doing, we must always bear in mind that in the mathematical definition of the First Law, equation (5.1a)

$$dU = đq - đw \tag{5.1a}$$

đq and đw are signed, algebraic quantities complying with the sign conventions defined on pages 27 and 56:

- If heat is gained by the system from the surroundings, đq is positive.
- If heat is lost from the system to the surroundings, đq is negative.
- If work is done by the system on the surroundings, đw is positive.
- If work is done on the system by the surroundings, đw is negative.

5.9 The First Law in an isolated system

By definition, an isolated system cannot exchange matter, heat or work with the surroundings, and so, for any infinitesimal change in an isolated system

$$đq_{isol} = đw_{isol} = 0$$

where the subscript $_{isol}$ serves as a reminder that we are considering changes in an isolated system.

Hence, for any infinitesimal change in an isolated system

$$dU_{isol} = 0$$

and for any finite change in an isolated system

$$\Delta U_{isol} = 0$$

This implies that the internal energy U_{isol} of an isolated system can never change, but remains constant for all time. This is a statement of energy conservation: since the only true isolated system is the universe as a whole, then the First Law can be expressed as

> **The internal energy U of the universe is constant: energy can be neither created nor destroyed.**

5.10 The First Law in a closed system – the adiabatic change, $đq = 0$

An adiabatic change is one that does not exchange heat, and so $đq = 0$. Consequently, for a closed system undergoing a small or infinitesimal adiabatic change

$$dU_{adi} = -đw_{adi}$$

and for a finite change

$$\Delta U_{adi} = -\{_1w_2\}_{adi} \tag{5.11}$$

where, even though the change ΔU in the state function U is independent of the path, the subscript $_{adi}$ acts as a reminder that, in this particular discussion, we are considering adiabatic paths only.

Hence, if work is done by the system on the surroundings, $đw_{adi}$ and $\{_1w_2\}_{adi}$ are positive and so $-đw_{adi}$ and $-\{_1w_2\}_{adi}$ are negative, implying that the internal energy U_{adi} of the system decreases, and the temperature T of the system will fall. Conversely, if the surroundings do work on the system, $đw_{adi}$ and $\{_1w_2\}_{adi}$ are negative and so $-đw_{adi}$ and $-\{_1w_2\}_{adi}$ are positive, implying that the internal energy U_{adi} of the system increases, and the temperature T of the system will rise.

5.11 The First Law in a closed system – the isochoric change, $dV = 0$

An isochoric change is one at constant volume, implying that $dV = 0$. For a closed system, and if the only work that can be performed is P, V work of expansion, then, for a small or infinitesimal isochoric change

$$đw_V = P_{ex}\, dV = 0$$

and so

$$dU_V = đq_V \tag{5.12a}$$

and for a finite change

$$\Delta U_V = \{_1q_2\}_V \tag{5.12b}$$

where the subscript $_V$ serves as a reminder that the change is at constant volume.

For any change in state at constant volume, then the corresponding change in internal energy ΔU_V has the special property of being equal to the heat transferred between the system and the surroundings. If, however, the change does not take place at constant volume, then ΔU still has a value, but it will not equal the heat transferred during the change. However, because internal energy is a state function, then that value of ΔU is equal to the quantity of heat that would have been transferred, had the change taken place at constant volume.

Equations (5.12a) and (5.12b) are important, for they tell us that if heat energy đq_V is transferred into a system at constant volume, then all of that energy is transformed into an increase in internal energy dU_V, resulting in a rise dT_V in the system's temperature. Conversely if heat energy is lost from a system at constant volume, then đq_V is negative, resulting in a corresponding decrease in the system's internal energy (dU_V is negative) and a fall in the system's temperature (dT_V is also negative).

5.12 The heat capacity at constant volume, C_V

5.12.1 The definition of C_V

For any given system, the ratio of the infinitesimal gain (or loss) of heat đq_V at constant volume to the corresponding infinitesimal increase (or decrease) in temperature dT_V is known as the **heat capacity at constant volume** C_V

$$C_V = \frac{đq_V}{dT_V} \tag{5.13a}$$

from which

$$đq_V = C_V \, dT_V \tag{5.13b}$$

But, according to equation (5.12a), for a change at constant volume

$$dU_V = đq_V \tag{5.12a}$$

hence, using equation (5.13b)

$$dU_V = C_V \, dT_V \tag{5.14a}$$

and so

$$C_V = \frac{đq_V}{dT_V} = \frac{dU_V}{dT_V} = \left(\frac{\partial U}{\partial T}\right)_V \tag{5.14b}$$

where the partial derivative $(\partial U/\partial T)_V$ has exactly the same meaning as the ratio dU_V/dT_V, but is mathematically more informative, for if we know the equation-of-state that defines the internal energy U of a system of constant mass as a function $U(T, V)$ of the temperature T and the volume V, then we can determine $(\partial U/\partial T)_V$, and therefore C_V, by differentiation.

Reference to page 103 will show that the term $(\partial U/\partial T)_V$ has already appeared in the chapter, in equation (5.8b)

$$dU = \left(\frac{\partial U}{\partial T}\right)_V dT + \left(\frac{\partial U}{\partial V}\right)_T dV \tag{5.8b}$$

Using equation (5.14b), we can now write equation (5.8b) as

$$dU = C_V dT + \left(\frac{\partial U}{\partial V}\right)_T dV \tag{5.15}$$

Equation (5.15) is of general applicability to any closed system of constant mass.

The heat capacity C_V helps us answer questions such as

- "By how much does the temperature of a system change when it exchanges a given amount of heat with its surroundings at constant volume?" and
- "If I want to increase the temperature of this system, at constant volume, by a given amount, how much heat energy do I need?"

Hence, the term 'capacity': if C_V is relatively large, then, according to equation (5.13a)

$$C_V = \frac{đq_V}{dT_V} \tag{5.13a}$$

we see that a relatively large amount of heat $đq_V = C_V\, dT$ is required to raise the temperature of the system, at constant volume, by a given amount dT, implying that the system has the 'capacity' to 'absorb' large amounts of energy for a small temperature increase; conversely, if C_V is relatively small, then the same temperature change dT can be achieved with much less heat, implying that the system's capacity to absorb heat is correspondingly less.

Equation (5.14b)

$$C_V = \left(\frac{\partial U}{\partial T}\right)_V \tag{5.14b}$$

which is the formal mathematical definition of C_V, offers a rather different, but closely related, interpretation: if C_V is relatively large, then a given increase ∂T_V in temperature, at constant volume, corresponds to a relatively large increase ∂U_V in the system's internal energy; If C_V is relatively small, then a given temperature increase ∂T_V corresponds to a relatively small increase ∂U_V in the system's internal energy.

Equation (5.14b) refers to a closed system of constant mass, usually defined in terms of the mole number n; for a system of 1 mol, the internal energy of the system is the molar internal energy $\boldsymbol{U}$, and the corresponding value of C_V, usually symbolised as $\boldsymbol{C_V}$, is defined as

$$\boldsymbol{C_V} = \left(\frac{\partial \boldsymbol{U}}{\partial T}\right)_V \tag{5.16}$$

$\boldsymbol{C_V}$ is known as the **molar heat capacity at constant volume**. For a system of n mol, C_V and $\boldsymbol{C_V}$ are related as

$$C_V = n\boldsymbol{C_V} = n\left(\frac{\partial \boldsymbol{U}}{\partial T}\right)_V = \left(\frac{\partial (n\boldsymbol{U})}{\partial T}\right)_V = \left(\frac{\partial U}{\partial T}\right)_V \tag{5.17}$$

Values for $\boldsymbol{C_V}$ for some selected gases are shown in Table 5.1.

As can be seen, the (monatomic) inert gases all have the same value for $\boldsymbol{C_V}$, 12.5 J K^{-1} mol^{-1}, even though they range in molecular weight from 0.004 kg mol^{-1} (helium) to 0.131 kg mol^{-1} (xenon); the diatomic gases have a higher value of $\boldsymbol{C_V}$, at about 21 J K^{-1} mol^{-1}; with CH$_4$ and CO$_2$ having higher values still. Also, in general, $\boldsymbol{C_V}$ is not a constant, but varies both with temperature and pressure.

For reactions involving gases, isochoric changes require that the reaction takes place in a vessel of constant volume, manufactured of a material strong enough to withstand any pressures resulting from the reaction. If the reaction results in the creation of gases, the pressure inside the reaction vessel will increase, so the material must prevent the vessel from exploding; if the

Table 5.1 Values of the molar heat capacity at constant volume, C_V, for selected gases

	C_V J K^{-1} mol^{-1}		C_V J K^{-1} mol^{-1}
He (g)	12.5	H$_2$ (g)	20.5
Ne (g)	12.5	N$_2$ (g)	20.8
Ar (g)	12.5	O$_2$ (g)	21.1
Kr (g)	12.5	CO$_2$ (g)	28.9
Xe (g)	12.5	NH$_3$ (g)	28.0

Source: The company, Air Liquide, has a comprehensive on-line resource, http://encyclopedia.airliquide.com/, containing much information about gases, with data corresponding to the IUPAC recommended temperature of 298.15 K, but at a pressure of 1 atm = 1.013 bar, rather than 1.000 bar. In practice, this pressure difference is immaterial, and the data shown in this table may safely to taken as complying with the IUPAC recommendations.

reaction consumes gases and transforms them into, for example, solids or liquids (as, for example, happens when cooling causes a gas to condense into a liquid), then the pressure inside the vessel is likely to fall below atmospheric pressure, and so the vessel needs to be protected against implosion.

Many reactions, of course, don't involve gases, but take place in solution, and so the volume of the corresponding systems are usually unchanged (or very nearly so) as a result of a chemical reaction. Many biochemical reactions that take place within living cells, and that do not involve gaseous reactants or products, may therefore be regarded as isochoric.

5.12.2 The finite change ΔU_V at constant volume

From the definition of C_V

$$C_V = \left(\frac{\partial U}{\partial T}\right)_V \tag{5.14b}$$

Then, for a finite change in state at constant volume, we may integrate equation (5.14b) as

$$\int_{U_1}^{U_2} dU = U_2 - U_1 = \Delta U = \int_{T_1}^{T_2} C_V dT \tag{5.18}$$

For systems in general, C_V is itself a function of temperature, $C_V(T)$, and so the integral on the right-hand side of equation (5.18) needs to be evaluated according to the specific equation that defines the function $C_V(T)$ in any particular circumstance. If, however, C_V can be regarded as a constant over the temperature range from T_1 to T_2, then equation (5.18) becomes

$$\Delta U = C_V(T_2 - T_1) \qquad \text{at constant volume} \tag{5.19}$$

In practice, the assumption that C_V is a constant is reasonable, but there is one particular circumstance for which C_V actually is a constant – for a system of an ideal gas.

5.12.3 C_V, Rumford, and Joule

The definition of C_V completes the story for the experiments of Rumford and Joule. As we saw on page 52, for the Joule experiment, the work done by the two masses, falling through a finite distance h, is $2\,Mgh$ J. For the system comprising the water, there is no heat exchange, and work equal to $2\,Mgh$ J is done on the system, implying that $\Delta U = 0 + 2\,Mgh$. If we use equation (5.19), and assume that C_V for water is constant, then

$$\Delta U = C_V(T_2 - T_1) = 2\,Mgh$$

Joule knew M, g, h, C_V and T_1, and he measured the temperature T_2, the temperature to which the water was raised by the action of the falling masses. Joule's units of measure were those of the mid-19th century – what they actually were does not matter, the important point is at that time, the units of measure for heat, as used for the term $C_V(T_2 - T_1)$ were different from those used for work, as applied to the term $2\,Mgh$, for when Joule was performing his experiments, heat and work were regarded as totally different phenomena, and were not perceived as different manifestations of the same concept – the concept we now call energy. The result of Joule's work was to establish that the ratio $[C_V(T_2 - T_1)]/[2\,Mgh]$ is a constant – the "mechanical equivalent of heat".

As regards Rumford's cannon, the work done by the drill on the cannon, w_{drill}, causes an increase $\Delta U = 0 + w_{\text{drill}}$ in the internal energy U of the cannon. If C_V is the total heat capacity of the bronze of the cannon, then equation (5.19) implies that the temperature of the cannon will rise by $\Delta T = \Delta U/C_V = w_{\text{drill}}/C_V$. As indeed happened.

5.12.4 C_V for an ideal gas

Equation (5.15)

$$dU = C_V dT + \left(\frac{\partial U}{\partial V}\right)_T dV \qquad (5.15)$$

is a general equation, and applies to all closed systems of constant mass. As we saw on page 97, however, the internal energy U of an ideal gas is a function of the temperature T only, and so, for an ideal gas

$$\left(\frac{\partial U}{\partial V}\right)_T = 0$$

implying that equation (5.15) becomes

$$dU = C_V dT$$

or

$$C_V = \frac{dU}{dT} \qquad (5.20a)$$

which, for molar quantities, is

$$C_V = \frac{dU}{dT} \tag{5.20b}$$

Equations (5.20a) and (5.20b) apply to an ideal gas only, and are expressed as total derivatives dU/dT and dU/dT, rather than as partial derivatives $(\partial U/\partial T)_V$ and $(\partial U/\partial T)_V$, since the internal energy U is a function of the temperature T only, and depends on no other variables.

5.12.5 A molecular interpretation of C_V for an ideal gas

As we discussed on page 95, the internal energy U of any system is fundamentally attributable to the energy of each molecule within the system, and, as we saw on page 96, for an ideal monatomic gas, the average molecular energy $\langle \mathcal{E} \rangle$ is given by equation (3.19) as

$$\langle \mathcal{E} \rangle = \frac{3}{2} k_B T \tag{3.19}$$

from which we derived

$$U = \frac{3}{2} RT \tag{5.2}$$

and

$$\frac{dU}{dT} = \frac{3}{2} R \tag{5.3}$$

But according to equation (5.20b), for an ideal gas

$$C_V = \frac{dU}{dT} \tag{5.20b}$$

and so, for a monatomic ideal gas, we can combine the microscopically derived equation (5.3) with the macroscopically derived equation (5.20b) to give

$$C_V = \frac{3}{2} R \tag{5.21}$$

Given that $R = 8.314$ J K^{-1} mol^{-1}, then $3R/2 = 12.5$ J K^{-1} mol^{-1}. This number appeared a few pages ago: reference to Table 5.1 will show that the value of C_V for each of helium, neon, argon, krypton and xenon is ...12.5 J K^{-1} mol^{-1}. This is not a coincidence. The derivation of equation (5.3) was based on the assumption of a monatomic ideal gas, and each of helium, neon, argon, krypton and xenon are certainly monatomic. Also, as 'inert' gases, these gases have only minimal inter-molecular interactions, and although each atom occupies a finite volume, the behaviour of these gases at room temperature and atmospheric pressure is about as close to ideal as is possible.

But what about ideal gases that are not monatomic? Reference to Table 5.1, for example, shows that diatomic gases such as hydrogen, nitrogen and oxygen have values of C_V of about 21 J K^{-1} mol^{-1}, with higher values for the poly-atomic molecules of carbon dioxide and ammonia.

Very briefly, there is an explanation at the atomic level. As we discussed on page 95, a monatomic molecule possesses kinetic energy as it moves through three-dimensional space. If the molecular kinetic energy associated with each dimension in space is $k_B T/2$, then motion in three dimensions corresponds to an energy of $3k_BT/2$, which, for 1 mol of atoms gives $3RT/2$, as is consistent with equation (5.2).

As well as possessing translational kinetic energy, a polyatomic molecule also has rotational energy as it tumbles in space, and vibrational energy as the constituent atoms vibrate with respect to one another along their respective bonds. To determine how these different modes of energy change as external energy is supplied is an exercise in quantum physics, and beyond our scope. One result, however, is of interest here, and that relates to a system of diatomic molecules at a typical room temperature, and at pressures of around 1 atmosphere. Like a monatomic gas, at any temperature T, each molecule has translational kinetic energy in three dimensions, which accounts for an energy of $3k_BT/2$. In addition, an input of external energy at temperatures of around 300 K can be converted into the rotation of the molecule about each of two axes perpendicular to the interatomic bond (but not rotation about the intermolecular axis), and each of these two rotational modes accounts for a further $k_BT/2$ each; furthermore, at 'normal' temperatures, vibrational modes of movement are little influenced by additional heat. At any temperature T the average molecular energy therefore comprises two components: a contribution $3k_BT/2$, corresponding to the kinetic energy of translation, plus $2 \times k_BT/2$ for the two rotational modes, giving a total of $5k_BT/2$ per molecule, and $5RT/2$ per mole. This suggests that, for a diatomic ideal gas, $C_V = 5R/2 = 20.8$ K^{-1} mol^{-1} – close, as shown in Table 5.1, to the values of 20.5 J K^{-1} mol^{-1} for hydrogen, and 21.1 K^{-1} mol^{-1} for oxygen, and equal to the value of 20.8 J K^{-1} mol^{-1} for nitrogen, the least reactive molecule of this trio, and so the most likely to show ideal behaviour.

5.12.6 Is C_V a state function or a path function?

So far in this book, all functions have been classified either as state functions – such as mass M, mole number n and volume V – or path functions, of which we have defined two, heat $đq$ and work $đw$.

From equation (5.14b)

$$C_V = \left(\frac{\partial U}{\partial T}\right)_V \quad (5.14b)$$

C_V is defined as the ratio of two infinitesimal changes ∂U_V and ∂T_V in the state functions U and V, so this seems to suggest that C_V is itself a state function. However, equation (5.13a)

$$C_V = \frac{đq_V}{dT_V} \quad (5.13a)$$

expresses C_V as the ratio between a path function, $đq_V$ and an infinitesimal change dT_V in a state function T, so perhaps the situation is rather less clear.

As we saw on page 5, state functions are properties of a system at equilibrium, and can be measured at a point in time, whereas (see page 24) path functions can be measured only as a system changes state, and have no meaning at a point in time. Accordingly, we can measure a system's volume V only when the system is in equilibrium, when that state 'has' a volume; but heat $đq$ is measureable only when a system changes from one equilibrium state to another, and has no meaning at equilibrium – the system does not 'have' heat.

According to equation (5.13a), the physical measurement of C_V requires the measurement of two changes – đq_V and the corresponding value of dT_V – which can happen only as a system changes state, and not whilst a system is in equilibrium. Furthermore, one of these measurements is of đq_V, a path function which has no meaning at equilibrium. Does this imply that C_V is therefore also a path function?

A counter-argument, however, can be made from the definition of C_V according to equation (5.14b)

$$C_V = \left(\frac{\partial U}{\partial T}\right)_V \tag{5.14b}$$

which is a partial derivative. Fundamental to the mathematics of the calculus is the recognition that, for a general function $y(x)$ of a single variable x, the ratio dy/dx has a value, and a meaning, even when both dy and dx have values of zero; also, although each of dy and dx can be physically measured only when a system undergoes a small change, the ratio dy/dx has a real meaning at an instant, even when the values dy and dx are both infinitesimally small, and even when both are zero. So, if we were to carry out an experiment in which we keep the volume of a system constant, and measure the values of the internal energy U at various temperatures T, we can then plot a graph with U as the vertical axis and T as the horizontal axis. At any point on that graph, we can draw a tangent, and measure the slope. This is the value of $(\partial U/\partial T)_V$ at that point, 'the' value of C_V, corresponding to a specific equilibrium value of U and a corresponding specific equilibrium value of T.

This resolves the state function/path function dilemma. Even though C_V is defined as the ratio of two changes, the essence of calculus, and the definition of C_V as $(\partial U/\partial T)_V$, implies that C_V is a property that has true meaning at a single point in time, and so represents an equilibrium property of a system. C_V is therefore, unambiguously, a state function. And, as a state function, C_V for a given mass of material can be expressed as a function of any other two intensive state functions, so we may define the molar heat capacity $\boldsymbol{C_V}$ of any substance as $\boldsymbol{C_V} = \boldsymbol{C_V}(T, P)$ – as we noted on page 108, in general, $\boldsymbol{C_V}$ varies with both temperature and pressure.

One further point about the definition of C_V, as expressed by equation (5.14b)

$$C_V = \left(\frac{\partial U}{\partial T}\right)_V \tag{5.14b}$$

in general, and (5.20a)

$$C_V = \frac{dU}{dT} \tag{5.20a}$$

for an ideal gas.

Mathematically, the differential form of equations (5.14b) and (5.20a) implies that the infinitesimal quantities ∂U_V, dU, ∂T_V and dT can all be uniquely defined at all temperatures – as indeed is a requirement to allow the integration by which equation (5.19) was derived from equation (5.14b). The infinitesimal changes of state associated with equations (5.14b) and (5.20a), and the finite change associated with the integral equation (5.19), must therefore all be quasistatic – this being the temperature equivalent of the quasistatic change we discussed on page 30, which related to pressure.

5.13 The First Law in a closed system – the isothermal change, $dT = 0$, for an ideal gas

As we have seen, the internal energy U of an ideal gas is a function of temperature only, and so, for any isothermal change in an ideal gas

$$dU = 0$$

Since

$$dU = đq - đw \qquad (5.1a)$$

then, for an isothermal change in an ideal gas

$$đq_T = đw_T \qquad (5.22)$$

implying that

- if work is done by the system ($đw_T$ is positive), then an equivalent amount of heat is transferred into the system by the surroundings ($đq_T$ is positive); likewise, if heat is transferred into the system, then work is done by the system, and
- if work is done on the system by the surroundings ($đw_T$ is negative), then an equivalent amount of heat is lost by the system to the surroundings ($đq_T$ is negative); likewise, if heat is lost by the system, then work is done on the system.

For any system other than an ideal gas, a change at constant temperature does *not* imply that $dU = 0$, and the First Law

$$dU = đq - đw \qquad (5.1a)$$

has to be applied in that form, according to the specific situation.

5.14 The First Law in a closed system – the isobaric change, $dP = 0$

An isobaric change is one that takes place at constant pressure P. Since

$$dU_P = đq - đw \qquad (5.1a)$$

then, for a closed system that performs only P, V work

$$dU_P = đq_P - P_{ex}\, dV$$

If the external pressure P_{ex} constant, we may write

$$dU_P + P_{ex}\, dV = đq_P$$

and if the change is quasistatic, and also frictionless (as this change must be since the only work performed is P, V work of expansion), then the pressure P of the system is equal to the external pressure P_{ex}, implying that

$$dU_P + P\, dV = đq_P \qquad (5.23)$$

Equation (5.23) is now expressed in terms of variables relating only to the system, and states that the heat $đq_P$ gained by, or lost from, a system during any quasistatic change in state at constant pressure P is equal to the change in the system's internal energy dU_P plus the product $P\,dV$.

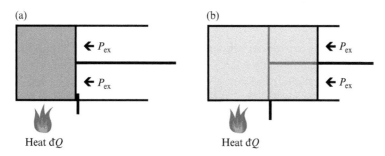

Figure 5.2 Heat supplied at constant volume, and at constant pressure. Figure 5.2(a) shows a system of an ideal gas held at a constant volume, and Figure 5.2(b) shows an otherwise identical system, but operating at constant pressure. The same amount of heat $đQ$ is supplied to both systems, so raising the temperatures of both systems.

To explore the significance of equation (5.23), consider Figure 5.2, which compares two systems of an ideal gas, one being heated at constant volume, and the other at constant pressure. Before the heat is applied, both systems are in identical states, with the same temperatures, pressures and volumes.

When a defined quantity $đQ$ of heat is supplied to the system shown in Figure 5.2(a) at constant volume, the gas heats up. Since this change is at constant volume, we can invoke the special case we explored on page 106 for an isochoric change, where equation (5.12a) shows that the change in internal energy dU_V is equal to the quantity of heat $đQ$

$$dU_V = đQ \tag{5.24}$$

Consider now the system shown in Figure 5.2(b), which started in exactly the same state as the system shown in Figure 5.2(a), but operates at constant pressure. When the same quantity $đQ$ of heat is supplied to this system, it heats up, and, at the same time, the increasing pressure causes the piston to move to the right, against the external pressure P_{ex}. The system in Figure 5.2(b) therefore does P, V work of expansion against the surroundings, and, if the change is quasistatic and frictionless, then equation (5.23) applies, with $đq_P = đQ$

$$dU_P + P\,dV = đQ \tag{5.25}$$

The quantity $đQ$ of heat is the same for both of these changes, so equations (5.24) and (5.25) can be combined to give

$$dU_P + P\,dV = dU_V \tag{5.26}$$

For an expansion, dV is positive, and P is necessarily positive, so a consequence of equation (5.26) is that

$$dU_V > dU_P \tag{5.27}$$

THE FIRST LAW OF THERMODYNAMICS

The inequality (5.27) implies that, for a given quantity đQ of heat transferred into any system, the resulting change dU in internal energy must always be greater for a system changing at constant volume, as compared to an identical system changing at constant pressure.

If the system is an ideal gas, then since the internal energy U of an ideal gas is a function of temperature only, the inequality (5.14) implies

$$dT_V > dT_P \qquad (5.15)$$

where dT_V represents the increase in temperature for the system at constant volume, and dT_P the increase in temperature for the system at constant pressure. The two systems, however, started at the same temperature, and so the inequality (5.15) implies that the final temperature of the system is higher for a change at constant volume, as compared to a change at constant pressure.

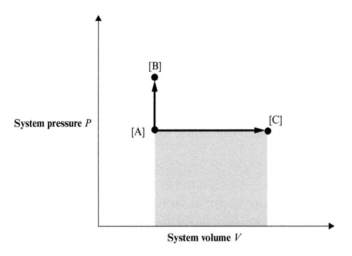

Figure 5.3 A P, V diagram for heat supplied at constant volume, and at constant pressure. A system originally in state [A] receives a quantity of heat đQ, and changes state at constant volume to state [B]. No work is done, and so all the energy đQ can be used to raise the system's temperature. If the same amount of heat đQ is supplied to a system which changes state at constant pressure from state [A] to state [C], some of the energy is required to enable the system to do P, V work of expansion, as shown by the shaded area. The energy available to raise the temperature of the system is therefore đQ − PdV. The temperature of state [B] is therefore greater than the temperature of state [C].

These results – that, for a given amount of heat đQ, $dU_V > dU_P$ and $dT_V > dT_P$ – can readily be understood in physical terms, as represented in the P, V diagram shown in Figure 5.3. The heat đQ supplied to any system is a form of energy, energy that is transferred into the system from the surroundings. If the system is constrained to maintain its volume, then all of that energy can be used to raise the system's temperature. But if the system is operating at constant pressure, some of that energy is required to do P, V work of expansion PdV against the atmosphere, so rather less energy, đQ − PdV, is available to increase the system's temperature. As a consequence, $dT_V > dT_P$.

5.15 Reversible and irreversible paths

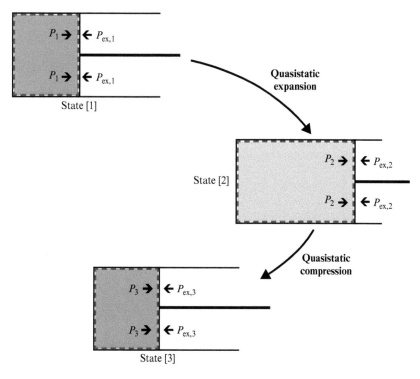

Figure 5.4 A frictionless quasistatic adiabatic expansion, followed by a frictionless quasistatic adiabatic compression, returns both the system and the surroundings to their exact original states.

Figure 5.4 shows the familiar system of an ideal gas within a cylinder, such that the boundary of the system is the interior of the cylinder, and the inside surface of the piston. The surroundings comprise the cylinder, the piston, and the external atmosphere, which we assume to be an ideal gas of vast extent.

Let's allow the system to expand, along a frictionless, quasistatic, adiabatic path, from an initial state $[1]_{sys}$ to a second state $[2]_{sys}$. The system is then compressed, once again along a frictionless, quasistatic, adiabatic path, from state $[2]_{sys}$ to a final state $[3]_{sys}$, such that the volume of the system in state $[3]_{sys}$ is equal to the original volume in state $[1]_{sys}$. Simultaneously, the surroundings are firstly compressed from a state $[1]_{sur}$ to a second state $[2]_{sur}$, and then expand back to $[3]_{sur}$, in which the volume of the surroundings is the same as in the original state.

Initially, the pressure P_1 exerted by the system on the inside surface of the piston is equal to the pressure P_{ex} exerted by the surroundings on the outer surface of the piston. The piston remains stationary, and both the system and the surroundings are in equilibrium. When the external pressure is then reduced, the pressure difference causes the piston to move frictionlessly to the right in Figure 5.4, such that the system expands from an initial volume V_1 to a final volume V_2. Since the path is quasistatic, all the state functions of both the system and the surroundings are well defined throughout, and so the system does P, V work of expansion against the surroundings given by

$$P, V \text{ work done by system on the surroundings} = \int_{V_1}^{V_2} P\, dV$$

Simultaneously, from the 'standpoint' of the surroundings

$$P, V \text{ work done by the surroundings on the system} = -\int_{V_1}^{V_2} P\, dV$$

Similarly, during the compression from V_2 to V_1,

$$P, V \text{ work done by system on the surroundings} = \int_{V_2}^{V_1} P\, dV = -\int_{V_1}^{V_2} P\, dV$$

where the sign of the second integral is reversed, corresponding to the switching of the lower and upper limits of integration. Also, from the 'standpoint' of the surroundings, for the compression from V_2 to V_1,

$$P, V \text{ work done by the surroundings on the system} = \int_{V_1}^{V_2} P\, dV$$

Since the changes are all adiabatic, there are no heat exchanges, and so all values of đq are zero. The key thermodynamic variables associated with these two changes, which together constitute a thermodynamic cycle, are shown in Table 5.2. As can be seen, this table is in three sections, the first for the system, the second for the surroundings, and the third for the sum of the system and the surroundings, which together constitute the universe.

The entries within Table 5.2 contain no surprises. Since the boundary between the system and the surroundings is adiabatic, there are no heat exchanges between the system and the surroundings; and because the paths followed are quasistatic, all the state functions of both the system and the surroundings are well-defined throughout, so that all work done can be expressed in terms of the state functions P and V of the system. Also, all changes within the surroundings are equal in magnitude, and opposite in sign, to the corresponding changes in the system, and nothing is 'lost' across the system boundary. The First Law is obeyed throughout.

By assumption, the system comprises an ideal gas, and so the internal energy U_{sys} is a function of temperature only. The sign of ΔU_{sys}, as shown in Table 5.2, therefore indicates whether the temperature of the system increases or decreases. As can be seen, as a result of the initial adiabatic expansion, ΔU_{sys} is negative, and so the system cools; during the subsequent adiabatic compression, ΔU_{sys} is positive, and the system heats up. Importantly, as shown on

Table 5.2 Thermodynamic data corresponding to the changes depicted in Figure 5.4

For the system

	System expansion	System compression	Total cycle
P, V work done by system on surroundings, $(w_{P,V})_{sys}$	$+\int_{V_1}^{V_2} P\,dV$	$-\int_{V_1}^{V_2} P\,dV$	0
Heat gained by system from surroundings, $(q)_{sys}$	0	0	0
$\Delta U_{sys} = (q)_{sys} - (w_{P,V})_{sys}$	$-\int_{V_1}^{V_2} P\,dV$	$+\int_{V_1}^{V_2} P\,dV$	0
ΔT_{sys}	Negative	Positive	0

For the surroundings

	Compression of surroundings	Expansion of surroundings	Total cycle
P, V work done by surroundings on system, $(w_{P,V})_{sur}$	$-\int_{V_1}^{V_2} P\,dV$	$+\int_{V_1}^{V_2} P\,dV$	0
Heat gained by surroundings from system, $(q)_{sur}$	0	0	0
$\Delta U_{sur} = (q)_{sur} - (w_{P,V})_{sur}$	$+\int_{V_1}^{V_2} P\,dV$	$-\int_{V_1}^{V_2} P\,dV$	0
ΔT_{sur}	Positive	Negative	0

For both the system and the surroundings

	State [1] to State [2]	State [2] to State [3]	Total cycle
P, V work done by the universe, $(w_{P,V})_{uni}$	0	0	0
Heat gained by the universe, $(q)_{uni}$	0	0	0
$\Delta U_{uni} = (q)_{uni} - (w_{P,V})_{uni}$	0	0	0
ΔT_{uni}	0	0	0

Table 5.2, for the whole cycle, $\Delta U_{sys} = 0$, implying that, for the whole cycle $\Delta T_{sys} = 0$. The final temperature of the system is therefore equal to the initial temperature.

An important feature of this sequence of changes is that the volume of the system in the final state [3], is deliberately chosen to be equal to the volume in the initial state, state [1]. Now, we have just shown that the temperature of the system in the final state is equal to that of the original state, and we know that, in the absence of effects such as those of gravity, motion and the like (see page 48), any pure substance has two and only two independent state functions. Given that we have proven that the volume and temperature of the initial and final states of the system are unchanged, then the pressure of the final state [3] must also be equal to the pressure in the initial state [1].

This demonstrates that, as a result of the entire frictionless, adiabatic, cycle, the final state of the system is identical to the initial state, as represented in the P, V diagram shown in Figure 5.5.

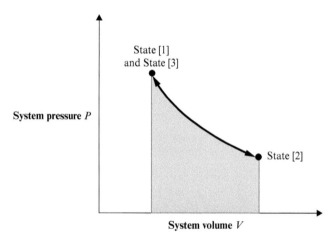

Figure 5.5 A P, V diagram corresponding to the states of the system depicted in Figure 5.4. The system starts in state [1]. The quasistatic, adiabatic, expansion follows the path shown by the solid arrow to state [2]. The subsequent quasistatic, adiabatic, compression retraces exactly the same path back to state [3], which is identical to state [1]. The quasistatic, adiabatic path from state [1] to state [2] is a reversible path, as is the quasistatic, adiabatic path from state [2] to state [3]. The shaded area represents the work done by the system on the surroundings during the expansion, and by the surroundings on the system during the compression. The whole cycle from state [1] to state [2], and back from state [2] to state [1], comprises only two reversible paths - paths which are coincident and opposite - and so this is an example of the rare situation in which $\oint dw = 0$ (see page 39).

Since the surroundings comprise the cylinder, the piston, and also (by assumption) the ideal gas of the external atmosphere, we cannot equate ΔU_{sur} to the change in the internal energy of the ideal gas of the external atmosphere alone – we must also take account of changes in the internal energies of the cylinder and the piston too. But as shown in Table 5.2, for the whole cycle, no heat is exchanged either into or out of the surroundings. Since, for the whole cycle, there is no source of heat, the temperature of the surroundings as a whole in the final state must equal the temperature in the initial state, as verified by the fact that $\Delta U_{sur} = 0$. Also, throughout the change, the volumes and pressures associated with the cylinder and piston themselves remain constant, and the ideal gas comprising the rest of the surroundings returned to its original volume and temperature, and hence pressure.

Collectively, this all implies that the surroundings too have returned to their original state, enabling us to verify that, for the frictionless, quasistatic, adiabatic cycle as a whole, *both* the system *and* the surroundings return *exactly* to their original states.

Now, that may seem perfectly obvious; which indeed it might be. But it is significant, as illustrated in Figure 5.6.

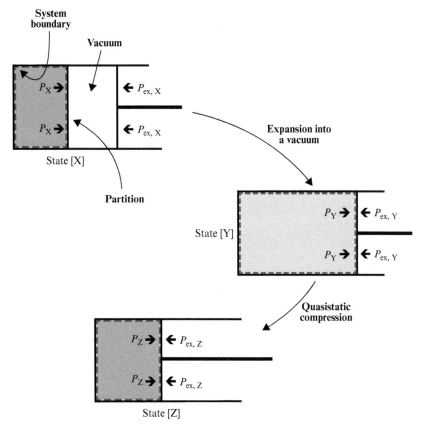

Figure 5.6 An adiabatic expansion into a vacuum, followed by a quasistatic adiabatic compression, does not return either the system or the surroundings to their original states.

In Figure 5.6, the ideal gas within the system in state [X] has exactly the same pressure, volume and temperature as the ideal gas in the system in state [1] as shown in Figure 5.4. In Figure 5.4, however, the system was bounded by the piston, whereas in Figure 5.6, the system is bounded by a partition, on the far side of which is a vacuum, itself bounded by the cylinder and a frictionless piston.

If the partition shown in Figure 5.6 is broken, the ideal gas expands adiabatically into the vacuum, to reach state [Y] – a process known as **free expansion** or **Joule expansion**. Since the gas in the system is expanding against a vacuum, there is no external pressure to overcome, and so no P, V work of expansion is done. Furthermore, the change is adiabatic, and so no heat is exchanged with the surroundings. Applying the First Law to the system

$$dU_{sys} = đq_{sys} - đw_{sys}$$

then, since both $đq_{sys}$ and $đw_{sys}$ are zero, dU_{sys} must also be zero. Given that we know that the internal energy U of an ideal gas is a function of temperature only, this implies that the temperature of the system is unchanged, and that the temperature T_Y of the system in state [Y] is the same as the temperature T_X of the system in the original state [X]: $T_Y = T_X$.

For the surroundings, no work is done either by the system on the surroundings or by the surroundings on the system; nor is there any heat exchange. There is therefore no change in the internal energy of the surroundings, and no change in temperature.

Given that the system expanded adiabatically from state [X] to state [Y], if we wish to return the system from state [Y] back to its original state [X], it seems that a good way to do this would be by means of an adiabatic compression. So, let's now compress the system frictionlessly, adiabatically and quasistatically so that the final volume is made equal to the original volume V_X. This return path, however, cannot be the same as the original path: the original path was an expansion into a vacuum, but since the vacuum has now been filled, the return path requires that the piston must be pushed back by the surroundings against the internal pressure. To achieve this, P, V work of compression needs to be done on the system by the surroundings, implying that, for the return motion of the piston, from the system's 'point of view', $đw_{sys}$ is non-zero, and negative. Since the compression is adiabatic, $đq_{sys}$ is zero, and so, applying the First Law

$$dU_{sys} = đq_{sys} - đw_{sys}$$
$$= - đw_{sys}$$

Since $đw_{sys}$ is a negative number, dU_{sys} is positive, implying that the temperature of the final state is different from, and higher than, the original temperature T_X of the original state [X]. The state that the system reaches as a result of the adiabatic compression is therefore *different* from the original state [X], so let's designate the final state as state [Z]. We have just shown that $T_Z > T_X$; in addition, if we apply the ideal gas law $PV = nRT$ to both the initial state [X] and the final state [Z], then

$$\frac{P_X V_X}{T_X} = \frac{P_Z V_X}{T_Z}$$

and so

$$P_Z = \frac{P_X T_Z}{T_X}$$

Since we know that $T_Z > T_X$, then $P_Z > P_X$, implying that both the pressure, and the temperature, of the final state [Z] are higher than the pressure and the temperature of the initial state [X], as shown in Figure 5.7.

The key thermodynamic variables for these changes are all shown in Table 5.3: as can be seen, the adiabatic compression from state [Y] to state [Z] results in a simultaneous heating of the system and a corresponding cooling of the surroundings.

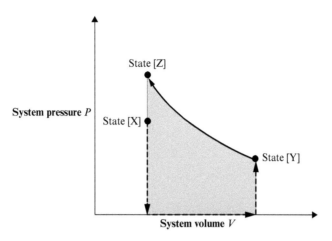

Figure 5.7 This is a P, V diagram for the system depicted in Figure 5.6. The system starts in state [X], and expands adiabatically against a vacuum to state [Y]. This change follows the non-quasistatic path approximated by the dashed line. The area between the dashed line and the horizontal axis is zero, as is consistent with the fact that no work is done when a gas expands against a vacuum. In returning the system frictionlessly, quasistatically and adiabatically from state [Y] to a state which has a volume equal to the volume of the initial state, the system follows the solid line to state [Z]. State [Z], as required, has the same volume as state [X], but the pressure is higher, as is the temperature. The work done by the surroundings on the system for the path from state [Y] to state [Z] is represented by the shaded area. No adiabatic path from state [Y] to state [X] exists, and so the adiabatic path from state [X] to state [Y] is therefore irreversible.

What does all this mean? In a word – and an important word too – *irreversibility*. Fundamentally, the change of state from state [X] to state [Y], in which the gas in the system expanded into a vacuum, is **irreversible**, in that gases naturally and spontaneously expand into a vacuum, as happens when the partition in Figure 5.6 is broken. But gases never, just never, spontaneously change from filling the entire space available, as in state [Y], to a state in which the gas is bunched up on the left-hand side, with a vacuum to the right, as in state [X], but without the partition. Expansion of a gas into a vacuum is an example of a change that is **unidirectional** – a change that naturally happens in one direction only, and not in reverse: other examples are heat flow down a temperature gradient (hot things spontaneously cool down, whereas cool things don't 'suddenly' warm up), the mixing of many fluids (milk stirred into tea doesn't subsequently separate out), and any number of chemical reactions go in only one direction (once ignited, octane immediately burns in oxygen to give carbon dioxide and water; a mixture of the corresponding quantities of carbon dioxide and water will never, spontaneously, react to give octane and molecular oxygen).

Irreversibility and unidirectionality are universal features of the world in which we live: there are many changes that take place in only one direction. And some deeper insight into irreversibility and unidirectionality can be obtained from Figures 5.4, 5.5, 5.6 and 5.7, and the corresponding entries in Tables 5.2 and 5.3.

With reference to Figure 5.4, and the corresponding entries in Table 5.2, we see that the second frictionless, quasistatic, adiabatic path from state [2] to state [3] reversed, precisely, the original adiabatic change from state [1] to state [2] for both the system and the surroundings, such that the final states $[3]_{sys}$ and $[3]_{sur}$ are identical to the original states $[1]_{sys}$ and $[1]_{sur}$. The original change from state [1] to state [2] is therefore known as a **reversible change**, as achieved by following a **reversible path**.

THE FIRST LAW OF THERMODYNAMICS

Table 5.3 Thermodynamic data corresponding to the changes depicted in Figure 5.6

For the system

	System expansion into a vacuum	System compression	Total cycle
P, V work done by system on surroundings, $(w_{P,V})_{sys}$	0	$-\int_{V_1}^{V_2} P\,dV$	$-\int_{V_1}^{V_2} P\,dV$
Heat gained by system from surroundings, $(q)_{sys}$	0	0	0
$\Delta U_{sys} = (q)_{sys} - (w_{P,V})_{sys}$	0	$+\int_{V_1}^{V_2} P\,dV$	$+\int_{V_1}^{V_2} P\,dV$
ΔT_{sys}	0	Positive	Positive

For the surroundings

	No change	Expansion of surroundings	Total cycle
P, V work done by surroundings on system, $(w_{P,V})_{sur}$	0	$+\int_{V_1}^{V_2} P\,dV$	$+\int_{V_1}^{V_2} P\,dV$
Heat gained by surroundings from system, $(q)_{sur}$	0	0	0
$\Delta U_{sur} = (\Delta q)_{sur} - (\Delta w_{P,V})_{sur}$	0	$-\int_{V_1}^{V_2} P\,dV$	$-\int_{V_1}^{V_2} P\,dV$
ΔT_{sur}	0	Negative	Negative

For both the system and the surroundings

	State [X] to State [Y]	State [Y] to State [Z]	Total cycle
P, V work done by the universe, $(w_{P,V})_{uni}$	0	0	0
Heat gained by the universe, $(q)_{uni}$	0	0	0
$\Delta U_{uni} = (q)_{uni} - (w_{P,V})_{uni}$	0	0	0
ΔT_{uni}	0	0	0

In contrast, the adiabatic change from state [Y] to state [Z], as depicted in Figure 5.7 and as described mathematically in Table 5.3, *did not* return the system, or the surroundings, to their original states, even though the final volume V_Z was deliberately made equal to the original volume V_X. The first adiabatic path, from state [X] to state [Y], *cannot* be reversed such that both the system and the surroundings are returned to their original states.

Any path that cannot be reversed so that *both* the system *and* the surroundings are returned to their exact original states is known – unsurprisingly – as an irreversible path, and the corresponding change of state is known as an irreversible change. Furthermore, if any thermodynamic cycle contains just one irreversible step, however small, the cycle as a whole must be regarded as irreversible.

Why is the path from state [X] to [Y] irreversible? The reason is this: the path from state [X] to state [Y] was not quasistatic. Because the ideal gas expanded into a vacuum, the change from state [X] to state [Y] did not take place through an infinite sequence of equilibrium states in which all the system's state functions are well-defined – rather, the gas expanded suddenly, there was turbulence, and during the change, the values of the system's pressure and volume were not well-defined, as indicated by the dashed lines in the P, V diagram shown in Figure 5.7. For a change to be reversible, it must be quasistatic – and since the path actually followed from state [X] to state [Y] wasn't quasistatic, it must be irreversible.

As we have just demonstrated, the path actually taken from state [X] to state [Y], as shown by the dashed line, is irreversible. This is true. It is false, however, to make a statement of the form "The path from state [X] to state [Y] is irreversible, therefore state [X] cannot be reached from state [Y]". At first sight, this statement seems reasonable, and even more so when Figure 5.7 (correctly) shows that a quasistatic adiabatic compression of state [Y] results in state [Z], not state [X]. In fact, the original state [X] can be reached from state [Y] by any number of paths, of which these are two possibilities

- From state [Y], reach state [Z] by a quasistatic adiabatic compression as shown in Figure 5.7. As discussed in the earlier narrative, state [Z] has the same volume as the original state [X], but a higher temperature and pressure than the original state [X]. But if state [Z] is then cooled at constant volume, the temperature and pressure of the system will both be reduced, until state [X] is reached. In terms of the P, V diagram shown in Figure 5.7, this final step would trace a path, vertically downwards, from state [Z] to state [X], closing the 'top left hand corner', with the exchange of heat from the system to the surroundings.
- Another path from state [Y] to state [X] is based on the observation that, as once again discussed in the narrative, the temperature T_Y of the system in state [Y] is the same as the temperature T_X of the system in the original state [X]. An isothermal – rather than an adiabatic – compression of state [Y] will therefore reach the original state [X].

State [X] can therefore be reached from state [Y] by a number of different paths – *but not by retracing, exactly, the original, irreversible, path from [X] to [Y] in the opposite direction*. The important point is this: irreversibility is a property of paths, not states. For any system, any two states [X] and [Y] are mutually accessible by many paths. One particular category of path from state [X] to state [Y], known as a reversible path, is characterised by being quasistatic and free from all friction (and similar 'dissipative' effects) – characteristics which imply that it is possible to retrace the path 'backwards', so returning both the system and the surroundings to their exact original states. Any path from state [X] to state [Y] which does not have both of these characteristics is necessarily irreversible, implying that state [X] cannot be reached from state [Y] by retracing the original path 'backwards'; if it is necessary to reach state [X] from state [Y], then a totally different path must be followed.

As we shall see when we discuss the Second Law, the distinction between reversible and irreversible paths is of great significance. We now know that a necessary condition for a path to be reversible is that it must be quasistatic, and that all non-quasistatic paths must be irreversible. In the following sections, we explore two other phenomena which relate to irreversibility: firstly, mixing, and then friction.

5.16 Mixing

Figure 5.8 shows a system comprising two compartments of equal volume. Each compartment contains 0.5 mol of the same ideal gas, at equal temperatures, but at different pressures P_1 and P_2, such that $P_1 > P_2$. Although the gases are in fact identical, let's assume that the molecules of the gas in the left-hand compartment can be distinguished in some way from those in the

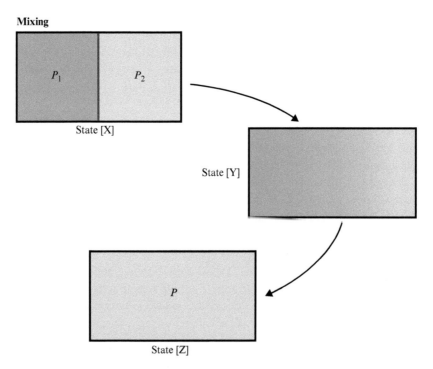

Figure 5.8 Mixing. State [X] comprises two compartments of equal volume, separated by a partition. Each compartment contains 0.5 mol of the same ideal gas, at the same temperature, but at different pressures P_1 and P_2, such that $P_1 > P_2$, as represented by the different shading. If the partition is removed, the gases mix, passing through a series of non-equilibrium states, any one of which is represented by state [Y]. Ultimately, the system reaches a final equilibrium state [Z], in which the gases are totally mixed, at pressure $P = (P_1 + P_2)/2$.

right-hand compartment – perhaps those on the left are 'blue', and those on the right, 'red'. When the partition is removed, the pressure difference $P_1 - P_2$, which is positive from left to right, will cause the gas originally in the left compartment to flow to the right. But at the same time as the 'blue' molecules originally on the left move to the right, we will notice that some of the 'red' molecules originally on the right move to the left, even though the original pressure difference $P_1 - P_2$ acts from left to right. Initially, the number of 'blue' molecules moving to right will be greater than the number of 'red' molecules moving to the left. For a time, the system will move through a continuum of turbulent, non-equilibrium, states during which the two gases mix – collectively represented by the state [Y] in Figure 5.8 – until an equilibrium state [Z] is reached, in which both gases are evenly distributed throughout the total volume available, with a final, stable, system pressure $P = (P_1 + P_2)/2$.

The turbulence we observe immediately after the partition is removed indicates that the system is not in equilibrium, implying that the path from state [X] to state [Z] is not quasistatic, and hence irreversible. This irreversibility is totally in accord with our experience, for such a change is unidirectional: a pressure difference drives a non-quasistatic, irreversible, flow of molecules down the pressure gradient until the pressure equilibrates and the molecules are evenly mixed, just as a temperature difference will drive a non-quasistatic, irreversible, flow of heat down a temperature gradient until the temperature equilibrates - in which context, it is instructive to compare Figure 5.8 to Figure 1.2. As we know, if a system is in thermal equilibrium, we never observe that it suddenly becomes hotter in one-half and cooler in the other; similarly, a system in which 'blue' and 'red' molecules are evenly mixed never suddenly becomes one in which all the 'blue' molecules are bunched up on the left, and the 'red' ones on the right.

In this example, we have assumed that there is an initial pressure gradient $P_1 - P_2$, that drives the mixing. But that is not the whole story. Something deeper is happening too – as is readily appreciated by considering what would happen if the pressures on each side of the partition in Figure 5.8 are the same, so that there is no pressure difference. If the pressures are indeed the same, when the partition is removed, *we will observe exactly the same mixing effect*: the 'blue' molecules will diffuse to the right, and the 'red' ones to the left, until the gases are evenly mixed. The process might take place rather more slowly than in the presence of an initial pressure difference, but it will take place nonetheless. And during the mixing process, the volume actually occupied by the 'blue' molecules, which was well-defined before the partition was removed, becomes ill-defined, as is the pressure associated with 'blue'; likewise for 'red'.

At the microscopic level, the process is therefore non-quasistatic, but at the macroscopic level, the system appears to be in equilibrium: the total volume of the system remains unchanged, as does the pressure, and hence the temperature too. And if the molecules were no longer 'blue' and 'red', we would not be able to distinguish one molecule from another, to track where any particular molecule is going, or to identify where it came from. Macroscopically, we would just observe 1 mol of an ideal gas, in equilibrium, with a volume V, a pressure P, and at temperature T. Yet, at a molecular level, the system is in a turmoil of molecular motion, a molecular motion that is the fundamental basis of mixing.

Now, if the two sides of the system shown in Figure 5.8 were formed from a solid, we would not observe mixing. Mixing as depicted in Figure 5.8 takes place only for fluids – liquids and especially gases. The basic difference between a solid and a fluid is that, in a fluid, the molecules have much more freedom of movement, and so the mixing depicted in Figure 5.8 is attributable to the independent motions of each of the molecules within the gas. Once the partition has been removed, those molecules originally on the left have the opportunity, and freedom, to move to the right, and some, randomly, will do so; simultaneously, those molecules originally on the right have the opportunity, and freedom, to move to the left, and some, randomly, will do so. At the molecular level, the process of mixing results from the random motion of each molecule, with the final result, in which the gases are fully mixed, being the aggregate consequence of these individual random motions. For the process to be reversed, at the molecular level, this would require all the 'blue' molecules in the mixed gas to move into the left-hand side of the container, whilst, at the same time, all the 'red' molecules will need to move into the right-hand side. Since 1 mol of an ideal gas contains some 6×10^{23} molecules, all of which are moving randomly, the likelihood that a total of 6×10^{23} 'blue' and 'red' molecules will happen to co-ordinate their independent random movements in just the 'right' way is vanishing small. The probability of such an event is zero – it just won't happen.

Our discussion of macroscopic, non-quasistatic, unidirectional, irreversible mixing, and what is happening at a molecular level, makes a connection between, on the one hand, the macroscopically observed phenomena of unidirectionality and irreversibility and, on the other, the microscopic probability of particular types of behaviour at the molecular level – a connection which will be developed further on in Chapter 11.

Molecular irreversibility – of which physical mixing is one example, and chemical reactions are another – is a second cause, alongside the failure to follow a quasistatic path, of irreversibility. The next section examines a third – friction.

5.17 Friction

5.17.1 Static friction and dynamic friction

All the examples of gases expanding in cylinders fitted with pistons that we have examined so far in this book have assumed the absence of friction – indeed the words 'frictionless piston' have been used so often, it's easy to stop noticing them. In this section, we'll explore what happens when friction is present – as is inevitably the case in all real systems, even when its effects are minimised by using, for example, lubricating oils.

As we saw in Chapter 2 (see page 28), there are two types of friction – **static friction** that stops two surfaces from sliding over one another, and **dynamic friction** (sometimes known as kinetic friction) which acts when two surfaces are moving against one another. We'll explore the effect of static friction shortly (see page 140); for the moment we will focus on dynamic friction.

As verified by every-day experience, and first studied rigorously by Rumford and Joule (see pages 51 and 52), dynamic friction has three key characteristics.

- Firstly, dynamic friction is a force which always opposes motion, which is why **frictional work** needs to be done overcome it.

- Secondly, the performance of frictional work inevitably creates heat – known as **frictional heat** – such that all the work done against dynamic friction is instantaneously and completely converted into heat.
- Thirdly, the relationship between frictional work and frictional heat is unidirectional, in that frictional work is inevitably converted into frictional heat, but no-one has ever observed the conversion of frictional heat back into work: with reference to Count Rumford's observations, the work done by the drill makes the cannon hot, but heating the cannon will never drive the drill.

The purpose of this section is therefore to gain a deeper understanding of frictional work, and the corresponding frictional heat.

5.17.2 Friction is asymmetrical

As we know well, the two path functions, work and heat, are phenomena that take place at the boundary between a system and its surroundings, as a change of state takes place. Also, in accordance with the First Law, and the principle of the conservation of energy, no energy 'gets lost' as the result of any change in state – any reduction in the internal energy of a system is associated with a corresponding increase in the internal energy of the surroundings; likewise any increase in the internal energy of a system is associated with a corresponding reduction in the internal energy of the surroundings.

In all the examples given so far in this book, this principle of 'what the system loses is gained by the surroundings' has been applied to each of the components of work and heat individually – reference, for example, to the entries in Tables 5.2 and 5.3 will show that the 'work done by the system on the surroundings' is equal to the negative of the 'work done by the surroundings on the system'; likewise any 'heat gained by the system from the surroundings' is equal to the negative of any 'heat gained by the surroundings from the system'. All of the flows of work and heat have been 'symmetrical' in the sense that work done by the system maps on to work done on the surroundings, and heat lost from the system maps on to heat gained by the surroundings.

All our examples, however, have also intentionally excluded friction, so let us now allow for friction, and see how this can be incorporated into the First Law. Figure 5.9 therefore shows the familiar system of an expanding gas within a cylinder fitted with a piston, but now there is dynamic (but not static) friction between the rim of the piston and the internal surface of the cylinder. The boundary of the system is the interface between the gas within the cylinder, along the interior surface of the cylinder and the internal face of the piston: this implies that the dynamic friction between the rim of the piston and the interior surface of the cylinder takes place within the surroundings, not within the system.

Suppose for the moment that the system boundary, as defined in Figure 5.9 is adiabatic, that the external pressure P_{ex} is zero, and that the piston is held in position at point A by a peg. If this peg is removed, the pressure P within the cylinder drives the piston to the right, where it is stopped by a peg at point B. In the *absence of friction*, since the expansion is against a vacuum, the system does no work on the surroundings, so đw_{sys} = 0; nor is there any exchange of heat, and so đq_{sys} = 0.

THE FIRST LAW OF THERMODYNAMICS

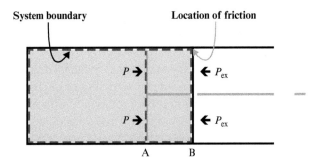

Figure 5.9 Friction. The system boundary is on the inside of both the cylinder and the piston. As the gas expands, and the piston moves from left to right from **A** to **B**, the system does two kinds of work - P, V work of expansion against the external pressure, and also frictional work in pushing the piston against the force of dynamic friction. The dynamic friction always opposes the motion of the piston, and acts around the rim of the piston, on the inside of the cylinder, in the surroundings. The work done against friction is necessarily positive, whether the piston is moving to the right or to the left.

For the system, according to the First Law we therefore have

$$dU_{sys} = đq_{sys} - đw_{sys} = 0$$

and for the surroundings

$$dU_{sur} = đq_{sur} - đw_{sur} = 0$$

Furthermore, $dU_{sys} + dU_{sur} = 0$, and so energy is conserved for the universe.

We now return the piston to position A, but when the piston is allowed to move to the right a second time, let us suppose that *there is now some dynamic friction*, acting between the rim of the piston and the interior surface of the cylinder, opposing the motion of the piston. Since the external pressure P_{ex} is zero, the system does no P, V work of expansion against the surroundings, but the system does have to do work against friction to make the piston move. If we represent this frictional the work as $đw_{fric}$, then $đw_{sys} = đw_{fric}$.

Since the boundary of the system is on the inside of both the piston and the cylinder, the point of contact between the piston and the cylinder is in the surroundings. So it is within the surroundings, not the system, that the friction takes place, and it is within the surroundings that the heating effect of friction arises. Since, by assumption, the system boundary is adiabatic, no frictional heat can 'leak back' into the system, and so $đq_{sys} = 0$. Accordingly, for the system, which performs frictional work

$$dU_{sys} = đq_{sys} - đw_{sys} = 0 - đw_{fric} = - đw_{fric}$$

Since it is the system that does the work to overcome the friction, $đw_{fric}$, is positive, and so dU_{sys} is negative – the system cools.

In the surroundings, in principle, work $đw_{fric}$ is done on the surroundings by the system, and so $đw_{sur} = - đw_{fric}$, where the – sign is required to transform the positive quantity $đw_{fric}$ into a negative quantity so as to comply with the sign convention that 'work done on' is negative. The system boundary is adiabatic, and so $đq_{sur} = 0$. For the surroundings, we therefore have

$$dU_{sur} = đq_{sur} - đw_{sur} = 0 - (- đw_{fric}) = đw_{fric} \tag{5.28}$$

Since đw_{fric} is positive, dU_{sur} is positive, and so the surroundings get hotter – this being the effect of frictional heat. Furthermore, dU_{sys} + dU_{sur} = – đw_{fric} + đw_{fric} = 0, so verifying the conservation of energy.

The analysis so far has been symmetrical, for we have equated the frictional work đw_{sys} = đw_{fric} done by the system against friction with the negative of the work đw_{sur} done on the surroundings by the system. But there is an alternative – and perhaps more realistic – way of describing what is happening in the surroundings.

Although the system 'experiences' work – the system really has to do work to push the piston against the dynamic friction – the surroundings 'experience' heat: the wall of the cylinder gets hot, immediately. Furthermore, an observer, taking measurements of the cylinder, and without any knowledge of what was happening, would detect an increase in the temperature of the cylinder wall, exactly as if heat were being directly transferred across the system boundary.

So, rather than describing the action of friction as one in which the frictional work đw_{fric} done by the system against friction is set equal to the work done on surroundings by the system, an alternative is to set the frictional work đw_{fric} done by the system equal to the frictional heat đq_{fric} = đq_{sur} gained by the surroundings. Furthermore, since the piston moves against a vacuum, no P, V work is done on the surroundings, and so đw_{sur} = 0.

We now have

đq_{sur} = đq_{fric} = đw_{fric}

and

đw_{sur} = 0

implying that

$$dU_{sur} = đq_{sur} - đw_{sur} = đq_{fric} - 0 = đw_{fric} \tag{5.29}$$

Comparison of equations (5.28) and (5.29) shows that they are identical: dU_{sur} = đw_{fric}, and since đw_{fric} is positive, dU_{sur} is positive, and the surroundings get hotter. That's frictional heat. But they were derived in different ways: equation (5.28) was based on the symmetry of 'work done by' = – 'work done on', and the increase in temperature attributable to friction is a result of the increase in dU_{sur}; in contrast, equation (5.29) 'breaks the symmetry', and equates the work done by the system, đw_{fric}, directly to the heat gained by the surroundings, đq_{sur} = đq_{fric}, and it is đq_{fric} that then drives the increase in temperature of the surroundings, and is the cause of the increase in dU_{sur}.

Given these two rather different interpretations, which is 'right'? Both give the same, final, correct result, but our view is that the second interpretation – the interpretation that underpins equation (5.29), and invokes the direct conversion of frictional work đw_{fric} into frictional heat đq_{fric} – is preferable, for three reasons.

Firstly, in equating the frictional work đw_{fric} directly to the frictional heat đq_{fric}, the frictional heat đq_{fric} is an intrinsic part of the analysis, and immediately corresponds to an increase in temperature. This is a more appropriate description of what actually takes place.

Secondly, consider how we would describe what happens if the system boundary were not situated around the interior of the cylinder and piston as shown in Figure 5.9, but some distance away from the cylinder and the piston. In this case, 'the system' would comprise the gas within the cylinder, the piston, the cylinder itself, and some external space too. The motion of the piston therefore takes place totally within the (now bigger) system; there is no heat exchange between the system and the (now further away) surroundings; there is no work done across the (more distant) boundary, but as the piston moves to the right in Figure 5.9, frictional work still is done, and frictional heat is still generated, but both happen entirely within the system.

We would now have, for the 'bigger' system

$$dU_{sys} = đq_{sys} - đw_{sys} = 0$$

But within the system work $đw_{fric}$ will have been done against friction, frictional heat $đq_{fric}$ will have been generated, and the wall of the cylinder will have become hotter. Equating $đw_{fric}$ directly to $đq_{fric}$ implies that

$$dU_{sys} = đq_{fric} - đw_{fric} = 0$$

which has real meaning: because $dU_{sys} = 0$, energy is conserved, which must be right; furthermore, since work is done by the system against friction, $đw_{fric}$ must be positive, implying that $đq_{fric}$ must be positive too, and of the same magnitude as $đw_{fric}$ – so verifying that frictional heat makes the system hotter, and also validating Joule's 'mechanical equivalent of heat'.

But it is the third reason that is the most compelling. By equating $đw_{fric} = đq_{fric}$, we are 'breaking the symmetry' of 'work done by' = – 'work done on'. The introduction of this asymmetry runs deep, especially since our experience tells us that the asymmetry is not just about equating $đw_{fric} = đq_{fric}$, it is also about recognising that this is a one-way, unidirectional, relationship: as Count Rumford showed, it is natural to convert frictional work into frictional heat, but the reverse process is impossible – frictional heat can never be converted into frictional work.

Friction is fundamentally an asymmetrical process. This asymmetry is of great significance, and is intrinsically related to unidirectionality and irreversibility. Also, as we shall see, especially in Chapters 8, 9 and 11, it relates to two other types of asymmetry: firstly, the asymmetry associated with the spontaneity of change, and the observation that spontaneous changes only go one way, and secondly, the asymmetry of time – time runs only forwards, and not backwards.

5.17.3 The mathematics of friction

Returning once more to the system depicted in Figure 5.9, suppose that the surroundings are now no longer a vacuum, but the atmosphere, so that there is a non-zero external pressure P_{ex} acting on the right-hand side of the piston. Let us also assume that there is no static friction, but there is dynamic friction acting between the rim of the piston and the inner surface of the cylinder. As the gas within the system expands, the system does two kinds of work – P, V work of expansion against the external pressure P_{ex}, and also the work required to push the piston against the dynamic friction, which opposes the piston's motion.

In accordance with the usual definition of work, we can express the work done against dynamic friction as

Work done against dynamic friction = dynamic frictional force overcome
× distance moved

The nature of the dynamic frictional force depends on the specific circumstances, but, in many situations, this force can be expressed in terms of the product of a positive constant φ and the instantaneous velocity of the piston $dx/d\tau$ (where we use the symbol τ for time, so as to avoid any confusion with the symbol t for centigrade temperature)

$$\text{Dynamic frictional force opposing motion} = \varphi \frac{dx}{d\tau} \qquad (5.30)$$

One implication of this relationship is that if there is no motion, then $dx/d\tau = 0$, implying that the dynamic frictional force is zero, as makes intuitive sense; also, the greater the velocity $dx/d\tau$, the stronger the opposing dynamic frictional force.

If the piston moves against friction through a distance dx, then the work $đw_{\text{fric}}$ done by the system against dynamic friction is given by

$$đw_{\text{fric}} = \varphi \frac{dx}{d\tau} dx \qquad (5.31)$$

So, for a finite change in which the piston moves from an initial position x_1 to a subsequent position x_2, the total work done by the system against dynamic friction is

$$w_{\text{fric}} = \varphi \int_{x_1}^{x_2} \frac{dx}{d\tau} dx \qquad (5.32)$$

It so happens that it is more informative to express the work done by the system against dynamic friction rather differently, by using what appears to be something of a mathematical 'trick'. If we express dx as

$$dx = \frac{dx}{d\tau} d\tau$$

which is quite valid, since $d\tau/d\tau$ is necessarily equal to one, then we may write

$$đw_{\text{fric}} = \varphi \frac{dx}{d\tau} \frac{dx}{d\tau} d\tau = \varphi \left(\frac{dx}{d\tau}\right)^2 d\tau \qquad (5.33)$$

Hence, for a finite change, in which the piston moves from an initial position x_1, which it occupies at time τ_1, to a subsequent position x_2, which it occupies at time τ_2, the total work done against dynamic friction during this change in state can be expressed as an integral over time as

$$w_{\text{fric}} = \varphi \int_{\tau_1}^{\tau_2} \left(\frac{dx}{d\tau}\right)^2 d\tau \qquad (5.34)$$

In equations (5.33) and (5.34), the frictional factor φ is, by definition, positive, and $(dx/d\tau)^2$, as a square, must also always be positive, whether the piston is moving to the right in Figure 5.9 (in which case $dx/d\tau$ is positive) or *moving in the reverse direction, to the left* (in which case $dx/d\tau$ is negative). The integrand in equation (5.34) is therefore *always positive*, under all conditions, and can never be negative. The value of the integral is therefore always positive, and, as an integral over time (which can be considered as a series of sequential additions), the value of this integral *must increase with time*: the longer the motion takes place, and the greater the value of the upper time limit τ_2, the greater the amount of work that is done against dynamic friction, even when the piston moves just to the left, just to the right, or to-and-fro.

Equations (5.32)

$$w_\text{fric} = \varphi \int_{x_1}^{x_2} \frac{dx}{d\tau} dx \qquad (5.32)$$

and (5.34)

$$w_\text{fric} = \varphi \int_{\tau_1}^{\tau_2} \left(\frac{dx}{d\tau}\right)^2 d\tau \qquad (5.34)$$

express the same quantity – the total work done by the system against dynamic friction – but the form of equation (5.34) makes it more obvious that this is always a positive quantity, and that, in overcoming dynamic friction, the system must *always* perform work: work is inevitably done *by* a system against friction; friction can never do work *on* a system. This is the mathematical demonstration of irreversibility, unidirectionality and asymmetry.

And, as observed so carefully by Rumford and Joule, this work is inevitably, and immediately, converted into heat – heat that is released into the surroundings, whether the piston is moving outwards or inwards. And because equations (5.32) and (5.34) can never be negative, it can never be possible for work to be done on the system by frictional heat: frictional work can be converted into heat, but frictional heat can never be converted back into work.

The fact that all the frictional work is converted, in its entirety, into frictional heat is a manifestation of the First Law and the principle of the conservation of energy – indeed, as we saw on page 53, it was the research by Rumford and Joule that identified work and heat as equivalent, but apparently different, realisations of the deeper, unifying, concept of energy. The unidirectionality of change, however – the empirical fact that frictional work is transformed into heat, but that frictional heat can never be transformed into work – is not addressed by the First Law, which is concerned with quantities, rather than directionality. As we shall see, the issue of directionality is central to the Second Law, the basic principles underlying which we are exploring here.

5.18 Friction and irreversible paths

Figure 5.10 shows the familiar system of an ideal gas within a cylinder, such that the boundary of the system is the interior of the cylinder, and the inside surface of the piston. Let's further suppose that the system boundary is perfectly adiabatic, so that no heat can pass between the system and the surroundings: this might be achieved, for example, by enclosing the gas within a very thin 'balloon', which can be expanded and compressed, made from a material that acts as

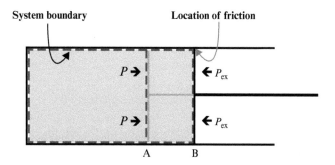

Figure 5.10 Friction and irreversible paths. The piston moves quasistatically and adiabatically from position A to position B, and back again. Dynamic friction is present for the expansion as the piston moves from position A to position B, but not for the return path. The path in which friction is present is irreversible.

a perfect thermal insulator. Also, we need to assume that the material from which the 'balloon' is made is not only perfectly adiabatic, but also has the property that no work is required to expand the 'balloon', nor does the 'balloon' do any work as it contracts. The surroundings comprise the 'balloon', the material of the cylinder and piston, and the atmosphere beyond, which is assumed to be an ideal gas of very large volume.

In this example, we will assume that dynamic friction is present between the rim of the piston and the interior of the cylinder only for the path in which the gas in the system expands, driving the piston from left to right in Figure 5.10; when the piston moves to the left, the motion is frictionless. Also, we'll assume that the only friction present is dynamic friction, which acts to oppose the left-to-right motion of the piston, and that there is no static friction – an assumption we will relax in the next section.

The system undergoes a quasistatic adiabatic expansion from state $[1]_{sys}$ to state $[2]_{sys}$, followed by a quasistatic adiabatic compression from state $[2]_{sys}$ to state $[3]_{sys}$, such that the volume in state $[3]_{sys}$ is the same as the volume in state $[1]_{sys}$.

When the external pressure is reduced by an infinitesimal amount, the absence of static friction permits the pressure difference to cause the piston to move at once – there is no 'stickiness'. As the piston moves to the right in Figure 5.10, the system does P, V work of expansion $(đw_{P,V})_{sys}$ against the surroundings; since the path is quasistatic, this work may be expressed as the integral of $P\,dV$ from an initial volume V_1, corresponding to position A in Figure 5.10, to a final volume V_2, corresponding to position B. In addition, the system must also do work to overcome the dynamic friction, as given by equation (5.33) for an infinitesimal change

$$đw_{fric} = \varphi \left(\frac{dx}{d\tau}\right)^2 d\tau \tag{5.33}$$

and, for a finite change, by the time integral defined by equation (5.34)

$$\text{Total work } w_{fric} \text{ done by system against dynamic friction} = \varphi \int_{\tau_1}^{\tau_2} \left(\frac{dx}{d\tau}\right)^2 d\tau \tag{5.34}$$

This work is immediately converted into heat in the surroundings – heat that can't 'leak' back into the system by virtue of the adiabatic boundary.

During the return step, the surroundings do P, V work of expansion $(đw_{P,V})_{sur} = P\,dV$ against the pressure in the system, which corresponds to P, V work of compression $(đw_{P,V})_{sys} = -P\,dV$ performed by the surroundings on the system. Since we are assuming that this step is frictionless, no work needs to be done to overcome dynamic friction.

These changes may all be summarised in Table 5.4, which shows all the thermodynamic variables, for the system, the surroundings, and the universe, for the quasistatic adiabatic cycle in which dynamic friction is present for the system expansion, but not the compression.

The key feature of Table 5.4 – the feature that distinguishes Table 5.4 from both Table 5.2 and Table 5.3 – is the presence of the quantity w_{fric}, which represents the work done against dynamic friction, and also the corresponding amount of frictional heat.

When the gas in the system expands, driving the piston to the right in Figure 5.10, the system does the work against the dynamic friction. Significantly, the frictional work, and the corresponding frictional heat, is always associated with a positive sign, and never a negative sign – unlike the P, V work of expansion, which can take either sign depending on whether this work represents work done *by* (positive sign) or work done *on* (negative sign) the system or the surroundings. Also, in accordance with the discussion on page 131, and the derivation of equation (5.29), the frictional work $(w_{fric})_{sys} = w_{fric}$ done by the system during the expansion step is equated directly to the frictional heat $(q_{fric})_{sur}$ generated within the surroundings.

As can be seen in Table 5.4, for the complete cycle, the overall change in internal energy ΔU_{sys} for the system is equal to the quantity $-w_{fric}$, where w_{fric} is given by the time integral

$$\text{Total work } w_{fric} \text{ done by system against dynamic friction} = \varphi \int_{\tau_1}^{\tau_2} \left(\frac{dx}{d\tau}\right)^2 d\tau \quad (5.34)$$

As we saw on page 133 in our discussion of equation (5.34), the frictional factor φ is, by definition, positive, and $(dx/d\tau)^2$, as a square, is necessarily positive too, implying that w_{fric} is always positive. Since $\Delta U_{sys} = -w_{fric}$, ΔU_{sys} must always be negative, implying that the system *must* have a final temperature $[T_3]_{sys}$ *lower* than the original temperature $[T_1]_{sys}$ – the system has cooled down as a result of the energy expended in overcoming the dynamic friction.

For the surroundings, $\Delta U_{sur} = w_{fric}$, implying the final temperature $[T_3]_{sur}$ is higher than the original temperature $[T_1]_{sur}$ – the surroundings have become hotter as a result of the generation of frictional heat.

For the complete cycle, neither the system, nor the surroundings, return to their identical original states. The complete cycle is therefore irreversible. But the complete cycle comprises two paths: the first path, from state [1] to state [2], which is adiabatic, quasistatic, and associated with dynamic friction; and the second path from state [2] to state [3], which is adiabatic, quasistatic and frictionless. As we demonstrated on pages 117 to 121, this second path – which is adiabatic, quasistatic and frictionless – is reversible, implying that it is the first path – the one that involves friction – that has introduced the irreversibility.

Although this result has been demonstrated for a quasistatic adiabatic cycle, the result that the presence of friction along any path makes that path irreversible is generally true, and of

Table 5.4 Key thermodynamic variables for the system and the surroundings corresponding to the quasistatic, adiabatic, cyclic change depicted in Figure 5.8 in which friction is present. In this table, w_{fric} is the value of the time integral defined by equation (5.34).

For the system

	System expansion	System compression	Total cycle
P, V work done by system on surroundings, $(w_{P,V})_{sys}$	$+\int_{V_1}^{V_2} P\,dV$	$-\int_{V_1}^{V_2} P\,dV$	0
Work done by system against friction $(w_{fric})_{sys}$	$+w_{fric}$	0	$+w_{fric}$
Heat gained by system from surroundings, $(q)_{sys}$	0	0	0
Heat generated within system from friction, $(q_{fric})_{sys}$	0	0	0
$\Delta U_{sys} = (q + q_{fric})_{sys}$ $-(w_{P,V} + w_{fric})_{sys}$	$-\int_{V_1}^{V_2} P\,dV - w_{fric}$	$+\int_{V_1}^{V_2} P\,dV$	$-w_{fric}$
ΔT_{sys}	Negative	0	Negative

For the surroundings

	Compression of surroundings	Expansion of surroundings	Total cycle
P, V work done by surroundings on system, $(w_{P,V})_{sur}$	$-\int_{V_1}^{V_2} P\,dV$	$+\int_{V_1}^{V_2} P\,dV$	0
Work done by surroundings against friction, $(w_{fric})_{sur}$	0	0	0
Heat gained by surroundings from system, $(q)_{sur}$	0	0	0
Heat generated within surroundings from friction, $(q_{fric})_{sur}$	$+w_{fric}$	0	$+w_{fric}$
$\Delta U_{sur} = (q + q_{fric})_{sur}$ $-(w_{P,V} + w_{fric})_{sur}$	$w_{fric} + \int_{V_1}^{V_2} P\,dV$	$-\int_{V_1}^{V_2} P\,dV$	$+w_{fric}$
ΔT_{sur}	Positive	0	Positive

(continued)

Table 5.4 Continued

For both the system and the surroundings			
	State [1] to State [2]	State [2] to State [3]	Total cycle
P, V work done by the universe, $(w_{P,V})_{uni}$	0	0	0
Work done by, and within, the universe against friction, $(w_{fric})_{uni}$	$+w_{fric}$	0	$+w_{fric}$
Heat gained by the universe, $(q)_{uni}$	0	0	0
Heat generated within the universe from friction, $(q_{fric})_{uni}$	$+w_{fric}$	0	$+w_{fric}$
$\Delta U_{uni} = (q + q_{fric})_{uni} - (w_{P,V} + w_{fric})_{uni}$	0	0	0
ΔT_{uni}	0	0	0

great importance. The distinction between reversible and irreversible paths is fundamental to the Second Law, and so, to summarise - for a path to be reversible, the path

- must take place quasistatically, through an infinite sequence of equilibrium steps, and
- must be totally free from friction, and (as discussed further shortly) all other 'dissipative' effects, such as electrical resistance and magnetic hysteresis.

Any path that does not comply with both of these conditions, simultaneously, is necessarily irreversible. And a moment's thought will verify that *all* real paths are irreversible: reversible paths are a helpful theoretical model, but the real world is irreversible.

Before we explore some of the implications of irreversibility, we need to reflect on some of the assumptions used in establishing the concept of irreversibly, as illustrated in Figures 5.4, 5.5, 5.6, 5.7, 5.8, 5.9 and 5.10 and as expressed mathematically in Tables 5.2, 5.3 and 5.4.

One assumption was that all the changes were adiabatic, as achieved, for example, by the presence of the adiabatic 'balloon'. The 'balloon' was an 'invention' to ensure an adiabatic boundary, and the reason for doing this was to keep the discussion of heat 'tidy'. Because the various changes were assumed to be adiabatic, no heat can cross the boundary between the system and the surroundings, thereby ensuring that all the heat resulting from the work done against dynamic friction originated within the surroundings, and stayed there. If the boundary were not adiabatic, keeping track of the heat flows would have been more clumsy, but the result would have been the same – in the presence of friction, it is impossible to return both the system and the surroundings to their exact original states.

A second assumption was that the system comprised an ideal gas, and that the gaseous component of the surroundings was also ideal (remembering that the surroundings also include the 'balloon', the cylinder and the piston). The key purpose of assuming that the system was an ideal gas was to ensure that there was a direct relationship between the change ΔU in the internal energy of the system or the surroundings, and the corresponding change ΔT in the temperature of the system or the surroundings. Once again, the assumption of the ideal gas can be relaxed without changing the overall result – frictional heat is still produced, and the temperature of the surroundings must rise.

A third assumption was that many of the changes were quasistatic. This served two purposes: firstly, as a basis for introducing the concept of reversibility, in that all reversible paths are necessarily quasistatic; and secondly since it enabled us to quantify any P, V work in terms of an integral of the form $\int P\,dV$, as expressed in terms of state functions relating only to the system. In the following section, we will explore what happens when a change is not quasistatic, and, as we shall see, this makes real, non-quasistatic, paths 'even more' irreversible then their theoretical counterparts. And a fourth assumption was the absence of static friction – an assumption that will also be relaxed in the next section.

The fifth assumption, fundamental to the concept of irreversibility, is that frictional work is instantaneously converted into frictional heat, but that frictional heat can never be converted back into work. This assumption is justified on two grounds: firstly, empirical evidence, such as that codified by Count Rumford and James Joule, and secondly, the expression of frictional work as a necessarily positive quantity given by equation (5.33)

$$đw_{\text{fric}} = \varphi \left(\frac{dx}{d\tau}\right)^2 d\tau \qquad (5.33)$$

and the corresponding time integral for a finite change

$$\text{Total work } w_{\text{fric}} \text{ done by system against dynamic friction} = \varphi \int_{\tau_1}^{\tau_2} \left(\frac{dx}{d\tau}\right)^2 d\tau \qquad (5.34)$$

As we have demonstrated several times, this integral can only be positive, and never negative, proving that work can only be done *by* the system against friction, and never by friction on the system. This, combined with the principle of the conservation of energy as stated by the First Law, implies that frictional work is converted in its entirety into frictional heat, but that frictional heat can never be converted into frictional work.

The generation of frictional heat from frictional work is one, very frequently encountered, example of a number of phenomena which can also be expressed as the time integral of a necessarily positive quantity, much like equation (5.34), two others being

- electrical resistance
- magnetic hysteresis.

Collectively, these are known as **dissipative effects** because they all dissipate other forms of energy, such as electrical energy and magnetic energy, irreversibly, as heat.

And there is yet another assumption – and this one is 'hidden' in equations (5.33) and (5.34): the assumption that the time increment $d\tau$ is always positive. If $d\tau$ might ever have a negative sign, then equations (5.33) and (5.34) would compute to a negative number, even when φ and $(dx/d\tau)^2$ are positive, resulting in a situation in which frictional heat could in fact generate frictional work. Now $d\tau = \tau_{x+dx} - \tau_x$, where time τ_{x+dx} is the time at which the piston reaches a given position $x + dx$, and time τ_x is the time at which the piston was in position x. Time τ_{x+dx} is therefore later than time τ_x, and so $d\tau = \tau_{x+dx} - \tau_x$ must always be positive. This assumes, however, that *time runs forwards*. If – startlingly – time were to run *backwards*, then τ_{x+dx} would be *less* than τ_x, in which case $d\tau$ would be negative. As a consequence, equations (5.33) and (5.34) would give a negative result, and frictional heat would spontaneously be converted into work.

In reality, of course, time runs only forwards, not backwards – at least in the universe with which we are familiar! And, as a consequence, frictional work spontaneously generates frictional heat, rather than frictional heat generating frictional work. As we shall see when we explore the Second Law of Thermodynamics, this asymmetry in time – the reality that time naturally runs forwards rather than backwards – is directly linked to the fact that all real processes are, thermodynamically, irreversible, and that spontaneous changes, such as the flow of heat down a temperature gradient rather than up one, and many chemical reactions, are unidirectional.

As we have seen, a reversible path is necessarily both quasistatic and friction free. We now examine what happens when a non-quasistatic path is followed in the presence of friction – and, as we shall shortly see, in practice, the presence of friction prevents a path from being quasistatic.

5.19 Real paths

We will now examine real paths – paths that do not follow an infinite quasistatic sequence of equilibrium states; paths that involve some form of molecular irreversibility; paths that take place in the presence of both static and dynamic friction. Any one of these implies that the corresponding path is necessarily irreversible, and they are often present in combination. In practice, all real paths are irreversible, and we all live in an irreversible world.

To further our understanding, let us return to our system of an ideal gas within a cylinder, but the piston in the cylinder is no longer perfectly frictionless: there is now both static and dynamic friction between the piston and the inside surface of the cylinder. As usual, the initial volume of the gas within the cylinder is V_1, and the pressure is P; the pressure external to the piston is $P_{ex} = P$. And as usual, the external pressure is reduced by an infinitesimal amount dP. The pressure P inside the cylinder is now greater than the pressure $P_{ex} - dP$ outside, and so there is a net force on the piston. If the piston is frictionless, this force causes the piston to move against the external pressure, and the system does work $P_{ex}\,dV = P\,dV$ on the surroundings.

But if there is static friction between the piston and the cylinder, then the effect of the pressure difference between the internal pressure P and the external pressure $P - (P_{ex} - dP)$ is … nothing – the piston does not move, for static friction causes 'stickiness'. To make the piston move, and to overcome the static friction, the external pressure has to be reduced by a more substantial amount, until the pressure difference is great enough to for the piston, suddenly, to move. So, rather than the smooth, continuous motion shown by the frictionless piston, a piston subject to static friction moves in a sequence of 'jerks', as illustrated in Figure 5.11. And, given, the suddenness of each 'jerk', the gas inside the cylinder is subject to turbulence, and so the sequence of states through which the gas passes cannot be described as equilibrium states – as indicated by the dashed lines in Figure 5.11.

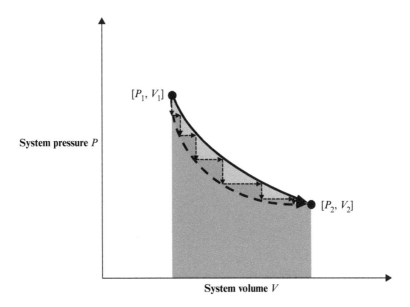

Figure 5.11 Reversible and irreversible expansion. The solid line represents a reversible expansion of an ideal gas from $[P_1, V_1]$ to $[P_2, V_2]$, during which the system does reversible work w_{rev} on the surroundings, shown by the total area under the solid line. An irreversible change takes place in 'jerks', as approximated by the intermediate, stepped, dashed line, or perhaps more smoothly as represented by the lower dashed line. The irreversible work w_{irrev} done by the system on the surroundings is represented by the area under either dashed line, both of which are necessarily less than the area under the solid line. During the irreversible change, energy that would have resulted in work during an equivalent reversible change is dissipated as frictional heat, implying that $w_{rev} > w_{irrev}$.

The 'jerkiness' of the movement of the piston is not the only consequence of friction. Another is the fact that work needs to be done to overcome the dynamic frictional force that opposes the motion of the piston. This implies that, as the piston moves, two forms of work are being done – P, V work of expansion against the external pressure, and also work against the dynamic friction. Accordingly, in the presence of friction, the First Law must be written as

$$dU = đq - (đw_{P,V} + đw_{fric}) \tag{5.35}$$

where $đw_{P,V}$ is the P, V work done by the system on the surroundings, and $đw_{fric}$ is the work done by the system against friction.

And, as we know, this frictional work $đw_{fric}$ is always a positive number, and is necessarily dissipated, immediately, as heat.

Let us now consider what happens when, once the system has reached state $[2] \equiv [P_2, V_2]$ in Figure 5.11, the external pressure is increased, and the gas inside the cylinder is compressed. As we have seen, increasing the pressure by a small amount has no effect, for the 'stickiness' attributable to static friction allows the piston to withstand modest pressure differences. But as the external pressure increases, a point is reached at which the external pressure is sufficiently high to push the piston back, and to overcome the static friction - the piston then 'jerks' inwards, and the gas becomes compressed, as indicated in Figure 5.12.

Eventually, the piston reaches its original position, with the volume of the gas returning to its original value V_1, but by tracing a path represented approximately (given that the intermediate

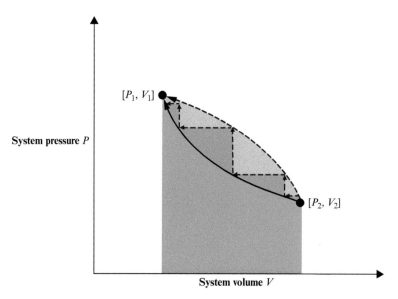

Figure 5.12 Reversible and irreversible compression. The solid line represents a reversible compression of an ideal gas from $[P_2, V_2]$ to $[P_1, V_1]$, during which the surroundings do reversible work w_{rev} on the system, represented by the total area under the solid curve. An irreversible change takes place in 'jerks', as approximated by the upper, stepped, dashed line, or perhaps more smoothly as represented by the top-most dashed line. The irreversible work w_{irrev} done on the system by the surroundings is represented by the total area under either dashed line, both of which are necessarily greater than the area under the solid line. As explained in the text, these areas are associated with negative signs, and it is therefore still true that

$$w_{rev} > w_{irrev}.$$

states are not equilibrium states) by one of the dashed lines shown in Figure 5.12 – a dashed line that necessarily lies above the precise path that a reversible change would have taken. Furthermore, during the compression, the surroundings do P, V work of compression on the system, as well as work against dynamic friction which is instantaneously converted into heat.

There is another important implication of the presence of friction too. As can be seen from Figure 5.11, the **reversible work** w_{rev} done by an expanding gas is represented by the area under the full line, and the **irreversible work** w_{irrev} done by the gas in expanding between the same initial and final states is represented by the area under one of the dashed lines. Graphically, these areas are different, and it is evident that

$$w_{rev} > w_{irrev} \tag{5.36}$$

This result was anticipated on page 28, and although Figure 5.11 represents a particular case, the inequality (5.36) is universally true, for the difference đw_{rev} − đw_{irrev} or $w_{rev} - w_{irrev}$ represents the energy dissipated as frictional heat.

The situation is similar for compression, as shown in Figure 5.12. For a reversible change, the compression follows the path shown by the solid line, and since the change is a compression, the surroundings do reversible work w_{rev} on the system. The magnitude of this reversible work is represented by the area under the solid line: let us assume that this magnitude is, say, 50 J. But, since the system is being compressed and work w_{rev} is being done on the system by the surroundings, w_{rev} is a negative quantity, and so, in this case $w_{rev} = -50$ J.

If friction is present, the external pressure has to be increased to overcome the 'stickiness', and so all irreversible paths are above the reversible path, as illustrated in Figure 5.12. Furthermore, work has to be done against the dynamic friction – work which is dissipated as frictional heat. The irreversible work w_{irrev} done by the surroundings on the system therefore corresponds to the total area between either of the dashed lines and the horizontal axis – both areas being greater than the area corresponding to the reversible path. Suppose, then, that, that the magnitude of this work is, say, 80 J. Algebraically, as was the case for the reversible compression, the irreversible work w_{irrev} done on the system by the surroundings is a negative quantity, implying that $w_{irrev} = -80$ J.

A moment's thought will verify that the negative number $-50 = w_{rev}$ is *larger* than the negative number $-80 = w_{irrev}$ (think, for example, of the x axis of a graph, in which numbers to the right are greater than numbers to the left – on which side of –50 is –80?). Hence we may write

$$w_{rev} > w_{irrev} \tag{5.36}$$

for a compression, as was the case for expansion.

The inequality (5.36) has three, very important, implications.

Firstly, as we have demonstrated, when the system does work on the surroundings, then since $w_{rev} > w_{irrev}$, the maximum work that the system can perform as a result of any change in state is the reversible work w_{rev}. Since all real paths are irreversible, the work w_{irrev} performed by the system is always less than the theoretical maximum, w_{rev}.

Secondly, consider the case in which a change in state of a system is associated with work being done on the system by the surroundings. The inequality $w_{rev} > w_{irrev}$ remains valid, but in this case, the signs associated with w_{rev} and w_{irrev} are now negative. As illustrated in Figure 5.12, the reversible work $w_{rev} = -50$ J is indeed a larger number, algebraically, than the irreversible work, $w_{irrev} = -80$ J. If, however, we think of the magnitudes of these quantities and not their algebraic signs, then the situation is the other way around: when work is being done on a system during any change in state, then the *magnitude* of the reversible work is the *minimum* work that needs to be done on the system: for any real change, the magnitude of the irreversible work that needs to be done on the system is inevitably a greater amount than the reversible work.

In both cases – work done by the system, and work done on the system – the unifying concept is that of efficiency: the reversible change is always more efficient than the corresponding irreversible change. If the system does the work, then, for any given change in state, more work is done reversibly than irreversibly; if the surroundings have to do work on the system, then the least amount of work that needs to be done is the reversible work – if the work being done is irreversible, then the surroundings have to 'work harder'.

The third implication is the consequence of invoking the First Law, which must apply to both a reversible, and an irreversible, change. Since, for any a system, the internal energy U is a state function, any change dU is a property of the initial and final states only, and independent of the path. Hence, for any change in state taking place either reversibly or irreversibly, if đq_{rev} represents the **reversible heat** flow, and đq_{irrev} the **irreversible heat** flow,

$$dU = đq - đw = đq_{rev} - đw_{rev} = đq_{irrev} - đw_{irrev} \tag{5.1a}$$

THE FIRST LAW OF THERMODYNAMICS

implying that

$$đq_{rev} - đq_{irrev} = đw_{rev} - đw_{irrev}$$

so if

$$đw_{rev} > đw_{irrev} \tag{5.37}$$

then

$$đw_{rev} - đw_{irrev} > 0$$

implying that

$$đq_{rev} - đq_{irrev} > 0$$

and therefore that

$$đq_{rev} > đq_{irrev} \tag{5.38}$$

Expression (5.38) states that the heat gained by a system during a reversible change between any two states is necessarily greater than the heat gained by the system for any irreversible change between the same two states. This appears to be rather inconsequential, for, as an *in*equality, the fact that, 17, say, is greater than, say, 5 does not appear to be very interesting: 17 is also greater than 13, 4 and −6. Furthermore, surely science is all about equations – from Newton's Laws to Einstein's theory of General Relativity – not about inequalities? Yes, much of science is about equations. But there are two examples where inequalities are of paramount importance. One is Heisenberg's 'uncertainty principle' in quantum mechanics, which, briefly, sets a limit to what we can measure. And the second is expression (5.38)

$$đq_{rev} > đq_{irrev} \tag{5.38}$$

As we shall see in Chapter 9, the inequality (5.38) is arguably one of the most important expressions in the whole of science – and it is underpinned by the asymmetry we first discussed on page 132.

EXERCISES

1. Complete the blank entries in the following table. All quantities are in J.

dU	dq	dw	dU − dq + dw
250		−150	
−50	100		
	−100	−150	
−250	−100		
	100	100	
150		−150	
100		0	
0	0		

2. If the system corresponding to the data in this table is one mole of an ideal gas, for which $C_V = 12.5 \text{ J K}^{-1}\text{mol}^{-1}$, calculate the final temperature of the system in each case, assuming that the initial temperature is 300 K.

3. Define, precisely, a 'reversible' path. How does a reversible path differ from a quasistatic path? And from an irreversible path? Explain why all reversible paths are necessarily quasistatic, but not all quasistatic paths are necessarily reversible. Give an example of a path which is both quasistatic and reversible. Give an example of a path which is both quasistatic and irreversible. Give an example of a path which is both non-quasistatic and irreversible.

4. Do question 9 in the exercises to Chapter 2 again.

6 Enthalpy and thermochemistry

 Summary

The extensive state function **enthalpy** H is defined as

$$H = U + PV \tag{6.4}$$

and the extensive state function the **heat capacity at constant pressure** C_P as

$$C_P = \left(\frac{\partial H}{\partial T}\right)_P \tag{6.8a}$$

For any change in a system at constant pressure, the change ΔH_P is equal to the heat $\{_1q_2\}_P$ reversibly exchanged between the system and the surroundings.

All chemical reactions are changes in state from the original reactants to the final products. For a chemical reaction taking place at constant pressure, if the enthalpy change ΔH is negative, the reaction produces heat, and is called **exothermic**. An **endothermic** reaction is one for which ΔH is positive, implying that the reaction absorbs heat.

Although many chemical reactions that take place spontaneously are exothermic, the statements "all spontaneous reactions are exothermic" and "if a reaction is exothermic, and $\Delta H < 0$, then it will be spontaneous" are false: a true criterion for spontaneity will be met in Chapter 13.

Reference tables of thermodynamic data are based on chemical **standard states** (see Table 6.4 on page 172) which specify the conditions under which the corresponding data apply.

Hess's law of constant heat formation states that

The heat released or absorbed in any chemical reaction depends solely on the initial reactants and final products, and is independent of the path taken during the reaction.

This statement is a direct consequence of the fact that the heat released or absorbed in any chemical reaction taking place at constant pressure is the corresponding change ΔH in enthalpy, and since enthalpy is a state function, the change ΔH is independent of the path.

The **standard enthalpy change** $\Delta_r H^\ominus$ for any reaction – in which the reactants in their standard states react to form the products in their standard states, with the reaction assumed to go to completion – is given by

$$\Delta_r H^\ominus = \sum_{\text{Products}} n_{\text{prod}} \Delta_f H_{\text{prod}}^\ominus - \sum_{\text{Reactants}} n_{\text{reac}} \Delta_f H_{\text{reac}}^\ominus \tag{6.21}$$

in which n_i represents the mole number of component i as determined by the stoichiometry of the reaction, and $\Delta_f H_i^\ominus$ represents the **standard molar enthalpy of formation** of a reactant or a product – the enthalpy change associated with the formation of 1 mol of the molecule in its standard state from its component elements, in their standard states.

Modern Thermodynamics for Chemists and Biochemists. Dennis Sherwood and Paul Dalby.
© Oxford University Press 2018. Published 2018 by Oxford University Press.

In general, the enthalpy change ΔH for any reaction taking place at constant pressure will vary according to the temperature at which the reaction takes place. The most generalised expression for this variation is **Kirchhoff's first equation**

$$\left(\frac{\partial(\Delta_r H)}{\partial T}\right)_P = \Delta C_P \tag{6.33}$$

in which

$$\Delta C_P = \sum_{\text{products, } j} n_j\, C_P(j) - \sum_{\text{reactants, } i} n_i\, C_P(i) \tag{6.28}$$

where, as in equation (6.21), n_i represents the mole number of component i, and $C_P(i)$ is the corresponding molar heat capacity at constant pressure.

6.1 Man's most important technology

What do you think is man's most important technology? The mobile phone? Perhaps. The internet? Maybe. Our vote goes to fire. Fire, of course, is a natural phenomenon: if lightning strikes dry grass on a prairie, or wood in a forest, fire can then erupt. But when, thousands of years ago, our ancestors discovered how to make fire deliberately – perhaps by striking a flint, or from the friction between two wood surfaces when rubbed together – and then how to control it, fire became a technology. And with what an impact!

Firstly, the use of fire for heat – by lighting a fire, and keeping it going, man could create heat to keep living spaces warm, and so protect against cold weather. Secondly, the use of fire for light – light that enabled man to 'stretch' the natural day-light into the hours of darkness. In this context, let's remember that the electric light bulb, as we would recognise one today, was first demonstrated in 1879 - up until that time, other than some electric prototypes, light could be obtained only from fire: fire from burning natural oils, such as olive oil (oil lamps date back to ancient times), waxes and fats (in candles), or gases (from about the early 1800s). Thirdly, the use of fire for cooking. Man is the only animal that cooks food, not only making it taste better, but also making it easier to digest. By cooking, early man was able to get more nutrition from a given mass of food, and would require less time to digest it, allowing more time for doing things other than hunting, gathering, eating and digesting. And there is even a theory that the enhanced nutrition resulting from cooking enabled more rapid development of the brain, so spurring human development (see *Catching Fire: How Cooking made us Human*, by Richard Wrangham, Basic Books, New York, 2009). And fourthly, the use of fire to drive machines, such as steam engines, and then steam turbines, which in turn provide electricity. Heat, light, cooking and the energy to drive machines: all the result of fire.

Underpinning all these examples is a particular type of chemical reaction: an oxidation reaction in which certain reactants – materials such as coal, natural gas (usually methane) or a hydrocarbon such as octane - combine with oxygen to give certain products, such that, as the reaction takes place, a considerable amount of energy, often in the form of heat, is produced. When suitably harnessed, this energy can then be used for any number of purposes from cooking, to the generation of electricity. Ultimately, the energy for all these different applications originates from the energy produced as a result of the breaking and reforming of chemical bonds during an oxidation reaction.

A particular type of oxidation reaction – one of quite literally vital importance – is the oxidation of sugars, such as glucose, within biological organisms during the process known as respiration. It is the energy released from this reaction that keeps us all alive: energy that is used directly as heat to maintain our body temperature; energy, in the form of chemical energy, that is temporarily 'trapped' in, primarily, the molecule known as ATP, chemical energy that can then be used to drive other biochemical reactions, such as the synthesis of proteins, when the ATP subsequently decomposes.

Oxidation reactions, such as the burning of a fuel, are not the only type of chemical reaction that generates heat: many other types of reaction create heat too, and there are also some chemical reactions that absorb, rather than create, heat. The interplay between the energy stored in chemical bonds, and the energy released into, or absorbed from, the surroundings as chemical reactions take place is of fundamental importance, and so the purpose of this chapter is to explore the principles underpinning the association between heat and chemical reactions – a branch of thermodynamics known as **thermochemistry**.

6.2 Enthalpy

To set the scene, we need to return to our discussion on pages 114 and 116 of the First Law

$$dU = đq - đw \tag{5.1a}$$

in the context of a system undergoing a change at constant pressure. This is of particular relevance given the fact that many chemical and biochemical reactions take place under conditions open to the pressure of the atmosphere, a pressure which may be considered to be constant over the time during which those reactions take place.

For a system which performs only P, V work against an external pressure P_{ex}, and where there are no dissipative effects (see page 139), equation (5.1a) may be written as

$$dU = đq - P_{ex}\, dV$$

If the change in state follows a reversible path, then this is by definition quasistatic. The pressure P of the system therefore equals the external pressure P_{ex} throughout the change, and so

$$dU = đq_{rev} - P\, dV$$

and

$$dU + P\, dV = đq_{rev} \tag{6.1}$$

where the term $đq_{rev}$ reminds us that the heat exchange refers to a reversible path.

For a change at constant pressure, P is constant, and equation (6.1) may then be written as

$$dU + P\, dV = \{đq_{rev}\}_P \qquad \text{at constant pressure} \tag{6.2}$$

where $\{đq_{rev}\}_P$ is the heat gained by the system (if positive), or lost from the system (if negative), for a reversible change at constant pressure.

Because the change is at constant pressure, dP must be zero, and so the product VdP is zero too. According to the rules of mathematics, given any equation, if you add a quantity to one side of the equation, then the equation remains true if you add the same quantity to the other side. So, let us add zero to both sides of equation (6.2), but we'll express 'zero' on the left-hand side of equation (6.2) as the expression VdP, which happens to equal zero when the pressure P is constant, and on the right-had side as the more familiar symbol 0

$$dU + P\,dV + V\,dP = \{đq_{rev}\}_P + 0 \qquad \text{at constant pressure}$$

But

$$P\,dV + V\,dP = d(PV) \qquad \text{always}$$

and so

$$dU + d(PV) = \{đq_{rev}\}_P \qquad \text{at constant pressure}$$

hence

$$d(U + PV) = \{đq_{rev}\}_P \qquad \text{at constant pressure} \qquad (6.3)$$

Now, the internal energy U, pressure P and volume V of a system are all state functions – and so the quantity $U + PV$ must also be a state function too, implying that the cyclic integral of the quantity $U + PV$ must be zero:

$$\oint d(U + PV) = 0$$

Since the quantity $U + PV$ is a state function, we may define a new state function H as

$$H = U + PV \qquad (6.4)$$

then

$$dH_P = d(U + PV)_P = \{đq_{rev}\}_P \qquad \text{at constant pressure} \qquad (6.5a)$$

and for a finite change at constant pressure

$$\Delta H_P = (H_2 - H_1)_P = \Delta(U + PV)_P = \{q_{rev}\}_P \qquad \text{at constant pressure} \qquad (6.5b)$$

The state function H is known as **enthalpy**, and its significance is a consequence of equation (6.5): for a system which can perform P, V work only, the change in enthalpy ΔH_P for any change in state taking place at constant pressure is equal to the reversible heat gained or lost by the system during that change. As already noted, this is important because a huge number of changes in real systems, especially normal chemical and biological systems, take place at constant pressure – the pressure of the atmosphere – and perform P, V work only. So, if we want to know something about the heat exchanged during many real processes, then this implies we are dealing with enthalpy, and enthalpy changes. Enthalpy H is an extensive state function; the corresponding intensive state function, the **molar enthalpy** H, for a single mole of material, may be defined as

$$H = U + PV \qquad (6.6)$$

Since enthalpy is a state function, then, just as any system at equilibrium has a pressure P, a temperature T and an internal energy U, it also has an enthalpy H. Also, just as any change in state from any initial equilibrium state to any final equilibrium state will be associated with a change in pressure ΔP, a change in temperature ΔT, and a change in internal energy ΔU, there is also a change in enthalpy ΔH too. What's special about enthalpy is that *if the change in state is at constant pressure, then the corresponding change in enthalpy ΔH_P is equal to the reversible heat transferred between the system and the surroundings* – which, as we shall see in this chapter, is a quantity of considerable importance. If, however, the change does not take place at constant pressure, then ΔH will still have a value, but this value will not equal the reversible heat transferred during the change. However, because enthalpy is a state function, then that value of ΔH is equal to what the reversible heat transfer would have been, had the change taken place at constant pressure.

As we now know from equations (6.5)

$$dH_P = \{đq_{rev}\}_P \tag{6.5a}$$

$$\Delta H_P = (H_2 - H_1)_P = \{q_{rev}\}_P \tag{6.5b}$$

the change dH_P or ΔH_P equates to the heat exchanged, at constant pressure, for a reversible change in the system. Reversible changes, however, are theoretical abstractions; furthermore, the expression (5.38)

$$đq_{rev} > đq_{irrev} \tag{5.38}$$

tells us that the heat exchanged for a real, irreversible, change is always less than the heat exchanged for the corresponding reversible change. Combining equations (6.5) with the inequality (5.38), we may write

$$dH_P = \{đq_{rev}\}_P > \{đq_{irrev}\}_P \tag{6.6a}$$

and

$$\Delta H_P = (H_2 - H_1)_P = \{q_{rev}\}_P > \{q_{irrev}\}_P \tag{6.6b}$$

In reality, all real changes are necessarily irreversible, and, in the case of chemical reactions, the irreversibility can arise from, for example, a departure from the theoretical quasistatic path (as might happen if there is a very sudden, and substantial, change in local pressure or volume, as is likely during an explosion) or the presence of dissipative effects (such as those associated with the turbulent mixing of liquids and gases, rather than the dynamic friction of pistons moving within cylinders).

Equations (6.6a) and (6.6b) imply that the actual heat exchange $\{q_{irrev}\}_P$ associated with any change at constant pressure will be less than the theoretical maximum $\{q_{rev}\}_P$ corresponding to the appropriate value of dH or ΔH. dH and ΔH therefore represent the theoretical maximum available heat that can be obtained from any thermodynamic change – and any chemical reaction – taking place at constant pressure. In practice, however, in many situations, especially in the realms of chemistry and biochemistry, the difference between the actual heat exchange and the theoretical maximum as defined by dH or ΔH is either small, or just doesn't matter. Values of dH and ΔH for chemical reactions are therefore commonly interpreted as equal to the quantity of heat that can actually be measured, even though, in principle, these are theoretical maxima.

6.3 The mathematics of H

6.3.1 H as a function $H(T, P)$ of T and P, and the definition of C_P

For a closed system of constant mass, the enthalpy H can be expressed as a function of temperature T and pressure P as $H(T, P)$. Accordingly,

$$dH = \left(\frac{\partial H}{\partial T}\right)_P dT + \left(\frac{\partial H}{\partial P}\right)_T dP \tag{6.7}$$

which is an equation we introduced in Chapter 4 as equation (4.13). In this equation, $(\partial H/\partial T)_P$ represents the increase ∂H_P in the enthalpy attributable to an increase ∂T_P for a change at constant pressure. We have also seen, from equation (6.5a), that ∂H_P is equal to the reversible heat $\{đq_{\text{rev}}\}_P$ exchanged between the system and the surroundings for a change at constant pressure, so we now define the **heat capacity at constant pressure**, C_P, as

$$C_P = \left(\frac{\partial H}{\partial T}\right)_P = \frac{\{đq_{\text{rev}}\}_P}{dT_P} \tag{6.8a}$$

from which

$$dH_P = \{đq_{\text{rev}}\}_P = C_P \, dT_P \tag{6.8b}$$

The definition of C_P, equation (6.8a) implies that equation (6.7) may be written as

$$dH = C_P \, dT + \left(\frac{\partial H}{\partial P}\right)_T dP \tag{6.9}$$

This definition of C_P, which applies to constant pressure, is very similar to the definition of C_V as

$$C_V = \left(\frac{\partial U}{\partial T}\right)_V = \frac{đq_V}{dT_V} \tag{5.13a) and (5.14b}$$

which applies at constant volume. In the definition of C_P, the heat term $\{đq_{\text{rev}}\}_P$ is explicitly shown as a reversible heat exchange: as we saw on page 148, the assumption of reversibility was required in the development of the definition of H so that the work $đw$ could be represented as $P \, dV$. In the definition of C_V (see page 107), we did not need, at that point, to assume reversibility, but the discussion on page 113 did point out that the definition of C_V makes the implicit assumption that the change of state is quasistatic. We now see that the assumption is in fact stronger: the change not just quasistatic, but reversible.

For 1 mol of a pure substance, we may define the **molar heat capacity at constant pressure** C_P as

$$C_P = \frac{đq_P}{dT_P} = \left(\frac{\partial H}{\partial T}\right)_P \tag{6.10}$$

Some values for C_p for some selected materials at 1 bar and 298.15 K are shown in Table 6.1.

ENTHALPY AND THERMOCHEMISTRY

Table 6.1 Values of C_P for selected materials, at 1 bar and 298.15 K

	C_P J K^{-1} mol^{-1}		C_P J K^{-1} mol^{-1}
He (g)	20.8	H$_2$ (g)	28.8
Ne (g)	20.8	N$_2$ (g)	29.1
Ar (g)	20.8	O$_2$ (g)	29.4
Kr (g)	20.8	CO$_2$ (g)	37.1
Xe (g)	20.8	NH$_3$ (g)	35.1
C$_2$H$_5$OH (l)	112	CCl$_4$ (l)	131
Al (s)	24.2	Fe(s)	25.1
Cu (s)	24.4	Pb(s)	26.8
SiO$_2$ (s)	44.4	NaCl (s)	50.5

Source: *CRC Handbook of Chemistry and Physics*, 96th edn, 2015, Section 4, pages 4–124, and Section 5, pages 5–4 to 5–42.

Like C_V, and $\boldsymbol{C_V}$, C_P and $\boldsymbol{C_P}$ are state functions (see also page 113). And, as a state function, C_P for a given mass of material can be expressed as a function of any other two intensive state functions, for example, the temperature T and pressure P, and so C_P may be written as $C_P = C_P(T, P)$. The temperature variation of $\boldsymbol{C_P}$ is discussed further on pages 195 to 197, and some values for $\boldsymbol{C_P}$ for water at different temperatures are shown in Table 6.2.

Table 6.2 Values of C_p for water, at 1 bar and a variety of temperatures

	T K	C_p J K^{-1} mol^{-1}
H$_2$O (s)	173.15	24.8
H$_2$O (s)	273.15	37.8
H$_2$O (l)	273.16	76.0
H$_2$O (l)	298.15	75.3
H$_2$O (l)	372.75	75.9
H$_2$O (g)	373	37.4
H$_2$O (g)	500	35.7
H$_2$O (g)	1000	41.3

Source: *CRC Handbook of Chemistry and Physics*, 96th edn, 2015, Section 6, pages 6–1 to 6–4, and 6–12.

6.3.2 H and C_P for an ideal gas

By definition, for any system

$$H = U + PV \tag{6.4}$$

If the system is an ideal gas, then we also know that

$$PV = nRT \tag{3.3}$$

implying that, for an ideal gas

$$H = U + nRT$$

For a system of an ideal gas of constant mass, n is a constant; furthermore, we also know that the internal energy U of a constant mass of an ideal gas is a function of the temperature T only (see page 97). Since both the terms U and nRT depend only on the temperature T, it follows that the enthalpy H of an ideal gas is also a function of temperature only. This implies that C_P for a fixed mass of an ideal gas may be written as a total derivative

$$C_P = \frac{dH}{dT} \tag{6.11}$$

rather than as a partial derivative $(\partial H/\partial T)_P$.

Given that, for an ideal gas

$$H = U + nRT$$

then, for 1 mol of material, $n = 1$ and so

$$\boldsymbol{H = U + RT}$$

Differentiating with respect to the temperature T

$$\frac{dH}{dT} = \frac{dU}{dT} + R \tag{6.12}$$

But

$$\frac{dH}{dT} = C_P \tag{6.11}$$

and, for an ideal gas

$$\frac{dU}{dT} = C_V \tag{5.16}$$

and so equation (6.11) becomes, for an ideal gas

$$\boldsymbol{C_P = C_V + R} \tag{6.13}$$

As we saw on page 111, for an ideal gas, C_V is a constant, independent of pressure and temperature, and so equation (6.13) implies that C_P for an ideal gas is also constant. Furthermore, for a monoatomic ideal gas, we showed on page 111 that $C_V = 3R/2$, and so equation (6.13) predicts that, for a monatomic gas, $C_P = 5R/2 = 3 \times 8.314/2 = 20.8$ J K^{-1} mol^{-1} – as indeed is validated by the values for C_P for the inert gases, as shown in Table 6.1.

6.4 Endothermic and exothermic reactions

So, after all this theory, let's do some chemistry! Consider a generalised chemical reaction

A → B

taking place at constant pressure, to completion, as written. Regarding the reactants A as an initial thermodynamic state, and the products B as a final thermodynamic state, then the corresponding thermodynamic change in state is that in which an initial state [1], comprising 1 mol of A and 0 mol of B, changes by virtue of a chemical reaction to a final state [2], comprising 0 mol of A and 1 mol of B. If the enthalpy of the initial state is H_1, and that of the final state H_2, then the enthalpy change ΔH is related to the heat $\{q_{rev}\}_P$ – or more simply q_P – exchanged (in principle reversibly) between the system and the surroundings during the constant pressure change as

$$\Delta H = H_2 - H_1 = \{q_{rev}\}_P = q_P \qquad (6.5b)$$

Depending on what A and B actually are, the reaction can

- absorb heat from the surroundings (ΔH and q_P are both positive)
- release heat to the surroundings (ΔH and q_P are both negative), or
- be heat-neutral ($\Delta H = q_P = 0$).

The third case, in which $\Delta H = q_P = 0$, is theoretically possible, but rare; the first two cases cover the overwhelming majority of chemical and biochemical reactions, and are given special names:

- an **endothermic** reaction *absorbs* heat, and ΔH is *positive* ($H_2 > H_1$)
- an **exothermic** reaction *releases* heat, and ΔH is *negative* ($H_2 < H_1$).

The words 'endothermic' and 'exothermic' are not part of our everyday vocabulary, and are easily muddled – one way of remembering which is which is to associate 'exothermic' with the much more frequently used word 'exit', which is all about leaving and going away – so in an exothermic reaction, heat exits and is lost, and ΔH is negative.

Some examples of endothermic reactions are

- all **phase changes** from a more condensed **phase** to a less condensed phase – solid to liquid, liquid to gas
- photosynthesis, in which light energy is used by green plants to synthesise sugar molecules from CO_2 and H_2O
- the important industrial process used for the manufacture of cement, in which, at temperatures in excess of about 850 °C, limestone decomposes into calcium oxide and carbon dioxide

$$CaCO_3 \text{ (s)} \rightarrow CaO \text{ (s)} + CO_2 \text{ (g)} \qquad \Delta H = 179 \text{ kJ}$$

And some examples of exothermic reactions are

- all **phase changes** from a less condensed **phase** to a more condensed phase – gas to liquid, liquid to solid
- respiration, in which sugars are decomposed into CO_2 and H_2O, so exothermically releasing the energy that was 'trapped' in the endothermic reactions of photosynthesis
- the burning of fuels, in which the energy released is used, for example, to heat spaces, or to power engines – for example, the combustion of (liquid) octane to form carbon dioxide and (liquid) water

$$C_8H_{18}\,(l) + \frac{25}{2}\,O_2\,(g) \rightarrow 8\,CO_2\,(g) + 9\,H_2O\,(l) \qquad \Delta H = -5471\,\text{kJ}$$

As is illustrated in the examples of the decomposition of $CaCO_3$ and the combustion of C_8H_{18}, the value of the enthalpy change ΔH for a chemical reaction is usually expressed in units of J, kJ or MJ for the reaction as written: so, for the combustion of octane, 5471 kJ of heat are released for each mol of C_8H_{18} oxidised, using 25/2 = 12.5 mols of oxygen. Since enthalpy is an extensive state function, enthalpy changes will depend on the amounts of materials involved, and so if the combustion of C_8H_{18} is represented by the more conventional chemical equation

$$2\,C_8H_{18}\,(l) + 25\,O_2\,(g) \rightarrow 16\,CO_2\,(g) + 18\,H_2O\,(l) \qquad \Delta H = -10{,}942\,\text{kJ}$$

in which all the **stoichiometric coefficients** are whole numbers, then the enthalpy change is $-10{,}942$ kJ, twice that for the reaction written with the fractional coefficient 25/2 for oxygen.

Sometimes, the enthalpy change for a reaction is expressed in units of J mol^{-1}, kJ mol^{-1} or MJ mol^{-1}, where the 'per mol', mol^{-1}, has been added. For a reaction such as

$$CaCO_3\,(s) \rightarrow CaO\,(s) + CO_2\,(g) \qquad \Delta H = 179\,\text{kJ mol}^{-1}$$

this is fine, for the reactant, $CaCO_3$, and also each of the products CaO and CO_2, are all expressed in quantities of 1 mol, and so the enthalpy change of 179 kJ is appropriate to the decomposition of 1 mol of $CaCO_3$, to the production of 1 mol of CaO, and also to the production of 1 mol of CO_2. For a more complex reaction, however, such as the combustion of octane

$$C_8H_{18}\,(l) + \frac{25}{2}\,O_2\,(g) \rightarrow 8\,CO_2\,(g) + 9\,H_2O\,(l) \qquad \Delta H = -5471\,\text{kJ}$$

then 1 mol of octane (l) reacts with 25/2 mols of oxygen (g), to produce 8 mols of CO_2 (g) and 9 mols of H_2O (l). To express the enthalpy change associated with this reaction as -5471 kJ mol^{-1} is potentially unclear – to which molecule does 'per mol', mol^{-1}, relate? In this case, it might be considered 'obvious' that the molecule being referred to is octane, but what might be 'obvious' to one person might not be so 'obvious' to another, especially as reactions become increasingly complex. To avoid this problem, this book will adopt the convention that the enthalpy change associated with any chemical reaction will refer to the reaction as written, and will be expressed in units of J, kJ or MJ, as, for example

$$2\,C_8H_{18}\,(l) + 25\,O_2\,(g) \rightarrow 16\,CO_2\,(g) + 18\,H_2O\,(l) \qquad \Delta H = -10{,}942\,\text{kJ}$$

If the unit of measure is expressed as J mol^{-1}, kJ mol^{-1} or MJ mol^{-1}, then the accompanying text will make explicitly clear what 'mol^{-1}' refers to.

One further important point is that the enthalpy of a reaction depends on the phase of the reactants: so, in the combustion of octane, as defined by the reaction formulae just quoted, liquid octane reacts with gaseous oxygen to produce gaseous carbon dioxide and liquid water. If we are interested in the reaction which results in water vapour rather than liquid water

$$C_8H_{18}\,(l) + \frac{25}{2}\,O_2\,(g) \rightarrow 8\,CO_2\,(g) + 9\,H_2O\,(g) \qquad \Delta H_? = ?\,\text{kJ}$$

then the enthalpy change ΔH is different, for we need to take into account the enthalpy associated with the phase change

$$H_2O\,(l) \rightarrow H_2O\,(g) \qquad \Delta H = 44\,\text{kJ}$$

As a phase change from a more condensed state (liquid water) to a less condensed state (water vapour), this is an endothermic reaction, for which, in this case, the appropriate value of ΔH is 44 kJ for each mol of liquid water transformed into water vapour.

To determine the unknown enthalpy change for the reaction which results in water vapour, we may invoke the fact that enthalpy is a state function, which implies that the enthalpy change for any reaction depends only on the initial and final states, and is independent of the path taken. Figure 6.1 therefore shows how the combustion of octane to result in water vapour can be represented as two steps – the first burning the gaseous octane to result in liquid water, and the second to convert the liquid water into water vapour:

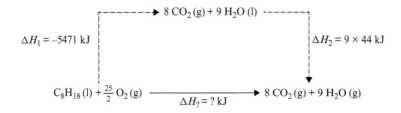

Figure 6.1 How the fact that enthalpy is a state function can be used to determine ΔH for a reaction. The enthalpy change $\Delta H_?$ for the reaction associated with the solid arrow is equal to the sum of the enthalpy changes, ΔH_1 and ΔH_2 for the two reactions associated with the dotted arrows: $\Delta H_? = \Delta H_1 + \Delta H_2 = -5471 + 9 \times 44 = -5075$ kJ

The enthalpy change ΔH_1 for the first step, burning 1 mol of liquid octane to produce 9 mols of liquid water, is −5471 kJ, and the enthalpy change ΔH_2 for the vaporisation of 9 mols of liquid water is $9 \times 44 = 396$ kJ. The enthalpy change for the reaction in which liquid octane is burnt to form carbon dioxide and water vapour is therefore $-5471 + 396 = -5075$ kJ per mol of octane oxidised.

$$C_8H_{18}\,(l) + \frac{25}{2} O_2\,(g) \rightarrow 8\,CO_2\,(g) + 9\,H_2O\,(g) \qquad \Delta H = -5075\,\text{kJ}$$

The study of enthalpy changes associated with chemical reactions is known as **thermochemistry**, as discussed in detail later in this chapter.

6.5 Enthalpy, directionality and spontaneity

Since enthalpy is a state function, the change ΔH in the enthalpy of a system depends only on the enthalpies of the initial and final states of the system. So, for a change in state in which some reactants, collectively represented by A, react together to form some products, collectively represented by B, under given conditions of temperature and pressure

$$A \rightarrow B$$

then we may express the corresponding enthalpy change $\Delta H_{A \rightarrow B}$ as

$$\Delta H_{A \rightarrow B} = H_B - H_A$$

Suppose now that the direction of the reaction, under the same conditions of temperature and pressure, is reversed

B → A

The enthalpy change $\Delta H_{B \to A}$ for this reverse reaction is, by definition, given by

$$\Delta H_{B \to A} = H_A - H_B = -(H_B - H_A) = -\Delta H_{A \to B}$$

So, if we know the enthalpy change $\Delta H_{A \to B}$ for any reaction, then the enthalpy change $\Delta H_{B \to A}$ for the reverse reaction has the same magnitude as the enthalpy change of the original reaction, but the opposite sign.

As an example, if we consider the exothermic reaction in which octane is oxidised as

$$C_8H_{18} \text{ (l)} + \frac{25}{2} O_2 \text{ (g)} \to 8\, CO_2 \text{ (g)} + 9\, H_2O \text{ (l)} \qquad \Delta H = -5471 \text{ kJ}$$

for which the enthalpy change is −5471 kJ per mol of octane burnt, then the reverse reaction

$$8\, CO_2 \text{ (g)} + 9\, H_2O \text{ (l)} \to C_8H_{18} \text{ (l)} + \frac{25}{2} O_2 \text{ (g)} \qquad \Delta H = 5471 \text{ kJ}$$

will be endothermic, with an enthalpy change of +5471 kJ per mol of octane produced.

This reverse reaction, however, is distinctly odd: it is most unlikely that a mixture of gaseous carbon dioxide and liquid water will react together, and completely, to form octane and molecular oxygen. In contrast, however, the original reaction – the combustion of octane – happens instantaneously once the octane is ignited by a spark, as happens in every octane-burning petrol engine. However, if this reverse reaction were to happen, then the First Law of Thermodynamics tells us that the corresponding enthalpy change is 5471 kJ. As this example illustrates, one of the beneficial applications of the First Law is that it can be used to determine important properties – such as enthalpy changes – of reactions that do not take place in reality, but in which, for whatever reasons, we might be interested.

A deeper issue raised here concerns the *spontaneity* of the chemical reaction – the combustion of octane in oxygen happens naturally, and is spontaneous; the reverse reaction, in which carbon dioxide and water form octane and oxygen, is decidedly 'unnatural' and 'not spontaneous' – this being another example of a spontaneous, unidirectional, change like those we encountered on page 123. Since the 'natural' combustion reaction is exothermic, whilst the reverse 'unnatural' reaction is endothermic, this could trigger the thought that "reactions which are exothermic are spontaneous, but reactions which are endothermic aren't". This, however, cannot be true, for there are many examples of spontaneous endothermic reactions – take, for instance, the spontaneous change of ice at 283 K (approximately 10°C) into liquid water, which, as a phase change from a more condensed state (ice) to a less condensed state (water) is necessarily endothermic.

The question "why are some reactions spontaneous, and others aren't?" is very important, but it is one that the First Law of Thermodynamics does not answer. But, as we shall see in Chapters 9, 11 and 13, the Second Law of Thermodynamics does.

6.6 The difference $\Delta H - \Delta U$

Since, by definition

$$H = U + PV \tag{6.4}$$

then, for a finite change from state [1] to state [2]

$$\Delta H = \Delta U + \Delta(PV) \tag{6.14}$$

and therefore

$$\Delta H - \Delta U = (P_2 V_2 - P_1 V_1)$$

For a given change in any system, the difference $\Delta H - \Delta U$ between the values of the change ΔH in the system's enthalpy, and the change ΔU in the system's internal energy, is therefore given by the quantity $\Delta(PV) = (P_2 V_2 - P_1 V_1)$. Now, for an isothermal change in an ideal gas, we know that

$$PV = \text{constant}$$

so that

$$\Delta(PV) = 0$$

implying that, for an isothermal change in an ideal gas

$$\Delta H_T = \Delta U_T$$

But we also know that the internal energy U of an ideal gas is a function of temperature T only, from which we infer that, for an isothermal change

$$\Delta U_T = 0$$

and therefore that $\Delta H_T = 0$

For an isothermal change for an ideal gas, both the change ΔU_T in the internal energy U and the change ΔH_T in the enthalpy H are zero; furthermore, the fact that $\Delta H_T = 0$ confirms the conclusion we drew on page 153 – that, for an ideal gas, enthalpy, like internal energy, is a function of temperature only.

An isothermal change for an ideal gas is a special, hypothetical, case, in which $\Delta(PV) = 0$. There are, however, many real changes for which

$$P_2 V_2 \approx P_1 V_1$$

Implying that

$$\Delta H \approx \Delta U$$

This is the case for many changes that do not involve gases, such as those that take place wholly in the solid and liquid states, and in solution.

As an example, consider the phase change in which one mol of solid water (ice) changes into one mol of liquid water at a temperature of 273.15 K, 0 °C, and a pressure of 1 bar (10^5 Pa). This is an endothermic reaction, absorbing 6.01 kJ:

$$H_2O\,(s) \rightarrow H_2O\,(l) \qquad\qquad \Delta H = 6.01 \text{ kJ mol}^{-1}$$

Given that the molar volumes of ice and water at 273.15 K are 1.96×10^{-5} m^3 mol^{-1} and 1.80×10^{-5} m^3 mol^{-1}, respectively, then, at a constant pressure of 1 bar

$$\Delta(PV) = (PV)_{water} - (PV)_{ice} = 10^5 \times (1.80 - 1.96) \times 10^{-5} \text{ J mol}^{-1} = -0.16 \text{ J mol}^{-1}$$

This value of $\Delta(PV)$, -0.16 J mol^{-1}, is some four orders magnitude smaller than the enthalpy change $\Delta H = 6.01$ kJ mol^{-1} for the phase transition, demonstrating that, for the phase change from ice to liquid water, ΔH and ΔU are very nearly equal.

When gases are involved, however, the difference between ΔH and ΔU becomes more evident, as illustrated by the example of the vapourisation of liquid water to steam, which takes place endothermically at 373.15 K, 100 °C, and a pressure of 10^5 Pa

$$H_2O \text{ (l)} \rightarrow H_2O \text{ (g)} \qquad\qquad\qquad \Delta H = 41 \text{ kJ mol}^{-1}$$

The molar volume of water at 373.15 K is 1.90×10^{-5} m^3 mol^{-1} (that's about 5% greater than the value of 1.80×10^{-5} m^3 mol^{-1} at 273.15 K, as a result of the thermal expansion of water), and that of steam at 373.15 K is $3.01 \times 10^{-2} = 3010 \times 10^{-5}$ m^3 mol^{-1}, and so, at a constant pressure of 10^5 Pa

$$\Delta(PV) = (PV)_{steam} - (PV)_{water}$$
$$\Delta(PV) = 10^5 \times (3010 - 1.90) \times 10^{-5} \text{ J mol}^{-1} = 3 \text{ kJ mol}^{-1}$$

Now

$$\Delta H = \Delta U + \Delta(PV)$$

and since $\Delta H = 41$ kJ mol^{-1}, then

$$41 = \Delta U + 3$$

and so

$$\Delta U = 38 \text{ kJ mol}^{-1}$$

For the phase transition from water to steam at 373.15 K and a pressure of 10^5 Pa 1 ($\approx$ 1 bar), ΔH (41 kJ mol^{-1}) and ΔU (38 kJ mol^{-1}) therefore differ by some 8%.

A more general case is that of the gas phase reaction represented by

$$aA + bB \rightarrow cC + dD$$

where the reactants, A and B, and the products, C and D, are all gases, which – for simplicity of analysis – are all assumed to be ideal.

We know that

$$PV = nRT$$

is universally true for all ideal gases, in which n is the total number of mols of gas present; furthermore, since an ideal gas is one in which the molecules occupy no volume, and experience no mutual interactions, then this law is valid not only for a gas of a single chemical type of molecule, but also, as we saw on page 49 in our discussion of Dalton's law of partial pressures, for mixtures of ideal gases. Therefore

$$(PV)_{products} = (c + d) RT$$

and

$$(PV)_{\text{reactants}} = (a + b)RT$$

hence

$$\Delta(PV) = [(c + d) - (a + b)]RT$$

The quantity $[(c+d)-(a+b)]$ is the difference between the total mole numbers of the products and the total mole numbers of the reactants, as determined by the stoichiometric coefficients of the reaction $aA + bB \rightarrow cC + dD$ as written. If we express this as Δn,

$$\Delta n = [(c + d) - (a + b)] \tag{6.15}$$

then

$$\Delta(PV) = \Delta nRT$$

and

$$\Delta H = \Delta U + \Delta nRT \tag{6.16}$$

As an example, consider the exothermic combustion of gaseous octane (see exercise 3)

$$C_8H_{18}\,(g) + \frac{25}{2}\,O_2\,(g) \rightarrow 8\,CO_2\,(g) + 9\,H_2O\,(g) \qquad \Delta H = -5116\,\text{kJ}$$

taking place at 298 K (about 25 °C, rather higher that the 'normal' room temperature of about 20 °C), and 10^5 Pa.

For this reaction

$$\Delta n = (8 + 9) - \left(1 + \frac{25}{2}\right) = \frac{7}{2}$$

At a temperature of 298 K, and using a value of 8.31 J K^{-1} mol^{-1} for R, then

$$\Delta nRT = \frac{7}{2} \times 8.31 \times 298\,\text{J} = 9\,\text{kJ}$$

And since $\Delta H = -5116$ kJ, and assuming ideal behaviour, then

$$\Delta U = \Delta H - \Delta nRT = -5116 - 9 = -5125\,\text{kJ}$$

As we have just seen, for the reaction written as $aA + bB \rightarrow cC + dD$, Δn is defined as

$$\Delta n = [(c + d) - (a + b)] \tag{6.15}$$

in which the stoichiometric coefficients a, b, c and d are necessarily positive. These stoichiometric coefficients are different from the corresponding **stoichiometric numbers**, which are the signed, algebraic, quantities $-a$, $-b$, $+c$ and $+d$, in which the stoichiometric coefficients of all products are associated with a positive sign, and the stoichiometric coefficients of all reactants with a negative sign. Stoichiometric numbers therefore depend on the direction in which a chemical equation is written – for example, if the reaction $aA + bB \rightleftharpoons cC + dD$ is reversible, then the stoichiometric *numbers* for the forward reaction $aA + bB \rightarrow cC + dD$, are the set $-a$, $-b$, $+c$ and $+d$, but for the reverse reaction $cC + dD \rightarrow aA + bB$, the stoichiometric numbers are $+a$, $+b$, $-c$ and $-d$. Both the forward, and the reverse, reactions, however, have the same stoichiometric *coefficients*, +a, +b, +c and +d. In general, stoichiometric coefficients are more convenient to use, and so this book will always define reactions, and equations such as

$$\Delta n = [(c + d) - (a + b)] \tag{6.15}$$

in terms of stoichiometric coefficients, and *not* stoichiometric numbers.

6.7 Phase changes

If we measure the temperature of ice as it is gently heated, at normal atmospheric pressure, and at a constant rate, from an initial temperature, of, say, $-20\ °C$ towards $0\ °C$, the temperature rises at a steady rate. But when the temperature of the ice reaches $0\ °C$, and the ice begins to melt, the temperature of the ice-water mixture remains at $0\ °C$, even though heat continues to be supplied to the system at a constant rate – only when all the ice has melted does the temperature of the water once more begin to rise.

This experiment demonstrates that the phase change from ice to water requires heat, this being just one instance of the general principle that melting a solid into a liquid is an endothermic reaction. For 1 mol of solid, the enthalpy change associated with this reaction is known as the **molar enthalpy of fusion** $\Delta_{fus}H$ (an older term being the molar latent heat of fusion), as measured at the melting point at a pressure of 1 bar

$$H_2O\ (s) \rightarrow H_2O\ (l) \qquad\qquad \Delta_{fus}H = 6.01\ kJ\ mol^{-1}$$

Similarly, the molar **enthalpy of vaporisation** $\Delta_{vap}H$ is the enthalpy change accompanying the transition of 1 mol of a substance from liquid to gas, as measured at the boiling point at a pressure of 1 bar

$$H_2O\ (l) \rightarrow H_2O\ (g) \qquad\qquad \Delta_{vap}H = 40.65\ kJ\ mol^{-1}$$

Table 6.3 shows some representative enthalpies of phase changes for fusion and vaporisation.

All phase changes from more condensed states to less condensed states are endothermic, representing the need to break the forces of attraction that hold condensed states together. Conversely, all phase changes from less condensed states to more condensed states are exothermic, representing the energy released with those forces of attraction form the associated condensed state bonds. An important consequence of the heat released as a result of a phase transition from a less, to a more, condensed state happens naturally when water vapour condenses to liquid water in the atmosphere – under appropriate atmospheric conditions, the heat released by this condensation is the source of the energy of thunderstorms, cyclones, tornados and hurricanes.

Sublimation – the direct transition from solid to gas, without passing through the liquid phase, as associated with the **molar enthalpy of sublimation** $\Delta_{sub}H$ – is not a common phenomenon under normal conditions of atmospheric pressure. However, if iodine crystals are gently heated, they sublimate into a distinctive, violet-coloured, gas; solid carbon dioxide ("dry ice") sublimates at room temperature, and the resulting dense fumes are used, for example, in theatres to dowse the stage in 'smoke'; ice too can sublimate, resulting in the 'disappearance' of snow and ice on very cold days, without the formation of water:

$$H_2O\ (s) \rightarrow H_2O\ (g) \qquad\qquad \Delta_{sub}H = 51.1\ kJ\ mol^{-1}$$

Table 6.3 Some values of $\Delta_{fus}H$ and $\Delta_{vap}H$ for phase changes for some selected materials at the stated temperatures

Selected elements at 1 atm pressure

	Fusion		Vaporisation	
	T_{fus} K	$\Delta_{fus}H$ kJ mol^{-1}	T_{vap} K	$\Delta_{vap}H$ kJ mol^{-1}
H_2	14	0.12	20	0.90
N_2	63	0.71	77	5.57
O_2	54	0.44	90	6.82
S	388	1.72	718	45
Br_2	266	10.6	332	30.0
I_2	387	15.5	458	41.6
Al	933	10.7	2792	294
Hg	234	2.30	630	59.1

Selected inorganic molecules at 1 atm pressure

	Fusion		Vaporisation	
	T_{fus} K	$\Delta_{fus}H$ kJ mol^{-1}	T_{vap} K	$\Delta_{vap}H$ kJ mol^{-1}
H_2O	273	6.01	373	40.65
HCl	159	2.00	118	16.15
HI	222	2.87	238	19.76
NH_3	195	5.66	240	23.33

Selected organic molecules at 1 atm pressure

	Fusion		Vaporisation	
	T_{fus} K	$\Delta_{fus}H$ kJ mol^{-1}	T_{vap} K	$\Delta_{vap}H$ kJ mol^{-1}
CH_4	91	0.94	112	8.19
C_6H_6	279	9.87	353	30.7
$CHCl_3$	210	9.5	334	29.2
C_2H_5OH	159	4.93	351	38.6
CH_3COOH	290	11.73	391	23.7
Phenol	314	11.51	455	45.7

Source: *CRC Handbook of Chemistry and Physics*, 96th edn, 2015, Section 6, pages 6–127 to 6–154. Note that the reference pressure is 1 atm = 1.01325 bar.

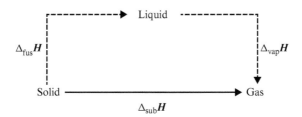

Figure 6.2 The molar enthalpies of fusion, vaporisation and sublimation.

As illustrated in Figure 6.2, we may invoke the fact that enthalpy is a state function to determine the relationship between the enthalpies of fusion, $\Delta_{fus}H$, vaporisation $\Delta_{vap}H$ and sublimation $\Delta_{sub}H$. For a given mass of any pure material, there are two paths from a solid to a gas: one path, shown by the solid line, is directly, by sublimation; an alternative is indirectly, firstly from solid to liquid, and subsequently from liquid to gas, as shown by the dashed line. Hence, in general

$$\Delta_{sub}H = \Delta_{fus}H + \Delta_{vap}H \tag{6.17}$$

and so, if we know any two of $\Delta_{fus}H$, $\Delta_{vap}H$ and $\Delta_{sub}H$, we can compute the third. But with care. If we refer, for example, to Table 6.1, we see that, for water, $\Delta_{fus}H = 6.01$ kJ mol^{-1}, and $\Delta_{vap}H = 40.65$ kJ mol^{-1}, suggesting, according to equation (6.16) that $\Delta_{sub}H = 46.66$ kJ mol^{-1}. It isn't – as we have just seen, $\Delta_{sub}H = 51.1$ kJ mol^{-1}. The discrepancy arises from the fact that, as shown in Table 6.1, $\Delta_{fus}H = 6.01$ kJ mol^{-1} is the value as measured at the normal melting point of ice at 1 atm pressure, which is a temperature of 273 K, whereas $\Delta_{vap}H = 40.65$ kJ mol^{-1} is the value as measured at the normal boiling point of water at 1 atm pressure, 373 K. Since enthalpy is a function of both temperature and pressure, the value of $\Delta_{vap}H$ at 273 K is likely to be different from its value at 373 K – as indeed it is: $\Delta_{vap}H$ at 273 K is in fact 45.05 kJ mol^{-1} (as will be demonstrated on page 194). If we now use equation (6.16) using consistent data for 273 K, $\Delta_{fus}H\,(273) + \Delta_{vap}H\,(273) = 6.01 + 45.05 = 51.06$ kJ mol^{-1}, which rounds to the 'right' answer, 51.1 kJ mol^{-1}, which also refers to 273 K.

This example illustrates an important point: it's very easy to make an inadvertent mistake by adding and subtracting numbers found in reference tables, when the reference data applies to specific conditions which are different from the conditions of current interest. So be alert to that – and especially to the fact that all tables contain data measured at a given temperature, which is either stated explicitly or (we hope!) can be assumed to be 298.15 K = 25 °C. If the data you need corresponds to a different temperature, then an adjustment need to be made – which, as we shall see on pages 190 to 200, is not difficult, provided you remember to do it!

6.8 Measuring enthalpy changes – calorimetry

Phase changes are physical changes, rather than chemical reactions, so we now turn to the enthalpy changes associated with reactions in which reagents, under appropriate conditions, react chemically to give the corresponding products. Experimentally, enthalpy changes associated with chemical reactions can be measured using a **calorimeter**. Calorimeters are of different types, but all work on the same principle. The chemical reaction of interest takes place under controlled conditions, and the heat generated (or absorbed) by the reaction is transferred either to (or from) a known quantity of a 'working substance' (usually water) of known molar heat capacity. If the temperature of the working substance is T_1 before the reaction, and T_2 afterwards, then the quantity of heat q_w absorbed by (or released from) the working substance (hence the subscript $_w$) during the reaction is

$$q_w = n\,C_P\,(T_2 - T_1)$$

where C_P is the molar heat capacity at constant pressure of the working substance, and n is its quantity, in mols. If $T_2 > T_1$, then q_w is positive, and the working substance gains heat; if $T_2 < T_1$, then q_w is negative, and the working substance loses heat. If the transfer of heat from the reaction to the working substance is efficient, and no heat is lost, then the quantity q_w of heat gained by (or lost from) the working substance is equal to the quantity of hear resulting from (or absorbed by) the chemical reaction – heat equal to the reaction's ΔH if carried out at constant pressure. Hence

$$\Delta H = -q_w$$

where the negative sign recognises that heat gained (or lost) by the working substance corresponds to heat generated from (or absorbed by) the reaction. So, if the temperature of the working substance rises, q_w is positive, implying that ΔH is negative corresponding to an exothermic reaction; likewise, if the result of the reaction is to cool the working substance, q_w is negative, implying that ΔH is positive.

For reactions involving combustion, it is often safer to carry out the reaction within a strong sealed container, known as a **bomb calorimeter**. The reaction therefore takes place not at constant pressure but at constant volume, implying that the heat q_w is equal $-\Delta U$, the negative of the change ΔU in the internal energy, rather than to $-\Delta H$, the negative of the change ΔH in the enthalpy. To determine the reaction's ΔH, the experimentally measured value of ΔU then needs to be adjusted using equation (6.14)

$$\Delta H = \Delta U + \Delta(PV) \tag{6.14}$$

As we saw on page 158, if the reaction involves no gases, then $\Delta(PV) \approx 0$, and so $\Delta H \approx \Delta U$. Many combustion reactions, however, do involve gases, either as the products of the combustion, or as the reactants too. For a reaction in which the final temperature is ultimately equal to the initial temperature, and assuming that the various gases involved approximate to ideal gases, then $\Delta(PV) = \Delta nRT$, where Δn is the net change in mole number, as given by equation (6.15).

In practice, accurate calorimetry requires considerable experimental care, especially to ensure that all heat is accounted for correctly (remembering, for example, that the vessel used for the experiment might also absorb or release heat, as well as the working substance). In principle, though, calorimetry is straight-forward, and is based soundly on the fundamental concept underpinning the First Law – namely, that energy, in this case in the form of heat, can be neither created nor destroyed.

6.9 Calculating enthalpy changes – Hess's law

A requirement for successful calorimetry is that the reaction of interest goes to completion, without any side products. Some reactions do not go naturally to completion, but reach a dynamic stable equilibrium mixture containing measureable quantities of both reactants and products – as, for example, happens when hydrogen and gaseous iodine react to form gaseous hydrogen iodide, which itself decomposes back to hydrogen and iodine, as represented by the reversible reaction

$$H_2\,(g) + I_2\,(g) \rightleftharpoons 2\,HI\,(g)$$

Furthermore, some other reactions, although apparently simple on paper, are much more complex in practice: as an example, consider the reaction between graphite and oxygen to produce carbon monoxide, which in principle is written as

$$2\,C\,(graphite,\,s) + O_2\,(g) \rightarrow 2\,CO\,(g)$$

but the reality is that some CO_2 is inevitably formed at the same time.

One of the benefits of the First Law of Thermodynamics is that the identification of enthalpy as a state function enables the enthalpies of 'pure' reactions to be calculated, even if the corresponding reactions do not take place, as written, in reality. As an example, the enthalpy change ΔH for the full combustion of 1 mol of graphite to carbon dioxide, at 298.15 K and 1 bar pressure, can be determined by calorimetry as $\Delta H = -393.5$ kJ

$$C\,(graphite,\,s) + O_2\,(g) \rightarrow CO_2\,(g) \qquad\qquad \Delta H = -393.5\,\text{kJ}$$

Also, calorimetry can determine the enthalpy change for the oxidation of carbon monoxide to carbon dioxide, at the same conditions of temperature and pressure, as

$$CO\,(g) + \tfrac{1}{2}\,O_2\,(g) \rightarrow CO_2\,(g) \qquad\qquad \Delta H = -283.0\,\text{kJ}$$

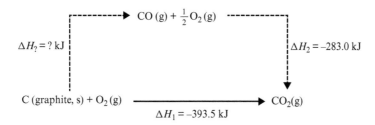

Figure 6.3 An example of the use of Hess's law of constant heat formation.

Figure 6.3 illustrates that, in principle, it is possible to change from an initial state of graphite and gaseous oxygen to a final state of gaseous carbon dioxide by two paths: firstly, directly (as shown by the solid line), and secondly, indirectly via an intermediate state of carbon monoxide (as shown by the dashed lines). Since enthalpy is a state function, the enthalpy change for the reaction in which graphite is totally oxidised to carbon dioxide must be independent of the

ENTHALPY AND THERMOCHEMISTRY

path between those two states. The enthalpy change ΔH_1 associated with the direct reaction must therefore equal the sum $\Delta H_? + \Delta H_2$ associated with the indirect path

$$\Delta H_1 = \Delta H_? + \Delta H_2$$

$$\therefore -393.5 = \Delta H_? - 283.0$$

$$\therefore \Delta H_? = -110.5 \text{ kJ}$$

This method of calculating the enthalpy changes associated with chemical reactions was first elucidated by the Swiss-born chemist Germain Hess in 1840, before the First Law, and the concept of enthalpy, had been established. So, rather than referring to enthalpy changes, **Hess's law of constant heat formation** refers to heat:

> The heat released or absorbed in any chemical reaction depends solely on the initial reactants and final products, and is independent of the path taken during the reaction.

Another example. What is the enthalpy change associated with the formation of 1 mol of propane, C_3H_8 (g) from elemental carbon (graphite, s), and gaseous hydrogen?

To answer this question, we can use this information

C_3H_8 (g) + 5 O_2 (g) → 3 CO_2 (g) + 4 H_2O (g) $\hspace{2cm}$ $\Delta H = -2044$ kJ

C (graphite, s) + O_2 (g) → CO_2 (g) $\hspace{2cm}$ $\Delta H = -393.5$ kJ

H_2 (g) + $\frac{1}{2} O_2$ (g) → H_2O (g) $\hspace{2cm}$ $\Delta H = -241.8$ kJ

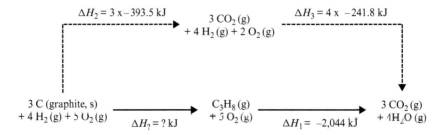

Figure 6.4 Another example of how to solve a thermochemical puzzle.

Figure 6.4 shows how this information may be fitted together to solve the thermochemical puzzle. Starting with the appropriate stoichiometric quantities of elemental carbon, hydrogen and oxygen, we can 'carry out' the hypothetical reaction of combining the carbon (graphite, s) with hydrogen (g) to form ethane (g), with the oxygen (g) as yet not being used. We can then use that oxygen (g) to burn the ethane (g) to form carbon dioxide (g) and water (g). These two reactions are shown by the solid path – the overall result being to convert 3 mols of carbon (graphite, s), 4 mols of hydrogen and 5 mols of oxygen into 3 mols of carbon dioxide, and 4 mols of water (g). The alternative path, shown by the dashed lines, forms, firstly, 3 mols of

CO_2, and secondly, 4 moles of H_2O (g). The initial states and final states of both paths are identical, and so the total enthalpy changes along each path are the same:

$$\Delta H_? + \Delta H_1 = \Delta H_2 + \Delta H_3$$

and so

$$\Delta H_? - 2044 = 3 \times -393.5 + 4 \times -241.8$$

giving

$$\Delta H_? = -103.7 \text{ KJ}$$

Thermochemical puzzles of this type are favourite exam questions. When tackling a problem, it's important to remember that every problem is designed to 'work': no 'impossible' problem will ever be set, and you will always be given all the information you need – the 'trick' is to assemble the information you are given in just the right way. To help do this, get a pencil and some paper, and draw some quick sketches of how the reactions might work – and don't worry about throwing things away and scratching things out, for it's unlikely you'll get the sequences right first time.

The starting point is *always* the reaction with the unknown enthalpy change – write that reaction down first. Then it's a matter of sketching out possible answers to these four questions:

- What other ways might there be of producing the target products?
- What ways might there be of producing the required reactants?
- What other reactions are there for which the reactants of the 'target reaction' might also be reactants?
- What other reactions are there for which the products of the 'target reaction' might themselves be reactants?

Not all of these questions will apply in any particular case, but together, they should help you solve any thermochemical problem. So, in Figure 6.4, the 'target' reaction was

$$3 \text{ C (graphite, s)} + 4 \text{ H}_2 \text{ (g)} \rightarrow \text{C}_3\text{H}_8 \text{ (g)} \qquad \Delta H_? = ?$$

with the reactants being elemental carbon (graphite, s) and hydrogen (g), and the product, ethane (g). Since the reactants are elements, the second question – "What ways might there be of producing the required reactants?" – is not applicable; but the third and fourth questions are: the reactants, carbon and hydrogen, can, with oxygen, take part in other reactions, for example, the formation of carbon dioxide and water; and ethane is the reactant in its own combustion. We then see that the information provided is all relevant: the enthalpy changes for the formation of CO_2 and H_2O, and the combustion of C_3H_8. So, with perhaps a few false starts, a diagram like the one shown in Figure 6.4 can be drawn, and the problem is well on its way to being solved.

But not quite yet: there are two other concepts to bear in mind.

Firstly, it is important to ensure that the stoichiometry is correct throughout the pathways being analysed. Enthalpy is an extensive state function, and so depends on the quantities involved. This implies that if the enthalpy change for the reaction written as

$$CO\,(g) + \frac{1}{2}\,O_2\,(g) \rightarrow CO_2\,(g)$$

is − 283.5 kJ, then the enthalpy change associated with the reaction written as

$$2\,CO\,(g) + O_2\,(g) \rightarrow 2\,CO_2\,(g)$$

is 2 × − 283.5 = − 567 kJ. You will therefore need to take considerable care to ensure that the numbers used to solve any given problem are right for the given stoichiometry. So, once again with reference to Figure 6.4, we need to multiply the enthalpy change given for the reaction

$$C\,(graphite,\,s) + O_2\,(g) \rightarrow CO_2\,(g) \qquad \Delta H = -393.5\,kJ$$

by a factor of 3, and that for the reaction

$$H_2\,(g) + \frac{1}{2}\,O_2\,(g) \rightarrow H_2O\,(g) \qquad \Delta H = -241.8\,kJ$$

by a factor of 4 to get the stoichiometry right for the combustion of 1 mol of ethane, C_3H_8.

Secondly, take care to note the phases of the reactants and products. As we saw on pages 155 and 156, and in Figure 6.1, the enthalpy change for the reaction

$$C_8H_{18}\,(l) + \frac{25}{2}\,O_2\,(g) \rightarrow 8\,CO_2\,(g) + 9\,H_2O\,(l) \qquad \Delta H = -5471\,kJ$$

is different from the enthalpy change for the reaction

$$C_8H_{18}\,(l) + \frac{25}{2}\,O_2\,(g) \rightarrow 8\,CO_2\,(g) + 9\,H_2O\,(g) \qquad \Delta H = -5075\,kJ$$

by virtue of the enthalpy change associated with the vaporisation of water

$$H_2O\,(l) \rightarrow H_2O\,(g) \qquad \Delta H = 44\,kJ$$

whereby

$$-5075 = -5471 + 9 \times 44$$

And one further note on the use of Hess's law: it's important to remember that there is an implicit assumption that the temperature is constant for all of the reactions involved: if some reactions take place at a temperature which is different from that for other reactions, then these temperature changes need to be taken into account, as will be described on pages 190 to 200.

6.10 Chemical standards

6.10.1 What are 'standards'?

When observing natural phenomena, and making measurements, it is very helpful if the scientific community uses well-defined, and universally acknowledged, units-of-measure: by doing so, statements such as "the mass of the system is 3.2 kg" can be understood, unambiguously, by scientists and engineers, and indeed the general public, around the world. So, by international agreement, the 'standard' unit-of-measure for mass is acknowledged as the kilogram, as the standard unit-of-measure for length is acknowledged as the metre. Both of these 'standards' are arbitrary, in that there is no fundamental science associated with the kilogram or the metre – indeed, the kilogram and the metre are not the only ways by which quantities of matter and length can be measured: for many years, British scientists used the 'pound' for mass, and the 'foot' for length. What the unit-of-measure actually *is* does not matter; what *does* matter is that the international community agrees to adopt the same convention, and that the 'standards' chosen are relevant, convenient, and easy-to-use. So, for example, the kilogram and the metre are both 'human-scale' units, and many of the activities we carry out are easily measured in these units.

Units-of-measure are just one aspect of internationally agreed standards: two others are

- the definition of the precise state of the system to which a particular measurement refers
- the reference level of measurement, which is usually expressed as the definition of a zero point for a given scale.

The familiar statement "pure water, at a pressure of 1.01325 bar = 1 atm, boils at a temperature of 100 °C" is an example of all three of these attributes of standards.

Firstly, the unit of measurement of temperature is specified as the °C, and reference is made to two, different, units of measurement of pressure, the bar and the atm.

Secondly, the definition of the precise state of the system is specified by the phrases "pure water" and "at a pressure of 1.01325 bar". If water is not pure, but contains, for example, some dissolved substance such as common salt, then – as every cook knows, and as we shall show in Chapter 18 – the boiling point of water is higher than 100 °C. Similarly, if water is heated at a pressure lower than 1 bar, it will boil at a lower temperature. So the boiling point of 100 °C applies only to a system of pure water, at a pressure of 1 bar – this being a particular example of the more general case that measurements of thermodynamic data must be accompanied by a statement of the conditions that define, completely, the state of the system for which any measurement is made.

The third point – concerning the reference level – is not directly mentioned, but is an attribute of the Celsius scale, as explicitly identified by the symbol °C. The reference level of any measurement scale is the level against which all other measurements are compared. The **reference level** is therefore a measurement, usually defining the zero of the corresponding scale, and the state of the system to which that measurement corresponds is the **reference state**. So, for example, the original reference level of the Celsius scale of temperature measurement, defined as 0 °C, is the temperature of the reference state, an equilibrium mixture of pure liquid water and pure ice, at a pressure of 1.01325 bar. The choices of reference level and reference state for the Celsius scale are arbitrary – an alternative, for example, might be to define the reference level of 0 °C as the boiling point of liquid ammonia at a pressure of 0.9 bar. The key issue is convenience: the water reference level is much more convenient for widespread use than the ammonia reference level, and the reference state of a water-ice mixture at normal atmospheric pressure is much more accessible than the state of boiling ammonia at 0.9 bar.

With these thoughts in mind, we now introduce the **conventional standards** used in thermodynamics, as recommended by the International Union of Pure and Applied Chemistry (IUPAC). Full details can be found on pages 1427 and 1438 of the IUPAC 'Gold Book' (*Compendium of Chemical Terminology, Gold Book*, version 2.3.3, published by IUPAC, 2014), which in turn references pages 53 and 54 of the 2nd edition of the 'Green Book' (*Quantities, Units and Symbols in Physical Chemistry*, published by IUPAC, 1993 – corresponding to pages 61 and 62 of the 3rd edition, published in 2007).

6.10.2 Units-of-measure

The units-of-measure in thermodynamics comply with SI standards, such as the kg for mass; the J for energy, work and heat; and the K for temperature. The number of particles in a chemical system is measured in mol, defined by reference to the Avogradro constant $N_A = 6.022141 \times 10^{23}$ mol^{-1}: a system of 3 mol of molecular oxygen O_2 therefore contains $3 \times 6.022141 \times 10^{23}$ oxygen molecules, and a system of 10^{-5} mol of aqueous hydrogen ions H^+ (aq) contains $6.022141 \times 10^{23} \times 10^{-5} = 6.022141 \times 10^{18}$ ions.

The SI unit of pressure, the Pascal (Pa), is defined as the pressure exerted by a force 1 Newton over an area of 1 m^2 : 1 N m^{-2} = 1 Pa. For reactions, and measurements, in the laboratory, the Pa is an inconveniently small unit, and so, for many years, the internationally accepted unit of pressure was the 'atmosphere', corresponding to the pressure naturally exerted at sea level by the earth's atmosphere, such as to support a column of mercury in a barometer 0.760 m high. This pressure is equal to 1.01325×10^5 Pa, and so the thermodynamic standard pressure is defined as 10^5 Pa, which is sensibly close to 'normal' atmospheric pressure, and convenient to use. The standard 10^5 Pa could be designated as 10^5 Pa, or 10^2 kPa, or 10^{-1} MPa, using the well-known 'kilo' or 'mega' prefixes, but in the absence of a prefix for 10^5, the international community has adopted the unit known as the bar, such that 1 bar = 10^5 Pa.

The SI unit of length is the metre, m, implying that the SI unit of volume is the m^3. At a laboratory scale, one cubic metre is rather large, and so the smaller volume, the cubic decimetre, symbolised as dm^3, is still in widespread use, even though it is not an 'official' SI unit. Note that 1 dm^3 = $(10^{-1}$ m$)^3$ = 10^{-3} m^3 = $(10$ cm$)^3$ = 10^3 cm^3 = 1 litre l.

6.10.3 Measures of mixtures

Many systems, especially systems in which chemical reactions take place, are composed not of pure materials but of mixtures – and, whilst a chemical reaction is taking place, the composition of the reaction mixture is changing. Central to much chemistry is therefore the measurement of the quantities of various components within a mixture at any time. The most general, and fundamental, measure of the relative quantity of any component within a mixture is the **mole fraction**, x_i, this being a dimensionless number defined as the ratio of the total number of moles n_i of component i to the total number $\sum_i n_i$ of moles of all components in the mixture

$$x_i = \frac{n_i}{\sum_i n_i} \tag{6.18}$$

Some mixtures – for example, many mixtures of many gases, and mixtures of miscible liquids such as water and ethanol – are such that the mole fractions of the various components are of similar magnitudes. For other mixtures, however, the mole fraction of one component is overwhelmingly larger than that of any other, in which case we speak of a **solution** of one or more **solutes** (those components with small mole fractions) within a **solvent** (that component with the much larger mole fraction). As an example, consider a solution of 0.1 mol of the water-soluble sugar glucose dissolved in 1 dm³ of pure water. The mass of 1 dm³ of pure water is 1 kg, and since the molecular weight of water is 0.018 kg mol^{-1}, a mass of 1 kg of water corresponds to the quantity 1/0.018 = 55.56 mol. Ignoring the (very small) ionisation of water (which we shall discuss in detail in Chapter 17), the resulting system is a mixture of pure water and dissolved glucose, such that the mole fraction x_{H_2O} of water is 55.56/(55.56+0.1) = 0.9982, and the mole fraction x_{glu} of glucose is 0.1/(55.56 + 0.1) = 0.0018. Since $x_{H_2O} \gg x_{glu}$, we refer to this system as a solution of the solute, glucose, within the solvent, pure water.

For solutes in solution, two measures of quantity, other than the mole fraction, are frequently used

- The **molality** b_i, defined as the ratio of the number of moles n_i of solute i to the total mass M_S of the solvent S

$$b_i = \frac{n_i}{M_S} \tag{6.19}$$

The dimensions of the molality b_D are therefore mol kg^{-1}.

- The **molar concentration** [I], defined as the ratio of the total number of moles n_i of component i to the total volume V of the solution

$$[I] = \frac{n_i}{V} \tag{6.20}$$

The dimensions of [I] are mol m^{-3}, or mol dm^{-3} (this being the same as mol l^{-1}), depending on the choice of the unit of measure for the volume V. An alternative name for the molar concentration [I], but only if measured in units of mol dm^{-3}, is the **molarity**, M_i (not to be confused with the molality b_i!).

For aqueous solutions at normal laboratory temperatures, the mass of 1 dm³ of pure water is exactly, or very nearly, 1 kg. Accordingly, if n_i mol of a substance i are dissolved in 1 kg of pure water, the molality is n_i mol kg^{-1}; similarly, the molar concentration is either 10^3 n_i mol m^{-3}, or n_i mol dm^{-3} – from which we see that, for aqueous solutions, the value of the molality is equal to the value of the molar concentration as measured in mol dm^{-3}.

6.10.4 Standard states

As we saw on page 169 in connection with the boiling point of water, to gain a complete understanding of any thermodynamic measurement (in that case, the temperature at which the water boils), we need to know the conditions of the state of the corresponding system (in that case, pure water at a pressure of 1.01325 bar). Since temperature is an intensive state function, the boiling point does not depend on how much water in present, expressed either in terms of the mass, in kg, or the number of moles n. But for more general thermodynamic data – such as a system's enthalpy – we usually need to know the quantity of matter present, as well as the temperature and pressure at which the measurement is made.

If thermodynamic data is to be tabulated, and made available for reference, that presents a problem: there are so many combinations of quantity, temperature and pressure that it would be impossible to make all the measurements, let alone have all these measurements available as reference data.

To overcome this problem, reference tables publish thermodynamic data for reactions in which specific quantities of reactants and products react under specific conditions of temperature and pressure, from which the corresponding data under any other conditions – for example, at a different temperature or a different pressure – can then be computed using appropriate formulae (as will be introduced during the course of this book).

The "specific quantity" is usually 1 mol, and the "specific conditions" are known as **standard states**, as shown in Table 6.4.

Table 6.4 IUPAC standard states. See page 61 of the IUPAC 'Green Book', 3rd edn, 2007.

Reactant or product	Standard form	Standard pressure $P^\ominus$
Solid	The most stable form of the pure solid at the stated temperature, but if not stated, 298.15 K = 25 °C.	1 bar
Liquid	The most stable form of the pure liquid at the stated temperature, but if not stated, at 298.15 K = 25 °C.	1 bar
Gas – ideal	The pure gas at the stated temperature, but if not stated, at 298.15 K = 25 °C.	1 bar
Gas – real	The hypothetical state of the pure gas at the standard pressure, exhibiting ideal behaviour, and at the stated temperature, but if not stated, at 298.15 K = 25 °C.	1 bar
Solvent in solution	The most stable form of the pure solvent at a pressure of 1 bar, and at the stated temperature, but if not stated, at 298.15 K = 25 °C.	1 bar
Solute in an ideal solution	The state at unit concentration (or unit molality) at a pressure of 1 bar, and at the stated temperature, but if not stated, at 298.15 K = 25 °C.	1 bar
Solute in a real solution	The hypothetical state at unit molality, behaving like the infinitely dilute solution, at a pressure of 1 bar, and at the stated temperature, but if not stated, at 298.15 K = 25 °C.	1 bar

Several features of Table 6.4 merit explanation, notably

- The standard state of a material depends on the phase of the material, distinguishing between solids, liquids, ideal and real gases, solvents, and solutes in solution.
- The standard pressure is defined as 1 bar = 10^5 Pa, which is slightly less than standard atmospheric pressure (1 atm = 1.01325 bar, and 1 bar = 0.9869 atm). The standard pressure is represented by the symbol $P^\ominus$.
- The standard states defined in Table 6.4 do *not* specify 'the' standard temperature at which measurements are made, for the (good) reason that, if the pressure is specified

(as it is, at 1 bar), then certain reactions take place 'naturally' at different temperatures – to take a specific example, at a pressure of 1 bar, the reaction in which ice changes phase to liquid water takes place at 273.15 K, whereas the reaction in which liquid water changes phase to water vapour takes place at 373.15 K. Reference tables therefore identify the temperature, represented by the symbol T_s, at which a given standard data value applies, using the convention that, if a temperature is not explicitly stated, then T_s is 298.15 K = 25 °C.

The mention in Table 6.4 of the 'most stable form' recognises that, at a pressure of 1 bar and a temperature of 298.15 K, certain substances can exist in more than one form as **allotropes** – carbon, for example, can exist as graphite or diamond at many different temperatures and pressures. Given any such variety, one form is always the most stable, and so it is that form that defines the standard – with the single exception of phosphorus, for which the standard form is accepted as the white allotrope, even though the red allotrope is in fact the more stable form.

As can be seen from Table 6.4, the standard state for a real gas refers to "the hypothetical state of the pure gas … exhibiting ideal behaviour". The issue being addressed here is the fact that no real gas shows ideal behaviour, yet many of the useful mathematical relationships are derived from an ideal gas model. In order to make the mathematics applicable to real gases with the minimum of modification, it is desirable to choose the standard state of a real gas as that state in which the gas would in principle show ideal behaviour, as we will discuss in more detail in Chapter 22.

For solutes in solution, Table 6.4 defines two standards – one for **ideal solutions**, and one for **real solutions**. As we will see in more detail in Chapter 16, an ideal solution is the liquid solution equivalent of an ideal gas – a solution of a solute in a solvent which obeys simple laws – and, for an ideal solution, the standard of unit molar concentration, 1 mol m^{-3} or 1 mol dm^{-3}, is the analogue of the standard pressure of 1 bar for an ideal gas. Real solutions, however, like real gases, obey more complex laws, and for very similar reasons: reasons attributable to, for example, the volume occupied by each solute molecule or **ion**, and – very importantly – the effect of inter-molecular or inter-ion interactions, which are more significant in a liquid than in a gas as a result of the closer mutual proximity, especially for charged ions in solution. As a solution becomes increasingly dilute, the solute molecules or ions become on average further apart, and so interact with each other progressively more weakly. The behaviour of a real solution therefore becomes progressively closer to the ideal as it becomes more dilute. **Infinite dilution** is therefore the hypothetical state in which the solute molecules have no mutual interaction at all: for a real solution, infinite dilution is the analogue of zero pressure for a real gas. And although infinite dilution is indeed a hypothetical state, measurements associated with infinite dilution can be estimated by the extrapolation of experimental results obtained from real solutions for progressively greater dilutions. For simplicity, much of the discussion of solutions in this book will relate to ideal solutions; real solutions will be explored in Chapter 22, which will also explain why the standard state for a real solution explicitly refers to the molality as the measure of quantity of dissolved solute, rather than the molar concentration.

One further note on standard states, and – unfortunately – a note of potential confusion. Table 6.4 identifies the standard states as recommended by IUPAC, but

- not all reference tables comply with IUPAC recommendations, and
- some reference tables contain data from measurements made in the past, at, say, 1 atm pressure and 298.15 K = 25 °C (in fact, prior to 1982, IUPAC recommended this combination as the standard, rather than the current standard of 1 bar and 298.15 K).

In many practical situations, this variability doesn't matter, but in principle, it does – it is preferable for 'standards' to be uniquely defined, universally accepted, and unambiguous, and for reference tables to be mutually consistent. The reality, however, is that 'standard' data may be found in reference sources at either of the two pressures of 1 bar and 1 atm, combined with any one of the temperatures of 293.15 K = 20 °C, 298.15 K = 25 °C, and (rather more rarely) 288.15 K = 15 °C or 273.15 K = 0 °C. To muddle matters even further, it sometimes happens that different sources, both claiming to use the same standards, quote different numbers for ostensibly the same quantity. So, when using reference sources, take care, especially in relation to what 'standard' is being used. This book, unless explicitly stated otherwise, will comply with the IUPAC recommendations of a standard pressure of $P^\ominus$ = 1 bar, with measurements made at T_s = 298.15 K = 25 °C; also, as far as possible, we will quote thermodynamic data from a single reference source – the 96th edition of the *CRC Handbook of Chemistry and Physics*, published by CRC Press in 2015, which (for the most part!) complies with the IUPAC recommendations.

6.10.5 Reference levels

As we noted on page 169, all measurements are made by comparison to a reference level that specifies the baseline of the measurement scale. In most cases, the baseline represents the zero of the scale, but, for example, the Fahrenheit scale of temperature uses 32 °F as the reference level. This is very rare, and so for the rest of this book, we will assume that the reference level defines the zero point.

In some cases, the definition of the meaning of 'zero' is unambiguous and physically absolute: so, for example, if we measure the quantity of a particular material in terms the mole number n, then n = 0 means that no material is present, and a volume of 0 m^3 also implies absence. There are, however, many contexts in which the identification of 'zero' is not so obvious. One example is the measurement of time: when an Olympic athlete runs the 100 m sprint, the reference level, defining the zero of time, is the instant the official fires the starting pistol; whilst for other measurements of time – for example, the date on which the race takes place – we use a different, but just as arbitrary, reference level, the start of the month. We all know that 'time' did not 'start' at the instant the athlete leaves the blocks, or on the first day of the month: rather, we all agree to use whatever reference level is convenient for any particular circumstance.

A second example of an arbitrary reference level is the measurement of length. In geography, the height of a mountain is conventionally measured by reference to the average sea level; in a laboratory, the position of a mass m kg will be measured in terms of the height h = 1.5 m above the laboratory bench. If the laboratory bench is itself 106 m above average sea level, then, using average sea level as the reference level, the height h' of the mass is 107.5 m. Any measurement h using the bench reference level is related to the measurement h' of the same system using the sea reference level such that

$$h' = h + r \quad \text{m}$$

in which the constant r, which in this particular case is 106 m, is known as the **reference level shift**, this being the height of the laboratory bench above sea level, corresponding to $h = 0$.

If we measure the gravitational potential energy E of the mass as mgh J (where g is the acceleration due to gravity), then that measurement too has an arbitrary reference level, as implied by the reference level used for the measurement of the height h. So, if the mass m is 2 kg, and the height h_1, using the laboratory bench as the reference level, is 1.5 m, then, since $g = 9.81$ m sec^{-2}, the gravitational potential energy E_1 is measured as $E_1 = mgh_1 = 29.43$ J. But if we use the sea as the reference level, then the height h_1' of the mass will, correspondingly, be measured as 107.5 m above average sea level, and the gravitational potential energy E_1' would be measured as $E_1' = mgh_1' = 2109.15$ J. Physically, the system of a specific mass m, at a specific position within the laboratory, is identical for both measurements, yet one measurement gives the gravitational potential energy as 29.43 J, and the other, as 2109.15 J. Which measurement is 'right'?

This question – "which measurement is 'right'?" – assumes that there is indeed a 'right' answer, and that is it possible to measure *the* gravitational potential energy of the mass. In fact, there is no 'right' answer, for the mass does not have a uniquely defined, absolute, gravitational potential energy – all measurements of gravitational potential energy are arbitrary, and the value of any measurement depends on which reference level is chosen.

There is, however, a subtlety. Consider the value of the gravitational potential energy of the mass when positioned at $h_2 = 0.3$ m above the laboratory bench, corresponding to a height $h_2' = 106.3$ m above average sea level. Using the laboratory bench reference level, $E_2 = mgh_2 = 5.886$ J, and using average sea level as the reference level, $E_2' = mgh_2' = 2085.61$ J. If the mass then falls from $h_1 = 1.5$ m to $h_2 = 0.3$ m, using the bench as the reference level, then the corresponding change ΔE in potential energy is therefore $\Delta E = E_2 - E_1 = 5.89 - 29.43 = -23.54$ J. Using the sea as the reference level, the mass falls from $h_1' = 107.5$ m to $h_2' = 106.3$ m, and the corresponding change $\Delta E' = E_2' - E_1' = 2085.61 - 2109.15 = -23.54$ J. As can be seen, provided that the same units of measurement (kg, m, sec, J) are used in each case, $\Delta E = \Delta E'$, for the difference in the reference levels cancels out.

This may be proven mathematically for

$$\Delta E = E_2 - E_1 = mgh_2 - mgh_1 = mg(h_2 - h_1)$$

and

$$\Delta E' = E_2' - E_1' = mgh_2' - mgh_1' = mg(h_2' - h_1')$$

But we know that, in terms of the reference level shift r

$$h' = h + r$$

and so

$$h_2' - h_1' = (h_2 + r) - (h_1 + r) = h_2 - h_1$$

implying that $\Delta E = \Delta E'$ regardless of the reference level shift, and indeed regardless of the reference levels themselves.

Gravitational potential energy is one example of a measurement of a form of energy; the thermodynamic internal energy U is another, and the same principle applies – it is impossible to measure 'absolute' values of U. Since, by definition, the enthalpy $H = U + PV$, there are also no 'absolute' values of H. Measurements of thermodynamic data, as published in reference tables, are therefore made using internationally agreed standards for reference levels, which will be introduced and defined in the appropriate contexts throughout this book. But just as the values of ΔE and $\Delta E'$, as discussed in the previous paragraph, are equal, and independent of the reference level, values of ΔU and ΔH for chemical reactions are also independent of the reference level. As we shall see throughout this book, a change in a thermodynamic variable, such as $\Delta H = H_2 - H_1$, is in general much more significant than the absolute measurements of H_1 or H_2 by themselves.

6.10.6 Standard states in practice

Standard states define the key features of a system of interest so that 'standard systems' can be replicated, and used as universally accepted reference states for the measurement of thermodynamic data. It's important that the standard states are useful in practice, and correspond to states of systems that are identical, or close to, the practical circumstances in which measurements take place, as exemplified by the choice of 1 bar as the standard pressure, very close to the ambient pressure in most laboratories. The choice of the standard states, however, is one of mutual agreement, based on practical convenience, rather than fundamental science: the standard pressure, for example, could be agreed as 1.7 bar, but this would be very inconvenient – scientists would need to carry out experiments in specially-designed pressure cabinets to make standard measurements. If, however, circumstances change, then it is possible to redefine the standards accordingly; furthermore, if different circumstances are important and relevant to a different community, then it is possible to define another set of standards, and have both available for use by different communities.

This second situation actually does arise. The standard states defined in Table 6.4 have a particular implication, which is not obvious from Table 6.4 itself. For a solute in solution, the standard state is defined as 'unit concentration', this being, for example, 1 mol dm^{-3}. In general, that's fine – we all know what that means, and solutions containing unit molar concentration of a particular solute can, for many solutes, be prepared and measured. For one particular solute, however, unit molar concentration has a particular significance – if the solute is the hydrogen ion H$^+$(aq), then the standard molar concentration [H$^+$] = 1 mol dm^{-3} corresponds to pH 0 (by definition, pH = $-\log_{10}$ [H$^+$] = $-\log_{10}$ [1] = 0). For many normal laboratory and industrial contexts, having a standard associated with pH 0 presents no problems, but in biochemistry, it does: pH 0 is highly acid, and far from the conditions of pH 7 at which (or close to which) much biochemistry takes place. As a consequence, biochemists have adopted the **biochemical standard state**, defined in relation to pH 7, which is different from the **conventional standard states**, defined in relation to pH 0, as in Table 6.4. We'll discuss the biochemical standard state in more detail in Chapter 23; until then, we will be using the conventional standard states as defined in Table 6.4.

6.10.7 Reaction categories

Table 6.5 Common reaction categories

Name	Description	Enthalpy change
Fusion	A reaction in which a pure substance changes phase from solid to liquid.	$\Delta_{fus}H$
Vaporisation	A reaction in which a pure substance changes phase from liquid to gas.	$\Delta_{vap}H$
Sublimation	A reaction in which a pure substance changes phase from solid to gas.	$\Delta_{sub}H$
Formation	A (possibly hypothetical) reaction in which a molecule is formed from its constituent elements.	$\Delta_{f}H$
Combustion	A reaction defining the combustion of a molecule into defined products – usually those resulting from complete combustion in excess oxygen.	$\Delta_{c}H$
Solution	A reaction in which 1 mol of a solute is dissolved in a solvent to infinite dilution.	$\Delta_{sol}H$
Dilution	A reaction in which a specified amount of solvent is added to a solution.	$\Delta_{dil}H$
General chemical reaction	A generalised chemical reaction in which specific reactants form specific products, and that does not fall within any of the other categories.	$\Delta_{r}H$ or ΔH

It is helpful to group similar types of reaction together into certain categories, as shown in Table 6.5. Each reaction category is associated with a corresponding enthalpy change represented as $\Delta_{xyz}H$, where the subscript $_{xyz}$ refers to the appropriate reaction category, as shown in the third column of the table. If the reaction is such that the reactants start in their respective standard states, and the products are formed in their respective standard states, and the reaction takes place isothermally at the specified temperature (or, if not specified, at a temperature assumed to be 298.15 K = 25°C), then the corresponding enthalpy change is referred to as a **standard enthalpy of [the appropriate] reaction**, and is represented as $\Delta_{xyz}H^{\ominus}$ (as will be used in this book) or $\Delta_{yxz}H^{\circ}$.

6.10.8 Standards vary with temperature

Consider once more to the combustion of octane

$$C_8H_{18}\,(l) + \frac{25}{2}O_2\,(g) \rightarrow 8\,CO_2\,(g) + 9\,H_2O\,(g) \qquad \Delta_c H = -5075\,kJ$$

If all the products and reactants are in their standard states, as defined by Table 6.4, then, as we have just seen, the enthalpy change ΔH associated with this reaction is the standard enthalpy of combustion of octane, designated as $\Delta_c H^{\ominus}$.

As we noted on page 172, however, although all chemical standards are defined for a specific standard pressure $P^\ominus$ – the IUPAC recommendation being 1 bar – they do *not* define any specific temperature. This implies that the standard enthalpy change $\Delta_r H^\ominus$ associated with any reaction varies with temperature. We draw attention to this because sometimes the use of the word 'standard' causes some confusion, especially when (mis)taken to mean 'constant value associated with'. In fact, in relation to chemistry, 'standard' does not mean 'constant value under all conditions' – it means 'with reference to standard states (as defined in Table 6.4), at the standard pressure (if complying with IUPAC recommendations) $P^\ominus$ = 1 bar, and at the temperature as stated'.

Mathematically, we know that the enthalpy of a given mass of material is a function of two other state functions, and those most often associated with enthalpy are temperature and pressure: accordingly, H may be represented as $H(T, P)$. As a consequence, the enthalpy change $\Delta_r H$ associated with any reaction is also a function of temperature and pressure, implying that $\Delta_r H$ is in fact $\Delta_r H(T, P)$. But if we are referring to standard enthalpy changes $\Delta_r H^\ominus$, by definition, the pressure is fixed at the standard pressure $P^\ominus$, and so, for a reaction of fixed stoichiometry, the standard enthalpy change $\Delta_r H^\ominus$ is a function of temperature only, implying that $\Delta_r H^\ominus$ is more correctly written as $\Delta_r H^\ominus(T)$.

Tables of reference data therefore specify the temperature corresponding to the values of $\Delta H^\ominus$ presented. As we shall see on page 193, if we know, either from reference data or by experiment, the value of $\Delta H^\ominus(T_1)$ at one particular temperature T_1, then knowledge of the behaviour of partial derivatives of the form $(\partial H/\partial T)_P = C_P$ for the reactants and the products enables us to compute the value of $\Delta H^\ominus(T_2)$ at any other temperature T_2.

6.11 Standard enthalpies of formation

In general, the enthalpy change associated with any reaction depends not only on the nature of the reaction, but also on the specific values of the (constant) pressure, and temperature, at which the reaction takes place. In principle, this could imply that reference data would need to be tabulated, and made available, for all reactions under all conditions. This would clearly require an enormous amount of work to compile the data, and the resulting tables would be huge. So, to avoid this problem, reference data is available for the **standard enthalpies of formation** of a large number of molecules, these being the enthalpy changes associated with the (often hypothetical) reaction in which 1 mol of the molecule of interest, in its standard state, is formed isothermally from the appropriate quantities of the molecule's constituent elements, also in their standard states:

Constituent elements in standard states → 1 mol of molecule of interest in standard state $\quad \Delta H = \Delta_f H^\ominus$

The formal symbol for the standard enthalpy of formation for any molecule is $\Delta_f H^\ominus(T)$, where T represents the temperature to which the standard state refers; if this is omitted, then the assumption is that the standard temperature is 298.15 K = 25 °C. Expressing the standard enthalpy of formation as $\Delta_f H^\ominus(T)$ also reminds us, as we have just seen, that standard enthalpies are functions of temperature, but not of pressure, since the pressure is fixed at the standard pressure P°, this being 1 bar in compliance with IUPAC guidelines. Also, the use of the bold font indicates that $\Delta_f \boldsymbol{H}^\circ(T)$ is a molar quantity.

In principle, for the reaction

$$\text{Constituent elements in standard states} \rightarrow \text{1 mol of molecule of interest in standard state} \qquad \Delta H = \Delta_f \boldsymbol{H}^\circ$$

$\Delta_f \boldsymbol{H}^\circ$ is given by

$$\Delta_f \boldsymbol{H}^\circ = \boldsymbol{H}^\circ(\text{molecule of interest}) - \sum_{\text{elements}} \boldsymbol{H}^\circ(\text{constituent element})$$

in which $\boldsymbol{H}^\circ$(molecule of interest) represents the absolute molar enthalpy of the molecule of interest in its standard state, and $\boldsymbol{H}^\circ$(constituent element) represents the absolute molar enthalpy of each of the constituent elements, in their appropriate standard states. Since, as we discussed on page 176, it is impossible to measure absolute enthalpies, we need a reference level, and so the reference level for the measurement of enthalpy is defined such that $\boldsymbol{H}^\circ$(constituent element) = 0 J mol^{-1} for all elements, where the reference state is the standard state. Note that this applies both to elements that, in their standard states, are atoms, such as C(s), and also to those that are molecules, such as O_2(g). As a consequence, 'the' molar enthalpy of any molecule in its standard state, $\boldsymbol{H}^\circ$(molecule of interest), is equal to that molecule's standard molar enthalpy of formation $\Delta_f \boldsymbol{H}^\circ$. Furthermore, if the 'molecule of interest' is itself an element, then $\Delta_f \boldsymbol{H}^\circ$ (element) = $\boldsymbol{H}^\circ$(element) = 0 J, for all elements.

Note that the assignment $\boldsymbol{H}^\circ$(constituent element) = 0 J for all elements does not imply that each element 'has' zero enthalpy. In an absolute sense, the enthalpy of an element is in principle attributable to the enthalpies of all the subatomic particles from which the element is formed. These subatomic particles will certainly be associated with a non-zero 'absolute' value of internal energy U, and since $H = U + PV$, a non-zero 'absolute' value of H too. What that value might be could well be of interest to a nuclear physicist, but to a laboratory chemist, $\boldsymbol{H}^\circ$(constituent element) = 0 J is a convenient, and very practical, reference level for measurement.

Also, since enthalpy is a state function, the enthalpy change for the reverse reaction, in which 1 mol of a molecule in its standard state is decomposed into its constituent elements in their standard states, is the negative $-\Delta_f \boldsymbol{H}^\circ$ of the standard enthalpy of formation $\Delta_f \boldsymbol{H}^\circ$

$$\text{1 mol of molecule of interest in standard state} \rightarrow \text{Constituent elements in standard states} \qquad \Delta H = -\Delta_f \boldsymbol{H}^\circ$$

Some examples of standard enthalpies of formation are shown in Table 6.6.

Scanning this table, which contains data for only a few molecules, we see that the numerical values for $\Delta_f \boldsymbol{H}^\circ$ extend over a broad range, with most values negative, but some positive. Furthermore, there appear to be some meaningful patterns: for example, large negative values of $\Delta_f \boldsymbol{H}^\circ$, such as those of $CaCO_3$ (s) ($\Delta_f \boldsymbol{H}^\circ$ = −1208 kJ mol^{-1}) and SiO_2 (s, quartz) ($\Delta_f \boldsymbol{H}^\circ$ = −911 kJ mol^{-1}) seem to be associated with very stable molecules; molecules with positive values of $\Delta_f \boldsymbol{H}^\circ$, such as acetylene, C_2H_2 (g) ($\Delta_f \boldsymbol{H}^\circ$ = +227 kJ mol^{-1}), and nitric oxide, NO (g) ($\Delta_f \boldsymbol{H}^\circ$ = +91 kJ mol^{-1}), are more reactive. It can, however, be misleading to interpret values of $\Delta_f \boldsymbol{H}^\circ$ in terms of general 'reactivity': as an example, consider ammonium chloride, NH_4Cl (s), for which $\Delta_f \boldsymbol{H}^\circ$ is −314 kJ mol^{-1}, a 'reasonably' negative number, from which we

Table 6.6 Some standard enthalpies of formation $\Delta_f H^\ominus$

All elements in their standard states: $\Delta_f H^\ominus = 0 \text{ J mol}^{-1}$.

Selected inorganic molecules at a temperature of $298.15 \text{ K} = 25\,°\text{C}$.

	kJ mol^{-1}		kJ mol^{-1}
H_2O (l)	−285.8	H_2S (g)	−20.6
H_2O (g)	−241.8	NO (g)	+91.3
HCl (g)	−92.3	NO_2 (g)	+33.2
HBr (g)	−36.3	NH_3 (g)	−45.9
HI (g)	+26.5	CO (g)	−110.5
SO_2 (g)	−296.8	CO_2 (g)	−393.5

	kJ mol^{-1}		kJ mol^{-1}
AgCl (s)	−127.0	$CaCO_3$ (s, calcite)	−1207.6
AgBr (s)	−100.4	Fe_2O_3 (s)	−824.2
AgI (s)	−61.8	SiO_2 (s, quartz)	−910.7
CaO (s)	−634.9	NaCl (s)	−411.2
$Ca(OH)_2$ (s)	−985.2	NH_4Cl (s)	−314.4

Selected organic molecules at a temperature of $298.15 \text{ K} = 25\,°\text{C}$.

	kJ mol^{-1}		kJ mol^{-1}
CH_4 (g)	−74.6	C_6H_6 (l)	+49.1
C_2H_6 (g)	−84.0	CH_3OH (l)	−239.2
C_2H_4 (g)	+52.4	C_2H_5OH (l)	−277.6
C_2H_2 (g)	+227.4	CH_3COOH (l)	−484.3
C_3H_8 (g)	−103.8	$CHCl_3$ (l)	−134.1

Source: *CRC Handbook of Chemistry and Physics*, 96th edn, 2015, Section 5, pages 5–4 and 5–42.

might infer that ammonium chloride is 'reasonably' stable. In fact, ammonium chloride very easily decomposes into ammonia and hydrogen chloride. This apparent 'paradox' is readily resolved, for the reaction by which ammonium chloride decomposes

NH_4Cl (s) → NH_3 (g) + HCl (g)

is very different from the (hypothetical) reaction

½ N_2 (g) + 2H_2 (g) + ½ Cl_2 (g) → NH_4Cl (s)

to which $\Delta_f H^\ominus$ refers. The observed 'decomposition' of ammonium chloride is not a decomposition to its constituent elements, but rather to two other molecules, ammonia and hydrochloric acid. So beware about mapping values of $\Delta_f H^\ominus$ onto 'stability' or 'reactivity' – as we shall see in Chapter 13, there is a much more reliable thermodynamic indicator of 'reactivity' than $\Delta_f H^\ominus$.

The importance of standard enthalpies of formation is that they enable us to use Hess's law to calculate the enthalpy change associated with almost any reaction in which reactants in their standard states react to form the corresponding products in their standard states. And if the reaction of interest takes place under conditions different from the standard ones, then, as we shall show on page 193, we can use a formula to convert the standard enthalpy change to the enthalpy change at some other temperature or pressure.

Dealing firstly with reactions taking place under standard conditions, consider, as an example, the generalised chemical reaction

$$n_A \text{ A} + n_B \text{ B} \rightarrow n_C \text{ C} + n_D \text{ D} \qquad \Delta_r H^\ominus_? = ?$$

in which n_A mol of A react with n_B mol of B to yield n_C mol of C and n_D mol of D, and we use the symbol Δ_r to indicate that this is a general chemical reaction. Let us assume that the reaction goes to completion, so that, before the reaction, the quantities of C and D are zero, and on completion of the reaction, the quantities of A and B are zero; let us further assume that all reagents are in their standard states at the start of the reaction, as are the products at the end of the reaction. Our problem is to determine the enthalpy change for this reaction, which will be a standard enthalpy change $\Delta_r H^\ominus_?$.

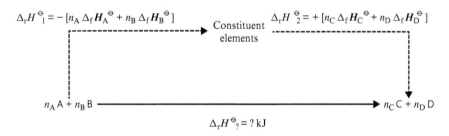

Figure 6.5 Using standard enthalpies of formation. Note that $\Delta_r H^\ominus_1$ is the total standard enthalpy of decomposition for each of the reactants, whilst $\Delta_r H^\ominus_2$ is the total standard enthalpy of formation for each of the products.

Figure 6.5 shows a hypothetical pathway in which the reaction takes place in two steps: the first in which n_A mol of A and n_B mol of B are decomposed into their elements, as associated with a standard enthalpy change $\Delta_r H^\ominus_1$; the second in which these elements are recombined to form n_C mol of C and n_C mol of D, as associated with a standard enthalpy change $\Delta_r H^\ominus_2$.

Enthalpy is a state function, and so, applying Hess's law, we may write

$$\Delta_r H^\ominus_? = \Delta_r H^\ominus_1 + \Delta_r H^\ominus_2$$

Taking firstly component A, we know that the standard enthalpy of formation $\Delta_f H_A^\ominus$ of A is defined as

Constituent elements in standard states $\rightarrow$ 1 mol of molecule of interest in standard state $\qquad \Delta_r H = \Delta_f H^\ominus$

and so, for the decomposition of n_A mol of A under standard conditions, we may therefore write

n_A mol of A in standard state $\rightarrow$ Constituent elements in standard states $\qquad \Delta_r H = -n_A \Delta_f H_A^\ominus$

Likewise, for the decomposition of n_B mol of B under standard conditions

n_B mol of B in standard state $\rightarrow$ Constituent elements in standard states $\qquad \Delta_r H = -n_B \Delta_f H_B^\ominus$

We now assume that the mixture of A and B is ideal, which, as explained on page 8, implies that state functions are linearly additive. Accordingly, the total standard enthalpy change $\Delta_r H^\ominus{}_1$ associated with the simultaneous decomposition of n_A mol of A and n_B mol of B into their respective constituent elements is given by

$$\Delta_r H^\ominus{}_1 = -[n_A \Delta_f H_B^\ominus + n_B \Delta_f H_B^\ominus]$$

Similarly, for the formation of n_C mol of C and n_D mol of D from their respective constituent elements under standard conditions, and once again assuming an ideal mixture

$$\Delta_r H^\ominus{}_2 = [n_C \Delta_f H_C^\ominus + n_D \Delta_f H_D^\ominus]$$

Since, by Hess's law

$$\Delta_r H^\ominus{}_? = \Delta_r H^\ominus{}_1 + \Delta_r H^\ominus{}_2$$

then

$$\Delta_r H^\ominus{}_? = [n_C \Delta_f H_C^\ominus + n_D \Delta_f H_D^\ominus] - [n_A \Delta_f H_A^\ominus + n_B \Delta_f H_B^\ominus]$$

giving a general expression for the **standard enthalpy of reaction** $\Delta_r H^\ominus$ as

$$\Delta_r H^\ominus = \sum_{\text{Products}} n_{\text{prod}} \Delta_f H_{\text{prod}}^\ominus - \sum_{\text{Reactants}} n_{\text{reac}} \Delta_f H_{\text{reac}}^\ominus \qquad (6.21)$$

Equation (6.21) applies to any reaction, and can be used to find the corresponding standard enthalpy change, provided that the standard enthalpies of formation all the reactants $_{\text{reac}}$ and products $_{\text{prod}}$ are known, either by experiment or from reference literature. Note also that, in equation (6.21), the standard enthalpy of formation $\Delta_f H^\ominus$ for each of the reactants and products is a signed quantity, that may positive (for an endothermic formation reaction, such as that for acetylene, C_2H_2 (g), for which $\Delta_f H^\ominus = 226.7$ kJ mol^{-1}) or negative (for an exothermic formation reaction, such as that for methane CH_4 (g), for which $\Delta_f H^\ominus = -74.6$ kJ mol^{-1}).

One further aspect of equation (6.21) merits attention, and that concerns the units of measure associated with the resulting computation of $\Delta_r H^\ominus$ for the given reaction. Standard enthalpies of formation $\Delta_f H^\ominus$ are always expressed in units of J mol^{-1}, kJ mol^{-1}, or, on occasion, MJ mol^{-1}, for, by definition, $\Delta_f H^\ominus$ refers to the formation of 1 mol of a substance in its standard state. Since the stoichiometric coefficients n_{prod} and n_{reac} in equation (6.21) represent the appropriate number of moles of each product and reactant, then the various terms

$n_{prod}\Delta_f H_{prod}^{\ominus}$ and $n_{reac}\Delta_f H_{reac}^{\ominus}$ have dimensions of J, kJ or MJ. The units of measure for $\Delta_r H^{\ominus}$ as computed using equation (6.21) are therefore J, kJ or MJ. $\Delta_r H^{\ominus}$, however, is still an extensive state function, the value of which will depend on the relevant quantities of material: so, for example, if the all the stoichiometric coefficients n_{prod} and n_{reac} are doubled (or, say, halved), the chemical equation will still balance, but the value of $\Delta_r H^{\ominus}$ as computed using equation (6.21) will be doubled (or halved). The numerical value of any particular $\Delta_r H^{\ominus}$ must therefore relate to a specific chemical equation, with specifically designated stoichiometric coefficients. Some sources express $\Delta_r H^{\ominus}$ for a particular reaction using units such as kJ mol^{-1}, even though it might not be clear what 'mol^{-1}' might mean: as we discussed on page 155, this book will use the convention that values of $\Delta_r H^{\ominus}$ which are not explicitly expressed per mol will refer to the value of $\Delta_r H^{\ominus}$ as associated with the molar quantities defined by the stoichiometry of the corresponding reaction, as written.

If we apply equation (6.21) to a reaction in which elements, in their standard states, naturally take part, for example

$$C\,(s) + O_2\,(g) \rightarrow CO_2\,(g) \qquad\qquad \Delta_r H^{\ominus} = \Delta_r H^{\ominus}_?$$

we have

$$\Delta_r H^{\ominus}_? = \sum_{\text{Products}} n_{prod}\Delta_f H_{prod}^{\ominus} - \sum_{\text{Reactants}} n_{reac}\Delta_f H_{reac}^{\ominus}$$

$$= \Delta_f H^{\ominus}\,CO_2\,(g) - [\Delta_f H^{\ominus}\,C\,(s) + \Delta_f H^{\ominus}\,O_2\,(g)] \qquad (6.21)$$

But $\Delta_f H^{\ominus}\,CO_2\,(g)$ is itself defined by the equation

$$C\,(s) + O_2\,(g) \rightarrow CO_2\,(g) \qquad\qquad \Delta_r H^{\ominus} = \Delta_f H^{\ominus}\,CO_2\,(g)$$

Hence

$$\Delta_f H^{\ominus}\,CO_2\,(g) = \Delta_f H^{\ominus}\,CO_2\,(g) - [\Delta_f H^{\ominus}\,C\,(s) + \Delta_f H^{\ominus}\,O_2\,(g)]$$
$$\therefore \Delta_f H^{\ominus}\,C\,(s) + \Delta_f H^{\ominus}\,O_2\,(g) = 0$$

which is totally consistent with the definition of the reference level for enthalpy (see page 179), for which

$$\Delta_f H^{\ominus}\,C\,(s) = \Delta_f H^{\ominus}\,O_2\,(g) = 0\,\text{J mol}^{-1}$$

and similarly for all other elements in their standard states.

If a reaction takes place involving an element not in its standard state, then we must ensure we take account of the enthalpy associated with the reaction

element in standard state → element in non-standard state

So, as an example, if we are studying a reaction using the non-standard red allotrope of phosphorus, rather than the standard white form, we must remember that the standard enthalpy of formation of red phosphorus is −17.6 kJ mol^{-1}

$$P\,(s,\,\text{white}) \rightarrow P\,(s,\,\text{red}) \qquad\qquad \Delta_f H^{\ominus} = -17.6\,\text{kJ mol}^{-1}$$

6.12 Ionic enthalpies

So far, we've discussed enthalpy changes associated with reactions involving only elements or whole molecules. Much chemistry, however, takes place in solution, and involves ions. We now examine the enthalpy change associated with a reaction in which element in its standard state either gains or loses one or more electrons to form an aqueous ion, for example

$$\tfrac{1}{2} Cl_2 (g) + e^- \rightarrow Cl^- (aq)$$

and,

$$Na (s) \rightarrow Na^+ (aq) + e^-$$

where (aq) – for 'aqueous' – indicates that the corresponding entity (in the current cases, ions) are in solution, and e^- is an electron.

Consider, then, the reaction

$$\tfrac{1}{2} H_2 (g) + \tfrac{1}{2} Cl_2 (g) \rightarrow H^+ (aq) + Cl^- (aq) \qquad \Delta_r H^\circ_? = ?$$

in which all entities are in their standard states. The standard enthalpy change $\Delta_r H^\circ_?$ for this reaction may be calculated using equation (6.1) as

$$\Delta_r H^\circ_? = [\Delta_f H^\circ \; H^+ (aq) + \Delta_f H^\circ \; Cl^- (aq)] - \tfrac{1}{2} [\Delta_f H^\circ \; H_2 (g) + \Delta_f H^\circ \; Cl_2 (g)]$$

But since H_2 (g) and Cl_2 (g) are elements in their standard states

$$\Delta_f H^\circ \; H_2 (g) = \Delta_f H^\circ \; Cl_2 (g) = 0$$

and so

$$\Delta_r H^\circ_? = \Delta_f H^\circ \; H^+ (aq) + \Delta_f H^\circ \; Cl^- (aq)$$

We now define a second reference level, that the standard enthalpy of formation of an aqueous proton – a hydrogen ion – is designated as zero at all temperatures:

$$\Delta_f H^\circ \; H^+ (aq) = 0 \quad \text{at all temperatures}$$

This then enables us to equate the enthalpy change $\Delta_r H^\circ_?$ for the reaction

$$\tfrac{1}{2} H_2 (g) + \tfrac{1}{2} Cl_2 (g) \rightarrow H^+ (aq) + Cl^- (aq) \qquad \Delta_r H^\circ_? = ?$$

to the standard ionic enthalpy of formation of the Cl^- (aq) ion,

$$\Delta_r H^\circ_? = \Delta_f H^\circ \; Cl^- (aq)$$

Experiment shows that this enthalpy change is -167.2 kJ mol^{-1}; Table 6.7 contains some further data.

As an example of the use of the data in Table 6.7, we can determine the standard enthalpy change $\Delta_r H^\circ$ for the reaction

$$Na^+ (aq) + Cl^- (aq) \rightarrow NaCl (s) \qquad \Delta_r H = \Delta_r H^\circ_?$$

Table 6.7 Standard enthalpies of formation $\Delta_f H^\ominus$ for some selected aqueous ions at a temperature of 298.15 K = 25 °C

	kJ mol^{-1}		kJ mol^{-1}
H$^+$ (aq)	0	OH$^-$ (aq)	−230.0
Na$^+$ (aq)	−240.1	Cl$^-$ (aq)	−167.2
K$^+$ (aq)	−252.4	Br$^-$ (aq)	−121.6
Ag$^+$ (aq)	+105.6	I$^-$ (aq)	−55.2
Ca^{2+} (aq)	−542.8	H$_2$PO$_4^-$ (aq)	−1296.3
Ba^{2+} (aq)	−537.6	S^{2-} (aq)	+33.1
Cu^{2+} (aq)	+64.8	CO$_3^{2-}$ (aq)	−677.1
Zn^{2+} (aq)	−153.9	SO$_4^{2-}$ (aq)	−909.3

Source: *CRC Handbook of Chemistry and Physics*, 96th edn, 2015, Section 5, pages 5–66 and 5–67.

According to equation (6.21)

$$\Delta_r H^\ominus = \sum_{\text{Products}} n_{\text{prod}} \Delta_f H_{\text{prod}}^\ominus - \sum_{\text{Reactants}} n_{\text{reac}} \Delta_f H_{\text{reac}}^\ominus$$

$$= \Delta_f H^\ominus \text{ NaCl (s)} - [\Delta_f H^\ominus \text{ Na}^+ \text{(aq)} + \Delta_f H^\ominus \text{ Cl}^- \text{(aq)}] \tag{6.21}$$

Reference to Table 6.6 shows that $\Delta_f H^\ominus$ NaCl (s) = −411.2 kJ mol^{-1}, whilst Table 6.7 shows $\Delta_f H^\ominus$ Na$^+$ (aq) to be −240.1 kJ mol^{-1}, and $\Delta_f H^\ominus$ Cl$^-$ to be −167.2 kJ mol^{-1}, implying that

$$\Delta_r H^\ominus_? = -411.2 - [-240.1 - 167.2] = -3.9 \text{ kJ}$$

Hence

Na$^+$ (aq) + Cl$^-$ (aq) → NaCl (s) $\Delta_r H^\ominus = -3.9$ kJ

6.13 Bond energies

We have seen how the standard enthalpy $\Delta_c H^\ominus$ associated with a chemical reaction such as the combustion of methane

$$\text{CH}_4 \text{ (g)} + 2\, \text{O}_2 \text{ (g)} \rightarrow \text{CO}_2 \text{ (g)} + 2\, \text{H}_2\text{O (l)} \qquad \Delta_c H^\ominus = -890.3 \text{ kJ mol}^{-1}$$

can be computed from standard enthalpies of formation $\Delta_f H^\ominus$ as

$$\Delta_r H^\ominus = \sum_{\text{Products}} n_{\text{prod}} \Delta_f H_{\text{prod}}^\ominus - \sum_{\text{Reactants}} n_{\text{reac}} \Delta_f H_{\text{reac}}^\ominus \tag{6.21}$$

which in this case becomes

$$\Delta_c H^\ominus = [\Delta_f H^\ominus \text{ CO}_2 \text{ (g)} + 2\, \Delta_f H^\ominus \text{ H}_2\text{O (l)}] - [\Delta_f H^\ominus \text{ CH}_4 \text{ (g)} + 2\, \Delta_f H^\ominus \text{ O}_2 \text{ (g)}]$$

Using the data from Table 6.6, we therefore have

$$\Delta_c H^\ominus = [-393.5 + 2 \times (-285.8)] - [-74.6 + 2 \times 0] = -890.5 \text{ kJ mol}^{-1}$$

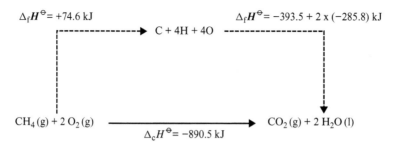

Figure 6.6 The use of standard enthalpies of formation $\Delta_f H^\ominus$ implies breaking reactants into their constituent elements, and then recombining those elements into the required products.

This calculation is based on the principle underpinning the concept of the standard enthalpy of formation $\Delta_f H^\ominus$ – namely, the decomposition of the reactants into their constituent elements, followed by the recombination of those elements into the required products, as shown in Figure 6.6.

At the molecular level, of course, the reactants – methane and oxygen – don't just 'break up' into carbon, hydrogen and oxygen, nor do the products – carbon dioxide and water – just 'form' from the bare elements. What actually happens is complicated, but necessarily involves the breaking of the four C–H bonds in the methane, and the O=O double bond in the oxygen, followed by the formation of the two C=O bonds in the carbon dioxide, and the two O-H bonds in the water. In general, breaking chemical bonds requires energy, and making them releases energy, so the overall net result of all this, in the case of the oxidation of methane at constant pressure (in which some energy is required for P, V work), is an enthalpy change of -890.5 kJ mol^{-1}. This raises an interesting question – is it possible to attribute standard enthalpy changes to the breaking and making of individual chemical bonds? For if we could do this, then we could write

$$\Delta_c H^\ominus \, [CH_4\,(g) + 2O_2(g)] = \Delta H^\ominus \,[\text{breaking 4 C–H bonds and 2 O=O bonds}]$$

$$+ \Delta H^\ominus \,[\text{forming 2 C=O bonds and 4 O–H bonds}]$$

as illustrated in Figure 6.7.

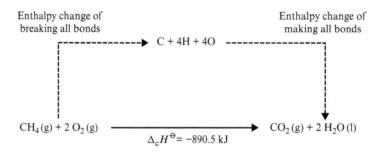

Figure 6.7 Can we analyse the standard enthalpies of formation of molecules into the standard enthalpies associated with the breaking and making of the appropriate chemical bonds?

More generally, if we had standard tables of the standard enthalpy changes of all types of chemical bond, then this would enable us to compute the standard enthalpy change for any chemical reaction.

The idea of identifying the standard enthalpy changes of breaking and making chemical bonds is very appealing, but in practice, there is a significant practical problem – careful measurement shows that the enthalpy change associated with the breaking of, say, a single C–H bond is not a constant, but depends on the specific reaction, and on what other atoms are bonded (or indeed not bonded) to the other three valencies of any particular carbon atom. As an example, the standard enthalpy change associated with the breaking of the first C–H bond in methane CH_4 is about $+ 430$ kJ mol^{-1}

$$CH_4 \rightarrow CH_3 + H \qquad \Delta H^{\ominus} = + 430 \text{ kJ mol}^{-1}$$

whereas the standard enthalpy change assciated with the breaking of the C–H bond in chloroform $CHCl_3$ is about $+ 380$ kJ mol^{-1}

$$CHCl_3 \rightarrow CCl_3 + H \qquad \Delta H^{\ominus} = + 380 \text{ kJ mol}^{-1}$$

Furthermore, the standard enthalpy change for the breaking of each successive C–H bond in methane is different, as shown by this (somewhat approximate) data

$$CH_4 \rightarrow CH_3 + H \qquad \Delta H^{\ominus} = + 430 \text{ kJ mol}^{-1}$$
$$CH_3 \rightarrow CH_2 + H \qquad \Delta H^{\ominus} = + 470 \text{ kJ mol}^{-1}$$
$$CH_2 \rightarrow CH + H \qquad \Delta H^{\ominus} = + 420 \text{ kJ mol}^{-1}$$
$$CH \rightarrow C + H \qquad \Delta H^{\ominus} = + 340 \text{ kJ mol}^{-1}$$

(data from https://srd.nist.gov/NSRDS/NSRDS-NBS31.pdf).

Accordingly, the 'Gold Book' of the International Union of Pure and Applied Chemistry (IUPAC) defines two types of 'bond energy' – in fact, 'bond enthalpy' – the **bond dissociation energy**, D^o, and the **mean bond energy**, ϵ, or $\Delta H^{\ominus}$ (X–Y):

> The bond dissociation energy, D^o, is the enthalpy change, usually measured at a temperature of 298 K, associated with the breaking of 1 mol of a given bond of some specific molecular entity, so creating two free radicals.

> The mean bond energy, ϵ, or $\Delta H^{\ominus}$(X–Y), is the average value of the gas-phase bond dissociation energies, usually at a temperature of 298 K, for all X–Y bonds of the same type within the same chemical species.

So, for methane, the bond dissociation energy of the first C–H bond is the enthalpy associated with the reaction

$$CH_4 \text{ (g)} \rightarrow CH_3^{\bullet} \text{ (g)} + H^{\bullet} \text{ (g)} \qquad D^o = + 430 \text{ kJ mol}^{-1}$$

where the symbol $^{\bullet}$ indicates that one of the two electrons originally shared in the now-broken covalent C – H bond remains associated with the $CH_3^{\bullet}$ radical, and the other with the $H^{\bullet}$ radical. This is in contrast to the formation of an ion pair of a CH_3^- ion and an H^+ ion, in which the electron pair stays with the CH_3^- entity, and the H^+ entity is electron free.

ENTHALPY AND THERMOCHEMISTRY

The mean bond energy ϵ of the C–H bond in methane is different: it does not refer to the enthalpy change associated with the breaking of any particular C–H bond; rather, it is the arithmetic average of the four bond dissociation energies associated with the breaking of each of methane's four C–H bonds

$$CH_4 (g) \rightarrow CH_3^{\bullet} (g) + H^{\bullet} (g) \qquad D^o = +431 \text{ kJ mol}^{-1}$$
$$CH_3^{\bullet} (g) \rightarrow CH_2^{\bullet\bullet} (g) + H^{\bullet} (g) \qquad D^o = +473 \text{ kJ mol}^{-1}$$
$$CH_2^{\bullet\bullet} (g) \rightarrow CH^{\bullet\bullet\bullet} (g) + H^{\bullet} (g) \qquad D^o = +422 \text{ kJ mol}^{-1}$$
$$CH^{\bullet\bullet\bullet} (g) \rightarrow C^{\bullet\bullet\bullet\bullet} (g) + H^{\bullet} (g) \qquad D^o = +337 \text{ kJ mol}^{-1}$$

so that $\Delta H^{\ominus}(C-H)$, using methane as the base molecule, is about $+416$ kJ mol^{-1}.

Bond dissociation energies, and mean bond energies, are defined with respect to breaking bonds, and all values are positive, implying that bond breaking is in general endothermic, and that energy needs to be put into the system for bonds to be broken. Bond formation is the reverse of bond breaking, implying that the sign of the enthalpy change is reversed

$$CH_3^{\bullet} (g) + H^{\bullet} (g) \rightarrow CH_4 (g) \qquad \Delta H^{\ominus} = -431 \text{ kJ mol}^{-1}$$

and the enthalpy change is written as $\Delta H^{\ominus}$ rather than D^o, which, by definition, refers to the bond-breaking reaction only.

Bond formation is in general exothermic, releasing heat, and the overall heat exchange associated with any given reaction is the algebraic sum of the (positive) enthalpy change associated with bond breaking, plus the (negative) enthalpy change associated with bond formation. If this results in a positive number (say, bond breaking = 300 kJ mol^{-1}, bond formation = –200 kJ mol^{-1}, total = +100 kJ mol^{-1}), then the reaction is endothermic; if the result is a negative number (say, bond breaking = 300 kJ mol^{-1}, bond formation = –450 kJ mol^{-1}, total = –150 kJ mol^{-1}), then the reaction is exothermic.

Table 6.8 gives some representative values of bond dissociation energies, and Table 6.9, mean bond energies.

Table 6.8 Bond dissociation energies D^o for some selected bonds, as shown by the – symbol, at 1 bar and 298.15 K

	kJ mol^{-1}		kJ mol^{-1}
H – H	+ 436	H$_3$C – H	+ 439
H – Cl	+ 431	H$_5$C$_6$ – H	+ 466
H – NH$_2$	+ 450	(HC $\equiv$ C) – H	+ 558
H – OH	+ 497	H$_3$C – CH$_3$	+ 377
F – F	+ 159	H$_3$C – OH	+ 385
Cl – Cl	+ 243	H$_3$C – Cl	+ 350

Source: *CRC Handbook of Chemistry and Physics*, 96th edn, 2015, Section 9, pages 9–65 to 9–79.

The implied accuracy of the mean bond energies shown in Table 6.9 is somewhat misleading, in that these numbers are averages, and are indicative only. That said, and if we associate the magnitude of the mean bond energy with bond strength (in that the greater the mean bond

BOND ENERGIES

Table 6.9 Mean bond energies ϵ (X–Y) for some selected bonds

	kJ mol^{-1}		kJ mol^{-1}
H – H	+ 436	N – H	+ 391
C – H	+ 416	N – N	+ 160
C – C	+ 356	C = C	+ 598
C – N	+ 285	C $\equiv$ C	+ 813
C – O	+ 336	O = O	+ 498
C – Cl	+ 327	N $\equiv$ N	+ 946
O – H	+ 467	C $\equiv$ N	+ 866
O – O	+ 146	C $\equiv$ O	+ 1073

Source: https://chem.libretexts.org/Textbook_Maps/General_Chemistry_Textbook_Maps/Map%3A_Chem-PRIME_(Moore_et_al.)/15Thermodynamics%3A_Atoms%2C_Molecules_and_Energy/15.09%3A_Bond_Enthalpies

energy, the greater the energy required to break the bond into the constituent atoms, hence the 'stronger' the bond), then the N – N and O – O covalent bonds are relatively weak, with energies of around 150 kJ mol^{-1}; the C – X bond has an intermediate energy of around + 350 kJ mol^{-1}; the H – H and O – H bonds are relatively strong at around + 450 kJ mol^{-1}; double bonds are stronger than single bonds; and triple bonds stronger than double bonds. This last point might, at first sight, appear to be confusing, for acetylene H – C $\equiv$ C – H is certainly more reactive than ethylene, H$_2$ – C = C – H$_2$, which in turn is more reactive than ethane H$_3$ – C – C – H$_3$ – yet, according to the data in Table 6.9, the C $\equiv$ C is stronger than the C = C bond, which is in turn stronger than the C – C bond. The apparent paradox is easily resolved, for the numbers in Table 6.9 refer to the energy required to break a given bond in its entirety into the corresponding elements, and it is true that considerably more energy is required to break a C $\equiv$ C bond completely than a C – C bond. When acetylene reacts, however, the C $\equiv$ C is rarely broken down to carbon – more likely a reaction will transform the C $\equiv$ C bond to either a C = C bond, or a C – C bond.

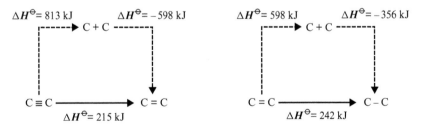

Figure 6.8 The formation of C = C from C $\equiv$ C, and C–C from C = C.

As can be seen in Figure 6.8, and using the mean bond energy data from Table 6.9, the energy required to break a C $\equiv$ C bond into a C = C bond is + 215 kJ mol^{-1}, rather less than the + 242 kJ mol^{-1} required to break a C = C bond into a C – C bond, and significantly less than the C – C mean bond energy of + 356 kJ mol^{-1}. This indicates that less energy is required to break a C≡C bond into a C=C bond, than to break a C=C bond into a C−C bond, than to break a C−C bond itself – as is consistent with the observed fact that acetylene is more reactive than ethylene, which itself is more reactive than ethane.

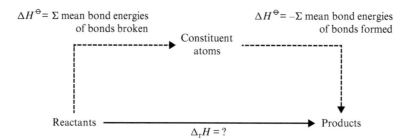

Figure 6.9 The general use of mean bond energies.

Figure 6.9 shows how mean bond energies can be used to find the approximate value of the standard enthalpy change for any reaction

$\Delta_r H^\ominus = \Sigma$ [mean bond energies of all bonds in reactants that are broken]

$\quad\quad - \Sigma$ [mean bond energies of all bonds in products that formed]

In principle, this is useful if we wish to estimate the enthalpy change associated with a reaction for which no other helpful information is available, for example, if we are hypothesising a reaction that does not take place in practice, or if we wish to explore a product that does not yet exist. In practice, however, the approximate nature of mean bond energy data implies that any calculated results could be associated with relatively large errors, and so alternative methods, such as the use of modern computer models, which can take account of the details of the specific chemical context in which any given reaction takes place, in general give better results.

6.14 The variation of ΔH with temperature

6.14.1 H as a function $H(T, P)$ of T and P

As we know, for a given mass of a pure material, the enthalpy H is a function $H(T, P)$ of T and P, and so the enthalpy change ΔH associated with any reaction will, in general, be different under different conditions of both temperature and pressure. In many real situations, the pressure is effectively constant, and so the purpose of this section is to show how we can calculate the enthalpy change $\Delta_r H(T_2)$ for any reaction at any temperature T_2 from knowledge of the enthalpy change $\Delta_r H(T_1)$ for that reaction at a specific temperature T_1.

There is one important case in which we know that the pressure is constant, and that's when a reaction takes place under standard conditions, with all reagents and products in their standard states. The pressure is necessarily constant at 1 bar, and the enthalpy change is the standard enthalpy change $\Delta_r H^\ominus$, which, for a given mass, is a function of temperature only $\Delta_r H^\ominus(T)$.

THE VARIATION OF ΔH WITH TEMPERATURE

To determine how enthalpy changes vary with temperature, we return to the fundamental definition of a change dH in the enthalpy of any system of constant mass, as given by equation (6.9)

$$dH = C_P\, dT + \left(\frac{\partial H}{\partial P}\right)_T dP \tag{6.9}$$

At constant pressure dP is zero, and so

$$dH = C_P\, dT$$

implying that, for a finite change from an initial temperature T_1 to a final specific temperature T_2

$$\int_{H(T_1)}^{H(T_2)} dH = H(T_2) - H(T_1) = \int_{T_1}^{T_2} C_P\, dT \tag{6.22}$$

This relationship is very simple: it equates the change in the enthalpy of a system resulting from a change in temperature to the integral of the corresponding heat capacity C_P, over the appropriate temperature range. Let's now see how we can apply equation (6.22) to some real situations.

6.14.2 ΔH at different temperatures

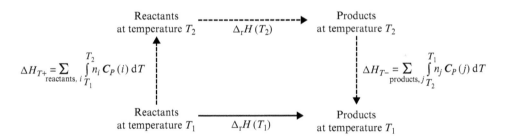

Figure 6.10 Enthalpy changes at different temperatures. Reactants are firstly heated from temperature T_1 to temperature T_2; the reaction takes place at temperature T_2; and the products are then cooled from temperature T_2 to temperature T_1.

Figure 6.10 shows how Hess's law can be applied to the calculation of the enthalpy change $\Delta_r H\,(T_2)$ for a reaction taking place at temperature T_2 given knowledge of the enthalpy change $\Delta_r H\,(T_1)$ for the same reaction taking place at temperature T_1, under conditions of constant pressure throughout. Since enthalpy is a state function, the enthalpy change $\Delta_r H\,(T_1)$ for the reaction at temperature T_1 must equal

- the enthalpy change ΔH_{T+} associated with the heating of all reactants from temperature T_1 to temperature T_2, plus
- the enthalpy change $\Delta_r H\,(T_2)$ for the reaction at temperature T_2, plus
- the enthalpy change ΔH_{T-} associated with the cooling of all reactants from temperature T_2 to temperature T_1

so that

$$\Delta_r H(T_1) = \Delta H_{T+} + \Delta_r H(T_2) + \Delta H_{T-}$$

giving

$$\Delta_r H(T_2) = \Delta_r H(T_1) - [\Delta H_{T+} + \Delta H_{T-}] \qquad (6.23)$$

The enthalpy changes $\Delta H_{T+}(i)$ associated with the heating of n_i mol of any the reactant i can be determined using equation (6.22)

$$\int_{H(T_1)}^{H(T_2)} dH = H(T_2) - H(T_1) = \int_{T_1}^{T_2} C_P \, dT \qquad (6.22)$$

which, given that for any component i, $C_P = n_i \, C_P(i)$, can be expressed as

$$\Delta H_{T+}(i) = n_i \int_{T_1}^{T_2} C_P(i) \, dT \qquad (6.24)$$

For ideal gases, C_P is a constant, and, for most real substances, C_P can, as a first approximation, be regarded as a constant, especially over relatively small temperature ranges. If we can regard each $C_P(i)$ as a constant, then, for any reactant i

$$\Delta H_{T+}(i) = n_i \int_{T_1}^{T_2} C_P(i) \, dT = n_i C_P(i) \int_{T_1}^{T_2} dT = n_i \, C_P(i) \, (T_2 - T_1) = n_i \, C_P(i) \, \Delta T$$

Assuming ideal conditions and the linear additivity of all state functions, the total enthalpy change ΔH_{T+} associated with the heating of all reactants i from temperature T_1 to temperature T_2 is therefore given by

$$\Delta H_{1+} = \sum_{\text{reactants, } i} n_i \, C_P(i) \Delta T \qquad (6.25)$$

The products are cooled from temperature T_2 to temperature T_1, and so, for each product j, we may use an exactly similar argument, such that equation (6.24) becomes

$$\Delta H_{T-}(j) = n_j \int_{T_2}^{T_1} C_P(j) \, dT = - n_j \int_{T_1}^{T_2} C_P(j) \, dT$$

The sign of the integral changes when the limits are reversed, and since $T_2 > T_1$, the resulting negative sign indicates cooling, and a drop in temperature. Assuming as before that $C_P(j)$ is a constant for all products, and that the mixture is ideal, then

$$\Delta H_{T-} = - \sum_{\text{products, } j} n_j \, C_P(j) \Delta T \qquad (6.26)$$

Equation (6.25) for ΔH_{T+} and (6.26) for ΔH_{T-} can now be used in equation (6.23) to give

$$\Delta_r H(T_2) = \Delta_r H(T_1) - \left[\sum_{\text{reactants, } i} n_i C_P(i) \Delta T - \sum_{\text{products, } j} n_j C_P(j) \Delta T \right]$$

$$= \Delta_r H(T_1) + \left[\sum_{\text{products, } j} n_j C_P(j) \Delta T - \sum_{\text{reactants, } i} n_i C_P(i) \Delta T \right]$$

and so

$$\Delta_r H(T_2) = \Delta_r H(T_1) + \left[\sum_{\text{products, } j} n_j C_P(j) - \sum_{\text{reactants, } i} n_i C_P(i) \right] \Delta T \tag{6.27}$$

We now define ΔC_P as

$$\Delta C_P = \sum_{\text{products, } j} n_j C_P(j) - \sum_{\text{reactants, } i} n_i C_P(i) \tag{6.28}$$

where we note that, in contrast to each of the molar heat capacities $C_P(i)$ and $C_P(j)$, ΔC_P is in general not a molar quantity. Equation (6.27) then becomes

$$\Delta_r H(T_2) = \Delta_r H(T_1) + \Delta C_P \Delta T \tag{6.29}$$

Under the assumptions of constant pressure, constant values for all the C_Ps, and an ideal mixture, equation (6.29) allows us to calculate the enthalpy change $\Delta_r H(T_2)$ for any reaction at any temperature T_2, given knowledge of the enthalpy change $\Delta_r H(T_1)$ at a known temperature T_1, and also data concerning the heat capacities at constant pressure for each of the reactants and products, so enabling the calculation of ΔC_P according to equation (6.28). Although that list of assumptions appears to be burdensome, in practice, they apply under many real conditions, and, in practice, equation (6.29) is both useful – and simple,

Reference to the various examples in this chapter will show that the magnitudes of $\Delta_r H$ for many chemical reactions are of the order 100 kJ mol^{-1} = 10^5 J mol^{-1}. In contrast, as shown in Table 6.1, values of C_P are typically in the range 10 to 10^2 J K^{-1} mol^{-1}, implying that values of ΔC_P (in which numbers of similar order-of-magnitude are subtracted from one another) are likely to be of order 10 J K^{-1} mol^{-1}. For temperature changes measured in up to small number of hundreds of degrees, ΔT will be of order 10^2 K, and so the product $\Delta C_P \Delta T$ will be of order 10^3 K – considerably smaller than typical magnitudes of $\Delta_r H$ of order 10^5 J mol^{-1}. This indicates that although, for any reaction $\Delta_r H$ changes with temperature, it does so relatively slowly.

6.14.3 A worked example

As an example of the application of equation (6.29)

$$\Delta_r H(T_2) = \Delta_r H(T_1) + \Delta C_P \Delta T \tag{6.29}$$

we return to the discussion on page 163, in which we established the relationship between the molar enthalpies of fusion $\Delta_{\text{fus}} H$, vaporisation $\Delta_{\text{vap}} H$, and sublimation $\Delta_{\text{sub}} H$, as

$$\Delta_{\text{sub}} H = \Delta_{\text{fus}} H + \Delta_{\text{vap}} H \tag{6.17}$$

In the case of water, we noted that, at the normal boiling point of water, close to 373 K, $\Delta_{vap}H$ (373) is 40.65 kJ mol^{-1}, a value which is readily found in standard tables. In applying equation (6.17), we needed to use $\Delta_{vap}H$ (273), which we quoted as 45.05 kJ mol^{-1} (see page 163). We can now verify this value using equation (6.29).

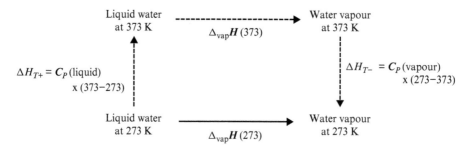

Figure 6.11 A practical application of equation (6.29).

Figure 6.11 is structurally identical to Figure 6.10, but applies to the vaporisation of water

water, liquid → water, vapour

at two different temperatures, T_1 = 273 K and T_2 = 273 K.

If we assign $\Delta_r H(T_1)$ as $\Delta_{vap}H$ (273), and $\Delta_r H(T_2)$ as $\Delta_{vap}H$ (373), then

$$\Delta_r H(T_2) = \Delta_r H(373) = \Delta_{vap}H(373) = 40.65 \text{ kJ mol}^{-1}$$

Reference to Table 6.2 will show that, for water,

C_P (liquid, 273.15 K) = 76.0 J K^{-1} mol^{-1}

C_P (vapour, 373 K) = 37.4 J K^{-1} mol^{-1}

and if we assume that the values of C_P are constant over the temperature range from 273 K to 373 K, then

$$\Delta C_P = C_P \text{(vapour)} - C_P \text{(liquid)} = 37.4 - 76.0 = -38.6 \text{ J K}^{-1} \text{ mol}^{-1} = 0.0386 \text{ kJ K}^{-1} \text{ mol}^{-1}$$

Since ΔT = 373 − 273 = 100 K, equation (6.29) becomes

$$40.65 = \Delta_r H(T_1) + (-0.0386) \times 100$$

hence

$$\Delta_r H(T_1) = 40.65 + 3.86 = 44.51 \text{ kJ mol}^{-1}$$

This calculation implies that $\Delta_{vap}H$ (273) = 44.51 kJ mol^{-1}, sensibly close to the true value of 45.05 kJ mol^{-1}: the discrepancy is primarily attributable to the fact (as suggested by the data for liquid water in Table 6.2), that the values of C_P for liquid water and water vapour are not constant over the temperature range 273 K to 373 K. In many practical circumstances, this variation does not introduce a material error, but sometimes we need to be more accurate, so let's now see how any variations of C_P with temperature may be taken into account.

6.14.4 The variation of C_P with temperature

In deriving equation (6.29), we assumed that all the molar heat capacities C_P at constant pressure are constants, as is the case for an ideal gas (see page 153), and, by approximation, acceptable in many practical circumstances. In reality, however, and especially over larger temperature ranges, molar heat capacities at constant pressure (and also at constant volume) vary with temperature, and this needs to be taken in to account. For any given material, the temperature variability of C_P is usually expressed as a polynomial, either of the form

$$C_P = a + bT + cT^2 + dT^3 + eT^4 \tag{6.30}$$

or in a slightly different form known as the **Shomate Equation**

$$C_P = A + BT + CT^2 + DT^3 + \frac{E}{T^2} \tag{6.31}$$

where, for any given substance, a, b, c…A, B, C…are empirically determined temperature-independent constants which apply over a broad temperature range, and all five terms are required for any given computation. Some values for the coefficients a, b, c… A, B, C…for some selected gases are shown in Tables 6.10 and 6.11.

Table 6.10 The temperature variation of $C_P = a + bT + cT^2 + dT^3 + eT^4$ for some selected gases

	a	$b \times 10^3$	$c \times 10^6$	$d \times 10^9$	$e \times 10^{12}$
	J K^{-1} mol^{-1}	J K^{-1} mol^{-1}	J K^{-1} mol^{-1}	J K^{-1} mol^{-1}	J K^{-1} mol^{-1}
Ar (g)	20.79	0	0	0	0
H$_2$ (g)	23.97	30.61	−64.19	57.54	−17.71
N$_2$ (g)	29.42	−2.170	0.5820	13.05	−8.231
O$_2$ (g)	30.18	−14.92	54.71	−50.00	14.88
CO$_2$ (g)	27.10	11.27	124.9	−197.4	87.80
NH$_3$ (g)	35.24	−35.05	169.7	−176.8	63.27
H$_2$O (g)	36.54	−34.80	116.8	−130.0	52.55

Source: *The Properties of Gases and Liquids*, Bruce E. Poling, John M. Prausnitz and John P. O'Connell, 5th edn, McGraw-Hill (2001), pages A–35 to A–46.

If we take account of the temperature-dependence of C_P, then equations (6.25) and (6.26) are not valid, and we have to return to equation (6.24) for $\Delta H_{T+}(i)$

$$\Delta H_{T+}(i) = n_i \int_{T_1}^{T_2} C_P(i)\, dT \tag{6.24}$$

and compute the integral for each reactant i, according to the specific behaviour of the corresponding molar heat capacity $C_P(i)$. Likewise, we can compute $\Delta H_{T-}(j)$

$$\Delta H_{T-}(j) = n_j \int_{T_2}^{T_1} C_P(j)\, dT = -n_j \int_{T_1}^{T_2} C_P(j)\, dT$$

for each product j, from which we can determine $\Delta_r H\,(T_2)$ according to equation (6.23)

$$\Delta_r H\,(T_2) = \Delta_r H\,(T_1) - [\Delta H_{T+} + \Delta H_{T-}] \tag{6.23}$$

Table 6.11 The Shomate temperature variation of $C_P = A + BT + CT^2 + DT^3 + E/T^2$ for some selected gases

	A J K^{-1} mol^{-1}	$B \times 10^3$ J K^{-1} mol^{-1}	$C \times 10^6$ J K^{-1} mol^{-1}	$D \times 10^9$ J K^{-1} mol^{-1}	$E \times 10^{-6}$ J K^{-1} mol^{-1}
Ar (g)	20.79	0	0	0	0
H$_2$ (g)	33.07	−11.36	11.43	−2.773	−0.1586
N$_2$ (g)	28.99	1.854	−9.647	16.63	0
O$_2$ (g)	31.32	−20.24	57.87	−36.51	−0.007
CO$_2$ (g)	25.00	55.19	−33.69	7.948	−0.1366
NH$_3$ (g)	20.00	49.77	−15.38	1.921	0.1892
H$_2$O (g)	30.09	6.833	6.793	−2.534	0.0821

Source: NIST (National Institute of Standards and Technology) Chemistry WebBook, http://webbook.nist.gov/chemistry/.

Where

$$\Delta H_{T+} = \sum_{\text{reactants, } i} \Delta H_{T+}(i) = \sum_{\text{reactants, } i} n_i \int_{T_1}^{T_2} C_P(i)\, dT$$

and

$$\Delta H_{T-} = \sum_{\text{products, } j} \Delta H_{T-}(j) = -\sum_{\text{products, } j} n_j \int_{T_1}^{T_2} C_P(j)\, dT$$

so that

$$\Delta_r H(T_2) = \Delta_r H(T_1) - \left[\sum_{\text{reactants, } i} n_i \int_{T_1}^{T_2} C_P(i)\, dT - \sum_{\text{products, } j} n_j \int_{T_1}^{T_2} C_P(j)\, dT \right]$$

$$= \Delta_r H(T_1) + \left[\sum_{\text{products, } j} n_j \int_{T_1}^{T_2} C_P(j)\, dT - \sum_{\text{reactants, } i} n_i \int_{T_1}^{T_2} C_P(i)\, dT \right]$$

Mathematically, summing the integrals is the same as integrating the sums, implying that

$$\sum_{\text{products, } j} n_j \int_{T_1}^{T_2} C_P(j)\, dT - \sum_{\text{reactants, } i} n_i \int_{T_1}^{T_2} C_P(i)\, dT = \int_{T_1}^{T_2} \left[\sum_{\text{products, } j} n_j C_P(j) - \sum_{\text{reactants, } i} n_i C_P(i) \right] dT$$

But we have defined ΔC_P as

$$\Delta C_P = \sum_{\text{products, } j} n_j C_P(j) - \sum_{\text{reactants, } i} n_i C_P(i) \tag{6.28}$$

which is still valid, even if any $C_P(i)$ or $C_P(j)$ is a function of temperature of the form shown in equations (6.30) or (6.31). We may therefore write

$$\int_{T_1}^{T_2} \left[\sum_{\text{products},j} n_j C_P(j) - \sum_{\text{reactants},i} n_i C_P(i) \right] dT = \int_{T_1}^{T_2} \Delta C_P \, dT$$

and so

$$\Delta_r H(T_2) = \Delta_r H(T_1) + \int_{T_1}^{T_2} \Delta C_P \, dT \tag{6.32}$$

Equation (6.32) is a more general form of equation (6.29)

$$\Delta_r H(T_2) = \Delta_r H(T_1) + \Delta C_P \Delta T \tag{6.29}$$

which, as we have seen, applies only if all the molar heat capacities $C_P(i)$ and $C_P(j)$ are constant.

6.14.5 Kirchhoff's equations

Equation (6.32) may be written as

$$\Delta_r H(T_2) - \Delta_r H(T_1) = \int_{T_1}^{T_2} \Delta C_P \, dT$$

which itself may be written as a partial derivative as

$$\left(\frac{\partial (\Delta_r H)}{\partial T} \right)_P = \Delta C_P \tag{6.33}$$

where the subscripts P serve as a reminder that the pressure is constant. Equation (6.33), which is the most general expression of how any enthalpy change $\Delta_r H$ varies with temperature, is known as **Kirchhoff's first equation**.

Kirchhoff's first equation has just been established by expressing the integral equation (6.32) as a partial differential equation (6.33): in fact, there is an alternative approach, based on the fundamental definition of C_P as

$$\left(\frac{\partial H}{\partial T} \right)_P = C_P \tag{6.8}$$

For a system comprising n_A mol of A, we may therefore write

$$\left(\frac{\partial H_A}{\partial T} \right)_P = n_A C_P(A)$$

and for a system comprising n_B mol of B

$$\left(\frac{\partial H_B}{\partial T} \right)_P = n_B C_P(B)$$

ENTHALPY AND THERMOCHEMISTRY

For an ideal mixture, all state functions are linearly additive and so

$$\left(\frac{\partial H_A}{\partial T}\right)_P + \left(\frac{\partial H_B}{\partial T}\right)_P = \left(\frac{\partial (H_A + H_B)}{\partial T}\right)_P = n_A C_P(A) + n_B C_P(B) \quad (6.34)$$

Likewise, for an ideal mixture of n_C mol of C and n_D mol of D

$$\left(\frac{\partial H_C}{\partial T}\right)_P + \left(\frac{\partial H_D}{\partial T}\right)_P = \left(\frac{\partial (H_C + H_D)}{\partial T}\right)_P = n_C C_P(C) + n_D C_P(D) \quad (6.35)$$

Subtracting (6.34) from equation (6.35), we have

$$\left(\frac{\partial (H_C + H_D)}{\partial T}\right)_P - \left(\frac{\partial (H_A + H_B)}{\partial T}\right)_P = [n_C C_P(C) + n_D C_P(D)] - [n_A C_P(A) + n_B C_P(B)]$$

and so

$$\left(\frac{\partial [(H_C + H_D) - (H_A + H_B)]}{\partial T}\right)_P = [n_C C_P(C) + n_D C_P(D)] - [n_A C_P(A) + n_B C_P(B)]$$

which is in fact Kirchhoff's first equation

$$\left(\frac{\partial (\Delta_r H)}{\partial T}\right)_P = \Delta C_P \quad (6.33)$$

but derived from first principles, the fundamental definition of C_P.

Likewise, for changes at constant volume, and from the definition of C_V as

$$\left(\frac{\partial U}{\partial T}\right)_V = C_V \quad (5.14b)$$

we may derive **Kirchhoff's second equation**

$$\left(\frac{\partial (\Delta U)}{\partial T}\right)_V = \Delta C_V \quad (6.36)$$

6.15 The variation of $\Delta_r H^\circ$ with temperature

A particularly useful special case of the principles presented in the last section is their application to standard enthalpies, in which all reactants and products are in their standard states, and the reaction takes place at the constant pressure of 1 bar. As we saw on page 172, reference sources provide tables of thermodynamic data as associated with standard states, at a specified temperature, usually 298.15 K = 25 °C.

So, for the generalised reaction

$$aA + bB \rightarrow cC + dD$$

if we use equation (6.21)

$$\Delta_r H^\circ = \sum_{\text{Products}} n_{\text{prod}} \Delta_f H_{\text{prod}}^\circ - \sum_{\text{Reactants}} n_{\text{reac}} \Delta_f H_{\text{reac}}^\circ \quad (6.21)$$

and data such as that shown in Table 6.1 to compute the standard enthalpy change $\Delta_r H^\circ$ for that reaction, then the result gives us $\Delta_r H^\circ$ for that reaction taking place, to completion as written, at 298.15 K, $\Delta_r H^\circ$ (298.15).

If we wish to determine the standard enthalpy change $\Delta_r H^\circ(T)$ for that reaction, but at a temperature T other than 298.15 K, then, if the C_Ps of all the reagents, and all the products, may be regarded as constant, we may use equation (6.29)

$$\Delta_r H(T_2) = \Delta_r H(T_1) + \Delta C_P \Delta T \qquad (6.29)$$

such that

$$\Delta_r H^\circ(T_1) = \Delta_r H^\circ (298.15)$$
$$\Delta_r H^\circ(T_2) = \Delta_r H^\circ(T)$$

and $T_1 = 298.15$ K, giving

$$\Delta_r H^\circ(T) = \Delta_r H^\circ(298.15) + \Delta C_P (T - 298.15) \qquad (6.35)$$

Alternatively, if any of the C_Ps vary with temperature, we may use equation (6.32) in the form

$$\Delta_r H^\circ(T) = \Delta_r H^\circ(298.15) + \int_{298.15}^{T} \Delta C_P \, dT \qquad (6.34)$$

As an example, consider the reaction

$$C_2H_4 \, (g) + H_2 \, (g) \rightarrow C_2H_6 \, (g)$$

taking place under standard conditions of 1 bar, and at 298.15 K. The standard enthalpy change $\Delta_r H^\circ(298.15)$ for this reaction may be calculated from equation (6.21)

$$\Delta_r H^\circ = \underbrace{\sum n_{prod} \Delta_f H_{prod}^\circ}_{\text{Products}} - \underbrace{\sum n_{reac} \Delta_f H_{reac}^\circ}_{\text{Reactants}}$$

$$= [\Delta_f H^\circ \, C_2H_6 \, (g)] - [\Delta_f H^\circ \, C_2H_4 \, (g) + \Delta_f H^\circ \, H_2 \, (g)] \qquad (6.21)$$

Reference to Table 6.6 shows that

$$\Delta_f H^\circ \, C_2H_6 \, (g) = -84.0 \, \text{kJ mol}^{-1}$$
$$\Delta_f H^\circ \, C_2H_4 \, (g) = +52.4 \, \text{kJ mol}^{-1}$$
$$\Delta_f H^\circ \, H_2 \, (g) = 0 \, \text{kJ mol}^{-1}$$

and so, for the hydrogenation of ethylene at 1 bar and 298.15 K

$$\Delta_r H^\circ = [-84.0] - [52.4 + 0] = -136.4 \, \text{kJ mol}^{-1}$$

For a reaction taking place at 200 °C = 473.15 K, we can use equation (6.35)

$$\Delta_r H^\circ(T) = \Delta_r H^\circ(298.15) + \Delta C_P (T - 298.15) \qquad (6.35)$$

to estimate the standard enthalpy change $\Delta H^\circ(473.15)$, assuming that all the C_Ps are constant as

$C_P\ C_2H_6\ (g) = 52.5\ J\ K^{-1}\ mol^{-1}$

$C_P\ C_2H_4\ (g) = 42.9\ J\ K^{-1}\ mol^{-1}$

$C_P\ H_2\ (g) = 28.8\ J\ K^{-1}\ mol^{-1}$

so that

$$\Delta C_P = \sum_{\text{products},\ j} n_j C_P(j) - \sum_{\text{reactants},\ i} n_i C_P(i) \qquad (6.28)$$

$= [C_P\ C_2H_6\ (g)] - [C_P\ C_2H_4\ (g) + C_P\ H_2\ (g)]$

$= [52.5] - [42.9 + 28.8]$

$= -19.2\ J\ K^{-1}\ mol^{-1}$

$= -0.0192\ kJ\ K^{-1}\ mol^{-1}$

Using equation (6.25), we now have

$$\Delta_r H^\circ\ (473.15) = -127.0 - 0.0192\ (473.15 - 298.15) = -127.0 - 3.4 = -130.4\ kJ\ mol^{-1} \qquad (6.36)$$

We therefore see that $\Delta_r H^\circ\ (473.15) = -130.4$ kJ mol^{-1} is different from $\Delta_r H^\circ\ (298.15) = -127.7$ kJ mol^{-1}, but not by much. This is in accordance with the inference we made on page 193: although in principle the value of $\Delta_r H^\circ\ (T)$ is different from that of $\Delta_r H^\circ\ (298.15)$, in practice, for many reactions, ΔC_P is substantially smaller than ΔH, and so the difference $\Delta_r H^\circ(T) - \Delta_r H^\circ\ (298.15)$ is relatively small, especially if the temperature difference $(T - 298.15)$ is small too.

6.16 Flames and explosions

In this final section in this chapter, we shall use the principles of thermochemistry to analyse two rather special types of reaction – flames and explosions.

Flames and explosions are reactions which take place very fast. In a flame, the reaction is highly exothermic, and the energy released not only gives the warmth of the fire, but also excites the atoms and molecules in the immediate vicinity of the reaction to such an extent as to cause electronic transitions resulting in the emission of light – the flickering yellows of the excited gases, and the reds of the embers. In an explosion, reactants, usually in a condensed state such as a solid or a gas, react very quickly to form gaseous products. Since a given mass of a gas occupies a volume some 10^3 times greater than the volume occupied by the same mass of the corresponding solid (see, for example, pages 63 and 159), the sudden expansion of the product gases creates the explosive effect.

A schematic thermodynamic representation of a flame is shown in Figure 6.12. Overall, some reactants at, say, 298 K and atmospheric pressure undergo an exothermic reaction to generate products, at atmospheric pressure, but at a much higher temperature $T_?$. Figure 6.12 breaks this down into two hypothetical steps: firstly, a reaction, at constant pressure and at 298 K, in which the reactants are transformed into the products, and secondly, a process in

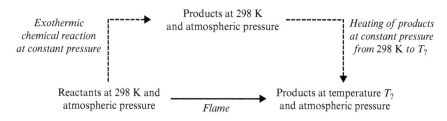

Figure 6.12 A flame.

which the energy released during the reaction is used to heat the products from 298 K to the final temperature $T_?$.

As an example, consider the combustion of acetylene, as defined by the reaction

$$C_2H_2\,(g) + \frac{5}{2}O_2\,(g) \rightarrow 2\,CO_2\,(g) + H_2O\,(g) \qquad \Delta_cH^\circ = \Delta H^\circ_?$$

as happens in an oxy-acetylene torch. From Table 6.6, we have

$\Delta_fH^\circ\ C_2H_2\,(g) = +227.4\,\text{kJ mol}^{-1}$

$\Delta_fH^\circ\ O_2\,(g) = 0\,\text{kJ mol}^{-1}$

$\Delta_fH^\circ\ CO_2\,(g) = -393.5\,\text{kJ mol}^{-1}$

$\Delta_fH^\circ\ H_2O\,(g) = -241.8\,\text{kJ mol}^{-1}$

We now use equation (6.21)

$$\Delta_rH^\circ = \sum_{\text{Products}} n_{\text{prod}} \Delta_f H_{\text{prod}}^\circ - \sum_{\text{Reactants}} n_{\text{reac}} \Delta_f H_{\text{reac}}^\circ \qquad (6.21)$$

to compute $\Delta_rH^\circ = \Delta_cH^\circ$ for the combustion of acetylene as

$\Delta_cH^\circ = [2 \times (-393.5) + 1 \times (-241.8)] - [1 \times (227.4) + 5/2 \times (0)] = -1256.2\,\text{kJ mol}^{-1}$

and, as expected, the burning of acetylene is strongly exothermic.

As we saw on page 150, for a change at constant pressure, the enthalpy change ΔH is equal to the (reversible) heat $\{q_{\text{rev}}\}_P$ exchanged between the system and the surroundings

$$\Delta H = H_2 - H_1 = \{q_{\text{rev}}\}_P \qquad (6.5b)$$

If we assume that 1256.2 kJ are available to heat the products of the burning of acetylene, then our task is to estimate the resulting temperature. To do this, we need to compute how much heat is required to raise the temperature of the reaction products, 2 mol of CO_2 (g) and 1 mol of H_2O (g), from 298 K to some temperature $T_?$. Using values for C_P for CO_2 (g) and H_2O (g) as shown in Tables 6.1 and 6.2

$C_P\ CO_2\,(g) = 37.1\,\text{J K}^{-1}\,\text{mol}^{-1}$

$C_P\ H_2O\,(g) = 37.4\,\text{J K}^{-1}\,\text{mol}^{-1}$

and assuming that these values are both valid, and constant over a wide temperature range, then the total heat required to raise the temperature of 2 mol of CO_2 (g), and 1 mol of H_2O (g), from 298 K to $T_?$, at constant pressure, is given by

Heat required to raise temperature of products from 298 K to $T_?$

$$= \sum_{\text{products}} n_{\text{prod}}(C_P)_{\text{prod}} \Delta T = 2 \times 37.1 \times (T_? - 298) + 1 \times 37.4 \times (T_? - 298)$$

$$= 111.6 \times (T_? - 298) \quad \text{J per mol } C_2H_2 \text{ burnt}$$

Assuming that all the heat generated by the exothermic combustion of 1 mol of C_2H_2 (g) – namely, 1256.2 kJ = 1.26×10^6 J – is used to raise the temperature of the products, then

$$1.26 \times 10^6 = 111.6 \times (T_? - 298)$$

hence

$$T_? = 11{,}588 \text{ K}$$

suggesting that the maximum temperature that could be reached by an oxy-acetylene flame is approximately a roasting 11,600 K.

In fact, the actual temperature is significantly less than this estimate – about 3300 K. Why is there such a difference?

The explanation lies in the various assumptions made in our theoretical computations – assumptions such as that

- the reaction

$$C_2H_2 \text{ (g)} + \frac{5}{2}O_2 \text{ (g)} \rightarrow 2\,CO_2(g) + H_2O\,(g)$$

proceeds, to completion, as written, and with no other products
- all the energy represented by the enthalpy change of this reaction is used to heat the products, and no energy is used in any other way
- the molar heat capacities at constant pressure of the products are constant over the temperature range
- and, importantly, the assumption embodied in equation (6.5b)

$$\Delta H = H_2 - H_1 = \{q_{\text{rev}}\}_P \tag{6.5b}$$

that the change in state is reversible.

Although these assumptions are reasonable in the context of demonstrating the principles of how to use thermodynamic data to address thermochemical questions, in reality

- at such high temperatures, the reaction is much more diverse and complex, with the formation, for example, of radicals too
- much energy is used in other ways, for example as light, as sound, and as heat lost to the surroundings, implying that the amount of energy available for heating the reaction products is much less than ΔH
- the molar heat capacities are not constant, and
- a flame is very far from a reversible system, and so the change in enthalpy $\Delta_c H^\circ$, which equates to the reversible heat $\{q_{\text{rev}}\}_P$, will inevitably be greater than the actual heat $\{q_{\text{irrev}}\}_P$, as implied by equation (6.6b)

$$\Delta H = H_2 - H_1 = \{q_{\text{rev}}\}_P > \{q_{\text{irrev}}\}_P \tag{6.6b}$$

A more sophisticated analysis could allow for all these factors - and others too – so giving a more accurate value of the temperature. These refinements, however, are just that – refinements – refinements that certainly improve the accuracy of the computations, but the fundamental principles of Hess's law, of the state function concept, and of the First Law, remain.

Explosions, too, can be analysed thermodynamically, as illustrated in Figure 6.13.

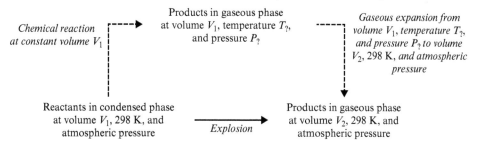

Figure 6.13 An explosion.

An explosion is a chemical reaction in which reactants, usually in a condensed state such as a solid or a liquid at atmospheric pressure, very rapidly transform into gaseous products. Furthermore, if the reaction is exothermic, the heat produced raises the temperature of the reactants, as well as having other effects, such as creating light – and a big bang! At atmospheric pressure, these gaseous products would occupy a volume much larger than the condensed materials from which they were formed, but at the instant of their formation, they occupy essentially that, same, small, original volume. The pressure associated with the newly-formed gaseous products is therefore very high – much higher than the ambient atmospheric pressure – and it is the very rapid expansion of the product gases, so as to reach atmospheric pressure, that is experienced as a violent explosion.

Whilst the reaction is taking place, and shortly thereafter, the system is violently not in equilibrium, implying that the state functions of any intermediate state are not well-defined, and that the change is far from being reversible. That said, we know that the system's pressure during the explosion will be very high, and the purpose of the thermodynamic analysis is to estimate how high this pressure might be.

To do this, we break the real explosion down into two hypothetical steps

- an initial chemical reaction in which the reactants form products at constant volume V_1, implying that the gaseous products are highly compressed, and therefore at some very high pressure $P_?$, as well as at a higher temperature $T_?$, followed by
- the expansion of the initially highly compressed gaseous products from their initial volume V_1 to whatever volume V_2 is appropriate at a pressure of 1 bar and a temperature of 298 K.

as shown in Figure 6.13.

As an example, consider the explosion resulting from the decomposition of nitroglycerine

$$C_3H_5(ONO_2)_3\,(l) \rightarrow 3\,CO_2\,(g) + \frac{5}{2}H_2O\,(g) + \frac{3}{2}N_2\,(g) + \frac{1}{4}O_2\,(g)$$

which, as may be inferred from the equation, is not a combustion reaction, for it does not require external oxygen.

Here is some relevant thermodynamic data

$\Delta_f H^\circ$ $C_3H_5(ONO_2)_3$ (l) = -364.0 kJ mol^{-1}

$\Delta_f H^\circ$ CO_2 (g) = -393.5 kJ mol^{-1}

$\Delta_f H^\circ$ H_2O (g) = -241.8 kJ mol^{-1}

$\Delta_f H^\circ$ N_2 (g) = 0 kJ mol^{-1}

which can be used in equation (6.21) to enable us to compute the standard enthalpy change $\Delta_r H^\circ$ for the decomposition of nitroglycerine

$$\Delta_r H^\circ = \sum_{\text{Products}} n_{\text{prod}} \Delta_f H_{\text{prod}}^\circ - \sum_{\text{Reactants}} n_{\text{reac}} \Delta_f H_{\text{reac}}^\circ \tag{6.21}$$

$$= [3 \times -393.5 + 2.5 \times -241.8] - [-364.0]$$

$$= -1421 \text{ kJ mol}^{-1}$$

As we know, for any reaction, ΔH

$$\Delta H = H_2 - H_1 = \{q_{\text{rev}}\}_P \tag{6.5b}$$

represents the amount of heat $\{q_{\text{rev}}\}_P$ (reversibly) exchanged at constant pressure. The first (hypothetical) step in Figure 6.13, however, is a reaction at constant volume, not at constant pressure, and so, to determine the heat exchange at constant volume, we need to use not ΔH, but ΔU

$$\Delta U = q_V \tag{5.12a}$$

Now, as we saw on page 158, ΔH and ΔU are related by equation (6.14)

$$\Delta H = \Delta U + \Delta(PV) \tag{6.14}$$

which, for a chemical reaction in which reactants in their standard states react to form products in their standard states, becomes

$$\Delta H^\circ = \Delta U^\circ + [(PV)^\circ]_{\text{Products}} - [(PV)^\circ]_{\text{Reactants}} \tag{6.37}$$

In the case of the decomposition of nitroglycerine, the standard state of the reactant is a liquid at 298.15 K and a pressure of 1 bar, and the standard states of the products are gases, also at 298.15 K and a pressure of 1 bar.

If we assume that the products behave as ideal gases, then, for each product

$$PV = nRT$$

Where n is the mole number of the appropriate product, as given by the stoichiometry of the decomposition of nitroglycerine, R is the universal gas constant, 8.314 J K^{-1} mol^{-1}, and T is 298.15 K. For the products resulting from the decomposition of 1 mol of nitroglycerine, and using the appropriate stoichiometric coefficients, we therefore have

$$[(PV)^\circ]_{\text{Products}} = (3 + 2.5 + 1.5 + 0.25) \times 8.314 \times 298.15 \text{ J} = 18.0 \text{ kJ}$$

FLAMES AND EXPLOSIONS

The reactant, nitroglycerine, has a molecular weight of 0.227 kg mol^{-1}, and a density of 1.6×10^3 kg m^{-3}, implying that the volume of 1 mol of nitroglycerine is given by

volume of 1 mol of nitroglycerine = mass of 1 mol/density = $0.227/(1.6 \times 10^3)$ m^3

$$= 1.4 \times 10^{-4} \text{ m}^3$$

For a pressure of 1 bar = 1.01325×10^5 Pa, then

$$[(PV)^{\ominus}]_{\text{Reactants}} = 1.01325 \times 10^5 \times 1.4 \times 10^{-4} \text{ J} = 14.2 \text{ J}$$

In accordance with our discussion on pages 158 and 159, since the nitroglycerine is a liquid, and the products are gases, $[(PV)^{\ominus}]_{\text{Products}} = 18{,}000$ J is very much greater than $[(PV)^{\ominus}]_{\text{Reactants}} = 14.2$ J, and so

$$[(PV)^{\ominus}]_{\text{Products}} - [(PV)^{\ominus}]_{\text{Reactants}} \approx [(PV)^{\ominus}]_{\text{Products}} = 18 \text{ kJ}$$

and since, in general

$$\Delta H^{\ominus} = \Delta U^{\ominus} + [(PV)^{\ominus}]_{\text{Products}} - [(PV)^{\ominus}]_{\text{Reactants}} \tag{6.37}$$

then, for the decomposition of nitroglycerine, for which $\Delta_r H^{\ominus} = -1421$ kJ mol^{-1}

$$\Delta U^{\ominus} = -1421 + 18 = -1403 \text{ kJ mol}^{-1}$$

According to equation (5.12b)

$$\Delta U = q_V \tag{5.12b}$$

the 1403 kJ of energy generated by the decomposition of 1 mol of nitroglycerine can, assuming no energy losses, be used to increase the temperature of the reactants from the initial temperature of 298 K to some, as yet unknown, temperature $T_?$.

We can, however, estimate $T_?$, given knowledge of the molar heat capacities C_V at constant volume, for each of the reactants. Reference to tables of thermodynamic data (and see also page 109) will confirm that

C_V CO_2 (g) = 28.9 J K^{-1} mol^{-1}

C_V H_2O (g) = 27.5 J K^{-1} mol^{-1}

C_V N_2 (g) = 20.8 J K^{-1} mol^{-1}

C_V O_2 (g) = 21.1 J K^{-1} mol^{-1}

For n mol of any substance of molar heat capacity C_V, the temperature ΔT rise attributable to the absorption of a quantity of heat $q_V = \Delta U$ at constant volume is given by equation (5.19)

$$\Delta U = q_V = C_V (T_2 - T_1) = nC_V \Delta T \tag{5.19}$$

implying that

$$\Delta T = \frac{q_V}{nC_V}$$

So, for the absorption of 1403 kJ of heat by the products of the decomposition of nitroglycerine

$$n\,C_V = (3 \times 28.9) + (2.5 \times 27.5) + (1.5 \times 20.8) + (0.25 \times 21.1) = 192$$

and so

$$\Delta T = \frac{q_V}{nC_V} = 1.403 \times 10^6/192 = 7307 = T_? - 298$$

$$\therefore \quad T_? \approx 7600 \text{ K}$$

$T_?$ is our estimate of the instantaneous temperature of the decomposition products of nitroglycerine: as we saw in our discussion of the temperature of the oxy-acetylene flame, this is an upper limit – many factors, such as other reaction products, energy losses and the like, will cause the actual heat used to raise the products' temperature to be less than the estimate obtained by equating this quantity with ΔU. And an explosion is about as irreversible a change as one could imagine!

It is, of course, quite an intellectual leap to assume that the immediate decomposition of nitroglycerine can be thought of as taking place at constant volume. In reality, in any measurable time, there will be nothing constant about an explosion, so, the estimate of 7600 K is sure to be significantly higher than the actual temperature, as indeed it is – the actual temperature is around 5000 K to 5500 K.

Nonetheless, let us continue to estimate the instantaneous pressure $P_?$. We now follow the second (hypothetical) path shown in Figure 6.12, in which the product gases at the original volume V_1, the as-yet unknown pressure $P_?$, and the newly-estimated temperature of 7600 K expands to the final state of volume V_2, temperature 298 K, and pressure 1 bar. To do this, we invoke the ideal gas law for each of the product gases

$$PV = nRT$$

implying that, for each of the product gases

$$\frac{P_? V_1}{7,600} = \frac{1 \times V_2}{298}$$

$$\therefore \quad P_? = \frac{7,600}{298} \times \frac{V_2}{V_1} = 25.5 \times \frac{V_2}{V_1} \text{ bar}$$

In this expression, V_1 is the volume of the original liquid nitroglycerine, which we calculated as $1.4 \times 10^{-4} \text{ m}^3 \text{ mol}^{-1}$ (see page 205). V_2 is the volume of the decomposition products, comprising a total of 7.25 mol of gases for each mol of nitroglycerine. Since each mol of gas occupies 0.0224 m^3,

$$V_2 = 7.25 \times 0.0224 \text{ m}^3 = 0.16 \text{ m}^3$$

and so

$$P_? = 25.5 \times \frac{V_2}{V_1} = 25.5 \times \frac{0.16}{1.4 \times 10^{-4}} = 30{,}000 \text{ bar}$$

Our estimate of the instantaneous pressure of the explosion is a staggering 30,000 – yes, 30,000 – bar! As with all our estimates, this is very much a theoretical maximum, but maybe not as over-the-top as one might think: the real pressure is around 20,000 bar.

EXERCISES

1. Define enthalpy, H. How does the enthalpy H differ from the internal energy U? Are there any conditions in which a change in a system can be such that $\Delta H = \Delta U$? If so,

what are they? If a change is such that $\Delta H \neq \Delta U$, what are the conditions for which that $\Delta H > \Delta U$? And $\Delta H < \Delta U$?

2. Define 'exothermic' and 'endothermic'. For each of the following changes in state, identify whether the change is always exothermic, always endothermic, or either exothermic or endothermic depending on the conditions:
   - the melting of ice
   - the burning of paper
   - water falling over a waterfall
   - the evaporation of ether
   - the formation of water droplets in the atmosphere.

3. Given thermochemical data for octane as

$$C_8H_{18}\,(l) + \frac{25}{2}\,O_2\,(g) \to 8\,CO_2\,(g) + 9\,H_2O\,(g) \qquad \Delta H = -5075\,\text{kJ}$$

$$C_8H_{18}\,(l) \to C_8H_{18}\,(g) \qquad \Delta H = 41\,\text{kJ}$$

verify the data on page 160, namely

$$C_8H_{18}\,(g) + \frac{25}{2}\,O_2\,(g) \to 8\,CO_2\,(g) + 9\,H_2O\,(g) \qquad \Delta H = -5116\,\text{kJ}$$

4. Explain, in physical terms, why C_P is, in principle, always greater than C_V. Under what conditions are C_P and C_V approximately equal?

5. Set up a spreadsheet to calculate C_P according to the equation

$$C_P = a + bT + cT^2 + dT^3 + eT^4 \qquad (6.28)$$

Use the data in Table 6.10 to compute C_P for each of the eight gases in the table.

6. Set up a spreadsheet to calculate C_P according to the Shomate equation

$$C_P = A + BT + CT^2 + DT^3 + \frac{E}{T^2} \qquad (6.29)$$

Use the data in Table 6.11 to compute C_P for each of the eight gases in the table.

7. What is a 'standard state'? Why are standard states important?

8. What is the difference between molality and molarity?

9. Distinguish between a 'reference state' and a 'reference level'.

10. In the morning, the sun warms the earth. So, consider an area of land along the coast, so that, over a given time, the sun's energy transfers an amount of heat $\{q_{rev}\}_P$ to the land, and an equal amount $\{q_{rev}\}_P$ to the sea.
    - Using the data in Tables 6.1 and 6.2, give a sensible estimate of C_P (land) and C_P (sea), respectively.
    - Write an expression for the increase in temperature ΔT (land) of the land, attributable to the absorption of the heat $\{q_{rev}\}_P$, and a similar expression for ΔT (sea).
    - Which of the following three possibilities is true: ΔT (land) $> \Delta T$ (sea), ΔT (land) $= \Delta T$ (sea), or ΔT (land) $< \Delta T$ (sea)?
    - What happens to the air directly above the land, and above the sea?
    - Which of the following three possibilities is true: there is no wind; the wind blows from the sea onto the land; the wind blows from the land onto the sea?
    - In the evening, both the land and the sea lose heat to the atmosphere. Describe, as fully as you can, what happens to the wind.

7 Ideal gas processes – and two ideal gas case studies too

 Summary

See Table 7.2 on page 229.

7.1 What this chapter is about

The purpose of this chapter is to examine some applications of the First Law to processes involving ideal gases so as to derive formulae for, firstly, an important relationship between the molar heat capacity C_V at constant volume, and the molar heat capacity C_P at constant pressure. We next examine three equations which are universally true for all changes in state for an ideal gas, regardless of the path taken, and then, for each of the four main types of path

- the isochoric change, at constant volume, in which $dV = 0$
- the isobaric change, at constant pressure, in which $dP = 0$
- the isothermal change, at constant temperature, in which $dT = 0$
- the adiabatic change, in which no heat is exchanged between the system and the surroundings, and so $đq = 0$

we will establish expressions for

- the work ($đw$ for an infinitesimal change and $\{_1w_2\}$ for a finite change) done by the system to the surroundings, or on the system by the surroundings, and
- the heat ($đq$ for an infinitesimal change and $\{_1q_2\}$ for a finite change) lost from the system to the surroundings, or gained by the system from the surroundings.

Few new concepts will be introduced, and some of the material will be familiar – we trust, however, that it is useful to bring all the key concepts relating to ideal gases together. There's also quite a lot of mathematics, so all the key results are summarised in the table on page 229. Although the mathematics does require some attention (and also familiarity with differentiation and integration), an important feature of this chapter is that all the formulae are derived strictly and rigorously from first principles – first principles that, once you are comfortable with the material, you will be able to use with confidence in situations which are not explicitly covered here.

The final sections present two 'case studies', in which many of the individual fundamental principles explored in the earlier part of the chapter are brought together to demonstrate how they can be applied within a more complex context. The first case study concerns a thermodynamic cycle of considerable importance, known as the 'Carnot cycle', and the second is a system in which a piston oscillates to-and-fro just like the pendulum of a clock. The case studies are quite demanding, and the second, in particular, is quite mathematical. That said, each shows how a step-by-step application of a limited number of basic concepts can be used to tackle even the most complex problems.

7.2 Ideal gases

As we have already seen (see page 17), an ideal gas is one whose molecules:

- occupy zero volume
- are spherical
- undergo elastic collisions, and
- have no mutual interactions, however far apart or close together.

Collectively, these conditions imply that:

- the entire volume of the system is available to the molecules, and no space is excluded
- all the energy of the molecules is represented by the kinetic energy of translational motion, and no other modes of motion, such as rotational and vibrational motion, are present
- all molecules can move totally freely, and are not inhibited by any inter-molecular attractions or repulsions.

Although no real gas is truly ideal, the behaviour of many real systems approximates well to the theoretical ideal model, especially at low pressures; furthermore, the lighter mono-atomic inert gases such as helium, neon and argon maintain almost-ideal behaviour over a much greater range of conditions then poly-atomic polar molecules such as water vapour and ammonia. And – of course – the ideal gas model has no relevance to the condensed states of matter, liquids and solids.

Mathematically, for n mol of an ideal gas, the state functions pressure P, volume V, and temperature T, are related according to the equation-of-state

$$PV = nRT \tag{3.3}$$

for all values of P, V, and T. An implication of equation (3.3) is that, for a fixed quantity of n mol, if any two of P, V, and T are defined, the third is determined. This implies that a fixed mass of an ideal gas has two, and only two, independent state functions. In this regard, however, two of an ideal gas's state functions are 'special': as we have noted several times (but not yet proven – see pages 97 and 643), the internal energy U of an ideal gas is a function of temperature only; furthermore, since, by definition

$$H = U + PV \tag{6.4}$$

which, for an ideal gas, may be written as

$$H = U + nRT$$

then the enthalpy H of an ideal gas too is a function of temperature only. Although T, U and H are all different state functions, knowledge of any two of these does not specify the state of the system; to define the state of a system requires the specification of either

- simultaneous values of P and V, or
- the value of either P or V, and the simultaneous value of any one of T, U or H.

7.3 Some important assumptions

As well as assuming that all systems are composed of ideal gases, throughout this chapter, we will be making these additional assumptions too:

- the only work done by, or on, the system is P,V of expansion or compression, and
- unless otherwise explicitly stated, all changes are quasistatic, and
- there are no dissipative effects in general, and no friction in particular.

These last two assumptions together imply that, unless otherwise stated, all changes are reversible.

7.4 C_V and C_P for ideal gases

The molar heat capacities at constant volume C_V and constant pressure C_P were introduced in Chapter 5 (see pages 107 to 113), and Chapter 6 (see pages 151 any 153), respectively; in this section, we develop a deeper understanding.

7.4.1 C_V and U

We know that the internal energy U of an ideal gas is a function of temperature T only. In simpler language, that means that, as the temperature changes, the internal energy changes; and *vice versa*, as the internal energy changes, the temperature changes. This implies that the *rate of change* of internal energy with respect to temperature – the ratio of how much the internal energy changes for a given change in temperature – must exist, and must have a value other than zero. Mathematically, this rate of change is expressed by the total differential dU/dT, implying that, for n mol of an ideal gas, we may write

$$\frac{dU}{dT} = n\,C_V \tag{7.1a}$$

or

$$dU = n\,C_V\,dT \tag{7.1b}$$

where C_V is a positive constant, independent of all thermodynamic variables, known, as we saw on page 108, as the molar heat capacity at constant volume.

For finite change in state, we need to integrate equation (7.1b) as

$$\int_{U_1}^{U_2} dU = U_2 - U_1 = n\,C_V \int_{T_1}^{T_2} dT$$

and, since n and C_V are constants, then

$$\Delta U = n \, C_V \, \Delta T \tag{7.1c}$$

or

$$U_2 - U_1 = n \, C_V (T_2 - T_1) \tag{7.1d}$$

As we saw on page 113, integrating equation (7.1b) makes an implicit assumption – an important assumption, and an assumption that we would hardly ever notice, but an assumption that is present nonetheless – that $C_V = dU/dT$ is well-defined throughout the change from state [1] to state [2]. Mathematically, this requires that the instantaneous values of both U and T are well-defined throughout the change, which implies that the change must take place through a sequence of infinitesimal equilibrium states, along a quasistatic path. Embedded into the integration of equation (7.1b) to equations (7.1c) and (7.1d) is therefore the assumption that the change in state is quasistatic, that friction and other dissipative effects are absent, and that the change is therefore reversible.

The 'hidden' assumption of reversibility is important in connection with the First Law

$$dU = đq - đw \tag{5.1a}$$

which, using equation (7.1b), we may write as

$$n \, C_V \, dT = đq_{\text{rev}} - đw_{\text{rev}}$$

If the system of interest can perform P,V work only, then all the work $đw$ is necessarily associated with the expansion or compression of the system of interest, which implies a change in volume. If we limit our discussion to changes at constant volume, then $đw = P \, dV$ must be zero and

$$dU_V = đq_V \tag{5.12a}$$

and so, for an infinitesimal change at constant volume

$$dU_V = n \, C_V \, dT = đq_V \tag{7.2a}$$

and

$$n \, C_V = \frac{đq_V}{T} \tag{7.2b}$$

For a finite change

$$\Delta U_V = n \, C_V \, \Delta T = \{_1q_2\}_V \tag{7.2c}$$

and

$$n \, C_V = \frac{\{_1q_2\}_V}{\Delta T} \tag{7.2d}$$

Equations (7.2) imply that the value of dU or ΔU for any change is equal to the heat that would be exchanged, had that change been at constant volume. This – perhaps somewhat clumsy – statement recognises that U is a state function, and that the value of dU or ΔU for any change depends only on the initial and final states, and is independent of the path. So, for any change from state [1] to state [2], dU or ΔU is a specific value. If that change happens to take place *along a path in which the volume does not change*, then the value of dU or ΔU is equal to the heat exchanged along that path; but if the path between states [1] and [2] is not at constant volume throughout, dU or ΔU will still have the same value, but this will in general be different from the actual heat exchange that took place along the path followed. But that value of dU or ΔU will equal what the heat exchange *would have been* had the path been one of constant volume.

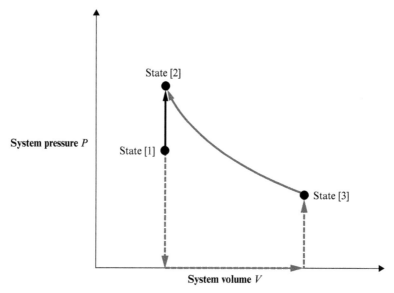

Figure 7.1 The heat exchange for two different paths from state [1] to state [2]. For the first, direct, constant volume path from [1] to [2], no work is done, heat is gained by the system from the surroundings, and so $\Delta U_{1\rightarrow 2,V} = n\,C_V\,\Delta T = \{_1q_2\}_V$. The change through state [3] is adiabatic throughout, and so there is no heat exchange. No work is done for the expansion into a vacuum from state [1] to state [3], but work $\{_3w_2\}_{adi}$ is done on the system by the surroundings for the change from state [3] to state [2]. The overall change $\Delta U_{1\rightarrow 2}$ is independent of the path, and so $\Delta U_{1\rightarrow 2,V} = \Delta U_{1\rightarrow 3} + \Delta U_{3\rightarrow 2}$, implying that $\{_1q_2\}_V = -\{_3w_2\}_{adi}$.

An example should make that quite clear, so consider Figure 7.1, which shows two different paths for a change in state from state [1] to state [2] for an ideal gas. The first path is direct, at constant volume, from state [1] to state [2]; the second path is indirect, following an irreversible expansion into a vacuum to state [3], followed by a reversible adiabatic compression to state [2].

The change from state [1] to state [2] is at constant volume, and, as shown in Figure 7.1, results in an increase in pressure of the ideal gas. According to the ideal gas law

$$PV = nRT \tag{3.3}$$

an increase in pressure ΔP at constant volume V implies an increase in temperature ΔT given by

$$\Delta T = \frac{V \Delta P}{nR}$$

Since no work is done during this change, this increase in temperature must be attributable to the flow of heat $\{_1q_2\}_V$ from the surroundings into the system. This flow of heat increases the internal energy U of the ideal gas according to equation (7.2c)

$$\Delta U_{1\rightarrow 2,V} = n\,C_V\,\Delta T = \{_1q_2\}_V \tag{7.2c}$$

where the subscript $_{1\to 2,V}$ reminds us that the change directly from state [1] to state [2] is at constant volume, and the value of $\Delta U_{1\to 2,V}$ is, as expected, equal to the actual quantity of heat transferred into the system from the surroundings.

Consider now the path from state [1], through state [3], to state [2], as shown in Figure 7.1. The change from state [1] to state [3], following the dashed path, is an irreversible adiabatic expansion into a vacuum. As we saw on pages 121 and 122, the system's volume V increases, and the pressure V decreases (as depicted in Figure 7.1), but since no work is done, and no heat is exchanged, $\Delta U_{1\to 3}$ is zero, and so the temperature of the system in state [3] is the same as the temperature of the original state [1].

The system is now compressed, reversibly and adiabatically, following the path shown in Figure 7.1 from state [3] to state [2], which has the same volume as state [1], but, as discussed on pages 121 to 123, a higher pressure and temperature. The compression requires work $\{_3w_2\}_{adi}$ to be done on the system by the surroundings, and, in accordance with the First Law,

$$\Delta U_{3\to 2} = -\{_3w_2\}_{adi}$$

where the sign of $\{_3w_2\}$ is negative, implying that $\Delta U_{3\to 2}$ is positive. Furthermore, since U is a state function, the change ΔU between any two states is independent of the path, and so

$$\Delta U_{1\to 2} = \Delta U_{1\to 2,V} \text{ directly}$$

and

$$= \Delta U_{1\to 3} + \Delta U_{3\to 2} \text{ indirectly through path [3]}$$

implying that

$$\{_1q_2\}_V = 0 - \{_3w_2\}_{adi}$$

Physically, this result captures the fact that, during the direct change from state [1] to state [2] along the constant volume path, the system gained heat $\{_1q_2\}_V$ from the surroundings, but no work was done; in contrast, during the indirect change from state [1] to state [2] through state [3] along the adiabatic path, there was no heat exchange between the system and the surroundings, but the surroundings did work $\{_3w_2\}_{adi}$ (which is negative) on the system.

In both cases, however, there is a change ΔU in the internal energy U of the system. Since U is a state function, the value of ΔU for any change between a given initial state and a given final state is independent of the path taken, and so $\Delta U_{1\to 2}$ is the same for both the direct and indirect paths. What's different is the attribution of the value of $\Delta U_{1\to 2}$ to flows of heat, or the performance of work, or a mixture of both, for any particular path.

In the example shown in Figure 7.1, the change directly from state [1] to state [2], followed a path at constant volume throughout, and so the value of $\Delta U_{1\to 2} = \Delta U_{1\to 2,V}$ is equal to the value of $\{_1q_2\}_V$, heat actually transferred from the surroundings into the system. $\Delta U_{1\to 2}$ is therefore totally attributable to a flow of heat, and there was no performance of work. The second path, however, from state [1] to state [3] and then to state [2], was adiabatic, and so there is no flow of heat, but work $\{_3w_2\}_{adi}$ was done on the system during the adiabatic compression from state [3] to state [2]. In this case, the value of $\Delta U_{1\to 2}$ is totally attributable to work.

The paths actually shown on Figure 7.1 were deliberately chosen as two particular, special, cases of a path from state [1] to state [2], the first being 'all heat, no work', and the second 'all work, no heat'. As can be readily imagined, there are any number of other paths from state [1] to state [2], each composed of any number of steps i, and each associated with a value for the total heat $\{_1q_2\}_{tot}$ exchanged, and a corresponding value for the total work $\{_1w_2\}_{tot}$ performed, during that change, where

$$\{_1q_2\}_{tot} = \sum_i q_i$$

and

$$\{_1w_2\}_{tot} = \sum_i w_i$$

in which q_i is the amount of heat gained by, or lost from, the system during step i, and w_i is the corresponding amount of work done by, or on, the system during the same step i. For the same two states [1] and [2], each path will have its own corresponding values of $\{_1q_2\}_{tot}$ and $\{_1w_2\}_{tot}$, but, in accordance with the First Law, these values will always have the same difference given by

$$\Delta U_{1\to 2} = \{_1q_2\}_{tot} - \{_1w_2\}_{tot}$$

This is another illustration of the fact that, for any change in state, there will always be a corresponding change in all state functions (even if that change evaluates to zero), including a change ΔU in the internal energy U. If the change takes place along a path of constant volume throughout, then $\{_1w_2\}_{tot}$ will be zero, and $\Delta U_{1\to 2,V}$ equates to a flow of heat $\{_1q_2\}_V$ as given by equation (7.2c)

$$\Delta U_{1\to 2,V} = n\, C_V\, \Delta T = \{_1q_2\}_V \tag{7.2c}$$

But if the change in state does not take place along a path of constant volume throughout, then there will still be a value for $\Delta U_{1\to 2}$, and there may, or may not, be a flow of heat $\{_1q_2\}_{tot}$; furthermore, the values of $\Delta U_{1\to 2}$ and $\{_1q_2\}_{tot}$ will, in general, be different.

7.4.2 C_p and H

By definition, the enthalpy H for any system is given by

$$H = U + PV \tag{6.4}$$

For an ideal gas, this becomes

$$H = U + nRT \tag{7.3}$$

Showing that, since U is a function of T only, then H must be a function of T only. Consequently, for n mol of an ideal gas, we may define a quantity dH/dT as

$$\frac{dH}{dT} = n\, C_P \tag{7.4a}$$

or

$$dH = n\, C_P\, dT \tag{7.4b}$$

where C_P is a positive constant, independent of all thermodynamic variables, known, as we saw on page 151, as the molar heat capacity at constant pressure.

For finite change in state, equation (7.4b) becomes

$$\Delta H = n\, C_P\, \Delta T \tag{7.4c}$$

or

$$H_2 - H_1 = n\, C_p (T_2 - T_1) \tag{7.4d}$$

We now invoke the First Law

$$dU = đq - đw \tag{5.1a}$$

For a system that performs P, V work only

$$đw = P_{ex}\, dV$$

and for a reversible change in which the external pressure P_{ex} is always equal to the internal pressure P, V

$$dU = đq_{rev} - P\, dV$$

As we saw on page 149, for a change at constant pressure P, dP is zero, and so

$$dU_P = đq_{rev,P} - P\, dV - V\, dP$$

implying that

$$dU_P + P\, dV + V\, dP = đq_{rev,P}$$

giving

$$d(U + PV)_P = đq_{rev,P}$$

Since

$$U + PV = H \tag{6.4}$$

Then, for a change at constant pressure

$$dH_P = đq_{rev,P} \tag{6.8b}$$

Combining equation (6.8b) with equation (7.4b), we have

$$dH_P = đq_{rev,P} = n\, C_P\, dT \tag{7.5a}$$

and so

$$n\, C_P = \frac{đq_{rev,P}}{T} \tag{7.5b}$$

For finite changes

$$\Delta H_P = \{_1 q_2\}_{rev,P} = n\, C_P\, \Delta T \tag{7.5c}$$

and

$$n\, C_P = \frac{\{_1 q_2\}_{rev,P}}{\Delta T} \tag{7.5d}$$

For a change taking place reversibly and at constant pressure throughout, the value of the change in enthalpy ΔH_P is equal to the quantity of heat $\{_1q_2,_{\text{rev}}\}_P$ exchanged between the system and the surroundings. If the change takes place at constant pressure throughout, but not reversibly, then, as discussed on page 150, since

$$\Delta H_P = \{_1q_2\}_{\text{rev}, P} > \{_1q_2\}_{\text{irrev}, P}$$

then ΔH_P is greater than, and the theoretical maximum of, the quantity of heat $\{_1q_2,_{\text{irrev}}\}_P$ actually exchanged. And if the change does not take place at constant pressure, then the actual quantity of heat $\{_1q_2\}$ exchanged will, in general, be different from the value of ΔH for that change.

7.4.3 The difference $C_P - C_V$

We have seen that, for an ideal gas,

$$H = U + nRT \tag{7.3}$$

Differentiating with respect to temperature T

$$\frac{dH}{dT} = \frac{dU}{dT} + nR$$

But by definition, for an ideal gas

$$\frac{dU}{dT} = n\,C_V \tag{7.1a}$$

and

$$\frac{dH}{dT} = n\,C_P \tag{7.4a}$$

from which we see that, for an ideal gas

$$n\,C_P = n\,C_V + nR$$

and so, for an ideal gas

$$C_P - C_V = R \tag{7.6}$$

Equation (7.6) presents the somewhat surprising result that, for any ideal gas, the difference $C_P - C_V$ between the molar heat capacities at constant pressure and at constant volume is equal to the universal gas constant R.

7.4.4 Real values of C_V and C_P

Table 7.1 shows values for C_V and C_P for some selected molecules – some gases, some liquids and some solids.

As can be seen, at 298.15 K, C_V for helium is 12.5 J K^{-1}mol^{-1} and C_P is 20.8 J K^{-1}mol^{-1}. The difference $C_P - C_V$ is 20.8 − 12.5 = 8.3 J K^{-1}mol^{-1}, totally consistent with R = 8.314 J K^{-1}mol^{-1}. This holds for all the monatomic gases, and, very closely, for the other gases shown too – with the exception of H$_2$O (g) at 373.15 K, for which $C_P - C_V$ = 9.4 J K^{-1}mol^{-1}. At this temperature,

Table 7.1 Molar heat capacities of selected elements and compounds.

	T K	C_V J K^{-1} mol^{-1}	C_P J K^{-1} mol^{-1}	$C_P - C_V$ J K^{-1} mol^{-1}
He (g)	298.15	12.5	20.8	8.3
Ne (g)	298.15	12.5	20.8	8.3
Ar (g)	298.15	12.5	20.8	8.3
Kr (g)	298.15	12.5	20.8	8.3
Xe (g)	298.15	12.5	20.8	8.3
H_2 (g)	298.15	20.5	28.8	8.3
N_2 (g)	298.15	20.8	29.1	8.3
O_2 (g)	298.15	21.1	29.4	8.3
CO_2 (g)	298.15	28.9	37.1	8.2
NH_3 (g)	298.15	28.0	35.1	7.1
C_2H_5OH (l)	298.15	112	112	0
CCl_4 (l)	298.15	131	131	0
Al (s)	298.15	24.2	24.2	0
Cu (s)	298.15	24.4	24.4	0
Fe (s)	298.15	25.1	25.1	0
Pb (s)	298.15	26.8	26.8	0
SiO_2 (s)	298.15	44.6	44.6	0
NaCl (s)	298.15	50.5	50.5	0
H_2O (s)	173.15	24.8	24.8	0
H_2O (s)	273.15	37.8	37.8	0
H_2O (l)	273.16	76.0	75.9	−0.1
H_2O (l)	298.15	75.3	75.3	0
H_2O (l)	372.75	67.9	75.9	8.0
H_2O (g)	373.15	28.0	37.4	9.4
H_2O (g)	773	27.1	35.7	8.6
H_2O (g)	1,273	32.9	41.3	8.5

Sources as Tables 5.1, 6.1 and 6.2, with additional values of C_V for liquid and gaseous water from *CRC Handbook of Chemistry and Physics*, 96th Edition, 2015, Section 6, pages 6-1 to 6-4. C_V for the other condensed state materials has been set equal to C_P.

water vapour is on the point of condensing into liquid water, and so the assumptions associated with an ideal gas are at their least valid, so we would expect a departure from the ideal value of $C_P - C_V$ And, as can be seen, at higher temperatures, the value of $C_P - C_V$ of even water vapour approaches that of an ideal gas.

For the condensed states of matter – liquids and solids – the values of C_V and C_P for any particular substance are, very closely, equal, with values of around 25 J K^{-1}mol^{-1} for the metals (recognising, once again, the wide range of atomic weights from aluminium, 27, through copper, 64, to lead, 207), and considerably higher values for the organic liquids. The theory explaining why a particular material has a particular value of C_V or C_P is a question of condensed matter physics; the fact that C_V and C_P are equal for the condensed states of matter is readily explained, as we saw on pages 158 to 161, from the definition of enthalpy as

$$H = U + PV \tag{5.4}$$

For any change of state

$$\Delta H = \Delta U + \Delta(PV)$$

For a condensed state of matter, for most changes, especially those over a small temperature differential, $\Delta(PV) \approx 0$ and so $\Delta H \approx \Delta U$, which in turn implies that $C_P \approx C_V$.

7.5 Three formulae that apply to all ideal gas changes

There are three formulae that are universal, and apply to all changes for an ideal gas, irrespective of the nature of the path.

The first is the ideal gas law

$$PV = nRT \tag{3.3}$$

which, given values for any two of P, V and T, enables the third variable to be determined.

The two other formulae are

$$\Delta U = n\, C_V\, \Delta T \tag{7.2c}$$

and

$$\Delta H = n\, C_P\, \Delta T \tag{7.4c}$$

These two formulae relate to changes ΔU and ΔH in the state functions U and H for a change in state from any state $[1] \equiv [P_1, V_1, T_1]$ to any state $[2] \equiv [P_2, V_2, T_2]$. Both ΔU and ΔH therefore depend only on the initial and final states, and are independent of the path taken.

The nature of the path, however, does determine the heat $đq$ or $\{_1q_2\}$ exchanged between the system and the surroundings, and the work $đw$ or $\{_1w_2\}$ done by the system on the surroundings, or on the system by the surroundings. The next four sections therefore determine $đq$, $\{_1q_2\}$, $đw$ and $\{_1w_2\}$ for each of the four main types of path:

- the isochoric path, at constant volume, such that $dV = 0$
- the isobaric path, at constant pressure, such that $dP = 0$
- the isothermal path, at constant temperature, such that $dT = 0$
- the adiabatic path, with no heat exchange, such that $đq = 0$.

7.6 The isochoric path, $dV = 0$

An isochoric path, at constant volume, implies that $dV = 0$. If only P,V work against an external pressure P_{ex} takes place, then, by definition, at constant volume

$$đw_V = P_{ex}\, dV = 0$$

and so, for a finite change

$$\{_1w_2\}_V = 0$$

According to the First Law

$$dU = đq - đw \tag{5.1a}$$

then, since $\{đw\}_V = 0$

$$đq_V = dU$$

But,

$$dU = n\,C_V\,dT$$

hence

$$đq_V = n\,C_V\,dT$$

and

$$\{_1q_2\}_V = n\,C_V\,\Delta T$$

7.7 The isobaric path, $dP = 0$

Since, in general

$$dU = đq - đw \tag{5.1a}$$

then, for an infinitesimal reversible change at constant pressure

$$dU = đq_{rev,P} - đw_{rev,P}$$

If the system can perform only P,V work of expansion

$$đw_{rev,P} = P\,dV$$

and so, for a finite change

$$\{_1w_2\}_{rev,P} = P\,\Delta V$$

Also, according to the ideal gas law

$$PV = nRT \tag{3.3}$$

then, at constant pressure P

$$P\,dV = nR\,dT$$

and

$$P\,\Delta V = nR\,\Delta T$$

implying that

$$đw_{rev,P} = P\,dV = nR\,dT$$

$$\{_1w_2\}_{rev,P} = P\,\Delta V = nR\,\Delta T$$

To determine $đq_{rev,P}$, we use the expression for $đw_{rev,P}$ in the First Law as

$$dU = đq_{rev,P} - nR\,dT$$

which, combined with equation (7.1b),

$$dU = n\,C_V\,dT \qquad (7.1b)$$

gives

$$đq_{rev,P} = n\,C_V\,dT + nR\,dT = n\,(C_V + R)\,dT$$

But we know that

$$C_P - C_V = R \qquad (7.6)$$

and so

$$đq_{rev,P} = n\,C_P\,dT$$

Furthermore, since

$$dH = n\,C_P\,dT \qquad (7.4c)$$

then

$$đq_{rev,P} = n\,C_P\,dT = dH$$

and, for a finite change,

$$\{_1q_2\}_{rev,P} = n\,C_P\,\Delta T = \Delta H$$

7.8 The isothermal path, $dT = 0$

For a change along an isothermal path at a given temperature T, $dT = 0$. Since, for an ideal gas, U and H are functions of temperature only, then if $dT = 0$, then $dU = 0$ and $dH = 0$.

To determine $đw$, our starting point is the fundamental definition of $đw_{rev}$ as

$$đw_{rev} = P\,dV$$

The ideal gas law

$$PV = nRT \qquad (3.3)$$

is universally true, and so, in general

$$đw_{rev} = \frac{nRT}{V}\,dV$$

If the temperature T is constant throughout an isothermal reversible change from any state $[1] \equiv [P_1, V_1, T_1]$ to any state $[2] \equiv [P_2, V_2, T_1]$, then the quantity nRT is constant, and so this expression can be integrated as

$$\{_1w_2\}_{rev,T} = nRT \int_{V_1}^{V_2} \frac{dV}{V} = nRT \ln \frac{V_2}{V_1} = nRT \ln \frac{P_1}{P_2} \qquad (7.7)$$

To determine đq, we return to the First Law

$$dU = đq - đw \qquad (5.1a)$$

Since, for an isothermal change $dU = 0$, then

$$đq_{rev,T} = đw_{rev,T} = P\,dV = nR\,dT$$

and

$$\{_1q_2\}_{rev,T} = \{_1w_2\}_{rev,T} = nRT \int_{V_1}^{V_2} \frac{dV}{V} = nRT \ln \frac{V_2}{V_1} = nRT \ln \frac{P_1}{P_2} \qquad (7.7)$$

If the system expands, $V_2 > V_1$, and $P_2 < P_1$, implying that $\{_1w_2\}_{rev}$ and $\{_1q_2\}_{rev}$ are both positive: during an isothermal expansion of an ideal gas, the gas therefore does work on the surroundings, and also gains an equivalent amount of heat from the surroundings, as given by equation (7.7). Conversely, if the system is compressed, $V_2 < V_1$, and $P_2 > P_1$, implying that $\{_1w_2\}_{rev}$ and $\{_1q_2\}_{rev}$ are both negative: during an isothermal compression of an ideal gas, the surroundings do work on the gas, and the gas releases an equivalent amount of heat to the surroundings, also in accordance with equation (7.7).

As we saw on page 144, for a real, irreversible, change

$$đw_{rev} > đw_{irrev} \qquad (5.37)$$

and

$$đq_{rev} > đq_{irrev} \qquad (5.38)$$

Equation (7.7) therefore represents theoretical maxima for the work $\{_1w_2\}_{irrev}$ actually done on, or by, the system, and the corresponding exchange of heat $\{_1q_2\}_{irrev}$.

7.9 The adiabatic path, đ$q = 0$

7.9.1 Adiabatic work đw_{adi}

During an adiabatic change, no heat is exchanged between the system and the surroundings, and đq is necessarily zero. Under these circumstances, the First Law

$$dU = đq - đw \qquad (5.1a)$$

becomes

$$dU_{adi} = -đw_{adi} \qquad (7.8)$$

A consequence of this equation is that, if a system does adiabatic work on the surroundings, đw_{adi} is positive, and so dU_{adi} is negative. If the system is an ideal gas, U is a function of temperature only, and so if dU is negative, the gas must cool. Conversely, as we saw on page 213, if the surroundings do adiabatic work on the ideal gas system, $đw_{adi}$ is negative, dU is positive, and the gas will warm.

Since

$$dU = n\,C_V\,dT \qquad (7.1b)$$

then, for an infinitesimal adiabatic change

$$dU_{adi} = n\,C_V\,dT_{adi} = -đw_{adi} \qquad (7.8)$$

and so, for a finite change

$$\{_1w_2\}_{rev,\,adi} = -n\,C_V\,\Delta T_{adi} \qquad (7.9)$$

As discussed on page 211, when C_V is used to quantify the amount of heat or work associated with any change, that change is necessarily assumed to be quasistatic; the same applies to the use of C_P too. Furthermore, in the absence of friction and all other dissipative effects, this implies the change is reversible, hence the used of the subscripts $_{rev,\,adi}$ in equation (7.9).

Equation (7.9) is an expression for the work associated with a reversible adiabatic change for an ideal gas, and so, in accordance with our discussion on page 143, where we showed that

$$đw_{rev} > đw_{irrev} \qquad (5.37)$$

equation (7.9) represents a theoretical maximum.

For an adiabatic expansion, the system does work on the surroundings, and so $\{_1w_2\}_{rev,\,adi}$ is positive. In equation (7.9), n and C_V are necessarily positive, implying that $\Delta T_{adi} = T_2 - T_1$ must be negative. The final temperature T_2 of the system must therefore be lower than the initial temperature T_1 - as a result of an adiabatic expansion, the system cools. Conversely, during a compression, the surroundings do work on the system $\{_1w_2\}_{rev,\,adi}$ is negative, $T_2 > T_1$, and the system becomes warmer.

The key results for a finite reversible adiabatic change are therefore

$$\{_1q_2\}_{adi} = 0$$

$$\{_1w_2\}_{adi} = -n\,C_V\,\Delta T_{adi} \qquad (7.9a)$$

$$\Delta U_{adi} = -n\,C_V\,\Delta T_{adi} \qquad (7.9b)$$

$$\Delta H_{adi} = -n\,C_P\,\Delta T_{adi} \qquad (7.10)$$

Reference to page 212 will show that the expression $n\,C_V\,\Delta T$ - with a positive sign rather than a negative sign - arose in connection with the heat exchanged by an ideal gas changing state at constant volume:

$$\Delta U_{1\to 2,V} = n\,C_V\,\Delta T = \{_1q_2\}_V \qquad (7.2c)$$

Is this just a coincidence?

No, it isn't - as illustrated in Figure 7.2, which is very similar to Figure 7.1, and which shows a reversible isothermal expansion from state [1] $\equiv [P_1, V_1, T_1]$ to state [3] $\equiv [P_3, V_3, T_1]$, followed by a reversible adiabatic compression from state [3] to state [2] $\equiv [P_2, V_1, T_2]$, such that the volumes in states [1] and [2] are identical, but with $P_2 > P_1$, and $T_2 > T_1$.

THE ADIABATIC PATH, $dq = 0$

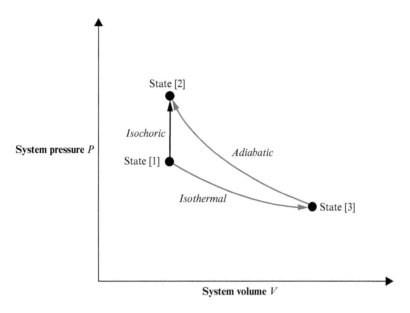

Figure 7.2 The heat exchange for an isochoric path is equal to the work done by the equivalent adiabatic path. State [2] has the same volume as state [1], and a higher temperature and pressure. All paths are reversible, and the path from state [1] to state [2] is isochoric, from state [1] to state [3] isothermal, and from state [3] to state [2] adiabatic.

Since U is a state function, with reference to the states shown in Figure 7.2,

$$\Delta U_{1 \to 2,\, V} = \Delta U_{1 \to 3,\, \text{iso}} + \Delta U_{3 \to 2,\, \text{adi}}$$

The path from state [1] to state [3] is isothermal, and so $\Delta U_{1 \to 3,\, \text{iso}} = 0$, hence

$$\Delta U_{1 \to 2,\, V} = \Delta U_{3 \to 2,\, \text{adi}}$$

Now, using the First Law for the isochoric path from state [1] to state [2]

$$\Delta U_{1 \to 2,\, V} = \{_1 q_2\}_V - \{_1 w_2\}_V$$

but at constant volume, $\{_1 w_2\}_V = 0$, and so

$$\Delta U_{1 \to 2,\, V} = \{_1 q_2\}_V$$

Also, for the adiabatic path from state [2] to state [3]

$$\Delta U_{3 \to 2,\, \text{adi}} = \{_3 q_2\}_{\text{adi}} - \{_3 w_2\}_{\text{adi}}$$

and for an adiabatic path, $\{_3 q_2\}_{\text{adi}} = 0$, implying that

$$\Delta U_{3 \to 2,\, \text{adi}} = - \{_3 w_2\}_{\text{adi}}$$

So, since

$$\Delta U_{1 \to 2,\, V} = \Delta U_{3 \to 2,\, \text{adi}}$$

then

$$\{_1 q_2\}_V = - \{_3 w_2\}_{\text{adi}} = n\, C_V\, \Delta T$$

7.9.2 An important equation - $PV^\gamma = $ constant

As we saw in Chapter 1, the earliest equation of state to be established was Boyle's Law

$$PV = \text{constant} = c_{\text{iso}} \qquad (1.2)$$

which applies to a fixed mass of an ideal gas undergoing an isothermal change at temperature T, such that the constant, c_{iso}, on the right hand side of this equation is nRT. Equation (1.2) gives a relationship for the simultaneous values of P and V for an isothermal change, which raises the question of whether there might be another equation giving the simultaneous values of P and V for an adiabatic change. In fact, there is:

$$PV^\gamma = \text{constant} = c_{\text{adi}} \qquad (7.11)$$

where the constant, c_{adi}, in this equation is different from the Boyle's Law constant, c_{iso}, and the exponent γ is equal to the ratio between the molar heat capacities at constant pressure and constant volume

$$\gamma = \frac{C_P}{C_V} \qquad (7.12)$$

It may be that equations (7.11) and (7.12) are familiar, and even if they are, it is likely that they were originally presented as 'givens'. So we take the opportunity here to derive equations (7.11) and (7.12) from first principles.

Our starting point is one of the equations we have just derived, equation (7.8)

$$dU_{\text{adi}} = n\,C_V\,dT_{\text{adi}} = -\dj w_{\text{adi}} \qquad (7.8)$$

For a reversible change, we know that $\dj w = P\,dV$, and so for a reversible adiabatic change, we may write

$$n\,C_V\,dT_{\text{rev, adi}} = -\{P\,dV\}_{\text{rev, adi}} \qquad (7.13)$$

We'll return to equation (7.13) shortly; for the moment, let's take the definition of enthalpy

$$H = U + PV \qquad (5.4)$$

from which, on taking differentials, we may write

$$dH = dU + P\,dV + V\,dP$$

Applying the First Law for dU we have

$$dH = \dj q - \dj w + P\,dV + V\,dP$$

For an adiabatic change $\dj q = 0$, and for a reversible change $\dj w = P\,dV$. So, for a change which is both reversible and adiabatic

$$dH_{\text{rev, adi}} = \{V\,dP\}_{\text{rev, adi}}$$

Now, by definition

$$dH = n\,C_P\,dT \qquad (7.4b)$$

and so

$$n\,C_P\,dT_{\text{rev, adi}} = \{V\,dP\}_{\text{rev, adi}} \qquad (7.14)$$

We now have two equations which relate to a reversible, adiabatic change for an ideal gas

$$n\,C_V\,dT_{\text{rev, adi}} = -\{P\,dV\}_{\text{rev, adi}} \tag{7.13}$$

and

$$n\,C_P\,dT_{\text{rev, adi}} = \{V\,dP\}_{\text{rev, adi}} \tag{7.14}$$

In these two equations, $dT_{\text{rev, adi}}$ is common, and so from equation (7.13),

$$dT_{\text{rev, adi}} = -\frac{\{P\,dV\}_{\text{rev, adi}}}{n\,C_V}$$

and substituting for $dT_{\text{rev, adi}}$ in equation (7.14), we have

$$-n\,C_P\frac{\{P\,dV\}_{\text{rev, adi}}}{n\,C_V} = \{V\,dP\}_{\text{rev, adi}}$$

which simplifies to

$$-\frac{C_P}{C_V}\frac{dV}{V} = \frac{dP}{P} \tag{7.15}$$

where, for clarity, we've dropped the $_{\text{rev, adi}}$ suffix: we just have to remember that equation (7.15) includes three, important, assumptions – that the system of interest is an ideal gas, that the change in state is adiabatic, and that the change in state is reversible.

For an ideal gas, C_P and C_V are constants, and so for any change from any state $[1] \equiv [P_1, V_1, T_1]$ to any other state $[2] \equiv [P_2, V_2, T_2]$, we may integrate equation (7.15) as

$$-\frac{C_P}{C_V}\int_{V_1}^{V_2}\frac{dV}{V} = \int_{P_1}^{P_2}\frac{dP}{P}$$

and so

$$-\frac{C_P}{C_V}\ln\left(\frac{V_2}{V_1}\right) = \ln\left(\frac{P_2}{P_1}\right) \tag{7.16}$$

The ratio C_P/C_V is conventionally represented by the symbol γ

$$\gamma = \frac{C_P}{C_V} \tag{7.12}$$

allowing us to express equation (7.16) as

$$-\gamma \ln\left(\frac{V_2}{V_1}\right) = \ln\left(\frac{P_2}{P_1}\right) \tag{7.17}$$

Given that $\ln(x/y) = \ln(x) - \ln(y)$, equation (7.17) may be rewritten as

$$-\gamma \ln V_2 + \gamma \ln V_1 = \ln P_2 - \ln P_1$$

and so

$$\ln P_1 + \gamma \ln V_1 = \ln P_2 + \gamma \ln V_2$$

But $\gamma \ln V_1 = \ln V_1^{\gamma}$, implying that

$$\ln P_1 + \ln V_1^{\gamma} = \ln P_2 + \ln V_2^{\gamma}$$

hence

$$\ln P_1 V_1^\gamma = \ln P_2 V_2^\gamma$$

giving the result that, for an ideal gas undergoing a reversible adiabatic change from any state $[1] \equiv [P_1, V_1, T_1]$ to any other state $[2] \equiv [P_2, V_2, T_2]$

$$P_1 V_1^\gamma = P_2 V_2^\gamma \tag{7.18}$$

where

$$\gamma = \frac{C_P}{C_V} \tag{7.12}$$

Equation (7.18), often expressed in the form

$$PV^\gamma = constant = c_{adi} \tag{7.11}$$

is the adiabatic equivalent of Boyle's Law

$$PV = constant = c_{iso} \tag{1.2}$$

for a reversible isothermal change for an ideal gas. Importantly, the *constant* c_{adi} in equation (7.13) is different from the *constant* c_{iso} in equation (1.2): c_{iso} is of course nRT.

7.9.3 Adiabatic work dw_{adi} - another derivation

We have already shown that the work $\{_1w_2\}_{adi}$ done by or on a system of an ideal gas during a reversible adiabatic change is given by

$$\{_1w_2\}_{adi} = -n C_V \Delta T \tag{7.9a}$$

We also know that,

$$\{_1w_2\} = \int_{V_1}^{V_2} P \, dV$$

but this integral can only be evaluated if we have knowledge of the functional relationship between P and V for any particular path. For an isothermal change, Boyle's Law $PV = constant = c_{iso}$ is valid, and this is how we established, on page 31, the work done by an ideal gas for a reversible isothermal path.

Now that we have established that

$$PV^\gamma = c_{adi} \tag{7.11}$$

for a reversible adiabatic path, we can use equation (7.11) as an alternative way of determining the work $\{_1w_2\}_{rev, adi}$ done on, or by, an ideal gas undergoing a reversible adiabatic change from any state $[1] \equiv [P_1, V_1, T_1]$ to any other state $[2] \equiv [P_2, V_2, T_2]$.

So, given that, by definition

$$\{_1w_2\}_{rev, adi} = \int_{V_1}^{V_2} P \, dV$$

then, using equation (7.13)

$$PV^\gamma = \text{constant} = c_{adi} \tag{7.11}$$

we may write

$$\{_1w_2\}_{\text{rev, adi}} = \int_{V_1}^{V_2} \frac{c_{adi}}{V^\gamma} dV = c_{adi} \int_{V_1}^{V_2} V^{-\gamma} dV$$

The algebra gets rather messy, but this is a standard integral with the result

$$\{_1w_2\}_{\text{rev, adi}} = \frac{c_{adi}}{-\gamma + 1} \left(V_2^{-\gamma+1} - V_1^{-\gamma+1} \right) \tag{7.19}$$

Since $\gamma = C_P/C_V$, γ is always greater than 1 (see equation (7.6)), and so equation (7.19) is better expressed as

$$\{_1w_2\}_{\text{rev, adi}} = \frac{c_{adi}}{\gamma - 1} \left(\frac{1}{V_1^{\gamma-1}} - \frac{1}{V_2^{\gamma-1}} \right) \tag{7.20}$$

According to equation (7.11)

$$P_1 V_1^\gamma = c_{adi} = P_2 V_2^\gamma \tag{7.11}$$

so that

$$\{_1w_2\}_{\text{rev, adi}} = \frac{1}{\gamma - 1} \left(\frac{P_1 V_1^\gamma}{V_1^{\gamma-1}} - \frac{P_2 V_2^\gamma}{V_2^{\gamma-1}} \right) = \frac{1}{\gamma - 1} (P_1 V_1 - P_2 V_2) \tag{7.21}$$

Now, for an ideal gas, the ideal gas law

$$PV = nRT \tag{3.3}$$

is true under all conditions, and so

$$P_1 V_1 = nRT_1$$

and

$$P_2 V_2 = nRT_2$$

enabling us to express equation (7.21) as

$$\{_1w_2\}_{\text{rev, adi}} = \frac{nR}{\gamma - 1} (T_1 - T_2) \tag{7.21}$$

But since

$$\gamma = \frac{C_P}{C_V}$$

then

$$\gamma - 1 = \frac{C_P}{C_V} - 1 = \frac{C_P - C_V}{C_V}$$

But, for an ideal gas

$$C_P - C_V = R \tag{7.6}$$

and so
$$\gamma - 1 = \frac{R}{C_V}$$
thereby simplifying equation (7.21) to
$$\{_1w_2\}_{\text{rev, adi}} = nC_V(T_1 - T_2) = -nC_V\Delta T \tag{7.9a}$$
which is exactly the same as the result we derived (much more easily!) on pages 221 and 222.

7.9.4 Some other adiabatic relationships

The adiabatic equation
$$PV^\gamma = \text{constant} = c_{\text{adi}} \tag{7.11}$$
can be combined with the ideal gas law
$$PV = nRT \tag{3.3}$$
to derive a number of other relationships between the state functions $[P_1, V_1, T_1]$ and $[P_2, V_2, T_2]$ of any two states which are mutually accessible by a reversible adiabatic path.

So, for example, from the ideal gas law, we have
$$\left(\frac{P_1V_1}{T_1}\right)^\gamma = \left(\frac{P_2V_2}{T_2}\right)^\gamma$$
$$\therefore \frac{P_1^{\gamma-1}}{P_2^{\gamma-1}}\left(\frac{P_1V_1^\gamma}{P_2V_2^\gamma}\right) = \frac{T_1^\gamma}{T_2^\gamma}$$

But for an adiabatic change
$$P_1V_1^\gamma = P_2V_2^\gamma$$
$$\therefore \frac{P_1}{P_2} = \left(\frac{T_1}{T_2}\right)^{\frac{\gamma}{\gamma-1}} \tag{7.22}$$

and since
$$\frac{P_1}{P_2} = \left(\frac{V_2}{V_1}\right)^\gamma$$

then
$$\frac{V_2}{V_1} = \left(\frac{T_1}{T_2}\right)^{\frac{1}{\gamma-1}} \tag{7.23}$$

Finally, given that
$$\gamma = \frac{C_P}{C_V} \tag{7.12}$$

and
$$C_P - C_V = R \tag{7.6}$$
then
$$\frac{\gamma}{\gamma - 1} = \frac{C_P}{R}$$
and
$$\frac{1}{\gamma - 1} = \frac{C_V}{R}$$
implying that
$$\frac{P_1}{P_2} = \left(\frac{T_1}{T_2}\right)^{\frac{\gamma}{\gamma-1}} = \left(\frac{T_1}{T_2}\right)^{\frac{C_P}{R}} \tag{7.22}$$
and
$$\frac{V_2}{V_1} = \left(\frac{T_1}{T_2}\right)^{\frac{1}{\gamma-1}} = \left(\frac{T_1}{T_2}\right)^{\frac{C_V}{R}} \tag{7.23}$$

7.10 The key ideal gas equations

A summary of the key equations for the main processes associated with ideal gases is shown in Table 7.2.

Table 7.2 Summary of ideal gas processes. This table shows the key thermodynamic variables for a change in state, from state [1] ≡ [P_1, V_1, T_1] to state [2] ≡ [P_2, V_2, T_2], for an ideal gas, along four different reversible paths.

	Isochoric	Isobaric	Isothermal	Adiabatic
Condition	Constant volume $dV = 0$	Constant pressure $dP = 0$	Constant temperature $dT = 0$	No heat exchange $đq = 0$
Equation	$\frac{P}{T} = \frac{nR}{V}$	$\frac{V}{T} = \frac{nR}{P}$	$PV = nRT$	$PV^\gamma = c_{adi}$
Work done by system $\{_1w_2\}$	0	$P\Delta V$ $=$ $nR\Delta T$	$nRT \ln(V_2/V_1)$ $=$ $nRT \ln(P_1/P_2)$	$-nC_V \Delta T$ $=$ $-\frac{1}{\gamma - 1} \Delta(PV)$
Heat gained by system $\{_1q_2\}$	$nC_V \Delta T$	$nC_P \Delta T$	$nRT \ln(V_2/V_1)$ $=$ $nRT \ln(P_1/P_2)$	0
ΔU	$nC_V \Delta T$	$nC_V \Delta T$	0	$nC_V \Delta T$
ΔH	$nC_P \Delta T$	$nC_P \Delta T$	0	$nC_P \Delta T$
Other relationships	$\Delta U = \{_1q_2\}_V$	$\Delta H = \{_1q_2\}_P$	$\{_1q_2\}_T = \{_1w_2\}_T$	$\frac{P_2}{P_1} = \left(\frac{V_1}{V_2}\right)^\gamma$ $= \left(\frac{T_2}{T_1}\right)^{\frac{\gamma}{\gamma-1}}$
	$PV = nRT$ $\gamma = C_P/C_V$ and $C_P - C_V = R$			

7.11 Case Study: The Carnot cycle

7.11.1 The four changes in state that constitute a Carnot cycle

In Chapter 2, on page 37, we introduced the concept of the 'thermodynamic cycle' – a sequence of steps such that the system returns to its original state. Amongst the conclusions drawn were two important results concerning the definitions of state functions and path functions: for any complete thermodynamic cycle, if a function X is a state function, then

$$\oint dX = 0 \tag{2.6}$$

whereas if a function f is a path function, then in general

$$\oint đf \neq 0 \tag{2.8}$$

Equation (2.6) is universally true; the inequality (2.8) has a single exception, for which the cyclic integral will equal zero – the case in which a system changes reversibly from one state to another along a particular path, and then returns to the original state by retracing the original path exactly, but in the opposite direction. This is an extremely rare situation, and so the inequality (2.8) can be taken as a reliable test for a path function.

A thermodynamic cycle of considerable importance is one in which as ideal gas – as might be contained within a cylinder fitted with a frictionless piston – starts in an initial state [1], passes through three intermediate states [2], [3] and [4], and then returns to the original state [1], with the each of the corresponding paths defined as:

- from state [1] to state [2]: reversible isothermal expansion
- from state [2] to state [3]: reversible adiabatic expansion
- from state [3] to state [4]: reversible isothermal compression
- from state [4] to state [1]: reversible adiabatic compression.

This cycle, known as the Carnot cycle, is depicted in the P, V diagram shown in Figure 7.3.

During the initial isothermal expansion, as we saw on page 221, the system does work on, and gains heat from, the surroundings. If the temperature of both the system and surroundings is T_H, then we may designate state [1] as $[P_1, V_1, T_H]$, and state [2] as $[P_2, V_2, T_H]$, where $P_2 < P_1$ and $V_2 > V_1$.

The adiabatic expansion from state [2] to state [3] cools the system down to a colder temperature T_C (see page 222); work is done by the system on the surroundings; there is no heat exchange; and we may designate state [3] as $[P_3, V_3, T_C]$, where $P_3 < P_2$, $V_3 > V_2$ and $T_C < T_H$.

The isothermal compression, at the colder temperature T_C, from state [3] to state [4], requires the surroundings to do work on, and gain heat from, the system. Defining state [4] as $[P_4, V_4, T_C]$, then $P_4 > P_3$ and $V_4 < V_3$.

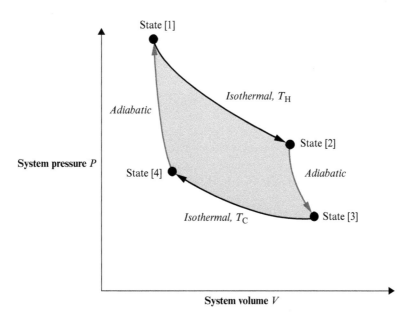

Figure 7.3 The Carnot cycle. Starting in state [1], an ideal gas undergoes a sequence of reversible changes of state, through states [2], [3] and [4], returning to state [1].

During the final step, the adiabatic compression from state [4] back to the initial state [1], the surroundings do work on the system, and, since no heat is exchanged, this causes the temperature of the system to rise from the colder temperature T_C to the hotter temperature T_H. Also, we may relate the state functions associated with states [3] and [4] as $P_1 > P_4$ and $V_1 < V_4$.

Summarising the cycle as a whole:

- Reversible work is done on, or by, the system in each of the four steps, as indicated by the areas between the appropriate line segment shown on the P, V diagram and the horizontal axis …
- …but the total reversible work done by the system on the surroundings is greater than the total reversible work done on the system by the surroundings …
- …so that, for the entire cycle, the net reversible work $\{W\}_{rev}$ done by the system on the surroundings is indicated by the shaded area in the centre of the P, V diagram …
- …from which it is evident that, the greater the temperature difference $T_H - T_C$, and so the wider the gap between the upper and lower isothermals, the greater the amount of net reversible work the system does.
- Heat is exchanged reversibly during only two of the four steps…
- …the first isothermal step, from state [1] to state [2] at a hotter temperature T_H, during which the system absorbs heat $\{_1q_2\}_{rev,H}$ from the surroundings, where $\{_1q_2\}_{rev,H}$ will be a positive quantity, and …

- ...the second isothermal step, from state [3] to state [4] at a colder temperature T_C, during which the system loses heat $\{_3q_4\}_{rev,C}$ to the surroundings, where $\{_3q_4\}_{rev,C}$ will be a negative quantity.

7.11.2 How much reversible work does the system perform on each cycle?

There are two ways to determine the value of $\{W\}_{rev}$, the net reversible work done by the system on the surroundings over a complete cycle. One way is to compute the work associated with each path, using the formulae for isothermal and adiabatic work as shown in Table 7.2, and then adding them together. That process is reliable, but laborious. The second way is to use the First Law

$$dU = đq - đw \qquad (5.1a)$$

as applied to a complete cycle

$$\oint dU = \oint đq - \oint đw$$

recognising that, since the internal energy U is a state function, then

$$\oint dU = 0$$

and so

$$\oint đq = \oint đw \qquad (5.7)$$

Equation (5.7) applies to cycles that are both reversible and irreversible. For the complete Carnot cycle, each step is reversible, and we may equate $\oint đw = \{W\}_{rev}$. Also, since heat is exchanged only during the isothermal steps, equation (5.7) becomes

$$\oint đq_{rev} = \{_1q_2\}_{rev,H} + \{_3q_4\}_{rev,C}$$

Reference to the 'isothermal' column of Table 7.2 gives

$$\{_1q_2\}_{rev,H} = nRT_H \ln \frac{V_2}{V_1} \qquad (7.24)$$

and

$$\{_3q_4\}_{rev,C} = nRT_C \ln \frac{V_4}{V_3} \qquad (7.25)$$

implying that

$$\oint đq_{rev} = nRT_H \ln \frac{V_2}{V_1} + nRT_C \ln \frac{V_4}{V_3} \qquad (7.26)$$

In equation (7.26), $V_2 > V_1$, and so the first term is positive, which is expected, since the system gains heat from the surroundings during the reversible isothermal expansion from state [1] to state [2]. The reversible isothermal compression from state [3] to state [4] implies that $V_4 < V_3$, and so the second term in equation (7.26) is negative, as is also expected since the system loses heat to the surroundings. Equation (7.26) is therefore more naturally written as

$$\oint đq_{rev} = nRT_H \ln \frac{V_2}{V_1} - nRT_C \ln \frac{V_3}{V_4}$$

and so the total net reversible work $\{W\}_{rev}$ done by the system on the surroundings for a complete cycle is given by

$$\oint đw_{rev} = nRT_H \ln \frac{V_2}{V_1} - nRT_C \ln \frac{V_3}{V_4} = \{W\}_{rev} \qquad (7.27)$$

7.11.3 An important relationship between the state volumes

Equation (7.27) expresses the total net reversible work $\{W\}_{rev}$ done by the system on the surroundings for a complete cycle on terms of the ratios V_2/V_1 and V_3/V_4 of the volumes associated with each 'end' of each isothermal cycle.

Our purpose now is to establish an important relationship between these ratios, which emerges from the facts that the volumes V_1, V_2, V_3 and V_4 are not just arbitrary numbers: rather, V_1 and V_2 are 'linked' by virtue of the 'hot' isothermal expansion, and V_3 and V_4 are linked by virtue of the 'cold' isothermal compression; furthermore, V_2 and V_3 are linked by the adiabatic expansion, and V_4 and V_1 by the adiabatic compression. The mathematics is about to get a little clumsy, but each step is very straight-forward, and based on principles that are already familiar.

For the reversible isothermal expansion

$$P_1 V_1 = P_2 V_2$$

and so

$$\frac{P_2}{P_1} = \frac{V_1}{V_2}$$

For the reversible isothermal compression

$$P_3 V_3 = P_4 V_4$$

and so

$$\frac{P_3}{P_4} = \frac{V_4}{V_3}$$

For the two adiabatic paths

$$P_2 V_2^{\gamma} = P_3 V_3^{\gamma}$$

and

$$P_1 V_1^{\gamma} = P_4 V_4^{\gamma}$$

where γ, as usual, is the ratio C_P/C_V.

If we divide the first adiabatic equation by the second, we obtain

$$\frac{P_2}{P_1} \left(\frac{V_2}{V_1}\right)^{\gamma} = \frac{P_3}{P_4} \left(\frac{V_3}{V_4}\right)^{\gamma}$$

from which the ratios between the pressures can be eliminated using the isothermal equations as

$$\frac{V_1}{V_2}\left(\frac{V_2}{V_1}\right)^\gamma = \frac{V_4}{V_3}\left(\frac{V_3}{V_4}\right)^\gamma$$

giving

$$\left(\frac{V_2}{V_1}\right)^{\gamma-1} = \left(\frac{V_3}{V_4}\right)^{\gamma-1} \tag{7.28}$$

and so

$$\frac{V_2}{V_1} = \frac{V_3}{V_4}$$

Equation (7.28) is a rather unexpected result, which allows us to simplify expression (7.27) for the total net reversible work $\{W\}_{\text{rev}}$ as

$$\{W\}_{\text{rev}} = nR(T_H - T_C)\ln\frac{V_2}{V_1} = nR(T_H - T_C)\ln\frac{V_3}{V_4} \tag{7.29}$$

7.11.4 How much work do we get for a given quantity of heat?

Each complete cycle returns the system to its original state, so the cycle can continue indefinitely. And for each complete cycle, the system performs net reversible work $\{W\}_{\text{rev}}$ on the surroundings. For this to happen, the system absorbs heat $\{_1q_2\}_{\text{rev, H}}$ from the surroundings during the first reversible isothermal expansion at the hotter temperature T_H, and discards heat $\{_3q_4\}_{\text{rev, C}}$ to the surroundings at the colder temperature T_C.

By the First Law

$$\{_1q_2\}_{\text{rev, H}} = \{W\}_{\text{rev}} + \{_3q_4\}_{\text{rev, C}}$$

and so the heat $\{_1q_2\}_{\text{rev, H}}$ absorbed at the hotter temperature T_H is the fundamental source of the net work $\{W\}_{\text{rev}}$. If we are designing a system to do as much work as possible, then, for any given heat input $\{_1q_2\}_{\text{rev, H}}$, we would wish the corresponding value of $\{W\}_{\text{rev}}$ to be as large as possible, with $\{_3q_4\}_{\text{rev, C}}$ as small as possible.

One measure of the extent to which a given amount of heat $\{_1q_2\}_{\text{rev, H}}$ is converted to work $\{W\}_{\text{rev}}$, rather than being discarded as heat $\{_3q_4\}_{\text{rev, C}}$, is the ratio η, defined as

$$\eta = \frac{\{W\}_{\text{rev}}}{\{_1q_2\}_{\text{rev}}} \quad \text{or} \quad \eta = \frac{\{W\}_{\text{rev}}}{\{_1q_2\}_{\text{rev}}} \times 100 \tag{7.30}$$

η represents the system's **efficiency** at converting heat into work, numerically expressed as the fraction, or percentage, of the reversible heat input $\{_1q_2\}_{\text{rev, H}}$ converted into reversible work $\{W\}_{\text{rev}}$. The maximum possible value of η is 1, or 100%, when all the heat is converted into work, and the minimum value is 0, or 0%, when none of the heat is converted.

Now

$$\{W\}_{\text{rev}} = nR(T_H - T_C)\ln\frac{V_2}{V_1} = nR(T_H - T_C)\ln\frac{V_3}{V_4} \tag{7.29}$$

and Table 7.2 tells us that, for the reversible isothermal expansion from state [1] to state [2]

$$\{_1q_2\}_{\text{rev, H}} = nRT_H \ln \frac{V_2}{V_1} \tag{7.24}$$

and so

$$\eta = \frac{T_H - T_C}{T_H} \tag{7.31}$$

Equation (7.31) shows that the efficiency of a system undergoing a Carnot cycle depends only on the temperatures of the two heat reservoirs – one at a hotter temperature T_H, and one at a colder temperature T_C – between which it operates. One inference from equation (7.28) is that the greater the temperature difference $T_H - T_C$, the greater the efficiency; and a second is that the system can only achieve the theoretical maximum efficiency of 1, or 100%, when T_C is zero, the absolute zero of temperature.

7.11.5 A new state function

There is one further important result to derive. Our starting point is equation (7.24) for the reversible heat $\{_1q_2\}_{\text{rev, H}}$ gained by the system during the isothermal expansion

$$\{_1q_2\}_{\text{rev, H}} = nRT_H \ln \frac{V_2}{V_1} \tag{7.24}$$

implying that

$$\frac{\{_1q_2\}_{\text{rev, H}}}{T_H} = nR \ln \frac{V_2}{V_1} \tag{7.32}$$

Likewise, for the reversible heat $\{_3q_4\}_{\text{rev, C}}$ lost from the system during the isothermal compression

$$\{_3q_4\}_{\text{rev, C}} = nRT_C \ln \frac{V_4}{V_3} \tag{7.25}$$

and so

$$\frac{\{_3q_4\}_{\text{rev, C}}}{T_L} = nR \ln \frac{V_4}{V_3} \tag{7.33}$$

Adding equations (7.32) and (7.33)

$$\frac{\{_1q_2\}_{\text{rev, H}}}{T_H} + \frac{\{_3q_4\}_{\text{rev, C}}}{T_C} = nR \ln \frac{V_2}{V_1} + nR \ln \frac{V_4}{V_3} = nR \ln \left(\frac{V_2 V_4}{V_1 V_3} \right) \tag{7.34}$$

But

$$\frac{V_2}{V_1} = \frac{V_3}{V_4} \tag{7.28}$$

implying that $(V_2 V_4 / V_1 V_3) = 1$ and that $\ln(V_2 V_4 / V_1 V_3) = 0$.

Equation (7.34) then becomes

$$\frac{\{_1q_2\}_{\text{rev, H}}}{T_H} + \frac{\{_3q_4\}_{\text{rev, C}}}{T_C} = 0 \tag{7.35}$$

Now, $\{_1q_2\}_{rev,H}$ is the reversible heat exchanged between the system and the surroundings at temperature T_H, and $\{_3q_4\}_{rev,C}$ is the reversible heat exchanged between the system and the surroundings at temperature T_C. Both of these terms are of the general form

$$\frac{\text{Quantity of reversible heat exchanged between the system and the surroundings}}{\text{Temperature at which that heat is exchanged}}$$

so, if the heat exchanged at any temperature is infinitesimally small, each term in equation (7.35) is of the mathematical form

$$\frac{đq_{rev}}{T}$$

For a complete Carnot cycle, the only heat flows are $\{_1q_2\}_{rev,H}$ and $\{_3q_4\}_{rev,C}$, and so equation (7.35) represents a total for the complete cycle. This can be expressed as a cyclic integral

$$\oint \frac{đq_{rev}}{T} = \frac{\{_1q_2\}_{rev,H}}{T_H} + \frac{\{_3q_4\}_{rev,C}}{T_C} = 0$$

or more simply

$$\oint \frac{đq_{rev}}{T} = 0 \tag{7.36}$$

As we saw at the start of this section, any function X is a state function if

$$\oint dX = 0 \tag{2.6}$$

We also noted that the cyclic integral of a path function can be zero, under the exceptional circumstance that the cycle consists only of a sequence of paths such that the 'return' paths retrace the 'outward' paths in their totality. A Carnot cycle does not do this, and so equation (7.33) has identified a new state function, expressed as $đq_{rev}/T$ – a story which we will continue in Chapter 9.

7.12 Case Study: The thermodynamic pendulum

This final section of this chapter presents an extended example utilising much of the content of this chapter, and also the latter part of Chapter 5, which explored the effect of friction in real systems. The concept here is that of the 'thermodynamic pendulum' – a piston that oscillates to-and-fro in a closed system, just as a mechanical pendulum swings to-and-fro under a clock.

7.12.1 Simple harmonic motion

A mechanical pendulum comprises a mass m kg, suspended by a string of length l m from a fixed point. If the mass is stationary, the string will be perfectly vertical, and the system is at rest, in equilibrium. Suppose, however, that an external agent (such as a person) displaces the mass to the right, so that the string is no longer vertical but at an angle θ_{max} to the vertical. If the agent continues to hold the mass, the system will remain at the non-equilibrium angle θ_{max}, but if the agent lets go, the mass will fall, and, constrained by the string, will move in an arc. But rather than stopping at the equilibrium position, where the angle of displacement θ is zero, the momentum of the moving mass will cause the mass to swing through the equilibrium position, and continue moving to the left. When the mass reaches the point at which the angle to the vertical is $-\theta_{max}$ (where the negative sign indicates that this angle is to the left of the vertical), the mass will stop momentarily, and then swing back to the right, through the equilibrium position, returning to its originally displaced position angle θ_{max}. The cycle will then start again, and the mass will swing to and fro. If the system is perfectly frictionless – which implies no friction at the point at which the string is suspended, and no friction as the mass moves through the air – then this to-and-fro motion will last for ever.

This is an example of *simple harmonic motion*, a form of motion in which a mass m oscillates to and fro about an equilibrium position. Mathematically, this motion can be described by an equation of the form

$$x(\tau) = x_{max} \cos 2\pi f \tau \tag{7.37}$$

where $x(\tau)$ is the displacement of the mass m from its equilibrium position at time τ, x_{max} is the maximum displacement, and f is the frequency of the oscillation – the number of complete oscillations (right to left and back to the right again) per second. The frequency f is related to the periodic time τ_{cycle} for one complete cycle of oscillation as

$$\tau_{cycle} = \frac{1}{f} \tag{7.38}$$

and to a variable usually written as ω – also somewhat confusingly known as the frequency – by the equation

$$\omega = 2\pi f$$

implying that equation (7.37) becomes

$$x(\tau) = x_{max} \cos \omega \tau \tag{7.39}$$

Equation (7.39) is important, for it not only tells us the position $x(\tau)$ of the mass m at any time τ, and the velocity at any time τ

$$\frac{dx}{d\tau} = -x_{max} \omega \sin \omega \tau$$

but also the acceleration at any time τ

$$\frac{d^2x}{d\tau^2} = -x_{max} \omega^2 \cos \omega \tau$$

Furthermore, equation (7.39)

$$x(\tau) = x_{max} \cos \omega \tau$$

implies that

$$\frac{d^2x}{d\tau^2} = -\omega^2 x \tag{7.40}$$

Equation (7.40) is an expression for the acceleration of a mass m exhibiting simple harmonic motion. But we know another equation for acceleration – Newton's Second

Law of Motion, which states that a **force** F, acting on a mass m, results in an acceleration $d^2x/d\tau^2$ as

$$F = m \frac{d^2x}{d\tau^2}$$

from which

$$\frac{d^2x}{d\tau^2} = \frac{F}{m}$$

In the specific case of simple harmonic motion, we may therefore write

$$\frac{F}{m} = -\omega^2 x$$

giving

$$F = -m\omega^2 x$$

If the mass m of the system is constant, and recognising that the frequency ω of simple harmonic motion is also constant, we may then write

$$F = -kx \tag{7.41}$$

where k is a constant such that

$$k = m\omega^2$$

from which

$$\omega = \sqrt{\frac{k}{m}} \tag{7.42}$$

Equations (7.39), (7.40) and (7.41)

$$x(\tau) = x_{max} \cos \omega\tau = x_{max} \cos \sqrt{\frac{k}{m}}\tau \tag{7.39}$$

$$\frac{d^2x}{d\tau^2} = -\omega^2 x = -\frac{k}{m}x \tag{7.40}$$

$$F = -kx = m\frac{d^2x}{d\tau^2} \tag{7.41}$$

in which k, ω, m and x_{max} are constants, are different, equivalent, mathematical descriptions of a system in which the magnitude of a force F, acting on a mass m, is proportional to the (negative of the) displacement $x(\tau)$ of the mass from its equilibrium position, resulting in simple harmonic motion. It turns out that this is a widespread phenomenon – the oscillation of a pendulum is one example, others are the bobbing up-and-down of a mass on a spring, the vibrations of a musical tuning fork, and the behaviour of many electrical circuits.

The importance of this discussion is that if a study of the forces operating on a system shows that it is possible to express the time-dependent behaviour of that system in terms of an equation of the form (7.40) or (7.41), then the system will oscillate with simple harmonic motion according to equation (7.37); and, *vice versa*, if we observe a system oscillating with simple harmonic motion, then we may infer that the forces are acting in accordance with equations

(7.40) and (7.41). This is what we are about to demonstrate for a thermodynamic system, the 'thermodynamic pendulum'…

…but just before we do that, we need to explore one other aspect of a frictionless pendulum – what happens to the energy of the system.

So, consider the pendulum displaced to its maximum extent, and at an angle θ_{max} to the vertical, the mass m is at a vertical height $l\,(1 - \cos\theta_{max})$ above its equilibrium position. At this point, the mass m has gravitational potential energy (relative to the equilibrium position) of $mgl(1 - \cos\theta_{max})$, where g is the acceleration attributable to gravity – energy that was originally introduced into the system by the agent that displaced the pendulum away from its equilibrium position at the outset. When the mass is released, it swings in an arc, and loses height. The resultant loss in gravitational potential energy corresponds to an increase in the kinetic energy of the mass as it moves faster, and as the mass moves through the equilibrium position, the gravitational potential energy reduces to zero, and all the system's energy is kinetic. Throughout the motion – provided that it is perfectly frictionless – the total energy of the system, the sum of the gravitational potential energy and the kinetic energy, remains constant, as required by the principle of conservation of energy, and the First Law of Thermodynamics.

7.12.2 The thermodynamic pendulum – oscillation

Let's now explore a rather different example of simple harmonic motion – the 'thermodynamic pendulum'.

Figure 7.4(a) shows a perfectly adiabatic cylinder, which contains a perfectly fitting piston of mass m kg, at its mid-point x_0 from the left-hand end of the cylinder, and n mol of an ideal gas on each side of the piston. The whole system is in equilibrium, with the piston at rest, and with the gas on each side of the piston of the cylinder having a pressure P_0, temperature T_0, volume V_0, and internal energy U_0, implying that the total internal energy of the gas in the entire system is $2U_0$.

Suppose that the piston is made of metal, and is displaced, frictionlessly, to the right to a position x_{max} by using a magnet, held outside the cylinder. To move the piston in this way will require some energy, for work has to be done by the agent moving the magnet to compress the gas on the right-hand side.

As shown in Figure 7.4, when the piston is displaced to the right, the gas to the right is compressed, whilst the gas to the left has expanded. As a consequence, the pressure P_R to the right is increased above the original equilibrium pressure P_0, whilst the pressure P_L to the left is reduced below the original equilibrium pressure P_0. The resulting pressure difference $P_R - P_L$ across the piston therefore gives rise to a force on the piston from right to left. At the outset, the piston is displaced by the external magnet to its extreme position x_{max}, and the pressure difference $P_R - P_L$ across the piston is at a maximum $(P_R - P_L)_{max}$. Whilst the external magnet is held in place, the corresponding force is resisted, and the piston will remain at its extreme position; but if the magnet is removed, the force attributable to the pressure difference will cause the piston to move from right to left in Figure 7.4. Since the piston has a mass m, the piston gains momentum as it moves from right to left towards the equilibrium position x_0, and – because of this momentum – the piston

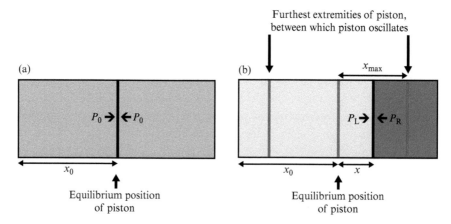

Figure 7.4 **The thermodynamic pendulum.** Figure 7.4(a) shows a cylinder, of total length $2x_0$, with a piston of mass m kg situated at rest at its mid-point. Both halves of the cylinder contain equal volumes V_0 of n mol of an ideal gas. The system is in equilibrium, such that the temperature of the gas on both sides of the piston is T_0 K, and the pressure on both sides of the piston is P_0 Pa. The system is perfectly adiabatic, so no heat can flow through the cylinder walls, or through the piston.
Figure 7.4(b) shows the piston displaced to the right by a distance x from the mid-point. The gas to the right has been compressed, and exerts an instantaneous pressure P_R from right to left in the diagram, whilst the gas to the left has expanded, exerting a pressure P_L, from left to right in the diagram. With the position of the piston to the right as shown on the diagram, since $P_R > P_L$, there is a net force on the piston from right to left, causing the piston to accelerate to the left. As it does so, the gas to the left becomes compressed, whilst the gas to the right expands. Assuming that the maximum displacement x_{max} of the piston, on either side of the central equilibrium position, is relatively small, then, if the piston is perfectly frictionless, the piston will oscillate between the two extreme positions (on the right and left, respectively), just like a pendulum.

does not stop at the equilibrium position, but continues to move to the left. However, as the piston moves leftwards past the equilibrium position, the gas to the left becomes progressively more compressed, and the gas to the right more expanded, so creating a steadily increasing pressure difference from left to right, in opposition to the motion of the piston. The piston therefore slows down, until it eventually, instantaneously, stops, at which point it will begin to move from left to right. If we assume that there is no friction, the piston will oscillate indefinitely, just like a perfect pendulum. Intuitively, this is readily understood. Here is the analysis that proves it.

Consider the forces on the piston when it is positioned at an arbitrary distance x from the mid-point. Suppose that the area of the cylinder is A_0, and that the pressure on the piston from the (compressed) ideal gas on the right-hand side is P_R. This exerts a force on the piston of $P_R A_0$ N, from right to left in the diagram. As the same time, the (expanded) ideal gas on the left-hand side has a pressure P_L, exerting a force on the piston of $P_L A_0$ N, from left to right in the diagram. In the absence of friction, the net force $F_{R \to L}$, from right to left in the diagram is given by

$$F_{R \to L} = P_R A_0 - P_L A_0 = (P_R - P_L) A_0 \tag{7.43}$$

The system is perfectly adiabatic, and so, if the volumes of the gas on each side of the piston are V_L and V_R, then, by reference to the original pressures P_0 and volumes V_0, we may write

CASE STUDY: THE THERMODYNAMIC PENDULUM

$$P_0 V_0^\gamma = P_L V_L^\gamma = P_R V_R^\gamma$$

and so equation (7.43) becomes

$$F_{R \to L} = \left[\frac{P_0 V_0^\gamma}{V_R^\gamma} - \frac{P_0 V_0^\gamma}{V_L^\gamma} \right] A_0 = P_0 A_0 \left[\frac{V_0^\gamma}{V_R^\gamma} - \frac{V_0^\gamma}{V_L^\gamma} \right] \qquad (7.44)$$

Now, with reference to Figure 7.4, we know that

$$V_0 = A_0 x_0$$

so we can replace A_0 in equation (7.30) by V_0/x_0, giving

$$F_{R \to L} = \frac{P_0 V_0}{x_0} \left[\frac{V_0^\gamma}{V_R^\gamma} - \frac{V_0^\gamma}{V_L^\gamma} \right]$$

We also know that the gas within the cylinder is ideal, so that

$$P_0 V_0 = nRT_0$$

implying that

$$F_{R \to L} = \frac{nRT_0}{x_0} \left[\frac{V_0^\gamma}{V_R^\gamma} - \frac{V_0^\gamma}{V_L^\gamma} \right] \qquad (7.45)$$

To simplify the term in square brackets, we note from Figure (7.4) that

$$V_L = A_0 (x_0 + x)$$
$$V_R = A_0 (x_0 - x)$$

and

$$V_0 = A_0 x_0$$

implying that we may write equation (7.45) as

$$F_{R \to L} = \frac{nRT_0}{x_0} \left[\frac{x_0^\gamma}{(x_0 - x)^\gamma} - \frac{x_0^\gamma}{(x_0 + x)^\gamma} \right]$$

Now

$$\frac{x_0^\gamma}{(x_0 + x)^\gamma} = \frac{1}{(1 + x/x_0)^\gamma} = \left(1 + \frac{x}{x_0}\right)^{-\gamma}$$

hence

$$F_{R \to L} = \frac{nRT_0}{x_0} \left[\left(1 - \frac{x}{x_0}\right)^{-\gamma} - \left(1 + \frac{x}{x_0}\right)^{-\gamma} \right]$$

If x is small compared to x_0, then the binomial theorem tells us that, to a good approximation

$$\left(1 + \frac{x}{x_0}\right)^{-\gamma} = 1 - \gamma \frac{x}{x_0}$$

and

$$\left(1 - \frac{x}{x_0}\right)^{-\gamma} = 1 + \gamma \frac{x}{x_0}$$

implying that

$$\left(1 - \frac{x}{x_0}\right)^{-\gamma} - \left(1 + \frac{x}{x_0}\right)^{-\gamma} = 2\gamma \frac{x}{x_0}$$

$F_{R \to L}$ may therefore be expressed as

$$F_{R \to L} = \frac{nRT_0}{x_0} \left(2\gamma \frac{x}{x_0}\right)$$

from which

$$F_{R \to L} = \frac{2\gamma nRT_0}{x_0^2} x \tag{7.46}$$

Now, in equation (7.46), γ, n, R, T_0 and x_0, are all constants, and so if we define the constant k as

$$k = \frac{2\gamma nRT_0}{x_0^2} \tag{7.47}$$

then equation (7.46) becomes

$$F_{R \to L} = k x \tag{7.48}$$

Now, the force $F_{R \to L}$ imparts an acceleration from *right* to *left*, which is the direction in which the displacement x decreases. Mathematically, we need to determine the force $F_{L \to R}$ which acts in the direction in which x increases, as given by

$$F_{L \to R} = -F_{R \to L} = -k x \tag{7.49}$$

Since, according to Newton's Second Law of Motion

Force = mass × acceleration

then, for a piston of mass m, if we express the acceleration as the second order derivative of distance, x, with respect to time τ, we have

$$F_{L \to R} = -kx = m \frac{d^2 x}{d\tau^2}$$

from which

$$\frac{d^2 x}{d\tau^2} = -\frac{k}{m} x \tag{7.50}$$

Equation (7.50) is the standard differential equation for simple harmonic motion, such as the motion of a pendulum, or that of a weight bobbing up and down on the end of spring. In the current case, the piston in the cylinder will oscillate to-and-fro, according to equation (7.39),

$$x(\tau) = x_{max} \cos \omega \tau \tag{7.39}$$

which gives the position $x(\tau)$ of the piston at any time τ. In this equation, x_{max} is the maximum extremity of the oscillation, the frequency of oscillation, ω, is given by

$$\omega^2 = \frac{k}{m} = \frac{2\gamma nRT_0}{mx_0^2} \tag{7.51}$$

CASE STUDY: THE THERMODYNAMIC PENDULUM

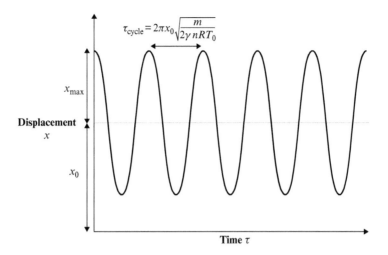

Figure 7.5 The adiabatic pendulum. If a piston of the mass m is displaced by a distance x_{max} from its equilibrium position x_0 within a frictionless adiabatic cylinder, it will oscillate indefinitely. The pressures, volumes and temperatures on each side of the piston will oscillate with the same frequency. Throughout the motion, the sum of the internal energies of the gases on both sides of the piston, and the kinetic energy of the piston, is a constant, in accordance with the First Law.

and, and as shown in Figure 7.5, the time τ_{cycle} for one complete cycle of oscillation is

$$\tau_{cycle} = 2\pi x_0 \sqrt{\frac{m}{2\gamma nRT_0}}$$

7.12.3 The thermodynamic pendulum – energy

We now determine the energy of the system, and how the energy changes with time. In this system, when the piston is displaced by a distance x from the equilibrium position x_0, the total energy of the system is the sum of the internal energies U_L and U_R of the gases on each side of the piston, plus the kinetic energy U_K of the piston itself.

Taking firstly the kinetic energy U_K of the piston, this is given by

$$U_K = \frac{1}{2} m v^2 \tag{7.52}$$

where m is the mass of the piston, and v is the velocity of the piston at position x. The velocity v can be determined by differentiating equation (7.39) with respect to time τ as

$$v = \frac{dx}{d\tau} = -\omega x_{max} \sin \omega \tau$$

hence

$$U_K = \frac{1}{2} m \omega^2 x_{max}^2 \sin^2 \omega \tau$$

But

$$\omega^2 = \frac{2\gamma nRT_0}{mx_0^2} \tag{7.51}$$

and so

$$U_K = \frac{\gamma nRT_0}{x_0^2} x_{max}^2 \sin^2 \omega\tau \tag{7.53}$$

To calculate the internal energies U_L and U_R of the gases on each side of the piston, we can use the First Law

$$dU = đq - đw \tag{5.1a}$$

remembering that the system is perfectly adiabatic, implying that $đq$ is zero throughout. If we consider firstly the gas to the right of the piston, when the piston has been displaced a distance x to the right, this gas has been compressed from an initial volume $V_0 = A_0 x_0$ to a volume $V_R = A_0 (x_0 - x)$. During this compression, the piston moves frictionlessly to the right, and does adiabatic work $đw_R$ on the gas on the right. If the system is acting reversibly, then we may write

$$dU_R = - đw_R = - P_R \, dV_R$$

Since the work is reversible and adiabatic, we know that, throughout the change, the simultaneous values of the pressure P and volume V are such that

$$P_0 V_0^\gamma = PV^\gamma$$

implying that

$$dU_R = - P_0 V_0^\gamma \frac{dV}{V^\gamma}$$

For a finite change, and assuming that the initial internal energy of the gas on the right hand side of the cylinder is U_0, then

$$\Delta U_R = U_R - U_0 = - P_0 V_0^\gamma \int_{V_0}^{V_R} \frac{dV}{V^\gamma} = - P_0 V_0^\gamma \int_{V_0}^{V_R} V^{-\gamma} dV$$

and so

$$U_R = U_0 - P_0 V_0^\gamma \left[\frac{V_R^{-\gamma+1} - V_0^{-\gamma+1}}{-\gamma + 1} \right]$$

Since $\gamma > 1$, this expression is more informatively written as

$$U_R = U_0 + \frac{P_0 V_0^\gamma}{(\gamma - 1)} \left[\frac{1}{V_R^{\gamma-1}} - \frac{1}{V_0^{\gamma-1}} \right] \tag{7.54}$$

Since $V_0 > V_R$, the expression in square brackets in equation (7.54) is necessarily positive, implying that $U_R > U_0$. The temperature of the gas on the right has therefore increased, as is consistent with an adiabatic compression.

We now calculate the corresponding change in internal energy ΔU_L for the gas in the left-hand side of the cylinder, which expands from a volume V_0 to a volume V_L. Since the change is adiabatic, $đq = 0$, and, as before, we have

$$\Delta U_L = U_L - U_0 = - P_0 V_0^\gamma \int_{V_0}^{V_L} \frac{dV}{V^\gamma} = - P_0 V_0^\gamma \int_{V_0}^{V_L} V^{-\gamma} dV$$

giving

$$U_L = U_0 - P_0 V_0^\gamma \left[\frac{V_L^{-\gamma+1} - V_0^{-\gamma+1}}{-\gamma + 1} \right]$$

and

$$U_L = U_0 + \frac{P_0 V_0^\gamma}{(\gamma - 1)} \left[\frac{1}{V_L^{\gamma-1}} - \frac{1}{V_0^{\gamma-1}} \right] \tag{7.55}$$

In equation (7.55), $V_0 < V_L$, and so the expression in square brackets in equation (7.55) is necessarily negative, implying that $U_L < U_0$. The temperature of the gas on the left has therefore decreased, as is consistent with an adiabatic expansion.

To find the total internal energy U_G for the gas, we now add equations (7.54) and (7.55):

$$U_G = U_L + U_R = U_0 + \frac{P_0 V_0^\gamma}{(\gamma - 1)} \left[\frac{1}{V_L^{\gamma-1}} - \frac{1}{V_0^{\gamma-1}} \right] + U_0 + \frac{P_0 V_0^\gamma}{(\gamma - 1)} \left[\frac{1}{V_R^{\gamma-1}} - \frac{1}{V_0^{\gamma-1}} \right]$$

$$= 2U_0 + \frac{P_0 V_0^\gamma}{(\gamma - 1)} \left[\frac{1}{V_L^{\gamma-1}} + \frac{1}{V_R^{\gamma-1}} - \frac{2}{V_0^{\gamma-1}} \right]$$

$$= 2U_0 + \frac{P_0 V_0^\gamma}{(\gamma - 1)} \left[V_L^{-(\gamma-1)} + V_R^{-(\gamma-1)} - 2 V_0^{-(\gamma-1)} \right] \tag{7.56}$$

The term in square brackets in equation (7.56) can be simplified by using the relationships

$$V_L = A_0 (x_0 + x) = A_0 x_0 \left(1 + \frac{x}{x_0}\right) = V_0 \left(1 + \frac{x}{x_0}\right)$$

and

$$V_R = A_0 (x_0 - x) = A_0 x_0 \left(1 - \frac{x}{x_0}\right) = V_0 \left(1 - \frac{x}{x_0}\right)$$

Equation (7.56) therefore becomes

$$U_G = 2U_0 + \frac{P_0 V_0^\gamma}{(\gamma - 1)} V_0^{-(\gamma-1)} \left[\left(1 + \frac{x}{x_0}\right)^{-(\gamma-1)} + \left(1 - \frac{x}{x_0}\right)^{-(\gamma-1)} - 2 \right]$$

$$= 2U_0 + \frac{P_0 V_0}{(\gamma - 1)} \left[\left(1 + \frac{x}{x_0}\right)^{-(\gamma-1)} + \left(1 - \frac{x}{x_0}\right)^{-(\gamma-1)} - 2 \right]$$

$$= 2U_0 + \frac{nRT_0}{(\gamma - 1)} \left[\left(1 + \frac{x}{x_0}\right)^{-(\gamma-1)} + \left(1 - \frac{x}{x_0}\right)^{-(\gamma-1)} - 2 \right] \tag{7.57}$$

If, for convenience, we represent $-(\gamma - 1)$ as α, then we may use the binomial theorem as

$$\left(1 + \frac{x}{x_0}\right)^{-(\gamma-1)} = \left(1 + \frac{x}{x_0}\right)^\alpha = 1 + \alpha \frac{x}{x_0} + \frac{\alpha(\alpha - 1)}{2} \left(\frac{x}{x_0}\right)^2 + \ldots$$

and

$$\left(1 - \frac{x}{x_0}\right)^{-(\gamma-1)} = \left(1 - \frac{x}{x_0}\right)^{\alpha} = 1 - \alpha \frac{x}{x_0} + \frac{\alpha(\alpha-1)}{2}\left(\frac{x}{x_0}\right)^2 + \ldots$$

If we assume that x is considerably smaller than x_0, then, to a very good approximation

$$\left[\left(1 + \frac{x}{x_0}\right)^{-(\gamma-1)} + \left(1 - \frac{x}{x_0}\right)^{-(\gamma-1)} - 2\right] = \alpha(\alpha-1)\left(\frac{x}{x_0}\right)^2$$

But since

$$\alpha = -(\gamma - 1)$$

then

$$\alpha(\alpha - 1) = \gamma(\gamma - 1)$$

and so

$$\left[\left(1 + \frac{x}{x_0}\right)^{-(\gamma-1)} + \left(1 - \frac{x}{x_0}\right)^{-(\gamma-1)} - 2\right] = \gamma(\gamma-1)\left(\frac{x}{x_0}\right)^2$$

Equation (7.57) therefore becomes

$$U_G = 2U_0 + \frac{nRT_0}{(\gamma-1)} \gamma(\gamma-1)\left(\frac{x}{x_0}\right)^2$$

from which

$$U_G = 2U_0 + \frac{\gamma nRT_0}{x_0^2} x^2 \tag{7.58}$$

Equation (7.58) defines how the total internal energy U_G of the gas varies as the piston moves to-and-fro, as measured by the displacement parameter x. When the piston is at its maximum displacement, $x = x_{max}$, at which point

$$U_G = 2U_0 + \frac{\gamma nRT_0}{x_0^2} x_{max}^2 \tag{7.59}$$

Now, when the system was initially in equilibrium, with the piston at rest at the central position x_0, the gas on each side of the piston was associated with an internal energy equal to U_0, so the total internal energy of the system was $2U_0$. In order to start the oscillation, the piston had to be displaced (perhaps by a magnet) to its extremity x_{max} to the right, and this required the introduction of energy, in order to push against the gas on the right-hand side. Equation (7.59) tells us that the amount of energy that had to be introduced into the system to achieve this is the second term

$$\text{Energy introduced into the system to start the oscillation} = \frac{\gamma nRT_0}{x_0^2} x_{max}^2$$

Equation (7.58) tells us how the total internal energy of the gas U_G in the system varies with the displacement parameter x. Since we also know that x varies with time according to equation (7.39)

$$x(\tau) = x_{max} \cos \omega \tau \tag{7.39}$$

then the total internal energy U_G of the gas will vary with time t as

$$U_G = 2U_0 + \frac{\gamma n R T_0}{x_0^2} x_{max}^2 \cos^2 \omega \tau \tag{7.60}$$

Now equation (7.53) defined the kinetic energy U_K at time τ of the piston as

$$U_K = \frac{\gamma n R T_0}{x_0^2} x_{max}^2 \sin^2 \omega \tau \tag{7.53}$$

implying that the total energy U of the whole system as time τ is

$$U = U_K + U_G$$

and so

$$U = \frac{\gamma n R T_0}{x_0^2} x_{max}^2 \sin^2 \omega \tau + 2U_0 + \frac{\gamma n R T_0}{x_0^2} x_{max}^2 \cos^2 \omega \tau$$

$$= 2U_0 + \frac{\gamma n R T_0}{x_0^2} x_{max}^2 \sin^2 \omega \tau + \frac{\gamma n R T_0}{x_0^2} x_{max}^2 \cos^2 \omega \tau$$

$$= 2U_0 + \frac{\gamma n R T_0}{x_0^2} x_{max}^2 \left(\sin^2 \omega \tau + \cos^2 \omega \tau \right)$$

But since

$$\sin^2 \omega \tau + \cos^2 \omega \tau = 1$$

for all values of τ, this implies that

$$U = 2U_0 + \frac{\gamma n R T_0}{x_0^2} x_{max}^2 \tag{7.61}$$

Now, every term in equation (7.61) is constant, and there is no dependence on time τ. The total energy of the system is therefore constant throughout time. Furthermore, the structure of equation (7.61) is expressed as the sum of two terms. The first term, $2U_0$, represents the total internal energy U of the gas when the system is in equilibrium, with the piston at rest at position x_0, and with the gas on each side associated with an internal energy U_0. And, as we have seen, the second term represents the energy introduced into the system by moving the piston from its equilibrium position x_0 to its maximum displacement x_{max}, so enabling the oscillations to start. But once the oscillations have started, they go on for ever.

7.12.4 Introducing friction

The discussion so far has assumed that the to-and-fro motion of the piston is frictionless. If we now introduce dynamic friction (see page 128), then this breaks the conditions for reversibility (see page 134), implying that the system is now behaving irreversibly. If we further assume that all other conditions remain unchanged, then, despite the fact that the system is now irreversible, it is still quasistatic (see page 30), so enabling us to continue to use equations such as

$$PV^\gamma = constant = c_{adi}$$

to calculate the work associated with compressing and expanding the gases in the cylinder, and the corresponding changes in the system's internal energy.

Mathematically, however, the introduction of dynamic friction requires us to modify equation (7.50) for simple harmonic motion

$$\frac{d^2x}{d\tau^2} = -\frac{k}{m}x \qquad (7.50)$$

to

$$\frac{d^2x}{d\tau^2} = -\frac{\varphi}{m}\frac{dx}{d\tau} - \frac{k}{m}x \qquad (7.62)$$

In equation (7.62), the additional term represents dynamic friction as a force proportional to the velocity $dx/d\tau$ of the piston (see page 133), with the negative sign indicating that this force always opposes the motion. The parameter φ represents the strength of the frictional force, such that the more 'sticky' the friction, the greater the value of φ, and the stronger the force resisting the motion. Also, the inverse proportionality to the mass m captures the fact that the heavier the mass, the lower the corresponding acceleration attributable to any given force. This all makes intuitive sense:

- if the piston is not moving, then the velocity $dx/d\tau$ is zero, and no frictional force is evident
- dynamic friction takes effect only as the piston moves, and always acts against the motion – if the piston is moving to the right, $dx/d\tau$ is a positive quantity and so $(-\varphi/m)(dx/d\tau)$ is negative, acting to the left, but if the if the piston is moving to the left, $dx/d\tau$ is a negative quantity and so $(-\varphi/m)(dx/d\tau)$ is positive, acting to the right
- and if there is no friction at all, φ is zero, and equation (7.62) becomes (7.50), the equation for 'pure' simple harmonic motion.

Reference to a standard mathematics text will verify that if, as is often the case

$$\frac{k}{m} > \frac{\varphi^2}{4m^2}$$

then the solution to equation (7.62) is

$$x(\tau) = x_{max}\, e^{-\frac{\varphi}{2m}\tau} \cos\left[\sqrt{\left(\frac{k}{m} - \frac{\varphi^2}{4m^2}\right)}\,\tau\right] \qquad (7.63)$$

which is less intimidating than it might look.

Firstly, note that if $\varphi = 0$, corresponding to the absence of dynamic friction, equation (7.63) becomes

$$x(\tau) = x_{max} \cos\sqrt{\left(\frac{k}{m}\right)}\,\tau$$

which, as expected, is the standard equation for simple harmonic motion, equation (7.39), with $\omega^2 = k/m$.

Secondly, note that equation (7.63) comprises three elements, all multiplied together

- A constant x_{max}, which defines the position of the mass at time $\tau = 0$, this being the maximum displacement of the mass from its equilibrium position.
- A cosine function, which oscillates with a frequency given by $\sqrt{\left(\dfrac{k}{m} - \dfrac{\varphi^2}{4m^2}\right)}$ this being somewhat less than the frictionless frequency $\sqrt{(k/m)}$, implying that the piston moves rather more slowly when dynamic friction is present ($\varphi > 0$) than when it is absent ($\varphi = 0$) – as makes intuitive sense.
- A factor $\exp-(\varphi/2m)\tau$, an exponential term that becomes progressively smaller with time, so that the stronger the dynamic friction, the greater the value of the parameter φ, and the more quickly this exponential term decays.

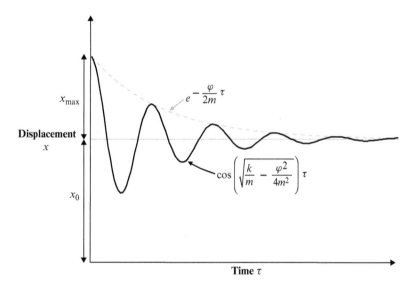

Figure 7.6 The effect of dynamic friction. If there is dynamic friction between the piston and the inner surface of the cylinder, then the motion of the piston is 'damped'. The energy dissipated by frictional heat is lost from the system, and so the amplitude of the oscillation becomes progressively smaller, until the piston stops. The system spontaneously and irreversibly moves from its initial, unstable, non-equilibrium state to its final, stable, equilibrium state.

The behaviour of this equation is shown in Figure 7.6, which shows how the displacement x of the piston from its equilibrium position varies over time. As can be seen, the piston oscillates, but the amplitude of each successive oscillation becomes progressively smaller, until such time as the piston stops, at (or very close to) the equilibrium position. The frequency of the oscillation is given by complex-looking square root

$$\sqrt{\left(\dfrac{k}{m} - \dfrac{\varphi^2}{4m^2}\right)}$$

and the progressive decrease in the magnitude of each successive oscillation is given by the exponential term $\exp -(\varphi/2m)\tau$, which depends on the dynamic frictional force, represented by the term φ: the stronger the frictional force, the larger the value of φ, and the more quickly the oscillations die away; conversely, if the frictional force is weak, φ is relatively small, and so the oscillations diminish slowly. And if there is no dynamic friction at all, $\varphi = 0$, and the oscillations continue for ever. For non-zero values of φ, the motion is of the general form shown in Figure 7.6, and is known as **damped harmonic motion**.

For completeness, we note here that if

$$\frac{k}{m} < \frac{\varphi^2}{4m^2}$$

then the system does not oscillate; rather it comes to rest at the equilibrium position either directly, or by returning to the equilibrium position after a single pass through it. This happens when the magnitude of the frictional factor φ is relatively large, corresponding to a strong, rather than a weak, frictional force. For the remainder of this discussion, we will assume that the magnitude of the frictional factor φ is sufficiently low so that the piston oscillates, as shown on Figure 7.6.

Let's now examine the energy within this system. At any time τ, between the start of the motion of the piston and the eventual cessation of the motion as a result of the cumulative effect of the dynamic friction, the piston will be displaced a distance $x(\tau)$ from the equilibrium position x_0, as given by equation (7.63)

$$x(\tau) = x_{max}\, e^{-\frac{\varphi}{2m}\tau} \cos\left[\sqrt{\left(\frac{k}{m} - \frac{\varphi^2}{4m^2}\right)}\right]\tau \tag{7.63}$$

for the appropriate value of τ. For simplicity, we will write as equation (7.63) as

$$x(\tau) = x_{max}\, e^{-\Phi\tau} \cos \Omega\tau \tag{7.64}$$

where

$$\Phi = \frac{\phi}{2m}$$

and

$$\Omega = \sqrt{\left(\frac{k}{m} - \frac{\phi^2}{4m^2}\right)}$$

At time τ, the piston has an instantaneous velocity $dx/d\tau$, which can be determined from equation (7.64) as

$$\frac{dx}{d\tau} = -x_{max}\, \Omega\, e^{-\Phi\tau} \sin \Omega\tau - x_{max} \Phi\, e^{-\Phi\tau} \cos \Omega\tau$$

$$= -x_{max}\, e^{-\Phi\tau}(\Omega \sin \Omega\tau - \Phi \cos \Omega\tau) \tag{7.65}$$

That's all rather clumsy, but do not be concerned: as we shall shortly see, the important aspect of this expression is not its detailed form, but its structure as the (negative) sum of two terms, one a sin function, the other a cos function, both damped by the negative exponential $\exp(-\Phi\tau)$.

The energy associated with the system at time τ comprises three components:

- The energy associated with the P,V work of the compression and expansion of the gas on each side of the piston, as given by equation (7.58)

$$U_G = 2U_0 + \frac{\gamma n R T_0}{x_0^2} x^2 \qquad (7.58)$$

where $2U_0$ is the total initial internal energy of the gas on both sides of the piston before the original displacement of the piston, and the effect of friction is embedded within the behaviour of x (the parameter φ is not explicitly shown, but is present in equation (7.63) for x, and the derivation of equation (7.58), for a reversible change, is also valid for a quasistatic irreversible change).

- The kinetic energy of the piston, given by

$$\text{Kinetic energy of piston} = \frac{1}{2} m \left(\frac{dx}{d\tau} \right)^2 \qquad (7.52)$$

where the velocity $dx/d\tau$ is given by equation (7.65).

- The energy attributable to the work done in overcoming friction, which is the total work done against the frictional force over the time from time 0 to time τ.

This last item, the energy attributable to the work done in overcoming dynamic friction, was examined in Chapter 5, where, on page 133, we showed that the total work done against dynamic friction from time 0 to any time T is given by

$$\text{total work done against dynamic friction} = = \varphi \int_0^T \left(\frac{dx}{d\tau} \right)^2 d\tau \qquad (5.34)$$

The total energy of the system at any time T is therefore given by the sum of these three components

$$\text{Total energy} = 2U_0 + \frac{\gamma n R T_0}{x_0^2} x^2 + \frac{1}{2} m \left(\frac{dx}{d\tau} \right)^2 + \varphi \int_0^T \left(\frac{dx}{d\tau} \right)^2 d\tau \qquad (7.66)$$

This expression is quite scary, but we can understand how it behaves by exploring each term:

- The first term, $2U_0$ is the total initial internal energy of the system in its original equilibrium state, and remains constant over time.
- The second term represents the energy associated with the compression and expansion of the gas on each side of the piston. When the piston is originally displaced by the external magnet, this action introduced energy into the system corresponding to the value of this term with $x = x_{max}$, plus, as we will see shortly, some energy to overcome friction. Over time, this term behaves as x^2, which, according to equation (7.63), is a $\cos^2$ function, damped by a declining exponential. This term therefore oscillates over time, with a steadily declining amplitude.
- The third term represents the kinetic energy of the piston, and, over time, behaves as $(dx/d\tau)^2$. Equation (7.65) for $dx/d\tau$ is messy, but it comprises the sum of a cos

function and a sin function, all multiplied by a negative exponential. $(dx/d\tau)^2$ will therefore be a function of $\cos^2$, $\sin^2$ and the product $\cos \sin$, all multiplied by a negative exponential which, by virtue of being squared, declines all the faster. What's important is that – whatever the details – this term will oscillate with declining amplitude.

- The fourth term is different. The frictional parameter φ is positive, and this multiplies the time integral of $(dx/d\tau)^2$, which, as the square of $dx/d\tau$, is also necessarily positive, whether $dx/d\tau$ is positive (in physical terms, the piston is moving to the right), or negative (the piston is moving to the left). And, as a time integral of a positive quantity, this fourth term does not oscillate – rather, it progressively increases with time, until a time is reached when $dx/d\tau = 0$ (the piston comes to rest), and the value of the integral stays at a constant positive value. Furthermore, right at the start, when the piston is moved by the external agent from its original position x_0, the value of this integral, over the short time over which this initial motion happens, represents the work done against friction by the external agent.

Overall, then, the behaviour of the total energy of the system over time, as described mathematically by equation (7.66)

$$\text{Total energy} = 2U_0 + \frac{\gamma n R T_0}{x_0^2} x^2 + \frac{1}{2} m \left(\frac{dx}{d\tau}\right)^2 + \varphi \int_0^T \left(\frac{dx}{d\tau}\right)^2 d\tau \qquad (7.66)$$

can be interpreted physically as

- An initial equilibrium state, with the piston stationary at its equilibrium position x_0. The total energy of the system is $2U_0$, the sum of the internal energies U_0 of the gas on each side of the piston. All other terms in equation (7.66) are zero.
- An initial event then takes place in which external agent, say a magnet, displaces the piston to the right to a position x_{max}. In so doing this external agent does work on the system equal to

  Initial work done on system by external agent

$$= \frac{\gamma n R T_0}{x_0^2} x_{max}^2 + \varphi \int_0^{\tau_1} \left(\frac{dx}{d\tau}\right)^2 d\tau$$

where the first term represents the work done by the external agent on compressing the gas to the right, 'helped' by the expansion of the gas on the left, and the second term represents the work done against friction over the time taken for this initial displacement to take place, from time 0, as specified by the lower limit of the integral, to time τ_1, as specified by the upper limit of the integral. The value of this integral is a positive number, and represents energy dissipated as heat generated by the frictional work. Since the system is adiabatic, this heat cannot escape. The temperatures of the gases on both sides of the system will therefore be somewhat higher than these temperatures would have been in the absence of friction.

- When the piston is released, it will execute damped simple harmonic motion, oscillating about the equilibrium position x_0, with a progressively reducing amplitude, as represented by Figure 7.6. During this time, the two central terms in equation (7.66) oscillate, as energy exchanges between the kinetic energy of the piston, and the energy associated with the expansion and compression of the gases on each side of the piston. In the absence of dynamic friction, as shown on page 247, the term proportional to x^2 (the second term, associated with the energy of expansion and compression) behaves as a $\cos^2$ function, whilst the term proportional to $(dx/d\tau)^2$ (the third term, associated with the kinetic of the piston) behaves as a $\sin^2$ function. These two terms are therefore 'out-of-phase': when either one is at a maximum, the other is at a minimum. But their sum remains constant. With dynamic friction present, the mathematics is more complicated, but these two terms still oscillate, with the sum of the two terms progressively diminishing, until the piston stops, when both terms become, and remain, zero. This is because, with dynamic friction present, the last term, attributable to the work done against friction, *does not oscillate*, but just *gets bigger and bigger with time* – until the piston stops, after which the last term remains constant. The work done against friction, as represented by the last term, is dissipated as heat, causing the gases to get hotter.

- Ultimately, the piston will stop at, or very close to, the equilibrium position x_0 – theoretically, it should stop exactly at the equilibrium position, but if the friction causes some 'stickiness', the actual final position might be slightly to one side or the other, with the stickiness of the static friction countering the corresponding pressure differential. At that time, and thereafter, the second and third, oscillating, terms in equation (7.66) become, and remain, zero, and the final term is a constant, so that the total energy of the system is

$$\text{Total energy} = 2U_0 + \varphi \int_0^{T_r} \left(\frac{dx}{d\tau}\right)^2 d\tau \tag{7.67}$$

where the upper limit T_r of the integral is the time at which the piston comes to rest. And as a result of the heat produced by the frictional work, the temperature of the gases is higher than it would have been in the absence of friction.

Overall, since, according to the First Law, energy is conserved, the total energy of the system, once the piston has been displaced, as given by equation (7.66)

$$\text{Total energy} = 2U_0 + \frac{\gamma n R T_0}{x_0^2} x^2 + \frac{1}{2} m \left(\frac{dx}{d\tau}\right)^2 + \varphi \int_0^T \left(\frac{dx}{d\tau}\right)^2 d\tau \tag{7.66}$$

is constant.

In the absence of dynamic friction, the system oscillates for ever, with the energy associated with the compression and expansion of the gases varying with the kinetic energy of the piston, such that the sum remains constant. But in the presence of dynamic friction, the final term, the time integral, gets bigger and bigger. This acts to 'suck' energy away from the oscillation, and dissipate it as heat.

From a thermodynamic point of view, damped harmonic motion can be understood in terms of the energy dissipated as heat in overcoming dynamic friction. In the absence of dynamic friction, as we have seen, once the piston is originally displaced, it will oscillate for ever. In the presence of dynamic friction, however, some of the energy that would propel the piston from, say, the right-hand extreme position to the left-hand extreme position is required to do work to overcome friction. The energy associated with this work is instantaneously and necessarily dissipated as frictional heat, and so is not available to compress the gas on the left-hand side of the cylinder. As the piston slows down on the left-hand side, it cannot travel quite as far as it would have done had no energy been lost due to dynamic friction. The piston therefore comes to rest on the left-hand side at a point closer to the central equilibrium position than the theoretical, friction-free, position. And, as the piston then moves the right, the same phenomenon occurs, and yet more energy is lost to frictional heat. Over time, the oscillations diminish, until the piston finally stops – as illustrated in Figure 7.6.

One further important point. The frictionless piston corresponds to a reversible process, whereas the motion of the piston in the presence of dynamic friction is an example of a real, irreversible, process – a process that moves, spontaneously, from an initial, non-equilibrium, state (with the piston originally displaced by an external agent to its extreme position) to a final, equilibrium, state (the piston at rest at the centre of the cylinder, with the pressure equal on both sides). That's what all non-equilibrium real systems do – they move unidirectionally from a non-equilibrium state to an equilibrium state. And, as we shall see in Chapter 8, this observation is central to the Second Law.

EXERCISES

Some exercises to explore isothermal and adiabatic changes for an ideal gas.

1. Figure 7.7 represents three states of a given mass of ideal gas, such that states [1] and [2] have the same volume, implying that a change from state [1] to state [2] is isochoric. Suppose that the states have been labelled as state [1] $\equiv [P_1, V_1, T_1]$, state [2] $\equiv [P_2, V_1, T_1]$, and state [3] $\equiv [P_3, V_3, T_1]$. Why must this labelling be incorrect? What must be the (single) correct label for state [2]? What are the three, different, possible labels for state [3]?

2. Once again with reference to Figure 7.7, suppose that a narrative associated with the diagram were to read "Consider a change from state [1] to state [2] along two different paths. The first path is a reversible isochoric change directly from state [1] to state [2]; the second path is a reversible adiabatic expansion from state [1] to state [3], followed by a reversible adiabatic compression from state [3] to state [2]". Why is this narrative incorrect? Prove your answer to this question mathematically.

3. If a change from state [1] to state [3] takes place along a path corresponding to expansion against a vacuum, what is the value of T_3? If the change from state [1] to state [3] had taken place reversibly, how would that change have been described?

4. For an arbitrary point on the P,V diagram corresponding to state X $\equiv [P, V, T]$, show that the slope $(\partial P/\partial V)_T$ of an isotherm at that point is given by $-nRT/V^2$, and that

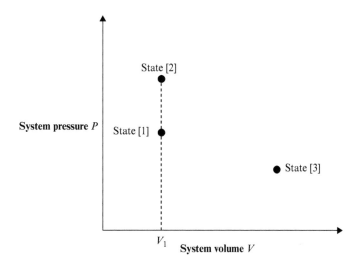

Figure 7.7 Three states of an ideal gas, such that the volumes of states [1] and [2] are the same.

the slope $(\partial P/\partial V)_{\text{adi}}$ of an adiabat at the same point is $(C_P/C_V)(nRT/V^2)$. Does the isotherm, or the adiabat, have the steeper slope? Draw a diagram showing how, in general, an isotherm and an adiabat pass through the same point.

5. In Figure 7.7, for a given state [1], suppose that the path from state [1] to an arbitrary state [X] is isothermal. How many states [X] are there that satisfy this condition? Plot these points on a copy of Figure 7.7. Suppose further, for a given state [2], that the path from state [2] to an arbitrary state [Y] is adiabatic. How many states [Y] might there be? Plot these points too. Suppose now that state [3] in Figure 7.7 satisfies both conditions simultaneously – state [3] is on an isotherm with respect to state [1], and also on an adiabat with respect to state [2]. How many states [3] are there that satisfy this condition?

6. With these thoughts in mind, re-read pages 212 and 222, with particular reference to the interpretation of Figures 7.1. and 7.2.

Some exercises based on the case study of the Carnot cycle

7. Redraw Figure 7.3 for a reversible Carnot cycle, labelling the four system states. For each of the four steps of the reversible Carnot cycle, identify the work done by the system, or on the system, as an area, distinguishing clearly between work done 'by' and work done 'on'.

8. For the reversible cycle from state [1], through state [2], then states [3] and [4], and back to state [1], is net work done by the system, or on the system? What is the area on the diagram that represents this work?

9. For the reversible cycle in the opposite direction, from state [1], to state [4], and then through states [3] and [2] back to state [1], is net work done by the system, or on the system? What is the area on the diagram that represents this work? How does your answer to this question compare to your answer to question 8?

10. For the cycle in the direction [1] → [2] → [3] → [4], suppose that the work that takes place for the change from state [1] to state [2] is irreversible, but the remaining three

steps are reversible. On your diagram, draw a dashed line representing the approximate, irreversible, path from state [1] to state [2]. Is your dashed line positioned above, or below, the solid line representing the reversible path from state [1] to state [2]? Why? If the reversible work associated with the change from state [1] to state [2] is $\{_1w_2\}_{rev}$, and the irreversible work $\{_1w_2\}_{irrev}$, what is the relationship between $\{_1w_2\}_{rev}$ and $\{_1w_2\}_{irrev}$?

11. For the cycle in the direction $[1] \to [2] \to [3] \to [4]$, suppose that all steps are irreversible. Draw dashed lines representing the four approximate paths accordingly. For each of the four steps, what is the relationship between the corresponding quantities of reversible and irreversible work? For the whole cycle, what is the relationship between the total reversible work $\oint dw_{rev}$ and the total irreversible work $\oint dw_{irrev}$?

12. Consider the step $[1] \to [2]$, an isothermal expansion at a temperature T_H. What is the change ΔU in internal energy? Why? If the change takes place reversibly, what is the relationship between the reversible work $\{_1w_2\}_{rev}$ and the reversible heat flow $\{_1q_2\}_{rev}$? If this step is irreversible, what is the relationship between $\{_1w_2\}_{irrev}$ and $\{_1q_2\}_{irrev}$? Given the relationship between $\{_1w_2\}_{rev}$ and $\{_1w_2\}_{irrev}$ derived from exercise 10, what is the relationship between $\{_1q_2\}_{rev,H}$ and $\{_1q_2\}_{irrev,H}$? And – bearing in mind that the temperature T_H is constant for an isothermal change – what is the relationship between $\{_1q_2\}_{rev,H}/T_H$ and $\{_1q_2\}_{irrev,H}/T_H$?

13. For a complete reversible cycle, write down an expression for $\oint (dq_{rev}/T)$ as the sum of four terms of the form $\{_iq_j\}_{rev}/T$ for each of the four steps. What is the value of $\oint (dq_{rev}/T)$? Why? Suppose now that the steps $[1] \to [2]$ and $[3] \to [4]$ are irreversible, so that the cycle is irreversible as a whole. Write down a similar expression of four terms for $\oint (dq_{irrev}/T)$. What is the relationship between $\oint (dq_{rev}/T)$ and $\oint (dq_{irrev}/T)$? As a result, show that $0 > \oint (dq_{irrev}/T)$ – this being known, as we shall see in Chapter 9, as the 'Clausius inequality'.

Hurricanes

14. Suppose that the pressure of the atmosphere at sea level is P_0, and at an altitude h m above sea level, the pressure is P_h. Which of the following three statements is true: $P_0 < P_h, P_0 = P_h, P_0 > P_h$? Why?

15. A volume V_0 of air at sea level, at pressure P_0, has a temperature T_0. Suppose that the volume rises within the atmosphere to a height h m above sea level, corresponding to a volume V_h, a pressure P_h, and a temperature T_h. Assuming that this change of state is adiabatic, is V_h greater than, equal to, or less than V_0? Why? Is T_h greater than, equal to, or less than T_0? Why? Analyse this change in terms of the First Law, with particular reference to the work done on, or by, the volume of air.

16. In general, warmer air can 'carry' a larger volume of water vapour than cooler air. If the warm air at sea level is saturated with water vapour, what happens to that water vapour as the air rises in the atmosphere? What would an observer, on the surface of the earth, see happening in the atmosphere, and what is that observer likely to experience?

17. When water changes phase between the liquid state and the vapour state, the corresponding change in enthalpy is the enthalpy of vaporisation $\Delta_{vap}H$. If the change in state is from vapour to liquid, is $\Delta_{vap}H$ positive or negative?

18. The formation of rain in a cloud corresponds to a phase change for water from vapour to liquid. If the corresponding enthalpy change is positive, what is the source of

that enthalpy? If the corresponding enthalpy change is negative, what happens to the enthalpy released?

19. A hurricane is a natural phenomenon in which warm, saturated, air rises from the sea, associated with heavy rain fall, and violently swirling winds. With your answers to questions 14 to 18 in mind, explain a hurricane in thermodynamic terms.

20. "A hurricane is an example of a Carnot cycle, operating in the atmosphere between a hot reservoir, the sea at temperature T_0, and a cold reservoir, the upper atmosphere at temperature T_h." Discuss.

The thermodynamic pendulum

21. The case study on the thermodynamic pendulum assumed that the changes in state were adiabatic. Derive the isothermal equivalents of equations (7.51) and (7.61). Comparing the adiabatic and isothermal results, what do you notice?

8 Spontaneous changes

Summary

Spontaneous changes in isolated systems are **unidirectional** and **irreversible**, and proceed from non-equilibrium states to equilibrium states.

8.1 A familiar picture

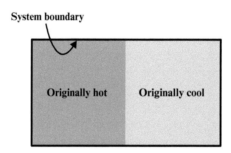

Figure 8.1 A non-equilibrium system. Initially, the solid block on the left is at a higher temperature than the solid block on the right. The system is not in equilibrium. Over time, the originally hotter block becomes cooler, and the originally cooler one becomes hotter, until both attain the same temperature, at which point the system achieves equilibrium. This is an example of a spontaneous, unidirectional and irreversible change in state.

Figure 8.1 is familiar: it is identical to Figure 1.2, and shows a system comprising two solid blocks made of the same quantity of the same material. The block on the left is originally at a temperature T_L; the block on the right is originally at a temperature T_R, such that $T_L > T_R$. Since an intensive state function, temperature, is different in the two halves of the system, the system is not in equilibrium.

Let us now suppose that the system boundary is adiabatic so that no heat can be lost, and that the two blocks are in thermal contact along their adjacent surfaces. An observer will then detect that, over time, the originally hotter block on the left becomes cooler, and the originally cooler block on the right becomes hotter, until they both achieve the same temperature T, at which point the system achieves thermal and thermodynamic equilibrium indefinitely.

Familiar territory indeed, and we may use the First Law to calculate the final equilibrium temperature T. Since the system boundary is adiabatic, there is no heat flow into, or out of, the system; furthermore, since the system is solid, then, for relatively small changes in temperature, we can assume that there is no change in volume, and so there is no work done by, or on, the system. The internal energy U of the system is unchanged, and so, for the system as a whole, ΔU is zero.

The total internal energy U of the system, however, is an extensive state function, and so is the sum of the internal energies U_L and U_R of the left- and right-hand blocks

$$U = U_L + U_R$$

and so

$$\Delta U = \Delta U_L + \Delta U_R = 0 \tag{8.1}$$

Each block undergoes a change at constant volume. If we assume that the molar heat capacity at constant volume C_V for the material from which each blocks is made is constant over the temperatures under consideration, then we may use equation (5.19)

$$\Delta U = C_V(T_2 - T_1) \quad \text{at constant volume} \tag{5.19}$$

to relate the change in internal energy ΔU for each block to the corresponding change in temperature ΔT. The left-hand block, comprising n mol of material, cools from temperature T_L to temperature T, and so $\Delta T = T - T_L$ implying that

$$\Delta U_L = n\,C_V\,\Delta T = n\,C_V\,(T - T_L) \tag{8.2}$$

Since the left-hand block cools, $T < T_L$, and so ΔU_L is negative – the left-hand block loses heat, as expected.

For the right-hand block, which also comprises n mol of the same material, and changes from temperature T_R to temperature T, we have

$$\Delta U_R = n\,C_V\,\Delta T = n\,C_V\,(T - T_R) \tag{8.3}$$

For the right-hand block, $T > T_R$, and so ΔU_R is positive, verifying that the right-hand block gains heat.

Combining equations (8.1), (8.2) and (8.3),

$$\Delta U_L + \Delta U_R = n\,C_V\,(T - T_L) + n\,C_V\,(T - T_R) = 0$$

and so

$$T = \frac{T_L + T_R}{2} \tag{8.4}$$

Substituting for T in equations (8.2) and (8.3)

$$\Delta U_L = nC_V\,\frac{T_R - T_L}{2} \tag{8.5}$$

and

$$\Delta U_R = nC_V\,\frac{T_L - T_R}{2} \tag{8.6}$$

Since $T_L > T_R$, ΔU_L is negative, and ΔU_R is positive, as expected. Also, adding equations (8.5) and (8.6)

$$\Delta U_L + \Delta U_R = 0$$

so confirming that the First Law is upheld.

8.2 Spontaneity, unidirectionality and irreversibility

The change of state illustrated in Figure 8.1 is, as we have just shown, fully compliant with the First Law, but it is also characterised by three further features: the change is

- spontaneous,
- undirectional, and
- irreversible.

These words are important, so let us explore their meaning rather more deeply.

A **spontaneous** change is one that takes place naturally, of its own accord, and without the intervention of an external agent. Certainly, the change illustrated in Figure 8.1 takes place as a result of the temperature gradient $T_L - T_R$; its presence drives the change; and once the temperature of the system equilibrates, and the temperature gradient becomes zero, the change stops. All of this, however, happens within the system itself, and is not attributable to any external events.

The change is **unidirectional** in that the originally hotter block becomes cooler, and the originally cooler block becomes hotter, until both blocks achieve the same temperature, after which no further change takes place. It is totally beyond anyone's experience to observe a system of this type, at equilibrium, spontaneously to change, so that the left-hand half becomes hotter, and the right-hand half cooler. It just doesn't happen. But there is a deeper inference: the example of the two blocks at different temperatures is just a particular case of a system moving unidirectionally from a non-equilibrium state to an equilibrium state.

The change is **irreversible**, in that, once the change has happened, and the final equilibrium state has been obtained, it is impossible – in the literal sense of that word, just not possible – for the change to be reversed, following the original path 'backwards', to restore the original states. As we saw in our discussions of irreversibility on pages 123 to 126, this does not imply that it is impossible to restore the original states at all, for it is perfectly possible to take an item, in thermal, and thermodynamic, equilibrium with its surroundings, and cool it down – but to do so requires the system to take a different path.

As an example, consider what happens when an item – say some milk – is taken out of a refrigerator, into a room. Originally, the temperature of the milk is lower than the temperature of the room, and the situation is just like that depicted in Figure 8.1: a cooler body, the milk, is in thermal contact with a warmer body, the room. The milk warms up, and – in principle – the room cools down, but since the room is much bigger than the milk, as we saw in our discussion of the thermometer on page 45, the change in temperature for the room is negligible, and the milk achieves room temperature. For the milk, this is a natural, spontaneous, and unidirectional change. It is also irreversible, in that, to cool the milk down again, we need to return it to the refrigerator – a complex, carefully designed machine, which uses energy in the form of electricity to achieve the cooling we require. The action is certainly not the reverse of the natural and spontaneous process by which the milk warmed up in the room.

The change of state for the system illustrated in Figure 8.1 is indeed irreversible; but let's suppose, just for a moment, that the change of state could be reversed. In so doing, a quantity of heat, equal in magnitude to the value of ΔU in equations (8.5) and (8.6), would need to be transferred from one block to the other. The quantity of heat lost by one block would be gained in equal amount by the other, no energy would be gained or lost overall, and so the First Law would be upheld. The First Law does not distinguish between the natural, spontaneous, irreversible change and the 'unnatural', reverse, change – both comply.

Our experience, however, is that spontaneous changes are real, and that they are unidirectional and irreversible. Since the direction of change is not addressed by the First Law, we must look elsewhere for an explanation. And to set the scene for doing just that, let's take a deeper look at those three fundamental features of many real changes: spontaneity, unidirectionality and irreversibility.

8.3 Some more examples of spontaneous, unidirectional, and irreversible changes

Heat flow down a temperature gradient is just one example of a very general phenomenon – the tendency of non-equilibrium systems to move spontaneously, unidirectionally and irreversibly to equilibrium states.

8.3.1 The expansion of a gas into a vacuum

Another instance is the system we examined on page 121, and in Figure 5.6 – the adiabatic expansion of a gas into a vacuum. As soon as a gas, originally constrained within a given volume, is able to expand into a vacuum, it will do so, spontaneously (without external cause), unidirectionally (from a non-equilibrium to an equilibrium state) and irreversibly (as discussed on page 123).

8.3.2 Mixing

A second instance is mixing: milk stirred into tea or coffee spontaneously, unidirectionally and irreversibly forms a homogeneous mixture. The milk does not spontaneously separate out, this being an example of molecular irreversibility, as discussed on page 128. We note, however, three situations in which mixing appears to be reversible. The first can happen to milk in tea or coffee: if the milk degrades, or otherwise 'goes off', it can happen that the tea or coffee transforms into a rather nasty liquid, in which the products of the now-degraded milk separate from the long-cold coffee. This is not an example of reversibility; rather, it is an example of (complex) chemical change.

A second situation is that of the mixing of say, water and olive oil, or olive oil and vinegar. If water and olive oil, or vinegar and olive oil, are vigorously shaken, it appears as if the resulting fluid is a homogeneous mixture. If, however, the mixture is left to stand, after quite a short time, the mixture will separate – certainly spontaneously, and in fact, though not obviously, unidirectionally and irreversibly – into separate layers of water or vinegar below the less dense olive oil. This (real) example of 'unmixing' seems to contradict the principle of molecular irreversibility; in fact, this is not the case, for the original 'mixing' was not mixing at the molecular level, in which molecules of water, vinegar and olive oil mutually surround one another randomly. Rather, the olive oil is dispersed within the water or vinegar in the form of internally homogeneous micro-droplets, which subsequently coalesce to form an easily visible layer.

And finally we note a class of chemical reactions, of which the most dramatic is the **Belousov–Zhabotinsky reaction**, in which an originally homogeneous mixture develops 'stripes' which intensify and then fade, intensify and then fade again, and continue to 'oscillate', as vividly shown, for example, on https://www.youtube.com/watch?v=PpyKSRo8Iec. In essence, this reaction is the chemical analogue of the 'thermodynamic pendulum' discussed on pages 236 to 253, a reaction which slowly 'swings to and fro' through a sequence of non-equilibrium states, until, after considerable time, the stable equilibrium state is reached (for a detailed explanation, see example, http://www.scholarpedia.org/article/Belousov-Zhabotinsky_reaction).

8.3.3 Phase changes

A further instance of spontaneity, unidirectionality and irreversibility is the melting of ice at, say, 293 K, about 20 °C, at 1 atm pressure: at that temperature and pressure, ice melts to form liquid water spontaneously, unidirectionally and irreversibly, in that liquid water will never change into ice under those conditions. If, however, the conditions are changed – suppose, for example, that the temperature is reduced to say, 253 K, about –20 °C – then liquid water will spontaneously, unidirectionally and irreversibly change into ice. At any given temperature, the appropriate change – solid to liquid at higher temperatures, liquid to solid at lower temperatures – complies with the required trio of features of being spontaneous, unidirectional and irreversible, as a non-equilibrium state changes to an equilibrium state.

This example illustrates an important point: for any given system, the stable, equilibrium state is not necessarily an absolute, but may be contingent on the system's conditions. So, at a pressure of 1 atm, and at temperatures between 273.15 K and 373.15 K, the stable, equilibrium state of water is as a liquid, and other forms of water – ice or water vapour – will spontaneously, unidirectionally and irreversibly change into liquid water accordingly. But at 1 atm pressure, and temperatures above 373.15 K, the stable, equilibrium form of water is water vapour, and below 273.15 K, the stable equilibrium form is ice. At any given temperature, however, a non-equilibrium form will always change into the appropriate equilibrium form spontaneously, unidirectionally and irreversibly.

8.3.4 Chemical reactions

The dependence of the equilibrium state on external conditions is of particular importance as regards chemical, and, in particular, biochemical reactions. For some chemical reactions, the stable equilibrium state is a single state, over a very wide range of conditions – so, for example, if aqueous silver nitrate is mixed with any aqueous halide, such as sodium iodide

$AgNO_3$ (aq) + NaI (aq) → AgI (s) + $NaNO_3$ (aq)

silver iodide is instantaneously precipitated, and no matter how long we might wait, and even if the aqueous solution is boiling, solid silver iodide will never dissolve to any meaningful extent into aqueous solution. This reaction goes to completion as written, with the final, stable equilibrium state being one in which essentially all the silver nitrate is converted into silver iodide. Likewise, when paper is ignited, it immediately burns, consuming all the paper, and producing carbon dioxide, water vapour, and ash – products that can never react together to form paper.

A contrasting example is the exothermic reaction at room temperature of gaseous ammonia and hydrogen chloride to form solid ammonium chloride

$NH_3\,(g) + HCl\,(g) \rightarrow NH_4Cl\,(s)$

If, however, solid aluminium chloride is gently heated, it decomposes endothermically to form ammonia and hydrogen chloride

$NH_4Cl\,(s) \rightarrow NH_3\,(g) + HCl\,(g)$

At any given temperature, both reactions are taking place simultaneously, as expressed by the reversible reaction

$NH_3\,(g) + HCl\,(g) \rightleftharpoons NH_4Cl\,(s)$

with the reaction to the right, forming solid ammonium chloride, predominant at lower temperatures, and the reverse reaction, forming ammonia and hydrogen chloride, predominant at higher temperatures. It is therefore possible, at intermediate temperatures, for a system of NH_3 (g), HCl (g) and NH_4Cl (s) to be in equilibrium, and with measureable quantities of NH_3 (g), HCl (g) and NH_4Cl (s) all present: the reaction has not gone to completion, but is in a state of dynamic chemical equilibrium, with both products and reactants co-existing.

We'll examine dynamic chemical equilibrium in some detail in Chapter 14; as we shall see on pages 449 to 455, the existence of dynamic chemical equilibrium, and the stable co-existence of both reactants and products, does not contradict, but is totally consistent with, the fundamental principles of spontaneously, unidirectionally and irreversibly.

8.4 Spontaneity and causality

Central to our discussion is the concept of the 'spontaneous change': a change in the state of a system, which – to quote the words used on page 260 – "takes place naturally, of its own accord, and without the intervention of an external agent". A spontaneous change is therefore one which happens 'by itself', and so is not 'caused by' anything – it just happens. This 'just happening' is totally consistent with our experience: hot things cool down 'all by themselves', and silver iodide precipitates as soon as a solution of silver nitrate is mixed with a solution of sodium iodide. Certainly, a hotter body has to be in contact with a cooler one in order for the change in temperature to happen, and the aqueous silver nitrate needs to be mixed with the sodium iodide solution, but these initial events create the conditions in which the changes can happen, and are not the cause of the changes themselves: once the conditions are present, the changes happen naturally, spontaneously, 'of their own accord', without being caused by anything else.

If we observe a change in state of a system, and believe it to be spontaneous, then all possible causality must be absent. Let's take a moment, then, to examine more deeply just what causality is, how causality can be recognised, and hence how the absence of causality can be verified.

Fundamentally, spontaneity is attributable to the absence of causality: conversely, if it is possible to make a statement of the form "[Event A] is the cause of [event B]", then [event B] – whatever [event B] might be – is not spontaneous, but the result of [event A] having happened beforehand.

Two aspects of this are important. The first concerns time. For [event A] to be attributed as the cause of [event B], then [event A] must happen first: it makes no sense to say [event A] is the cause of [event B] if [event A] occurs after [event B]. Our experiences of time, and of spontaneity, are therefore intimately connected.

The second important aspect concerns the attribution of causality, as opposed to coincidence. Events happen everywhere, all the time, and some events will occur before others. But the fact that [event A] precedes [event B] does not imply that [event A] is the cause of [event B], even if the [event A] and [event B] happen often, and in sequence. The apparent correlation between [event A] and [event B] might be mere coincidence.

To take a trivial example: one evening, a child might notice that a bus stops outside her house, and, a few minutes later, a street light is lit. The child is curious, and, the following day, notices once again that a few minutes after the bus stops, the street light comes on. Being a careful observer, each evening, over a period of several weeks, the child takes notes of the time at which the bus arrives, and the time at which the street light comes on. She reviews her data, and observes that, on each of forty successive days, the time interval between the arrival of the bus and the illumination of the street light is not less 3.0 minutes, and not more than 3.5 minutes, with a mean value of 3.3 minutes. [Event A], the arrival of the bus, consistently precedes [event B], the illumination of the street light, and so the child could, quite convincingly, claim "because the bus arrives, the street light comes on". We know, of course, that this is coincidence – the bus might be following a timetable, to arrive outside the child's house at, say, 6.27 pm in the early evening, and, at that time of year, the street lights, independently, might be timed to come on at 6:30 pm. The child, however, remains convinced.

Attributing causality is problematical, and the anecdote of the child is a metaphor for much scientific activity, in which careful observations are made of how [this] might influence [that] in the hope of gaining a greater understanding of causality.

A moment's thought will verify that, if a particular [cause] is believed to drive a given [effect], then this requires that the [effect] is within a defined system, and that the corresponding [cause] is not within the system, but in the surroundings: if both the [cause] and [effect] are within the same system, then the change, as a whole, would be regarded as being spontaneous, without external cause. If this is accepted, then it implies that causality can be attributed only if something happens *across the boundary* between a system and its surroundings, something that connects an event, the [cause], originating in the surroundings, with a subsequent consequence, the [effect], within the system.

This throws the focus of our attention onto the boundary between a system and the surroundings: to attribute causality, we must be able to identify 'something' that crosses the boundary, that connects the [cause], which takes place within the surroundings, with the [effect], that takes place within the system. What, then, might that 'something' be?

In general, there are three categories of 'something' that might cross the boundary between a system's surroundings and the system itself, two of which are familiar, and one less so:

- a flow of matter
- a flow of energy
- a flow of information.

So – at the risk of stretching credibility – this is how each of these might work for the example of the child watching the bus and the street light. The system in this case is the street light, and the surroundings are everywhere else, including the road along which the bus drives, and the bus stop. Although quite implausible, it is possible that, as the bus arrives at the bus stop, a mechanism could be triggered which causes a small metal bar to roll over the boundary between the surroundings and the system, such that the metal slots into an electrical circuit, so causing the light to come on. An event has happened in the surroundings (the arrival of the bus), which has caused a flow of matter across the boundary (the rolling of the metal bar), so completing the electrical circuit, causing the consequent effect (the illumination of the street light).

Slightly more plausible – but still unlikely! – might be the situation in which, when the bus stops, the weight of the bus compresses a gas in a cylinder under the road. If the cylinder is connected by a pipe to the base of the light, the increased pressure of the gas within the cylinder might perform P, V of expansion, pushing a piston and closing an electrical switch, causing the light to illuminate. In this case, if the system boundary is the inner surface of the piston, there is a flow of energy, in the form of work, across the system boundary, rather than a flow of matter. A hydraulic linkage of pipes and pistons is rather clumsy: a neater alternative is to drive a flow of energy in the form of electricity along a cable, as could happen if the weight of the bus were to trigger a switch, sending an electric current to the light.

Flows of energy and matter across system boundaries are very familiar, and the definitions of closed and open make explicit reference to these flows.

Perhaps slightly less familiar is the flow of information. In our example, this would be the situation in which there is a human operator, at the base of the street light, and within the street light system, who can observe the surroundings. When the operator sees the bus arrive, the operator switches the light on – but in order to do this, the operator needs to know that the bus has indeed arrived. How does the operator know? Because the operator receives appropriate information, which the operator might do personally (be seeing), or indirectly (by being told of the arrival, by means of some sort of message). In this case, it is the availability of information that is key – so building a link between information and thermodynamics, which we shall develop further in due course (see pages 346 and 347).

In any of these three mechanisms, or related ones, were to operate, then the switching on of the street light would indeed be caused by the arrival of the bus. This example, of course, is both hypothetical and unrealistic – but the point illustrated by the example is valid nonetheless. In order to establish causality, three conditions must be fulfilled:

- the [cause], which takes place in the surroundings, must take place earlier in time than the [effect], which takes place within the system

- something – in the form of matter, energy or information – must cross the boundary between a system and its surroundings, after the [cause] has been initiated, and before the [effect] starts
- the overall 'story', linking the [cause] to the [effect] by means of the matter, energy or information flowing across the boundary, must be plausible.

8.5 The significance of the isolated system

To summarise.

Causality can be attributed only if matter, energy or information crosses the boundary between a system and its surroundings, such that we can reliably say that a given cause in the surroundings triggered the flow of matter, energy or information across the boundary, so giving rise to the consequent effect within the system – an effect we detect as a change in the state of the system.

As we have seen, however, attributing cause and effect is fraught with difficulty: within any given system, lots of different things might be happening, at different times; many other things might be happening beforehand in the surroundings; and there might be all sorts of flows of matter, energy or information across the system boundary. Sorting all this out, and distinguishing reliably between coincidence and causality, is genuinely difficult.

Except in one particular circumstance. Suppose that very careful observation of the system boundary detects absolutely nothing at all – no flow of matter, no flow of energy, no flow of information. Suppose further that we do indeed observe a change of state within the system. Since the observed change of state within the system happens with no flow of any form across the system boundary, then surely that change must be truly spontaneous. This is verified by reference to all the examples given at the start of this chapter: in each of the cases of the flow of heat down a temperature gradient, the expansion of a gas into a vacuum, the melting of ice and the precipitation of silver iodide, nothing at all happened at, or across, the system boundary, so we may state, with confidence, that each of these changes in state took place spontaneously.

Those examples are particular cases, but by no means rare or unusual ones: our real world is full of spontaneous, unidirectional and irreversible changes. And there is one class of system in which *every* change *must* be spontaneous. As we saw on page 57, by definition, an isolated system has no activity at the boundary, ever. It therefore follows that any change within an isolated system must, always, be spontaneous – and not only spontaneous, but also unidirectional and irreversible too, taking the system from a non-equilibrium to an equilibrium state.

And with that thought in mind, the stage is now set to introduce the Second Law.

EXERCISE

1. What do you understand by the word 'spontaneous'? You observe a change in a system. How would you determine whether the change was 'spontaneous' or not?

9 The Second Law of Thermodynamics

Summary

The Second Law of Thermodynamics may be stated as

A spontaneous change proceeds, unidirectionally and irreversibly, from a non-equilibrium to an equilibrium state. For any system, we may define an extensive state function S, the entropy of the system, such that the change dS in the system's entropy for any change in state is given by

$$dS = \frac{dq_{rev}}{T} \tag{9.1}$$

where dq_{rev} is the quantity of heat exchanged, reversibly, at temperature T, during that change in state.

When equation (9.1) is combined with the result that, for all irreversible changes

$$dq_{rev} > dq_{irrev} \tag{5.38}$$

then, for any change from an initial state [1] to a final state [2]

$$\Delta S > \int_{state\ 1}^{state\ 2} \frac{dq_{irrev}}{T} \tag{9.9}$$

If the inequality (9.9) is applied to a thermodynamic cycle, we derive the **Clausius inequality**

$$0 \geqslant \oint \frac{dq}{T} \tag{9.11}$$

where the equals sign applies to a reversible change, corresponding to a reversible flow of heat dq_{rev}, and the greater-than sign holds for a real, irreversible change, corresponding to an irreversible flow of heat dq_{irrev}.

For an isolated system, a change from an initial state [1] to a final state [2] will be spontaneous if $\Delta S = S_2 - S_1 > 0$; if $\Delta S = S_2 - S_1 = 0$, then states [1] and [2] are in equilibrium with each other; if $\Delta S = S_2 - S_1 < 0$, then state [1] will not spontaneously change to state [2], but state [2] will spontaneously change to state [1].

Since the universe as a whole is necessarily an isolated system, the first of these three cases – $\Delta S = S_2 - S_1 > 0$ – implies that the entropy of the universe continuously increases.

For a change from an initial state [1] to a final state [2] within a closed or open system, we must take account of the entropy change ΔS_{sys} within the surroundings. The criterion for spontaneity is now $\Delta S_{sys} + \Delta S_{sur} > 0$.

The First and Second Laws can be combined to give two important equations, both of which are expressed only in terms of state functions of the system:

$$dU = T\,dS - P\,dV \qquad (9.20a)$$

and

$$dH = T\,dS + V\,dP \qquad (9.21a)$$

Entropy is related to the molar heat capacities C_V and C_P as

$$\left(\frac{\partial S}{\partial T}\right)_V = \frac{nC_V}{T} \qquad (9.23)$$

and

$$\left(\frac{\partial S}{\partial T}\right)_P = \frac{nC_P}{T} \qquad (9.26)$$

A special case of equation (9.1) relates to a phase change, for which

$$\Delta_{\text{ph}}S = \frac{\Delta_{\text{ph}}H}{T_{\text{ph}}} \qquad (9.43)$$

where $\Delta_{\text{ph}}S$ is the entropy change associated with a change of phase for 1 mol of a pure substance, taking place at a temperature T_{ph}, and associated with a molar enthalpy change of $\Delta_{\text{ph}}H$.

For n mol of a pure substance being heated at constant pressure from a lower temperature T_1 to a higher temperature T_2, then the corresponding change ΔS in the entropy of the system is given by

$$\Delta S_P = n \int_{T_1}^{T_{\text{ph1}\to 2}} C_P(\text{phase 1})\,\frac{dT}{T}$$

$$+ n\,\frac{\Delta_{\text{ph1}\to 2}H}{T_{\text{ph1}\to 2}} + n \int_{T_{\text{ph1}\to 2}}^{T_{\text{ph2}\to 3}} C_P(\text{phase 2})\,\frac{dT}{T} \qquad (9.46)$$

$$+ n\,\frac{\Delta_{\text{ph2}\to 3}H}{T_{\text{ph2}\to 3}} + n \int_{T_{\text{ph2}\to 3}}^{T_2} C_P(\text{phase 3})\,\frac{dT}{T}$$

where the terms $\Delta_{\text{ph}}H/T_{\text{ph}}$ allow for the possibility of phase changes at intermediate temperatures. Since the molar heat capacity at constant pressure C_P is necessarily positive, equation (9.46) implies that the entropy of any pure material increases with temperature.

An inference from equation (9.46) is the **Third Law of Thermodynamics** as stated as

At the absolute zero of temperature, the entropy of all pure perfectly crystalline solids is zero.

9.1 The Second Law

A spontaneous change proceeds, unidirectionally and irreversibly, from a non-equilibrium to an equilibrium state. For any system, we may define an extensive state function S, the entropy of the system, such that the change dS in the system's entropy for any change in state is given by

$$dS = \frac{đq_{rev}}{T} \tag{9.1}$$

where $đq_{rev}$ **is the quantity of heat exchanged, reversibly, at temperature** T, **during that change in state.**

Like the definition of the First Law given on page 94, this statement of the Second Law comprises three elements:

- Firstly, a statement based on empirical observation – that spontaneous changes, as discussed in the previous chapter, are unidirectional and irreversible, and proceed from a non-equilibrium to an equilibrium state.
- Secondly, the identification of an extensive state function – in this case, a system's entropy S.
- A mathematical definition of dS for any change in state – in this case, as the ratio between the reversible heat $đq_{rev}$ exchanged during that change in state, and the temperature T at which that heat exchange takes place.

The **entropy** S of any system is an extensive state function; for a system of n mol of material, the corresponding intensive function, the **molar entropy** S, is defined as $S = S/n$. Also, in accordance with equation (9.1), entropy is measured in units of J K^{-1}.

The new extensive state function entropy, S, is by no means 'obvious', in the sense that the state functions pressure P, volume V, and temperature T, are 'obvious'. Furthermore, the definition of entropy – or rather the definition of the infinitesimal change dS in entropy as the ratio $đq_{rev}/T$ – is positively obscure, and so merits some explanation.

9.2 Entropy – a new state function

Although the definition of entropy as

$$dS = \frac{đq_{rev}}{T} \tag{9.1}$$

seems to be not only obscure, but also 'out-of-the-blue', let's begin our discussion of entropy by pointing out that perhaps equation (9.1) is rather less of a surprise than it might, as first sight, appear. Also, we note that we identified that the quantity $đq_{rev}/T$ is a state function in our discussion of the Carnot Cycle on page 236 – and this is indeed the 'classical' proof: we present a rather simpler approach here.

If we rewrite equation (9.1) as

$$đq_{rev} = T\, dS \tag{9.2}$$

we see that a path function, heat $đq_{rev}$, is expressed as the product of an intensive state function, temperature T, and the change dS in an extensive function, the entropy S. We've seen an example of this structure before, many times:

$$đw_{rev} = P\, dV \tag{2.9}$$

in which a path function, work đw_{rev}, is expressed as the product of an intensive state function, pressure P, and the change dV in an extensive function, the volume V. Equation (9.2) is therefore structurally familiar, and its existence was hinted at on page 56.

Also, just as the First Law does not define the internal energy U, but rather the infinitesimal change dU as

$$dU = đq - đw \tag{5.1a}$$

so the Second Law does not define the entropy S, but rather the infinitesimal change dS as

$$dS = \frac{đq_{rev}}{T} \tag{9.1}$$

The context here is that of an infinitesimal change in state in a closed system of constant mass. As a result, any state function X might change by an amount dX – so, for example, the volume might change by dV, the internal energy by dU, the entropy by dS. During that change in state, if the system can perform only P, V work of expansion, work đ$w = P_{ex}$dV might be performed by the system on the surroundings, or on the system by the surroundings; furthermore, there might be an exchange of heat đq, across the system boundary, between the system and the surroundings. If the change in state is reversible, then it is necessarily quasistatic, and as we saw on page 29, the external pressure P_{ex} of the surroundings may be taken as being equal to the internal pressure P of the system, allowing us to express the reversible work đw_{rev} as

$$đw_{rev} = P\,dV \tag{2.9}$$

Similarly, if there is a reversible flow of heat đq_{rev} across the system boundary, then the external temperature T_{ex} of the surroundings will equal the internal temperature T of the system, and so the ratio đq_{rev}/T will have a value, a value which determines the change dS in the system's entropy, in accordance with equation (9.1)

$$dS = \frac{đq_{rev}}{T} \tag{9.1}$$

In equations (9.1) and (9.2), the heat flow đq_{rev} is the heat flow associated with a reversible change in state. As we saw on pages 140, reversible changes are a theoretical abstraction, rather than a reality, for all real changes are necessarily irreversible. We have also seen, however, that consideration of the theoretical abstraction can be very useful in throwing light on physical reality – as will surely become apparent in this chapter's discussion. For the time being, we will therefore continue to consider reversible changes and the corresponding value of đq_{rev}; real, irreversible, changes will be addressed shortly.

Empirically, for any change in state, we now see that it is possible (at least in principle) to measure the reversible heat flow đq_{rev} between a system and the surroundings for any change, and the temperature T of the system at which that heat exchange took place. We can therefore compute the ratio đq_{rev}/T accordingly. It is by no means self-evident, however, that the ratio so computed represents a change in a state function of the system, so we will now demonstrate that the ratio đq_{rev}/T is indeed the differential of a state function.

Our starting point is the First Law for a closed system of constant mass, which states

$$dU = đq - đw \tag{5.1a}$$

Since the internal energy U of the system is a state function, the value of any change dU depends only on the initial and final states of the system, and is independent of the path. If the path happens to be reversible, we may write

$$dU = đq_{rev} - đw_{rev}$$

and since, if the only form of work is P, V work of expansion

$$đw_{rev} = P\,dV \tag{2.9}$$

$$\therefore \quad đq_{rev} = dU + P\,dV$$

Now, on page 210, we established that, for n mol of an ideal gas,

$$dU = n\,C_V\,dT \tag{7.1b}$$

and so

$$đq_{rev} = n\,C_V\,dT + P\,dV$$

If we now divide throughout by T, then

$$\frac{đq_{rev}}{T} = nC_V\frac{dT}{T} + \frac{P\,dV}{T} \tag{9.3}$$

We may now use the ideal gas law

$$PV = nRT \tag{3.3}$$

to replace P/T by nR/V giving

$$\frac{đq_{rev}}{T} = nC_V\frac{dT}{T} + nR\frac{dV}{V} \tag{9.4}$$

Since, for an ideal gas, n, C_V and R are all constants, then for any cycle, we can integrate equation (9.4) as

$$\oint \frac{đq_{rev}}{T} = nC_V \oint \frac{dT}{T} + nR \oint \frac{dV}{V} \tag{9.5}$$

Now, for a change in state from any initial state [1] to any other state [2]

$$\int_{T_1}^{T_2} \frac{dT}{T} = \ln\frac{T_2}{T_1}$$

But, for a complete cycle, the final state [2] is the same as the initial state [1], implying that $T_2 = T_1$, hence

$$\oint \frac{dT}{T} = \ln\frac{T_1}{T_1} = 0$$

Similarly

$$\oint \frac{dV}{V} = \ln\frac{V_1}{V_1} = 0$$

and so equation (9.5) becomes

$$\oint \frac{đq_{rev}}{T} = 0 \tag{9.6}$$

Equation (9.6) is important. As we saw on pages 37 and 38, any function for which the cyclic integral is zero *must be a state function*. Equation (9.6) therefore tells us that the quantity $đq_{rev}/T$ must be the differential of a state function, a state function that we may assign the symbol S, and the name entropy

$$\frac{đq_{rev}}{T} = dS \tag{9.1}$$

The discussion from equation (9.3) to equation (9.6) demonstrates that the quantity $đq_{rev}/T$ is indeed a state function, but it is not a general proof, but a special case relating to an ideal gas. But although the key result that demonstrates that $đq_{rev}/T$ is a state function, equation (9.6), was derived for a system comprising an ideal gas, it turns out the result – that $đq_{rev}/T$ is a state function – is true for all systems, not just an ideal gas. A more general discussion is given in Chapter 10, but for the purposes of this chapter, and our general understanding of the Second Law, we trust the demonstration just given will be sufficient.

9.3 The Clausius inequality

We have demonstrated that the quantity $đq_{rev}/T$ is a state function

$$\frac{đq_{rev}}{T} = dS \tag{9.1}$$

and that

$$\oint \frac{đq_{rev}}{T} = 0 \tag{9.6}$$

We now invoke a result from page 144, where we proved that, for any change in state, the reversible heat exchange $đq_{rev}$ is always greater than the heat exchange $đq_{irrev}$ for any irreversible path between the same two states

$$đq_{rev} > đq_{irrev} \tag{5.38}$$

If we divide both sides of equation (5.38) by T

$$\frac{đq_{rev}}{T} > \frac{đq_{irrev}}{T} \tag{9.7}$$

then for a finite change from state [1] to state [2],

$$\int_{state\,1}^{state\,2} \frac{đq_{rev}}{T} > \int_{state\,1}^{state\,2} \frac{đq_{irrev}}{T} \tag{9.8}$$

where the integral on the left refers to a change from state [1] to state [2] along a reversible path, and that on the right to an irreversible path between the same two states.

But, by definition

$$\int_{\text{state 1}}^{\text{state 2}} \frac{đq_{\text{rev}}}{T} = \int_{S_1}^{S_2} dS = S_2 - S_1 = \Delta S$$

and so

$$\Delta S > \int_{\text{state 1}}^{\text{state 2}} \frac{đq_{\text{irrev}}}{T} \qquad (9.9)$$

Also, returning to equation (9.7) and integrating around a complete cycle

$$\oint \frac{đq_{\text{rev}}}{T} > \oint \frac{đq_{\text{irrev}}}{T}$$

but we also know that

$$\oint \frac{đq_{\text{rev}}}{T} = 0$$

implying that

$$0 > \oint \frac{đq_{\text{irrev}}}{T} \qquad (9.10)$$

Combining equation (9.6) with the inequality (9.10), we derive

$$0 \geqslant \oint \frac{đq}{T} \qquad (9.11)$$

where $đq$ represents any flow of heat, with the equality applying to a reversible change for which $đq = đq_{\text{rev}}$, and the inequality applying to an irreversible change for which $đq = đq_{\text{irrev}}$.

The inequality (9.11) is one of the most important expressions in science. It is known as the **Clausius inequality**, after the nineteenth century German scientist Rudolf Clausius, who was one of the giants of the development of thermodynamics. As hinted at on page 144, and as we are about to see, it is the Clausius Inequality – the general expression (9.11), and its irreversible companion, expression (9.10) – that unlocks spontaneity, unidirectionality and irreversibility.

9.4 Real changes

Consider any system that undergoes a change in state, from an initial state [1] to a final state [2]. States [1] and [2] are equilibrium states, and so all the state functions – the pressure P, the temperature T, the enthalpy H, and the entropy S – will all have well-defined, single values in state [1] and also in state [2]. Focusing on the entropy S, expression (9.9) tells us that, for any real, irreversible, change in state from state [1] to state [2]

$$\Delta S > \int_{\text{state 1}}^{\text{state 2}} \frac{đq_{\text{irrev}}}{T} \qquad (9.9)$$

At first sight, that does not appear to be very informative: inequality (9.9) simply tells us that the entropy change $\Delta S = S_2 - S_1$ for any given change in state is greater than another number. Well, 23 is greater than 9, greater than 2, and greater than –6 too. So what?

As it happens, so a lot.

One important aspect of the inequality (9.9) is that it is real. Because it includes the irreversible heat flow đq_{irrev}, it must apply to real-world situations, rather than to the theoretical abstractions associated with reversible paths. And a second important aspect of the inequality (9.9) is that it is perfectly general, applying to any system.

So let us apply expression (9.9) to a change that takes place within an isolated system. Because the system is isolated, there is no exchange of matter, and no exchange of energy, across the system boundary. Hence, for any change in an isolated system đq_{irrev} must be zero. But, according to expression (9.9)

$$\Delta S > \int_{state\ 1}^{state\ 2} \frac{đq_{irrev}}{T} \tag{9.9}$$

hence, for any real change in an isolated system

$$\Delta S = S_2 - S_1 > 0$$

and so

$$S_2 > S_1 \tag{9.12}$$

Expression (9.12) tells us that, for any change in an isolated system, the entropy S_2 of the final state [2] is necessarily greater than the entropy S_1 of the initial state [1]; it also implies that, if state [1] is in thermodynamic equilibrium with state [2], then $\Delta S = 0$ and so $S_2 = S_1$.

At this point, we need to refer to page 266, where is says "*It therefore follows that any change within an isolated system must, always, be spontaneous*". With reference to the inequality (9.12), this implies that a spontaneous – and hence unidirectional and irreversible – change in an isolated system is therefore associated with an increase in the system's entropy. Looked at the other way around, if we know, or can compute, the entropy S_X of the any state [X] of an isolated system, and the entropy S_Y of any other state [Y], then we now know that, if $S_X < S_Y$, then state [X] will spontaneously change to state [Y] *but not the other way around* - state [Y] will never spontaneously change to state [X]. The change from state [X] to state [Y] is spontaneous, unidirectional and irreversible. And it is the entropy of each state that tells us so.

Although barriers which are perfectly impermeable to both matter and energy are theoretical rather than real, many real barriers are sufficiently impermeable to allow the corresponding systems to act as isolated systems, and we shall examine some examples shortly.

But there is one system which is truly isolated. The universe as a whole – if only because we can choose some zone far enough out in space, wherever we might like, in which to 'draw' the boundary. If we apply inequality (9.12) to the universe, we derive the statement

The entropy of the universe necessarily increases

which is another way of stating the Second Law.

What about changes in open or closed systems, for which dq_{irrev} is likely to be non-zero? Since entropy is an extensive state function, then the entropy S_{uni} of the universe is the sum of the entropy S_{sys} of the system and the entropy S_{sur} of the surroundings

$$S_{uni} = S_{sys} + S_{sur}$$

So for any change of state of the system, and the corresponding change of state in the surroundings

$$\Delta S_{uni} = \Delta S_{sys} + \Delta S_{sur}$$

If the change of state within the system is spontaneous, then

$$\Delta S_{uni} = \Delta S_{sys} + \Delta S_{sur} > 0 \tag{9.13}$$

The inequality (9.13) tells us that, for a change in an open or closed system to be spontaneous, we must investigate the entropy change ΔS_{sur} for the surroundings, as well as the entropy change ΔS_{sys} of the system: knowledge of the entropy change ΔS_{sys} within an open or closed system alone does not allow us to determine whether or not a change is spontaneous. If, for example, ΔS_{sys} has a value, say, of +30 J K^{-1}, then, since this is a positive value, it might be thought that the corresponding change is not spontaneous. And if the corresponding value for the entropy change ΔS_{sur} of the surroundings is, say, -10 J K^{-1}, then the total entropy change for the universe ΔS_{uni} is +20 J K^{-1}, and the change is indeed not spontaneous. But if ΔS_{sur} is -80 J K^{-1}, then ΔS_{uni} is -50 J K^{-1}, and the change is spontaneous. This suggests that if we wish to understand spontaneity in closed and open systems, we need to keep track of entropy changes not only within the system, but also within the surroundings too. This is true, but, as we shall see in the Chapter 13, there is a powerful, and much more easily applied, way of understanding spontaneity within a closed system by reference to state functions of the system alone.

That completes the key elements of the 'story' of the Second Law. It's important, and deep – so you may wish to re-read the last few pages before continuing. When you are ready to continue, you will find that the next section will focus on some familiar applications, showing how entropy changes can be calculated, and how they can be interpreted in terms of verifying that some changes are, or are not, spontaneous. We then derive a very important equation combining the First and Second Laws into a single expression, followed by some applications.

9.5 Two examples

9.5.1 Heat flow down a temperature gradient

As a first example of putting equation (9.1) into practice, and of how to compute entropy changes, consider the familiar system shown in Figure 9.1 of two blocks, each of n mol of the same material, such that the block on the left is originally at a temperature T_L, and the block on the right at a temperature T_R, such that $T_L > T_R$. As we know, the system, which initially is not in equilibrium, changes so that the block on the left cools whilst the block on the right

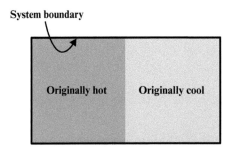

Figure 9.1 Heat flow down a temperature gradient.

warms up, until both blocks have the same temperature T. We also know that this change is spontaneous – our task here is to prove it.

The system boundary is adiabatic, and so no heat is exchanged between the system and the surroundings; furthermore, there is no exchange of matter across the boundary, which therefore allows the system to behave as an isolated system. Our prediction is that the entropy of the system, once both blocks have attained the final temperature T, is greater than the entropy of the initial state of the system, with the left-hand block at temperature T_L, and the right-hand block at temperature T_R.

To test this prediction, we need to compute the entropy change associated with the heat flow between the two blocks. Since entropy is an extensive function, the entropy S_{sys} of the whole system is the sum of the entropies S_L and S_R of the two blocks

$$S_{sys} = S_L + S_R$$

and so

$$\Delta S_{sys} = \Delta S_L + \Delta S_R$$

For the block on the left, according to equation (9.1)

$$dS_L = \left(\frac{\dj q_{rev}}{T}\right)_L \tag{9.1}$$

If we regard the heat flow from the left hand block to the right-hand block to be reversible, then, as we saw on page 259, the heat flow can be assumed to take place at constant volume, and so we can use equation (5.14a)

$$dU_V = n\, C_V\, dT_V = \dj q_V \tag{5.14a}$$

where C_V is the molar heat capacity at constant volume for the material from which the block is made.

Expressing (5.14a) as

$$\dj q_{rev} = n\, C_V\, dT$$

then

$$dS_L = nC_V\, \frac{dT}{T}$$

and so, for a finite change from an initial temperature T_L to a final temperature T

$$\Delta S_L = nC_V \int_{T_L}^{T} \frac{dT}{T} = nC_V \ln \frac{T}{T_L}$$

Similarly, for the right-hand block

$$\Delta S_R = nC_V \int_{T_R}^{T} \frac{dT}{T} = nC_V \ln \frac{T}{T_R}$$

and so

$$\Delta S_{sys} = \Delta S_L + \Delta S_R$$

$$= nC_V \ln \frac{T}{T_L} + nC_V \ln \frac{T}{T_R}$$

$$= nC_V \ln \frac{T^2}{T_L T_R}$$

Now, in Chapter 8, on page 259, we showed that

$$T = \frac{T_L + T_R}{2} \tag{8.4}$$

from which

$$\Delta S_{sys} = nC_V \ln \frac{(T_L + T_R)^2}{4 T_L T_R} \tag{9.14}$$

Since the system boundary is adiabatic, and no matter passes between the system and the surroundings, the system can be considered to be isolated. Accordingly, if ΔS_{sys}, as expressed by equation (9.14), is positive, the change in state is spontaneous; if equation (9.14) is negative, then it isn't. What, then, is the sign of the expression in equation (9.14)?

Given that n and C_V are necessarily positive, the sign of the expression in equation (9.14) is determined by the term $\ln[(T_L + T_R)^2 / 4T_L T_R]$, which is positive if $(T_L + T_R)^2 > 4T_L T_R$, but negative if $(T_L + T_R)^2 < 4T_L T_R$.

Now, we know that, in the initial state

$$T_L > T_R$$

giving

$$T_L - T_R > 0$$

and so

$$(T_L - T_R)^2 > 0$$

implying that $\quad T_L^2 - 2T_L T_R + T_R^2 > 0$

If we now add $4 T_L T_R$ to both sides

$$T_L^2 + 2 T_L T_R + T_R^2 > 4 T_L T_R$$

and so

$$(T_L + T_R)^2 > 4 T_L T_R$$

from which

$$\Delta S_{sys} = nC_V \ln \frac{(T_L + T_R)^2}{4T_L T_R} > 0 \tag{9.15}$$

Since both T_L and T_R are necessarily positive, the entropy change ΔS_{sys} is also positive, and so the corresponding change in state – heat flow down a temperature gradient – happens spontaneously.

One final point: if $T_L = T_R$, the system has a uniform temperature throughout, and so is in thermodynamic equilibrium. In this equilibrium case,

$$(T_L + T_R)^2 = (2T_L)^2 = 4T_L^2 = 4T_L T_R$$

and so equation (9.14) becomes

$$\Delta S_{sys} = nC_V \ln \frac{(T_L + T_R)^2}{4T_L T_R} = nC_V \ln(1) = 0$$

This demonstrates that a condition for two states [1] and [2] of an isolated system to be in thermodynamic equilibrium is that $\Delta S_{sys} = 0$.

9.5.2 Adiabatic expansion into a vacuum

Our second example is the adiabatic expansion of an ideal gas into a vacuum – another change of state which we know is spontaneous. Our analysis here will be based on the results of the discussion in Chapter 5 (see pages 121 to 126), with Figure 9.2 being a simplified version of Figure 5.6.

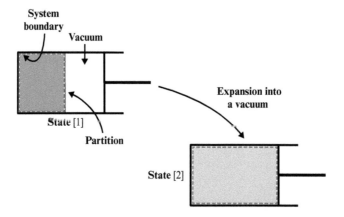

Figure 9.2 Expansion of an ideal gas into a vacuum. An ideal gas in an initial state [1] ≡ [P_1, V_1, T_1] expands into a vacuum to reach a final state [2] ≡ [$P_2, V_2, T_2 = T_1$].

The system comprises an ideal gas that expands adiabatically and irreversibly from state [1] to state [2]. The system boundary is adiabatic, and there is no flow of matter across the boundary, implying that the system behaves as an isolated system. Since the expansion is against a vacuum, no work is done either by, or on, the system and so đw = 0; furthermore, the adiabatic path implies that đq = 0, where both đq and đw correspond to the values of the heat, and work, associated with the real, irreversible change.

The change in internal energy $dU = đq - đw = 0$; also, since the gas is ideal, and U is a function of temperature only (see page 97), then $dT = 0$. The final temperature T_2 is therefore equal to the initial temperature T_1.

To calculate the entropy change we need to compute

$$dS = \frac{đq_{rev}}{T} \tag{9.1}$$

Now, because the change is adiabatic, there is an immediate temptation to say "because the change is adiabatic, $đq_{rev} = 0$, therefore $dS = \Delta S = 0$". Tempting; but wrong.

The error is that $đq_{rev}$ in equation (9.1) is the *reversible* heat flow associated with the change from the initial to the final state – the heat that is exchanged along a fully reversible path between the required states. This fully reversible path is hypothetical, and especially so in this case, for the real path, the path that was actually followed, is (inevitably) irreversible, and (in this particular case) adiabatic. So, for the actual path, the heat exchange $đq_{irrev} = 0$, and the corresponding value of $đq_{rev}$ remains to be determined. There is, however, one clue, for we know, according to the inequality (5.38), that

$$đq_{rev} > đq_{irrev} \tag{5.38}$$

and so, since $đq_{irrev} = 0$, then $đq_{rev}$ must be some positive number.

In fact, this information alone is all we need to determine whether the change is spontaneous or not, for, according to equation (9.1)

$$dS = \frac{đq_{rev}}{T} \tag{9.1}$$

then, since the temperature T is necessarily positive, and, given that we have just shown that $đq_{rev}$ is positive, then any infinitesimal dS must be positive, as must a finite difference ΔS. The entropy change ΔS is therefore positive, so the change must be spontaneous, as expected.

That argument is valid, but it does not quantify what the entropy change ΔS actually is. To do this, we do have to determine $đq_{rev}$ – which requires us to identify the appropriate reversible change. The key here is to recognise that the temperature of the final state is the same as that of the original state, so a reversible isothermal change will take the system from the initial state $[1] \equiv [P_1, V_1, T_1]$ to the correct final state $[2] \equiv [P_2, V_2, T_1]$.

For an isothermal change of an ideal gas, $dU = 0$, and, according to the First Law

$$dU = đq_{rev} - đw_{rev} = 0 \tag{5.1a}$$
$$\therefore đq_{rev} = đw_{rev} = P\,dV$$

Equation (9.1) therefore becomes

$$dS = \frac{đq_{rev}}{T} = \frac{P\,dV}{T} \tag{9.16}$$

For an ideal gas

$$PV = nRT \tag{3.3}$$

and so, for a finite change, ΔS may be quantified as

$$\Delta S = \int_{\text{state 1}}^{\text{state 2}} \frac{đq_{\text{rev}}}{T} = \int_{V_1}^{V_2} \frac{P\,dV}{T} = nR \int_{V_1}^{V_2} \frac{dV}{V} = nR \ln \frac{V_2}{V_1} \tag{9.17}$$

The change in state is an expansion, and so $V_2 > V_1$, implying that $\ln(V_2/V_1) > 0$. Since n and R are necessarily positive, equation (9.17) verifies that ΔS is positive. Adiabatic expansion into a vacuum is therefore spontaneous, as expected. Also, if $V_2 = V_1$, the original state does not change at all, the system remains in equilibrium, and, according to equation (9.17), $\Delta S = 0$.

9.6 The First and Second Laws combined

According to the First Law

$$dU = đq - đw = 0 \tag{5.1a}$$

This law is universally true, and applies to all systems, and all paths between any initial state and any final state. So, for a real, irreversible, change between any two states

$$dU = đq_{\text{irrev}} - đw_{\text{irrev}} \tag{9.18}$$

and, for a hypothetical, reversible, change between the same two states

$$dU = đq_{\text{rev}} - đw_{\text{rev}} \tag{9.19}$$

Now, from the definition of entropy, if dS is the corresponding entropy change for that change in state, then

$$đq_{\text{rev}} = T\,dS \tag{9.1}$$

and for a system that performs only P, V work of expansion

$$đw_{\text{rev}} = P\,dV \tag{2.9}$$

so that equation (9.18) now becomes

$$dU = T\,dS - P\,dV \tag{9.20a}$$

or

$$T\,dS = dU + P\,dV \tag{9.20b}$$

Equations (9.20a) and (9.20b) bring the First and Second Laws together into a pair of equivalent equations – equations that apply to all systems, and all paths between any initial and final states. Furthermore, since all the variables in equations (9.20a) and (9.20b) are state functions, the equations apply both to hypothetical, reversible, paths as well as to real, irreversible, paths.

A further important equation – an equation that relates the change dH in enthalpy to the change dS in entropy – can be derived from the definition of enthalpy as

$$H = U + PV \tag{6.4}$$

from which

$$dH = dU + P\,dV + V\,dP$$

Given that

$$dU = T\,dS - P\,dV \tag{9.20a}$$

we may substitute equation (9.20a) for dU and so

$$dH = T\,dS - P\,dV + P\,dV + V\,dP$$

implying that

$$dH = T\,dS + V\,dP \tag{9.21a}$$

and

$$T\,dS = dH - V\,dP \tag{9.21b}$$

Equations (9.21a) and (9.21b) combine the First and Second Laws in the context of the enthalpy H, which is appropriate to conditions of constant pressure; equations (9.20a) and (9.20b), which are expressed in terms of the internal energy U, are especially appropriate to conditions of constant volume.

Also, equations (9.20a), (9.20b), (9.21a) and (9.21b) are particularly helpful in enabling us to compute entropy changes without having to determine $đq_{rev}$ first. So, taking the example of the adiabatic expansion of an ideal gas into a vacuum, as discussed on pages 121 to 126, and 278 to 280, since we know that, for this change, $dU = 0$, then equation (9.20b) immediately tells us that, in this case

$$T\,dS = P\,dV$$

and so, as we saw on page 280, for a finite change from state [X] to state [Y]

$$\Delta S = \int_{state\ X}^{state\ Y} dS = \int_{V_X}^{V_Y} \frac{P\,dV}{T} = nR \int_{V_X}^{V_Y} \frac{dV}{V} = nR \ln \frac{V_Y}{V_X} \tag{9.17}$$

Also, for the system shown in Figure 9.1, which depicts the flow of heat down a temperature gradient, we know, as discussed on page 276, that the change for each block, and for the system as a whole, is at constant volume, and so $dV = 0$. Equation (9.20b), therefore becomes

$$T\,dS = dU$$

Reference to page 107 will show that, for a general system of constant mass

$$dU = C_V dT + \left(\frac{\partial U}{\partial V}\right)_T dV \tag{5.15}$$

and so at constant volume

$$T\,dS = dU = C_V dT = n\,\boldsymbol{C_V} dT$$

The left-hand block changes temperature from T_L to T, and so, for the left-hand block

$$\Delta S_L = \int_{\text{initial state}}^{\text{final state}} dS = nC_V \int_{T_L}^{T} \frac{dT}{T} = nC_V \ln \frac{T}{T_L}$$

and similarly for the right-hand block

$$\Delta S_R = \int_{\text{initial state}}^{\text{final state}} dS = nC_V \int_{T_R}^{T} \frac{dT}{T} = nC_V \ln \frac{T}{T_R}$$

giving the same results as those we derived on page 277.

9.7 The mathematics of entropy

9.7.1 S as a function $S(T, V)$ of T and V

Invoking the now-familiar general principle that

> In the absence of gravitation, motion, electric, magnetic and surface effects, a given mass of a pure substance has two, and only two, independent state functions.

then, for a closed system of constant mass, the entropy S may be expressed as a function of any other two state functions, for example, the temperature T and the volume V. Accordingly, if $S = S(T, V)$, then

$$dS = \left(\frac{\partial S}{\partial T}\right)_V dT + \left(\frac{\partial S}{\partial V}\right)_T dV \tag{9.22}$$

Also, from equation (9.20b)

$$T\,dS = dU + P\,dV \tag{9.20b}$$

and so, dividing throughout by $T\,dT$

$$\frac{dS}{dT} = \frac{(dU + P\,dV)}{T\,dT}$$

At constant volume, $dV = 0$, giving

$$\left(\frac{\partial S}{\partial T}\right)_V = \frac{1}{T}\left(\frac{\partial U}{\partial T}\right)_V$$

The molar heat capacity at constant volume C_V is defined as

$$C_V = \left(\frac{\partial U}{\partial T}\right)_V \tag{5.16}$$

and for a system of n mol of constant mass,

$$\left(\frac{\partial U}{\partial T}\right)_V = n\left(\frac{\partial U}{\partial T}\right)_V$$

THE MATHEMATICS OF ENTROPY

and so

$$\left(\frac{\partial S}{\partial T}\right)_V = \frac{nC_V}{T} \tag{9.23}$$

An important inference from equation (9.23), attributable to the facts that n, C_V and T are necessarily positive, is that $(\partial S/\partial T)_V$ must always be positive – at constant volume, the entropy of any system must increase with temperature.

Using equation (9.23) to substitute nC_V/T for $(\partial S/\partial T)_V$ in equation (9.22), we obtain

$$dS = nC_V \frac{dT}{T} + \left(\frac{\partial S}{\partial V}\right)_T dV \tag{9.24a}$$

In equation (9.24a), the partial derivative $(\partial S/\partial V)_T$ represents the change ∂S_T in the system's entropy attributable to a change ∂V_T in the system's volume, at constant temperature – by how much the entropy of the system changes if it is allowed to expand. Given that entropy is not as familiar a state function as, say, pressure, volume, or temperature, partial derivatives that involve entropy are not so easy to interpret. However, as we shall see on page 639, it is possible to prove that, for any closed system, $(\partial S/\partial V)_T = (\partial P/\partial T)_V$, which is rather more accessible – it represents the increase ∂P_V in pressure attributable to an increase ∂T_V in temperature at constant volume. An alternative expression of equation (9.24a) is therefore

$$dS = nC_V \frac{dT}{T} + \left(\frac{\partial P}{\partial T}\right)_V dV \tag{9.24b}$$

9.7.2 S as a function $S(T, P)$ of T and P

Since the entropy S of any closed system may be represented as a function of any two other state functions, we now choose the temperature T and the pressure P. So, if $S = S(T, P)$, we have

$$dS = \left(\frac{\partial S}{\partial T}\right)_P dT + \left(\frac{\partial S}{\partial P}\right)_T dP \tag{9.25}$$

Equation (9.21b) states

$$T\,dS = dH - V\,dP \tag{9.21b}$$

Dividing throughout by $T\,dT$

$$\frac{dS}{dT} = \frac{(dH - V\,dP)}{T\,dT}$$

At constant pressure, $dP = 0$, and so

$$\left(\frac{\partial S}{\partial T}\right)_P = \frac{1}{T}\left(\frac{\partial H}{\partial T}\right)_P$$

From the definition of the molar heat capacity at constant pressure, C_P

$$C_P = \left(\frac{\partial H}{\partial T}\right)_P \tag{6.10}$$

implying that for a system of n mol of constant mass,

$$\left(\frac{\partial H}{\partial T}\right)_P = n\left(\frac{\partial \overline{H}}{\partial T}\right)_P$$

and so

$$\left(\frac{\partial S}{\partial T}\right)_P = \frac{nC_P}{T} \tag{9.26}$$

As we saw in relation to $(\partial S/\partial T)_V$ in equation (9.23), the right-hand side of equation (9.26) must always be positive, implying that $(\partial S/\partial T)_P$ is always positive too – at constant pressure, the entropy of any system must increase with temperature.

If we use equation (9.26) to substitute nC_P/T for $(\partial S/\partial T)_P$ in equation (9.25), we derive

$$dS = nC_P\frac{dT}{T} + \left(\frac{\partial S}{\partial P}\right)_T dP \tag{9.27a}$$

In equation (9.27a), the partial derivative $(\partial S/\partial P)_T$ represents the change ∂S_T in the system's entropy attributable to a change ∂P_T in the system's pressure, at constant temperature – by how much the entropy of the system changes if it is compressed. Once again, that's not a very obvious concept, but, as we shall see on page 639, it is possible to prove that, for any closed system, $(\partial S/\partial P)_T = -(\partial V/\partial T)_P$, this being the negative of the increase ∂V_P in volume attributable to an increase ∂T_P in temperature at constant pressure – which is a measure of how much a system expands when it is heated. An alternative form of equation (9.27a) is therefore

$$dS = nC_P\frac{dT}{T} - \left(\frac{\partial V}{\partial T}\right)_P dP \tag{9.27b}$$

9.7.3 S as a function $S(P, V)$ of P and V

If we choose to express the entropy S as a function $S(P, V)$ of the pressure P and volume V, then

$$dS = \left(\frac{\partial S}{\partial P}\right)_V dP + \left(\frac{\partial S}{\partial V}\right)_P dV \tag{9.28}$$

Equation (9.28) contains two more partial derivatives relating to entropy, the first representing the rate of change of entropy with respect to pressure at constant volume, and the second representing the rate of change of entropy with respect to volume at constant pressure.

To find a replacement for $(\partial S/\partial P)_V$, our starting point is the mathematical identity that

$$\left(\frac{\partial S}{\partial P}\right)_V = \left(\frac{\partial S}{\partial T}\right)_V \left(\frac{\partial T}{\partial P}\right)_V$$

Reference to page 283 will show that

$$\left(\frac{\partial S}{\partial T}\right)_V = \frac{nC_V}{T} \tag{9.23}$$

and so

$$\left(\frac{\partial S}{\partial P}\right)_V = \frac{nC_V}{T}\left(\frac{\partial T}{\partial P}\right)_V \tag{9.29}$$

Similarly, for a replacement for $(\partial S/\partial V)_P$, since

$$\left(\frac{\partial S}{\partial V}\right)_P = \left(\frac{\partial S}{\partial T}\right)_P \left(\frac{\partial T}{\partial V}\right)_P$$

and

$$\left(\frac{\partial S}{\partial T}\right)_P = \frac{nC_P}{T} \tag{9.26}$$

then

$$\left(\frac{\partial S}{\partial V}\right)_P = \frac{nC_P}{T}\left(\frac{\partial T}{\partial V}\right)_P \tag{9.30}$$

Using equations (9.29) and (9.30) to substitute for $(\partial S/\partial P)_V$ and $(\partial S/\partial V)_P$ in equation (9.28) gives

$$dS = \frac{nC_V}{T}\left(\frac{\partial T}{\partial P}\right)_V dP + \frac{nC_P}{T}\left(\frac{\partial T}{\partial V}\right)_P dV \tag{9.31}$$

In equation (9.31), $(\partial T/\partial P)_V$ represents the increase in temperature ∂T_V attributable to an increase in the system pressure ∂P_V at constant volume; $(\partial T/\partial V)_P$ represents the increase in temperature ∂T_P attributable to an expansion ∂V_P at constant pressure.

9.8 Entropy changes for an ideal gas

9.8.1 Three equivalent equations

Equations (9.24a), (9.27a) and (9.31)

$$dS = nC_V\frac{dT}{T} + \left(\frac{\partial S}{\partial V}\right)_T dV \tag{9.24a}$$

$$dS = nC_P\frac{dT}{T} + \left(\frac{\partial S}{\partial P}\right)_T dP \tag{9.27a}$$

$$dS = \frac{nC_V}{T}\left(\frac{\partial T}{\partial P}\right)_V dP + \frac{nC_P}{T}\left(\frac{\partial T}{\partial V}\right)_P dV \tag{9.31}$$

all apply to closed systems of a single pure substance, and they are all equivalent: each may be used under all circumstances, and the choice of which is used in any particular circumstance depends on which pair of the three variables P, V and T is most relevant. To illustrate how these equations can be used, we now apply them to calculate the entropy change ΔS for the specific system of n mol of an ideal gas, undergoing a change from an initial state $[1] \equiv [P_1, V_1, T_1]$ to a final state $[2] \equiv [P_2, V_2, T_2]$

9.8.2 ΔS for an ideal gas as a function of T and V

We deal firstly with equation (9.24a)

$$dS = nC_V\frac{dT}{T} + \left(\frac{\partial S}{\partial V}\right)_T dV \tag{9.24a}$$

and to find a replacement for the term $(\partial S/\partial V)_T$, we use equation (9.20b), which contains both dS and dV

$$T\,dS = dU + P\,dV \tag{9.20b}$$

Dividing throughout by $T\,dV$, and at constant temperature T, we therefore have

$$\left(\frac{\partial S}{\partial V}\right)_T = \frac{1}{T}\left(\frac{\partial U}{\partial V}\right)_T + \frac{P}{T} \tag{9.32}$$

Equation (9.32) is general, and applies to all closed systems. In the special case of a system of an ideal gas, firstly, we know that the internal energy U is a function of temperature only, and so, as we saw on page 110, $(\partial U/\partial V)_T = 0$. Furthermore, for an ideal gas,

$$PV = nRT \tag{3.3}$$

and so

$$\frac{P}{T} = \frac{nR}{V}$$

implying that equation (9.32) becomes

$$\left(\frac{\partial S}{\partial V}\right)_T = \frac{nR}{V}$$

For an ideal gas, equation (9.24a) therefore becomes

$$dS = nC_V\frac{dT}{T} + nR\frac{dV}{V} \tag{9.33}$$

For a finite change from an initial state $[1] \equiv [P_1, V_1, T_1]$ to a final state $[2] \equiv [P_2, V_2, T_2]$, we therefore have

$$\Delta S = \int_{\text{state 1}}^{\text{state 2}} dS = \int_{T_1}^{T_2} nC_V\frac{dT}{T} + \int_{V_1}^{V_2} nR\frac{dV}{V}$$

For an ideal gas, n and C_V are constants and so

$$\Delta S = \int_{\text{state 1}}^{\text{state 2}} dS = nC_V\int_{T_1}^{T_2}\frac{dT}{T} + nR\int_{V_1}^{V_2}\frac{dV}{V}$$

Hence, for any change for an ideal gas

$$\Delta S = nC_V \ln\frac{T_2}{T_1} + nR\ln\frac{V_2}{V_1} \tag{9.34}$$

9.8.3 ΔS for an ideal gas as a function of T and P

Turning now to equation (9.27a)

$$dS = nC_P\frac{dT}{T} + \left(\frac{\partial S}{\partial P}\right)_T dP \tag{9.27a}$$

to find a replacement for $(\partial S/\partial P)_T$, we use equation (9.21b)

$$T\,dS = dH - V\,dP \tag{9.21b}$$

which contains both dS and dP.

Dividing throughout by $T\,dP$, and at constant temperature T

$$\left(\frac{\partial S}{\partial P}\right)_T = \frac{1}{T}\left(\frac{\partial H}{\partial P}\right)_T - \frac{V}{T} \tag{9.35}$$

Like equation (9.32), equation (9.35) is general, and if we apply it to an ideal gas, firstly, $(\partial H/\partial P)_T = 0$ since, as we saw on page 153, the enthalpy of an ideal gas is a function of temperature only, and secondly, given that

$$PV = nRT \tag{3.3}$$

then

$$\frac{V}{T} = \frac{nR}{P}$$

Accordingly, for n mol of an ideal gas, equation (9.35) becomes

$$\left(\frac{\partial S}{\partial P}\right)_T = -\frac{nR}{P}$$

and equation (9.27a) simplifies to

$$dS = nC_P\frac{dT}{T} - nR\frac{dP}{P} \tag{9.36}$$

For a finite change from an initial state $[1] \equiv [P_1, V_1, T_1]$ to a final state $[2] \equiv [P_2, V_2, T_2]$, we therefore have

$$\Delta S = \int_{\text{state 1}}^{\text{state 2}} dS = \int_{T_1}^{T_2} nC_P\frac{dT}{T} - \int_{P_1}^{P_2} nR\frac{dP}{P}$$

For an ideal gas, n and C_P are constants and so

$$\Delta S = \int_{\text{state 1}}^{\text{state 2}} dS = nC_P \int_{T_1}^{T_2} \frac{dT}{T} - nR \int_{P_1}^{P_2} \frac{dP}{P}$$

Hence, for any change for an ideal gas

$$\Delta S = nC_P \ln \frac{T_2}{T_1} - nR \ln \frac{P_2}{P_1} \tag{9.37}$$

9.8.4 ΔS for an ideal gas as a function of P and V

We now turn now to equation (9.31)

$$dS = \frac{nC_V}{T}\left(\frac{\partial T}{\partial P}\right)_V dP + \frac{nC_P}{T}\left(\frac{\partial T}{\partial V}\right)_P dV \tag{9.31}$$

Equation (9.31) is general for any closed system of constant mass; for an ideal gas

$$PV = nRT \tag{3.3}$$

and so

$$\left(\frac{\partial T}{\partial P}\right)_V = \frac{V}{nR}$$

implying that

$$\frac{1}{T}\left(\frac{\partial T}{\partial P}\right)_V = \frac{V}{nRT} = \frac{1}{P}$$

Similarly

$$\left(\frac{\partial T}{\partial V}\right)_P = \frac{P}{nR}$$

and

$$\frac{1}{T}\left(\frac{\partial T}{\partial V}\right)_P = \frac{P}{nRT} = \frac{1}{V}$$

and so equation (9.31) becomes, for an ideal gas

$$dS = nC_V \frac{dP}{P} + nC_P \frac{dV}{V} \tag{9.38}$$

For a finite change from an initial state $[1] \equiv [P_1, V_1, T_1]$ to a final state $[2] \equiv [P_2, V_2, T_2]$, we therefore have

$$\Delta S = \int_{\text{state 1}}^{\text{state 2}} dS = \int_{P_1}^{P_2} nC_V \frac{dP}{P} + \int_{V_1}^{V_2} nC_P \frac{dV}{V}$$

For an ideal gas, n, C_V and C_P are all constants and so

$$\Delta S = \int_{\text{state 1}}^{\text{state 2}} dS = nC_V \int_{P_1}^{P_2} \frac{dP}{P} + nC_P \int_{V_1}^{V_2} \frac{dV}{V}$$

Hence, for any change for an ideal gas

$$\Delta S = nC_V \ln \frac{P_2}{P_1} + nC_P \ln \frac{V_2}{V_1} \tag{9.39}$$

9.8.5 Three more equivalent equations

Those last few pages included quite a lot of mathematics, but everything was based on just three basic principles:

- firstly, that the entropy S of any closed system of a constant mass of a single pure substance is a function of two, and only two, other state functions, so enabling us to write expressions of the form

$$dS = \left(\frac{\partial S}{\partial P}\right)_V dP + \left(\frac{\partial S}{\partial V}\right)_P dV \tag{9.28}$$

- secondly, the definitions of C_V and C_P as

$$C_V = \left(\frac{\partial U}{\partial T}\right)_V \tag{5.16}$$

and

$$C_P = \left(\frac{\partial H}{\partial T}\right)_P \tag{6.10}$$

- and thirdly, for the applications to an ideal gas, the ideal gas law

$$PV = nRT \tag{3.3}$$

Given these principles, everything else was a matter of mathematical manipulation, which required some care, but is not intrinsically difficult.

As a result, we derived three equations for the change ΔS in the entropy of n mol of an ideal gas, undergoing a finite change in state from an initial state $[1] \equiv [P_1, V_1, T_1]$ to a final state $[2] \equiv [P_2, V_2, T_2]$:

$$\Delta S = nC_V \ln \frac{T_2}{T_1} + nR \ln \frac{V_2}{V_1} \tag{9.34}$$

$$\Delta S = nC_P \ln \frac{T_2}{T_1} - nR \ln \frac{P_2}{P_1} \tag{9.37}$$

and

$$\Delta S = nC_V \ln \frac{P_2}{P_1} + nC_P \ln \frac{V_2}{V_1} \tag{9.39}$$

Since the gas is ideal, the ideal gas law

$$PV = nRT \tag{3.3}$$

is always true, and so

$$\frac{V_2}{V_1} = \frac{T_2}{T_1} \frac{P_1}{P_2}$$

Equation (9.34) may therefore be written as

$$\Delta S = nC_V \ln \frac{T_2}{T_1} + nR \ln \frac{T_2}{T_1}\frac{P_1}{P_2} = n(C_V + R) \ln \frac{T_2}{T_1} + nR \ln \frac{P_1}{P_2} \tag{9.40}$$

But for an ideal gas, reference to page 153 will verify that

$$C_P = C_V + R \tag{6.13}$$

and so equation (9.40) becomes

$$\Delta S = nC_P \ln \frac{T_2}{T_1} - nR \ln \frac{P_2}{P_1} \tag{9.37}$$

which is identical to equation (9.37).

Similarly, since $PV = nRT$ is always valid for an ideal gas

$$\frac{T_2}{T_1} = \frac{P_2}{P_1} \frac{V_2}{V_1}$$

and so substituting for T_2/T_1 in equation (19.37)

$$\Delta S = nC_P \ln \frac{P_2}{P_1} \frac{V_2}{V_1} - nR \ln \frac{P_2}{P_1} = n(C_P - R) \ln \frac{P_2}{P_1} + nC_P \ln \frac{V_2}{V_1}$$

But

$$C_P = C_V + R \tag{6.13}$$

and so

$$\Delta S = nC_V \ln \frac{P_2}{P_1} + nC_P \ln \frac{V_2}{V_1}$$

which is identical to equation (9.39).

The three equations

$$\Delta S = nC_V \ln \frac{T_2}{T_1} + nR \ln \frac{V_2}{V_1} \tag{9.34}$$

$$\Delta S = nC_P \ln \frac{T_2}{T_1} - nR \ln \frac{P_2}{P_1} \tag{9.37}$$

and

$$\Delta S = nC_V \ln \frac{P_2}{P_1} + nC_P \ln \frac{V_2}{V_1} \tag{9.39}$$

are therefore equivalent.

9.9 Entropy changes at constant pressure

Most chemical and biochemical processes take place at constant pressure, and so the behaviour of a system's entropy, as the temperature varies at constant pressure, is of especial importance. So, expressing S as $S(T, P)$, reference to pages 283 and 284 will show that a change dS in the entropy of a closed system of constant mass of a pure substance is given by

$$dS = \left(\frac{\partial S}{\partial T}\right)_P dT + \left(\frac{\partial S}{\partial P}\right)_T dP \tag{9.25}$$

and

$$dS = nC_P \frac{dT}{T} + \left(\frac{\partial S}{\partial P}\right)_T dP \tag{9.27a}$$

At constant pressure $dP = 0$, and so equation (9.27a) becomes

$$dS = nC_P \frac{dT}{T} \quad \text{at constant pressure} \tag{9.40}$$

hence, for a change in state from state $[1] \equiv [S_1, T_1]$ to state $[2] \equiv [S_2, T_2]$

$$\int_{S_1}^{S_2} dS = S_2 - S_1 = \Delta S_P = n \int_{T_1}^{T_2} C_P \frac{dT}{T} \tag{9.41}$$

If C_P can be assumed to be a constant over the temperature range from T_1 to T_2, then equation (9.41) integrates to

$$S_2 - S_1 = \Delta S_P = nC_P \ln \frac{T_2}{T_1} \tag{9.42}$$

If C_P is not a constant, then equation (9.41) needs to take account of the specific behaviour of C_P as a function of temperature, as discussed on page 194, for example according to equation (6.28)

$$C_P = a + bT + cT^2 + dT^3 + eT^4 \tag{6.28}$$

or the Shomate Equation

$$C_P = A + BT + CT^2 + DT^3 + \frac{E}{T^2} \tag{6.29}$$

9.10 Phase changes

There is, however, one particular circumstance in which C_P is not constant over a given temperature range, and that is when that temperature range spans a phase transition, such as the melting of a solid into a liquid, or the vaporisation of a liquid into a gas.

Figure 9.3 shows a schematic representation of how the molar enthalpy of a representative pure substance changes, at constant pressure, as the temperature rises. Towards the left, the material is a solid, the molar enthalpy rises as the temperature increases, and the slope $(\partial H/\partial T)_P$ of the curve at any point represents the value of C_P for the solid at that temperature. Towards the centre, the material is a liquid, and once again, the molar enthalpy rises, with the slope of the curve at any point representing the corresponding value of C_P for the liquid. At the melting point, the enthalpy 'jumps' from a lower value to a higher value, the magnitude of this jump being the molar enthalpy change associated with the phase change from solid to liquid, an enthalpy change known, as we saw on page 161, as the molar enthalpy of fusion

$$\text{solid} \to \text{liquid} \qquad \qquad \Delta H = \Delta_{\text{fus}} H$$

Given that the solid to liquid phase transition is a change in state, as well as being associated with a molar enthalpy change $\Delta_{\text{fus}} H$, it will also be associated with a molar entropy change $\Delta_{\text{fus}} S$, which can readily be determined from equation (9.1)

$$dS = \frac{\text{d}q_{\text{rev}}}{T} \tag{9.1}$$

Since, at the temperature T_{fus} at which the phase change takes place, the change in state is reversible, then, at constant pressure,

$$\text{d}q_{\text{rev},P} = \Delta_{\text{fus}} H$$

and so

$$\Delta_{\text{fus}} S = \frac{\Delta_{\text{fus}} H}{T_{\text{fus}}} \tag{9.43}$$

So, for the melting of 1 mol of ice into liquid water at a pressure of 1 bar and a temperature of 273 K

$$H_2O\,(s) \to H_2O\,(l) \qquad\qquad \Delta_{\text{fus}} H = 6.01 \text{ kJ mol}^{-1}$$

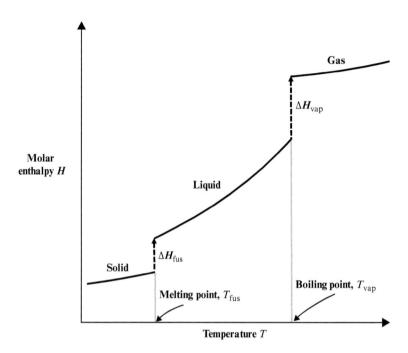

Figure 9.3 Schematic representation of the temperature variation of the molar enthalpy of a pure substance through two phase changes. At the melting point, T_{fus}, the enthalpy 'jumps' by the molar enthalpy of fusion ΔH_{fus}, and at the boiling point, T_{vap}, by the enthalpy of vaporisation ΔH_{vap}. The slope of the enthalpy-temperature graph at any point is the value of C_P at the corresponding temperature, and if C_P is constant over any temperature range, then that portion of the enthalpy-temperature graph is a straight upward-sloping line. The value of C_P is well-defined whilst the material is a solid, a liquid or a gas, but at the point of a phase change, the value of C_P, theoretically, becomes infinite.

implying that

$$\Delta_{fus}S = \frac{6{,}010}{273} = 22.0 \text{ J K}^{-1} \text{ mol}^{-1}$$

Similarly, for the phase change

liquid → vapour $\hspace{4cm} \Delta H = \Delta_{vap}H$

$$\Delta_{vap}S = \frac{\Delta_{vap}H}{T_{vap}} \tag{9.44}$$

and so, for the vaporisation of 1 mol of liquid water at a pressure of 1 bar and a temperature of 373 K

$H_2O\,(l) \rightarrow H_2O\,(g)$ $\hspace{4cm} \Delta_{vap}H = 40.65 \text{ kJ mol}^{-1}$

we have

$$\Delta_{vap}S = \frac{40{,}650}{373} = 109.0 \text{ J K}^{-1} \text{ mol}^{-1}$$

Some representative values for the molar entropies associated with some phase changes are shown in Table 9.1.

PHASE CHANGES

Table 9.1 Some thermodynamic data for phase changes

Selected elements at 1 atm pressure

	Fusion			Vaporisation		
	$\Delta_{fus}H$ kJ mol^{-1}	T_{fus} K	$\Delta_{fus}S$ J K^{-1} mol^{-1}	$\Delta_{vap}H$ kJ mol^{-1}	T_{vap} K	$\Delta_{vap}S$ J K^{-1} mol^{-1}
H$_2$	0.12	14	8.57	0.90	20	45.0
N$_2$	0.71	63	11.27	5.57	77	72.3
O$_2$	0.44	54	8.15	6.82	90	75.8
S	1.72	388	4.43	45	718	62.7
Br$_2$	10.6	266	39.9	30.0	332	90.4
I$_2$	15.5	387	40.1	41.6	458	90.8
Al	10.7	933	11.5	294	2792	105.3
Hg	2.30	234	9.83	59.1	630	93.8

Selected inorganic compounds at 1 atm pressure

	Fusion			Vaporisation		
	$\Delta_{fus}H$ kJ mol^{-1}	T_{fus} K	$\Delta_{fus}S$ J K^{-1} mol^{-1}	$\Delta_{vap}H$ kJ mol^{-1}	T_{vap} K	$\Delta_{vap}S$ J K^{-1} mol^{-1}
H$_2$O	6.01	273	22.0	40.65	373	109.0
HCl	2.00	159	12.6	16.15	118	136.9
HI	2.87	222	12.9	19.76	238	83.0
NH$_3$	5.66	195	29.0	23.33	240	97.2

Selected organic compounds at 1 atm pressure

	Fusion			Vaporisation		
	$\Delta_{fus}H$ kJ mol^{-1}	T_{fus} K	$\Delta_{fus}S$ J K^{-1} mol^{-1}	$\Delta_{vap}H$ kJ mol^{-1}	T_{vap} K	$\Delta_{vap}S$ J K^{-1} mol^{-1}
CH$_4$	0.94	91	10.3	8.19	112	73.1
C$_6$H$_6$	9.87	279	35.4	30.7	353	87.0
CHCl$_3$	9.5	210	45.2	29.2	334	87.4
C$_2$H$_5$OH	4.93	159	31.0	38.6	351	110.0
CH$_3$COOH	11.73	290	40.5	23.7	391	60.6
Phenol	11.51	314	36.7	45.7	455	100.4

Source: *CRC Handbook of Chemistry and Physics*, 96th edn, 2015, Section 6, pages 6-127 to 6-154, for $\Delta_{fus}H$, T_{fus}, $\Delta_{vap}H$ and T_{vap}, and using equations (9.43) and (9.44) to compute the corresponding values of $\Delta_{fus}S$ and $\Delta_{vap}S$. Note that the reference pressure is 1 atm = 1.01325 bar.

As shown in Table 9.1, for a transition from a more condensed state to a less condensed state, the corresponding phase change enthalpies are positive – the reactions are endothermic. The corresponding entropy changes are therefore also positive, implying that the entropy of a liquid is greater than the entropy of the corresponding solid, and that the entropy of a gas is greater than the entropy of the corresponding liquid. Also, for all the materials shown in the table, $\Delta_{vap}S > \Delta_{fus}S$, and this is generally true for all materials, as will be discussed further on pages 343 and 344; note also what is known as **Trouton's rule**, which states that $\Delta_{vap}S$ often takes a value of between 85 J K^{-1} mol^{-1} and 95 J K^{-1} mol^{-1}, as is the case for a number of largely non-polar liquids.

Physically, at a phase transition from solid to liquid, heat supplied to the solid is not used to raise the material's temperature; rather, the heat is used as the enthalpy of fusion, transforming the solid into a liquid. At the melting point, C_P therefore has no meaning; mathematically, since by definition

$$C_P = \left(\frac{\partial H}{\partial T}\right)_P \tag{6.10}$$

then, at the melting point, $\partial H_P = \Delta_{fus}H$ and $\partial T_P = 0$, and so the value of C_P theoretically becomes infinite, as shown by the vertical 'jump' in Figure 9.3. Likewise, at the boiling point, $\partial H_P = \Delta_{vap}H$ and $\partial T_P = 0$, giving rise to another discontinuity, as also shown in Figure 9.3.

This 'breakdown' in the value of C_P through a phase transition implies that we cannot use equation (9.41)

$$\Delta S_P = n \int_{T_1}^{T_2} C_P \frac{dT}{T} \tag{9.41}$$

if the temperature range from T_1 to T_2 spans a phase change. If in fact there is a phase transition at some temperature T_{ph} between T_1 and T_2, then to compute the total change ΔS_P in entropy at constant pressure for the entire temperature range, then we must express ΔS_P as the sum of three components

$$\Delta S_P = \Delta S_P \text{ (phase 1)} + \Delta S_P \text{(phase transition)} + \Delta S_P \text{ (phase 2)}$$

We can use then equation (9.41) for each phase, and an equation of the form of equation (9.43) for the phase transition, as

$$\Delta S_P = n \int_{T_1}^{T_{ph}} C_P\text{(phase 1)}\frac{dT}{T} + n\frac{\Delta_{ph}H}{T_{ph}} + n \int_{T_{ph}}^{T_2} C_P\text{(phase 2)}\frac{dT}{T} \tag{9.45}$$

In principle, equation (9.45) enables us to calculate the entropy change for any system of a constant mass of a pure substance, undergoing a heating process at constant pressure, from any lower temperature T_1 to any higher temperature T_2, within in which there is a phase change from, say, a solid to a liquid; if there is more than one phase change over that temperature range, equation (9.45) can be extended as

$$\Delta S_P = n \int_{T_1}^{T_{\text{ph1} \to 2}} C_P(\text{phase 1}) \frac{dT}{T}$$

$$+ n \frac{\Delta_{\text{ph1} \to 2} H}{T_{\text{ph1} \to 2}} + n \int_{T_{\text{ph1} \to 2}}^{T_{\text{ph2} \to 3}} C_P(\text{phase 2}) \frac{dT}{T} \qquad (9.46)$$

$$+ n \frac{\Delta_{\text{ph2} \to 3} H}{T_{\text{ph2} \to 3}} + n \int_{T_{\text{ph2} \to 3}}^{T_2} C_P(\text{phase 3}) \frac{dT}{T}$$

where $T_{\text{ph1} \to 2}$ and $T_{\text{ph2} \to 3}$ are the temperatures at which the phase transitions take place, and $\Delta_{\text{ph1} \to 2} H$ and $\Delta_{\text{ph2} \to 3} H$ are the molar enthalpy changes for the corresponding phase transitions. The general behaviour of equation (9.46), for one mole of a pure substance, is illustrated in Figure 9.4. As can be seen, at constant pressure, the entropy of the system increases with temperature, as expected from equation (9.26); the graph also shows the discontinuous 'jumps' attributable to the phase changes – these discontinuities being similar to those shown in Figure 9.3.

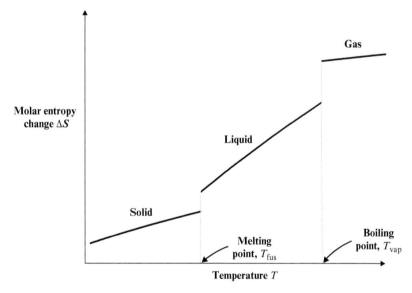

Figure 9.4 Schematic representation of the temperature variation of the molar entropy of a pure substance through two phase changes. The graph shows how, for a pure material at constant pressure, $\Delta S = S(T) - S(T_0)$ varies with the temperature T, where $S(T_0)$ is the molar entropy of the system at an arbitrary reference temperature T_0. Since $S(T_0)$ is a constant, $S(T)$ increases with temperature, with discontinuous 'jumps' at each phase transition.

Over a range of low temperatures from, say, a very low temperature T_1 to a somewhat higher temperature T, most materials will be solid. If C_P can be regarded as constant over the temperature range T_1 to T, then the first term in equation (9.46) integrates to $n\, C_P \ln(T/T_1)$. But as T_1 becomes lower, experiment shows that C_P steadily decreases, tending towards zero as the temperature approaches absolute zero. To evaluate the first term in equation (9.46) requires us to know how C_P for the system of interest varies with temperature as the temperature gets progressively closer to absolute zero. In general, there are two ways of doing this: either to use a rather complex equation, derived by the Dutch physicist and Nobel Prize Winner Peter Debye, or to estimate the area under an experimentally determined graph of C_P/T as the vertical axis and T as the horizontal axis, or of C_P as the vertical axis and $\ln T$ as the horizontal axis – a matter we will explore further in Chapter 11.

A final note on phase changes. As illustrated in Figure 9.3, and as we discussed on page 291, a phase change for a pure material is, in principle, accompanied by a 'jump' $\Delta_{ph1 \to 2} H$, in molar enthalpy, taking place at a unique temperature $T_{ph1 \to 2}$. The value of $C_P = (\partial H/\partial T)_P$ at this temperature is therefore infinite, corresponding to the slope of the vertical line defining the step. However, for some phase, or phase-like, changes – for example, the melting of non-crystalline materials such as polymers and glasses - the change $\Delta_{ph1 \to 2} H$ in molar enthalpy H is not a sharp step, but rather a sigmoid curve, as illustrated in Figure 9.5(a), for which the temperature derivative, $(\partial H/\partial T)_P = C_P$, has the distinctive shape shown in Figure 9.5(b).

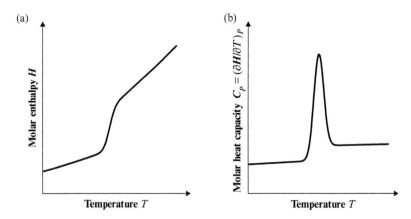

Figure 9.5 The melting of a non-crystalline solid. Unlike the phase transitions shown in Figures 9.3 and 9.4, for which the change in molar enthalpy at a phase boundary is a step, for the melting of a non-crystalline solid, the molar enthalpy - temperature graph has a sigmoid shape of the form illustrated in (a). The corresponding temperature dependence of C_P is shown in (b). The 'sharper' the phase transition, the more vertical the central part of (a), ultimately resulting in a step; likewise, the 'sharper' the phase transition, the narrower, and higher, the peak in (b), ultimately resulting in an infinitely high 'spike'.

An important phase-like change that shows the behaviour depicted in Figure 9.4 is the unfolding of a globular protein, as we shall examine in detail in Chapter 25.

9.11 The Third Law of Thermodynamics

We now explore one further feature concerning equation (9.46), which is a general expression for the entropy change ΔS_P of any system undergoing a change in state, at constant pressure, from an initial temperature T_1 to a final temperature T_2

$$\Delta S_P = S(T_2) - S(T_1)$$

Equation (9.46), as written, assumes that $T_2 > T_1$, and that the system is undergoing a heating process, in principle, from a solid, through a liquid, and finally to a gaseous, state. If the change in state is one of cooling, then the same equation applies, subject to changing the (positive) signs associated with the enthalpies of the phase changes to negative signs.

Assuming that the change of interest is one of heating, in which the final temperature T_2 is greater than the initial temperature T_1, then inspection of each of the terms in equation (9.46) will show that they are all intrinsically positive: C_P is by definition positive at all temperatures; each of the enthalpy changes is also positive, corresponding to a change from a more, to a less, condensed state; and the temperature T is also necessarily positive, as is the temperature increment dT. The result of equation (9.46) must therefore be positive, for all systems, and at all temperatures: provided that $T_2 > T_1$, then $S(T_2) > S(T_1)$ always: the entropy of all systems must increase with temperature.

Suppose now that T_1 is the absolute zero of temperature, 0 K. Since any other temperature must be greater, then, for all systems, the entropy $S(T)$ at any temperature (other than absolute zero) must be greater than $S(0)$, the entropy of that system at absolute zero:

$$S(T) > S(0)$$

This raises the question "What is the entropy $S(0)$ of any system at absolute zero?". As we shall see in Chapter 12, one statement of the Third Law of Thermodynamics is

> **At the absolute zero of temperature, the entropy of all pure perfectly crystalline solids is zero.**

So, if, at absolute zero, the system of interest takes the form of a pure crystalline solid – as many chemical systems do – then $S(0) = 0$, implying that $S(T) > 0$, at all temperatures T. The entropy of any system is therefore necessarily positive, under all conditions; furthermore, the entropy increases with temperature.

9.12 T, S diagrams

Finally in this chapter, we note the T, S diagram – a graph in which the vertical axis is the temperature T, and the horizontal axis the corresponding entropy S, for any system. For a reversible change between an initial equilibrium state [1] and a final equilibrium state [2], since the path takes place through a sequence of intermediate equilibrium states, the corresponding values of T and S are well-defined throughout, implying that the change can be represented as a solid line on the diagram; furthermore, since

$$đq_{rev} = T\,dS \tag{9.1}$$

then the area underneath the curve represents the total reversible heat $\{_1q_2\}_{rev}$ exchanged between the system and the surroundings during the change in state.

As an example, Figure 9.6 shows a T, S diagram for the reversible Carnot Cycle discussed on pages 230 to 236.

As we saw on page 230, the Carnot Cycle is a thermodynamic cycle comprising four steps

- from state [1] to state [2]: reversible isothermal expansion, at a hotter temperature T_H

- from state [2] to state [3]: reversible adiabatic expansion
- from state [3] to state [4]: reversible isothermal compression, at a cooler temperature T_C
- from state [4] to state [1]: reversible adiabatic compression.

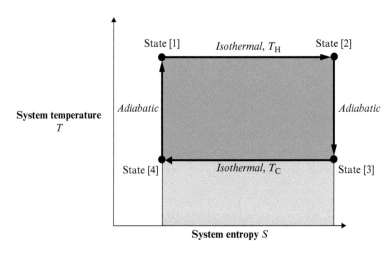

Figure 9.6 A T, S diagram for the reversible Carnot cycle. The central, darker-shaded, area represents both the net heat gained by the system from the surroundings, and also the net work done by the system on the surroundings. This diagram is the counterpart to the P, V diagram shown as Figure 7.3.

These four states are shown in Figure 9.6. Since the paths from state [1] to state [2], and from state [3] to state [4] are isothermal, these paths are represented by horizontal lines at a hotter temperature T_H, and a cooler temperature T_C, respectively. The paths from state [2] to state [3], and from state [4] back to state [1], are both adiabatic and reversible, implying that $đq_{rev} = 0 = T\,dS$. Both of these paths are therefore at constant entropy, and are shown in Figure 9.6 as vertical lines.

The total area between the upper horizontal line and the horizontal axis (the sum of the two shaded areas) is equal to the total heat transferred into the system from the surroundings during the isothermal expansion from state [1] to state [2], and the heat lost from the system to the surroundings during the isothermal compression from state [3] to state [4] is represented by the light blue area. Since the two adiabatic steps exchange no heat, the corresponding areas are zero.

For the entire cycle, the net heat gained by the system from the surroundings is shown by the dark area bounded by the cycle. Mathematically, this area represents $\oint đq$, and since, according to the First Law

$$\oint đq = \oint đw \tag{5.7}$$

this area also represents the net work done by the system on the surroundings.

Figure 9.6 represents a particular case, the T, S diagram for a reversible Carnot Cycle. If, however, a T, S diagram is drawn for any reversible thermodynamic cycle, then it is always the case that

- The total area between a curve tracing the reversible path between any two states represents the total reversible heat exchanged between the system and the surroundings during that change, this being a gain in heat by the system from the surroundings if the entropy increases (moving from left to right in the diagram), and a loss in heat from the system to the surroundings if the entropy decreases (moving from right to left in the diagram).
- The total area enclosed by the cycle is equal to the net quantity of reversible heat exchanged between the system and the surroundings, this being a net gain of heat by the system for a clockwise cycle, and a net loss for an anticlockwise cycle...
- ...and the area is also equal to the net reversible work done by the system on the surroundings (for a clockwise cycle), or done by the surroundings on the system (for an anticlockwise cycle).

For a real cycle, the corresponding irreversible flows of work and heat are subject to the inequalities

$$đw_{rev} > đw_{irrev} \tag{5.37}$$

and

$$đq_{rev} > đq_{irrev} \tag{5.38}$$

implying that the central area represents the corresponding theoretical maxima.

T, S diagrams are counterparts to P, V diagrams, and are especially useful in engineering; also, given the mathematical symmetry between the terms $P\,dV$ and $T\,dS$ – both are of the form (intensive state function) x d(extensive state function) – the variables P and V are known as **conjugate variables**, as are the variables T and S.

EXERCISES

1. Find a friend whom you trust – and who is patient! Sit down with your friend, and, over a cup of tea or coffee, explain the concept of 'entropy', so that your friend really understands the term, and its significance.

2. An exam question reads: "Calculate the entropy change for a gas expanding adiabatically into a vacuum, from an initial volume V_1 to a final volume V_2." A candidate's answer is "The change is adiabatic, and so there is no flow of heat. Since $dS = đq/T$, $dS = 0$.". The candidate is awarded 1 mark out of 10. What should the candidate have written to be awarded 10 marks?

3. Starting from the definition of enthalpy H, derive equation (9.21a)

$$dH = T\,dS + V\,dP \tag{9.21a}$$

Is equation (9.21a) universally true, and applicable to all systems under all conditions? If not, what are the assumptions and limitations?

4. "Yes," said Sam, "of course temperature is a measure of how hot a system is. But more fundamentally, temperature is in fact the rate of change of enthalpy with respect to entropy at constant pressure". Is it?

5. An ideal gas changes state from an initial state $[1] \equiv [P_1, V_1, T_1]$ to a final state $[2] \equiv [P_2, V_2, T_2]$. The change is real, and therefore irreversible; the change is not isothermal, and so $T_1 \neq T_2$; the change is not adiabatic, and so $P_1 V_1^\gamma \neq P_2 V_2^\gamma$: the change is therefore totally general. Spend a few minutes thinking through, and then making some notes on, how you would calculate the entropy change $\Delta S_{1 \to 2}$, and how you intend to solve the central problem – the fact that entropy can be calculated only for reversible paths, but the path of interest is irreversible. This question does not require any formulae or mathematics – it's about thinking through the 'strategy' for solving what is, in fact, quite a difficult problem. Don't rush it... and when you've finished, try the following questions, which will guide you through the process ....

6. Show that it is *always* possible to change from any arbitrary state $[1] \equiv [P_1, V_1, T_1]$ to any other arbitrary state $[2] \equiv [P_2, V_2, T_2]$ by a sequence of two reversible steps: the first, a reversible isothermal change at temperature T_1 from state $[1] \equiv [P_1, V_1, T_1]$ to an intermediate state $[3] \equiv [P_3, V_3, T_1]$, followed by a reversible adiabatic change from the intermediate state $[3] \equiv [P_3, V_3, T_1]$ to the final state $[2] \equiv [P_2, V_2, T_2]$. Show this two-step path on a P, V diagram. Is this the only reversible path between state [1] and state [2], or are there any others? If there are others, what are they? Show these other paths (if any) on a P, V diagram.

7. For the process described in question 6 – a reversible isothermal change from state $[1] \equiv [P_1, V_1, T_1]$ to state $[3] \equiv [P_3, V_3, T_1]$, followed by a reversible adiabatic change from the intermediate state $[3] \equiv [P_3, V_3, T_1]$ to the final state $[2] \equiv [P_2, V_2, T_2]$ – show that the intermediate state [3] is a single, unique, state such the pressure P_3 is given by

$$P_4 = P_1 \left(\frac{T_2}{T_1}\right)^{\frac{\gamma}{\gamma-1}}$$

and the volume by

$$V_3 = V_2 \left(\frac{T_2}{T_1}\right)^{\frac{1}{\gamma-1}}$$

8. Representing the entropy change from state [X] to state [Y] as $\Delta S_{X \to Y}$, calculate $\Delta S_{1 \to 3}$ and $\Delta S_{3 \to 2}$, and show that

$$\Delta S_{1 \to 2} = R \ln \left(\frac{V_2}{V_1}\right) \left(\frac{T_2}{T_1}\right)^{\frac{1}{\gamma-1}} = R \ln \left(\frac{P_1}{P_2}\right) \left(\frac{T_2}{T_1}\right)^{\frac{\gamma}{\gamma-1}}$$

Why does this result depend only on state functions associated with states [1] and [2], and is independent of the state functions associated with state [3]?

9. If the change from state [1] to state [2] is an expansion, which of the following three statements is true: $V_1 > V_2$; $V_1 = V_2$; or $V_1 < V_2$? Why? And these three: $P_1 > P_2$; $P_1 = P_2$; or $P_1 < P_2$? Why? And these three: $T_1 > T_2$; $T_1 = T_2$; or $T_1 < T_2$? Why? As a result, prove that $\Delta S_{1 \to 2} > 0$. What does the answer to the last question imply about the change from state [1] to state [2]?

10. If the change from state [1] to state [2] is a compression, which of the following three statements is true: $V_1 > V_2$; $V_1 = V_2$; or $V_1 < V_2$? Why? And these three: $T_1 > T_2$;

$T_1 = T_2$; or $T_1 < T_2$? Why? As a result, prove that $\Delta S_{1\to 2} < 0$. What does the answer to the last question imply about the change from state [1] to state [2]?

11. Question 6 suggested that state [2] may be reached from state [1] via an intermediate state [3] such that the path from state [1] to state [3] is reversible and isothermal, and the path from state [3] to state [2] is reversible and adiabatic. This is valid, but not the only possibility – consider, for example, a different intermediate state [4] such that the path from the initial state $[1] \equiv [P_1, V_1, T_1]$ to state $[4] \equiv [P_4, V_4, T_2]$ is reversible and adiabatic, and the path from state $[4] \equiv [P_4, V_4, T_2]$ to the final state $[2] \equiv [P_2, V_2, T_2]$ is reversible and isothermal. Show this path on the P, V diagram you drew for question 6, and prove that

$$V_4 = V_1 \left(\frac{T_1}{T_2}\right)^{\frac{1}{\gamma-1}}$$

$$P_4 = P_1 \left(\frac{T_2}{T_1}\right)^{\frac{\gamma}{\gamma-1}}$$

and

$$\Delta S_{1\to 2} = R \ln \left(\frac{V_2}{V_1}\right) \left(\frac{T_2}{T_1}\right)^{\frac{1}{\gamma-1}} = R \ln \left(\frac{P_1}{P_2}\right) \left(\frac{T_2}{T_1}\right)^{\frac{\gamma}{\gamma-1}}$$

Why is this the same result as that derived in question 8?

12. If the change from state [1] to state [2] is an expansion, prove that $\Delta S_{1\to 2} > 0$; similarly, if the change from state [1] to state [2] is a compression, prove that $\Delta S_{1\to 2} < 0$.

13. Suppose that a real, irreversible change from state [1] to state [2] is followed by a second, real, irreversible change from state [2] back to state [1]. What is the change in entropy for the complete cycle? Suppose now that this cycle is analysed as a sequence of four reversible steps: an isothermal expansion from state [1] to state [3], followed by an adiabatic expansion from state [3] to state [2] (as in questions 6, 7, 8 and 9); then an isothermal compression from state [2] to state [4], followed by an adiabatic compression from state [4] back to state [1] (as in questions 10, 11 and 12 in reverse). Compute the overall entropy change for the cycle as the sum of $\Delta S_{1\to 2}$ (from question 8) and $\Delta S_{2\to 1}$ (as derived from question 11). Draw the four-step cycle on a P, V diagram, and compare your sketch with Figure 7.3.

14. In questions 6 to 10, the intermediate state [3] is reached from the initial state [1] by a reversible isothermal change, and itself is linked to the final state [2] by a reversible adiabatic path. Suppose now that the intermediate state [3] is itself an arbitrary state, which cannot be directly reached isothermally from state [1], or directly linked adiabatically to state [2]. Considering therefore a state [X], between states [1] and [3], and a state [Y] between states [3] and [2], such that the path between states [1] and [2] is the four-step sequence $[1] \to [X] \to [3] \to [Y] \to [2]$, with the paths $[1] \to [X]$ and $[3] \to [Y]$ being reversible and isothermal, and the paths $[X] \to [3]$ and $[Y] \to [2]$ reversible and adiabatic. Compute the entropy changes $\Delta S_{1\to X}$, $\Delta S_{X\to 3}$, $\Delta S_{3\to Y}$, $\Delta S_{Y\to 2}$, $\Delta S_{1\to 3}$, $\Delta S_{3\to 2}$ and $\Delta S_{1\to 2}$. Do you see the pattern?

15. Suppose further that the real, irreversible, path from state [1] to state [2] takes place through an infinite number of intermediate states [i], [j], [k]… such that each state [j]… is linked to the preceding state [i]… by an infinitesimal reversible isothermal path followed by an infinitesimal reversible adiabatic path. What is the infinitesimal entropy change $dS_{i \to j}$? And the total entropy change

$$\Delta S_{1 \to 2} = \int_{\text{state 1}}^{\text{state 2}} dS \ ?$$

10 Clausius, Kelvin, Planck, Carathéodory and Carnot

 Summary

The fact that spontaneous changes in isolated systems are unidirectional and irreversible, and proceed from non-equilibrium states to equilibrium states, is a fundamental truth. One such spontaneous, unidirectional and irreversible change is the natural flow of heat down a temperature gradient: when two bodies, at different temperatures, are in thermal contact, the originally hotter body will become cooler, and the originally cooler body will become warmer, until both bodies achieve the same, intermediate, temperature, and the whole system achieves equilibrium. The reverse change – in which two bodies in thermal equilibrium spontaneously change, such that one becomes hotter and the other cooler – just never happens.

The **Clausius statement of the Second Law**,

It is impossible to create a device that operates in a cycle and has no effect other than the transfer of heat from a cooler to a warmer body.

and the **Kelvin-Planck statement**,

It is impossible to construct a device that operates in a cycle and has no effect other than the performance of work and the exchange of heat with a single reservoir.

are, fundamentally, variants of the much simpler statement "heat flows spontaneously down a temperature gradient, not up it". Importantly, however, both of these statements relate the flow of heat to the possibility, or otherwise, of using the spontaneous flow of heat down a temperature gradient to do useful work – just like the flow of water down a gravitational gradient can drive a mill wheel or a turbine.

So, in simpler language, the Clausius statement affirms that heat will spontaneously flow down a temperature gradient, but if we wish to push heat up a temperature gradient, then we have to do work on the system. The Kelvin-Planck statement makes the additional point that, if we wish to harness the heat flow down a temperature gradient to do useful work, such as to drive an engine, then some heat of the heat that flows into the engine from a hot reservoir must flow out of the system into a cold reservoir. The heat necessarily flowing out of the system into the cold reservoir therefore cannot be used for work – implying that some of the heat that flows into the engine from the hot reservoir must flow out to the cold one. No engine, however well designed, can transform all the energy flowing into the engine into useful work: some energy must necessarily be discarded, or 'wasted'. This then sets a limit on the efficiency of engines, as studied in depth by Carnot in his examination of the **Carnot engine**.

The **Carathéodory statement of the Second Law**,

Two adiabats cannot cross.

is rather more abstract, but, as will be explained on pages 321 to 333, it is essentially another aspect of the same fundamental truth that heat spontaneously flows 'downhill', not 'uphill'.

10.1 The Clausius statement of the Second Law

10.1.1 An impossible machine

The development of the Second Law during the mid-nineteenth century is one of the great achievements of science. The identification of entropy as a state function, and of the Clausius Inequality, shed light on one of the world's true mysteries – why do so many changes only go one way? Because entropy increases; that's why. And it is because the Clausius Inequality is just that – an inequality, rather than an equation – that our scientific understanding of our world provides an explanation of a fundamental asymmetry: the asymmetry between the 'right' direction in which many changes happen and the 'wrong' direction; the asymmetry between 'cause' and 'effect'; the asymmetry between 'then' and 'now'; the asymmetry that defines the flow of time.

And as well as identifying the inequality that now bears his name, amongst many other things, Rudolf Clausius is attributed with stating both the law of the conservation of energy ("The energy of the universe is constant") as well as the principle that the entropy of the universe increases ("The entropy of the world tends to a maximum"). He also expressed the Second Law using these words

> It is impossible to create a device that operates in a cycle and has no effect other than the transfer of heat from a cooler to a warmer body.

That's the kind of scientific law that needs to be read three times; in fact, it's right, but it does need some explanation. It's also somewhat strange in that it states what doesn't happen, rather than what does – "it is impossible ..." is rather harder to think about than more direct statements of physical laws such as "the orbit of each planet is an ellipse, with the sun at one focus".

The essence of what is now known as the **Clausius statement of the Second Law of Thermodynamics** is our experience that heat spontaneously flows from hotter bodies to cooler ones, and not in the other direction. This is indeed our common experience; but is this because our experience is inevitably limited? Might it be possible to devise some type of machine (that's a "device that operates in a cycle") that can actually achieve this? If we were clever enough, if we had the right materials, could we design a machine that really did result in moving heat up a temperature gradient?

However attractive this concept might be, the Clausius Statement states that such a machine not only doesn't exist, it can never exist. It is an absolute truth that heat spontaneously flows down a temperature gradient, unidirectionally and irreversibly, and no machine, however designed, can reverse that process.

The nature of the 'impossible machine' merits some consideration, for the Clausius statement defines it has having two key features. The first is that it "operates in a cycle". What this means is that the machine goes through a sequence of processes, returns to its original state, and then repeats the same sequence – the "cycle" – again and again. As we saw on page 61, this is far more effective than operating just once. Accordingly, throughout this chapter, 'machine' will always imply a device that operates in a cycle.

The second feature is defined by the words "device…that has no effect other than…". The significance of this is the distinction Clausius is making between a machine that *only* allows heat to flow from a cooler body to a warmer one – which is the type of machine that Clausius identifies as being impossible to build – and a machine that allows heat to flow from a cooler to a warmer body *but with something else happening too*. What might that 'something else' be? Well, given that this book is about thermodynamics, we don't have to think too hard to identify a possible candidate: the discussion so far has focused on heat; if work can somehow be involved too, then maybe we *can* design a machine that can move heat up a temperature gradient. Importantly, Clausius is not denying the possibility of heat moving up a temperature gradient: what he is denying is the possibility that heat can move up a gradient *alone, apparently of its own accord, spontaneously*. And Clausius's choice of words hints at the possibility that heat might be able to move up a temperature gradient if a particular condition is fulfilled – for example, if work is performed as well. Let's see …

Figure 10.1 illustrates the distinction between a machine that Clausius identifies as being impossible to build, and a machine that could, in principle, work.

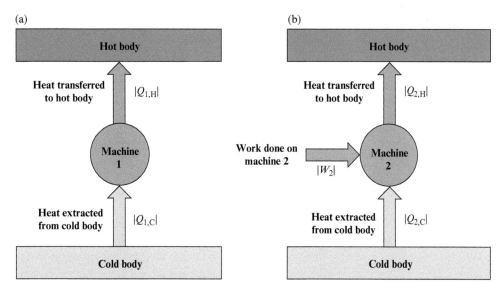

Figure 10.1 An impossible machine, and a possible one. Machine 1 extracts heat $|Q_{1,C}|$ from the cold body, and transfers heat $|Q_{1,H}|$ to a hot body. According to the Clausius Statement, such a machine is impossible. Machine 2, in contrast, is possible. Like machine 1, it extracts heat $|Q_{2,C}|$ from a cold body and transfers heat $|Q_{2,C}|$ to a hot body; but in addition, work $|W_2|$ is done on the machine.

Machine 1, as shown in Figure 10.1(a), extracts heat from a cold body, and transfers heat to a hot body, and nothing else happens. According to the Clausius Statement, this machine cannot exist. Machine 2, as shown in Figure 10.1(b), is very similar, but there is one, vital, difference. As well as extracting heat from the cold body and transferring heat to the hot body, the surroundings are doing work on the machine too. If Machine 1 were 'invisible', or just very small, then an external observer would detect heat flowing, apparently spontaneously, up a temperature gradient – an observation that is contrary to all experience. In contrast, an external observer of Machine 2 would detect not only the flow of heat up the temperature gradient, but also some work being done on the machine by the surroundings, work that is acting to 'push' the heat up the temperature gradient. This is, in fact, how a refrigerator works: it is a machine that extracts heat from its interior (the 'cold body'), and transfers it to the local area in which the refrigerator is situated (the 'hot body'). But it is only able to do so by virtue of a clever mechanism, powered by an engine, which does work, work powered by the electricity fed into the refrigerator from the nearby electrical socket.

10.1.2 The First Law and machines operating in a cycle

We can apply the First Law to both machines, where, with reference to Figure 10.1, the 'system' is the machine; the system boundary is represented by the line defining the machine, and the 'surroundings' are everything outside the system, as defined by the machine. In particular, both the hot and the cold bodies are in the surroundings, as is the source of the work that is done on Machine 2. All flows of heat and work therefore cross the system boundary.

As we know, the First Law may be expressed as

$$dU = đq - đw \tag{5.1a}$$

For a complete thermodynamic cycle, as we saw on page 101, the First Law becomes

$$\oint dU = \oint đq - \oint đw$$

this being relevant, in that, in accordance with the Clausius Statement, the machine to which he refers "operates in a cycle". Since the internal energy U is a state function,

$$\oint dU = 0$$

and so

$$\oint đq = \oint đw \tag{5.7}$$

Any complete cycle is inevitably associated with a number of individual steps – the minimum number being two (the first from 'here' to 'there', and the second back from 'there' to 'here'), and the maximum being any number at all, as long as the last step returns the system to its original state. During any individual step, in general, some heat $đq$ will be exchanged between the system and the surroundings, and some work $đw$ will be done on the surroundings by the system, or by the surroundings on the system. $\oint đq$ represents the sum of all the individual flows of heat $đq$, both into the system and out of the system, for the complete cycle, and so represents the *net heat exchange between the system and the surroundings*. If $\oint đq$ is a positive number for any given cycle, then the overall effect of that cycle is for the system to gain that amount of heat from the surroundings; if $\oint đq$ is negative, then, during a complete cycle, the system loses that amount of heat to the surroundings. Similarly, $\oint đw$ represents the *net work done by, or on, the system* as the result of a complete cycle: if $\oint đw$ is positive, the system does net work on the surroundings, if $\oint đw$ is negative, then the surroundings do net work on the system. In physical terms, equation (5.7) states that for *any* thermodynamic cycle, the net gain of heat by the system from the surroundings during one complete cycle must equal the net work done by the system on the surroundings over that same cycle (if both $\oint đq$ and $\oint đw$ are positive); or, if both $\oint đq$ and $\oint đw$ are negative, then the net heat lost by the system to the surroundings is equal to the net work done by the surroundings on the system over that same cycle. In rather simpler terms, this means that, for any cycle, "what goes in (as heat or work), must come out (as the equivalent amount of work or heat)", and no energy can get "stuck" in the middle!

10.1.3 A (temporary) change in our sign convention

Up to this point, as defined on page 56, the symbol q has been used to represent a quantity of heat exchanged between a system and the surroundings, such that q is positive if the system gains heat, and negative if the system loses heat. Similarly, the symbol w has been used for work, a positive value representing work done by a system on the surroundings, and a negative value representing work done on a system by the surroundings. According to this convention, both q and w are algebraic, signed, quantities, and so the same symbols can take either positive or negative values, depending on the circumstances. As a consequence, the symbol itself does not make explicit the direction of the flow of heat, or whether work is done 'by' or 'on' – the direction has to be inferred from other information.

In many circumstances, the sign convention used so far is very useful, by virtue of its general applicability. In this chapter, however, there is a considerable benefit in making it absolutely explicit whether, in any given circumstances, heat is gained by, or lost from, a system, and whether work is done by, or done on, a system. Accordingly, in this chapter, we will use the symbol $|Q|$ to represent a *quantity* of heat, and $|W|$ to represent a *quantity* of work. Importantly, $|Q|$ and $|W|$ are inherently and necessarily always positive numbers, representing magnitudes, and can *never* be negative. The associated directionality will therefore be identified explicitly by a preceding sign:

- $|Q|$, or $+|Q|$, represents heat gained by a system
- $-|Q|$ represents heat lost by a system
- $|W|$, or $+|W|$, represents work done by a system
- $-|W|$ represents work done on a system.

10.1.4 The First Law and the impossible machine

We now apply the First Law to the cycles performed by the 'impossible' Machine 1 shown in Figure 10.1(a). Since there are only two heat flows associated with a complete cycle of Machine

1's operation, using the new sign convention, we may express the net heat exchange for one complete cycle as the difference

$$\oint dq = |Q_{1,C}| - |Q_{1,H}|$$

where $|Q_{1,C}|$ is the quantity of heat gained by Machine 1, extracted from the cold body, and $|Q_{1,H}|$ is the quantity of heat lost by the Machine 1, and transferred into the hot body. Since nothing else happens

$$\oint dw = 0$$

and so

$$|Q_{1,C}| = |Q_{1,H}|$$

All the heat $|Q_{1,C}|$ extracted from the cold body is therefore transferred, as heat $|Q_{1,H}|$, to the hot body. No heat is gained or lost during this process, and so the First Law is upheld. As a result of the process, however, Machine 1 transfers quantity $|Q_{1,C}| = |Q_{1,H}|$ of heat from a cooler to a warmer body, and, according the Clausius Statement, this machine cannot exist.

Applying exactly the same principles to Machine 2

$$\oint dq = |Q_{2,C}| - |Q_{2,H}|$$

and, since a quantity of work $|W_2|$ is done on the system by the surroundings

$$\oint dw = -|W_2|$$

Since, for one cycle of the operation of Machine 2

$$\oint dq = \oint dw$$

then

$$|Q_{2,C}| - |Q_{2,H}| = -|W_2|$$

and so

$$|Q_{2,C}| + |W_2| = |Q_{2,H}| \tag{10.1}$$

Since all the quantities in equation (10.1) are necessarily positive, $|Q_{2,H}|$ must be greater than $|Q_{2,C}|$ - more heat is transferred into the hot body than was extracted from the cold body, the additional amount of heat being attributable to the conversion of the work $|W_2|$ done on the system into heat. Once again, the First Law is upheld. But, as vividly demonstrated by the experiments of James Joule, the conversion of work into heat is a natural process, and so the action of Machine 2 is in accord with normal experience. Whereas Machine 1, which has no effect other than transferring heat from a cooler to a warmer body, does not exist, Machine 2 – which moves heat up a temperature gradient but in addition has work performed on it – is real, and can exist.

THE KELVIN-PLANCK STATEMENT OF THE SECOND LAW

One further point about equation (10.1). Comparison of equation (10.1) with Machine 2, as depicted in Figure 10.1(b) will show that $|Q_{2,C}|$ and $|W_2|$ are associated with flows *into* Machine 2, and $|Q_{2,H}|$ is associated with a flow *away from* Machine 2. Equation (10.1) therefore equates the total inflows to the total outflows, which is, in essence, what the conservation of energy, and the First Law, is all about. Equation (10.1) can therefore be inferred directly from Figure 10.1(b), which is why this chapter uses the more explicit sign convention. This principle is generally true: even if the 'machine' in a diagram such as Figure 10.1(b) has many more inflows of heat or work, and many more outflows, for any single cycle of the machine's operation, the total of the inflows of both heat and work will always equal to the total of the outflows.

10.2 The Kelvin–Planck statement of the Second Law

10.2.1 Another impossible machine

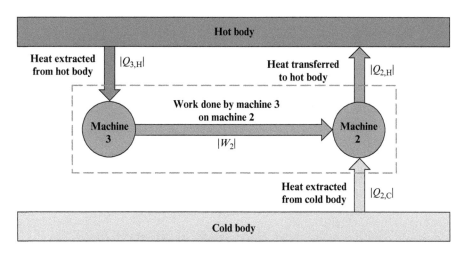

Figure 10.2 Another machine that doesn't exist. Machine 2 extracts heat from the cold body and transfers heat to the hot body, but since work is done on this machine, Machine 2 is not deemed impossible by the Clausius statement, and so can exist. The work done on Machine 2 is derived from Machine 1, which creates this work by extracting heat from the hot body. The First Law is upheld throughout, but the 'composite machine', as defined by the system boundary shown as the dashed line, can be shown by the Clausius statement to be impossible. Since Machine 2 can exist, Machine 3 must be impossible.

Figure 10.2 shows two machines, each of which operates in a cycle, such that the cycles are synchronised. Machine 2, on the right, is identical to Machine 2 as shown in Figure 10.1(b), which, as we have just demonstrated, can exist. Also, the flows of heat and work associated with Machine 2 comply with equation (10.1)

$$|Q_{2,C}| + |W_2| = |Q_{2,H}| \tag{10.1}$$

On the left-hand side of Figure 10.2 is a new machine, Machine 3, which can extract heat from the hot body and convert it into work: in essence, it is an engine, in that all engines convert a source of heat, such as burning coal or fuel, into work, such as the motion of a locomotive or a car. A particular feature of the design of Machine 3, however, is that the amount of work it performs on each cycle is just the right amount of work needed to drive one cycle of Machine 2, as indicated by the horizontal arrow.

Applying the First Law

$$\oint dq = \oint dw$$

to Machine 3, we have

$$|Q_{3,H}| = |W_2| \qquad (10.2)$$

as verified by applying the 'what goes in must come out' rule to Machine 3 as depicted on the left-hand side of Figure 10.2. If we use equation (10.2) to substitute $|Q_{3,H}|$ for $|W_2|$ in equation (10.1), we derive

$$|Q_{2,C}| + |Q_{3,H}| = |Q_{2,H}|$$

implying that

$$|Q_{2,C}| = |Q_{2,H}| - |Q_{3,H}| \qquad (10.3)$$

As we have stated, the two machines work in synchrony, so let's imagine that they are 'very small' – so small that an external observer cannot distinguish between them, nor can the external observer detect the flow of work from Machine 3 to Machine 2. The external observer can, however, detect the flows of heat between the 'composite-machine' and the hot and cold bodies, these being flows that cross the boundary of the newly-defined 'composite system', as shown by the dashed line in Figure 10.2.

Accordingly, with reference to the flows of heat associated with the 'composite machine' as shown in Figure 10.2, there is a single inflow of heat $|Q_{2,C}|$ from the cold body, and a net exchange of heat with the hot body given by

Net heat transferred from 'composite machine' to hot body = $|Q_{2,H}| - |Q_{3,H}|$

But according to equation (10.3)

$$|Q_{2,H}| - |Q_{3,H}| = |Q_{2,C}| \qquad (10.3)$$

implying that

Net heat transferred from 'composite machine' to hot body = $|Q_{2,C}|$

We have now demonstrated that, as a result of one cycle of the 'composite machine'

- a quantity $|Q_{2,C}|$ of heat is extracted from the cold body, and
- an equal quantity $|Q_{2,C}|$ of heat is transferred into the hot body, and ...
- ... nothing else happens.

The 'composite machine' therefore "operates in a cycle and has no effect other than the transfer of heat from a cooler to a warmer body", and so is deemed impossible by the Clausius Statement. But since we know that the Machine 2 component of the 'composite machine' can exist, the problem must lie with Machine 3 – Machine 3 must be another form of 'impossible machine', so that the combination of an 'impossible machine 3' with a 'real machine 2' makes the 'composite machine' impossible.

10.2.2 Can heat be converted in its entirety into work?

Why might Machine 3 be 'impossible'? Machine 3 is in fact very simple: it just extracts heat from the hot body, and converts all that heat into work. Maybe that's the problem: the conversion of *all* the heat into work. This raises an important question: is it possible for an engine to convert all the heat it absorbs into work, or must some heat be 'wasted'?

In fact, as we shall prove shortly, some heat must be 'wasted', or, to put in another way, it is impossible to construct an engine that can convert all its heat input into work. This concept is known as the **Kelvin–Planck statement of the Second Law of Thermodynamics**, which is often expressed as

> It is impossible to construct a device that operates in a cycle and has no effect other than the performance of work and the exchange of heat with a single reservoir.

Like the Clausius statement, the Kelvin-Planck statement – named after William Thomson, later Lord Kelvin, and Max Planck – is expressed as a negative; also, like the Clausius Statement, the words are very carefully chosen.

The essence of the Kelvin-Planck Statement is that you can't design a machine which takes in heat from an outside body (that's the "single reservoir") and converts all that heat into work. But …

- you can design a machine that takes in heat, but performs no work (the machine just gets hot)
- you can design a machine *on which work is done*, and exchanges heat with just one external body (this being a description of, for example, the heat produced by frictional work), and
- you can design a machine that does indeed perform work – but for this to happen, the machine *must exchange heat with at least two reservoirs*: a hot one, which acts as a source of energy, and a cold one, to which some heat is necessarily discarded. Without releasing heat to that second, cold, reservoir, the machine just won't work.

10.2.3 A machine that does exist

The Kelvin-Plank statement is illustrated in Figure 10.3, which shows a new machine, Machine 4, replacing the impossible machine, Machine 3, shown on the left-hand side of Figure 10.2. Whereas the impossible machine, Machine 3, only extracted heat $|Q_{3,H}|$ from the hot body, and performed work $|W_2|$ on Machine 2, the new machine, Machine 4, additionally transfers heat $|Q_{4,C}|$ to the cold body.

As before, the work $|W_2|$ performed by Machine 4 on each cycle is exactly the right amount required for each cycle performed by Machine 2. So, for Machine 2, equation (10.1) still holds

$$|Q_{2,C}| + |W_2| = |Q_{2,H}| \tag{10.1}$$

For Machine 4, and from the associated inflows and outflows shown in Figure 10.3, the First Law implies that

$$|Q_{4,C}| + |W_2| = |Q_{4,H}| \tag{10.4}$$

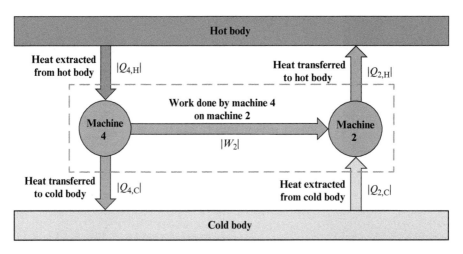

Figure 10.3 This machine can exist. Machine 2 in this diagram is identical to Machine 2 in Figure 10.2; Machine 4, however, differs from Machine 3 in that, in addition to extracting heat $|Q_{4,H}|$ from the hot body and performing work $|W_2|$ on Machine 2, this machine transfers heat $|Q_{4,C}|$ to the cold body.

Equations (10.1) and (10.4) can be combined by eliminating $|W_2|$ to give

$$|Q_{2,C}| - |Q_{4,C}| = |Q_{2,H}| - |Q_{4,H}| \tag{10.5}$$

For the 'composite machine' formed from the combined operation of Machine 4 and Machine 3, the exchange of work $|W_2|$ is 'hidden' within the boundary of the 'composite machine', and so is 'invisible' to an external observer, who detects only

- the extraction of a net amount of heat $|Q_{2,C}| - |Q_{4,C}|$ from the cold reservoir, and
- the transfer of net amount of heat $|Q_{2,H}| - |Q_{4,H}|$ into the hot reservoir, such that
- the two net quantities, $|Q_{2,C}| - |Q_{4,C}|$ and $|Q_{2,H}| - |Q_{4,H}|$, according to equation (10.5), are equal.

Now if the net quantities $|Q_{2,C}| - |Q_{4,C}|$ and $|Q_{2,H}| - |Q_{4,H}|$, which we know are equal, are also positive, then the external observer would detect a 'composite machine' which has no effect other than the transfer of heat from a cooler to a warmer body – a situation which, by the Clausius statement, must be impossible.

But there is also the possibility that the net quantities $|Q_{2,C}| - |Q_{4,C}|$ and $|Q_{2,H}| - |Q_{4,H}|$ are *negative*. In this case, the observer would actually be detecting the loss of a net amount of heat $|Q_{4,H}| - |Q_{2,H}|$ from the hot body, and the simultaneous gain of an equal amount of heat $|Q_{4,C}| - |Q_{2,C}|$ by the cold body. The observer would simply be detecting the flow of heat down a temperature gradient, which is, of course, a perfectly normal, every-day experience.

The 'composite machine' shown in Figure 10.3 therefore can exist, but only on the condition that the net quantities $|Q_{2,C}| - |Q_{4,C}|$ and $|Q_{2,H}| - |Q_{4,H}|$ are negative, or, alternatively, that the net quantities $|Q_{4,C}| - |Q_{2,C}|$ and $|Q_{4,H}| - |Q_{2,H}|$ are positive.

Given that all the quantities $|Q_{4,C}|$, $|Q_{2,C}|$, $|Q_{4,H}|$ and $|Q_{2,H}|$ are, by definition, positive, is there a way to determine whether or not the two differences $|Q_{4,C}| - |Q_{2,C}|$ and $|Q_{4,H}| - |Q_{2,H}|$ are necessarily positive?

Yes, there is. Suppose that $|Q_{2,C}| = |Q_{2,H}| = 0$. Under these circumstances, the two differences $|Q_{4,C}| - |Q_{2,C}|$ and $|Q_{4,H}| - |Q_{2,H}|$ must be positive, always. But in reality, what does the condition that $|Q_{2,C}| = |Q_{2,H}| = 0$ actually mean? That's easy. It means that Machine 2 on the right-hand side of Figure 10.3 is 'switched off', such that Machine 4 is working on its own. Alternatively, if Machine 2 were not there at all, Machine 4 could exist by itself, as indeed it does: a machine that extracts heat from a hot body, performs work, and discards heat to a cold body, exists. We call them 'engines'.

10.3 Heat engines and heat pumps

A device which performs work, using heat as the source of energy, is known as a **heat engine**, examples being the steam engine and the petrol engine. All heat engines operate on the same principle, as illustrated in Figure 10.4(a), in which heat $|Q_{E,H}|$ drawn from a hot body, such as a fire, is used to perform external work $|W_E|$, whilst simultaneously discarding heat $|Q_{E,C}|$ to a cold body, often the engine's surroundings. And, as we saw on page 311, in accordance with the Kelvin-Planck Statement, this machine can exist only if $|Q_{E,C}| > 0$.

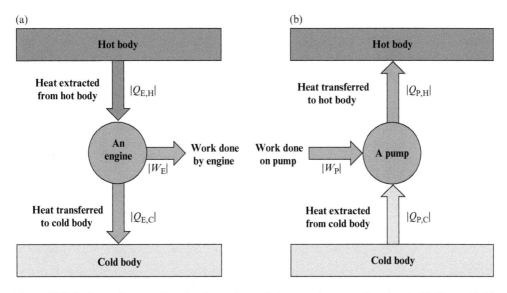

Figure 10.4 Engines and pumps. All engines that use heat as their source of energy work on the principle illustrated in (a). Heat $|Q_{E,H}|$ is transferred from a hot body (for example, a fire) into the engine, which performs external work $|W_E|$ whilst simultaneously discarding heat $|Q_{E,C}|$ to a cold body. Importantly, $|Q_{E,C}|$ must always be greater than zero, and can never equal zero. According to the First Law, $|Q_{E,H}| = |W_E| + |Q_{E,C}|$, and so it is always the case that $|W_E| < |Q_{E,H}|$. This implies that a given amount $|Q_{E,H}|$ of heat can never be converted in its entirety into work $|W_E|$ – some heat, $|Q_{E,C}|$, must always be lost to the 'cold body', which, in many real cases, comprises the system's surroundings. On the right, the flows of heat and work are the other way about: the 'pump' extracts heat $|Q_{P,C}|$ from a cold body, and, if work $|W_P|$ is performed on the pump, then heat $|Q_{P,H}|$ is transferred – or 'pumped' – into the hot body. For the pump, $|Q_{P,C}| + |W_P| = |Q_{P,H}|$. This machine is identical to that shown in Figure 10.1(b).

According to the First Law

$$|Q_{E,H}| = |W_E| + |Q_{E,C}|$$

and so $|Q_{E,H}|$ is always greater than $|W_E|$ – it is *impossible* to transform a given quantity of heat $|Q_{E,H}|$ in its entirety into work: some heat must *always* be wasted as an amount of heat $|Q_{E,C}|$ discarded into some form of cold body. The essence of the Kelvin-Planck statement of the Second Law is that the quantity $|Q_{E,C}|$ of heat that an engine must discard is inevitably greater than zero. As a consequence, the amount of work $|W_E|$ that can be 'captured' by an engine from its source of heat must be less than the amount of heat $|Q_{E,H}|$ supplied to the engine, leading to the concept of the **efficiency** η_E of the engine, defined, either as a number between 0 and 1

$$\eta_E = \frac{|W_E|}{|Q_{E,H}|} = \frac{|Q_{E,H}| - |Q_{E,C}|}{|Q_{E,H}|} \tag{10.6a}$$

or as a percentage, from 0 % to 100 %

$$\eta_E = \frac{|W_E|}{|Q_{E,H}|} \times 100 = \frac{|Q_{E,H}| - |Q_{E,C}|}{|Q_{E,H}|} \times 100 \tag{10.6b}$$

We introduced the concept of efficiency in Chapter 7 (see page 234), in the case study of the Carnot cycle – a story which we shall continue shortly. For the moment, let's note that the Kelvin-Planck statement demands that $|Q_{E,C}| > 0$, implying that the efficiency η_E of any engine can never equal 1 or 100%. Given that the ultimate efficiency 1 or 100% is unattainable, that leaves open the question of how closely this limit might actually be approached - so equations (10.6) pose a continuous challenge to engineers as to how to design ever more efficient engines.

Returning to Figure 10.4, we see, on the right-hand side, a representation of a machine which is identical to the engine shown on the left-hand side, subject to the flows of heat and work being the other way around: heat is drawn in from a cold body, work is done on the machine, and heat is transferred into the hot body. This machine 'pumps' heat up a temperature gradient, but since it is a machine, operating in a cycle, that transfers heat from a cooler to a warmer body, *and in addition* has work performed on it, this machine is not 'impossible' according to the Clausius Statement, it can exist – indeed it is identical to the machine depicted in Figure 10.1(b). This machine, not surprisingly, is known as a **heat pump**, and is the basis of all refrigeration.

For a heat pump, the definition of the efficiency η_P is mathematically identical to that for a heat engine

$$\eta_P = \frac{|W_P|}{|Q_{P,H}|} = \frac{|Q_{P,H}| - |Q_{P,C}|}{|Q_{P,H}|} \tag{10.6c}$$

or as a percentage, from 0 % to 100 %

$$\eta_P = \frac{|W_P|}{|Q_{P,H}|} \times 100 = \frac{|Q_{P,H}| - |Q_{P,C}|}{|Q_{P,H}|} \times 100 \tag{10.6d}$$

For a heat pump, however, the meaning of the efficiency is slightly different from the meaning as applied to a heat engine: for a heat pump, the efficiency is the fraction, or percentage, of heat pumped into the hot reservoir which is attributable to work. However, if there is no cold reservoir at all, it is still possible to convert work totally into heat – that's Count Rumford and his cannon once again – in which case, technically, $|Q_{P,C}|$ in equations (10.6c) and (10.6d) is zero, and the efficiency of the 'pump' is 1 or 100%.

10.4 Irreversibility

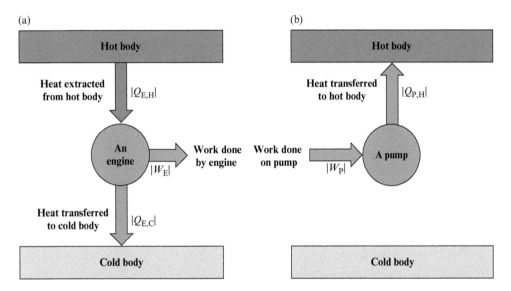

Figure 10.5 Irreversibility. On the right-hand side, (b), the flow of heat $|Q_{P,C}|$ into the pump from the cold body is zero. The pump, however, still exists, demonstrating that it is possible to convert a given amount of work $|W_P|$ in its entirety into heat $|Q_{P,H}|$. But as shown by the engine on the left, it is not possible for a given amount of heat $|Q_{E,H}|$ to be converted in its entirety into work. This is a fundamental example of irreversibility.

Figure 10.5 is very similar to Figure 10.4 – the sole difference is that the flow of heat $|Q_{P,C}|$ from the cold body into the pump is zero. An observer of the pump would detect a machine that

- exchanges heat $|Q_{P,H}|$ with a single body, the hot one
- has work $|W_P|$ performed on it.

This is not a machine that only transfers heat from a cooler to a warmer body, and so is not 'impossible' in terms of the Clausius Statement. Furthermore, this is not a machine that performs work and exchanges heat with only a single reservoir – it is a machine *on which work is performed*, and exchanges heat with a single reservoir – and so is 'allowed' according to the Kelvin-Planck Statement. The machine depicted in Figure 10.5(b) therefore can exist – but it has a special property: according to the First Law

$$|W_P| = |Q_{P,H}|$$

implying that the work $|W_P|$ done on the machine is converted, in its entirety, into heat. A machine that converts work into heat is, of course, totally in line with experience – it is, once again, a representation of Count Rumford's cannon or, better, James Joule's apparatus (see Figure 3.3 on page 52).

But suppose that the quantity of heat $|Q_{P,H}|$ – attributable, perhaps, to the paddles in James Joule's apparatus – is now used as the heat source $|Q_{E,H}| = |Q_{P,H}|$ for the heat engine shown in Figure 10.5(a). Since no engine can be 100% efficient, the heat $|Q_{E,H}| = |Q_{P,H}|$ supplied as the energy source to the engine *cannot* be converted in its entirety into work, and some heat $|Q_{E,C}|$ must be discarded to the cold body. $|W_E|$ must therefore be less than $|W_P|$.

The comparison of the two machines depicted in Figure 10.5 shows that the pump on the right can convert a given amount of work $|W_P|$ in its entirety into an equal quantity of heat $|Q_{P,H}|$, but this amount of heat $|Q_{P,H}|$, when used by the engine on the left, can only produce an amount of work $|W_E|$ *less than* $|W_P|$ – even if the machine is operating at the maximum possible efficiency. This is a manifestation of irreversibility – for the concept of irreversibility is at the very heart of the Clausius and Kelvin-Planck statements.

This irreversibility can be demonstrated by a 'thought experiment', based on Figure 10.5. Let's imagine that the two hot bodies are in fact a single hot body, and that this single hot body is at some given temperature, higher than the temperature of the cold body. Some heat $|Q_{E,H}|$ flows from the 'hot body' into the engine, causing the hot body to cool down a little. The heat $|Q_{E,H}|$ is sufficient to drive one complete cycle of the engine, which discards some heat $|Q_{E,C}|$ into the cold body, so warming it a little, and performing some work $|W_E| = |Q_{E,H}| - |Q_{E,C}|$.

As a result of the first cycle of the engine, the hot body has become somewhat cooler, the cold body has become somewhat warmer, and the engine has performed some work. The natural heat flow down a temperature gradient has been harnessed so as to produce some work, just as the flow of water under gravity can be harnessed by a water wheel to power, say, the milling of grain.

Suppose now that the quantity of work $|W_E|$ performed by the engine on the left-hand side of Figure 10.5 is used to power one cycle of the pump shown on the right hand side – as might (in principle!) happen if the work $|W_E|$ performed by the engine were used to lift the weights in Joule's apparatus. As the weights then drop - and assuming that no energy is lost in the mechanics – then the quantity of work $|W_E|$ can be converted in its entirety into heat $|Q_{P,H}|$. The overall effect on the hot body of the combined single cycle of both the engine and the pump is therefore to lose a quantity of heat $|Q_{E,H}| = |W_E| + |Q_{E,C}|$ to the engine, and then to gain heat $|Q_{P,H}| = |W_E|$ from the pump. On each cycle of both machines acting together in this way, the heat $|Q_{P,H}|$ gained by the hot body from the pump must therefore *always be less* than the heat $|Q_{E,H}|$ lost from the hot body to the engine, the net loss being $|Q_{E,H}| - |Q_{P,H}| = |Q_{E,C}|$, precisely the quantity of heat gained by the cool body.

Exactly the same events happen on each combined cycle of both machines. And over time, the originally hot body becomes progressively cooler, and the originally cold body becomes progressively hotter, until both bodies achieve the same equilibrium temperature, and both machines stop. This, of course, is the familiar spontaneous, unidirectional, and irreversible, flow of heat down a temperature gradient. So, although the wording used in both the Clausius and the Kelvin-Planck statements is somewhat obscure – all that reference to "devices operating in cycles", to the "exchange of heat with this-or-that body" and to "no effect other than [whatever]" – the essence of both statements is the reality that spontaneous changes proceed unidirectionally and irreversibly from non-equilibrium to equilibrium states.

10.5 A graphical interpretation

Let's now take a deeper look at how a heat engine might work, and - in particular – at how heat might be exchanged between the engine and either of the two heat reservoirs between which the engine operates.

Fundamental to the engine's operation is its ability to perform P, V work of expansion: the expansion of a gas in a cylinder can be harnessed to drive a piston, and if the piston can be returned to its original position, as discussed on pages 61, then the cycle can be repeated indefinitely. For this to happen, a cylinder made of a suitable material, and containing a suitable gas (the **working substance**), must be in thermal contact with an external source of heat (the 'hot body' or 'hot reservoir'), so that the heat flow across the (necessarily non-adiabatic) system boundary will cause the pressure of the gas inside the cylinder to increase. The pressure difference between the gas inside the cylinder and the gas in the surroundings will, as we know, drive the piston against the external pressure, so allowing the system to perform work. In accordance with the First Law, during the expansion – the 'power stroke' – heat is absorbed by the system from the surroundings, and work is done by the system on the surroundings.

In order for the system to operate in a cycle, the piston needs to be returned to its original position. This requires the gas inside the cylinder to be compressed, implying – once again in accordance with the First Law – that during the 'return stroke', work is done on the system by the surroundings, and that heat is lost by the system to the surroundings.

For the engine to be useful, the work done by the engine during the power stroke needs to be greater than the work done on the system during the return stroke, so we assume for the moment that this is indeed the case. And as this brief description makes quite clear, for one complete cycle of the engine's operation, and the overall performance of work, there are two, different, exchanges of heat – the first in which the system gains heat for the power stroke, the second in which the system loses heat during the return stroke. This is very much in line with the implications of the Kelvin-Planck Statement as regards the necessity of a real engine to exchange heat with two reservoirs, not one. Yes, this is 'very much in line', but not totally so: at the moment, our description of a real engine demonstrates that there are necessarily two exchanges of heat, but we haven't yet demonstrated that this also requires two different reservoirs. To demonstrate this, we need to capture the working of the engine graphically, using a P, V diagram.

To do this, we need to make three assumptions that enable us to compile a suitable diagram:

- firstly, that the working substance within the system is an ideal gas
- secondly, that all changes take place along thermodynamically reversible paths
- thirdly, that all exchanges of heat between the system and the surroundings are isothermal.

The assumption that the gas within the cylinder is ideal is not restrictive – many real gases approximate to ideal behaviour. The assumption that all changes are thermodynamically reversible, as described on pages 31 to 33, has the benefit that all state functions are well-defined throughout all changes in state, and so can be depicted precisely on a P, V diagram. Real changes are not so well-behaved, and we must always bear in mind that any computations relating to the reversible work done, or reversible heat exchanged, represent theoretical maxima, always greater than any corresponding real, irreversible values. But as we shall see, the assumption of reversibility does not have any adverse influence on the discussion. And finally, the assumption of isothermal changes makes the explanation clear, without losing generality.

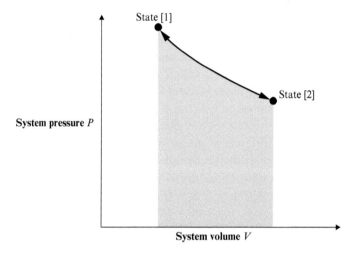

Figure 10.6 Isothermal expansion and compression at the same temperature. The power stroke is a reversible isothermal expansion at a given temperature T from state [1] to state [2], and the return stroke is a reversible isothermal compression, at the same temperature T, back to the original state [1]. For the cycle, the engine does no net work, and there is no net heat exchange with a single external reservoir.

Suppose, then, that an engine is in contact with a heat source at a temperature T. If the initial state of the engine is state [1], then the power stroke is a reversible isothermal expansion to some state [2], as shown in Figure 10.6. During this expansion, the engine does work $|_1W_2|_{\text{rev}}$, as represented by the shaded area in Figure 10.6, and absorbs heat $|_1Q_2|_{\text{rev}} = |_1W_2|_{\text{rev}}$ from the heat source.

Let us now assume that the return stroke takes place from state [2] to the original state [1], once again reversibly and isothermally, at the same temperature T. The system traces the same path, but in the reverse direction. During the return stroke, the surroundings do work $-|_2W_1|_{\text{rev}}$ on the system, but the quantity $|_2W_1|_{\text{rev}}$ of this work is exactly equal to the quantity $|_1W_2|_{\text{rev}}$ of the work done by the system on the surroundings during the power stroke - as is vividly demonstrated in Figure 10.6, in which the area under the curve defining the power stroke is obviously equal to the area under the curve defining the return stroke. Similarly, the quantity of heat $|_2Q_1|_{\text{rev}}$ lost by the system to the surroundings during the return stroke is equal to the quantity of heat $|_1Q_2|_{\text{rev}}$ gained by the system from the surroundings during the power stroke.

For the cycle as a whole, the engine depicted in Figure 10.6 performs no net work, and exchanges no net heat – which is, without a doubt, rather dull. But the engine whose operation is depicted in Figure 10.6 has a particular feature: both the expansion and the compression take place isothermally at the same temperature – and because of this, the return stroke is identical to the power stroke, but in the opposite direction. If we want our engine to perform work during the complete cycle, then the area under the line in the P, V diagram representing the return stroke *must be less than* the area under the line representing the power stroke. And for this to happen, then the line in the P, V diagram representing the return stroke *must be different from, and lower than,* the line representing the power stroke: that's the *only* way in which the area associated with the return stroke, and hence the work done on the system during the return stroke, can be less than the area associated with the power stroke, and hence the work done by the system during the power stroke, so ensuring that, over a complete cycle, the engine does net work on the surroundings.

But if the return stroke is to follow a reversible isothermal path lower in the P, V diagram than that of the power stroke, then the return stroke *must take place at a lower temperature than the power stroke*. The engine therefore must exchange heat with two reservoirs – a hot reservoir which provides the energy for the power stroke, and a cold reservoir for the return stroke: exactly in accordance with the Kelvin-Planck statement.

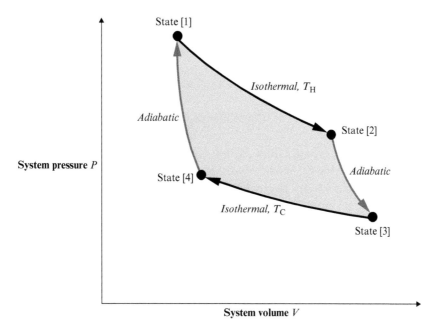

Figure 10.7 Schematic representation of an engine that does work. The power stroke is a reversible isothermal expansion, at a given hotter temperature T_H, from state [1] to state [2], and the return stroke is a reversible isothermal compression from state [3] to state [4] at a colder temperature T_C. The change from state [2] to state [3] is a reversible adiabatic expansion, and that from state [4] back to the original state [1] is a reversible adiabatic compression. For a complete cycle, the engine does net work on the surroundings as represented by the shaded area, and exchanges heat with two reservoirs.

Figure 10.7 shows a schematic representation of an engine that actually does perform work on each cycle. The power stroke, from state [1] to state [2], is a reversible isothermal expansion at a given, hotter, temperature T_H, exactly as in Figure 10.6. The return stroke in Figure 10.7, however, is different from the return stroke in Figure 10.6: rather than retracing the original path in the reverse direction, the return stroke from state [3] to state [4] takes place, reversibly and isothermally, at a colder temperature T_C.

How, though, does the engine change state so that the temperature of the ideal gas working substance falls from T_H to T_C? That's the path from state [2] to state [3] – a reversible adiabatic expansion, during which no heat is exchanged between the system and its surroundings, but the system does perform some P, V work of expansion on the surroundings, resulting in the required fall in temperature (see page 222). Similarly, the path from state [3] back to the original state [1] is a reversible adiabatic compression, during which the surroundings do work on the system, and the temperature rises from T_C to T_H, allowing the cycle to repeat.

During the complete cycle, the engine

- performs net work $|W|_{rev}$ on the surroundings, as represented by the shaded area in the centre of the diagram
- receives heat $|Q_H|_{rev}$ at a hotter temperature T_H from an external, hot, reservoir
- discards heat $|Q_C|_{rev}$ at a colder temperature T_C to an external, cool, reservoir
- such that, for the whole cycle, the (positive) net work done by the system on the surroundings equals the (positive) difference between the quantity of heat absorbed by the system from the hot reservoir and the quantity of heat discarded by the system to the cool reservoir: $|W|_{rev} = |Q_H|_{rev} - |Q_C|_{rev}$

This is totally in accordance with the Kelvin-Planck statement: it is a device that operates in a cycle, performs work, and exchanges heat with two reservoirs.

It is also totally consistent with the Clausius statement. To verify this, consider what happens when the process depicted in Figure 10.7 happens in reverse – as is possible, since all the changes are reversible. The change from the initial state [1] to state [4] is an adiabatic expansion. There is no heat exchange, but the system does work on the surroundings, resulting in a fall in temperature from T_H to T_C. The change from state [4] to state [3] is an isothermal expansion at the colder temperature T_C, in which work is done by the system on the surroundings, and heat is absorbed into the system from the cool reservoir. There then follows an adiabatic compression, in which work is done on the system, the temperature of which rises to T_H. The final step, from state [2] to state [1] is an isothermal compression at the hotter temperature T_H in which the surroundings do work on the system (the magnitude of which is the area between the line from state [2] to state [1] and the horizontal axis), and the system releases an equivalent amount of heat to the hot reservoir.

During the complete cycle, the system

- has net work $|W|_{rev}$ performed on it by the surroundings, as represented by the shaded area in the centre of the diagram
- absorbs heat $|Q_C|_{rev}$ at a colder temperature T_C from an external, cool, reservoir
- releases heat $|Q_H|_{rev}$ at a hotter temperature T_H to an external, hot, reservoir
- such that, for the whole cycle, the (positive) net work done by the surroundings on the system equals the (positive) difference between the quantity of heat released by the system into the hot reservoir and the quantity of heat absorbed by the system to the cool reservoir: $|W|_{rev} = |Q_H|_{rev} - |Q_C|_{rev}$

The overall result of this is that work is done on the system, with the effect of transferring heat from a cooler body to a warmer body – the system is therefore acting as a heat pump. The operation of this system is fully in accord with the Clausius statement: the transfer of heat is not the sole action of the system – in addition, work is done on the system by the surroundings.

Figure 10.7 is, in fact, a visual representation of the Clausius Statement. The transfer of heat from a cooler to a warmer body is represented visually in Figure 10.7 by the upper and lower isothermal paths. These two paths must be at different 'heights' in the diagram, for they represent different temperatures. Any device that operates in a cycle and transfers heat from a cooler body to a warmer body must follow some form of closed path in the P, V diagram – a closed path which must include both the lower and the upper isothermal paths. This closed path must therefore define some non-zero area; an area that represents the work that is required to be done on the device to enable the heat transfer to take place. It is simply impossible to 'connect' the lower and the upper paths without, at the same, time, defining a central area, and so the P, V diagram shown in Figure 10.7 vividly illustrates that the transfer of heat up a temperature gradient, of its own accord, and without any work, cannot happen. As indeed the Clausius Statement states very clearly.

10.6 The Carathéodory statement

Figure 10.7 introduces two reversible adiabatic paths to link the two isothermal paths, so enabling a complete cycle to take place with the two isothermal paths operating at different temperatures. These two adiabatic steps are required since the (upper) isothermal path from state [1] to state [2] is necessarily separated from the (lower) isothermal path from state [3] to state [4] – in rather different language, two different isothermal paths can never cross.

This can best be explained by reference to Figure 10.8, which is a P, V diagram for an ideal gas. States [1] and [2] have the same volume, with state [2] at a higher pressure than state [1]. According to the ideal gas law $PV = nRT$, the temperature T_2 of state [2] must be higher than the temperature T_1 of state [1]. An isotherm through state [1] connects all other states with a temperature of T_1, and an isotherm through state [2] connects all other states with a temperature of T_2. There can therefore be no state [3] which has the same temperature as both state [1] and state [2], and so the lines shown in Figure 10.8 that connect state [3] to state [1] and state [3] to state [2] cannot both be isotherms. Either one can be an isotherm, but the other cannot. Two isotherms therefore cannot cross.

It should be noted that an implicit assumption is that Figure 10.8 relates to system composed of a gas only, and that there is no change of phase during any of the processes depicted. If there is a change in phase – for example from a gas to a liquid – then there is a discontinuity in the volume, for the volume of a liquid at the vaporisation temperature is very much less than the same mass of the corresponding gas. The path showing the behaviour of the gas then 'stops' at the pressure and volume at which the phase change occurs, leaving a gap in the diagram to the point at which the pressure and volume correspond to the liquid (see Figure 15.10).

We have just demonstrated that, within any phase, two isotherms cannot cross, hence the need for the two adiabatic paths shown in Figure 10.7. The same is true for adiabats, for there

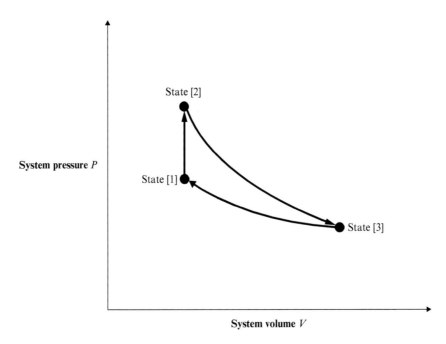

Figure 10.8 Two isotherms cannot cross. For an ideal gas, the temperature T_2 of state [2] must be higher than the temperature T_1 of state [1]. There is no point, anywhere on the diagram, corresponding to a state [3] which has the same temperature as both states [1] and [2], and so two isotherms cannot cross. The same is true of two adiabats.

is another statement of the Second Law which you might come across: it is attributable to the Greek mathematician Constantin Carathéodory, and is often stated as

Two adiabats cannot cross.

This statement is quite a long way from more familiar concepts such as spontaneity, unidirectionality and irreversibility, but it too can be explained, once again by reference to Figure 10.8, but interpreting Figure 10.8 rather differently.

As before, the system is an ideal gas, and state [2] has a higher pressure, and therefore a higher temperature, than state [1]. For a reversible path from state [1] to state [2] at constant volume, there is a heat flow $\{_1q_2\}_V$ across the boundary, from the surroundings into the system, given by

$$n\,C_V \Delta T_V = \{_1q_2\}_V \tag{7.2c}$$

where $\Delta T_V = T_2 - T_1$ is the temperature change, at constant volume, for the change from state [1] to state [2].

Let us assume that the path from state [2] to state [3], as shown in Figure 10.8, is a reversible adiabatic expansion. Work is done by the system on the surroundings, and the temperature drops to, say, T_3. Let us further assume that the path from state [3] back to state [1] is a reversible adiabatic compression, implying that work is done on the system by the surroundings, and the temperature of the system increases to T_1. At first sight, this appears to be quite plausible, for there is no apparent reason why the temperature T_3, which is necessarily less than T_2 by virtue of the adiabatic expansion, might not also be less than T_3, so allowing for a temperature increase from T_3 to T_1 attributable to the adiabatic compression.

There is, however, a problem. For the complete cycle from state [1], to state [2], then to state [3], and back to state [1], the system does net work on the surroundings, as indicated by the approximately triangular area in the middle of the diagram. But since the paths from state [2] to state [3], and from state [3] to state [1], are both adiabatic, the only path which involves an exchange of heat is the constant volume path from state [1] to state [2]. The cycle is therefore one in which the system does work, and exchanges heat with only a single reservoir – a cycle deemed by the Kelvin-Planck statement as impossible. This implies that it is impossible for both paths through state [2] to be adiabatic: one may be, but not both – a situation which we have met before (see pages 123 to 126, and also page 223). Carathéodory's statement rather less intuitively obvious statement that two adiabats cannot cross is therefore totally consistent with the other forms of expressing the Second Law.

10.7 Carnot engines and Carnot pumps

Take another look at Figure 10.7 (page 319), and compare it with Figure 7.3 (page 231). The two figures are identical are identical. Figure 10.7 is in fact a representation of a very special thermodynamic cycle, known as the Carnot cycle, first studied by the French scientist-engineer, Sadi Carnot. In 1824, aged 28, Carnot published a work entitled *Reflections on the Motive Power of Fire*, so laying one of the most important foundation stones of thermodynamics, on which people such as Clausius and Kelvin later built. The centre-piece of *Reflections* is an analysis of the thermodynamic cycle that now bears his name – although a diagram of the form of Figure 10.7 does not appear in his original work!

The Carnot cycle was explored in depth in the case study in Chapter 7 (see pages 230 to 236). As depicted in Figures 7.3 and 10.7, the cycle operates so that the system performs net work $|W_E|_{\text{rev}}$ $(= |\{W\}_{\text{rev}}|$, as expressed in the symbols used in Chapter 7) on the surroundings, gains heat $|Q_{E,H}|_{\text{rev}} = |\{_1q_2\}_{\text{rev,H}}|$ from the hot reservoir at temperature T_H, and discards heat $|Q_{E,C}|_{\text{rev}} = |\{_3q_4\}_{\text{rev,C}}|$ to the cold reservoir at temperature T_C, so acting as a heat engine. It is helpful to draw on the results of Chapter 7 here, but to do so, we need to note that the variables used in Chapter 7 are in accordance with the original sign convention, and so can represent both positive and negative quantities, whereas the variables used in this chapter represent necessarily positive magnitudes only. Bearing that in mind, we may write, for a reversible **Carnot heat engine** operating between a hot reservoir at temperature T_H and a cold reservoir at temperature T_C

$$\{_1q_2\}_{\text{rev,H}} = |Q_{E,H}|_{\text{rev}} = nRT_H \ln \frac{V_2}{V_1} \tag{7.24}$$

$$\{_3q_4\}_{\text{rev,C}} = |Q_{E,C}|_{\text{rev}} = nRT_C \ln \frac{V_4}{V_3} \tag{7.25}$$

and

$$|W_E|_{\text{rev}} = |Q_{E,H}|_{\text{rev}} - |Q_{E,C}|_{\text{rev}} \tag{10.7}$$

Since each of the four steps that constitute a Carnot cycle are thermodynamically reversible, the cycle can be driven the other way around, implying that the system acts not as a heat engine but as a heat pump, in which the surroundings do net work $|W_P|_{rev}$ on the system, the system extracts heat $|Q_{P,C}|_{rev}$ from the cold reservoir, and transfers heat $|Q_{P,H}|_{rev}$ into the hot reservoir. If the **Carnot heat pump** operates between a hot reservoir at temperature T_H and a cold reservoir at temperature T_C, and passes through the same states as the corresponding reversible Carnot heat engine, then

$$|Q_{P,H}|_{rev} = nRT_H \ln \frac{V_2}{V_1} \tag{10.8}$$

$$|Q_{P,C}|_{rev} = nRT_C \ln \frac{V_4}{V_3} \tag{10.9}$$

and

$$|W_P|_{rev} = |Q_{P,H}|_{rev} - |Q_{P,C}|_{rev} \tag{10.10}$$

These two modes of operation of a Carnot cycle are illustrated schematically in Figure 10.9, and as P, V diagrams in Figure 10.10.

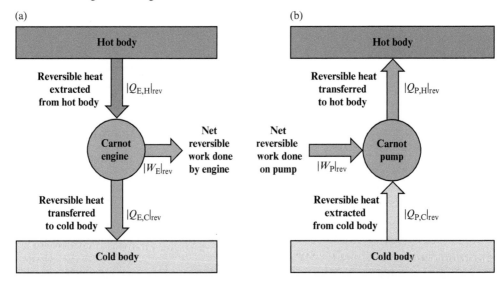

Figure 10.9 A reversible Carnot heat engine (a), and a reversible Carnot heat pump (b). If the temperatures of the hot and cold reservoirs are the same for both the engine and the pump, then all the corresponding energy flows are equal: $|Q_{E,H}|_{rev} = |Q_{P,H}|_{rev}$, $|Q_{E,C}|_{rev} = |Q_{P,C}|_{rev}$, and $|W_E|_{rev} = |W_P|_{rev}$. As a consequence the efficiencies of the two systems are the same.

Comparison of the set of equations (7.24), (7.25) and (10.7) for the heat engine with the corresponding set of equations (10.8), (10.9) and (10.10) for the heat pump shows that the corresponding flows of energy are identical. If we use the generalised definition of the efficiency η

$$\eta = \frac{|W|}{|Q_H|} = \frac{|Q_H| - |Q_C|}{|Q_H|} \tag{10.6a}$$

CARNOT ENGINES AND CARNOT PUMPS

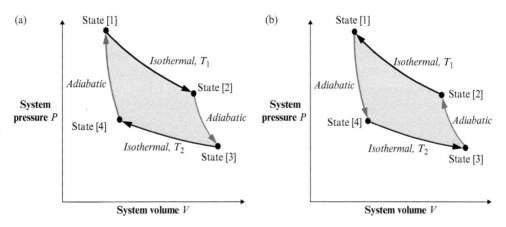

Figure 10.10 *P, V* diagrams for a reversible Carnot heat engine (a), and a reversible Carnot heat pump (b). The Carnot engine follows the sequence [1] → [2] → [3] → [4] → [1], during which the system performs net work on the surroundings, extracts heat from the hot reservoir, and discards heat to the cold reservoir. The Carnot pump follows the reverse sequence [1] → [4] → [3] → [2] → [1], during which the surroundings perform net work on the system, whilst the system extracts heat from the cold reservoir and transfers heat into the hot reservoir.

then the efficiency η^E_{rev} of a reversible Carnot heat engine is equal to the efficiency η^P_{rev} of a reversible Carnot heat pump which operates between the same two temperatures. And, according to equation (7.31)

$$\eta^E_{rev} = \eta^P_{rev} = \frac{T_H - T_C}{T_H} \tag{7.31}$$

the efficiencies η^E_{rev} and η^P_{rev} depend *only* on the temperatures T_H and T_C of the reservoirs between which the engine, or the pump, operates. This is known as **Carnot's theorem**.

Equations (10.6a) and (7.31) can therefore be combined as

$$\eta^E_{rev} = \eta^P_{rev} = \frac{T_H - T_C}{T_H} = \frac{|Q_H| - |Q_C|}{|Q_H|} \tag{10.11}$$

The statement that the efficiency of an engine operating according to a reversible Carnot cycle depends only on the temperatures of the corresponding hot and cold reservoirs appears, at first sight, to be rather dull. But a moment's thought will indicate that, far from being dull, it is most surprising – perhaps by virtue of what that statement implies, rather than explicitly states. The implications concern what the efficiency of the engine *does not* depend on – for if the efficiency depends only on the temperatures of the reservoirs, then the efficiency does not depend on *anything else at all*. So, for example, the efficiency does not depend on the nature of the working substance – the gas that is expanding and contracting; and – rather startlingly – it doesn't depend on the actual design of the engine. As long as the engine can follow a reversible Carnot cycle, then the efficiency is a given, and there is no 'better' design or 'worse' one: the efficiency is determined by the basic science of thermodynamics, not by the ingenuity of a design engineer.

In practice, of course, it is impossible to design any engine that will operate according to a Carnot cycle: each of the four paths are reversible, yet all real paths are necessarily irreversible.

Design engineers therefore still have the every opportunity to design engines which approach reversibility by, for example, reducing friction, and we shall look at some of the implications of real irreversibility shortly; for the moment, let's carry out a 'thought experiment' to explore what might happen if a very imaginative design engineer did in fact invent an engine that really is more efficient than a reversible Carnot engine operating between the same two temperatures.

The 'thought experiment' involves a 'test engine' – this being the newly-designed, high-efficiency engine – coupled to a reversible Carnot heat engine, working not as an engine, but as a reversible Carnot heat pump. As shown in Figure 10.11, the test engine and the Carnot pump are operating between the same two external heat reservoirs, and are coupled together such that their cycles are synchronised to ensure that the quantity of work done by the test engine on each cycle is just the right amount to drive one cycle of the Carnot pump.

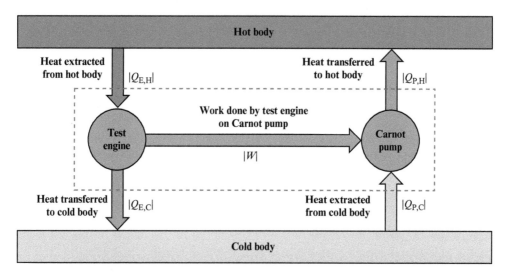

Figure 10.11 What happens if the test engine can operate at a higher efficiency than the reversible Carnot heat pump? The test engine and the Carnot pump operate between the same temperature reservoirs, such that the work $|W|$ done by one cycle of the test engine is just sufficient to drive one cycle of the Carnot pump.

If we represent the efficiency of the reversible Carnot heat pump by η^P_{rev}, and the efficiency of the test engine by η^E (where the subscript $_{rev}$ has been omitted, since we don't know whether the test engine is thermodynamically reversible or not), then the hypothesis we wish to test is that $\eta^E > \eta^P_{rev}$.

The analysis will draw on the general definition of the efficiency η, as

$$\eta = \frac{|W|}{|Q_H|} = \frac{|Q_H| - |Q_C|}{|Q_H|} \tag{10.6a}$$

which also implies that

$$|W| = \eta \, |Q_H| \tag{10.12}$$

and

$$|Q_C| = |Q_H|(1 - \eta) \tag{10.13}$$

Over one cycle of the test engine, the test engine absorbs a quantity $|Q_{E,H}|$ of heat from the hot reservoir, performs a quantity $|W|$ of work, and discards a quantity $|Q_{E,C}|$ of heat to the cold reservoir. According to equations (10.12) and (10.13)

$$|Q_{E,H}| = \frac{|W|}{\eta^E} \tag{10.14}$$

and

$$|Q_{E,C}| = |Q_{E,H}|\,(1 - \eta^E) = \frac{|W|}{\eta^E}(1 - \eta^E) \tag{10.15}$$

Over the corresponding cycle of the reversible Carnot pump, the test engine does a quantity $|W|$ of work on the pump, whilst the pump absorbs a quantity $|Q_{P,C}|$ of heat from the cold reservoir and transfers a quantity $|Q_{P,H}|$ of heat into the hot reservoir. Using equations (10.12) and (10.13) once again

$$|Q_{P,H}| = \frac{|W|}{\eta^P_{rev}} \tag{10.16}$$

and

$$|Q_{P,C}| = |Q_{P,H}|\,(1 - \eta^P_{rev}) = \frac{|W|}{\eta^P_{rev}}(1 - \eta^P_{rev}) \tag{10.17}$$

Subtracting equation (10.14) from equation (10.16)

$$|Q_{P,H}| - |Q_{E,H}| = \frac{|W|}{\eta^P_{rev}} - \frac{|W|}{\eta^E}$$

and so

$$|Q_{P,H}| - |Q_{E,H}| = \frac{|W|}{\eta^P_{rev}\eta^E}(\eta^E - \eta^P_{rev}) \tag{10.18}$$

Equation (10.18) looks rather complicated, but it carries a message. On the left-hand side, the difference $|Q_{P,H}| - |Q_{E,H}|$ represents the net amount of heat exchanged between the hot reservoir and the combined system of the test engine and the reversible Carnot heat pump. If this quantity is positive, $|Q_{P,H}| > |Q_{E,H}|$, then, for each complete cycle of the combined machine's operation, more heat is transferred into the hot reservoir by the Carnot pump than is extracted by the test engine, and so the hot reservoir becomes hotter; if this quantity is zero, $|Q_{P,H}| = |Q_{E,H}|$, and so the temperature of the hot reservoir remains constant; if this quantity is negative, then $|Q_{P,H}| < |Q_{E,H}|$, implying that, for each complete cycle, the engine extracts more heat from the hot reservoir than the Carnot pump returns, and so the hot reservoir becomes cooler.

The sign of the expression on the left-hand side of equation (10.17) therefore has physical significance, which takes us to the right-hand side, where we see that the variables $|W|$, η^P_{rev} and η^E are necessarily positive. The sign of the difference $|Q_{P,H}| - |Q_{E,H}|$ is therefore totally determined by the sign of the difference $\eta^E - \eta^P_{rev}$. According to our hypothesis, $\eta^E > \eta^P_{rev}$, and so the difference $|Q_{P,H}| - |Q_{E,H}|$ is positive. As a result of the combined machine's operation, there is a net inflow of heat into the hot reservoir: the hot reservoir gets hotter.

We can determine what happens, simultaneously, to the cold reservoir by subtracting equation (10.14) from equation (10.16)

$$|Q_{P,C}| - |Q_{E,C}| = \frac{|W|}{\eta^P_{rev}}(1 - \eta^P_{rev}) - \frac{|W|}{\eta^E}(1 - \eta^E)$$

implying that

$$|Q_{P,C}| - |Q_{E,C}| = \frac{|W|}{\eta^P_{rev}\eta^E}(\eta^E - \eta^P_{rev}) \tag{10.19}$$

Like equation (10.18), equation (10.19) is rather cluttered, but it too has meaning. If $|Q_{P,C}| > |Q_{E,C}|$ more heat is extracted from the cold reservoir by the Carnot pump than is received from the test engine, and so the cold reservoir becomes cooler; if $|Q_{P,C}| = |Q_{E,C}|$, the system is balanced, and the temperature of the cold reservoir remains constant; if $|Q_{P,C}| < |Q_{E,C}|$, more heat is discarded into the cold reservoir by the test engine than is extracted by the Carnot pump, and the cold reservoir becomes hotter. Since, by hypothesis, $\eta^E > \eta^P_{rev}$, then equation (10.19) tells us that the difference $|Q_{P,C}| - |Q_{E,C}|$ is necessarily positive. As a result of the combined machine's operation, more heat is extracted from the cold reservoir by the Carnot pump than is discarded by the test engine, and so the cold reservoir becomes cooler.

For any one cycle of the combined operation of the test engine and the reversible Carnot heat pump, equations (10.18) and (10.19) – whose right-hand sides are in fact identical - apply simultaneously, implying that the combined machine

- does no net work, but
- makes the hot reservoir hotter, whilst
- making the cold reservoir cooler.

This combined machine therefore has no effect other than the transfer of heat from a cooler to a warmer body, and so is deemed impossible according to the Clausius statement.

A combined machine, as shown in Figure 10.11 therefore cannot exist. But the reversible Carnot heat pump on the right can exist, and so the 'impossible' component must be the machine on the left, the test engine. Since the key feature of the test engine is the assumption about its efficiency – namely, that the efficiency of the test engine is greater than that of the reversible Carnot heat pump – then this assumption must be false.

The conclusion that this implies is important: *no engine* - however well designed, no matter what working material might be used, no matter what the nature of the cycle might be – can ever be more efficient than a reversible Carnot heat engine, operating between the same two temperatures. As regards efficiency – in terms of the efficient conversion of heat into work – a machine operating a Carnot cycle represents the 'perfect' engine.

10.8 Real engines

The analyses of the reversible Carnot heat engine and pump shows that the Carnot cycle – a cycle of four thermodynamically reversible steps, two isothermal and two adiabatic – defines a machine of the maximum possible efficiency η^E_{rev}, as given by equation (10.11)

$$\eta_{\text{rev}}^{\text{E}} = \frac{T_{\text{H}} - T_{\text{C}}}{T_{\text{H}}} = \frac{|Q_{\text{H}}| - |Q_{\text{C}}|}{|Q_{\text{H}}|} \tag{10.11}$$

Provided that the cold reservoir is not at the absolute zero of temperature (of which more in Chapter 12), T_{C} must be greater than zero, implying that the efficiency of even a 'perfect' Carnot engine or pump, can never reach the ideal of 1 or 100%. Even by using the most efficient possible process, heat cannot be converted in its entirety into work: some heat must always be discarded to a cold reservoir. In – perhaps more complicated – words, it is impossible to construct a device that operates in a cycle that has no effect other than the performance of work and the exchange of heat with a single reservoir. The inferences drawn from the study of the Carnot cycle are therefore totally consistent with the Kelvin-Planck statement of the Second Law.

The Carnot cycle tells us that even the most 'perfect' engine – an engine that works totally frictionlessly through a sequence of infinitesimal reversible steps – can never be 100% efficient. What about a real engine – an engine with friction, an engine that does not work infinitesimally, a real engine that is (thermodynamically) irreversible?

In fact, the Carnot cycle, and the analysis of the super-efficient 'test engine' depicted in Figure 10.11, give insight into real engines too. Let us imagine that the 'test engine' on the left of Figure 10.11 is a real engine – say a steam engine – one with friction and that runs irreversibly. It too will have some efficiency (we hope!) $\eta_{\text{irrev}}^{\text{E}}$ in that some heat $|Q_{\text{E,H}}|$ will be applied to engine (from, say, burning coal), and the engine will perform some work $|W|$ (say, to power a locomotive).

Let us now carry out another 'thought experiment', in which the work $|W|$ performed by the real engine is used, as in Figure 10.11, to drive a reversible Carnot heat pump. We know that the efficiency of the Carnot pump is the theoretical maximum, and so the hypothesis we are testing now is that the efficiency of the real engine is less than the efficiency of the Carnot pump, namely, that $\eta_{\text{irrev}}^{\text{E}} < \eta_{\text{rev}}^{\text{P}}$.

The analysis on pages 323 to 328 is identical, so let us revisit equations (10.18) and (10.19)

$$|Q_{\text{P,H}}| - |Q_{\text{E,H}}| = \frac{|W|}{\eta_{\text{rev}}^{\text{P}} \eta_{\text{irrev}}^{\text{E}}} (\eta_{\text{irrev}}^{\text{E}} - \eta_{\text{rev}}^{\text{P}}) \tag{10.18}$$

$$|Q_{\text{P,C}}| - |Q_{\text{E,C}}| = \frac{|W|}{\eta_{\text{rev}}^{\text{P}} \eta_{\text{irrev}}^{\text{E}}} (\eta_{\text{irrev}}^{\text{E}} - \eta_{\text{rev}}^{\text{P}}) \tag{10.19}$$

Since the hypothesis we are testing is that $\eta_{\text{irrev}}^{\text{E}} < \eta_{\text{rev}}^{\text{P}}$, then the term $(\eta_{\text{irrev}}^{\text{E}} - \eta_{\text{rev}}^{\text{P}})$ is necessarily negative. From equation (10.18), $|Q_{\text{P,H}}|$ will therefore be less than $|Q_{\text{E,H}}|$, implying that the real engine takes more heat from the hot reservoir than the Carnot pump replaces. The hot reservoir therefore cools. And from equation (10.19), $|Q_{\text{P,C}}|$ will therefore be less than $|Q_{\text{E,C}}|$, implying that the Carnot pump extracts less heat from the cold reservoir that the real engine discards. The cool reservoir therefore becomes hotter.

For any one cycle of the combined operation of the real engine and the reversible Carnot heat pump, the combined machine

- does no net work, but
- makes the hot reservoir cooler, whilst
- making the cold reservoir hotter.

An observer would detect a hot reservoir getting cooler, and a cold reservoir getting hotter, with some work being done by a machine along the way. This, of course, is the totally natural, real, flow of heat down a temperature gradient, and so the combined machine must exist.

Real engines therefore must be such that their efficiency is necessarily *less than* the corresponding, maximally efficient, Carnot engine or Carnot pump: in the real world

$$\eta^E_{irrev} < \eta^P_{rev} = \eta^E_{rev} \tag{10.20}$$

There is, in principle, a third case, that in which $\eta^E_{irrev} = \eta^P_{rev}$. In this instance, the temperatures of the two reservoirs would stay constant, just as if the reversible Carnot heat pump were being driven by a reversible Carnot heat engine. The 'test engine' would therefore be totally indistinguishable from a reversible Carnot engine, and so would be deemed a Carnot engine. For all real engines, the inequality (10.20) must be true.

Not all systems, however, are engines or pumps. How can the inequality (10.20) be applied to changes in general systems? To answer this question, we will draw on an observation made on page 124: that any process that contains just a single irreversible step is, as a whole, necessarily irreversible. So let us imagine a 'not-quite' Carnot heat engine in which the return isothermal path at temperature T_C, from state [3] to state [4] in Figure 10.10(a), is irreversible. For this cycle, heat $|Q_H|$ is absorbed reversibly from the hot reservoir at temperature T_H, but the net work done by the engine for a complete cycle must be represented as $|W|_{irrev}$, recognising the irreversibility of the return isothermal path.

For the irreversible engine

$$\eta_{irrev} = \frac{|W_{irrev}|}{|Q_H|}$$

and for a reversible Carnot heat engine operating between the same two temperatures, and absorbing the same quantity of heat $|Q_H|$ from the hot reservoir

$$\eta_{rev} = \frac{|W_{rev}|}{|Q_H|}$$

According to the inequality (10.20)

$$\eta_{rev} > \eta_{irrev} \tag{10.20}$$

from which

$$\frac{|W_{rev}|}{|Q_H|} > \frac{|W_{irrev}|}{|Q_H|}$$

and so

$$|W|_{rev} > |W|_{irrev}$$

If the change is infinitesimal

$$đw_{rev} > đw_{irrev}$$

which is an inequality we have seen before. As discussed on pages 143 and 144, it is the inequality that represents unidirectionality and irreversibility, and it is the inequality that underpins the Clausius Inequality (see page 272) and the principle of the increase in entropy – truly fundamental concepts that are all consequences of the analysis of the Carnot cycle.

EXERCISE

1. "I understand what Carathéodory is saying, and why two adiabats can't cross," said Alex. "If they did, you could design a machine which follows two adiabatic paths and one isothermal path – a machine which would operate in a cycle, performing work, and exchanging heat with just a single reservoir, during the isothermal step. This machine would contravene the Kelvin-Planck Statement, and so can't exist." Sam thought about that for a moment, and then replied, "Yes, that makes sense. But suppose a machine operated in a cycle formed from two isothermal steps and a single adiabatic step. This machine would perform work and exchange heat with two reservoirs – and so would not contravene any of the statements." Is Sam right?

2. Here is a (hypothetical!) conversation between James Joule and Count Rumford...

    James: "I've just been looking at a book on thermodynamics, and I read on page 316 that you can convert work, in its entirety, into heat – as your cannon and my falling weights prove. But on page 311 it says that you can't convert heat in its entirety into work."

    Benjamin: "That makes sense to me. No matter how much I heat the water in your apparatus, that will never lift your weights."

    James: "Yes. And no matter how much I heat your cannon, that will never cause your drill to rotate."

    Benjamin: "Wait a moment, though. Suppose that the heat generated in your apparatus from the action of your paddle is used to boil some of the water as steam. The steam can then power an engine that can lift the weights. That way, the heat from your apparatus can be used to lift the weights."

    James: "Mmm. I see what you mean. And the heat in your cannon, as generated by the friction from your drill, could boil some water, and the steam used to power an engine – which can then drive the drill."

    Benjamin: "So it *is* possible to lift your weights, and to drive my drill, from the heat they each produce."

    James: "Yes! The book must be wrong!"

    Benjamin: "Indeed! It must!"

    Discuss.

11 Order, information and time

> **Summary**
>
> Entropy is a measure of *disorder*: the more disordered a system, the higher the entropy; the more ordered, the lower the entropy.
>
> In the context of thermodynamics, order and disorder have very precise meanings, and bridge the macroscopic standpoint of thermodynamics with the microscopic standpoint of statistical mechanics. The central concept is that of a **microstate** – a complete description of the instantaneous locations, and momenta, of all the microscopic particles within a system. Macroscopically, the system exhibits a **macrostate**, which can be measured in terms of the system's state functions. The 'bridge' between observable, macroscopic, phenomena and the microscopic behaviour at the particle level is the realisation that any particular macrostate can correspond to very many different microstates. So, for example, a gas may be in macroscopic thermodynamic equilibrium with a given pressure and temperature, but the particles within the system can have many different locations, and momenta, all changing over time, but still resulting in the same, single, macroscopically observable state.
>
> The number of microstates than correspond to a given macrostate is that macrostate's **multiplicity** W, which is related to the systems entropy S according to **Boltzmann's equation**
>
> $$S = k_B \ln W \qquad (11.2)$$
>
> where k_B is Boltzmann's constant, equal to R/N_A.
>
> The natural, spontaneous, increase in entropy of all systems is associated with our perception of the passage of time, and the unidirectionality associated with thermodynamic irreversibility is associated with our perception that time runs only 'forwards', and not 'backwards'. The association of entropy with disorder also impacts on other, widely different fields, such as information theory, economics, and even the behaviour of high-performing teams.

11.1 What this chapter is about

This chapter broadens our understanding of entropy beyond the boundaries of chemical reactions, and flows of heat and work. As we shall see, entropy has relevance to a wide variety of contexts, and provides insights into phenomena as diverse as the tangling of the cables of laptop chargers, and the success of high-performing teams.

11.2 Why do things get muddled?

We have all had the experience that a room, which was tidy yesterday, is now a mess. Or that the contents of that drawer have become muddled. And why is the cable for my laptop charger always tangled? As far as I know, no-one has come into my room and deliberately thrown everything around; no-one has intentionally mixed the contents of that drawer up; no-one has spent ten minutes tying that cable into knots. Yet these things have all happened, 'by themselves'. Or – using the language of thermodynamics – spontaneously, unidirectionally and irreversibly. Yes, these changes can all be reversed, but not by retracing their original paths – if I want to tidy the room, sort the drawer, or untangle the cable, I have to do a lot of work.

What all these events have in common is a transition from a highly *ordered* state – the tidy room, the well-organised drawer, the neatly-coiled cable – to a *disordered* state – the messy room, the muddled drawer, the tangled cable. These disordered states arise spontaneously, and although the corresponding systems (the room, the drawer, the cable) are parts of larger systems (my house, my office, my brief case), these are all instances of spontaneous changes within the universe as a whole. Now, the Second Law tells us that all spontaneous changes in isolated systems are accompanied by an increase in entropy, and our experience shows us that many spontaneous changes are associated with to a transition from order to disorder. Is there a link between entropy and disorder? Does a change from an ordered state to a disordered one always correspond to an increase in entropy? And, conversely, does an increase in a system's entropy imply a change from a relatively well-ordered state to one of disorder?

The answer to all these questions is "yes". There is indeed a very strong link between disorder and entropy, a link which we will explore in this chapter – an exploration which begins with an examination of what we mean, in a thermodynamic context, by the words 'order' and 'disorder'.

11.3 Order and disorder

So, as an example of 'order', consider a system of five coloured pencils positioned side-by-side, in the sequence of the visible spectrum, with the blue pencil on the left, then green, yellow and orange, and finally the red pencil on the right. If we know the pattern that determines the 'rule' for the order – in this case, the sequence of colours that defines visible light – we can find any particular pencil immediately: we know that the orange pencil will be the fourth from the left, between the yellow pencil and the red one. Even when it is dark, and we are unable to distinguish the colours, we still know where each pencil is. If, however, the pencils are in a highly disordered state, we will have to examine every position to discover the orange pencil – or, if it is dark, take each pencil individually into the light.

The 'right' order for the pencils is uniquely specified – blue, green, yellow, orange, red. Any other sequence – say, green, red, yellow, orange, blue – is 'wrong'. A moment's thought will show that the number of 'wrong', disordered, sequences is much greater than the number of 'right' ones, for we can scramble the pencils in all manner of different ways, but there is only one 'right', ordered, sequence. If only one sequence is 'right', how many 'wrong' ones are there? And if the pencils are sequenced randomly, what is the likelihood, or probability, that they will be 'right'?

To answer these questions, we firstly need to determine the total number of all possible sequences, both 'right' and 'wrong'. Given 5 pencils, there are 5 ways of choosing the first one (say, yellow). For each of these 5 original choices, there are 4 pencils left, and so there are 4 ways of choosing the second pencil (say, red); then 3 ways of choosing the third; 2 ways of choosing the fourth; and finally there is only 1 pencil remaining. The total number of possibilities is therefore the product $5 \times 4 \times 3 \times 2 \times 1$ – written more simply as 5! (known as "factorial 5"). A moment's arithmetic gives the result that 5! = 120, showing that there are a total of 120 possible sequences, of which only 1 is 'right', and the other 119 'wrong'. The number of disordered states (119) is, as expected, significantly greater than the number of states in which the pencils are in precisely the 'right' order (1); looked at in a rather different way, if the pencils are in a random order, then there is 1 chance in 120 that the pencils are 'just right', and 119 chances in 120 that the pencils are 'wrong'.

This introduces the concept of probability: 1 chance in 120 corresponds to a probability of 1/120 = 0.008333, or around 0.8%; 119 chances in 120 corresponds to a probability of 5039/5040 = 0.991667, or around 99.2%. So, if 120 people are each holding a bag containing five differently coloured pencils, and if each person is invited to take the pencils out one-by-one, then we should expect that just one person would take all the five pencils out in the 'right' order. Similarly, if we were watching a film, lasting 10 hours (600 minutes), such that, each minute, the pencils are shown in a different random sequence, then, as the different configurations are displayed, we would expect to see the pencils in the 'right' order on just five occasions (1/120 × 600 = 5), and then for only one minute, with an average interval between these 'special' events of 2 hours; furthermore, 5 such 'special' events are likely to take place within a total elapsed time of 10 hours.

Another interpretation of probability concerns the likelihood that the pencil on the left is blue. In the highly ordered state, this probability is 1, corresponding to the certainty that the blue pen is on the left, in accordance with the colour sequence of the spectrum. If the 5 pencils are disordered, however, the probability that the pencil on the left is blue falls to 1/5 = 0.2 = 20%.

This example demonstrates the general principle that ordered states are associated with a low probability of occurrence; conversely, disordered states are associated with a high probability of occurrence.

11.4 Macrostates and microstates

We now introduce two new terms – *macrostate* and *microstate*:

- A *macrostate* is a macroscopically observable state of a system, such as 'coloured pencils in the right order', 'coloured pencils in the wrong order' – and, more scientifically, 'a system not in thermodynamic equilibrium' or 'a system in thermodynamic equilibrium'.
- A *microstate* is a specific configuration, at a particular time, of the component parts from which a system is formed – for example, the specific order of the coloured pencils, or the exact arrangement of the molecules within a gas.

In our examples of the pencils, the room, the drawer and the cable, macrostates will be defined using simple words or phrases such as 'tidy', 'untidy', 'neat', 'muddled', 'tangled', 'ordered' and 'disordered'; for thermodynamic systems, macrostates are defined by the values of the familiar macroscopic state functions such as mass, volume, pressure, temperature, internal energy, enthalpy and entropy. The corresponding microstates, however, require much more information, and much longer descriptions. So to define a specific microstate for the pencils, we must describe the sequence of colours; for the room, we would need to define the precise location of the furniture, the magazines, and everything else in the room; for the drawer, we would specify the exact position and orientation of each item; for the cable, would describe exactly how the cable is configured; for a thermodynamic system, we need to know the precise locations and movements of each individual molecule.

Let us focus on the desk drawer for a moment, and suppose that we have just finished organising the contents of the drawer so that all the ball-point pens are stacked tidily side-by-side on the left, the scraps of paper with scribbled notes are in the middle, the paper-clips are at the back, and the five coloured pencils are on the right, ordered in spectrum sequence, from blue to red. The macrostate of the drawer is 'well-organised', and the corresponding microstate has all the components of the drawer 'just so'. We now close the drawer, pause, and open it again. It's very likely that the act of closing the drawer and then opening it will have caused the contents of the drawer to have moved. The components of the drawer are now in a different configuration – a different microstate – and the macrostate is no longer as well-organised as it was, but is now a little 'muddled'. If the drawer is opened and closed a few more times, the pens, paper, paper-clips and pencils will now be strewn throughout the drawer, resulting in a truly 'muddled' macrostate, the precise configuration of which can be captured in a photograph. We now open and close the drawer several times again, and take another photograph. Without any doubt, this second photograph will be different from the first one: the microstate defined by the exact positions of the drawer's contents will be different each time we open the drawer, and although these microstates are indeed different, they all correspond to the same macrostate, 'muddled'. Whereas the macrostate 'well-organised' corresponds to just one microstate, with all the pens, paper, paper-clips and pencils 'just so', the macrostate 'muddled' corresponds to a large number of possible microstates, each with the pens, paper, paper-clips and pencils in different positions, all of which are 'muddled'.

The action of opening and closing the drawer 'jiggles' the contents, so creating the circumstances in which the contents of the drawer can access a range of microstates. And in doing this, we know from experience that we can open and close the drawer a huge number of times – thereby allowing the system to access all the allowable microstates – but we will never open it to find that the contents have returned to their original 'well-organised' state. Furthermore, if we open the drawer at random, it is vastly more probable that we will find the drawer to be 'muddled', rather than 'well-organised'.

That's all very obvious, and fully in accord with our every-day experience. It's also a powerful analogy for what happens at a microscopic level to a gas in a cylinder, as illustrated in Figure 11.1.

Figure 11.1(a) shows a system comprising a cylinder with a central partition, with the same number of molecules of an ideal gas on each side, at a given temperature T. All the molecules are identical, but let us suppose that we can distinguish the molecules on the left ('dark') from

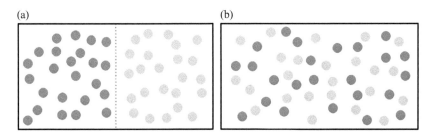

Figure 11.1 Macrostates and microstates. (a) shows a cylinder, with a central partition, and with the same number of molecules of an ideal gas on each side. If we assume that the molecules are distinguishable, (a) shows a configuration in which all the 'dark' molecules are to the left, and all the 'pale' molecules are to the right. The corresponding macrostate is highly ordered, rather like the 'well-organised' drawer. If the partition is removed, the thermal motion of the molecules causes them to mix as shown in (b), rather like the 'muddled' drawer. The corresponding macrostate is much more disordered; furthermore, it is extremely unlikely that the gas shown in (b) will spontaneously assume a configuration of the type shown in (a).

those on the right ('pale'). We know with certainty that all the 'dark' molecules are on the left, and all the 'pale' molecules are on the right, and so the probability that any molecule on the left is 'dark' is 1, and that it is 'pale' is 0; similarly, the probability that any molecule on the right is 'dark' is 0, and that it is 'pale' is 1. The macrostate illustrated in Figure 11.1(a) is therefore highly ordered, rather like the 'well-organised' drawer.

If the partition is removed, the entire volume of the cylinder is available to all the molecules. Thermal motion, consistent with the temperature T, causes those molecules originally on the left to move to the right, and those originally on the right to move to the left: as we saw on page 126, the gas will spontaneously mix. At any instant, the precise configuration of all the molecules represents the corresponding instantaneous microstate, and, since the molecules can all move at random throughout the entire cylinder, the microstate changes very quickly: the action of thermal motion is therefore a natural 'jiggling' process, rather like the opening and closing the drawer, enabling the contents of the system to access all possible microstates. If we observe the system over any period of time, we will detect the macrostate (or macrostates) corresponding to the microstates that the system accesses during the time of our observation. If, over that time, those microstates all correspond to the same macrostate, then we will see a single, stable, macrostate; a macrostate that has macroscopic state functions that are constant over time, and – for intensive state functions – have the same value at all points within the system. This will therefore be an equilibrium macrostate. Given that any real system will contain many molecules – say, multiples of the Avogadro constant, 6×10^{23}, each of which is moving very fast – then the number of microstates being accessed over any finite time – say, minutes – is truly huge. Each of these microstates, however, corresponds to a single macrostate – the equilibrium state at temperature T – and each of these microstates necessarily has the same internal energy U, the internal energy of the equilibrium macrostate.

Once the gas has reached an equilibrium state, with the molecules distributed evenly throughout the cylinder, the probability that any molecule on the left is 'dark' is 0.5, and that it is 'pale' is 0.5; similarly, the probability that any molecule on the right is 'dark' is 0.5, and that it is 'pale' is 0.5. Before the partition was removed, we knew with certainty the identity of any molecule in either half-cylinder; now, with the partition removed, we can only guess.

Compared to the macrostate illustrated in Figure 11.1(a), the macrostate illustrated in Figure 11.1(b) is much more disordered, rather like the 'muddled' drawer, or the disordered coloured pencils.

11.5 Three important principles

The examples of the pencils and the gas illustrate three important general principles about macrostates and microstates:

- Any single macrostate is likely to correspond to a number of different microstates, all of which will have the same internal energy U.
- The number of microstates that corresponds to the same macrostate (sometimes referred to as that macrostate's **multiplicity**) is a measure of that macrostate's degree of order – the smaller the number of microstates, the higher the order; the greater the number of microstates, the lower the order (and hence the greater the disorder).
- If a system can access a large number of microstates, then the macrostate which we are most likely to observe will be that macrostate which has the highest probability of occurrence – namely, that macrostate which corresponds to the largest number of microstates, that macrostate with the highest multiplicity.

11.6 Measuring disorder

Our discussion so far illustrates the importance of the number of microstates corresponding to a single macrostate – the macrostate's multiplicity – so let's represent this number by the variable W (or, alternatively, Ω). And, in accordance with the second principle just mentioned, as well as being a measure of the number of microstates corresponding to a given macrostate, W is also a direct measure of a system's *dis*order: the greater the value of a macrostate's multiplicity W, the more disordered the system.

To show how W can be computed, consider an ideal gas, at a given temperature, which can occupy a cylinder of a fixed total volume, as shown in Figure 11.2.

For simplicity, suppose that the gas comprises ten identical molecules, each individually identifiable – hence the numbers in Figure 11.2. As we discussed in Chapter 1 (see page 17), and Chapter 3 (see page 67), each molecule has kinetic energy, as determined by the gas's temperature, and each molecule can move at random through the entire volume of the cylinder. At any time, all 10 molecules will be distributed within the cylinder, with some molecules on the left of the mid-line, and some on the right. As time evolves, the numbers on the left and the right will change (subject to their sum always equalling ten), and if we can 'freeze' time, and then examine the system, this will define the system's instantaneous microstate.

Figure 11.2(a) shows one such possible microstate, in which seven molecules are on the left, and three on the right. At the instant at which the configuration shown in Figure 11.2(a) happens, each molecule will be located at a particular position, and will be travelling with a given velocity. The description of this microstate therefore requires, for each molecule, the specification of six numbers – three representing the instantaneous position, and three more

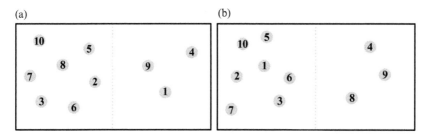

Figure 11.2 How many microstates correspond to a given macrostate? Both (a) and (b) show an ideal gas of 10 molecules, at a given temperature, within a cylinder of a fixed total volume. The macrostate of interest – the mass of gas in each half-cylinder – is determined only by the number of molecules within either side of the cylinder, and not by any other parameters such as an individual molecule's exact location or velocity. Both (a) and (b) therefore represent the same macrostate [7, 3], with seven molecules in the left-hand half, and three in the right-hand half. Microstate (a) is different from the microstate (b), in that in (a), molecule 8 is on the left, and molecule 1 is on the right, whereas in (b) molecule 8 is on the right, and molecule 1 is on the left.

defining the instantaneous velocity. Overall, for a system of 10 molecules, the description of any microstate will require $6 \times 10 = 60$ numbers; for a more realistic system, say, one comprising $1 \, \text{mol} = 6 \times 10^{23}$ molecules, the quantity of data required, $6 \times 6 \times 10^{23}$ numbers, is immense.

Let us consider the distribution of mass within each half-cylinder, to the left and to the right of the mid-line. The total volume of the system is fixed, and we are assuming a constant temperature; furthermore, the total mass of the gas is constant at $10m$, where m is the mass of each molecule. If, at a particular instant, there are seven molecules on the left and three on the right, then the mass in the left-hand half is $7m$, and the mass in the right-hand half is $3m$. Since the mass in each half depends only on the number of molecules in each half, and not on other parameters such as each molecule's specific location or velocity, then the distribution of mass can be fully described using only three numbers: the mass m of each molecule, and the number pair [7, 3], representing the number of molecules to the left and the right respectively. Furthermore, the same distribution of mass – the same macrostate [7, 3] – can result from a number of different microstates, as illustrated in Figure 11.2.

This raises the question "what is the value of W_i – the number of different microstates – that corresponds to any given macrostate i?". To answer this question, let us first consider an extreme case, the macrostate [10, 0], in which all 10 molecules are on the left. This macrostate corresponds to just a single microstate, with all the molecules on the left: remember that the mass of gas on the left does not depend on the locations and velocities of the individual molecules, it depends only on the number of molecules. And if there are 10 molecules in total, there is only a single microstate with all of them on the left – $W[10, 0] = 1$.

How about macrostate [9, 1]? A moment's thought will verify that there are 10 different microstates that correspond to this single macrostate: the first has only molecule 1 on the left, and all the others on the right; the second with only molecule 2 ... and the tenth with only molecule 10. $W[9, 1]$ therefore equals 10.

The number of microstates associated with the macrostate [8, 2] is less easy to guess. Rather, we need to use a formula, which is related to what mathematician's know as the 'binomial theorem': for a total of 10 molecules, the total number of microstates corresponding to the macrostate [8, 2] is given by 10!/(8! × 2!), where 10! = 10 × 9 × 8 × 7 × 6 × 5 × 4 × 3 × 2 × 1. 8! cancels in the nominator and denominator, and so 10!/(8! × 2!) = (10 × 9)/2 = 45. The total number of microstates W[8, 2] corresponding to the macrostate [8, 2] is therefore 45.

Similarly, macrostate [7, 3] corresponds to 10!/(7! × 3!) microstates; [6, 4] to 10!/(6! × 4!); [5, 5] to 10!/(5! × 5!); [4, 6] to 10!/(4! × 6!); [3, 7] to 10!/(3! × 7!); [2, 8] to 10!/(2! × 8!); [1, 9] to 10!/(1! × 9!); and [0, 10] to 10!/(0! × 10!) – where mathematicians define 0! to be equal to 1. The actual numbers corresponding to these computations are shown in Table 11.1.

Table 11.1 The number of microstates corresponding to a given macrostate for a system of 10 molecules, as illustrated in Figure 11.1, and the associated probabilities

Macrostate i	Number W_i of corresponding microstates	Probability P_i of macrostate
[10, 0]	1	0.0009766
[9, 1]	10	0.009766
[8, 2]	45	0.04395
[7, 3]	120	0.1172
[6, 4]	210	0.2051
[5, 5]	252	0.2461
[4, 6]	210	0.2051
[3, 7]	120	0.1172
[2, 8]	45	0.04395
[1, 9]	10	0.009766
[0, 10]	1	0.0009766

As can be seen from Table 11.1, the macrostate with the highest value of W is [5, 5] – the state in which the molecules are evenly distributed on both sides. Furthermore, the values of W for the symmetrical macrostates [X, Y] and [Y, X] are equal – as makes intuitive sense. Also, the total number of possible microstates for all macrostates is 2^{10} = 1024, where the exponent, 10, equals the total number of molecules spread across the 2 compartments.

If all microstates are equally probable – as is the case for the molecules of an ideal gas, at a given temperature, moving randomly – then we may define the **mathematical probability** P_i of a given macrostate i as

mathematical probability P_i of macrostate i

$$= \frac{\text{Number of microstates for macrostate } i}{\text{Total number of possible microstates}} = \frac{W_i}{\sum_i W_i} \quad (11.1)$$

The probabilities of each macrostate are also given in Table 11.1 – as can be seen, state [5, 5] has the highest probability, 0.2461, meaning that, if the system were to be observed on 1000 occasions, on 246 of those, the system would be seen to be in state [5, 5]. Although this state is the most probable, states [6, 4] and [4, 6] are not far behind – these would each be observed on 205 of the 1000 observations. The states [10, 0] and [0, 10], with all the molecules in one half or the other, however, are decidedly improbable, with either of these states being observed only about once in every 1000 observations. Looked at rather differently, it is $0.2461/(2 \times 0.0009766) = 126$ times more likely that we would observe the $[5, 5]$ state rather than either $[10, 0]$ or $[0, 10]$.

For a gas of 20 molecules, there are a total of 2^{20} possible microstates. The most probable macrostate is $[10, 10]$, this being some 10^5 times more probable than either $[20, 0]$ or $[0, 20]$. For a gas of 100 molecules, the total number of microstates is 2^{100}, and the most probable macrostate is $[50, 50]$, some 5×10^{28} times more probable than either $[100, 0]$ or $[0, 100]$. If we take a realistic situation, say, a gas containing 1 mol of molecules – that's about 6×10^{23} molecules – there are a total of $2^{6 \times 10^{23}}$ possible microstates. The most probable macrostate has 0.5 mol in each half-cylinder, this macrostate being a factor of some $2^{6 \times 10^{23}} \approx 10^{2 \times 10^{23}}$ times more likely than having all the molecules in one-half of the cylinder. This number is unimaginably large – the number of seconds in a *billion* years is some 3×10^{16}, and to put a billion years into context, the age of the Earth has been estimated at about 4.5 billion years, and the age of the universe, back to Big Bang, at around 13.8 billion years, around 4×10^{17} seconds. 10^{17} – that's 1 followed by 17 zeros – is, in human terms, a big number, but compared to $10^{2 \times 10^{23}}$ – that's 1 followed by 200,000,000,000,000,000,000,000 zeros – it is infinitesimal! So, we can very assuredly say that it just will never happen that the air in the room in which you are now sitting will spontaneously assemble in just one-half of that room!

11.7 What happens when a gas expands into a vacuum

This analysis of macrostates, and the corresponding microstates, gives us a deeper insight into what happens when a gas expands into a vacuum, and also of the correspondence between entropy and disorder.

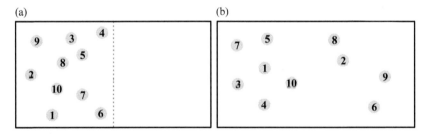

Figure 11.3 Expansion into a vacuum. In (a), all molecules (of which only 10 are shown) are constrained within the left-hand half of the cylinder, and the system is in thermodynamic equilibrium. When the partition is broken, the molecules are able to move through the entire volume, and all possible microstates within the whole cylinder are now accessible to the system. (b) shows one such microstate, [6, 4]. Macroscopically, we observe the most probable macrostate – that macrostate which has the maximum number of corresponding microstates.

Figure 11.3(a) represents a gas of n molecules confined to the left-hand half of a cylinder, with the right-hand half being a vacuum. Within the left-hand side of the cylinder, the molecules are moving randomly, and so are able to access all possible microstates within the half-cylinder, but since the partition is present, there are no microstates accessible to the system in which any molecules are on the right-hand side. At any instant, the positions and velocities of the molecules on the left-hand side take on particular values, corresponding to the instantaneous microstate, and as time evolves, these positions and values change, resulting in different microstates. If, however, the system is in thermodynamic equilibrium, all the state functions have well-defined values which are constant over time, implying that all these different microstates correspond to a single macrostate 'the gas at thermodynamic equilibrium within the left-hand half of the cylinder at a given volume, pressure and temperature'.

If we now remove the partition, suddenly, the entire volume of the whole cylinder is accessible to the gas. Macroscopically, we observe an initial non-equilibrium state as the gas expands into the vacuum; then, after a short time, equilibrium is restored with the gas evenly filling the whole volume. Microscopically, as soon as the partition is removed, molecules which were previously constrained within the left-hand side of the cylinder can now move to the right. At the instant that the partition is removed, the system is in the microstate $[n, 0]$; then, when only a single molecule has moved into the right-hand side of the cylinder, the system is in the microstate $[n-1, 1]$, and, as time evolves, it passes, very quickly, through states such as $[n-2, 2]$, $[n-3, 3]$ and so on. Since the molecules are all moving randomly, the system has access to all possible microstates, from its original state $[n, 0]$, to the other possible extreme state $[0, n]$, and all intermediate states. Furthermore, since the molecules are all in motion, these microstates are continuously changing, randomly and very rapidly.

We know, however, that many different microstates can correspond to the same macrostate. What, then, does a macroscopic observer detect? Initially, turmoil – as the microstates change, a series of non-equilibrium macrostates will be observed, and it is not possible to define any of the state functions, for these will change over time and have different values in different parts of the system. Quite soon, however, the gas distributes itself evenly throughout the whole volume, and state functions such as pressure become well-defined, and stable both over time, and at different points within the system. Macroscopically, we describe this as an equilibrium state, but microscopically, individual molecules are still moving randomly, and the underlying microstates are still changing, randomly and very rapidly. This microscopic activity corresponds to macroscopic equilibrium, implying that all these different microstates must correspond to the same macrostate, 'the system at equilibrium'. The equilibrium state is the state with the largest number of corresponding microstates, and the highest probability: the state $[n/2, n/2]$ in which the molecules are evenly distributed throughout the total volume.

11.8 The Boltzmann equation

We can now bring together three apparently different concepts:

- The observation that spontaneous changes in isolated systems take place unidirectionally, from non-equilibrium states to equilibrium states, as is in accordance with our every-day experience, and as stated by the Second Law (see page 268).
- The principle of the increase in entropy, as established by the Clausius inequality (see page 272).
- The correspondence between microstates and macrostates, such that equilibrium macrostates are those with the greatest probability, as expressed in terms of the number of microstates that can represent the same macrostate.

ORDER, INFORMATION AND TIME

These three concepts are in fact just different aspects of the same phenomenon, and were unified firstly by Ludwig Boltzmann around 1872, and then, in 1901, by Max Planck, into a single equation

$$S = k_B \ln W \tag{11.2}$$

In this equation, now known as **Boltzmann's equation**,

- S is the entropy of any isolated system in thermodynamic equilibrium (the system must be in equilibrium, so that S is well-defined).
- k_B (alternatively k) is Boltzmann's constant, which we introduced in equation (3.17). By definition, k_B is equal to R/N_A, the ratio of the ideal gas constant R to the Avogadro constant N_A (the number of molecules per mol of material), and has a numerical value of 1.381×10^{-23} J K^{-1}.
- W is macrostate's multiplicity – the number of microstates corresponding to the macrostate of the system for which S is the entropy, assuming that all microstates are equally likely. In the context of Boltzmann's equation, W is sometimes referred to as the macrostate's **thermodynamic probability**. Mathematically, as we saw in Table 11.1, probabilities are necessarily less than 1, and since values of W are necessarily greater than 1, and often vastly so, to avoid potential confusion, we will refer to W as the macrostate's multiplicity.

Boltzmann's equation is the mathematical relationship linking the multiplicity W of a given equilibrium (macro)state of an isolated system, and that state's entropy S. According to this equation, the greater the multiplicity of that equilibrium (macro)state, the greater the value of W, and so the greater the value of the entropy S. Since W is also a measure of disorder, Boltzmann's equation represents the direct link between entropy and disorder that we anticipated at the start of this chapter.

If an isolated system changes state from (macro)state $[1] \equiv [S_1, W_1]$ to (macro)state $[2] \equiv [S_2, W_2]$, then we may use equation (11.1) to evaluate the corresponding change ΔS in entropy as

$$\Delta S = S_2 - S_1 = k_B \ln W_2 - k_B \ln W_1 = k_B \ln \frac{W_2}{W_1} \tag{11.3}$$

If $W_2 > W_1$, then $\ln(W_2/W_1) > 0$, and $\Delta S > 0$, implying that the change from state [1] to state [2] is spontaneous. This corresponds to a change in which the number of microstates corresponding to state [2] is greater than the number of microstates corresponding to state [1], which is just a different way of saying that a change from order (as represented by state [1]) to disorder (state [2]) is spontaneous – as our experience of untidy rooms, muddled drawers and tangled cables bears witness.

Conversely, if $W_2 < W_1$, then $\ln(W_2/W_1) < 0$, and $\Delta S < 0$, implying that the change from state [1] to state [2], from disorder to order, is not spontaneous.

Equation (11.2) can be compared to equation (9.17)

$$\Delta S = \int_{\text{state 1}}^{\text{state 2}} \frac{\mathchar'26\mkern-12mu d q_{\text{rev}}}{T} = \int_{V_1}^{V_2} \frac{P \, dV}{T} = nR \int_{V_1}^{V_2} \frac{dV}{V} = nR \ln \frac{V_2}{V_1} \tag{9.17}$$

which, as we saw on page 280, gives the change ΔS in entropy corresponding to the adiabatic expansion of an ideal gas into a vacuum from an original volume V_1 to a final volume V_2. Since the change is adiabatic, no heat crosses the system boundary; furthermore, no work is done in expanding against a vacuum; together, these conditions imply that this change is, in effect, within an isolated system.

Equations (9.17) and (11.3) are therefore describing the same phenomenon, and so

$$\Delta S = k_B \ln \frac{W_2}{W_1} = nR \ln \frac{V_2}{V_1}$$

If we assume that there is 1 mol of gas, then $n = 1$, and if the gas expands to twice its original volume, then $V_2 = 2V_1$, and so

$$\Delta S = k_B \ln \frac{W_2}{W_1} = R \ln 2$$

Noting that, by definition, $k_B = R/N_A$, we now have

$$\Delta S = \frac{R}{N_A} \ln \frac{W_2}{W_1} = R \ln 2$$

and so

$$\ln \frac{W_2}{W_1} = N_A \ln 2 = \ln 2^{N_A}$$

Hence

$$\frac{W_2}{W_1} = 2^{N_A} \approx 2^{6 \times 10^{23}} \tag{11.4}$$

$2^{6 \times 10^{23}}$ is that (very!!!) large number we met on page 340, showing that W_2, the number of microstates corresponding to the macrostate in which the gas fills the entire volume $V_2 = 2V_1$ of the cylinder, is unimaginably larger than W_1, the number of microstates corresponding to the macrostate in which the gas is entirely within the half-cylinder V_1. As soon the partition is removed, all microstates, throughout the entire volume of the cylinder, are instantaneously accessible – and equation (11.4) tells us that the macrostate in which the gas fills the entire volume is $2^{6 \times 10^{23}}$ times more likely to be observed than the macrostate in which all the molecules are bunched up in the left-hand half of the cylinder. As soon as the gas has the opportunity to expand into the vacuum, it will indeed do so, spontaneously, unidirectionally, and – without any doubt whatsoever – irreversibly.

One further interpretation of equation (11.3)

$$\Delta S = k_B \ln \frac{W_2}{W_1} \tag{11.3}$$

is in the context of the data on the molar entropy changes associated with phase changes, as shown in Table 9.1. As we saw on page 294, for all substances, both the molar entropy change $\Delta_{fus}S$ associated with a phase change from a solid to the corresponding liquid, and the molar entropy change $\Delta_{vap}S$ associated with a phase change from a liquid to the corresponding gas, are always positive. If we represent the order of the solid, liquid and gaseous states by W_s, W_l, and W_g, respectively, then, in terms of equation (11.3), the fact that $\Delta_{fus}S > 0$ implies that $W_l > W_s$; likewise, since $\Delta_{vap}S > 0$ too, then $W_g > W_l$. This implies that a solid is more ordered than a liquid, and that a liquid is more ordered than a gas – as we now know from our molecular understanding of each of these states. Furthermore, since for any given material, $\Delta_{vap}S > \Delta_{fus}S$, then W_g is much larger in comparison to W_l than W_l is larger than W_s. Physically, this implies that the increase in disorder corresponding to the transition from a liquid to a gas is substantially greater than the increase in disorder that arises when a solid melts into a liquid. This too is totally consistent with our molecular knowledge: solids are highly ordered, often as crystals; liquids are rather less ordered, but molecules are still quite close together; gases are much more disordered, with the molecules moving randomly. Structurally, a liquid is 'closer' to the corresponding solid than a gas is to the corresponding liquid.

11.9 Maxwell's demon

In 1867, James Clerk Maxwell, another giant in the development of thermodynamics – and physics in general too – proposed a 'thought experiment' which appeared to contravene the Second Law.

Maxwell imagined two cylinders, each containing the same number of molecules of a monatomic ideal gas, at a constant temperature throughout, and in thermodynamic equilibrium. The cylinders are connected, so gas molecules can move from one cylinder to the other, in both directions; thermal motion ensures that the gases are well mixed. The connection between the cylinders contains an atomic-scale door, which can be opened and closed so quickly that, in principle, when the door is open, only a single molecule can pass from one cylinder to the other. The door is totally frictionless, and requires no work to operate.

The gas in both cylinders is at a constant temperature T, implying, as we saw on page 70, that the molecules throughout the system have an average kinetic energy $\langle \mathcal{E} \rangle$ given by

$$\langle \mathcal{E} \rangle = \frac{3}{2}k_B T \tag{3.19}$$

$\langle \mathcal{E} \rangle$, however, is an average – some molecules have higher energies, and so are moving at faster speeds, and some have lower energies, and so are moving at slower speeds. Maxwell then hypothesised the existence of a "being with faculties so sharpened" so as to be able to identify the speed at which each molecule is moving. So, when the "being" sees a molecule in the left-hand cylinder moving towards the right at less than the average speed, it opens the door to allow that molecule to move into the right-hand cylinder; but if a molecule in the left-hand cylinder approaches the door at greater than the average speed, the door remains shut. Similarly, if a molecule in the right-hand cylinder approaches the door at greater than the average speed, the "being" opens the door to allow it to enter the left-hand cylinder, but the door remains closed to molecules in the right-hand side moving at less than the average speed.

Over time, as result of the "being's" selective opening and shutting of the door, the left-hand side will contain rather more molecules with higher speeds, and hence higher energies, whilst the right-hand side will have rather more molecules with lower speeds and correspondingly lower energies. The average energy for the entire system remains unchanged, but the distribution of energy between the cylinders is now uneven, as illustrated in Figure 11.4. And since equation (3.19) implies that the higher the average molecular energy, the higher the

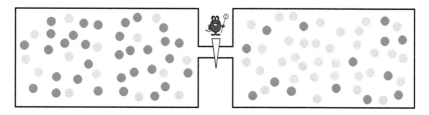

Figure 11.4 Maxwell's demon. The connection between two cylinders contains a door, which a 'demon' opens to allow 'pale' (slow) molecules to move from left to right, and 'dark' (fast) molecules from right to left. Over time, the 'pale' (slow) molecules will accumulate on the right, and the 'dark' (fast) molecules on the left, implying that a system that was originally at a constant temperature changes into a system with a higher temperature on the left, and a lower temperature on the right, without the performance of work – so breaking the Second Law.

temperature, the temperature of the left-hand cylinder will now be higher than the original temperature T, and the temperature in the right-hand cylinder will be lower than the original temperature T.

An external observer, who cannot detect the presence of the "being" or the door, would see an isolated system, originally 'in one piece' and at a uniform equilibrium temperature T, apparently spontaneously changing into a system of two halves, one at a higher temperature $T_L > T$, and the other at a lower temperature $T_R < T$. Physically, this is the flow of heat up a temperature gradient, and the corresponding change ΔS in entropy is given by the negative of the entropy change for heat flowing down the same temperature gradient. Reference to equation (9.15) therefore implies that, for heat flow up a temperature gradient

$$\Delta S = -nC_V \ln \frac{(T_L + T_R)^2}{4T_L T_R}. \tag{11.4}$$

As we saw on pages 277 and 278, the term $(T_L + T_R)^2/4T_L T_R$ is necessarily greater than 1. ΔS in equation (11.4) is therefore negative, corresponding to a decrease in entropy. It therefore appears that the action of Maxwell's "being" causes a spontaneous heat flow up a temperature gradient in an isolated system, accompanied by a decrease in entropy, so blowing the Second Law to pieces – causing Lord Kelvin to refer to Maxwell's "being" as "Maxwell's demon", a name that has stuck ever since.

Maxwell was no believer in demons, and he was a truly superb physicist and thinker. In hypothesising his "being", Maxwell is asking whether it might, in principle, be possible to design a machine, perhaps operating on a molecular level (Maxwell didn't know about nanotechnology!), which could contravene the Second Law. And if such a machine cannot be designed, what is the corresponding proof of impossibility?

Maxwell's 'thought experiment' stimulated debate and enquiry for decades. Perhaps the assumption that no work is required to operate the door is false, and that the amount of work required is sufficient to drive the heat up the gradient, like a Carnot heat pump; perhaps the demon needs 'feeding' from time to time, so as to be able to operate the door, and not be dead – in which case, the system isn't isolated, but has a flow of energy across the system boundary from the surroundings to the demon in the form of food; in more modern terms, this would be a flow of electrical energy, or perhaps light, as required to drive a nanomachine.

The most satisfactory resolution of the demon dilemma recognises that the door can be operated correctly only if the demon has the appropriate *information* – at any time, the demon must be able to measure the speed of an approaching atom, and then use that information to take a decision as to whether to open the door or not. If the gathering, storage and use of information – and indeed the discarding of old information which is no longer required (perhaps to make room for storing newly acquired information) – might be linked, in some way, to entropy, then perhaps there is a solution to the demon dilemma. The processing of the required information might then be associated with an increase in entropy greater than the decrease given by equation (11.4), resulting in a total entropy change which is positive, so saving the Second Law.

This raises a powerful question: is it possible to extend the concept of entropy far beyond heat, work, and gases in cylinders to the much more abstract realm of "information"? The answer is "yes", and entropy plays a central role in information theory – of which we can give only the most superficial overview here.

Let us therefore return to the system of five coloured pencils in a drawer. If someone asks "how many coloured pencils are there in the drawer?", I might answer "I don't know" – indicating that I have no information about the state of the system, and that there could, in principle, be any number of pencils. My knowledge of the system is very poor, and my uncertainty about the system is very high. Alternatively, if I respond "there are five pencils", my knowledge is higher, and my uncertainty lower. Considerable uncertainty still remains, however: if I know that there are five pencils, there are still 5! = 120 possible ways in which those pencils might be arranged, all (as far as I know) of equal probability. Additional information might enable me to say "there are five pencils, and the blue one is on the left", so increasing my knowledge, reducing my uncertainty, and reducing the number of remaining possible arrangements to 4! = 20.

This throws a new, and different light, on the parameter W, the number of microstates that correspond to a single macrostate. If the system is highly ordered (the five pencils in the 'right' sequence), then the number W of microstates corresponding to that highly ordered macrostate is small, perhaps just 1; and, according to Boltzmann's equation

$$S = k_B \ln W \qquad (11.2)$$

the entropy S of the system is correspondingly low. For this system, however, we know exactly where each pencil is located, and so our knowledge is correspondingly high. Conversely, if the system is highly disordered, then the number W of corresponding microstates is large, and – according to Boltzmann's equation – the entropy S is correspondingly large too. But since this system is disordered, we don't have much information about the location of a particular pencil, and our knowledge about the system is low.

An important quantity in information theory is the 'Hartley entropy' $\mathbb{H}$ (where we use a different font to distinguish the Hartley entropy $\mathbb{H}$ from a system's enthalpy H), defined as

$$\mathbb{H} = \log_2 |\mathbb{M}| \qquad (11.5)$$

in which $|\mathbb{M}|$ is the number of different ways in which a given message can be transmitted using a given number of a specific set of symbols, such as letters of an alphabet, or the dots and dashes of Morse code. |𝕄| is the information scientist's equivalent of the thermodynamicist's W, the number of microstates that correspond to a given macrostate.

Equations (11.2) and (11.5) are clearly of a similar form, and so the concept of entropy extends well beyond the realms of heat, work, engines and thermodynamics into the world of information, communications and computing. And to return to Maxwell and his demon – information theorists can now demonstrate that the increase in entropy associated with the use of the information that the demon needs to do his job more than compensates for the entropy reduction associated with the creation of the temperature gradient. The Second Law is saved.

11.10 Entropy and time

Let's consider once more our desk drawer – the drawer that becomes increasingly muddled each time it is opened. Suppose that you are watching a film in which the drawer is opened and closed many times. That's probably not a very exciting film, but imagine that, as you watch the film, you see the drawer becoming progressively tidier, so that as the film ends, the drawer is 'just so'.

Your initial reaction to the film is likely to be mild bewilderment: drawers 'naturally' become untidy, and so a film in which a drawer becomes progressively tidier 'all by itself' makes no sense. But then you realise that there is an explanation – the film is running backwards.

This is, of course, a simple illustration of the connection between our perception of order, our intuitive understanding of the 'natural' increase in disorder, and the passage of time. The direction of time's 'arrow' – from the past, through the present, into the future – is inextricably linked to our awareness of the natural transition from ordered states to disordered ones, which we now appreciate more deeply in the context of the Second Law, the continuing increase in the universe's entropy. Furthermore, the fact that time flows only forwards, and not backwards, is in essence a manifestation of thermodynamic irreversibility.

Every human being has a natural understanding of entropy, even though most have never heard of the Second Law: our perception of the passage of time is based upon an intuitive recognition of the increase in entropy associated with irreversible, spontaneous, changes. For change is fundamental to the measurement of time: if the position of the earth relative to the sun, if the positions of the hands on a clock, if the location of grains of sand in an hour glass did not change, we would not be conscious of the passage of time. And the direction in which time flows is determined by how entropy changes – if you were transported to another universe in which entropy increases as time flows, you would observe a continuous tendency for order to change to disorder, from which you would conclude that time in this new universe runs in the same direction as time in our familiar universe. But if, in the new universe, you sense a progressive increase in order, you would conclude that time would be running backwards.

11.11 Thermoeconomics

Much of the focus of thermodynamics is the study of systems pertinent to science and engineering – chemical reactions, biochemical processes and engines – and of the flows of heat and work associated with, for example, the expansion of gases. There are, however, a host of other important systems, not least the economic system of industry, finance and that other kind of work, the kind performed by humans.

For many years, the question "might thermodynamics, and concepts such as internal energy and entropy, have any application to economics?" has been under consideration, and especially so since the global economic crash of 2008 created considerable, and widespread, doubt concerning the validity of the conventional approach to economic theory. Hence '**thermoeconomics**', the hybrid created by blending together thermodynamics and economics.

Thermoeconomics is controversial, and very much in its infancy: no-one has yet resolved, for example, precisely what entropy means in an economic context. That said, it cannot be denied that much economic activity requires energy, and has the effect of increasing order, and hence reducing entropy. To take a single example: the manufacture of a car requires the assembly, in just the right configuration, of all the right component parts: the right engine has to be in the right place, as do the right wheels and the right seats. The number of microstates corresponding to the required macrostate – a car that works, and is safe – is just one, and so the entropy of a car is very low. Assembling that car did not happen by accident, spontaneously; rather, all the components had to be made, and then the assembly had to take place, with everything happening in the right sequence at the right time. Much energy is expended, and much money is required too. Does the flow of money play a role in economics rather like the flow of heat in thermodynamics? And, given that entropy is a non-conservative function, such that the entropy of the universe increases, is one (if not the key) goal of economics to do something similar – to have a continuing increase in the wealth (if not the well-being) of society?

11.12 Organodynamics

Let's now take a very different context: a high-performing team. We all recognise that high-performing teams are very special – we get excited watching them, let alone being part of one, and we sense that a high-performing team somehow achieves more than could possibly be achieved by the same people, carrying out similar activities by themselves.

The most obvious examples of high-performing teams relate to sport – champion football teams such as Real Madrid, Barcelona, Manchester United and Bayern Munich immediately come to mind, in addition to the national teams of the countries of your choice. But there are other examples too – the team that changes the wheels on Formula 1 racing cars is as well-choreographed as the most talented troupe of dancers; an orchestra, a jazz band, and many other musical ensembles are further examples, with the additional feature that, whereas sport is usually win-lose (even Barcelona can lose to Bayern Munich), with music, everyone wins, from the players to the audience, from the conductor to the composer. Many businesses and enterprises can also give rise to high-performing teams, from the team that works together to deliver a project, to the medical teams that perform near-miracles in emergencies.

What, then, are the key differences between a community of, say, n people 'doing their own thing', and those same n people acting as a high-performing team? That's a question that is talked about at length in many organisations, in their training courses, and at business schools. And it's a question that has some very familiar answers, such as

- members of a high-performing team have a shared goal that they collectively strive to achieve, whereas individuals have their own, often different, and possibly conflicting, goals
- members of a high-performing team communicate effectively with each other, whereas individuals may never communicate with each other
- members of a high-performing team are usually strongly motivated, whereas individuals may find it hard to maintain motivation.

and many others.

These answers are all true – but there is a further truth about high-performing teams that is much less often noticed. High-performing teams show a high degree of order, maintained over long periods of time: high-performing teams are associated with a low entropy, and keep that entropy low throughout the team's existence.

It's not common to use the word 'entropy' in associated with teams, so let's explore that for a moment. But rather than thinking about entropy, let's firstly think about order. Take, for example, the Formula 1 wheel-changing team. In the description just above, we used the phrase 'well-choreographed', for it is as if each member of the team is executing the most exquisite dance: everyone knows exactly where to be in relation to everyone else, and in relation to the car and its wheels; everyone makes exactly the right movements at the right times; each individual is in total harmony with everyone else. At any instant, for the wheel to be changed correctly in the minimum possible time, the number of microstates (as defined, for example, by the exact position and action of each team member) for the corresponding macrostate (what is happening to the car at that time) is likely to be just 1, with everyone doing exactly the right thing. At that instant, the 'system' is very highly ordered, and has a very low entropy. Furthermore, the sequence of macrostates over time, each corresponding to a single, different, microstate, requires the degree of order for the team as a whole to be continue to be very high indeed: the entropy must be kept low, and the order high, from the instant the car arrives until the moment it drives away.

Much the same can be said of an orchestra: at any instant, each player is playing exactly the right note, at the right time, at the right volume, and for the right duration, and continues to do so throughout the whole piece. The level of order is, once again, extraordinarily high, and it just never happens that, say, the trumpet player suddenly decides to 'do his own thing', and play a series of random, unordered, notes as loudly as possible.

Sport too, is highly ordered. In a football match, there are 22 players on the pitch, as well as the referee, and the touch judges close by. The movements, both on and off the ball, of not only all the players but also of the officials too, are highly correlated – people are not just randomly running around.

High-performing teams are therefore characterised by a high degree of order – and hence a low entropy. And they maintain the order high, and the entropy low, over long periods of time.

Which is 'interesting': the maintenance of low entropy over time is an 'unnatural act' in that it seems to go against the Second Law. According to the Second Law, entropy increases over time, and it is the 'natural' state of the world for order to degrade into chaos. The 'natural' behaviour of a high-performing team is therefore for the entropy to increase as time passes: the degree of order should decrease, and the team should progressively 'fall apart'.

We have all probably experienced this, either directly, or by observation – what started off as a great team degenerates into bickering, surliness and non-co-operation. The Second Law tells us that this is the 'natural way' things are, so we shouldn't be surprised when it happens. What should surprise us is when the team stays together, and achieves even more – for this appears to be going against 'natural law'.

Yes, the Second Law does tell us that systems progress naturally from order to chaos. But it also tells us how to stop that from happening, and how to maintain, and even increase, order over time. The 'trick' lies in the definition of the 'system': entropy does increase – but only in an *isolated* system. For closed and open systems, systems that allow energy to cross their boundaries, it is perfectly within 'the laws of physics' for the system's entropy to be kept at a low value, and for order to be maintained within the system, provided that there is a more-than-compensating increase in entropy in the surroundings. And one way of achieving that is to ensure that there is a continuous flow of energy into, and through, the non-isolated system – an energy flow that does not result in an increase in the system's temperature, but an energy flow that maintains the system's low entropy.

Every living thing on the planet is doing exactly this: living things are open systems, exchanging matter and energy with their surroundings. For green plants, the energy from the sun is the fundamental external source which maintains the 'order' within the plant, and keeps the plant alive; for animals, the external energy source is food, which, when combined chemically with atmospheric oxygen (in the process called respiration), produces the energy that keeps you, and me, alive. And when that energy flow stops, the plant, or the animal, dies.

Where, then, does the 'energy flow' come from for a high-performing team? Several places. From within each individual team member, from a sense of commitment and motivation. From the enthusiasm of the crowd. And, most importantly, from the team leader. The key role of the leader is to keep the energy flow into, and through, the team, keeping spirits up when the going is tough; encouraging; motivating; driving; supporting. Indeed, if you think of the words we commonly use to describe great leaders, they are words like 'enthusiastic', 'motivating', 'untiring', 'charismatic', 'visionary', 'exciting', 'energising'. These are all words about energy. And if 'thermodynamics' is the science of how to get useful work out of machines, maybe there is a science of **organodynamics** - how to get useful work out of organisations!

EXERCISES

1. Write an essay entitled "Thermodynamics, and the arrow of time".
2. Imagine that you are observing a number of activities such as
    - People walking across a city square.
    - Players moving on a football pitch, as the game progresses.
    - Members of an orchestra playing a symphony.

    How might you define the 'entropy' of each of these three 'systems'? What variables would you measure accordingly? How would you test the hypothesis that a high-performing football team, as determined, for example, by winning a league, has a sustainably lower entropy than teams that finished the season towards the bottom of the league?

12 The Third Law of Thermodynamics

Summary

There are a number of different statement of the **Third Law of Thermodynamics**, including

At the absolute zero of temperature, the entropy of all pure perfectly crystalline solids is zero.

At the absolute zero of temperature, the entropy change associated with any isothermal change is zero.

It is impossible to reach the absolute zero of temperature in a finite number of operations.

According to the first statement, if a pure substance forms a perfectly crystalline solid at the absolute zero of temperature, then we may calculate the entropy $S_P(\mathbb{T})$ of any pure substance at any temperature $\mathbb{T}$ from the equation

$$S_P(\mathbb{T}) = n \int_0^{T_{\text{fus}}} C_p(\text{solid}) \frac{dT}{T}$$

$$+ n \frac{\Delta_{\text{fus}} H}{T_{\text{fus}}} + n \int_{T_{\text{fus}}}^{T_{\text{vap}}} C_p(\text{liquid}) \frac{dT}{T} \qquad (12.2)$$

$$+ n \frac{\Delta_{\text{vap}} H}{T_{\text{vap}}} + n \int_{T_{\text{vap}}}^{\mathbb{T}} C_p(\text{gas}) \frac{dT}{T}$$

assuming that the pressure P is constant throughout. Entropies calculated in this way are known as **absolute entropies** or **Third Law entropies**.

The relationship between entropy and order can be used in the process known as **adiabatic demagnetisation** to cool a paramagnetic salt down to very low temperatures, approaching absolute zero.

12.1 The Third Law

We mentioned the Third Law of Thermodynamics very briefly at the end of Chapter 9; the purpose of this chapter is to examine it in more detail. Like the Second Law, which may be

Modern Thermodynamics for Chemists and Biochemists. Dennis Sherwood and Paul Dalby.
© Oxford University Press 2018. Published 2018 by Oxford University Press.

expressed in a number of different ways (see, for example, pages 267, 274, 304, 311 and 322), the Third Law can also take different forms, for example:

> At the absolute zero of temperature, the entropy of all pure perfectly crystalline solids is zero.

or

> At the absolute zero of temperature, the entropy change associated with any isothermal change is zero.

or

> It is impossible to reach the absolute zero of temperature in a finite number of operations.

The Third Law differs from the First and Second Laws in that it does not introduce a new state function: rather, it is concerned with the behaviour of systems as they approach the absolute zero of temperature.

12.2 Absolute entropies

As the temperature of a gas of a pure substance decreases at constant pressure, a temperature is reached at which we observe that the gas transforms into a liquid; as the temperature is reduced even more, we see the liquid will transform into a solid, often a crystal. We also know that as the temperature decreases, the internal energy U decreases, as does the average molecular kinetic energy. Accordingly, at a molecular level – which we can't see of course, but can only infer – as a gas, the molecules are moving fast, and throughout the total volume available; as a liquid, the molecules are moving more slowly, and over a more restricted distance; and as a solid, molecular motion reduces to vibration about a particular location as defined by the crystal lattice. Furthermore, we know (see pages 295) that as the temperature decreases, the entropy also decreases, implying that the progressive reduction in molecular movement is associated with a progressive reduction in the system's entropy.

This, of course, is totally in accordance with our understanding of the relationship between entropy and order, as discussed on pages 343 and 344. A gas, in which the molecules can move randomly over macroscopic distances, is far more disordered than a crystal, in which all the molecules are arranged in specific positions, as described by the crystal lattice. The entropy of a gas is therefore greater than the entropy of the same substance as a solid.

The physical differences between a gas, a liquid and a solid can also be interpreted in terms of Boltzmann's equation

$$S = k_B \ln W \tag{11.2}$$

As the motions of the molecules become progressively restricted as the system cools down through from a gas, through a liquid, to a crystalline solid, the number W of microstates that correspond to any given macrostate diminishes, and so the entropy S reduces. For a gas, any given molecule can be anywhere, order is low, and so W and S are large; in a crystalline solid,

ABSOLUTE ENTROPIES

any given molecule is constrained to be vibrating about its lattice position, order is high, and so W and S are both low.

As W decreases, S decreases, and ultimately, when $W = 1$, then Boltzmann's equation implies that $S = 0$. Furthermore, since W cannot be less than 1 (having fewer than 1 microstate for a given macrostate makes no sense), the situation in which $W = 1$ and $S = 0$ represents a limit, beyond which it is impossible to go.

$W = 1$ corresponds to a state in which the molecules in the system are all 'just so', rather like the contents of the highly-ordered drawer (see page 335) – physically, this implies that the molecules are all at their exact lattice positions, the crystal is perfectly ordered, and the molecules have stopped vibrating, and so have zero kinetic energy. If we now invoke equation (3.19)

$$\langle \mathcal{E} \rangle = \frac{3}{2} k_B T \tag{3.19}$$

for the average kinetic energy $\langle \mathcal{E} \rangle$ of a molecule, we see that $\langle \mathcal{E} \rangle$ is zero when $T = 0$ K. At the absolute zero of temperature, molecular motion ceases, the crystalline solid is perfectly ordered, $W = 1$ and $S = 0$: at absolute zero, the entropy of a perfectly crystalline solid is zero.

The importance of this statement is that it defines a reference level for the measurement of the entropy of any pure substance. If, for example, the entropy of any pure substance at any given temperature T and pressure is P is denoted as $S_P(T)$, then the entropy change at constant pressure ΔS_P for a change in state of that substance from temperature T_1 to temperature T_2 is

$$\Delta S_P = S_P(T_2) - S_P(T_1)$$

But if T_1 is absolute zero, and if the material can be considered to be a perfectly crystalline solid at that temperature, then $S_P(T_1) = 0$, and so

$$S_P(T_2) = \Delta S_P \tag{12.1}$$

We now invoke equation (9.46), which defines ΔS_P for any change in state at constant pressure from any temperature T_1 to any other temperature T_2 for n mol of any material:

$$\Delta S_P = n \int_{T_1}^{T_{\text{ph1}\to 2}} C_p \text{ (phase 1)} \frac{dT}{T}$$

$$+ n \frac{\Delta_{\text{ph1}\to 2} H}{T_{\text{ph1}\to 2}} + n \int_{T_{\text{ph1}\to 2}}^{T_{\text{ph2}\to 3}} C_p \text{ (phase 2)} \frac{dT}{T} \tag{9.46}$$

$$+ n \frac{\Delta_{\text{ph2}\to 3} H}{T_{\text{ph2}\to 3}} + n \int_{T_{\text{ph2}\to 3}}^{T_2} C_p \text{ (phase 3)} \frac{dT}{T}$$

So, if T_1 is 0 K, and T_2 any other non-zero, arbitrary, temperature $\mathbb{T}$, then equation (9.46) may be combined with equation (12.1) to give

$$\Delta S_P = S_P(\mathbb{T}) - S_P(0) = S_P(\mathbb{T})$$

and so

$$S_P(\mathcal{T}) = n \int_0^{T_{fus}} C_p(\text{solid}) \frac{dT}{T}$$

$$+ n \frac{\Delta_{fus}H}{T_{fus}} + n \int_{T_{fus}}^{T_{vap}} C_p(\text{liquid}) \frac{dT}{T} \qquad (12.2)$$

$$+ n \frac{\Delta_{vap}H}{T_{vap}} + n \int_{T_{vap}}^{\mathcal{T}} C_p(\text{gas}) \frac{dT}{T}$$

Equation (12.2), in principle, allows the entropy $S_P(\mathcal{T})$ of any pure substance to be calculated at any temperature $\mathcal{T}$, given knowledge of the temperature dependency of the various heat capacities (so allowing the integrals to be determined), and also the values of the parameters T_{fus}, $\Delta_{fus}H$, T_{vap}, and $\Delta_{vap}H$ for the two phase transitions from solid to liquid, and from liquid to gas.

For the liquid and gas phases, as we saw on pages 192 and 195, to a first approximation, C_P may be regarded as a constant, and so the corresponding integrals can be expressed in the form $n\, C_P \ln(T_2/T_1)$; for a more accurate determination, C_P may be expressed (see page 195) as a polynomial such as

$$C_P = a + bT + cT^2 + dT^3 + eT^4 \qquad (6.28)$$

and the integrals for each term computed accordingly.

For the solid phase, however, C_P is not a constant, especially as the temperature cools – as can be seen from Figure 12.1, which shows typical experimental data for a crystalline, insulating, solid:

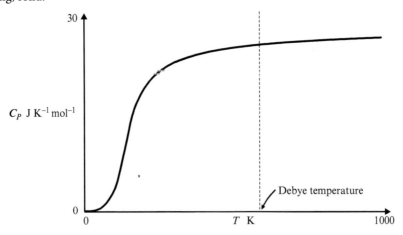

Figure 12.1 C_P, as a function of temperature. This graph is not based on specific data for a specific substance; rather, it captures the generalised behaviour of C_P for a typical crystalline, insulating, solid, for which the shape of the curve, close to absolute zero, is proportional to T^3. For a solid that conducts electricity, the low temperature shape is linear, proportional to T.

As can be seen, at absolute zero, C_P is zero; C_P then follows an S-shaped curve until a point at which C_P stabilises, thereafter rising gently with temperature. The theory behind this curve was first examined by Albert Einstein in 1907, soon after his very famous work on special relativity, and subsequently enhanced by the Dutch Nobel Prize Winner, Peter Debye, in 1912. The actual shape depends on the substance, but the general shape shown in Figure 12.1 is representative for solids which do not conduct electricity at very low temperatures; for conducting solids, the behaviour at very low temperatures is linear, proportional to T, rather than curved, proportional to T^3, as shown.

Debye identified a special temperature, now known as the Debye temperature, T_D or Θ_D, which has a particular significance in his theoretical analysis, and may be interpreted as the temperature at which the curve flattens out, as indicated in Figure 12.1. So, for example, for a metal such as lead, T_D is 105 K, and C_P stabilises at about 27 J K^{-1} mol^{-1}; for the much lighter metal aluminium, T_D is 428 K, and C_P stabilises at a somewhat lower value, about 24 J K^{-1} mol^{-1}; for the very different material, diamond, T_D is around 2,230 K, and C_P approaches 25 J K^{-1} mol^{-1} (compared with its value of about 6 J K^{-1} mol^{-1} at 298 K).

In practice, the entropy represented by equation (12.2) can be computed theoretically by carrying out the required integrations using appropriate formulae, or estimated using experimental data, and measuring the area under either the curve of C_P/T as the vertical axis, and T as the horizontal axis, or the curve of C_P as the vertical axis, and $\ln T$ as the horizontal axis. Such entropies, which invoke the Third Law statement that the entropy of the substance at absolute zero is zero, are known as Third law entropies, or absolute entropies.

An assumption associated with the computation of the absolute entropy for any pure substance is that it forms a perfectly crystalline solid at absolute zero. All substances, other than helium, are solid at very low temperatures and at a pressure of 1 bar (even hydrogen is a solid at temperatures below about 14 K), and many pure substances, but not all, are crystalline, often perfectly so. Helium remains a liquid at ultra-low temperatures, and at 1 bar pressure, but does solidify with a crystalline structure at temperatures below about 1 K and at pressures above 25 bar; and some pure substances form irregular 'glass-like' non-crystalline structures, and so are not 'perfectly' ordered. That said, standard reference tables list the standard molar entropies $S^\ominus$ of many compounds, these being the absolute entropies, usually as at 298.15 K, for a variety of elements and compounds in their standard states (see page 172), as shown in Table 12.1.

Two key features of Table 12.1 are

- All values are necessarily positive. This is a direct consequence of the structure of equation (12.2), and the intrinsically positive nature of all of the terms, and fully in accordance with the discussion on page 284.
- Unlike, for example, the tables of standard molar enthalpies of formation $\Delta_f H^\ominus$ (Table 6.6), in which the entries for all elements are zero, this table shows that each element has a non-zero standard molar entropy $S^\ominus$. This is attributable to the adoption of the convention that the molar enthalpies and free energies of formation of all elements in their standard states are zero, a convention that does not apply to standard entropies: rather, the standard molar entropy of any substance – including all elements – is determined according to equation (12.2), with $T = T^\ominus = 283.15$ K.

As regards specific values, diamond stands out as having a particularly low standard molar entropy (2.4 J K^{-1} mol^{-1}) which suggests it has a very highly ordered structure, even at 298.15 K.

THE THIRD LAW OF THERMODYNAMICS

Table 12.1 Some standard molar entropies $S^\ominus$

Selected elements a temperature of 298.15 K = 25 °C.			
	J K^{-1} mol^{-1}		J K^{-1} mol^{-1}
He (g)	126.2	C (s, diamond)	2.4
Xe (g)	169.7	Si (s)	18.8
H$_2$ (g)	130.7	Li (s)	29.1
N$_2$ (g)	191.6	Na (s)	51.3
O$_2$ (g)	205.2	Al (s)	28.3
Cl$_2$ (g)	223.1	Cu (s)	33.2
Br$_2$ (l)	152.2	Pb (s)	64.8
Selected inorganic compounds at a temperature of 298.15 K = 25 °C.			
	J K^{-1} mol^{-1}		J K^{-1} mol^{-1}
H$_2$O (l)	70.0	H$_2$S (g)	205.8
H$_2$O (g)	188.8	NO (g)	210.8
HCl (g)	186.9	NO$_2$ (g)	240.1
HBr (g)	198.7	NH$_3$ (g)	192.8
HI (g)	206.6	CO (g)	197.7
SO$_2$ (g)	248.2	CO$_2$ (g)	213.8
	J K^{-1} mol^{-1}		J K^{-1} mol^{-1}
AgCl (s)	96.3	CaCO$_3$ (s, calcite)	91.7
AgBr (s)	107.1	Fe$_2$O$_3$ (s)	87.4
AgI (s)	115.5	SiO$_2$ (s, quartz)	41.5
CaO (s)	38.1	NaCl (s)	72.1
Ca(OH)$_2$ (s)	83.4	NH$_4$Cl (s)	94.6
Selected organic compounds at a temperature of 298.15 K = 25 °C.			
	J K^{-1} mol^{-1}		J K^{-1} mol^{-1}
CH$_4$ (g)	186.3	C$_6$H$_6$ (l)	173.4
C$_2$H$_6$ (g)	229.1	CH$_3$OH (l)	126.8
C$_2$H$_4$ (g)	219.3	C$_2$H$_5$OH (l)	160.7
C$_2$H$_2$ (g)	200.9	CH$_3$COOH (l)	159.8
C$_3$H$_8$ (g)	270.3	CHCl$_3$ (l)	201.7

Source: *CRC Handbook of Chemistry and Physics*, 96th edn, 2015, Section 5, pages 5–4 to 5–42.

The highest values (in this table) are those of the heavier gases propane (270.3 J K^{-1} mol^{-1}), sulphur dioxide (248.2 J K^{-1} mol^{-1}) and nitrogen dioxide (240.1 J K^{-1} mol^{-1}). Amongst the liquids, water has the lowest standard molar entropy (70.0 J K^{-1} mol^{-1}), as associated with the order resulting from the **hydrogen bonds** between adjacent molecules – an effect that can be identified, albeit less strongly, in the comparison between the entropies, on the one hand, of methanol (126.8 J K^{-1} mol^{-1}), ethanol (160.7 J K^{-1} mol^{-1}) and acetic acid (159.8 J K^{-1} mol^{-1}), which can form hydrogen bonds, and, on the other, chloroform (201.7 J K^{-1} mol^{-1}), which cannot.

As was the case for standard molar enthalpies (see equation (6.21), page 182), for any reaction

Reactants → Products

we can calculate the corresponding standard entropy change $\Delta S^{\ominus}$

$$\Delta S^{\ominus} = \sum_{\text{Products}} n_{\text{prod}} S_{\text{prod}}^{\ominus} - \sum_{\text{Reactants}} n_{\text{reac}} S_{\text{reac}}^{\ominus} \qquad (12.3)$$

where n_{prod} and n_{reac} are the mole numbers for the products and reactants, as determined by the stoichiometry of the reaction.

12.3 Approaching absolute zero

Cryogenics, the science of low temperatures, is a blend of physics, engineering, and also ingenuity, for it is not at all obvious *how* it might be possible to reduce the temperature of a system progressively lower towards absolute zero. The temperature of a system can be reduced using, in essence, a Carnot heat pump (see page 314), whereby a device has work performed on it, so enabling heat to be extracted from the body that is to be cooled, and transferred to a warmer body. Devices that do this under 'normal' conditions – such as those used within domestic refrigerators - usually achieve this by exploiting the enthalpy of vaporisation of a special fluid, called a refrigerant. In essence, the refrigerant, in liquid form and at a high pressure, is allowed to expand quickly against a low pressure, causing the liquid to vaporise. As we saw on page 154, the phase transition from liquid to gas is endothermic, and the heat required is taken from the interior of the refrigerator, which cools. The gas now flows out of the refrigerator, and is compressed back into a liquid, so releasing the heat, originally extracted from the interior of the refrigerator, into the surroundings. This operation of this form of heat pump therefore requires that the refrigerant can exist in both liquid and gaseous form at different stages within a single cycle – but at very cold temperatures, most materials are solids, and a cooling process based on this method just won't work.

Very low temperatures are therefore achieved by a very different process: a process that directly uses the relationships between entropy and disorder, and between entropy and temperature.

The relationship between entropy and disorder that is of particular relevance here concerns the behaviour of certain types of molecule in response to a magnetic field. Certain materials – such as crystals of the salts cerium magnesium nitrate and gadolinium sulphate – are known as **paramagnetic**, whereby they are attracted by magnetic fields, and themselves can act as small magnets.

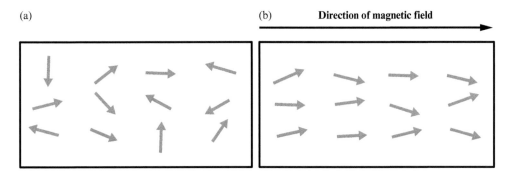

Figure 12.2 Paramagnetism. Both (a) and (b) represent the structures of a crystal formed from molecules which behave as magnetic dipoles. In (a), there is no external magnetic field, and so the molecular dipoles are oriented randomly. The crystal as a whole therefore shows no overall magnetic properties. In (b), the presence of an external magnetic field tends to align the molecular dipoles in the direction of the field. The crystal as a whole therefore act as a small magnet. The structure in (a) is more disordered than the structure in (b), and hence has a higher entropy.

As illustrated in Figure 12.2, paramagnetic materials are composed of molecules, each of which behaves as a 'molecular magnet', known as a 'dipole'. These 'molecular magnets', however, are relatively weak, and so in the absence of an external magnetic field, the dipoles are arranged randomly, so that the material as a whole has no magnetic properties. In the presence of a magnetic field, however, the dipoles align, implying that the molecular magnetic fields mutually reinforce each other, so conferring magnetic properties on the macroscopic material. If, however, the external magnetic field is removed, the thermal motion of the molecules causes the molecular orientations to randomise once more, so the macroscopic magnetic properties are lost – this being in contrast to **ferromagnetic** materials, which maintain their magnetism, once magnetised, even if the external magnetic field is removed. 'Permanent' magnets are therefore necessarily ferromagnetic, not paramagnetic.

Comparison of Figures 12.2(a) and 12.2(b) shows that the structure in which the dipoles are randomised, as in Figure 12.2(a), is more disordered than the structure in which the dipoles are aligned, as in Figure 12.2(b). At any given temperature, the randomised structure therefore has a greater entropy than the ordered structure, implying that the alternate turning on, and off, of an external magnetic field can 'flip' a paramagnetic material, at any temperature, between a lower entropy state (external magnetic field 'on') and a higher entropy state (external magnetic field 'off').

The entropies of the magnetised, and unmagnetised, states of a paramagnetic crystal, as a function of temperature, are shown in Figure 12.3, where the upper curve represents the entropy of the unmagnetised state, and the lower curve, the entropy of the magnetised state. In general, entropy increases with temperature (see equation (12.2), and the discussions on pages 284 and 295), and so both curves slope upwards as the temperature increases; furthermore, at any given temperature, the entropy of the unmagnetised state is greater than that of the magnetised state at the same temperature.

To demonstrate how we can use changes in entropy to result in cooling, consider a system comprising a crystal of a pure paramagnetic material, at a very low temperature, in a suitable vessel surrounded by a flow of liquid helium, at the same low temperature. The thermal contact

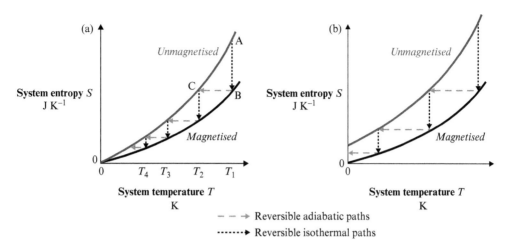

Figure 12.3 Adiabatic demagnetisation. Both (a) and (b) represent the behaviour of the entropy of a paramagnetic crystal as a function of temperature. In general, entropy increases with temperature, and, as shown, at any temperature, the entropy of the unmagnetised state (upper curve) is greater than the entropy of the magnetised state (lower curve). The sequence of reversible isothermal and adiabatic changes shown in (a) can approach absolute zero, but never quite reach it; the sequence shown in (b) appears to reach absolute zero, but breaks the Third Law – so this sequence is impossible.

between the vessel and the flow of liquid helium can be controlled so that the vessel can either be thermally isolated from the liquid helium, or in thermal equilibrium with the helium, in which case the temperature within the vessel can be maintained constant, at the temperature of the liquid helium.

The vessel is placed within a magnetic field, which can be switched on and off, but if the magnetic field changes from its initial value to its final value relatively slowly – meaning, in this case, at a rate slower than molecular motions, which is quite feasible – then the change in state caused by the application of the external magnetic field is, in essence, reversible.

If the magnetic field is off, the system is in a demagnetised state, for example, state A shown in Figure 12.3(a). If the magnetic field is switched on, the action of the field tends to align the molecular dipoles, and, in causing the dipoles to re-orient, the external magnetic field does work on the crystal. Furthermore, as just described, the change in the magnetic field strength takes place relatively slowly, and so the change in state of the paramagnetic material may be regarded as reversible.

Applying the First Law, which is universally true, to the system of the paramagnetic crystal, we may write

$$dU = đq_{rev} - đw_{rev} \qquad (5.1a)$$

and by definition

$$đq_{rev} = TdS \qquad (9.1)$$

and so

$$dU = TdS - đw_{rev} \qquad (12.4)$$

If the magnetic field is applied whilst thermal contact is maintained with the flowing liquid helium, the change of state is isothermal. As we know (see page 102), for a constant mass of any pure substance, the internal energy U is a function of any two other state functions, so for the system of the paramagnetic crystal, let us choose the volume V and the temperature T. Given that the crystal is a solid, and that any temperature changes under consideration are small, we may assume that the volume V remains constant; furthermore, because the change is isothermal, the temperature T is constant too. The internal energy U therefore remains constant, and so, for this change, $dU = 0$, and so equation (12.4) becomes

$$0 = TdS - đw_{rev}$$

implying that, for this isothermal magnetisation

$$TdS = đw_{rev} \tag{12.5}$$

In equation (12.5), $đw_{rev}$ represents the reversible work done by the system on the surroundings. In most examples given in this book, the work in question has been P, V work associated with the expansion of a gas; in this case, however, there is no P, V work of expansion – rather, the work is 'magnetic work' associated with the alignment of magnetic dipoles within the crystal. For the dipoles to become aligned, work has to be done on the system by the surroundings, and so $đw_{rev}$ is negative. Since, for an isothermal change, the temperature T is both constant and necessarily positive, equation (12.5) therefore implies that dS is negative: as a result of the isothermal magnetisation, the entropy of the paramagnetic crystal decreases, and the system achieves state B in Figure 12.2(a). As Figure 12.2(a) shows, the change of state from state A to state B is one in which the entropy of the system falls at constant temperature.

Suppose now that two conditions change together. Firstly, the system is thermally isolated from the surrounding liquid helium, implying that any change of state is necessarily adiabatic. Secondly, the magnetic field is reduced to zero quite slowly, so that the corresponding change of state is as reversible as possible. Because the crystal is suitably paramagnetic rather than ferromagnetic, thermal motion – even at a very low temperature – causes the molecules to randomise their orientations, so increasing the system's entropy. But just as the application of the external magnetic field during the change from state A to state B did work *on* the system, switching off the magnetic field may be thought of as a process in which the system does work on the surroundings. Once again, according to the First Law

$$dU = đq_{rev} - đw_{rev} \tag{5.1a}$$

then, if the system does work, $đw_{rev}$ is positive. But since the system is now thermally isolated so that the change in state is adiabatic, $đq_{rev}$ must be zero, implying that dU is negative. Also, we know that, for n mol of any material undergoing a change at constant volume

$$dU = nC_V \, dT \tag{5.17}$$

where C_V is the molar heat capacity at constant volume of the paramagnetic crystal. C_V is necessarily positive, and so if dU is negative, then dT is negative too: the system must cool. Furthermore, since $đq_{rev} = 0$, then $dS = đq_{rev}/T = 0$, and so this change in state is one in which the entropy stays constant.

This is shown in Figure 12.3(a) as the change from state B to state C – a reversible, adiabatic, demagnetisation, which has the effect of reducing the temperature from T_1 to T_2 at constant entropy. Overall, as a result of the initial reversible isothermal change from state A to state B, followed by the reversible adiabatic change from state B to state C, the system has 'moved down' the unmagnetised curve from state A to state C, with a reduction in both entropy, and importantly, temperature.

So, if the sequence of isothermal and adiabatic changes is repeated, the system follows the 'ladder' shown in Figure 12.3(a), with the temperature getting closer and closer to absolute zero.

This process, known as **adiabatic demagnetisation**, was first suggested, independently, by Peter Debye in 1926, and by William Giauque in 1927; then, in 1933, Giauque announced the achievement of a then-record-low temperature of 0.53 K, using gadolinium sulphate as the paramagnetic material. Since then, the 'record' has continued to be broken, and the method has evolved from the reduction of entropy resulting from the randomisation of previously aligned molecular magnetic dipoles in paramagnetic crystals to the similar phenomenon associated with magnetic dipoles in atomic nuclei. At the time of writing, the temperature of some 10^5 rubidium atoms has been lowered to about 5×10^{-11} K by a team of researchers at Stanford University, and, at a rather different scale, a research team in Italy has reduced the temperature of a 400 kg block of copper – that's, for comparison, about 4×10^{27} atoms - to 6×10^{-3} K.

5×10^{-11} K is very close to absolute zero, but not quite there. Is absolute zero possible to reach?

No.

For two reasons. Firstly, the successive steps of adiabatic demagnetisation reduce the temperature, step-by-step, by a given fraction – so, given a starting temperature T_1, the next temperature reached is, say, αT_1 (where $\alpha < 1$), then $\alpha^2 T_1 \ldots$ to $\alpha^n T_1$. For any particular of α, $\alpha^n T_1$ will always be some positive, non-zero, value. And even if α is not a constant for each step, but a series of values $\alpha_1, \alpha_2, \alpha_3, \ldots$, any function of the form $\alpha_1 \alpha_2 \alpha_3 T_1$ will still remain positive.

The second reason is illustrated in Figure 12.3(b), which shows a 'thought experiment' in which a succession of isothermal magnetisations and adiabatic demagnetisations does result in a state for which the system's temperature is at absolute zero. For this to happen, there must be a 'gap' between the lower 'ends' of the curves that represents the entropy of the demagnetised, and magnetised, states of the paramagnetic material. An implication of this is that there is a change of state, taking place isothermally at absolute zero, from the magnetised state to the demagnetised state, which is accompanied by a finite, non-zero, entropy change, as represented by the 'gap'. Such a change of state, however, is contrary to the Third Law, as stated on page 351, in the form

> **At the absolute zero of temperature, the entropy change associated with any isothermal change is zero.**

Such a change of state is forbidden, and so no such 'gap' can exist. The two entropy curves cannot remain separated as suggested in Figure 12.3(b), but must converge, as shown in Figure 12.3(a). This convergence therefore implies the Third Law as stated as

THE THIRD LAW OF THERMODYNAMICS

> It is impossible to reach the absolute zero of temperature in a finite number of operations.

Furthermore, according to the statement of the Third Law as

> At the absolute zero of temperature, the entropy of all pure perfectly crystalline solids is zero.

then, since the paramagnetic material may be assumed to be a perfectly crystalline solid, then the point at which the two entropy curves converge is precisely at the origin – the point at which zero entropy corresponds to the absolute zero of temperature, as shown in Figure 12.3(a).

EXERCISES

1. State the Third Law of Thermodynamics in three different forms.
2. Define 'absolute entropy'.
3. What is 'adiabatic demagnetisation'? How does it enable the absolute zero of temperature to be approached?

PART 3

Free energy, spontaneity, and equilibrium

13 Free energy

 Summary

The extensive state function the **Gibbs free energy** G is defined as

$$G = H - TS \qquad (13.10a)$$

For a closed system undergoing a change in state from state [1] to state [2] under conditions of constant temperature and pressure, and if the system can perform only P,V work of expansion, then equation (13.10a) becomes

$$\Delta G = \Delta H - T\Delta S \qquad (13.13)$$

If $\Delta G < 0$, the change is known as **exergonic**, and that change is spontaneous; if $\Delta G = 0$, then states [1] and [2] are in equilibrium with each other; if $\Delta G > 0$, the change is known as **endergonic**, and state [1] will not spontaneously change to state [2], but state [2] will spontaneously change to state [1].

A change in state in which ΔH is the dominant factor is known as an **enthalpy-driven** change; if $T\Delta S$ is the dominant factor, the change is **entropy-driven**.

For any system, G may be expressed as a function $G(P, T)$ of the pressure P and temperature T. For a system of a single pure substance that can perform P, V work of expansion only, the change dG associated with the changes dP and dT is given by

$$dG = V\,dP - S\,dT \qquad (13.16a)$$

from which the pressure variation of G at constant temperature is described by the equation

$$\left(\frac{\partial G}{\partial P}\right)_T = V \qquad (13.20)$$

In general, equation (13.20) may be integrated as

$$\Delta G_T = G_2 - G_1 = \int_{P_1}^{P_2} V\,dP \qquad (13.22)$$

which can be evaluated on a case-by-case basis according to the relevant equation-of-state.

So, for example, for an ideal gas at any temperature T, if the Gibbs free energy of the gas in its standard state is $G^\ominus$ at the standard pressure of $P^\ominus$, then the Gibbs free energy G at any other pressure P is given by

$$G = G^\ominus + nRT \ln \frac{P}{P^\ominus} \qquad (13.47b)$$

Modern Thermodynamics for Chemists and Biochemists. Dennis Sherwood and Paul Dalby.
© Oxford University Press 2018. Published 2018 by Oxford University Press.

The temperature variation of G at constant pressure is described most fundamentally by the **Gibbs-Helmholtz equation**

$$\left(\frac{\partial (G/T)}{\partial T}\right)_P = -\frac{H}{T^2} \tag{13.28}$$

The **Helmholtz free energy** A is defined as

$$A = U - TS \tag{13.34}$$

The Helmholtz free energy plays the same role for closed systems at constant temperature and volume as the Gibbs free energy G plays for closed systems at constant temperature and pressure. In all equations containing G and H there is an equivalent equation containing A and U.

If a system changes state from an initial state [1] to a final state [2] at constant temperature and pressure, and can perform some other form of reversible work $\{_1w_2\}_{\text{other,rev}}$ in addition to P, V work of expansion, then

$$\{_1w_2\}_{\text{other,rev}} = -\Delta G \tag{13.43b}$$

Since the reversible work a system can perform is always greater than the irreversible work, equation (13.39b) states that $-\Delta G$ is the maximum work a system can perform, in addition to P, V work of expansion, when undergoing a change at constant temperature and pressure. $-\Delta G$ is therefore a measure of the **maximum available work**, the maximum amount of energy that can be practically harnessed from a change of state taking place under the stated conditions.

An important consequence of equation (13.43b) is that if work is done *on* a system, then $\{_1w_2\}_{\text{other,rev}}$ has a negative sign, implying that $-\Delta G$ has a positive sign: a non-spontaneous change in state can therefore be driven by performing the appropriate amount of work on the system. This then leads to the concept of **coupled systems** – two systems coupled together so that the energy released by a spontaneous change in one system is harnessed so as to drive a non-spontaneous change in the other.

For any chemical reaction, if the reactants and products are all in their standard states, then

$$\Delta_r G^\circ = \sum_{\text{Products}} n_{\text{prod}} \Delta_f G_{\text{prod}}^\circ - \sum_{\text{Reactants}} n_{\text{reac}} \Delta_f G_{\text{reac}}^\circ \tag{13.45}$$

where n_i represents the mole number of component i as determined by the stoichiometry of the reaction, and $(\Delta_f G^\circ)_i$ represents the **standard molar Gibbs free energy of formation** of a reactant or a product – the change in the Gibbs free energy associated with the formation of 1 mol of the molecule in its standard state from its component atoms.

Chemical reactions inevitably involve mixtures of pure components, and the composition of the mixture changes as the reaction progresses. The Gibbs free energy G of a mixture is therefore a function $G(P, T, n_A, n_B, \ldots)$ of the pressure P, the temperature T, and also the mole numbers n_i of each of the components. The **chemical potential** $\mu_i(P, T)$ of any pure component i in a mixture is defined as

$$\mu_i(P, T) = \left(\frac{\partial G}{\partial n_i}\right)_{P,T,n_j \neq i} = G_i(P, T) \tag{13.55}$$

and, for a mixture, equation (13.16a)

$$dG = V\,dP - S\,dT \tag{13.16a}$$

which applies to a single pure substance, becomes

$$dG = V\,dP - S\,dT + \sum_i \mu_i\,dn_i \qquad (13.60)$$

Equation (13.60) is the **fundamental equation of chemical thermodynamics**.

The 'Fourth Law' of Thermodynamics is

Under any given conditions, of all the states that are accessible to a system, the system will occupy the state of lowest Gibbs free energy.

13.1 Changes in closed systems

13.1.1 Entropy changes in the system and in the surroundings

The Second Law tells us that the entropy of an isolated system inevitably increases. Conceptually, the universe as a whole is an isolated system; more practically, a sealed adiabatic container approximates closely to an isolated system, but most changes – especially those relevant to chemical and biological systems – take place in closed or open systems. The purpose of this chapter is therefore to examine how the principle of the increase of entropy can provide very useful insight into the behaviour of closed systems. This will identify not only how to predict whether or not a given chemical or biochemical reaction is spontaneous, but will also enrich our understanding of chemical equilibrium, and many chemical processes.

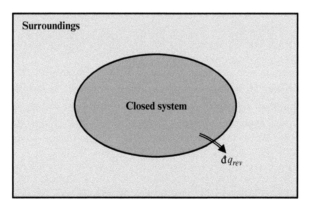

Figure 13.1 A spontaneous change in a closed system. The closed system and the surroundings together comprise the universe. Initially, the system and the surroundings are in equilibrium at a pressure P and a temperature T. An infinitesimal spontaneous change then takes place within the system, and so $dS_{sys} + dS_{sys} > 0$. Since the change takes place at constant temperature, the corresponding heat exchange between the system and the surroundings is reversible.

Figure 13.1 shows a closed system, and the surroundings, which collectively constitute the universe. Initially, the system and the surroundings are in equilibrium, at the same pressure P and the same temperature T. Suppose that a spontaneous change happens within the system. Since the system is closed, it is likely to exchange energy with the surroundings, and so the spontaneous change within the system will cause a flow of heat or work – or indeed both – across the system boundary, so causing a corresponding change in the surroundings.

Both the system and the surroundings will experience a change in state, and, if the change is assumed to be infinitesimal, we should therefore expect a change in entropy dS_{sys} within the system, and a corresponding change in entropy dS_{sur} within the surroundings. Since entropy is an extensive function, the change within the system, and the corresponding change within the surroundings, together cause a change in entropy $dS_{uni} = dS_{sys} + dS_{sur}$ in the entropy of the universe. If the change within the system is spontaneous, than the change within the universe is spontaneous too, and so, in accordance with the Second Law

$$dS_{uni} = dS_{sys} + dS_{sur} > 0 \tag{13.1}$$

13.1.2 A condition for spontaneity within a closed system

Suppose that the spontaneous change within the system is associated with an infinitesimal loss of heat from the system to the surroundings. Since the change is infinitesimal, during the change of state, the system and the surroundings may be regarded as being at the same temperature $T_{sys} = T_{sur} = T$, and so the flow of heat is reversible. Using the original sign convention (see page 56), in which symbols such as đq represent signed, algebraic quantities, then the (reversible) heat lost by the system to the surroundings may be written as $\{đq_{rev}\}_{sys}$, which is a negative quantity. The surroundings gain heat $\{đq_{rev}\}_{sur}$, which will be a positive quantity, and since no heat is 'lost' over the boundary

$$\{đq_{rev}\}_{sur} = -\{đq_{rev}\}_{sys}$$

For the surroundings, according to the definition of entropy

$$dS_{sur} = \frac{\{đq_{rev}\}_{sur}}{dT_{sur}} = -\frac{\{đq_{rev}\}_{sys}}{dT_{sys}} \tag{9.1}$$

An infinitesimal change, as well as taking place at constant temperature, may also be regarded as taking place at constant pressure. As we saw on page 150, for a system operating at constant pressure, and that performs P,V work of expansion only, the infinitesimal reversible heat exchange $\{đq_{rev}\}_P$ may be expressed in terms of the infinitesimal change dH_{sys} in the system's enthalpy as

$$dH_{sys} = \{đq_{rev}\}_{sys, P} \tag{6.5a}$$

Under these conditions, equation (9.1) therefore becomes

$$dS_{sur} = -\frac{dH_{sys}}{T_{sys}} \tag{13.2}$$

and so we may write the inequality (13.1) as

$$dS_{sys} - \frac{dH_{sys}}{T_{sys}} > 0 \tag{13.3}$$

All the variables in equation (13.3) relate only to the system, and so the subscript $_{sys}$ may be dropped; furthermore, equation (13.3) can be rearranged, so that, for a spontaneous change in a closed system

$$dH - T\,dS < 0 \tag{13.4}$$

13.2 Spontaneous changes in closed systems

13.2.1 Four important, but not unduly restrictive assumptions

The inequality (13.4)

$$dH - T\,dS < 0 \tag{13.4}$$

is important, for it gives a condition for a spontaneous change in a closed system that operates at constant temperature and pressure, and that performs P, V work of expansion only. Furthermore, because the condition is expressed in terms of state functions of the system alone, and makes no reference to the surroundings, or indeed the universe, then this condition can be applied by reference only to the system: rather than having to determine the entropy change for the universe as a whole, we can focus on variables of the system. So, if a change of state within a system is associated with a value of dH which is less than the corresponding value of $T\,dS$, then $dH - T\,dS$ is negative, and so the change is spontaneous; if dH is greater than the corresponding value of $T\,dS$, then $dH - T\,dS$ is positive, and so the change is not spontaneous. And if the value of dH is equal to the value of $T\,dS$, then $dH - T\,dS$ is zero, implying that there is no change at all – the system is in thermodynamic equilibrium.

In deriving the inequality (13.4), four assumptions were made:

- The first is that the system is assumed to be closed, exchanging energy in the form of heat and work across the system boundary, but not matter.
- The second assumption was that of constant temperature, which ensured that the heat flow $đq_{rev}$ across the boundary was reversible, as well as enabling us to use T_{sys} for the temperature at which the heat was exchanged.
- The third assumption was that of constant pressure.
- The fourth was that the system performs only P, V work of expansion, and no other forms of work.

These third and fourth assumptions have a particular significance in that they enabled us to replace $đq_{rev}$ by dH_{sys} as shown by equation (6.5a).

These four assumptions are somewhat 'hidden' in the inequality (13.4), but they are there, and they need to be kept in mind when the inequality (13.4) is used. As it happens, many chemical, and especially biochemical, processes take place under conditions which comply, either precisely or well enough, with the conditions of being a closed system, operating at constant temperature and constant pressure, and performing P, V work only: in practice, as we shall see, the inequality (13.4) has widespread applicability. Also, if a situation arises in which one any one or more of these conditions is not fulfilled, then an understanding of the fundamental principles enable us to modify matters accordingly – as we shall see, for example, on pages 621 to 623, where we examine some systems that perform other forms of work, such as electrical work, as well as P, V work.

13.2.2 In which direction does a spontaneous change take place?

For a closed system undergoing a finite change in state at constant temperature and pressure, and performing P, V work only, inequality (13.4) tells us that this (finite) change will be spontaneous if

$$\Delta H - T\Delta S < 0 \tag{13.5a}$$

or if

$$T > \frac{\Delta H}{\Delta S} \tag{13.5b}$$

So, consider any change

state [1] $\rightarrow$ state [2]

associated with a particular enthalpy change $\Delta H = H_2 - H_1$ and a corresponding entropy change $\Delta S = S_2 - S_1$, at a given temperature T. If the simultaneous values of ΔH, ΔS and T are such that the inequality (13.5a) $(\Delta H - T\Delta S) < 0$ is fulfilled, then state [1] will change spontaneously into state [2]. State [2] will therefore be the stable, equilibrium, state of the system at temperature T. If, however, the values are such that $(\Delta H - T\Delta S) > 0$, then the change from state [1] to state [2] will not be spontaneous – but the reverse reaction, from state [2] to state [1] will. Why so?

The reason is readily explained by considering the reverse reaction

state [2] $\rightarrow$ state [1]

also taking place at temperature T, for which the enthalpy change $\Delta_{bak}H$ (where the subscript bak refers to the reverse reaction) must have the same magnitude as the enthalpy change ΔH for the forward reaction, but the opposite sign

$$\Delta_{bak}H = -\Delta H$$

and similarly

$$\Delta_{bak}S = -\Delta S$$

so that

$$\Delta_{bak}H - T\Delta_{bak}S = -(\Delta H - T\Delta S) \tag{13.6}$$

Equation (13.6) implies that when $(\Delta H - T\Delta S)$ for a particular change in state at constant temperature and pressure is positive, then $(\Delta_{bak}H - T\Delta_{bak}S)$ for the reverse change, under the same conditions, is negative. Accordingly, when the condition for spontaneity for the forward reaction is not fulfilled – that is when $(\Delta H - T\Delta S)$ is positive – then the condition for spontaneity for the reverse reaction is fulfilled. So, for any change in state from state [1] to state [2], at constant pressure and at temperature T, for which $(\Delta H - T\Delta S)$ is positive, state [2] will spontaneously change into state [1], and state [1] is the stable, equilibrium, state at that temperature.

There is a third possibility too, as happens when the simultaneous values of ΔH, ΔS and T are such that $(\Delta H - T\Delta S) = 0$. Under these conditions, neither the forward reaction from state [1] to state [2], nor the reverse reaction from state [2] to state [1], are spontaneous. Physically, this implies that states [1] and [2] are in equilibrium.

To summarise: for any of fixed mass system that may exist in either of two states [1] and [2], the inequalities (13.5a) and (13.5b)

$$\Delta H - T\Delta S < 0 \tag{13.5a}$$

$$T > \frac{\Delta H}{\Delta S} \tag{13.5b}$$

determine, at any temperature T and at constant pressure, how that system will behave:

- if $(\Delta H - T\Delta S) < 0$, implying that $T > \Delta H/\Delta S$, state [1] will spontaneously change into state [2], and state [2] will be stable
- if $(\Delta H - T\Delta S) > 0$, implying that $T < \Delta H/\Delta S$, state [2] will spontaneously change into state [1], and state [1] will be stable
- if $(\Delta H - T\Delta S) = 0$, implying that $T = \Delta H/\Delta S$, the two states [1] and [2] will be in equilibrium.

13.2.3 Spontaneous changes, and the signs of ΔH and ΔS

As we have just shown, for any change in state

state [1] → state [2]

in a closed system at a given pressure, and at constant temperature T, then if

$$\Delta H - T\Delta S < 0 \tag{13.5a}$$

the reaction is spontaneous in the direction shown.

The values of ΔH and ΔS associated with any particular change will be algebraic, signed quantities – so, for example, for endothermic chemical reactions ΔH is positive, whereas for exothermic reactions, ΔH is negative.

Given that the temperature T is necessarily positive, there are four possible combinations of the signs of ΔH and ΔS:

- ΔH positive, combined with ΔS negative
- ΔH negative, combined with ΔS positive
- ΔH negative, combined with ΔS negative
- ΔH positive, combined with ΔS positive

The inequality (13.5a) can be used to give us an insight into whether or not the change in state is spontaneous for each of these possibilities.

13.2.4 ΔH positive, ΔS negative

So, firstly, if the change in state is such that ΔH is positive, and ΔS is negative, then since the temperature T is necessarily positive, the quantity $(\Delta H - T\Delta S)$ will always be greater than zero, and the inequality (13.5a) can never be fulfilled. The reaction from state [1] to state [2] can therefore never be spontaneous, but the reverse reaction from state [2] to state [1] is spontaneous.

13.2.5 ΔH negative, ΔS positive

If, however, the change of state is such that ΔH is negative, and ΔS is positive, then the quantity $(\Delta H - T\Delta S)$ is necessarily less than zero, and the inequality (13.5a) will always be fulfilled. The reaction from state [1] to state [2] will therefore always be spontaneous. This explains why many exothermic chemical reactions – for which ΔH is negative – are spontaneous, especially so if the reaction is associated with an increase in the entropy of the system.

13.2.6 ΔH and ΔS have the same sign

Not all exothermic reactions, however, are spontaneous, and the inequality (13.5a) offers an explanation for this too. If an exothermic reaction (ΔH is negative) is associated with a negative value of ΔS, then whether or not the reaction from state [1] to state [2] is spontaneous depends on the relative magnitudes of ΔH and $T\Delta S$ at a given temperature T: if the reaction is very exothermic, so that $(\Delta H - T\Delta S) < 0$ even though ΔS is positive, then the reaction will be spontaneous; but if ΔS is sufficiently negative, then $(\Delta H - T\Delta S)$ could be a positive quantity even if the reaction is exothermic, and ΔH is negative.

Similarly, if ΔH is positive, and ΔS is positive too, then $(\Delta H - T\Delta S)$ can be either positive or negative, depending on the specific values. We therefore see that if ΔH and ΔS have *different signs*, then the spontaneity of the change

state [1] $\rightarrow$ state [2]

is unambiguous:

- If ΔH is positive and ΔS is negative, $(\Delta H - T\Delta S)$ is always positive, the change of state as written is not spontaneous, and state [1] is the stable state.
- If ΔH is negative and ΔS is positive, $(\Delta H - T\Delta S)$ is always negative, the change of state as written is spontaneous, and state [2] is the stable state.

But if ΔH and ΔS have the *same signs*, then whether or not the reaction as written is spontaneous depends on the actual value of $(\Delta H - T\Delta S)$ under any given conditions. One of these conditions is the temperature T at which the change is taking place, and if, for any change, ΔH and ΔS have the same signs, then it is always possible to define a temperature T_{eq} such that

$$T_{eq} = \frac{\Delta H}{\Delta S} \tag{13.7}$$

in which case

$$\Delta H - T_{eq}\Delta S = 0$$

This implies that T_{eq} is the temperature at which states [1] and [2] are in equilibrium. Since T_{eq} is necessarily positive, equation (13.7) requires that both ΔH and ΔS have the same sign – either both positive, or both negative; if ΔH and ΔS have different signs, then T_{eq} has no meaning.

If the values of ΔH and ΔS may both be assumed to constant over a temperature range around T_{eq}, then we may use equation (13.7) to replace ΔS by $\Delta H/T_{eq}$ in the inequality (13.5a) giving, as a condition for the spontaneous change of state from state [1] to state [2]

$$\Delta H - T \frac{\Delta H}{T_{eq}} < 0$$

and so

$$\frac{\Delta H}{T_{eq}} (T_{eq} - T) < 0 \qquad (13.8)$$

Since T_{eq} is always positive, the sign of the expression $\Delta H(T_{eq} - T)/T_{eq}$ – and so whether this expression is less than, or greater than, zero – depends on the sign of ΔH, which is a thermodynamic property of the change being studied, and also on sign of the term $(T_{eq} - T)$, which depends on the temperature T at which the change is taking place. This dependence on temperature implies that whether or not state [1] changes spontaneously to state [2], or state [2] to state [1], depends on the temperature T.

Overall, for a change in which both ΔH and ΔS are positive, equation (13.7) defines a temperature T_{eq} such that

- when $T > T_{eq}$, the inequality (13.8) is fulfilled, and so state [1] will spontaneously change to state [2], implying that state [2] is the stable state at temperatures above T_{eq}; but
- when $T < T_{eq}$, then the inequality (13.8) is not fulfilled, and so state [2] will spontaneously change to state [1], implying that state [1] is the stable state at temperatures above T_{eq}.

If, however, ΔH and ΔS are both negative, then

- if $T > T_{eq}$, then the inequality (13.8) is not fulfilled, and so state [2] will spontaneously change to state [1], implying that state [1] is the stable state at temperatures above T_{eq}; but
- if $T < T_{eq}$, then the inequality (13.8) is not fulfilled, and so state [1] will spontaneously change to state [2], implying that state [2] is the stable state at temperatures above T_{eq}.

13.2.7 A worked example of spontaneity – phase changes

To make all that real, we now give a worked example of a frequently encountered change in state that depends on the temperature: a phase transition, such as that from, say, ice to liquid water, or from liquid water to water vapour. So, if state [1] is 1 mol of pure ice at a pressure of 1 bar, and state [2] is 1 mol of pure water at the same pressure, then for the phase change

$$H_2O\ (s) \rightarrow H_2O\ (l)$$

reference to Table 9.1 will show that $\Delta_{fus}H$ is 6.01 kJ mol^{-1} and $\Delta_{fus}S$ is 22.0 J K^{-1} mol^{-1}, both of which are positive. Applying equation (13.7) shows that ice and liquid water will be in equilibrium at $T_{eq} = 6010/22 = 273$ K – this being the temperature we immediately recognise as the familiar melting point of ice. Assuming that the values of $\Delta_{fus}H$ and $\Delta_{fus}S$ remain constant over a range of temperatures, we may now apply the inequality (13.8) for changes of phase that take place at 1 bar pressure and at other temperatures T, different from T_{eq}. When $T > T_{eq}$, then $T_{eq} - T < 0$, and so ice will spontaneously melt to form liquid water – as indeed corresponds to our experience at temperatures higher than 273.15 K (and less than 373.15 K, at which temperature water boils).

Alternatively, once again at 1 bar pressure, when $T < T_{eq}$, then $T_{eq} - T > 0$, and so liquid water will spontaneously become ice: at temperatures lower than 273.15 K, liquid water spontaneously freezes.

Similarly, for the vaporisation of water

$$H_2O\,(l) \rightarrow H_2O\,(g)$$

Table 9.1 shows that $\Delta_{vap}H$ is 40.65 kJ mol^{-1} and $\Delta_{vap}S$ is 109 J K^{-1} mol^{-1}. Applying equation (13.7), liquid water and water vapour will be in equilibrium at $T_{eq} = 40,650/109 = 373$ K – the familiar temperature at which water boils. Above this temperature, and at 1 bar pressure, liquid water will spontaneously vaporise; below this temperature (and above 273.15 K), and once again at 1 bar pressure, water vapour will spontaneously condense to liquid water.

13.3 Gibbs free energy

As we have seen, a criterion for a spontaneous change in a closed system is inequality (13.4)

$$dH - T\,dS < 0 \qquad (13.4)$$

This criterion has the benefit that it is defined in terms of state functions of the system only, and makes no reference to the surroundings; however, as expressed, it requires knowledge of three different quantities, dH, T and dS. The purpose of this section is to discover a new state function which enables us to define a criterion for a spontaneous change in a closed system in terms of a single quantity.

To do this, we recall that the inequality (13.4) contains the 'hidden' assumptions (see page 369) of constant temperature and constant pressure. Since the temperature is constant, dT is zero, and so the product $S\,dT$ is necessarily zero too, enabling us to modify the inequality (13.4) as

$$dH - T\,dS - S\,dT < 0$$

which may be re-written mathematically as

$$dH - d(TS) < 0$$

and so

$$d(H - TS) < 0 \qquad (13.9a)$$

A mathematical implication of the inequality (13.9a) is that the quantity $(H - TS)$ is a perfect differential. $(H - TS)$ must therefore represent a state function (see pages 12, 37 and 38), so let us define a new state function G as

$$G = H - TS \qquad (13.10a)$$

G is known as the **Gibbs free energy** (or **free energy**) in honour of the American chemist Josiah Willard Gibbs. G is an extensive state function; the intensive equivalent per mole of material, **G**, the **molar Gibbs free energy**, is defined as

$$\mathbf{G} = G/n = \mathbf{H} - T\mathbf{S} \tag{13.10b}$$

The key importance of the Gibbs free energy is a consequence of inequality (13.9a)

$$d(H - TS) = dG < 0 \tag{13.9a}$$

and the finite change equivalent

$$\Delta G < 0 \tag{13.9b}$$

Both the inequalities (13.9a) and (13.9b) are expressed in terms of a single state function G of the system, and are very compact criteria of the spontaneity of change within a closed system, performing P,V work of expansion only, and at constant temperature and pressure throughout. Furthermore, dG and ΔG are determinants of thermodynamic equilibrium too: if no change is happening, then the system is in thermodynamic equilibrium, and so the inequalities (13.9a) and (13.9b) become equalities as

$$dG = 0 \tag{13.11a}$$

and

$$\Delta G = 0 \tag{13.11b}$$

In general, any state of any system will be associated with values for all state functions – the familiar pressure P, temperature T, and volume V; the rather less familiar enthalpy H and entropy S; and the newly-defined Gibbs free energy G. And, for any system undergoing any change of state

$$\text{state [1]} \rightarrow \text{state [2]}$$

those state functions will all change accordingly. If the system is closed, and undergoes a particular form of change of state characterised by constant temperature and pressure, and the performance of P, V work only, then the corresponding value of $\Delta G = G_2 - G_1$ has a special significance as an indicator of equilibrium or spontaneity:

- If ΔG is zero, then no change takes place, and states [1] and [2] are in equilibrium.
- If ΔG is negative, the change from state [1] to state [2] will be spontaneous.
- If ΔG is positive, the change from state [1] to state [2] will be not be spontaneous, but since the reverse change

  $$\text{state [2]} \rightarrow \text{state [1]}$$

  will be associated with a value of ΔG of the same magnitude, but of opposite sign, then the change in this reverse direction will be spontaneous.

13.4 The significance of the non-conservative function

Entropy has a particular, and rather unexpected, property: the quantity of entropy is not conserved, and, as the Second Law tells us, the entropy of the universe increases. This is unusual: the scientific laws we are used to are conservation laws, such as the law of the conservation of mass, or of momentum, or of energy. So the fact that entropy is not conserved, but steadily increases, is rather odd, and raises questions such as "where does entropy come from?" – as if entropy is a type of 'substance', which might be thrown into the universe from some, as yet undiscovered, 'cosmic volcano'.

However, as we saw in Chapter 11, entropy is not a substance; rather it is a more abstract concept, associated with order and the randomisation of microstates – so entropy doesn't have to 'come from' anywhere. Our purpose here is therefore to build on our discussions in Chapter 11, and to explore an especially important consequence of the fact that entropy is indeed not conserved.

Consider, then, a 'thought experiment' in which two observers – let's call one William (for Lord Kelvin), and the other Rudolf (for Clausius) – are discussing how they might identify a criterion for spontaneous change. Let's further assume that William is within what he regards as a 'system', and that, from William's point-of-view, Rudolf is in the surroundings; Rudolf, however, is within his system, so that, from Rudolf's point-of-view, William is in Rudolf's surroundings. The key point here is that neither William nor Rudolf have a 'special' view – each of their observations are equivalent. One further note on context: the systems within which William and Rudolf make their respective observations cannot be isolated, for if either were, then one observer would see nothing happen. Let's therefore assume that the boundary allows the transfer of either heat or work, so that a change observed by William in 'his' system will give rise to a corresponding change within Rudolf's system. Both William and Rudolf are therefore in their respective closed systems.

Let's eavesdrop on their conversation...

> *William:* We are seeking to define a criterion for spontaneous change. I suggest that this implies that we need to identify something we can each measure – a change in a state function – that we will both agree corresponds to a spontaneous change.
>
> *Rudolf:* Yes, that makes sense to me. So what about internal energy U? Suppose I observe a change for which $\Delta U = 0$, might that be a criterion for a spontaneous change?
>
> *William:* Mmm... I'm not sure... If $\Delta U = 0$, then nothing much is happening. That can hardly be a criterion for change. What about a change in which $\Delta U < 0$? Suppose that any change for which $\Delta U < 0$ is spontaneous, and any change for which $\Delta U > 0$ isn't. We know, from the First Law, that
>
> $$dU = đq - đw \qquad (5.1a)$$
>
> so, let's assume, for the moment that I observe a change that takes place without any work, so $đw = 0$. If my system loses heat, $đq$ is negative, and so $\Delta U < 0$. Perhaps that might be a criterion of spontaneity?
>
> *Rudolf:* I understand what you mean, but I think there's a problem. If you observe a change that loses heat, that heat has to go somewhere – it can't just disappear. Any heat you lose, I gain, in equal quantity. There's no work across our mutual boundary, and so, from my standpoint, $\Delta U > 0$. We both observe the same change, which could indeed be spontaneous. But you measure a negative value for your ΔU, whilst I measure a positive value for my ΔU. We will both observe a change, which you will interpret as spontaneous, and I will interpret as not spontaneous. We can't both be right...

William: Yes, that's a good point ... Well, what about enthalpy? Many exothermic reactions – burning paper, for example – are spontaneous, and so might a criterion of spontaneity be $\Delta H > 0$?

Rudolf: I think we'll hit the same problem. At constant pressure, ΔH equals the heat produced by your exothermic reaction. The heat you lose, I gain, so when you observe an exothermic change, with a positive value of ΔH, I observe an endothermic change, with an equal, negative, value of ΔH. Once again, we won't agree.

William: There's clearly a fundamental issue here. For any conservative function, any change in my system corresponds to an equal and opposite change in yours, and, of course, *vice versa*.

Rudolf: That must be right. No conservative function can ever be a criterion for a spontaneous change. The only solution has to be a non-conservative function – a function which can increase, or decrease, for both of us, simultaneously, for the same change.

William: That must lead to entropy. It is quite possible for me to observe a change in my system for which the change in entropy is, say, positive...

Rudolf: ...and, given that entropy is not conserved, it is also possible for me to observe an increase in entropy too. And if I do, we will both agree that a given change is associated with an increase in entropy ...

William: ...and so we will both agree that the change is spontaneous

As William and Rudolf have just agreed, only a non-conservative function can serve as a criterion of spontaneity of change. Entropy is one such function; another is the Gibbs free energy G, for it includes the entropy S in its definition

$$G = H - TS \tag{13.10a}$$

The fact that the Gibbs free energy G is not conserved might suggest that referring to G as 'free energy' is misleading, for a fundamental attribute of all other forms of energy – notably, in thermodynamics, the internal energy U and the enthalpy H, and more widely, gravitational potential energy, kinetic energy and the rest – is the principle of the conservation of energy: indeed, the First Law is based entirely on that principle. G is therefore not 'energy' in the fundamental sense; nor, for that matter, is G 'free'. Rather, 'Gibbs free energy' is simply the name given to the quantity $H - TS$, and its significance is that it is *not* conserved.

13.5 Enthalpy- and entropy-driven reactions

The definition of the Gibbs free energy

$$G = H - TS \tag{13.10a}$$

is the starting point for the derivation of a very important equation. For an infinitesimal change at constant temperature (explicitly), and constant pressure (implicitly), we may take differentials giving

$$dG = dH - T\,dS \tag{13.12}$$

and, for a finite change, once again at constant temperature and pressure throughout

$$\Delta G = \Delta H - T\,\Delta S \tag{13.13}$$

Equation (13.13) is a more generalised version of the inequality (13.5a)

$$\Delta H - T\Delta S < 0 \tag{13.5a}$$

and is one of the keystones of practical thermodynamics.

As we saw on pages 371 and 372, the signs, and relative magnitudes, of ΔH and ΔS determine whether any change in state is spontaneous in the forward direction, or the reverse direction, or is in equilibrium – a matter of particular importance to chemical and biochemical processes. As indicated by equation (13.13), a reaction associated with a large negative value of ΔH, a strongly exothermic reaction, will not necessarily be spontaneous: if the corresponding value of ΔS is positive, then the term $T\Delta S$ (which necessarily increases with temperature) might be more positive than ΔH is negative, with the result that ΔG is positive. One way of tipping the balance towards making ΔG negative might be to lower the temperature of the reaction – provided that this does not have other adverse consequences, not least the freezing of water if the reaction takes place in aqueous solution. Similar considerations indicate that reactions associated with large positive values of ΔS, especially those that take place at higher temperatures, are more likely to be spontaneous, even if ΔH is positive.

We shall consider the interactions between the signs and magnitudes of ΔG, ΔH and ΔS further on pages 390 to 396, but, as will become evident in chapters 14, 24 and 25, in general

- the value of ΔG, and hence the spontaneity, of many laboratory reactions, especially in organic chemistry, tends to be determined by ΔH
- the value of ΔG, and hence the spontaneity, of many natural biochemical reactions tends to be determined by ΔS.

For any reaction, if ΔG is primarily determined by ΔH, the reaction is said to be **enthalpy-driven**; if ΔG is primarily determined by ΔS, the reaction is said to be **entropy-driven**.

13.6 The mathematics of G

13.6.1 G as a function $G(P,T)$ of P and T

Given the definition of G as

$$G = H - TS \tag{13.10a}$$

we may take differentials, giving

$$dG = dH - T\,dS - S\,dT$$

But since

$$H = U + PV \tag{6.4}$$

then

$$dH = dU + P\,dV + V\,dP$$

and so

$$dG = dU + P\,dV + V\,dP - T\,dS - S\,dT$$

Now, according to the First Law

$$dU = đq - đw \tag{5.1a}$$

and so

$$dG = đq - đw + P\,dV + V\,dP - T\,dS - S\,dT \tag{13.14}$$

Equation (13.14) is a complete, and correct, expression for dG under all conditions, and can be used as a starting point for the determination of dG in any particular situation. So, for example, if the change in state takes place thermodynamically reversibly, then

$$dG = đq_{rev} - đw_{rev} + P\,dV + V\,dP - T\,dS - S\,dT \tag{13.15}$$

But from the definition of entropy

$$đq_{rev} = T\,dS$$

and if the system can perform only P,V work of expansion

$$đw_{rev} = P\,dV$$

then equation (13.15) simplifies to

$$dG = V\,dP - S\,dT \tag{13.16a}$$

which, for molar quantities, becomes

$$d\mathbf{G} = \mathbf{V}\,dP - \mathbf{S}\,dT \tag{13.16b}$$

Equations (13.16a) and (13.16b) are important, and they will be referred to on many occasions as we develop the theory underlying many chemical and biological applications of thermodynamics. And although they were derived assuming a reversible change, since they are expressed in terms of state functions alone, they apply to all changes, whether reversible or not.

These equations also embody some significant mathematical properties of G – properties that are attributable to a fact we already know: the fact that, for n mol of a pure substance, and in the absence of gravitation, motion, electric, magnetic and surface effects (as is consistent with the assumption just made that the system can perform only P, V work of expansion), any one state function is a mathematical function of only two other state functions (see page 48). G is a state function, and can therefore be expressed as a mathematical function of the pressure P and the temperature T as

$$G = G(P, T) \tag{13.17}$$

Since this expresses G as a function of two variables, P and T, an infinitesimal change dG may be written using partial derivatives as

$$dG = \left(\frac{\partial G}{\partial P}\right)_T dP + \left(\frac{\partial G}{\partial T}\right)_P dT \tag{13.18}$$

where the partial derivatives $(\partial G/\partial P)_T$ and $(\partial G/\partial T)_P$ – although somewhat abstract-looking and even intimidating – have a real, physical meaning: $(\partial G/\partial P)_T$ describes how the Gibbs free energy G of a system changes with pressure, whilst the temperature is held constant, and

$(\partial G/\partial T)_P$ describes how the Gibbs free energy G of a system changes with temperature, whilst the pressure is held constant.

Equation (13.18) is mathematically correct, and universally true, and may be compared to equation (13.16a), which is physically correct, and also universally true. Furthermore, the dG on the left-hand side of each equation is the same quantity, implying that the right-hand sides must be the same too:

$$dG = V\,dP - S\,dT = \left(\frac{\partial G}{\partial P}\right)_T dP + \left(\frac{\partial G}{\partial T}\right)_P dT \tag{13.19}$$

It must therefore be the case that

$$\left(\frac{\partial G}{\partial P}\right)_T = V \tag{13.20}$$

and

$$\left(\frac{\partial G}{\partial T}\right)_P = -S \tag{13.21}$$

These equations tell us that the rate-of-change of a system's Gibbs free energy with respect to pressure, at constant temperature, is equal to that system's instantaneous volume; and the rate-of-change with respect to temperature, at constant pressure, is the negative of the system's instantaneous entropy. These results are decidedly not obvious ... but, as we shall shortly see, they are very helpful to us in determining how a system's Gibbs free energy changes with pressure and temperature. In particular, given that the volume V of all systems must be positive, equation (13.20) tells us that, at constant temperature, G must increase as the pressure P increases; likewise, since the entropy S of all systems must also be positive, equation (13.20) tells us that, at constant temperature, G must decrease as the temperature T increases.

13.6.2 How G varies with P at constant T

Taking matters one step further, consider a change of state, from an initial pressure P_1 to a final pressure P_2, at constant temperature. The corresponding change $\Delta G_T = G_2 - G_1$ can be determined by integrating equation (13.20) as

$$\Delta G_T = G_2 - G_1 = \int_{P_1}^{P_2} V\,dP \tag{13.22}$$

Equation (13.22) can also be derived from equation (13.16a)

$$dG = V\,dP - S\,dT \tag{13.16a}$$

under conditions of constant temperature, for which $dT = 0$.

To evaluate the integral of equation (13.22), we need to know the equation-of-state – the explicit relationship between V and P – for the system being studied. For a solid or a liquid, to a good approximation, the volume V is constant over a small change in pressure, and so $\Delta G_T \approx V\Delta P$; for an ideal gas, the relevant equation-of-state is the ideal gas law

$$PV = nRT$$

and so, since T is constant

$$\Delta G_T = G_2 - G_1 = \int_{P_1}^{P_2} V\, dP = nRT \int_{P_1}^{P_2} \frac{dP}{P} \qquad (13.23a)$$

hence

$$\Delta G_T = nRT \ln \frac{P_2}{P_1} = nRT \ln \frac{V_1}{V_2} \qquad (13.23b)$$

and, more generally

$$G(P, T) = nRT \ln P + \text{constant} \qquad (13.23c)$$

in which the constant is the constant of integration.

Equation (13.23b) allows us to calculate the change ΔG_T in the Gibbs free energy for any ideal gas undergoing an isothermal change at temperature T from any initial state $[P_1, V_1]$ to any final state $[P_2, V_2]$; for a non-ideal gas, we can use equation (13.22) provided that we have knowledge of the appropriate equation-of-state.

13.6.3 How G varies with T at constant P

In principle, we can determine the change ΔG_P for a change of state for any system, from an initial temperature T_1 to a final temperature T_2, at constant pressure, in an exactly analogous way, by computing the integral form of equation (13.21) as

$$\Delta G_P = G_2 - G_1 = -\int_{T_1}^{T_2} S\, dT \qquad (13.24)$$

or, identically, from equation (13.16a)

$$dG = V\, dP - S\, dT \qquad (13.16a)$$

under the condition that $dP = 0$

To do this in practice, however, requires the use of an equation-of-state in which the entropy S is expressed as a function of the temperature T – just as we used the ideal gas law to express the volume V as a function of the pressure P in the previous section. Unfortunately, in this book, we have not defined an equation-of-state expressing S as a function of T, and so we cannot integrate equation (13.24) directly. So, to establish how G varies with T, we need to approach the problem rather differently, but still making use of equation

$$\left(\frac{\partial G}{\partial T}\right)_P = -S \qquad (13.21)$$

Our starting point is the definition of G as

$$G = H - TS \qquad (13.10a)$$

From which

$$-S = \frac{G}{T} - \frac{H}{T} \qquad (13.25)$$

According to equation (13.21), $(\partial G/\partial T)_P = -S$, and so equation (13.25) may be written as

$$\left(\frac{\partial G}{\partial T}\right)_P = \frac{G}{T} - \frac{H}{T} \qquad (13.26)$$

Taking the term G/T by itself, if we differentiate G/T with respect to temperature at constant pressure, then, according to the usual rules for differentiating a quotient

$$\left(\frac{\partial (G/T)}{\partial T}\right)_P = \frac{1}{T}\left(\frac{\partial G}{\partial T}\right)_P - \frac{G}{T^2} \qquad (13.27)$$

Using equation (13.26) to substitute for $(\partial G/\partial T)_P$ in equation (13.27) gives

$$\left(\frac{\partial (G/T)}{\partial T}\right)_P = \frac{1}{T}\left(\frac{G}{T} - \frac{H}{T}\right) - \frac{G}{T^2}$$

which simplifies to

$$\left(\frac{\partial (G/T)}{\partial T}\right)_P = -\frac{H}{T^2} \qquad (13.28)$$

Equation (13.28) is known as the **Gibbs-Helmholtz equation**, and defines how the ratio G/T varies with the temperature T. For a finite change in state at constant pressure from an initial state [1] to a final state [2], equation (13.28) becomes

$$\int_{\text{state 1}}^{\text{state 2}} d\left(\frac{G}{T}\right) = -\int_{T_1}^{T_2} \frac{H}{T^2} dT$$

If we assume that the enthalpy H is constant over the temperature range from T_1 to T_2, then these integrals evaluate to

$$\frac{G_2}{T_2} - \frac{G_1}{T_1} = H\left(\frac{1}{T_2} - \frac{1}{T_1}\right) \qquad (13.29)$$

Given values for G_1, H, T_1 and T_2, we can use equation (13.29) to compute G_2, the Gibbs free energy of the system at temperature T_2 – subject to the assumption that both H and P are constant. We have therefore solved the problem of discovering how the Gibbs free energy G of a system changes with temperature at constant pressure.

As already noted, equation (13.29) assumes that the relevant enthalpy H is constant over the temperature range from T_1 to T_2. In principle, this assumption is false, for we know that H does vary with temperature, as defined by the heat capacity C_P at constant pressure

$$\left(\frac{\partial H}{\partial T}\right)_P = C_P \qquad (6.8a)$$

But if the enthalpy change $\Delta H = C_P \Delta T$ over the temperature range $\Delta T = T_2 - T_1$ is small compared to the corresponding change ΔG in Gibbs free energy, then the assumption that H is constant over that range is acceptable. As we shall see on page 448, in general, H varies with temperature much more slowly than G, implying that, in most practical circumstances, H can be assumed to be constant, and so equation (13.29) may be used with confidence.

13.6.4 G as a function $G(V,T)$ of V and T

As we have seen,

$$\left(\frac{\partial G}{\partial P}\right)_T = V \tag{13.20}$$

Mathematically, however,

$$\left(\frac{\partial G}{\partial P}\right)_T = \left(\frac{\partial G}{\partial V}\right)_T \left(\frac{\partial V}{\partial P}\right)_T$$

and so

$$\left(\frac{\partial G}{\partial V}\right)_T \left(\frac{\partial V}{\partial P}\right)_T = V$$

implying that

$$\left(\frac{\partial G}{\partial V}\right)_T = \frac{V}{(\partial V/\partial P)_T} = V \left(\frac{\partial P}{\partial V}\right)_T \tag{13.30}$$

The quantity $(\partial V/\partial P)_T$ represents the infinitesimal change ∂V_T in volume of a system attributable to an infinitesimal increase ∂P_T in pressure at constant temperature T. For all known materials, an increase in pressure results in a decrease in volume, and so $(\partial V/\partial P)_T$ is always negative. The volume V of any system, however, is necessarily positive, and so, according to equation (13.30), the quantity $(\partial G/\partial V)_T$ is always negative – as the volume V of any system increases at constant temperature, the Gibbs free energy G decreases. Volume increases of a system at constant temperature are therefore spontaneous; similarly, at any given pressure and temperature, a system will occupy the maximum available volume. For gases, this maximum volume is determined by the size of the container; for condensed phases – liquids and solids – this maximum volume is determined by the balance, at any given temperature, between the thermal motion of the molecules (which acts to drive neighbouring molecules apart) and the strength of inter-molecular forces (which act to hold neighbouring molecules together).

The functional dependence $G(V, T)$ on V and T can be determined from equation (13.30) if the equation-of-state of the system is known, from which $(\partial P/\partial V)_T$ can be determined. So, for an ideal gas, we know that the equation-of-state is

$$PV = nRT \tag{3.3}$$

and so

$$\left(\frac{\partial P}{\partial V}\right)_T = -\frac{nRT}{V^2}$$

from which

$$\left(\frac{\partial G}{\partial V}\right)_T = -V\frac{nRT}{V^2} = -\frac{nRT}{V}$$

This expression can be integrated to give

$$\Delta G_T = -nRT \ln \frac{V_2}{V_1} \tag{13.31a}$$

or the more general result

$$G(V, T) = -nRT \ln V + \text{constant} \qquad (13.31b)$$

Equation (13.31a) is an expression for the change ΔG_T in the Gibbs free energy of an ideal gas corresponding to a change in volume from V_1 to V_2 at constant temperature T. If this is an expansion, $V_2 > V_1$, and so ΔG_T is negative, as expected.

A rather more complex equation-of-state is the van der Waals equation

$$\left(P + \frac{an^2}{V^2}\right)(V - nb) = nRT \qquad (3.6)$$

which, as we saw on page 50, is an equation-of-state for a gas whose molecules occupy a finite volume of b m^3 mol^{-1}, and experience an inter-molecular attractive force as determined by the parameter a. Equation (3.6) can be re-arranged as

$$P = \frac{nRT}{V - nb} - \frac{an^2}{V^2}$$

from which

$$\left(\frac{\partial P}{\partial V}\right)_T = -\frac{nRT}{(V - nb)^2} + \frac{2an^2}{V^3}$$

Since

$$\left(\frac{\partial G}{\partial V}\right)_T = V \left(\frac{\partial P}{\partial V}\right)_T \qquad (13.30)$$

then, for a van der Waals gas

$$\left(\frac{\partial G}{\partial V}\right)_T = V\left(-\frac{nRT}{(V - nb)^2} + \frac{2an^2}{V^3}\right) = -\frac{nRTV}{(V - nb)^2} + \frac{2an^2}{V^2} \qquad (13.32)$$

Now, since

$$\frac{V}{(V - nb)^2} = \frac{1}{V - nb} + \frac{nb}{(V - nb)^2}$$

then

$$\left(\frac{\partial G}{\partial V}\right)_T = -\frac{nRT}{V - nb} - \frac{n^2 bRT}{(V - nb)^2} + \frac{2an^2}{V^2}$$

which integrates, in general, as

$$G(V, T) = -nRT \ln(V - nb) + \frac{n^2 bRT}{(V - nb)} - \frac{2an^2}{V} + \text{constant} \qquad (13.33)$$

At large volumes, $V \gg nb$, and so $\ln(V - nb) \approx \ln V$; furthermore, when V is large, $\ln V$ is also very much greater than both the second and the third terms in equation (13.33). Under these conditions, equation (13.33) therefore approximates equation (13.31b), demonstrating that, at large volumes, the volume-dependent behaviour of a van der Waals gas approximates to that of an ideal gas. At small volumes, and especially as the volume tends towards multiples of the quantity nb – which represents the physical volume actually occupied by the molecules of the van der Waals gas – equation (13.33) behaves very differently from equation (13.31b).

As we shall see in Chapter 15, this difference in behaviour helps explain why real gases become liquids as they are progressively compressed at constant temperature.

13.7 Helmholtz free energy

The definition of the Gibbs free energy

$$G = H - TS \tag{13.10a}$$

hints at the existence of another state function, in which the internal energy U replaces the enthalpy H

$$A = U - TS \tag{13.34}$$

Given that U, T, and S are state functions, A too is a state function, known as the **Helmholtz free energy**. Like the Gibbs free energy G, the Helmholtz free energy A is an extensive state function; the intensive equivalent per mole of material, the **molar Helmholtz free energy** A, is defined, as usual, as A/n. Also, we note that to avoid ambiguity, the term 'free energy' is used as an abbreviation only for the Gibbs free energy G, and never for Helmholtz free energy A, which is always referred to by its full name.

We have seen that the Gibbs free energy is of especial importance in relation to changes of state at constant temperature and constant pressure. The relevance of constant pressure can be traced to the discussion, earlier in this chapter, leading to equation (13.2), where, on page 368, the assumption of constant pressure allowed us to replace $đq_{rev}$ in the expression for the entropy change for the surroundings by dH (equation (6.5a)).

Had we been considering changes for a closed system at constant volume, rather than at constant pressure, then, in replacing $đq_{rev}$, we would have used a variant of equation (5.12a)

$$dU_{sys} = \{đq_{rev}\}_{sys,\,V} \tag{5.12a}$$

rather than equation (6.5a)

$$dH_{sys} = \{đq_{rev}\}_{sys,\,P} \tag{6.5a}$$

The internal energy U would then have appeared in all subsequent equations, rather than the enthalpy H, with the proviso that the corresponding results would apply at constant temperature and constant volume, rather than constant temperature and constant pressure.

The two free energies – that named after Gibbs, G, and that named after the German physicist, Hermann von Helmholtz, A – play similar roles, and obey identical equations. As state functions, any system in an equilibrium state will have values for both A and G, and any change in state, under any conditions, will result in values for the changes ΔA and ΔG. If the change takes place at constant temperature and constant volume, then the corresponding value of ΔA provides insight into the spontaneity, or otherwise, of the change; if the change takes place at constant temperature and constant pressure, then those same insights are obtained from the corresponding value of ΔG. In practice, changes at constant temperature and pressure are more frequently observed than those at constant temperature and volume, and so, in practice, ΔG plays a rather more prominent role than ΔA.

13.8 The mathematics of A

13.8.1 A as a function $A(V,T)$ of V and T

Just as the Gibbs free energy G is most naturally a function $G(P, T)$ of P and T, then the Helmholtz free energy A is most naturally a function $A(V, T)$ of V and T. From the definition of A

$$A = U - TS \tag{13.34}$$

then

$$dA = dU - T\,dS - S\,dT = đq_{rev} - đw_{rev} - T\,dS - S\,dT$$

By definition, $đq_{rev} = T\,dS$, and for a system that does P, V work of expansion only $đw_{rev} = P\,dV$, and so

$$dA = -P\,dV - S\,dT \tag{13.35}$$

Following exactly the same logic as we used in connection with equation (13.19), equation (13.35) implies that

$$\left(\frac{\partial A}{\partial V}\right)_T = -P \tag{13.36}$$

and

$$\left(\frac{\partial A}{\partial T}\right)_V = -S \tag{13.37}$$

Equations (13.36) and (13.37) define the rate-of-change of A with respect to volume, at constant temperature, and with respect to temperature, at constant volume.

So, just as we saw on pages 372 to 382 for G, the corresponding results for A are

- for a change in volume at constant temperature for a solid or liquid, $\Delta A = -P\Delta V$
- for a change in volume, or pressure, at constant temperature for an ideal gas

$$\Delta A_T = \frac{nR}{T}\ln\frac{V_1}{V_2} = \frac{nR}{T}\ln\frac{P_2}{P_1}$$

- for a change in temperature at constant volume

$$\left(\frac{\partial (A/T)}{\partial T}\right)_V = -\frac{U}{T^2} \tag{13.38}$$

13.9 Maximum available work

The inequalities (13.9a) and (13.9b)

$$dG < 0 \tag{13.9a}$$

$$\Delta G < 0 \tag{13.9b}$$

are very important, for they provide us with a criterion of spontaneity defined in terms of a state function of the system of interest alone: rather than having to deal with both the system and the surroundings in order to determine the entropy of the universe, we can focus on a well-defined property of the system. The use of changes in the Gibbs free energy G as a criterion of spontaneity, however, depends on the simultaneous compliance of the change in state with four conditions: the change in state from state [1] to state [2] must

- take place within a closed system ...
- ...at constant temperature ...
- ...and at constant pressure ...
- ...whilst performing only P, V work of expansion.

If these 'big four' conditions are simultaneously fulfilled, then we may correctly infer that if the corresponding value of $\Delta G = G_2 - G_1$ is negative, then that change in state will be spontaneous; alternatively, if ΔG is positive, then the change in state will not be spontaneous (but the reverse change from state [2] to state [1] will be); and if ΔG is zero, the system is in equilibrium.

Not all changes in state, however, comply simultaneously with all four of these conditions: some changes take place in open systems; some changes take place over a range of temperatures; some changes are not at constant pressure (for example, changes at constant volume, for which, as we have just seen, the Helmholtz free energy A is the equivalent of the Gibbs free energy G at constant pressure); and some changes involve the performance of forms of work other than P, V work of expansion against the atmosphere – for example, mechanical work associated with driving some form of machine, or the electrical work required to move charged particles through electrical fields, as happens in batteries and electrical circuits.

But since G is a state function, there will be a value of ΔG for any change of state, whether that change of state complies with all of the 'big four' conditions or not; the interpretation of what that change ΔG means, however, will depend on the circumstances. So, if the 'big four' conditions apply, then a negative (or positive) value ΔG provides information about spontaneity, but if any one of the conditions does not apply, then ΔG will still have a value, but its interpretation might simply be as a number, or perhaps of something else of interest, other than spontaneity.

One such circumstance is when a closed system can do work in addition to P,V work of expansion, perhaps electrical work, or driving some form of machine. To demonstrate how we may use 'first principles' to examine this, we start with the definition of G as

$$G = H - TS \tag{13.10}$$

Taking differentials, as we saw on page 379,

$$dG = đq - đw + P\,dV + V\,dP - T\,dS - S\,dT \tag{13.14}$$

Equation (13.14) is a complete, and correct, expression for dG under all conditions, and can be used as a starting point for the determination of dG in any particular situation. At constant temperature, $dT = 0$, and at constant pressure, $dP = 0$, and so if both of these conditions hold simultaneously

$$dG = đq - đw + P\,dV - T\,dS \tag{13.39}$$

If the change in state takes place thermodynamically reversibly, then

$$dG = đq_{rev} - đw_{rev} + P\,dV - T\,dS \qquad (13.40)$$

But from the definition of entropy

$$đq_{rev} = T\,dS$$

and so equation (13.17) becomes

$$dG = -đw_{rev} + P\,dV \qquad (13.41)$$

For a closed system that performs P,V work of expansion only, then

$$đw_{rev} = P\,dV$$

and so, under this additional condition

$$dG = 0 \qquad (13.42)$$

This result might be something of a surprise, for it tells us that, for a closed system undergoing a change at constant temperature and pressure, and that can perform reversible P, V work of expansion only, then dG is zero. This therefore seems to contradict the inequality (13.9a) which states

$$dG < 0 \qquad (13.9a)$$

and which also applies to a closed system undergoing a change at constant temperature and pressure, and that can perform P, V work of expansion only.

In fact, there is no contradiction: both equation (13.42) and the inequality (13.9a) are true – or rather, true in the context of the appropriate conditions. Reference to the text at the top of this page will show that the derivation of equation (13.39) from equation (13.40) stated "If the change in state takes place thermodynamically reversibly, then ...", thereby assuming thermodynamic reversibility for the change in state as a whole, as regards both heat and work. In contrast, as verified by reference to page 368, the derivation of equation (13.3), which was an important step in the argument leading to the definition of the Gibbs free energy, and the subsequent derivation of the inequality (13.9a), assumed only the reversibility of the flow of heat, but *not the reversibility of the flow of work*. The inequality (13.9a) therefore applies to all real, irreversible, changes of state at constant temperature and pressure, whereas equation (13.42) applies only to theoretical, reversible, changes of state, once again at constant temperature and pressure. And since thermodynamically reversible changes take place through a sequence of equilibrium states, then the equality expressed by equation (13.19) is totally consistent with the criterion for equilibrium that for an infinitesimal change $dG = 0$, and, for a finite change, $\Delta G = 0$.

Returning now to equation (13.41)

$$dG = -đw_{rev} + P\,dV \qquad (13.41)$$

let us suppose that the system performs some reversible P,V work of expansion $đw_{exp, rev}$, for example, against the atmosphere, and also some other form of reversible work $đw_{other, rev}$, such as driving a machine, lifting a weight, or enabling the flow of an electric current. The total reversible work $đw_{rev}$ done by the system is therefore the sum of two components

$$đw_{rev} = đw_{exp,\,rev} + đw_{other,\,rev}$$

Equation (13.41) therefore becomes

$$dG = -(đw_{exp,\,rev} + đw_{other,\,rev}) + P\,dV \tag{13.42}$$

But, for the reversible P,V work of expansion

$$đw_{exp,\,rev} = P\,dV$$

implying that

$$đw_{other,\,rev} = -dG \tag{13.43a}$$

Hence, for a finite change from state [1] to state [2]

$$\{_1w_2\}_{other,\,rev} = -\Delta G \tag{13.43b}$$

If the change in state is spontaneous, then dG and ΔG will be negative, implying that $đw_{other,\,rev}$ and $\{_1w_2\}_{other,\,rev}$ will be positive, corresponding to a situation in which the system does work, of this other form, on the surroundings.

Equation (13.43b) show that, for a finite change in state at constant temperature and pressure, during which the system performs both P, V work of expansion against the atmosphere, and also some other form of work, then the quantity $-\Delta G$ is equal to the quantity of reversible work, other than, and in addition to, P, V work of expansion, that the system can do on the surroundings. Since we know that, in general

$$đw_{rev} > đw_{irrev} \tag{5.37}$$

then the quantity of work represented by $-\Delta G$ is a theoretical maximum; the actual amount of 'other' work that the system performs is necessarily less than this upper limit, as defined by $-\Delta G$.

This provides another interpretation for ΔG for any change in the state of a closed system at constant temperature and constant pressure: the value of $-\Delta G$ represents the maximum work the system can perform – a quantity of work known as the **maximum available work** – other than the P, V work of expansion that the system necessarily does against the atmosphere.

That might sound rather abstract, but it is very real, and very important. Consider any system – a chemical system, an industrial system, a biological system – that undergoes a change in state from state [1] to state [2], at constant pressure and temperature. That change of state is associated with changes in each of the system's state functions, including a change ΔG in the Gibbs free energy. Suppose that we ask the question "what is the maximum amount of energy, associated with this change of state, apart from, and in addition to, any P, V work of expansion that the system does against the atmosphere, that can be harnessed for whatever purpose we might wish?" The answer to that question is $-\Delta G$: $-\Delta G$ represents the maximum available work – the maximum amount of energy that can actually be used to power a machine, or to make an electric current flow. In rather looser language, the energy represented by $-\Delta G$ might be thought of as being 'freely available' – hence the term 'free energy'.

In the last few paragraphs, the words "apart from, and in addition to, any P, V work of expansion that the system does against the atmosphere" are rather clumsy, and might appear superfluous. These words, however, make an important distinction between the total maximum work that the system performs, and the work represented by $-\Delta G$, for these are *not* the same. When a system undergoes a change of state at constant pressure, the total maximum amount of work that the system can perform is the sum of the (reversible) P, V work of expansion = $P\,dV$, usually done against the atmosphere, plus any 'other' (reversible) work, which is equal to $-\Delta G$. The maximum total amount of work that the system performs, and that can in principle be captured for other purposes, is therefore the sum $P\,dV - \Delta G$, which is in practice an addition when ΔG is negative. In many real situations, however, the P, V work of expansion is not actually harnessed, even though it could be harnessed if we wished: this work is not necessarily 'wasted', for it could be used should we choose to do so. $-\Delta G$ is therefore not the maximum amount of work that the system can in principle perform; rather it is the maximum practical amount of work obtainable from the system "apart from, and in addition to, any P, V work of expansion that the system does against the atmosphere".

Exactly similar conclusions, in relation to changes in state of a closed system at constant temperature and constant volume, can be drawn for the Helmholtz free energy A. Starting with the definition of equation (13.34)

$$A = U - TS \qquad (13.34)$$

if we firstly take differentials, and then apply conditions of constant temperature ($dT = 0$), constant volume ($dV = 0$, which also implies that there is no P, V work of expansion), and thermodynamic reversibility, then the quantity $-\Delta A$, the negative of the change in the Helmholtz free energy, will be shown to be equal to the maximum available work associated with any change in state under the assumed conditions of constant temperature and constant volume. So, just as $-\Delta G$ represents the maximum energy that can be captured from a system changing state at constant temperature and constant pressure, $-\Delta A$ represents the maximum energy that can be captured from a system changing state at constant temperature and constant volume.

13.10 How to make non-spontaneous changes happen

In the last section, we saw that, for a system of constant mass undergoing a change of state at constant temperature and pressure

state [1] → state [2]

then the corresponding value of $-\Delta G_{1\to 2}$ represents the maximum energy that can practically be harnessed from the change in state, as represented by equation (13.43b)

$$\{_1w_2\}_{\text{other, rev}} = -\Delta G \qquad (13.43b)$$

For a spontaneous change, ΔG is negative, implying that $-\Delta G$ is positive. $\{_1w_2\}_{\text{other, rev}}$ therefore represents work done by the system on the surroundings, during the corresponding change in state. A change in state associated with a negative value of ΔG is known as **exergonic**, and since, in general, $\Delta G = \Delta H - T\Delta S$, large negative values of ΔG are often associated with large negative values of ΔH – exergonic processes are often strongly exothermic. And, according to equation (13.43b), an exergonic change, associated with a large, negative, value of ΔG can, potentially, do a large quantity of work $\{_1w_2\}_{\text{other, rev}}$, other than P, V work of expansion, on the surroundings.

There is, however, another interpretation of equation (13.43b). So far, we have taken the case in which $\{_1w_2\}_{\text{other, rev}}$ is positive, corresponding to a change in which the system does work on the surroundings. So suppose, alternatively, that $\{_1w_2\}_{\text{other, rev}}$ is negative, implying that the change is one in which the surroundings do work on the system. Equation (13.43b) is still valid, but if $\{_1w_2\}_{\text{other, rev}}$ is now negative, then ΔG must be positive. A change in state for which ΔG is positive is known as **endergonic**, and there's something else we know about such as change too: if ΔG is positive, then that change is not spontaneous. Or, more precisely, not spontaneous of its own accord. For what we have now shown, from equation (13.43b) as applied to a case on which $\{_1w_2\}_{\text{other, rev}}$ is negative, is that a change in state which is not spontaneous of its own accord can be *made to happen* if a quantity of work $\{_1w_2\}_{\text{other, rev}}$ is done *on* the system, such that $\{_1w_2\}_{\text{other, rev}}$ is (at least) equal to the value of ΔG for that change.

This consequence of equation (13.43b) is of great significance, for it tells us that a change in state that is normally not spontaneous can be made to happen by a suitable injection of external energy: if the amount $\{w\}_{\text{other, rev}}$ of external energy 'pumped' into the system is equal to (or greater than), the value of ΔG for a particular change in state, then that change will take place, even though the change is not spontaneous of its own accord. This explains why non-spontaneous changes happen – non-spontaneous changes such as keeping a room tidy (to take a trivial example!), and non-spontaneous changes such as protein synthesis (a decidedly non-trivial example!). Non-spontaneous actually changes happen because they are not taking place of their own accord, by themselves: they are taking place in the context of other changes, in which systems are coupled together, so that a spontaneous change in one system, which 'releases' available energy, expressed as $\{_1w_2\}_{\text{other, rev}}$, is linked to another system, in which that energy is used to drive a reaction which would otherwise be non-spontaneous.

13.11 Coupled systems...

To develop the concept of **coupled systems**, consider a system, at a given temperature and pressure, undergoing a change in state

state [1] → state [2]

for which $\Delta G_{1\to 2}$ is negative. Since $\Delta G_{1\to 2} < 0$, this change of state is exergonic, and will be spontaneous of its own accord.

Consider also a second system, at the same temperature and pressure, undergoing a change of state

state [3] → state [4]

for which $\Delta G_{3\to 4}$ is positive. Since $\Delta G_{3\to 4} > 0$, this change of state is endergonic, and will not be spontaneous of its own accord.

Suppose now that both of these changes of state happen simultaneously, and are coupled together within the same 'super-system', which we can represent as

$$\left.\begin{array}{l}\text{state [1]} \rightarrow \text{state [2]} \\ \text{state [3]} \rightarrow \text{state [4]}\end{array}\right\}$$

If the resulting mixture can be considered to be ideal (see page 8), then all state functions are additive. We may then express the change ΔG_{coup} of the Gibbs free energy for the coupled system as

$$\Delta G_{coup} = \Delta G_{1 \rightarrow 2} + \Delta G_{3 \rightarrow 4}$$

The sign of ΔG_{coup} is determined by the sign of $\Delta G_{1 \rightarrow 2} + \Delta G_{3 \rightarrow 4}$. Our hypothesis, however, is that $\Delta G_{1 \rightarrow 2} < 0$ and that $\Delta G_{3 \rightarrow 4} > 0$, and so there are two different possibilities:

- If the magnitude of $\Delta G_{1 \rightarrow 2}$ is less than the magnitude of $\Delta G_{3 \rightarrow 4}$, then $\Delta G_{1 \rightarrow 2} + \Delta G_{3 \rightarrow 4}$ is positive, and the combined {state [1] + state [3]} will not spontaneously change to the combined {state [2] + state [4]}.
- But if $\Delta G_{1 \rightarrow 2}$ is more negative than $\Delta G_{3 \rightarrow 4}$ is positive, then ΔG_{coup} is negative, and the combined {state [1] + state [3]} will spontaneously change to the combined {state [2] + state [4]}.

In this second case, an observer will therefore see a change

state [3] $\rightarrow$ state [4]

that is known not to be spontaneous actually to be taking place spontaneously, which presents a puzzle, for it seems to be breaking the Second Law – especially if the observer does not notice, or pay attention to, the coupled change

state [1] $\rightarrow$ state [2]

which is spontaneous.

In fact, this is not a puzzle at all. The two changes are coupled in the sense that the change

state [3] $\rightarrow$ state [4]

will not happen spontaneously, by itself. But if this change is coupled to the change

state [1] $\rightarrow$ state [2]

then the overall change {state [1] + state [3]} to {state [2] + state [4]} is accompanied by a value ΔG_{coup} which is negative. So the coupled changes in state are, together, spontaneous, and everything is totally consistent with the Second Law.

There are two conditions which must be satisfied for this to happen. Firstly, the quantity of Gibbs free energy 'released', corresponding to $-\Delta G_{1 \rightarrow 2}$, must at least equal the 'requirement' for free energy $\Delta G_{3 \rightarrow 4}$ – which may be expressed mathematically as $\Delta G_{1 \rightarrow 2} + \Delta G_{3 \rightarrow 4} < 0$. Secondly, the two systems must be suitably coupled so that the energy 'released' by the 'donor' system can be 'harvested' by the 'acceptor' system.

The concept of coupled systems is very important. If we have a non-spontaneous change in state for which $\Delta G_{3 \rightarrow 4} > 0$, then we can make that change happen by coupling it with a simultaneous change in state associated with a value of $\Delta G_{1 \rightarrow 2}$ sufficiently negative so that

$\Delta G_{1\to 2}+\Delta G_{3\to 4} < 0$. Likewise, if we observe an apparently 'impossible' change in state actually taking place, then we know that it is coupled with a perhaps less obvious change in state so that together, once again, $\Delta G_{1\to 2} + \Delta G_{3\to 4} < 0$.

13.12 ... an explanation of frictional heat, and tidying a room ...

To make this discussion more concrete, consider a system undertaking a real change in state

state [1] → state [2]

which, under ordinary conditions of temperature and pressure, is exergonic, and so is associated with a value of negative value of $\Delta G_{1\to 2}$. This change in state is therefore spontaneous, and is associated with a maximum practical release of energy given by equation (13.43b)

$$-\Delta G_{1\to 2} = \{_1w_2\}_{\text{other, rev}} \tag{13.43b}$$

For a real, irreversible, change, if we represent the actual amount of energy released as $\{_1w_2\}_{\text{other, irrev}}$, then

$$-\Delta G_{1\to 2} = \{_1w_2\}_{\text{other, rev}} > \{_1w_2\}_{\text{other, irrev}}$$

Consider further a second change of state

state [3] → state [4]

associated with a positive value of $\Delta G_{3\to 4}$. This endergonic change of state is not spontaneous, and will take place only if external energy is transferred into the system as represented by equation (13.43b)

$$-\{_3w_4\}_{\text{other, rev}} = \Delta G_{3\to 4} \tag{13.43b}$$

In this case, $-\{_3w_4\}_{\text{other, rev}}$ represents the reversible work that needs to be performed on the system to drive the endergonic reaction. In practice, the work that actually needs to be done on the system to drive the endergonic change from state [3] to state [4] will be real, irreversible, work rather than theoretical, reversible, work. Since in general

$$đw_{\text{rev}} > đw_{\text{irrev}} \tag{5.37}$$

then, if we change the signs, the inequality 'flips' giving

$$-đw_{\text{rev}} < -đw_{\text{irrev}}$$

Equation (13.43b) for the endergonic, non-spontaneous, real change therefore becomes

$$-\{_3w_4\}_{\text{other, irrev}} > -\{_3w_4\}_{\text{other, rev}} = \Delta G_{3\to 4} \tag{13.43c}$$

Suppose now that the two systems are coupled, so that the exergonic, spontaneous, change from state [1] to state [2] drives the endergonic, non-spontaneous, change from state [3] to state [4]. The essence of the coupling between the two systems is that the irreversible work $\{_1w_2\}_{\text{other, irrev}}$ done by the exergonic change from state [1] to state [2] is the source of, and at least equal to, the irreversible work $-\{_3w_4\}_{\text{other, irrev}}$ needed to drive the endergonic change from state [3] to state [4]. For the two coupled systems, we may therefore state that

$$\{_1w_2\}_{\text{other, irrev}} \geqslant -\{_3w_4\}_{\text{other, irrev}}$$

But since

$$-\Delta G_{1\to 2} = \{_1w_2\}_{\text{other, rev}} > \{_1w_2\}_{\text{other, irrev}}$$

and

$$-\{_3w_4\}_{\text{other, irrev}} > -\{_3w_4\}_{\text{other, rev}} = \Delta G_{3\to 4}$$

then these three last expressions can be combined as

$$-\Delta G_{1\to 2} = \{_1w_2\}_{\text{other, rev}} > \{_1w_2\}_{\text{other, irrev}} \geqslant -\{_3w_4\}_{\text{other, irrev}} > -\{_3w_4\}_{\text{other, rev}} = \Delta G_{3\to 4}$$

which simplifies to

$$-\Delta G_{1\to 2} > \Delta G_{3\to 4}$$

This captures the reality that, in practice, not all of the energy $-\Delta G_{1\to 2}$ released by the exergonic change can be used to drive the endergonic change – some energy must be wasted, which is, of course, the Second Law in yet another form. But if there is enough 'surplus', then the exergonic change will drive the endergonic one.

Now, for the endergonic, non-spontaneous, change, we know that

$$\Delta G_{3\to 4} = \Delta H_{3\to 4} - T\Delta S_{3\to 4}$$

Let's now consider what happens within the system undergoing this endergonic change. This system receives an input in energy, driving the endergonic change, for which $\Delta G_{3\to 4} = \Delta H_{3\to 4} - T\Delta S_{3\to 4}$. If the change happens to be such that $T\Delta S_{3\to 4}$ is small compared to $\Delta H_{3\to 4}$, then $\Delta G_{3\to 4} \approx \Delta H_{3\to 4}$. Since $\Delta G_{3\to 4} > 0$, then $\Delta H_{3\to 4} > 0$ and the system's enthalpy will increase. But, for any system changing state at constant pressure, we know that

$$\Delta H = C_P \Delta T \tag{6.8b}$$

An increase ΔH in enthalpy therefore corresponds to an increase ΔT in temperature. The input in energy therefore causes the system's temperature to increase. So, for example, when frictional work is done on a system – which is a real instance of an energy injection into a system – the system gets hotter.

But this is not the only possibility. Given that

$$\Delta G_{3\to 4} = \Delta H_{3\to 4} - T\Delta S_{3\to 4}$$

an alternative is that the enthalpy change $\Delta H_{3\to 4}$ is small, but that $-T\Delta S_{3\to 4}$ is large. Since external energy is being transferred into the system, $\Delta G_{3\to 4}$ is positive, and so $-T\Delta S_{3\to 4}$ must be positive too, implying that $\Delta S_{3\to 4}$ must be negative. As a result of the input of energy, the order of the system increases: to tidy that room, we must expend energy.

Injecting energy into a system can therefore have two, very different outcomes: the system could get hotter, or the system could become more ordered. Both could of course happen simultaneously in various combinations, but the recognition of the two extreme possibilities – either getting hotter or becoming more ordered – is of general interest, and of particular importance in one, very significant, context …

13.13 …and the maintenance of life

…for this is the explanation of almost all living processes. Most living processes create order – the synthesis of a protein, for example, creates a vastly more ordered product, the protein, as compared to its original components, the corresponding number of individual amino acids. As we saw in Chapter 11, an increase in order is associated with a decrease in entropy – so much of biochemistry is associated with reactions for which $\Delta S < 0$, often associated with relatively small values of ΔH such that $\Delta G > 0$. These biochemical reactions therefore 'shouldn't happen' – as indeed they don't, in isolation. For these reactions to happen, they must be coupled with one or more simultaneous reactions associated with a sufficiently negative value of ΔG so that the overall value of ΔG for the two (or more) coupled reactions is less than 0.

This is the role provided, in particular (but not exclusively), by ATP. For the reaction

$$ATP + H_2O \rightarrow ADP + phosphate$$

$\Delta G = -30.5$ kJ mol^{-1}, and it is this 'release' of Gibbs free energy that 'drives' much of biochemistry, such as the synthesis of proteins.

Coupled reactions are important at all scales. The entire earth, for example, has, maintains, and increases order, especially in relation to all the living processes. So, if, in terms of our discussion in this, and the previous, section, state [3] represents [the earth as at now] and state [4] represents [the earth tomorrow], then ΔG for this change of state is positive. The earth, however, is coupled to the system of the sun, and so state [1] represents [the sun as at now] and state [2] represents [the sun tomorrow]. Fortunately, ΔG for this change is strongly negative (think, for example, of ΔH), so for the coupled system [earth + sun], ΔG is negative. Life therefore continues.

13.14 Climate change and global warming – the 'big picture'

At the time of writing, there is a conference taking place in Paris on climate change and global warming. At the opening yesterday, 30th November 2015, the largest-ever assembly of world leaders – including the leaders of some 150 countries – stated their individual, and collective, commitment to making real progress as regards the most significant threat to the planet. There are many books on climate change, some very large indeed (for example, as mentioned in the references), but since climate change and global warming is about heat, and the temperature of the earth, we include some thoughts here – for the fundamentals of global warming are all about the thermodynamics of coupled systems.

One system is man, and what man does. Human beings, individually, do lots of different things, but collectively, taking a 'big picture' view, they do one big thing. Human beings create order. Much of human activity is about making things and changing things. So, agriculture takes a natural ecosystem, and transforms it into ordered paddy fields of rice, or grasslands for raising cattle – a farm being a much more ordered state than the wilderness it replaces. Manufacturing takes raw materials and transforms them into automobiles and computers. Building houses requires that rocks are hacked out of a quarry, or mud is taken that from a river bank, to form orderly structures. Yes, there are some activities of man that create disorder – of which war is one. But, for the most part, man creates order, and decreases entropy. So, the for system

Man (doing things) → Man (doing more things)

ΔS is negative. For the things that man does – making things, building things – the corresponding ΔH for the 'thing' itself is often rather small, implying that man's activities are dominated by the negative value of ΔS, and that ΔG is correspondingly positive. For these things to happen, there needs to be a coupled system for which ΔG is sufficiently negative for the combined value of ΔG for the two coupled systems is to be negative overall.

For millennia, the coupled system was the human being, or another animal, who did work, so that the work performed by the living system, and the corresponding negative value of ΔG, drove the farming, the making, the building. Then man discovered another source of energy that could be used: heat, produced originally from burning wood, then coal, gas and oil. In these combustion reactions, ΔG is strongly negative, primarily as a result of correspondingly large negative values of ΔH: exothermic combustion reactions were harnessed, either directly (by means of, for example, a steam engines or oil-burning internal combustion engines) or indirectly (by using steam to create electricity), to drive the agricultural machinery, to power industry, to help build huge structures. The strongly enthalpy-driven systems are therefore coupled with the strongly entropy-reducing systems of human activity.

And as long as man continues to do things, man will continue to reduce entropy, and will continue to required energy accordingly.

The global warming problem is a by-product. Exothermic combustion reactions of wood, coal, gas and oil release carbon dioxide into the atmosphere. This carbon dioxide, alongside other 'greenhouse gases', collectively act as an ever-more-effective thermal 'blanket', progressively slowing down the rate at which the earth radiates heat energy back into space, so causing the global temperature to rise. This in turn has a number of consequences: the melting of polar ice, so increasing sea level and causing flooding; thermal expansion of the sea, which also drives the sea level up; and also an increase in the incidence of violent weather, as the planet dissipates energy through means other than radiation to the atmosphere. For the human population, these are all bad news. Rising sea levels destroys habitable and agricultural land, and violent storms create destruction.

The solution to the global warming problem can never lie in a significant reduction of the use of energy. Of course, we can be more efficient in energy usage, and that is a good thing to do. But, very fundamentally, man's creation of order will never stop, and thermodynamics tells us that this will always require energy. So the solution to the global warming problem is primarily elsewhere: to develop technologies which provide energy, so that those exothermic changes can still be coupled to man's entropy-reducing activities, but without the by-product of atmospheric carbon dioxide, and other deleterious materials and effects. And, very urgently, to develop technologies that can extract carbon dioxide, and other 'greenhouse gases', directly from the atmosphere so that the current global thermal disequilibrium can be stabilised.

13.15 Standard Gibbs free energies

On page 179, we defined the standard enthalpy of formation $\Delta_f H^\ominus$ for any molecule as the enthalpy change associated with the formation of 1 mol of that molecule in its standard state from the corresponding constituent elements in their standard states

$$\text{Constituent elements in standard states} \rightarrow \text{1 mol of molecule of interest in standard state} \qquad \Delta H = \Delta_f H^\ominus$$

We then showed how the First Law, expressed as Hess's Law of Constant Heat Formation, enabled us to calculate the standard enthalpy change $\Delta_r H^\ominus$ for any reaction

Reactants $\rightarrow$ Products

as

$$\Delta_r H^\ominus = \sum_{\text{Products}} n_{\text{prod}} \Delta_f H_{\text{prod}}^\ominus - \sum_{\text{Reactants}} n_{\text{reac}} \Delta_f H_{\text{reac}}^\ominus \qquad (6.21)$$

Similarly, we may define the **standard Gibbs free energy of formation** $\Delta_f G^\ominus$ for any molecule as the change in G associated with the formation of 1 mol of that molecule in its standard state from the corresponding constituent elements in their standard states

$$\text{Constituent elements in standard states} \rightarrow \text{1 mol of molecule of interest in standard state} \qquad \Delta G = \Delta_f G^\ominus$$

For this reaction

$$\Delta_f G^\ominus (\text{molecule}) = G^\ominus (\text{molecule}) - \sum_{\text{elements}} G^\ominus (\text{element}) \qquad (13.44a)$$

In equation (13.44a), $G^\ominus$(molecule) represents the absolute molar Gibbs free energy of the molecule of interest in its standard state, and $G^\ominus$ (element) represents the absolute molar Gibbs free energy of any constituent element in its standard state. However, as we saw on page 176, it is impossible to measure absolute enthalpies, and since, by definition, $G = H - TS$, it is therefore impossible to measure absolute Gibbs free energies. So, just as we defined a reference level for H by assigning $H^\ominus$ (element) = 0 J mol^{-1} for all elements in their standard states, we define the reference level for G such that $G^\ominus$ (element) = 0 J mol^{-1} for all elements in their standard states, so that the reference state is the standard state. The value of the summation in equation (13.44a) is therefore zero, and so $\Delta_f G^\ominus$ for any molecule may be regarded as 'the' molar Gibbs free energy $G^\ominus$ of that molecule in its standard state,

$$\Delta_f G^\ominus (\text{molecule}) = G^\ominus (\text{molecule}) \qquad (13.44b)$$

Also, if the molecule itself is an element, $\Delta_f G^\ominus$ (element) = $G^\ominus$ (element) = 0 J mol^{-1} for all elements in their standard states. Building on the discussion on page 179 in connection with the definition of the reference level for $H^\ominus$, the definition of the reference level $G^\ominus$ (element) = 0 J mol^{-1} for all elements does not imply that all elements 'have' zero $G^\ominus$; furthermore, since all elements are composed from different numbers of protons, neutrons, electrons and all the

Table 13.1 Some standard molar Gibbs free energies of formation $\Delta_f G^\ominus$

All elements in their standard state: $\Delta_f G^\ominus = 0$ kJ mol^{-1}.

Selected inorganic molecules at a temperature of 298.15 K = 25°C.

	kJ mol^{-1}		kJ mol^{-1}
H_2O (l)	−237.1	H_2S (g)	−33.4
H_2O (g)	−228.6	NO (g)	+87.6
HCl (g)	−95.3	NO_2 (g)	+51.3
HBr (g)	−53.4	NH_3 (g)	−16.4
HI (g)	+1.7	CO (g)	−137.2
SO_2 (g)	−300.1	CO_2 (g)	−394.4

	kJ mol^{-1}		kJ mol^{-1}
$AgCl$ (s)	−109.8	$CaCO_3$ (s, calcite)	−1,129.1
$AgBr$ (s)	−96.9	Fe_2O_3 (s)	−742.2
AgI (s)	−66.2	SiO_2 (s, quartz)	−856.3
CaO (s)	−603.3	$NaCl$ (s)	−384.1
$Ca(OH)_2$ (s)	−897.5	NH_4Cl (s)	−202.9

Selected organic molecules at a temperature of 298.15 K = 25°C.

	kJ mol^{-1}		kJ mol^{-1}
CH_4 (g)	−50.5	C_6H_6 (l)	+124.5
C_2H_6 (g)	−32.0	CH_3OH (l)	−166.6
C_2H_4 (g)	+68.4	C_2H_5OH (l)	−174.8
C_2H_2 (g)	+209.9	CH_3COOH (l)	−389.9
C_3H_8 (g)	−23.4	$CHCl_3$ (l)	−73.7

Selected aqueous ions at a temperature of 298.15 K = 25°C, and at a standard concentration of 1 mol dm^{-3}.

	kJ mol^{-1}		kJ mol^{-1}
H^+ (aq)	0	OH^- (aq)	−157.2
Na^+ (aq)	−261.9	Cl^- (aq)	−131.2
K^+ (aq)	−283.3	Br^- (aq)	−104.0
NH_4^+ (aq)	−79.3	NO_3^- (aq)	−111.3
Ca^{2+} (aq)	−553.6	SO_4^{2-} (aq)	−744.5
Cu^{2+} (aq)	+65.5	PO_4^{3-} (aq)	−1,018.7

Source: *CRC Handbook of Chemistry and Physics*, 96th edn, 2015, Section 5, pages 5–4 to 5–42 and 5–66 to 5.67.

other sub-atomic particles, it is most unlikely that, in an absolute sense, all elements have the same value of $G^\standardstate$, let alone a value of zero. In essence, the assignment of $G^\standardstate$ (element) = 0 J mol^{-1} for all elements identifies a reference level for each element, and so rather than there being a single reference level, there are as many reference levels as there are elements. And although that sounds rather confusing, in fact, everything works in practice.

Table 13.1 shows some values of the standard Gibbs free energy of formation $\Delta_f G^\standardstate$ for some selected molecules and ions.

Note that, in Table 13.1, for the hydrogen ion H$^+$(aq), the value $\Delta_f G^\standardstate(H^+) = 0$ J mol^{-1} is the definition of the reference level for the measurement of $\Delta_f G^\standardstate$ for all ions in solution, as discussed on page 184 in connection with the assignment $\Delta_f H^\standardstate(H^+) = 0$ J mol^{-1}.

By analogy with the discussion on pages 181 and 182 in connection with the standard enthalpy $\Delta_r H^\standardstate$ of a generalised chemical reaction, for any reaction

Reactants → Products

in which n_{reac} mol of each reactant, of standard molar Gibbs free energy of formation $\Delta_f G_{\text{reac}}^\standardstate$, form n_{prod} mol of each product, of standard molar Gibbs free energy of formation $\Delta_f G_{\text{prod}}^\standardstate$, the **standard Gibbs free energy of reaction** $\Delta_r G^\standardstate$ is given by

$$\Delta_r G^\standardstate = \sum_{\text{Products}} n_{\text{prod}} \Delta_f G_{\text{prod}}^\standardstate - \sum_{\text{Reactants}} n_{\text{reac}} \Delta_f G_{\text{reac}}^\standardstate \tag{13.45}$$

Computations involving Gibbs free energies are very similar to those involving enthalpies, and all the methods of thermochemistry (see Chapter 6) can be applied, with G replacing H. Also, as we saw in our discussion of $\Delta_r H^\standardstate$ on 181 and 182, whereas the units of measure of $\Delta_f G^\standardstate$ are kJ mol^{-1}, the units of $\Delta_r G^\standardstate$, as computed using equation (13.45), are kJ, for the molar quantities defined by the stoichiometry of the corresponding reaction, as written.

13.16 Gibbs free energies at non-standard pressures

In general, as we know, the Gibbs free energy G of any system, and the change $\Delta_r G$ for any reaction, are functions $G(P, T)$ and $\Delta_r G(P, T)$ of the pressure P and temperature T, and so will vary with both pressure and temperature; since the standard quantities $G^\standardstate$ and $\Delta_r G^\standardstate$ are defined at a fixed standard pressure $P^\standardstate$, they are functions $G^\standardstate(T)$ and $\Delta_r G^\standardstate(T)$ of temperature.

We will deal with temperature variations on pages 444 to 448; here, our current purpose is to show how to determine a value of G or $\Delta_r G$ at any pressure P. Since reference tables provide data relating to standard states, which, according to IUPAC guidelines, relate to measurements and reactions taking place at a standard pressure $P^\standardstate$ of 1 bar, then to determine values of G and $\Delta_r G$ at any other pressure, consider the change of state, at constant temperature, for n mol of any pure material

Material in standard state at 1 bar → Material at pressure P

for which the change ΔG in Gibbs free energy is given by

$\Delta G = G$ (Material at pressure P) $- G^\standardstate$

FREE ENERGY

To determine ΔG, with reference to page 380, we invoke equation (13.22)

$$\Delta G = G_2 - G_1 = \int_{P^\ominus}^{P} V \, dP \qquad (13.22)$$

If the material is in a condensed state – a solid or a gas – then the volume V varies only very slowly with pressure: many condensed materials are in essence incompressible. For a condensed state, V in equation (13.22) can therefore be regarded as a constant, especially over relatively small changes in volume, and so, for a condensed state, at the temperature for which the standard state is defined, the value of G at a pressure P different from the standard pressure $P^\ominus$ is given by

$$G = G^\ominus + V(P - P^\ominus) \qquad (13.46a)$$

If P is measured in bar, then $P^\ominus = 1$ bar, and so

$$G = G^\ominus + V(P - 1) \qquad (13.46b)$$

If the material is an ideal gas at temperature T, as we saw on page 381,

$$\Delta G = G - G^\ominus = nRT \int_{P^\ominus}^{P} \frac{dP}{P} = nRT \ln \frac{P}{P^\ominus} \qquad (13.47a)$$

implying that the Gibbs free energy G of n mol of an ideal gas at temperature T, but at a non-standard pressure P, is given by

$$G = G^\ominus + nRT \ln (P/P^\ominus) \qquad (13.47b)$$

If P is measured in bar, and the standard pressure $P^\ominus = 1$ bar, this becomes

$$G = G^\ominus + nRT \ln P \qquad (13.47c)$$

Also, for 1 mol of a pure ideal gas, the molar Gibbs free energy G is given by

$$\mathbf{G = G^\ominus + RT \ln (P/P^\ominus)} \qquad (13.47d)$$

and, since $P^\ominus = 1$ bar

$$\mathbf{G = G^\ominus + RT \ln P} \qquad (13.47e)$$

Equations (13.47) are important, and we shall use them in Chapter 14 to determine a condition for chemical equilibrium for gas phase reactions (see page 416), and in Chapter 16 in connection with reactions in solution (see page 511). Also, note that although the pressure P is measured in units of bar, the quantity P shown in equations (13.47c) and (13.47e) is dimensionless (as is required for the ln term), since in these equations P is standing for the ratio $P/P^\ominus$, where $P^\ominus = 1$ bar.

13.17 The Gibbs free energy of mixtures

13.17.1 From pure substance to mixture

So far, this book has focused on the thermodynamics of pure substances, for the most part ideal gases. Chemical reactions, however, whether in the laboratory, in industry, or in living systems, inevitably involve mixtures: mixtures of various reagents reacting to form mixtures of products. In this section, we therefore extend our understanding of the thermodynamics beyond pure substances to mixtures of pure substances (which, in fact, also covers mixtures of impure substances too, for, ultimately, an impure substance is a mixture of the corresponding pure substance with unwanted substances called 'contaminants', these being other pure substances or themselves mixtures . . .).

Our starting point is to recall that, in the absence of gravitation, motion, electric, magnetic and surface effects, a fixed mass of any pure substance has only two independent state functions (see page 48). So, for example, the Gibbs free energy G (pure substance) of a given mass of a pure substance can be expressed, mathematically, as a function $G(P, T)$ of the pressure P and temperature T

$$G \text{ (pure substance)} = G(P, T) \tag{13.17}$$

from which, as we saw on page 380, we established that

$$dG \text{ (pure substance)} = V\,dP - S\,dT = \left(\frac{\partial G}{\partial P}\right)_T dP + \left(\frac{\partial G}{\partial T}\right)_P dT \tag{13.19}$$

implying that, for a pure substance

$$\left(\frac{\partial G}{\partial P}\right)_T = V \tag{13.20}$$

and

$$\left(\frac{\partial G}{\partial T}\right)_P = -S \tag{13.21}$$

These expressions all relate to 'a given mass' of material, but we haven't specified what that mass actually is. If the mass is 1 mol, the extensive functions G, V and S become the corresponding intensive molar quantities $\mathbf{G}$, $\mathbf{V}$ and $\mathbf{S}$ and so

$$\mathbf{G} \text{ (pure substance)} = \mathbf{G}(P, T) \tag{13.48}$$

and

$$d\mathbf{G} \text{ (pure substance)} = \mathbf{V}\,dP - \mathbf{S}\,dT = \left(\frac{\partial \mathbf{G}}{\partial P}\right)_T dP + \left(\frac{\partial \mathbf{G}}{\partial T}\right)_P dT \tag{13.49}$$

For a system comprising n mol of a pure substance, the Gibbs free energy $G(n\text{ mol})$ of the whole system is related to the molar quantity $\mathbf{G}$ as

$$G(n\text{ mol}) = n\,\mathbf{G}(P, T)$$

and for a system comprising, say, $2n$ mol of the same pure substance, the Gibbs free energy G $(2n$ mol$)$ of the bigger whole system is similarly

$$G(2n \text{ mol}) = 2n\, \mathbf{G}(P, T) = n\, \mathbf{G}(P, T) + n\, \mathbf{G}(P, T)$$

This indeed obvious; but it is a key step leading towards an important result. Suppose that the 'bigger' system is not composed of $2n$ mol of the same pure substance, but of n_A mol of a pure substance A, and n_B mol of a pure substance B, both at the same pressure P and temperature T. Following the same logic as we have just used, the Gibbs free energy $G(n_A$ mol A and n_B mol B$)$ of the mixture of pure A and pure B is given by

$$G(n_A \text{ mol A and } n_B \text{ mol B}) = n_A\, \mathbf{G}_A(P, T) + n_B\, \mathbf{G}_B(P, T) \tag{13.50}$$

Equation (13.50), however, makes a 'hidden' assumption: in determining the free energy of a mixture by the linear addition of the Gibbs free energies of each component, we are, in fact, assuming that the mixture shows ideal behaviour, and, as we saw on pages 17 and 173, that there are, amongst other things, no molecular interactions. We shall accept this assumption for the moment, and deal with the behaviour of non-ideal systems later (see Chapter 22).

Suppose, then, that the mixture is formed from two ideal gases, with gas A exerting a partial pressure p_A, and gas B, a partial pressure p_B. If the system is at temperature T, then equation (13.47d)

$$\mathbf{G} = \mathbf{G}^{\ominus} + RT \ln(P/P^{\ominus}) \tag{13.47d}$$

enables us to express $\mathbf{G}_A(P, T)$ and $\mathbf{G}_B(P, T)$ as

$$\mathbf{G}_A(P, T) = \mathbf{G}_A^{\ominus} + RT \ln(p_A/P^{\ominus})$$

and

$$\mathbf{G}_B(P, T) = \mathbf{G}_B^{\ominus} + RT \ln(p_B/P^{\ominus})$$

The Gibbs free energy $G(P, T)$ of the mixture is therefore

$$G(P, T) = n_A \mathbf{G}_A^{\ominus} + n_A RT \ln(p_A/P^{\ominus}) + n_B \mathbf{G}_B^{\ominus} + n_B RT \ln(p_B/P^{\ominus}) \tag{13.51}$$

An important inference from equations (13.50) and (13.51) – and an inference which remains true even if some adjustment needs to be made to these equations to take account of non-ideal behaviour – is that the total Gibbs free energy of a mixture of pure substances depends on the composition of the mixture, as specified by the mol numbers n_A and n_B. So, even if the pressure P and temperature T both remain constant, the Gibbs free energy $G(P, T)$ of the mixture will change if the composition changes. *This is the essence of chemistry*. A chemical reaction is what happens when reagents transform into products – or, using the language of the current section, what happens as the composition of a mixture changes. As a reaction takes place, the relative proportions of the reactants and products changes – sometimes very quickly, sometimes more slowly – and, as this happens, the instantaneous value of the system's total Gibbs free energy $G(P, T)$ changes too. Whilst this is happening, the inequality

$$\Delta G < 0 \tag{13.9b}$$

remains true: if the reaction can move from one state to another such that the Gibbs free energy of the system as a whole becomes smaller, then that change will be spontaneous – as we shall explore in detail in the next chapter.

13.17.2 The chemical potential

Mathematically, to represent the Gibbs free energy G of a mixture, we must recognise this dependence on composition, and so, for a two-component system at a given pressure P and temperature T, the Gibbs free energy G of the whole system is not solely a function of the two intensive state functions P and T, but of a total of four state functions, the two intensive state functions P and T, and also the two extensive state functions n_A and n_B, which define the precise composition of the mixture.

So, for a mixture of two pure components

G (two-component mixture) $= G(P, T, n_A, n_B)$

and for a mixture of an arbitrary number of pure components

G (general mixture) $= G(P, T, n_A, n_B, \ldots, n_i, \ldots)$

Accordingly, for a two-component mixture, the Gibbs free energy G of the mixture (where we now drop the parenthesis (mixture) for clarity) may be expressed mathematically as

$G = G(P, T, n_A, n_B)$

In this expression, G is a function of four variables, P, T, n_A and n_B, and so if we differentiate this equation, we have

$$dG = \left(\frac{\partial G}{\partial P}\right)_{T, n_A, n_B} dP + \left(\frac{\partial G}{\partial T}\right)_{P, n_A, n_B} dT + \left(\frac{\partial G}{\partial n_A}\right)_{P, T, n_B} dn_A + \left(\frac{\partial G}{\partial n_B}\right)_{P, T, n_A} dn_B \quad (13.52)$$

Equation (13.52) looks very complicated – in fact, it is quite straight-forward. The (apparent) complexity is attributable to the fact the Gibbs free energy G of the mixture can change as a result of independent changes to any one of four different variables – P, T, n_A and n_B. Each of the partial derivatives therefore represents the incremental change in G attributable to an incremental change in each of these four variables independently, whilst all of the other three variables are constant. The partial derivatives $(\partial G/\partial P)_{\ldots}$ and $(\partial G/\partial T)_{\ldots}$ are already familiar (see page 379); the two new partial derivatives $(\partial G/\partial n_A)_{\ldots}$ and $(\partial G/\partial n_B)_{\ldots}$ specify the incremental change in G for any incremental change in the mole numbers n_A and n_B, as will happen as a reaction progresses, and the composition of the mixture changes.

For a mixture of an arbitrary number n of components, G is a function of $n + 2$ variables, $G = G(P, T, n_A, n_B, \ldots, n_i, \ldots)$, and the differential dG is given by

$$dG = \left(\frac{\partial G}{\partial P}\right)_{T, n_i \ldots} dP + \left(\frac{\partial G}{\partial T}\right)_{P, n_i \ldots} dT + \sum_i \left(\frac{\partial G}{\partial n_i}\right)_{P, T, n_{j \neq i}} dn_i \quad (13.53a)$$

in which the summation is over each component, and $(\partial G/\partial n_i)_{\ldots}$ is the partial derivative of G with respect to the mole number n_i, holding the pressure, the temperature, and the mole numbers of all other components constant. This quantity is known as the **chemical potential** of component i in the mixture, represented by the symbol $\mu_i(P, T)$

$$\left(\frac{\partial G}{\partial n_i}\right)_{P,T,n_j \neq i} = \mu_i(P, T) \tag{13.54}$$

As can be seen, for any pure material i, the chemical potential $\mu_i(P, T)$ is a function of the pressure P and temperature T.

An alternative representation of equation (13.53a) is therefore

$$dG = \left(\frac{\partial G}{\partial P}\right)_{T,n_i...} dP + \left(\frac{\partial G}{\partial T}\right)_{P,n_i...} dT + \sum_i \mu_i(P, T)\, dn_i \tag{13.53b}$$

Equations (13.53a) and (13.53b) apply to all mixtures - and applies to pure substances too. So, for a fixed quantity - say n_A mol - of a pure substance A, $dn_A = 0$, and equations (13.53a) and (13.53b) become

$$dG = \left(\frac{\partial G}{\partial P}\right)_T dP + \left(\frac{\partial G}{\partial T}\right)_P dT$$

which is identical to equation (13.19).

Furthermore, if the system comprises only n_A mol of a single component A, the total Gibbs free energy $G(P, T, n_A)$ of the system is

$$G(P, T, n_A) = n_A \mathbf{G}_A(P, T)$$

and if the pressure P and the temperature T are constant,

$$\left(\frac{\partial G}{\partial n_A}\right)_{P,T} = \mathbf{G}_A(P, T) = \mu_A(P, T) \tag{13.55}$$

Equation (13.55) is important, for it tells us that, at any pressure P and temperature T, the chemical potential $\mu_i(P, T)$ of any pure substance i is equal to the molar Gibbs free energy $\mathbf{G}_i(P, T)$ of that pure substance i, at the same pressure P and temperature T. Although $\mu_i(P, T)$ is a molar quantity, it is not conventionally represented in a bold font; also, as a molar quantity, $\mu_i(P, T)$ - which is often abbreviated to μ_i or simply μ - is an intensive state function.

13.17.3 Three important properties of the chemical potential μ

The identification that, for a pure substance, the chemical potential μ equals the molar Gibbs free energy $\mathbf{G}$ implies that every property of $\mathbf{G}$, and every equation in which $\mathbf{G}$ appears, also has an equivalent using μ. Of those we have seen so far, three are of particular importance.

Firstly, if a pure substance is in its standard state, usually at a pressure of 1 bar, and at a stated temperature, we may define the **standard chemical potential** $\mu^{\ominus}$ as equal to $\mathbf{G}^{\ominus}$.

Secondly, just as $\mathbf{G}$ varies with pressure as

$$\mathbf{G} = \mathbf{G}^{\ominus} + RT \ln(P/P^{\ominus}) \tag{13.47d}$$

then μ varies with pressure as

$$\mu = \mu^{\ominus} + RT \ln(P/P^{\ominus}) \tag{13.56}$$

in which $P^{\ominus}$ is the standard pressure, 1 bar.

Thirdly, consider a system composed at any instant of n_A mol of A and n_B mol of B that can react chemically, so that the mole numbers n_A and n_B change over time. If we assume ideal conditions, and therefore that the Gibbs free energies of the two components can be added, then, at any instant, the total Gibbs free energy G of the mixture, at any pressure and temperature, is given by

$$G = n_A G_A + n_B G_B = n_A \mu_A + n_B \mu_B$$

If the temperature and pressure are constant, we can differentiate, giving

$$dG = n_A\,d\mu_A + n_B\,d\mu_B + \mu_A\,dn_A + \mu_B\,dn_B \tag{13.57}$$

If we now apply equation (13.53b)

$$dG = \left(\frac{\partial G}{\partial P}\right)_{T,n_i\ldots} dP + \left(\frac{\partial G}{\partial T}\right)_{P,n_i\ldots} dT + \sum_i \mu_i(P,T)\,dn_i \tag{13.53b}$$

which is generally true, to a two-component mixture at constant pressure and temperature, $dP = dT = 0$, and so

$$dG = \mu_A\,dn_A + \mu_B\,dn_B \tag{13.58}$$

Equations (13.57) and (13.58) are describing the same quantity dG, which implies that for any two-component system of substances A and B that mutually react, then, at all times

$$n_A d\mu_A + n_B\,d\mu_B = 0$$

and, more generally, for a system of i components

$$\sum_i n_i\,d\mu_i = 0 \tag{13.59}$$

Equation (13.59) is known as the **Gibbs-Duhem equation**, named after Josiah Willard Gibbs, and the French chemist Pierre Duhem. Its significance is that it puts a constraint on changes to the chemical potentials of the various components in the mixture: in a two-component mixture, for example, any increase $d\mu_A$ for component A is accompanied by a corresponding decrease $d\mu_B = -(n_B/n_A)\,d\mu_B$ for component B.

Furthermore, for any reaction, the Gibbs-Duhem condition is not the only constraint: another is imposed by the stoichiometry of the reaction. So, consider the simple case of the reaction represented by

$$A \rightarrow 2B$$

in which one molecule of A decomposes into two molecules of B – a real example being the decomposition of $N_2O_4(g)$ into $NO_2(g)$

$$N_2O_4(g) \rightarrow 2\,NO_2(g)$$

For each molecule of A that decomposes, 2 molecules of B are produced. So, consider a small period of time during which, say 25 molecules of A decompose. dn_A is therefore -25. Simultaneously, 50 molecules of B are produced, and so dn_B is 50. dn_A and dn_B are therefore mutually related as

$$-2\,dn_A = dn_B$$

and so the changes dn_A and dn_B cannot take arbitrary values, but are constrained such that

$$2\,dn_A + dn_B = 0$$

13.17.4 The fundamental equation of chemical thermodynamics

Equation (13.53b)

$$dG = \left(\frac{\partial G}{\partial P}\right)_{T,n_j...} dP + \left(\frac{\partial G}{\partial T}\right)_{P,n_j...} dT + \sum_i \mu_i\, dn_i \tag{13.53b}$$

can be combined with (slightly modified versions of) equations (13.20) and (13.21)

$$\left(\frac{\partial G}{\partial P}\right)_{T,n_j...} = V \tag{13.20}$$

$$\left(\frac{\partial G}{\partial T}\right)_{P,n_j...} = -S \tag{13.21}$$

to give what is known as the **fundamental equation of chemical thermodynamics**

$$dG = V\,dP - S\,dT + \sum_i \mu_i\, dn_i \tag{13.60}$$

Equation (13.60) is a more generalised version of equation (13.16a)

$$dG = V\,dP - S\,dT \tag{13.16a}$$

which applies to a fixed quantity of a single pure substance. Furthermore, equation (13.60) also applies to systems in which the quantity of material changes – as represented by a change dn_i in any mole number n_i – as can happen, for example, when matter flows into or out of the system of interest, across the system boundary. Equation (13.60) is therefore valid in open systems, and relaxes the constraint we introduced on page 104 of limiting our analysis to closed systems only, for which $dn_i = 0$.

13.17.5 μ and the First Law

In this final section of this chapter, we will derive an expression for the First Law of Thermodynamics that incorporates the chemical potential μ, and so extends the First Law to open systems.

Our starting point is equation (13.60)

$$dG = V\,dP - S\,dT + \sum_i \mu_i\, dn_i \tag{13.60}$$

which applies to any system containing a mixture of pure materials, as defined by the corresponding mole numbers n_i. We now invoke the Gibbs-Duhem equation

$$\sum_i n_i\, d\mu_i = 0 \tag{13.59}$$

implying that equation (13.60) may be expressed as

$$dG = V\,dP - S\,dT + \sum_i \mu_i\,dn_i + \sum_i n_i\,d\mu_i$$

and so

$$dG = V\,dP - S\,dT + d\sum_i \mu_i n_i \qquad (13.61)$$

Given the definitions of G as

$$G = H - TS \qquad (13.10)$$

and of H

$$H = U + PV \qquad (6.4)$$

then

$$G = U + PV - TS$$

from which

$$dG = dU + P\,dV + V\,dP - T\,dS - S\,dT \qquad (13.62)$$

Equations (13.61) and (13.62) are general, and can be applied to the same system. Accordingly, we can equate the right-hand side of equation (13.62) to the right-hand side of equation (13.61), giving

$$dU + P\,dV + V\,dP - T\,dS - S\,dT = V\,dP - S\,dT + d\sum_i \mu_i n_i$$

implying that

$$dU = T\,dS - P\,dV + d\sum_i \mu_i n_i \qquad (13.63)$$

Equation (13.63) can be compared to equation (9.20a)

$$dU = T\,dS - P\,dV \qquad (9.20a)$$

which applies to a closed system of constant mass. Equation (13.63) is a more general expression, and is the mathematical expression of the combination of the First and Second Laws for an open system in which the composition can change, as happens when at least one value of dn_i is non-zero. Equation (13.63) also implies that a change $d\mu_i$ in the chemical potential of at least one component i can also cause a change dU in the system's internal energy U.

Equation (13.63) can be expressed in terms of heat $đq$ and work $đw$: for a system that undergoes a reversible change, and can perform P, V work only

$$T\,dS = đq_{rev} \qquad (9.1)$$

and

$$P\,dV = đw_{rev} \qquad (2.9)$$

implying that equation (13.63) becomes

$$dU = đq_{rev} - đw_{rev} + d\sum_i \mu_i n_i \tag{13.64}$$

and more generally

$$dU = đq - đw + d\sum_i \mu_i n_i \tag{13.65}$$

Equation (13.65) can be compared to the statement of the First Law given in Chapter 4 as

$$dU = đq - đw \tag{5.1a}$$

which applies to a closed system of constant mass; equation (13.61) is the form of the First Law that applies to open systems.

Furthermore, since

$$H = U + PV \tag{6.4}$$

then

$$dH = dU + d(PV)$$

and so

$$dH = đq - đw + d(PV) + d\sum_i \mu_i n_i \tag{13.66}$$

For a reversible change in a system that performs P, V work of expansion, $đq_{rev} = T\,dS$ and $đw_{rev} = P\,dV$ and so

$$dH = T\,dS + V\,dP + d\sum_i \mu_i n_i \tag{13.67}$$

Equations (13.66) and (13.67) are the open-system expressions for changes in enthalpy.

13.18 The 'Fourth Law' of Thermodynamics

We conclude this chapter with a statement of a principle that is so important that it might even merit being known as the 'Fourth Law'

> Under any given conditions, out of all the states that are accessible to a system, the system will occupy the state of lowest Gibbs free energy.

As a first example, consider an ideal gas, within a container of volume V, at a temperature T, and exerting a pressure P on the container wall. Since the molecules within the gas are moving randomly, it is (theoretically) possible that, at any given temperature T, the gas could occupy a volume V_X, necessarily smaller than V and so within the container, and at a correspondingly higher pressure P_X. Many such states are possible, for there are a huge number of volumes $V_X < V$. All these states are accessible to the gas, for all obey the equation-of-state $P_X V_X = nRT$ – yet observation of the gas will show that the gas occupies one specific state, of volume V, this being the maximum volume available to the gas, as defined by the boundaries of the container.

For this gas, the Gibbs free energy G_X of any accessible state X is given by $G_X = H_X - TS_X$. Since the gas is ideal and the temperature T is constant, the enthalpies H_X of all these states are the same, implying that any differences in G_X are totally attributable to differences in the entropies S_X. As we saw in Chapter 9, for an ideal gas

$$dS = nC_V \frac{dT}{T} + nR \frac{dV}{V} \tag{9.33}$$

from which, if the temperature is constant, $dT = 0$, and so $dS = nR\,(dV/V)$. Since n, R and V are necessarily positive, any increase dV in volume is associated with an increase dS in the system's entropy, implying that the entropy is maximised when the volume is maximised. According to the definition of the Gibbs free energy G as $G = H - TS$, given that T and H are constant, when S is a maximum – as it is when the gas occupies the maximum volume available – then G is a minimum. Which is precisely what the 'Fourth Law' states – the system occupies the state of minimum G.

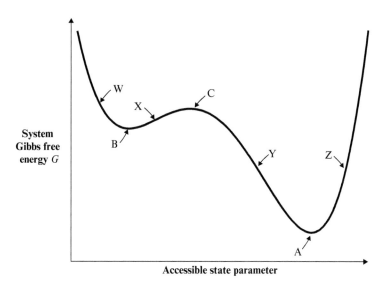

Figure 13.2 The 'Fourth Law' of Thermodynamics. The horizontal axis identifies all the states that are potentially accessible to the system under any given conditions, and the vertical axis represents the corresponding Gibbs free energy G. State A represents a stable equilibrium; state B, a metastable equilibrium; and state C, an unstable equilibrium. In general, the system will move spontaneously from any initial state to the state of lowest Gibbs free energy that is accessible. If the 'energy hill', whose 'summit' is the unstable state C, is relatively 'high', then the system can be 'trapped' in the metastable state B.

Figure 13.2 represents these ideas graphically. The horizontal axis represents a parameter that identifies all the states that are potentially accessible to a system under any given conditions, and the vertical axis represents the corresponding Gibbs free energy G. Suppose that the system happens to be in state $Y \equiv [G_Y]$, and can, in principle, change into any other state in the diagram. If the system were to change to state $C \equiv [G_C]$, then the corresponding value of $\Delta G_{Y \to C} = G_C - G_Y$ is positive, and so the transition from state Y to state C is not spontaneous; however, if the system were to change to state $A \equiv [G_A]$, then, since $\Delta G_{Y \to A} = G_A - G_Y$ is negative, the system will spontaneously change from state Y to state A. This applies to all states Y between state C and state A, and also to all states Z to the right of state A. And once the system is in state A, any change from state A to any other state – including both large changes and incremental changes – must be associated with a positive value of ΔG, and so cannot happen spontaneously; furthermore, if state A is disturbed, for example, by a thermal fluctuation, then the system will return to state A. State A is therefore known as a **stable equilibrium state**.

Likewise, to the left of Figure 13.2, all states W will spontaneously change to state B, as will all states X between state B and state C. State B is also a minimum on the curve, but at a higher value of G than the minimum associated with state A. $\Delta G_{B \to A} = G_A - G_B$ is therefore negative, and so state B should change spontaneously to state A. For this to happen, however, state B must pass through the intermediate state C, which, as can be seen from Figure 13.2, is 'uphill': $\Delta G_{B \to C} = G_C - G_B$ is positive, and so the change from state B to state C is not spontaneous, even though the overall change from state B to state A is.

The existence of state C, and the 'height' of the associated 'energy hill', determines whether or not state A is accessible from state B. If the 'height' of the 'energy hill' $\Delta G_{B \to C} = G_C - G_B$ is relatively small, it may be possible for thermal fluctuations to enable some molecules to 'cross the barrier', and so change from state B to state A; furthermore, if the 'height' can be 'lowered', for example by the use of a catalyst, state B can then spontaneously change to state A. If state A is potentially accessible from state B under suitable circumstances, state B is known as a **metastable equilibrium state**, 'metastable' because state B is stable over measureable periods of time, but will spontaneously move, if suitably disturbed, to the stable equilibrium state A. An example of a metastable equilibrium state is a **supercooled gas** – as can sometimes arise when a real gas is compressed to a combination of pressure and temperature at which the gas 'should' liquefy, but remains as a gas. As we shall discuss on page 474, the transformation of a gas to a liquid requires a process known as **nucleation**, in which micro-droplets of liquid are formed, around which the bulk of the gas can then condense. In terms of Figure 13.1, the supercooled gas is represented by state B, the corresponding liquid by state A, and the micro-droplets by state C. Under normal conditions, impurities in the gas, and the (microscopic) roughness of the vessel in which the gas is contained, reduce the height of the 'energy hill' at state C, and the gas (in state B) condenses directly into the liquid (state A). If, however, the gas is very pure, and the container has very smooth walls, the gas can be compressed to form the metastable supercooled state, which is maintained until the system is disturbed, for example by shaking, or by scratching the interior surface of the container, so causing the gas to liquefy instantaneously.

In some systems, the value of $\Delta G_{B \to C} = G_C - G_B$ is relatively large, and the 'energy hill' correspondingly 'high', so that state A is not readily accessible from state B. State B is therefore 'trapped', and can remain stable for long periods of time, even though state B does not correspond to the lowest possible value of G. As an example, the form of carbon known as diamond has a higher Gibbs free energy than an equal mass of graphite, yet no-one has witnessed a diamond spontaneously changing into graphite – diamond is a 'trapped' state which can last a very long time indeed!

At both the stable equilibrium state A, and the metastable equilibrium state B, the slope of the tangent to the G curve, as measured mathematically by the first derivative, is zero. Furthermore, the concave shape of the curve at states A and B implies that any fluctuation around state A or state B is accompanied by an incremental increase in G, and so, if disturbed, the system returns to state A or to state B respectively. Mathematically, the second derivatives at states A and B are positive. At state C, the top of the 'energy hill', the slope of the tangent to the G curve is also zero, but the convex shape implies that any fluctuation moves away from state C, either to state A or to state B. State C is therefore known as an **unstable equilibrium**, and is characterised mathematically by a point on the G curve for which the first derivative is zero, and the second derivative is negative.

Throughout the rest of this book, we will invoke the 'Fourth Law' on many occasions. The study of chemical equilibrium, for example, is about the identification of the characteristics of 'state A' (as in Figure 13.1) in terms of the composition of the equilibrium mixture (as discussed in Chapters 14 and 16); phase changes (Chapter 16) are concerned with the transition from one equilibrium state (say, a liquid) to another (say, a gas) which has a lower value of G; and, as we shall see in Chapter 25, the three-dimensional equilibrium configuration of a protein is determined as that specific configuration, out of all possible configurations, that has a minimum value of G. In principle, the problem of determining the stable configuration of a protein is exactly the same as that of establishing that an ideal gas will occupy the maximum possible volume – the mathematics, however, is rather harder!

EXERCISES

1. Define:
    - Gibbs free energy
    - Helmholtz free energy
    - Exergonic
    - Endergonic
    - Available work
    - Maximum available work
    - Chemical potential
    - Stable equilibrium state
    - Metastable equilibrium state
    - Unstable equilibrium state.
2. Are all exothermic reactions exergonic? Are all exergonic reactions exothermic? Under what conditions, if any, can an exergonic reaction be endothermic? And under what conditions can an endothermic reaction be exergonic?
3. Comment on the validity, or otherwise, of the following statements, all of which relate to a system undergoing a spontaneous change:
    - $\Delta H < 0$ can never be used as a criterion of spontaneity.
    - If $\Delta S > 0$, the change is spontaneous.
    - All spontaneous changes must obey the condition that $\Delta G < 0$.
    - ΔA has no importance as regards spontaneity.
4. Under what conditions is equation (13.13)

$$\Delta G = \Delta H - T\,\Delta S \tag{13.13}$$

    valid?

5. What do the terms 'enthalpy-driven reaction' and 'entropy-driven reaction' mean? Give some examples. A reaction is enthalpy-driven at a pressure of 1 bar and at 300 K. Will the same reaction be enthalpy-driven under all other conditions of pressure and temperature?

6. Derive equation (13.16a)

$$dG = V\,dP - S\,dT \tag{13.16a}$$

and identify precisely, and completely, the conditions under which it is valid.

7. In response to the exam question "How does the Gibbs free energy G vary with the temperature T?", a candidate writes "Since $G = G^\circ + RT \ln P$, G varies linearly with T, and so, at constant pressure, a graph of G against T will be a straight line of slope $R \ln P$." The candidate is awarded 2 marks out of 10. Is the student's answer plausible? Why? Write an answer that would be awarded 10 marks.

8. Explain the concept of 'coupled systems', and illustrate how an otherwise non-spontaneous reaction can be made to happen.

9. In physical terms, what do the equations

$$dU = đq - đw + d\sum_i \mu_i n_i \tag{13.65}$$

and

$$dH = T\,dS + V\,dP + d\sum_i \mu_i n_i \tag{13.67}$$

mean?

10. "All metastable states are, at least to some extent, 'trapped'." Discuss, with examples.

14 Chemical equilibrium and chemical kinetics

Summary

Consider a chemical reaction in which gaseous reagents A and B react to give gaseous products C and D, according to the chemical equation

$$aA + bB \rightarrow cC + dD$$

At any time whilst the reaction is taking place, the system is a mixture of the four pure components A, B, C and D, and, as the reactants A and B change into the products C and D, the composition of the mixture changes. At any instant, the Gibbs free energy G of the mixture is a function $G(P, T, n_A, n_B, ...)$ of the pressure P, the temperature T, and also the instantaneous mole numbers n_i of each of the components. Since the mole numbers n_i change as the reaction progresses, the Gibbs free energy G of the mixture will also change, until it reaches a minimum, at which point the reaction mixture is in equilibrium, and the corresponding mole numbers are those that define the equilibrium mixture. If the mole numbers of the reagents in the equilibrium mixture are very small, the reaction essentially goes to completion; if the mole numbers of the products in the equilibrium mixture are very small, the reaction doesn't take place spontaneously in the direction from reagents to products as written, but the reverse reaction is spontaneous; if the equilibrium mixture contains measureable quantities of A, B, C and D, we say that the reaction is reversible.

For any gas phase reaction, at any temperature T, the equilibrium mixture is defined by

$$\Delta_r G_p^{\ominus} = -RT \ln K_p \tag{14.13a}$$

in which

$$\Delta_r G_p^{\ominus} = (c\Delta_f G_{p,C}^{\ominus} + d\Delta_f G_{p,D}^{\ominus}) - (a\Delta_f G_{p,A}^{\ominus} + b\Delta_f G_{p,B}^{\ominus}) \tag{14.9}$$

The **pressure thermodynamic equilibrium constant** K_p is given by

$$K_p = \frac{(p_{C,eq}/P_C^{\ominus})^c \, (p_{D,eq}/P_D^{\ominus})^d}{(p_{A,eq}/P_A^{\ominus})^a \, (p_{B,eq}/P_B^{\ominus})^b} = \frac{p_{C,eq}^c \, p_{D,eq}^d}{p_{A,eq}^a \, p_{B,eq}^b} \tag{14.12}$$

where $p_{i,eq}$ is the partial pressure of component i in the equilibrium mixture.

Alternatively, we can define the equilibrium mixture in terms of molar concentrations [I], defining the (molar) **concentration thermodynamic equilibrium constant**, K_c, as

$$K_c = \frac{([C]_{eq}/[C]^{\ominus})^c \, ([D]_{eq}/[D]^{\ominus})^d}{([A]_{eq}/[A]^{\ominus})^a \, ([B]_{eq}/[B]^{\ominus})^b} = \frac{[C]_{eq}^c \, [D]_{eq}^d}{[A]_{eq}^a \, [B]_{eq}^b} \tag{14.21}$$

such that

$$K_p = K_c RT^{\Delta n} \tag{14.22a}$$

Modern Thermodynamics for Chemists and Biochemists. Dennis Sherwood and Paul Dalby.
© Oxford University Press 2018. Published 2018 by Oxford University Press.

The equilibrium constant K_p varies with temperature according to the **van't Hoff equation**

$$\frac{d \ln K_p}{dT} = \frac{\Delta_r H^\circ}{RT^2} \qquad (14.31a)$$

Equation (14.31a) is a quantitative description of how K_p varies with temperature, and is a mathematical form of a particular instance of the qualitative statement known as **Le Chatelier's principle**

If an external constraint is applied to a system, the system adjusts itself so as to oppose that constraint.

So, with reference to equation (14.31a), if the temperature increases ($dT > 0$), and if the reaction is endothermic ($\Delta_r H^\circ > 0$), then since R and T are necessarily positive, then $d \ln K_p$ must be positive. An increase in temperature therefore causes an increase in $\ln K_p$ - and therefore in K_p - so driving the reaction to the right. Since the reaction is, by assumption, endothermic, this absorbs heat. An increase in the system's temperature therefore causes the system to respond by absorbing heat - which is precisely what Le Chatelier's principle, in general, states.

Chemical kinetics is the study of how quickly, or slowly, a chemical reaction takes place, as measured by the **rate of reaction**. When a chemical system is macroscopically in equilibrium, microscopically, the rate $v_f(\tau)$ of the forward reaction equals the rate $v_r(\tau)$ of the reverse reaction, where (τ) highlights that these rates vary with time τ. For a reversible reaction of the form

$$aA + bB \rightleftharpoons cC + dD$$

$v_f(\tau)$ and $v_r(\tau)$ can be expressed in terms of the instantaneous molar concentrations $[I(\tau)]$, or partial pressures $p_i(\tau)$, of the components as

$$v_f(\tau) = k_f[A(\tau)]^a [B(\tau)]^b = k_f'[p_A(\tau)]^a [p_B(\tau)]^b \qquad (14.34a)$$

and

$$v_r(\tau) = k_r[C(\tau)]^c [D(\tau)]^d = k_r'[p_C(\tau)]^c [p_D(\tau)]^d \qquad (14.34b)$$

in which k_f, k_f', k_r and k_r' are known as **reaction rate constants**.

Thermodynamics meets kinetics when, at equilibrium, $v_f(\tau) = v_r(\tau)$ and

$$\frac{k_f}{k_r} = \frac{[C(\tau_{eq})]^c [D(\tau_{eq})]^d}{[A(\tau_{eq})]^a [B(\tau_{eq})]^b} = \frac{[C]_{eq}^c [D]_{eq}^d}{[A]_{eq}^a [B]_{eq}^b} = K_c \qquad (14.37d)$$

and

$$\frac{k_f'}{k_r'} = \frac{[p_C(\tau_{eq})]^c [p_D(\tau_{eq})]^d}{[p_A(\tau_{eq})]^a [p_B(\tau_{eq})]^b} = \frac{p_{C,eq}^c \, p_{D,eq}^d}{p_{A,eq}^a \, p_{B,eq}^b} = K_p \qquad (14.37e)$$

The rate constants k_f, k_f', k_r and k_r' depend on temperature, in accordance with the **Arrhenius equation**

$$k(T) = A e^{-\frac{E_a}{RT}} \qquad (14.38)$$

in which $k(T)$ is a generalised temperature-dependent reaction rate constant; E_a is the reaction **activation energy**; and A is the **frequency factor** or **pre-exponential factor**.

Equation (14.21) explains why some reactions for which $\Delta G < 0$, and which should therefore be spontaneous, don't actually happen at a given temperature T. If, at that temperature, the activation energy E_a is large, then the term $\exp(-E_a/RT)$ is small, and the corresponding reaction rate $k(T)$ is also

CHEMICAL REACTIONS

small. The reaction will therefore proceed only very slowly, if at all, even though, thermodynamically, the reaction is spontaneous.

If the temperature T is increased, the reaction rate $k(T)$ will also increase, possibly very substantially, and so the thermodynamic equilibrium composition will be reached more quickly; that equilibrium composition, however, is also temperature-dependent, according to the van't Hoff equation, equation (14.31a).

14.1 Chemical reactions

So far in this book, we have been dealing with changes in state of the form

state [1] → state [2]

and the associated changes in a variety of state functions, as represented by, for example, ΔP, ΔV, ΔT, ΔU, ΔH, ΔS and ΔG. One of the (usually unstated) assumptions has been that state [1] changes to state [2] completely, and that an initial state [1], which existed 'earlier', is replaced, some time 'later', in its entirety, by state [2], which then remains stable over time. So, previous examples have included an ideal gas in half-a-cylinder (state [1]) which expands into a vacuum to fill the cylinder (state [2]), or two blocks at different temperatures (state [1]) that mutually change their temperatures so that the temperature becomes uniform (state [2]). Our analysis of spontaneity took this further by identifying that spontaneous changes are those in which state [1] is an unstable non-equilibrium state, which changes unidirectionally and irreversibly into state [2], a stable equilibrium state.

In chemistry, a reaction written as

$AgNO_3$ (aq) + NaI (aq) → AgI (s) + $NaNO_3$ (aq)

suggests that there is an initial state [1], comprising 1 mol of $AgNO_3$ (aq), 1 mol of NaI (aq), 0 mol of AgI (s), and 0 mol of $NaNO_3$ (aq), such that the initial mixture of $AgNO_3$ (aq) and NaI (aq) reacts to form a final state [2], comprising 0 mol of $AgNO_3$ (aq), 0 mol of NaI (aq), 1 mol of AgI (s), and 1 mol of $NaNO_3$ (aq). In this reaction, state [1], the mixture of only $AgNO_3$ (aq) and NaI (aq), is the initial, unstable, non-equilibrium state, and state [2], the mixture of only AgI (s) and $NaNO_3$ (aq), is the final, stable, equilibrium state. This chemical reaction is not only spontaneous, unidirectional and irreversible, but it also 'goes to completion' as written, in that all the reagents on the left-hand side react in their totality to form the products shown on the right-hand side, and the reverse reaction

AgI (s) + $NaNO_3$ (aq) → $AgNO_3$ (aq) + NaI (aq)

just doesn't happen.

Many chemical reactions, however, behave differently, in that an initial state [1] comprising a mixture of a mol of A and b mol of B do indeed react spontaneously, but the final equilibrium state is not a mixture of C and D only, but rather a mixture of all four compounds A, B, C and D, where the relative proportions of A, B, C and D in the equilibrium mixture (at any given temperature and pressure) are well-defined. Furthermore, at the same temperature and pressure, an initial mixture of c mol of C and d mol of D will react spontaneously to yield the *same* equilibrium mixture of A, B, C and D as resulted from the initial state of a mol of A and b mol of B.

This implies that the 'forward' reaction

$$aA + bB \rightarrow cC + dD$$

and the 'reverse' reaction

$$cC + dD \rightarrow aA + bB$$

are taking place simultaneously, and that the reaction-pair is best represented as the reversible reaction

$$aA + bB \rightleftharpoons cC + dD$$

At any time, the system comprises a mixture of the pure components A, B, C and D, and as the reaction progresses, the quantities of these components mutually change, as determined by the stoichiometry of the reaction. So, at time $\tau = 0$, the system comprises a mol of A, b mol of B, 0 mol of C and 0 mol of D, which we may represent as $[a, b, 0, 0]$; if the reaction goes to completion, at time $\tau = T$, the final state of the system is $[0, 0, c, d]$; and if, at some intermediate time τ, the system instantaneously contains $n_i(\tau)$ moles of component i, then the instantaneous state of the system at that time may be represented as $[n_A(\tau), n_B(\tau), n_C(\tau), n_D(\tau)]$.

If A, B, C and D are all ideal gases, then, at any time τ, each component i exerts a corresponding partial pressure $p_i(\tau)$, where this partial pressure changes over time, as the quantity of component i in the reaction mixture changes as the reaction progresses. As we saw on page 400, at any given temperature T, we can use equation (13.47d)

$$\mathbf{G} = \mathbf{G}^{\ominus} + RT \ln \frac{P}{P^{\ominus}} \tag{13.47d}$$

to express the instantaneous value $G_i(\tau)$ of the Gibbs free energy of component i at time τ in terms of the corresponding mole number $n_i(\tau)$, partial pressure $p_i(\tau)$, and standard molar Gibbs free energy $\mathbf{G}_i^{\ominus}$ as

$$G_i(\tau) = n_i(\tau)\,\mathbf{G}_i(\tau) = n_i(\tau)\,\mathbf{G}_i^{\ominus} + n_i(\tau)\,RT \ln \frac{p_i(\tau)}{P^{\ominus}}$$

The standard pressure $P_i^{\ominus} = 1$ bar for all components i, and so we may write

$$G_i(\tau) = n_i(\tau)\,\mathbf{G}_i(\tau) = n_i(\tau)\,\mathbf{G}_i^{\ominus} + n_i(\tau)\,RT \ln p_i(\tau)$$

and it is this less cluttered 'shorthand' form that we will use in this section.

Since the system, and the mixture, is ideal, the total Gibbs free energy $G_{\text{sys}}(\tau)$ of the system at any time τ is the linear sum of the instantaneous Gibbs free energies of each of the components

$$G_{\text{sys}}(\tau) = G_A(\tau) + G_B(\tau) + G_C(\tau) + G_D(\tau)$$
$$= n_A(\tau)\,\mathbf{G}_A(\tau) + n_B(\tau)\,\mathbf{G}_B(\tau) + n_C(\tau)\,\mathbf{G}_C(\tau) + n_D(\tau)\,\mathbf{G}_D(\tau)$$

and so

$$\begin{aligned}G_{\text{sys}}(\tau) = &\; n_A(\tau)\,\mathbf{G}_A^{\ominus} + n_A(\tau)\,RT \ln p_A(\tau) \\ &+ n_B(\tau)\,\mathbf{G}_B^{\ominus} + n_B(\tau)\,RT \ln p_B(\tau) \\ &+ n_C(\tau)\,\mathbf{G}_C^{\ominus} + n_C(\tau)\,RT \ln p_C(\tau) \\ &+ n_D(\tau)\,\mathbf{G}_D^{\ominus} + n_D(\tau)\,RT \ln p_D(\tau)\end{aligned}$$

This equation is rather messy, but it is, in principle, very simple: the total Gibbs free energy G_{sys} of the reaction mixture is expressed as $G_{sys}(\tau)$ so as to recognise that this changes over time as the reaction progresses; $G_{sys}(\tau)$ is expressed as a linear sum of the instantaneous Gibbs free energies $G_i(\tau)$ of each component i; and each $G_i(\tau)$ depends on the molar Gibbs free energy $G_i^{\ominus}$ of component i, the quantity of component i in the mixture at time τ, as determined by the mole number $n_i(\tau)$, and the corresponding partial pressure $p_i(\tau)$.

Furthermore, as we saw on page 397, for any component i, $G_i^{\ominus}$ – 'the' molar Gibbs free energy of that component in its standard state – is in fact equal to that component's molar Gibbs free energy of formation $\Delta_f G_i^{\ominus}$, enabling us to define $G_{sys}(\tau)$ as

$$G_{sys}(\tau) = n_A(\tau)\,\Delta_f G_A^{\ominus} + n_A(\tau)\,RT \ln p_A(\tau)$$
$$+ n_B(\tau)\,\Delta_f G_B^{\ominus} + n_B(\tau)\,RT \ln p_B(\tau)$$
$$+ n_C(\tau)\,\Delta_f G_C^{\ominus} + n_C(\tau)\,RT \ln p_C(\tau)$$
$$+ n_D(\tau)\,\Delta_f G_D^{\ominus} + n_D(\tau)\,RT \ln p_D(\tau) \tag{14.1}$$

always bearing in mind that each term $p_i(\tau)$ is in fact 'shorthand' for the ratio $p_i(\tau)/P_i^{\ominus}$.

14.2 How $G_{sys}(\tau)$ varies over time

Equation (14.1) is the fundamental equation that describes how all chemical reactions take place. In its current form, however, is it rather uninformative, so the purpose of this section is to examine how equation (14.1) behaves. Our analysis here will apply to a gas phase reaction, in which all the components A, B, C and D are assumed to be ideal gases, undergoing a reaction at a constant temperature T and a constant pressure of 1 bar, as is consistent with normal laboratory conditions. The analysis will be extended to reactions taking place in solution in Chapter 16, and non-ideal conditions will be addressed in Chapter 22.

The analysis that follows is mathematical, and, though not difficult, is quite complicated. We trust our approach is intelligible; those less confident in mathematics might wish to skim-read the equations – but do pay attention to the diagrams, and the accompanying descriptions.

The central feature of the analysis is to determine the state of the system at some time τ, by which time a fraction $\xi(\tau)$ of the original a moles of A have reacted. At time $\tau = 0$, the reaction has not yet started, and so no A has reacted, implying that $\xi(\tau) = 0$; if the reaction has gone to completion by time $\tau = T$, then all the originally present A has reacted, and $\xi(\tau) = 1$; at some intermediate time τ, the number of moles of A that have reacted, and have therefore 'disappeared', is given by $a\,\xi(\tau)$, and the remaining number $n_A(\tau)$ of as-yet unreacted moles of A is therefore $a\,(1 - \xi(\tau))$.

Given the stoichiometry of the gas phase reaction

$aA + bB \rightleftharpoons cC + dD$

then a moment's thought will verify that

$$n_A(\tau) = a\,(1 - \xi) \tag{14.2a}$$
$$n_B(\tau) = b\,(1 - \xi) \tag{14.2b}$$
$$n_C(\tau) = c\,\xi \tag{14.2c}$$
$$n_D(\tau) = d\,\xi \tag{14.2d}$$

in which, to reduce the clutter, ξ stands for $\xi(\tau)$.

We now derive some expressions for each of the instantaneous partial pressures $p_i(\tau)$. To do this, we remember that each component A, B, C and D is ideal, implying that each component obeys Dalton's law of partial pressures (see page 48), such that, for each component i

$$p_i(\tau) = n_i(\tau) \frac{RT}{V(\tau)} \tag{3.3}$$

where $V(\tau)$ is the instantaneous volume of the total system (which will, in general, change over time since the reaction takes place at a constant pressure of 1 bar).

Adding the partial pressures $p_i(\tau)$ for all the components

$$\sum_i p_i(\tau) = \frac{RT}{V(\tau)} \sum_i n_i(\tau) = P \tag{14.3}$$

where P is the total pressure of the system, which is maintained constant throughout the reaction.

A consequence of equation (14.3) is that

$$\frac{RT}{V(\tau)} = \frac{P}{\sum_i n_i(\tau)}$$

and so, substituting for $RT/V(\tau)$ in equation (3.3), we derive an expression for the instantaneous partial pressures $p_i(\tau)$ in terms of the instantaneous mole numbers $n_i(\tau)$

$$p_i(\tau) = \frac{n_i(\tau)}{\sum_i n_i(\tau)} P \tag{14.4}$$

The ratio $n_i(\tau)/\sum_i n_i(\tau)$ represents the number of moles of component i within the mixture relative to the total number of moles of all components. This ratio is known as the **mole fraction**, $x_i(\tau)$ of component i, which we met in Chapter 6

$$x_i(\tau) = \frac{n_i(\tau)}{\sum_i n_i(\tau)} \tag{6.18}$$

Reference to equations (14.2) will verify that

$$\sum_i n_i(\tau) = (a+b) + \xi[(c+d) - (a+b)] = (a+b) + \xi \Delta n \tag{14.5}$$

where, as we saw on page 160

$$\Delta n = [(c+d) - (a+b)] \tag{6.15}$$

Using equations (14.2), (14.4) and (14.5), we may therefore express the partial pressures $p_i(\tau)$ as

$$p_A(\tau) = \frac{n_A(\tau)}{\sum_i n_i(\tau)} P = \frac{a(1-\xi)}{(a+b) + \xi \Delta n} P \tag{14.6a}$$

$$p_B(\tau) = \frac{n_B(\tau)}{\sum_i n_i(\tau)} P = \frac{b(1-\xi)}{(a+b) + \xi \Delta n} P \tag{14.6b}$$

$$p_C(\tau) = \frac{n_C(\tau)}{\sum_i n_i(\tau)} P = \frac{c\xi}{(a+b) + \xi \Delta n} P \tag{14.6c}$$

$$p_D(\tau) = \frac{n_D(\tau)}{\sum_i n_i(\tau)} P = \frac{d\xi}{(a+b) + \xi \Delta n} P \tag{14.6d}$$

In equations (14.2) and (14.6), a, b, c, d and Δn are all constants, determined by the stoichiometry of the chemical reaction

$$aA + bB \rightleftharpoons cC + dD$$

under study, and P is the constant pressure (often 1 bar) at which reaction takes place. The only variable is therefore ξ – the fraction of molecules of A that have reacted over the time from time 0 to time τ. Equations (14.2) and (14.6) can now be used to substitute for the corresponding variables $n_i(\tau)$ and $p_i(\tau)$ in equation (14.1) for the total Gibbs free energy G_{sys} of the reaction mixture. When we do this, the result will contain only a single variable ξ – all the other parameters (the stoichiometric mole numbers a, b, c and d, and the various standard Gibbs free energies G_i°) are, for any given reaction, constants. G_{sys} can therefore be written as $G_{sys}(\xi)$, emphasising that the value of G_{sys} (at any given temperature T, and at a pressure of 1 bar) depends on the variable ξ, which is a measure of how far the reaction has progressed, this being the **extent-of-reaction**, from $\xi = 0$ (the reaction hasn't yet started) to $\xi = 1$ (the reaction has gone to completion to the right as written).

When we actually make the substitutions of equations (14.2) and (14.6) into equation (14.1), the result looks hugely complicated – but, in fact, it isn't: it is the sum of a number of terms, corresponding to each of the components taking part in the reaction, and the only variable is ξ – all the other parameters are constants. Take a deep breath:

$$\begin{aligned} G_{sys}(\xi) = {} & a(1-\xi)\Delta_f G_A^{\circ} + a(1-\xi)RT \ln \frac{a(1-\xi)P}{(a+b)+\xi\Delta n} \\ & + b(1-\xi)\Delta_f G_B^{\circ} + b(1-\xi)RT \ln \frac{b(1-\xi)P}{(a+b)+\xi\Delta n} \\ & + c\xi \Delta_f G_C^{\circ} + c\xi RT \ln \frac{c\xi P}{(a+b)+\xi\Delta n} \\ & + d\xi \Delta_f G_D^{\circ} + d\xi RT \ln \frac{d\xi P}{(a+b)+\xi\Delta n} \end{aligned} \tag{14.7}$$

It is quite impossible to look at equation (14.7) and gain an intuitive understanding of how it behaves. But it is possible to capture equation (14.7) in a spreadsheet, and explore how, for any given reaction, $G_{sys}(\xi)$ changes as ξ changes from 0 to 1. When this is done, the results are quite surprising: although the details are different depending on which reaction is chosen, the generalised behaviour of equation (14.7) is such that there are only three types of result: two of which are shown in Figure 14.1.

In Figure 14.1(a), the system Gibbs free energy $G_{sys}(\xi)$ continually decreases as the reaction progresses from the initial state $[a, b, 0, 0]$ of pure A and B to a final state $[0, 0, c, d]$ of pure C and D. Because the curve is downwards sloping along its entire length, the value of $G_{sys}(\xi)$ for

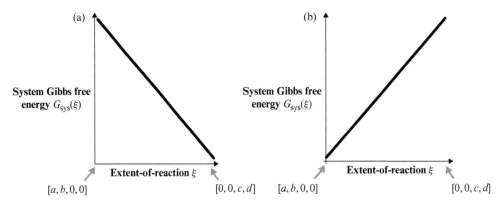

Figure 14.1 Reactions that go to completion. These diagrams show two (of three) possible behaviours of equation (14.7) for the total Gibbs free energy $G_{sys}(\xi)$ of the chemical system corresponding to the reaction $aA + bB \to cC + dD$. Figure (a) corresponds to the case in which the reaction $aA + bB \to cC + dD$ spontaneously goes to completion in the direction as written; figure(b) is the case in which the reverse reaction $cC + dD \to aA + bB$ goes to completion. See also Figure 14.2 for the third behaviour of $G_{sys}(\xi)$.

any value of ξ will always be slightly greater than the value of $G_{sys}(\xi + \delta\xi)$, corresponding to the reaction which has progressed by an incremental amount $\delta\xi$. The change from state $[\xi]$ to state $[\xi + \delta\xi]$ therefore corresponds to a value of $\Delta G = G_{sys}(\xi + \delta\xi) - G_{sys}(\xi)$ which is always negative. This incremental change is therefore spontaneous for all values of ξ from 0 to 1, implying that the overall reaction proceeds spontaneously to completion in the direction as written, as happens, for example, for the reaction

$$CH_4 \text{ (g)} + 2O_2 \text{ (g)} \to CO_2 \text{ (g)} + 2H_2O \text{ (g)}$$

The graph shown in Figure 14.1(a) is essentially linear; this is a particular case of a more general, downwards sloping, but gently curved, behaviour.

Figure 14.1(b) illustrates the opposite case: a reaction in which $G_{sys}(\xi)$ increases continuously as the reaction progresses. Any change from a state $[\xi]$ to a neighbouring state $[\xi + \delta\xi]$ therefore corresponds to a value of $\Delta G = G_{sys}(\xi + \delta\xi) - G_{sys}(\xi)$ which is always positive, and so will not happen spontaneously as written – but the reverse reaction is spontaneous. This corresponds to a reaction such as

$$CO_2 \text{ (g)} + 2H_2O \text{ (g)} \to CH_4 \text{ (g)} + 2O_2 \text{ (g)}$$

As with Figure 14.1(a), the linear form of the graph shown in Figure 14.1(b) is a particular case of a more general, upwards sloping, gently curved, behaviour.

So, to develop matters further, consider a generalised gas phase reaction

$$aA + bB \rightleftharpoons cC + dD$$

for which the total Gibbs free energy of the reactants is $a\, \Delta_f G_A^\circ + b\, \Delta_f G_B^\circ$, and of the products, $c\, \Delta_f G_C^\circ + d\, \Delta_f G_D^\circ$. If

$$c\, \Delta_f G_C^\circ + d\, \Delta_f G_D^\circ < a\, \Delta_f G_A^\circ + b\, \Delta_f G_B^\circ$$

then is the reaction spontaneous, to completion to the right as written? And if

$$c\,\Delta_f\mathbf{G_C}^\circ + d\,\Delta_f\mathbf{G_D}^\circ > a\,\Delta_f\mathbf{G_A}^\circ + b\,\Delta_f\mathbf{G_B}^\circ$$

is the reaction to the left spontaneous to completion?

In fact, the answers to both of these questions is "no", for two different reasons. Firstly, the total Gibbs free energy of the initial state $[a, b, 0, 0]$ is not $a\,\Delta_f\mathbf{G_A}^\circ + b\,\Delta_f\mathbf{G_B}^\circ$; likewise, the total Gibbs free energy of the final state $[0, 0, c, d]$ is not $c\,\Delta_f\mathbf{G_D}^\circ + d\,\Delta_f\mathbf{G_D}^\circ$. Both of the initial and final states are mixtures, and so we need to use equation (14.7) to determine $G_{sys}(\xi)$ for the two states $\xi = 0$, corresponding to the initial state $[a, b, 0, 0]$, and $\xi = 1$, corresponding to final state $[0, 0, c, d]$. Accordingly,

$$G_{sys}(\xi = 0) \equiv G_{sys}[a, b, 0, 0]$$

$$= a\,\Delta_f\mathbf{G_A}^\circ + a\,RT\ln\frac{aP}{a+b} + b\,\Delta_f\mathbf{G_B}^\circ + b\,RT\ln\frac{bP}{a+b}$$

and

$$G_{sys}(\xi = 1) \equiv G_{sys}[0, 0, c, d]$$

$$= c\,\Delta_f\mathbf{G_D}^\circ + c\,RT\ln\frac{cP}{c+d} + d\,\Delta_f\mathbf{G_D}^\circ + d\,RT\ln\frac{dP}{c+d}$$

In the equations for $G_{sys}(\xi = 0)$ and $G_{sys}(\xi = 1)$, the ratios of the form $a/(a+b)$ in each of the ln terms are all less than 1, and so all the ln terms are negative. As a result, $G_{sys}(0)$ is less than the sum $a\,\Delta_f\mathbf{G_A}^\circ + b\,\Delta_f\mathbf{G_B}^\circ$, and $G_{sys}(1)$ is less than the sum $c\,\Delta_f\mathbf{G_C}^\circ + d\,\Delta_f\mathbf{G_D}^\circ$. This is a specific instance of the entropy effect of mixing: the total entropy of a mixture is always greater than the sum of the entropies of the pure components, for the mixture is intrinsically less ordered: since, by definition, $G = H - TS$, an increase in S drives a decrease in G. The comparison of the initial and final states for the gaseous mixture must therefore be made using the more complex equations for $G_{sys}(\xi = 0)$ and $G_{sys}(\xi = 1)$, rather than the 'unmixed' expressions $a\,\Delta_f\mathbf{G_A}^\circ + b\,\Delta_f\mathbf{G_B}^\circ$ and $c\,\Delta_f\mathbf{G_C}^\circ + d\,\Delta_f\mathbf{G_D}^\circ$.

But even if $G_{sys}(1) \equiv G_{sys}[0, 0, c, d] < G_{sys}(0) \equiv G_{sys}[a, b, 0, 0]$, this does not imply that the reaction proceeds to the right, to completion, even though the corresponding value of $\Delta G = G_{sys}(1) - G_{sys}(0)$ is negative. To predict the behaviour of a reaction by comparing $G_{sys}(1)$ to $G_{sys}(0)$ is misleading, since it makes the implicit assumption that $G_{sys}(0) \equiv G_{sys}[a, b, 0, 0]$ and $G_{sys}(1) \equiv G_{sys}[0, 0, c, d]$ are the only states available to the system. The initial state of the system is certainly $G_{sys}(0) \equiv G_{sys}[a, b, 0, 0]$, but before the system can reach $G_{sys}(1) \equiv G_{sys}[0, 0, c, d]$, it must pass through all other states $G_{sys}(\xi)$. If there is an intermediate state which has a value of G lower that both $G_{sys}(0)$ and $G_{sys}(1)$, then the system will equilibrate at that state, as shown in Figure 14.2.

Figure 14.2 illustrates a chemical system in which the total Gibbs free energy $G_{sys}(0)$ of the state $[a, b, 0, 0]$, is greater than $G_{sys}(1)$ of the state $[0, 0, c, d]$, rather like the endpoints of the case shown in Figure 14.1(a). In Figure 14.2, however, there is an intermediate point at which $G_{sys}(\xi)$ is at a minimum, lower than both $G_{sys}(0) \equiv G_{sys}[a, b, 0, 0]$ and $G_{sys}(1) \equiv G_{sys}[0, 0, c, d]$. This represents a point of **chemical equilibrium**, at which all components A, B, C and D are simultaneously present in the reaction mixture, at a

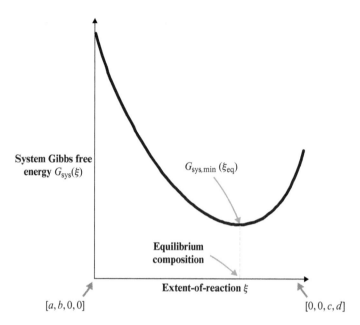

Figure 14.2 Chemical equilibrium. This diagrams shows the third possible behaviour of equation (14.7) for the total Gibbs free energy $G_{sys}(\xi)$ of the chemical system corresponding to the reaction $aA + bB \rightarrow cC + dD$. In this case, the reaction does not go to completion in either direction, but achieves a stable equilibrium state in which all four components A, B, C and D are simultaneously present at a specific composition, corresponding to the equilibrium extent-of-reaction ξ_{eq}.

specific composition, corresponding to the **equilibrium extent-of-reaction** ξ_{eq}, and so the reaction is indeed reversible

$$aA + bB \rightleftharpoons cC + dD$$

The presence of a stable chemical equilibrium can be inferred, intuitively, from Figure 14.2. At time $\tau = 0$, the reaction has not yet started; the system is in the state $[a, b, 0, 0]$, corresponding to the left-most point on the graph shown in Figure 14.2, $\xi = 0$, and in this state $G_{sys}(\xi) = G_{sys}(0)$. A neighbouring state, just to the right of the initial state, corresponds to a positive, small, value of ξ, and since the curve slopes downwards, the Gibbs free energy $G_{sys}(\xi)$ of the system in this state is less than the original Gibbs free energy $G_{sys}(0)$. For the change from the initial state to the current state, $\Delta G = G_{sys}(\xi) - G_{sys}(0) < 0$, implying that the change from the initial state to the current state is spontaneous. As the reaction proceeds, the system moves from left to right in Figure 14.2, and, as long as the slope of the curve at the current state is negative, the incremental reaction, from left to right as written, will be spontaneous. The reaction will therefore continue to take place spontaneously from left to right as written until the system reaches the state corresponding to the minimum of the $G_{sys}(\xi)$ curve, at which $\xi = \xi_{eq}$. Since any further reaction, to a point further to the right, will take the system from a minimum value of $G_{sys}(\xi)$ to a higher value, the corresponding change will not be spontaneous – the system will stabilise at the point at which $G_{sys}(\xi)$ is a minimum, corresponding to the equilibrium point $\xi = \xi_{eq}$.

Exactly the same analysis can be carried out for the reverse reaction, in which the starting state is $[0, 0, c, d]$, corresponding to $\xi = 1$. The reverse reaction corresponds to moving from

CHEMICAL EQUILIBRIUM

right to left in Figure 14.2, such that the reverse reaction will take place spontaneously 'going down the slope', until the equilibrium composition is reached. And once the system has reached its equilibrium composition, any small change, in either reaction direction, increases $G_{sys}(\xi)$, and so the system will spontaneously revert to the equilibrium state.

Given that $\xi = \xi_{eq}$ has a specific, single, value, we may use equations (14.2) to determine the corresponding mole numbers

$$n_A(eq) = a(1 - \xi_{eq}) \tag{14.8a}$$
$$n_B(eq) = b(1 - \xi_{eq}) \tag{14.8b}$$
$$n_C(eq) = c\,\xi_{eq} \tag{14.8c}$$
$$n_D(eq) = d\,\xi_{eq} \tag{14.8d}$$

which collectively define the equilibrium composition of the reaction mixture.

To make that concrete, suppose that the reaction under study is of the form

$$A + 3B \rightleftharpoons 2C + 4D$$

so that $a = 1$, $b = 3$, $c = 2$ and $d = 4$, and suppose further that $\xi_{eq} = 0.6$. If the initial state of the reaction is $[a, b, 0, 0] \equiv [1, 3, 0, 0]$, then equations (14.8) imply that the reaction will reach equilibrium with a mixture of A, B, C and D such that the mole numbers of each component at equilibrium are

$$n_A(eq) = a(1 - \xi_{eq}) = 1(1 - 0.6) = 0.4$$
$$n_B(eq) = b(1 - \xi_{eq}) = 3(1 - 0.6) = 1.2$$
$$n_C(eq) = c\,\xi_{eq} = 2 \times 0.6 = 1.2$$
$$n_D(eq) = d\,\xi_{eq} = 4 \times 0.6 = 2.4$$

14.3 Chemical equilibrium

It is remarkable that equation (14.7) exhibits only three generic types of behaviour, as shown in Figures 14.1(a), 14.1(b) and 14.2; even more remarkable is that the fact that this result holds for *all* chemical reactions, no matter how many reactants and products there are, and no matter what the stoichiometry might be. So, given any chemical reaction, how can we determine which behaviour the reaction shows, at any given temperature (the pressure is assumed to be 1 bar under all circumstances)? We have already seen (see page 421) that the 'obvious' method of testing the sign of the quantity $(c\,\Delta_f G_C^{\ominus} + d\,\Delta_f G_D^{\ominus}) - (a\,\Delta_f G_A^{\ominus} + b\,\Delta_f G_B^{\ominus})$ will certainly indicate the generalised direction of the reaction (if this quantity is negative, the reaction will in general proceed to the right as written; if negative, to the left), but this test cannot identify the possibility of an intermediate equilibrium state. So how can the possibility of an equilibrium state be determined?

The answer to this question is embedded in Figure 14.2: if the reaction does have an intermediate state, this will correspond to a minimum in the graph of $G_{sys}(\xi)$, implying that if we can find the minimum value of $G_{sys}(\xi)$, then the value of $\xi = \xi_{eq}$ at this minimum defines the equilibrium composition.

There are two ways of discovering the equilibrium state of any chemical reaction. The first is to use a spreadsheet to compute the values of $G_{sys}(\xi)$, using an equation of the form of equation (14.7) as the specification of the spreadsheet cell contents, as appropriate to the reaction under study. If the columns in the spreadsheet represent each of the terms in equation (14.7), and the rows represent successive values of ξ from 0 to 1, using increments of, say, 0.01, then the column sum for any given row gives the corresponding value of $G_{sys}(\xi)$. The set of values of $G_{sys}(\xi)$ can then be studied, or, better, plotted as a graph, so establishing the behaviour of $G_{sys}(\xi)$, and the existence of an equilibrium state if one exists. This is how the authors created the graphs in Figures 14.1(a), 14.1(b) and 14.2, using the data shown in Table 14.1.

Table 14.1 Parameters used for the computation of $G_{sys}(\xi)$. The table shows the parameters used in the spreadsheet computations of $G_{sys}(\xi)$, using equation (14.7), at total pressure P = 1 bar = 10^5 Pa, resulting in the graphs shown in Figures 14.1(a), 14.1(b) and 14.2.

Reaction $aA + bB \rightleftharpoons cC + dD$	T K	a	b	c	d	$\Delta_f G_A^\circ$ kJ mol^{-1}	$\Delta_f G_B^\circ$ kJ mol^{-1}	$\Delta_f G_C^\circ$ kJ mol^{-1}	$\Delta_f G_D^\circ$ kJ mol^{-1}
$CH_4 + 2O_2 \rightleftharpoons CO_2 + 2H_2O$	298	1	2	1	2	−50.5	0	−394.4	−228.6
$CO_2 + 2H_2O \rightleftharpoons CH_4 + 2O_2$	298	1	2	1	2	−394.4	−228.6	−50.5	0
$N_2 + 3H_2 \rightleftharpoons 2NH_3$	1000	1	3	2	0	0	0	−16.4	–

The second way to determine the equilibrium state is by using mathematics: if equation (14.7) if differentiated with respect to ξ, then the value of $dG_{sys}(\xi)/d\xi$ at any value of ξ is equal to the slope of the tangent to the $G_{sys}(\xi)$ curve at that point. The equilibrium point, corresponding to $\xi = \xi_{eq}$, is the value of ξ at which the slope of the tangent is zero, and so $dG_{sys}(\xi)/d\xi = 0$.

In principle, differentiating equation (14.7) with respect to ξ is straight-forward; in practice, given the number of terms in equation (14.7), it is rather clumsy, and somewhat lengthy. We present the full analysis here: as with the derivation of equation (14.7) itself, the analysis may be skim-read – the key result is the surprisingly simple equation (14.10).

Equation (14.7) can be expressed as

$$G_{sys}(\xi) = G_A(\xi) + G_B(\xi) + G_C(\xi) + G_D(\xi)$$

in which

$$G_A(\xi) = a(1-\xi)\Delta_f G_A^\circ + a(1-\xi)RT \ln \frac{a(1-\xi)P}{(a+b)+\xi \Delta n}$$

$$= a(1-\xi)\Delta_f G_A^\circ + a(1-\xi)RT \ln [a(1-\xi)P] - a(1-\xi)RT \ln [(a+b)+\xi \Delta n]$$

$$G_B(\xi) = b(1-\xi)\Delta_f G_B^\circ + b(1-\xi)RT \ln \frac{b(1-\xi)P}{(a+b)+\xi \Delta n}$$

$$= b(1-\xi)\Delta_f G_B^\circ + b(1-\xi)RT \ln [b(1-\xi)P] - b(1-\xi)RT \ln [(a+b)+\xi \Delta n]$$

CHEMICAL EQUILIBRIUM

$$G_C(\xi) = c\xi \Delta_f \mathbf{G_C}^{\ominus} + c\xi RT \ln \frac{c\xi P}{(a+b)+\xi \Delta n}$$

$$= c\xi \Delta_f \mathbf{G_C}^{\ominus} + c\xi RT \ln [c\xi P] - c\xi RT \ln [(a+b)+\xi \Delta n]$$

$$G_D(\xi) = d\xi \Delta_f \mathbf{G_D}^{\ominus} + d\xi RT \ln \frac{d\xi P}{(a+b)+\xi \Delta n}$$

$$= d\xi \Delta_f \mathbf{G_D}^{\ominus} + d\xi RT \ln [d\xi P] - d\xi RT \ln [(a+b)+\xi \Delta n]$$

As can be seen, the forms of $G_A(\xi)$ and $G_B(\xi)$ are identical, using the mole numbers a and b, respectively; likewise, the forms of $G_C(\xi)$ and $G_D(\xi)$ are identical, using the mole numbers c and d.

Differentiating $G_{sys}(\xi)$ with respect to ξ, we may write

$$\frac{dG_{sys}(\xi)}{d\xi} = \frac{dG_A(\xi)}{d\xi} + \frac{dG_B(\xi)}{d\xi} + \frac{dG_C(\xi)}{d\xi} + \frac{dG_D(\xi)}{d\xi}$$

and, taking firstly the expression for $G_A(\xi)$, and differentiating each term in turn, we have

$$\frac{dG_A(\xi)}{d\xi} = -a\, \Delta_f \mathbf{G_A}^{\ominus} - a\, RT \ln [a(1-\xi)P] - a\, RT + a\, RT \ln [(a+b)+\xi \Delta n]$$

$$- \frac{a(1-\xi)\Delta n\, RT}{(a+b)+\xi \Delta n}$$

Similarly

$$\frac{dG_B(\xi)}{d\xi} = -b\, \Delta_f \mathbf{G_B}^{\ominus} - b\, RT \ln [b(1-\xi)P] - b\, RT + b\, RT \ln [(a+b)+\xi \Delta n]$$

$$- \frac{b(1-\xi)\Delta n\, RT}{(a+b)+\xi \Delta n}$$

$$\frac{dG_C(\xi)}{d\xi} = c\, \Delta_f \mathbf{G_C}^{\ominus} + c\, RT \ln [c\xi P] + c\, RT - c\, RT \ln [(a+b)+\xi \Delta n]$$

$$- \frac{c\xi \Delta n\, RT}{(a+b)+\xi \Delta n}$$

and

$$\frac{dG_D(\xi)}{d\xi} = d\, \Delta_f \mathbf{G_D}^{\ominus} + d\, RT \ln [d\xi P] + d\, RT - d\, RT \ln [(a+b)+\xi \Delta n]$$

$$- \frac{d\xi \Delta n\, RT}{(a+b)+\xi \Delta n}$$

To determine $dG_{sys}(\xi)/d\xi$, we need to add these last four expressions, each of which contains five terms. Adding the last of these terms

$$-\frac{a(1-\xi)\Delta n\, RT}{(a+b)+\xi \Delta n} - \frac{b(1-\xi)\Delta n\, RT}{(a+b)+\xi \Delta n} - \frac{c\xi \Delta n\, RT}{(a+b)+\xi \Delta n} - \frac{d\xi \Delta n\, RT}{(a+b)+\xi \Delta n}$$

$$= -\frac{\Delta n RT}{(a+b)+\xi \Delta n}[a(1-\xi) + b(1-\xi) + c\xi + d\xi]$$

$$= -\frac{\Delta n RT}{(a+b)+\xi \Delta n}[(a+b) + \xi\{(c+d)-(a+b)\}]$$

$$= -\frac{\Delta n RT}{(a+b)+\xi \Delta n}[(a+b) + \xi \Delta n] = -\Delta n\, RT$$

Summing the third terms in each of the expressions for $dG_i(\xi)/d\xi$, we obtain

$$-a\,RT - b\,RT + c\,RT + d\,RT = RT[(c+d) - (a+b)] = \Delta n\,RT$$

The overall sum of the third terms and the fifth terms is therefore zero.

We now have

$$\frac{dG_{sys}(\xi)}{d\xi} = \frac{dG_A(\xi)}{d\xi} + \frac{dG_B(\xi)}{d\xi} + \frac{dG_C(\xi)}{d\xi} + \frac{dG_D(\xi)}{d\xi}$$

$$= -(a\,\Delta_f G_A^\circ + b\,\Delta_f G_B^\circ) + (c\,\Delta_f G_C^\circ + d\,\Delta_f G_D^\circ)$$

$$- a\,RT \ln \frac{a(1-\xi)P}{(a+b) + \xi\,\Delta n} - b\,RT \ln \frac{b(1-\xi)P}{(a+b) + \xi\,\Delta n}$$

$$+ c\,RT \ln \frac{c\xi P}{(a+b) + \xi\,\Delta n} + d\,RT \ln \frac{d\xi P}{(a+b) + \xi\,\Delta n}$$

But we also know that the partial pressures p_i may be expressed as

$$p_A = \frac{a(1-\xi)P}{(a+b) + \xi\,\Delta n} \tag{14.6a}$$

$$p_B = \frac{b(1-\xi)P}{(a+b) + \xi\,\Delta n} \tag{14.6b}$$

$$p_C = \frac{c\xi P}{(a+b) + \xi\,\Delta n} \tag{14.6c}$$

$$p_D = \frac{d\xi P}{(a+b) + \xi\,\Delta n} \tag{14.6d}$$

Hence

$$\frac{dG_{sys}(\xi)}{d\xi} = -(a\,\Delta_f G_A^\circ + b\,\Delta_f G_B^\circ) + (c\,\Delta_f G_C^\circ + d\,\Delta_f G_D^\circ)$$

$$- a\,RT \ln p_A - b\,RT \ln p_B + c\,RT \ln p_C + d\,RT \ln p_D$$

$$= (c\,\Delta_f G_C^\circ + d\,\Delta_f G_D^\circ) - (a\,\Delta_f G_A^\circ + b\,\Delta_f G_B^\circ)$$

$$+ RT \ln p_C^c + RT \ln p_D^d - RT \ln p_A^a - b\,RT \ln p_B^b$$

and so

$$\frac{dG_{sys}(\xi)}{d\xi} = (c\,\Delta_f G_C^\circ + d\,\Delta_f G_D^\circ) - (a\,\Delta_f G_A^\circ + b\,\Delta_f G_B^\circ) + RT \ln \frac{p_C^c\,p_D^d}{p_A^a\,p_B^b}$$

Now, on page 399, we defined the standard Gibbs free energy change $\Delta_r G^\circ$ for any reaction as

$$\Delta_r G^\circ = \sum_{\text{Products}} n_{\text{prod}} \Delta_f G_{\text{prod}}^\circ - \sum_{\text{Reactants}} n_{\text{reac}} \Delta_f G_{\text{reac}}^\circ \tag{13.45}$$

If we apply this expression to the reaction of current interest

$$aA + bB \rightleftharpoons cC + dD$$

we see that, for this reaction

$$\Delta_r G^\circ = (c\,\Delta_f G_C^\circ + d\,\Delta_f G_D^\circ) - (a\,\Delta_f G_A^\circ + b\,\Delta_f G_B^\circ)$$

Since all of the components in this reaction are ideal gases, each of the standard molar Gibbs free energies of formation $\Delta_f G_i^\circ$ refers to the standard state of an ideal gas, which, as shown in Table 6.4, is the pure gas at the standard pressure $P_i^\circ = 1$ bar. To act as a reminder that we are dealing with ideal gases, we will use the notation $\Delta_r G_p^\circ$ for $\Delta_r G^\circ$, and $\Delta_f G_{p,i}^\circ$ for $\Delta_f G_i^\circ$, noting that, in this context, the lower case subscript p implies "with reference to the ideal gas standard state defined as $P_i^\circ = 1$ bar" and is used in preference to the upper case subscript P to avoid potential confusion with "at constant pressure". Using this notation, we may therefore write

$$\Delta_r G_p^\circ = (c\,\Delta_f G_{p,C}^\circ + d\,\Delta_f G_{p,D}^\circ) - (a\,\Delta_f G_{p,A}^\circ + b\,\Delta_f G_{p,B}^\circ) \tag{14.9}$$

giving

$$\frac{dG_{sys}(\xi)}{d\xi} = \Delta_r G_p^\circ + RT \ln \frac{p_C^c\, p_D^d}{p_A^a\, p_B^b} \tag{14.10}$$

which is the result we are seeking.

As we saw on page 424, when $dG_{sys}(\xi)/d\xi = 0$, $G_{sys}(\xi)$ has a minimum value, and the system is in equilibrium. Accordingly, at equilibrium

$$0 = \Delta_r G_p^\circ + RT \ln \frac{p_{C,eq}^c\, p_{D,eq}^d}{p_{A,eq}^a\, p_{B,eq}^b} \tag{14.11}$$

in which the partial pressures p_i of each component i correspond to those partial pressures $p_{i,eq}$ exerted by each component as present in the equilibrium mixture. Since, at any given temperature, the quantity of each component in the equilibrium mixture is uniquely defined, we may define a (temperature dependent) constant, K_p, as

$$K_p = \frac{p_{C,eq}^c\, p_{D,eq}^d}{p_{A,eq}^a\, p_{B,eq}^b}$$

Since each quantity $p_{i,eq}$ is the ratio $p_{i,eq}/P_i^\circ$, K_p is more explicitly written as

$$K_p = \frac{p_{C,eq}^c\, p_{D,eq}^d}{p_{A,eq}^a\, p_{B,eq}^b} = \frac{(p_{C,eq}/P^\circ)^c\,(p_{D,eq}/P^\circ)^d}{(p_{A,eq}/P^\circ)^a\,(p_{B,eq}/P^\circ)^b} \tag{14.12}$$

Equation (14.11) can therefore be written as

$$0 = \Delta_r G_p^\circ + RT \ln K_p$$

implying that

$$\Delta_r G_p^\circ = -RT \ln K_p \tag{14.13a}$$

and

$$K_p = \exp(-\Delta_r G_p^\circ / RT) \tag{14.13b}$$

in which

$$\Delta_r G_p^\circ = (c\,\Delta_f G_{p,C}^\circ + d\,\Delta_f G_{p,D}^\circ) - (a\,\Delta_f G_{p,A}^\circ + b\,\Delta_f G_{p,B}^\circ) \tag{14.9}$$

CHEMICAL EQUILIBRIUM AND CHEMICAL KINETICS

Note that, in all equations involving K_p, all quantities shown as p_i are 'shorthand' for the ratio $p_i/P_i^{\ominus}$, where $P_i^{\ominus}$ is the standard pressure for the pure component i, namely 1 bar. Since $p_i/P_i^{\ominus} = p_i/1 = p_i$, writing $p_i/P_i^{\ominus}$ adds unnecessary clutter – hence the form of equations (14.10), (14.11) and (14.12). Sometimes, however, the ratio $p_i/P_i^{\ominus}$ is important, and so the 'hidden' $P_i^{\ominus}$ should not be forgotten.

14.4 The pressure thermodynamic equilibrium constant K_p ...

Given that equation (14.12)

$$K_p = \frac{p_{C,eq}^c \, p_{D,eq}^d}{p_{A,eq}^a \, p_{B,eq}^b} = \frac{(p_{C,eq}/P_C^{\ominus})^c \, (p_{D,eq}/P_D^{\ominus})^d}{(p_{A,eq}/P_A^{\ominus})^a \, (p_{B,eq}/P_B^{\ominus})^b} \tag{14.12}$$

is expressed in terms of the partial pressures of the reactants and products, K_p is known as the **pressure thermodynamic equilibrium constant** for the reaction taking place at constant pressure and at a temperature T K.

If we happened to write the reaction with, say, doubled stoichiometric coefficients $2a$, $2b$, $2c$ and $2d$, as

$$2aA + 2bB \rightleftharpoons 2cC + 2dD$$

then the equation would still balance chemically, and the equilibrium partial pressures would still be p_A, p_B, p_C and p_D. The equilibrium constant, however, would (using the 'shorthand' form) be written as

$$\frac{p_{C,eq}^{2c} \, p_{D,eq}^{2d}}{p_{A,eq}^{2a} \, p_{B,eq}^{2b}} = \left(\frac{p_{C,eq}^c \, p_{D,eq}^d}{p_{A,eq}^a \, p_{B,eq}^b}\right)^2 = K_p^2$$

The quantity shown on the left-hand side of this expression is still a constant, but it is a different constant from that defined by equation (14.12), which uses the single stoichiometric coefficients a, b, c and d. The value of K_p as computed using equation (14.12) therefore depends on the stoichiometric coefficients used to define the chemical reaction, even though the equilibrium point is a matter of physical and chemical reality, which has nothing to do with how a chemical reaction might be written down. In order to avoid any ambiguity, there is a general convention that values of K_p are computed from equation (14.12) using stoichiometric coefficients corresponding to the smallest integers which can be used to express the reaction – in particular, no stoichiometric coefficient can be a fraction.

To gain an insight into the meaning of equations (14.12), (14.13a) and (14.13b)

$$K_p = \frac{p_{C,eq}^c \, p_{D,eq}^d}{p_{A,eq}^a \, p_{B,eq}^b} = \frac{(p_{C,eq}/P_C^{\ominus})^c \, (p_{D,eq}/P_D^{\ominus})^d}{(p_{A,eq}/P_A^{\ominus})^a \, (p_{B,eq}/P_B^{\ominus})^b} \tag{14.12}$$

$$\Delta_r G_p^{\ominus} = -RT \ln K_p \tag{14.13a}$$

$$K_p = \exp(-\Delta_r G_p^{\ominus}/RT) \tag{14.13b}$$

let's assume, for simplicity, that the mole numbers of the reactants and the products are all one, corresponding to the reversible chemical reaction

A + B ⇌ C + D

in which case, using the 'shorthand' version,

$$K_p = \frac{p_{C,eq}\ p_{D,eq}}{p_{A,eq}\ p_{B,eq}}$$

Physically, at equilibrium, the system comprises a stable mixture of the reactants A and B, and the products C and D, with each component i exerting its own partial pressure p_i, this being the pressure that would be exerted by component i if that component were to occupy the entire volume available to the whole system. For any given volume, the greater the partial pressure of component i, the greater the amount of component i present; conversely, the lower the partial pressure of component i, the lesser the amount of component i present. p_i is therefore a measure of the quantity of component i in the equilibrium mixture.

So, if K_p – which is necessarily always positive – is a large number, say, thousands, then there are considerably more products in the equilibrium mixture than reactants; if K_p is very large – say, tens of thousands, there are hardly any reactants present at all – in practical terms, the forward reaction

A + B → C + D

goes to completion.

Conversely, if K_p is a small number, say, 10^{-3}, then there are considerably more reactants in the equilibrium mixture than products, and if K_p is very small (10^{-6}, or smaller), there are hardly any products present at all – in practical terms, A and B are stable, and don't react, whereas a mixture of C and D will spontaneously react to form A and B, so it is the reverse reaction

C + D → A + B

that goes to completion.

If K_p is in the range, say, 0.1 to 10, then the product of the partial pressures of the reactants will be of a similar magnitude to the product of the partial pressures of the products, and so A, B, C and D will be present in measureable quantities in the equilibrium mixture.

This analysis demonstrates that although equations (14.12) and (14.13b)

$$K_p = \frac{p_{C,eq}\ p_{D,eq}}{p_{A,eq}\ p_{B,eq}} \qquad (14.12)$$

$$K_p = \exp(-\Delta_r G_p^{\circ}/RT) \qquad (14.13b)$$

both refer to 'equilibrium', 'equilibrium' does not imply that both the reactants and the products are present in the equilibrium mixture in measureable quantities. Rather, every chemical reaction will be associated with an equilibrium constant K_p – even reactions which, in practice, go to completion. 'Equilibrium' is a more general concept, meaning 'stable state' - which might be a state in which the reaction has gone to completion, with the reactants transformed into products (corresponding a to a very large value of K_p); it might mean that nothing happens at all, and that the reactants are stable (K_p very small); it might mean that the equilibrium state contains a measureable mixture of both reactants and products (K_p has an intermediate value, say, between 10^{-3} and 10^3).

14.5 ... and the meaning of $\Delta_r G_p^{\ominus}$

Equation (14.13b)

$$K_p = \exp(-\Delta_r G_p^{\ominus}/RT) \tag{14.13b}$$

relates K_p to the value of $\Delta_r G_p^{\ominus}$, the standard change in Gibbs free energy for the reversible gas phase reaction

$$aA + bB \rightleftharpoons cC + dD$$

taking place at the temperature T for which the standard states apply. As we saw on page 427, $\Delta_r G_p^{\ominus}$ is defined as

$$\Delta_r G_p^{\ominus} = (c\, \Delta_f \mathbf{G}_{p,C}^{\ominus} + d\, \Delta_f \mathbf{G}_{p,D}^{\ominus}) - (a\, \Delta_f \mathbf{G}_{p,A}^{\ominus} + b\, \Delta_f \mathbf{G}_{p,B}^{\ominus}) \tag{14.9}$$

and values of $\Delta_f \mathbf{G}_{p,i}^{\ominus}$ all refer to the standard state of the corresponding ideal gas as tabulated in reference works. $\Delta_r G_p^{\ominus}$ can therefore easily be determined for any reaction by looking up the appropriate numbers. What, though, does the resulting value of $\Delta_r G_p^{\ominus}$ mean? And the sign of $\Delta_r G_p^{\ominus}$ too?

From a numerical standpoint, $\Delta_r G_p^{\ominus}$ is the arithmetic difference $G_{\text{final}} - G_{\text{initial}}$ between the Gibbs free energies of two states

- a final state comprising c mol of C, in its standard state, alongside, but not mixed with, d mol of D, in its standard state, such that $G_{\text{final}} = c\, \Delta_f \mathbf{G}_{p,C}^{\ominus} + d\, \Delta_f \mathbf{G}_{p,D}^{\ominus}$
- an initial state comprising a mol of A, in its standard state, alongside, but not mixed with, b mol of B, in its standard state, such that $G_{\text{initial}} = a\, \Delta_f \mathbf{G}_{p,A}^{\ominus} + b\, \Delta_f \mathbf{G}_{p,B}^{\ominus}$.

This is somewhat hypothetical, in that reactions rarely take place under conditions in which all the reactants, and all the products, are in their standard states – for gases, a partial pressure of 1 bar for all components, or, as we shall see in the next chapter, for reactions in solution, a molar concentration of 1 mol dm^{-3}. Furthermore, both G_{initial} and G_{final} are the sums of Gibbs free energies of formation of the pure reactants and products respectively, with no recognition of mixing. And we must always bear in mind that $\Delta_r G_p^{\ominus}$ relates solely to an initial state of pure reactants, and a final state of pure products, and does not, in itself, give any information about any intermediate states.

The importance of $\Delta_r G_p^{\ominus}$ – and a feature that does link the value of $\Delta_r G_p^{\ominus}$ to a very significant intermediate state, the equilibrium state – is equation (14.13b)

$$K_p = \exp(-\Delta_r G_p^{\ominus}/RT) \tag{14.13b}$$

For any reaction, the value of $\Delta_r G_p^{\ominus}$ can be used to compute the corresponding value of the pressure, equilibrium constant K_p, the magnitude of which gives us direct information on whether the equilibrium lies towards the reactants or the products (as regards the reaction as written), and the relative composition of the reactants and products in the equilibrium mixture: to use some non-scientific language, the value of $\Delta_r G_p^\ominus$ enables us to answer the questions "does the reaction 'go' or not?", and "how far does it go?".

So, firstly, the sign of $\Delta_r G_p^\ominus$ immediately gives you important information as regards K_p: with reference to equation (14.13b), if $\Delta_r G_p^\ominus$ is negative, then K_p is necessarily greater than 1, and the equilibrium will be towards the right of the reaction as written, whereas if $\Delta_r G^\ominus$ is positive, then $K_p < 1$, and the equilibrium will be towards the left. Secondly, the larger the magnitude of $\Delta_r G_p^\ominus$, the further towards the right, or left, that equilibrium will be. So numerically, for a reaction taking place at 298.15 K, a value of $\Delta_r G_p^\ominus$ of −35 kJ corresponds to a value of K_p of 1.4×10^6, and so reactions with values of $\Delta_r G_p^\ominus$ progressively more negative than −35 kJ will correspond to reactions with equilibria progressively further towards the products; likewise, a value of $\Delta_r G_p^\ominus$ of + 35 kJ corresponds to a value of K_p of 7.4×10^{-7}, and so reactions with values of $\Delta_r G_p^\ominus$ progressively more positive than +35 kJ will correspond to reactions with equilibria progressively further towards the reactants.

One further point about the pressure equilibrium constant K_p and $\Delta_r G_p^\ominus$, in relation to the graph of $G_{sys}(\xi)$ as shown in Figure 14.2. Although the minimum value of $G_{sys}(\xi)$ identifies the existence of an equilibrium state, the value of the pressure equilibrium constant K_p cannot be directly deduced from the graph itself, nor does the graph specify the corresponding value of $\Delta_r G_p^\ominus$. What the graph shown in Figure 14.2 does identify directly is the equilibrium value ξ_{eq} of the extent-of-reaction parameter ξ, which corresponds to the point on the horizontal axis ξ at which the graph of $G_{sys}(\xi)$ is at a minimum. ξ_{eq} is not the same as K_p, but knowledge of ξ_{eq} does allow K_p to be (rather messily) computed from a combination of equation (14.12)

$$K_p = \frac{p_{C,eq}^c \, p_{D,eq}^d}{p_{A,eq}^a \, p_{B,eq}^b} \tag{14.12}$$

and equations (14.6), using the equilibrium value ξ_{eq}

$$p_{A,eq} = \frac{a(1-\xi_{eq})P}{(a+b) + \xi_{eq}\Delta n} \tag{14.6a}$$

$$p_{B,eq} = \frac{b(1-\xi_{eq})P}{(a+b) + \xi_{eq}\Delta n} \tag{14.6b}$$

$$p_{C,eq} = \frac{c\,\xi_{eq}\,P}{(a+b) + \xi_{eq}\Delta n} \tag{14.6c}$$

$$p_{D,eq} = \frac{d\,\xi_{eq}\,P}{(a+b) + \xi_{eq}\Delta n} \tag{14.6d}$$

Furthermore, the minimum value of $G_{sys}(\xi)$, as indicated by the point on the vertical axis G_{sys} corresponding to the minimum of the curve, has no particular significance, and – in particular – is *not* related to $\Delta_r G_p^\ominus$: $\Delta_r G_p^\ominus$ is calculated from equation (14.9)

$$\Delta_r G_p^\ominus = (c\,\Delta_f G_{p,C}^\ominus + d\,\Delta_f G_{p,D}^\ominus) - (a\,\Delta_f G_{p,A}^\ominus + b\,\Delta_f G_{p,B}^\ominus) \tag{14.9}$$

in which each of the standard molar Gibbs free energies of formation $\Delta_f G_i^\circ$ is determined from tables of reference data, such as Table 13.1, and, once $\Delta_r G_p^\circ$ has been computed, K_p can then be calculated as

$$K_p = \exp(-\Delta_r G_p^\circ / RT) \tag{14.13b}$$

14.6 Non-equilibrium systems

The last several pages have examined gas phase chemical equilibrium in some detail, so we now turn to behaviour of systems away from equilibrium. To do this, we return to equation (14.10)

$$\frac{dG_{\text{sys}}(\xi)}{d\xi} = \Delta_r G_p^\circ + RT \ln \frac{p_C^c \, p_D^d}{p_A^a \, p_B^b} \tag{14.10}$$

As we have seen, when the extent-of-reaction $\xi = \xi_{\text{eq}}$, $dG_{\text{sys}}(\xi)/d\xi = 0$, $G_{\text{sys}}(\xi)$ has a minimum value, the system is in equilibrium, each of the partial pressures p_i corresponds to those exerted by the components as present in the equilibrium mixture, and we may define the pressure equilibrium constant K_p as

$$K_p = \frac{p_{C,\text{eq}}^c \, p_{D,\text{eq}}^d}{p_{A,\text{eq}}^a \, p_{B,\text{eq}}^b} = \frac{(p_{C,\text{eq}}/P_C^\circ)^c \, (p_{D,\text{eq}}/P_D^\circ)^d}{(p_{A,\text{eq}}/P_A^\circ)^a \, (p_{B,\text{eq}}/P_B^\circ)^b} \tag{14.12}$$

At all values of ξ other than $\xi = \xi_{\text{eq}}$, the system is in a non-equilibrium state, with each component i exerting the corresponding partial pressure p_i, and so let us define the **pressure mass action ratio** Γ_p (sometimes called the **pressure reaction quotient** Q_p) as

$$\Gamma_p = \frac{p_C^c \, p_D^d}{p_A^a \, p_B^b} = \frac{(p_C/P_C^\circ)^c \, (p_D/P_D^\circ)^d}{(p_A/P_A^\circ)^a \, (p_B/P_B^\circ)^b} \tag{14.14}$$

Mathematically, as well as being (as usual) a function of the temperature T, the pressure mass action ratio Γ_p is function $\Gamma_p(\xi)$ of the extent-of-reaction ξ, and has a value for all values of ξ from $\xi = 0$ to $\xi = 1$. A special case arises when $\xi = \xi_{\text{eq}}$, for which $\Gamma_p(\xi_{\text{eq}}) = K_p$. Also, as we saw on page 428 in relation to K_p, the terms shown in equation (14.14) as p_i are in fact ratios p_i/P_i°, and so Γ_p, like K_p, is necessarily dimensionless.

If we now define the **reaction Gibbs free energy** $\Delta_r G$ as

$$\Delta_r G = \frac{dG_{\text{sys}}(\xi)}{d\xi} \tag{14.15}$$

then, for gas phase reactions, we may use the notation $\Delta_r G_p$, and equation (14.10) becomes

$$\Delta_r G_p = \Delta_r G_p^\circ + RT \ln \frac{p_C^c \, p_D^d}{p_A^a \, p_B^b} \tag{14.16a}$$

which, using the definition of Γ_p as given by equation (14.14), may be written as

$$\Delta_r G_p = \Delta_r G_p^{\ominus} + RT \ln \Gamma_p \tag{14.16b}$$

Now, since

$$K_p = \exp(-\Delta_r G_p^{\ominus}/RT) \tag{14.13b}$$

equation (14.16b) can be written as

$$\Delta_r G_p = -RT \ln K_p + RT \ln \Gamma_p \tag{14.16c}$$

or

$$\Delta_r G_p = RT \ln \frac{\Gamma_p}{K_p} \tag{14.16d}$$

We now have a cluster of equivalent equations

$$\frac{dG_{sys}(\xi)}{d\xi} = \Delta_r G_p^{\ominus} + RT \ln \frac{p_C^c \, p_D^d}{p_A^a \, p_B^b} \tag{14.10}$$

$$\Delta_r G_p = \Delta_r G_p^{\ominus} + RT \ln \frac{p_C^c \, p_D^d}{p_A^a \, p_B^b} \tag{14.16a}$$

$$\Delta_r G_p = \Delta_r G_p^{\ominus} + RT \ln \Gamma_p \tag{14.16b}$$

$$\Delta_r G_p = -RT \ln K_p + RT \ln \Gamma_p \tag{14.16c}$$

$$\Delta_r G_p = RT \ln \frac{\Gamma_p}{K_p} \tag{14.16d}$$

all of which describe chemical systems away from equilibrium, once again remembering, as always, that each p_i represents the ratio $p_i/P_i^{\ominus}$.

To show how these equations can be used, let's examine equation (14.16d) in particular. For any given reaction

$$aA + bB \rightleftharpoons cC + dD$$

taking place at a constant pressure of 1 bar, and at a given temperature T, the pressure equilibrium constant K_p is a constant given by

$$\Delta_r G_p^{\ominus} = -RT \ln K_p \tag{14.13a}$$

in which

$$\Delta_r G_p^{\ominus} = (c \, \Delta_f G_{p,C}^{\ominus} + d \, \Delta_f G_{p,D}^{\ominus}) - (a \, \Delta_f G_{p,A}^{\ominus} + b \, \Delta_f G_{p,B}^{\ominus}) \tag{14.9}$$

For any given reaction, the standard molar Gibbs free energies of formation $\Delta_f G_{p,i}^{\ominus}$ can be determined from reference tables, and so K_p can be computed.

Suppose we are interested in the behaviour of the chemical system at a chosen extent-of-reaction $\xi = 0.2$, corresponding to a state in which 20% of the reactants have reacted. For any reaction pressure P (usually 1 bar), we may now use this value of ξ to compute the corresponding partial pressures p_i

CHEMICAL EQUILIBRIUM AND CHEMICAL KINETICS

$$p_A = \frac{a(1-\xi)P}{(a+b) + \xi \Delta n} \tag{14.6a}$$

$$p_B = \frac{b(1-\xi)P}{(a+b) + \xi \Delta n} \tag{14.6b}$$

$$p_C = \frac{c\xi P}{(a+b) + \xi \Delta n} \tag{14.6c}$$

$$p_D = \frac{d\xi P}{(a+b) + \xi \Delta n} \tag{14.6d}$$

from which we may calculate Γ_p as

$$\Gamma_p = \frac{p_C^c \, p_D^d}{p_A^a \, p_B^b} \tag{14.14}$$

We can now compare the calculated value of Γ_p with the known value of K_p, which will result in one of three cases, each relating to equation (14.16d)

$$\Delta_r G_p = RT \ln \frac{\Gamma_p}{K_p} \tag{14.16d}$$

- Case 1: $\Gamma_p = K_p$, implying that $\Gamma_p/K_p = 1$ and $\Delta_r G_p = 0$
- Case 2: $\Gamma_p < K_p$, implying that $\Gamma_p/K_p < 1$ and $\Delta_r G_p < 0$
- Case 3: $\Gamma_p > K_p$, implying that $\Gamma_p/K_p > 1$ and $\Delta_r G_p > 0$

The significance of these three cases is best appreciated by recalling the definition of $\Delta_r G$ as

$$\Delta_r G = \frac{dG_{sys}(\xi)}{d\xi} \tag{14.15}$$

which, for a gas phase reaction, we write as $\Delta_r G_p$. An insightful way of interpreting $\Delta_r G_p$ is as a measure of the slope of the tangent to the $G_{sys}(\xi)$ curve for any value of ξ, as illustrated in Figure 14.3.

As is evident from Figure 14.3, case 1, $\Gamma_p = K_p$, corresponds to the equilibrium state; case 2, $\Gamma_p < K_p$, corresponds to a non-equilibrium state in which the reaction will spontaneously proceed to the right, as written, towards the equilibrium state; case 3, $\Gamma_p > K_p$, corresponds to a non-equilibrium state in which the reaction will spontaneously proceed to the left, as written, also towards the equilibrium state. For any given mixture of components A, B, C and D, which can react according to the reversible reaction

$$aA + bB \rightleftharpoons cC + dD$$

the ratio Γ_p/K_p therefore specifies the direction in which the reaction will proceed.

One important note: the symbol Δ is being used in a very special way in the notation $\Delta_r G$, as defined by equation (14.15)

$$\Delta_r G = \frac{dG_{sys}(\xi)}{d\xi} \tag{14.15}$$

NON-EQUILIBRIUM SYSTEMS

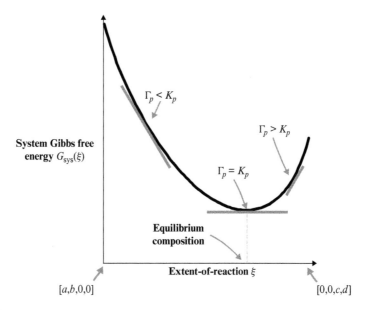

Figure 14.3 Non-equilibrium systems. For any chemical reaction, the pressure equilibrium constant K_p, is, as its name suggests, a constant, and the mass action ratio Γ_p varies according to the extent-of-reaction ξ. For any value of ξ, the slope of the tangent to the $G_{sys}(\xi)$ curve is given by $\Delta_r G_p = RT \ln(\Gamma_p/K_p)$.

and as used in equations (14.16)

$$\Delta_r G_p = \Delta_r G_p^{\circ} + RT \ln \frac{p_C^c \, p_D^d}{p_A^a \, p_B^b} \tag{14.16a}$$

$$\Delta_r G_p = \Delta_r G_p^{\circ} + RT \ln \Gamma_p \tag{14.16b}$$

$$\Delta_r G_p = -RT \ln K_p + RT \ln \Gamma_p \tag{14.16c}$$

$$\Delta_r G_p = RT \ln \frac{\Gamma_p}{K_p} \tag{14.16d}$$

Up to this point, the symbol Δ has been used to denote a difference between any final state and any initial state: $\Delta H = H_2 - H_1$, $\Delta S = S_2 - S_1$ and so on. $\Delta_r G$, however, does *not* represent the difference in Gibbs free energy between any two states; rather, it represents the slope of the $G_{sys}(\xi)$ curve at the point corresponding to that value of ξ at which the various components are exerting the partial pressures given in the ln term in equation (14.16a), so determining the corresponding value of Γ_p.

One further point about the pressure mass action ratio Γ_p, which is a direct consequence of its definition as

$$\Gamma_p = \frac{p_C^c \, p_D^d}{p_A^a \, p_B^b} \tag{14.14}$$

As we know, each term p_i in equation (14.14) is more strictly written as the ratio p_i/P_i°, and so if it happens that each component i is in its standard state, then $p_i = P_i^{\circ}$ and each of the ratios $p_i/P_i^\circ = 1$, implying that the standard pressure mass action ratio $\Gamma_p^\circ = 1$. Since $\ln 1 = 0$, if all the components of a reaction are in their standard states, then equations (14.16a) and (14.16b) imply that the corresponding value of $\Delta_r G_p = \Delta_r G_p^\circ = (c\, \Delta_f G_{p,C}^\circ + d\, \Delta_f G_{p,D}^\circ) - (a\, \Delta_f G_{p,A}^\circ + b\, \Delta_f G_{p,B}^\circ)$.

14.7 Another equilibrium constant, K_x ...

Returning now to equilibrium states, K_p, as defined by equation (14.12)

$$K_p = \frac{p_{C,eq}^c\, p_{D,eq}^d}{p_{A,eq}^a\, p_{B,eq}^b} = \frac{(p_{C,eq}/P_C^\circ)^c\, (p_{D,eq}/P_D^\circ)^d}{(p_{A,eq}/P_A^\circ)^a\, (p_{B,eq}/P_B^\circ)^b} \tag{14.12}$$

is expressed in terms of the partial pressures of the various reactants and products. Now, at any time during a reaction taking place at a constant pressure P, the partial pressure p_i exerted by component i is related to the corresponding mole fraction x_i according to equation (14.6)

$$p_i = x_i\, P \tag{14.6}$$

Note that, since equation (14.6) is derived directly from Dalton's law of partial pressures, the term p_i represents a true pressure, and is not 'shorthand' for the ratio p_i/P_i° that appears in all the equations for G_i – such as equations (13.47e) and (14.1) – and all the equations for K_p and Γ_p. At equilibrium

$$p_{i,eq} = x_{i,eq}\, P$$

and if we divide both sides of this equation by P_i°, the standard pressure of component i, and also multiply the right-hand side by $x_i^\circ/x_i^\circ = 1$, then

$$\frac{p_{i,eq}}{P_i^\circ} = \frac{x_{i,eq}}{x_i^\circ}\, \frac{x_i^\circ P}{P_i^\circ}$$

enabling us to write equation (14.12) for K_p as

$$K_p = \frac{(x_{C,eq}/x_C^\circ)^c\, (x_{D,eq}/x_D^\circ)^d\, (x_C^\circ P/P_C^\circ)^c\, (x_D^\circ P/P_D^\circ)^d}{(x_{A,eq}/x_A^\circ)^a\, (x_{B,eq}/x_B^\circ)^b\, (x_A^\circ P/P_A^\circ)^a\, (x_B^\circ P/P_B^\circ)^b}$$

Now, the standard pressure P_i° has the same value for all components i, which we may write as P°; similarly, all the standard mole fractions x_i° have the same value x°, and so

$$K_p = \frac{(x_{C,eq}/x_C^\circ)^c\, (x_{D,eq}/x_D^\circ)^d}{(x_{A,eq}/x_A^\circ)^a\, (x_{B,eq}/x_B^\circ)^b} \left(\frac{x^\circ P}{P^\circ}\right)^{[(c+d)-(a+b)]}$$

But since

$$\Delta n = [(c+d) - (a+b)] \tag{6.15}$$

and, according to Table 6.4, the standard mole fraction $x^\circ = 1$, then

$$K_p = \frac{(x_{C,eq}/x_C^\circ)^c\, (x_{D,eq}/x_D^\circ)^d}{(x_{A,eq}/x_A^\circ)^a\, (x_{B,eq}/x_B^\circ)^b} \left(\frac{P}{P^\circ}\right)^{\Delta n}$$

At equilibrium, each mole fraction $x_{i,\,\text{eq}}$ is a specific number, and so we may define a new constant K_x as

$$K_x = \frac{(x_{C,\text{eq}}/x_C^{\ominus})^c \, (x_{D,\text{eq}}/x_D^{\ominus})^d}{(x_{A,\text{eq}}/x_A^{\ominus})^a \, (x_{B,\text{eq}}/x_B^{\ominus})^b} \tag{14.17a}$$

or, in the 'shorthand' form

$$K_x = \frac{x_{C,\text{eq}}^c \, x_{D,\text{eq}}^d}{x_{A,\text{eq}}^a \, x_{B,\text{eq}}^b}$$

According to Table 6.4, the standard pressure $P^{\ominus} = 1$ bar, giving

$$K_p = K_x \, P^{\Delta n} \tag{14.18a}$$

and

$$K_x = K_p \, P^{-\Delta n} \tag{14.18b}$$

K_x is known as the **mole fraction thermodynamic equilibrium constant**, and since each mole fraction x_i is dimensionless, K_x, like K_p, is dimensionless. Furthermore, in equations (14.18), the terms $P^{\Delta n}$ and $P^{-\Delta n}$ are dimensionless too, despite the appearance that the dimensions are $\text{Pa}^{\Delta n}$ or $\text{Pa}^{-\Delta n}$: as can be seen from the derivation of these equations, P is 'shorthand' for the ratio $x^{\ominus}(P/P^{\ominus})$.

In equations (14.18), if Δn is zero, then $K_x = K_p$; furthermore, if a reaction takes place under normal atmospheric conditions, $P \approx 1$ bar, and so the numerical values of K_p and K_x will be equal, or nearly so, for all values of Δn. However, for reactions taking place at pressures other than 1 bar (for example, an industrial process at, say, 100 bar), the numerical values of K_p and K_x may differ by orders of magnitude, according to the value of Δn. In this context, it's important to remember that K_p and K_x are just different measures of the same phenomenon: for any given system, even if K_p and K_x have different values, the actual system is the same.

Also, for non-equilibrium mixtures, we may define the (dimensionless) **mole fraction mass action ratio** Γ_x (or **mole fraction reaction quotient** Q_x) as

$$\Gamma_x = \frac{(x_C/x_C^{\ominus})^c \, (x_D/x_D^{\ominus})^d}{(x_A/x_A^{\ominus})^a \, (x_B/x_B^{\ominus})^b} = \frac{x_C^c \, x_D^d}{x_A^a \, x_B^b} \tag{14.19}$$

14.8 ... and a third equilibrium constant, K_c

A third measure of the quantity of component i within a mixture – in addition to the partial pressure p_i, and the mole fraction x_i – is the molar concentration $[I]$, defined as the number of moles n_i of component i present within the total system volume V

$$[I] = \frac{n_i}{V} \tag{6.20}$$

For a mixture of ideal gases, the equilibrium molar concentration $[I]_{\text{eq}}$ of component i within the equilibrium reaction mixture is therefore related to the corresponding equilibrium mole number $n_{i,\,\text{eq}}$, and the total system volume V, as $[I]_{\text{eq}} = n_{i,\,\text{eq}}/V$. Also, within the mixture of ideal gases, each component i obeys the ideal gas law, and, for an equilibrium mixture, Dalton's law of partial pressures tells us that

$$p_{i,\text{eq}} = n_{i,\text{eq}} \frac{RT}{V}$$

and so, for each component i within an equilibrium mixture

$$p_{i,\text{eq}} = n_{i,\text{eq}} \frac{RT}{V} = \frac{n_{i,\text{eq}}}{V} RT = [I]_{\text{eq}} RT \qquad (14.20)$$

where each p_i is a true partial pressure, and not the ratio $p_i/P_i^{\ominus}$.

We now divide both sides of equation (14.20) by the standard pressure $P_i^{\ominus}$ as

$$\frac{p_{i,\text{eq}}}{P_i^{\ominus}} = [I]_{\text{eq}} \frac{RT}{P_i^{\ominus}}$$

and multiply the right-hand side by $[I]^{\ominus}/[I]^{\ominus} = 1$, giving

$$\frac{p_{i,\text{eq}}}{P_i^{\ominus}} = \frac{[I]_{\text{eq}}}{[I]^{\ominus}} \frac{RT[I]^{\ominus}}{P_i^{\ominus}}$$

so allowing us to express equation (14.12) for K_p

$$K_p = \frac{p_{C,\text{eq}}{}^c \, p_{D,\text{eq}}{}^d}{p_{A,\text{eq}}{}^a \, p_{B,\text{eq}}{}^b} = \frac{(p_{C,\text{eq}}/P_C^{\ominus})^c \, (p_{D,\text{eq}}/P_D^{\ominus})^d}{(p_{A,\text{eq}}/P_A^{\ominus})^a \, (p_{B,\text{eq}}/P_B^{\ominus})^b} \qquad (14.12)$$

in terms of molar concentrations as

$$K_p = \left(\frac{([C]_{\text{eq}}/[C]^{\ominus})^c \, ([D]_{\text{eq}}/[D]^{\ominus})^d}{([A]_{\text{eq}}/[A]^{\ominus})^a \, ([B]_{\text{eq}}/[B]^{\ominus})^b} \right) \left(\frac{([C]^{\ominus}/P_C^{\ominus})^c \, ([D]^{\ominus}/P_D^{\ominus})^d}{([A]^{\ominus}/P_A^{\ominus})^a \, ([B]^{\ominus}/P_D^{\ominus})^b} \right) RT^{[(c+d)-(a+b)]}$$

Since all components i have the same standard molar concentration $[I]^{\ominus}$, and the same standard pressure $P^{\ominus}$, then

$$K_p = \left(\frac{([C]_{\text{eq}}/[C]^{\ominus})^c \, ([D]_{\text{eq}}/[D]^{\ominus})^d}{([A]_{\text{eq}}/[A]^{\ominus})^a \, ([B]_{\text{eq}}/[B]^{\ominus})^b} \right) \left(\frac{[I]^{\ominus} RT}{P^{\ominus}} \right)^{[(c+d)-(a+b)]}$$

According to the standards defined in Table 6.4, $[I]^{\ominus} = 1$ mol dm^{-3}, $P^{\ominus} = 1$ bar, and, at any given temperature T, the term RT is a constant. For any given reaction of defined stoichiometry, we may therefore define the (molar) **concentration thermodynamic equilibrium constant**, K_c, as

$$K_c = \frac{([C]_{\text{eq}}/[C]^{\ominus})^c \, ([D]_{\text{eq}}/[D]^{\ominus})^d}{([A]_{\text{eq}}/[A]^{\ominus})^a \, ([B]_{\text{eq}}/[B]^{\ominus})^b} \qquad (14.21a)$$

or, in the more commonly-used form in which $[I]_{\text{eq}}$ is used as 'shorthand' for the ratio $[I]_{\text{eq}}/[I]^{\ominus}$

$$K_c = \frac{[C]_{\text{eq}}{}^c \, [D]_{\text{eq}}{}^d}{[A]_{\text{eq}}{}^a \, [B]_{\text{eq}}{}^b} \qquad (14.21b)$$

implying that

$$K_p = K_c \, RT^{\Delta n} \qquad (14.22a)$$

and

$$K_c = K_p RT^{-\Delta n} \tag{14.22b}$$

In equations (14.22a) and (14.22b), the term RT is a dimensionless number since it is being used as 'shorthand' for the dimensionless ratio $[I]^{\ominus}RT/P^{\ominus}$; also, like K_p and K_x, K_c is dimensionless.

Likewise, the (molar) **concentration mass action ratio** Γ_c (or **concentration reaction quotient** Q_c), for non-equilibrium mixtures is defined as

$$\Gamma_c = \frac{([C]/[C]^{\ominus})^c \, ([D]/[D]^{\ominus})^d}{([A]/[A]^{\ominus})^a \, ([B]/[B]^{\ominus})^b} = \frac{[C]^c \, [D]^d}{[A]^a \, [B]^b} \tag{14.23}$$

Once again, in equations (14.21) for K_c, and (14.23) for Γ_c, each term $[I]$ is more correctly written as the ratio $[I]/[I]^{\ominus}$, where $[I]^{\ominus}$ is the standard molar concentration of component i, defined in Table 6.4 as either 1 mol m^{-3} or 1 mol dm^{-3}. K_c and Γ_c, like K_p, K_x, Γ_p and Γ_x, are therefore dimensionless; furthermore, if all components are in their standard states, $\Gamma_c^{\ominus} = 1$.

K_p, K_x and K_c are all measures of the same physical phenomenon: the relative quantities of the various components in an equilibrium mixture – or, for Γ_p, Γ_x and Γ_c, in a non-equilibrium mixture – and the choice of which measure to use depends on convenience. For gases, partial pressures and mole fractions are often used, and, as we shall see in the next chapter, for reactions in solution, mole fractions and molar concentrations tend to be preferred. Numerically, as we have seen, at a pressure of 1 bar, $K_p = K_x$ and $\Gamma_p = \Gamma_x$; and if a gas phase reaction is such that $\Delta n = 0$, then, according to equation (14.22), $RT = 1$, and so $K_p = K_c$ and $\Gamma_p = \Gamma_c$. If, however, $\Delta n \neq 0$, then, since the magnitude of RT is approximately 2,500 at a temperature of 298 K, the magnitudes of K_p and K_c can be very different.

14.9 Two worked examples – methane and ammonia

As an example of how these relationships, and the key equations, work in practice, consider firstly the combustion of methane

$$CH_4 \, (g) + 2O_2 \, (g) \rightarrow CO_2 \, (g) + 2H_2O \, (g)$$

Reference to Table 13.1 gives the following data at 298.15 K and a pressure of 1 bar

$\Delta_f G_p^{\ominus} \, CH_4 \, (g) = -50.5 \text{ kJ mol}^{-1}$

$\Delta_f G_p^{\ominus} \, O_2 \, (g) = 0 \text{ kJ mol}^{-1}$

$\Delta_f G_p^{\ominus} \, CO_2 \, (g) = -394.4 \text{ kJ mol}^{-1}$

$\Delta_f G_p^{\ominus} \, H_2O \, (g) = -228.6 \text{ kJ mol}^{-1}$

According to the general equation (13.45)

$$\Delta_r G^{\ominus} = \sum_{\text{Products}} n_{\text{prod}} \Delta_f \mathbf{G}_{\text{prod}}^{\ominus} - \sum_{\text{Reactants}} n_{\text{reac}} \Delta_f \mathbf{G}_{\text{reac}}^{\ominus} \tag{13.45}$$

then, for the gas phase combustion of methane

$$\Delta_r G_p^{\ominus} = [(-394.4) + 2 \times (-228.6)] - [(-50.5) + 2 \times (0)] = -801.1 \text{ kJ}$$

If we now use this value of $\Delta_r G_p^\ominus$ in equation (14.13b),

$$K_p = \exp(-\Delta_r G_p^\ominus / RT) \tag{14.13b}$$

then, at 298.15 K

$$K_p = 2.26 \times 10^{140}$$

Also,

$$\Delta n = (1 + 2) - (1 + 2) = 0$$

and so

$$RT^{-\Delta n} = (8.314 \times 298.15)^0 = 1$$

giving, from equation (14.22b)

$$K_c = K_p \, RT^{-\Delta n} = 2.26 \times 10^{140}$$

For the combustion of methane, K_p and K_c are equal, and of the order of 10^{140} – an unimaginably large number. This shows that the numerators in equations (14.12)

$$K_p = \frac{p_{C,eq}{}^c \, p_{D,eq}{}^d}{p_{A,eq}{}^a \, p_{B,eq}{}^b} \tag{14.12}$$

and (14.21b)

$$K_c = \frac{[C]_{eq}{}^c \, [D]_{eq}{}^d}{[A]_{eq}{}^a \, [B]_{eq}{}^b} \tag{14.21b}$$

both of which are driven by the quantity of products in the equilibrium mixture, are vastly larger than the denominators, which are driven by the quantity of reactants in the equilibrium mixture. The combustion of methane is certainly a chemical reaction that goes to completion!

A second example is the formation of ammonia from elemental hydrogen and nitrogen – a reaction which is commercially important as the Haber–Bosch Process

$$N_2(g) + 3\,H_2(g) \rightarrow 2\,NH_3(g)$$

for which (see Table 13.1), at 298.15 K and a pressure of 1 bar

$$\Delta_f G_p^\ominus \, N_2(g) = 0 \, \text{kJ mol}^{-1}$$
$$\Delta_f G_p^\ominus \, H_2(g) = 0 \, \text{kJ mol}^{-1}$$
$$\Delta_f G_p^\ominus \, NH_3(g) = -16.4 \, \text{kJ mol}^{-1}$$

Using equation (13.45) once again

$$\Delta_r G^\ominus = \sum_{\text{Products}} n_{\text{prod}} \Delta_f G_{\text{prod}}^\ominus - \sum_{\text{Reactants}} n_{\text{reac}} \Delta_f G_{\text{reac}}^\ominus \tag{13.45}$$

we have

$$\Delta_r G_p^\ominus = [2 \times (-16.4)] - [(0) + (0)] = -32.8 \, \text{kJ}$$

Since

$$K_p = \exp(-\Delta_r G^\ominus / RT) \tag{14.13b}$$

then at 298.15 K

$$K_p = 5.6 \times 10^5$$

For the synthesis of ammonia,

$$\Delta n = (2) - (1 + 3) = -2$$

and so

$$RT^{-\Delta n} = (8.314 \times 298.15)^2 = 6.145 \times 10^6$$

giving, from equation (14.22b)

$$K_c = K_p \, RT^{-\Delta n} \tag{14.22b}$$
$$K_c = K_p \, RT^{-\Delta n} = 3.45 \times 10^{12}$$

The synthesis of ammonia is very different from the combustion of methane: firstly, the magnitudes of K_p are vastly different – some 2×10^{140} for methane, as compared to approximately 6×10^5 for ammonia; secondly, $K_p = K_c$ for methane, but for ammonia, $K_p \approx 6 \times 10^5$, whereas $K_c \approx 3 \times 10^{13}$.

Although the magnitude of K_p for ammonia, some 6×10^5, is a number that, from a 'human' point of view is quite large, it's not obvious what that means in terms of the real, physical, composition of the N_2, H_2, NH_3 equilibrium mixture. This is important industrially: the objective of the ammonia synthesis process is to produce ammonia from nitrogen and hydrogen, and so the purer the final product the better – any residual unreacted nitrogen and hydrogen are, in essence, contaminants. So what actually is the level of the 'contaminants' of an equilibrium mixture with a value of $K_p \approx 6 \times 10^5$? Given that K_p is a ratio of products to reactants, it might be thought that, if $K_p \approx 6 \times 10^5$, then for every 6×10^5 molecules of product, there is 1 residual molecule of reactant, implying that the unreacted hydrogen and nitrogen is around 1 part per 6×10^5 – say, approximately 2 parts per million. If this is the case, then the extent-of-reaction ξ_{eq} is approximately 1 minus 2 parts per million, namely about 0.999998. Is this intuitive hunch correct?

To test this, we start with what we know: in general

$$K_p = \frac{p_{C,eq}^c \, p_{D,eq}^d}{p_{A,eq}^a \, p_{B,eq}^b} \tag{14.12}$$

and for the ammonia synthesis reaction

$$N_2(g) + 3\,H_2(g) \rightarrow 2\,NH_3(g)$$

for which $a = 1$, $b = 3$, $c = 2$, $d = 0$, $\Delta n = [(c + d) - (a + b)] = -2$, and $K_p \approx 6 \times 10^5$ implying that

$$K_p = \frac{p_{NH_3,eq}^2}{p_{N_2,eq} \, p_{H_2,eq}^3} = 6 \times 10^5 \tag{14.24}$$

Our task is to determine ξ_{eq}, the equilibrium extent-of-reaction. We now invoke equations (14.6a), (14.6b) and (14.6c), with $P = 1$ bar

$$p_{A,eq} = p_{N_2,eq} = \frac{a(1-\xi_{eq})P}{(a+b) + \xi_{eq}\Delta n} = \frac{1-\xi_{eq}}{4-2\xi_{eq}} \tag{14.6a}$$

$$p_{B,eq} = p_{H_2,eq} = \frac{b(1-\xi_{eq})P}{(a+b) + \xi_{eq}\Delta n} = \frac{3(1-\xi_{eq})}{4-2\xi_{eq}} \tag{14.6b}$$

$$p_{C,eq} = p_{NH_3,eq} = \frac{c\xi_{eq}P}{(a+b) + \xi_{eq}\Delta n} = \frac{2\xi_{eq}}{4-2\xi_{eq}} \tag{14.6c}$$

We may use these values in equation (14.24) to give

$$K_p = \frac{[2\xi_{eq}/(4-2\xi_{eq})]^2}{[(1-\xi_{eq})/(4-2\xi_{eq})][3(1-\xi_{eq})/(4-2\xi_{eq})]^3}$$

$$= \left(\frac{\xi_{eq}}{(2-\xi_{eq})}\right)^2 \left(\frac{2(2-\xi_{eq})}{(1-\xi_{eq})}\right) \left(\frac{2^3(2-\xi_{eq})^3}{3^3(1-\xi_{eq})^3}\right) = \frac{16}{27} \frac{\xi_{eq}^2(2-\xi_{eq})^2}{(1-\xi_{eq})^4} \tag{14.25}$$

In equation (14.25), ξ_{eq} is the extent-of-reaction corresponding to the equilibrium mixture, the fraction of the reactant molecules that have reacted. If we now define a variable ζ

$$\zeta = 1 - \xi \tag{14.26}$$

then, physically, ζ represents the fraction of reactants that have not yet reacted, the fraction of reactants still remaining at any stage during the reaction. Using equation (14.26), equation (14.25) becomes

$$K_p = \frac{16}{27} \frac{(1-\zeta_{eq})^2(1+\zeta_{eq})^2}{\zeta_{eq}^4} = \frac{16}{27} \frac{[(1-\zeta_{eq})(1+\zeta_{eq})]^2}{\zeta_{eq}^4}$$

and so

$$K_p = \frac{16}{27} \frac{(1-\zeta_{eq}^2)^2}{\zeta_{eq}^4} \tag{14.27}$$

Given that $K_p \approx 6 \times 10^5$, we expect that ξ_{eq} is quite, if not very, close to 1, and so $\zeta_{eq} = 1 - \xi_{eq}$ will be correspondingly small, close to 0. In equation (14.27), it is likely that, to a good approximation $(1 - \zeta_{eq}) \approx 1$ and to an even better approximation, $(1 - \zeta_{eq}^2) \approx 1$. Under these approximations, equation (14.27) becomes

$$K_p \approx \frac{16}{27\zeta_{eq}^4} = \frac{16}{27(1-\xi_{eq})^4} \tag{14.28a}$$

and so

$$\zeta_{eq}^4 \approx \frac{16}{27K_p} = (1-\xi_{eq})^4 \tag{14.28b}$$

If $K_p \approx 6 \times 10^5$, then

$$\zeta_{eq}^4 \approx \frac{16}{27 \times 6 \times 10^5} = (1-\xi_{eq})^4$$

giving

$$\zeta_{eq} \approx 0.03$$

and therefore

$$\xi_{eq} \approx 0.97$$

This analysis indicates that that, at a temperature of 298.15 K, and a pressure of 1 bar, for the synthesis of ammonia for which $K_p \approx 6 \times 10^5$, the equilibrium extent of-reaction ξ_{eq} is approximately 0.97, implying that about 97% of the nitrogen and hydrogen reactants have in fact reacted, and that 3% remains unreacted. So the equilibrium mixture probably contains rather more unreacted reactants that one might intuitively infer from the value of $K_p \approx 6 \times 10^5$. Although this analysis involved a number of approximations, we note that the approximations are, in fact, very reasonable: using the spreadsheet to simulate equation (14.7), and that was the source of the graphs shown in Figures 13.2 and 13.3, the minimum of the $G_{sys}(\xi)$ curve was found to correspond to a value of $\xi_{eq} = 0.97$.

Equation (14.28a) can be used to determine the value of K_p that needs to be exceeded for any given equilibrium extent-of-reaction. So, for example, if we require the ammonia synthesis equilibrium mixture to correspond to a value of $\xi_{eq} = 0.999$ (corresponding to unreacted reactants at about 1 part per thousand), then K_p needs to be at least about 6×10^{11}; a value of $\xi_{eq} = 0.999999$ (corresponding to unreacted reactants at about 1 part per million), then K_p needs to be at least about 6×10^{23}.

This therefore raises the question "if K_p is about 6×10^5 at 298 K, at what temperature does K_p have a value of, say, 10^{12}?".

To answer this question, it is very tempting to set $K_p = 10^{12}$ in equation (14.13b), with $\Delta_r G_p^\circ = -16.4 \times 2$ kJ (the factor of 2 is required given the stoichiometry of the reaction as written) and $R = 8.314$ J K^{-1} mol^{-1}, giving a temperature of about 143 K. Algebraically, that works: given those numbers for K_p, $\Delta_r G_p^\circ$ and R, equation (14.13b) does indeed compute the temperature as 145 K. Scientifically, however, that's the wrong answer. And the reason is attributable to the assumptions that underpin the two variables in equation (14.13b), $\Delta_r G_p^\circ$ and T.

By definition, $\Delta_r G_p^\circ$ is given by equation (13.45)

$$\Delta_r G^\circ = \sum_{\text{Products}} n_{\text{prod}} \Delta_f G_{\text{prod}}^\circ - \sum_{\text{Reactants}} n_{\text{reac}} \Delta_f G_{\text{reac}}^\circ \qquad (13.45)$$

in which all reactants and products are in their ideal gas standard states, and T is the temperature to which those standard states apply. Reference to the definitions of standard states as shown in Table 6.4 shows that the value of $\Delta_r G_p^\circ = -16.4$ kJ per mol of NH$_3$ corresponds to a temperature T of 298.15 K; if a different temperature is used, then the value of $\Delta_r G_p^\circ$ will be different too. Use of equation (14.13b) requires that the values of $\Delta_r G_p^\circ$ and T are *mutually consistent*: the value of $\Delta_r G_p^\circ$ must be the appropriate value for the corresponding value of T. That was the case when we computed $K_p = 5.6 \times 10^5$ at 298.15 K, but if we change the temperature, then we must change the value of $\Delta_r G_p^\circ$ as well – it is not valid to set $K_p = 10^{12}$, to keep $\Delta_r G_p^\circ = -16.4$ kJ per mol of NH$_3$, and then use equation (14.13b) to compute a temperature.

This therefore creates a problem: if we wish to determine the temperature at which $K_p \approx 10^{12}$, how do we do it? We can't use equation (14.13b) 'backwards' in the form

$$T = -\frac{\Delta_r G_p^{\ominus}}{R \ln K_p} \tag{14.13c}$$

for if we set K_p to any particular value, say 10^{12}, and, given that we know that $R = 8.314$ J K^{-1} mol^{-1}, we're still left with a single equation with two unknowns, $\Delta_r G_p^{\ominus}$ and T. Equation (14.13c) therefore doesn't help us determine directly how the equilibrium constant K_p varies with temperature T: we have to solve this problem in a different way – a way we will define in the next section.

14.10 How the equilibrium constant varies with temperature

To solve the problem of how the equilibrium constant K_p for the generalised ideal gas reversible reaction

$$aA + bB \rightleftharpoons cC + dD$$

varies with the temperature T, we return to equation (14.13a)

$$\Delta_r G_p^{\ominus} = -RT \ln K_p \tag{14.13a}$$

in which $\Delta_r G_p^{\ominus}$ is the value of the change in Gibbs free energy for the reaction taking place with the reactants and products in their standard states at a constant pressure of 1 bar, and at a constant temperature T.

If we write equation (14.13a) as

$$\frac{\Delta_r G_p^{\ominus}}{T} = -R \ln K_p \tag{14.13d}$$

we may then differentiate with respect to temperature, at constant pressure, giving

$$\left(\frac{\partial (\Delta_r G_p^{\ominus}/T)}{\partial T}\right)_P = -R \left(\frac{\partial \ln K_p}{\partial T}\right)_P \tag{14.29a}$$

In this equation, $\Delta_r G_p^{\ominus}$ refers to the change in Gibbs free energy for a reaction in which the reactants and the products are in their standard states. Since the definition of standard states specifies a pressure of 1 bar, the reaction necessarily takes place at constant pressure, and so both $\Delta_r G_p^{\ominus}$ and K_p vary only with temperature. Equation (14.29a) is therefore more correctly written using total derivatives rather than partial derivatives as

$$\frac{d(\Delta_r G_p^{\ominus}/T)}{dT} = -R \frac{d \ln K_p}{dT} \tag{14.29b}$$

The left-hand side of equation (14.29b) is very similar to the left-hand side of the Gibbs-Helmholtz equation, equation (13.28)

$$\left(\frac{\partial (G/T)}{\partial T}\right)_P = -\frac{H}{T^2} \tag{13.28}$$

subject to two differences.

One difference is the use of a total derivative in equation (14.29b), in contrast to the use of a partial derivative at constant pressure in equation (13.28), but that has just been explained; the second concerns the presence in equation (14.29b) of a change $\Delta_r G_p^\circ$ in standard ideal gas Gibbs free energies, in contrast to the general value G of the Gibbs free energy associated with a given state of a system, as in equation (13.28).

To resolve this second difference, consider a general change in state, from an initial state [A], associated with an enthalpy H_A and a Gibbs free energy G_A, to a final state [B], associated with an enthalpy H_B and a Gibbs free energy G_B. Since $\Delta G = G_B - G_A$, then $\Delta G/T = G_B/T - G_A/T$ and so

$$\left(\frac{\partial (\Delta G/T)}{\partial T}\right)_P = \left(\frac{\partial (G_B/T)}{\partial T}\right)_P - \left(\frac{\partial (G_A/T)}{\partial T}\right)_P$$

Using the Gibbs-Helmholtz equation, equation (13.28), for these two terms, we have

$$\left(\frac{\partial (\Delta G/T)}{\partial T}\right)_P = -\frac{H_B}{T^2} + \frac{H_A}{T^2} = -\frac{(H_B - H_A)}{T^2} = -\frac{\Delta H}{T^2}$$

and so, for a generalised chemical reaction

$$\left(\frac{\partial (\Delta_r G/T)}{\partial T}\right)_P = -\frac{\Delta_r H}{T^2}$$

Accordingly, if the reaction takes place under standard conditions, which implies constant pressure

$$\frac{d(\Delta_r G^\circ/T)}{dT} = -\frac{\Delta_r H^\circ}{T^2} \tag{14.30}$$

Equation (14.30) applies to all standard states; it also resolves the differences between the left-hand sides of equations (14.29b) and (13.28), and implies that, for any general gas phase reaction

$$\frac{d \ln K_p}{dT} = \frac{\Delta_r H^\circ}{RT^2} \tag{14.31a}$$

In equation (14.31a), $\Delta_r H^\circ$ represents the standard enthalpy change for the reaction for which K_p is the pressure equilibrium constant, as defined by equation (6.21)

$$\Delta_r H^\circ = \sum_{\text{Products}} n_{\text{prod}} \Delta_f H_{\text{prod}}^\circ - \sum_{\text{Reactants}} n_{\text{reac}} \Delta_f H_{\text{reac}}^\circ \tag{6.21}$$

in which $\Delta_f H_i^\circ$ is the standard molar enthalpy of formation of component i, which, for an ideal gas, corresponds to the pure component i at the standard pressure $P^\circ = 1$ bar.

Equation (14.31a) is known as the **van't Hoff equation**, or **van't Hoff isochore**, named after the Dutch chemist who won the very first Nobel prize for chemistry in 1901, and it specifies how the equilibrium constant K_p varies with the temperature T at constant pressure.

An alternative form of equation (14.31a) is

$$d \ln K_p = -\frac{\Delta_r H^\circ}{T} d\left(\frac{1}{T}\right) \tag{14.31b}$$

from which we see that, if the equilibrium constant K_p is measured at different temperatures T, then a plot of $\ln K_p$ as the vertical axis against $1/T$ as the horizontal axis will yield a straight line of slope $-\Delta_r H^\circ/R$, as illustrated in Figure 14.4, so allowing $\Delta_r H^\circ$ to be estimated without any calorimetry.

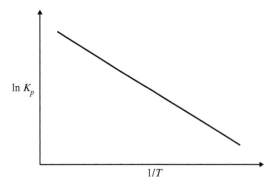

Figure 14.4 **The van't Hoff equation.** A plot of $\ln K_p$ against $1/T$ results in an approximately straight line of slope $-\Delta_r H^\circ/R$.

Fitting experimental measurements of $\ln K_p$ to a straight line assumes that $\Delta_r H^\circ$ is constant over the temperature range measured; in fact, $\Delta_r H^\circ$ is not constant (see equation 6.33 on page 197), but in practice the assumption that $\Delta_r H^\circ$ is constant is usually acceptable.

We note that values of $\Delta_r H^\circ$ determined experimentally using equations 14.31 (a) or 14.31(b), rather than by calculation or by using other methods such as calorimetry, are known as **van't Hoff enthalpies** $\Delta_r H^\circ_{vH}$.

Equations (14.31a) and (14.31b) may be integrated as

$$\int_{K_p(T_1)}^{K_p(T_2)} d\ln K_p = \int_{T_1}^{T_2} \frac{\Delta_r H^\circ}{RT^2} dT = -\int_{T_1}^{T_2} \frac{\Delta_r H^\circ}{R} d\left(\frac{1}{T}\right)$$

giving

$$\ln K_p(T_2) - \ln K_p(T_1) = \frac{\Delta_r H^\circ}{R}\left(\frac{1}{T_1} - \frac{1}{T_2}\right)$$

and

$$\ln \frac{K_p(T_2)}{K_p(T_1)} = \frac{\Delta_r H^\circ}{R}\left(\frac{T_2 - T_1}{T_1 T_2}\right) \tag{14.32}$$

If $K_p(T_1)$ at temperature T_1 is known, then knowledge of $\Delta_r H^\circ$ enables us to use equation (14.32) to estimate $K_p(T_2)$ at temperature T_2, so solving the problem encountered on page 444. So, returning to the ammonia synthesis equation

$$N_2 (g) + 3 H_2 (g) \rightleftharpoons 2 NH_3 (g)$$

we have seen, on page 441, that K_p at 298.15 K is 5.6×10^5, and reference to Table 6.6 shows that $\Delta_r H^\circ$ at that temperature is -45.9 kJ per mol of NH_3. We can now use equation (14.32) to determine the temperature T_2 at which K_p is 10^{12} as

$$\ln\left(\frac{10^{12}}{5.6 \times 10^5}\right) = -\frac{2 \times 45,900}{8.314}\left(\frac{T_2 - 298.15}{298.15 \times T_2}\right)$$

which gives a value of $T_2 = 215$ K.

This example shows that a relatively small change in temperature (from about 300 K to about 215 K) can have a relatively large impact on the equilibrium constant (from about 5.6×10^5 to around 10^{12}) – a point which will be explored in more detail in the next section. Before we do that, however, we need to draw attention to an assumption that we made, but did not state, when we integrated equations (14.31a) and (14.31b) to derive equation (14.32). What we didn't highlight at that time – though we do now – is the assumption that, over the temperature range over which the integration is performed, the standard enthalpy change $\Delta_r H^{\circ}$ remains constant. If this were not the case – if $\Delta_r H^{\circ}$ varied with temperature over that range – then the integration is not valid. Reference to page 197, however, will show that we know that $\Delta_r H^{\circ}$ *does* vary with temperature: that's what equation (6.33) tells us

$$\left(\frac{\partial (\Delta_r H)}{\partial T}\right)_P = \Delta C_P \tag{6.33}$$

This point is even more relevant if you re-read pages 443 and 444, where the whole reason that we could not use equation

$$K_p = \exp(-\Delta_r G_p^{\circ}/RT) \tag{14.13b}$$

to determine how K_p varies with temperature was precisely because $\Delta_r G_p^{\circ}$ is *not* constant, but varies with temperature. So, if equation (14.13b) is ruled out because $\Delta_r G_p^{\circ}$ is not constant over a given temperature range, how is it that we can use equation

$$\ln \frac{K_p(T_2)}{K_p(T_1)} = \frac{\Delta_r H^{\circ}}{R}\left(\frac{T_2 - T_1}{T_1 T_2}\right) \tag{14.32}$$

when we know, according to equation (6.33), that $\Delta_r H^{\circ}$ also varies with temperature?

The answer is that equation (14.32) is not 'right' – it is an approximation, based on the assumption that $\Delta_r H^{\circ}$ is constant over the temperature range T_1 to T_2 – or, rather more pragmatically, that any variation in $\Delta_r H^{\circ}$ is small, and considerably smaller than the corresponding increase in $\Delta_r G_p^{\circ}$. The key issue here is the difference between the magnitude of the rate of change of $\Delta_r H^{\circ}$ with temperature, as defined in general by equation (6.33)

$$\left(\frac{\partial (\Delta_r H)}{\partial T}\right)_P = \Delta C_P \tag{6.33}$$

and the magnitude of the rate of change of $\Delta_r G^{\circ}$ with temperature, as defined in general by equation (13.21)

$$\left(\frac{\partial G}{\partial T}\right)_P = -S \tag{13.21}$$

as applied to a reaction, in the form

$$\left(\frac{\partial \Delta_r G}{\partial T}\right)_P = -\Delta_r S$$

As an example, for the ammonia synthesis reaction

$$N_2(g) + 3 H_2(g) \rightarrow 2 NH_3(g)$$

taking place at 298.15 K, reference to Table 6.1 shows that

$$\Delta C_P = 2 \times 35.1 - (29.1 + 3 \times 28.8) = -45.3 \text{ J K}^{-1}$$

and reference to Table 12.1 gives

$$\Delta_r S^\ominus = 2 \times 192.8 - (191.6 + 3 \times 130.7) = -198.1 \text{ J K}^{-1}$$

In comparing ΔC_P to $\Delta_r S^\ominus$, it's the magnitude that is important rather than the sign, and, as can be seen, ΔC_P is some four to five times smaller than $\Delta_r S^\ominus$. According to equations (6.33) and (13.21), these values therefore imply that $\Delta_r H^\ominus$ changes considerably more slowly with temperature than $\Delta_r G^\ominus$. This is a specific example of a more general result, the consequence of which is to validate the assumption that $\Delta_r H^\ominus$ is reasonably constant, implying that the van't Hoff equation, equation (14.31a) and (14.31b), can, in most cases, be integrated to give equation (14.32).

14.11 Le Chatelier's principle

The example of the synthesis of ammonia

$$N_2(g) + 3 H_2(g) \rightleftharpoons 2 NH_3(g)$$

showed that K_p at a temperature of 298.15 K is about 5.6×10^5, increasing to about 10^{12} at a temperature of about 215 K. An *increase* in the equilibrium constant – implying that the composition of the equilibrium mixture is enriched in ammonia – has been achieved by a *decrease* in temperature. The reaction is also exothermic, for reference to Table 6.6 shows that $\Delta_r H^\ominus$ at 298.15 K is -45.9 kJ per mol of NH_3, implying that as the forward reaction takes place, heat is produced.

As our analysis of the behaviour of K_p has shown, if the temperature of the system is decreased, the system spontaneously responds by shifting its equilibrium point to the right, so producing more heat, as if the system 'wants' to counteract the reduction in temperature by creating heat. The system, of course, does not 'want' anything – we cannot attribute human feelings to a chemical reaction in a test tube. Rather, this is an example of how a chemical system naturally, and spontaneously, responds to a change in circumstances.

More formally, this is an example of what is known as Le Chatelier's principle, named for the French chemist who articulated it in 1884:

> If an external constraint is applied to a system, the system adjusts itself so as to oppose that constraint.

This principle predicts that, for any reaction, a decrease in temperature – whereby the 'external constraint' is cooling the system – will trigger a response in which the system opposes this 'constraint' by creating heat. For the synthesis of ammonia, the forward reaction is exothermic, and so a decrease in the system temperature will shift the reaction to the right, producing an equilibrium mixture richer in ammonia; likewise, an increase in temperature will drive the reaction backwards, so that the equilibrium mixture is poorer in ammonia.

Changes in pressure can also drive similar self-opposing effects. For the ammonia reaction written as

$$N_2(g) + 3 H_2(g) \rightleftharpoons 2 NH_3(g)$$

four molecules of reactant combine to form two molecules of product, so, if the volume of the system is fixed, the pressure exerted by a mixture of pure reactants will be twice that of a system comprising a mixture of pure product. So, if the pressure is increased at constant temperature, the system can self-adjust by producing more ammonia; likewise, the system can counteract a reduction in pressure by the decomposition of ammonia into nitrogen and hydrogen.

The examples of the ammonia reaction are specific, but the principle holds very widely.

One final point about the ammonia reaction. Ammonia is commercially important: it is the raw material for many chemical processes such as the production of agricultural fertilisers and nitric acid. A commercial company wishing to produce ammonia on an industrial scale, and using nitrogen and hydrogen as the raw materials, will therefore seek to run the ammonia synthesis reaction

$$N_2(g) + 3\,H_2(g) \rightleftharpoons 2\,NH_3(g)$$

under conditions that maximise the production of $NH_3(g)$, so that the equilibrium mixture is rich in $NH_3(g)$, and has only a very low concentration of unreacted $N_2(g)$ and $H_2(g)$, which, in this context, are contaminants. The commercial ideal is therefore to run the reaction so that the equilibrium is as far to the right as possible, with a correspondingly large value of K_p.

Given the stoichiometry, and the fact that the reaction is exothermic, this suggests a combination of low temperature, and high pressure. Both of these are possible, but the combination can be expensive; lowering the temperature, however, creates a different problem. In general, chemical reactions are faster at higher temperatures, and slower at lower ones – so the lower the temperature, the slower the reaction. From a business point of view, although lowering the temperature for the synthesis of ammonia gives an equilibrium mixture rich in ammonia, the benefit of an equilibrium mixture rich in product, and low in contaminants, can be considerably offset if it takes a long time for that equilibrium to be reached. Might a reaction which runs more quickly, but requires some further processing to remove unwanted hydrogen and nitrogen, be more commercially viable than a reaction that runs much more slowly, but yields a higher quality initial product? Defining the optimal conditions for an industrial process is therefore not just a matter of temperature and pressure; it is also a matter of how fast the reaction runs.

This highlights one of the limitations of thermodynamics: thermodynamics addresses questions such as "what is the equilibrium mixture of a chemical reaction under given conditions?" and "if those conditions change, what is the new equilibrium mixture?"; thermodynamics, however, cannot answer the question "how quickly, or indeed slowly, is the thermodynamic equilibrium state reached?". This is a matter of chemical kinetics. So, let's move beyond thermodynamics to kinetics …

14.12 Thermodynamics meets kinetics

Equations (14.12)

$$K_p = \frac{p_{C,eq}{}^c\, p_{D,eq}{}^d}{p_{A,eq}{}^a\, p_{B,eq}{}^b} \tag{14.12}$$

and (14.21b)

$$K_c = \frac{[C]_{eq}^c \, [D]_{eq}^d}{[A]_{eq}^a \, [B]_{eq}^b} \tag{14.21b}$$

define the equilibrium constants K_p and K_c of the reversible gas phase reaction

$$aA + bB \rightleftharpoons cC + dD$$

taking place at a constant pressure and temperature, in terms of thermodynamic state functions, the partial pressures p_A, p_B, p_C and p_D, or the molar concentrations [A], [B], [C], and [D], of the reactants and the products, each of which is well-defined at equilibrium. Macroscopically, the equilibrium mixture of the reactants A and B, and the products C and D, remains stable; microscopically, however, individual molecules are moving at random throughout the available volume, they are colliding with one another, and – as a consequence – they are reacting. If, for simplicity, we assume that $a = b = c = d = 1$, then a collision between one molecule of A and one molecule of B could produce one molecule of C and one molecule of D; similarly, a collision between one molecule of C and one molecule of D could result in the formation of one molecule of A and one molecule of B. Even though, macroscopically, the system is in thermodynamic equilibrium, microscopically, the two reactions

$$A + B \rightarrow C + D$$

and

$$C + D \rightarrow A + B$$

are still taking place.

Given the dynamic turmoil happening microscopically, how is it possible for the system to be stable macroscopically?

The answer to this question introduces the concept of the **rate of reaction** – a measure of how fast a reaction takes place, as determined, for example, by assessing, for any reaction

$$A + B \rightarrow C + D$$

how much C or D is produced over any given time, and how much A or B is consumed over that time. The study of how fast chemical reactions take place is known as **chemical kinetics**, and, as an example, consider the 'forward' reaction

$$A + B \rightarrow C + D$$

Suppose that an experimental study, using the methods of chemical kinetics, measures the rate of this reaction, under given conditions of temperature and pressure, as v_f; similarly, let v_r represent the rate of the reverse reaction

$$C + D \rightarrow A + B$$

at the same temperature and pressure.

If, at any time, the rate v_f of the forward reaction is greater than the rate v_r of the reverse reaction, then, over a short period of time, there will be a net increase in the products, C and D, and a net decrease in the reactants A and B. Similarly, if at any time $v_f < v_r$, then over a short period of time, there will be a net decrease in the products, C and D, and a net increase in the reactants A and B. Suppose, however, the two rates are equal, so that $v_f = v_r$: as fast as 'new' products are formed by the forward reaction, they are consumed by the reverse reaction. This is known as a state of **microscopic dynamic equilibrium**, or, more simply, **dynamic equilibrium**. Over any short period of time, although both the forward and the reverse reactions are taking place microscopically, macroscopically, the quantities of A, B, C and D all remain unchanged – the system is in chemical, as well as thermodynamic, equilibrium. So, for a reversible reaction

$$aA + bB \rightleftharpoons cC + dD$$

in both chemical and thermodynamic equilibrium, two conditions are simultaneously fulfilled: kinetically

$$v_f = v_r \tag{14.33}$$

and thermodynamically

$$K_p = \frac{p_{C,eq}{}^c \, p_{D,eq}{}^d}{p_{A,eq}{}^a \, p_{B,eq}{}^b} \tag{14.12}$$

and

$$K_c = \frac{[C]_{eq}{}^c \, [D]_{eq}{}^d}{[A]_{eq}{}^a \, [B]_{eq}{}^b} \tag{14.21b}$$

Equation (14.33) must be linked in some way to equations (14.12) and (14.21b), for all are describing, in rather different ways, the same phenomenon. To determine the link, we need to express the reaction rates v_f and v_r in terms of parameters relevant to the chemical reaction in question, ideally using thermodynamic state functions. To do this, consider a generalised gas phase reversible reaction of the form

$$aA + bB \rightleftharpoons cC + dD$$

For the forward reaction to take place, molecules of A and B need to be in direct contact – the molecules must collide. We would therefore expect that the greater the number of molecular collisions that take place per second, the greater the rate of reaction. Furthermore, at any given temperature, the number of molecular collisions per second is likely to be greater when there are more, rather than fewer, of the appropriate molecules within any given volume. So, consider a gas mixture, of total volume V and at a given temperature T, which contains, at any time, n_i molecules of component i, at a molar concentration n_i/V. If the gas mixture is ideal, we may use Dalton's law of partial pressures (see page 48) to define the partial pressure p_i of component i as

$$p_i = n_i \frac{RT}{V}$$

Since the reaction requires molecules to come into direct contact, it is likely that the higher the molar concentration n_i/V, and hence the higher the corresponding partial pressure p_i, the greater the likelihood of a reaction. This is indeed the case. Chemical kinetics is a big subject in its own right (see, for example, the references), so we will quote only the most important results here. For a gas phase reaction

$$aA + bB \rightarrow cC + dD$$

taking place at a given temperature and pressure, the instantaneous rate $v_f(\tau)$ of this reaction, at any time τ after the reaction has started, is given by

$$v_f(\tau) = k_f [A(\tau)]^a [B(\tau)]^b \qquad (14.34a)$$

where k_f is a (temperature dependent) constant, known as the **reaction rate constant**, and $[A(\tau)]$ and $[B(\tau)]$ are the molar concentrations of the reactants, as determined by the quantities of A and B present at time τ.

An alternative, and equivalent, expression for $v_f(\tau)$ is equation (14.34b)

$$v_f(\tau) = k_f' [p_A(\tau)]^a [p_B(\tau)]^b \qquad (14.34b)$$

in which $v_f(\tau)$ is related to a temperature-dependent rate constant k_f' (different from the rate constant k_f), and $p_A(\tau)$ and $p_B(\tau)$ are the partial pressures attributable to the quantities of A and B present at time τ.

As we saw on page 438, at any time τ, for any component i

$$p_A(\tau) = [I(\tau)] RT$$

and so, for this particular reaction, the rate constants k_f and k_f' are related as

$$k_f = k_f' RT^{(a+b)}$$

The rate of this reaction, $v_f(\tau)$, is the instantaneous rate, at time τ. The meaning of this is evident on consideration of what it actually happening during the reaction, and the form of equations (14.34a) and (14.34b). At any time τ, during the course of the reaction, the system comprises a number $n_A(\tau)$ of molecules of A, with a molar concentration $[A(\tau)]$, exerting a partial pressure $p_A(\tau)$, reacting with a number $n_B(\tau)$ of molecules of B, with a concentration $[B(\tau)]$, exerting a partial pressure $p_B(\tau)$. At that instant, the rate $v_f(\tau)$ of the reaction is given by equations (14.34a) and (14.34b). A short time later, at time $\tau + \delta\tau$, because the reaction has been taking place, there are now fewer molecules of both A and B. If the volume V remains unchanged, the reduction in the number of molecules A and B implies that the concentrations $[A(\tau+\delta\tau)]$ and $[B(\tau+\delta\tau)]$, and the corresponding partial pressures $p_A(\tau+\delta\tau)$ and $p_B(\tau+\delta\tau)$, are lower than their earlier values $[A(\tau)]$ and $[B(\tau)]$, and $p_A(\tau)$ and $p_B(\tau)$. Equations (14.34a) and (14.34b) still apply, but the resulting rate $v_f(\tau+\delta\tau)$ is now lower than the rate $v_f(\tau)$, as measured earlier. During the course of the reaction, the variables $[A]$, $[B]$, p_A, p_B, and v_f all change with time – as indeed makes intuitive sense in that all reactions start quite fast (at which time v_f is a positive number), slow down as the reaction progresses (v_f becomes steadily smaller), and then eventually stop (v_f drops to zero) when all the reactants have been consumed, as illustrated in Figure 14.5.

Physically, equations (14.24a) and (14.24b) are an instantaneous view of the reaction; mathematically, they are differential equations (the word 'rate' implies that $v_f(\tau)$ is a time derivative), which can be solved under a variety of conditions, as described in the texts on chemical kinetics. For our current purposes, however, equations (14.24a) and (14.24b) are very helpful, so let us return to the forward gas phase reaction

$$aA + bB \rightarrow cC + dD$$

THERMODYNAMICS MEETS KINETICS

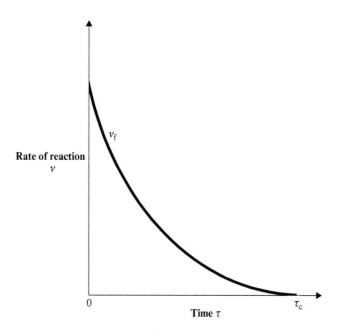

Figure 14.5 The rate of a chemical reaction. This diagram shows a schematic representation of how, at a given temperature and pressure, the rate of a generalized gas phase chemical reaction $aA + bB \rightarrow cC + dD$ varies over time. At time $\tau = 0$, only the reactants are present, and the initial rate of the reaction $v_f(0)$ is given by $k_f [A(0)]^a [B(0)]^b$ or $k_f'[p_A(0)]^a[p_B(0)]^b$. As time progresses, the reactants are consumed, and the corresponding partial pressures decrease. $v_f(\tau)$ therefore also decreases. At time $\tau = \tau_c$, when the reaction reaches completion, there is no A or B left: $[A(\tau_c)]$, $[B(\tau_c)]$, $p_A(\tau_c)$, and $p_B(\tau_c)$, are all zero, implying that $v_f(\tau_c)$ is then zero.

for which, at any time τ, and at any given temperature and pressure, the rate $v_f(\tau)$ of the reaction may be written as

$$v_f(\tau) = k_f [A(\tau)]^a [B(\tau)]^b = k_f' [p_A(\tau)]^a [p_B(\tau)]^b \tag{14.34}$$

Similarly, for the reverse reaction

$$cC + dD \rightarrow aA + bB$$

at the same temperature and pressure

$$v_r(\tau) = k_r [C(\tau)]^c [D(\tau)]^d = k_r'[p_C(\tau)]^c [p_D(\tau)]^d \tag{14.35}$$

where the rate constants k_r and k_r' for the reverse reaction are not necessarily the same as the rate constants k_f and k_f' for the forward reaction.

For the reversible reaction written as

$$aA + bB \rightleftharpoons cC + dD$$

and assuming that we start at time $\tau = 0$ with a mol of A, b mol of B, 0 mol of C, and 0 mol of D, then, as A and B are consumed, the rate $v_f(\tau)$ of the forward reaction decreases with time. The rate of $v_r(\tau)$ of the reverse reaction however, increases with time, as C and D are produced. At any time τ, both the forward and the reverse reactions are taking place at rates in accordance with equations (14.34) and (14.35), each driven by the instantaneous values of the molar concentrations, $[A(\tau)]$, $[B(\tau)]$, $[C(\tau)]$ and $[D(\tau)]$, and the corresponding partial pressures $p_A(\tau)$, $p_B(\tau)$, $p_C(\tau)$ and $p_D(\tau)$.

As time τ increases, $v_f(\tau)$ decreases from some initial value towards zero; simultaneously, $v_r(\tau)$ increases from zero. At some particular time $\tau = \tau_{eq}$, a state will be reached for which $v_f(\tau_{eq}) = v_r(\tau_{eq})$ – as illustrated in Figure 14.6.

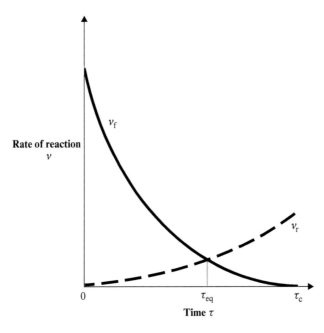

Figure 14.6 Kinetic and thermodynamic equilibrium. This diagram shows a schematic representation of how, at a given temperature and pressure, the forward and reverse rates of a generalized gas phase reversible chemical reaction $aA + bB \rightleftharpoons cC + dD$ varies over time. The rate $v_f(\tau)$ of the forward reaction decreases over time (solid line), whilst simultaneously the rate $v_r(\tau)$ of the reverse reaction increases over time (dashed line). At time $\tau = \tau_{eq}$, the two rates are equal, at which point the forward and reverse reactions are in kinetic equilibrium. The corresponding values of the molar concentrations $[A(\tau_{eq})]$, $[B(\tau_{eq})]$, $[C(\tau_{eq})]$, and $[D(\tau_{eq})]$, and the partial pressures $p_A(\tau_{eq})$, $p_B(\tau_{eq})$, $p_C(\tau_{eq})$, and $p_D(\tau_{eq})$, are those of the quantities of A, B, C and D that define the thermodynamic equilibrium mixture.

Since at time $\tau = \tau_{eq}$, the rates of the forward and reverse reaction are equal, then

$$v_f(\tau_{eq}) = v_r(\tau_{eq}) \tag{14.36}$$

Using equations (14.34) and (14.35), equation (14.36) implies that, at time $\tau = \tau_{eq}$

$$k_f [A(\tau_{eq})]^a [B(\tau_{eq})]^b = k_r [C(\tau_{eq})]^c [D(\tau_{eq})]^d$$

and

$$k_f' [p_A(\tau_{eq})]^a [p_B(\tau_{eq})]^b = k_r' [p_C(\tau_{eq})]^c [p_D(\tau_{eq})]^d$$

and so

$$\frac{k_f}{k_r} = \frac{[C(\tau_{eq})]^c [(D(\tau_{eq})]^d}{[A(\tau_{eq})]^a [B(\tau_{eq})]^b} = \frac{[C]_{eq}^c [D]_{eq}^d}{[A]_{eq}^a [B]_{eq}^b} \tag{14.37a}$$

and

$$\frac{k'_f}{k'_r} = \frac{[p_C(\tau_{eq})]^c\ [p_D(\tau_{eq})]^d}{[p_A(\tau_{eq})]^a\ [p_B(\tau_{eq})]^b} = \frac{p_{C,eq}{}^c\ p_{D,eq}{}^d}{p_{A,eq}{}^a\ p_{B,eq}{}^b} \qquad (14.37b)$$

Because k_f, k_r, k'_f and k'_r are constants, the ratios k_f/k_r and k'_f/k'_r must be constants too, implying that the ratios of molar concentrations, and of partial pressures, on the right hand sides of equations (14.37a) and (14.37b) must also be constant, and independent of time. One possibility might be that each of $[A(\tau_{eq})]$, $[B(\tau_{eq})]$, $[C(\tau_{eq})]$ and $[D(\tau_{eq})]$, and $p_A(\tau_{eq})$, $p_B(\tau_{eq})$, $p_C(\tau_{eq})$ and $p_D(\tau_{eq})$, varies independently over time, such that the ratios defined by equations (14.37a) and (14.37b) happen to remain constant; a far simpler explanation is that each of $[A(\tau_{eq})]$, $[B(\tau_{eq})]$, $[C(\tau_{eq})]$ and $[D(\tau_{eq})]$, and $p_A(\tau_{eq})$, $p_B(\tau_{eq})$, $p_C(\tau_{eq})$ and $p_D(\tau_{eq})$, are themselves constant. If this is the case, then since each of the molar concentrations, and the corresponding partial pressures, is a thermodynamic state function, the fact that each of these state functions is constant over time implies that $[A(\tau_{eq})]$, $[B(\tau_{eq})]$, $[C(\tau_{eq})]$ and $[D(\tau_{eq})]$, and $p_A(\tau_{eq})$, $p_B(\tau_{eq})$, $p_C(\tau_{eq})$ and $p_D(\tau_{eq})$, are those concentrations, and partial pressures, which define the state of thermodynamic equilibrium, which takes us back to equations (14.12) and (14.21b)

$$\frac{k_f}{k_r} = \frac{[C(\tau_{eq})]^c\ [(D(\tau_{eq})]^d}{[A(\tau_{eq})]^a\ [B(\tau_{eq})]^b} = \frac{[C]_{eq}{}^c\ [D]_{eq}{}^d}{[A]_{eq}{}^a\ [B]_{eq}{}^b} = K_c \qquad (14.37d)$$

and (14.12)

$$\frac{k'_f}{k'_r} = \frac{[p_C(\tau_{eq})]^c\ [p_D(\tau_{eq})]^d}{[p_A(\tau_{eq})]^a\ [p_B(\tau_{eq})]^b} = \frac{p_{C,eq}{}^c\ p_{D,eq}{}^d}{p_{A,eq}{}^a\ p_{B,eq}{}^b} = K_p \qquad (14.37e)$$

Thermodynamics meets kinetics!

14.13 The Arrhenius equation

Let us consider once again the gas phase reaction

$aA + bB \rightarrow cC + dD$

As we have seen, at any time τ the rate $v_f(\tau)$ of this reaction, taking place at a given temperature and pressure, may be expressed as

$$v_f(\tau) = k_f\,[A(\tau)]^a\ [B(\tau)]^b \qquad (14.34a)$$

The rate constant k_f can be determined by carrying out a series of experiments, each with different quantities of A and B, and hence different molar concentrations $[A(\tau)]$ and $[B(\tau)]$. If the temperature and pressure are maintained constant for all these experiments, then the rate $v_f(\tau)$ of the reaction will be found to be proportional to the product $[A(\tau)]^a\ [B(\tau)]^b$, and the constant of proportionality will be the rate constant k_f.

If, however, a similar series of experiments is carried out at the same pressure but a different temperature, the rate $v_f(\tau)$ of the reaction will still be found to be proportional to the product $[A(\tau)]^a\ [B(\tau)]^b$, but the constant of proportionality will take a different value, such that the higher the temperature T, the greater the value of k_f, and the faster the reaction takes place. The 'constant' in equation (14.34a) is in fact dependent on temperature, and so is more correctly written as $k_f(T)$.

Measurements of the values of the generalised reaction rate constant $k(T)$ at different temperatures T show that, for many chemical reactions, the relationship

$$k(T) = A\, e^{-\frac{E_a}{RT}} \qquad (14.38)$$

is a good description, in which the term E_a is known as the reaction's **activation energy**, as expressed in J mol^{-1}, and A is the **frequency factor** or **pre-exponential factor**, the units of measurement of which depend on the nature of the reaction (for details on this, please refer to a text on kinetics). Equation (14.38) is known as the **Arrhenius equation**, after the Swedish chemist and Nobel prize winner, Svante Arrhenius.

It has been found experimentally that, over quite wide ranges of temperature, A and E_a are constants for any given chemical reaction. Assuming that A and E_a are indeed constants, we can then infer from equation (14.38) how $k(T)$ varies with temperature. At low temperatures, T is small, E_a/RT is large, and so $\exp(-E_a/RT) = 1/\exp(E_a/RT)$ is small. At low temperatures, $k(T)$ is therefore small, and the corresponding reaction will run slowly. As the temperature increases, E_a/RT continuously decreases, implying that $\exp(-E_a/RT) = 1/\exp(E_a/RT)$ continuously increases, such that at very high temperatures, E_a/RT approaches zero, and $\exp(-E_a/RT) = 1/\exp(E_a/RT)$ approaches the value of 1. The maximum possible value of $k(T)$, as expressed in equation (14.38), is therefore A. Overall, equation (14.28) states that $k(T)$ steadily increases as T increases, so confirming the every-day observation that most chemical reactions go faster at higher temperatures.

Physically, for reactions involving just the decomposition of a single molecule, A is related to the frequency of molecular vibration – this (crudely!) being an indication of the tendency of a single molecule to shake itself apart; for reactions involving multiple molecules, A is a measure of the frequency of molecular collisions, this being sensibly relevant, since reactant molecules need to be in contact for a reaction to take place. Not all molecular vibrations, or collisions, however, result in a reaction, and so the probability that a vibration, or a collision, actually results in a reaction is recognised by the exponential term $\exp(-E_a/RT)$. In this expression, the activation energy E_a is a measure of the energy that needs to be injected into the system for the reaction to take place – in general, before the chemical bonds of the reaction products can be formed, the bonds of the reactants need to be broken, and, as we saw on page 188, breaking bonds is usually endothermic, requiring energy. The microscopic process associated with the breaking of the bonds within the reactants, followed by the formation of the bonds of the products, can be represented diagrammatically, as shown in Figure 14.7.

The horizontal axis in Figure 14.7 represents the progress of a reaction from pure reactants (towards the left) to pure products (towards the right), with the central part of the diagram representing intermediate states through which the reaction might pass. In essence, the horizontal axis is time, but for many chemical reactions, the time scales are very fast, measured in nano, if not pico, seconds. Also, the use of solid lines might seem to imply that the intermediate states have well-defined, single valued, energies – in practice, this is unlikely to be the case, so the diagram is a representation of what is happening as the reaction progresses, rather than a map of reality. What's important is the presence of the central 'hill', as we shall explain.

At the start of a reaction, the reactants have a Gibbs free energy G_R, and when the reaction has gone to completion, the products have a Gibbs free energy G_P. If $G_P < G_R$, as shown in

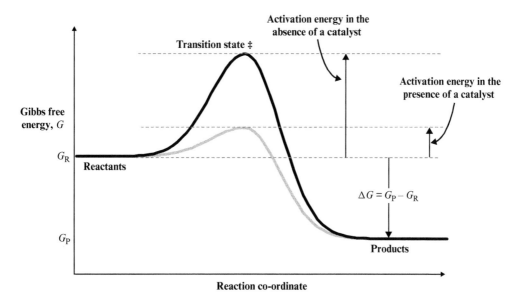

Figure 14.7 Activation energy. The horizontal axis, the 'reaction coordinate', represents the progress of a reaction from reactants to products at a given temperature T. In this diagram, $G_P < G_R$, and so $\Delta G_P = G_P - G_R$ is negative, implying that the reaction from products to reactants is thermodynamically spontaneous. The black line shows a representation of the progress of the 'normal' reaction. The grey line corresponds to a catalysed reaction, in which the catalyst acts to lower the required activation energy. The zone at the 'top of the hill' is known as the activated complex, and the apex as the transition state.

Figure 14.7, then the change $\Delta G = G_P - G_R$ is negative, and the reaction is thermodynamically spontaneous.

Before the reaction can actually take place, however, the chemical bonds in the reactants must be broken, so creating the situation in which the chemical bonds of the products can be formed. As we saw on page 188, in general, the breaking of chemical bonds is endothermic – energy has to be put into the system to break molecules apart. This is represented by the 'uphill' slope of the left-hand side of the 'hill' in the centre of Figure 14.7. When sufficient energy has been injected, the system's Gibbs free energy has risen to the 'top of the hill', this zone being known as the **activated complex**, and the apex as the **transition state**, often represented by the symbol ‡. Thereafter, the system 'rolls down the slope' on the right of the 'hill', releasing energy (corresponding to the usually exothermic process of bond formation), until the final state has been reached, the thermodynamically stable state of the products.

Figure 14.7 captures the essence of the blending of thermodynamics and kinetics. Thermodynamics is primarily concerned with the equilibrium states on the left and on the right, whilst kinetics is concerned with the process of 'climbing the activation hill', the mechanisms that take place at the molecular level as the bonds and broken and reformed, and the time it all takes.

So, consider a reaction such as

Reactants → Products

which, in principle, goes to completion, as shown in Figure 14.7. If ΔG at any given temperature and pressure for this reaction is negative, then, once again in principle, the reactants will change, spontaneously, into the products. Thermodynamically, this is true, but thermodynamics takes no account of time: at some point within an infinite amount of time, the reaction will happen. Waiting an infinite amount of time, however, isn't very convenient: for a reaction to take place in a reasonable amount of time, not only must the thermodynamics imply spontaneity, but the kinetics must be favourable too, so that a sufficient number of reactant molecules can 'climb the activation hill', and so 'fall down' the other side.

The key determinant is the 'height' of the 'activation hill', as determined by the magnitude of the activation energy E_a – or, better, by the comparison between E_a and the value of RT at a given temperature T. Physically, kT is the average kinetic energy of a single molecule at temperature T, and so RT is a measure of the average kinetic energy of 1 mol of molecules. If a reaction is such that E_a is of a similar magnitude to RT, then there is enough thermal energy to enable the 'average molecule' to surmount the 'activation hill', and we would expect, intuitively, that, at that temperature, the reaction would take place spontaneously, at a detectable rate. This condition corresponds to $E_a/RT \approx 1$, implying that the exponential term $\exp(-E_a/RT)$ in equation (14.38) is of order 0.4. If RT is considerably larger than E_a, the average molecular thermal energy is greater than the activation energy, and so we would expect the reaction to be very fast: in this case $E_a/RT < 1$, possibly significantly so, and $\exp(-E_a/RT)$ approaches 1. A third situation is that for which RT is very much less than E_a so that $E_a/RT > 1$, or indeed $\gg 1$. Very few molecules will have sufficient thermal energy to reach the transition state at the 'top of the hill', and so we would expect the reaction to be very slow, or not to take place at all. As the temperature drops, RT decreases relative to E_a, the ratio E_a/RT becomes progressively larger, and $\exp(-E_a/RT)$ ever smaller: the lower the temperature, the slower the reaction.

To make that discussion more concrete, suppose that T is room temperature, say 298 K, implying that $RT \approx 2.5$ kJ mol^{-1}. If

- $E_a = 2.5$ kJ mol^{-1}, $E_a/RT = 1$, and $\exp(-E_a/RT) = 0.37$
- $E_a = 0.25$ kJ mol^{-1}, $E_a/RT = 0.1$, and $\exp(-E_a/RT) = 0.90$
- $E_a = 25$ kJ mol^{-1}, $E_a/RT = 10$, and $\exp(-E_a/RT) = 5 \times 10^{-5}$
- $E_a = 250$ kJ mol^{-1}, $E_a/RT = 100$, and $\exp(-E_a/RT) = 4 \times 10^{-44}$ – this being such a small number – from a practical standpoint, zero – that the reaction doesn't take place at all.

As the ratio E_a/RT increases, $\exp(-E_a/RT)$ decreases very quickly, showing that this ratio is very significant as regards the likelihood of a reaction actually taking place. For the combustion of a fuel such as octane, E_a is some 180 kJ mol^{-1}, corresponding to a value of $\exp(-E_a/RT)$ of about 3×10^{-32}, so you can be sure that the fuel in a car's petrol tank will remain stable, and won't burst into flames spontaneously! In this regard, it's worth noting that the combustion of octane is accompanied by a Gibbs free energy change of some $-5,300$ kJ mol^{-1}, a very large negative value, implying that, once the combustion of octane is initiated, as achieved by the spark from the car's spark plug, the reaction is indeed spontaneous.

Equation (14.38)

$$k(T) = A\, e^{-\frac{E_a}{RT}} \tag{14.38}$$

suggests two strategies that can be adopted if we wish to increase the speed of a chemical reaction. The first is by increasing the temperature T – if the (absolute) temperature is doubled, from, say $T = 300$ K to $2T = 600$ K, and if we assume that both A and E_a remain constant over that temperature range, then the rate of reaction increases by a factor of $[\exp(-E_a/2RT)/\exp(-E_a/RT)] \approx \exp(E_a/2RT) \approx \exp(E_a/5000)$. If E_a is, say, 180 kJ mol^{-1}, $\exp(180000/5000) \approx 4 \times 10^{15}$ – a very big number – so this has a very significant effect.

This explains why many reactions – especially exothermic combustions – are self-sustaining. If a reaction can be initiated by raising the temperature of the system by a suitable amount (for example, by putting a match to paper), then, as soon as the reaction starts, the local release of heat, attributable to the exothermic enthalpy change, can raise the temperature even more, so enabling the reaction to go even faster, so creating even more exothermic energy...

The second strategy is to reduce the activation energy E_a, so making the 'activation hill' lower: halving the activation energy has exactly the same numerical effect as doubling the (absolute) temperature. Hence the importance of catalysts: as shown by the grey line in Figure 14.7, a catalyst acts to lower the activation energy, at a given temperature, of the reaction being catalysed, and this is very important. Catalysts allow a reaction at a given temperature to go faster, as happens in all living systems, where enzymes play an essential role in ensuring that biochemical reactions taking place at an appropriate rate at the organism's normal temperature. The use of catalysts in industrial processes also has huge economic importance, in allowing many manufacturing activities to take place at a lower temperature than would otherwise be necessary, so saving energy costs, and giving environmental benefits too.

Returning to the synthesis of ammonia

$N_2(g) + 3 H_2(g) \rightleftharpoons 2 NH_3(g)$

it was the discovery of a suitable catalyst – originally osmium, but soon replaced by much cheaper iron-based materials – that transformed a laboratory process into an industry that produces some 140 million tonnes a year, from plants around the world that each manufacture 3000 or more tonnes each day. In accordance with Le Chatelier's principle, the forward reaction is favoured by higher pressure, but although Le Chatelier's principle suggests that a low temperature should be used, the reaction at low temperatures is far too slow. Accordingly, the operating conditions for the industrial process are typically a pressure of 100 to 200 bar, and a temperature of between 400 K and 700 K, at which the catalysed reaction is sufficiently fast, even if the thermodynamic equilibrium mixture is not ideal.

14.14 The overall effect of temperature on chemical reactions

We conclude this chapter with a summary of how a change in temperature affects a general ideal gas chemical reaction

$a\text{A} + b\text{B} \rightleftharpoons c\text{C} + d\text{D}$

There are two key factors, one thermodynamic, the other kinetic.

The, first, thermodynamic, factor determines the equilibrium composition, as defined by the equilibrium constant K_p

$$K_p = \frac{p_{C,eq}{}^c \, p_{D,eq}{}^d}{p_{A,eq}{}^a \, p_{B,eq}{}^b} \tag{14.12}$$

Since K_p changes with temperature, as described by the van't Hoff equation

$$\frac{d \ln K_p}{dT} = \frac{\Delta_r H^\circ}{RT^2} \quad (14.31a)$$

then, if $d(\ln K_p)/dT$ is positive, as the temperature T increases, so does $\ln K_p$, and hence K_p itself. An increase in temperature therefore drives the reaction to the right, such that the equilibrium mixture will become progressively richer in products; conversely, a decrease in temperature favours the reactants. Alternatively, if $d(\ln K_p)/dT$ is negative, then as the temperature T increases, $\ln K_p$ and K_p both decrease, driving the reaction to the left, and favouring reactants over products; conversely, a decrease in temperature drives the reaction to the right, towards the products. As can be seen from the van't Hoff equation, equation (14.31a), since the temperature T is necessarily always positive, the sign of $d(\ln K_p)/dT$ is determined by the sign of ΔH°:

- If a reaction is such that ΔH° is positive – if the reaction is endothermic – then an increase in temperature will drive the reaction towards the products; likewise, a decrease in temperature will favour the reactants.
- If a reaction is such that ΔH° is negative – if the reaction is exothermic – then an increase in temperature will drive the reaction towards the reactants; likewise, a decrease in temperature will favour the products.

Both of these statements are in accordance with Le Chatelier's principle.

The second, kinetic, factor concerns the rate at which the reaction takes place, as described by the Arrhenius equation

$$k(T) = Ae^{-\frac{E_a}{RT}} \quad (14.38)$$

As the temperature of the system increases, the average energy of a molecule within the system increases, and so an increasingly greater number of molecules are able to 'climb the activation energy hill'. The reaction therefore takes place faster. And since, for any given reaction, the activation energy E_a, the temperature T, and the ideal gas constant R are always necessarily positive, as the temperature T increases, the ratio E_a/RT becomes progressively smaller, implying that $\exp(-E_a/RT)$ becomes progressively bigger. An increase in temperature therefore causes the rate of the reaction to increase under all circumstances.

If the reaction is endothermic, then an increase in temperature increases the rate of the reaction (by virtue of the kinetic effect), and also shifts the equilibrium towards the products. If, however, the reaction is exothermic – as is the case for the synthesis of ammonia – then an increase in temperature makes the reaction run faster, but the resulting equilibrium is towards the left, favouring the reactants over the products.

14.15 A final thought

As a conclusion to this chapter, we note that, of all the equations in thermodynamics, probably the single most important is equation (14.13a)

$$\Delta_r G_p^\circ = -RT \ln K_p \quad (14.13a)$$

As we have seen, equation (14.13a) relates the equilibrium constant K_p for the reversible ideal gas reaction

$$aA + bB \rightleftharpoons cC + dD$$

to the standard Gibbs free energy change $\Delta_r G^\ominus$, as given by

$$\Delta_r G_p^\ominus = (c\,\Delta_f G_{p,C}^\ominus + d\,\Delta_f G_{p,D}^\ominus) - (a\,\Delta_f G_{p,A}^\ominus + b\,\Delta_f G_{p,B}^\ominus) \tag{14.9}$$

in which each of the standard molar Gibbs free energies $G_i^\ominus$ is defined by reference to the standard state for ideal gases, this being (as shown in Table 6.4) the pure gaseous component i at a pressure of 1 bar, and a stated temperature T.

Systems in equilibrium, or in a steady state, are, in reality, very common, and arise in many physical, chemical and biochemical contexts – for example, the equilibrium between different phases of the same substance, the equilibrium of chemical reactions, and the equilibrium exhibited by the stable state of an enzyme in a living cell. The importance of equation (14.13a) is that it is a particular case – as associated with an ideal gas phase chemical reaction – of a much wider set of equations all of which have the form

$$\Delta G^\ominus = -RT \ln K \tag{14.39}$$

where K is an equilibrium constant which specifies the equilibrium composition of a given system, and $\Delta G^\ominus$ is the corresponding change in standard Gibbs free energy, defined in terms of the standard states appropriate to the equilibrium in question (for example, for a chemical reaction taking place in solution, the appropriate standard states are defined in terms of unit molar concentrations, as shown in Table 6.4). Wherever there is an equilibrium, there is a corresponding equation of this form, and we will meet many examples in the following chapters.

EXERCISES

1. Define:
    - Extent-of-reaction
    - Gibbs free energy of reaction $\Delta_r G_p$
    - Standard Gibbs free energy of reaction $\Delta_r G_p^\ominus$
    - Pressure equilibrium constant, K_p
    - Mole fraction equilibrium constant, K_x
    - Concentration equilibrium constant, K_p
    - Pressure mass action ratio Γ_p
    - Mole fraction mass action ratio Γ_x
    - Concentration mass action ratio Γ_c
    - Rate of reaction
    - Reaction rate constant.
2. Explain Le Chatelier's principle, and show how it predicts what happens when
    - heat is added to a system of a liquid at its boiling point
    - how the equilibrium mixture of a gas phase reaction varies with pressure.

3. In response to the exam question "How does an increase in temperature affect a chemical reaction?", a candidate responds "As the temperature increases, the average molecular energy increases, and this makes the reaction go faster." The candidate is awarded 1 mark out of 10. Is the candidate's response plausible? Write an answer that would be awarded 10 marks.

4. This is an exercise which requires use of a spreadsheet. Its purpose is to explore a generalised reaction of the form $aA + bB \rightarrow cC + dD$, and to generate graphs of the form of Figure 14.2. This is achieved by using the spreadsheet to compute $G_{sys}(\xi)$, for successive values of the extent-of-reaction parameter ξ, according to equation (14.7). This will enable you to see how the equilibrium composition of a chemical reaction changes as the various parameters change.

So, at the top of the spreadsheet, define the system's constants: the stoichiometric coefficients a, b, c and d, and the corresponding standard molar Gibbs free energies G_A°, G_B°, G_C° and G_D° – you can use, for example, data taken from Table 14.1. You will also need a value for the temperature T (= 298.15 K, so as to be consistent with the values of each G_i°), the system pressure P (say, the standard pressure $P^\circ = 10^5$ Pa), and $R = 8.314$ J K^{-1}mol^{-1}.

Next, incorporate a cell to calculate $\Delta n = [(c + d) - (a + b)]$; a second cell to compute $\Delta_r G_p^\circ$, as defined by equation (14.9); and a third cell for K_p, as given by equation (14.13b).

The main part of the spreadsheet calculates $G_{sys}(\xi)$ according to equation (14.7). Accordingly, set up a column for the extent-of-reaction ξ for values from $\xi = 0$ to $\xi = 1$ in increments of 0.01, corresponding to 100 rows. Successive columns can then be used to compute each of the terms on the right hand side of equation (14.7) for each value ξ, so allowing the appropriate columns to be summed to give $G_{sys}(\xi)$. To make the spreadsheet easier to code (and also to de-bug!), you might find it helpful to use several columns for each of the components: so, for component B, one column might compute $b(1 - \xi)$; the next column, $b(1 - \xi)G_B^\circ$; the succeeding column, the quantity $b(1 - \xi)/[(a + b) + \xi \Delta n]$; then the ln term (remembering that the pressure P is in fact 'shorthand' for the ratio P/P°); and finally a column for the sum that defines $G_B(\xi)$. Similar clusters of columns can be used to compute $G_A(\xi)$, $G_C(\xi)$ and $G_D(\xi)$, with a final column for $G_{sys}(\xi) = G_A(\xi) + G_B(\xi) + G_C(\xi) + G_D(\xi)$.

You can then define a line graph, with values of ξ for the x axis, and values of $G_{sys}(\xi)$ for the y axis; in addition, you can define line graphs for each of $G_A(\xi)$, $G_B(\xi)$, $G_C(\xi)$ and $G_D(\xi)$ independently. You may also wish to display the value of K_p on each graph as a legend.

As you vary the parameters – a, b, c and d; G_A°, G_B°, G_C° and G_D°; P and T – you will generate many different graphs, but all should be of the type shown in Figure 14.2, or, if the equilibrium composition happens to be far to the left or far to the right, as shown in Figure 14.1. The shape of any graph, and the value of ξ defining the equilibrium composition, should be in accord with the corresponding value of K_p: if $\sim 10^{-4} < K_p < \sim 10^4$, the graph should be U-shaped; if $K_p < \sim 10^{-4}$, the graph is upward sloping; if $K_p > \sim 10^4$, the graph is downward sloping; if $K_p = 1$, the graph is U-shaped, with the minimum at the half-way point $\xi = 0.5$.

Also, make sure you explore how the same reaction varies with the system pressure P at constant temperature, and that you can interpret (using, for example, Le Chatelier's principle) why changing the pressure changes the equilibrium composition; also, predict how you think the equilibrium composition is likely to change for a given change in pressure (in which direction, and approximately by how much), then change the pressure – and see if your prediction was right!

You can also explore how the reaction varies with the temperature T at constant pressure, but take care! In doing this, there's an implicit assumption. What is it?

PART 4
Chemical applications

15 Phase equilibria

Summary

Two phases, X and Y, are in equilibrium when

$$G_X = G_Y \qquad (15.4a)$$

or

$$\mu_X = \mu_Y \qquad (15.4b)$$

Any pure material in a condensed phase is in equilibrium with that pure material in the vapour phase, and the pressure exerted by that vapour is known as the **equilibrium saturated vapour pressure** p.

For any phase change, a change dP_{ex} in the external pressure results in a change dT_{ph} in the phase transition temperature T_{ph} as given by the **Clapeyron equation**

$$\frac{dP_{ex}}{dT_{ph}} = \frac{\Delta_{ph}H}{T_{ph}\Delta_{ph}V} \qquad (15.7)$$

For the liquid $\rightleftharpoons$ vapour and solid $\rightleftharpoons$ vapour phase changes, we may use a special case of the Clapeyron equation, known as the **Clausius-Clapeyron equation**

$$\frac{d\ln p}{dT_{vap}} = \frac{\Delta_{vap}H}{RT_{ph}^2} \quad \text{or} \quad \frac{d\ln p}{dT_{sub}} = \frac{\Delta_{sub}H}{RT_{sub}^2} \qquad (15.8)$$

15.1 Vapour pressure

Figure 15.1 is a representation of a 'Torricelli barometer', which can be constructed by filling a tube, about 1 m in length, with mercury, and then inverting the tube in a bath of mercury as shown. When the tube is vertical, the mercury in the tube will rise about 0.76 m above the surface of the mercury in the surrounding reservoir. The mercury column is supported by the action of atmospheric pressure on the surrounding mercury, such that the pressure P_{atm} exerted by atmosphere on the open surface of the mercury is equal to the pressure $h\rho g$ attributable to the mass of mercury within the column. Since variations in atmospheric pressure P_{atm} change the height h of the mercury column, the instrument acts as a barometer.

Immediately after the instrument is constructed, the space inside the tube, above the mercury column, is a vacuum, the 'Torricellian vacuum'. But very soon, a few atoms of mercury evaporate, so the space is no longer a void, but contains mercury vapour. Assuming that the external atmospheric pressure remains stable, and that the system is at a constant temperature, then, macroscopically, within the vertical column, all the mercury – in both liquid and gaseous

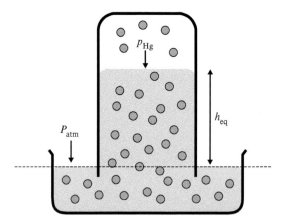

Figure 15.1 A representation of a Torricelli barometer at room temperature. Atmospheric pressure maintains a column of mercury, above which there is a vacuum – but not a total void: a few mercury atoms have evaporated, and they exert a small pressure, the vapour pressure p_{Hg}. Since the pressure attributable to the column of mercury of height h_{eq} m is $h_{eq}\rho g$ (where ρ is the density of mercury, and g is the acceleration due to gravity), then $P_{atm} = h_{eq}\rho g + p_{Hg}$. The height h_{eq} m is therefore not a direct measure of the atmospheric pressure P_{atm}, but rather of the difference $P_{atm} - p_{Hg}$. Since $P_{atm} \gg p_{Hg}$, if measurements of high accuracy are not required, the effect of the vapour pressure of mercury is usually ignored, and h_{eq} m is taken as a measure of P_{atm}.

form – is in thermodynamic equilibrium. Microscopically, however, as we know from the discussion on page 70, all the mercury atoms are moving with an average energy determined by the temperature T. We also know, from Figures 3.10 and 3.11, that, at any temperature T, there is a distribution of molecular energies, so that some molecules have a greater energy than others, and that a few molecules have quite high energies.

Within the liquid, forces such as those attributable to van der Waals attraction act to bind the mercury atoms loosely together, so that the mercury maintains its liquid state. Within the bulk of the mercury, these forces act in all directions, but at the upper surface, the intermolecular attractive forces are asymmetric, acting 'downwards' into the liquid, rather than 'upwards' into the space above. As a consequence, a mercury atom close to the surface that happens to have sufficient energy to overcome the 'downward' forces of intermolecular attraction might be able to 'escape' – a phenomenon we know as **evaporation**. Similarly, a mercury atom that had previously evaporated and is now in the gaseous state might impact the surface of the liquid, and become 'trapped', so condensing back into the liquid state. At any given temperature T, some liquid molecules are evaporating into the gaseous state, whilst some gaseous molecules are condensing into the liquid state, and – as we saw on page 451 – when the rates of these two processes are equal, the system is in microscopic dynamic equilibrium, and also macroscopic thermodynamic equilibrium.

The liquid → vapour transition is a reversible endothermic reaction

Liquid ⇌ Vapour $\hspace{6em} \Delta H = \Delta_{vap} H$

such that at any temperature T, some molecules are in the liquid state, and some in the gaseous state. At any instant, those molecules that are in the gaseous state exert a pressure, known as the **vapour pressure**, and when the liquid ⇌ vapour system is in equilibrium, we refer to the **equilibrium saturated vapour pressure**, or just **saturated vapour pressure**, usually represented by the lower case italic symbol p. As shown in Figure 15.1, the vapour pressure p_{Hg} exerted by the mercury vapour within the Torricellian vacuum 'pushes down' on the mercury column, implying that the original height h of the mercury column is reduced such that the equilibrium height h_{eq} is given by $P_{atm} = h_{eq}\rho g + p_{Hg}$. The height h_{eq} as observed is therefore a measure of $P_{atm} - p_{Hg}$, rather than P_{atm}.

For any given liquid at any given temperature, the higher the equilibrium saturated vapour pressure, the greater the number of molecules in the gas phase, from which we infer that that it has been relatively 'easy' for molecules to evaporate. In physical terms, we would expect that the 'ease' of molecular evaporation will be influenced by two factors:

- firstly, the lighter the molecule, the smaller the amount of energy required for evaporation, and so the higher the vapour pressure, and
- secondly, the stronger the intermolecular forces within the liquid, the greater the amount of energy required for evaporation, and so the lower the vapour pressure.

These predictions are verified by experimental data. So, as an example of the first factor, both di-isopropyl ether ($C_3H_7OC_3H_7$, molecular weight 0.102 kg mol^{-1}) and di-ethyl ether ($C_2H_5OC_2H_5$, molecular weight 0.074 kg mol^{-1}) are non-polar liquids at 298 K, and so the intermolecular forces within the liquids are weak. At that temperature, the saturated vapour pressure of the lighter molecule, di-ethyl ether is 72 kPa, whereas the saturated vapour pressure of the heavier molecule, di-isopropyl ether, is 20 kPa. As regards the second factor, the strong polar nature of water 'holds' the molecules within the liquid phase, and so the saturated vapour pressure has a low value of 3.2 kPa at 298 K, even though the molecular weight of water is 0.018 kg mol^{-1}, much less than diethyl ether; mercury is a special case because it is a liquid composed of very heavy atoms (molecular weight 0.201 kg mol^{-1}) which are mutually interacting as within a metal, and so has an exceptionally low saturated vapour pressure of about 0.3 Pa (not kPa) at 298 K.

As with the generalised reversible gas phase reaction which we examined on pages 423 to 430, at any temperature T, the equilibrium between the liquid and gas phases will be characterised by an equilibrium constant which specifies the relative quantities of liquid and vapour in the equilibrium mixture; furthermore, we should expect that equilibrium constant to change as a result of a change in the temperature T.

To determine the qualitative nature of the change, we may invoke Le Chatelier's principle (see page 448), which tells us that any system will respond by counteracting the effect of any external disturbance. If the temperature T of the system decreases, the system will therefore change so as to produce heat. Since the liquid $\rightarrow$ vapour reaction is endothermic, the equilibrium position will move towards the liquid. Conversely, if the system is heated, the system will respond by absorbing heat, and so the equilibrium position will move to the vapour. Vapour pressures therefore increase with temperature. Intuitively, this is exactly we would expect: as a liquid is heated, not only does the average molecular energy increase, but as we saw in Figure 3.11, relatively more molecules have relatively higher energies, and so more molecules can evaporate.

The discussion so far has focused on the liquid $\rightleftharpoons$ vapour phase transition, and the corresponding vapour pressure, but exactly similar considerations apply to the solid $\rightleftharpoons$ vapour phase transition: at any temperature T, some molecules 'escape' from the solid phase into the vapour phase, where they exert a corresponding vapour pressure. Since, however, the molecules in a solid are more tightly bound than in the corresponding liquid, considerably more energy is required to 'escape' from a solid than a liquid. According to the Boltzmann distribution, as illustrated in Figures 3.10 and 3.11, even in a solid, at any given temperature, there will be some molecules with sufficient energy to evaporate, but significantly fewer than from a liquid at the same temperature. In general, the vapour pressure associated with a solid is therefore considerably less than the vapour pressure associated with the corresponding liquid.

The general tendency of a solid or a liquid to evaporate is referred to by the term **volatility** – at any given temperature, a more volatile substance evaporates more readily than a less volatile substance, and has a correspondingly higher vapour pressure. A non-volatile substance has a very low tendency to evaporate, and has a correspondingly low vapour pressure, and so many solids can be described as non-volatile, especially in comparison to liquids.

15.2 Vapour pressure and external pressure

The analysis so far has focused on systems comprising a pure liquid or solid in contact with a vacuum; in practice, this is very rare – most liquids or solids are in contact with the atmosphere, which exerts atmospheric pressure on the liquid or solid surface. Evaporation from the surface still takes place, however, for evaporation is fundamentally attributable to the distribution of energy within the molecules of the solid or the liquid, as determined by the system temperature. On evaporating, the vapour molecules mix with the gases of the atmosphere, with each component within the resulting mixture exerting the appropriate partial pressure. As an example, for (liquid) water subject to an external pressure P of 1 atm = 101.325 kP, the equilibrium saturated vapour pressure for the water molecules is about 3 kP, alongside nitrogen, with a partial pressure of about 78 kP, and oxygen at 21 kP.

Furthermore, just as molecules of the liquid can pass between the liquid and the atmosphere, molecules of nitrogen, oxygen and carbon dioxide present in the atmosphere can also exchange, and dissolve into the liquid. This implies that the liquid is no longer pure. We will ignore this effect for the moment, and continue to assume that the liquid remains pure – we explore the properties of solutions in the next chapter.

Also, since atmospheric pressure acts as a 'weight' on the surface of the liquid, this 'squeezes' the liquid, and 'helps' molecules 'escape' and evaporate. At any temperature, rather more molecules evaporate under external atmospheric pressure as compared to a vacuum, and so the vapour pressure is somewhat increased – in general, higher external pressures raise vapour pressures, and lower external pressures depress them.

It might be thought that there is a counter-argument: doesn't the external pressure 'force' the vapour molecules back into the liquid, so causing the vapour pressure to decrease? This argument, in fact, is false in general, and in particular for an ideal gas. Since one of the assumptions of an ideal gas is that there are no intermolecular interactions, it follows that the molecules of the vapour do not 'know' that any other gases are present, and so the vapour molecules cannot be 'forced' back into the liquid. And even if the gas mixture is not ideal, any 'forcing back' effect is extremely weak.

15.3 The Gibbs free energy of phase changes

To explore the thermodynamics of phase changes in more detail, consider a system in which a liquid and the corresponding vapour are in thermodynamic equilibrium at a given temperature T K, and equilibrium saturated vapour pressure p Pa, as represented by the reversible reaction

Liquid $\rightleftharpoons$ Vapour $\hspace{4cm} \Delta H = \Delta_{vap}H$

As we shall see, the analysis we are about to carry out applies to all phase transitions, of which the liquid $\rightleftharpoons$ vapour change is one example. So, rather than referring specifically to 'liquid' and 'vapour', we will now use the more general terms 'phase X' and 'phase Y'

Phase X $\rightleftharpoons$ Phase Y $\hspace{4cm} \Delta H = \Delta_{ph}H$

in which, for any transition, phase X is the more condensed phase, and phase Y, the less condensed phase, and the corresponding enthalpy change $\Delta_{ph}H$ is positive.

If the molar Gibbs free energies of each phase are G_X and G_Y, then the total Gibbs free energy G_{sys} of the system is given by

$$G_{sys} = n_X \, G_X + n_Y \, G_Y \tag{15.1}$$

where n_X and n_Y are the mole numbers for each phase, corresponding to the equilibrium mixture. On differentiation, we have

$$dG_{sys} = n_X \, dG_X + n_Y \, dG_Y + G_X \, dn_X + G_Y \, dn_Y \tag{15.2}$$

As we know (see page 48), in the absence of gravitation, motion, electric, magnetic and surface effects, a fixed mass of a pure substance has only two independent state functions. Since the temperature, and the pressure, of the system are constant, these two independent state functions are determined, implying that all other state functions are fixed. Both G_X and G_Y are state functions, and so G_X and G_Y must therefore be fully defined, implying that dG_X and dG_Y must both be zero. Furthermore, the system is in equilibrium, and so dG_{sys} must also be zero, implying that equation (15.2) simplifies to

$$0 = G_X \, dn_X + G_Y \, dn_Y \tag{15.3}$$

Since the system is of constant mass, then

$n_X + n_Y =$ constant

and so

$dn_X + dn_Y = 0$

therefore

$dn_X = - dn_Y$

which implies that equation (15.3) becomes

$$G_X = G_Y \tag{15.4a}$$

Note that equation (15.4a) represents the equality of the molar Gibbs free energy G_i of each phase, and not the *standard* molar Gibbs free energy $G_i^\ominus$. This is important, for $G_i^\ominus$ is defined for the standard pressure $P^\ominus = 1$ bar, and can vary only with temperature; G_i, on the other hand, is simply the Gibbs free energy of a specific quantity, 1 mol, of component i, and so can vary with both temperature and pressure – as we shall see, it is the pressure variation that is of particular significance.

Also, as we saw on page 404, at any pressure P and temperature T, the molar Gibbs free energy G is equal to the chemical potential μ

$$G_X(P, T) = \mu_X(P, T) \tag{13.55}$$

and so equation (15.4a) can be expressed in terms of the chemical potentials μ_X and μ_Y of each phase as,

$$\mu_X = \mu_Y \tag{15.4b}$$

Equations (15.4a) and (15.4b) are important, since they define the conditions for the equilibrium between the two phases: their molar Gibbs free energies – and hence their chemical potentials – must be equal. Much of this chapter will examine a variety of real situations in which different phases are in equilibrium, each of which is a different application of the very simple-looking equations (15.4a) and (15.4b). Given the equivalence of the molar Gibbs free energy G and the chemical potential μ, any equation expressed in terms of G has a corresponding form expressed in terms of μ, and so to avoid undue repetition, this chapter will express equations using G.

15.4 Melting and boiling

Here is a summary of the key facts we know about phase changes, all of which apply to a system comprising a constant quantity of 1 mol of a pure material, at constant external pressure P_{ex}, for example, 1 bar.

- At low temperatures, the material is a solid of volume V_s.
- At a specific temperature T_{fus}, the system changes in its entirety from a solid to a liquid of volume V_l.
- At a specific temperature $T_{vap} > T_{fus}$, the system changes in its entirety from a liquid to a gas of volume V_g.
- At T_{fus}, according to equation (15.4a), the molar Gibbs free energies of the solid and the liquid are equal, and so $G_s = G_l$.
- Likewise, at T_{vap}, the molar Gibbs free energies of the liquid and the gas are equal, $G_l = G_g$.

We also know, according to equation (13.21), that, for any system, the slope $(\partial G/\partial T)_P$ of a graph of G against T at constant pressure P is given by

$$\left(\frac{\partial G}{\partial T}\right)_P = -S \tag{13.21}$$

MELTING AND BOILING

The entropy S of any system is necessarily positive, and so the slope defined by equation (13.21) will always be negative. Furthermore, S increases with temperature, with discontinuities at T_{fus} and T_{vap}, as shown in Figure 9.4. This implies that a graph of the molar Gibbs free energy G as a function of temperature will be a generally downwards-sloping curve, formed from three segments. The first segment will show G_s for the solid, and will form a shallow-sloping curve, corresponding to S_s, from low temperatures to T_{fus}; between T_{fus} and T_{vap}, the system is a liquid, and the G_l curve will be somewhat steeper, corresponding to $S_l > S_s$; above T_{vap}, the system is a gas, and G_g will be even steeper, corresponding to $S_g > S_l$.

These experimental observations, combined with the intuitive interpretation of the behaviour of equation (13.21), imply that, for any pure material at constant pressure, the graph of G as a function of temperature must be as illustrated in Figure 15.2.

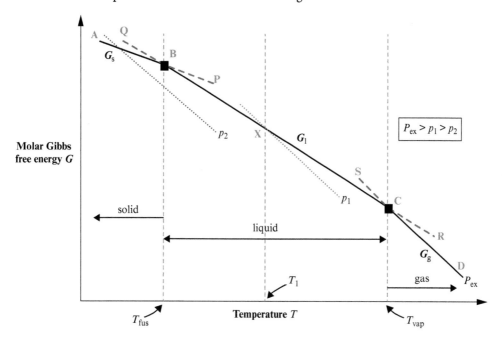

Figure 15.2 Phase changes and G. As the temperature T of a pure material increases at a constant pressure P_{ex}, the material changes phase, firstly from solid to liquid at T_{fus}, at which point $G_s = G_l$, and then from liquid to gas at T_{vap}, where $G_l = G_g$. Within any phase, G reduces smoothly as T increases, such that the slope of the curve $(\partial G/\partial T)_P = -S$ at each point. At each phase boundary, however, the discontinuity in S implies that there is a corresponding discontinuity in the slope $(\partial G/\partial T)_P$, hence the 'kinks' at states **B** and **C**.

The mathematics associated with the curves shown in Figure 15.2 is discussed in the last section of this chapter, but we don't need the mathematics to interpret what happens physically. So, as the material is heated, the system follows the G_s curve from state **A** on the left towards the right. When $T = T_{fus}$, the system achieves state **B**, where the steeper G_l curve cuts across the shallower G_s curve.

If the system were to remain solid at a temperature $T > T_{fus}$, corresponding to a state somewhere along the line-segment **BP**, then $G_s > G_l$, implying that $G_l - G_s < 0$. This negative change in the system Gibbs free energy implies that the system would spontaneously change from solid to liquid. States along the line segment **BP** are therefore unstable, and so, to the right of state **B**, the system follows the line segment **BC**. This explains melting, and the fact that, during melting, the system changes in its entirety from solid to liquid – in general, as a solid is heated, at state **B**, at which $T = T_{\text{fus}}$, the solid and the liquid are in equilibrium, but as soon as T exceeds T_{fus}, even if only by a very small amount, any quantity of solid is unstable as compared to the liquid. Likewise, for all states along the line segment **BQ**, for which $T < T_{\text{fus}}$, $G_s < G_l$, implying that $G_s - G_l < 0$ and that the system spontaneously changes from liquid to solid: when a liquid freezes, the system changes in its entirety from liquid to solid. The fundamental principle is that, at all temperatures, the system adopts the minimum molar Gibbs free energy **G**.

As the temperature increases further, at $T = T_{\text{vap}}$, the G_g curve cuts across the G_l curve, and, in accordance with the principle of minimizing **G**, as soon as T exceeds T_{vap}, the liquid turns in its entirety into a gas. Similarly, if a gas at a temperature $T > T_{\text{vap}}$ is cooled, at $T = T_{\text{vap}}$, the gas and the liquid are in equilibrium, but as soon as T drops below T_{vap}, the gas liquefies.

There are, however, two subtleties. Suppose, for example, that a gas is cooled, following the G_g curve from right to left. As the gas approaches T_{vap} from the right, along the line segment **DC**, the temperature T is always greater than T_{vap}, and $G_g < G_l$. All the system is therefore a gas, as expected. As the system is cooled through T_{vap}, suppose that – for whatever reason – the liquid state cannot be reached. The system will then follow the line segment **CS**, remaining as a gas, even though $G_g > G_l$. Such a phenomenon can arise, and is known as a **supercooled gas**. As we saw on page 410, supercooled states can exist because the formation of a liquid requires the presence of suitable **nuclei** which allow micro-droplets of liquid to form. These nuclei then act as centres about which further condensation can occur, and the gas liquefies. The process of forming the very first nuclei, known as **nucleation**, is facilitated by the presence of small impurities, such as dust particles, or by contact of the gas with the (microscopically) rough surface of a container. In most circumstances, nucleation happens 'naturally' within the system as soon as $T = T_{\text{vap}}$, and the system follows the path **DCB**. If, however, the gas is very pure, and contains no dust particles, and is stored within a container with very smooth walls, then, as T falls below T_{vap}, nucleation might not happen, and so a supercooled state, along the line segment **CS**, can arise. If, however, the system is disturbed – for example, by scratching the container wall – then nucleation takes place instantaneously, and the gas liquefies at once, changing state from a point on the higher-**G** line segment **CS** to the corresponding point, at the same temperature, on the lower-**G** line segment **CB**. Similarly, states along the line segment **CR** represent a **superheated liquid**. Nucleation is also a feature of the liquid → solid change, and so the line segment **BQ** represents a **supercooled liquid**; the line segment BP represents a **superheated solid**, but in practice, superheated solids are very rare.

The second subtlety concerns the effect of pressure on a solid, a liquid and a gas. For n mol of a pure material, at any given temperature T, the volume V_s of a solid is well-defined, and since solids are largely incompressible, this volume is constant for even quite large changes in the external pressure P_{ex}. Similarly, when the solid melts, the volume V_l of the corresponding liquid is largely independent of the external pressure. But when the liquid changes into a gas, the gas can occupy any volume V_g, and exert any pressure P_g. Furthermore, if we assume that the gas is ideal, then the (partial) pressure P_g exerted by the gas is not influenced by any external pressure P_{ex}, as exerted, for example, by the atmosphere: since one of the assumptions of an ideal gas is that there are no molecular interactions, if a material, on heating, changes from a liquid to a gas, then that gas does not 'know' that the atmosphere is present, and so exerts a partial pressure P_g as determined by the volume V_g available to it, in accordance with the ideal gas law $P_g V_g = nRT$ at the appropriate temperature T.

The significance of this is that, for the solid, at any external pressure P_{ex} and temperature T, the molar Gibbs free energy G_s can take only a single value, and so the $G_s(T)$ curve is a unique line, as shown by the line segment AB in Figure 15.2. Similarly, for the liquid, $G_l(T)$ is uniquely defined, as shown by the line segment **BC**. But for the gas, at any temperature T, the molar Gibbs free energy is *not* uniquely defined, but (assuming the gas is ideal) obeys equation (13.47b)

$$G_g(P, T) = G^\ominus + RT \ln(P/P^\ominus) \tag{13.47b}$$

in which the notation $G_g(P, T)$ emphasises that the molar Gibbs free energy of the gas depends on both the (partial) pressure P exerted by the gas (this being independent of any external pressure P_{ex}, such as that exerted by the atmosphere) and the temperature T, and also on $G^\ominus$, the standard molar Gibbs free energy of the pure gas at the standard pressure $P^\ominus$.

The implication of equation (13.47b) is illustrated in Figure 15.2. As can be seen, for the solid phase, $G_s(T)$ is represented by single line **AB**, and, for the liquid phase, $G_l(T)$ is represented by single line **BC**. For the gaseous phase, the line segment **CD** represents $G_g(P_{ex}, T)$, this being the behaviour of equation (13.47b) for the specific case of $P = P_{ex}$; in addition, however, Figure 15.2 shows two dotted lines, labelled p_1 and p_2, which correspond to the behaviour of equation (13.47b) at two different pressures such that $p_1 < p_2 < P_{ex}$. These two dotted lines are just two specific instances – in fact, there are an infinite number of 'dotted lines' right across Figure 15.2, each corresponding to a different value of the gas pressure p.

As shown in Figure 15.2, the $G_g(P, T)$ curve corresponding to $P = p_1$ intersects the $G_l(T)$ curve at $T = T_1$ (note that the subscript $_1$ in T_1 and p_1 refers to 'one', and is not $_l$ for 'liquid'). What does this mean? The fact that the two curves cross must mean that the molar Gibbs free energy $G_l(T_1)$ of the liquid at $T = T_1$ equals the molar Gibbs free energy $G_g(p_1, T_1)$ of the corresponding gas, also at $T = T_1$, and with the gas exerting a partial pressure p_1. The two phases are therefore in equilibrium, and so p_1 *must be the equilibrium saturated vapour pressure* above the liquid at $T = T_1$. By exactly the same reasoning, once again referring to Figure 15.2, p_2 is the equilibrium saturated vapour pressure above the solid at the temperature corresponding to the point at which the p_2 curve crosses the $G_s(T)$ curve for the solid.

As we saw on page 472, the (negative) slope of the G_g curve is always greater than that of the G_s and G_l curves. Accordingly, at any temperature T, there will always be one $G_g(P, T)$ curve that will cross any G_s and G_l curve, showing that there will always be an equilibrium between a solid or a liquid and the corresponding vapour at a some corresponding equilibrium saturated vapour pressure p. This implies that the condensed phases of the solid or the liquid are always associated with a vapour, and that the solid and vapour, or liquid and vapour, must co-exist in equilibrium. If, however, a material is present as a gas – except on the point of liquefaction – the gas is always present by itself, and not associated with a drop of liquid, or a speck of solid. This too can be explained by reference to Figure 15.2: when the material is naturally in the gas phase, corresponding to any point along the line segment **CD** (or further to the right), there is no intersection with the $G_s(T)$ curve or the $G_l(T)$ curve. These curves are uniquely defined, and are shown as the line segments **AB** and **BC** respectively. When extrapolated to the right, to temperatures greater than T_{vap}, the resulting solid and liquid states would have values of **G** far greater than the corresponding value of G_g as shown by the line segment **CD**. If a solid or a liquid were present at some temperature $T > T_{vap}$, then that solid or liquid would instantaneously vaporise to achieve a state with a much lower **G**.

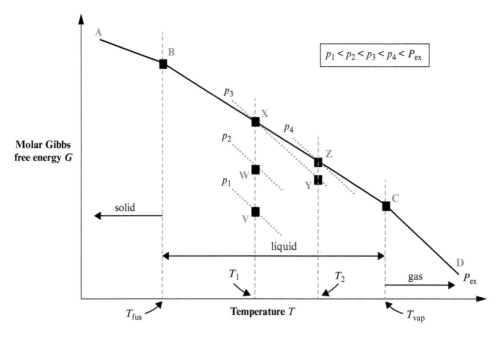

Figure 15.3 Unstable and stable states. A system of a liquid and vapour at state **V** is unstable, and the liquid will vaporize, passing through state **W**, until equilibrium is reached at state **X**, with the vapour exerting the equilibrium saturated vapour pressure p_3. If the system is then heated, state **Y** cannot be reached, and so the system achieves state **Z**, corresponding to a higher equilibrium saturated vapour pressure p_4. At state **C**, the liquid boils.

Figure 15.3 takes matters one step further. Consider a system comprising a liquid at a temperature T_1, and subject to an external pressure P_{ex}, corresponding to state **X**. Suppose further that the vapour pressure above the liquid happens to be p_1, so that the vapour phase is represented by state **V**. As can be seen, at a pressure p_1, $G_g(p_1, T_1) < G_l(T_1)$, implying that $G_g(p_1, T_1) - G_l(T_1) < 0$, and so the liquid will spontaneously vaporise. If the newly-produced vapour is not extracted, but allowed to accumulate, this increases the vapour pressure to p_2 corresponding to state **W**. $G_g(p_2, T_1)$ of the vapour in state **W** is still at a lower value than $G_l(T_1)$, and so further evaporation takes place until the vapour pressure rises to p_3, such that $G_g(p_3, T_1) = G_l(T_1)$, corresponding to state **X**. At this point, $G_g(p_3, T_1) - G_l(T_1) = 0$, and so the liquid and the vapour are in equilibrium. No further evaporation takes place, and p_3 is the equilibrium saturated vapour pressure at $T = T_1$.

Suppose now that the temperature of the system rises to T_2. Given that state **X** is on the intersection of two **G** curves, the system could potentially change either to state **Y**, following the path **XY**, or to state **Z**, corresponding to the path **XZ**. As shown in Figure 15.3, state **Y** has a lower molar Gibbs free energy than state **Z**, suggesting that the system will follow path **XY**.

State **Y**, however is a gas, and so in order to move from the liquid-vapour equilibrium state **X** to the gas-only state **Y**, the entire system has to vaporise. Suppose, then, that a micro-bubble of vapour is created within the bulk of the liquid. If the liquid is in, or very close to, state **X**, then the vapour pressure within that micro-bubble will be equal, or very close to, p_3 Pa. Since this micro-bubble is within the bulk of the liquid, at a distance of, say, h m below the surface, then the external pressure on the micro-bubble will be $P_{ex} + h\rho g$ Pa, where P_{ex} is the external pressure on the surface of the liquid, and $h\rho g$ is the hydrostatic pressure attributable to a height h m of liquid of density ρ kg m^{-3}, and g is the acceleration due to gravity. As shown in Figure 15.3, $p_3 < P_{ex}$, and so $p_3 < P_{ex} + h\rho g$, implying that the internal pressure p_3 within the micro-bubble will be less than the pressure $P_{ex} + h\rho g$ on the outside of the micro-bubble. The micro-bubble must therefore collapse. Any micro-bubble that might form is therefore unstable, and so the bulk of the liquid must remain as a liquid.

The path **XY** is not possible, and state **Y** is therefore inaccessible from state **X**. Accordingly, as the liquid is heated to $T = T_2$, the system follows the path **XZ**, becoming a hotter liquid, with a higher equilibrium saturated vapour pressure p_4.

With further heating, the liquid becomes progressively hotter, and the equilibrium saturated vapour pressure progressively greater, as the system follows the path from state **Z** towards state **C**. At state **C**, the system temperature T has risen to T_{vap}, and the saturated vapour pressure $p = P_{ex}$ Pa. If we now re-consider what might happen within the bulk of the liquid, then suppose, as before, that a micro-bubble forms at a depth h m below the surface. The pressure on the inside of the micro-bubble is now $p = P_{ex}$ Pa, and the pressure on the outside is $P_{ex} + h\rho g$ Pa. If $P_{ex} \gg h\rho g$, then the pressures on both sides of the micro-bubble are equal, and so – importantly – the micro-bubble does not collapse. Furthermore, if the liquid temperature is slightly higher than T_{vap}, $p > P_{ex}$, and so (subject to overcoming forces of surface tension, which are seeking to keep the micro-bubble small) the micro-bubble will grow. This process will be happening throughout the liquid, as the liquid phase changes, in bulk, to the gas phase. Bubbles form, and grow, throughout the liquid, which becomes turbulent – a process we know as boiling. And, once all the liquid has become a gas, further heating follows the path **C** to **D**, and beyond, corresponding to a progressively hotter gas.

As this description shows, boiling – the phenomenon in which a liquid changes in bulk to a gas – takes place when the vapour pressure p within the interior of the liquid is greater than the quantity $P_{ex} + h\rho g$. This clearly varies with the quantity of liquid, as represented by the parameter h, but for many liquids under normal conditions, $P_{ex} \gg h\rho g$, and so the condition for boiling becomes $p > P_{ex}$. So, for example, in a laboratory, we may be boiling some water, which may have a depth of, say, 0.1 m. The hydrostatic pressure $h\rho g$ attributable to a column of $h = 0.1$ m is about 9.81×10^2 Pa, which is substantially less than the pressure of the atmosphere $P_{ex} = 1.01 \times 10^5$ Pa, and so boiling takes place when $p = P_{ex}$.

Since there are many circumstances in which the hydrostatic effect can in practice be ignored, this leads to the rule-of-thumb that "boiling takes place when the saturated vapour pressure of a liquid equals that of the external environment". This rule-of-thumb, however, can be misleading, as exemplified by the Torricelli barometer. As shown in Figure 15.1, when the barometer is first formed, the space above the mercury is a vacuum. Immediately, some vaporisation takes place (rather like state **V** in Figure 15.3), until equilibrium is reached (state **X** in Figure 15.3). This creates a system in which the equilibrium saturated vapour pressure equals the pressure of the external environment, for the external pressure *is* the equilibrium saturated vapour pressure. If the criterion for boiling is "$p = P_{ex}$", then the mercury in the barometer should boil – which, clearly, it doesn't. This apparent paradox is easily resolved by using the (correct) rule that boiling takes place when the saturated vapour pressure p within the bulk of the liquid is equal to $P_{ex} + h\rho g$. Since, even just below the uppermost surface of the mercury, $h > 0$, then if $P_{ex} = p$, the condition for boiling cannot be fulfilled, and so the mercury does not boil away. Something else is happening too – see Exercise 6!

15.5 Changing the external pressure P_{ex}

The discussion over the preceding pages has examined the behaviour of the molar Gibbs free energy G as a function of temperature at a constant external pressure P_{ex}; in this section, we explore what happens to the solid-liquid-gas system as P_{ex} changes. The key relationship in this context is the molar version of equation (13.20)

$$\left(\frac{\partial G}{\partial P}\right)_T = V \tag{13.20}$$

which states that if the pressure increases by an amount ∂P_T at constant temperature, then the corresponding change ∂G_T in the system's molar Gibbs free energy is equal to the system's instantaneous molar volume V. This has an immediate implication – since for all materials, $V > 0$, ∂G_T is also necessarily always greater than 0, and so an increase ∂P_T in the external pressure P_{ex} causes an increase ∂G_T in the system's molar Gibbs free energy. In terms of Figures 15.2 and 15.3, if P_{ex} increases, each of the three line segments **AB**, **BC** and **CD** moves upwards as indicated schematically in Figure 15.4; likewise, if P_{ex} decreases, each of the three line segments **AB**, **BC** and **CD** moves downwards.

Furthermore, since we know that, for almost all materials, the molar volume V increases with temperature, then, for any given change ∂P_T in the external pressure P_{ex}, the corresponding change ∂G_T is progressively greater as the temperature increases. If the external pressure P_{ex} increases, then for each of the line segments **AB**, **BC** and **CD**, the 'right-hand' part (corresponding to higher temperatures) will rise by a larger amount than the 'left-hand' part (corresponding to lower temperatures), implying that each line segment becomes 'flatter'. Likewise, if the external pressure P_{ex} decreases, then each of the line segments becomes steeper.

Increasing P_{ex} therefore displaces the three line segments **AB**, **BC** and **CD** upwards, as well as making each line segment flatter, but the magnitude of these effects differs for each of the three phases. In general, the upwards displacement ∂G_T at any temperature T is, as we have seen, given by equation (13.20) as $V \, \partial P_T$, and so for the solid phase, the displacement $\partial G_{T,s}$ is $V_s \, \partial P_T$; likewise, for the liquid phase, the displacement $\partial G_{T,l}$ is $V_l \, \partial P_T$. For most materials, $V_l > V_s$, but not substantially so, and so $\partial G_{T,l}$ is in general a little greater than $\partial G_{T,s}$: in terms of Figure 15.4, the line segment **BC**, corresponding to the liquid, is displaced upwards rather more than the line segment **AB**, for the solid.

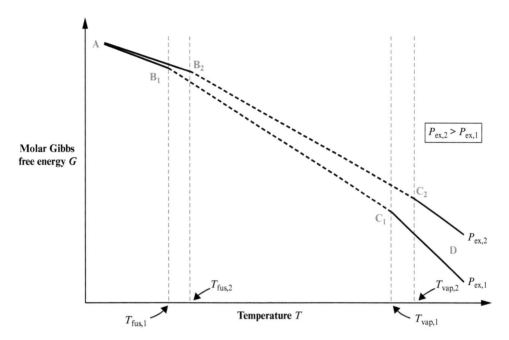

Figure 15.4 A schematic representation of phase changes at two different external pressures P_{ex}. The line $A \rightarrow B_1 \rightarrow C_1 \rightarrow D$ is the same as the line $A \rightarrow B \rightarrow C \rightarrow D$ shown in Figures 15.2 and 15.3: at an external pressure $P_{ex,1}$, the solid $\rightarrow$ liquid phase transition takes place at $T = T_{fus,1}$, corresponding to state B_1, and the liquid $\rightarrow$ gas phase transition takes place at $T = T_{vap,1}$, corresponding to state C_1. If the external pressure is increased to $P_{ex,2}$, the whole curve shifts upwards, but by different amounts for each phase; also, each line segment becomes flatter. At pressure $P_{ex,2}$, the solid $\rightarrow$ liquid phase transition takes place at state C_2, such that $T_{vap,2} > T_{vap,1}$; the solid $\rightarrow$ liquid phase transition takes place at state B_2, for which $T_{fus,2} > T_{fus,1}$, as is consistent with the assumption that $V_l < V_s$. For clarity, in this diagram, the upwards displacements, and slope changes, of each line segment are unrealistic, but they do indicate the general behaviour.

For the gas, however, V_g is substantially greater than both V_s and V_s – as an example, for 1 mol of water at a pressure of 1 bar, for ice at 223.15 K (−50°C), $V_s \approx 1.95 \times 10^{-5}$ m³; for water at 323.15 K (+50°C), $V_l \approx 1.82 \times 10^{-5}$ m³; but for steam at 423.15 K(+150°C), $V_g \approx 3.47 \times 10^{-2}$ m³. Water is a very unusual material in that $V_s > V_l$, but the numeric values of V_s and V_l are similar. For steam, however, V_g is approximately 2000 times greater than V_s and V_l. This implies that the magnitude of the displacement of the line segment CD is of the order of 2×10^3 greater than that of the line segments AB and BC.

All these effects are shown, in schematic form, in Figure 15.4, from which it can be seen that, when the external pressure increases, the boiling point T_{vap} increases, as does T_{fus} – but only if $V_s < V_l$; if the material is, for example, water, for which $V_s > V_l$, then T_{fus} decreases as P_{ex} increases, as shown in Figure 15.5(a).

Figure 15.5(b) shows another special case, which arises at a specific, low, external pressure. With reference to Figure 15.4, as P_{ex} decreases, the line $A \rightarrow B \rightarrow C \rightarrow D$ shifts downwards, and the line segments become steeper, especially the gas line segment CD. The liquid line segment BC becomes progressively shorter, and T_{vap} progressively decreases. Eventually, the

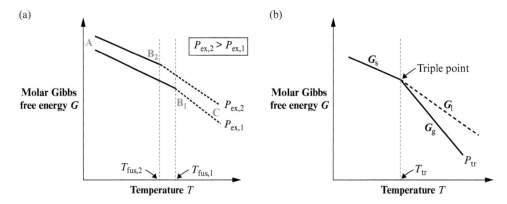

Figure 15.5 Two special cases. (a) The behaviour of a system, around T_{fus}, for a material, such as water, for which $V_s > V_l$. When $P_{ex,2} > P_{ex,1}$ the displacement of the solid phase line segment **AB** is greater than that of the liquid line segment **BC**, and so $T_{fus,2} < T_{fus,1}$: an increase in the external pressure P_{ex} causes the freezing point to decrease. (b) shows a case that arises at a particular low pressure at which G_s, G_l and G_g share a single, specific value, implying that all the three phases of solid, liquid and gas are simultaneously in mutual equilibrium. This is the 'triple point', corresponding to the triple-point pressure P_{tr} and triple-point temperature T_{tr}.

line segment **BC** shrinks to zero, and the G, T diagram takes the form shown in Figure 15.5(b), in which G_s, G_l and G_g all share a single, specific value, known as the **triple point**, at which the solid, liquid and gas are all in mutual equilibrium. This is a unique point for any material, defined at a single combination of the triple-point pressure, P_{tr}, and the triple-point temperature, T_{tr}: for water, P_{tr} = 611.7 Pa and T_{tr} = 273.16 K: in fact, the triple point of water is the reference state for the measurement of absolute temperatures according to the Kelvin temperature scale. At pressures below P_{tr}, with reference to Figure 15.5(b), the line segment representing G_g is positioned even further to the left. The liquid state no longer exists, and the only phase transition is between the solid and the gas.

A third special case, which is also associated with the 'disappearance' of the liquid state, occurs at high pressures. With reference to Figure 15.4, as we saw on page 478, when the external pressure P_{ex} increases, the line **A** → **B** → **C** → **D** shifts upwards, and the line segments become flatter, with the G_g curve becoming flatter more quickly than the G_l curve. Eventually, a pressure P_{crit} is reached at which the slope of the G_l curve equals that of the G_s curve. Mathematically, this implies that $(\partial G_g/\partial T)_P = -S_g = (\partial G_l/\partial T)_P = -S_l$; physically, this implies that the difference between the gas phase and the liquid phase, as represented by the discontinuity between S_l and S_g that occurs at $T = T_{vap}$, as shown in Figure 9.4, disappears. At external pressures greater than P_{crit}, there is no liquid state, and the system remains as a gas, condensing to a solid at low temperatures.

The state at which this happens is known as the **critical point**, the pressure P_{crit} is the **critical pressure**, and the temperature which corresponds to the last identifiable T_{vap} is the **critical temperature** T_{crit}: for water, P_{crit} = 2.2064 × 10^7 Pa, approximately 218 atm, and T_{crit} = 647.096 K. As we shall see in more detail on page 489, T_{crit} is the highest temperature at which a gas can be liquefied by increasing the system pressure: a gas at a temperature $T > T_{crit}$ can never become a liquid by exerting pressure alone.

15.6 The Clapeyron and Clausius-Clapeyron equations

15.6.1 The Clapeyron equation

As we know from Figure 15.4, as the external pressure P_{ex} changes, the temperature T_{ph} at which a phase change occurs also changes. Our purpose in this section is to quantify this, and to answer the question "If the external pressure P_{ex} changes by a defined amount dP_{ex}, what is the corresponding change dT_{ph} in the temperature of a phase change?"

Our starting point is the phase equilibrium condition

$$G_X = G_Y \tag{15.4a}$$

which applies to all phase changes under all conditions. Differentiating equation (15.4a) gives

$$dG_X = dG_Y \tag{15.5}$$

We now invoke equation (13.16b)

$$dG = V\,dP - S\,dT \tag{13.16b}$$

As we saw on pages 378 and 379, this equation is a direct consequence of the definition of the Gibbs free energy as $G = H - TS$, and applies to any closed system of constant mass that performs only P,V work of expansion. By definition, the molar Gibbs free energy G of any system relates to a system of constant mass; furthermore, a system of a liquid in equilibrium with its vapour at temperature T_{ph} and pressure P_{ex} is one in which the only form of work is P,V work of expansion. We may therefore apply equation (13.16b) to each of the infinitesimal changes dG_X and dG_Y in equation (15.5) giving

$$V_X\,dP_{ex} - S_X\,dT_{ph} = V_Y\,dP_{ex} - S_Y\,dT_{ph}$$

Rearranging this expression, we have

$$\frac{dP_{ex}}{dT_{ph}} = \frac{S_Y - S_X}{V_Y - V_X} = \frac{\Delta_{ph}S}{\Delta_{ph}V} \tag{15.6}$$

In equation (15.6), $\Delta_{ph}V$ is the difference between the molar volumes V of the two phases – for example, the difference between the volume of 1 mol of water vapour, and 1 mol of liquid water. $\Delta_{ph}S$ is the difference between the molar entropies S of the two phases, which, as shown on pages 291 and 292, is in general given by equation (9.43)

$$\Delta_{ph}S = \frac{\Delta_{ph}H}{T_{ph}} \tag{9.43}$$

where $\Delta_{ph}H$ is the molar enthalpy of the phase transition of interest, such as the molar enthalpy of fusion $\Delta_{fus}H$, the molar enthalpy of vaporisation $\Delta_{vap}H$, or the molar enthalpy of sublimation $\Delta_{sub}H$, and T_{ph} is the temperature at which the corresponding phase transition takes place.

Equation (15.6) therefore becomes

$$\frac{dP_{ex}}{dT_{ph}} = \frac{\Delta_{ph}H}{T_{ph}\Delta_{ph}V} \tag{15.7}$$

Equation (15.7) is known as the **Clapeyron equation**, and defines how changes in temperature and pressure affect phase changes. It is therefore the quantitative description that complements Le Chatelier's qualitative description that we discussed on page 448.

15.6.2 The liquid ⇌ vapour change

As an illustration of how the Clapeyron equation can be used, consider the liquid ⇌ vapour phase change, for which T_{ph} is T_{vap}, $\Delta_{ph}H$ is $\Delta_{vap}H$, and $\Delta_{ph}V = V_g - V_l$. Since, for all vaporisations, $V_g \gg V_l$, $\Delta_{ph}V \approx V_g$, and so equation (15.7) becomes

$$\frac{dP_{ex}}{dT_{vap}} = \frac{\Delta_{vap}H}{T_{vap}V_g}$$

This phase change is boiling, which, as we saw on page 477, takes place when the equilibrium saturated vapour pressure p is equal to $P_{ex} + h\rho g$, where $h\rho g$ is the hydrostatic pressure at a depth h within the bulk of the liquid. In many cases, $P_{ex} \gg h\rho g$ and so boiling takes place when $p = P_{ex}$, implying that $dp = dP_{ex}$. Accordingly, we may write

$$\frac{dp}{dT_{vap}} = \frac{\Delta_{vap}H}{T_{vap}V_g}$$

Furthermore, if the vapour can be approximated as an ideal gas,

$$pV_g = RT_{vap}$$

and so

$$\frac{dp}{dT_{vap}} = p\frac{\Delta_{vap}H}{RT_{vap}^2}$$

from which

$$\frac{1}{p}\frac{dp}{dT_{vap}} = \frac{\Delta_{vap}H}{RT_{vap}^2}$$

Since $d\ln p = dp/p$, we derive

$$\frac{d\ln p}{dT_{vap}} = \frac{\Delta_{vap}H}{RT_{vap}^2} \tag{15.8}$$

Equation (15.8) is known as the **Clausius-Clapeyron equation**, this being a special case, as applied to the phase change from liquid to vapour, of the more general Clapeyron equation, equation (15.7).

If we assume that $\Delta_{vap}H$ is constant over a change from $[p_1, T_1]$ to $[p_2, T_2]$, then equation (15.8) integrates to

$$\ln\frac{p_2}{p_1} = \frac{\Delta_{vap}H}{R}\left(\frac{1}{T_1} - \frac{1}{T_2}\right) \tag{15.9a}$$

In equation (15.9a), T_1 represents the boiling point $T_{vap,1}$ corresponding to the saturated vapour pressure p_1, and T_2 represents the boiling point $T_{vap,2}$ corresponding to the saturated vapour pressure p_2. Under most conditions, the vapour pressures p_1 and p_2 are equal to the external pressures $P_{ex,1}$ and $P_{ex,2}$, and so an alternative form of equation (15.9a) is

$$\ln \frac{p_2}{p_1} = \ln \frac{P_{ex,2}}{P_{ex,1}} = \frac{\Delta_{vap}H}{R}\left(\frac{1}{T_{vap,1}} - \frac{1}{T_{vap,2}}\right) \tag{15.9b}$$

15.6.3 The solid ⇌ vapour change

For sublimation, the direct change from the solid phase to the vapour phase, since $\Delta_{ph}V = V_{vap} - V_{sol} \approx V_{vap}$, the analysis is identical, subject to the replacement in equations (15.8) and (15.9a) of $\Delta_{vap}H$ by $\Delta_{sub}H$, and T_{vap} by T_{sub}, giving

$$\ln \frac{p_2}{p_1} = \ln \frac{P_{ex,2}}{P_{ex,1}} = \frac{\Delta_{sub}H}{R}\left(\frac{1}{T_{sub,1}} - \frac{1}{T_{sub,2}}\right) \tag{15.9c}$$

15.6.4 The solid ⇌ liquid change

For the solid ⇌ liquid phase change, however, the analysis is different, for two reasons: firstly, $\Delta_{ph}V = V_l - V_s$ will be a small number, very different from either V_l or V_s in that, for the condensed phases, V_l and V_s are nearly equal; secondly, there are no gases at all, and so the ideal gas equation – as was used for a phase change resulting in a vapour – has no relevance. Each case has to be treated separately, and we need to use the Clausius-Clapeyron equation in its original form

$$\frac{dP_{ex}}{dT_{ph}} = \frac{\Delta_{ph}H}{T_{ph}\Delta_{ph}V} \tag{15.7}$$

which, for the solid ⇌ liquid phase change becomes

$$\frac{dP_{ex}}{dT_{fus}} = \frac{\Delta_{fus}H}{T_{fus}\Delta_{fus}V} \tag{15.10a}$$

or, for small changes, in finite-difference form

$$\Delta P_{ex} = \frac{\Delta_{fus}H}{T_{fus}\Delta_{fus}V} \Delta T_{fus} \tag{15.10b}$$

In equations (15.10), both $\Delta_{fus}H$ and T_{fus} are necessarily positive, but $\Delta_{fus}V = V_l - V_s$ may be positive or negative, depending on whether the material expands on melting ($V_l > V_s$), or contracts ($V_l < V_s$). For most materials, $V_l > V_s$, and so an increase in external pressure ($\Delta P_{ex} > 0$) results in an increase in the melting point ($\Delta T_{ex} > 0$), as shown in Figure 15.4, but if $V_l < V_s$, as is the case for water, an increase in external pressure ($\Delta P_{ex} > 0$) results in an decrease in the melting point ($\Delta T_{ex} < 0$), as shown in Figure 15.5(a).

15.6.5 A worked example

As an example of how equation (15.9b) can be applied, consider the question "At what temperature does water boil if the external atmospheric pressure is 0.9 bar?". We know that water boils at 373.15 K at an external pressure of 1.01325 bar, implying that the equilibrium saturated vapour pressure of water is 1.01325 bar at 100 °C = 373.15 K. The question requires us to compute the temperature at which the equilibrium saturated vapour pressure of water is 0.9 bar, so causing the water to boil at that lower atmospheric pressure. From Table 9.1, we see that $\Delta_{vap}H$ = 40.65 kJ mol^{-1}; also, $T_{vap,1}$ = 373.15 K; p_1 = 1.01325 bar; and p_2 = 0.9 bar. If these numbers are entered into equation (15.9b), the resulting arithmetic yields the result $T_{vap,2}$ = 369.8 K = 96.7 °C, showing that the boiling point of water is lowered at lower pressures. This is in agreement with the inference we made on page 469 using Le Chatelier's principle: the higher the external pressure, the higher the temperature at which any liquid will vaporise; and, *vice versa*, the lower the external pressure, the lower the vaporisation temperature.

15.6.6 A graphical interpretation

Since

$$d\left(\frac{1}{T}\right) = -\frac{1}{T^2}dT$$

then an alternative way of expressing the Clausius-Clapeyron equation

$$\frac{d\ln p}{dT_{vap}} = \frac{\Delta_{vap}H}{RT_{vap}^2} \quad \text{or} \quad \frac{d\ln p}{dT_{sub}} = \frac{\Delta_{sub}H}{RT_{sub}^2} \tag{15.8}$$

is

$$d\ln p = -\frac{\Delta_{vap}H}{R}d\left(\frac{1}{T_{vap}}\right) \quad \text{or} \quad d\ln p = -\frac{\Delta_{sub}H}{R}d\left(\frac{1}{T_{sub}}\right)$$

These expressions imply that, if the saturated vapour pressure p for a liquid ⇌ vapour, or solid ⇌ vapour, equilibrium system is measured at different phase transition temperatures T_{ph}, then a plot of ln p as the vertical axis, against $1/T_{ph}$ as the horizontal axis, will in principle yield a straight line of slope $-\Delta_{vap}H/R$ or $-\Delta_{sub}H/R$, as shown in Figure 15.6, so allowing $\Delta_{vap}H$ and $\Delta_{sub}H$ to be estimated without having to do any calorimetry. Also, since $\Delta_{sub}H$ = $\Delta_{fus}H + \Delta_{vap}H$ (as is readily demonstrated from thermochemistry, as discussed on page 163), estimates of $\Delta_{vap}H$ and $\Delta_{sub}H$ allow the estimation of $\Delta_{fus}H$ too.

In practice, a graph of the type illustrated in Figure 15.6 can be determined by placing a small amount of liquid or solid in an environment in which the external pressure P_{ex} can be controlled. As the temperature of the system is raised, a temperature is reached at which the liquid boils, or the solid sublimes. This determines T_{vap} or T_{sub} corresponding to a given P_{ex} = p. P_{ex} can then be changed, and new values of T_{vap} or T_{sub} measured accordingly.

As shown in Figure 15.6, if the experimentally determined liquid ⇌ vapour and solid ⇌ vapour lines are extrapolated, they intersect at a point at which the equilibrium saturated vapour pressure for all three phases – solid, liquid and vapour – have the same value, implying that all three phases are in mutual equilibrium. This, as we saw on page 480, is the triple point, as characterized by the unique combination of pressure P_{tr} and temperature T_{tr} at which this phenomenon occurs.

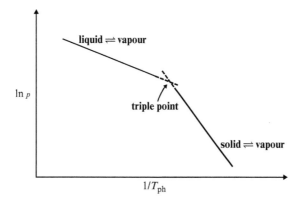

Figure 15.6 A graphical representation of the Clausius-Clapeyron equation. Experimental plots of $\ln p$ against $1/T_{ph}$ in principle yields a straight line of slope $-\Delta_{ph}H/R$. The slope of the solid $\rightleftharpoons$ vapour line is therefore $-\Delta_{sub}H/R$, and of the liquid $\rightleftharpoons$ vapour line, $-\Delta_{vap}H/R$. Given that $\Delta_{sub}H = \Delta_{fus}H + \Delta_{vap}H$, and that $\Delta_{ph}H$ is positive for all phase transitions from a more condensed to a less condensed state, $\Delta_{sub}H > \Delta_{fus}H$, and so the solid $\rightleftharpoons$ vapour line is steeper than the liquid $\rightleftharpoons$ vapour line. Any deviation from a straight line suggests that $\Delta_{ph}H$ is not constant over the corresponding temperature range. The point at which the extrapolated lines intersect is the triple point.

15.7 Phase changes, ideal gases and real gases

15.7.1 Ideal gases cannot liquefy...

This last section of this chapter explores the liquid $\rightleftharpoons$ vapour phase change for ideal gases and real gases, so as to enrich our understanding of the process of liquefaction.

The most important feature of the liquid $\rightleftharpoons$ vapour phase change for an ideal gas is that it cannot happen: an ideal gas is an ideal gas under all conditions of pressure, volume and temperature, and can never become a liquid. This is evident from the ideal gas law

$$P = \frac{nRT}{V} \tag{3.3}$$

which, as illustrated in Figure 15.7, maintains its P,V hyperbolic form under all circumstances, and never shows the discontinuity in volume, at constant temperature and pressure, that corresponds to the formation of a (low volume) liquid from the corresponding (high volume) gas.

The non-existence of the liquid phase is a consequence of the assumptions underlying the definition of an ideal gas (see page 17) – in particular, the assumption that, within an ideal gas, there are no intermolecular interactions. As described on page 70, at any temperature T, any molecule is associated with thermal energy, of the order of $k_B T$, and it is this energy that allows thermal motion. If there are no interactions between neighbouring molecules, then each molecule can move independently, in all directions, and the system is a gas. If, however, closely neighbouring molecules experience some form of mutually attractive force, this will tend to attract the molecules to one another, so counteracting the tendency of thermal energy to drive them apart. The average distance between any pair of molecules subject to a mutually attractive force will therefore be smaller than the corresponding distance between any pair of molecules not subject to such a force, suggesting that, at any given temperature and pressure,

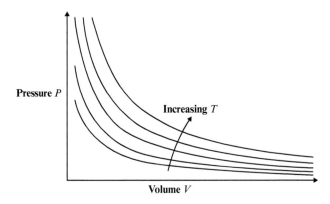

Figure 15.7 Ideal gases do not liquefy. At all temperatures T, from close to absolute zero to as high a temperature as you might imagine, the ideal gas law $PV = nRT$ predicts a smooth, hyperbolic, relationship between the pressure P and the volume V. There are no circumstances in which, at a given temperature and pressure, there is the discontinuity in volume associated with the formation of a liquid from the gas. Also, the slope $(\partial P/\partial V)_T$ is negative at all points, and is never zero or positive.

the volume occupied by a gas in which intermolecular attractive forces are present will be less than the volume occupied by an ideal gas at the same temperature and pressure.

This behaviour is illustrated in the central portion of Figure 15.8, which shows a P,V curve for an ideal gas at a given temperature T, and the corresponding P,V curve, at the same temperature, for a real gas which complies with the van der Waal equation-of-state

$$\left(P + \frac{an^2}{V^2}\right)(V - nb) = nRT \tag{3.6}$$

and which remains a gas at all pressures.

Starting towards the right of the diagram, at high volumes and low pressures, the average distance between any pair of molecules within the real gas is large, implying that the intermolecular forces are very weak. The real gas approximates to ideal behaviour, and, in terms of the van der Waals equation, at high volumes, $V \gg nb$ and $an^2/V^2 \ll P$, so that the van der Waals equation reduces to the ideal equation $PV = nRT$.

As we move to the left at constant temperature, the experimental observation is that the volume of a real gas at a given pressure is less than that of an ideal gas at the same pressure, as shown by the horizontal light grey line in Figure 15.8; similarly, the pressure exerted by a real gas occupying a given volume is less than that exerted by an ideal gas occupying the same volume, as indicated by the vertical light grey line. Both of these observations can be explained in terms of a short-range intermolecular attractive force which becomes progressively stronger as the average distance between any pair of molecules reduces, as caused by increasing compression. In the context of the van der Waals equation, the effect of intermolecular attractive forces is represented by the term an^2/V^2, which increases as the volume V (and hence the average intermolecular distance) decreases. Physically, these forces are attributable to the electrostatic attraction between the electric dipoles associated with the electron 'clouds' surrounding each molecule.

Towards the far left of Figure 15.8, however, we see that the situation reverses: at any given (very high) pressure, the volume occupied by a real gas is greater than that of the

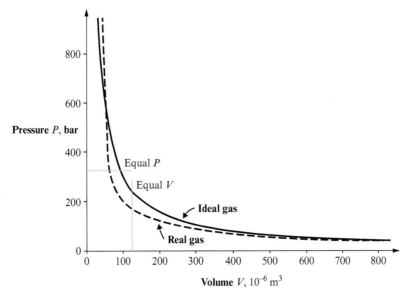

Figure 15.8 Real gases occupy a smaller volume than an ideal gas at the same temperature and pressure. The upper, solid, line shows the P,V curve for an ideal gas at a temperature T; the lower, dashed, line shows the behaviour, at the same temperature, of the same quantity of a real gas which complies with the van der Waals equation-of-state. At high volumes, and low pressures, towards the right, the behaviour of the real gas approximates to that of an ideal gas. At intermediate pressures, the volume occupied by a real gas at a given pressure is less than the volume occupied by an ideal gas at the same pressure (as shown by the horizontal light grey line); similarly, the pressure exerted by a real gas at a given volume is less than that exerted by an ideal gas at the same volume (as shown by the vertical light grey line). Both of these observations imply the existence of short-range attractive intermolecular forces. At very high pressures and low volumes, towards the far left, the volume occupied by the real gas at a given pressure is greater than that occupied by an ideal gas, implying the existence of very short-range intermolecular forces of repulsion. Note that the pressures on the vertical axis are very high: the volume associated with normal atmospheric pressure of ~1 bar corresponds to a point far to the right, beyond the graph shown.

corresponding ideal gas. Physically, this implies that, at very high pressures and very small volumes – that's when neighbouring molecules are extremely close together – a force comes into play that counteracts compression. This must be a very short-range intermolecular repulsion, which prevents neighbouring molecules from being pressed too closely together, as attributable, for example, to the electrostatic repulsion between the electron orbitals of neighbouring molecules. In the van der Waals equation, this very short-range force is primarily represented by the parameter b m^3 mol^{-1} which, physically, is a measure of the volume actually occupied by one mole of molecules. The volume b/N_A, where N_A is Avogadro's constant, therefore represents the volume occupied by a single molecule, and so denied to any neighbouring molecules. If the volume b/N_A approximates to a sphere, then the radius of this sphere is given by $(3b/4\pi N_A)^{1/3}$, and so there must be a very strong repulsive force which prevents neighbouring molecules from approaching within a mutual distance of approximately $2(3b/4\pi N_A)^{1/3}$.

These physical insights are verified mathematically from the van der Waals equation, most easily when rearranged as

$$P = \frac{nRT}{V - nb} - \frac{an^2}{V^2} \tag{15.11}$$

Firstly, if there are no intermolecular forces, $a = 0$, and if the molecules also occupy zero volume, $b = 0$. These two conditions comply with the assumptions defining an ideal gas, and the van der Waals equation becomes the ideal equation, $PV = nRT$.

Secondly, at any given temperature T and volume V, the larger the value of the parameter a, the lower the corresponding pressure P. Assuming for the moment that $b = 0$, then, if $a = 0$, the pressure P is that exerted by an ideal gas. But if $b = 0$ and $a > 0$ the system pressure is reduced by an amount $a(n/V)^2$, implying that a real gas exerts a lower pressure than an equivalent ideal gas. In terms of the molecular interpretation of pressure discussed on pages 17 and 18, this implies that the molecules of a real gas are in some way 'held back', and so don't collide with the walls of the container as forcefully as the molecules of an ideal gas – exactly as we would expect if there are forces of intermolecular attraction, and as shown by the central part of Figure 15.8.

Thirdly, if $a = 0$ and $b > 0$, at any given temperature T, a real gas occupying a volume V exerts a pressure $P = nRT/(V - nb)$, this being greater than the pressure nRT/V exerted by ideal gas occupying the same volume. Similarly, an ideal gas exerting a pressure P occupies a volume $V = nRT/P$, whilst a real gas exerting the same pressure P occupies a greater volume given by $nRT/P + nb$ – exactly as shown in the far left of Figure 15.8.

Collectively, the short-range attractive, and very short-range repulsive, intermolecular forces are known as **van der Waals forces**, and Table 15.1 shows the values of the van der Waals parameters a and b for a selection of real gases.

Table 15.1 Van der Waals parameters a and b for selected gases

	Molecular weight	a J m^3 mol^{-2}	b 10^{-6} m^3 mol^{-1}
Helium, He	4	0.0035	23.8
Neon, Ne	20	0.0208	16.72
Argon, Ar	40	0.1355	32.01
Krypton, Kr	84	0.2325	39.60
Xenon, Xe	131	0.4192	51.56
Radon, Rd	222	0.6601	62.39
Hydrogen, H_2	2	0.0245	26.51
Nitrogen, N_2	28	0.1370	38.7
Oxygen O_2	32	0.1382	31.86
Chlorine, Cl_2	71	0.6343	54.22
Water, H_2O (g)	18	0.5537	30.49
Carbon monoxide, CO	28	0.1472	39.48
Carbon dioxide, CO_2	44	0.3658	42.86
Methane, CH_4	16	0.2300	43.01
Pentane, C_5H_{12}	72	1.9130	145.1
Decane, $C_{10}H_{22}$	142	5.2880	305.1

Source: https://chem.libretexts.org/Reference/Reference_Tables/Atomic_and_Molecular_Properties/A8%3A_van_der_Waal's_Constants_for_Real_Gases

As intuitively expected, heavier – and hence larger – molecules have greater values of the molecular volume b (compare, for example, helium and radon, and methane and decane); a molecule (such as carbon dioxide) has a value of b greater than an atom (such as argon) of similar molecular weight; also, a in general increases with molecular weight too. In addition, there is an effect attributable to molecular polarity, which is most evident by comparing water, which is strongly polar, with, say, the non-polar molecule methane – although these molecules have similar molecular weights, the value of a for water is considerably greater than that for methane. This implies that the inter-molecular attractive forces between neighbouring water molecules are considerably stronger than the corresponding forces between neighbouring methane molecules, which is consistent with the electrostatic forces associated with polarity; furthermore, as is evident from the values of b, the water molecule is more compact than the methane molecule.

15.7.2 ...but real gases can

Figure 15.8 shows the behaviour of a real gas, which complies with the van der Waals equation-of-state, and which remains as a gas even when very highly compressed. If, however, a real gas is compressed at a lower temperature, then the intermolecular forces of attraction can cause the gas to condense into a liquid. The formation of a liquid can be examined mathematically from the van der Waals equation (15.11), when expressed as a cubic equation in V as

$$PV^3 - nV^2(bP + RT) + an^2V - abn^3 = 0 \tag{15.12}$$

The behaviour of equation (15.12), at a number of temperatures T, is shown in the P, V diagram, Figure 15.9.

For a real gas, at higher temperatures, the term $nRT/(V - nb)$ dominates the term $a(n/V)^2$, and so the P, V curve is a hyperbola, similar to that of an ideal gas; furthermore, when V is large, $V - nb \approx V$ and $a(n/V)^2$ is small, so that approximate ideal behaviour is shown at larger volumes even at lower temperatures.

However, as illustrated in Figure 15.9, for small values of V, and at lower temperatures, the P, V graph of a van der Waals gas shows the minimum and maximum characteristic of a cubic function. Furthermore, there is a specific temperature at which the P, V curve shows what is known as a **point of inflexion**, at which both the slope $(\partial P/\partial V)_T$ of the P, V curve, and also the second derivative $(\partial^2 P/\partial V^2)$, are zero. At all higher temperatures, the P, V curve is either close to a hyperbola, or 'kinked', such that the slope $(\partial P/\partial V)_T$ is negative, but never zero, for all volumes and pressures; at all lower temperatures, the P, V curve shows a local minimum and a maximum. Physically, this specific temperature is significant: at all higher temperatures, the system is a gas at all volumes and pressures; at all lower temperatures, the system can exist as a gas (at larger volumes and lower pressures, towards the right of the P, V curve), or as a liquid (at lower volumes and higher pressures, towards the left of the P, V curve). The temperature at which the P, V curve shows a point of inflexion, is, in fact, the critical temperature T_{crit} that we met on page 480, and T_{crit} defines the highest temperature at which a real gas can be liquefied by compression alone.

To determine the critical temperature T_{crit}, our starting point is the van der Waals equation

$$P = \frac{nRT}{V - nb} - \frac{an^2}{V^2} \tag{15.11}$$

PHASE EQUILIBRIA

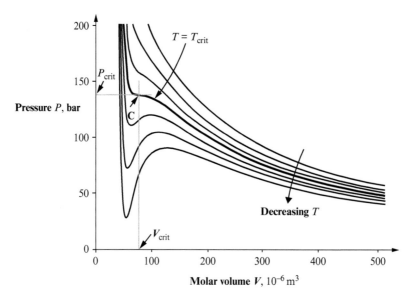

Figure 15.9 A P, V diagram illustrating the behaviour of one mole of a van der Waals gas at seven different temperatures. The van der Waals parameters for these graphs are $a = 0.23$ J m³ mol⁻² and $b = 2.5 \times 10^{-5}$ m³ mol⁻¹, at each of the seven temperatures of 380 K, 356 K, 340 K, 327.9 K = T_{crit}, 315 K, 300 K and 285 K. At larger volumes, and at higher temperatures, the van der Waals gas approximates to ideal behaviour. As the temperature decreases, the van der Waals curve deviates from the ideal hyperbolic curve, such that, at T_{crit}, the van der Waals curve shows a point of inflexion **C**, the critical point, at which both the slope $(\partial P/\partial V)_T$, and the second derivative $(\partial^2 P/\partial V^2)_T$, of the P,V curve are zero. At temperatures below T_{crit}, each P, V curve shows a minimum and a maximum. Note that the range of the horizontal axis is from zero to about $5V_{crit}$.

from which

$$\left(\frac{\partial P}{\partial V}\right)_T = -\frac{nRT}{(V-nb)^2} + \frac{2an^2}{V^3}$$

and

$$\left(\frac{\partial^2 P}{\partial V^2}\right)_T = \frac{2nRT}{(V-nb)^3} - \frac{6an^2}{V^4}$$

For a point of inflexion, both

$$\left(\frac{\partial P}{\partial V}\right)_T = -\frac{nRT}{(V-nb)^2} + \frac{2an^2}{V^3} = 0 \tag{15.13}$$

and

$$\left(\frac{\partial^2 P}{\partial V^2}\right)_T = \frac{2nRT}{(V-nb)^3} - \frac{6an^2}{V^4} = 0 \tag{15.14}$$

implying, from equation (15.13), that

$$\frac{nRT_{crit}}{(V_{crit}-nb)^2} = \frac{2an^2}{V_{crit}^3} \tag{15.15}$$

PHASE CHANGES, IDEAL GASES AND REAL GASES

and, from equation (15.14), that

$$\frac{2nRT_{crit}}{(V_{crit} - nb)^3} = \frac{6an^2}{V_{crit}^4} \tag{15.16}$$

where T_{crit} is the critical temperature, and V_{crit} is the corresponding critical volume.

Dividing equation (15.15) by equation (15.16)

$$\frac{(V_{crit} - nb)}{2} = \frac{V_{crit}}{3}$$

from which

$$V_{crit} = 3nb \tag{15.17}$$

implying that the molar critical volume $V_{crit} = 3b$. Reference to Table 15.1 shows that a representative value for V_{crit} is of the order of $100 \times 10^{-6} \times 10^{-6}$ m^3 = 0.1 litre.

Substituting $3nb$ for V_{crit} in equation (15.15) gives

$$\frac{nRT_{crit}}{4n^2b^2} = \frac{2an^2}{27n^3b^3}$$

and so

$$T_{crit} = \frac{8a}{27Rb} \tag{15.18}$$

Equations (15.16) and (15.17) may now be used in the van der Waals equation, equation (15.11), to give the value of the critical pressure P_{crit} as

$$P_{crit} = \frac{a}{27b^2} \tag{15.19}$$

The three equations (15.17), (15.18) and (15.19) therefore define the molar critical volume V_{crit}, critical temperature T_{crit}, and critical pressure P_{crit} of a van der Waals gas in terms of the parameters a and b; conversely, experimental measurements of V_{crit}, T_{crit}, and P_{crit} allow the parameters a and b to be estimated. So, for example, for 1 mol of a van der Waals gas for which $a = 0.23$ J m^3 mol^{-2} and $b = 2.5 \times 10^{-5}$ m^3 mol^{-1} (although these values are hypothetical, they are quite consistent with those in Table 15.1), $V_{crit} = 7.5 \times 10^{-5}$ m^3, $T_{crit} = 327.9$ K, and $P_{crit} = 1.363 \times 10^7$ Pa = 136.3 bar, as shown in Figure 15.9.

As can be seen by comparing the left-hand sides of Figures 15.7, 15.8 and 15.9, the van der Waals curves are much steeper at low volumes than the ideal curves: this corresponds to the physical reality that, whereas gases can be compressed relatively easily, liquids are much less compressible. Mathematically, as shown on page 89, a material's isothermal compressibility may be expressed in terms of the partial derivative $(\partial V/\partial P)_T$, which represents the change in volume ∂V_T attributable to a change in the external pressure ∂P_T. For all known materials, when ∂P_T is positive (corresponding to an increase in pressure), ∂V_T is inevitably negative (the system is compressed), implying that $(\partial V/\partial P)_T$ is always negative. A large (negative) value of $(\partial V/\partial P)_T$ therefore implies that that the system is easily compressed, and a small (negative) value implies that the system is difficult to compress. Since the slope of the P, V curve is $(\partial P/\partial V)_T = 1/(\partial V/\partial P)_T$, a very steep P, V curve, as on the left of Figure 15.9, implies that $(\partial P/\partial V)_T$ has a large (negative) value, corresponding to a small (negative) value for $(\partial V/\partial P)_T$, confirming that liquids are difficult to compress.

Physically, as we have seen, the critical temperature T_{crit} is the highest temperature at which a van der Waals gas can be liquefied by applying pressure alone, so reducing the volume, corresponding to moving along any isotherm at $T < T_{crit}$ from the right of Figure 15.9 towards the left. At the critical temperature T_{crit}, condensation of the gas into a liquid becomes thermodynamically favourable, and the vapour will spontaneously self-aggregate into liquid droplets that appear as a fog, at which point what was a system of a single gaseous phase becomes a system of two phases: the gas, and the newly-formed liquid droplets. The pressure drops accordingly, re-establishing thermodynamic equilibrium between the vapour phase and the liquid phase. Any further decrease in volume leads to a sharp rise in pressure because, as we have seen, the liquid is much less compressible than the gas phase.

15.7.3 The Maxwell construction

To examine the liquefaction of the gas, and the liquid ⇌ vapour phase transition is more detail, consider Figure 15.10(a), which shows a P, V curve for one mole of a van der Waals gas at a temperature $T < T_{crit}$, over a (very narrow) range of volumes from about $0.5 V_{crit} \approx 50 \times 10^{-6}$ m^3 (on the left) to about $2.5 V_{crit} \approx 250 \times 10^{-6}$ m^3 (on the right).

If we observe what happens as the van der Waals gas is steadily compressed isothermally at a temperature $T < T_{crit}$, then, as the molar volume V decreases, the system pressure P will increase according to the van der Waals equation, following the isotherm shown in Figure 15.10(a) from right to left, passing from state **J** through state **K** to state **L**. Under normal conditions, when the system is at state **L**, the system volume can suddenly 'jump', at constant pressure, from the volume V_L to the much smaller volume V_M, and we observe that the system has now become an equilibrium mixture between the van der Waals gas and the corresponding liquid. Then, as the volume of the now fully liquid system decreases below V_M, the system pressure rises very quickly, as shown towards the left of the graph in Figure 15.10(a).

The real behaviour of the system therefore follows the van der Waals equation in the gas phase, for the line segment from state **J** to state **L**, and also in the liquid phase, for the line segment from **M** towards state **N**. Between state **L** and state **M**, however, the system departs from the van der Waals equation, 'jumping' directly from state **L** to state **M**, as represented by the straight line segment **LM**.

With reference the P, V curve shown in Figure 15.10(a), the line segment **LM** fulfils three conditions:

- the system temperature T is the same for both state **L** and state **M**, implying that state **L** and state **M** are on the same isotherm; and
- the system pressure P is the same for both state **L** and state **M**, implying that the line segment **LM** is horizontal; and also (and much less obviously)
- the area of the shaded region **LXY** is equal to the area of the shaded region **YZM**.

There are many pairs of states that satisfy the first two conditions, for many horizontal lines can be drawn across the diagram at pressures P between the local pressure minimum represented by state **Z** and the local pressure maximum represented by state **X**. But for any given isotherm at any temperature $T < T_{crit}$, there is only one horizontal line segment which fulfils the third, equal-area, condition, which is known as the **Maxwell construction**. The line segment

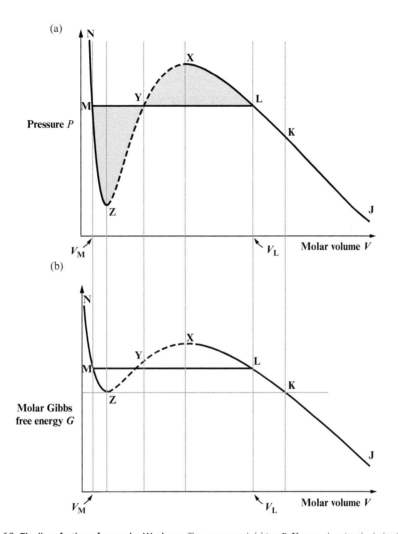

Figure 15.10 The liquefaction of a van der Waals gas. The upper graph (a) is a P, V curve showing the behaviour of the van der Waals equation at a temperature $T < T_{crit}$ at small molar volumes V; the lower graph shows the corresponding behaviour of the molar Gibbs free energy G. As the gas is compressed at constant temperature from state **J**, it follows the P, V curve from right to left, through state **K** until it reaches state **L**. The molar volume of the system can then 'jump', at constant pressure, from V_L to the considerably smaller volume V_M. This corresponds to the liquefaction of the gas: at state **L**, the system is a gas; at state **M**, the system in a liquid. Further compression follows the P, V curve from state **M** towards the left: the steep rise of the P, V curve towards state **N** represents the incompressibility of the liquid. As shown in the lower graph (b), the phase transition from **L** to **M** corresponds to $\Delta G = 0$. Also, as shown in the upper graph, the line **LM** complies with the 'Maxwell construction' such that the two shaded regions **LXY** and **YZM** have equal areas. The line segment **LX** represents a supercooled gas, and the line segment **MZ**, a superheated liquid. States along the line segment **XYZ**, shown dashed, are physically impossible, for there are no known materials for which $(\partial P/\partial V)_T$ and $(\partial G/\partial V)_T$ are positive. Note that the range of the horizontal axis is very small, straddling V_{crit} from about 0.5 $V_{crit} \approx 50 \times 10^{-6}$ m³ (on the left) to about 2.5 $V_{crit} \approx 250 \times 10^{-6}$ m³ (on the right).

LM is therefore unique, from a specific state **L**, between states **K** and **X**, to a specific state **M**, between states **Z** and **N**.

To gain a deeper insight as to why the states **L** and **M** are unique, and why the areas of the shaded regions **LXY** and **YZM** are equal, we need to consider the behaviour of the molar Gibbs free energy **G** of the system as the system is compressed. As we saw on page 384, the Gibbs free energy of a van der Waals gas, expressed as a function $G(V, T)$ of volume V and temperature T, is given by

$$G(V, T) = -nRT \ln(V - nb) + \frac{n^2 bRT}{(V - nb)} - \frac{2an^2}{V} + \text{constant} \tag{13.33}$$

The constant of integration may be determined by reference to the standard state, which, for a real gas – as shown in Table 6.4 – is the hypothetical state of the real gas, acting as an ideal gas, at the standard pressure $P^{\ominus}$. For an ideal gas, $a = b = 0$, and under standard conditions, $V = nRT/P^{\ominus}$, and so equation (13.33) reduces to

$$G^{\ominus} = -nRT \ln\left(\frac{nRT}{P^{\ominus}}\right) + \text{constant}$$

giving

$$\text{constant} = n\mathbf{G}^{\ominus} + nRT \ln\left(\frac{nRT}{P^{\ominus}}\right)$$

where $\mathbf{G}^{\ominus}$ is the standard molar Gibbs free energy of the real gas. Equation (13.33) therefore becomes

$$G(V, T) = n\mathbf{G}^{\ominus} + nRT \ln\left(\frac{nRT}{P^{\ominus}(V - nb)}\right) + \frac{n^2 bRT}{(V - nb)} - \frac{2an^2}{V} \tag{15.20}$$

For a given mass of a specific real gas at a constant temperature, the parameters n, a, b, T, $\mathbf{G}^{\ominus}$ and R are all constants, and so equation (15.20) is a function of the volume V only, as illustrated in Figure 15.10(b) for $n = 1$, and a chosen temperature $T < T_{\text{crit}}$.

As the system is compressed from state **J**, the system pressure P is represented in Figure 15.10(a), and the corresponding molar Gibbs free energy **G** in Figure 15.10(b). As can be seen, the graphs of the pressure P and the molar Gibbs free energy **G** have rather similar shapes, and the maxima and minima of both curves, states **X** and **Z**, are at the same molar volumes V (but note that state **Y**, which lies on the line segment **LM** shown in Figure 15.10(a), does *not* lie on the line segment **LM** shown in Figure 15.10(b), but is somewhat above it).

The special relevance of the molar Gibbs free energy curve relates to the condition for phase equilibrium between any two phases X and Y that we met on page 471, where we showed that two phases are in equilibrium when the molar Gibbs free energies **G** of each phase are equal:

$$\mathbf{G}_X = \mathbf{G}_Y \tag{15.4a}$$

An implication of equation (15.4a) is that the change $\Delta \mathbf{G}$ in the molar Gibbs free energy between the two phases X and Y in equilibrium must be such that $\Delta \mathbf{G} = \mathbf{G}_Y - \mathbf{G}_X = 0$. An equilibrium phase transition must therefore correspond to a horizontal line in Figure 15.10(b).

So, as the van der Waals gas is compressed from state **J**, the pressure *P* follows the upper curve of Figure 15.10(a), and the molar Gibbs free energy *G* tracks the lower curve of Figure 15.10(b). At state **J**, the molar Gibbs free energy *G* has only a single value for all molar volumes *V*, and so no phase transition is possible. However, when the gas approaches point **K**, there are now two possible states of the system with the same value of *G*: state **K**, the current state of the system, and state **Z**, to the left. A change in state from state **K** to state **Z** fulfils the condition that $\Delta G = 0$, suggesting that a phase change is potentially possible. Reference to Figure 15.10(a), however, shows that a change in state from state **K** to state **Z** must be associated with a drop in system pressure *P*. A condition for the phase change, however, is that it takes place at a constant pressure, and so a phase transition from state **K** to state **Z** cannot happen.

As the gas is compressed further from state **K** towards state **X**, Figure 15.10(b) shows that, in general, there are now two other states which have the same molar Gibbs free energy *G* – one on the left-hand branch of the curve, to the left of state **Z** on the line segment **ZN**, and the second on the line segment **XZ**, which is shown dashed.

Suppose that a phase change takes place from some state along the line segment **KX** to a corresponding state along the line segment **XZ**. As can be seen from Figure 15.10(a), wherever the final state might be along the line segment **XZ**, the slopes $(\partial P/\partial V)_T$ and $(\partial G/\partial V)_T$ of both graphs are positive. All known substances, however, are such that both $(\partial P/\partial V)_T = 1/(\partial V/\partial P)_T$ and $(\partial G/\partial V)_T$ are necessarily *negative* at all volumes: all substances decrease in volume under increasing pressure, and, as we saw on page 383, the Gibbs free energy *G* also necessarily decreases as the volume *V* increases. This implies that there can be no real states along the line segment **XZ** – they just do not exist, which is why the line segment **XZ** is shown dashed. All states along the line segment **XZ** are therefore inaccessible, and there can be no phase transitions from any state on the line segment **KX** to any state on the line segment **XZ**.

In contrast, a phase change from a state on the right-hand line segment **KX** to the left-hand line segment **ZN** is possible, for all states on the line segment **ZN** are associated with negative values of both $(\partial P/\partial V)_T$ and $(\partial G/\partial V)_T$, and so all states on that line segment can exist. For a phase change actually to take place, however, three conditions must be fulfilled simultaneously: the phase change must take place

- at a constant temperature *T*, implying that the two phase states are on the same isotherm, and
- at a constant pressure *P*, implying that the two phase states must be on the same horizontal on the *P, V* diagram shown in Figure 15.10(a), and
- at a constant value of the molar Gibbs free energy *G*, implying that the two phase states must be on the same horizontal on the *G,V* diagram shown in Figure 15.10(b).

Under normal circumstances, these three conditions are fulfilled simultaneously *only* for states **L** and **M**, and so the van der Waals gas can be compressed, as a gas, from state **J**, and through state **K** – but at state **L**, a phase change takes place from the gaseous phase state **L** to the liquid phase state **M**. Likewise, a liquid allowed to expand from state **N** will follow the isotherm to state **M**, at which point the liquid will vaporise to form a gas in state **L**.

The first two of these conditions correspond to the first two conditions shown on page 492; the third condition here, however, is expressed in terms of the molar Gibbs free energy **G** and is, as we shall now show, equivalent to the third condition shown in page 492, which was expressed in terms of the equality of the areas of the two shaded regions **LXY** and **YZM** in Figure 15.10(a).

To prove that the areas of the two shaded regions **LXY** and **YZM** are equal, we invoke the fundamental equation for dG

$$dG = V\,dP - S\,dT \tag{13.16a}$$

which at constant temperature becomes

$$dG = V\,dP$$

Adding, and subtracting, $P\,dV$, we may write

$$dG = V\,dP + P\,dV - P\,dV$$
$$= d(PV) - P\,dV$$

which integrates for any change from state $[1] \equiv [P_1, V_1]$ to state $[2] \equiv [P_2, V_2]$ as

$$\int_{G_1}^{G_2} dG = \int_{P_1 V_1}^{P_2 V_2} d(PV) - \int_{V_1}^{V_2} P\,dV$$

or

$$\Delta G = \Delta(PV) - \int_{V_1}^{V_2} P\,dV \tag{15.21}$$

Equation (15.21) is general, so we can now apply it to the phase change from state **L** to state **M**, for which $V_1 = V_L$, $V_2 = V_M$ and $\Delta G = 0$, giving

$$0 = P_M V_M - P_L V_L - \int_{V_L}^{V_M} P\,dV \tag{15.22}$$

in which P_M is the pressure corresponding to V_M and P_L is the pressure corresponding to V_L. Reference to Figure 15.10(a) will show that, for the phase transition, $P_M = P_L$, and if we represent this pressure as P_{ph}, then equation (15.22) becomes

$$P_{ph}(V_M - V_L) = \int_{V_L}^{V_M} P\,dV \tag{15.23}$$

Since $V_M < V_L$, the difference $V_M - V_L$ is negative; if this difference is expressed as the positive quantity $V_L - V_M$, then the limits of the integral need to be exchanged, giving

$$P_{ph}(V_L - V_M) = \int_{V_M}^{V_L} P\,dV \tag{15.24}$$

Equation (15.24) may be interpreted by reference to Figure 15.10(a): the term $P_{ph}(V_L - V_M)$ represents the rectangular area between the line segment **LXY** and the horizontal axis, and the integral represents the area under the P, V curve from state **M**, through states **Z**, **Y** and **X**, to state **L**. Mathematically, equation (15.24) states that these two areas are equal; geometrically, Figure 15.10(a) shows that this can be true only if the area of the shaded region **LXY** is equal to that of the shaded area **YZM** – hence the Maxwell construction.

One further aspect of Figure 15.10 concerns the states represented in the gas phase by the line segment **LX**. The slopes $(\partial P/\partial V)_T$ and $(\partial G/\partial V)_T$ along this line segment are both negative, and so states in this region can in principle exist: they can only occur, however, if the gas is compressed beyond state **L** without becoming a liquid. This is, in fact, possible, and states along the line segment **LX** are supercooled (alternatively, supersaturated) gas states, as described on page 474; likewise, states along the line segment **MZ** represent superheated liquid states.

As we saw on page 410, supercooled gases and superheated liquids are known as metastable states, which can last for relatively long periods of time (minutes or longer), but, once disturbed (for example, by scratching the inside of the vessel), will immediately change phase – supercooled gas to liquid, superheated liquid to gas – at constant pressure. So, with reference to the Figure 15.7(a), a superheated liquid in state **Z** will change phase to a gas at a state close to state **J**. As shown in the Figure 15.10(b), the molar Gibbs free energy G_J of this gaseous state **J** is less than G_Z, implying that $\Delta G = G_J - G_Z$ for this change is negative, and therefore that this change is both spontaneous and irreversible. This is a specific case of the general truth that the phase transition from any metastable supercooled or superheated state is spontaneous and irreversible – in contrast to the phase change represented by the line segment **LM**, which is a reversible, equilibrium, change.

15.8 The mathematics of G_s, G_l and G_g

In this last section in this chapter, we present the mathematics associated with Figures 15.2, 15.3, 15.4 and 15.5, which each show a variety of graphs for each of G_s, G_l and G_g, as a function of the temperature T, for a variety of external pressures P_{ex}. These graphs were not derived mathematically, but rather on the basis of the observations listed as the bullet-points on page 472, combined with an intuitive understanding of the behaviour of equations (13.21)

$$\left(\frac{\partial G}{\partial T}\right)_P = -S \tag{13.21}$$

and (13.20)

$$\left(\frac{\partial G}{\partial P}\right)_T = V \tag{13.20}$$

In fact, $G_s(T)$, $G_l(T)$ and $G_g(T)$ – each as a function of temperature at constant pressure – can all be determined mathematically, by integrating equation (13.21), as

$$G(T) = -\int_0^T S_P(T)\,dT + f(P) \quad \text{at constant pressure} \tag{15.25}$$

Three aspects of equation (15.25) merit attention. Firstly, the upper limit of the integration is denoted as the arbitrary temperature variable $\mathbb{T}$ so as to distinguish this limit from the variable of integration T; and secondly, since entropy is itself a function of both pressure and temperature, the entropy variable, at constant pressure, is shown as $S_P(T)$. Thirdly, the constant of integration is not an absolute constant, but rather a function of pressure only, shown as $f(P)$.

The temperature dependence of entropy was explored in Chapters 9 and 12, and Chapter 12 discussed how the absolute, or third law, entropy $S_P(T)$ at any temperature T can be computed using equation (12.2); the resulting expression can then be used in equation (15.25) to determine the corresponding $G(T)$. As can be seen from equation (12.2), the calculation of $S_P(T)$ requires the evaluation of integrals of the general type $\int (C_P/T) dT$, and since C_P for any phase is usually not a constant, these integrals are most accurately evaluated using numerical methods of integration. If, however, we assume that C_P for a particular phase is a constant, at least over a specific temperature range, then an analytical expression for $G(T)$ can be determined, as we shall now show for the gas phase.

As can be seen from page 353, for a gas, the third law entropy $S_P(T)$, as computed from equation (12.2), comprises five terms: one integral for each of the three phases, plus two terms for entropy changes associated with the each of the changes of phase. For 1 mol of any given pure gas, at temperature T_{vap}, the first four terms in equation (12.2) equate to the molar third law entropy $S_P(T_{\text{vap}})$ of the gas phase, immediately after vaporization. Since $S_P(T_{\text{vap}})$ is a pure number, equation (12.2) may be expressed much more simply as

$$S_{P,g}(\mathbb{T}) = S_P(T_{\text{vap}}) + \int_{T_{\text{vap}}}^{\mathbb{T}} C_{P,g} \frac{dT}{T} \tag{15.26}$$

If we assume that $C_{P,g}$ is a constant over the range from T_{vap} to $\mathbb{T}$, then equation (15.26) integrates as

$$S_{P,g}(\mathbb{T}) = S_P(T_{\text{vap}}) + C_{p,g} \ln \frac{\mathbb{T}}{T_{\text{vap}}}$$

or, replacing the variable $\mathbb{T}$ by the more familiar variable T

$$S_{P,g}(T) = S_P(T_{\text{vap}}) + C_{P,g} \ln \frac{T}{T_{\text{vap}}} \tag{15.27}$$

in which the term $S_P(T_{\text{vap}})$ represents the contribution to $S_{P,g}(T)$ attributable to temperatures less than T_{vap}, and the term $C_{P,g} \ln(T/T_{\text{vap}})$, the contribution attributable to temperatures greater than T_{vap}.

This expression may now be used in equation (15.25) giving

$$G_g(\mathbb{T}) = -\int_0^{\mathbb{T}} S_P(T_{\text{vap}}) \, dT - \int_{T_{\text{vap}}}^{\mathbb{T}} C_{p,g} \ln \frac{T}{T_{\text{vap}}} \, dT + f(P)$$

in which the lower limit of the second integral is set at T_{vap}, since the integrand is defined only for $T > T_{vap}$. Once again, if we assume that $C_{P,g}$ is a constant, we have

$$G_g(\mathbb{T}) = -\mathbb{T} S_P(T_{vap}) - C_{P,g} \int_{T_{vap}}^{\mathbb{T}} \ln \frac{T}{T_{vap}} dT + f(P) \qquad (15.28)$$

The second term may be 'integrated by parts' as

$$\int_{T_{vap}}^{\mathbb{T}} \ln \frac{T}{T_{vap}} dT = \left[T \ln \frac{T}{T_{vap}} \right]_{T_{vap}}^{\mathbb{T}} - \int_{T_{vap}}^{\mathbb{T}} T \, d\left(\ln \frac{T}{T_{vap}} \right)$$

Since

$$d\left[\ln \frac{T}{T_{vap}} \right] = \left(\frac{T_{vap}}{T} \right) \left(\frac{1}{T_{vap}} \right) dT = \left(\frac{1}{T} \right) dT$$

then

$$\int_{T_{vap}}^{\mathbb{T}} \ln \frac{T}{T_{vap}} dT = \left[T \ln \frac{T}{T_{vap}} \right]_{T_{vap}}^{\mathbb{T}} - \int_{T_{vap}}^{\mathbb{T}} T \left(\frac{1}{T} \right) dT = \left[T \ln \frac{T}{T_{vap}} \right]_{T_{vap}}^{\mathbb{T}} - \int_{T_{vap}}^{\mathbb{T}} dT$$

giving

$$\int_{T_{vap}}^{\mathbb{T}} \ln \frac{T}{T_{vap}} dT = \left(\mathbb{T} \ln \frac{\mathbb{T}}{T_{vap}} - T_{vap} \ln \frac{T_{vap}}{T_{vap}} - (\mathbb{T} - T_{vap}) \right)$$

But $\ln(T_{vap}/T_{vap}) = \ln 1 = 0$, and so

$$\int_{T_{vap}}^{\mathbb{T}} \ln \frac{T}{T_{vap}} dT = \mathbb{T} \left(\ln \frac{\mathbb{T}}{T_{vap}} - 1 \right) + T_{vap}$$

Accordingly, equation (15.28) becomes

$$G_g(\mathbb{T}) = -\mathbb{T} S_P(T) - C_{P,g} \mathbb{T} \left(\ln \frac{\mathbb{T}}{T} - 1 \right) - C_{P,g} T_{vap} + f(P)$$

which may be expressed, using the more familiar variable T, as

$$G_g(T) = -T \left(C_{P,g} \left[\ln \frac{T}{T_{vap}} - 1 \right] + S_P(T_{vap}) \right) - C_{P,g} T_{vap} + f(P) \qquad (15.29)$$

Equation (15.29) is the result we have been seeking – an expression for the molar Gibbs free energy $G_g(T)$ of a gas, as a function of temperature $T > T_{vap}$, at constant pressure, assuming that $C_{P,g}$ is a constant.

By exactly similar reasoning, we may derive an expression for $G_l(T)$ for a liquid

$$G_l(T) = -T\left(C_{P,l}\left[\ln\frac{T}{T_{\text{fus}}} - 1\right] + S_P(T_{\text{fus}})\right) - C_{P,l}T_{\text{fus}} + f(P) \tag{15.30}$$

for temperatures T such that $T_{\text{vap}} > T > T_{\text{fus}}$. In this expression, $S_P(T_{\text{fus}})$ is the sum of the first two terms of equation (12.2), representing the molar third law entropy of the liquid phase, immediately after melting.

Likewise, for a solid

$$G_s(T) = -T\left(C_{P,s}\left[\ln\frac{T}{T_{\text{lo}}} - 1\right] + S_P(T_{\text{lo}})\right) - C_{P,s}T_{\text{lo}} + f(P) \tag{15.31}$$

at temperatures $T < T_{\text{fus}}$. In this equation, T_{lo} is a temperature below T_{fus}, but considerably greater than the Debye temperature (see page 354), and $S_P(T_{\text{lo}})$ is the molar third law entropy of the material at $T = T_{\text{lo}}$, taking into account the non-linear behaviour of $C_{P,s}$ at low temperatures, as shown in Figure 12.1.

Equations (15.29), (15.30) and (15.31) all assume that the appropriate C_P is a constant. In fact, as discussed on pages 195 and 196, each C_P is not a constant, but, within any phase, tends to increase rather slowly with temperature. These equations are therefore approximations, rather than 'the truth'; that said, the approximations are quite good, and all show the correct general behaviour, and equations of these forms were used to plot the graphs in Figures 15.2, 15.3, 15.4 and 15.5.

EXERCISES

1. Define:
   - Phase
   - Vapour pressure
   - Equilibrium saturated vapour pressure
   - Boiling
   - Van der Waals gas
   - Maxwell construction
   - Supercooled gas
   - Superheated liquid
   - Metastable state
   - Nucleation.

2. At a pressure of 1.01325 bar, and at all temperatures from 273.15 K to 373.15 K, water is a liquid, and is in equilibrium with its vapour. As the temperature increases within this range, the vapour pressure increases, but at all temperatures, there is an equilibrium between the two phases. At a pressure of 1.01325 bar, and at all temperatures above 373.15 K, water is a gas – but, at, say, 400 K, all the water is a gas, and there is no measureable liquid water. Why is it that, when the system is naturally a liquid (between 273.15 K and 373.15 K), then there is a liquid $\rightleftharpoons$ vapour equilibrium, but when the system is naturally a gas (above 373.15 K), there is no longer an equilibrium in that there is no liquid present?

3. Prove that the condition for equilibrium between two phases is $G_X = G_Y$.

4. Why can an ideal gas not liquefy? Why does a real gas liquefy?

5. This is a spreadsheet exercise to produce a diagram of the form of Figure 15.5 for a van der Waals gas, using the van der Waals equation-of-state, equation (15.11)

$$P = \frac{nRT}{V - nb} - \frac{an^2}{V^2} \qquad (15.11)$$

for $n = 1$.

At the top of the spreadsheet, define the van der Waals parameters a and b – as shown by Table 15.1, a is typically ~ 0.2 J m^3 mol^{-2}, and $b \sim 4 \times 10^{-5}$ m^3 mol^{-1}, Then code some cells using equations (15.20), (15.21) and (15.22) to compute the critical values V_{crit}, T_{crit} and P_{crit} – these values are important, for the cubic behaviour of the van der Waals equation is most evident at volumes and pressures around V_{crit} and P_{crit}, and at temperatures below T_{crit}. When $a = 0.2$ J m^3 mol^{-2}, $b = 40 \times 10^{-6}$ m^3 mol^{-1}, and $n = 1$, $V_{crit} = 120 \times 10^{-6}$ m^3, $P_{crit} = 4.6 \times 10^6$ Pa and $T_{crit} = 178.19$ K.

Given that $V_{crit} = 120 \times 10^{-6}$ m^3, set up a column (say column D) for values of V from 50×10^{-6} m^3 (in, say, cell D10) to 500×10^{-6} m^3 in increments of 5×10^{-6} m^3, corresponding to 90 rows (to cell D100). In the next column, code a/V^2 in cells E10 to E100.

Now leave a few blank columns, and then, at the head of the next column (say, cell G1), enter a value for $R = 8.314$ J K^{-1} mol^{-1}, and in cell G2, a value for $T = 150$ K (this value being below $T_{crit} = 178.19$ K). Then, in cells G10 to G100, code $RT/(V - b)$, using the value of V from column D. In cells H10 to H100, code P as column G – column E. Column H therefore contains values of the pressure of the van der Waals gas at different volumes, at a constant temperature of 150 K.

Leave a blank column, and then set up two columns I and J, like columns G and H, but using a different value of T, say, 160 K; likewise set up further pairs of columns at a range of temperatures up to, say, 200 K.

A multi-line graph, showing how the pressure varies with volume at each temperature, will be like Figure 15.5. Vary the temperature in small increments around $T_{crit} = 178.19$ K to observe the point of inflexion; also, if you set the parameter a to zero (keeping b non-zero), and also b to zero (keeping a non-zero), you will see the effect of each of these parameters independently. And if both a and b are set to zero simultaneously, the results will correspond to an ideal gas.

6. As noted on page 478, there is a second factor which contributes to the reason why the mercury in a Torricelli barometer does not boil. Thinking about what actually happens when further mercury atoms vapourise, and bearing in mind Figure 15.1, what is this second factor?

7. "I'm confused about 'evaporation' and 'boiling'." "Let me help. In fact, 'evaporation' and 'boiling' are just different words for exactly the same thing – the change in state from liquid to gas." "Ah... I see... well... that seems to be right..."

Do you agree that "that seems to be right"? How would you contribute to the conversation?

16 Reactions in solution

 Summary

An **ideal solution** is a solution of one or more solutes D within a solvent S such that each component i obeys **Raoult's law**

$$p_i = x_i p_i^* \tag{16.1}$$

in which p_i is the vapour pressure of component i above the solution; p_i^* is the vapour pressure, at the same temperature, associated with the pure component i; and x_i is the mole fraction of component i within the solution.

Ideal-dilute solutions obey **Henry's law**

$$p_i = x_i\, k_{H,i} \tag{16.2}$$

where $k_{H,i}$ is the **Henry's law constant**, and x_i is necessarily small.

The molar Gibbs free energy $\boldsymbol{G}_i$ of component i in an ideal solution is, in general, given by

$$\boldsymbol{G}_i \text{ (in solution)} = \boldsymbol{G}_i^{\ominus}\text{(pure)} + RT \ln (x_s/x_s^{\ominus}) \tag{16.7a}$$

in which the standard state is the pure component i.

For solutes D in a dilute ideal solution, two alternatives to equation (16.8) are

$$\boldsymbol{G}_D \text{ (in solution)} = \boldsymbol{G}_D^{\ominus}\text{(in solution, } b) + RT \ln (b_D/b_D^{\ominus}) \tag{16.9a}$$

where b_D is the molality of the solute D, and the standard state is that of unit molality, and

$$\boldsymbol{G}_D \text{ (in solution)} = \boldsymbol{G}_D^{\ominus}\text{(in solution)} + RT \ln ([D]/[D]^{\ominus}) \tag{16.10a}$$

in which [D] is the molar concentration of component D, and the standard state is that of unit molar concentration, 1 mol m^{-3}, or 1 mol dm^{-3}.

For a generalised chemical reaction

$$aA + bB \rightleftharpoons cC + dD$$

taking place within an ideal solution, and using standards states based on mole fractions

$$\Delta_r G_x = \Delta_r G_x^{\ominus} + RT \ln K_x \tag{16.13a}$$

where

$$K_x = \frac{(x_{C,eq}/x_C^{\ominus})^c\, (x_{D,eq}/x_D^{\ominus})^d}{(x_{A,eq}/x_A^{\ominus})^a\, (x_{B,eq}/x_B^{\ominus})^b} = \frac{x_{C,eq}^c\, x_{D,eq}^d}{x_{A,eq}^a\, x_{B,eq}^b} \tag{16.12a}$$

and

$$\Delta_r G_x^{\ominus} = (c\,\Delta_f G_{x,C}^{\ominus} + d\,\Delta_f G_{x,D}^{\ominus}) - (a\,\Delta_f G_{x,A}^{\ominus} + b\,\Delta_f G_{x,B}^{\ominus}) \tag{16.14a}$$

Modern Thermodynamics for Chemists and Biochemists. Dennis Sherwood and Paul Dalby.
© Oxford University Press 2018. Published 2018 by Oxford University Press.

varies with temperature according to

$$\frac{d \ln K_x}{dT} = \frac{\Delta_r H^\ominus}{RT^2} \qquad (16.20)$$

For dilute ideal solutions, and ideal-dilute solutions, equations (16.12a), (16.13a), (16.14a), and (16.20) – and indeed all other related equations – can be written in which either the molality b_i or the molar concentration [I] replaces the mole fraction x_i of component i, as associated with the appropriate standard state of either unit molality or unit molar concentration.

16.1 The ideal solution

16.1.1 Solvents, solutes and mole fractions

In this chapter, we will explore the thermodynamics of chemical reactions that take place in **solution**, this being of particular relevance to many biological systems. Solutions are by definition mixtures in which n_D mol of a **solute** D are dissolved within n_S mol of a **solvent** S. Unfortunately, the words 'solute' and 'solvent' both begin with the letter 'S', which risks ambiguity: we will therefore use S to represent the **S**olvent, and D to represent the **D**issolved solute. If, at the temperature of interest, the solute is a solid or a gas, and the solvent a liquid, we usually talk of a solution of the solute within the solvent; if both the solute and the solvent are liquids, and if the quantities of the components are similar, then the distinction between 'solvent' and 'solute' is blurred, and so we usually talk of a mixture, rather than a solution. It is also possible to consider other forms of mixtures and solutions, such as two gases, gases in liquids, and two solids (for example, metallic alloys): for the remainder of this book, we shall assume that a solution is the result of dissolving a solid, liquid or gas solute within a liquid solvent.

The relative quantities of the solvent S and the solute D can be specified in terms of their respective mole fractions x_S and x_D, defined according to equation (6.18) as

$$x_S = \frac{n_S}{n_S + n_D}$$

and

$$x_D = \frac{n_D}{n_S + n_D}$$

where n_S and n_D are the mole numbers of the solvent and solute at any time. For a pure substance, say the solvent S, $n_D = 0$, and so $x_S = 1$, whilst, for the solvent D, $x_D = 0$.

If the solution contains more than one solute, then the mole fraction x_i of component i is given by

$$x_i = \frac{n_i}{\sum_i n_i} \qquad (6.18)$$

in which the summation is over all components, including the solvent, and each x_i is necessarily less than. 1 As we shall see shortly, if a chemical reaction is taking place in solution, then the mole numbers n_i, and the corresponding mole fractions x_i, of each component i change over time as the reaction takes place, but at any instant, each n_i and x_i is well-defined.

From these definitions, for a single solute, it is always true that

$$x_S + x_D = 1 \qquad (6.18)$$

and for multiple solutes, likewise

$$\sum_i n_i = 1$$

A dilute solution is one in which there is much more solvent than solute: $n_S \gg n_D$, and $x_S \gg x_D$. Also, we note that the mole numbers n_S and n_D refer to the quantities of S and D in the solution, which may refer not only to complete molecules, but also to ions. So, for example, if $n_{Ca(NO_3)_2}$ mol of solid calcium nitrate, $Ca(NO_3)_2$, are dissolved in water, the $Ca(NO_3)_2$ molecule dissociates completely into one Ca^{2+} (aq) **cation** and two NO_3^- (aq) **anions**. Even though only a single entity, $Ca(NO_3)_2$, was introduced, there are two, different, solutes in the solution – the Ca^{2+} (aq) and NO_3^- (aq) ions – each with their own mole numbers $n_{Ca^{2+}}$ and $n_{NO_3^-}$. Furthermore, given that, on dissolving in water, each molecule of $Ca(NO_3)_2$ dissociates completely into one Ca^{2+} (aq) ion and two NO_3^- (aq) ions, the mole number $n_{Ca^{2+}}$ equals the mole number $n_{Ca(NO_3)_2}$; for the NO_3^- (aq) ion, however, the mole number $n_{NO_3^-}$ equals twice the mole number $n_{Ca(NO_3)_2}$. So although only n mol of solid calcium nitrate $Ca(NO_3)_2$ have been dissolved in the water, the total number of solute particles in solution is $n_{Ca^{2+}} + n_{NO_3^-} = 3n_{Ca(NO_3)_2}$.

16.1.2 Raoult's law

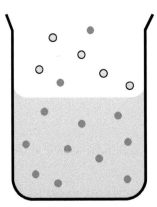

Figure 16.1 A representation of a solution. Molecules of the solute "dark" are dissolved in the solvent "light", in a vessel open to the atmosphere. Molecules of the solute and the solvent are also present in the gaseous phase, each exerting its own vapour pressure.

Figure 16.1 shows a representation of a solution of a solute, "dark", within a liquid solvent, "light", in a beaker open to the atmosphere.

Given that the solvent S is a liquid, at any temperature, some solvent molecules will be in the gaseous phase, as shown in Figure 16.1. These gaseous molecules will exert a pressure on the surface of the solution, in addition to the pressure attributable to the atmosphere. This additional pressure is the vapour pressure p_S, this being the partial pressure attributable to molecules of S within the S-enriched atmosphere above the solution – an atmosphere that we assume behaves as a mixture of ideal gases, including the gas attributable to the vapour S. Similarly, some molecules of the solute D will also be 'escaping' from the solution into the gaseous phase, exerting a D vapour pressure p_D. Not shown in Figure 16.1, however, are the molecules of nitrogen, oxygen and other atmospheric gases, which are present both in the space above the solution, and also dissolved within the solution – throughout this chapter, the presence of dissolved atmospheric gases will be ignored.

When the system is in equilibrium as a whole, two different equilibria co-exist: for the solute

D (in solution) $\rightleftharpoons$ D (vapour)

and for the solvent

S (in solution) $\rightleftharpoons$ S (vapour)

For the solvent, at any temperature, the equilibrium vapour pressure p_S above a solution is less than the equilibrium vapour pressure p_S^* at the same temperature for the pure liquid. This may be interpreted at a molecular level by considering the surface layer of the liquid: for the pure liquid, the entire surface is comprised of molecules of S, and, at any given temperature, they 'escape' at a given rate. For a solution, however, some of the surface molecules are those of the solvent D, so fewer molecules of S are present. At the same temperature, fewer molecules of S can therefore escape, and so the corresponding vapour pressure is lower.

An **ideal solution** is one for which, at any temperature T, the equilibrium vapour pressure p_i of each component i in the solution is directly proportional to the corresponding mole fraction x_i of the component i, with the constant of proportionality being equilibrium vapour pressure p_i^* of the pure component i, as

$$p_i = x_i p_i^* \tag{16.1}$$

Equation (16.1) is known as **Raoult's law**, after the French 19th century chemist François-Marie Raoult, and when $x_i = 1$, component i is in its pure state, and $p_i = p_i^*$. Although ideal solutions are theoretical abstractions, they play a very valuable role in the description of real solutions, similar to that played by the ideal gas as regards enabling us to approximate the behaviour of real gases.

So, for a two-component solution, for the solvent S

$$p_S = x_S p_S^*$$

where p_S^* is the vapour pressure at temperature T for the pure solvent S, usually a liquid. Simultaneously for the solute D, at the same temperature T

$$p_D = x_D p_D^*$$

where p_D^* is the vapour pressure at temperature T for the pure solute D. For any solution, since both x_S and x_D are necessarily less than 1, $p_S < p_S^*$ and $p_D < p_D^*$, as expected.

The solute D might be a liquid which is miscible with the solvent, in which case, it might not be obvious which liquid is the 'solvent', and which the 'solute', but this does not matter – equation (16.1) is true for all the components in an ideal solution or mixture. Very often, the solute is a solid, which, as we saw on page 470, is very likely to be considerably less volatile than the liquid solvent in which it is dissolved. In this case, p_D^*, the vapour pressure associated with the pure solid D, will be considerably less than p_S^*, the vapour pressure associated with the pure solvent S, and $p_D = x_D p_D^*$ will be even smaller – but even though the values of p_D and p_D^* might be small, for an ideal solution, the relationship $p_D = x_D p_D^*$ remains valid.

Raoult's law, for a two component mixture, is illustrated in Figure 16.2.

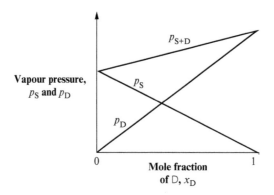

Figure 16.2 Raoult's law and the ideal solution. The graph shows the vapour pressures p_S and p_D for the two components in a mixture of S and D, as their respective mole fractions x_S and x_D change from $x_S = 1$ and $x_D = 0$, corresponding to pure S, on the left, to $x_S = 0$ and $x_D = 1$, corresponding to pure D, on the right. For all mole fractions, the vapour pressures of each component are proportional to the corresponding mole fraction, such that $p_S = x_S p_S^*$ and $p_D = x_D p_D^*$, in compliance with Raoult's law. The solution is therefore ideal, and the total vapour pressure above the solution is the sum $p_S + p_D$.

Figure 16.2 is important experimentally, for it is possible to measure, for any solution or mixture, all the variables x_S, x_D, p_S, p_D, p_S^* and p_D^*. The actual values of p_S and p_D can then be compared to the values $x_S p_S^*$ and $x_D p_D^*$ as predicted by Raoult's law, and if the measured values equal the predicted values, then the solution is ideal; if the actual values are either greater or less than the predicted values, then the solution is not ideal, as illustrated in Figure 16.3.

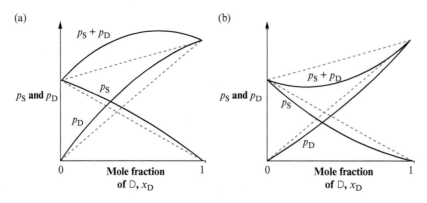

Figure 16.3 Deviations from Raoult's law, and non-ideal solutions. These two graphs illustrate non-ideal mixtures of two components S and D, such that $p_S \neq x_S p_S^*$ and $p_D \neq x_D p_D^*$. In (a), the vapour pressures are greater than as predicted by Raoult's law, and in (b), less than as predicted by Raoult's law.

If the actual vapour pressures are greater than as predicted by Raoult's law, this is known as a **positive deviation from Raoult's law**; if the actual vapour pressures are less than as predicted by Raoult's law, this is a **negative deviation from Raoult's law**. These deviations from Raoult's Law will be discussed further in Chapter 22.

16.1.3 A molecular interpretation of Raoult's law

To gain an understanding of the molecular conditions which need to be fulfilled for a solution to be considered ideal, consider Figure 16.4, which shows a representation of a small part of the surface of a solution, from which evaporation takes place.

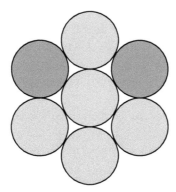

Figure 16.4 A representation of a small part of the surface of a two component solution. The molecules of the solvent S are lightly coloured and those of the solute D are darkly coloured. The mole fraction x_S of the solvent is 5/7, and the mole fraction x_D of the solute is 2/7.

Since the molecules in the solution are randomly distributed, the proportion of solvent molecules on the surface will be equal to the mole fraction of the solvent x_S; likewise, the proportion of solute molecules will be equal to the mole fraction of the solute x_D. These mole fractions are the key ratios in Raoult's law, and so an implication of the law is that the rate of evaporation of either the solvent or the solute is directly proportional to the number of the corresponding molecules on the surface, and is influenced by no other factors.

For a deeper insight, recall that evaporation requires that a molecule on the surface overcomes the intermolecular attractions within the liquid, and 'escapes'. The stronger the intermolecular attractions, the greater the energy required for evaporation, and so, at any given temperature, the fewer the number of molecules in the vapour, and the lower the corresponding vapour pressure. For a pure solvent S, the molecular interactions are of one type – the S $\cdots$ S interaction between any two neighbouring S molecules, where we are using the symbol $\cdots$ to represent an intermolecular force rather than a chemical bond. When a solute D is dissolved in the solvent S, however, any one S molecule experiences two, different, intermolecular interactions – S $\cdots$ S and S $\cdots$ D.

If the strength of the S $\cdots$ D interaction is equal to that of the S $\cdots$ S interaction, then any molecule of S on the surface of the solution is in exactly the same physical situation as a molecule of S on the surface of the pure solvent S. At any given temperature, the probability that a molecule of S on the surface of the solution has enough energy to evaporate is therefore exactly the same as that of a molecule of S on the surface of the pure solvent. But since the surface of the solvent has only a proportion x_S of solvent molecules present, the vapour pressure p_S above the solution will equal $x_S\, p_S{}^*$, $p_S{}^*$ being, as we know, the vapour pressure above the pure liquid at the same temperature, so explaining Raoult's law. Exactly the same argument can be applied to the solute, for which $p_D = x_D\, p_D{}^*$.

Suppose, however, that the S $\cdots$ D interaction is much stronger than the S $\cdots$ S interaction. In the solution, the more molecules of D that surround any surface molecule of S, the more that molecule of S is 'held back'. At any given temperature, the probability that any S molecule on the surface will have sufficient energy to 'escape' will be lessened, and, relative to the pure solvent, the corresponding vapour pressure will be lower for two reasons – firstly, because there are fewer S molecules on the surface, as determined by the mole fraction x_S, and secondly, because the probability of evaporation is reduced as a result of the strength of the S $\cdots$ D interaction. p_S will therefore be less than $x_S\, p_S{}^*$, increasingly so as the mole fraction x_D of the solute D increases. Raoult's law therefore does not apply for the solvent. Likewise, p_D will be less than $x_D\, p_D{}^*$, and Raoult's law will be broken for the solute too – as we shall explore in more detail in Chapter 22.

A third case is that in which the S ... D interaction is weaker than the S ... S interaction, in which case evaporation is enhanced, relative to the pure materials. p_S will be greater than $x_S\, p_S{}^*$, p_D will be greater than $x_D\, p_D{}^*$, and, once again, Raoult's law is broken for both the solvent and the solute.

Raoult's law therefore requires that the S ... S interaction is equal in all respects to the S ... D interaction, or, in practice, nearly so. A mixture of diethyl ether and dipropyl ether is therefore more likely to conform to ideal behaviour than a mixture of, say, water and sucrose.

The equality – or inequality – of the S $\cdots$ S and S $\cdots$ D interactions has three important, and experimentally observable, consequences.

Firstly, if the S $\cdots$ S and S ... D interactions are equal, the average separation between two molecules of S, and between one molecule of S and one molecule of D, will be identical. As a consequence, when a volume V_S of S is added to a volume V_D of D (which, for the purposes of this example, might best be thought of as another liquid), then the total volume V_{tot} of the mixed system will be the sum of the original volumes: $V_{tot} = V_S + V_D$. In contrast, if the S $\cdots$ D interaction is much stronger than the S $\cdots$ S interaction, then, on mixing, molecules of S and D will tend to be closer together, and we would expect that $V_{tot} < V_S + V_D$. This reduction in volume is exactly what happens when pure ethanol and pure water are mixed: as we saw on page 8, adding 1 m³ of ethanol to 1 m³ of water results in a volume of 1.92 m³, not 2 m³. An observable property of an ideal solution is therefore that

> There is no volume change on forming an ideal solution.

This is important, for the possibility that, for a non-ideal solution, $V_{tot} < V_S + V_D$ (or indeed that $V_{tot} > V_S + V_D$ if the S $\cdots$ D interaction is much weaker than the S $\cdots$ S interaction) breaks the 'rule' that extensive state functions are always additive. Note that this does not apply to all state functions of a non-ideal solution: for example, at a pressure of 1 bar at a temperature of 298.15 K, 1 m³ of ethanol has a mass of 798 kg, and 1 m³ of water a mass of 997 kg. When these are mixed, although the volume shrinks to 1.92 m³, mass is conserved, and $M_{tot} = M_S + M_D = 798 + 997 = 1795$ kg.

Secondly, in an ideal solution, the fact the S $\cdots$ S interaction equals the S $\cdots$ D interaction implies that there is no energy change on mixing, and so there is no evolution, or absorption, of heat. If two materials S and D are mixed with the result that heat is produced or absorbed, this indicates that the mixing is causing energy to be released or consumed at a molecular level, as associated with a reaction which we might write as

$$S + D \rightarrow S \cdots D \qquad \Delta H = \Delta_{mix} H$$

in which, as we have seen, S $\cdots$ D does not represent a new chemical entity, but rather the state resulting from the mixing of S and D, mixing that is associated with an enthalpy change $\Delta_{mix}H$. A familiar example of this is the substantial release of heat when pure sulphuric acid is mixed with pure water. A second observable property of an ideal solution is therefore that

There is no heat of mixing on forming an ideal solution.

Thirdly, as a solution becomes increasingly dilute – as the mole fraction x_D of the solute approaches zero whilst the mole fraction x_S of the solvent approaches one – the effect of any difference between the S $\cdots$ D molecular interaction and the S $\cdots$ S molecular interaction diminishes, and the solution progressively tends towards increasing ideal behaviour. Equally dilute solutions of different solutes, however, do not necessarily have the same degree of ideality – a mixture of two similar organic liquids is likely to result in a close-to-ideal solution, but a solution of the same dilution of an electrolyte in water is less likely to be ideal, as a result of the polar nature of the water molecule, and the long-range nature of electrostatic forces.

The more similar the molecules of the solvent S and the solute D, the more likely that a mixture of S and D will be an ideal solution. The extreme case of this is when S and D are the same – and so mixing a volume V_1 of, say, benzene with a volume V_2 of benzene does indeed result in an 'ideal solution' of a total volume of $V_1 + V_2$ of benzene-in-benzene. More realistically, a volume V_1 of benzene will mix with a volume V_2 of toluene to form an essentially ideal solution of total volume $V_1 + V_2$, with the 'solvent' being that component with the greater mole fraction within the solution, and the 'solute' that component with the lesser mole fraction. We would also expect both components to comply with Raoult's law at all mole fractions.

If the molecules of S and D are quite different, then we should expect deviations from Raoult's law. An extreme case is that in which S and D, though, for example, both liquids, are so different that they don't mix at all – for example, if a container containing olive oil and water are vigorously shaken, the two liquids will form a cloudy liquid which will soon, spontaneously, separate into two separate layers. The initial state, immediately after mixing, is not a true solution: rather, it is a mechanically-produced dispersion of small droplets, which soon coalesce to form the two different layers. Thermodynamically, this implies that the change $\Delta G = \Delta H - T \Delta S$ in Gibbs free energy associated with the change from the unstable dispersed-droplet state to the stable two-layer state must be negative. It so happens that the enthalpy change ΔH associated with the transition from many water-oil molecular interactions to the corresponding number of water-water and oil-oil molecular interactions is around 0 J mol^{-1}; ΔG is negative because ΔS is strongly positive when the oil and water separate. This appears to be counter-intuitive, for a state comprising two separate layers seems to be more ordered, and hence of lower entropy, than a fully mixed state. This macroscopic paradox is resolved at a microscopic level: in the olive oil-water dispersion, hydrogen bonds between the water molecules create ordered structures of water molecules surrounding oil molecules, whereas a homogeneous body of water is much more disordered at the molecular level. The entropy of the oil-water dispersion is therefore lower than that of the two layers, implying that ΔS is positive when the dispersion separates into two layers. ΔG is correspondingly negative, and so the separation into two layers is spontaneous – but increasingly less so as the temperature increases; water and olive oil mix better at higher temperatures.

16.1.4 Henry's law

A special case of a deviation from Raoult's law arises when the vapour pressure p_D of a solute D in a solution is found, experimentally, to be proportional to the mole fraction x_D, but the constant of proportionality is not the vapour pressure p_D^* of the pure component D, but rather a parameter $k_{H,D}$ that needs to be determined empirically

$$p_D = x_D k_{H,D} \tag{16.2}$$

This is known as **Henry's law**, and $k_{H,D}$ is the **Henry's law constant** for component D. Henry's Law is found to be valid for solutes, often gases, in dilute solutions – so-called **ideal-dilute solutions** – in which the mole fraction x_D is small, as shown in Figure 16.5.

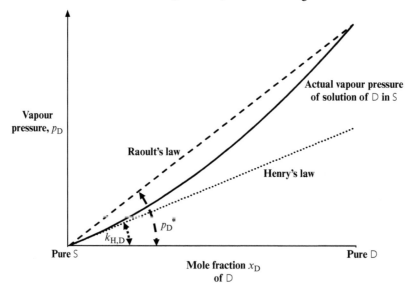

Figure 16.5 Raoult's law and Henry's law. The solid, curved, black line shows how the vapour pressure p_D of component D in a binary solution of a solute D dissolved in a solvent S (or a liquid mixture of D and S) varies with the mole fraction x_D of D from 0 (pure S) to 1 (pure D). If the solution were ideal, p_D would be given by Raoult's law as $p_D = x_D p_D^*$, as indicated by the black dashed line of slope p_D^*. The actual behaviour of p_D deviates negatively from ideal behaviour, and so the solution is non-ideal. However, at small mole fractions x_D, towards the left of the graph, the vapour pressure p_D is sensibly linear with respect to the mole fraction x_D, but with a constant of proportionality $k_{H,D}$, so that $p_D = x_D k_{H,D}$, where $k_{H,D}$ is the Henry's law constant for component D. See also Figure 22.2.

In Figure 16.5, the actual vapour pressure of the solution is less than the ideal, Raoult's law, vapour pressure for all mole fractions, implying that the Henry's law constant $k_{H,D}$ is smaller than the vapour pressure p_D^* of the pure solute D. This is a negative deviation from Raoult's law, as shown in Figure 16.5; positive deviations are also possible, and Henry's law can still apply, but with $k_{H,D} > p_D^*$. In all cases, however, Henry's law applies only to dilute, and very dilute, solutions.

A quick note on the term 'ideal-dilute solution'. As we have just seen, an 'ideal-dilute solution' is one that is necessarily dilute with respect to a solute D by virtue of the fact that the mole number $x_D \ll$ the mole number x_S, and ideal, in that the solution complies with Henry's law, but not Raoult's law. This is in contrast to a 'dilute ideal solution', which is an ideal solution by virtue of compliance with Raoult's law (which applies for all mole fractions x_D), and which happens to be dilute.

16.2 The Gibbs free energy of an ideal solution

16.2.1 Solvents, solutes and mole fractions

Consider an equilibrium ideal solution of n_D mol of a solute D dissolved in n_S mol of a solvent S. Since the solution is ideal, each of the components behaves independently, and so the molecules of S in the solution will be in equilibrium with molecules of S in the vapour above the solution

S (solvent in solution) ⇌ S (solvent as a vapour)

whilst, simultaneously, the molecules of D in the solution will be in equilibrium with molecules of D in the vapour above the solution

D (solute in solution) ⇌ D (solute as a vapour)

On page 471, we showed that a pure substance X, in liquid or solid form, is in equilibrium with its vapour when, at any given temperature, the molar Gibbs free energies of the two phases are equal

G_X (liquid or solid) = G_X (vapour) (15.4a)

So, for the solvent S, we may therefore write, for equilibrium at any given temperature

G_S (solvent in solution) = G_S (solvent as a vapour) (16.3a)

and, simultaneously, for the solute D

G_D (solute in solution) = G_D (solute as a vapour) (16.3b)

Equations (16.3a) and (16.3b), which are expressed in terms of molar Gibbs free energies G, are fundamental: all subsequent relationships in this chapter will be derived from them. As we saw on page 472, the chemical potential μ may be used as an alternative to G; in accordance with our practice in previous chapters, we will continue to use G here.

We now draw on a result we derived on page 400: that, at any temperature T, the molar Gibbs free energy G of an ideal gas, which exerts a pressure P bar, is given by

G (gas at pressure P) = $G^\ominus$ (gas at pressure $P^\ominus$) + $RT \ln (P/P^\ominus)$ (13.47d)

where $G^{\ominus}$ (gas at pressure $P^{\ominus}$) is the standard molar Gibbs free energy of the pure gas in its standard state at the standard pressure $P^{\ominus} = 1$ bar. If we apply equation (13.47d) to the vapour of the solvent S, which, at any temperature T, exerts a vapour pressure p_S, then

$$G_S \text{ (vapour at pressure } p_S) = G_S^{\ominus} \text{ (vapour at pressure } P^{\ominus}) + RT \ln (p_S/P^{\ominus}) \qquad (16.4)$$

Assuming that the solvent-vapour system obeys Raoult's Law, the vapour pressure p_S is related to the mole fraction x_S as

$$p_S = x_S p_S^* \qquad (16.1)$$

Dividing both sides of equation (16.1) by the standard pressure $P^{\ominus}$, and multiplying the right-hand side by the ratio of the standard mole fractions $x_S^{\ominus}/x_S^{\ominus} = 1$, we derive

$$\frac{p_S}{P^{\ominus}} = \frac{x_S}{x_S^{\ominus}} \left(\frac{x_S^{\ominus} p_S^*}{P^{\ominus}} \right)$$

Substituting the right-hand side of this equation for the ratio $p_S/P^{\ominus}$ in equation (16.4), we have

$$G_S \text{ (vapour at pressure } p_S) = G_S^{\ominus} \text{ (vapour at pressure } P^{\ominus}) \qquad (16.5)$$

$$+ RT \ln \frac{x_S}{x_S^{\ominus}} + RT \ln \frac{x_S^{\ominus} p_S^*}{P^{\ominus}}$$

In equation (16.5), the quantity G_S (vapour at pressure p_S) is the molar Gibbs free energy of the vapour of the pure solvent above the solution at any temperature T, which must be the same as the quantity represented as G_S (solvent as a vapour) in equation (16.3a)

$$G_S \text{ (solvent in solution)} = G_S \text{ (solvent as a vapour)} \qquad (16.3a)$$

Accordingly, equation (16.5) becomes

$$G_S \text{ (solvent in solution)} = G_S^{\ominus} \text{ (vapour at pressure } P^{\ominus})$$

$$+ RT \ln \frac{x_S}{x_S^{\ominus}} + RT \ln \frac{x_S^{\ominus} p_S^*}{P^{\ominus}} \qquad (16.6)$$

For any solvent S, as shown in Table 6.4, the standard pressure $P^{\ominus} = 1$ bar, and the standard mole fraction $x_S^{\ominus} = 1$; furthermore, at any temperature T, the vapour pressure p_S^* associated with the pure solvent is a constant, implying that the term $RT \ln (x_S^{\ominus} p_S^*/P^{\ominus})$ is also a constant. This term may therefore be combined with the constant $G_S^{\ominus}$ (vapour at pressure $P^{\ominus}$) to give a new standard, which we may write as $G_S^{\ominus}$ (pure solvent), such that

$$G_S \text{ (solvent in solution)} = G_S^{\ominus} \text{ (pure solvent)} + RT \ln (x_S/x_S^{\ominus})$$

or, more simply

$$G_S \text{ (in solution)} = G_S^{\ominus} \text{ (pure)} + RT \ln (x_S/x_S^{\ominus}) \qquad (16.7a)$$

The newly-introduced standard $G_S^{\ominus}$ (pure solvent) is the value of G_S (solvent in solution) corresponding to a mole fraction $x_S = 1$, for which $x_S/x_S^{\ominus} = 1$ and $\ln (x_S/x_S^{\ominus}) = 0$. When the mole fraction $x_S = 1$, the solvent S is the pure solvent, and since the pressure is the standard pressure $P^{\ominus} = 1$ bar, the solvent is in its standard state, hence the representation $G^{\ominus}$ (pure solvent).

An alternative form of equation (16.7a) is

$$G_S = G_S^\circ + RT \ln (x_S/x_S^\circ) \tag{16.7b}$$

in which it is tacitly understood that G_S refers to the solvent in solution, and G_S° refers to the pure solvent; also, since the standard mole fraction $x_S^\circ = 1$, we may use the 'shorthand' forms

$$G_S \text{ (in solution)} = G_S^\circ \text{ (pure)} + RT \ln x_S \tag{16.7c}$$

and

$$G_S = G_S^\circ + RT \ln x_S \tag{16.7d}$$

Equations (16.7) play the same role for an ideal solution as equations (13.47d) and (13.47e)

$$G = G^\circ + RT \ln (P/P^\circ) \tag{13.47d}$$
$$G = G^\circ + RT \ln P \tag{13.47e}$$

play for an ideal gas, and, just as equations (13.47d) and (13.47e) were the fundamental equations that were used to describe chemical equilibrium in the gas phase (see page 416 and pages 423 to 428), as we shall show shortly, equations (16.7) are the fundamental equations that are used to describe chemical equilibrium in solution.

For any material S, at any given temperature, the standard molar Gibbs free energy G_S° (pure) of the pure material S is a given number, which may be found in reference tables; equations (16.7) are therefore expressions which enables us to calculate the molar Gibbs free energy G_S (in solution) for that material when it is a component in an ideal solution at mole fraction x_S. Since the mole fraction x_S of any component in a solution is necessarily less than 1, G_S (in solution) will always be less than G_S° (pure). This implies that the change ΔG_S

$$\Delta G_S = G_S \text{ (in solution)} - G_S^\circ \text{ (pure)}$$

in the molar Gibbs free energy associated with the formation of a solution is necessarily negative: if a solvent can dissolve in a solute (thereby creating the state represented by the mole fraction x_S), then that change of state will be spontaneous – solid sugar will dissolve in water spontaneously. Also, although equation (16.7a) was derived by explicit reference to the solvent, it is a general equation which applies to any component within an ideal solution. And one small, but important, point: equations (16.7a) and (16.7b) require that $x_S > 0$, which, physically, means that there is at least a small amount of component S in the solution – since $\ln 0 = -\infty$, the equation doesn't work if there isn't any S there!

Equation (16.7a)

$$G_S \text{(in solution)} = G_S^\circ \text{(pure)} + RT \ln (x_S/x_S^\circ) \tag{16.7a}$$

uses the mole fraction x_S as the measure of the quantity of the S present in the solution, and refers to the pure component S as the standard state associated with G_S° (pure). This equation is true for both the solvent S and the solute D, and is appropriate when both S and D exist in a pure state, for which G_S° (pure) and G_D° (pure) can be measured and published in reference tables. Equation (16.7a) is therefore particularly applicable to the description of solutions for which the mole fraction x_S is an appropriate, and useful, measure of the quantity of S present, which implies that the various components in the solution are present in comparable quantities, for example, mixtures of miscible liquids, such as water and ethanol, in comparable amounts.

16.2.2 Dilute solutions, molalities and molar concentrations

There are some solutions, however, for which one, and often both, of these conditions may not apply. Solutions of dissolved ions, for example, do not comply with the condition that the ion exists in a pure state; also, many solutions are dilute, for which mole fractions are often a clumsy measure of quantity. As an example, consider a solution of 0.01 mol of NaCl, which, on being dissolved in 1 dm³ of pure water, dissociates into 0.01 mol of Na⁺ (aq) and 0.01 mol of Cl⁻ (aq) ions. The mass of 1 dm³ of pure water is 1 kg, and since the molecular weight of water is 0.018 kg mol⁻¹, a mass of 1 kg of water corresponds to $1/0.018 = 55.56$ mol. For this solution, and ignoring the (very sparse) ionisation of the water molecule (which is discussed in detail in the next chapter), the mole fractions of each component are given by

$$x_{H_2O} = \frac{n_{H_2O}}{n_{H_2O} + n_{Na^+} + n_{Cl^-}} = \frac{55.56}{55.56 + 0.01 + 0.01} = \frac{55.56}{55.58} = 0.9996$$

$$x_{Na^+} = \frac{n_{Na^+}}{n_{H_2O} + n_{Na^+} + n_{Cl^-}} = \frac{0.01}{55.56 + 0.01 + 0.01} = \frac{0.01}{55.58} = 0.0002$$

$$x_{Cl^-} = \frac{n_{Cl^-}}{n_{H_2O} + n_{Na^+} + n_{Cl^-}} = \frac{0.01}{55.56 + 0.01 + 0.01} = \frac{0.01}{55.58} = 0.0002$$

For many reactions in aqueous solution, especially those involving ions, our interest is in the dissolved solutes, the ions, rather than in the solute, the water, which may not be involved in the reaction at all, but rather provides the environment in which the reaction takes place. Since the mole fraction is very heavily weighted by the solvent, for dilute solutions, the mole fraction is often not a convenient measure of the quantities of the solutes. As we saw on page 171, however, there are two alternative measures relevant to solutions

- the **molality**, b_i, the ratio of the number of moles n_i of solute i to the total mass M_S of the solvent S

$$b_i = \frac{n_i}{M_S} \qquad (6.19)$$

and

- the **molar concentration** (or, more simply, **concentration**), [I], (sometimes, if measured in units of mol dm⁻³, called the **molarity** M), the ratio of the total number of moles n_i of component i to the total volume V of the solution

$$[I] = \frac{n_i}{V} \qquad (6.20)$$

To show how these measures can be used in expressions for the molar Gibbs free energy G_D (in solution) of a solute in solution, consider a dilute solution of a solute D in a solvent S. For the solute D, as well as the solvent S, equation (16.7a)

$$G_D \text{ (in solution)} = G_D^{\ominus} \text{ (pure)} + RT \ln (x_D/x_D^{\ominus}) \qquad (16.7a)$$

THE GIBBS FREE ENERGY OF AN IDEAL SOLUTION

is valid, in which

$$x_D = \frac{n_D}{n_S + n_D} \tag{6.18}$$

If the solution is dilute, $n_S \gg n_D$ and so

$$x_D \approx \frac{n_D}{n_S} = \frac{n_D M_S}{n_S M_S} = \frac{n_D M_S}{M_S} = b_D M_S \tag{16.8}$$

where M_S is the molar mass (the molecular weight in kg mol^{-1}) of the solvent, $M_S = n_S M_S$ is the actual mass of the solvent, and $b_D = n_D/M_S$ is the molality of the solute D as defined by equation (6.19).

Dividing both sides of equation (16.8) by the standard mole fraction $x_D^{\ominus} = 1$, and multiplying the right-hand side by the ratio $b_D^{\ominus}/b_D^{\ominus} = 1$, we derive

$$\frac{x_D}{x_D^{\ominus}} \approx \frac{b_D}{b_D^{\ominus}} \left(\frac{b_D^{\ominus} M_S}{x_D^{\ominus}} \right)$$

and so equation (16.7a) becomes

$$G_D \text{ (in solution)} = G_D^{\ominus} \text{ (pure)} + RT \ln \frac{b_D}{b_D^{\ominus}} + RT \ln \frac{b_D^{\ominus} M_S}{x_D^{\ominus}}$$

For any given solvent (which is usually pure water), the molar mass M_s is a constant, and so the (constant) last term may be combined with the standard $G_D^{\ominus}$ (pure) to give a new standard, which we may write as $G_D^{\ominus}$ (in solution, b), such that

$$G_D \text{ (in solution)} = G_D^{\ominus} \text{ (in solution, } b\text{)} + RT \ln (b_D/b_D^{\ominus}) \tag{16.9a}$$

The newly-defined standard $G_D^{\ominus}$ (in solution, b) is therefore the value of the molar Gibbs free energy G_D (in solution) of a solution of the solute D in a defined solvent at the standard molality when $b_D = b_D^{\ominus} = 1$ mol of D per kg of solute. An alternative, simpler, form of equation (16.9a) is

$$G_{b,D} = G_{b,D}^{\ominus} + RT \ln (b_D/b_D^{\ominus}) \tag{16.9b}$$

in which the subscript $_{b,D}$ both serves as a reminder that $G_{b,D}$ and $G_{b,D}^{\ominus}$ are measured by reference to the standard state of unit molality within a given solvent, as well as distinguishing them from G_D and $G_D^{\ominus}$, which reference the standard state of pure D.

Since the standard molality $b_D^{\ominus}$, as defined by the IUPAC standards specified in Table 6.4, is 1 mol kg^{-1} for all components D, equations (16.9a) and (16.9b) are often written as

$$G_D \text{ (in solution)} = G_D^{\ominus} \text{ (in solution, } b\text{)} + RT \ln b_D \tag{16.9c}$$

$$G_{b,D} = G_{b,D}^{\ominus} + RT \ln b_D \tag{16.9d}$$

Equations (16.9) express the molar Gibbs free energy of a solute D in a dilute ideal solution in terms of the molality b_D, and were derived from equations (16.7) by transforming the mole fraction x_D into the molality b_D; we may also transform the mole fraction x_D into the molar concentration [D] as

$$x_D \approx \frac{n_D}{n_S} = \frac{n_D/V}{n_S/V} = \frac{[D]}{n_S/V}$$

where V is the volume of the solution. Dividing both sides by the standard mole fraction $x_D^{\circ} = x^{\circ} = 1$, and multiplying the right-hand side by the ratio $[D]^{\circ}/[D]^{\circ} = 1$, we derive

$$\frac{x_D}{x^{\circ}} \approx \frac{[D]}{[D]^{\circ}} \left(\frac{[D]^{\circ}}{x^{\circ}(n_S/V)} \right)$$

Equation (16.7a)

$$G_D \text{ (in solution)} = G_D^{\circ} \text{ (pure)} + RT \ln (x_D/x^{\circ}) \tag{16.7a}$$

now becomes

$$G_D \text{ (in solution)} = G_D^{\circ} \text{ (pure)} + RT \ln \frac{[D]}{[D]^{\circ}} + RT \ln \frac{[D]^{\circ}}{x^{\circ}(n_S/V)}$$

For any given solution at any given temperature T, G_D° (pure) is a constant, as are the quantities V and n_S; furthermore, according to the definition of the standard states as given in Table 6.4, $[D]^{\circ} = 1$ mol m^{-3} (or mol dm^{-3}), and $x^{\circ} = 1$. The last term is therefore a constant, and so we may define another new standard state, which we write as G_D° (in solution, [D]), giving

$$G_D \text{ (in solution)} = G_D^{\circ} \text{ (in solution, [D])} + RT \ln ([D]/[D]^{\circ}) \tag{16.10a}$$

G_D° (in solution, [D]) is the value of the molar Gibbs free energy G_D (in solution) of a solution in which the molar concentration [D] of the solute D in a given solvent is the standard molar concentration $[D]^{\circ} = 1$ mol m^{-3} (or mol dm^{-3}). At the standard pressure $P^{\circ} = 1$ bar, as shown in Table 6.4, the solute D is in its standard state, hence the notation G_D° (in solution, [D]).

An alternative form of equation (16.10a) is

$$G_{c,D} = G_{c,D}^{\circ} + RT \ln ([D]/[D]^{\circ}) \tag{16.10b}$$

where the subscript $_{c,D}$ identifies that $G_{c,D}$ and $G_{c,D}^{\circ}$ are measured by reference to the standard state of unit molar concentration within a given solvent, and since the standard molar concentration $[D]^{\circ} = 1$ mol m^{-3} (or 1 mol dm^{-3}) for all components D, we also have the 'shorthand' equations

$$G_D \text{ (in solution)} = G_D^{\circ} \text{ (in solution, [D])} + RT \ln[D] \tag{16.10c}$$

and

$$G_{c,D} = G_{c,D}^{\circ} + RT \ln[D] \tag{16.10d}$$

Equations (16.10)

$$G_D \text{ (in solution)} = G_D^{\circ} \text{ (in solution, [D])} + RT \ln ([D]/[D]^{\circ}) \tag{16.10a}$$
$$G_{c,D} = G_{c,D}^{\circ} + RT \ln ([D]/[D]^{\circ}) \tag{16.10b}$$
$$G_D \text{ (in solution)} = G_D^{\circ} \text{ (in solution, [D])} + RT \ln[D] \tag{16.10c}$$
$$G_{c,D} = G_{c,D}^{\circ} + RT \ln[D] \tag{16.10d}$$

which are expressed in terms of molar concentrations [D], together with equations (16.9)

$$G_D \text{ (in solution)} = G_D^\ominus \text{ (in solution, } b) + RT \ln (b_D/b_D^\ominus) \quad (16.9a)$$

$$G_{b,D} = G_{b,D}^\ominus + RT \ln (b_D/b_D^\ominus) \quad (16.9b)$$

$$G_D \text{ (in solution)} = G_D^\ominus \text{ (in solution, } b) + RT \ln b_D \quad (16.9c)$$

$$G_{b,D} = G_{b,D}^\ominus + RT \ln b_D \quad (16.9d)$$

which are expressed in terms of the molality b_D, and also equations (16.7)

$$G_S \text{ (in solution)} = G_S^\ominus \text{ (pure)} + RT \ln (x_S/x_S^\ominus) \quad (16.7a)$$

$$G_S = G_S^\ominus + RT \ln (x_S/x_S^\ominus) \quad (16.7b)$$

$$G_S \text{ (in solution)} = G_S^\ominus \text{ (pure)} + RT \ln x_S \quad (16.7c)$$

$$G_S = G_S^\ominus + RT \ln x_S \quad (16.7d)$$

which are expressed in terms of the mole fraction x_D, form a 'trio' of sets of equations which all describe the molar Gibbs free energy of a component D within a dilute ideal solution. They are all equivalent, but they differ as regards the units-of-measure for the quantity of D present, and the corresponding standard $G^\ominus$. $G_D^\ominus$ (pure) = $G_D^\ominus$ is the molar Gibbs free energy of the pure component D, and is used in conjunction with the mole fraction x_D; $G_D^\ominus$ (in solution, b) = $G_{b,D}^\ominus$ is the molar Gibbs free energy of the component D in a solution in a defined solvent at unit molality, and is used with the molality b_D; $G_D^\ominus$ (in solution, [D]) = $G_{c,D}^\ominus$ is the molar Gibbs free energy of the component D in a solution in a defined solvent at unit molar concentration, and is used with the molar concentration [D].

Furthermore, equations (16.7), (16.9) and (16.10), which all apply to an ideal solution, are equivalent to equation (13.47d)

$$G = G^\ominus + RT \ln (P/P^\ominus) \quad (13.47d)$$

for the molar Gibbs free energy of an ideal gas, for which the standard $G^\ominus$ is the molar Gibbs free energy of the pure gas at a pressure of $P^\ominus$ = 1 bar.

16.3 Equilibria of reactions in solution

Equations (16.7), (16.9) and (16.10) enable us to determine the thermodynamic properties, at a temperature T and pressure P, of any chemical reaction, taking place in a solution, of the general form

$$a\text{A} + b\text{B} \rightleftharpoons c\text{C} + d\text{D}$$

under the assumption that the solution is ideal. Our assumption for this section is that the reactants A and B, and the products, C and D, are all in solution, and are all solutes – none is the solvent S. The solution therefore contains five components, but our focus is on the four

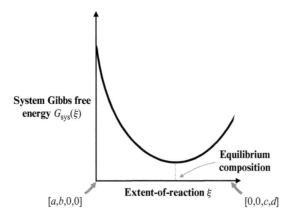

Figure 16.6 **The Gibbs free energy G_{sys} of a generalised chemical reaction taking place in solution.** This diagram is in essence identical to Figure 14.2, which shows G_{sys} for a generalised gas phase chemical reaction – but since the reaction depicted in Figure 14.2 is different from the reaction depicted here, the shapes of the U-shaped curves are somewhat different. See also Figure 14.1 for examples of reactions in which the equilibrium point is far to the right (the reaction in essence goes to completion as written), or the left (the reactants are stable).

reagents A, B, C and D: the solvent S is present, not as a reactant, but as the medium in which the reaction takes place. In most situations, the solvent S is water, and so the assumption we are making here is that molecular water is not involved in the reaction, either as a reagent or a product, which is true in many cases; the case in which water is indeed a reagent will be discussed in the next chapter.

Just as we saw on pages 417 to 423, which discussed a generalised gas phase reaction, for a generalised reaction taking place in solution, the total Gibbs free energy G_{sys} will vary as the composition of the components changes, as shown in Figure 16.6.

The solution is ideal, and so the molar Gibbs free energies of each of A, B, C and D, and of the solvent S, are additive, implying that the total Gibbs free energy $G_{sys}(\tau)$ of the reaction mixture at any instant τ is given by

$$G_{sys}(\tau) = G_A(\tau) + G_B(\tau) + G_C(\tau) + G_D(\tau) + G_S(\tau)$$
$$= n_A(\tau)\,\boldsymbol{G}_A(\tau) + n_B(\tau)\,\boldsymbol{G}_B(\tau) + n_C(\tau)\,\boldsymbol{G}_C(\tau) + n_D(\tau)\,\boldsymbol{G}_D(\tau) + n_S(\tau)\,\boldsymbol{G}_S(\tau)$$

where $G_i(\tau)$ is the instantaneous Gibbs free energy of component i. As we have just seen, for any component i, we may use any of equations (16.7), (16.9) or (16.10) to express $\boldsymbol{G}_i(\tau)$ in terms of the mole fraction, the molality or the molar concentration, and the corresponding standard states respectively; for the moment, we will examine the behaviour of the reaction mixture using the mole fraction equation

$$\boldsymbol{G}_i = \boldsymbol{G}_i^{\ominus} + RT\ln(x_i/x_i^{\ominus}) \tag{16.7b}$$

where $\boldsymbol{G}_i^{\ominus}$ refers to the standard state of the pure component i. Accordingly, at any time τ

$$G_i(\tau) = n_i(\tau)\,\boldsymbol{G}_i(\tau) = n_i(\tau)\,\boldsymbol{G}_i^{\ominus} + n_i(\tau)\,RT\ln(x_i(\tau)/x_i^{\ominus}) \tag{16.11}$$

As the reaction takes place, for each of the reagents A, B, C and D, the mole number n_i, and the corresponding mole fraction x_i, both change, implying that $G_i(\tau)$ changes accordingly. The solvent S, however, is not involved in the reaction, and so $n_S(\tau)$ remains constant; also, the solvent is likely to be present in considerable excess, implying that $x_S(\tau) \approx 1$ for all times τ. The term $G_S(\tau) = n_S(\tau)\,\mathbf{G}_S(\tau)$ is therefore constant, and so we may re-define $G_{\text{sys}}(\tau)$ as

$$G_{\text{sys}}(\tau) = G_A(\tau) + G_B(\tau) + G_C(\tau) + G_D(\tau)$$

which refers only to the reagents, and it is the behaviour of $G_{\text{sys}}(\tau)$ that determines the reaction equilibrium composition.

Now, on pages 417 to 423, we studied the general behaviour of a reaction involving ideal gases, and our starting point (see page 416) was to express the molar Gibbs free energy $G_i(\tau)$ of component i in the ideal gas mixture as

$$G_i(\tau) = n_i(\tau)\mathbf{G}_i(\tau) = n_i(\tau)\mathbf{G}_i^{\ominus} + n_i(\tau)RT\ln p_i(\tau) \tag{14.1a}$$

We then showed – see equations (14.6) – that the partial pressure p_i of each component is proportional to the corresponding mole fraction x_i. Equations (16.11) and (14.1a) are therefore, in essence, identical, and all the results for ideal gases, as explored on pages 417 to 432, apply to ideal solutions, subject to the mapping of ideal gas partial pressures p_i onto ideal solution mole fractions x_i, taking into account the corresponding difference in the standard states.

Accordingly, by analogy with equation (14.12), which defined the pressure equilibrium constant K_p in terms of pressure ratios $p_i/P_i^{\ominus}$, and as shown on page 437, we may define the **mole fraction thermodynamic equilibrium constant** K_x in terms of mole fraction ratios $x_i/x_i^{\ominus}$ as

$$K_x = \frac{(x_{C,\text{eq}}/x_C^{\ominus})^c\,(x_{D,\text{eq}}/x_D^{\ominus})^d}{(x_{A,\text{eq}}/x_A^{\ominus})^a\,(x_{B,\text{eq}}/x_B^{\ominus})^b} = \frac{x_{C,\text{eq}}^c\,x_{D,\text{eq}}^d}{x_{A,\text{eq}}^a\,x_{B,\text{eq}}^b} \tag{16.12a}$$

where the 'shorthand' term recognises the definition of the standard mole fraction $x_i^{\ominus} = 1$ for all pure components i.

K_x is associated with a corresponding standard Gibbs free energy change $\Delta_r G_x^{\ominus}$ such that

$$\Delta_r G_x^{\ominus} = -RT\ln K_x \tag{16.13a}$$

in which $\Delta_r G_x^{\ominus}$ is defined in terms of the standard molar Gibbs free energy of formation $\Delta_f \mathbf{G}_{x,i}^{\ominus}$ of each reactant i, as measured by reference to the standard state of unit mole fraction as

$$\Delta_r G_x^{\ominus} = (c\,\Delta_f \mathbf{G}_{x,C}^{\ominus} + d\,\Delta_f \mathbf{G}_{x,D}^{\ominus}) - (a\,\Delta_f \mathbf{G}_{x,A}^{\ominus} + b\,\Delta_f \mathbf{G}_{x,B}^{\ominus}) \tag{16.14a}$$

Equation (16.13a) is the condensed phase equivalent of the gas phase equation (14.13a)

$$\Delta_r G_p^{\ominus} = -RT\ln K_p \tag{14.13a}$$

and is an example of the point made on page 461 concerning the general nature of equations of the form $\Delta G^{\ominus} = -RT\ln K$.

For non-equilibrium systems, we may define the **mole fraction mass action ratio** Γ_x as

$$\Gamma_x = \frac{(x_C/x_C^{\ominus})^c\,(x_D/x_D^{\ominus})^d}{(x_A/x_A^{\ominus})^a\,(x_B/x_B^{\ominus})^b} = \frac{x_C^c\,x_D^d}{x_A^a\,x_B^b} \tag{16.15a}$$

such that

$$\Delta_r G_x = \Delta_r G_x^{\ominus} + RT\ln \Gamma_x \tag{16.16a}$$

As depicted in Figure 14.3 (see page 435), $\Delta_r G_x$, is equal to the instantaneous slope, $dG_{sys}(\xi)/d\xi$, of the $G_{sys}(\tau)$ curve, and we also have

$$\Delta_r G_x = -RT \ln K_x + RT \ln \Gamma_x \qquad (16.17a)$$

$$\Delta_r G_x = RT \ln \frac{\Gamma_x}{K_x} \qquad (16.18a)$$

Furthermore, we can use exactly the same methods as we used on pages 444 to 446 to derive other useful relationships such as

$$\frac{d(\Delta_r G_x^\ominus/T)}{dT} = -R \frac{d\ln K_x}{dT} \qquad (16.19)$$

$$\frac{d\ln K_x}{dT} = \frac{\Delta_r H^\ominus}{RT^2} \qquad (16.20)$$

and

$$\ln \frac{K_x(T_2)}{K_x(T_1)} = \frac{\Delta_r H^\ominus}{R}\left(\frac{T_2 - T_1}{T_1 T_2}\right) \qquad (16.21)$$

noting that equation (16.21) assumes that $\Delta_r H^\ominus$ is constant over the temperature range from T_1 to T_2.

16.4 Dilute solutions, molalities and molar concentrations

The previous section analysed the reaction $aA + bB \rightleftharpoons cC + dD$, taking place in an ideal solution, using the mole fraction x_i as the measure of the quantity of component i in the solution, with the corresponding standard state of the pure component i. All the results are general, the only assumptions being that the solvent is not a reagent, that the solution is ideal, and that the solvent S is present in excess. We will explore non-ideal solutions in Chapter 22; in this section, we extend our analysis of dilute ideal solutions in which the quantities, and hence mole numbers, of the reagents A, B, C and D are small in comparison to the quantity, and mole number, of the solvent S. We will also continue to assume that the solvent S is not a reagent.

For any component j in the solution, including the solvent S

$$x_j = \frac{n_j}{n_A + n_B + n_C + n_D + n_S}$$

For a dilute solution, $n_S \gg (n_A + n_B + n_C + n_D)$, and so

$$x_j \approx \frac{n_j}{n_S}$$

implying that, for the solvent, $x_S \approx n_S/n_S = 1$, as expected.

As we saw on pages 514 to 517, this approximation enables the molar Gibbs free energy G_i of any reagent i (but not the solvent S) to be expressed either as

$$G_i \text{ (in solution)} = G_i^\ominus \text{ (in solution, } b) + RT \ln (b_i/b_i^\ominus) \qquad (16.9a)$$

$$G_{b,i} = G_{b,i}^\ominus + RT \ln (b_i/b_i^\ominus) \qquad (16.9b)$$

where b_i is the molality, and the standard state is that of unit molality, or as

$$G_i \text{ (in solution)} = G_i^\ominus \text{ (in solution, [I])} + RT \ln ([I]/[I]^\ominus) \tag{16.10a}$$

$$G_{c,i} = G_{c,i}^\ominus + RT \ln ([I]/[I]^\ominus) \tag{16.10b}$$

where [I] is the molar concentration, and the standard state is that of unit molar concentration. The results of the previous section were all derived from the equation

$$G_i(\tau) = n_i(\tau) G_i(\tau) = n_i(\tau) G_i^\ominus + n_i(\tau) RT \ln (x_i(\tau)/x_i^\ominus) \tag{16.11}$$

but had we started with the equivalent equation

$$G_{b,i}(\tau) = n_i(\tau) G_{b,i}(\tau) = n_i(\tau) G_{b,i}^\ominus + n_i(\tau) RT \ln (b_i(\tau)/b_i^\ominus)$$

which uses molalities, we would have derived equations of the form of equations (16.12a) to (16.23), subject to the replacement of $x_i/x_i^\ominus$ by $b_i/b_i^\ominus$, and the use of unit molality standard states $G_{b,i}^\ominus$. Accordingly, we may define the **molality equilibrium constant** K_b as

$$K_b = \frac{(b_{C,eq}/b_C^\ominus)^c (b_{D,eq}/b_D^\ominus)^d}{(b_{A,eq}/b_A^\ominus)^a (b_{B,eq}/b_B^\ominus)^b} = \frac{b_{C,eq}^c \, b_{D,eq}^d}{b_{A,eq}^a \, b_{B,eq}^b} \tag{16.12b}$$

which is related to the corresponding $\Delta_r G_b^\ominus$ as

$$\Delta_r G_b^\ominus = - RT \ln K_b \tag{16.13b}$$

where

$$\Delta_r G_b^\ominus = (c \, \Delta_f G_{b,C}^\ominus + d \, \Delta_f G_{b,D}^\ominus) - (a \, \Delta_f G_{b,A}^\ominus + b \, \Delta_f G_{b,B}^\ominus) \tag{16.14b}$$

in which the standard molar quantities $\Delta_f G_{b,i}^\ominus$ refer to the standard state of unit molality $b_i^\ominus = 1$ mol kg^{-1}.

Also

$$\Gamma_b = \frac{(b_C/b_C^\ominus)^c (b_D/b_D^\ominus)^d}{(b_A/b_A^\ominus)^a (b_B/b_B^\ominus)^b} = \frac{b_C^c \, b_D^d}{b_A^a \, b_B^b} \tag{16.15b}$$

and

$$\Delta_r G_b = \Delta_r G_b^\ominus + RT \ln \Gamma_b \tag{16.16b}$$

$$\Delta_r G_b = - RT \ln K_b + RT \ln \Gamma_b \tag{16.17b}$$

$$\Delta_r G_b = RT \ln \frac{\Gamma_b}{K_b} \tag{16.18b}$$

Likewise, starting with the equation

$$G_{c,i}(\tau) = n_i(\tau) G_{c,i}(\tau) = n_i(\tau) G_{c,i}^\ominus + n_i(\tau) RT \ln ([I](\tau)/[I]^\ominus)$$

which is expressed in terms of the molar concentration [I(τ)], corresponding to a standard state of unit molar concentration, we derive the (molar) **concentration thermodynamic equilibrium constant** K_c as

$$K_c = \frac{([C]_{eq}/[C]^\ominus)^c ([D]_{eq}/[D]^\ominus)^d}{([A]_{eq}/[A]^\ominus)^a ([B]_{eq}/[B]^\ominus)^b} = \frac{[C]_{eq}^c \, [D]_{eq}^d}{[A]_{eq}^a \, [B]_{eq}^b} \tag{16.12c}$$

such that

$$\Delta_r G_c^\ominus = -RT \ln K_c \tag{16.13c}$$

and

$$\Delta_r G_c^\ominus = (c\Delta_f G_{c,C}^\ominus + d\Delta_f G_{c,D}^\ominus) - (a\Delta_f G_{c,A}^\ominus + b\Delta_f G_{c,B}^\ominus) \tag{16.14c}$$

in which the standard molar quantities $\Delta_f G_{c,i}^\ominus$ refer to the standard state of unit molar concentration $[I]^\ominus = 1$ mol m^{-3} (or 1 mol dm^{-3}).

Also, as is now familiar

$$\Gamma_c = \frac{([C]/[C]^\ominus)^c \, ([D]/[D]^\ominus)^d}{([A]/[A]^\ominus)^a \, ([B]/[B]^\ominus)^b} = \frac{[C]^c \, [D]^d}{[A]^a \, [B]^b} \tag{16.15c}$$

$$\Delta_r G_c = \Delta_r G_c^\ominus + RT \ln \Gamma_c \tag{16.16c}$$

$$\Delta_r G_c = -RT \ln K_c + RT \ln \Gamma_c \tag{16.17c}$$

$$\Delta_r G_c = RT \ln \frac{\Gamma_c}{K_c} \tag{16.18c}$$

16.5 Multi-state equilibria and chemical activity

The four sets of equations

$$K_p = \frac{(p_{C,eq}/P_C^\ominus)^c \, (p_{D,eq}/P_D^\ominus)^d}{(p_{A,eq}/P_A^\ominus)^a \, (p_{B,eq}/P_B^\ominus)^b} = \frac{p_{C,eq}^c \, p_{D,eq}^d}{p_{A,eq}^a \, p_{B,eq}^b} \tag{14.12}$$

$$\Delta_r G_p^\ominus = -RT \ln K_p \tag{14.13a}$$

$$\Delta_r G_p^\ominus = (c\,\Delta_f G_{p,C}^\ominus + d\,\Delta_f G_{p,D}^\ominus) - (a\,\Delta_f G_{p,A}^\ominus + b\,\Delta_f G_{p,B}^\ominus) \tag{14.9}$$

$$K_x = \frac{(x_{C,eq}/x_C^\ominus)^c \, (x_{D,eq}/x_D^\ominus)^d}{(x_{A,eq}/x_A^\ominus)^a \, (x_{B,eq}/x_B^\ominus)^b} = \frac{x_{C,eq}^c \, x_{D,eq}^d}{x_{A,eq}^a \, x_{B,eq}^b} \tag{16.12a}$$

$$\Delta_r G_x^\ominus = -RT \ln K_x \tag{16.13a}$$

$$\Delta_r G_x^\ominus = (c\,\Delta_f G_{x,C}^\ominus + d\,\Delta_f G_{x,D}^\ominus) - (a\,\Delta_f G_{x,A}^\ominus + b\,\Delta_f G_{x,B}^\ominus) \tag{16.14a}$$

$$K_b = \frac{(b_{C,eq}/b_C^\ominus)^c \, (b_{D,eq}/b_D^\ominus)^d}{(b_{A,eq}/b_A^\ominus)^a \, (b_{B,eq}/b_B^\ominus)^b} = \frac{b_{C,eq}^c \, b_{D,eq}^d}{b_{A,eq}^a \, b_{B,eq}^b} \tag{16.12b}$$

$$\Delta_r G_b^\ominus = -RT \ln K_b \tag{16.13b}$$

$$\Delta_r G_b^\ominus = (c\,\Delta_f G_{b,C}^\ominus + d\,\Delta_f G_{b,D}^\ominus) - (a\,\Delta_f G_{b,A}^\ominus + b\,\Delta_f G_{b,B}^\ominus) \tag{16.14b}$$

$$K_c = \frac{([C]_{eq}/[C]^\ominus)^c \, ([D]_{eq}/[D]^\ominus)^d}{([A]_{eq}/[A]^\ominus)^a \, ([B]_{eq}/[B]^\ominus)^b} = \frac{[C]_{eq}{}^c \, [D]_{eq}{}^d}{[A]_{eq}{}^a \, [B]_{eq}{}^b} \quad \quad (16.12c)$$

$$\Delta_r G_c^\ominus = -RT \ln K_c \quad \quad (16.13c)$$

$$\Delta_r G_c^\ominus = (c \, \Delta_f G_{c,C}^\ominus + d \, \Delta_f G_{c,D}^\ominus) - (a \, \Delta_f G_{c,A}^\ominus + b \, \Delta_f G_{c,B}^\ominus) \quad \quad (16.14c)$$

all describe the equilibrium of a generalised chemical reaction, taking place at the standard pressure $P^\ominus = 1$ bar, and at a given temperature T, but expressed in terms of different ratios for the measurement of reaction quantities, namely, mole fraction $x/x^\ominus$, molality $b/b^\ominus$, and molar concentration $[I]/[I]^\ominus$, each associated with the appropriate standard state. In deriving each of these sets of equations, the assumption was made that all the reagents are in the same physical state – ideal gases, pure liquids, or solutes in solution – so that all the terms in the corresponding equilibrium constants are of the same form: the terms in K_p are all of the form $p/P^\ominus$; in K_x, $x/x^\ominus$; in K_b, $b/b^\ominus$; and in K_c, $[I]/[I]^\ominus$.

Furthermore, as shown by the discussions on pages 415 to 417 and 511 to 517, all these expressions ultimately derive from an equation of the form

$$G_i = G_i^\ominus + RT \ln \frac{p_i}{P^\ominus} \quad \quad (13.47d)$$

in which the molar Gibbs free energy of any component i within the reaction mixture at any instant is expressed in terms of an appropriate standard molar Gibbs free energy, such as $G_i^\ominus$, and the quantity of that component at that instant, as measured by a corresponding ratio such as $p_i/P^\ominus$. As we have seen, different states – such as ideal gases or ideal solutions – are associated with different standard states and correspondingly different ratios for the measurement of quantities, and so equation (13.47d) may therefore be generalised to

$$G_i = G_i^\ominus + RT \ln a_i \quad \quad (16.19a)$$

where a_i stands for whatever quantity ratio is appropriate for component i – $p_i/P^\ominus$ for an ideal gas, $x_i/x^\ominus$ for a pure substance, and $b/b^\ominus$ or $[I]/[I]^\ominus$ for a solute in an ideal solution – and $G_i^\ominus$ refers to the corresponding state. We may also express equation (16.19a) in terms of the chemical potential μ (see page 404) such that

$$\mu_i = \mu_i^\ominus + RT \ln a_i \quad \quad (16.19b)$$

which, as we shall see in Chapter 22, is of particular importance when describing real, non-ideal, systems.

The term a_i is known as the **chemical activity**, or simply **activity**, of component i under any particular conditions, and so, for any generalised chemical reaction, we may write

$$K_r = \frac{a_{C,eq}{}^c \, a_{D,eq}{}^d}{a_{A,eq}{}^a \, a_{B,eq}{}^b} \quad \quad (16.12d)$$

$$\Delta_r G^\ominus = -RT \ln K_r \quad \quad (16.13d)$$

$$\Delta_r G^\ominus = (c \, \Delta_f G_C^\ominus + d \, \Delta_f G_D^\ominus) - (a \, \Delta_f G_A^\ominus + b \, \Delta_f G_B^\ominus) \quad \quad (16.14d)$$

along with

$$\Gamma_r = \frac{a_C^c \, a_D^d}{a_A^a \, a_B^b} \tag{16.15d}$$

$$\Delta_r G = \Delta_r G^\circ + RT \ln \Gamma_r \tag{16.16d}$$

$$\Delta_r G = -RT \ln K_r + RT \ln \Gamma_r \tag{16.17d}$$

$$\Delta_r G = RT \ln \frac{\Gamma_r}{K_r} \tag{16.18d}$$

As an example of the use of activities, consider the reaction in which solid metallic sodium is placed in contact with pure liquid water, producing gaseous hydrogen, and sodium and hydroxide ions in solution, as represented by

$$2\text{Na (s)} + 2\text{H}_2\text{O (l)} \rightarrow \text{H}_2\text{(g)} + 2\text{Na}^+\text{(aq)} + 2\text{ OH}^-\text{(aq)}$$

This reaction, in which the reagents are in different physical states, does not fit neatly into the framework described by K_p, K_x, K_b or K_c, with their corresponding values of $\Delta_r G_p^\circ$, $\Delta_r G_x^\circ$, $\Delta_r G_b^\circ$ and $\Delta_r G_c^\circ$. The reaction does, however, have an equilibrium constant, which we may write as K_r

$$K_r = \frac{a_{C,eq}^c \, a_{D,eq}^d}{a_{A,eq}^a \, a_{B,eq}^b} \tag{16.12d}$$

which will be large, since the equilibrium lies far to the right, associated with a (large, negative) value of $\Delta_r G^\circ$ as

$$\Delta_r G^\circ = -RT \ln K_r \tag{16.13d}$$

Given that the reactants are in different physical states, K_r is a generalised 'hybrid' equilibrium constant expressed in terms of whatever quantity ratios, as expressed by the activity a_i, are appropriate: the key requirement is that any given activity a_i, and the corresponding $\Delta_f G_i^\circ$, must both refer to the same standard state.

So, for the reaction of metallic sodium and water, for $\Delta_r G^\circ$ we have

$$\Delta_r G^\circ = [\Delta_f G^\circ \text{H}_2(g) + 2 \Delta_f G^\circ \text{Na}^+(aq) + 2 \Delta_f G^\circ \text{OH}^-(aq)]$$
$$- [2 \Delta_f G^\circ \text{Na}(s) + 2 \Delta_f G^\circ \text{H}_2\text{O}(l)]$$

in which the values for each $\Delta_f G^\circ$ can be found in reference tables.

For K_r

$$K_r = \frac{a_{eq}(\text{H}_2) \, (a_{eq}(\text{Na}^+))^2 \, (a_{eq}(\text{OH}^-))^2}{(a_{eq}(\text{Na}))^2 \, (a_{eq}(\text{H}_2\text{O}))^2}$$

we need the equilibrium activities $a_{i,eq}$ for each reactant. Accordingly, for the gaseous hydrogen $\text{H}_2(g)$, we may use the equilibrium partial pressure $p_{eq}\,\text{H}_2$

$$a_{eq}(\text{H}_2) = p_{eq}\,\text{H}_2/P^\circ\,\text{H}_2$$

and, assuming that the solution is ideal and also dilute, for the Na$^+$(aq) and OH$^-$(aq) ions, we may use molar concentrations as

$$a_{eq}(\text{Na}^+) = [\text{Na}^+]/[\text{Na}^+]^{\ominus}$$

and

$$a_{eq}(\text{OH}^-) = [\text{OH}^-]/[\text{OH}^-]^{\ominus}$$

For the molecular water, H$_2$O (l), we need to recognise that, as well as being a reactant, the water also serves as the solvent of the resulting solution. Reference to Table 6.4 will show that the standard state for solvents is the pure component, and so the appropriate quantity ratio is expressed in terms of the mole fraction,

$$a_{eq}(\text{H}_2\text{O}) = x_{\text{H}_2\text{O}}/x_{\text{H}_2\text{O}}^{\ominus}$$

for which the standard mole fraction, $x_{\text{H}_2\text{O}}^{\ominus} = 1$, corresponding, as required, to pure water. Furthermore, as the solvent, the water is present in vast excess, implying that the actual equilibrium mole fraction of water $x_{\text{H}_2\text{O}}$ is very close to 1, and so, to a very good approximation

$$a_{eq}(\text{H}_2\text{O}) = x_{\text{H}_2\text{O}}/x_{\text{H}_2\text{O}}^{\ominus} = 1$$

We now turn to $a_{eq}(\text{Na})$, the activity of the metallic sodium. So far in this book, we have made little reference to solids, and we have not derived an equation like the gas phase equation (13.47d)

$$\boldsymbol{G_i} = \boldsymbol{G_i}^{\ominus} + RT \ln(P_i/P^{\ominus}) \tag{13.47d}$$

for a solid. To do so, consider equation (16.10b)

$$\boldsymbol{G_{c,i}} = \boldsymbol{G_{c,i}}^{\ominus} + RT \ln ([\text{I}]/[\text{I}]^{\ominus}) \tag{16.10b}$$

which applies to a solute i in an ideal solution. In this equation, at a constant external pressure P and at any temperature T, the term $\boldsymbol{G_{c,i}}^{\ominus}$ is a constant, as are R, T and $[\text{I}]^{\ominus}$: the only variable is the molar concentration $[\text{I}]$, which is a measure of the quantity of component i present in the reaction mixture. Because the reaction is taking place in solution, the molecules (or ions) of component i are all randomly distributed within the solvent, alongside all the randomly distributed molecules (or ions) of the other reaction components. As a result of thermal motion, all the molecules (or ions) can move freely within the solvent, and the greater the number of molecules (or ions) of any component i that are distributed throughout the solution, as measured by the molar concentration $[\text{I}]$, the greater the likelihood of a reaction. Exactly similar reasoning applies if we use molality as the measure of quantity on a solution, mole fraction in the case of a liquid mixture, or partial pressure for a gas.

For a solid, however, the physical context is different. Suppose that 1 mol of a solid, in the form of a cube, is a reactant in a chemical reaction with a liquid, for example, the reaction of metallic iron with a dilute acid:

$$\text{Fe (s)} + \text{H}_2\text{SO}_4 \text{ (aq)} \rightarrow \text{Fe}^{2+} \text{ (aq)} + \text{SO}_4^{2-} \text{ (aq)} + \text{H}_2 \text{ (g)}$$

Suppose further that the liquid is enclosed in a vertical column, such that only the upper surface of the solid cube is in contact with the lower surface of the column of liquid. Physically, the reaction takes place on the surface of the solid, over the area of contact between the solid and liquid. Only the surface molecules of that area can react – those molecules within the mass of the solid, and within the bulk of the liquid, are not involved.

This is very different from the physical context of a reaction taking place in a mixture of ideal gases, a mixture of (necessarily miscible) liquids, or in an ideal solution, where all molecules are involved simultaneously – by contrast, for a reaction involving solids, or immiscible liquids, where the reaction can take place only on a contact surface, only those molecules on the appropriate surface are can react. This implies that an equation of the general form

$$G_{c,i} = G_{c,i}^{\circ} + RT \ln ([I]/[I]^{\circ}) \tag{16.10b}$$

cannot apply to reactions involving solids or immiscible liquids.

So, what does apply? In principle, we could introduce a term that defines the surface area where molecular contact takes place, but that becomes very complex. Accordingly, a much simpler approach is adopted: the forms of equations (16.19a) and (16.19b)

$$\mathbf{G}_i = \mathbf{G}_i^{\circ} + RT \ln a_i \tag{16.19a}$$

$$\mu_i = \mu_i^{\circ} + RT \ln a_i \tag{16.19b}$$

are maintained, and the convention is adopted that $a_i = 1$ for all solids, and for immiscible liquids. If, at any instant τ whilst the reaction is in progress, the number of moles of the solid component i present is $n_i(\tau)$, then, following the reasoning discussed on page 519, the contribution $G_i(\tau)$ attributable to component i to $G_{sys}(\tau)$ is

$$G_i(\tau) = n_i(\tau) \, \mathbf{G}_i(\tau) = n_i(\tau) \, \mathbf{G}_i^{\circ} + n_i(\tau) \, RT \ln a_i$$

But since, for a solid, $a_i = 1$, then $\ln a_i = 0$, and so

$$G_i(\tau) = n_i(\tau) \, \mathbf{G}_i(\tau) = n_i(\tau) \, \mathbf{G}_i^{\circ}$$

So, returning to the reaction

$$\text{Na (s)} + 2\text{H}_2\text{O (l)} \rightarrow \text{H}_2\text{(g)} + 2\text{Na}^+ \text{(aq)} + 2\text{ OH}^- \text{(aq)}$$

we have now ascertained that both a_{eq} (Na) = 1 and a_{eq} (H$_2$O) = 1 and a_{eq} (H$_2$O) = 1, and so K_r becomes

$$K_r = \frac{a_{eq}(\text{H}_2) \, (a_{eq}(\text{Na}^+))^2 \, (a_{eq}(\text{OH}^-))^2}{(a_{eq}(\text{Na}))^2 (a_{eq}(\text{H}_2\text{O}))^2}$$

$$= \left(p_{eq} \, \text{H}_2/P^{\circ} \, \text{H}_2\right) ([\text{Na}^+]/[\text{Na}^+]^{\circ})^2 \, ([\text{OH}^-]/[\text{OH}^-]^{\circ})^2$$

We will discuss activities a_i in more detail in Chapter 22, which deals with real, rather than ideal, systems: for the next several chapters, which use ideal systems as a model, we will, for the most part, be using the molar Gibbs free energy $\mathbf{G}_i$ and quantity ratios such as $[I]/[I]^{\circ}$, rather than the chemical potential μ_i and the chemical activity a_i.

16.6 Coupled reactions in solution

The discussion so far has been concerned with a generalised reaction $aA + bB \rightleftharpoons cC + dD$, taking place in an ideal solution with no other reactions happening, and this has very wide applicability. In biochemical contexts, however, many reactions take place in solution simultaneously, and an important class of reactions is the 'coupled system', which as we saw in Chapter 13 (see page 391), is a system in which an endergonic change

$$A \rightleftharpoons B \qquad\qquad \Delta_r G^\circ = \Delta_r G^\circ_{A \to B} > 0$$

is directly associated with an exergonic change

$$X \rightleftharpoons Y \qquad\qquad \Delta_r G^\circ = \Delta_r G^\circ_{X \to Y} < 0$$

such that $\Delta_r G^\circ_{A \to B} + \Delta_r G^\circ_{X \to Y} < 0$. In these expressions, A, B, X and Y do not necessarily represent single molecules, but could represent more complex mixtures; also, each $\Delta_r G^\circ$ is a general expression, and can apply to standards referring to the pure substance, unit molality or unit molar concentration – provided that both $\Delta_r G^\circ_{A \to B}$ and $\Delta_r G^\circ_{X \to Y}$ are based on the same standard.

Since $\Delta_r G^\circ_{A \to B} > 0$, the equilibrium for the reaction $A \rightleftharpoons B$, taking place by itself, is towards the left, and the greater the magnitude of $\Delta_r G^\circ_{A \to B}$, the lower the quantity of the product B in the $A \rightleftharpoons B$ equilibrium mixture. In contrast, because $\Delta_r G^\circ_{X \to Y} < 0$, the equilibrium for the reaction $X \rightleftharpoons Y$, also taking place by itself, is towards the right, and the greater the (negative) magnitude of $\Delta_r G^\circ_{X \to Y}$, the greater the quantity of the product Y in the $X \rightleftharpoons Y$ equilibrium mixture.

When these two reactions are coupled, and if $\Delta_r G^\circ_{A \to B} + \Delta_r G^\circ_{X \to Y} < 0$, then the equilibrium of the combined reaction

$$A + X \rightleftharpoons B + Y \qquad\qquad \Delta_r G^\circ = \Delta_r G^\circ_{A \to B} + \Delta_r G^\circ_{X \to Y} < 0$$

lies to the right, implying that the quantity of the product B in the $A + X \rightleftharpoons B + Y$ equilibrium mixture is possibly quite substantial, the more so the greater the (negative) magnitude of $\Delta_r G^\circ_{A \to B} + \Delta_r G^\circ_{X \to Y}$.

Operationally, the coupling of these reactions results in an apparently non-spontaneous reaction, $A \rightleftharpoons B$, becoming spontaneous; thermodynamically, this happens because the quantity $\Delta_r G^\circ_{X \to Y}$, the free energy released by the exergonic reaction $X \rightleftharpoons Y$, is more than sufficient to provide the free energy $\Delta_r G^\circ_{A \to B}$ required to drive the endergonic reaction $A \rightleftharpoons B$ towards B.

Coupled reactions are critical in biology. All biosynthetic reactions, such as the formation of proteins from amino acids, are endergonic. By themselves, these reactions are not spontaneous, and so they 'shouldn't happen'. The synthesis of a protein, however, is not a reaction that happens by itself: in the living cell, this reaction that is coupled to the exergonic hydrolysis of ATP (adenosine triphosphate) to ADP (adenosine diphosphate), such that the free energy released by the ATP hydrolysis drives the free energy requirement of protein synthesis. That's just one example, and there are many more.

The study of the flow of free energy through biological systems is known as **bioenergetics**, and will be explored in more detail in Chapter 24; here, we take the opportunity to examine further the thermodynamics of the generalised coupled system

$$\left. \begin{array}{l} A \rightleftharpoons B \\ X \rightleftharpoons Y \end{array} \right. \qquad\qquad \left. \begin{array}{l} \Delta_r G^\circ_{A \to B} = +50 \text{ kJ} \\ \Delta_r G^\circ_{X \to Y} = -60 \text{ kJ} \end{array} \right\}$$

in which we shall assume that A, B, X and Y are all solutes in an ideal dilute aqueous solution. As we have just seen, although A, B, X and Y are all written as single entities, they could represent more complex systems: 'A', for example, might represent the mixture 2C + 3D. Also, the reaction $A \rightleftharpoons B$, by itself, is endergonic ($\Delta_r G^\ominus_{A \to B}$ = +50 kJ); the reaction $X \rightleftharpoons Y$, by itself, is exergonic ($\Delta_r G^\ominus_{X \to Y}$ = −60 kJ); and the coupled reaction $A + X \rightleftharpoons B + Y$ is exergonic ($\Delta_r G^\ominus_{A \to B} + \Delta_r G^\ominus_{X \to Y}$ = +50 − 60 = −10 kJ).

Taking firstly the endergonic reaction $A \rightleftharpoons B$, if an initial n mol of A are dissolved in an aqueous solution of V dm^{-3}, then, at any point in the reaction corresponding to the extent-of-reaction parameter $\xi_{A \to B}$, the molar concentrations [A] and [B] of A and B are given by

$$[A] = \frac{n(1 - \xi_{A \to B})}{V}$$

and

$$[B] = \frac{n\xi_{A \to B}}{V}$$

If we represent the concentration equilibrium constant K_c for this reaction as $K_{A \to B}$, then, using equation (16.12c), we have

$$K_{A \to B} = \frac{[B]_{eq}}{[A]_{eq}} = \frac{n\xi_{A \to B,eq}/V}{n(1 - \xi_{A \to B,eq})/V} = \frac{\xi_{A \to B,eq}}{1 - \xi_{A \to B,eq}}$$

from which

$$\xi_{A \to B eq} = \frac{K_{A \to B}}{1 + K_{A \to B}} \qquad (16.20)$$

Furthermore, according to equation (16.13c)

$$\Delta_r G_c^\ominus = - RT \ln K_c \qquad (16.13c)$$

then, for the reaction $A \rightleftharpoons B$, and assuming that the value $\Delta_r G^\ominus_{A \to B}$ = +50 kJ is measured using unit molar concentrations as the standard, we have

$$K_{A \to B} = \exp\left(-\frac{50,000}{RT}\right) = 1.74 \times 10^{-9} \text{ at } 298.15 \text{ K}$$

and, from equation (16.20)

$$\xi_{A \to B, eq} = \frac{1.74 \times 10^{-9}}{1 + 1.74 \times 10^{-9}} = 1.74 \times 10^{-9}$$

$K_{A \to B}$ and $\xi_{A \to B,eq}$ are therefore very small numbers: for every 10^9 initial molecules of A, the equilibrium mixture contains only about 2 molecules of the product B. The equilibrium for the reaction $A \rightleftharpoons B$ therefore lies very far to the left: in practice, the reaction doesn't happen – A is stable.

An exactly similar analysis can be carried out for the reaction $X \rightleftharpoons Y$, for which $\Delta_r G^\ominus_{X \to Y}$ = −60 kJ, once again using unit molar concentrations as the standard, giving

$$K_{X \to Y} = \exp\left(\frac{60,000}{RT}\right) = 3.25 \times 10^{10} \quad \text{at } 298.15 \text{ K}$$

and
$$\xi_{X \to Y, eq} = \frac{3.25 \times 10^{10}}{1 + 3.25 \times 10^{10}} \approx 1$$

The reaction $X \rightleftharpoons Y$ therefore goes to completion.

We now consider the coupled reaction

$$A \rightleftharpoons B \qquad \Delta_r G^\ominus_{A \to B} = +50 \text{ kJ}$$
$$X \rightleftharpoons Y \qquad \Delta_r G^\ominus_{X \to Y} = -60 \text{ kJ}$$

which we may express as

$$A + X \rightleftharpoons B + Y \qquad \Delta_r G^\ominus_{A,X \to B,Y} = -10 \text{ kJ}$$

Using equation (16.13c), and representing the concentration equilibrium constant for the coupled reaction as K, we have

$$K = \exp\left(-\frac{10{,}000}{RT}\right) = 56.5 \text{ at } 298.15 \text{ K}$$

The equilibrium for the coupled reaction therefore lies towards the right, but much less so as compared to the reaction $X \rightleftharpoons Y$ taking place in isolation, for which the concentration equilibrium constant is 3.25×10^{10}. To determine just how far the equilibrium of the coupled reaction is to the right, we need to compute the corresponding equilibrium extent-of-reaction $\xi_{AX \to BY, eq}$, which, for simplicity, we will represent as ξ_{eq}.

Since

$$K = \frac{[B]_{eq}[Y]_{eq}}{[A]_{eq}[X]_{eq}}$$

and noting that, for the stoichiometry $A + X \rightleftharpoons B + Y$

$$[A]_{eq} = [X]_{eq} = \frac{n(1 - \xi_{eq})}{V}$$

and

$$[B]_{eq} = [Y]_{eq} = \frac{n\xi_{eq}}{V}$$

then

$$K = \frac{[B]_{eq}[Y]_{eq}}{[A]_{eq}[X]_{eq}} = \frac{\xi_{eq}^2}{(1 - \xi_{eq})^2}$$

which rearranges to the quadratic equation

$$\xi_{eq}^2(K - 1) - 2K\xi_{eq} + K = 0$$

with solutions

$$\xi_{eq} = \frac{2K \pm \sqrt{4K^2 - 4K(K-1)}}{2(K-1)} = \frac{K \pm \sqrt{K}}{K - 1}$$

Since, as we have seen, $K = 56.5$, then $\xi_{eq} = 1.15$ or 0.88. Physically, however, ξ_{eq} cannot be less than 0 or greater than 1, implying that the only feasible value of ξ_{eq} is 0.88.

This result shows that, for the coupled reaction $A + X \rightleftharpoons B + Y$, for every 100 initial molecules of A, the equilibrium mixture contains 12 molecules of the reactant A and 88 molecules of

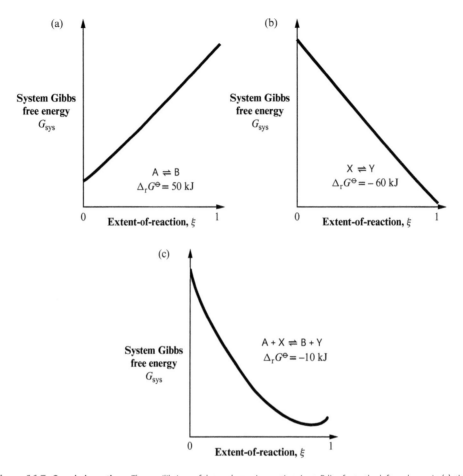

Figure 16.7 Coupled reactions. The equilibrium of the endergonic reaction A ⇌ B lies far to the left as shown in (a): this reaction is not spontaneous to the right, and A is stable. By contrast, the equilibrium of the exergonic reaction X ⇌ Y lies far to the right as shown in (b). The behaviour of the coupled reaction A + X ⇌ B + Y is shown in (c) – as can be seen, the equilibrium position is to the right, corresponding (as shown in the text) to an equilibrium extent-of-reaction ξ_{eq} = 0.88. Whereas the equilibrium mixture for the reaction A ⇌ B alone contains no measureable quantity of the product B, the equilibrium mixture for the coupled reaction A + X ⇌ B + Y contains 88 molecules of B for every 100 initial molecules of A.

the product B. By contrast, for the uncoupled reaction A ⇌ B, for every 10^9 initial molecules of A, the equilibrium mixture contains just 2 molecules of B. This verifies how coupling an endergonic reaction to an exergonic reaction can make apparently non-spontaneous reactions happen, without breaking the Second Law – as vividly illustrated in Figure 16.7.

EXERCISES

1. Define
   - solute
   - solvent

- mole fraction
- molality
- molarity
- molar concentration
- ideal solution
- dilute ideal solution
- ideal-dilute solution
- Raoult's law
- Henry's law
- coupled reactions.

2. What is the difference between a solution and a mixture?
3. What is the difference between Raoult's Law and Henry's Law?
4. A, B, C and D are pure liquids that react at 298.15 K as

$$A + 4B \rightleftharpoons 3C + D$$

The equilibrium composition comprises 0.05 mol A, 0.2 mol B, 2.85 mol C and 0.95 mol D, within a total volume of 6×10^{-3} m^3 and mass 4.3 kg. Assuming that the solution is ideal, at equilibrium, calculate
- the mole fraction of each component
- the molality of each component
- the molar concentration of each component
- the extent-of-reaction
- the mole fraction equilibrium constant
- the corresponding value of $\Delta_r G_x^\ominus$
- the molality equilibrium constant
- the corresponding value of $\Delta_r G_b^\ominus$
- the (molar) concentration equilibrium constant
- the corresponding value of $\Delta_r G_c^\ominus$.

Compare the three values of $\Delta_r G^\ominus$, and explain the result.

5. A, B, C and D are solutes that react in aqueous solution at 298.15 K as

$$A + 4B \rightleftharpoons 3C + D$$

At equilibrium, the molar concentrations of each component are 1.5×10^{-6} mol m^{-3} A, 6×10^{-6} mol m^{-3} B, 25.5×10^{-6} mol m^{-3} C and 8.5×10^{-6} mol m^{-3} D, within a total volume of 5×10^{-3} m^3. Assuming that the solution is ideal, at equilibrium, calculate
- the mole fraction of each component
- the molality of each component
- the molarity of each component
- the extent-of-reaction
- the mole fraction equilibrium constant
- the corresponding value of $\Delta_r G_x^\ominus$
- the molality equilibrium constant
- the corresponding value of $\Delta_r G_b^\ominus$
- the concentration equilibrium constant
- the corresponding value of $\Delta_r G_c^\ominus$

Compare the three values of $\Delta_r G^\ominus$, and explain the result.

6. Consider the generalised reaction

$$aA + bB \rightleftharpoons cC + dD$$

in which A, B, C and D are mutually miscible liquids. Starting from an expression for $G_{sys}(\tau)$ of a form similar to the equation on page 518, prove that equations 16.12a, 16.13a, 16.14a, 16.15a, 16.16a, 16,17a and 16.18a are valid for an ideal liquid mixture. Are equations 16.12b - 16.18b, and 16.12c - 16.18c also valid? If so, why? And if not, why not?

7. The oxidation of glucose is strongly exergonic

$$\text{Glucose} + 6O_2 \rightarrow 6CO_2 + 6H_2O \qquad \Delta_c G^\ominus = -2930 \text{ kJ}$$

Photosynthesis is the biochemical process, taking place in, for example, the leaves of green plants, in which glucose is synthesised from molecular carbon dioxide and water. This process happens naturally and spontaneously. From a thermodynamic standpoint – and without having to know any of the biochemistry – explain how this can happen.

8. The reaction

$$A \rightleftharpoons C + D$$

takes place in an ideal aqueous solution. From an initial state of an ideal solution of pure A, the equilibrium composition comprises molar concentrations of 10^{-7} mol dm^{-3} A, 10^{-2} mol dm^{-3} C, and 10^{-2} mol dm^{-3} D. Calculate the concentration equilibrium constant K_c, and also the ratio $[C]_{eq}/[A]_{eq}$. Suppose that some more D is added to the solution, bringing the molar concentration of D to 10^{-1} mol dm^{-3}. After the addition of more D, what is the concentration equilibrium constant K_c, and the ratio $[C]_{eq}/[A]_{eq}$? Even more D is now added, raising the molar concentration of D to 1 mol dm^{-3} – what is the concentration equilibrium constant K_c, and the ratio $[C]_{eq}/[A]_{eq}$? Interpret all your results in physical terms – what is happening?

9. The reaction

$$P + Q + H^+ (aq) \rightleftharpoons R + S$$

takes place in an ideal aqueous solution, and, as shown, consumes hydrogen $H^+(aq)$ ions. Write down an expression for the concentration equilibrium constant K_c. Suppose that $[H^+]_{eq} = 10^{-6}$ mol dm^{-3}, implying that the equilibrium pH $= -\log_{10}[H^+]_{eq} = $ pH 6; suppose further that the corresponding equilibrium concentration of component R is represented as $[R]_{eq}$ (pH 6). As we shall see in the next chapter, a **buffer solution** is a solution in which the molar concentration $[H^+]$ of hydrogen $H^+(aq)$ ions is controlled. If the reaction takes place within a buffer solution at constant pH 8, what is the effect on the reaction? If we represent the corresponding equilibrium concentration of component R as $[R]_{eq}$(pH 8), which of the following statements is true:

$[R]_{eq}$ (pH 8) = $[R]_{eq}$ (pH 6)
$[R]_{eq}$ (pH 8) < $[R]_{eq}$ (pH 6)
$[R]_{eq}$ (pH 8) > $[R]_{eq}$ (pH 6)?

Why?

Similarly, if the reaction takes place at pH 3, which statement is true:

$[R]_{eq}$ (pH 3) = $[R]_{eq}$ (pH 6)
$[R]_{eq}$ (pH 3) < $[R]_{eq}$ (pH 6)
$[R]_{eq}$ (pH 3) > $[R]_{eq}$ (pH 6)?

Questions 8 and 9 demonstrate an important characteristic of chemical equilibrium: the composition of the equilibrium mixture can be influenced by controlling the quantity of one, specific, reagent – the product D in question 8, and the reactant H^+(aq) in question 9. So, if we are especially interested in the product C as produced by the reaction A $\rightleftharpoons$ C + D, then the production of C can be increased by depleting D, so 'pulling' the reaction to the right (assuming that a sufficient quantity of A is available); similarly, if R is produced by the reaction P + Q + H^+(aq) $\rightleftharpoons$ R + S, then the production of R can be increased by decreasing the pH at which the reaction is buffered, so increasing the molar concentration of H^+ ions and 'pushing' the reaction to the right (assuming that sufficient quantities of P and Q are available). This 'mass action effect' has very considerable significance in biology, as we shall see in Chapters 23 and 24.

17 Acids, bases and buffer solutions

 Summary

The dissociation of the acid HA

$$HA \rightleftharpoons H^+ (aq) + A^- (aq)$$

has a corresponding **acid dissociation constant** K_a

$$K_a = \frac{[H^+]_{eq} [A^-]_{eq}}{[HA]_{eq}} \tag{17.12}$$

from which we derive the **Henderson-Hasselbalch equation**

$$pH = pK_a + \log_{10} \frac{[A^-]_{eq}}{[HA]_{eq}} \tag{17.16}$$

and, for weak acids (say, $K_a < 10^{-4}$), the **Henderson-Hasselbalch approximation**

$$pH \approx pK_a + \log_{10} \frac{[A^-]_{eq}}{[HA]_i} \tag{17.26}$$

Buffer solutions – solutions of a mixture of a weak acid HA and a corresponding salt XA, or of a weak base BOH and a corresponding salt BX – are able to maintain the pH stable, especially close to the pK_a of the weak acid HA (or the pK_b of the weak base BOH).

If a reaction, such as

$$aA + bB \rightleftharpoons cC + hH^+ (aq)$$

is taking place within a buffer solution, then the action of the buffer solution in holding the molar concentration [H⁺] constant affects the reaction's equilibrium. If the concentration equilibrium constant K_c for the reaction unconstrained by a buffer solution is

$$K_c = \frac{[C]_{eq}^c [H^+]_{eq}^h}{[A]_{eq}^a [B]_{eq}^b} \tag{17.48}$$

then, when the same reaction takes place in a solution buffered at $pH_{buf} = -\log_{10}[H^+]_{buf}$, the observed equilibrium constant $\Lambda (pH_{buf})$ is given by

$$\Lambda(pH_{buf}) = \frac{[C]_{eq}^c}{[A]_{eq}^a [B]_{eq}^b} \tag{17.49}$$

such that

$$K_c = \Lambda(pH_{buf}) [H^+]_{buf}^h = \Lambda(pH_{buf}) \, 10^{-h \, pH_{buf}} \tag{17.55}$$

Modern Thermodynamics for Chemists and Biochemists. Dennis Sherwood and Paul Dalby.
© Oxford University Press 2018. Published 2018 by Oxford University Press.

17.1 Dissociation

In this chapter, we will explore an especially important class of reactions which take place in aqueous solution – those that either produce, or consume, hydrogen H^+ (aq) ions, as happens, for example, when acetic acid, CH_3COOH, abbreviated to AcH, dissolves in water and ionises as

$$AcH \rightleftharpoons H^+ \text{ (aq)} + Ac^- \text{ (aq)}$$

In fact, single protons in solution, as implied by the symbol H^+ (aq), do not exist: rather, the proton released from the acid combines with another water molecule, so the reversible dissociation reaction is written more correctly as

$$AcH + H_2O \rightleftharpoons H_3O^+ \text{ (aq)} + A^- \text{ (aq)}$$

to form the **hydronium ion** H_3O^+ (aq), sometimes referred to as the **hydroxonium ion** or **oxonium ion**. For simplicity, this book will use H^+ (aq) or H^+ as a 'shorthand' for the more correct from H_3O^+ (aq).

The ionisation of acetic acid is a particular case of the dissociation, without further reaction, of a molecule XY as

$$XY \rightleftharpoons X + Y$$

where X and Y are usually oppositely charged ions in solution. In accordance with equation (16.12c)

$$K_c = \frac{[C]_{eq}^c \, [D]_{eq}^d}{[A]_{eq}^a \, [B]_{eq}^b} \tag{16.12c}$$

which applies to the generalised reaction $aA + bB \rightleftharpoons cC + dD$ taking place in an ideal solution, we may define a concentration equilibrium constant K_d for the dissociation of XY as

$$K_d = \frac{[X]_{eq} \, [Y]_{eq}}{[XY]_{eq}} \tag{17.1}$$

K_d is known as the **dissociation constant**, and a large value implies that the equilibrium lies far to the right, with the equilibrium composition largely the dissociated species X and Y; conversely, a small value of K_d implies that the equilibrium is lies far to the left, with the equilibrium composition being largely undissociated XY.

If, at equilibrium, a fraction ξ_{eq} of n mol of the XY molecules dissociate, then, for the stoichiometry $XY \rightleftharpoons X + Y$, the equilibrium mixture comprises $n(1 - \xi_{eq})$ moles of undissociated XY, $n\xi_{eq}$ moles of X and $n\xi_{eq}$ moles of Y. The molar concentrations of the various components of the equilibrium mixture are therefore given by

$$[X]_{eq} = \frac{n\xi_{eq}}{V} \tag{17.2a}$$

$$[Y]_{eq} = \frac{n\xi_{eq}}{V} \tag{17.2b}$$

and

$$[XY]_{eq} = \frac{n(1 - \xi_{eq})}{V} \tag{17.2c}$$

where V is the total volume of the solution. The form of equations (17.2) is specific to the stoichiometry $XY \rightleftharpoons X+Y$; expressions for other stoichiometries are easily established (see, for example, pages 441 and 442) – for clarity, we will continue here with the simple stoichiometry $XY \rightleftharpoons X + Y$.

Equations (17.1) and (17.2) can be combined to give

$$K_d = \frac{n}{V} \frac{\xi_{eq}^2}{(1 - \xi_{eq})} \tag{17.3}$$

which is a general equation from which K_d can be determined for any given value of ξ_{eq}. A particular, and important, case is that in which ξ_{eq} is small – say 0.01 or less – corresponding to a molecule XY that dissociates only weakly. In this case, $1 - \xi_{eq} \approx 1$, and so

$$K_d = \frac{n}{V} \xi_{eq}^2 \tag{17.4}$$

Now n/V represents the molar concentration of XY as would happen if it did not dissociate at all; also, if we started with n mol of pure XY, which we then add to water to form a solution of total volume V, then the quantity n/V is known. If we define the 'initial molar concentration' $[XY]_i$ as

$$[XY]_i = \frac{n}{V} \tag{17.5}$$

then

$$K_d \approx \frac{\xi_{eq}^2}{[XY]_i} \tag{17.6a}$$

and

$$\xi_{eq} \approx \sqrt{K_d [XY]_i} \tag{17.6b}$$

If $[XY]_i$ is, say, 1 mol dm^{-3}, then equations (17.6) imply that if $\xi_{eq} < 0.01$, $K_d < 10^{-4}$, so enabling us to quantify what 'weakly dissociated' means: XY is weakly dissociated if $K_d < 10^{-4}$.

17.2 The ionisation of pure water, pH and pOH

A very important example of a weakly dissociating molecule is water, which dissociates into two ions as

$$H_2O \rightleftharpoons H^+ (aq) + OH^- (aq)$$

$$\therefore K_w = \frac{[H^+]_{eq} [OH^-]_{eq}}{[H_2O]_{eq}} \tag{17.7}$$

in which (aq) for the ions is implied, and we use the special symbol K_w to refer specifically to the equilibrium ionisation of pure water. K_w is known as the **ionic product of water** (alternatively, the **ion-product of water**, or the **water ionisation constant**).

Since each molecule of H_2O which dissociates creates an ion-pair of one H^+ (aq) ion and one OH^- (aq) ion, the sum of the negative charges always equals the sum of the negative charges, so ensuring electrical neutrality, a condition known as **charge balance**. Also, the molar concentrations $[H^+]_{eq}$ and $[OH^-]_{eq}$ are necessarily equal, and are shown by experiment each to be 10^{-7} mol dm^{-3} at 298.15 K and a pressure of 1 bar.

As we saw on page 516, in equation (17.1), each molar concentration written as $[X]$ is 'shorthand' for the (dimensionless) ratio $[X]/[X]^{\circ}$, where $[X]^{\circ}$ is the corresponding standard. For the ions H^+ (aq) and OH^- (aq), which are in solution, reference to Table 6.4 will show that the standard state for solutes in solution is unit molar concentration, $[X]^{\circ} = 1$ mol dm^{-3}, at a pressure of 1 bar, and so $[H^+]_{eq}/[H^+]^{\circ} = [H^+]_{eq} = 10^{-7}$ mol dm^{-3}, and $[OH^-]_{eq}/[OH^-]^{\circ} = [OH^-]_{eq} = 10^{-7}$ mol dm^{-3}.

Since the mass of 1 dm^3 of pure water is 1 kg, and the molecular weight of water is 0.0180 kg mol^{-1}, the equilibrium molar concentration $[H_2O]_{eq}$ of pure water is $1/0.0180 = 55.56$ mol dm^{-3}. It therefore seems that the denominator in equation (17.7) is 55.56. In fact, this is not the case, for the denominator is the ratio $[H_2O]_{eq}/[H_2O]^{\circ}$, and – unlike the situation for the H^+ (aq) and OH^- (aq) ions – for pure water, $[H_2O]^{\circ}$ is *not* 1. The reason for this is attributable to the fact that, in this case, molecular water is not a dissolved *solute*, but rather the *solvent* in which the H^+ (aq) and OH^- (aq) ions are themselves dissolved. Reference to Table 6.4 shows that the standard state for a solvent is "the most stable form of the pure solvent" of that solvent – in this case, pure water. Since the molar concentration of pure water is 55.56 mol dm^{-3}, then $[H_2O]^{\circ} = 55.56$ mol dm^{-3}, and the ratio $[H_2O]_{eq}/[H_2O]^{\circ} = 55.56/55.56 = 1$. Accordingly, in equation (17.1), the term shown as $[H_2O]_{eq}$ takes the value 1, and equation (17.7), for pure water, becomes

$$K_w = [H^+]_{eq}[OH^-]_{eq} = 10^{-7} \times 10^{-7} = 10^{-14} \tag{17.8}$$

If we define

$$pH = -\log_{10}[H^+]_{eq} \tag{17.9}$$

and

$$pOH = -\log_{10}[OH^-]_{eq} \tag{17.10}$$

then, for pure water, pH = pOH = 7; furthermore, from equation (17.8)

$$\log_{10} K_w = \log_{10}[H^+]_{eq} + \log_{10}[OH^-]_{eq} = -14 \tag{17.11a}$$

and so, in general

$$pH + pOH = 14 \tag{17.11b}$$

In connection with the definitions of pH and pOH as equations (17.9) and (17.10), we note that, mathematically, logarithms can be taken only of pure numbers, and not of dimensioned quantities. In the definitions of pH and pOH, it might appear that $[H^+]_{eq}$ and $[OH^-]_{eq}$ are molar concentrations with dimensions mol dm^{-3}; in fact, they are the ratios $[H^+]_{eq}/[H^+]^{\circ}$, and $[OH^-]_{eq}/[OH^-]^{\circ}$, and so are dimensionless, as required. Furthermore, as we shall see in Chapter 22, the definition pH = $-\log_{10}[H^+]$, as in equation (17.9), is in fact an approximation – but a very good, and pragmatic, approximation, hence its widespread use.

Equation (17.8)

$$K_w = [H^+]_{eq}[OH^-]_{eq} = 10^{-14} \tag{17.8}$$

remains true for all equilibrium solutions in which H^+ (aq) and OH^- (aq) ions are present. For pure water, charge balance requires that the molar concentrations $[H^+]_{eq}$ and $[OH^-]_{eq}$ are equal, each being 10^{-7} mol dm^{-3}; but if other ions are present, the molar concentrations $[H^+]_{eq}$ and $[OH^-]_{eq}$ might change, subject to their product being constant, as determined by equation (17.8).

So, suppose that we add a compound HA to pure water. If the compound dissolves in its entirety, and ionises according to the reversible reaction

$$HA\ (aq) \rightleftharpoons H^+\ (aq) + A^-\ (aq)$$

then the solution comprises an equilibrium mixture of undissociated water molecules H_2O, undissociated HA molecules, H^+ (aq) ions, OH^- (aq) ions, and A^- (aq) ions. Charge balance requires that

$$[H^+]_{eq} = [OH^-]_{eq} + [A^-]_{eq}$$

implying that $[H^+]_{eq} > [OH^-]_{eq}$, yet it is still the case that the product $[H^+]_{eq}[OH^-]_{eq} = 10^{-14}$. This solution of HA in pure water therefore contains a greater concentration of H^+ (aq) ions, and a lesser concentration of OH^- (aq) ions, as compared to pure water, and although the solution also contains A^- (aq) ions, these do not 'count' as regards the solution's pH. The solution therefore becomes more acidic, with a pH < 7, and a pOH > 7.

Similarly, if we add a different compound BOH which ionises as

$$BOH\ (aq) \rightleftharpoons B^+\ (aq) + OH^-\ (aq)$$

then charge balance tells us that

$$[H^+]_{eq} + [B^+]_{eq} = [OH^-]_{eq}$$

and so $[H^+]_{eq} < [OH^-]_{eq}$. The solution therefore becomes more alkaline, with a pH > 7, and a pOH < 7.

The last few paragraphs explain why the addition of an acid HA lowers, or a base BOH raises, the pH of a solution, and so the rest of this chapter explores the important topic of the thermodynamics of acids and bases in more detail.

17.3 Acids

17.3.1 The acid dissociation constant K_a

Consider a solution of a **monoprotic acid** HA that releases a single hydrogen ion in accordance with the reversible reaction

$$HA\ (aq) \rightleftharpoons H^+\ (aq) + A^-\ (aq)$$

where, as we have seen, H^+ (aq) stands for H_3O^+ (aq). If we assume that the solution is ideal, we may define a concentration equilibrium constant, K_a, known as the **acid dissociation constant**, as

$$K_a = \frac{[H^+]_{eq}[A^-]_{eq}}{[HA]_{eq}} \tag{17.12}$$

in which (aq) is implied, from which we may define the parameter pK_a as

$$pK_a = -\log_{10} K_a \tag{17.13a}$$

or, alternatively,

$$K_a = 10^{-pK_a} \tag{17.13b}$$

In every-day language, we refer to a 'weak' acid as one which produces a relatively few hydrogen H^+ (aq) ions, corresponding to a low equilibrium molar concentration $[H^+]_{eq}$, in contrast to a 'strong' acid, which produces relatively more hydrogen H^+ (aq) ions, corresponding to a high equilibrium molar concentration $[H^+]_{eq}$. Equation (17.13b) allows us to quantify this distinction: the weaker the acid, the more the equilibrium of the reversible reaction HA (aq) $\rightleftharpoons$ H^+ (aq) + A^- (aq) lies to the left, and the lower the value of K_a, which can often take values orders of magnitude less than 1 – for example, acetic acid (the acid present in vinegar), for which $K_a = 10^{-4.5}$, and pK_a = 4.5. Weak acids therefore have large positive values of pK_a, and the weaker the acid, the greater the value of pK_a. This analysis is totally consistent with the general principles of dissociation as discussed on pages 535 and 536, where we established that weak dissociation is associated with values of the equilibrium dissociation fraction ξ_{eq} of less than about 0.01, corresponding to values of K_d smaller than about 10^{-4}.

The greater the value of ξ_{eq}, the greater the value of K_a, and the stronger the acid – examples being the strong mineral acids such as hydrochloric, sulphuric and nitric, which have values of K_a in the range 10 to 10^{10} or more, with corresponding values of pK_a in the range –1 to –10. Strong acids therefore have negative values of pK_a, and the stronger the acid, the more negative the value of pK_a. Note, however, that all but the most dilute solutions of these strong acids depart significantly from ideal behaviour, and so their values of K_a are not given by equation (17.6), and, in practice, are difficult to determine with accuracy.

Pure water is a particular case of a monoprotic acid, in that it ionises to produce H^+ (aq) ions as

$$H_2O \rightleftharpoons H^+ (aq) + OH^- (aq)$$

According to equation (17.12), K_a for water is given by

$$K_a = \frac{[H^+]_{eq}[OH^-]_{eq}}{[H_2O]_{eq}}$$

As we saw in page 537, $[H_2O]_{eq} = 1$, and so

$$K_a = [H^+]_{eq}[OH^-]_{eq}$$

But according to equation (17.8)

$$K_w = [H^+]_{eq}[OH^-]_{eq} = 10^{-14} \tag{17.8}$$

and so, for pure water

$$K_a = K_w = 10^{-14} \tag{17.14}$$

implying that pK_a for pure water is 14.

Polyprotic acids release more than one hydrogen ion, which can usually be represented as a distinct multi-step process of the form

$$H_2A \text{ (aq)} \rightleftharpoons H^+ \text{ (aq)} + HA^- \text{ (aq)}$$

followed by

$$HA^- \text{ (aq)} \rightleftharpoons H^+ \text{ (aq)} + A^{2-} \text{ (aq)}$$

each with its own acid dissociation constant K_{a1} and K_{a2} given by

$$K_{a1} = \frac{[H^+]_{eq}[HA^-]_{eq}}{[H_2A]_{eq}}$$

$$K_{a2} = \frac{[H^+]_{eq}[A^{2-}]_{eq}}{[HA^-]_{eq}}$$

So, for carbonic acid, we have

$$H_2CO_3 \text{ (aq)} \rightleftharpoons H^+ \text{ (aq)} + HCO_3^- \text{ (aq)} \qquad K_{a1} = 10^{-6.4}, pK_a = 6.4$$

and

$$HCO_3^- \text{ (aq)} \rightleftharpoons H^+ \text{ (aq)} + CO_3^{2-} \text{ (aq)} \qquad K_{a2} = 10^{-10.3}, pK_a = 10.3$$

showing that the first ionisation of carbonic acid is weak, and the second very much weaker. Some representative values of pK_a are shown in Table 17.1.

Table 17.1 Values of pK_a for some selected acids in aqueous solution at 298 K. The *greater* the value of pK_a, the *weaker* the acid. All entities are (aq).

	pK_a		pK_a
$HF \rightleftharpoons H^+ + F^-$	3.20	$H_2CO_3 \rightleftharpoons H^+ + HCO_3^-$ (pK_{a1})	6.35
$HCl \rightleftharpoons H^+ + Cl^-$	-6.2*	$HCO_3^- \rightleftharpoons H^+ + CO_3^{2-}$ (pK_{a2})	10.33
$HI \rightleftharpoons H^+ + I^-$	-8.6*	Formic $\rightleftharpoons H^+$ + Formate$^-$	3.75
$HNO_2 \rightleftharpoons H^+ + NO_2^-$	3.25	Acetic $\rightleftharpoons H^+$ + Acetate$^-$	4.76
$HNO_3 \rightleftharpoons H^+ + NO_3^-$	-1.37*	Oxalic $\rightleftharpoons H^+$ + Oxalate$^-$ (pK_{a1})	1.25
$H_3PO_4 \rightleftharpoons H^+ + H_2PO_4^-$ (pK_{a1})	2.16	Oxalate$^- \rightleftharpoons H^+$ + Oxalate^{2-} (pK_{a2})	3.81
$H_2PO_4^- \rightleftharpoons H^+ + HPO_4^{2-}$ (pK_{a2})	7.21	Pyruvic $\rightleftharpoons H^+$ + Pyruvate$^-$	2.39
$HPO_4^{2-} \rightleftharpoons H^+ + HPO_4^{3-}$ (pK_{a3})	12.32	Citric $\rightleftharpoons H^+$ + Citrate$^-$ (pK_{a1})	3.13
$H_2S \rightleftharpoons H^+ + HS^-$ (pK_{a1})	7.05	Citrate$^- \rightleftharpoons H^+$ + Citrate^{2-} (pK_{a2})	4.76
$HSO_4^- \rightleftharpoons H^+ + SO_4^{2-}$ (pK_{a2})	1.99	Citrate$^{2-} \rightleftharpoons H^+$ + Citrate^{3-} (pK_{a3})	6.40
$H_2O \rightleftharpoons H^+ + OH^-$	14.00	Benzoic $\rightleftharpoons H^+$ + Benzoate$^-$	4.20

Sources: *CRC Handbook of Chemistry and Physics*, 96th edn, 2015, Section 5, pages 5–91 to 5–102, and *, approximate values from *Lange's Handbook of Chemistry*, 15th edn, 1999, McGraw-Hill Inc., Section 8, Tables 8.7.

17.3.2 The Henderson-Hasselbalch equation

Returning to the definition of the acid dissociation constant K_a, equation (17.12),

$$K_a = \frac{[H^+]_{eq}[A^-]_{eq}}{[HA]_{eq}} \quad (17.12)$$

if we take logarithms to base 10, we derive

$$\log_{10} K_a = \log_{10}[H^+]_{eq} + \log_{10}\frac{[A^-]_{eq}}{[HA]_{eq}}$$

$$\therefore -\log_{10}[H^+]_{eq} = -\log_{10} K_a + \log_{10}\frac{[A^-]_{eq}}{[HA]_{eq}} \quad (17.15)$$

from which

$$pH = pK_a + \log_{10}\frac{[A^-]_{eq}}{[HA]_{eq}} \quad (17.16)$$

Equation (17.16) is known as the **Henderson-Hasselbalch equation**.

Since the dissociation reaction HA (aq) $\rightleftharpoons$ H$^+$(aq) + A$^-$(aq) takes place in solution, if we assume that the solution behaves ideally, then all of the results discussed in Chapter 16 apply. So, for example, the reaction will be associated with a reaction Gibbs free energy change $\Delta_r G_c$, and also a concentration equilibrium constant K_c given by

$$K_c = \frac{[H^+]_{eq}[A^-]_{eq}}{[HA]_{eq}} \quad (16.12c)$$

Comparison of equations (16.12c) and (17.12) confirms, as expected, that

$$K_c = K_a$$

The reaction is also associated with a standard molar Gibbs free energy change $\Delta_r G_c^{\circ}$, which in this case is calculated as

$$\Delta_r G_c^{\circ} = \Delta_f G_c^{\circ}(H^+) + \Delta_f G_c^{\circ}(A^-) - \Delta_f G_c^{\circ}(H_2O) \quad (16.14c)$$

for which the values of the standard molar Gibbs free energies of formation $\Delta_f G_{c,i}^{\circ}$ can be found in Table 13.1 (see page 398). Furthermore, since

$$\Delta_r G_c^{\circ} = -RT \ln K_c \quad (16.13c)$$

then, for the dissociation of an acid characterised by values of K_a and pK_a

$$\Delta_r G_c^{\circ} = -RT \ln K_a$$

Taking our discussion further, in general, equation (17.12)

$$K_a = \frac{[H^+]_{eq}[A^-]_{eq}}{[HA]_{eq}} \quad (17.12)$$

together with its logarithmic form, the Henderson-Hasselbalch equation, equation (17.16), apply to the reversible dissociation HA (aq) $\rightleftharpoons$ H$^+$ (aq) + A$^-$ (aq) of any monoprotic acid HA, as well as to any successive ionisation HA^{n-} (aq) $\rightleftharpoons$ H$^+$ (aq) + A$^{(n+1)-}$ (aq) of a polyprotic acid, and so have wide applicability. To gain additional insight into the behaviour of equations

ACIDS, BASES AND BUFFER SOLUTIONS

(17.12) and (17.16), consider a quantity of n mol of a pure acid, undissociated, acid HA which is then mixed with (if HA is naturally a liquid), or dissolved in (if HA is naturally a solid), pure water to form a solution of total volume V dm^3. The acid HA then dissociates, so that the solution contains an equilibrium mixture of H$^+$ (aq) and A$^-$ (aq) ions and undissociated, dissolved, HA molecules HA (aq).

The Henderson-Hasselbalch equation may be expressed in a rather different form in terms of the equilibrium fraction ξ_{eq} of dissociation. Since the dissociation of the acid HA is just one particular case of the dissociation of the generalised molecule XY, the analysis we carried out on pages 535 and 536 in connection with equations (17.1) to (17.3) can be applied to the dissociation of the acid HA such that

$$[H^+]_{eq} = \frac{n\xi_{eq}}{V} = [HA]_i \, \xi_{eq} \tag{17.17a}$$

$$[A^-]_{eq} = \frac{n\xi_{eq}}{V} = [HA]_i \, \xi_{eq} \tag{17.17b}$$

and

$$[HA]_{eq} = \frac{n(1 - \xi_{eq})}{V} = [HA]_i \, (1 - \xi_{eq}) \tag{17.17c}$$

in which $[HA]_i = n/V$, the molar concentration of the totally undissociated acid, this being the initial molar concentration at the instant the n mol of HA were added to the solution.

Equations (17.12) and (17.17) can be combined to give **Ostwald's dilution law**

$$K_a = [HA]_i \, \frac{\xi_{eq}^2}{(1 - \xi_{eq})} \tag{17.18a}$$

and its logarithmic equivalent

$$pK_a = -\log_{10}[HA]_i - \log_{10} \frac{\xi_{eq}^2}{(1 - \xi_{eq})} \tag{17.18b}$$

Also, if we add equations (17.17b) and (17.17c)

$$[HA]_{eq} + [A^-]_{eq} = [HA]_i \, (1 - \xi_{eq}) + [HA]_i \, \xi_{eq} = [HA]_i \tag{17.19}$$

we obtain a mathematical statement of the physical reality that the quantity of the moiety A is conserved: all the A that was originally present in the undissociated acid HA, as measured by the molar concentration $[HA]_i$, exists in the equilibrium mixture in the solution in the form of either the ion A$^-$ (aq), as measured by the molar concentration $[A^-]_{eq}$, or the remaining undissociated acid, as measured by the molar concentration $[HA]_{eq}$.

17.3.3 Weak acids

In equations (17.17) and (17.18), the equilibrium extent-of-reaction parameter ξ_{eq} can take on any value from 0 to 1. As shown by equation (17.17a), large values of ξ_{eq} correspond to a higher equilibrium concentration $[H^+]_{eq}$ of hydrogen H$^+$ (aq) ions, implying that the acid HA is strong; conversely, small values of ξ_{eq} correspond to a lower equilibrium concentration $[H^+]_{eq}$ of hydrogen H$^+$ (aq) ions, implying that the acid HA is weak. If ξ_{eq} is less than, say, 0.01, then $(1 - \xi_{eq}) \approx 1$, and equation (17.18a) becomes

$$K_a \approx [HA]_i \, \xi_{eq}^2 \qquad (17.20a)$$

and

$$pK_a \approx -\log_{10}[HA]_i - 2\log_{10} \xi_{eq} \qquad (17.20b)$$

So, if $[HA]_i = 1$ mol dm^{-3}, and $\xi_{eq} \approx 0.0100$, $K_a \approx 10^{-4}$ and $pK_a \approx 4$. Furthermore, since pK_a is a property of the acid HA, and therefore a constant in equations (17.20), if the solution of the acid becomes more dilute – say, $[HA]_i = 0.1$ mol dm^{-3} – then $\xi_{eq} \approx 10^{-1.5} \approx 0.0316$, implying that the acid becomes more dissociated as the solution becomes more dilute. This is the explanation of the name for equation (17.18a) as Ostwald's law of dilution.

If we now combine equation (17.17a)

$$[H^+]_{eq} = \frac{n\xi_{eq}}{V} = [HA]_i \xi_{eq} \qquad (17.17a)$$

with equation (17.20a) expressed as

$$\xi_{eq} \approx \sqrt{\frac{K_a}{[HA]_i}}$$

then, for a weak acid

$$[H^+]_{eq} \approx \sqrt{[HA]_i K_a} \qquad (17.21a)$$

implying that the pH of the equilibrium solution is given by

$$\text{pH} \approx \frac{1}{2}pK_a - \frac{1}{2}\log_{10}[HA]_i \qquad (17.21b)$$

For the example for which $[HA]_i = 1$ mol dm^{-3}, $K_a \approx 10^{-4}$ and $pK_a \approx 4$, then $[H^+]_{eq} \approx 10^{-2}$ mol dm^{-3}, and the pH of the equilibrium solution is approximately 2; if the acid is diluted so that $[HA]_i = 0.1$ mol dm^{-3}, then $[H^+]_{eq} \approx 10^{-2.5}$ mol dm^{-3}, and the equilibrium pH becomes 2.5. This shows that diluting the acid results in a lower molar concentration $[H^+]_{eq}$, and a corresponding higher (less acidic) pH, even though, as we have seen the acid is more disassociated (when $[HA]_i = 1$ mol dm^{-3}, $\xi_{eq} \approx 0.0100$, and when $[HA]_i = 0.1$ mol dm^{-3}, $\xi_{eq} \approx 0.0316$).

At first sight, this is odd: surely, if the acid is more dissociated, then more hydrogen H$^+$ (aq) ions are produced for the same initial quantity of undissociated acid HA, and therefore the molar concentration $[H^+]_{eq}$ should *increase*, with a corresponding *decrease* in pH. In fact, the first part of this statement is true: if the acid is more dissociated (corresponding to an increase in ξ_{eq} from ~0.0100 to ~0.0316), then, yes, more hydrogen H$^+$ (aq) ions are indeed produced from the same initial quantity of the acid HA – specifically, if there is 1 mol of the undissociated acid HA, then, when $\xi_{eq} \approx 0.0100$, the solution contains 0.01 mol of hydrogen H$^+$ (aq) ions, and when $\xi_{eq} \approx 0.0316$, the solution contains 0.0316 mol of hydrogen H$^+$ (aq) ions, more than three times as many. But at the same time, the acid has been diluted by a factor of 10 from 1 mol dm^{-3} to = 0.1 mol dm^{-3}. Overall, some three times the number of hydrogen H$^+$ (aq) ions are distributed over ten times the original volume. The molar concentration $[H^+]_{eq}$ is therefore reduced by a factor of approximately 3.16/10 = 0.316. This is indeed verified by the computations that, when $[HA]_i = 1$ mol dm^{-3}, $[H^+]_{eq} \approx 10^{-2}$ mol dm^{-3}, but on dilution such that $[HA]_i = 0.1$ mol dm^{-3}, $[H^+]_{eq} \approx 10^{-2.5}$ mol dm^{-3}, from which the ratio $10^{-2.5}/10^{-2} = 10^{-0.5} = 0.316$, as predicted.

In summary, equation (17.21b) is very useful for, as we have shown, it can be used to estimate the pH of a solution of a weak acid, given values for the acid's pK_a (which can be found in reference tables), the quantity of the pure acid (expressed as n mol), and the volume V dm^3 of the solution, which together, in accordance with equation (17.5), determine $[HA]_i$ as n/V. Figure 17.1 shows the general behaviours of the equilibrium extent-of-reaction ξ_{eq}, hydrogen ion concentration $[H^+]_{eq}$, and pH, for different initial molar concentrations $[HA]_i$, for a variety of weak acids of different pK_a.

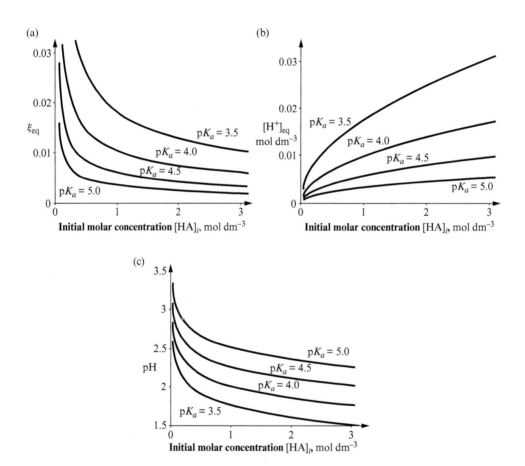

Figure 17.1 Weak acids. For the dissociation of a weak monoprotic acid HA according to the reversible reaction HA (aq) ⇌ H$^+$(aq) + HA$^-$ (aq), diagram (a) shows the behaviour of the equilibrium extent-of-reaction ξ_{eq}; diagram (b), the equilibrium hydrogen ion molar concentration $[H^+]_{eq}$; and diagram (c), the pH, as a function of the initial molar concentration $[HA]_i$, for four different values of pK_a: 3.5, 4.0, 4.5 and 5.0.

17.4 Bases

17.4.1 The base dissociation constant K_b

A **monohydroxic base** BOH releases a single hydroxyl ion OH$^-$ (aq) in solution, according to the reversible reaction

$$\text{BOH(aq)} \rightleftharpoons \text{B}^+ \text{(aq)} + \text{OH}^- \text{(aq)}$$

for which we may define the **base dissociation constant** K_b as

$$K_b = \frac{[\text{B}^+]_{eq}[\text{OH}^-]_{eq}}{[\text{BOH}]_{eq}} \tag{17.22}$$

Taking logarithms as before, we derive a variant on the Henderson-Hasselbalch equation, which applies to bases, as

$$\text{pOH} = \text{p}K_b + \log_{10} \frac{[\text{B}^+]_{eq}}{[\text{BOH}]_{eq}} \tag{17.23}$$

in which

$$\text{p}K_b = -\log_{10} K_b \tag{17.24a}$$

and

$$K_b = 10^{-\text{p}K_b} \tag{17.24b}$$

Just as pK_a is a measure of the strength or weakness of an acid (strong acids have low, even negative values of pK_a; weak acids have larger, positive values), pK_b is a measure of the strength or weakness of a base: strong bases dissociate relatively easily to produce relatively high concentrations [OH$^-$]$_{eq}$ of OH$^-$ (aq) ions, and so pK_b has a low, possibly negative, value; weak bases produce relatively low concentrations [OH$^-$]$_{eq}$ of OH$^-$ (aq) ions, and so pK_b has a higher, positive, value. Some representative values of pK_b are shown in Table 17.2.

17.4.2 Conjugate acids and bases

As we know, an acid is a molecule which, on being dissolved in pure water, makes the solution more acidic, as measured by a decrease in the pH of the solution, as compared to the value of pH = 7 for pure water. This happens because the acid molecule loses a hydrogen atom as a hydrogen ion H$^+$ (aq), as represented by the reversible reaction

$$\text{HA (aq)} \rightleftharpoons \text{H}^+ \text{(aq)} + \text{A}^- \text{(aq)}$$

As we saw on page 535, H$^+$ (aq) ions don't actually exist – H$^+$ (aq) is in fact 'shorthand' for the hydronium ion H$_3$O$^+$ (aq) – and so a more realistic description of the ionisation of the acid HA is

$$\text{HA (aq)} + \text{H}_2\text{O} \rightleftharpoons \text{H}_3\text{O}^+ \text{(aq)} + \text{A}^- \text{(aq)}$$

This reversible reaction highlights the essence of what an acid actually is, for it represents the forward reaction as the transfer of a hydrogen ion from a 'donor' molecule, HA, to an 'acceptor' molecule H$_2$O.

Table 17.2 Values of pK_b for some selected bases in aqueous solution at 298 K. The *greater* the value of pK_b, the *weaker* the base, and so the list is in order of decreasing base strength. All entities are (aq), except as explicitly noted.

	pK_b
Sodium: $2\,Na + 2\,H_2O \rightleftharpoons 2\,Na^+ + H_2\,(g) + 2\,OH^-$	−0.67*
Hydroxide ion: $OH^- + H_2O \rightleftharpoons H_2O + OH^-$	0.00
Lithium: $2\,Li + 2\,H_2O \rightleftharpoons 2\,Li^+ + H_2\,(g) + 2\,OH^-$	0.2*
Calcium: $Ca + 2\,H_2O \rightleftharpoons Ca^{2+} + H_2\,(g) + 2\,OH^-$	1.33*
Dimethylamine: $(CH_3)NH + H_2O \rightleftharpoons (CH_3)NH_2^+ + OH^-$	3.27
Methylamine: $CH_3NH_2 + H_2O \rightleftharpoons CH_3NH_3^+ + OH^-$	3.34
Ethylamine: $C_2H_5NH_2 + H_2O \rightleftharpoons C_2H_5NH_3^+ + OH^-$	3.35
Bicarbonate: $HCO_3^- + H_2O \rightleftharpoons H_2CO_3 + OH^-$	3.67
Ammonia: $NH_3\,(aq) + H_2O \rightleftharpoons NH_4^+ + OH^-$	4.75
Hydrazine: $N_2H_4 + H_2O \rightleftharpoons N_2H_5^+ + OH^-$	6.02
Pyridine: $C_5H_5N + H_2O \rightleftharpoons C_5H_5NH^+ + OH^-$	8.77
Aniline: $C_6H_5NH_2 + H_2O \rightleftharpoons C_6H_5NH_3^+ + OH^-$	9.13
Urea: $CO(NH_2)_2 + H_2O \rightleftharpoons CO(NH_2)_2\,H^+ + OH^-$	13.9

Sources: *CRC Handbook of Chemistry and Physics*, 96th edn, 2015, Section 5, pages 5-91 to 5-102, and *, approximate values from *Lange's Handbook of Chemistry*, 15th edn, 1999, McGraw-Hill Inc., Section 8, Tables 8.7.

A different situation, using ammonia as a specific example, is the reversible reaction that take place when ammonia is dissolved in water as

$$NH_3\,(aq) + H_2O \rightleftharpoons NH_4^+\,(aq) + OH^-\,(aq)$$

in which OH^- (aq) ions are created, implying that ammonia is a base. But, as this equation illustrates, in this case, the ammonia molecule is acting as the 'acceptor' of the hydrogen ion from the 'donor', water. We have already demonstrated that an acid is a hydrogen ion donor, and this example suggests that a base is a hydrogen ion acceptor. Furthermore, since every 'donation' requires 'acceptance', any reaction which involves an acid as the hydrogen ion donor necessarily involves a base as the corresponding hydrogen ion acceptor; likewise, every reaction which involves a base as a hydrogen ion acceptor necessarily involves an acid as the corresponding hydrogen ion donor – a concept known as the **Brønsted-Lowry theory**, after the Danish chemist, Johannes Brønsted, and the English chemist, Thomas Lowry, who independently formulated the theory in 1923. So for the forward reaction

$$HA\,(aq) + H_2O \rightleftharpoons H_3O^+\,(aq) + A^-\,(aq)$$

HA (aq) is the acid, and H_2O is the base, as is familiar. This reaction, however, is reversible, the reverse reaction being the transfer of a hydrogen ion from H_3O^+ (aq) to A^- (aq) – a reaction in which H_3O^+ (aq) acts as an acid, and A^- (aq) as the base. Similarly, for the forward reaction

$$NH_3\,(aq) + H_2O \rightleftharpoons NH_4^+\,(aq) + OH^-\,(aq)$$

NH$_3$ (aq) is the base, and H$_2$O the acid; and for the reverse reaction, NH$_4^+$ (aq) is the acid, and OH$^-$ (aq) the base. These two examples illustrate that water can act either as a base (in relation to HA), or as an acid (in relation to, for example NH$_3$): to describe this dual property, water is known as an **amphoteric** molecule.

There is a further feature of each of these reactions which merits attention: as we have seen, for the reversible reaction

$$\text{HA (aq)} + \text{H}_2\text{O} \rightleftharpoons \text{H}_3\text{O}^+ \text{(aq)} + \text{A}^- \text{(aq)}$$

HA (aq) is the acid in the forward reaction, whilst A$^-$ (aq) acts as the base in the reverse reaction; for the reversible reaction

$$\text{NH}_3 \text{(aq)} + \text{H}_2\text{O} \rightleftharpoons \text{NH}_4^+ \text{(aq)} + \text{OH}^- \text{(aq)}$$

NH$_3$ (aq) is the base for the forward reaction, whilst NH$_4^+$ (aq) acts as the acid in the reverse reaction. HA (aq) and A$^-$ (aq), and NH$_4^+$ (aq) and NH$_3$ (aq), are known as **conjugate pairs**, HA (aq) and NH$_4^+$ (aq) being the **conjugate acids** to A$^-$ (aq) and NH$_3$ (aq) as the **conjugate bases** respectively.

Furthermore, for the forward reaction

$$\text{HA (aq)} + \text{H}_2\text{O} \rightleftharpoons \text{H}_3\text{O}^+ \text{(aq)} + \text{A}^- \text{(aq)}$$

we know that

$$K_a = \frac{[\text{H}_3\text{O}^+]_{eq}[\text{A}^-]_{eq}}{[\text{HA}]_{eq}[\text{H}_2\text{O}]_{eq}} \tag{17.12}$$

For the conjugate base A$^-$ (aq), we may express the action of A$^-$ (aq) as a base by the reaction

$$\text{A}^- \text{(aq)} + \text{H}_2\text{O} \rightleftharpoons \text{HA (aq)} + \text{OH}^- \text{(aq)}$$

for which, using equation (17.22), we may write

$$K_b = \frac{[\text{HA}]_{eq}[\text{OH}^-]_{eq}}{[\text{A}^-]_{eq}[\text{H}_2\text{O}]_{eq}} \tag{17.22}$$

Multiplying the expressions for K_a, for the conjugate acid HA, and K_b, for the conjugate base A$^-$ (aq), together

$$K_a K_b = \frac{[\text{H}_3\text{O}^+]_{eq}[\text{A}^-]_{eq}}{[\text{HA}]_{eq}[\text{H}_2\text{O}]_{eq}} \frac{[\text{HA}]_{eq}[\text{OH}^-]_{eq}}{[\text{A}^-]_{eq}[\text{H}_2\text{O}]_{eq}} = \frac{[\text{H}_3\text{O}^+]_{eq}[\text{OH}^-]_{eq}}{[\text{H}_2\text{O}]_{eq}^2}$$

As we saw on page 537, [H$_2$O]$_{eq}$ is in fact [H$_2$O]$_{eq}$/[H$_2$O]$^{\ominus}$ = 1, and so, using equation (17.8)

$$K_a K_b = [\text{H}_3\text{O}^+]_{eq}[\text{OH}^-]_{eq} = K_w = 10^{-14} \tag{17.25a}$$

and

$$pK_a + pK_b = pK_w = 14 \tag{17.25b}$$

showing that the sum of pK_a (for any conjugate acid) and pK_b (for the corresponding conjugate base) is a constant, 14.

For the particular case of pure water

$$\text{H}_2\text{O} + \text{H}_2\text{O} \rightleftharpoons \text{H}_3\text{O}^+ \text{(aq)} + \text{OH}^- \text{(aq)}$$

molecular H_2O acts as the conjugate acid to the OH^- (aq) ion as the conjugate base. As we saw on page 540, pK_a for molecular water H_2O is 14, and so, according to equation (17.25b), pK_b for conjugate base, the OH^- (aq) ion, is 0. This may be verified by expressing the action of the OH^- (aq) ion as a base by the somewhat odd-looking reversible reaction

$$OH^-(aq) + H_2O \rightleftharpoons H_2O + OH^-(aq)$$

in which the H_2O molecule on the left acts as a hydrogen ion donor to the hydrogen ion acceptor OH^- (aq) on the left, which is therefore a base. Although this equation appears rather strange, it is correct, as verified by its equivalence to the reversible reaction for a generalised base A^- (aq)

$$A^-(aq) + H_2O \rightleftharpoons HA(aq) + OH^-(aq)$$

For the reaction OH^- (aq) $+ H_2O \rightleftharpoons H_2O + OH^-$ (aq),

$$K_b = \frac{[H_2O]_{eq}[OH^-]_{eq}}{[OH^-]_{eq}[H_2O]_{eq}} = 1$$

verifying that $pK_b = -\log_{10} K_b = 0$ for the base OH^- (aq).

Equation (17.25b)

$$pK_a + pK_b = pK_w = 14 \tag{17.25b}$$

has one, further, important implication. For any reversible reaction of the form

$$HA(aq) + H_2O \rightleftharpoons H_3O^+(aq) + A^-(aq)$$

HA (aq) is the conjugate acid, with an associated value of pK_a, and A^- (aq) is the conjugate base, with an associated value of pK_b, where the sum $pK_a + pK_b$ is a constant, 14. The entities HA (aq) and A^- (aq) are defined by the stoichiometry of the reaction, and if either one is known, the other is determined; likewise, if either the value of either pK_a or pK_b is known, the other is determined too. This implies that the entire system is determined by the stoichiometry of the reaction, and by the value of pK_a. Most standard reference tables therefore list only values of pK_a, usually over a range from about 0 to 14, leaving it to the user to compute $pK_b = 14 - pK_a$ as needed.

17.5 The Henderson-Hasselbalch approximation

As we saw on page 541, the Henderson-Hasselbalch equation

$$pH = pK_a + \log_{10} \frac{[A^-]_{eq}}{[HA]_{eq}} \tag{17.16}$$

is a general equation, which applies to all acids HA which ionise as

$$HA(aq) \rightleftharpoons H^+(aq) + A^-(aq)$$

to form a solution which can be approximated as ideal. If the initial concentration of the undissociated acid is $[HA]_i$, then, as we saw in connection with equation (17.19), since the quantity of the moiety A must be conserved

$$[HA]_i = [HA]_{eq} + [A^-]_{eq} \tag{17.19}$$

We also know that

$$[A^-]_{eq} = \frac{n\xi_{eq}}{V} \tag{17.17b}$$

and

$$[HA]_{eq} = \frac{n(1-\xi_{eq})}{V} \tag{17.17c}$$

where ξ_{eq} represents the fraction of the HA molecules dissociated in the equilibrium mixture. For a weak acid, ξ_{eq} is small, typically 0.01, implying that $(1 - \xi_{eq}) \approx 1$, and therefore that $[HA]_{eq} \gg [A^-]_{eq}$. Accordingly, since $[HA]_i = [HA]_{eq} + [A^-]_{eq}$, for a weak acid

$$[HA]_i \approx [HA]_{eq}$$

and so the Henderson-Hasselbalch equation becomes the **Henderson-Hasselbalch approximation**

$$pH \approx pK_a + \log_{10} \frac{[A^-]_{eq}}{[HA]_i} \tag{17.26}$$

The difference between the Henderson-Hasselbalch equation, equation (17.16), which contains the term $[HA]_{eq}$, and the approximation, equation (17.26), in which the corresponding term is $[HA]_i$, is subtle; in practice, however, it is important. Experimentally, $[HA]_{eq}$ is more difficult to measure than $[HA]_i$, for the amount of pure acid HA is usually known at the outset, and there is one particular experimental context in which this is of especial significance – the use of **buffer solutions**.

17.6 Buffer solutions

A buffer solution is an aqueous solution which tends to maintain a near-constant pH, even whilst a reaction is taking place which produces hydrogen ions H⁺ (aq) – so making the solution more acid and reducing the pH – or consuming hydrogen ions H⁺ (aq) – so making the solution more alkaline and increasing the pH. Physically, the solution is able either to convert hydrogen ions into some other form as soon as they are produced, or to act as a source of hydrogen ions as they consumed elsewhere. As a result, the molar concentration [H⁺] of free hydrogen ions in the solution remains reasonably constant, so maintaining the pH.

Biologically, buffer solutions are critical to the maintenance of life. Many enzymes are very sensitive to pH, and function correctly only over narrow ranges of pH – much cell metabolism takes place around pH 7, but some reactions take place at other pH values, notably mammalian digestion which requires acidic conditions. Living systems have evolved many highly sophisticated methods of holding a local environment at the required pH, even whilst the biochemical reactions that are taking place produce, or consume, hydrogen ions.

In the laboratory, buffer solutions can be formulated which maintain almost any particular pH, as required for the experiment of interest. Buffer solutions which maintain an acidic pH < 7 comprise a solution, of total volume V dm³, in pure water as the solvent, of a mixture of

- a known initial quantity, n_a mol, of a weak acid HA (such as acetic acid, CH_3COOH), and
- a known initial quantity, n_s mol, of a salt XA of that acid (such as sodium acetate CH_3COONa), where X represents a monovalent metallic element such as sodium.

Similarly, a buffer solution that maintains an alkaline pH > 7 comprises a solution, of total volume V dm³, in pure water as the solvent, of a mixture of

- a known initial quantity, n_a mol, of a weak base XOH (such as ammonia NH_3 dissolved in water, so forming NH_4OH), and
- a known initial quantity, n_s mol, of a salt XA of that weak base, such as ammonium chloride (NH_4Cl).

The analysis of acidic buffers and alkaline buffers is identical, and so, for clarity, the following discussion will refer to a weak acid-salt mixture: all of the equations, and inferences, apply equally as well to a weak base-salt mixture.

For a buffer solution formed from a weak acid HA and a corresponding salt XA, the salt XA ionises completely, and so the reaction

$$XA\ (aq) \rightarrow X^+\ (aq) + A^-\ (aq)$$

goes to completion to the right (hence the use of the symbol $\rightarrow$ rather than $\rightleftharpoons$). The molar concentration $[A^-]_s$ of A^- ions, attributable to the dissociation of the salt XA, is given by

$$[A^-]_s = \frac{n_s}{V} = [XA]_i \qquad (17.27)$$

where $[XA]_i = n_s/V$ is the instantaneous initial molar concentration of the salt XA, which, though hypothetical (in that the molecule XA dissociates totally on dissolving in the solution), is easily computed since both n_s and V are known.

The weak acid HA also ionises as

$$HA\ (aq) \rightleftharpoons H^+\ (aq) + A^-\ (aq)$$

but since the acid is weak, the equilibrium lies towards the left. Using equations (17.17), the corresponding molar concentrations, expressed in terms of the equilibrium extent-of-reaction parameter ξ_{eq}, are

$$[H^+]_a = \frac{n_a \xi_{eq}}{V} = [HA]_i\, \xi_{eq} \qquad (17.28a)$$

$$[A^-]_a = \frac{n_a \xi_{eq}}{V} = [HA]_i\, \xi_{eq} \qquad (17.28b)$$

and

$$[HA] = \frac{n_a(1-\xi_{eq})}{V} = [HA]_i\,(1-\xi_{eq}) \qquad (17.28c)$$

In these equations, $[HA]_i = n_a/V$, the instantaneous, hypothetical, but readily calculated initial molar concentration of the undissociated acid HA; $[H^+]_a$ and $[A^-]_a$ represent the equilibrium molar concentrations of H^+ (aq) and A^- (aq) ions attributable to the dissociation of the weak acid; and [HA] is the equilibrium molar concentration of the acid HA molecule, after dissociation.

As we have seen, on dissociation, the weak acid HA contributes hydrogen ions H⁺ (aq) to the solution; in addition, H⁺ (aq) ions are also contributed by the ionisation of water, and may be contributed, or consumed, by any other reactions that are taking place in the solution. For the moment, we will assume that both of these contributions are negligible – for the ionisation of water, for example, $[H^+] \approx 10^{-7}$ mol dm⁻³, compared (see page 543) to values of $[H^+] \approx 10^{-2}$ mol dm⁻³ or 10^{-3} mol dm⁻³ for a weak acid of $pK_a = 4$.

Within the solution, the mixture of ions equilibrates according to the fundamental definition of K_a

$$K_a = \frac{[H^+]_{eq}[A^-]_{eq}}{[HA]_{eq}} \tag{17.12}$$

where, for the buffer solution mixture (ignoring the ionisation of water and any other reactions), we may use equations (17.18a), (17.28b) and (17.27) to give

$$[H^+]_{eq} = [H^+]_a = [HA]_i \xi_{eq} \tag{17.29a}$$

$$[A^-]_{eq} = [A^-]_a + [A^-]_s = [HA]_i \xi_{eq} + [XA]_i \tag{17.29b}$$

and

$$[HA]_{eq} = [HA]_i (1 - \xi_{eq}) \tag{17.29c}$$

Accordingly,

$$K_a = \frac{[HA]_i \xi_{eq}([HA]_i \xi_{eq} + [XA]_i)}{[HA]_i(1 - \xi_{eq})} \tag{17.30}$$

Since the acid HA is, by assumption, weak, $\xi_{eq} \ll 1$, and it is always possible to ensure that the initial concentrations $[HA]_i$ and $[XA]_i$ are such that $[HA]_i \xi_{eq} \ll [XA]_i$. Accordingly, equation (17.26) can be approximated as

$$K_a \approx \frac{[HA]_i \xi_{eq} [XA]_i}{[HA]_i}$$

Assuming that the acid HA is the only source of hydrogen H⁺ (aq) ions, then according to equation (17.29a), the actual hydrogen ion molar concentration $[H^+] = [HA]_i \xi_{eq}$, and so

$$K_a \approx \frac{[H^+][XA]_i}{[HA]_i} \tag{17.31}$$

and in logarithmic form

$$\log_{10} K_a \approx \log_{10}[H^+] + \log_{10} \frac{[XA]_i}{[HA]_i}$$

giving

$$pK_a \approx pH - \log_{10} \frac{[XA]_i}{[HA]_i}$$

and

$$pH \approx pK_a + \log_{10} \frac{[XA]_i}{[HA]_i} \tag{17.32}$$

17.7 Why buffer solutions maintain constant pH

Equation (17.32) is a special case of the Henderson-Hasselbalch approximation, equation (17.26), as applied to buffer solutions, and offers the explanation as to why buffer solutions can maintain a constant pH. In equation (17.32), the term on the left-hand side is the pH of the buffer solution. On the right-hand side, pK_a is a property of the weak acid HA, and, at any given temperature, a constant; the term $[XA]_i = n_s/V$ is determined by the number n_s of moles of the salt XA dissolved in the volume V of the solution, and so is also a constant for a buffer solution of any defined composition; finally, the term $[HA]_i = n_a/V$ is determined by the number n_a of moles of the weak acid XA dissolved in the volume V of the solution, and so, for a buffer solution of any defined composition, is a constant too. Once the weak acid HA, a corresponding salt XA, and their molar quantities n_a and n_s have been determined – and all these can be chosen for any particular experiment – then all the terms on the right-hand side of equation (17.28) are constant, and so the pH of the buffer solution is constant, even if other reactions – such as enzyme-catalysed biochemical reactions – are taking place within the solution that produce or consume hydrogen H^+ (aq) ions.

In practice, if a buffer is formed from a weak acid HA associated with a given pK_a, and if the quantities of the acid HA and salt XA are chosen such that the molar concentrations $[XA]_i$ and $[HA]_i$ are equal, then equation (17.28) implies that the buffer solution will maintain the pH such that $pH = pK_a$; if the quantities are such that the ratio $[XA]_i/[HA]_i = 100$, then the buffer solution $pH = pK_a + 2$; if $[XA]_i/[HA]_i = 0.1$, then $pH = pK_a - 1$. In principle, equation (17.32) can be used to identify the ratio $[XA]_i/[HA]_i$ required to formulate a buffer maintaining any required pH from any HA, XA mixture. In practice, however, buffer solutions are most effective at controlling the pH when the required pH is reasonably close to the pK_a of the weak acid HA – as we shall see shortly, the discrepancy between theory and practice is attributable to the validity (or otherwise) of the assumptions associated with equation (17.32), which is (as is explicitly acknowledged) an approximation rather than a true equality.

The ability of a buffer solution to maintain a constant pH can be explained by considering what happens chemically in the solution as a reaction generates hydrogen ions. Because the buffer solution contains a large excess of A^- (aq) ions, attributable to the ionisation of the salt XA, as soon as any H^+ (aq) ions are produced, they are immediately converted into HA (aq) molecules according to the reversible reaction

$$H^+ \text{ (aq)} + A^- \text{ (aq)} \rightleftharpoons HA \text{ (aq)}$$

the equilibrium of which, since the acid XA is weak, is strongly to the right (when written as shown). As long as there are more than enough A^- (aq) ions present in the solution to react with all the H^+ (aq) ions produced by another reaction, then all these H^+ (aq) ions are transformed in HA (aq) molecules. But since the pH is determined only by hydrogen in the form of H^+ (aq) ions, rather than bound as HA (aq) molecules, the pH of the solution remains constant.

Suppose that another, perhaps enzymic, reaction is taking place in the solution that consumes H^+ (aq) ions. This displaces the weak acid reversible reaction

$$HA \text{ (aq)} \rightleftharpoons H^+ \text{ (aq)} + A^- \text{ (aq)}$$

to the right. The molecular form of the acid, HA (aq), however, is present in excess, and so HA (aq) molecules can then dissociate, so as to maintain equilibrium, with the result that the H^+ (aq) ions consumed by the enzymic reaction are replaced by the ionisation of HA (aq). Once again, the quantity of H^+ (aq) ions is maintained constant, as is the pH.

Buffer solutions therefore require the presence, simultaneously, of a means of both 'trapping' newly produced H^+ (aq) ions in molecular form, and also providing H^+ (aq) ions, as required, from a suitable source. In practice, this is achieved by making a solution comprising a mixture of a weak acid HA, and a salt of that acid, XA, with A^- (aq) ions contributed primarily by the salt XA as the H^+ (aq) ion 'trap', and the weak acid HA being the source. Furthermore, for the buffer solution to be effective, the quantities of both the weak acid HA, and the salt XA, need to be in excess of the quantities of H^+ (aq) ions likely to be produced or consumed. This last point explains why, for example, water is not an effective buffer: in principle, water is an acid, since it creates H^+ (aq) ions on dissociation, and since $pK_a = 14$, this acid is very weak indeed. Pure water therefore satisfies the first condition for a buffer solution, the presence of a weak acid, but it fails on the second: pure water contains no equivalent of the salt XA, and so does not have an excess of A^- (aq) ions, which, in this case, are OH^- (aq) ions.

17.8 How approximate is the Henderson-Hasselbalch approximation?

Experimentally, scientists need to create buffer solutions which have whatever pH is appropriate to any particular study, and to have confidence that the required pH will be maintained constant throughout the experiment, as hydrogen ions are produced or consumed by the reaction taking place within the buffer. The ability of a buffer to keep the pH constant is described by a property of the buffer solution known as the **buffer capacity**, and will discussed in the next section; in this section, we address the issue of how we can 'design' a buffer solution with the 'right' pH.

Suppose, for example, that we wish to create a buffer at pH 2. Equation (17.32)

$$pH \approx pK_a + \log_{10} \frac{[XA]_i}{[HA]_i} \tag{17.32}$$

suggests that it is possible to choose any weak acid HA, with a given pK_a, and a corresponding salt XA, and then choose the initial quantities $[HA]_i$ and $[XA]_i$ such that the resulting pH has the required value. So, if we choose a weak acid with $pK_a = 6$, and initial molar concentrations of $[XA]_i = 10^{-3}$ mol dm^{-3} and $[HA]_i = 10$ mol dm^{-3}, then $\log_{10}([XA]_i/[HA]_i) = -4$, and equation (17.32) predicts that the resulting pH will be $6 - 4 = 2$ as required.

In fact, equation (17.32) is an approximation, and so we should not expect that the pH as calculated from equation (17.32) will always correspond to the actual pH of a buffer solution formulated from a given choice of pK_a, $[HA]_i$ and $[XA]_i$,

The validity of equation (17.32) has been extensively examined, and, in general, the theoretical value of the pH as calculated from equation (17.32) is found to be a very good approximation (within about 1%) to the actual pH when

- the pK_a of the weak acid HA is between about 5 and 9, and
- the ratio $[XA]_i/[HA]_i$ is between about 0.1 and 10, and
- the total molar concentration of the weak acid and the corresponding salt (as measured by $[HA]_i + [XA]_i$) is neither very small, say, less than about 10^{-5} mol dm^{-3} (implying that the solution is very dilute, in which case the ionisation of the solvent, water, is significant), nor very large, say, greater than about 2 mol dm^{-3} (implying that the solution is far from ideal).

As a given buffer progressively deviates from these guidelines, there is an increasing discrepancy between the theoretical pH and the actual pH. Experimentally, it is therefore preferable to formulate a buffer from a weak acid with a pK_a within ± 1 of the desired pH, thereby ensuring that the ratio $[XA]_i/[HA]_i$ is between 0.1 and 10. In practice, this is usually possible, given the wide range of buffer materials available, just a few of which are shown in Table 17.3.

Table 17.3 Some materials suitable for buffer solutions at 298.15 K and 1 bar pressure

	pK_a
Acetic acid: $CH_3COOH \rightleftharpoons H^+ + CH_3COO^-$	4.756
Piperazine (1): $C_4H_{10}N_2.H_2^{2+} \rightleftharpoons H^+ + C_4H_{10}N_2.H^+$	5.333
MES: $(C_4H_8O)NH(C_2H_4SO_3) \rightleftharpoons H^+ + (C_4H_8O)N(C_2H_4SO_3)^-$	6.270
BES: $(C_2H_4OH)_2NH(C_2H_4SO_3) \rightleftharpoons H^+ + (C_2H_4OH)_2N(C_2H_4SO_3)^-$	7.187
Tris: $NH_2C(CH_2OH)_3.H^+ \rightleftharpoons H^+ + NH_2C(CH_2OH)_3$	8.072
Boric acid: $H_3BO_3 \rightleftharpoons H^+ + H_2BO_2^-$	9.237
Piperazine (2): $C_4H_{10}N_2.H^+ \rightleftharpoons H^+ + C_4H_{10}N_2$	9.731
CAPS: $(C_6H_{11})NH(C_3H_6SO_3H) \rightleftharpoons H^+ + (C_6H_{11})NH(C_3H_6SO_3)^-$	10.499

Source: *CRC Handbook of Chemistry and Physics*, 96th edn, 2015, Section 7, pages 7–26 to 7–28.

17.9 Buffer capacity

The key property of a buffer solution is its ability to maintain a stable pH in the presence of a reaction which produces or consumes hydrogen H^+ (aq) ions, or produces or consumes hydroxyl OH^- (aq) ions. If we wish to measure this property, we need to bring about a change which results in the production or consumption of H^+ (aq) or OH^- (aq) ions, and then measure the corresponding change in pH. In principle, the corresponding change in pH should be zero, for the purpose of the buffer is to maintain the pH constant. In practice, however, this ideal might not be achieved, and so the answer to the question "By how much does the pH of a real buffer solution change when H^+ (aq) or OH^- (aq) ions are produced or consumed?" is a good way of assessing how well a given buffer solution is achieving its intended purpose.

Suppose that a reaction is taking place in the buffer solution that creates a small number δn_H of H^+ (aq) ions, or that a source of δn_H H^+ (aq) ions is added to the solution (without materially changing the volume V dm^{-3} of the solution). If the solution were not acting as a buffer, the addition of δn_H H$^+$ (aq) ions would change the pH of the solution by a small amount $(\delta\text{pH})_{b-} = -\log_{10}(\delta n_H/V)$, where the negative sign of the log term recognises that an increase δn_H in H$^+$ (aq) ions makes the solution more acidic, so lowering the pH, and the subscript $_{b-}$ acts as a reminder that the δn_H H$^+$ (aq) ions are being added to a solution which is not acting as a buffer.

For a buffer solution comprising a mixture of the weak acid HA and a corresponding salt XA, however, many of the additional δn_H H$^+$ (aq) ions will react with the A$^-$ (aq) ions to produce undissociated HA (aq). As a consequence, only a fraction (say, α) of the δn_H H$^+$ (aq) ions originally added remain as H$^+$ (aq) ions, and so the actual change δpH in the buffer solution's pH is given by $(\delta\text{pH})_{b+} = -\log_{10}(\alpha \delta n_H/V) = -\log_{10}(\delta n_H/V) - \log_{10}\alpha$, where the subscript $_{b+}$ indicates 'in the presence of a buffer'. The term $-\log_{10}(\delta n_H/V)$ is the change $(\delta\text{pH})_{b-}$ in pH that would have happened had the solution not been a buffer, and so the change $(\delta\text{pH})_{b+}$ in pH for the buffer solution is therefore given by $(\delta\text{ pH})_{b+} = (\delta\text{ pH})_{b-} - \log_{10}\alpha$. Since the fraction α is necessarily positive, $(\delta\text{ pH})_{b+}$ is always less than $(\delta\text{ pH})_{b-}$: for a given quantity of H$^+$ (aq) ions added, the consequent change in pH for a buffer solution is always less than the change in pH for a similar solution that is not a buffer.

The conventional measure of a buffer solution's ability to resist changes in pH is known as the **buffer capacity** β, defined as

$$\beta = -\frac{1}{V}\frac{dn_H}{d\text{pH}} \tag{17.33a}$$

In equation (17.33a), dn_H represents an infinitesimal quantity of H$^+$ (aq) ions added to a solution of constant volume V, and d pH represents the corresponding change in the solution's pH. Also, as we have seen, when H$^+$ (aq) ions are added (implying that dn_H is positive), the solution becomes more acidic, and so d pH is negative: the negative sign in equation (17.33a) therefore ensures that the buffer capacity β, the units of which are m^{-3} or dm^{-3}, is positive.

An alternative definition of the buffer capacity is expressed in terms of the quantity dn_{OH} of hydroxyl OH$^-$ (aq) ions that might be added, if the system is made more alkaline rather than more acidic

$$\beta = +\frac{1}{V}\frac{dn_{OH}}{d\text{pH}} \tag{17.33b}$$

Equation (17.33b) has a positive sign, for when OH$^-$ (aq) ions are added (dn_{OH} is positive), the system becomes more alkaline, implying that d pH is positive. Equations (17.33a) and (17.33b) therefore ensure that the buffer capacity β is always positive.

A large value of β implies that a large change dn_H or dn_{OH} is required to bring about a given change d pH (say, 1 pH unit) in the system's pH, implying that the system is resistant to changes in pH; conversely, a small value of β implies that only a small change dn_H or dn_{OH} is required to bring about a change in the system's pH. The magnitude of β is therefore a measure of the system's ability to maintain the system's pH constant, hence the term 'buffer capacity'.

To determine an expression for β, consider a buffer solution of total volume V dm^{-3}, comprising n_a mol of a weak acid HA and n_s mol of a corresponding salt XA dissolved in pure water. As we have seen, we may define the (hypothetical) initial concentrations $[HA]_i$ and $[XA]_i$ of the weak acid and salt, respectively, as

$$[HA]_i = \frac{n_a}{V} \tag{17.5}$$

and

$$[XA]_i = \frac{n_s}{V} \tag{17.27}$$

so enabling us to define the total initial concentration $[Buf]_{tot}$ of the buffer solution as

$$[Buf]_{tot} = [HA]_i + [XA]_i = \frac{n_a}{V} + \frac{n_s}{V} = \frac{1}{V}(n_a + n_s) \tag{17.34}$$

We will assume that the system is in equilibrium, and so all molar concentrations $[\ldots]$ are the equilibrium molar concentrations $[\ldots]_{eq}$. The salt XA is totally ionised

$$XA(aq) \rightarrow X^+ (aq) + A^- (aq)$$

and so

$$[X^+] = [A^-] = [XA]_i \tag{17.35}$$

whilst the weak acid HA is only partially ionised

$$HA\,(aq) \rightleftharpoons H^+\,(aq) + A^-\,(aq)$$

such that

$$K_a \approx \frac{[H^+][A^-]}{[HA]_i} \tag{17.12}$$

The water also ionises

$$H_2O \rightleftharpoons H^+\,(aq) + OH^-\,(aq)$$

for which

$$K_w = [H^+][OH^-] = 10^{-14} \tag{17.8}$$

Suppose we now add n_H mol of some substance HQ which ionises totally as

$$HQ\,(aq) \rightarrow H^+\,(aq) + Q^-\,(aq)$$

HQ might be a strong acid which is added to the solution without changing the volume V; alternatively, HQ might represent the substrate of an enzyme catalysed reaction, taking place within the buffer solution, and acting as a source of H^+ (aq) ions. In this analysis, the purpose of HQ is to act as a source of additional H^+ (aq) ions within the buffer solution, so acting to make the buffer solution more acidic – an action which the buffer solution will seek to resist.

Since HQ is the only source of Q^- (aq) ions,

$$[Q^-] = [HQ]_i = \frac{n_H}{V} \tag{17.36}$$

Within the buffer solution, all these events – the total ionisation of XA and of HQ, and the partial ionisation of H_2O and of HA – all happen simultaneously, and all mutually equilibrate, such that the resulting value of $[H^+]$ determines the solution's pH. To determine what this pH actually is, we note that, since the equilibrium mixture is electrically neutral, the total quantity of positive charge must equal the total quantity of negative charge, and so

$$[H^+] + [X^+] = [OH^-] + [Q^-] + [A^-] \qquad (17.37)$$

We now make three changes to equation (17.37): using equation (17.35), we replace $[X^+]$ by $[XA]_i$; we use equation (17.8) to replace $[OH^-]$ by $K_w/[H^+]$; and we use equation (17.36) to replace $[Q^-]$ by n_H/V, giving

$$[H^+] + [XA]_i = \frac{K_w}{[H^+]} + \frac{n_H}{V} + [A^-] \qquad (17.38)$$

To develop equation (17.38) further, we now take into account the fact that all the atoms of the moiety A must be conserved: all the A atoms originally entering the solution from both the weak acid HA and the salt XA are either transformed into A^- (aq) ions, or remain within undissociated molecules of HA (aq), and so

$$[HA]_i + [XA]_i = [HA] + [A^-] = [Buf]_{tot} \qquad (17.39)$$

in which $[Buf]_{tot}$ was defined in equation (17.34). Accordingly,

$$[HA] = [Buf]_{tot} - [A^-]$$

Since

$$K_a = \frac{[H^+][A^-]}{[HA]} \qquad (17.12)$$

then

$$K_a = \frac{[H^+][A^-]}{[Buf]_{tot} - [A^-]} \qquad (17.40)$$

Equation (17.40) is an accurate expression for K_a, and avoids the simplifications associated with the buffer version of the Henderson-Hasselbalch approximation, equation (17.32), which sets $[HA] \approx [HA]_i$, and $[A^-] \approx [XA]_i$. Furthermore, if we rearrange equation (17.40) as

$$[A^-] = \frac{K_a[Buf]_{tot}}{K_a + [H^+]} \qquad (17.41)$$

then we may use this expression to replace $[A^-]$ in equation (17.38) as

$$[H^+] + [XA]_i = \frac{K_w}{[H^+]} + \frac{n_H}{V} + \frac{K_a[Buf]_{tot}}{K_a + [H^+]}$$

from which

$$\frac{n_H}{V} = [H^+] + [XA]_i - \frac{K_w}{[H^+]} - \frac{K_a[Buf]_{tot}}{K_a + [H^+]} \qquad (17.42)$$

In equation (17.42), the left-hand side refers to n_H, the number of moles of H^+ (aq) ions added to the buffer solution (see page 556 and equation (17.36)), and the volume V is a constant. On the right-hand side, the parameter K_w is a constant property of water, equal to 10^{-14}, and, for a buffer of defined composition, the parameters K_a, $[XA]_i$, and $[Buf]_{tot} = [HA]_i + [XA]_i$ are all constants too. The only variable on the right-hand side is therefore the molar concentration $[H^+]$.

ACIDS, BASES AND BUFFER SOLUTIONS

Accordingly, if we differentiate equation (17.42) with respect to $[H^+]$, we have

$$\frac{1}{V}\frac{dn_H}{d[H^+]} = 1 + \frac{K_w}{[H^+]^2} + \frac{K_a[\text{Buf}]_{tot}}{(K_a + [H^+])^2} \tag{17.43}$$

Now, mathematically

$$\frac{dn_H}{d[H^+]} = \left(\frac{dn_H}{d\,pH}\right)\left(\frac{d\,pH}{d[H^+]}\right)$$

and since

$$pH = -\log_{10}[H^+] = -\frac{1}{2.303}\ln[H^+] \tag{17.9}$$

then

$$\frac{d\,pH}{d[H^+]} = -\frac{1}{2.303}\frac{1}{[H^+]}$$

giving

$$\frac{dn_H}{d[H^+]} = -\frac{dn_H}{d\,pH}\frac{1}{2.303}\frac{1}{[H^+]}$$

Substituting this expression in equation (17.43), we have

$$-\frac{1}{V}\frac{dn_H}{d\,pH}\frac{1}{2.303}\frac{1}{[H^+]} = 1 + \frac{K_w}{[H^+]^2} + \frac{K_a[\text{Buf}]_{tot}}{(K_a + [H^+])^2}$$

and so

$$-\frac{1}{V}\frac{dn_H}{d\,pH} = 2.303\left([H^+] + \frac{K_w}{[H^+]} + \frac{K_a[\text{Buf}]_{tot}[H^+]}{(K_a + [H^+])^2}\right)$$

But since, according to the definition of the buffer capacity β

$$\beta = -\frac{1}{V}\frac{dn_H}{d\,pH} \tag{17.33a}$$

we therefore have

$$\beta = 2.303\left([H^+] + \frac{K_w}{[H^+]} + \frac{K_a[\text{Buf}]_{tot}[H^+]}{(K_a + [H^+])^2}\right) \tag{17.44}$$

Equation (17.44) is the result we have been seeking: an expression relating the buffer capacity β to the molar concentration $[H^+]$, and hence the buffer solution's pH. It's not intuitively obvious how equation (17.44) behaves, and so Figure 17.2 should help.

Figure 17.2 shows the behaviour of equation (17.44) for a buffer solution comprising 1.0 mol of a weak acid HA of pK_a 6, and 1.0 mol of a corresponding salt XA. As illustrated in Figure 17.2(a), the buffer capacity β peaks at β_{max} when $pH = 6 = pK_a$, showing that the buffer solution is most effective at maintaining a constant pH around the pK_a of the weak acid within the buffer solution. When $\beta = \beta_{max}$, $pH = 6$, and so, at this point, $[H^+] = 10^{-6}$ mol dm^{-3} = K_a, and so, from equation (17.44) we have

$$\beta_{max} = 2.303\left(K_a + \frac{K_w}{K_a} + \frac{K_a[\text{Buf}]_{tot}K_a}{(K_a + K_a)^2}\right) = 2.303\left(K_a + \frac{K_w}{K_a} + \frac{[\text{Buf}]_{tot}}{4}\right)$$

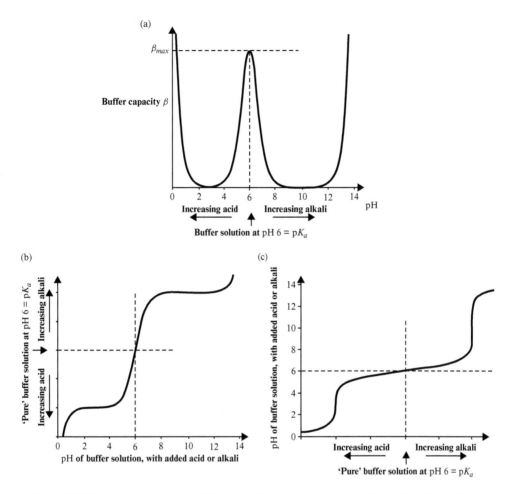

Figure 17.2 Buffer solutions and the buffer capacity. These graphs present three different illustrations of the key property of a buffer solution – the ability to maintain a constant **pH**. The buffer solution is the same in each case, comprising 1.0 mol of a weak acid **HA** of pK_a 6 and 1.0 mol of a corresponding salt **XA**. Graph (a) shows the buffer capacity β as defined by equation (17.44), the peak at **pH** 6 indicating that the buffer is maximally effective when **pH** = pK_a. Graphs (b) and (c) show how the **pH** of the buffer solution varies as more acid, or more alkali, is added to a solution originally at **pH** = pK_a = 6.

In this equation, $K_w = 10^{-14}$, and is a constant for all aqueous buffer solutions; K_a depends on the choice of the weak acid HA, but is typically in the range 10^{-3} to 10^{-9}, implying that the ratio K_w/K_a will be in the range 10^{-11} to 10^{-5}. The sum of the two terms $K_w/K_a + K_a$ is therefore in the range 10^{-3} to 10^{-5}, a number much smaller than 1. The quantity $[\text{Buf}]_{\text{tot}} = [\text{HA}]_i + [\text{XA}]_i$, however, is the sum of the initial molar concentrations of the weak acid HA and the salt XA which constitute the buffer solution, and typically has a value in the range $\sim$0.1 to $\sim$1 or 2 mol dm^{-3}, and certainly much greater than 10^{-3}. So, to a good approximation

$$\beta_{\max} \approx \frac{2.303}{4}[\text{Buf}]_{\text{tot}} = 0.576\,[\text{Buf}]_{\text{tot}} \qquad (17.45)$$

showing that the maximum buffer capacity β_{max} is proportional to the quantity $[Buf]_{tot}$ – 'heavier' buffers, which contain higher concentrations of their components HA and XA (up to, say, 2 mol dm^{-3}), are therefore more resilient to changes in pH than 'lighter' buffers (say, 0.1 mol dm^{-3}), with a lower total concentration $[Buf]_{tot} = [HA]_i + [XA]_i$.

Figure 17.2(a) also shows that the buffer capacity β increases as the pH of the buffer solution becomes very acid (pH towards 0), or alternatively very alkaline (pH towards 14). These values are far from the pK_a of the weak acid HA, and are also so acid or alkaline that the addition of more H$^+$ (aq) or OH$^-$ (aq) ions makes little difference. The sharply upward-sloping 'tails' at each side of Figure 17.2(a) are attributable to the first two terms in equation (17.44): the term $[H^+]$ dominates equation (17.44) at pH values approaching 0, whilst the term $K_w/[H^+]$ dominates at pH values approaching 14. The physical significance of these two 'tails' can be appreciated by setting $[Buf]_{tot}$ to a value of 0 in equation (17.44), implying that there is no buffer present, and that the 'solution' is pure water. The buffer capacity β then becomes the buffer capacity β_w of pure water, and equation (17.44) reduces to

$$\beta_w = 2.303 \left([H^+] + \frac{K_w}{[H^+]}\right) \tag{17.46}$$

The peak at pH 14 corresponds to pK_a for molecular water H$_2$O acting as an acid, and the peak at pH 0 to pK_b for the OH$^-$ (aq) ion, the corresponding conjugate base.

Figures 17.2(b) and 17.2(c) show how the pH of the buffer solution changes on the addition of H$^+$ (aq) ions (so making the buffer solution more acid), or OH$^-$ (aq) ions (more alkaline). As can be seen, relatively large amounts on ions need to be added to change the pH away from pH = 6 = pK_a; then, around pH 4 or pH 8, there is a rapid change to pH 2 or pH 12; thereafter, the pH changes more slowly.

Overall, Figure 17.2 confirms the two 'golden rules' concerning buffer solutions

- If you wish to control a solution to a particular pH, pH$_{target}$, choose a weak acid HA such that pK_a is as close to pH$_{target}$ as possible.
- Use higher, rather than lower, initial concentrations $[HA]_i$ and $[XA]_i$ of the weak acid HA and salt XA, so that $[Buf]_{tot} = [HA]_i + [XA]_i$ is sensibly large, usually in the range 0.1 – 2 mol dm^{-3}.

One final note on the buffer capacity β: if we ignore the buffer capacity of pure water, as represented by the terms $[H^+] + K_w/[H^+]$, equation (17.44) reduces to

$$\beta = 2.303 \frac{K_a[Buf]_{tot}[H^+]}{(K_a + [H^+])^2} \tag{17.47}$$

Equation (17.47) is known as the **Van Slyke equation**, after the American biochemist Donald Van Slyke, who developed this equation in connection with his research on the buffering of blood.

As can be seen from Figure 17.3, the behaviour of the Van Slyke equation is essentially identical to that of the more complex equation (17.44) within 2 or 3 units of pH on either side of pK_a – which, experimentally, is the important range – but does not have the upwards 'tails' at the extremes of pH.

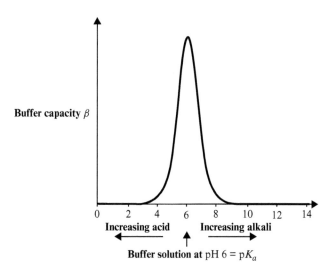

Figure 17.3 The Van Slyke equation.

17.10 Other reactions involving hydrogen ions

Our discussion so far has focused on acids, bases, buffer solutions, and the way in which buffer solutions maintain a constant pH. The context in which a buffer solution needs to maintain a constant pH arises when another reaction is taking place within the solution that generates, or consumes H$^+$ (aq) or OH$^-$ (aq) ions. This reaction might be the addition of an acid or base to the buffer solution, as happens, for example, in a titration; alternatively, this may be the result of a reaction taking place within the solution, such as the action of an enzyme. For the remainder of this chapter, we therefore explore the thermodynamics of this 'other' reaction, with particular reference to biochemical reactions, and the effect of the buffer solution on that reaction's equilibrium.

Our starting point is equation (16.13c)

$$\Delta_r G_c^\circ = -RT \ln K_c \tag{16.13c}$$

which, together with the definition of ΔG°

$$\Delta_r G_c^\circ = (c\Delta_f G_{c,C}^\circ + d\Delta_f G_{c,D}^\circ) - (a\Delta_f G_{c,A}^\circ + b\Delta_f G_{c,B}^\circ) \tag{16.14c}$$

may be applied to a reaction of the form

$aA + bB \rightleftharpoons cC + dD$

taking place in an ideal solution, at a pressure of 1 bar and at a given temperature, in which all components i are solutes. For any reaction of this type, we may use equation (16.13c) to calculate $\Delta_r G_c^\circ$ from data relating to the stoichiometry of the reaction, and the standard molar Gibbs free energies of formation $\Delta_f G_i^\circ$ of each component i, as published in reference works, with unit molar concentration as the standard state. Computing $\Delta_r G_c^\circ$, and the corresponding value of K_c is therefore very easy to do. And it is very informative too: as we saw on page 431, the sign of $\Delta_r G_c^\circ$ identifies the spontaneity and directionality of the reaction – if $\Delta_r G_c^\circ < 0$, the reaction will spontaneously take place from left to right as written, if $\Delta_r G_c^\circ > 0$, the reaction will spontaneously take place from right to left as written. Furthermore, the value of the concentration equilibrium constant K_c, as calculated from $\Delta_r G_c^\circ$ using equation (16.14c), specifies the composition of the equilibrium mixture as defined in terms of the equilibrium molar concentrations $[I]_{eq}$ as

$$K_c = \frac{[C]_{eq}^c \, [D]_{eq}^d}{[A]_{eq}^a \, [B]_{eq}^b} \qquad (16.12c)$$

In deriving the equation (16.12c), an implicit assumption was made that the components i are freely able to adjust their concentrations as the reaction progresses, in accordance with the chemistry of the reaction, and the corresponding thermodynamics. In very many situations, this is indeed the case, but there are particular contexts in which the concentration of one particular component is 'managed', in that there is a mechanism whereby that concentration can be controlled, independently of the reaction. This is of course what happens in a buffer solution, in which the combination of the weak acid HA and the corresponding salt XA 'manage' the pH of the solution, keeping the molar concentration $[H^+]$ of H^+ (aq) ions constant at, or very close to, the buffered molar concentration $[H^+]_{buf}$.

As we noted on page 549, this is of especial importance in biochemistry, for many biochemical reactions are very sensitive to pH. In general, enzymes have relatively narrow bands of pH within which they can catalyse their appropriate reactions: beyond those limits, the enzyme changes its configuration, becoming 'denatured', and can no longer function correctly. Most biochemical reactions take place at a neutral pH, at, or close to, pH 7, but some take place in more extreme conditions, such as the digestive enzymes of the stomach that catalyse reactions in the highly acidic conditions of around pH 2. Many biochemical systems therefore have often sophisticated mechanisms by which the pH of the reaction medium can be controlled so as to maintain the optimum pH, as appropriate to the particular biochemical reaction.

In general, an important consequence of buffering a reaction is to change the composition of the equilibrium mixture, as we will now show. To do this, consider a generalised reversible reaction

$$a\text{A} + b\text{B} \rightleftharpoons c\bar{\text{C}} + h\text{H}^+ \,(\text{aq})$$

taking place in an ideal solution, in which hydrogen ions H^+ (aq) are a product. By assumption, none of A, B or C are hydrogen ions, although they may themselves be ions; also, the solute is assumed to be water, which is not itself a reagent.

For this reaction, the concentration equilibrium constant K_c is given, as usual, by

$$K_c = \frac{[C]_{eq}^c \, [D]_{eq}^d}{[A]_{eq}^a \, [B]_{eq}^b} \qquad (16.12c)$$

which we choose to write as

$$K_c = \frac{[C]_{eq}^c}{[A]_{eq}^a \, [B]_{eq}^b} \times [H^+]_{eq}^h \qquad (17.48)$$

If we define the quantity Λ as

$$\Lambda = \frac{[C]_{eq}^{c}}{[A]_{eq}^{a} [B]_{eq}^{b}} \quad (17.49)$$

then

$$K_c = \Lambda [H^+]_{eq}^{h} \quad (17.50)$$

Physically, Λ represents the equilibrium ratio of all the reaction components, other than the hydrogen ion H^+ (aq).

Combining equation (16.13c)

$$\Delta_r G_c^\ominus = - RT \ln K_c \quad (16.13c)$$

with equation (17.50) we have

$$\Delta_r G_c^\ominus = - RT \ln \Lambda [H^+]_{eq}^{h}$$

implying that

$$\Delta_r G_c^\ominus = - RT \ln \Lambda - h RT \ln [H^+]_{eq} \quad (17.51)$$

Now, according to the rules for natural logarithms ln, to base e, and logarithms $\log_{10}$, to base 10

$$\ln[H^+] = 2.303 \log_{10}[H^+]$$

Furthermore, since, by definition

$$pH = - \log_{10}[H^+] \quad (17.9)$$

then equation (17.51) may be written as

$$\Delta_r G_c^\ominus = - RT \ln \Lambda + 2.303 \, h \, RT \, pH_{eq} \quad (17.52a)$$

where pH_{eq} is the pH of the equilibrium mixture, as determined by $[H^+]_{eq}$.

A special case of equation (17.52a) is when $h = 1$, for which, on rearranging

$$pH_{eq} = \frac{\Delta_r G_c^\ominus}{2.303 RT} + \frac{1}{2.303} \ln \Lambda$$

But since $\log_{10} \Lambda = \ln \Lambda / 2.303$, then, when $h = 1$

$$pH_{eq} = \frac{\Delta_r G_c^\ominus}{2.303 RT} + \log_{10} \Lambda$$

This expression is in fact the Henderson-Hasselbalch equation

$$pH = pK_a + \log_{10} \frac{[A^-]_{eq}}{[HA]_{eq}} \quad (17.16)$$

for a monoprotic acid $HA \rightleftharpoons H^+ + A^-$, for which $h = 1$ and $\Lambda = [A^-]/[HA]$; also, as we saw on page 541, $\Delta_r G_c^\ominus = - RT \ln K_a$ and so $pK_a = \Delta_r G_c^\ominus / 2.303 RT$.

In the general case, however, for any specific reaction, the left-hand side of equation (17.52a) is a constant, as defined by equation (16.14c)

$$\Delta_r G_c^\ominus = (c \, \Delta_f G_{c,C}^\ominus + d \, \Delta_f G_{c,D}^\ominus) - (a \, \Delta_f G_{c,A}^\ominus + b \, \Delta_f G_{c,B}^\ominus) \quad (16.14c)$$

Hence, if – for whatever reason – there is a change in the pH of the mixture in which the reaction takes place, the quantity Λ will change accordingly, as determined by equation (17.52a). Physically, this has an important consequence: the equilibrium ratio Λ of the non-hydrogen components of a reaction will vary according to the pH at which a given reaction takes place, and so Λ is more explicitly written as a function $\Lambda(\text{pH}_{\text{buf}})$ of the pH at which the reaction is buffered

$$\Delta_r G_c^{\circ} = -RT \ln \Lambda(\text{pH}_{\text{buf}}) + 2.303\, h\, RT\, \text{pH}_{\text{buf}} \tag{17.52b}$$

A particular case of equation (17.52b) is that at pH 0, for which

$$\Delta_r G_c^{\circ} = -RT \ln \Lambda(\text{pH}\,0)$$

But equation (16.13c) states that

$$\Delta_r G_c^{\circ} = -RT \ln K_c \tag{16.13c}$$

and so

$$K_c = \Lambda(\text{pH}\,0) \tag{17.53}$$

Equation (17.53) implies that, for a reaction producing hydrogen ions as a product, if $\Delta_r G_c^{\circ}$ is calculated from standard reference data for $\Delta_f G_c^{\circ}$ as

$$\Delta_r G_c^{\circ} = (c \Delta_f G_{c,C}^{\circ} + d\, \Delta_f G_{c,D}^{\circ}) - (a\, \Delta_f G_{c,A}^{\circ} + b \Delta_f G_{c,B}^{\circ}) \tag{16.14c}$$

and then used in equation (16.13c)

$$\Delta_r G_c^{\circ} = -RT \ln K_c \tag{16.13c}$$

to compute K_c, then that value of K_c is equal to the value of Λ (pH 0) – this being the equilibrium ratio of the non-hydrogen components when the reaction is constrained to take place at pH 0. This is an important, and perhaps somewhat surprising, result, for it is by no means obvious that the application of equations (16.14c) and (16.13c) to a reaction involving hydrogen ions has a built-in assumption that the resulting value of K_c corresponds to the reaction buffered at a constant pH of 0 – but this is indeed the case.

To verify this, consider the definition of K_c as

$$K_c = \frac{[C]_{\text{eq}}^c\, [D]_{\text{eq}}^d}{[A]_{\text{eq}}^a\, [B]_{\text{eq}}^b} \tag{16.12c}$$

under conditions in which the equilibrium state is at pH 0, corresponding to $[H^+]_{\text{eq}} = 10^0 = 1$ mol dm^{-3}. In this case

$$K_c = \frac{[C]_{\text{eq}}^c\, [H^+]_{\text{eq}}^h}{[A]_{\text{eq}}^a\, [B]_{\text{eq}}^b} = \frac{[C]_{\text{eq}}^c}{[A]_{\text{eq}}^a\, [B]_{\text{eq}}^b} = \Lambda\,(\text{pH}\,0)$$

This is of especial significance to biochemistry, for which conditions of pH 0 are hugely acid, often lethally so. Since most biochemical reactions take place at, or close to, neutral pH 7, the concentration equilibrium constant K_c, defined as

$$K_c = \frac{[C]_{\text{eq}}^c\, [D]_{\text{eq}}^d}{[A]_{\text{eq}}^a\, [B]_{\text{eq}}^b} \tag{16.12c}$$

which applies to a reaction taking place at pH 0, is of much less interest than the parameter Λ (pH 7)

$$\Lambda(\text{pH } 7) = \left(\frac{[C]_{eq}^c}{[A]_{eq}^a \, [B]_{eq}^b} \right)_{\text{pH7}} \tag{17.54}$$

which defines the composition of the equilibrium mixture, apart from the hydrogen ions, the molar concentration $[H^+]$ of which is fixed at pH 7, as determined by the appropriate buffer.

So, for the reaction

$$a\text{A} + b\text{B} \rightleftharpoons c\text{C} + h\text{H}^+ \,(\text{aq})$$

taking place in a buffered solution at any pH_{buf}, then, in accordance with equation (17.50)

$$K_c = \Lambda(\text{pH}_{\text{buf}}) \, [H^+]_{\text{buf}}^h = \Lambda(\text{pH}_{\text{buf}}) \, 10^{-h\,\text{pH}_{\text{buf}}} \tag{17.55}$$

At pH 7, $[H+]_{\text{buf}} = 10^{-7}$ mol dm^{-3}, implying that

$$K_c = 10^{-7h} \Lambda(\text{pH } 7) \tag{17.56a}$$

and so

$$\Lambda(\text{pH } 7) = 10^{7h} K_c \tag{17.56b}$$

The 'constrained', or 'forced', equilibrium ratio defined by $\Lambda(\text{pH } 7)$ – 'forced' in that the hydrogen ion concentration $[H^+]$ in the equilibrium mixture is constrained at 10^{-7}mol dm^{-3} – is therefore very much greater than 'the' equilibrium constant K_c, which applies at pH 0. As a consequence, the equilibrium mixture of A, B and C lies far further to the right, in favour of the product C, when the reaction takes place at pH 7, as compared to the equilibrium mixture at pH 0.

The discussion so far has explored the implications of a reaction

$$a\text{A} + b\text{B} \rightleftharpoons c\text{C} + h\text{H}^+ \,(\text{aq})$$

in which hydrogen ions $H^+(\text{aq})$ are a product; had we started with a reaction in which hydrogen ions H^+ (aq) are a reactant

$$a\text{A} + h\text{H}^+ \,(\text{aq}) \rightleftharpoons c\text{C} + d\text{D}$$

then the analysis would have been very similar, subject to K_c being defined as

$$K_c = \frac{[C]_{eq}^c \, [D]_{eq}^d}{[A]_{eq}^a \, [H^+]_{eq}^h}$$

and Λ as

$$\Lambda = \frac{[C]_{eq}^c \, [D]_{eq}^d}{[A]_{eq}^a}$$

As a consequence, if hydrogen ions H^+ (aq) are a reactant, the relationship between K_c and Λ is

$$K_c = \Lambda [H^+]_{eq}^{-h}$$

as compared to the relationship

$$K_c = \Lambda [H^+]_{eq}^h \tag{17.50}$$

which applies when hydrogen ions H⁺ (aq) are a product. If we allow the **stoichiometric number** h to be a signed, algebraic, quantity, which is positive when hydrogen ions H⁺ (aq) are a product, and negative when hydrogen ions H⁺ (aq) are a reactant, then equation (17.50) holds under all circumstances, as do all the consequences.

In particular, if hydrogen ions H⁺ (aq) are a reactant, and the equation is constrained to take place at pH 7, then equations (17.56)

$$K_c = 10^{-7h} \Lambda(\text{pH}7) \tag{17.56a}$$

$$\Lambda(\text{pH}7) = 10^{7h} K_c \tag{17.56b}$$

both remain valid, but, in this instance, the stoichiometric number h is negative. Accordingly, Λ (pH 7) is now very much smaller than K_c, implying that pH 7 favours the reactants rather than the products, as compared to pH 0.

17.11 The mass action effect

As we have just seen, for a reaction

$$a\text{A} + b\text{B} \rightleftharpoons c\text{C} + h\text{H}^+ \text{ (aq)}$$

in which hydrogen ions H⁺ (aq) are a product, the 'forced' equilibrium ratio Λ (pH 7) at pH 7 is significantly greater than the concentration equilibrium constant K_c

$$\Lambda(\text{pH}7) = 10^{7h} K_c \tag{17.56b}$$

Equation (17.51b) implies that, at pH 7, the reactants are favoured over the products, as compared to the same reaction taking place at pH 0.

In fact, this can be inferred from the reaction equation itself. If the reaction takes place at constant pH 7, corresponding to [H⁺] = 10^{-7} mol dm⁻³, as compared to constant pH 0, for which [H⁺] = 1 mol dm⁻³, then hydrogen ions need to be extracted from the system to keep the molar hydrogen ion concentration low. This continuous depletion of H⁺ (aq) 'pulls' the reaction to the right, increasing [C] whilst simultaneously reducing [A] and [B]. The resulting 'forced' equilibrium lies much further to the right, and the 'forced' equilibrium constant Λ(pH 7) is correspondingly significantly greater than the equilibrium constant at pH 0, K_c.

Similarly, if hydrogen ions H⁺ (aq) are a reactant

$$a\text{A} + h\text{H}^+ \text{ (aq)} \rightleftharpoons c\text{C} + d\text{D}$$

then equation (17.56b)

$$\Lambda(\text{pH}7) = 10^{7h} K_c \tag{17.56b}$$

still holds, but with h as a negative number. In this case, Λ (pH 7) is very much smaller than K_c, so that pH 7 favours the reactants over the products. This too can be inferred directly from the reaction equation: if H⁺ (aq) is maintained at the low molar concentration [H⁺] of 10^{-7} mol dm⁻³, as compared to the much higher molar concentration [H⁺] = 1 mol dm⁻³, corresponding to pH 0, then the depletion of H⁺ (aq) ions 'pulls' the reaction to the left as written, so favouring the reactants.

Both of these are examples of the general phenomenon known as the **mass action effect**, which refers to the influence of a particular component in a reaction mixture whose concentration is constrained or controlled in some particular way – for example, maintaining the molar concentration [H⁺] of hydrogen ions at a particular pH. In essence, what is happening is that the generalised equilibrium constant K_c

$$K_c = \frac{[C]_{eq}^c \, [D]_{eq}^d}{[A]_{eq}^a \, [B]_{eq}^b} \tag{16.12c}$$

is being 'forced' by holding one of the concentrations $[I]_{eq}$ at a chosen level.

And although this discussion has focussed on the concentration equilibrium constant K_c

$$K_c = \frac{[C]_{eq}^c \, [D]_{eq}^d}{[A]_{eq}^a \, [B]_{eq}^b} \tag{16.12c}$$

exactly the same analysis can be applied to non-equilibrium conditions as described by the concentration mass action ratio Γ_c

$$\Gamma_c = \frac{[C]^c \, [D]^d}{[A]^a \, [B]^b} \tag{16.15c}$$

17.12 Water as a reagent

The action of a buffer solution, holding the hydrogen ion concentration [H⁺] constant, is not the only way in which an equilibrium can be 'forced' – given the form of the concentration equilibrium constant K_c as

$$K_c = \frac{[C]_{eq}^c \, [D]_{eq}^d}{[A]_{eq}^a \, [B]_{eq}^b} \tag{16.12c}$$

then any individual molar concentration $[I]_{eq}$ can, in principle, be constrained, so 'forcing' the equilibrium composition of the other components. It is also possible that more than one molar concentration – say both $[A]_{eq}$ and $[D]_{eq}$ – might be constrained, so 'forcing' the equilibrium mixture defined by the ratio $[C]_{eq}^c/[D]_{eq}^d$. To explore this further, consider a reaction, taking place in an ideal solution, in which molecular water is itself a reactant, as well as the solvent.

This is a situation we have not yet discussed: so far in this book, all the reactions of the form

$$aA + cB \rightleftharpoons cC + dD$$

taking place in ideal solution, have been examined under the assumption that the components A, B, C and D are all solutes, with water as the solvent, and not participating in the reaction. Many reactions comply with this assumption, but there are some reactions in which molecular water H_2O – rather than the individual ions H^+ (aq) and OH^- (aq), or the ion pair H^+ (aq), OH^- (aq) – is itself a reagent, as well as the solvent, a notable example being the highly important biochemical reaction in which ATP is hydrolysed to ADP, according to the reaction represented as

$$ATP^{4-} \text{(aq)} + H_2O \text{(l)} \rightleftharpoons ADP^{3-} \text{(aq)} + HPO_4^{2-} \text{(aq)} + H^+ \text{(aq)}$$

ACIDS, BASES AND BUFFER SOLUTIONS

For the generalised reaction

$$aA + wH_2O \rightleftharpoons cC + dD$$

taking place in a dilute ideal solution, in which water is a reactant, it is likely that the components A (aq), C (aq) and D (aq) are present in molar concentrations of up to, say, 1 mol dm^{-3}, and quite possibly much less. The molar concentration [H$_2$O] of molecular water, however, is very much greater: 1 dm^{-3} of pure water weighs 1 kg, and since the molecular weight of water is 0.018 kg mol^{-1}, 1 dm^{-3} of pure water corresponds to 1/0.018 = 55.56 mol, implying that [H$_2$O] = 55.56 mol dm^{-3}, some nine orders of magnitude greater than the molar concentration [H$^+$] = 10^{-7} mol dm^{-3} of hydrogen ions at pH 7. This vast excess of molecular water will therefore exert a mass action effect, 'pushing' the reaction to the right if molecular water is a reactant, or to the left if a product.

Since water is both a reagent, and also the solvent, as discussed on page 524, the equilibrium constant K_r for this reaction is a 'hybrid' of the form

$$K_r = \frac{[C]_{eq}^c \, [D]_{eq}^d}{[A]_{eq}^a (x_{H_2O}/x_{H_2O}^{\ominus})^w} \tag{17.57}$$

As we saw on page 514, for a dilute solution, $x_{H_2O,eq} \approx 1$, and since $x_{H_2O}^{\ominus} = 1$, then the ratio $x_{H_2O}/x_{H_2O}^{\ominus} \approx 1$. As a consequence, the equilibrium constant K_r becomes

$$K_r = \frac{[C]_{eq}^c \, [D]_{eq}^d}{[A]_{eq}^a} \tag{17.58}$$

in which the term for water has 'dropped out'. This makes good sense: since water, as the solvent, is in excess, the molar concentration of water is in essence constant throughout the reaction, and so although the reaction is written chemically as

$$aA + wH_2O \rightleftharpoons cC + dD$$

the equilibrium constant that we are interested in is best expressed as

$$K_c = \frac{[C]_{eq}^c \, [D]_{eq}^d}{[A]_{eq}^a}$$

exactly as results from using $x_{H_2O,eq} = 1$ in the expression for the 'hybrid' concentration equilibrium constant, equation (17.57).

And for the hydrolysis of ATP, as expressed as

$$ATP^{4-} \text{ (aq)} + H_2O \text{ (l)} \rightleftharpoons ADP^{3-} \text{ (aq)} + HPO_4^{2-} \text{ (aq)} + H^+ \text{ (aq)}$$

we have a reaction in which not only is molecular water a reactant, but also hydrogen ions H$^+$ (aq) are a product. If this reaction takes place in a buffered solution at pH 7, then the equilibrium is being 'forced' by two constraints: the effect of water, and the fixing of the molar concentration [H+] at 10^{-7} mol dm^{-3}. This has considerable significance as regards the role played by ATP in biochemical systems, as we shall discuss further in Chapter 23.

EXERCISES

1. Define
   - acid
   - acid dissociation constant
   - weak acid
   - base
   - base dissociation constant
   - weak base
   - conjugate acid
   - conjugate base
   - ionic product of water
   - pH
   - pOH
   - pK_a
   - pK_b
   - pK_w
   - charge balance
   - buffer solution
   - buffer capacity.

2. Is water an acid, a weak acid, a base, or a weak base? Justify your answer.

3. Give some examples of conjugate acid-conjugate base pairs. Prove that, for any conjugate acid-conjugate base pair, $pK_a + pK_b = 14$.

4. Derive the Henderson-Hasselbalch equation. How does this differ from the Henderson-Hasselbalch approximation? How is the Henderson-Hasselbach equation related to Ostwald's dilution law?

5. An alkaline buffer solution comprises an aqueous mixture of a weak base XOH and a salt XA. Why must the acid be weak (as opposed to strong)? Why are the molar quantities of XOH and XA usually comparable? What is special about a buffer solution in which these molar quantities are equal?

6. Consider the reaction

   $$a\text{A} + h\text{H}^+ \text{(aq)} \rightleftharpoons c\text{C} + d\text{D}$$

   which takes place in an ideal aqueous solution. Defining

   $$\Lambda = \frac{[\text{C}]_{eq}^{\,c}}{[\text{A}]_{eq}^{\,a}\,[\text{B}]_{eq}^{\,b}} \tag{17.49}$$

   prove that

   $$K_c = \Lambda [\text{H}^+]_{eq}^{-h}$$

   and that

   $$\Delta_r G_c^\ominus = -RT \ln \Lambda - 2.303\, h\, RT\, \text{pH}_{eq}$$

   This equation is different from equation (17.52a)

   $$\Delta_r G_c^\ominus = -RT \ln \Lambda + 2.303\, h\, RT\, \text{pH}_{eq} \tag{17.52a}$$

Why? What does this analysis imply as regards the equilibrium composition for this reaction when buffered at different values of pH?

Explain why the value of K_c, as computed for this reaction from $\Delta_r G_c^{\circ}$, defines the equilibrium composition buffered at pH 0.

7. What is the mass action effect? In the human body, the metabolic pathway known as 'glycolysis' transforms glucose into pyruvic acid and the 'energy currency' ATP as

$$\text{glucose} \rightarrow 2 \text{ pyruvic acid} + 2 \text{ ATP}$$

The resulting pyruvate can follow several different pathways. One is 'the TCA cycle plus aerobic respiration', which may be represented as

$$2 \text{ pyruvic acid} + 6 \text{ O}_2 \rightarrow 6 \text{ CO}_2 + 6 \text{ H}_2\text{O} + 31 \text{ ATP}$$

and a second is the formation of lactic acid as

$$2 \text{ pyruvic acid} \rightarrow 2 \text{ lactic acid}$$

A runner, competing in a race such as the 100 m sprint, needs a sudden burst of energy, and rapidly consumes ATP. How does this affect these three metabolic processes? The molecular oxygen, as required in the second process, comes from the atmosphere, entering the body as a result of breathing, and being transported to cells through the blood stream. Although the athlete breathes harder and faster as the race takes place, and has an increased heart rate, the quantity of oxygen available within the cells can often become depleted. What happens as a result?

These processes will be explored further in Chapter 24.

18 Boiling points and melting points

Summary

The presence of a solute D, especially a non-volatile solute, within a solvent S causes the boiling point of the solution to be higher than that of the pure solvent, and the freezing point to be lower than that of the pure solvent. If the solution is assumed to be ideal, the **elevation of the boiling point** ΔT_{vap} is given by

$$\Delta T_{vap} = \left(\frac{R(T_{vap}{}^*)^2 M_S}{\Delta_{vap} H} \right) \frac{M_D}{M_S \, \mathbf{M}_D} \qquad (18.6)$$

and the **depression of the freezing point** ΔT_{fus} by

$$\Delta T_{fus} = \left(\frac{R(T_{fus}{}^*)^2 M_S}{\Delta_{fus} H} \right) \frac{M_D}{M_S \, \mathbf{M}_D} \qquad (18.13)$$

in which $\mathbf{M}_i$ is the molecular weight of component i, and M_i is the mass of component i within the solution.

For any solvent S, the **ebullioscopic constant** E_b is defined, with reference to equation (18.6), as

$$E_b = \frac{R(T_{vap}{}^*)^2 M_S}{\Delta_{vap} H} \qquad (18.7)$$

and the **cryoscopic constant** E_f is defined, with reference to equation (18.13), as

$$E_f = \frac{R(T_{fus}{}^*)^2 M_S}{\Delta_{fus} H} \qquad (18.14)$$

18.1 Non-volatile solutes

As we saw on page 505, at a given temperature T, for an ideal solution of a solute D dissolved in a liquid solvent S, the vapour pressure of the solvent p_S is given by Raoult's law as

$$p_S = x_S p_S{}^* \qquad (16.1)$$

where $p_S{}^*$ is the vapour pressure above the pure solvent at the same temperature. Likewise, for the solute D

$$p_D = x_D p_D{}^*$$

Total vapour pressure p_{tot} above the solution is therefore

$$p_{tot} = p_S + p_D = x_S p_S{}^* + x_D p_D{}^*$$

Modern Thermodynamics for Chemists and Biochemists. Dennis Sherwood and Paul Dalby.
© Oxford University Press 2018. Published 2018 by Oxford University Press.

If the solute D is non-volatile – for example, a solid – then the vapour pressure p_D^* above the pure substance D will be a small number, and certainly much smaller than p_S^*, the vapour pressure above the pure liquid solvent S. Furthermore, if the quantity of solute is small compared to the quantity of solvent – for example, if the solution is reasonably dilute – then the mole fraction x_D of the solute will be less (possibly much less) than the mole fraction x_S of the solvent. These two effects – a non-volatile solute in a reasonably dilute solution – combine so that the product $x_D p_D^*$ will be very much smaller than the product $x_S p_S^*$, implying that

$$p_{tot} \approx x_S p_S^*$$

Furthermore, since the mole fraction x_S of the solvent is necessarily less than 1, $p_{tot} < p_S^*$. For a reasonably dilute solution of a non-volatile solute in a liquid solvent, we may therefore draw two conclusions

- the total vapour pressure above the solution is attributable to the solvent alone, and
- at any given temperature, the total vapour pressure above the solution is less than the vapour pressure above the pure solvent.

Since, in general, the vapour pressure over any material rises with increasing temperature, a consequence of this second point is that the temperature of an ideal solution containing an non-volatile solute needs to be higher to achieve the same vapour pressure as that associated with the corresponding pure solvent.

18.2 Elevation of the boiling point

As we saw on page 477, a liquid boils when its vapour pressure p equals the pressure $P_{ex} + h\rho g$ within the bulk of the liquid, P_{ex} is the external pressure as exerted, for example, by the atmosphere, and $h\rho g$ is the hydrostatic pressure at a depth h m within the bulk of the liquid. In many circumstances, $P_{ex} \gg h\rho g$, and so the liquid boils when $p = P_{ex}$.

Consider, then, an ideal solution containing a non-volatile solute, open to the atmosphere, at a pressure $P_{ex} = 1$ bar. If the boiling point of the pure solute at 1 bar is T_{vap}^*, then this implies that p_S^* at T_{vap}^* is 1 bar; furthermore, since, for a solution, at any temperature, $p_S < p_S^*$, the vapour pressure p_S above a solution at temperature T_{vap}^* is some value less than p_S^*. Although the pure solvent boils at T_{vap}^*, an ideal solution containing a non-volatile solute does not – it is hot, but not yet boiling. To boil the solution requires a higher temperature: a temperature $T_{vap} > T_{vap^*}$, at which $p_S = 1$ bar, as illustrated in Figure 18.1.

For any given external pressure, the difference

$$\Delta T_{vap} = T_{vap} - T_{vap}^*$$

between the temperature T_{vap} at which the ideal solution boils, and T_{vap}^*, the temperature at which the pure solvent boils, is known as the **elevation of the boiling point**. ΔT_{vap} for any given solution can be determined by using the Clausius-Clapeyron equation in the form

$$\frac{d \ln p}{d T_{vap}} = \frac{\Delta_{vap} H}{R T_{vap}^2} \tag{15.8}$$

ELEVATION OF THE BOILING POINT

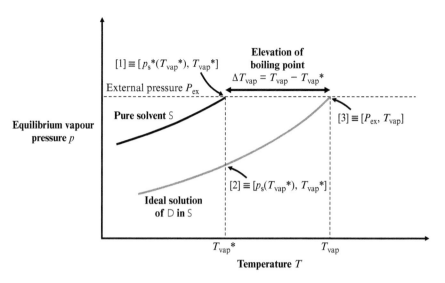

Figure 18.1 The elevation of the boiling point. The black line shows how the equilibrium saturated vapour pressure p_S^* of a pure solvent S increases with temperature. At the temperature T_{vap}^*, the vapour pressure $p_S^*(T_{vap}^*)$ equals the pressure P_{ex} of the external environment, and the pure solvent boils: this is state $[1] \equiv [p_S^*(T_{vap}^*), T_{vap}^*]$. The grey line shows the corresponding behaviour of the vapour pressure p_S of an ideal solution of a non-volatile solute within the solvent. At any temperature, $p_S = x_S p_S^*$, and since the mole fraction x_S is necessarily less than 1, p_S is always less than p_S^*, and so the grey line is always lower than the black line. For the solvent in the solution to boil, p_S must be become equal to the external pressure, which therefore requires the solution to be at a temperature $T_{vap} > T_{vap}^*$. The elevation of the boiling point ΔT_{vap} is the difference $T_{vap} - T_{vap}^*$. To determine ΔT_{vap}, we can use the integral form of the Clausius-Clapeyron equation between state $[2] \equiv [p_S(T_{vap}^*), T_{vap}^*]$ and state $[3] \equiv [P_{ex}, T_{vap}]$.

which, as we saw on page 482, is concerned with how changes in temperature T are related to changes in the vapour pressure p for liquid $\rightarrow$ vapour phase transitions for any system.

The integral form of equation (15.9a)

$$\ln \frac{p_2}{p_1} = \frac{\Delta_{vap} H}{R} \left(\frac{1}{T_1} - \frac{1}{T_2} \right) \tag{15.9a}$$

relates two states $[p_1, T_1]$ and $[p_2, T_2]$, so that if p_1, T_1, p_2 and $\Delta_{vap} H$ are known for any system, then T_2 can be determined.

The system of current interest is the ideal solution, as represented by the grey line in Figure 18.1, and we can apply the integral form of the Clausius-Clapeyron equation between an initial state $[2] \equiv [p_1, T_1] \equiv [p_S(T_{vap}^*), T_{vap}^*]$ and a final state $[3] \equiv [p_2, T_2] \equiv [P_{ex}, T_{vap}]$, and with $\Delta_{vap} H$ being the enthalpy of vaporisation of the pure solvent. Equation (15.9a) therefore becomes

$$\ln \frac{P_{ex}}{p_S(T_{vap}^*)} = \frac{\Delta_{vap} H}{R} \left(\frac{1}{T_{vap}^*} - \frac{1}{T_{vap}} \right) \tag{18.1}$$

But we also know that the pure solvent S boils at temperature T_{vap}^* when its vapour pressure $p_S^*(T_{vap}^*)$ is equal to P_{ex}, and so equation (18.1) becomes

$$\ln \frac{p_S^*(T_{vap}^*)}{p_S(T_{vap}^*)} = \frac{\Delta_{vap} H}{R} \left(\frac{1}{T_{vap}^*} - \frac{1}{T_{vap}} \right) \tag{18.2}$$

BOILING POINTS AND MELTING POINTS

The solution is ideal, and so we may now invoke Raoult's law

$$p_S = x_S p_S^* \tag{16.1}$$

In equation (16.1), the vapour pressures p_S^* of the pure solvent, and p_S of the component S within an ideal solution, refer to the same temperature. If this temperature is T_{vap}^*, corresponding to states [1] and [2] in Figure 18.1, then a particular case of equation (16.1) is

$$p_S(T_{vap}^*) = x_S p_S^*(T_{vap}^*)$$

implying that

$$\frac{p_S^*(T_{vap}^*)}{p_S(T_{vap}^*)} = \frac{1}{x_S}$$

and so equation (18.2) becomes

$$\ln \frac{1}{x_S} = \frac{\Delta_{vap} H}{R}\left(\frac{1}{T_{vap}^*} - \frac{1}{T_{vap}}\right) = \frac{\Delta_{vap} H}{R}\left(\frac{T_{vap} - T_{vap}^*}{T_{vap}^* T_{vap}}\right) \tag{18.3}$$

The difference $T_{vap} - T_{vap}^*$ is ΔT_{vap}, and will be quite small, implying that $T_{vap} \approx T_{vap}^*$, and so, to a good approximation, the product $T_{vap}^* T_{vap}$ can be expressed as $[T_{vap}^*]^2$. Equation (18.3) then becomes

$$\ln \frac{1}{x_S} = \frac{\Delta_{vap} H \Delta T_{vap}}{R(T_{vap}^*)^2} \tag{18.4}$$

Now

$$x_S = \frac{n_S}{n_S + n_D}$$

and so

$$\frac{1}{x_S} = \frac{n_S + n_D}{n_S} = 1 + \frac{n_D}{n_S}$$

For a dilute solution, n_S is very much greater than n_D, and so the ratio n_D/n_S will be much less than 1. Hence

$$\ln \frac{1}{x_S} = \ln\left(1 + \frac{n_D}{n_S}\right) \approx \frac{n_D}{n_S} = \frac{\Delta_{vap} H \, \Delta T_{vap}}{R(T_{vap}^*)^2} \tag{18.5}$$

The mole numbers n_D and n_S can be converted into masses of material M_D and M_S (both measured in units of kg) by reference to the appropriate molar masses $\bm{M}_D$ and $\bm{M}_S$ (the molecular weights measured in units of kg mol^{-1}) as

$$n_D = \frac{M_D}{\bm{M}_D}$$

and

$$n_S = \frac{M_S}{\bm{M}_S}$$

giving

$$\Delta T_{vap} = \left(\frac{R(T_{vap}^*)^2 \bm{M}_S}{\Delta_{vap} H}\right) \frac{M_D}{M_S \bm{M}_D} \tag{18.6}$$

Since all the terms on the right-hand side of equation (18.6) are necessarily positive, $\Delta T_{vap} = T_{vap} - T_{vap}^* > 0$, and $T_{vap} > T_{vap}^*$: dissolving a non-volatile solute in a solvent always raises the boiling point, as every cook who puts salt in water to nudge the boiling point up a little knows well. Furthermore, for a given solvent S, the term in parentheses in equation (18.6) is a property of that solvent, known as the **ebullioscopic constant**, E_b

$$E_b = \frac{R(T_{vap}^*)^2 M_S}{\Delta_{vap} H} \tag{18.7}$$

For any given experiment, the mass M_D of the solute, and the mass M_S of the solvent are known, and so equation (18.6) can be used to determine the molecular weight M_D of the solute molecule by measuring of the elevation of the boiling point, ΔT_{vap}.

Table 18.1 gives some values of E_b for a variety of solvents.

Table 18.1 Some values of the ebullioscopic constant E_b at 1 atm

	M_S kg mol^{-1}	T_{vap}^* K	$\Delta_{vap} H$ kJ mol^{-1}	E_b K kg mol^{-1}
Acetone	0.0581	329.3	29.1	1.80
Benzene	0.0781	353.2	30.72	2.64
Diethyl ether	0.0741	307.7	26.52	2.20
Ethanol	0.0461	351.4	38.56	1.23
Tetrachloromethane	0.1538	350.0	29.82	5.26
Water	0.0180	373.0	40.65	0.513

Source: *CRC Handbook of Chemistry and Physics*, 96th edn, 2015, pages 3–1 to 3–552, 4–98, 6–127 to 6–144, and 15–27.

18.3 Depression of the freezing point

On page 572, we saw that the vapour pressure p_S associated with an ideal solution of a non-volatile solute D within a solvent S is always less than the vapour pressure p_S^*, at the same temperature, associated with a pure solvent. That's why the grey line, representing the ideal solution, is below the black line, representing the ideal solution, in Figure 18.1, which shows what happens at higher temperatures, around the boiling point of the solvent.

At lower temperatures, p_S continues to be less than p_S^*, and Figure 18.2 illustrates what happens around the freezing point of the solvent.

Comparison of Figures 18.1 and 18.2 shows an important difference between boiling and freezing. Boiling takes place with respect to an external reference: when the vapour pressure of a liquid equals that of the external surroundings, the liquid boils, and that liquid can be a pure liquid or a solution. Freezing is different in that it takes place when the vapour pressure of the solid phase equals that of the liquid phase, at whatever the external pressure might be – hence the difference between Figures 18.1 and 18.2, and the absence of the external pressure reference in Figure 18.2.

In Figure 18.2, the black line shows how the vapour pressure p_S^* of pure S varies with temperature at a given external pressure. As can be seen, there are two segments: a steeper line for

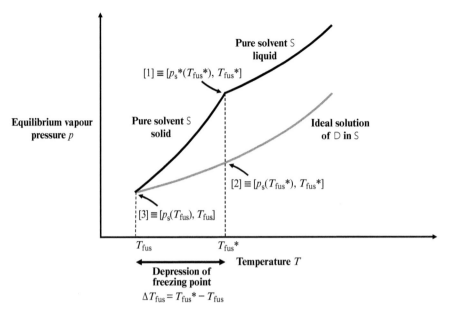

Figure 18.2 The depression of the freezing point. The black line shows how the equilibrium saturated vapour pressure p_S* of the solvent S, both as a pure solid, and as a pure liquid, varies with temperature at a given external pressure. The temperature at which the pure solvent melts is $T_{fus}*$, and at that temperature, the vapour pressures of the pure solid and pure liquid phases are equal, as shown by state $[1] \equiv [p_S*(T_{fus}*), T_{fus}*]$. The vapour pressure p_S of an ideal solution for which S is the solvent is the grey line, which is below the black line at all temperatures – in particular, at the temperature $T_{fus}*$ at which the pure liquid freezes, the solution is in state $[2] \equiv [p_S(T_{fus}*), T_{fus}*]$, and is still liquid. At the point at which the vapour pressure p_S of the ideal solution equals the vapour pressure p_S* of the pure solid, pure S solidifies from the ideal solution. This happens at a temperature T_{fus} lower than the freezing point $T_{fus}*$ of pure S, at state $[3] \equiv [p_S(T_{fus}), T_{fus}]$. The depression of the freezing point $\Delta T_{fus} = T_{fus}* - T_{fus}$ can be determined using the integral form of the Clausius-Clapeyron equation twice: firstly, between state $[1] \equiv [p_S*(T_{fus}*), T_{fus}*]$ and state $[3] \equiv [p_S(T_{fus}), T_{fus}]$; secondly, between state $[2] \equiv [p_S(T_{fus}*), T_{fus}*]$ and state $[3] \equiv [p_S(T_{fus}), T_{fus}]$.

S as a pure solid, and a shallower line for S as a pure liquid. These two lines intersect at the temperature $T_{fus}*$, the melting point, at which the two phases are in equilibrium. This defines state [1] in Figure 18.2, at which the vapour pressure of the pure liquid is $p_S*(T_{fus}*)$, a value equal to the vapour pressure over the pure solid at that temperature. The reason why the line representing pure S as a solid is steeper than that for the liquid can be appreciated from the general form Clausius-Clapeyron equation

$$\frac{d \ln p}{dT_{ph}} = \frac{\Delta_{ph} H}{RT_{ph}^2} \tag{15.8}$$

With reference to Figure 18.2, for the black line segment between state [3] and state [1], S is a pure solid, in equilibrium with its vapour, and so $\Delta_{ph} H$ is the molar enthalpy of sublimation $\Delta_{sub} H$; for the black line segment from state [3] towards higher temperatures, S is a pure liquid, in equilibrium with its vapour, and so $\Delta_{ph} H$ is the molar enthalpy of vaporisation $\Delta_{vap} H$. Since $\Delta_{sub} H = \Delta_{fus} H + \Delta_{vap} H$, and all three enthalpy changes are necessarily positive, then $\Delta_{sub} H > \Delta_{vap} H$. The slope of a graph of $\ln p$ against T is therefore greater for the solid than the liquid, and the same is true for a graph of p against T, as shown in Figure 18.2.

DEPRESSION OF THE FREEZING POINT

The grey line in Figure 18.2 shows the vapour pressure p_S for an ideal solution in which a non-volatile solute D is dissolved in the solvent S. As can be seen, the vapour pressure p_S is lower than the vapour pressure $p_S{}^*$ of pure S at the same temperature, and so the grey line is below the black line at all temperatures. In particular, at the freezing point $T_{fus}{}^*$ of the pure solvent, the solution is still liquid. As the temperature reduces, a temperature T_{fus} is reached at which the vapour pressure p_S of the solution equals the vapour pressure $p_S{}^*$ of pure solid S, at which point pure S will solidify from the solution. The freezing point of the solution is therefore lower than the freezing point of the pure solvent – this being known as the **depression of the freezing point**, as measured by $\Delta T_{fus} = T_{fus}{}^* - T_{fus}$.

We now apply the Clausius-Clapeyron equation twice: firstly, for the pure solid S, along the black line, from state $[1] \equiv [p_S{}^*(T_{fus}{}^*), T_{fus}{}^*]$ to state $[3] \equiv [p_S(T_{fus}), T_{fus}]$, and then secondly for the solution, along the grey line, from state $[2] \equiv [p_S(T_{fus}{}^*), T_{fus}{}^*]$ to state $[3] \equiv [p_S(T_{fus}), T_{fus}]$.

For the pure solid S, the change from state $[1] \equiv [p_S{}^*(T_{fus}{}^*), T_{fus}{}^*]$ to state $[3] \equiv [p_S(T_{fus}), T_{fus}]$ represents an equilibrium between pure solid S and S vapour, and so the appropriate form of the integrated Clausius-Clapeyron equation is that associated with sublimation

$$\ln \frac{p_2}{p_1} = \frac{\Delta_{sub}H}{R}\left(\frac{1}{T_1} - \frac{1}{T_2}\right) \tag{15.9c}$$

giving

$$\ln \frac{p_S{}^*(T_{fus}{}^*)}{p_S(T_{fus})} = \frac{\Delta_{sub}H}{R}\left(\frac{1}{T_{fus}} - \frac{1}{T_{fus}{}^*}\right) \tag{18.8}$$

For the ideal solution, the change from state $[2] \equiv [p_S(T_{fus}{}^*), T_{fus}{}^*]$ to state $[3] \equiv [p_S(T_{fus}), T_{fus}]$ is an equilibrium between liquid S in solution, and S vapour, and so we use the integrated Clausius-Clapeyron equation (15.9a) as associated with vaporisation

$$\ln \frac{p_2}{p_1} = \frac{\Delta_{vap}H}{R}\left(\frac{1}{T_1} - \frac{1}{T_2}\right) \tag{15.9a}$$

implying that

$$\ln \frac{p_S(T_{fus}{}^*)}{p_S(T_{fus})} = \frac{\Delta_{vap}H}{R}\left(\frac{1}{T_{fus}} - \frac{1}{T_{fus}{}^*}\right) \tag{18.9}$$

We now subtract equation (18.9) from (18.8) giving

$$\ln \frac{p_S{}^*(T_{fus}{}^*)}{p_S(T_{fus})} - \ln \frac{p_S(T_{fus}{}^*)}{p_S(T_{fus})} = \frac{\Delta_{sub}H}{R}\left(\frac{1}{T_{fus}} - \frac{1}{T_{fus}{}^*}\right) - \frac{\Delta_{vap}H}{R}\left(\frac{1}{T_{fus}} - \frac{1}{T_{fus}{}^*}\right)$$

and so

$$\ln \left(\frac{p_S{}^*(T_{fus}{}^*)}{p_S(T_{fus})}\right)\left(\frac{p_S(T_{fus})}{p_S(T_{fus}{}^*)}\right) = \frac{(\Delta_{sub}H - \Delta_{vap}H)}{R}\left(\frac{1}{T_{fus}} - \frac{1}{T_{fus}{}^*}\right) \tag{18.10}$$

Since

$$\Delta_{sub}H = \Delta_{fus}H + \Delta_{vap}H \tag{6.17}$$

equation (18.10) simplifies to

$$\ln \frac{p_S^*(T_{fus}^*)}{p_S(T_{fus}^*)} = \frac{\Delta_{fus}H}{R} \left(\frac{1}{T_{fus}} - \frac{1}{T_{fus}^*} \right) \quad (18.11)$$

In this equation, $p_S^*(T_{fus}^*)$ represents the vapour pressure of the pure solvent S, at its normal freezing point, T_{fus}^*, corresponding to state [1] in Figure 18.2; $p_S(T_{fus}^*)$ represents the vapour pressure of the ideal solution of S, at the same temperature, corresponding to state [2] in Figure 18.2. If the solution is ideal, these two vapour pressures are related by Raoult's law as

$$p_S(T_{fus}^*) = x_S p_S^*(T_{fus}^*) \quad (16.1)$$

and so equation (18.11) becomes

$$\ln \frac{1}{x_S} = \frac{\Delta_{fus}H}{R} \left(\frac{1}{T_{fus}} - \frac{1}{T_{fus}^*} \right) \quad (18.12)$$

Mathematically, equation (18.12) is identical in form to equation (18.3)

$$\ln \frac{1}{x_S} = \frac{\Delta_{vap}H}{R} \left(\frac{1}{T_{vap}^*} - \frac{1}{T_{vap}} \right) \quad (18.3)$$

and so, subject to the appropriate mapping of the different variables, the analysis on page 574 which led from equation (18.3) to equation (18.6)

$$\Delta T_{vap} = T_{vap} - T_{vap}^* = \left(\frac{R(T_{vap}^*)^2 M_S}{\Delta_{vap}H} \right) \frac{M_D}{M_S\, M_D} = E_b \frac{M_D}{M_S\, M_D} \quad (18.6)$$

can be used to transform equation (18.12) into

$$\Delta T_{fus} = T_{fus}^* - T_{fus} = \left(\frac{R(T_{fus}^*)^2 M_S}{\Delta_{fus}H} \right) \frac{M_D}{M_S\, M_D} = E_f \frac{M_D}{M_S\, M_D} \quad (18.13)$$

$\Delta T_{fus} = T_{fus}^* - T_{fus}$ is the depression of the freezing point, and E_f is a property of the solvent S, the **cryoscopic constant**

$$E_f = \frac{R(T_{fus}^*)^2 M_S}{\Delta_{fus}H} \quad (18.14)$$

some representative values for which are shown in Table 18.2.

Table 18.2 Some values of the cryoscopic constant E_f at 1 atm

	M_S kg mol^{-1}	T_{fus}^* K	$\Delta_{fus}H$ kJ mol^{-1}	E_f K kg mol^{-1}
Benzene	0.0781	278.6	9.87	5.07
Cyclohexanol	0.1002	299.1	1.78	42.2
Ethylene glycol	0.0621	260.5	9.96	3.11
Glycerol	0.0921	291.3	18.3	3.56
Phenol	0.0941	314.0	11.51	6.84
Toluene	0.0921	178.2	6.64	3.55
Water	0.0180	273.2	6.01	1.86

Source: *CRC Handbook of Chemistry and Physics*, 96th edn, 2015, pages 3–1 to 3–552, 4–98, 6–145 to 6–154, and 15–28.

The elevation of the boiling point ΔT_{vap}, and the depression of the freezing point ΔT_{fus}, are known as **colligative properties** of the solution, and their measurement offers a practical method for the determination of the molecular weight of a non-volatile solute. Given reference data for either the ebullioscopic constant E_b or the cryoscopic constant E_f for a particular solvent, and a known mass M_D of solute dissolved in a known mass of M_S of solvent, and assuming that the solution is ideal, then measurement of ΔT_{vap} or ΔT_{fus} enables us to use equation (18.7) or equation (18.14) as appropriate to estimate the molecular weight M_D of the solute.

EXERCISE

1. Describe, in physical terms, why a solution of a non-volatile solute elevates the boiling point, and depresses the freezing point, of the solvent. What happens if the solute is not 'non-volatile'?

19 Mixing and osmosis

Summary

Osmosis is the spontaneous flow of molecules of a solvent S, through a **semi-permeable membrane**, from a solution containing a lower concentration of a solute D into a solution containing a higher concentration of solute D, so diluting the solution of higher concentration. The **osmotic pressure** Π of a solution of a given molar concentration [D] is the pressure that needs to be exerted by the solution on the membrane to stop osmosis from the pure solvent S, as given by the **van't Hoff equation (osmosis)**

$$\Pi = RT \frac{n_D}{V} = RT \, [D] \tag{19.6}$$

19.1 Mixing

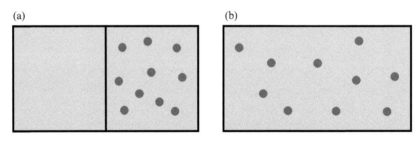

Figure 19.1 Mixing. When the partition is removed, the molecules of the solute spontaneously distribute evenly across the entire available volume.

Figure 19.1(a) shows a system of two compartments: on the left is a quantity of n_S mol of a pure liquid solvent S at pressure P and temperature T; on the right is a volume of an ideal solution of n_D mol of a solute D dissolved within n_S mol of the solvent S, at the same pressure P and temperature T.

With the central partition in place, each compartment is in equilibrium; when the partition is removed, the entire system volume is available to molecules of both the solvent S and the solute D. At the instant the partition is removed, the system is not in equilibrium: over time, the molecules spontaneously distribute themselves evenly across the entire available volume, as shown in Figure 19.1(b). This, of course, is mixing, as discussed in Chapter 8, and to explore the thermodynamics of mixing, our starting point is equation (16.7c) for the molar Gibbs free energy G_i of a component i in a solution

$$G_i \text{ (in solution)} = G_i^*(\text{pure}) + RT \ln x_i \tag{16.7c}$$

in which x_i is the mole fraction of component i in the solution, and $G_i^*(\text{pure})$ is the standard molar Gibbs free energy of the pure component, corresponding to $x_i = 1$.

At time 0, just before the partition is removed, for the solvent S in the left-hand compartment, $x_S = 1$ and so

$$G_S(\text{left}, 0) = G_S^*(\text{pure})$$

In the right-hand compartment, at time 0, n_S mol of solvent S are mixed with n_D mol of dissolved solvent D, and so

$$G_S(\text{right}, 0) = G_S^*(\text{pure}) + RT \ln x_S(\text{right}, 0) = G_S^*(\text{left}, 0) + RT \ln x_S(\text{right}, 0)$$

Since at time 0 the mole fraction $x_S(\text{right}, 0) = n_S/(n_S + n_D)$ is necessarily less than 1, $\ln x_S(\text{right}, 0)$ is negative, and so $G_S(\text{right},0)$ is less than $G_S(\text{left},0)$. This implies that, when the partition is removed, for the solvent S, the fundamental condition for equilibrium between two phases X and Y, equation (15.4a)

$$G_X = G_Y \tag{15.4a}$$

is not fulfilled. Removing the partition therefore creates a non-equilibrium state for the solvent.

Similarly, at time 0, for the solute D in the right-hand compartment

$$G_D(\text{right}, 0) = G_D^*(\text{pure}) + RT \ln x_D(\text{right}, 0)$$

and since there is no solute D on the left, $G_D(\text{left},0) = 0$. Once again, when the partition is removed, the condition for equilibrium is not fulfilled, and so this represents a non-equilibrium state for the solute.

The act of removing the partition suddenly makes all the volume of the system available to molecules of both the solvent S and the solute D, and thermal motion causes the molecules of both S and D to move across the now-removed boundary. Since the solvent S was already on both sides of the boundary, this motion is microscopic, and so might not be macroscopically detectable; for the solute D, however, the concentration of D towards the right decreases, whilst the concentration of D towards the left increases – a process we recognise as **diffusion**.

At any time τ after the partition has been removed, suppose that the number of molecules of the solute D that have moved to the left-hand half of the system is such that the instantaneous mole fraction of D in that half is $x_D(\text{left}, \tau)$; likewise, let the instantaneous mole fraction of D in the right-hand half be $x_D(\text{right}, \tau)$. The molar Gibbs free energy $G_D(\text{left}, \tau)$ of the solute on the left is therefore

$$G_D(\text{left}, \tau) = G_D^*(\text{pure}) + RT \ln x_D(\text{left}, \tau)$$

and on the right

$$G_D(\text{right}, \tau) = G_D^*(\text{pure}) + RT \ln x_D(\text{right}, \tau)$$

For the solute D to be in equilibrium, $G_D(\text{left},\tau)$ must equal $G_D(\text{right},\tau)$ for all times τ, implying that $x_D(\text{left},\tau)$ must equal $x_D(\text{right},\tau)$ for all times τ. The equilibrium state for the solute D is therefore one in which the mole fractions in each half of the system are the same. In fact, the condition is considerably stronger, for a moment's thought will show – using the same argument – that the molar Gibbs free energy G_D, and hence the corresponding mole fraction x_D, must have the same values for any volume, however small or large, within the overall volume to which molecules of the solute D have access. At equilibrium, the solute D must therefore be evenly distributed throughout the accessible volume; and exactly the same applies to the solvent S.

Mixing is therefore a spontaneous process, provided that the two components are miscible – which is by definition true for any two components that form an ideal solution.

19.2 Osmosis

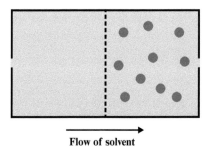

Figure 19.2 Osmosis. The barrier between the two halves of the system allows free flow of the solvent S, but not of the solute D. The molar Gibbs free energy G_S(left) of the solute on the left is that of the pure solvent G_S^*(pure), which will always be greater than the molar Gibbs free energy G_S(right) = G_S^*(pure) + $RT \ln x_S$ of the solute on the right. There will therefore be a continuous flow of solvent to the right – hence the gaps in the two walls, symbolising the continued availability of solvent on the left, and the presence of some reservoir on the right into which the solvent can flow.

Figure 19.2 is a variant of Figure 19.1(a), where the partition between the two halves of the system has been replaced by a barrier which has the special property that molecules of the solvent S are able to pass freely across the barrier, but the barrier blocks the movement of molecules of the solute D. Molecules of the solvent S therefore have access to the whole system, but molecules of the solute D are confined to the right-hand half. A barrier that exhibits this property for any given combination of a solvent S and a solute D is known as **a semi-permeable membrane** – as can be formed from a material which has pores large enough to permit the flow of a solvent such as water, but too small to allow the passage of a solute molecule of much larger size, such as a protein.

The solvent in the left-hand half is pure S, and so, at the standard pressure of 1 bar = 10^5 Pa, the molar Gibbs free energy G_S(left) is equal to G_S^*(pure). On the right-hand side, however, molecules of the solvent S are mixed with those of the solute D, and so the molar Gibbs free energy G_S(right) is equal to G_S^*(pure) + $RT \ln x_S$. Since the mole fraction x_S is always less than 1, G_S(left) is always greater than G_S(right), and solvent will continue to flow from left to right as long as there is a supply of solvent coming in from the left, and somewhere for the progressively more dilute solution to flow to on the right. In principle, this flow will last for ever, for there will always be molecules of the solute D on the right-hand side, however dilute the solution becomes, whilst there can never be any molecules of the solute D on the left-hand side.

This phenomenon is known as **osmosis**. Macroscopically, osmosis is recognised by a net flow of pure solvent across the semi-permeable membrane, from left to right as shown in Figure 19.2. Microscopically, however, solvent molecules are continuously diffusing in both directions, and so osmosis implies that, over any given time, more solvent molecules are diffusing from left to right than from right to left. To explain this, we can consider a model of molecular motion similar to that used in Chapter 1 in connection with gaining a deeper understanding of pressure (see pages 17 and 18). Because the compartment to the left contains the pure solvent, all the molecules striking the semi-permeable membrane on the left are those of the solvent, and they all pass through to the right. The compartment on the right, however, contains a mixture of the solvent and the solute, and so, over any given time, some solute molecules will strike the semi-permeable membrane and bounce back, whilst some solvent molecules will strike the membrane and pass through to the left. If the temperature and pressure are the same on both sides of the semi-permeable membrane, over any given time, the number of solvent molecules striking the left of the semi-permeable membrane is greater than the number of solvent molecules striking the right, and so there is a net flow of solvent molecules from left to right.

Osmosis has great importance biologically. All biological cells are surrounded by a **membrane** – for it is this membrane that defines the cell, and distinguishes the cell's 'inside' from the surrounding 'outside'. Cells contain many components, including a host of biochemically active agents dissolved in the cell's aqueous interior. Some of these are ions, such as Na^+, Ca^{2+} and Cl^-, or small molecules such as sugars; others are large molecules such as proteins. The interior of a cell is therefore a complex mixture, and the molar Gibbs free energy of the solvent, water, is correspondingly less than the molar Gibbs free energy of pure water. Furthermore, many cell membranes are semi-permeable, allowing the passage of water, and perhaps ions and small molecules, but keeping large molecules within the cell. As a consequence, if such a cell is in contact with water – for example, a cell of a plant root that might be in contact with water in the soil – osmosis will draw the external water into the cell. Osmosis is a key process for allowing the flow of water through biological systems.

19.3 Osmotic pressure

Figure 19.3 is a development of Figure 19.2: instead of having the pure solvent and the solution side-by-side, the two compartments are configured as a reservoir, open to the atmosphere, and a central column, with the semi-permeable membrane at the base of the column. Figure 19.3(a) is the initial, non-equilibrium state, just like Figure 19.2. Osmosis takes place as pure solvent molecules flow through the semi-permeable membrane into the solution, but in contrast to Figure 19.2 in which the solution flowed away to the right, in Figure 19.3(b), the result of the inflow of pure solvent into the solution is to cause the solution to occupy a greater volume within the column. As a result, the height of the solution in the column rises.

But unlike the situation depicted in Figure 19.2, in which the osmosis-driven flow of solvent into the solution continues, in principle, for ever, the system shown in Figure 19.3(b) is

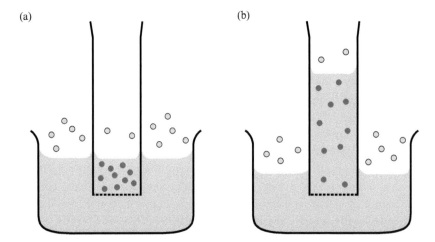

Figure 19.3 Osmosis – another perspective. Like the state shown in Figure 19.2, the state represented by Figure 19.3(a) is unstable: the system changes spontaneously to the state represented in Figure 19.3(b), in which solvent molecules pass by osmosis into the solution, diluting it, and increasing its volume. The resulting column of solution exerts additional pressure, downwards on the upper surface of the semi-permeable membrane, counteracting the flow of pure solvent, and achieving equilibrium.

in thermodynamic equilibrium: once the upper surface of the column of solution reaches a certain height, the flow of solvent stops. Why so?

As solvent molecules diffuse into the column, the height of the column steadily increases. As this happens, the pressure on the semi-permeable membrane also increases, one effect of which is to increase the rate at which solvent molecules within the column strike the semi-permeable membrane from above. Eventually, a pressure is reached at which, microscopically, the number of solvent molecules striking the membrane from above equals the number striking from below, at which point the net flow of solvent across the membrane falls to zero, the system achieves macroscopic equilibrium, and osmosis ceases.

To explain this thermodynamically, consider equation (16.7c)

$$G_i(\text{in solution}) = G_i^*(\text{pure}) + RT \ln x_i \tag{16.7c}$$

which explicitly shows that G_i for any component i in a solution depends on the mole fraction x_i, and the temperature T. What's less obvious is that G_i depends on the system pressure P too. As discussed on page 403, the Gibbs free energy of a component in a two-component mixture is in general a function $G(P, T, n_A, n_B)$ of the pressure P, temperature T and the mole numbers of each component n_A and n_B. The two variables n_A and n_B can be combined into a single variable, the mole fraction $x_A = n_A/(n_A + n_B)$, which is explicitly shown in equation (16.7c), as is the temperature T, but the pressure P has 'disappeared'. This has happened because, in our discussions so far of the thermodynamics of solutions, we have often used those familiar words "at constant pressure P and temperature T", and so the pressure has been assumed to be constant. Accordingly, it does not appear explicitly in equation (16.7c). But in the context of osmosis, as vividly illustrated in Figure 19.3(b), pressure is important, and equation (16.7c) needs to be amended accordingly.

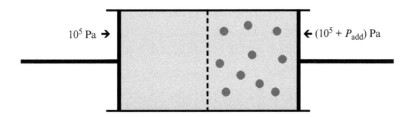

Figure 19.4 Osmotic pressure. The pure solvent, on the left of the semi-permeable membrane, is at a pressure of 1 bar = 10^5 Pa, and the solution on the right is at a higher pressure $(10^5 + P_{add})$ Pa. At a given temperature T, there will be a particular additional pressure $P_{add} = \Pi$ Pa at which the system is in equilibrium, and so G_S(pure solvent) = G_S(solution). The additional pressure Π Pa that is required to achieve this is known as the osmotic pressure.

Figure 19.4 is a variant of both Figure 19.2 and Figure 19.3. Compared with Figure 19.2, Figure 19.4 introduces two pistons, so that there is no inflow of solvent on the left, and no outflow of solution to the right. The system shown in Figure 19.4 is therefore closed, with the pistons on either side exerting pressure on the pure solvent from the left, and also on the solution from the right. The difference between Figure 19.4 and Figure 19.3 is primarily one of orientation – the whole system in Figure 19.4 is horizontal, and so there no effects attributable to the hydrostatic pressure exerted by the column of solution shown in Figure 19.3(b); rather, any additional pressure on the solution-side of the semi-permeable membrane is attributable to the pressure exerted by the right-hand piston.

As shown in Figure 19.4, the pressure exerted on the pure solvent by the left-hand piston is 1 bar = 10^5 Pa, and it is this pressure which is exerted by the pure solvent on the pure solvent-side surface of the semi-permeable membrane. If the piston on the right is free to move, as osmosis takes place, solvent molecules will flow through the semi-permeable membrane from left to right. The volume of the solution will therefore increase, and so this piston will move to the right.

Suppose, however, that the piston on the right exerts a pressure $(10^5 + P_{add})$ Pa, implying that the pressure exerted by the solution on the solution-side surface of the semi-permeable membrane is $(10^5 + P_{add})$ Pa. There is therefore a pressure difference of $(10^5 + P_{add}) - 10^5 = P_{add}$ Pa across the semi-permeable membrane from right to left in Figure 19.4, and we are assuming that the semi-permeable membrane can withstand this, and does not break. The effect of the additional pressure P_{add} is to oppose the flow of solvent molecules attributable to osmosis, and if P_{add} is high enough, osmosis can be stopped.

To calculate the additional pressure P_{add} that stops osmosis, we invoke equation (13.20)

$$\left(\frac{\partial G}{\partial P}\right)_{T,n_i\ldots} = V \tag{13.20}$$

which, as we saw on page 406, describes how the Gibbs free energy G of any system of volume V varies with the pressure P, holding constant both the temperature T and the system composition, as represented by the variables n_i. This equation can be applied to the right-hand compartment shown in Figure 19.4, which comprises a solution of n_D mol of a solute D dissolved in n_S mol of a solvent S. If the solution is ideal, all state functions are linearly additive,

MIXING AND OSMOSIS

and so equation (13.20) applies to both the solute D and the solvent S independently. So, for molar quantities, we may express equation (13.20) for the solute D as

$$\left(\frac{\partial G_D}{\partial P}\right)_{T,n_D,n_S} = V_D$$

and for the solvent S as

$$\left(\frac{\partial G_S}{\partial P}\right)_{T,n_D,n_S} = V_S$$

At constant temperature T and constant composition n_i, for any change from an initial pressure P_1 to a final pressure P_2, the expression

$$\left(\frac{\partial G_S}{\partial P}\right)_{T,n_D,n_S} = V_S$$

may be integrated as

$$\int_{G_1}^{G_2} dG_S = \int_{P_1}^{P_2} V_S \, dP \tag{19.1}$$

The left-hand side of this equation integrates to the change $\Delta G_S = G_2 - G_1$ in the molar Gibbs free energy of the solvent S; the value of the right-hand side depends on the equation-of-state that is appropriate to the system of interest. For an ideal gas, this is the ideal gas equation $PV = RT$, and the integral is the familiar $RT \ln (P_2/P_1)$. The current system of interest, however, is not an ideal gas but a liquid, and, assuming that the liquid is incompressible, the molar volume V_S of the solvent is constant, and so the right-hand side of equation (19.1) integrates to $V_S (P_2 - P_1)$ – as we saw on page 380.

If we apply equation (19.1) to the solvent S within the solution on the right hand-side of the semi-permeable membrane as shown in Figure 19.4, then, at a constant temperature T, and for a given mole fraction x_S of the solvent, we have

$$G_S(\text{in solution at pressure } (10^5 + P_{\text{add}}) \text{ Pa}) - G_S(\text{in solution at pressure} 10^5 \text{ Pa})$$
$$= V_S (10^5 + P_{\text{add}} - 10^5) = P_{\text{add}} V_S \tag{19.2}$$

Now although equation (16.7c)

$$G_i(\text{in solution}) = G_i^{\diamond}(\text{pure}) + RT \ln x_i \tag{16.7c}$$

does not explicitly show the pressure P as a variable, a value for the pressure has been assumed – the standard pressure $P^{\diamond} = 10^5$ Pa, this being the pressure referred to in the phrase "assuming constant pressure". Accordingly, for the solvent S, equation (16.7c), is more explicitly written as

$$G_S(\text{in solution at pressure } 10^5 \text{ Pa}) = G_S^{\diamond}(\text{pure}) + RT \ln x_S$$

implying that we may replace $G_S(\text{in solution at pressure } 10^5 \text{ Pa})$ in equation (19.2) by $G_S^{\diamond}(\text{pure}) + RT \ln x_S$ giving

$$G_S(\text{in solution at pressure } (10^5 + P_{\text{add}}) \text{ Pa}) = G_S^{\diamond}(\text{pure}) + RT \ln x_S + P_{\text{add}} V_S \tag{19.3}$$

Equation (19.3) is the equation we have been seeking – an enrichment of equation (16.7c) that takes pressure into account. As can be seen, the greater the additional pressure P_{add}, the greater G_S, and since the term $P_{add}\, V_S$ is necessarily positive, the value of $P_{add}\, V_S$ will counteract the necessarily negative value of $RT \ln x_S$.

For the system to achieve equilibrium, the molar Gibbs free energy of the solvent S in the solution on the right-hand side of the semi-permeable membrane, as given by equation (19.3), must equal the molar Gibbs free energy of the pure solvent S on the left-hand side. Since the pure solvent is at a pressure of 1 bar = 10^5 Pa, the pure solvent is in its standard state, and so the molar Gibbs free energy of the pure solvent is $G_S^{\ast}(\text{pure})$. Accordingly, equilibrium is reached at that particular additional pressure $P_{add} = \Pi$ for which

$$G_S^{\ast}(\text{pure}) = G_S^{\ast}(\text{pure}) + RT \ln x_S + \Pi\, V_S$$

implying that

$$\Pi\, V_S = -RT\, \ln x_S = RT \ln \frac{1}{x_S} = RT \ln \frac{n_S + n_D}{n_S}$$

and so

$$\Pi\, V_S = RT \ln \left(1 + \frac{n_D}{n_S}\right) \tag{19.4}$$

The pressure Π defined by equation (19.4) is known as the **osmotic pressure** of the solution. Since, by definition, the state for which Π is defined is an equilibrium state, there is no net flow of the solvent S across the semi-permeable membrane – the net flow of solvent, as depicted in Figure 19.4 has ceased, and osmosis has stopped. Microscopically, of course, solvent molecules continue to flow in both directions across the semi-permeable membrane; macroscopically, however, the system reaches thermodynamic equilibrium when these flow rates are equal. The osmotic pressure Π may therefore be interpreted as the pressure required to equalise these flows, so counteracting osmosis.

The expression for the osmotic pressure Π, equation (19.4), can be simplified for a dilute solution of the solute D, for which $n_D \ll n_S$, so that the ratio n_D/n_S is much less than 1. This implies that

$$\ln \left(1 + \frac{n_D}{n_S}\right) \approx \frac{n_D}{n_S}$$

and so equation (19.4) becomes

$$\Pi = RT \frac{n_D}{n_S V_S} \tag{19.5}$$

V_S is the molar volume of the pure solvent S, and is related to the volume V of the chamber on the right-hand side of the semi-permeable membrane as

$$V = n_S V_S + n_D V_D$$

where V_D is the molar volume of the solute D. As we have seen, for a dilute solution $n_S \gg n_D$ and so $V \approx n_S V_S$. Equation (19.5) therefore simplifies to

$$\Pi = RT \frac{n_D}{V} = RT\, [D] \tag{19.6}$$

where [D] = n_D/V, the molar concentration of the solute D within the chamber on the right-hand side of the semi-permeable membrane. Equation (19.6) presents a very simple relationship between the molar concentration [D] of the solute and the corresponding osmotic pressure Π, and is known as the **van't Hoff equation (osmosis)**, after Jacobus van't Hoff who, as we noted on page 445, was the recipient of the first Nobel Prize for chemistry in 1901 "in recognition of the extraordinary services he has rendered by the discovery of the laws of chemical dynamics and osmotic pressure in solutions".

19.4 Reverse osmosis

Returning once more to Figure 19.4, as we have seen, the osmotic pressure Π is the additional pressure P_{add}, above atmospheric pressure, that needs to be exerted on a solution to prevent osmosis, the flow of the pure solute into the solution. If, at any temperature, the pressure P_{add} actually exerted on the solution is greater than Π, then molecules of the pure solvent will flow from the solution into the pure solvent – from right to left in Figure 19.4. This is known as **reverse osmosis**, and it enables a pure solvent to be 'squeezed' out of a solution, as exploited in the industrially important industrial process of **desalination**, in which drinking water is produced from salty sea water.

Desalination requires that the pressure exerted on sea water is greater than the osmotic pressure, which is determined by sea water's ionic composition. Typically, 1 kg of sea water contains some 0.035 kg of various ions such as Na^+, K^+, Mg^{2+}, Ca^{2+}, Cl^-, SO_4^{2-} and HCO_3^-, of which Na^+ and Cl^- are the most abundant. For simplicity, let's assume that the ionic content of sea water is all Na^+ and Cl^-, and that each 1 kg of sea water contains 0.035 kg of NaCl. The molecular weight of NaCl is 0.0585 kg mol^{-1}, and so 0.035 kg of NaCl corresponds to 0.6 mol of NaCl for each 1 kg of sea water. Since 1 kg of sea water occupies approximately 10^{-3} m^3, the concentration of NaCl in sea water is approximately 6×10^2 mol m^{-3}. Each molecule of NaCl creates two particles, one Na^+ ion and one Cl^- ion, and so the concentration of ions in sea water is around 1.2×10^3 mol m^{-3} – this being the number required for [D] in the van't Hoff equation, equation (19.6). At $T = 298$ K, the osmotic pressure of sea water at that temperature is therefore given by

$$\Pi = RT[D] = 8.314 \times 298.14 \times 1.2 \times 10^3 = 3 \times 10^6 \text{ Pa}$$

Since 1 bar = 10^5 Pa, the osmotic pressure of sea water is the (probably surprisingly high) value of about 30 bar. Industrial plants that use reverse osmosis to produce drinking water from sea water therefore need to exert a continuous pressure in excess of 30 bar – in practice, pressures of between 40 bar and 80 bar are used.

19.5 A note on hydrostatic pressure

Before we leave our discussion of osmosis, we return to Figure 19.3(b), which represents an equilibrium state, for we now know that the hydrostatic pressure exerted by the vertical column of the solution provides the required osmotic pressure.

To calculate the hydrostatic pressure of the column, suppose that the height of this column, above the level of the semi-permeable membrane, is h m, and that the (constant) cross sectional area of the column is A m^2, giving the total volume of the solution within the column as hA m^3. For a solution of density ρ kg m^{-3}, the total weight of the solution in the column is therefore $hA\rho$ kg, which is associated with a gravitational force of $hA\rho g$ N, where g is the acceleration due to gravity, 9.81 m sec^{-2}. This gravitational force is exerted on the semi-permeable membrane, which also has an area of A m^2, and so the pressure exerted on the membrane by the solution in the column is given by $hA\rho g/A = h\rho g$ Pa.

In the context of equation (19.3)

$$G_S(\text{in solution at pressure}(10^5 + P_{add})\text{ Pa}) = G_S^{\circ}(\text{pure}) + RT\ln x_S + P_{add}V_S \qquad (19.3)$$

$P_{add} = h\rho g$, and so we may express the molar Gibbs free energy G_S(at membrane surface) of the solution at the surface of the semi-membrane as

$$G_S(\text{at membrane surface}) = G_S^{\circ}(\text{pure}) + RT\ln x_S + h\rho gV_S \qquad (19.7)$$

Equation (19.7) contains no surprises; but let's consider the value of molar Gibbs free energy of the solution not at the surface of the semi-permeable membrane, but somewhere else within the column of the solution – say, at a level half-way up the column, at distance $h/2$ m from the upper surface. The weight of the solution above that level is $hA\rho g/2$ kg, and so the pressure at that level is $h\rho g/2$ Pa. Using equation (19.3) once again, we now have

$$G_S(\text{half-way up column}) = G_S^{\circ}(\text{pure}) + RT\ln x_S + h\rho gV_S/2 \qquad (19.8)$$

Comparison of equations (19.7) and (19.8) shows that G_S (at membrane surface) and G_S (half-way up column) are different: G_S (at membrane surface) is greater than G_S (half-way up column) by the amount $h\rho gV_S/2$. This is one instance of a more general variability: $G_S(l)$ at any distance l m, as measured from the top of the column, is given by G_S° (pure) $+ RT\ln x_S + l\rho gV_S$, and so varies with l.

The system comprising the solution within the column is at equilibrium, yet we have a situation in which an intensive state function, the molar Gibbs free energy, is different at different levels l within the system; furthermore, a second intensive state function, the pressure, varies with l as the external pressure plus $l\rho g$.

The significance of this is that it breaks one of the fundamental 'rules' of thermodynamics – the rule, as described on pages 8 and 9, that a necessary condition for thermodynamic equilibrium is that the values of all intensive state functions are constant over time, and have the same values at all points within the system.

The example just described certainly complies with the first condition – the values of both the molar Gibbs free energy, and the pressure, are indeed constant over time. The problem lies with the second condition – the requirement that the values of any intensive state function are the same throughout the system. Perhaps this second requirement is 'softer' than the first, and can be relaxed in cases such as this – but surely such an explanation in unsatisfactory. A better explanation invokes another set of words that have been quoted many times throughout this book "in the absence of gravitation, motion, electric, magnetic and surface effects..." (see, for example, page 48). As is evident from the term $l\rho g$ – which defines the hydrostatic pressure attributable to a column of a material of density ρ kg m^{-3} and height l m – the presence of the factor g, the acceleration due to gravity, 9.81 m sec^{-2}, explicitly shows that hydrostatic pressure is a "gravitational effect". Gravitational effects, such as hydrostatic pressure, add complexity, and – from the standpoint of an introductory book such as this – are best 'wished away': which explains why the analysis of osmotic pressure, as shown in Figure 19.4, has the entire system at the same level, so that there is no reference to hydrostatic pressure, and there are no "gravitational effects".

EXERCISES

1. Define
   - semi-permeable membrane
   - osmosis
   - osmotic pressure
   - reverse osmosis.
2. In Figure 19.1(a), the right–hand compartment contains n_S mol of pure solvent S and n_D mol of solute D. What is the total Gibbs free energy of the whole system shown in Figure 19.1(a)? And also Figure 19.1(b)? Prove that mixing is spontaneous.
3. What is the fundamental physical process that underlies osmosis?
4. In Figure 19.4, suppose that both compartments contain some solute D of molar concentrations $[D]_L$ on the left, and $[D]_R$ on the right, such that $[D]_L < [D]_R$. Derive an expression, like equation (19.6), that defines the pressure P_{add} that stops osmosis.
5. Explain clearly how reverse osmosis can be used to produce drinking water from sea water.

20 Electrochemistry

Summary

If a rod of metallic zinc, an **electrode**, is placed in contact with a solution of Zn^{2+} (aq) ions, an **electrolyte**, a spontaneous **oxidation** reaction

$$Zn\ (s) \rightarrow Zn^{2+}\ (aq) + 2\ e^-$$

known as a **half-reaction**, takes place, in which metallic zinc dissolves into the solution, and a surplus of electrons remains on the metallic rod, conferring on the rod a negative electric charge, and the corresponding negative **electrical potential**. Likewise, when metallic copper is in contact with Cu^{2+} (aq) ions, a spontaneous **reduction** half-reaction

$$Cu^{2+}\ (aq) + 2\ e^- \rightarrow Cu\ (s)$$

takes place in which metallic copper is deposited on the rod, which becomes positively charged, and achieves a positive electrical potential. If the two solutions of Zn^{2+} (aq) ions and Cu^{2+} (aq) ions are in electrical contact – for example, if they are separated by a porous membrane – then, when the zinc and copper rods are connected by an electrical conductor, electricity flows through the conductor. Collectively, the zinc and copper **half-cells** constitute an **electrochemical cell**, and the fundamental source of the electricity is the **cell reaction**

$$Zn\ (s) + Cu^{2+}\ (aq) \rightarrow Zn^{2+}\ (aq) + Cu\ (s)$$

this being the combination of the oxidation half-reaction taking place at the cell **anode**, and the reduction half-reaction taking place at the cell **cathode**.

Each electrode has a corresponding **electrode potential**, E_{cat} at the cathode and E_{an} at the anode, and the **cell potential** E_{cell} is the difference $E_{cat} - E_{an}$ between the two electrode potentials. In everyday language, an electrochemical cell is an electrical battery, and the cell potential is the battery's voltage.

This zinc-copper electrochemical cell is a specific example of a more general phenomenon, in which redox reactions can be harnessed to drive an electric current.

An electrochemical cell's **electromotive force** $\mathbb{E}_{cell}$ is the electrical potential difference between the cell's electrodes when the cell is operating reversibly. This is the maximum electrical potential difference a cell with a given cell reaction can achieve, and $\mathbb{E}_{cell}$ is related to the change ΔG_{cell} in the Gibbs free energy associated with the cell reaction as

$$-\Delta_r G_{cell} = n_e F\ \mathbb{E}_{cell} \tag{20.8b}$$

where n_e is the number of electrons generated by the cell reaction (as determined by the reaction stoichiometry), and F is the magnitude of the electric charge carried by 1 mol of electrons.

Modern Thermodynamics for Chemists and Biochemists. Dennis Sherwood and Paul Dalby.
© Oxford University Press 2018. Published 2018 by Oxford University Press.

If the components of an electrochemical cell are in their standard states, the corresponding **standard electromotive force** $\mathbb{E}_{cell}^{\circ}$ is related to the generalised equilibrium constant K_r for the cell reaction as

$$\mathbb{E}_{cell}^{\circ} = \frac{RT}{n_e F} \ln K_r \tag{20.15}$$

If we generalise the reduction half-reaction

$$Cu^{2+} (aq) + 2 e^- \rightarrow Cu (s)$$

at a copper electrode as

$$\text{oxidised state} + n_e \, e^- \rightleftharpoons \text{reduced state}$$

then the corresponding reversible electrode potential $\mathbb{E}$ (ox,red) is given by the **Nernst equation**

$$\mathbb{E} (ox,red) = \mathbb{E}^{\circ} (ox,red) + \frac{RT}{n_e F} \ln \frac{[ox]}{[red]} \tag{20.19a}$$

20.1 Work and electricity

On page 52, we saw how, during the early 1840s, James Joule showed that a given amount of mechanical work, performed by masses falling under gravity, can be consistently converted into a fixed amount of heat, as measured by the increase in the temperature of a reservoir of water.

Around that time, however, Joule was carrying out a number of other experiments too – experiments concerning electricity, and what happens when an **electrical current** passes through a **conductor**, such as a copper wire. One of his key observations was that a flow of electricity, as driven by an electric battery, causes the wire to get hotter – so here was another phenomenon in which heat is produced without any fire, and produced not by mechanical work, but by the flow of electricity. To quantify the amount of heat produced by a known flow of electricity, Joule used exactly the same concept as the apparatus shown in Figure 3.3 – but rather than immersing a paddle, driven by mechanical energy, in water, Joule immersed an electrical conductor, driven by an electric battery, as illustrated in Figure 20.1.

Joule measured the electrical current I (measured in amps, A) flowing through the conductor, which was of known **electrical resistance** R (measured in ohms, Ω); Joule also knew the mass of the water, and calculated the corresponding quantity of heat produced. As a result, Joule established that the heat produced by the flow of a current through a conductor, over a period of time τ (sec), is given by

$$\text{Heat produced by the flow of electricity over time } \tau = I^2 \, R\tau \text{ J} \tag{20.1}$$

Equation (20.1), first published in 1840, is known as **Joule's first law**, and the heat produced is known as **Joule heating**. A key feature of equation (20.1) is that it is necessarily positive: as a square, I^2 is positive, the resistance R of all normal conducting materials – and certainly those used by Joule – is also positive, as is the measure τ of time (provided that time is running

WORK AND ELECTRICITY

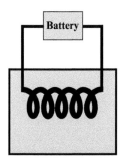

Figure 20.1 **A representation of Joule's electrical apparatus.** A source of electricity, such as a battery, operating at a voltage E V, drives an electric current I A through a conductor of electric resistance R Ω, immersed in a reservoir of water. Over time, the water becomes progressively hotter, demonstrating that the flow of electricity creates heat. Since that flow of electricity is driven by the battery, the battery must be doing work – just as the falling weights in Figure 3.3 do the work that results in the heating of the water by the paddles.

forwards!). This is very similar to the effect of friction, which also involves a squared term, as discussed on pages 133 and 134.

From his measurements of the heat produced by electricity, using the apparatus represented in Figure 20.1, and mechanically, as shown in Figure 3.3, Joule drew a momentous conclusion: the heat produced in the water from the flow of electricity is indistinguishable from the heat produced by the falling of the masses. When electricity flows through a conductor, is work performed, just like masses falling under gravity? Are work and heat essentially the same thing? The evidence that Joule obtained, and the questions he asked, were critical in establishing that heat and work, however performed, are both manifestations of the same underlying concept, energy.

Joule established that the flow of electricity performs work and can create heat, and since thermodynamics is all about the interplay between work, heat and energy, this suggests that electricity is within the scope of thermodynamics. So far in this book, however, we have stated "in the absence of gravitation, motion, electric, magnetic and surface effects ...", and so all electrical effects have been explicitly excluded. This will now be remedied: this purpose of this chapter is therefore to show how electricity can be incorporated into the formalism of chemical thermodynamics – as is relevant as regards, for example, the action of electric batteries (all batteries create electricity from chemical processes), and the chemistry and biochemistry of reactions in aqueous solution, which often involve electrically charged ions.

By the time Joule was working, there was considerable knowledge about electricity – Hans Christian Ørsted, in Denmark, and Michael Faraday, in England, had explored the links between electricity in magnetism in the 1810s and 1820s; reliable electric batteries were available, especially once the 'Daniell cell' had been invented by John Daniell in 1836; and in 1827, Georg Ohm published what we now know as 'Ohm's Law', relating the current I (A) flowing through a material of electric resistance R (Ω) as a result of an electrical potential difference E (measured in volts, V) as $E = IR$. What no-one knew then was that electricity is a physical flow of electrons within the interior of the conductor, and that heat is produced as a result of the 'internal friction' associated with this electron flow through the resisting medium of the molecular structure of the conductor. By analogy, just as the rotational motion of Count

Rumford's cannon drill created heat as a result of the friction between the drill and the internal surface of the bronze, the 'internal friction' attributable to the motion of electrons through the metallic structure creates the heat resulting from the flow of electricity. And like mechanical friction, the process of creating Joule heat from the flow of electricity is dissipative (see page 139), and irreversible: an electric current flowing through a conductor causes heating, but no amount of heating of the conductor will cause a flow of electricity.

At this point, for completeness, we note the Seebeck effect – a phenomenon in which a temperature difference drives an electric current. Superficially, the Seebeck effect might appear to be the reversal of Joule heating. In fact, it isn't – it's totally different. A requirement for the Seebeck effect is that the system comprises two conductors, each made of different materials, with each material at a different temperature – the electricity flow arises from the effect of the different temperatures on the energy levels of the electrons in the two different materials at their mutual junction. Joule heating takes place within a single conductor made of a single material as a result, as described, of 'internal friction'; and, importantly, Joule heating is truly dissipative and irreversible.

20.2 Electrical work and Gibbs free energy

Microscopically, we now know that the heat generated within the conductor is attributable to the work done against 'internal friction' by the flow of electrons. Macroscopically, where does the energy required to perform this work originate? There is only one possibility – from the chemical reaction taking place within the battery.

We now invoke a result from page 389: that the maximum work, other than P,V work of expansion, that can be performed by a system undergoing a change of state at constant pressure and temperature is given by

$$\{_1w_2\}_{\text{other,rev}} = -\Delta G \tag{13.43b}$$

where ΔG is the corresponding change in the Gibbs free energy of the system. Within the battery, a chemical reaction is taking place, and, as with all chemical reactions, this reaction will be associated with a reaction Gibbs free energy $\Delta_r G_{\text{bat}}$. This reaction, however, is rather special, in that it also has the result of driving an electric current. We'll examine the mechanism for this shortly; for the moment, the important point is that, if the reaction within the battery is spontaneous, and therefore associated with a negative value for $\Delta_r G_{\text{bat}}$, then $\{_1w_2\}_{\text{other,rev}}$ is positive. Consequently, as we saw on page 389, it is this quantity that determines the maximum amount of energy that can be transformed into the heat generated by the flow of electricity within the conductor.

We all know that batteries run down, and eventually stop. So if the maximum possible current a battery can generate is I_{max}, then the maximum heat that can be generated in the conductor is determined by equation (20.1) as

$$\text{Maximum heat} = I_{\text{max}}^2 \, R\tau \ \text{J} \tag{20.2}$$

Equation (13.43b) is derived from thermodynamics, whilst equation (20.2) is derived from the experimental observations of Joule. Both are describing the same phenomenon, in that it is the work done by the battery that results in the heat produced in the conductor: assuming that no work produced by the battery is 'lost', it must therefore be the case that

$$-\Delta_r G_{bat} = I_{max}^2 R\tau \qquad (20.3)$$

Equation (20.3) may be expressed rather differently, by combining it with Ohm's Law, which relates the current I A flowing through a conductor of resistance R Ω to the voltage E V across the conductor as

$$E = IR \qquad (20.4)$$

giving

$$-\Delta_r G_{bat} = E_{max} I_{max} \tau \qquad (20.5)$$

where E_{max} is the maximum voltage that the battery can deliver, so driving the maximum current I_{max}. In equation (20.5), each of E_{max} and I_{max} can be associated with either a positive or a negative sign, depending on how the potential difference across the conductor is defined (according to which terminal on the battery is designated 'positive' and which 'negative'), and the direction of current flow. The term $E_{max}I_{max}$ in equation (20.5), however, is simply an alternative to the term $I_{max}^2 R$ in equation (20.3), and since $I_{max}^2 R$ is necessarily positive, we must ensure that the product $E_{max} I_{max}$ is always positive too.

As the term 'electric current' suggests, any electric current I (A) can be represented in terms of the movement of a specific quantity Q (Coulomb, C) of electric charge, over a period of time τ (sec) as

$$I = \frac{Q}{\tau}$$

For the battery, if, over time τ, a maximum quantity of n_e(max) mol of electrons can flow through the external conductor, then the magnitude of the total charge transferred is n_e(max)F – where F, known as the **Faraday constant**, is the magnitude of the total charge associated with 1 mol of electrons, 9.6485×10^4 C mol^{-1} (noting that, in order to ensure that the product $E_{max}I_{max}$ is positive, we are using the magnitude of the charge carried, even though, formally, the charge on an electron is negative). I_{max} may therefore be expressed as

$$I_{max} = \frac{n_e(max) F}{\tau}$$

from which

$$I_{max} \tau = n_e(max) F$$

Equation (20.5) therefore becomes

$$-\Delta_r G_{bat} = n_e(max) F E_{max} \qquad (20.6)$$

We now have an expression, equation (20.6), linking the change in Gibbs free energy $\Delta_r G_{bat}$ for the reaction taking place within the battery to the maximum voltage E_{max} (necessarily expressed as a positive number) that the battery can generate, and the corresponding maximum number n_e mol of electrons the reaction can produce. Where, though, do the electrons come from?

20.3 Half-cells

To answer this question, we will examine how a chemical reaction within the battery produces electricity, and the underlying principles determining the voltage E_{max}. Our starting point is the phenomenon that spontaneously happens when a rod of pure metallic zinc is placed on contact with a solution of aqueous $ZnSO_4$, as shown in Figure 20.2(a).

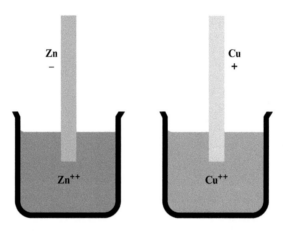

Figure 20.2 Two, separate, half-cells. If a rod of pure zinc metal, an electrode, is in contact with a solution of Zn^{2+} ions, an electrolyte, as in (a), the metallic rod becomes negatively charged. Similarly, a rod of pure copper metal, another electrode, in contact with a solution of Cu^{2+} ions, another electrolyte, becomes positively charged.

Before the zinc rod is inserted into the solution, the $ZnSO_4$ solution contains an equal number of oppositely charged, dissociated, Zn^{2+} (aq) and SO_4^{2-} (aq) ions, so that, although the solution contains charged ions, the solution is electrically neutral. When the zinc rod is placed in contact with the solution, a spontaneous reaction takes place between the surface of the metal and the aqueous solution, in which zinc atoms pass from the metal into the solution as Zn^{2+} (aq) ions, leaving the corresponding electrons within the metal rod

$$Zn\ (s) \rightarrow Zn^{2+}\ (aq) + 2\ e^-$$

The zinc rod therefore loses mass, and the surplus electrons within the metal rod confer a negative charge on the metal rod. An equal and opposite positive charge is created within the solution, attributable to the surplus of Zn^{2+} (aq) ions and so the overall electrical effect of the reaction is therefore to separate charges, with a net positive charge in the solution, known as the **electrolyte**, and a net negative charge on the metallic zinc rod, known as the **electrode**.

The system depicted in Figure 20.2(a) is called a zinc **half-cell**, and the corresponding reaction

$$Zn\ (s) \rightarrow Zn^{2+}\ (aq) + 2\ e^-$$

which describes what happens spontaneously when metallic zinc first has contact with an aqueous solution, is known as a **half-reaction** (or sometimes **half-cell reaction**). Since this half-reaction, in the direction as written, is one in which the (relatively) electron-rich metallic

Zn (s) (the reduced state) loses electrons to form the corresponding (relatively) electron-deficient Zn^{2+} (aq) ions (the oxidised state), this reaction is within the class of reactions known as oxidation.

Like all chemical reactions, this half-reaction is, in principle, reversible, the reverse reaction being

$$Zn^{2+} (aq) + 2\ e^- \rightarrow Zn\ (s)$$

in which Zn^{2+} (aq) ions recombine with electrons within the metallic zinc rod, depositing Zn(s) on the rod's surface. The system is therefore more correctly represented as reversible half-reaction as

$$Zn\ (s) \rightleftharpoons Zn^{2+} (aq) + 2\ e^-$$

This half-cell reaction is a particular case of the generalised reversible reaction

$$aA + bB \rightleftharpoons cC + dD$$

taking place in solution, as discussed in Chapter 16, and all the thermodynamic principles, and all the equations, can be applied to half-cell reactions too. So, under any given system conditions concerning, for example, temperature and pressure, a half-cell reaction will be associated with a standard reaction Gibbs free energy $\Delta_r G_c^{\circ}$, and a corresponding concentration equilibrium constant K_c such that

$$\Delta_r G_c^{\circ} = -RT \ln K_c \tag{16.16c}$$

We will explore the quantitative thermodynamics in more detail shortly (see page 621); for the moment, we note that since we are now dealing with entities such as positively charged Zn^{2+} (aq) ions in solution, and negatively charged electrons e^- on the metallic zinc rod, one of the 'system conditions' which is now relevant, but has not featured so far in this book, is the effect of electrostatic interactions, such as the repulsion between particles carrying similar charges, and the attraction between particles carrying opposite charges.

When the zinc rod is first placed into the $ZnSO_4$ solution, it is experimentally verifiable that the half-reaction

$$Zn\ (s) \rightarrow Zn^{2+} (aq) + 2\ e^-$$

takes place spontaneously, to the right, as shown. But as more Zn atoms pass into solution as Zn^{2+} (aq) ions, the zinc rod loses mass, whilst accumulating a surplus of electrons. As a result, the zinc rod becomes more negatively charged. Simultaneously, the formation of Zn^{2+} (aq) ions makes the solution progressively more positively charged. Over time, the increasing positive charge in the solution slows down the rate at which further Zn atoms ionise; simultaneously, the increasing negative charge on the metallic zinc rod attracts the Zn^{2+} (aq) ions out of solution, reforming metallic Zn atoms. As a result of these electrostatic interactions, the system therefore achieves equilibrium, at which point, as we saw on page 451, the rate of the forward ionisation half-reaction is equal to the rate of the reverse electron-capture half-reaction

$$Zn\ (s) \rightleftharpoons Zn^{2+} (aq) + 2\ e^-$$

Figure 20.2(b) shows another half-cell, in which metallic copper – another example of an electrode – is in contact with another electrolyte, a solution of CuSO$_4$, containing dissociated Cu^{2+} (aq) and SO$_4^{2-}$ (aq) ions. In this case, on contact with the Cu^{2+} (aq) ions in solution, the metallic copper reacts spontaneously in accordance with the half-reaction

$$Cu^{2+} \text{ (aq)} + 2\,e^- \rightarrow Cu \text{ (s)}$$

In contrast to the zinc half-reaction, the copper half-reaction, in the direction as written, is one in which the (relatively) electron-deficient Cu^{2+} (aq) ions (the **oxidised state**) gain electrons to form the corresponding (relatively) electron-rich Cu (s) (the **reduced state**), and is within the class of reactions known as **reduction**.

As a consequence of this half-reaction

- The Cu(s) atoms formed from the oxidation of the Cu^{2+} (aq) ions are deposited on the metallic copper rod, which therefore gains mass.
- Electrons are lost from the metallic copper rod, which therefore acquires a positive charge.
- Positively charged Cu^{2+} (aq) ions are withdrawn from the solution, which is therefore left with a surplus of SO$_4^{2-}$ (aq) ions. The solution therefore acquires a net negative charge.

An, just as we saw for the zinc half-cell, over time, the increasingly positive charge of the metallic copper rod acts to repel further positively charged Cu^{2+} (aq) ions, and so the copper half-reaction achieves an equilibrium state

$$Cu^{2+} \text{ (aq)} + 2\,e^- \rightleftharpoons Cu \text{ (s)}$$

with a separation of charge such that the metallic rod is positive, and the solution, negative.

The examples of the zinc and copper half-cells are two instances of a more general phenomenon: in principle, any element X, when in contact with an aqueous solution, will either, like zinc, spontaneously ionise

$$X \rightarrow X^{n+} \text{ (aq)} + n\,e^-$$

or, like copper, accept electrons

$$X^{n+} \text{ (aq)} + n\,e^- \rightarrow X$$

Because solid metals have an electronic structure in which individual electrons are only relatively loosely bound to the corresponding nuclei (as is macroscopically evident from the ability of metals to conduct electricity), half-reactions of this type for metals have relatively low activation energies, implying that the formation of ions (as for zinc), or the discharge of ions (as for copper), happens at a measureable rate at normal temperatures (see pages 455 to 459). Most half-cells are therefore metallic, but not all – as we shall see on page 614, an important half-cell uses gaseous hydrogen, associated with the half-reaction

$$2\,H^+ \text{ (aq)} + 2\,e^- \rightarrow H_2 \text{ (g)}$$

20.4 The electrochemical cell

The two half-cells shown in Figure 20.2 soon equilibrate, attributable, as we have seen, to the build-up of a positive charge in the solution for the zinc half-cell depicted in Figure 20.2(a), and of a negative charge in the solution for the copper half-cell shown in Figure 20.2(b). Suppose, however, that it were possible to avoid the build-up of charge within the solutions. If, for example, we could add negative ions to the solution for the zinc half-cell shown in Figure 20.2(a), then these would counteract the build-up of the positive charge, and the half-reaction

$$Zn\,(s) \;\rightarrow\; Zn^{2+}\,(aq) + 2\,e^{-}$$

could continue to the right, towards an equilibrium determined by the thermodynamics of the half-cell reaction, uninfluenced by electrostatic interactions. Likewise, if positive charges could be added to the copper half-cell shown in Figure 20.2(b), then the copper half-reaction

$$Cu^{2+}\,(aq) + 2\,e^{-} \;\rightarrow\; Cu\,(s)$$

could also continue towards its corresponding 'uninfluenced' thermodynamic equilibrium.

A moment's thought will verify that a zinc half-cell has a certain 'symmetry' with respect to a copper half-cell: the zinc half-cell creates a net positive charge in the $ZnSO_4$ solution, which could be helpful in neutralising the net negative charge in the $CuSO_4$ solution of the copper half-cell; likewise, the copper half-cell creates a net negative charge in the $CuSO_4$ solution, which could be helpful in neutralising the net positive charge in the $ZnSO_4$ solution. Might it be possible to link the two half-cells together, so that the two solutions neutralise each other?

Yes, it is. And the result is a useable electric battery, or **electrochemical cell**, known as the **Daniell cell**, as represented in Figure 20.3.

On the left is a zinc half-cell, comprising a metallic zinc electrode, in contact with a solution of $ZnSO_4$, which dissociates into Zn^{2+} (aq) and SO_4^{2-} (aq) ions; on the right is a copper half-cell comprising a metallic copper electrode, in contact with a solution of $CuSO_4$, which dissociates into Cu^{2+} (aq) and SO_4^{2-} (aq) ions. The two half-cells are connected externally by means of an electric conductor, such as a copper wire; internally, the electrolyte comprises the two solutions, which are in contact through a porous partition, which slows down the mixing of the two solutions, whilst allowing the flow of ions.

When two half-cells are connected together, collectively, they form a **Galvanic electrochemical cell**, or in more familiar terms, an electric battery. If an external conductor is connected between the cell's terminals, an electric current will flow through the conductor, so causing the heating studied by Joule. From an electrical point of view, the current I A through the external conductor is driven by the **electrical potential difference** E V between the cell's teminals, an electrical potential difference attributable to the presence of a net positive charge on the Cu electrode, and a net negative charge on the Zn electrode, charges which result from the chemical reactions taking place within the two half-cells. We'll take a deeper look at the chemistry of these reactions, and at what is happening within the half-cells, shortly: firstly, we need to clarify some technical terms.

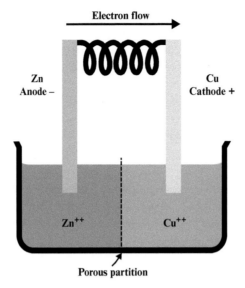

Figure 20.3 The Daniell cell. A zinc half-cell, on the left, is connected to a copper half-cell, on the right. Externally, the connection is by means of an electric conductor between the zinc and copper rods; within the cell, the two solutions are in contact through a porous partition. The negatively charged terminal (zinc), the anode, is conventionally shown on the left, and the positively charged terminal, the cathode (copper), is conventionally shown on the right. Electrons spontaneously flow from the anode to the cathode from left to right as shown, through the electric conductor, which becomes hot as a result of the 'internal friction'. Simultaneously, ions flow through the porous partition so as to maintain electrical neutrality throughout the electrolyte. By convention, the flow of the electric current through the external conductor is from the cathode to the anode, from right to left, this being opposite to the direction of the physical flow of electrons.

20.5 Anodes and cathodes ...

The electrode of any electrochemical cell is associated with a corresponding, and specific, half-reaction, and for the Daniell cell of Figure 20.3, the zinc electrode half-reaction is the oxidation

$$Zn\;(s) \rightarrow Zn^{2+}\;(aq) + 2\;e^-$$

whilst the copper electrode half-reaction is the reduction

$$Cu^{2+}\;(aq) + 2\;e^- \rightarrow Cu\;(s)$$

The two electrodes are distinguished by referring to one as the **anode**, and the other as the **cathode**, where these terms are defined by reference to the nature of the corresponding half-reaction

- The **anode** is the electrode at which the *oxidation* half-reaction takes place. Since the oxidation half-reaction produces electrons, the anode will acquire a *negative* charge. By convention, the anode is always depicted, or otherwise represented, to the left.
- The **cathode** is the electrode at which the *reduction* half-reaction takes place. Since the reduction half-reaction consumes electrons, the cathode will acquire a *positive* charge. By convention, the cathode is always depicted, or otherwise represented, to the right.

As a consequence of the charges on the anode and the cathode, negatively charged electrons will flow through the external conductor from the anode to the cathode, from left to right in Figure 20.3. Conventionally, the electric current I A is regarded as flowing through the external conductor from the positive electrode (the cathode) to the negative electrode (the anode), from right to left in Figure 20.3: the physical flow of electrons through the external conductor, and the conventional designation of the flow of electric current, are therefore in opposite directions.

The definitions of the anode and the cathode refer *only* to whether the corresponding half-reaction is an oxidation (at the anode) or a reduction (at the cathode), and *not* to the material, for the (good) reason that the same material can act as either an anode, or a cathode, depending on the circumstances. So, as we have seen, the zinc half-reaction is reversible

$$Zn\ (s) \rightleftharpoons Zn^{2+}\ (aq) + 2\ e^-$$

and it could happen that the actual reaction taking place under particular conditions is the reverse half-reaction

$$Zn^{2+}\ (aq) + 2\ e^- \rightarrow Zn\ (s)$$

This half-reaction is a reduction, and so, under these conditions, the zinc electrode is acting as a cathode.

Likewise, the copper half-reaction

$$Cu^{2+}\ (aq) + 2\ e^- \rightleftharpoons Cu\ (s)$$

is reversible, and, under certain circumstances, can run in reverse

$$Cu\ (s) \rightarrow Cu^{2+}\ (aq) + 2\ e^-$$

Since this half-reaction is an oxidation, the copper electrode is now acting as an anode.

Under all circumstances, the anode is always associated with an **oxidation half-reaction**, and the cathode with a **reduction half-reaction**. So, for a Galvanic Daniell electrochemical cell, the zinc electrode is the anode and the copper electrode is the cathode, but if a copper half-cell is coupled with a silver half-cell (formed from metallic silver in contact with Ag^+ (aq) ions), the copper electrode is the anode, and the silver electrode, the cathode.

20.6 …and the flow of ions and electrons

The 'content' of the battery, the electrolyte – which in the case of the Daniell cell shown on Figure 20.3 comprises the solution of $ZnSO_4$ on the left, and the solution of $CuSO_4$ on the right – plays a vital role on the operation of the cell. It is the flow of ions within the electrolyte, through the central porous partition, that fulfils two key functions: keeping the electrolyte electrostatically neutral, and completing the overall electric circuit.

Firstly, with reference to the Daniell cell, at the anode, Zn atoms migrate from the solid zinc rod into the $ZnSO_4$ solution as Zn^{2+} (aq) ions, so causing a local accumulation of positive charge; simultaneously, at the cathode, Cu^{2+} (aq) ions are being withdrawn from the solution and deposited as solid copper, creating a surplus of SO_4^{2-} (aq) ions. The contact between the anode and the cathode through the porous partition allows the SO_4^{2-} (aq) ions to flow from right to left in Figure 20.3, and Zn^{2+} (aq) ions from left to right, so maintaining electrical neutrality throughout the electrolyte, as required to solve the problem, discussed on page 599, of the build-up of charge.

Secondly, for the cell overall, negative charges, carried by electrons, flow through the external conductor from the anode to the cathode; simultaneously, within the electrolyte, the rightwards drift of positive ions, combined with the leftwards drift of negative ions, results in a net flow of negative charge from the cathode to the anode. The electric circuit, through both the external conductor and the electrolyte, is complete, corresponding (with reference to Figure 20.3) to a clockwise flow of negative charge – a flow ultimately attributable to the coupling of the half-reactions taking place spontaneously on the surfaces of the two metal electrodes, in contact with their respective electrolyte solutions.

20.7 The chemistry of the Daniell Cell

Let's now take a closer look at those two half-reactions. As we now know, at the zinc anode of the Daniell cell represented in Figure 20.3, surplus electrons accumulate in the metallic zinc rod as a result of the zinc half-reaction

$$Zn\ (s) \rightarrow Zn^{2+}\ (aq) + 2\ e^-$$

These electrons are able to flow through the metallic conductor to the metallic copper rod, where they are consumed by the copper half-reaction

$$Cu^{2+}\ (aq) + 2\ e^- \rightarrow Cu\ (s)$$

Overall, the zinc rod progressively dissolves into the $ZnSO_4$ solution, which becomes more concentrated, whilst, simultaneously, metallic copper is deposited on the copper rod, so depleting the $CuSO_4$ solution.

In the Daniell cell, the two half-reactions

$$Zn\ (s) \rightarrow Zn^{2+}\ (aq) + 2\ e^-$$

$$Cu^{2+}\ (aq) + 2\ e^- \rightarrow Cu\ (s)$$

happen simultaneously, and spontaneously, and may be described in terms of the overall **cell reaction**

$$Zn\ (s) + Cu^{2+}\ (aq) \rightarrow Zn^{2+}\ (aq) + Cu\ (s)$$

The corresponding electrochemical cell in which this reaction takes place may be represented as

$$Zn\ (s)\ |\ ZnSO_4\ (aq)\ \vdots\ CuSO_4\ (aq)\ |\ Cu(s)$$

which identifies all the components, with the symbol $\vdots$ indicating that the $ZnSO_4$ solution is separated from the $CuSO_4$ solution by a porous partition. In this representation, the anode, at which the oxidation half-reaction takes place, is always on the left; the cathode, at which the reduction half-reaction takes place, is always on the right.

Since the cell reaction couples together the reduction half-reaction

$$Cu^{2+} (aq) + 2\ e^- \rightarrow Cu\ (s)$$

which takes place at the cathode, with the oxidation half-reaction

$$Zn\ (s) \rightarrow Zn^{2+} (aq) + 2\ e^-$$

which takes place at the anode, the electrochemical cell as a whole forms a **redox system**, undergoing the **redox reaction**

$$Zn\ (s) + Cu^{2+} (aq) \rightarrow Zn^{2+} (aq) + Cu\ (s)$$

20.8 Currents, voltages and electrode potentials

As the electrons pass through the external conductor, they do work against the 'internal friction', so resulting in Joule heating. As we saw on page 595, the flow of electrons may be measured as the electric current I A, corresponding to the quantity Q C of charge flowing through the circuit over any time period τ sec, such that $I = Q/\tau$. From an electrical point-of-view, the flow of the current I A is the result of an electrical potential difference E V between the terminals of the cell. Fundamentally, this potential difference is attributable to the charges conferred on the cathode and anode as a result of the corresponding electrode half-reactions. We therefore define the **electrode potential** E_{el} of any electrode as the **electrical potential** attributable to the appropriate half-reaction, such that the electrical potential difference E of any Galvanic electrochemical cell is given by

$$E = E_{cat} - E_{an} \tag{20.7}$$

where E_{cat} is the **cathode potential**, and E_{an} is the **anode potential**. Since, as we saw on page 600, the cathode is always the positive terminal, E_{cat} is always greater than E_{an}, and so E, as defined by equation (20.7), is always positive.

The electrode potential is intrinsically associated with the electrode half-reaction. So, for a material X acting as a cathode, the corresponding half-reaction is the reduction

$$X^{n+} (aq) + n\ e^- \rightleftharpoons X\ (s)$$

for which the electrode potential may be represented as E_{el}, or more informatively, as E (oxidised state, reduced state), which in this case is $E(X^{n+}, X)$.

Also, this half-reaction is reversible, and will be associated with an equilibrium constant. If the half-reaction equilibrium lies far to the right, the equilibrium constant will be relatively large, and correspondingly large quantities of electrons will be consumed. The electrode will therefore have a high positive charge, and the corresponding electrode potential will have a large magnitude, and a positive sign.

Likewise, for a material X acting as an anode, the corresponding half-reaction is an oxidation, as generally represented by

$$X\ (s) \rightleftharpoons X^{n+} (aq) + n\ e^-$$

If the half-reaction equilibrium lies far to the right, the equilibrium constant will be relatively large, and correspondingly large quantities of electrons will accumulate on the pure material

X. The electrode will therefore have a high negative charge, and the corresponding electrode potential will have a large magnitude, and a negative sign.

The value of the electrode potential E_{el} is therefore directly attributable to the corresponding half-reaction, and thermodynamically, the electrode potential E_{el} of any electrode is an *intensive state function* of the system associated with the corresponding half-reaction. And as with all state functions, the electrode potential E_{el} is associated with a unit of measure, in this case, the volt, V; E_{el} is a signed, algebraic, quantity; and E_{el} will vary with other state functions that are associated with the appropriate half-reaction, such as the temperature and pressure at which the half-reaction is taking place.

20.9 A different type of Daniell cell

The Daniell cell shown in Figure 20.3, and represented as

$Zn\,(s) \mid ZnSO_4\,(aq) \vdots CuSO_4\,(aq) \mid Cu\,(s)$

provides electrical conductivity within the electrolyte by allowing direct contact between the solutions of $ZnSO_4$ and $CuSO_4$, through the central porous partition. It turns out, however, that when two different electrolytic solutions – such as solutions of $ZnSO_4$ and $CuSO_4$ – are in direct contact, a small electrical potential difference is spontaneously produced at the common boundary. In general, this **liquid junction potential** opposes the cell potential, and so the presence of a liquid junction potential implies that the cell potential actually delivered by cell is rather less than the theoretical maximum cell potential, as determined by the cell reaction. In designing an electrochemical cell, it is therefore advantageous to minimise, or if possible eliminate, any liquid junction potentials. Figure 20.4 therefore shows an alternative form of Daniell cell, in which the solutions of $ZnSO_4$ and $CuSO_4$ are not in direct contact through a porous partition as in Figure 20.3, but are connected electrically by means of a **salt bridge**, this being an inverted tube filled with an agar gel containing an ionic salt such as KCl or NaCl.

An electrochemical cell in which the electrolyte solutions are not in direct contact is represented as

$Zn\,(s) \mid ZnSO_4\,(aq) \mid\mid CuSO_4\,(aq) \mid Cu(s)$

where the symbol || replaces the symbol $\vdots$, as used to represent the porous partition. And as usual, the anode is on the left, and the cathode on the right.

The flow of ions across the salt bridge ensures electrical conductivity throughout the electrolyte, so enabling the flow of current, whilst maintaining electrical neutrality, and the presence of the agar gel is more effective than a porous membrane in preventing mixing. The electrochemical cell shown in Figure 20.4, however, has two boundaries – one at each end of the salt bridge – and so, at first sight, it might appear that the design shown in Figure 20.4, with two boundaries, and hence with two liquid junction potentials, would be less effective than the design shown in Figure 20.3, which has only a single boundary, at the porous partition. Figure 20.4 does indeed have two boundaries, and there are two liquid junction potentials – but because a negatively charged SO_4^{2-} (aq) ion flowing from right to left (in Figure 20.4) passes

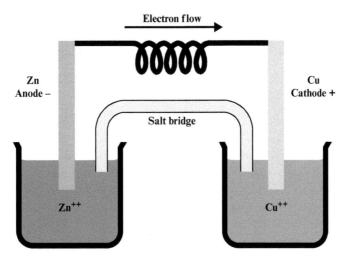

Figure 20.4 A Daniell cell with a salt bridge. Electrical conductivity throughout the electrolyte is achieved by means of a salt bridge – a tube filled with an agar gel containing an ionic salt.

firstly across a solution → bridge boundary, and then across a bridge → solution boundary, the two liquid junction potentials are of opposite signs. Furthermore, although the boundaries are different – that on the right is between the salt bridge and a solution of $CuSO_4$, whilst that on the left is between the salt bridge and a solution of $ZnSO_4$ – for these two solutions, the two liquid junction potentials are of nearly equal magnitude. As a result, from a practical point of view, they effectively cancel each other out. An electrochemical cell with a salt bridge therefore delivers a rather higher cell potential than the same cell with a porous partition.

20.10 Different types of electrode

The electrodes in the Daniell cell are both metals, and they are also metals that correspond to the positively charged ions in their respective electrolyte solutions: a Zn rod in contact with a solution of $ZnSO_4$ forms the anode, and a Cu rod in contact with a solution of $CuSO_4$ forms the cathode.

The purpose of the electrode is two-fold:

- firstly, to provide a conductor of free electrons, from the anode electrolyte to the external conductor, and then from the external conductor into the cathode electrolyte
- secondly, to permit the capture of electrons associated with the oxidation half-reaction at the anode, and the release of electrons associated with the reduction half-reaction at the cathode.

In the Daniell cell, both of these functions are performed by the zinc and copper rods. Many other metals act similarly, for example metallic lead, acting as the anode in a lead-acid storage battery, and metallic cadmium, acting as the anode in a rechargeable nickel-cadmium battery.

The required functions, however, can also be carried out by other forms of electrode, of which we mention three here. Firstly, graphite rods can act as electrodes since graphite conducts electricity, and can also mediate redox half-reactions on its surface. Graphite withstands high temperatures, which is useful in industrial processes, and offers the additional benefit of being inert, the significance of which is best appreciated in the context of the oxidation half-reaction

$$Zn\,(s) \rightarrow Zn^{2+}\,(aq) + 2\,e^-$$

associated with the zinc anode of a Daniell cell. As a consequence of this half-reaction, in which metallic zinc is oxidised to Zn^{2+} (aq) ions, electrons are created, which confer a negative charge on the zinc rod. Simultaneously, metallic zinc enters the solution as Zn^{2+} (aq) ions, implying that the zinc rod progressively loses mass as it dissolves. The material of the electrode, zinc metal, is therefore an active participant in the oxidation half-reaction, and during the oxidation, the material of the anode is consumed.

An oxidation half-reaction at a **graphite electrode**, however, is rather different. As an example, under the appropriate conditions (to be explained shortly), OH^- (aq) ions oxidise to molecular oxygen according to the half-reaction

$$4\,OH^-\,(aq) \rightarrow O_2\,(g) + 2\,H_2O + 4\,e^-$$

on the surface of a graphite anode, and graphite's ability to conduct electricity allows the resulting electrons to flow away to the external circuit. But, unlike the zinc rod which was an active participant in the oxidation half-reaction

$$Zn\,(s) \rightarrow Zn^{2+}\,(aq) + 2\,e^-$$

the graphite rod is not an active participant in the oxidation half-reaction

$$4\,OH^-\,(aq) \rightarrow O_2\,(g) + 2\,H_2O + 4\,e^-$$

Rather, the graphite acts more like a catalyst, enabling the half-reaction to take place, but not explicitly acting as a reagent or product. This **inert** property of a graphite electrode implies that it does not dissolve, and so is not consumed.

A further type of electrode, the **gas electrode**, is represented in the form of the half-cell represented in Figure 20.5.

Figure 20.5 shows a vessel of a solution containing H^+ (aq) ions, for example, a dilute solution of HCl in water, into which pure H_2 (g) is bubbled. Immersed in the solution is a **platinum electrode**, a small piece of metallic platinum, which has been specially treated by a process known as **platinisation**, in which pure, clean, platinum metal is coated with fine particles of platinum metal, known as platinum black. This creates a piece of pure platinum with a very much increased surface area, which can act as a catalyst of the reduction half-reaction

$$2\,H^+\,(aq) + 2\,e^- \rightarrow H_2\,(g)$$

Like a graphite electrode, the platinum in Figure 20.5 does not actively participate in the half-reaction; rather, the surface of the platinum enables the half-reaction to take place.

All the electrodes discussed so far involve the interaction of an element, in the form of a pure metal rod, or a gas, with a solution of the corresponding ions, in which electrons are exchanged between the element and the ions. A **redox electrode** is rather different: if an inert conductor such as platinised platinum is placed in a solution containing two different ions of

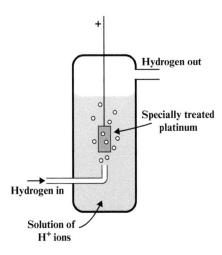

Figure 20.5 A gas electrode, represented as a half-cell. A gas, in this case hydrogen, is bubbled into a solution containing H⁺ (aq) ions, over a specially treated platinum foil, so forming an electrode.

the same element in different oxidation states, such as a mixture of Fe^{2+} (aq) and Fe^{3+} (aq), then the platinum mediates the redox half-reaction

$$Fe^{3+} (aq) + e^- \rightarrow Fe^{2+} (aq)$$

implying (for the reaction as shown) that the platinum rod will acquire a positive charge, and act as a cathode.

Although the term 'redox electrode' refers specifically to an electrode formed from an inert conductor in contact with a solution containing ions of the same element in different oxidation states as just described, a moment's thought will verify that all electrodes are directly involved in some form of redox reaction: by definition, a redox reaction is one in which electrons are transferred, and that is the fundamental basis of all electrochemistry. All electrodes therefore mediate redox reactions, and so, in that sense, all electrodes are, generically, redox electrodes; that said, the term 'redox electrode' is primarily used in the more restrictive sense as just defined.

20.11 Different types of electrochemical cell

A Daniell cell, as represented in Figure 20.3, is associated with the spontaneous cell reaction

$$Zn\,(s) + Cu^{2+}\,(aq) \rightarrow Zn^{2+}\,(aq) + Cu\,(s)$$

and generates a corresponding cell potential E_{Dan} accordingly. As we have seen, this is an example of a **Galvanic electrochemical cell**, named after the great eighteenth century Italian scientist, Luigi Galvani.

Suppose, however, that a Galvanic cell, such as a Daniell cell, is connected to another, external, electrochemical cell that can generate a cell potential E_{ex}. Suppose further that the Daniell cell and the external cell are connected so that the Daniell cell potential E_{Dan} is opposed to the external cell potential E_{ex} as shown in Figure 20.6.

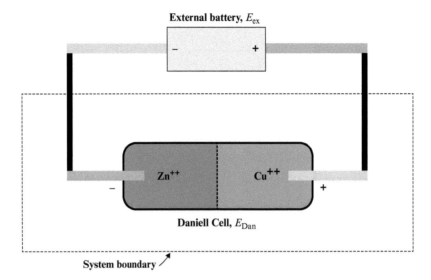

Figure 20.6 An electrolytic cell. An external battery of potential E_{ex} is connected in opposition to a Daniell cell, of cell potential E_{Dan}. If $E_{ex} > E_{Dan}$, cell reaction within the Daniell cell is forced to run in the reverse direction, and acts as an electrolytic cell.

Since the Daniell cell is in opposition to the external cell, the actual direction of flow of electrons through the circuit, and the corresponding conventional direction of flow of the electric current, depends on the relative magnitudes of E_{Dan} and E_{ex}.

If, for example, $E_{ex} = 0$, the external cell is not generating any voltage, and is just a conductor. The Daniell cell therefore operates as normal, driving a flow of electrons clockwise (with reference to Figure 20.6), and the corresponding conventional flow of the electric current is anti-clockwise. And, as we know, the electron flow results from the Daniell cell reaction

$$Zn\ (s) + Cu^{2+}\ (aq) \rightarrow Zn^{2+}\ (aq) + Cu\ (s)$$

in which metallic zinc dissolves from the anode, and metallic copper is deposited on the cathode. In this process, a spontaneous chemical reaction causes a flow of electricity.

Suppose that the external cell now generates a voltage E_{ex}, less than E_{Dan}. The effective cell potential of the Daniell cell is correspondingly reduced to $E_{Dan} - E_{ex}$, but since this quantity is greater than zero, electrons will still flow through clockwise the circuit, but the rate of flow will be reduced; similarly, an electric current will (conventionally) flow through the circuit anti-clockwise as before, but the magnitude of the current will be smaller.

Suppose now that the external cell generates a cell potential $E_{ex} > E_{Dan}$, perhaps substantially so. Metaphorically, the Daniell cell, which is 'pushing' electrons clockwise, is 'overcome' by the external cell, which is 'pushing' electrons anti-clockwise. As a result, there is a physical net flow of electrons anti-clockwise, corresponding to a conventional flow of an electric current clockwise. The reaction within the Daniell cell is therefore forced to take place in reverse

$$Zn^{2+}\ (aq) + Cu\ (s) \rightarrow Zn\ (s) + Cu^{2+}\ (aq)$$

with copper being dissolved, and zinc deposited.

When an electrochemical cell, which spontaneously operates as a Galvanic cell, is forced by an external voltage to run in reverse, the cell is now referred to as referred to as an **electrolytic electrochemical cell**. And, whereas a Galvanic cell creates an electric current from a normally spontaneous chemical reaction, the corresponding electrolytic cell uses an electric current to drive a chemical reaction in the reverse, non-spontaneous, direction.

For a Daniell cell, this forced reversal of the Galvanic cell reaction implies that the half-reaction at the zinc electrode of the Daniell electrolytic cell is the reduction

$$Zn^{2+} (aq) + 2\ e^- \rightarrow Zn\ (s)$$

whilst the half-reaction at the copper electrode is the oxidation

$$Cu\ (s) \rightarrow Cu^{2+} (aq) + 2\ e^-$$

If we maintain the definitions of the anode and the cathode, as given on page 600, as

- the anode is the location of the oxidation half-reaction, and
- the cathode is the location of the reduction half-reaction

then for the electrolytic Daniell cell, the anode is the copper electrode, and the cathode is the zinc electrode. This is the *other way around* as compared to the Galvanic Daniell cell, in which, as we saw in Figure 20.3, the anode is the zinc electrode, and the cathode, the copper electrode. This is a further example of the statement we made on page 601 that the designations 'anode' and 'cathode' stand for 'oxidation half-reaction' and 'reduction half-reaction', respectively, regardless of the material of the corresponding electrode.

Furthermore, for an electrolytic Daniell cell, the flow of electrons through the external circuit is from the copper terminal to the zinc terminal. The Zn terminal therefore now has a positive charge, and the Cu terminal a negative charge, implying that *the polarity of the terminals is reversed* for an electrolytic cell, as compared to the corresponding Galvanic cell.

In general, when a Galvanic cell operates as an electrolytic cell:

- The material of the electrode designated the cathode, and of the anode, reverses: so, for example, in a Galvanic Daniell cell, the anode is the zinc rod, and the cathode the copper rod; in an electrolytic Daniell cell, the anode is the copper rod and the cathode is the zinc rod.
- The polarity of the electrodes also reverses: in a Galvanic cell, the anode is the negative terminal, and the cathode positive, whereas in the corresponding electrolytic cell, the anode is the positive terminal and the cathode the negative terminal.
- But the fundamental definition of the anode as being associated with the oxidation half-reaction, and the cathode with the reduction half-reaction, is maintained.

Electrolytic cells are the basis of two important industrial processes, **electrolysis** and **electroplating**. Electrolysis is the process in which an external electrical source forces an electric current through a single electrolyte, driving a chemical reaction that, under normal conditions, is not spontaneous. So, consider the system shown in Figure 20.7, in which an external battery drives an electrolytic cell comprising two platinum electrodes, in contact with a dilute solution of an acid, such as H_2SO_4.

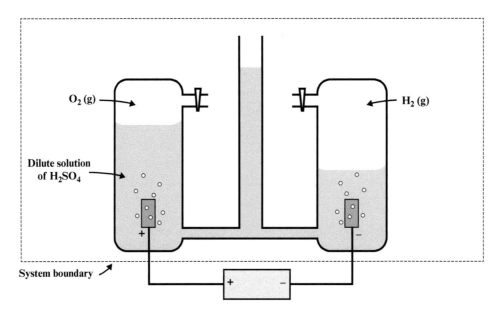

Figure 20.7 The electrolysis of water. An external battery drives an electric current through two platinum electrodes into a dilute solution of H_2SO_4, so forming an electrolytic cell, as shown within the dashed boundary. Water molecules are split into gaseous oxygen, which collects at the anode (which, for an electrolytic cell is the positive terminal on the left), and gaseous hydrogen, which collects at the cathode (which, for an electrolytic cell, is the negative terminal on the right)

When the electrodes are connected to a battery of a voltage of about 2 V, the space above the electrode attached to the battery's positive terminal collects a volume of oxygen gas, whilst the space above the electrode attached to the battery's negative terminal collects a double volume of hydrogen gas.

Pure water is a poor conductor of electricity, and the so some acid improves the conductivity considerably: the battery in Figure 20.7 therefore drives an electrolytic cell in which the electrolyte contains H^+ (aq), OH^- (aq) and SO_4^{2-} (aq) ions. The corresponding half-reactions are the reduction of hydrogen ions to gaseous hydrogen at the cathode on the right

$$2 H^+ (aq) + 2 e^- \rightarrow H_2 (g)$$

and (as we mentioned on page 606) the oxidation of hydroxyl ions to gaseous oxygen at the anode on the left

$$4 OH^- (aq) \rightarrow 2 O_2 (g) + 2 H_2O (g) + 4 e^-$$

Overall, the electrolytic cell reaction is

$$2 H_2O (l) \rightarrow H_2 (g) + 2 O_2 (g)$$

this being the splitting of water into hydrogen and oxygen. Under normal conditions, this is not spontaneous; electrolysis, however, can be used to drive this non-spontaneous reaction in the reverse direction.

Industrially, aluminium metal is manufactured by the electrolysis of molten alumina, Al_2O_3, in a process which consumes prodigious quantities of electricity; other industrial uses of electrolysis include, for example, the production of metallic sodium (from molten sodium hydroxide, or molten sodium chloride), magnesium (from molten magnesium chloride), and calcium (from, for example, molten calcium chloride); and of hydrogen, chlorine and sodium hydroxide from the electrolysis of an aqueous solution of sodium chloride (this being one of the most important uses of common salt).

Electroplating exploits the deposition of metallic ions on the cathode of an electrolytic cell. In general, for a metal M, the reduction half-reaction at the cathode may be represented as

$$M^{n+} (aq) + n\,e^- \rightarrow M\,(s)$$

and so if M^{n+} (aq) can be attracted to a suitable cathode, M(s) will be deposited as a fine layer.

Consider, then, an electrolytic variant of the Daniell cell, as shown in Figure 20.8.

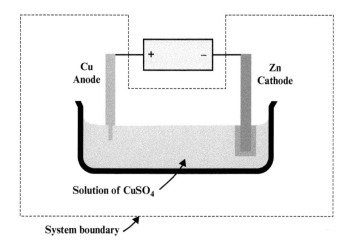

Figure 20.8 Electroplating. An external battery drives an electrolytic cell comprising a copper anode, a zinc cathode, and an electrolyte of an aqueous solution of $CuSO_4$. Solid copper Cu (s) from the anode dissolves as Cu^{2+} (aq) ions into the solution, and is deposited as a thin layer of Cu (s) on the surface of the cathode.

In Figure 20.8, an external battery drives an electrolytic cell, comprising a copper anode, a zinc cathode, and an electrolyte of $CuSO_4$ – unlike the Galvanic Daniell cell, the electrolyte of this cell contains no $ZnSO_4$, and so the only positively charged ion in the electrolyte is Cu^{2+} (aq).

The oxidation half-reaction at the anode is

$$Cu\,(s) \rightarrow Cu^{2+}\,(aq) + 2\,e^-$$

in which solid copper dissolves as Cu^{2+} (aq) ions. At the cathode, the reduction half-reaction is

$$Cu^{2+}\,(aq) + 2\,e^- \rightarrow Cu\,(s)$$

so depositing copper metal on the zinc cathode, for there are only Cu^{2+} (aq) ions in the solution. The overall result is to transfer solid copper from the anode to the cathode, such that the deposition on the cathode is in the form of a very thin, uniform, layer. This is the basis of electroplating, whereby cheaper metals can be covered with thin layers of more expensive metals, such as copper, chromium, silver or gold, so as to provide, for example, protection against corrosion, or to confer a more attractive appearance.

20.12 The electromotive force

As we have seen, a Galvanic cell produces a cell potential between its electrodes as the result of a spontaneous chemical reaction. In the case of the Daniell cell, the spontaneous cell reaction is

$$\text{Zn (s)} + \text{Cu}^{2+} \text{(aq)} \rightarrow \text{Zn}^{2+} \text{(aq)} + \text{Cu (s)}$$

The cell potential is present only whilst this reaction is taking place: as the reaction progresses from its initial, non-equilibrium state and progresses towards equilibrium, the cell potential continuously diminishes, until it falls to zero when the system is in chemical equilibrium. The rate at which the reaction takes place is, as we have seen, a matter of chemical kinetics rather than thermodynamics: clearly, a necessary condition for a useful electrical battery is that the cell reaction takes place slowly, so that the battery has a reasonable working life.

The maximum cell potential that the cell can generate is that corresponding to the state in which the cell reaction has just started – once the reaction has been initiated, some cell potential must be lost. Suppose, however, that a Daniell cell, which has a cell potential under any given conditions of E_{Dan}, is connected to another, external, cell, of cell potential E_{ex}, such that the two cell potentials oppose each other, as we saw in the representation of the electrolytic cell in Figure 20.6, and as illustrated, with a different caption, in Figure 20.9.

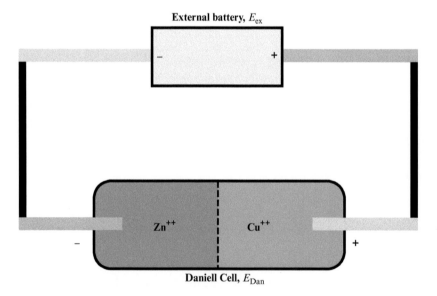

Figure 20.9 The electromotive force. A Daniell cell, which has a cell potential E_{Dan}, is connected, as shown, to another, external, cell whose cell potential E_{ex} is variable. If $E_{\text{ex}} = E_{\text{Dan}}$, there is no flow of electrons, and the current is zero. The cell potential of the Daniell cell is then the maximum possible, the electromotive force $I\!E$, and the Daniell cell is working reversibly.

On page 608, we saw that, when $E_{ex} < E_{Dan}$, the Daniell cell continues to operate as a Galvanic cell, but with an effective cell potential reduced to $E_{Dan} - E_{ex}$, and that when $E_{ex} > E_{Dan}$, the Daniell cell is no longer a Galvanic cell but an electrolytic cell.

A third case is that in which the opposing cell potential E_{ex} of the external cell is exactly equal to the cell potential E_{Dan} of the Daniell cell. There will then be no flow of electrons at all, and no current. If the external cell potential E_{ex} is now incrementally decreased, the Daniell cell reaction would take place as normal, with electrons flowing clockwise in Figure 20.8; if the external cell potential E_{cell} is incrementally increased, the Daniell cell reaction would be driven in reverse, and electrons would flow anti-clockwise.

This is very similar to the situation we explored on page 29, in which a gas at a pressure P within a cylinder expands against an external pressure P_{ex}, where we saw that the expansion of the gas is thermodynamically reversible when $P = P_{ex}$. For a system comprising an electrochemical cell, the cell potential E_{Dan} is the equivalent of the pressure P of the gas within the cylinder, and the cell potential E_{ex} of the external cell is the equivalent of the external pressure P_{ex}. The condition $E_{Dan} = E_{ex}$ is the equivalent of the condition $P = P_{ex}$, and implies that the Daniell cell is acting reversibly.

The cell potential E_{cell} of any Galvanic electrochemical cell acting reversibly is known as the cell's **electromotive force (emf)**, represented by the symbol $\mathbb{E}$. This represents the maximum potential the cell can achieve under any conditions of temperature and pressure, and, as we saw on page 143, when any system is acting reversibly, it performs the maximum possible work. If the cell is doing the maximum work, then this must correspond to the flow of the maximum possible number of electrons, which takes us back to equation (20.6)

$$-\Delta_r G_{bat} = n_e(\max) F \, \mathbb{E} \tag{20.6}$$

where $\Delta_r G_{bat}$ is the change in Gibbs free energy attributable to the chemical reaction taking place within the cell.

This demonstrates that for a Daniell cell, undergoing the cell reaction

$$\text{Zn (s)} + \text{Cu}^{2+} \text{(aq)} \rightarrow \text{Zn}^{2+} \text{(aq)} + \text{Cu (s)} \qquad \Delta G = \Delta_r G_{bat}$$

then

$$-\Delta_r G_{bat} = 2 F \, \mathbb{E}_{Dan} \tag{20.8a}$$

where $n_e(\max) = 2$ is determined by the stoichiometry of the cell reaction, this being a particular case of the more general equation, for any electrochemical cell which can generate an emf $\mathbb{E}_{cell}$

$$-\Delta_r G_{cell} = n_e F \, \mathbb{E}_{cell} \tag{20.8b}$$

Also, from equation (20.7)

$$E_{cell} = E_{cat} - E_{an} \tag{20.7}$$

we see that, when a cell is operating reversibly, such that $E_{cell} = \mathbb{E}_{cell}$, then each electrode is operating reversibly, at the corresponding **reversible electrode potentials** $\mathbb{E}_{cat}$ and $\mathbb{E}_{an}$

$$\mathbb{E}_{cell} = \mathbb{E}_{cat} - \mathbb{E}_{an} \tag{20.9}$$

Note that, if equation (20.8b) is written in the form

$$E_{cell} = E_{cat} - E_{an} = -\frac{\Delta_r G_{cell}}{n_e(\max) F} \tag{20.8c}$$

then E_{cell} can be seen to be the ratio of two extensive state functions, $\Delta_r G_{cell}$ and $n_e(\max)$, so verifying the statement made on page 604 that E is intensive state function.

Furthermore, if we attribute a value of $\Delta_r G$ to the individual half-reactions taking place at the cathode and the anode, equation (20.8b) can be generalised as

$$-\Delta_r G = n_e F E \tag{20.8d}$$

Equation (20.8d) implies that any reversible electrode potential E can be associated with a corresponding value of $\Delta_r G$, such that the greater the (positive) magnitude of E, the greater the (negative) value of $\Delta_r G$: a large (positive) value of E therefore implies a strongly spontaneous electrode reaction.

20.12.1 Standard reversible electrode potentials

Figure 20.10 shows a representation of an electrolytic cell consisting of a hydrogen gas electrode (see Figure 20.5) as the anode, and a generalised metal electrode as the cathode.

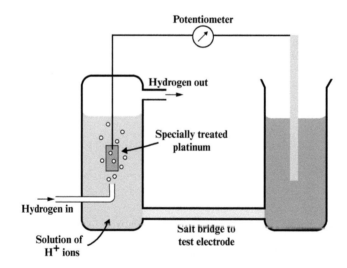

Figure 20.10 The standard hydrogen electrode. Hydrogen gas bubbles into a solution containing H⁺ (aq) ions, over a specially treated platinum foil, so forming an electrode. If the solution contains 1 mol m⁻³ of H⁺ (aq) ions, and the whole system is at a pressure of 1 bar, then this is known as a standard hydrogen electrode. A salt bridge connects the hydrogen electrode to a test electrode, also under standard conditions. If we assume that both electrodes are acting reversibly, then the potentiometer will show a reading of the standard cell emf $E^\circ = E_{cat}^\circ - E_{an}^\circ$. According to the definition of the reference level for the measurement of electrode potentials, the electrode potential of standard hydrogen electrode is zero, and so measurement of E_{cell}° equates to the standard electrode potential of the test electrode as E_{cat}° (if the hydrogen electrode acts as the anode, as shown in this diagram), or $-E_{an}^\circ$ (if the hydrogen electrode acts as the cathode).

Since, by assumption, the hydrogen gas electrode is acting as the anode, the corresponding half-reaction is the oxidation

$$H_2 (g) \rightarrow 2 H^+(aq) + 2 e^-$$

Using the representation E (oxidised state, reduced state), as on page 603, this half-reaction will be associated with a corresponding electrode potential E and if

- the solution is such that the concentration [H$^+$] of H$^+$ (aq) ions is 1 mol m^{-3} (or sometimes 1 mol dm^{-3})
- the pressure of the H$_2$ (g) is 1 bar, and
- we assume that the electrode is operating reversibly

then the **reversible hydrogen electrode** is in its standard state, and the corresponding electrode potential is the **standard reversible electrode potential** $E^{\ominus}$ (H$^+$, H$_2$). As just noted, this standard is defined in terms of a solution of H$^+$ (aq) ions at a molar concentration [H$^+$] of 1 mol dm^{-3}, corresponding to (the highly acidic) pH 0. Although this is suitable for much laboratory and industrial chemistry, it isn't for biochemistry, for many biochemical processes take place at or around pH 7. Accordingly, in Chapter 23, we will introduce a different standard, known as the 'biochemical standard', based on pH 7.

If the standard reversible hydrogen electrode is connected as the anode to a reversible 'test' electrode as the cathode, as shown in Figure 20.10, such that the 'test' solution is also in its standard state of a concentration of 1 mol m^{-3} of the appropriate ion, and at a pressure of 1 bar, then the two electrodes form a cell, the standard emf $E_{cell}^{\ominus}$ of which is given by equation (20.9) as

$$E_{cell}^{\ominus} = E^{\ominus} \text{(test)} - E^{\ominus} (\text{H}^+, \text{H}_2)$$

In practice, it is not possible to measure the absolute electrode potential E of a single electrode: the only measurements that can be made are those of a cell emf, E_{cell}, which is necessarily a difference of the form $E_{cat} - E_{an}$. If, however, we define the reference level for the measurement of electrode potentials such that $E^{\ominus}$ (H$^+$, H$_2$) = 0 V at all temperatures, then, for any test electrode acting as a cathode, as illustrated in Figure 20.10, the measured cell emf $E_{cell}^{\ominus}$ can be assigned as $E^{\ominus}$(test). This value can therefore be defined as the **standard electrode potential** of the test electrode, as tabulated in reference works. Some values of (near) standard reversible electrode potentials $E^{\ominus}$ for a variety of materials are given in Table 20.1 (the reference cited gives values at a pressure of 1 atm = 1.01325 bar, rather than at the standard pressure of 1 bar, but this difference in pressure does not, in practice, matter).

In Table 20.1, a positive standard reversible electrode potential implies that the corresponding electrode acts as the cathode with respect to a standard reversible hydrogen electrode as the anode of the Galvanic electrochemical cell formed from the two electrodes; a negative value implies that the corresponding electrode acts as the anode to the hydrogen electrode as the cathode. Also, since $E^{\ominus}$ is an intensive state function, its value for any given half-reaction is independent of the stoichiometry in which the half-reaction is expressed. So, for example, the half-reactions written as

$$\text{O}_2 \text{ (g)} + 4\text{ H}^+ \text{ (aq)} + 4\text{ e}^- \rightleftharpoons 2\text{ H}_2\text{O} \qquad E^{\ominus} = 1.229 \text{ V}$$

$$2\text{ O}_2 \text{ (g)} + 8\text{ H}^+ \text{ (aq)} + 8\text{e}^- \rightleftharpoons 4\text{ H}_2\text{O} \qquad E^{\ominus} = 1.229 \text{ V}$$

and

$$\tfrac{1}{2}\text{O}_2 \text{ (g)} + 2\text{ H}^+ \text{ (aq)} + 2\text{ e}^- \rightleftharpoons \text{H}_2\text{O} \qquad E^{\ominus} = 1.229 \text{ V}$$

Table 20.1 (Near) standard reversible electrode potentials. The table shows electrode potentials at a pressure of 1 atm = 1.01325 bar, rather than at 1 bar, and at a temperature of 298.15 K. In practice, this difference is pressure has no significant effect, and so the values shown in this table may reliably be used as standards. Gases are explicitly shown as (g) and liquids as (l); all ions are implicitly (aq), and all other elements are (s). Note that oxygen is associated with two standard reversible electrode potentials, one (O_2 (g), 2 H_2O) in acid conditions, the other (O_2 (g), 2 OH^-) in alkaline conditions.

Electrode	Half-reaction	Standard reversible electrode potential $E^\ominus$ V
Cl_2 (g), 2 Cl^-	Cl_2 (g) + e^- ⇌ 2 Cl^-	+ 1.36
O_2 (g), 2 H_2O	O_2 (g) + 4 H^+ + 4 e^- ⇌ 2 H_2O	+ 1.229
Ag^+, Ag	Ag^+ + e^- ⇌ Ag	+ 0.800
O_2 (g), 2 OH^-	O_2 (g) + 2 H^+ + 4 e^- ⇌ 2 OH^-	+ 0.401
Cu^{2+}, Cu	Cu^{2+} + 2 e^- ⇌ Cu	+ 0.342
2H^+, H_2 (g)	2 H^+ + 2 e^- ⇌ H_2 (g)	0
Pb^{2+}, Pb	Pb^{2+} + 2 e^- ⇌ Pb	− 0.126
Sn^{2+}, Sn	Sn^{2+} + 2 e^- ⇌ Sn	− 0.138
Ni^{2+}, Ni	Ni^{2+} + 2 e^- ⇌ Ni	− 0.257
Fe^{2+}, Fe	Fe^{2+} + 2 e^- ⇌ Fe	− 0.447
Zn^{2+}, Zn	Zn^{2+} + 2 e^- ⇌ Zn	− 0.762
Na^+, Na	Na^+ + e^- ⇌ Na	− 2.71
Ca^{2+}, Ca	Ca^{2+} + 2 e^- ⇌ Ca	− 2.87
K^+, K	K^+ + e^- ⇌ K	− 2.93
Li^+, Li	Li^+ + e^- ⇌ Li	− 3.04

Source: *CRC Handbook of Chemistry and Physics*, 96th edn, 2015, pages 5–79 to 5–88.

all have the same value for the intensive state function $E^\ominus$, but the corresponding values of the extensive function $\Delta_r G^\ominus = -n_e F E^\ominus$ are different, as − 474 kJ (n_e = 4), − 949 kJ (n_e = 8) and − 237 kJ (n_e = 2) respectively. As can be appreciated, these values of $\Delta_r G^\ominus$ are all large negative numbers, showing that the equilibrium lies far to the right – a mixture of molecular oxygen and H^+ (aq) ions is strongly reactive. Conversely, to split water to molecular oxygen requires a considerable quantity of free energy, 474 kJ per mol of O_2; as we shall see in Chapter 24, this is precisely what happens in the photosynthesis process within green plants, where the source of this free energy is sunlight.

With reference to Table 20.1, any pair of electrodes may be combined to form a cell, the standard reversible emf $E_{cell}^\ominus$ of which is given by equation (20.9) as

$$E_{cell}^\ominus = E_{cat}^\ominus - E_{an}^\ominus \tag{20.10}$$

For any pair of electrodes, the cathode is defined as the electrode with the more positive standard reversible electrode potential, and the anode as the electrode with the more negative standard reversible electrode potential – so ensuring, according to equation (20.10), that the standard emf $E_{cell}^\ominus$ of the corresponding cell is always positive. So, for example, a Daniell cell has copper as the cathode ($E_{cat}^\ominus$ = + 0.342 V), and zinc as the anode ($E_{an}^\ominus$ = − 0.762 V), giving a cell standard reversible emf $E_{cell}^\ominus$ = + 0.342 − (− 0.762) = 1.104 V. But when a

THE ELECTROMOTIVE FORCE

copper electrode is combined with a silver electrode, the silver electrode acts as the cathode ($\mathbb{E}_{cat}^{\ominus}$ = + 0.800 V), and the copper electrode as the anode ($\mathbb{E}_{an}^{\ominus}$ = + 0.342 V), giving a cell standard reversible emf $\mathbb{E}_{cell}^{\ominus}$ = + 0.800 − 0.342 = 0.458 V. This explains the comment made on page 601 that the designations 'cathode' and 'anode' are not properties of an individual electrode, but properties of a specific electrochemical cell. In general, however, those electrodes towards the bottom of Table 20.1 are more likely to act as cathodes, and those towards the top as anodes.

Also, as noted on pages 606 and 607, when a platinised platinum electrode is in contact with a solution containing two different ions of the same element in different oxidation states, such as a mixture of Fe^{2+} (aq) and Fe^{3+} (aq), then the platinum mediates the redox half-reaction

$$Fe^{3+} (aq) + e^{-} \rightarrow Fe^{2+} (aq)$$

so resulting in the corresponding **reversible redox electrode** potential. Table 20.2 shows some (near) standard reversible redox electrode potentials for a number of reactions, in which all reactants are in their standard state of a concentration of 1 mol dm^{-3} at the (not-quite standard) pressure of 1.01325 bar.

Table 20.2 (Near) standard reversible redox potentials $\mathbb{E}^{\ominus}$ of aqueous ions. The table shows redox potentials at a pressure of 1 atm = 1.01325 bar, rather than at 1 bar, and at a temperature of 298.15 K. All ions are (aq)

Electrode	Half-reaction	Standard reversible redox potential $\mathbb{E}^{\ominus}$ V
Co^{3+}, Co^{2+}	$Co^{3+} + e^{-} \rightleftharpoons Co^{2+}$	+ 1.92
Ce^{4+}, Ce^{3+}	$Ce^{4+} + e^{-} \rightleftharpoons Ce^{3+}$	+ 1.72
MnO_4^{-}, Mn^{2+}	$MnO_4^{-} + 8 H^{+} + 5 e^{-} \rightleftharpoons Mn^{2+} + 4 H_2O$	+ 1.51
$Cr_2O_7^{2-}$, Cr^{3+}	$Cr_2O_7^{2-} + 14 H^{+} + 6 e^{-} \rightleftharpoons 2 Cr^{3+} + 7 H_2O$	+ 1.36
O_2 (g), 2 H_2O	$O_2(g) + 4 H^{+} + 4 e^{-} \rightleftharpoons 2 H_2O$	+ 1.229
Hg^{2+}, Hg_2^{2+}	$2 Hg^{2+} + 2 e^{-} \rightleftharpoons Hg_2^{2+}$	+ 0.920
Fe^{3+}, Fe^{2+}	$Fe^{3+} + e^{-} \rightleftharpoons Fe^{2+}$	+ 0.771
MnO_4^{-}, MnO_4^{2-}	$MnO_4^{-} + e^{-} \rightleftharpoons MnO_4^{2-}$	+ 0.558
Cu^{2+}, Cu^{+}	$Cu^{2+} + e^{-} \rightleftharpoons Cu^{+}$	+ 0.153
Sn^{4+}, Sn^{2+}	$Sn^{4+} + 2 e^{-} \rightleftharpoons Sn^{2+}$	+ 0.151
H^{+}, H_2 (g)	$2 H^{+} + 2 e^{-} \rightleftharpoons H_2(g)$	0
Ti^{3+}, Ti^{2+}	$Ti^{3+} + e^{-} \rightleftharpoons Ti^{2+}$	− 0.369
Cr^{3+}, Cr^{2+}	$Cr^{3+} + e^{-} \rightleftharpoons Cr^{2+}$	− 0.407

Source: *CRC Handbook of Chemistry and Physics*, 96th edn, 2015, pages 5–79 to 5–88.

20.12.2 A note on standards, reference levels and conventions

The values for all the standard reversible electrode potentials shown in Tables 20.1 and 20.2 have been determined relative to the reference level $\mathbb{E}$ = 0 V, in accordance with the IUPAC standard that the reversible electrode potential of a standard hydrogen electrode is designated as 0 V at all temperatures. This choice of reference level is arbitrary, and a matter of international agreement – for example, the reference level might have been defined in relation to a standard silver electrode, such that the 'reference level according to the silver standard' is designated as, say, 12 V. Accordingly, since (according to Table 20.1) the silver electrode is 0.8 V more positive than the hydrogen electrode, then the value of the hydrogen electrode potential, as measured using the 'silver standard' would be 11.2 V; likewise, a nickel electrode, which, according to Table 20.1, is 1.057 V more negative than a silver electrode, would be measured as 12 – 1.057 = 10.943 V using the 'silver standard', this being 11.2 V more positive than the value – 0.257 V as measured using the hydrogen standard. In fact, the values of all standard electrode potentials, as measured using the 'silver standard', would be 11.200 V higher than the values using the hydrogen standard, as shown in Tables 20.1 and 20.2. The quantity 11.200 V is the reference level shift (see page 175), which enables any number measured in one standard to be converted into the corresponding number in the other standard.

To take two further examples, consider the copper electrode, which has a standard electrode potential of + 0.342 V on the hydrogen standard, but which would be + 0.342 + 11.200 = 11.542 V on the 'silver standard'; and also zinc, for which the corresponding numbers are – 0.762 (hydrogen standard), and – 0.762 + 11.200 = 10.438 V ('silver standard'). If we compare the standard emf of the Daniell cell, as measured according to the both standards, then, using the hydrogen standard, this emf is + 0.342 – (– 0.762) = + 1.104 V, and according to the 'silver standard', the emf would be measured as 11.542 – 10.438 = + 1.104 V – the *same* number. This illustrates the general point we made on page 175 – although the measurements associated with the values of the standard electrode potential are different according to the standard used, the *difference* between any two standard electrode potentials is the *same number*, regardless of the standard used, as long as the measurements are consistent, and are based on the same standard for any given pair of electrodes for which the cell emf is being measured.

As we shall see in Chapter 23, this has importance in biochemistry, for the biochemist community has adopted a different standard for the measurement of standard electrode potentials. The reason for this concerns an assumption somewhat 'hidden' in the definition of the standard reversible hydrogen electrode, which serves as the reference level for all the measurements shown in Tables 20.1 and 20.2. As noted on page 615, the standard reversible hydrogen electrode is defined in relation to a solution of hydrogen ions H^+ (aq) at unit molar concentration $[H^+] = 1$ mol dm^{-3}, corresponding to pH 0 (pH = $-\log_{10} [H^+] = -\log_{10} [1] = 0$), this being the reference state (which is a defined state of a system) associated with the definition of the reference level (which is a measurement) $E^\ominus = 0$ V. A solution at pH 0 is very acid, especially for most biochemical reactions which take place at, or close to, pH 7. Accordingly, biochemists adopt a different standard, based on pH 7, which – as we shall see on page 688 – affects the measurement of redox potentials which are relevant to biochemistry by a reference level shift of – 0.414 V.

One further convention is worth noting too. As can be seen from Tables 20.1 and 20.2, all standard electrode potentials are conventionally defined by reference to a reduction half-reaction, in which an oxidised state gains electrons to form a reduced state, as happens at the cathode of every electrochemical cell. All the entries in Tables 20.1 and 20.2 are of this form, for example

$$Cu^{2+} (aq) + 2 e^- \rightleftharpoons Cu (s)$$

or more generally

$$p\,\text{ox}^{q+} + (pq - rs)\,e^- \rightleftharpoons r\,\text{red}^{s+}$$

in which p mol of an oxidised component (or mixture) ox^{q+} react to form r mol of a reduced component (or mixture) red^{s+}. Simultaneously, a total quantity of $n_e = (pq - rs)$ electrons are consumed so as to balance the total positive charges associated with p mol of ox^{q+} on the left, and r mol of red^{s+} on the right. To allow for the possibility that, in any particular reaction, ox^{q+} or red^{s+} might carry a net negative, rather than positive, charge, the parameters q and s can be negative accordingly.

Since this reaction is a reduction, standard reversible electrode potentials can also be known as **standard reversible reduction potentials**, and represented as we saw on page 603 as $\mathbb{E}^\ominus$ (Cu^{2+}, Cu), or more generally $\mathbb{E}^\ominus$ (ox, red), where the sequence (ox, red) indicates the convention of expressing the corresponding half-reaction as an oxidised state being reduced.

20.13 Oxidising agents and reducing agents

As we saw on page 614, equation (20.8d)

$$-\Delta_r G = n_e\,F\,\mathbb{E} \tag{20.8d}$$

relates the reversible electrode potential $\mathbb{E}$ of any electrode to the value of $\Delta_r G$ attributable to the corresponding redox half-reaction, of the general form

$$p\,\text{ox}^{q+} + (pq - rs)\,e^- \rightleftharpoons r\,\text{red}^{s+}$$

If the electrode components are in their standard states, then equation (20.8d) becomes

$$-\Delta_r G^\ominus = n_e\,F\,\mathbb{E}^\ominus \tag{20.8e}$$

in which $\mathbb{E}^\ominus$ is the standard reversible reduction potential, and $\Delta_r G^\ominus$ the standard reaction Gibbs free energy change. Now, in Chapter 16, we established that, for any reaction taking place in an ideal solution, the general standard reaction Gibbs free energy change $\Delta_r G^\ominus$ is related to the corresponding equilibrium constant K_r as

$$\Delta_r G^\ominus = -RT \ln K_r \tag{16.13d}$$

in which the standard states are those appropriate to the various reagents.

Many redox reactions taking place at an electrode are particular instances of a reaction taking place in solution, and so, if we assume that the solution is ideal, and that the standard states are those of unit molar concentration, then the corresponding concentration equilibrium constant K_c (ox, red) is related to the appropriate values of $\mathbb{E}^\ominus$ and $\Delta_r G_c^\ominus$ as

$$n_e\,F\,\mathbb{E} = -\Delta_r G_c^\ominus = RT \ln K_c\,(\text{ox, red}) \tag{20.11a}$$

from which

$$K_c\,(\text{ox, red}) = \exp\left(\frac{n_e F\,\mathbb{E}^\ominus}{RT}\right) \tag{20.11b}$$

Equation (20.11b) implies that, for any redox reaction with a standard electrode potential E°, the more positive the value of E°, the greater the value of K_c (ox, red) for the corresponding redox reaction, and the further the equilibrium lies to the right; conversely, the more negative the value of E°, the smaller the value of K_c (ox, red) for the corresponding redox reaction, and the further the equilibrium lies to the left – as indeed discussed on page 603.

If we take, for example, the redox reaction

$$Co^{3+} + e^- \rightleftharpoons Co^{2+}$$

for which, according to Table 20.2, E° = 1.92 V, then we can infer that the equilibrium lies to the right, implying that Co^{2+} is a strong acceptor of electrons. By contrast, for the redox reaction

$$Li^+ + e^- \rightleftharpoons Li$$

for which, according to Table 20.1, E° = –3.04 V, the equilibrium lies to the left, implying that Li is a strong donor of electrons.

Strong acceptors of electrons are known as **oxidising agents** (or **oxidants**), in that their action is to seek to strip electrons from another moiety, thereby causing that moiety to become oxidised; similarly, strong donors of electrons are known as **reducing agents** (or **reductants**), in that their action is to seek to donate electrons to another moiety, thereby causing that moiety to become reduced. We may therefore draw the general conclusion that

> **The more positive E°, the stronger the oxidising agent.**
> **The more negative E°, the stronger the reducing agent.**

Reference to Tables 20.1 and 20.2 will show that they are ordered with the strongest oxidising agents at the top, and the strongest reducing agents at the bottom. In general, a half-reaction at any level in either Table will oxidise a half-reaction at any lower level; similarly, a half-reaction at any lower level in either table will reduce a half-reaction at any higher level in the table. To make that concrete: as an example, with reference to Table 20.2, consider the three half-reactions

$Fe^{3+} + e^- \rightleftharpoons Fe^{2+}$ $\qquad\qquad\qquad\qquad\qquad\qquad\qquad\qquad$ E° = 0.771 V

$Sn^{4+} + 2\,e^- \rightleftharpoons Sn^{2+}$ $\qquad\qquad\qquad\qquad\qquad\qquad\qquad\qquad$ E° = 0.151 V

$Ti^{3+} + e^- \rightleftharpoons Ti^{2+}$ $\qquad\qquad\qquad\qquad\qquad\qquad\qquad\qquad$ E° = – 0.369 V

Sn^{3+} therefore acts as an oxidising agent with respect to Ti^{2+}, according to the reaction

$$Sn^{4+} + 2\,Ti^{2+} \rightleftharpoons Sn^{2+} + 2\,Ti^{3+}$$

but Sn^{2+} acts as a reducing agent with respect to Fe^{3+} as

$$2\,Fe^{3+} + Sn^{2+} \rightleftharpoons Sn^{4+} + 2\,Fe^{2+}$$

Note that when an oxidising agent oxidises another moiety, the oxidising agent itself becomes reduced (Sn^{4+} oxidises Ti^{2+} to Ti^{3+}, whilst being reduced to Sn^{2+}); similarly, when a reducing agent reduces another moiety, the reducing agent itself becomes oxidised (Sn^{2+} reduces Fe^{3+} to Fe^{2+}, whilst being oxidised to Sn^{4+}).

20.14 The thermodynamics of electrochemical cells

In this section, we develop the general thermodynamics of electrochemical cells, taking as our example the familiar Galvanic Daniell cell. As we know, the anode oxidation half-reaction is

Zn (s) $\rightleftharpoons$ Zn^{2+} (aq) + 2 e$^-$

and the cathode reduction half-reaction is

Cu^{2+} (aq) + 2 e$^-$ $\rightleftharpoons$ Cu (s)

corresponding to the overall cell reaction

Zn (s) + Cu^{2+} (aq) $\rightleftharpoons$ Zn^{2+} (aq) + Cu (s)

In this reaction, the solid zinc rod gradually dissolves, and acts as an effectively infinite source of Zn^{2+} (aq) ions and electrons, whilst the solid copper rod acts as an effectively infinite sink for Cu^{2+} (aq) ions and electrons, and progressively gains mass as Cu^{2+} (aq) ions discharge and become deposited as atomic copper.

The Daniell cell reaction is a particular instance of the generalised reversible reaction

aA + bB $\rightleftharpoons$ cC + dD

which we studied in detail in the context of a mixture of ideal gases (see pages 417 to 439), and a reaction taking place in solution (pages 517 to 526). Accordingly, all the results presented on page 523 to 524 can be applied to the Daniell cell, for example

$$K_r = \frac{(a_{C,eq}/a_C^{\ominus})^c (a_{D,eq}/a_D^{\ominus})^d}{(a_{A,eq}/a_A^{\ominus})^a (a_{B,eq}/a_B^{\ominus})^b} = \frac{a_{C,eq}^c \, a_{D,eq}^d}{a_{A,eq}^a \, a_{B,eq}^b} \qquad (16.12d)$$

$$\Delta_r G^{\ominus} = -RT \ln K_r \qquad (16.13d)$$

$$\Delta_r G^{\ominus} = (c \, \Delta_f G_C^{\ominus} + d \, \Delta_f G_D^{\ominus}) - (a \, \Delta_f G_A^{\ominus} + b \, \Delta_f G_B^{\ominus}) \qquad (16.14d)$$

$$\Gamma_r = \frac{(a_C/a_C^{\ominus})^c (a_D/a_D^{\ominus})^d}{(a_A/a_A^{\ominus})^a (a_B/a_B^{\ominus})^b} = \frac{a_C^c \, a_D^d}{a_A^a \, a_B^b} \qquad (16.15d)$$

$$\Delta_r G = \Delta_r G^{\ominus} + RT \ln \Gamma_r \qquad (16.16d)$$

in which the standard states, and the activity ratios $a/a^{\ominus}$, correspond to the appropriate physical state of the component i.

So, for the Daniell cell reaction,

$$\Delta_r G^{\ominus} = [\Delta_f G^{\ominus} \text{Zn}^{2+} \text{(aq)} + \Delta_f G^{\ominus} \text{Cu (s)}] - [\Delta_f G^{\ominus} \text{Zn (s)} + \Delta_f G^{\ominus} \text{Cu}^{2+} \text{(aq)}]$$

in which the various standard molar Gibbs free energies of formation can be found in reference works, such as shown in Table 13.1, and

$$K_r = \frac{[\text{Zn}^{2+}]_{eq}/[\text{Zn}^{2+\ominus}](x_{Cu,eq}/x_{Cu}^{\ominus})}{[\text{Cu}^{2+}]_{eq}/[\text{Cu}^{2+\ominus}](x_{Zn,eq}/x_{Zn}^{\ominus})} = \frac{[\text{Zn}^{2+}]_{eq} \, x_{Cu,eq}}{[\text{Cu}^{2+}]_{eq} \, x_{Zn,eq}}$$

As can be seen K_r is a 'hybrid' (see page 524), as appropriate to the multi-state nature of the Daniell cell reaction: molar concentrations [I] are used for the ions, with the corresponding standard state $[I]^{\ominus} = 1$ mol dm^{-3}, and mole fractions x_i for the solids, with the standard state $x_i^{\ominus} = 1$, corresponding to the pure substance.

Continuing our analysis of the Daniell cell reaction further, at any instant whilst the cell reaction is in progress, the zinc anode will be associated with an instantaneous electrode potential E_{an}, and the copper cathode with an electrode potential E_{cat}, such that the overall Daniell cell emf E_{cell} is given by

$$E_{cell} = E_{cat} - E_{an} \tag{20.7}$$

For the purposes of this analysis, we will assume that the cell is working reversibly, and so the various electrode potentials E will be written as reversible electrode potentials $\mathbb{E}$.

As we saw on pages 434 and 435, and also page 520, and as illustrated in Figure 14.3, at any instant during any reaction, the slope of the $G_{sys}(\xi)$ curve is given by the quantity represented as $\Delta_r G$, as represented by equation (14.15)

$$\Delta_r G = \frac{dG_{sys}(\xi)}{d\xi} \tag{14.15}$$

from which

$$dG_{sys}(\xi) = \Delta_r G \, d\xi$$

In this expression, $dG_{sys}(\xi)$ is the infinitesimal change in the systems Gibbs free energy $G_{sys}(\xi)$ corresponding to an infinitesimal change $d\xi$ in the system's extent-of-reaction. But for an electrochemical cell, what does an 'infinitesimal change in the system's extent-of-reaction' actually mean?

A moment's thought will verify that, in real, physical, terms, for the Daniell cell undergoing the cell reaction as written, an infinitesimal change $d\xi$ in the system's extent-of-reaction corresponds to a flow of 2 $d\xi$ moles of electrons from the zinc anode to the copper cathode – as is explicit when the cell reaction is de-constructed into the two half-reactions

$$Zn\,(s) \rightleftharpoons Zn^{2+}\,(aq) + 2\,e^-$$

and the cathode reduction half-reaction

$$Cu^{2+}\,(aq) + 2\,e^- \rightleftharpoons Cu\,(s)$$

The flow of 2 $d\xi$ moles of electrons, through the external circuit, from the anode to the cathode, corresponds to a physical flow of charge equal to $2F\,d\xi$ C, where F, as we saw on page 595, is the charge associated with one mole of electrons, of magnitude 9.6485×10^4 C mol^{-1}. Since we are assuming that the cell is operating reversibly, these electrons move through an instantaneous potential difference $\mathbb{E}_{cell}$

$$\mathbb{E}_{cell} = \mathbb{E}_{cat} - \mathbb{E}_{an} \tag{20.9}$$

in which $\mathbb{E}_{cat}$ and $\mathbb{E}_{an}$ are the corresponding instantaneous reversible electrodes of the cathode and anode respectively. As we saw on pages 594 and 595, the flow of $2F\,d\xi$ C of charge up the potential gradient of $\mathbb{E}_{cell}$ V is associated with the performance of work equal to $2F\,d\xi\,\mathbb{E}_{cell}$ J. Furthermore, since the cell is assumed to be operating reversibly, this is the maximum possible work, which is equal to the negative of the corresponding change in the system Gibbs free energy $G_{sys}(\xi)$. This 'corresponding change in $G_{sys}(\xi)$' is precisely the quantity represented as $dG_{sys}(\xi)$ in equation (20.1), and so we may write

$$dG_{sys}(\xi) = -2F\,d\xi\,\mathbb{E}_{cell} = \Delta_r G\,d\xi$$

and so

$$\Delta_r G = -2F\,\mathbb{E}_{cell} = -2F(\mathbb{E}_{cat} - \mathbb{E}_{an})$$

The factor of 2 is attributable to the fact that, for the Daniell cell reaction, the stoichiometry of the reaction is such that two moles of electrons are created by the anode half-reaction for each mole of metallic zinc consumed; had the anode reaction been, for example, of the form

$$A(s) \rightleftharpoons A^{3+}(aq) + 3e^-$$

then, for an infinitesimal change $d\xi$ in the system's extent-of-reaction, the corresponding flow of charge would have been of $3F\,d\xi$ C. If we generalise the number of moles of electrons associated with the cell reaction as n_e, then

$$\Delta_r G = -n_e F\,\mathbb{E}_{cell}$$

But as well as being the slope of the $G_{sys}(\xi)$ curve, $\Delta_r G$ is also a variable in equation (16.6d)

$$\Delta_r G = \Delta_r G^\circ + RT \ln \Gamma_r \tag{16.16d}$$

which relates the instantaneous value of $\Delta_r G$ to instantaneous value of the mass action ratio Γ_r, and the standard Gibbs free energy $\Delta_r G^\circ$, implying that

$$\Delta_r G = -n_e F\,\mathbb{E}_{cell} = \Delta_r G^\circ + RT \ln \Gamma_r \tag{20.12}$$

where, for the Daniell cell reaction, $n_e = 2$.

For any given reaction taking place at a given temperature, the quantity $\Delta_r G^\circ$ is a constant, determined by the components taking part in the reaction, and the corresponding stoichiometry. Equation (20.12) is therefore a general expression for the relationship between the instantaneous reversible cell potential $\mathbb{E}_{cell}$ and the corresponding concentration mass action ratio Γ_r.

If all the components in the cell reaction are in their standard states, then, $\mathbb{E}_{cell}$ is the standard reversible cell emf $\mathbb{E}_{cell}^\circ$ and, as we saw on page 436, $\Gamma_r^\circ = 1$, implying that $\ln \Gamma_r^\circ = 0$. Equation (20.12) therefore becomes

$$\Delta_r G^\circ = -n_e F\,\mathbb{E}_{cell}^\circ \tag{20.13}$$

this being the standard state form of equation (20.8b)

$$-\Delta_r G_{cell} = n_e F\,\mathbb{E}_{cell} \tag{20.8b}$$

which we met on page 614.

Using equation (20.13) to replace $\Delta_r G^\circ$ in equation (20.12), we derive

$$-n_e F\,\mathbb{E}_{cell} = -n_e F\,\mathbb{E}_{cell}^\circ + RT \ln \Gamma_r$$

giving the important result

$$\mathbb{E}_{cell} = \mathbb{E}_{cell}^\circ - \frac{RT}{n_e F} \ln \Gamma_r \tag{20.14}$$

20.15 The Nernst equation

Equation (20.14) is known as the **Nernst equation**, in honour of the German chemist, Walther Nernst, who was instrumental in formulating the Third Law, and was awarded the Nobel Prize in chemistry in 1920 "in recognition of his work in thermochemistry". The Nernst equation is one of the key equations of the thermodynamics of electrochemical cells, for it relates the reversible cell emf $\mathbb{E}_{cell}$ for any reaction mixture, as expressed by the generalised mass action ratio Γ_r, to the standard reversible cell emf $\mathbb{E}_{cell}^{\ominus}$ – which, for any given temperature and pressure, is a specific constant, defined in relation to the standard states of each of the reaction components.

If we compare the Nernst equation

$$\mathbb{E}_{cell} = \mathbb{E}_{cell}^{\ominus} - \frac{RT}{n_e F} \ln \Gamma_r \tag{20.14}$$

which applies to electrochemical systems, with equation (16.19d)

$$\Delta_r G = \Delta_r G^{\ominus} + RT \ln \Gamma_r \tag{16.16d}$$

which applies to any chemical reaction, there is a striking correspondence, with the quantity $-n_e F \mathbb{E}$ being the electrochemical equivalent of the cell reaction Gibbs free energy change $\Delta_r G$.

A special case of the Nernst equation arises when the reaction mixture is at equilibrium, implying that the concentration mass action ratio Γ_r is equal to the redox reaction concentration equilibrium constant $K_r = K_r$ (ox, red). Also, when the system is in equilibrium, as shown in Figure 14.3, $\Delta_r G = 0$, and so, according to equation (20.12), $\mathbb{E}_{cell} = 0$, as we noted on page 613. Accordingly, at equilibrium, equation (20.14) becomes

$$0 = \mathbb{E}_{cell}^{\ominus} - \frac{RT}{n_e F} \ln K_r$$

and so

$$\mathbb{E}_{cell}^{\ominus} = \frac{RT}{n_e F} \ln K_r \tag{20.15}$$

Equation (20.15) is another of electrochemistry's important thermodynamic equations, relating the generalised equilibrium constant K_r of the cell reaction to the standard reversible cell emf $\mathbb{E}_{cell}^{\ominus}$.

A further application of the Nernst equation is best illustrated by reference once more to the Daniell cell reaction

$$Zn\,(s) + Cu^{2+}\,(aq) \rightleftharpoons Zn^{2+}\,(aq) + Cu\,(s)$$

as a specific case. Using the 'shorthand' form, for this reaction

$$\Gamma_r = \frac{[Zn^{2+}]\,x_{Cu}}{[Cu^{2+}]\,x_{Zn}}$$

then for the Daniell cell reaction, the Nernst equation, equation (20.14), may be written as

$$\mathbb{E}_{cell} = \mathbb{E}_{cell}^{\ominus} - \frac{RT}{2F} \ln \frac{[Zn^{2+}]\,x_{Cu}}{[Cu^{2+}]\,x_{Zn}}$$

where, for the Daniell cell, $n_e = 2$. Expressing the reversible cell emfs E_{cell} and $E_{cell}^{\ominus}$ as the difference between the corresponding reversible cathode and anode potentials, we have

$$E_{cat} - E_{an} = E_{cat}^{\ominus} - E_{an}^{\ominus} - \frac{RT}{2F} \ln \frac{[Zn^{2+}] x_{Cu}}{[Cu^{2+}] x_{Zn}} \tag{20.16}$$

Now

$$\ln \frac{[Zn^{2+}] x_{Cu}}{[Cu^{2+}] x_{Zn}} = \ln \frac{[Zn^{2+}]}{x_{Zn}} + \ln \frac{x_{Cu}}{[Cu^{2+}]} = \ln \frac{[Zn^{2+}]}{x_{Zn}} - \ln \frac{[Cu^{2+}]}{x_{Cu}}$$

and so equation (20.16) becomes

$$E_{cat} - E_{cat}^{\ominus} - \frac{RT}{2F} \ln \frac{[Cu^{2+}]}{x_{Cu}} = E_{an} - E_{an}^{\ominus} - \frac{RT}{2F} \ln \frac{[Zn^{2+}]}{x_{Zn}} \tag{20.17}$$

in which all the terms on the left refer to the copper cathode, and those on the right, to the zinc anode.

The fact that the left-hand side of equation (20.17) equals the right-hand side implies that each side is itself equal to some, as yet unknown, quantity X, and so we may write, for the copper cathode

$$E_{cat} - E_{cat}^{\ominus} - \frac{RT}{2F} \ln \frac{[Cu^{2+}]}{x_{Cu}} = X$$

and for the zinc anode

$$E_{an} - E_{an}^{\ominus} - \frac{RT}{2F} \ln \frac{[Zn^{2+}]}{x_{Zn}} = X$$

To determine X, consider the system in which all the components i are in their standard states. Recalling that all the terms [I] are in reality the ratios $[I]/[I]^{\ominus}$ (see page 516), then, under the assumption of standard states, $[I] = [I]^{\ominus}$. Accordingly, each of the terms $[I] = [I]/[I]^{\ominus} = [I]^{\ominus}/[I]^{\ominus} = 1$, implying that, under the assumption of standard states, both the ln terms become $\ln 1 = 0$; likewise, $x_i/x^{\ominus} = 1$. Furthermore, once again under the assumption of standard states, $E_{cat} = E_{cat}^{\ominus}$ and $E_{an} = E_{an}^{\ominus}$, and so X = 0.

The key inference of equation (20.17) is that, for the cathode

$$E_{cat} = E_{cat}^{\ominus} + \frac{RT}{2F} \ln \frac{[Cu^{2+}]}{x_{Cu}} \tag{20.18a}$$

and for the anode

$$E_{an} = E_{an}^{\ominus} + \frac{RT}{2F} \ln \frac{[Zn^{2+}]}{x_{Zn}} \tag{20.18b}$$

A feature of equations (20.18a) and (20.18b) is that the expressions in the numerator of the ln terms – Zn^{2+} (aq) and Cu^{2+} (aq) respectively – are oxidised with respect to the reduced states, Zn (s) and Cu (s), in the corresponding denominators.

Two equivalent, generalised, forms of equations (20.18) are therefore

$$E(ox, red) = E^{\ominus}(ox, red) + \frac{RT}{n_e F} \ln \frac{[ox]}{[red]} \tag{20.19a}$$

and

$$\mathbb{E}(\text{ox, red}) = \mathbb{E}^{\ominus}(\text{ox, red}) - \frac{RT}{n_e F} \ln \frac{[\text{red}]}{[\text{ox}]} \tag{20.19b}$$

Equations (20.19) relate to a half-cell reaction, and are equivalent to equation (20.14)

$$\mathbb{E}_{\text{cell}} = \mathbb{E}_{\text{cell}}^{\ominus} - \frac{RT}{n_e F} \ln \Gamma_r \tag{20.14}$$

which relates to an overall cell reaction. The term **Nernst equation** is therefore often used to refer to equations (20.19) as well as to equation (20.14).

Note that both the Nernst equations (20.19) refer to a half-reaction written in the specific direction in which the oxidised state is the reactant, the reduced state the product, and electrons are consumed

$$\text{oxidised state} + n_e\, e^- \rightleftharpoons \text{reduced state}$$

as indicated in the notation $\mathbb{E}(\text{ox, red})$, which implies a redox reaction from an oxidised state to a reduced state. This corresponds to a reduction reaction as takes place at a cathode, which is always, by convention, the 'positive' terminal of any electrochemical cell, and corresponds to the convention stated on pages 603 and 609. For a reaction written in this direction, $\Gamma_r = [\text{red}]/[\text{ox}]$, so ensuring the equivalence of equation (20.14) with equations (20.19).

Furthermore, as we saw on page 621, Γ_r for a generalised reaction

$$aA + bB \rightleftharpoons cC + dD$$

takes the form

$$\Gamma_r = \frac{(a_C/a_C^{\ominus})^c\,(a_D/a_D^{\ominus})^d}{(a_A/a_A^{\ominus})^a\,(a_B/a_B^{\ominus})^b} = \frac{a_C^c\, a_D^d}{a_B^a\, a_B^b}$$

in which $a_i/a^{\ominus}$ represents the activity ratio associated with the state of the component i (such as $[I]/[I]^{\ominus}$), corresponding to the appropriate standard state (such as unit molar concentration). Accordingly, for the generalised redox reaction

$$\text{oxidised state} + n_e\, e^- \rightleftharpoons \text{reduced state}$$

in the Nernst equations (20.19a) and (20.19b)

$$\mathbb{E}(\text{ox, red}) = \mathbb{E}^{\ominus}(\text{ox, red}) + \frac{RT}{n_e F} \ln \frac{[\text{ox}]}{[\text{red}]} \tag{20.19a}$$

and

$$\mathbb{E}(\text{ox, red}) = \mathbb{E}^{\ominus}(\text{ox, red}) - \frac{RT}{n_e F} \ln \frac{[\text{red}]}{[\text{ox}]} \tag{20.19b}$$

[ox] is shorthand for the term $(a_A/a_A^{\ominus})^a\,(a_B/a_B^{\ominus})^b$, corresponding to the redox (half-)reaction's reactants, as in the denominator of the expression for Γ_r; likewise, [red] is shorthand for the term $(a_C/a_C^{\ominus})^c(a_D/a_D^{\ominus})^d$, corresponding to the redox (half-)reaction's products, as in the numerator of the expression for Γ_r.

Since, by convention, all redox reactions are written as reduction reactions

$$\text{oxidised state} + n_e\, e^- \rightleftharpoons \text{reduced state}$$

for example, for the hydrogen electrode

$$2\,H^+\,(aq) + 2\,e^- \rightleftharpoons H_2\,(g)$$

then it might be thought that

$$[red] = \frac{p\,H_2}{P^{\ominus}H_2}$$

for the gaseous hydrogen product, using the partial pressure p as the quantity measure, and that, for the reactants

$$[ox] = ([H^+]/[H^+]^{\ominus})^2 ([e^-]/[e^-]^{\ominus})^2$$

in which the term $[e^-]/[e^-]^{\ominus}$ recognises the electrons in the redox half-reaction.

In fact, this expression for [ox] is incorrect, for the electron terms $[e^-]$ and $[e^-]^{\ominus}$ do not appear in expressions based on the Nernst equation. The reason for the omission of the terms $[e^-]$ and $[e^-]^{\ominus}$ concerns the fact that redox half-reactions never take place by themselves, but always in pairs, such that the electrons produced by the reduction reaction (at the cathode) are immediately consumed in the oxidation reaction (corresponding to the anode). The electrons never accumulate, and so can never be measured as a quantity state function, such as the molar concentration or the mole fraction – the only way they can be measured is in terms of the current flowing through the external circuit, which is a path function. Terms of the form $[e^-]$ and $[e^-]^{\ominus}$ therefore have no meaning – in contrast to the very real meaning of terms such as $[H^+]$ and $p\,H_2$ which represent the molar concentration of H^+ (aq) ions and the partial pressure of gaseous hydrogen H_2 (g) at any instant, both of which can be measured, and both of which vary as the reaction takes place.

Likewise, the 'standard molar Gibbs free energy of formation of the electron $\Delta_f G^{\ominus}$ (e^-)' does not exist (or, in practice, is zero), and so does not appear in any formulae for $\Delta_r G^{\ominus}$ for redox half-reactions. So, for the hydrogen electrode

$$\Delta_r G^{\ominus} = \Delta_f G^{\ominus}\,H_2(g) - 2\Delta_f G^{\ominus} H^+\,(aq)$$

and *not*

$$\Delta_r G^{\ominus} = \Delta_f G^{\ominus}\,H_2(g) - [2\Delta_f G^{\ominus} H^{2+}\,(aq) + 2\,\Delta_f G^{\ominus}(e^-)]$$

So, as a further example, consider the redox half-reaction

$$Cr_2O_7^{2-} + 14\,H^+ + 6\,e^- \rightleftharpoons 2\,Cr^{3+} + 7\,H_2O$$

the oxidised form is the reactant mixture $Cr_2O_7^{2-} + 14\,H^+$, and the reduced form is the product mixture $2\,Cr^{3+} + 7\,H_2O$, implying that the Nernst equation (20.19a) is written as

$$\mathbb{E}(Cr_2O_7^{2-},\,Cr^{3+}) = \mathbb{E}^{\ominus}(Cr_2O_7^{2-},\,Cr^{3+}) + \frac{RT}{6F}\ln\frac{[Cr_2O_7^{2-}][H^+]^{14}}{[Cr^{3+}]^2\,[x_{H_2O}]^7}$$

In this expression, all the standard molar concentrations $[I]^{\ominus} = 1$ mol dm^{-3}, and, as we saw on page 525, when water is the solvent as well as a reactant, the term x_{H_2O}, which is 'shorthand' for the ratio $x_{H_2O}/x_{H_2O}^{\ominus}$, is equal to 1. The Nernst equation therefore becomes

$$\mathbb{E}(Cr_2O_7^{2-},\,Cr^{3+}) = \mathbb{E}^{\ominus}(Cr_2O_7^{2-},\,Cr^{3+}) + \frac{RT}{6F}\ln\frac{[Cr_2O_7^{2-}][H^+]^{14}}{[Cr^{3+}]^2}$$

20.16 Redox reactions

As we saw on page pages 606 and 607, if a platinised platinum electrode is placed in an aqueous solution containing a mixture of Fe^{2+} (aq) and Fe^{3+} (aq) ions, the electrode will exhibit a reversible electrode potential E (Fe^{3+}, Fe^{2+}) attributable to the reduction half-reaction

$$Fe^{3+} \text{ (aq)} + e^- \rightleftharpoons Fe^{2+} \text{ (aq)}$$

taking place at the platinum's surface; if the two ions are present in their standard concentrations of 1 mol m^{-3}, then, at a pressure of 1 bar, the standard reversible electrode potential $E^{\ominus}$ (Fe^{3+}, Fe^{2+}) is 0.771 V at 298.15 K.

Similarly, a platinised platinum electrode in contact with an aqueous solution containing Sn^{2+} (aq) and Sn^{4+} (aq) ions will generate a reversible electrode potential attributable to the oxidation half-reaction

$$Sn^{4+} \text{ (aq)} + 2\,e^- \rightleftharpoons Sn^{2+} \text{ (aq)}$$

such that the standard reversible electrode potential $E^{\ominus}$(Sn^{4+}, Sn^{2+}) is 0.151 V at 298.15 K. If the two solutions are connected by a salt bridge, the iron-solution electrode (which has the higher standard electrode potential) will act as a cathode to the tin-solution electrode as the anode to form a Galvanic electrochemical cell with a standard reversible emf $E_{cell}^{\ominus}$ given by

$$E_{cell}^{\ominus} = E_{cat}^{\ominus} - E_{an}^{\ominus} = E^{\ominus}(Fe^{3+}, Fe^{2+}) - E^{\ominus}(Sn^{4+}, Sn^{2+}) = 0.771 - 0.151 = 0.620 \text{ V}$$

The corresponding cell reaction is

$$Sn^{2+} \text{ (aq)} + 2\,Fe^{3+} \text{ (aq)} \rightleftharpoons 2\,Fe^{2+} \text{ (aq)} + Sn^{4+} \text{ (aq)}$$

in which the Sn^{2+} (aq) ions are oxidised to Sn^{4+} (aq) ions, whilst simultaneously, the Fe^{3+} (aq) ions are reduced to Fe^{2+} (aq) ions, this being an example of a redox reaction. Accordingly, the reaction taking place within an electrochemical cell, all aspects of the thermodynamic analysis shown on pages 621 to 627 apply, including equation (20.15)

$$E_{cell}^{\ominus} = \frac{RT}{n_e F} \ln K_r \tag{20.15}$$

and the Nernst equation expressed as equation (20.14)

$$E_{cell} = E_{cell}^{\ominus} - \frac{RT}{n_e F} \ln \Gamma_r \tag{20.14}$$

Suppose now that a solution of, say, $FeCl_3$ is added to a solution of $SnCl_2$. The Fe^{3+} (aq) ions will react with the Sn^{2+} (aq) ions according to the reaction

$$SnCl_2 \text{ (aq)} + 2\,FeCl_3 \text{ (aq)} \rightleftharpoons 2\,FeCl_2 \text{ (aq)} + SnCl_4 \text{ (aq)}$$

or, more perceptively

$$Sn^{2+} \text{ (aq)} + 2\,Fe^{3+} \text{ (aq)} \rightleftharpoons 2\,Fe^{2+} \text{ (aq)} + Sn^{4+} \text{ (aq)}$$

This is of course the electrochemical cell reaction, but in this case, there are no electrodes, and no electrochemical cell. *But the same reaction still takes place*, for the reaction does not 'know' whether or not there are any electrodes, and whether or not any electric current is being driven through an external circuit. The thermodynamics of the 'real electrochemical' and 'hypothetical electrochemical' reactions are identical, and it is this insight that enables us to apply the concepts of the thermodynamics of electrochemical cells in general, and the Nernst equation in particular, to a huge number of redox reactions in which 'electricity' is not explicitly involved, but where electrons are exchanged in redox processes – as is especially important in many biological processes, as we shall see in Chapter 24.

As an example of how the Nernst equation can be used in practice, let us consider the reaction

$$Sn^{2+} (aq) + 2\ Fe^{3+} (aq) \rightleftharpoons 2\ Fe^{2+} (aq) + Sn^{4+} (aq)$$

which is of practical significance as a means of assaying Fe^{3+} (aq) ions by titration. The objective of the titration is to measure the quantity of Fe^{3+} (aq) ions by reducing them all to Fe^{2+} (aq) ions. For the titration to useful, the equilibrium of this reaction should be as far to the right as possible, implying that the equilibrium constant should be a large number – say, at least 10^{12} – so our goal is to determine the equilibrium constant of the titration reaction at a pressure of 1 bar, and a temperature of 298.15 K.

If this reaction were to take place in an electrochemical cell, as we have seen, the iron-solution would act as the cathode, and the tin-solution as the anode, and the Nernst reaction in the form of equation (20.14)

$$\mathbb{E}_{cell} = \mathbb{E}_{cell}^{\ominus} - \frac{RT}{n_e F} \ln \Gamma_r \qquad (20.14)$$

would apply, where, at any instant whilst the reaction is taking place

$$\Gamma_r = \frac{[Fe^{2+}]^2 [Sn^{4+}]}{[Fe^{3+}]^2 [Sn^{2+}]} = \Gamma_c$$

in which the quantities of all components are expressed as molar concentrations, each associated with the standard state $[I]^{\ominus} = 1$ mol dm^{-3}.

The correct value for n_e is best determined by firstly identifying the reactant which has the lowest stoichiometric coefficient in the cell reaction as written

$$Sn^{2+} (aq) + 2\ Fe^{3+} (aq) \rightleftharpoons 2\ Fe^{2+} (aq) + Sn^{4+} (aq)$$

which, in this case, is Sn^{2+} (aq), for which the corresponding half-reaction is

$$Sn^{2+} (aq) + 2\ e^{-} \rightleftharpoons Sn^{4+} (aq)$$

The coefficient for the electron e^- is 2, implying that $n_e = 2$. Note that, had the stoichiometry of the cell reaction been of the form, say

$$2\ Sn^{2+} (aq) + ... \rightleftharpoons ... + 2\ Sn^{4+} (aq)$$

such that Sn^{2+} (aq) is still the reactant with the lowest stoichiometric coefficient, then the corresponding half-reaction would be

$$2\ Sn^{2+} (aq) + 4\ e^{-} \rightleftharpoons 2\ Sn^{4+} (aq)$$

and n_e would be 4.

As we saw on page 624, the condition for equilibrium for an electrochemical cell is that $\mathbb{E}_{cell} = 0$, and so, at equilibrium, the various components are in their equilibrium compositions, giving

$$0 = \mathbb{E}_{cell}^{\ominus} - \frac{RT}{2F} \ln \frac{[Fe^{2+}]_{eq}^{2}\,[Sn^{4+}]_{eq}}{[Fe^{3+}]_{eq}^{2}\,[Sn^{2+}]_{eq}}$$

Hence

$$\mathbb{E}_{cell}^{\ominus} = \frac{RT}{2F} \ln \frac{[Fe^{2+}]_{eq}^{2}\,[Sn^{4+}]_{eq}}{[Fe^{3+}]_{eq}^{2}\,[Sn^{2+}]_{eq}} = \frac{RT}{2F} \ln K_c \qquad (20.20)$$

where the concentration equilibrium constant K_c is

$$K_c = \frac{[Fe^{2+}]_{eq}^{2}\,[Sn^{4+}]_{eq}}{[Fe^{3+}]_{eq}^{2}\,[Sn^{2+}]_{eq}}$$

This calculation is valid for the reaction, whether or not there are any 'real' electrochemical aspects such as electrodes and currents; accordingly, equation (20.20) for K_c is the result we seek for the titration. From Table 20.2, we see that $E^{\ominus}$ (Fe^{3+}, Fe^{2+}) = 0.771 V and $E^{\ominus}$ (Sn^{4+}, Sn^{2+}) = 0.151 V, giving a value for $\mathbb{E}_{cell}^{\ominus}$ of 0.771 − 0.151 = 0.620 V; R = 8.314 J K^{-1} mol^{-1}, F = 9.6485 × 10^4 C mol^{-1} and so K_c is of the order of 10^{21}. This is indeed a large number, much greater than 10^{12}, and so the reaction does indeed go to completion. The titration is therefore effective.

One further point about the reaction

$$Sn^{2+}\,(aq) + 2\,Fe^{3+}\,(aq) \rightleftharpoons 2\,Fe^{2+}\,(aq) + Sn^{4+}\,(aq)$$

in the context of the data in Table 20.2. In this reaction, the Sn^{2+} (aq) ion has acted as a reducing agent with respect to the Fe^{2+} (aq) ion; conversely, the Fe^{3+} (aq) ion has acted as an oxidising agent with respect to the Sn^{2+} (aq) ion. Reference to Table 20.2 will show that the reversible redox potential $E^{\ominus}$ (Fe^{3+}, Fe^{2+}) is more positive than the reversible redox potential $E^{\ominus}$ (Sn^{4+}, Sn^{2+}). As we saw on page 620, this is an instance of a general rule:

> The more positive the redox potential, the stronger the oxidising agent.
> The more negative the redox potential, the stronger the reducing agent.

Accordingly, the Co^{3+}, Co^{2+} system is therefore a very powerful oxidising agent, and the Cr^{3+}, Cr^{2+} system a very powerful reducing agent. Furthermore, equation (20.20) can be used to determine the minimum value of the standard emf of a redox cell for the corresponding reaction to be used as a titration. For a titration to be successful, the reaction must, from a practical point of view, go to completion: so let's propose that this requires a minimum value of K_c of about 10^{12}. If $n_e = 1$, then at 298.15 K, and taking the standard values of R = 8.314 J K^{-1} mol^{-1}, F = 9.6485 × 10^4 C mol^{-1}, then equation (20.20) implies that $\mathbb{E}_{cell}^{\ominus}$ must be greater than about 0.53 V. Table 20.2 can therefore be used to ensure that the difference between the standard reversible redox potentials of the reagents exceeds this threshold value.

One final thought on the Nernst equation

$$\mathbb{E}_{cell} = \mathbb{E}_{cell}^{\ominus} - \frac{RT}{n_e F} \ln \Gamma_r \qquad (20.14)$$

which, as we have seen, applies not only to electrochemical cells, but to all redox reactions. If we compare the Nernst equation to the general equation

$$\Delta_r G = \Delta_r G^\ominus + RT \ln \Gamma_r \qquad (16.16d)$$

the parallels are clear: for electrochemistry and redox reactions, the quantity $-n_e F \, \mathbb{E}_{cell}$ maps onto $\Delta_r G$, and $-n_e F \, \mathbb{E}_{cell}^\ominus$ onto $\Delta_r G^\ominus$. With this mapping in mind (and, importantly, the sign change), all of the inferences and interpretations associated with equation (16.16d) and the quantities $\Delta_r G$ and $\Delta_r G^\ominus$, in relation to general reactions taking place in solution, as discussed in Chapter 16, therefore apply to redox reactions, for which the corresponding thermodynamic measurements are $\mathbb{E}_{cell}$ and $\mathbb{E}_{cell}^\ominus$.

EXERCISES

1. Define
   - Joule heating
   - half-reaction
   - oxidation reaction
   - reduction reaction
   - electrochemical cell
   - cell reaction
   - Daniell cell
   - salt bridge
   - electrolyte
   - electrode
   - anode
   - cathode
   - Galvanic electrochemical cell
   - electrolytic electrochemical cell
   - electrolysis
   - electroplating
   - electrode potential
   - reversible electrode potential
   - standard electrode
   - standard reversible electrode potential
   - electromotive force
   - redox reaction
   - redox electrode
   - standard reversible redox potential
   - oxidising agent
   - reducing agent.

2. For each of the following pairs of electrodes, define the corresponding electrochemical cell – specifying which electrode is the anode, and which the cathode – and calculate the corresponding cell standard emf $\mathbb{E}^\ominus{}_{cell}$
   - calcium, silver
   - chlorine, iron
   - lithium, zinc

- hydrogen, oxygen (acidic)
- nickel, tin
- Ce^{4+}, hydrogen
- Fe^{3+}, MnO_4^-
- Cu^{2+}, Hg^{2+}

3. Two biochemical redox half-reactions, which are important in respiration, are

$$NAD^+ + 2\,e^- + H^+ \rightleftharpoons NADH \qquad E^\circ(NAD^+, NADH) = 0.094\text{ V}$$
$$FAD + 2\,e^- + 2\,H^+ \rightleftharpoons FADH_2 \qquad E^\circ(FAD, FADH_2) = 0.195\text{ V}$$

Calculate the value of E°, and the corresponding values of $\Delta_r G_c^\circ$ and K_c for each of the two redox reactions

$$O_2 + 2\,NADH + 2\,H^+ \rightleftharpoons 2\,H_2O + 2\,NAD^+$$

and

$$O_2 + 2\,FADH_2 + 2\,H^+ \rightleftharpoons 2\,H_2O + 2\,FAD^+$$

4. "The flow of electrons through an external conductor connecting the electrodes of a Galvanic electrochemical cell is just like the flow of heat between a hot body and a cold body. Both are non-equilibrium processes that take place spontaneously and unidirectionally." Discuss.

21 Mathematical round up

Summary

Four key equations

$$dU = T\,dS - P\,dV \qquad \text{in which } U = U(S, V) \tag{9.20a}$$

$$dH = T\,dS + V\,dP \qquad \text{in which } H = H(S, P) \tag{9.21a}$$

$$dG = V\,dP - S\,dT \qquad \text{in which } G = G(P, T) \tag{13.16a}$$

$$dA = -P\,dV - S\,dT \qquad \text{in which } A = A(V, T) \tag{13.35}$$

The four **Maxwell relations**

$$\left(\frac{\partial T}{\partial V}\right)_S = -\left(\frac{\partial P}{\partial S}\right)_V \tag{21.6a}$$

$$\left(\frac{\partial T}{\partial P}\right)_S = \left(\frac{\partial V}{\partial S}\right)_P \tag{21.6b}$$

$$\left(\frac{\partial V}{\partial T}\right)_P = -\left(\frac{\partial S}{\partial P}\right)_T \tag{21.6c}$$

$$\left(\frac{\partial P}{\partial T}\right)_V = \left(\frac{\partial S}{\partial V}\right)_T \tag{21.6d}$$

The **chain rule**

$$\left(\frac{\partial X}{\partial Y}\right)_Z \left(\frac{\partial Y}{\partial Z}\right)_X \left(\frac{\partial Z}{\partial X}\right)_Y = -1 \tag{21.10}$$

The two **thermodynamic equations-of-state**

$$\left(\frac{\partial U}{\partial V}\right)_T = T\left(\frac{\partial P}{\partial T}\right)_V - P \tag{21.12}$$

$$\left(\frac{\partial U}{\partial P}\right)_T = -T\left(\frac{\partial V}{\partial T}\right)_P - P\left(\frac{\partial V}{\partial P}\right)_T \tag{21.13}$$

The difference $C_P - C_V$

$$C_P - C_V = -T\left(\frac{\partial V}{\partial T}\right)_P^2 \left(\frac{\partial P}{\partial V}\right)_T \tag{21.28}$$

A **throttling process** is the expansion of a gas through a porous plug, from a higher pressure to a lower pressure, whilst maintaining the system's enthalpy constant. The **Joule-Thomson coefficient** μ_{JT} is defined as

$$\mu_{JT} = \left(\frac{\partial T}{\partial P}\right)_H = \frac{1}{C_P}\left[T\left(\frac{\partial V}{\partial T}\right)_P - V\right] \tag{21.37}$$

Modern Thermodynamics for Chemists and Biochemists. Dennis Sherwood and Paul Dalby.
© Oxford University Press 2018. Published 2018 by Oxford University Press.

For an ideal gas, $\mu_{JT} = 0$ at all temperatures and pressures; for a real gas, in general, $\mu_{JT} > 0$ at lower pressures, and so the gas will become cooler as a result of the throttling process ($\partial P_H < 0$ and so $\partial T_H < 0$), but at higher pressures, $\mu_{JT} < 0$, and the gas will become warmer ($\partial P_H < 0$ and so $\partial T_H > 0$). For a system operating at a given enthalpy, the temperature at which $\mu_{JT} = 0$ is known as the **inversion temperature**, and the curve that connects the inversion temperatures at different enthalpies is known as the **inversion curve**.

The **compressibility factor** Z is defined as

$$Z = \frac{PV}{RT} \tag{21.38}$$

For an ideal gas, $Z = 1$ under all conditions of pressure, volume and temperature. For a real gas, Z varies with pressure (and volume and temperature too), and so the behaviour of Z for a real gas under given conditions, and the extent to which measures of Z are close to 1, indicates the extent to which the real gas does, or does not, approximate to ideal behaviour.

21.1 The fundamental functions

In this chapter, we bring together the main mathematical results developed so far, and introduce some further relationships.

The *extensive state functions* we have discussed, and their corresponding units of measure, are

- the mole number n mol defining the quantity of a given pure material in a system
- the mass M kg of the system
- the volume V m^3 (or dm^{-3}) occupied by the system
- the internal energy U J
- the heat capacity at constant volume C_V J K^{-1} mol^{-1}
- the enthalpy H J
- the heat capacity at constant pressure C_P J K^{-1} mol^{-1}
- the entropy S J K^{-1}
- the Gibbs free energy G J
- the Helmholtz free energy A J
- the electric charge Q C

The *intensive state functions* are

- the temperature T K
- the pressure P Pa
- the partial pressure p Pa
- the density ρ kg m^{-3}
- the molality b mol kg^{-1}
- the molar concentration [I] mol m^{-3} (or mol dm^{-3})
- the chemical potential μ J mol^{-1}
- the electrode potential E V

and

- all extensive functions per mol – **M, V, U, C_V, H, C_P, S, G, A, Q, E**

We have also studied in detail two *path functions*

- heat đq J
- work đw J

The extensive state functions n, V, U, S and Q, and the intensive state functions P, T, μ and E are all 'primary', in contrast to the extensive state functions H, G and A which are defined by reference to 'primary' state functions as

$$H = U + PV \tag{6.4}$$
$$G = H - TS \tag{13.10a}$$
$$A = U - TS \tag{13.34}$$

21.2 Pure substances

As first stated on page 48, in the absence of gravitation, motion, electrical, magnetic and surface effects, a fixed quantity of n mol of a pure substance has two, and only two, independent state functions. Furthermore, for a pure substance, all extensive state functions are linearly proportional to n. For most thermodynamic contexts, these restrictions are not a problem, but there are two circumstances in particular which are beyond this scope

- open systems in which the quantity of matter changes, such as 'flow-through' systems in chemical engineering
- systems in which electrical effects arise, as discussed explicitly in Chapter 20, in connection with electrochemistry.

If the conditions defined on page 48 are fulfilled, then, mathematically, any state function X of a fixed quantity of a pure material can be expressed as function of any two other state functions Y and Z in the form $X(Y, Z)$ such that

$$dX = \left(\frac{\partial X}{\partial Y}\right)_Z dY + \left(\frac{\partial X}{\partial Z}\right)_Y dZ \tag{21.1}$$

The quantities represented by the generalised partial derivatives $(\partial X/\partial Y)_Z$ will have a particular meaning, as is best illustrated by the examples of the four differentials dU, dH, dG and dA, as applied to a closed system of a fixed mass of n mol, and which can perform only P, V work of expansion

$$dU = T\,dS - P\,dV \quad \text{in which } U = U(S, V) \tag{9.20a}$$
$$dH = T\,dS + V\,dP \quad \text{in which } H = H(S, P) \tag{9.21a}$$
$$dG = V\,dP - S\,dT \quad \text{in which } G = G(P, T) \tag{13.16a}$$
$$dA = -P\,dV - S\,dT \quad \text{in which } A = A(V, T) \tag{13.35}$$

These four equations are the basis of the mathematics of thermodynamics, and all have the form of equation (21.1) from which we may identify that

$$\left(\frac{\partial U}{\partial S}\right)_V = T \quad \text{and} \quad \left(\frac{\partial U}{\partial V}\right)_S = -P \tag{21.2a and b}$$

$$\left(\frac{\partial H}{\partial S}\right)_P = T \quad \text{and} \quad \left(\frac{\partial H}{\partial P}\right)_S = V \tag{21.3a and b}$$

$$\left(\frac{\partial G}{\partial P}\right)_T = V \quad \text{and} \quad \left(\frac{\partial G}{\partial T}\right)_P = -S \tag{21.4a and b}$$

$$\left(\frac{\partial A}{\partial V}\right)_T = -P \quad \text{and} \quad \left(\frac{\partial A}{\partial T}\right)_V = -S \tag{21.5a and b}$$

21.3 Heat capacities

Equation (9.20a)

$$dU = T\,dS - P\,dV \tag{9.20a}$$

is based on expressing U as a function $U(S, V)$ of S and V. This is not the only possibility: in principle, U may be expressed as a function of any other two state functions – for example, as a function $U(T, V)$ of T and V, in which case, we may write

$$dU = \left(\frac{\partial U}{\partial T}\right)_V dT + \left(\frac{\partial U}{\partial V}\right)_T dV \tag{5.8b}$$

The quantity $(\partial U/\partial T)_V$ has a special significance: it is the extensive state function C_V, the heat capacity at constant volume

$$C_V = \left(\frac{\partial U}{\partial V}\right)_T \tag{5.14b}$$

Mathematically, C_V is the quantity defined by equation (5.14b); physically – and hence its importance – it represents the quantity of heat $\{dq_{\text{rev}}\}_V$ that is required to flow into a closed system of fixed mass, at constant volume, in order to raise the temperature of the system by a given amount dT – for example, the amount of heat required to raise the temperature of 1 mol of a pure substance by 1 K. Although it is not explicitly obvious from equation (5.14b), C_V is itself a function of temperature $C_V(T)$.

The physical interpretation of the second quantity in equation (5.8b), $(\partial U/\partial V)_T$, which describes the change ∂U_T attributable to a change ∂V_T at constant temperature, is less obvious: a point to which we will return shortly.

A similar approach can be taken to equation (9.21a)

$$dH = T\,dS + V\,dP \tag{9.21a}$$

which is based on expressing H as a function $H(S, P)$ of S and P. Alternatively, we can express H as a function $H(T, P)$ of T and P, giving

$$dH = \left(\frac{\partial H}{\partial T}\right)_P dT + \left(\frac{\partial H}{\partial P}\right)_T dP \tag{6.7}$$

from which we define the extensive state function, the heat capacity at constant pressure C_P as

$$C_P = \left(\frac{\partial H}{\partial T}\right)_P \tag{6.8a}$$

Mathematically, C_P is the quantity defined by equation (6.8a); physically, it represents the quantity of heat $\{dq_{rev}\}_P$ that is required to flow, reversibly, into a closed system of fixed mass, at constant pressure, in order to raise the temperature of the system by a given amount dT – for example, the amount of heat required to raise the temperature of 1 mol of a pure substance by 1 K. Also, like C_V, C_P is a function of temperature $C_p(T)$.

21.4 The Maxwell relations

Mathematically, for any function X that can be expressed as function of two other state functions Y and Z in the form $X(Y, Z)$, we know that we can express dX as

$$dX = \left(\frac{\partial X}{\partial Y}\right)_Z dY + \left(\frac{\partial X}{\partial Z}\right)_Y dZ \tag{21.1}$$

In general, if X is a function of both Y and Z, $X(Y,Z)$, then we should expect the quantities $(\partial X/\partial Y)_Z$ and $(\partial X/\partial Z)_Y$ themselves to be (different) functions of Y and Z – and hence themselves differentiable with respect to Y and Z, so generating (new) functions which can be represented as

$$\left(\frac{\partial}{\partial Z}\left(\frac{\partial X}{\partial Y}\right)_Z\right)_Y \quad \text{and} \quad \left(\frac{\partial}{\partial Y}\left(\frac{\partial X}{\partial Z}\right)_Y\right)_Z$$

To make that real – and using an example that has nothing to do with thermodynamics – suppose that

$$X = Y^2 Z^3$$

in which X is a function of both Y and Z, $X(Y, Z)$, as required. Accordingly,

$$\left(\frac{\partial X}{\partial Y}\right)_Z = 2\,Y\,Z^3 \quad \text{and} \quad \left(\frac{\partial X}{\partial Z}\right)_Y = 3\,Y^2\,Z^2$$

and we see that $(\partial X/\partial Y)_Z$ and $(\partial X/\partial Z)_Y$ are indeed new, and different, functions of both Y and Z.

If we now differentiate $(\partial X/\partial Y)_Z$ with respect to Z at constant Y, we obtain

$$\left(\frac{\partial}{\partial Z}\left(\frac{\partial X}{\partial Y}\right)_Z\right)_Y = 6\,Y\,Z^2$$

Likewise, differentiating $(\partial X/\partial Z)_Y$ with respect to Y at constant Z

$$\left(\frac{\partial}{\partial Y}\left(\frac{\partial X}{\partial Z}\right)_Y\right)_Z = 6\,Y\,Z^2$$

from which we see that

$$\left(\frac{\partial}{\partial Z}\left(\frac{\partial X}{\partial Y}\right)_Z\right)_Y = \left(\frac{\partial}{\partial Y}\left(\frac{\partial X}{\partial Z}\right)_Y\right)_Z$$

This equality is not an accident of the example we chose: mathematically, for any function $X = X(Y, Z)$ of two variables Y and Z, provided that the function is what mathematicians describe as 'continuous' (meaning that the function behaves smoothly, and has no sudden jumps – such as those shown in Figure 9.3), then is it always true that

$$\left(\frac{\partial}{\partial Z}\left(\frac{\partial X}{\partial Y}\right)_Z\right)_Y = \left(\frac{\partial}{\partial Y}\left(\frac{\partial X}{\partial Z}\right)_Y\right)_Z = \frac{\partial^2 X}{\partial Y \partial Z}$$

This apparently abstract result can be applied, with benefit, to thermodynamics. If we take, in the first instance, the relationship

$$dU = T\,dS - P\,dV = \left(\frac{\partial U}{\partial S}\right)_V dS + \left(\frac{\partial U}{\partial V}\right)_S dV \tag{9.20a}$$

we have

$$\left(\frac{\partial U}{\partial S}\right)_V = T \quad \text{and} \quad \left(\frac{\partial U}{\partial V}\right)_S = -P$$

and so

$$\left(\frac{\partial}{\partial V}\left(\frac{\partial U}{\partial S}\right)_V\right)_S = \left(\frac{\partial T}{\partial V}\right)_S \quad \text{and} \quad \left(\frac{\partial}{\partial S}\left(\frac{\partial U}{\partial V}\right)_S\right)_V = -\left(\frac{\partial P}{\partial S}\right)_V$$

from which we infer that

$$\left(\frac{\partial T}{\partial V}\right)_S = -\left(\frac{\partial P}{\partial S}\right)_V$$

A consequence of equation (9.20a)

$$dU = T\,dS - P\,dV \tag{9.20a}$$

is therefore that

$$\left(\frac{\partial T}{\partial V}\right)_S = -\left(\frac{\partial P}{\partial S}\right)_V$$

and a similar analysis can be carried out for all of the fundamental equations of thermodynamics. Taking each of

$$dU = T\,dS - P\,dV \tag{9.20a}$$
$$dH = T\,dS + V\,dP \tag{9.21a}$$
$$dG = V\,dP - S\,dT \tag{13.16a}$$
$$dA = -P\,dV - S\,dT \tag{13.35}$$

in turn, then

$$\left(\frac{\partial T}{\partial V}\right)_S = -\left(\frac{\partial P}{\partial S}\right)_V \tag{21.6a}$$

$$\left(\frac{\partial T}{\partial P}\right)_S = \left(\frac{\partial V}{\partial S}\right)_P \tag{21.6b}$$

$$\left(\frac{\partial V}{\partial T}\right)_P = -\left(\frac{\partial S}{\partial P}\right)_T \tag{21.6c}$$

$$\left(\frac{\partial P}{\partial T}\right)_V = \left(\frac{\partial S}{\partial V}\right)_T \tag{21.6d}$$

These last four equations are collectively known as the **Maxwell relations**, and they are useful, as will shortly be seen, in changing an equation which contains a rather less familiar quantity such as $(\partial S/\partial V)_T$ into the more amenable expression $(\partial P/\partial T)_V$ – especially if the system under investigation has a known equation-of-state, such as ideal gas for which

$$PV = nRT \tag{3.3}$$

implying that

$$\left(\frac{\partial S}{\partial V}\right)_T = \left(\frac{\partial P}{\partial T}\right)_V = \frac{nR}{V}$$

The Maxwell relations have a 'symmetry' that makes them easier to remember

- The pairs of conjugate variables P and V, and S and T, are always diagonally opposite.
- The variable in the denominator of one side is the variable that is held constant on the other.
- The four equations are in two 'upside-down' pairs, more obviously shown by this sequence

$$\left(\frac{\partial T}{\partial V}\right)_S = -\left(\frac{\partial P}{\partial S}\right)_V \tag{21.6a}$$

$$\left(\frac{\partial V}{\partial T}\right)_P = -\left(\frac{\partial S}{\partial P}\right)_T \tag{21.6c}$$

and

$$\left(\frac{\partial T}{\partial P}\right)_S = \left(\frac{\partial V}{\partial S}\right)_P \tag{21.6b}$$

$$\left(\frac{\partial P}{\partial T}\right)_V = \left(\frac{\partial S}{\partial V}\right)_T \tag{21.6d}$$

- The sign is positive when the variables P and S are on different sides, and negative when P and S are on the same side. The negative sign is consistent with the physical reality that when the pressure P increases at constant temperature or volume – implying that ∂P_T and ∂P_V are positive – then the order of the system increases, implying that ∂S_T and ∂S_V are negative.

21.5 The chain rule

A further useful mathematical result is the so-called 'chain rule', which applies to any mathematical entity $X(Y, Z)$ that is a function of two variables Y and Z. As an example, for n mol of an ideal gas, the pressure P can be expressed as a function of temperature T and volume V according to the ideal gas law as

$$P(T, V) = \frac{nRT}{V} \tag{21.7a}$$

The same system, however, can be described by defining the volume V as a function of temperature T and pressure P as

$$V(T, P) = \frac{nRT}{P} \tag{21.7b}$$

Similarly, the temperature T can be expressed as a function of the pressure P and volume V as

$$T(P, V) = \frac{PV}{nR} \tag{21.7c}$$

These three equations are all describing the same system, but are different representations, according to which two state functions are chosen to be the independent variables that appear on the right-hand side.

In more general terms, for any function $X(Y, Z)$ of two variables Y and Z, there will always be two associated functions, $Y(X, Z)$ and $Z(X, Y)$. So, for the function $X(Y, Z)$, we may write

$$dX = \left(\frac{\partial X}{\partial Y}\right)_Z dY + \left(\frac{\partial X}{\partial Z}\right)_Y dZ \tag{21.1}$$

Likewise, for the related function $Y(X, Z)$

$$dY = \left(\frac{\partial Y}{\partial X}\right)_Z dX + \left(\frac{\partial Y}{\partial Z}\right)_X dZ \tag{21.8}$$

Equation (21.8) may be rearranged as

$$dX \left(\frac{\partial Y}{\partial X}\right)_Z = dY - \left(\frac{\partial Y}{\partial Z}\right)_X dZ$$

and since

$$\left(\frac{\partial X}{\partial Y}\right)_Z = 1 / \left(\frac{\partial Y}{\partial X}\right)_Z$$

then

$$dX = \left(\frac{\partial X}{\partial Y}\right)_Z \left(dY - \left(\frac{\partial Y}{\partial Z}\right)_X dZ\right)$$

giving

$$dX = \left(\frac{\partial X}{\partial Y}\right)_Z dY - \left(\frac{\partial X}{\partial Y}\right)_Z \left(\frac{\partial Y}{\partial Z}\right)_X dZ \tag{21.9}$$

Equations (21.1) and (21.8) – and hence (21.9) – all refer to the same system, and so dX in equation (21.9) is the same as dX in equation (21.1), hence

$$\left(\frac{\partial X}{\partial Y}\right)_Z dY + \left(\frac{\partial X}{\partial Z}\right)_Y dZ = \left(\frac{\partial X}{\partial Y}\right)_Z dY - \left(\frac{\partial X}{\partial Y}\right)_Z \left(\frac{\partial Y}{\partial Z}\right)_X dZ$$

and so

$$dZ = -\left(\frac{\partial X}{\partial Y}\right)_Z \left(\frac{\partial Y}{\partial Z}\right)_X \left(\frac{\partial Z}{\partial X}\right)_Y dZ$$

from which

$$\left(\frac{\partial X}{\partial Y}\right)_Z \left(\frac{\partial Y}{\partial Z}\right)_X \left(\frac{\partial Z}{\partial X}\right)_Y = -1 \qquad (21.10)$$

The elegant result shown in equation (21.10) is the **chain rule**, which takes a more concrete form when expressed in terms of the variables P, V and T of a thermodynamic system

$$\left(\frac{\partial P}{\partial V}\right)_T \left(\frac{\partial V}{\partial T}\right)_P \left(\frac{\partial T}{\partial P}\right)_V = -1 \qquad (21.11)$$

For an ideal gas, and with reference to equations (21.7)

$$\left(\frac{\partial P}{\partial V}\right)_T = -\frac{nRT}{V^2}$$

$$\left(\frac{\partial V}{\partial T}\right)_P = \frac{nR}{P}$$

$$\left(\frac{\partial T}{\partial P}\right)_V = \frac{V}{nR}$$

so verifying that

$$\left(-\frac{nRT}{V^2}\right)\left(\frac{nR}{P}\right)\left(\frac{V}{nR}\right) = -\frac{nRT}{PV} = -1$$

After all that mathematics, we will now use the results in the context of four applications

- the "thermodynamic equations-of-state"
- the difference $C_P - C_V$
- the Joule-Thomson coefficient
- the compressibility factor.

21.6 The thermodynamic equations-of-state

If we divide equation (9.20a)

$$dU = T\,dS - P\,dV \qquad (9.20a)$$

by dV at constant T, we have

$$\left(\frac{\partial U}{\partial V}\right)_T = T\left(\frac{\partial S}{\partial V}\right)_T - P$$

We may now use the Maxwell relation (21.6d)

$$\left(\frac{\partial P}{\partial T}\right)_V = \left(\frac{\partial S}{\partial V}\right)_T \tag{21.6d}$$

giving

$$\left(\frac{\partial U}{\partial V}\right)_T = T\left(\frac{\partial P}{\partial T}\right)_V - P \tag{21.12}$$

Mathematically, the term on the left-hand side of equation (21.12), $(\partial U/\partial V)_T$, represents the rate of change of U with respect to V at constant T; physically, this may be interpreted as a measure of how much the internal energy of a system changes when the system expands at constant temperature.

An important feature of equation (21.12) is that it expresses $(\partial U/\partial V)_T$ in terms of T and P, and also $(\partial P/\partial T)_V$, which is readily determined if an appropriate equation-of-state is known (such as the ideal gas law) for the system of interest. Given that equation (21.12) is of general applicability to any system of constant mass, equation (21.12) is known as the **first thermodynamic equation-of-state** – the second thermodynamic equation-of-state being the equivalent expression for $(\partial U/\partial P)_T$, which we can derive by starting, once again, with equation (9.20a)

$$dU = T\,dS - P\,dV \tag{9.20a}$$

If we divide throughout by dP at constant temperature, we have

$$\left(\frac{\partial U}{\partial P}\right)_T = T\left(\frac{\partial S}{\partial P}\right)_T - P\left(\frac{\partial V}{\partial P}\right)_T$$

We may now use the Maxwell relation (21.6c)

$$\left(\frac{\partial V}{\partial T}\right)_P = -\left(\frac{\partial S}{\partial P}\right)_T \tag{21.6c}$$

which allows us to substitute for $(\partial S/\partial P)_T$, giving

$$\left(\frac{\partial U}{\partial P}\right)_T = -T\left(\frac{\partial V}{\partial T}\right)_P - P\left(\frac{\partial V}{\partial P}\right)_T \tag{21.13}$$

and this is the **second thermodynamic equation-of-state**.

Equations (21.12) and (21.13) apply to any system of constant mass, and so, for an ideal gas of n mol

$$PV = nRT \tag{3.3}$$

In respect of equation (21.12), we therefore have

$$\left(\frac{\partial P}{\partial T}\right)_V = \frac{nR}{V} = \frac{P}{T} \tag{21.14}$$

and for equation (21.13)

$$\left(\frac{\partial V}{\partial T}\right)_P = \frac{nR}{P} = \frac{V}{T} \tag{21.15}$$

and

$$\left(\frac{\partial V}{\partial P}\right)_T = -\frac{nRT}{P^2} = -\frac{V}{P} \tag{21.16}$$

These results can be incorporated into equation (21.12)

$$\left(\frac{\partial U}{\partial V}\right)_T = T\left(\frac{\partial P}{\partial T}\right)_V - P \tag{21.12}$$

giving

$$\left(\frac{\partial U}{\partial V}\right)_T = T\left(\frac{P}{T}\right) - P = 0 \tag{21.17}$$

and into equation (21.13)

$$\left(\frac{\partial U}{\partial P}\right)_T = -T\left(\frac{\partial V}{\partial T}\right)_P - P\left(\frac{\partial V}{\partial P}\right)_T \tag{21.13}$$

with the result

$$\left(\frac{\partial U}{\partial P}\right)_T = -T\left(\frac{V}{T}\right) - P\left(\frac{V}{P}\right) = 0 \tag{21.18}$$

Equations (21.17) and (21.18) show that, for an ideal gas

$$\left(\frac{\partial U}{\partial V}\right)_T = \left(\frac{\partial U}{\partial P}\right)_T = 0$$

The internal energy U of a given mass of an ideal gas therefore does not vary with volume or with pressure. It does, however, vary with temperature as

$$\left(\frac{\partial U}{\partial T}\right)_V = C_V \tag{5.14b}$$

Since an ideal gas is totally described by the ideal gas law

$$PV = nRT \tag{3.3}$$

from which, for a given quantity of n mol of gas, the only variables that determine the gas's behaviour are P, V and T. Given that we have shown that U depends on T, but does not depend on either P or V, this now proves an assertion we have made many times in this book – that the internal energy of a given mass of an ideal gas is a function of temperature only.

A similar result – that the enthalpy of a given mass of an ideal gas is a function of temperature only – can be established from first principles by a very similar process, but this time starting from equation (9.21a)

$$dH = T\,dS + V\,dP \tag{9.21a}$$

from which, by dP at constant T gives

$$\left(\frac{\partial H}{\partial P}\right)_T = T\left(\frac{\partial S}{\partial P}\right)_T + V$$

We now use the Maxwell relation (21.6c)

$$\left(\frac{\partial V}{\partial T}\right)_P = -\left(\frac{\partial S}{\partial P}\right)_T \tag{21.6c}$$

and so

$$\left(\frac{\partial H}{\partial P}\right)_T = -T\left(\frac{\partial V}{\partial T}\right)_P + V$$

On page 642, we showed that, for an ideal gas

$$\left(\frac{\partial V}{\partial T}\right)_P = \frac{nR}{P} = \frac{V}{T} \tag{21.15}$$

implying that

$$\left(\frac{\partial H}{\partial P}\right)_T = -T\left(\frac{V}{T}\right) + V = 0$$

Also, starting once more with equation (9.21a)

$$dH = T\,dS + V\,dP \tag{9.21a}$$

and dividing by dV at constant T

$$\left(\frac{\partial H}{\partial V}\right)_T = T\left(\frac{\partial S}{\partial V}\right)_T + V\left(\frac{\partial P}{\partial V}\right)_T$$

Substituting for $(\partial S/\partial V)_T$ using the Maxwell relation (21.6d)

$$\left(\frac{\partial P}{\partial T}\right)_V = \left(\frac{\partial S}{\partial V}\right)_T \tag{21.6d}$$

gives

$$\left(\frac{\partial H}{\partial V}\right)_T = T\left(\frac{\partial P}{\partial T}\right)_V + V\left(\frac{\partial P}{\partial V}\right)_T$$

For an ideal gas

$$\left(\frac{\partial P}{\partial T}\right)_V = \frac{nR}{V} = \frac{P}{T} \tag{21.14}$$

and, inverting equation (21.16)

$$\left(\frac{\partial P}{\partial V}\right)_T = -\frac{P}{V} \tag{21.16}$$

from which

$$\left(\frac{\partial H}{\partial V}\right)_T = T\frac{P}{T} - V\frac{P}{V} = 0$$

For an ideal gas, both $(\partial H/\partial P)_T$ and $(\partial H/\partial V)_T$ are therefore zero, and since $(\partial H/\partial T)_P = C_P$, the enthalpy H of a given mass of an ideal gas is a function of temperature only.

As it happens, there is an easier way of demonstrating that the enthalpy H of an ideal gas is a function of temperature only, but this simpler proof – which we used on page 153 – is not from first principles, but reliant on having firstly proved that the internal energy U of an ideal gas is a function of temperature only. Since, by definition

$$H = U + PV \tag{6.4}$$

then, for an ideal gas

$$PV = nRT \tag{3.3}$$

and so, for an ideal gas

$$H = U + nRT$$

For an ideal gas of constant mass, n is a constant, as is R, and so the term nRT depends on the temperature T only. And if U depends on the temperature T only, then H must also depend only on temperature.

21.7 The difference $C_P - C_V$

In this section, we derive a general expression for the difference $C_P - C_V$. We start with the definition of C_P

$$C_P = \left(\frac{\partial H}{\partial T}\right)_P \tag{6.8a}$$

and equation (9.21a) for dH

$$dH = T\,dS + V\,dP \tag{9.21a}$$

from which, dividing by dT at constant P gives

$$\left(\frac{\partial H}{\partial T}\right)_P = T\left(\frac{\partial S}{\partial T}\right)_P = C_P \tag{21.19}$$

Similarly

$$\left(\frac{\partial U}{\partial T}\right)_V = C_V \tag{5.14b}$$

and since

$$dU = T\,dS - P\,dV \tag{9.20a}$$

then, dividing by dT at constant V gives

$$\left(\frac{\partial U}{\partial T}\right)_V = T\left(\frac{\partial S}{\partial T}\right)_V = C_V \tag{21.20}$$

Equations (21.19) and (21.20) therefore imply that

$$C_P - C_V = T\left(\frac{\partial S}{\partial T}\right)_P - T\left(\frac{\partial S}{\partial T}\right)_V \tag{21.21}$$

If we express the system's entropy S as a function $S(T,P)$ of T and P, then

$$dS = \left(\frac{\partial S}{\partial T}\right)_P dT + \left(\frac{\partial S}{\partial P}\right)_T dP \qquad (21.22)$$

Similarly, expressing the system's entropy S as a function $S(T,V)$ of T and V, we have

$$dS = \left(\frac{\partial S}{\partial T}\right)_V dT + \left(\frac{\partial S}{\partial V}\right)_T dV \qquad (21.23)$$

For any given system, equations (21.22) and (21.23) are describing the same change dS, and so

$$\left(\frac{\partial S}{\partial T}\right)_P dT + \left(\frac{\partial S}{\partial P}\right)_T dP = \left(\frac{\partial S}{\partial T}\right)_V dT + \left(\frac{\partial S}{\partial V}\right)_T dV$$

from which

$$\left(\frac{\partial S}{\partial T}\right)_P dT - \left(\frac{\partial S}{\partial T}\right)_V dT = \left(\frac{\partial S}{\partial V}\right)_T dV - \left(\frac{\partial S}{\partial P}\right)_T dP$$

Multiplying throughout by T

$$T\left(\frac{\partial S}{\partial T}\right)_P dT - T\left(\frac{\partial S}{\partial T}\right)_V dT = T\left(\frac{\partial S}{\partial V}\right)_T dV - T\left(\frac{\partial S}{\partial P}\right)_T dP$$

from which, using equation (21.20), we derive

$$(C_P - C_V)\, dT = T\left(\frac{\partial S}{\partial V}\right)_T dV - T\left(\frac{\partial S}{\partial P}\right)_T dP$$

We can substitute for $(\partial S/\partial V)_T$ and $(\partial S/\partial P)_T$ using the Maxwell relations (21.6d) and (21.6c)

$$\left(\frac{\partial P}{\partial T}\right)_V = \left(\frac{\partial S}{\partial V}\right)_T \qquad (21.6d)$$

$$\left(\frac{\partial V}{\partial T}\right)_P = -\left(\frac{\partial S}{\partial P}\right)_T \qquad (21.6c)$$

and so

$$(C_P - C_V)\, dT = T\left(\frac{\partial P}{\partial T}\right)_V dV + T\left(\frac{\partial V}{\partial T}\right)_P dP \qquad (21.24)$$

For any system of constant mass, the temperature T can be expressed as a function $T(V,P)$ of V and P and so we may write

$$dT = \left(\frac{\partial T}{\partial V}\right)_P dV + \left(\frac{\partial T}{\partial P}\right)_V dP$$

Multiplying this equation throughout by $(C_P - C_V)$

$$(C_P - C_V)\, dT = (C_P - C_V)\left(\frac{\partial T}{\partial V}\right)_P dV + (C_P - C_V)\left(\frac{\partial T}{\partial P}\right)_V dP \qquad (21.25)$$

The quantity $(C_P - C_V)\, dT$ on the left-hand sides of equations (21.24) and (21.25) are, for any given system, the same, and so the coefficients of the corresponding terms dV and dP on the right-hand sides must be equal. Hence, for the dV term

$$(C_P - C_V)\left(\frac{\partial T}{\partial V}\right)_P = T\left(\frac{\partial P}{\partial T}\right)_V$$

from which

$$C_P - C_V = T\left(\frac{\partial P}{\partial T}\right)_V \left(\frac{\partial V}{\partial T}\right)_P \tag{21.26}$$

and for the dP term

$$(C_P - C_V)\left(\frac{\partial T}{\partial P}\right)_V = T\left(\frac{\partial V}{\partial T}\right)_P$$

giving

$$C_P - C_V = T\left(\frac{\partial V}{\partial T}\right)_P \left(\frac{\partial P}{\partial T}\right)_V \tag{21.27}$$

Equations (21.26) and (21.27) are identical, and represent a general expression for the difference $(C_P - C_V)$ for any system of constant mass. If we now invoke the chain rule, equation (21.11)

$$\left(\frac{\partial P}{\partial V}\right)_T \left(\frac{\partial V}{\partial T}\right)_P \left(\frac{\partial T}{\partial P}\right)_V = -1 \tag{21.11}$$

then

$$\left(\frac{\partial P}{\partial T}\right)_V = -\left(\frac{\partial P}{\partial V}\right)_T \left(\frac{\partial V}{\partial T}\right)_P$$

and so we may substitute for $(\partial P/\partial T)_V$ in equation (21.27), giving

$$C_P - C_V = -T\left(\frac{\partial V}{\partial T}\right)_P^2 \left(\frac{\partial P}{\partial V}\right)_T \tag{21.28}$$

Equation (21.28) is an alternative to equation (21.27), but provides more insight: since T and $(\partial V/\partial T)_P^2$ are necessarily positive, the sign of the difference $(C_P - C_V)$ is determined by the sign of $(\partial P/\partial V)_T$. $(\partial P/\partial V)_T$, however, is not just a mathematical abstraction: it represents a very real phenomenon – a phenomenon most readily interpreted from the reciprocal expression $(\partial V/\partial P)_T$, which, in physical terms, represents the change ∂V_T in the volume of a system of constant mass resulting from a change ∂P_T in pressure, at constant temperature. This is a measure of how easy (or difficult) it is to compress the system, and, as we know, gases are much more easily compressed than a solid or a liquid, and so the magnitude of $(\partial V/\partial P)_T$ will be relatively large for a gas, and relatively small for a liquid or solid. Furthermore, for every known substance, an increase in pressure ∂P_T causes a *decrease* in volume ∂V_T – all materials become smaller under increasing pressure. This implies that $(\partial V/\partial P)_T$ – and hence $(\partial P/\partial V)_T$ – is universally negative, and so, according to equation (21.28) the difference $(C_P - C_V)$ is universally positive: C_P is always larger than C_V.

Furthermore, equation (21.28) implies that the difference $(C_P - C_V)$ is proportional to the temperature T, and will in general increase with increasing temperature, subject to the behaviour of the other variables $(\partial V/\partial T)_P^2$ and $(\partial P/\partial V)_T$. As T approaches absolute zero, however, the difference $(C_P - C_V)$ becomes ever smaller, such that $C_P = C_V$ when $T = 0$.

The term $(\partial V/\partial T)_P$ also has physical meaning: it represents the change in volume ∂V_P of a system of constant mass resulting from a change in temperature ∂T_P at constant pressure, and so is a measure of thermal expansion. Furthermore, since the density ρ of the system is the ratio of the system's mass M to its volume V

$$\rho = \frac{M}{V}$$

then, if we differentiate with respect to T at constant P, for a given mass M

$$\left(\frac{\partial \rho}{\partial T}\right)_P = -\frac{M}{V^2}\left(\frac{\partial V}{\partial T}\right)_P$$

If $(\partial V/\partial T)_P = 0$, then $(\partial \rho/\partial T)_P = 0$ too, implying that the density ρ of the system is at a minimum or maximum value as regards changes in temperature. As an example, water reaches its maximum density at about 277 K = 4 °C, at which point $(\partial \rho/\partial T)_P = 0$, implying that $(\partial V/\partial T)_P = 0$. According to equation (21.27), when $(\partial V/\partial T)_P = 0$, $C_P = C_V$, and so this will be the case for water at 277 K = 4 °C.

21.8 The Joule-Thomson coefficient

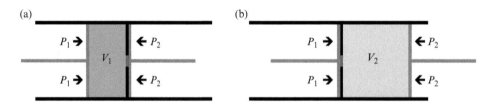

Figure 21.1 "**Throttling**". A gas, initially of volume V_1, as shown in (a), is forced, adiabatically, at a constant pressure P_1 through a narrow porous plug against a constant opposing pressure P_2, and subsequently occupies a volume V_2, as shown in (b).

Figure 21.1 shows what is known as a **throttling process**, in which a gas, originally occupying a volume V_1, is forced, adiabatically, at a constant pressure P_1 through a porous plug, against a constant opposing pressure P_2, subsequently occupying a volume V_2.

The process can be analysed as taking place in two steps:

- an adiabatic compression, at constant pressure P_1, from volume V_1 to a very small volume V_{small}, within the porous plug, followed by
- an adiabatic expansion, at constant pressure P_2, from volume V_{small} to volume V_2.

In the first, compression, step, the work done on the gas is given by

$$\text{Work done on gas during compression} = \int_{V_1}^{V_{small}} P\,dV = P_1\,(V_{small} - V_1)$$

Since $V_1 > V_{small}$, this quantity is approximately equal to $-P_1 V_1$, and necessarily negative, as is consistent with work being done on the gas.

In the second, expansion, step, the work done by the gas is given by

$$\text{Work done by gas during expansion} = \int_{V_{small}}^{V_2} P\,dV = P_2(V_2 - V_{small})$$

Since $V_2 > V_{small}$, this quantity is positive, as is consistent with work being done by the gas. The net work done by the gas is therefore

$$\text{Net work done by gas} = P_2(V_2 - V_{small}) + P_1(V_{small} - V_1)$$
$$= P_2 V_2 - P_1 V_1 + V_{small}(P_1 - P_2)$$

If V_{small} is indeed small compared to V_1 and V_2, then the term $V_{small}(P_1 - P_2)$ can be ignored, and so

$$\text{Net work done by gas} = P_2 V_2 - P_1 V_1 \tag{21.29}$$

According to the First Law, for a system of constant mass

$$dU = đq - đw \tag{5.1a}$$

Since the throttling process is adiabatic, $đq = 0$, and so for a finite change

$$\Delta U = -\{_1 w_2\}$$

where $\{_1 w_2\}$ is the net work done by the system, which in this case is given by equation (21.29). We may therefore write

$$\Delta U = U_2 - U_1 = -(P_2 V_2 - P_1 V_1)$$

and so

$$U_2 + P_2 V_2 = U_1 + P_1 V_1$$

But since, by definition

$$H = U + PV \tag{6.4}$$

then, for the adiabatic throttling process

$$H_2 = H_1$$

Adiabatic throttling therefore conserves enthalpy.

But what has happened to temperature of the gas? Is the temperature T_2 of the final state higher, or lower, than the temperature T_1 of the initial state?

To answer these questions concerning temperature, we need to determine the Joule-Thomson coefficient μ_{JT} (not to be confused with the chemical potential μ), defined as

$$\mu_{JT} = \left(\frac{\partial T}{\partial P}\right)_H \tag{21.30}$$

which specifies the change in temperature ∂T_H resulting from a change in pressure ∂P_H at constant enthalpy.

If we express the enthalpy H of a system of constant mass as a function $H(P, T)$ of pressure P and temperature T, then

$$dH = \left(\frac{\partial H}{\partial T}\right)_T dT + \left(\frac{\partial H}{\partial P}\right)_P dP \tag{21.31}$$

Given that

$$C_P = \left(\frac{\partial H}{\partial T}\right)_P \tag{6.8a}$$

then equation (21.31) becomes

$$dH = C_P\, dT + \left(\frac{\partial H}{\partial P}\right)_T dP \tag{21.32}$$

For a change at constant enthalpy, dH is zero on the left-hand side of equation (21.32), and if we divide throughout by dP and rearrange, then, in accordance with equation (21.29), we have

$$\left(\frac{\partial T}{\partial P}\right)_H = -\frac{1}{C_P}\left(\frac{\partial H}{\partial P}\right)_T = \mu_{JT} \tag{21.33}$$

We now seek a replacement for the term $(\partial H/\partial P)_T$. Starting from equation (9.21a)

$$dH = T\, dS + V\, dP \tag{9.21a}$$

we divide by dP at constant temperature T, giving

$$\left(\frac{\partial H}{\partial P}\right)_T = T\left(\frac{\partial S}{\partial P}\right)_T + V \tag{21.34}$$

Reference to page 639 will show that there is a Maxwell relation

$$\left(\frac{\partial V}{\partial T}\right)_P = -\left(\frac{\partial S}{\partial P}\right)_T \tag{21.6c}$$

and so equation (21.34) becomes

$$\left(\frac{\partial H}{\partial P}\right)_T = -T\left(\frac{\partial V}{\partial T}\right)_P + V \tag{21.35}$$

Physically, the term $(\partial V/\partial T)_P$ represents by how much a system expands (that's ∂V_P) for a small rise ∂T_P in temperature, at constant pressure – a property which readily measureable, and related to the system's **volumetric expansion coefficient**, α_V, defined (see exercise 4.8) as

$$\alpha_V = \frac{1}{V}\left(\frac{\partial V}{\partial T}\right)_P \tag{21.36}$$

We can now use equation (21.35) to substitute for $(\partial H/\partial P)_T$ in equation (21.33), giving

$$\mu_{JT} = \left(\frac{\partial T}{\partial P}\right)_H = \frac{1}{C_P}\left[T\left(\frac{\partial V}{\partial T}\right)_P - V\right] \tag{21.37}$$

Equation (21.37) is the standard expression for the Joule-Thomson coefficient μ_{JT}. Since C_P is necessarily positive, the sign of μ_{JT} depends on the sign of the term in square brackets, which can be positive or negative according to the gas involved, and the specific conditions of temperature and pressure. So, suppose that, in Figure 21.1, $P_1 > P_2$, implying that ΔP is negative. Since $\mu_{JT} = (\partial T/\partial P)_H$, then, if μ_{JT} is positive, the corresponding value of ΔT must be negative, and the gas will cool down as a result of the throttling process. If, however, the parameters in the square bracket in equation (21.36) are such that μ_{JT} is negative, then ΔT will be positive, and the gas will warm up.

For an ideal gas, as we saw on page 642,

$$\left(\frac{\partial V}{\partial T}\right)_P = \frac{nR}{P} = \frac{V}{T} \tag{21.15}$$

and so

$$\mu_{JT} = \frac{1}{C_P}\left[T\left(\frac{\partial T}{\partial P}\right)_P - V\right] = \frac{1}{C_P}\left[T\frac{V}{T} - V\right] = 0$$

under all conditions, implying that the temperature of an ideal gas remains constant as a result of a constant-enthalpy throttling process. This is consistent with the fact that the enthalpy of an ideal gas is a function of temperature only.

For real gases, however, the Joule-Thomson coefficient μ_{JT} takes different values under different conditions, as illustrated in the T, P **diagram** shown in Figure 21.2, the horizontal axis of which is the pressure P, and the vertical axis, the temperature T. Each curve represents a different **isenthalp**, a set of simultaneous values of P and T such that the enthalpy in constant. In general, any single isenthalp is a rather flattened inverted U-shape, and the slope of any curve at any point represents the instantaneous value of $(\partial T/\partial P)_H = \mu_{JT}$. As can be seen, in general, along any single isenthalp, at lower pressures, $\mu_{JT} = (\partial T/\partial P)_H$ is positive, implying that

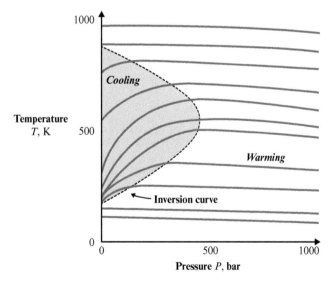

Figure 21.2 A representation of isenthalps for a real gas. Each continuous line – an isenthalp – links points of constant enthalpy, and so the slope at any point is a measure of the Joule-Thomson coefficient μ_{JT}. Since the throttling process is a change from a higher pressure to a lower pressure, ∂P_H is negative. When the slope μ_{JT} of an isenthalp is positive, the gas will cool as the pressure falls; when the slope μ_{JT} of an isenthalp is negative, the gas will become warmer as the pressure falls. The curve joining the points at which $\mu_{JT} = 0$ – the maxima of the isenthalps – is known as the inversion curve for the gas.

a result of the throttling process, during which the pressure falls ($\partial P_H < 0$), the gas becomes cooler ($\partial T_H < 0$); at higher pressures, $\mu_{JT} = (\partial T/\partial P)_H$ is negative, implying that a result of the throttling process, the gas becomes hotter ($\partial T_H > 0$). At the maximum, $\mu_{JT} = 0$, and, for any given isenthalp, the temperature at which this happens is known as the **inversion temperature**, T_{inv}. The **inversion curve** connects the inversion temperatures across all isenthalps, so defining the zones of temperature-pressure combinations at which a throttling process results either in cooling or heating. Industrially, the zone of cooling is important for refrigeration, and also for the processes used to liquefy gases such as ammonia, oxygen and nitrogen.

21.9 The compressibility factor

As we have just seen, for an ideal gas, the Joule-Thomson coefficient μ_{JT} is zero under all conditions, and so an isenthalp, as plotted on a graph such as that shown in Figure 21.2, would be a horizontal straight line; in contrast, for a real gas, μ_{JT} is non-zero, and an isenthalp is a curve. For any real gas, the values of μ_{JT}, and the shape of any isenthalp, therefore indicates how closely the gas approximates to, or deviates from, ideal behaviour.

Another parameter which can readily distinguish between ideal and real gases is the **compressibility factor Z**, defined as

$$Z = \frac{PV}{RT} \tag{21.38}$$

For an ideal gas, $Z = PV/RT = 1$ under all conditions of pressure, volume and temperature. Consider, however, a real gas that complies with the van der Waals equation

$$\left(P + \frac{an^2}{V^2}\right)(V - nb) = nRT \tag{3.6}$$

from which

$$P = \frac{nRT}{(V - nb)} - \frac{an^2}{V^2}$$

giving, for $n = 1$

$$Z = \frac{PV}{RT} = \frac{V}{(V - b)} - \frac{a}{VRT} \tag{21.39}$$

Equation (21.39) directly implies that Z varies with both the temperature T, and molar volume V; furthermore, since the van der Waals equation, equation (3.6), shows how the pressure P varies with temperature and volume, Z for a real gas will vary with pressure too, as illustrated in Figure 21.3.

Figure 21.3 shows how Z varies, as a function of pressure, at different temperatures for a van der Waals gas. Note that the horizontal pressure axis covers a very broad range, up to 600 bar, and so normal laboratory pressures are to the far left; the lowest, V-shaped, graph corresponds to a temperature just above the critical temperature T_{crit} (see pages 489 to 492). As can be seen, at low pressures, $Z \approx 1$ at all temperatures, and at higher temperatures, the behaviour of Z becomes 'flatter' over the pressure range.

THE COMPRESSIBILITY FACTOR 653

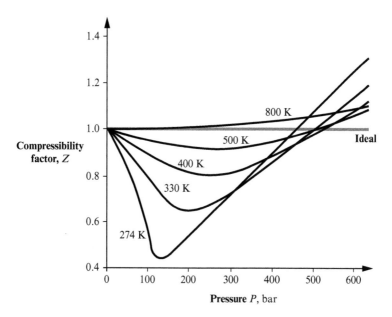

Figure 21.3 The compressibility factor Z for a van der Waals gas. The graphs show how the compressibility factor Z varies with pressure for a number of different temperatures. The lowest curve corresponds to a temperature just above the critical temperature T_{crit}. For an ideal gas, $Z = 1$ at all pressures and temperatures.

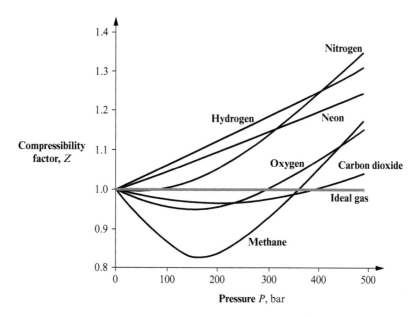

Figure 21.4 The compressibility factor Z for some selected real gases at 300 K.

Figure 21.4 shows how Z varies, at the same temperature, for some selected real gases. The range of pressures is again broad, and all the gases shown approach ideal behaviour at low pressures. Deviations from ideal behaviour, however, are evident at pressures greater than a few tens of bar, even for the light gases hydrogen and neon.

EXERCISES

1. Define
   - throttling process
   - isenthalp
   - Joule-Thomson coefficient
   - inversion temperature
   - inversion curve
   - compressibility factor.

2. This exercise uses a spreadsheet to create a graph of the compressibility factor Z for a van der Waals gas, of the type shown in Figure 21.3. Define three cells for constants: $R = 8.314$ J K^{-1} mol^{-1}, and the two van der Waals constants a and b (say, 0.23 J m^3 mol^{-2} and 3.9×10^{-5} m^3 mol^{-1} respectively, these being the parameters for argon). Then define three cells for V_{crit}, T_{crit} and P_{crit}, as defined by equations (15.17), (15.18) and (15.19) – as will be seen, for the parameters a and b as suggested, $T_{crit} = 210.17$ K.

   Leave a few blank rows, and then set up a column (say, column C) for V, with 100 rows, from 1.5×10^{-5} m^3, in increments of 1.0×10^{-5} m^3, with a final value of 1.05×10^{-3} m^3.

   At the top of column E, enter a value for T, which needs to be greater than T_{crit}, so as to ensure that the system is always a gas – so a value of, say, 230 K is suitable. Then, within column E, for each value of V, as defined by column C, compute P according to the van der Waals equation

   $$P = \frac{RT}{(V-b)} - \frac{a}{V^2} \tag{15.11}$$

   Now use column F to compute, for each value of V, the corresponding value of the compressibility factor Z

   $$Z = \frac{PV}{RT} \tag{21.39}$$

   Successive pairs of columns can be used to compute values of P and Z for successively greater values of T, for example, 300 K, 400 K, 500 K, and 800 K.

   To display the results, use a multi-line X-Y scatter format, with the X values corresponding to the pressure P, and the Y values to the compressibility factor Z. Explore what happens at different temperatures (above T_{crit}), and for different values of the parameters a and b – including $a = 0$, $b = 0$, not forgetting $a = b = 0$. (The values used for Figure 21.3 are $a = 0.23$ J m^3 mol^{-2} and $b = 3.3 \times 10^{-5}$ m^3 mol^{-1}.)

22 From ideal to real

Summary

Ideal gases and ideal solutions are theoretical abstractions, and studying them offers two benefits:

- since real gases and solutions behave in a manner more complex than ideal gases and solutions, the analysis of ideal systems is simpler, and easier to learn
- in many situations, real systems approximate well enough to the theoretical ideal, and so the results derived from the analysis of ideal systems can be applied in practice.

Although this second point is of practical value, circumstances inevitably arise in which the ideal analysis fails, and we need to take account of real behaviour. One way of approaching this is to develop appropriate theories from first principles, and this is especially the case for the study of the condensed phases of solids and liquids. An alternative is to attempt, as far as possible, to preserve the mathematical structures derived from ideal systems, and modify the parameters actually used.

For gases and solutions, this second approach is viable, and so

- For gases, the empirically-determined **fugacity** f replaces the ideal gas pressure P, or partial pressure p_i, as

$$\mathbf{G}_{\text{real}} = \mathbf{G}_{\text{real}}^\ominus + RT \ln \left(f/P^\ominus \right) \qquad (22.1\text{a})$$

and

$$\mu_{\text{real}} = \mu_{\text{real}}^\ominus + RT \ln \left(f/P^\ominus \right) \qquad (22.1\text{b})$$

where the **fugacity coefficient** ϕ is defined as

$$f = \phi P \qquad (22.3\text{a})$$

- For solvents in solution, the empirically-determined **mole fraction activity** $a_{x,i}$ replaces the mole fraction ratio $x_i/x_i^\ominus$ as

$$\mathbf{G}_i \text{ (in solution)} = \mathbf{G}_i^\ominus \text{(pure)} + RT \ln a_{x,i} \qquad (22.6\text{a})$$

and

$$\mu_i \text{ (in solution)} = \mu_i^\ominus \text{(pure)} + RT \ln a_{x,i} \qquad (22.6\text{b})$$

where the **mole fraction activity coefficient** $\gamma_{x,i}$ is defined such that

$$a_{x,i} = \gamma_{x,i} \left(x_i/x_i^\ominus \right) \qquad (22.5)$$

Modern Thermodynamics for Chemists and Biochemists. Dennis Sherwood and Paul Dalby.
© Oxford University Press 2018. Published 2018 by Oxford University Press.

- For solutes in solution, the empirically-determined **molal activity** $a_{b,i}$ replaces the molality ratio $b_i/b_i^{\ominus}$ as

$$G_{i,b} \text{ (in solution)} = G_{i,b}^{\ominus} \text{ (in solution)} + RT \ln a_{b,i} \tag{22.8a}$$

and

$$\mu_{i,b} \text{ (in solution)} = \mu_{i,b}^{\ominus} \text{ (in solution)} + RT \ln a_{b,i} \tag{22.8b}$$

where the **molal activity coefficient** $\gamma_{b,i}$ is defined as

$$a_{b,i} = \gamma_{b,i} \, (b_i/b_i^{\ominus}) \tag{22.9}$$

- The **ionic strength**, I, is a measure of the overall molality of ions within a solution, but weighted according to the square of the charge number z_i, as

$$I = \frac{1}{2} \sum_i z_i^2 \left(\frac{b_i}{b_i^{\ominus}}\right) \tag{22.20}$$

In general, the greater the ionic strength of a solution, the more the solution deviates from ideal behaviour.

22.1 Real gases and fugacity

As we saw in Chapter 21, the Joule-Thomson coefficient of an ideal gas is zero, and so the isenthalps of an ideal gas, if plotted in Figure 21.2, would all be horizontal straight lines at the appropriate system temperature. Likewise, for an ideal gas, the compressibility factor Z is equal to 1 under all conditions, but for a real gas, Z varies with pressure and temperature as shown in Figures 21.3 and 21.4. Figures 21.2, 21.3 and 21.4 therefore serve as a reminder that real systems deviate from ideal behaviour, and so, in this chapter, we explore how the theories developed for ideal systems throughout this book form the basis of describing the behaviour of real systems. Since the behaviour of real systems can be very complex, this chapter presents only an overview, building a bridge towards further study.

The thermodynamics of solids has been discussed only in passing in this book, primarily because solids play only a limited role in chemistry and biochemistry; for further information, the reader is referred to texts on condensed matter physics.

The behaviour of gases, however, has played a central role, and the ideal gas has featured in almost every chapter. One of the key results for an ideal gas was equation (13.47b)

$$G = G^{\ominus} + nRT \ln (P/P^{\ominus}) \tag{13.47b}$$

which for 1 mol of gas becomes

$$G = G^{\ominus} + RT \ln (P/P^{\ominus}) \tag{13.47d}$$

or, expressed in terms of the chemical potential μ

$$\mu = \mu^{\ominus} + RT \ln (P/P^{\ominus}) \tag{13.56}$$

These equations define how the molar Gibbs free energy G, and the chemical potential μ, of an ideal gas at a constant temperature T varies with the pressure P, in terms of the corresponding standard $G^{\ominus}$ or $\mu^{\ominus}$, and the standard pressure $P^{\ominus}$. Reference to Chapters 13, 14 and 16 will show that this equation underpinned the analysis of gas phase chemical equilibrium, of the equilibrium constant, and of how chemical equilibrium changes with temperature; this equation was also the key to understanding vapour pressure, the equilibrium between a solvent and its vapour, and the properties of ideal solutions.

As we discussed in Chapter 15 (see pages 489 to 492), real gases deviate from ideal behaviour, and so equation (13.47b) – and all the results that stem from it – ultimately break down. This presents quite a problem, and raises the question of whether we need to develop a host of different results, suitable for real systems. In principle, the answer to this question is "yes", and much academic research can be done accordingly. Pragmatically, however, there is a different answer – an answer that enables us to keep the form of all the equations derived on the assumption of ideal behaviour, but in which the system pressure P is replaced by a property – a new intensive state function – known as the **fugacity** f.

Operationally, using the fugacity f is very simple: wherever there is an equation, based on the assumption of an ideal gas, which includes the pressure P, or partial pressure p, then, for a real gas, the same equation is used, but replacing the pressure P or partial pressure p by the fugacity f. So, for example, since the molar Gibbs free energy G_{ideal} of an ideal gas at pressure P, and at a constant temperature T, is expressed as

$$G_{\text{ideal}} = G^{\ominus} + RT \ln(P/P^{\ominus}) \tag{13.47d}$$

then, for a system comprised of a real gas, we can express the molar Gibbs free energy G_{real} as

$$G_{\text{real}} = G^{\ominus} + RT \ln(f/P^{\ominus}) \tag{22.1a}$$

or, in terms of the chemical potential μ_{real}

$$\mu_{\text{real}} = \mu^{\ominus} + RT \ln(f/P^{\ominus}) \tag{22.1b}$$

Similarly, an equilibrium constant of the form

$$K_p = \frac{p_{C,\text{eq}}{}^c \, p_{D,\text{eq}}{}^d}{p_{A,\text{eq}}{}^a \, p_{B,\text{eq}}{}^b} \tag{14.12}$$

becomes

$$K_p = \frac{f_{C,\text{eq}}{}^c \, f_{D,\text{eq}}{}^d}{f_{A,\text{eq}}{}^a \, f_{B,\text{eq}}{}^b} \tag{22.2}$$

Comparison of equation (22.1b) with equation (16.19b)

$$\mu_i = \mu_i^{\ominus} + RT \ln a_i \tag{16.19b}$$

shows that the ratio $f_i/P^{\ominus}$ equals the activity a_i; also, since $P^{\ominus} = 1$ bar for both ideal and real gases, the activity a_i is numerically equal to the fugacity f_i. Importantly, the fugacity f - which has the dimensions of pressure, Pa or bar - is *not* the actual pressure P exerted by the real gas; rather, the fugacity f is a 'scaled' or 'modified' pressure, related to the actual pressure P as

$$f = \phi P \tag{22.3}$$

In equation (22.3), ϕ, the **fugacity coefficient**, is a dimensionless quantity that can be determined empirically, as can the associated value of f, as will shortly be shown in Figure 22.1.

Also, if we combine equations (22.1b) and (22.3a), we have

$$\mu_{real} = \mu^\ominus + RT \ln (\phi P/P^\ominus)$$

$$= \mu^\ominus + RT \ln(P/P^\ominus) + RT \ln \phi$$

$$= \mu_{ideal} + RT \ln \phi \qquad (22.1c)$$

When $\phi = 1$, $f = P$, and equations (22.1) all revert to their ideal forms: a value of ϕ approaching 1 therefore implies that a real system is approximating to ideal behaviour.

Physically, the value of ϕ is determined by the intermolecular forces within the gas. In general, at low pressures, the molecules of a real gas are relatively far apart. Any intermolecular forces are therefore weak, implying, as we saw in Chapter 15, that most real gases approach ideal behaviour at low pressures and large volumes (see page 486). The molar free energy G_{real} of a real gas, at a given temperature T and a given low pressure P, is therefore equal to, or very nearly equal to, the Gibbs free energy G_{ideal} of an ideal gas at the same temperature T and pressure P, implying that equation (22.1a) is the same as equation (13.47d), that $f \approx P$, and that $\phi \approx 1$.

As a real gas is compressed, the molecules are forced closer together, and the attractive intermolecular forces become progressively stronger. As a result, a real gas at a given temperature and volume exerts a lower pressure than an ideal gas at the same temperature and volume, as illustrated in the central part of Figure 15.8 (see page 486). If the volume of the real gas is held constant, then the only way to increase the system pressure to that exerted by the equivalent ideal gas is to increase its temperature – to supply energy as heat. This implies that the molar Gibbs free energy G_{real} of a real gas at a given temperature T and intermediate pressure P is less than the molar Gibbs free energy G_{ideal} of the equivalent ideal gas, at the same temperature and pressure. At much higher pressures, as shown at the far left of Figure 15.4, a given volume of a real gas exerts a higher pressure than a corresponding ideal gas, and so G_{real} is now greater than G_{ideal}. The behaviours of G_{ideal} and G_{real} for a van der Waals gas, over a wide range of pressures, are shown in Figure 22.1 – note that the temperature of the gas is greater than the critical temperature T_{crit}, so that, as explained on page 489, the gas remains as a gas at all pressures, and does not liquefy.

Point A in Figure 22.1 shows the molar Gibbs free energy G_{real} (P) of a real gas at a given temperature, and at a pressure P. As can be seen, point A, on the real curve, has the same molar Gibbs free energy as point C on the ideal curve, corresponding to a pressure f exerted by the ideal gas. Points A and C are therefore such that G_{real} $(P) = G_{ideal}$ (f), and it is this relationship that defines the fugacity f: physically, the fugacity f of a real gas at a real pressure P is equal to the pressure of the ideal gas of the same molar Gibbs free energy at the same temperature. The value of the fugacity coefficient ϕ can also be inferred from Figure 22.1 as the ratio f/P, or from the length of the line segment AB: this represents the difference $G_{ideal} - G_{real}$, which, according to equation (22.1c) equals $- RT \ln \phi$. For points A, B and C, $f < P$ and $\phi < 1$; moving to the right of Figure 22.1, at pressures greater than the pressure at which the two graphs cross, $f > P$ and $\phi > 1$, with $f = P$ and $\phi = 1$ at the cross-over point.

Figure 22.1 also shows the definition of the standard state, and the standard molar Gibbs free energy $G^\ominus$, for a real gas. At the standard pressure $P^\ominus$, the molar Gibbs free energy of the

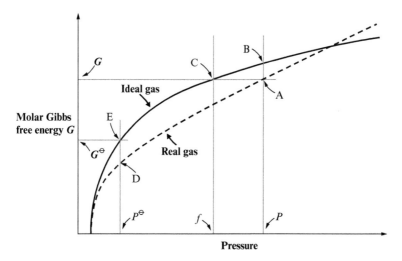

Figure 22.1 Fugacity. The solid line represents the molar Gibbs free energy G_{ideal} (P) for an ideal gas as a function of pressure, at a given temperature T. The dashed line shows the corresponding values G_{real} (P) for a real gas which behaves according to the van der Waals equation-of-state, at the same temperature T. The real gas remains as a gas at all pressures, and so the temperature T is higher than the critical temperature T_{crit}. At low pressures, to the left, the behaviour of the real gas approaches ideal behaviour; at intermediate pressures, G_{real} $(P) < G_{ideal}$ (P), as illustrated by points A and B; at very high pressures, to the right, G_{real} $(P) > G_{ideal}$ (P). For any given pressure P, there is a unique pressure, the fugacity f, such that G_{real} $(P) = G_{ideal}$ (f), as shown, for example, by points A and C. The standard molar Gibbs free energy $G_{real}^{\ominus}$ for the real gas is the value of G for the corresponding ideal gas at pressure $P = P^{\ominus}$, as defined by point E, and not at the actual value associated with point D. Note that, to illustrate the distinction between points D and E, the graph as shown has $P^{\ominus}$ much further to the right than its true value of 1 bar. See also the P, V diagram in Figure 15.4, which is based on the same data, and which also shows that a real gas exhibits near-ideal behaviour at low pressures, and that there is a cross-over point at high pressure.

real gas is defined by point D, and that of the corresponding ideal gas by point E: it is point E that defines the standard molar Gibbs free energy $G^{\ominus}$, and the standard chemical potential $\mu^{\ominus}$, of the real gas.

22.2 Real solutions and activity

22.2.1 Measuring quantities as mole fractions

As we discussed in Chapter 16 (see pages 522 and 523), the quantity of any component i in a solution can be measured in terms of the mole fraction x_i, the molality b_i or the molar concentration $[I]$. If the quantities of the components are comparable (for example, a mixture of liquids), or for the component that is present in significantly greater quantity – as is usually the case for the solvent – then the mole fraction x_i is usually the preferred measure. Accordingly, the molar Gibbs free energy G_i (in solution) of the component i in an ideal solution, at temperature T, is given by:

$$G_i \text{ (in solution)} = G_i^{\ominus}\text{(pure)} + RT \ln (x_i/x_i^{\ominus}) \qquad (16.7a)$$

or, in terms of the chemical potential μ

$$\mu_i \text{ (in solution)} = \mu_i^{\ominus}\text{(pure)} + RT \ln (x_i/x_i^{\ominus})$$

FROM IDEAL TO REAL

where G_i°(pure) is the standard molar Gibbs free energy of the pure component i, and μ_i°(pure), the standard chemical potential. Reference to pages 511 and 512 will show that equation (16.7a) was derived using Raoult's law

$$p_i = x_i p_i^* \qquad (16.1)$$

in which p_i is the vapour pressure of component i above the solution, and p_i^* is the vapour pressure of the pure component i, both at temperature T, and it is the validity of Raoult's law – or not – that determines whether, in practice, a real solution can be regarded as ideal.

As we saw on page 507, an important feature of an ideal solution is the equivalence of intermolecular forces between all components within the solution. So, consider a solution of a dissolved solute D within a solvent S, within which there are three different intermolecular forces: D $\cdots$ D between solute molecules, S $\cdots$ S between solvent molecules, and D $\cdots$ S between molecules of the solute and the solvent. If all these forces are equivalent, all molecules of D and S experience the same local environment, identical to that experienced in their pure states. Raoult's Law is therefore upheld – as, for example, for a solution (or rather mixture) of pure toluene, $C_6H_5CH_3$, and pure benzene, C_6H_6, as represented in Figure 22.2(a).

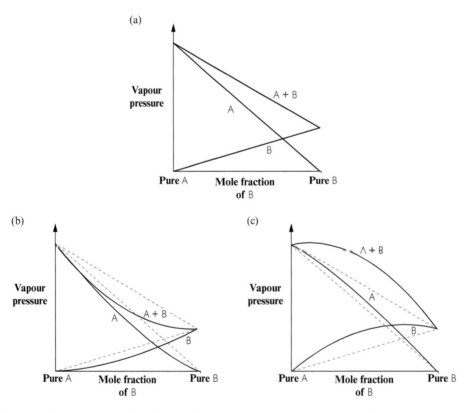

Figure 22.2 Deviations from Raoult's law. Each graph shows how the vapour pressure of a binary solution (or liquid mixture) of A and B varies with the composition of the solution, from pure A (on the left) to pure B (on the right). (a) represents an ideal solution that obeys Raoult's law; (b), a non-ideal solution with a negative deviation; and (c), a non-ideal solution with a positive deviation.

If, however, the intermolecular force D ··· S is stronger than both D ··· D and S ··· S, then neighbouring molecules of S and D and will tend to 'hold each other back', preventing them from 'escaping' from the liquid phase to the vapour phase, thereby reducing the vapour pressures of both D and S. In this case, Raoult's law is broken, and each p_i will be less than the ideal value $x_i p_i^*$. Any solution for which $p_i < x_i p_i^*$ is said to exhibit a **negative deviation from Raoult's law**, as represented in Figure 22.2(b), examples being a solution (or mixture) of chloroform, $CHCl_3$, and acetone, CH_3COCH_3, and solutions of ions in water in general.

A third case is that in which both the 'similar' intermolecular forces D ··· D and S ··· S are stronger than the 'dissimilar' intermolecular force D ··· S. In this case, molecules of both S and D 'prefer' to be surrounded by similar molecules, and will tend to 'exclude' a dissimilar molecule. This enables molecules of both D and S to 'escape' more readily, and so each p_i will be greater than the ideal value $x_i p_i^*$. This once again breaks Raoult's Law – as exemplified by a solution of carbon disulphide, CS_2, in acetone, CH_3COCH_3. Any solution for which $p_i > x_i p_i^*$ is said to exhibit a **positive deviation from Raoult's law**, as represented in Figure 22.2(c).

For any real solution, it is always possible to measure the mole fraction x_i of component i, the corresponding vapour pressure p_i (real), and the vapour pressure p_i^* of the pure component i (for which $x_i = 1$). If, for a variety of mole fractions x_i, measurements of p_i (real) are such that Raoult's Law

$$p_i \text{ (real)} = x_i p_i^*$$

is verified, then the solution is behaving ideally. If this is not the case, then the solution is not ideal, but the ratio p_i (real)$/p_i^*$ still has a value, which we define as the **mole fraction activity** $a_{x,i}$

$$a_{x,i} = \frac{p_i \text{ (real)}}{p_i^*} \quad (22.4)$$

For an ideal solution, the ratio p_i (real)$/p_i^* = x_i$, whilst for a real solution, p_i (real)$/p_i^* = a_{x,i}$, such that $a_{x,i} < x_i$ if the solution shows a negative deviation from Raoult's law, and $a_{x,i} > x_i$ for a solution showing positive deviation.

The mole fraction activity $a_{x,i}$ therefore acts as a 'scaled' or 'modified' mole fraction, and is related to the actual mole fraction x_i as

$$a_{x,i} = \gamma_{x,i} (x_i/x_i^\ominus) \quad (22.5)$$

where $\gamma_{x,i}$ is known as the **mole fraction activity coefficient** for the component i, with $\gamma_{x,i} < 1$ for negative deviations from Raoult's law, and $\gamma_{x,i} > 1$ for positive deviations.

Just as the fugacity f replaces the pressure P or partial pressure p to enable the thermodynamic equations derived for real gases to be applied to real gases, then the thermodynamic equations for real solutions are derived from the corresponding ideal solution equation by replacing the mole fraction x_i by the activity $a_{x,i}$, or the quantity $\gamma_{x,i} (x_i/x_i^\ominus)$. So, for example

$$G_i \text{ (in solution)} = G_i^\ominus \text{(pure)} + RT \ln a_{x,i} \quad (22.6a)$$

$$\mu_i \text{ (in solution)} = \mu_i^\ominus \text{(pure)} + RT \ln a_{x,i} \quad (22.6b)$$

$$G_i \text{ (in solution)} = G_i^{\circ}\text{(pure)} + RT \ln \gamma_{x,i}\, (x_i/x_i^{\circ}) \tag{22.7a}$$

$$\mu_i \text{ (in solution)} = \mu_i^{\circ}\text{(pure)} + RT \ln \gamma_{x,i}\, (x_i/x_i^{\circ}) \tag{22.7b}$$

As implied by their formal names, the symbols for both the mole fraction activity $a_{x,i}$ and the mole fraction activity coefficient $\gamma_{x,i}$ include the subscript x to indicate that these parameters are associated with the use of the mole fraction x_i as the measure of the quantity of component i present in the solution, and the corresponding definition of the standard state as that of pure i. Furthermore, as is evident by comparison with equation (16.19b)

$$\mu_i = \mu_i^{\circ} + RT \ln a_i \tag{16.19b}$$

$a_{x,i}$ is a particular case of the generalised chemical activity a_i.

From the 'perspective' of a solvent S, as the quantity of any dissolved solute D becomes smaller, the solution becomes progressively more dilute, and progressively more ideal. As this happens, the mole fraction x_S of the solvent approaches 1, $a_{x,S}$ approaches x_S/x_S°, and $\gamma_{x,S}$ approaches 1. In general, for any component i in a non-ideal solution, $\gamma_{x,i} \neq 1$, and the greater the deviation of $\gamma_{x,i}$ from 1, the greater the degree of non-ideality of the solution.

22.2.2 Ions, molalities and molar concentrations

For ions in solution, the situation is rather more complex. As we saw on page 510, an ideal-dilute is one for which, a solute D obeys Henry's Law

$$p_D = x_D\, k_{H,D} \tag{16.2}$$

where the vapour pressure of the dissolved solute D is proportional to the mole fraction x_D, but the constant of proportionality is an empirically determined constant $k_{H,D}$ rather than the vapour pressure p_D^* of pure D, as is the case for solutes obeying Raoult's Law. Furthermore, if the solution is one in which ions are dissolved in water – as is the case for many chemical and biochemical reactions – then the charged ions have strong interactions with one another, with the polar water molecules, and also with the $H^+(aq)$ and $OH^-(aq)$ ions resulting from the dissociation of the H_2O molecule. Since one of the fundamental properties of an ideal system – whether gas or solution – is that the entities within the system have no mutual interactions, any aqueous solution of ions will deviate from ideal behaviour, approaching ideal behaviour only when the solution is very dilute.

In general, ionic solutions are therefore non-ideal, and the higher the concentration of ions, the greater the departure from ideal behaviour. Furthermore, the standard states G_D°(pure) or μ_D°(pure), as used in equations (22.6) and (22.7), and as used in association with the mole fraction x_D as the measure of the quantity of D in solution, make little sense, since it is impossible to have a system composed of a pure ion, for which $x_D = 1$. On pages 515 and 516, we saw that there are two other measures of quantity in a solute D in solution

- the molality $b_D = n_D/M_S$ mol kg^{-1}, where n_D is the number of moles of D dissolved within a total mass M_S of the solvent S, and
- the molar concentration $[D] = n_D/V$ mol m^{-3} or mol dm^{-3}, in which n_D is the number of moles of D in a solution of total volume V m^3 or dm^3.

REAL SOLUTIONS AND ACTIVITY

For ideal-dilute solutions, and for real solutions that approximate to ideal behaviour, the molar concentration [D] is the more commonly used measure, which is why all the equations relating to solutions, as used so far in this book, have referred to molar concentrations, notably equation (16.10a)

$$G_D \text{ (in solution)} = G_D^\circ \text{(in solution, [D])} + RT \ln ([D]/[D]^\ast) \qquad (16.10a)$$

and its chemical potential equivalent

$$\mu_D \text{ (in solution)} = \mu_D^\circ \text{(in solution)} + RT \ln ([D]/[D]^\ast)$$

where the standard states, G_D°(in solution) and μ_D°(in solution), are defined in terms of solutions of unit molar concentration, [D] = 1 mol m^{-3} or 1 mol dm^{-3}.

For real solutions that do not approximate to ideal behaviour, however, the use of the molar concentration $[D] = n_D/V$ as the measure of quantity of the solute D in the solution has a drawback. As we saw on page 508, when a volume V_1 of one liquid is added to a volume V_2 of another liquid, the result – if the liquids are miscible, and the mixture ideal – is a liquid of total volume $V_{tot} = V_1 + V_2$. But if the solution is not ideal, then it is possible that $V_{tot} < V_1 + V_2$. For non-ideal solutions, volumes are not necessarily additive, and so measures of quantity based on measures of volume, such as the molar concentration $[D] = n_D/V$, are less useful.

The alternative is to use the molality $b_D = n_D/M_S$ as the measure of quantity, the benefit being that both mole numbers n and masses M are additive for both ideal, and non-ideal, solutions. Accordingly, for solutes D in solution, instead of adjusting the ideal equations

$$G_D \text{ (in solution)} = G_D^\circ \text{(in solution, [D])} + RT \ln ([D]/[D]^\ast) \qquad (16.10a)$$

$$\mu_D \text{ (in solution)} = \mu_D^\circ \text{(in solution)} + RT \ln ([D]/[D]^\ast)$$

to real behaviour, we use the equivalent ideal equations expressed in terms of molalities b_D

$$G_D \text{ (in solution)} = G_D^\circ \text{(in solution, } b\text{)} + RT \ln (b_D/b_D^\ast) \qquad (16.9a)$$

$$\mu_D \text{ (in solution)} = \mu_D^\circ \text{(in solution, } b\text{)} + RT \ln (b_D/b_D^\ast)$$

for which the standard states G_D°(in solution) and μ_D°(in solution) are defined in terms of solutions of unit molality $b_D^\ast$ = 1 mol per kg of solvent.

So, for a solute D in a real solution, to preserve the form of the ideal equations, the **molal activity** $a_{b,D}$ replaces the molality b_D as

$$G_D \text{ (in solution)} = G_D^\circ \text{(in solution)} + RT \ln a_{b,D} \qquad (22.8a)$$

and

$$\mu_D \text{ (in solution)} = \mu_D^\circ \text{(in solution)} + RT \ln a_{b,D} \qquad (22.8b)$$

The molal activity $a_{b,D}$ is related to the molality b_D as

$$a_{b,D} = \gamma_{b,D} (b_D/b_D^\ast) \qquad (22.9)$$

where $\gamma_{b,D}$ is an empirically determined parameter known as the **molal activity coefficient**, giving, for a solvent D in a real solution

$$G_D \text{ (in solution)} = G_D^\circ \text{(in solution)} + RT \ln \gamma_{b,D} (b_D/b_D^\ast) \qquad (22.10a)$$

and

$$\mu_D \text{ (in solution)} = \mu_D^{\ominus} \text{(in solution)} + RT \ln \gamma_{b,D} (b_D/b_D^{\ominus}) \quad (22.10b)$$

From the 'perspective' of a solute D, the more dilute the solution – and hence the smaller the value of b_D – the more the solution approaches ideal behaviour. With reference to equation (22.9), this implies that as b_D tends towards zero, $a_{b,D}$ tends towards the ideal ratio $b_D/b_D^{\ominus}$, and $\gamma_{b,D}$ therefore tends toward 1. Ultimately – in principle – the real solution shows true ideal behaviour (implying that $\gamma_{b,D} = 1$ and that $a_{b,D} = b_D/b_D^{\ominus}$) when $b_D = 0$, this hypothetical state being known, as we saw on page 173, as a solution at infinite dilution.

The definition of the standard state for a real solution involves a particular subtlety. For an ideal solution, the standard state corresponding to equations (16.9)

$$\mathbf{G}_D \text{ (in solution)} = \mathbf{G}_D^{\ominus} \text{(in solution)} + RT \ln (b_D/b_D^{\ominus}) \quad (16.9)$$

is that for which the molality $b_D^{\ominus} = 1$ mol per kg of a given solvent (usually water), in which case $\mathbf{G}_D = \mathbf{G}_D^{\ominus}$ and $\mu_D = \mu_D^{\ominus}$. For a real solution, however, and especially for a solution of ions, a molality of $b_D = 1$ mol kg^{-1} could well be quite far from the conditions under which Henry's Law is applicable, and one in which there are considerable intermolecular and inter-ionic interactions – all of which are highly non-ideal, and so quite unsatisfactory as regards the definition of a 'standard'. Page 61 of the 3rd edition (2007) of the IUPAC 'Green Book' therefore defines the standard state of a real solution as "the hypothetical state of the dissolved solute D at a molality of 1 mol kg^{-1}, at the standard pressure, and behaving like the infinitely dilute solution". The reference to the "infinitely dilute solution" recognises that real solutions of a solute D dissolved in a solvent S approach ideal behaviour as they become progressively more dilute, which, as we have seen, happens as the molality b_D tends towards zero. But since zero molality, corresponding to infinite dilution, is no longer a solution of D in S, this is necessarily a "hypothetical state". In practice, however, useable data for this standard can be obtained by extrapolating experimental data obtained from solutions containing measureable molalities b_D.

22.2.3 A note on the definition of pH

On page 537, we presented the very widely-used definition of pH as

$$\text{pH} = -\log_{10} [\text{H}^+] \quad (17.9)$$

Equation (17.9) is expressed in terms of the molar concentration [H$^+$] of H$^+$ (aq) ions in the system of interest, this being, as we saw on page 537, 'shorthand' for the ratio [H$^+$]/[H$^+$]$^{\ominus}$, in which [H$^+$]$^{\ominus} = 1$ mol dm^{-3}. In fact, the definition recommended by the International Union of Pure and Applied Chemistry is

$$\text{pH} = -\log_{10} a_{b,\text{H}+} = -\log_{10} \gamma_{b,\text{H}+} (b_{\text{H}+}/b_{\text{H}+}^{\ominus}) \quad (22.11)$$

as expressed in terms of the molal activity $a_{b,\text{H}+}$ of the H$^+$ (aq) ions, which is itself determined by the corresponding molal activity coefficient $\gamma_{b,\text{H}+}$, and the molality ratio $b_{\text{H}+}/b_{\text{H}+}^{\ominus}$.

At 298.15 K, the density of pure water is 997.05 kg m^{-3}, implying that 1 dm^{-3} of pure water weighs 0.99705 kg. Suppose, then, that a volume of 1 dm^{-3} contains $n_{\text{H}+}$ mol of H$^+$ (aq) ions –

perhaps, say, 10^{-8} mol – so that $[H^+] = n_{H+}$ mol dm^{-3}, and $-\log_{10}[H^+] = -\log_{10} n_{H+}$. The molality b_{H+} of the same system, however, is given by $b_{H+} = n_{H+}/0.99705 = 1.00296\, n_{H+}$ mol kg^{-1}. By definition, the standard molality $b_{H+}^{\ominus} = 1$ mol kg^{-1}, and, for a dilute solution of H^+ (aq) ions (as is the case for a molar concentration of 10^{-8} mol dm^{-3}), $\gamma_{b,H+}$ may be assumed to be equal to 1. Accordingly, $a_{b,H+} = \gamma_{b,H+}(b_{H+}/b_{H+}^{\ominus}) = 1.00296\, n_{H+}$, and so

$$-\log_{10} a_{b,H+} = -\log_{10} 1.00296\, n_{H+} = -\log_{10} n_{H+} - \log_{10} 1.00296$$

implying that

$$-\log_{10} a_{b,H+} = -\log_{10} n_{H+} - 0.00128$$

Accordingly, if $n_{H+} = 10^{-8}$ mol and we use the definition of pH as $-\log_{10}[H^+]$, then pH = $-\log_{10} n_{H+}$ = 8 exactly. But if we use the IUPAC definition that pH = $-\log_{10} a_{b,H+}$, then pH = 8 – 0.00128 = 7.99872. In essence, the difference between these two values represents a change in the reference level of measurement, such that, at 298.15 K, measurements of pH made using the 'molar concentration' scale are 0.00128 greater than measurements made using the 'molality' scale. In practice, of course, this difference is very small, and so, pragmatically, the definition of pH as

$$pH = -\log_{10}[H^+] \qquad (17.9)$$

is quite acceptable; in fact, however, equation (17.9) is an approximation of the 'right' expression

$$pH = -\log_{10} a_{b,H+} = -\log_{10} \gamma_{b,H+}(b_{H+}/b_{H+}^{\ominus}) \qquad (22.11)$$

The difference between the two values, –0.00128, is fundamentally attributable to the density of pure water, which defines the volume of 1 kg of water, so forming the 'link' between a measure of quantity based on the volume of a solvent (such as the molar concentration, mol dm^{-3}), and a measure based on the corresponding mass (such as the molality, mol kg^{-1}). At any given temperature, the density of pure water is a specific number, and so the difference between the pH as measured using molar concentrations, and the pH as measured using molalities, is a constant for all dilute solutions of H^+ (aq) ions. At 298.15 K, this number is – 0.00128; at other temperatures, this value will be different, as determined by the density of pure water at the appropriate temperature.

22.2.4 The ionic strength

A further feature of a solution of ions is that of **charge balance**: a solution at equilibrium must be electrically neutral, and so within the solution, the sum of the positive charges must equal the sum of the negative charges. This implies that ions can never exist by themselves, and so when a molecule XY dissociates in a real solution as

$$XY\,(aq) \rightleftharpoons X^+\,(aq) + Y^-\,(aq)$$

then the molalities b_{X+} of X^+ (aq) and b_{Y-} of Y^- (aq) must be equal.

According to equation (22.10b), for X^+ (aq) in the real solution

$$\mu_{X^+} = \mu_{X^+}^{\diamond} + RT \ln \gamma_{b,X^+} (b_{X^+}/b_{X^+}^{\diamond})$$

$$= \mu_{X^+}^{\diamond} + RT \ln (b_{X^+}/b_{X^+}^{\diamond}) + RT \ln \gamma_{b,X^+} \qquad (22.12)$$

If, however, the solution were ideal, then the ideal equation

$$\mu_D \text{ (in solution)} = \mu_D^{\diamond} \text{(in solution)} + RT \ln (b_D/b_D^{\diamond})$$

would apply, in the form

$$\mu_{X^+} \text{ (in ideal solution)} = \mu_{X^+}^{\diamond} + RT \ln (b_{X^+}/b_{X^+}^{\diamond})$$

Equation (22.12) therefore becomes

$$\mu_{X^+} = \mu_{X^+} \text{ (in ideal solution)} + RT \ln \gamma_{b,X^+} \qquad (22.13a)$$

Exactly the same applies to the anion Y^- (aq), and so

$$\mu_{Y^-} = \mu_{Y^-} \text{ (in ideal solution)} + RT \ln \gamma_{b,Y^-} \qquad (22.13b)$$

Since both X^+ (aq) and Y^- (aq) are present in equal amounts in the real solution, the total molar Gibbs free energy G_{tot} of the solution is given by

$$G_{tot} = G_{X^+} + G_{Y^-} = \mu_{X^+} + \mu_{Y^-}$$

$$= \mu_{X^+} \text{ (in ideal solution)} + RT \ln \gamma_{b,X^+}$$

$$+ \mu_{Y^-} \text{ (in ideal solution)} + RT \ln \gamma_{b,Y^-}$$

$$= \mu_{tot} \text{ (in ideal solution)} + RT \ln (\gamma_{b,X^+} \gamma_{b,Y^-}) \qquad (22.14)$$

Equation (22.14) shows that the difference between the molar Gibbs free energy of a real solution of ions, and the corresponding ideal solution, is the term $RT \ln (\gamma_{b,X^+} \gamma_{b,Y^-})$, which, at any given temperature T, depends only on the product of the molal activity coefficients γ_{b,X^+} and γ_{b,Y^-} of the positive and negative ions.

Experimentally, because ions can never be present by themselves, γ_{b,X^+} and γ_{b,Y^-} cannot be measured independently. It is, however, possible to measure the product $\gamma_{b,X^+} \gamma_{b,Y^-}$, and so we define the **mean molal activity coefficient** $\gamma_{b,\pm}$ as

$$\gamma_{b,\pm}^2 = \gamma_{b,X^+} \gamma_{b,Y^-} \qquad (22.15)$$

If we assume that each ion contributes equally to the non-ideality of the real solution, then equation (22.15) implies that we may write

$$\gamma_{b,\pm} = \gamma_{b,X^+} = \gamma_{b,Y^-} \qquad (22.16)$$

and so equations (22.13) become

$$\mu_{X^+} = \mu_{X^+} \text{ (in ideal solution)} + RT \ln \gamma_{b,\pm} \qquad (22.17a)$$

and

$$\mu_{Y^-} = \mu_{Y^-} \text{ (in ideal solution)} + RT \ln \gamma_{b,\pm} \qquad (22.17b)$$

giving equation (22.14) as

$$G_{\text{tot}} = \mu_{\text{tot}} \text{ (in ideal solution)} + 2RT \ln \gamma_{b,\pm} \tag{22.18}$$

In physical terms, the parameter $\gamma_{b,\pm}$ arises from the electrostatic interactions between ions, and there is a considerable body of theory which seeks to express $\gamma_{b,\pm}$ in terms of fundamental properties of ions, and the solutions in which they are dissolved. The details will be found elsewhere: we present here just the two main results.

For a dilute solution of a cation X^{z+} and an anion Y^{z-}, the **Debye-Hückel limiting law** states that

$$\log_{10} \gamma_{b,\pm} = -A|z_+z_-|\sqrt{I} \tag{22.19}$$

in which

- A is a cluster of fundamental constants (including, for example, the Boltzmann constant k_B, the Avogadro constant N_A, and π), as well as the temperature T; for aqueous solutions at 298.15 K, A evaluates to 0.509
- z_+ is the charge number for the cation ($z_+ = 1$ for Na^+, 2 for Ca^{2+}, …)
- z_- is the charge number for the anion ($z_- = -1$ for Cl^-, -2 for SO_4^{2-}, …)
- the modulus signs $|\ldots|$ are used to ensure that the product z_+z_- is always a positive number
- I is known as the solution's **ionic strength**, a dimensionless number given by

$$I = \frac{1}{2} \sum_i z_i^2 \left(\frac{b_i}{b_i^{\circ}}\right) \tag{22.20}$$

In equation (22.20), the summation goes over all the ions i in a given solution, each of which has a charge number z_i. The term z_i^2 is necessarily positive for both cations and anions, and a cation such as Fe^{3+} ($z_i = 3$) contributes 9 times as much as an anion such as Cl^- ($z_i = -1$). The term b_i represents the molality of ion i, and b_i° is the standard molality for ion i, 1 mol per kg of solvent at infinite dilution (see page 664).

There are some definitions of ionic strength that use the molar concentration ratio $[I]/[I]^{\circ}$ in equation (22.20), rather than the ratio of molalities b_i/b_i°. As we discussed on page 663, however, for real solutions, molalities b_i are preferred to molar concentrations $[I]$ as measures of quantity.

The ionic strength I is a measure of both the quantity of ions present in a given non-ideal solution (by virtue of the summation over the mole fractions b_i), and also the strength of electrostatic interactions (by virtue of the weighting of each b_i by the corresponding z_i^2), and is an important characteristic of any ionic solution. Broadly speaking, the greater the ionic strength I of a given solution, the greater the departure from ideal behaviour. Furthermore, I is necessarily positive, and it is the positive square root that is used in equation (22.19).

As we have noted, equation (22.19)

$$\log_{10} \gamma_{b,\pm} = -A|z_+z_-|\sqrt{I} \tag{22.19}$$

allows the mean concentration-based activity coefficient $\gamma_{b,\pm}$ to be computed for any ionic solution for which the charge numbers z_i and molalities b_i are known, and gives reliable results only for low molalities; for more concentrated solutions, various other expressions have been proposed, such as the **extended Debye-Hückel law**

$$\log_{10} \gamma_{b,\pm} = - \frac{A|z_+ z_-|\sqrt{I}}{1 + B\mathring{a}\sqrt{I}} \tag{22.21}$$

In equation (22.22), each of the terms A, z_+, z_- and I has the same meaning as in equation (22.19); the parameter $\mathring{a}$ represents the average **effective diameter** of the ion pair for which $\gamma_{b,\pm}$ is being estimated, and B is another constant, which has a value of 3.281×10^9 m^{-1} for aqueous solutions at 298.15 K.

As can be appreciated from equations (22.19) and (22.21), $\log_{10} \gamma_{b,\pm}$ is always negative, implying that $\gamma_{b,\pm}$ is always less than 1. This is consistent with the statement made on page 661 that all solutions of ions show negative deviations from Raoult's law, as is expected from the presence of attractive electrostatic forces between the dissolved ions and the water molecules.

22.3 A final thought

Overall, when dealing with real systems, one approach we could adopt is to develop new theories, keeping the equations relating to ideal systems in mind, but more as intellectual curiosities rather than of real applicability. And this is the approach which is (correctly) adopted by research scientists, who seek a deep understanding. An alternative, more pragmatic, approach is to preserve the equations relating to ideal systems, and adapt them, as little as possible, to comply with observed, real behaviour – always remembering that, sooner or later, especially at more extreme conditions of temperature and pressure, this will fail.

The concepts of fugacity, mole fraction activity and molal activity are therefore best regarded as adaptations of the ideal concepts of partial pressure, mole fraction and molality, replacing them in the ideal equations so that the ideal equations can continue to apply to real systems.

EXERCISES

1. Define
    - fugacity
    - fugacity coefficient
    - activity
    - activity coefficient
    - mean molal activity coefficient
    - ionic strength.
2. Determine values for the fugacities, and the fugacity coefficients, for a real gas in accordance with the following high-pressure measurements.

Pressure, bar	G ideal, kJ	G real, kJ	Fugacity, bar	Fugacity coefficient
100	13.78	13.30		
150	14.99	14.23		
200	15.85	14.85		
250	16.52	15.27		
300	17.06	15.62		
350	17.52	15.95		
400	17.93	16.24		
450	18.26	16.56		
500	18.59	16.83		
550	18.85	17.10		
600	19.14	17.37		
650	19.39	17.65		
700	19.59	17.89		
750	19.81	18.16		
800	19.96	18.38		
850	20.12	18.60		
900	20.29	18.88		
950	20.48	19.15		
1000	20.67	19.50		

3. What are the conditions that determine whether an activity coefficient is greater than, or less than, one? According to the Debye-Hückel limiting law, which of the following statements, all relating to the mean molal activity coefficient $\gamma_{b,\pm}$, is true?
   - $\gamma_{b,\pm}$ is always greater than one.
   - $\gamma_{b,\pm}$ is sometimes greater than one, and sometimes less than one.
   - $\gamma_{b,\pm}$ is always less than one.

   Why?

4. This is a spreadsheet exercise to reproduce Figure 22.1. This is done in two steps:
   - Firstly, the calculation of P, for increasing values of V, for a van der Waals gas, using equation (15.11)

$$P = \frac{nRT}{V - nb} - a\left(\frac{n}{V}\right)^2 \quad (15.11)$$

   - Secondly, the calculation of G_{real}, for increasing values of V, for a van der Waals gas, using equation (15.20)

$$G(V, T) = nG^\ominus + nRT \ln \frac{RT}{P^\ominus(V - nb)} + \frac{n^2 bRT}{(V - nb)} - \frac{2an^2}{V} \quad (15.20)$$

   for $n = 1$.

   For any given value of V, equation (15.11) generates the corresponding value of P, whilst equation (15.20) gives the corresponding value of G. Pairs of values of P and G that

correspond to the same value of V can then be selected, and used as the X and Y values respectively in an X-Y scatter diagram.

To do this, set up a spreadsheet, defining the constants $R = 8.314$ J K^{-1} mol^{-1}, $T = 298.15$ K, $n = 1$, $a = 0.23$ J m^3 mol^{-2}, $b = 3.0 \times 10^{-5}$ m^3 mol^{-1}, $P^{\ominus} = 10^5$ Pa, and $G^{\ominus} = 1{,}000$ J mol^{-1}.

Then define a column for V starting from a value of 3.1×10^{-5} m^3 (this being a small amount, 10^{-6} m^3 mol^{-1}, greater than the van der Waals parameter b) to around 8.6×10^{-4} m^3 in increments of 1×10^{-6} m^3, which requires some 830 rows. Subsequent columns can then be used to calculate P, from equation (15.11), and G_{real}, from equation (15.20), using intermediate columns as required for the various terms in the equations. You will also need a column for the ideal gas value of G_{ideal}, which corresponds to values of $a = b = 0$ in equation (15.20).

Within the spreadsheet, there will now be columns for V, P, G_{real} and G_{ideal}, in which each row gives corresponding values of P, G_{real} and G_{ideal}, for each value of V. To reproduce Figure 22.1, choose 'X-Y scatter with smooth lines' from the 'chart type' menu, and identify the X variable as the column with values of P, and, for the first graph, identify the Y variable as the column with values of G_{real}. For the second graph, identify the X variable as the column with values of P, and identify the Y variable as the column with values of G_{ideal}.

# PART 5
Biochemical applications

23 The biochemical standard state

Summary

Many biochemical processes take place in environments buffered at, or close to, pH 7, corresponding to a molar concentration [H⁺] of H⁺(aq) ions of 10^{-7} mol dm⁻³. The conventional standard state (see Table 6.4) for ions in solution is a molar concentration of 1 mol dm⁻³, corresponding, for H⁺(aq) ions, to pH 0. A highly acid environment of pH 0 is inappropriate as a reference state for biochemical systems, and so the **biochemical standard** defines the standard state for the hydrogen ion as a molar concentration of 10^{-7} mol dm⁻³. Although this appears to be only a very simple change, it has some significant effects, and many of the familiar data items, such as the concentration equilibrium constant, take different values when measured using the biochemical standard as compared to the conventional standards. To distinguish between measurements made under the two standards, those made under the biochemical standard are known as **transformed**, and are represented by the symbol ′.

The most important differences between measurements made under the two standards are captured in the following equations

- $[H^+]^{\ominus\prime} = 10^{-7}$ mol dm⁻³ (pH 7) $\hspace{4cm}$ (23.1a)

- $G'(H^+) = G(H^+) + 7\,RT \ln 10$ $\hspace{4cm}$ (23.10)

- $\mathbb{E}^{\ominus\prime}(\text{ox, red}) = \mathbb{E}^{\ominus}(\text{ox, red}) + \dfrac{7RT}{F} \ln 10$ $\hspace{3cm}$ (23.13)

- $\Delta_f G^{\ominus\prime}(H_2) = 14\,RT \ln 10$ $\hspace{2cm}$ J per mol H_2 $\hspace{2cm}$ (23.17)

- For any molecule of the form $X_x Y_y H_h$

    $\Delta_f G^{\ominus\prime}(X_x Y_y H_h) = \Delta_f G^{\ominus}(X_x Y_y H_h) + 7hRT \ln 10$ $\hspace{2cm}$ (23.22)

- For a generalised reaction of the form

    $a\text{AH}_p + b\text{BH}_q \rightleftharpoons c\text{CH}_r + d\text{DH}_s + h\text{H}^+$

    taking place in an ideal solution

    $\Delta_r G^{\ominus\prime} = \Delta_r G^{\ominus} - 7h\,RT \ln 10$ $\hspace{4cm}$ (23.29)

    $K_c' = 10^{7h} K_c$ $\hspace{6cm}$ (23.31b)

Modern Thermodynamics for Chemists and Biochemists. Dennis Sherwood and Paul Dalby.
© Oxford University Press 2018. Published 2018 by Oxford University Press.

23.1 Thermodynamics and biochemistry

One of the most exciting current contexts for the application of thermodynamics is biochemistry, embracing, for example, cellular metabolism, macromolecules, drug design, and the increasingly important interdisciplinary domain known as 'synthetic biology', which combines chemistry, engineering and biology to explore how biological systems behave, and might be built.

The following chapters therefore show how thermodynamics can help us understand biochemical systems, looking at three 'levels': the level of the individual reaction, in which energy is released or consumed; the level of the macromolecule; and the level of the macromolecular assembly.

The central theme of the next chapter is an exploration of how energy flows through biological systems – bioenergetics. As vividly discussed in Erwin Schrödinger's book *What is Life?*, a key characteristic of all living systems is the creation and maintenance of order. As we know from the Second Law, maintaining a state of low entropy is 'unnatural', in that spontaneous changes are necessarily associated with an increase in entropy. If, however, a spontaneous exergonic ($\Delta G < 0$) process can be coupled with a non-spontaneous endergonic ($\Delta G > 0$) process, then, as we saw on page 527, the overall result can reduce the entropy of a system locally, whilst ensuring that entropy increases overall. This principle underpins many metabolic pathways – sequences, or cycles, of reactions in which exergonic and endergonic processes are coupled, using molecules such as ATP and NADH as 'energy currencies'.

All biological macromolecules have a specific three-dimensional shape – each individual shape being essential for the corresponding function, such as the catalytic property of an enzyme. Given that any protein is a linear chain of perhaps several hundred amino acids, the number of theoretical possible three-dimensional configurations is huge – but the actual three-dimensional structure of each protein in the living cell is unique to that specific protein, and always the same for all molecules of that protein. How does the 'correct' configuration emerge from all of these possibilities? Chapter 25 discusses macromolecular configurations, drawing primarily on the fundamental thermodynamic principle that stable equilibrium configurations are those that have the minimum Gibbs free energy, in accordance with the 'Fourth Law' (see page 408).

Another feature of biological systems is that they exhibit a sequence of progressively more complex levels of aggregation: amino acids combine to form individual protein chains; individual protein chains cluster together to form multi-chain complexes (for example, haemoglobin, the oxygen-carrying protein in blood, comprises four individual chains); protein complexes combine to form components such as ribosomes or virus particles... and so on to cell organelles, cells and organisms. How sub-assemblies aggregate to form higher-level structures is a topic of active current research, key aspects of which are explored in our final, forward-looking, chapter, *Thermodynamics today – and tomorrow*.

To set the scene, the current chapter looks at biochemical systems in general, and deals with an important aspect of the application of thermodynamics to biochemical systems, the definition of the biochemical standard state.

23.2 A note on standards

The central theme of this chapter is 'the biochemical standard state'. As we shall see shortly, this is a particular standard state that is different from the corresponding standard state defined according to the conventional IUPAC standards which we introduced in Chapter 6. To put this change into context, we note here the general features of standards, and how there are used.

As we saw on pages 168 and 169, 'standards' play three different roles:

- to define units-of-measurement, such as the Joule and the kelvin
- to define standard states, these being carefully specified states of a system, which can serve as reference states, for which measurements can be made, and the corresponding data published in reference tables
- to define reference levels for measurement, usually expressed as the zero point of a scale.

Although these three roles are quite different, and distinguishable, they are inter-related: so, for example, as shown in Table 20.1 and 20.2, for the measurement of the standard reversible electrode potentials and redox potentials, the reference level $\mathbb{E}^\ominus = 0$ V is defined as the standard reversible electrode potential of a hydrogen electrode, in which molecular hydrogen $H_2(g)$ in its standard state at the standard pressure $P^\ominus = 1$ bar is in equilibrium with an ideal solution of hydrogen ions $H^+(aq)$ in their standard state, this being a solution at a molar concentration $[H^+] = 1$ mol dm^{-3}, and also at the standard pressure $P^\ominus = 1$ bar. Having then determined the reference level for measurement, the values of $\mathbb{E}^\ominus$ for other ions in ideal solution in their standard states can be determined, and published as reference data, such as the values shown in Table 20.1 – where, for example, we see that $\mathbb{E}^\ominus(O_2\,(g), 2\,H_2O) = 1.229$ V.

An important feature of standards is that they are arbitrary: energy, for example, can be measured in units-of-measure other than J; the standard state of an ideal solution of ions could be defined as a molar concentration other than 1 mol dm^{-3}; and the reference level for the standard reversible electrode potentials of an ion in an ideal aqueous solution could – as we saw on pages 617 to 619 - be defined in relation to, say, a standard silver electrode rather than a standard hydrogen electrode. This 'arbitrariness' does not present a problem: as long as we are all using an agreed set of standards consistently, and that all measurements are made in the same units, and relative to the same reference levels, everything works.

What's important about standards is that they are straight-forward to use in practice, and are agreed by the international community. So, in principle, it is possible to define the standard pressure $P^\ominus$ as, say, 1.2 bar, but since this is an 'unnatural' pressure, it would be very hard to use in practice. If, however, a particular branch of chemistry were to be developed in which a pressure of 1.2 bar is the norm, then it would make sense to change the standard pressure from 1 bar to 1.2 bar accordingly.

The central issue addressed by the introduction of the biochemical standard state relates to the fact that one of the conventional IUPAC standards is inconvenient to use for biochemistry. The key change concerns a particular standard state: as we shall see, this has a number of important consequences, which we shall now discuss in detail.

23.3 The biochemical standard state

A significant feature of biochemical systems is that they are real: most biochemical reactions usually take place in solution, and under conditions in which the ionic strength is relatively high. As a consequence, biochemical reactions do not, in general, comply with ideal conditions. As we saw in Chapter 22, for any solute i, we should therefore use equations expressed in terms of the chemical potential μ_i and the molal activity $a_{b,i}$, such as

$$\mu_i = \mu_i^\ominus + RT \ln a_{b,i} \tag{22.8b}$$

rather than the molar Gibbs free energy G_i, and the molar concentration [I]

$$G_i = G_i^\ominus + RT \ln([I]/[I]^\ominus) \tag{16.10a}$$

Strictly speaking, this is true; in practice, however, G_i and [I], and the assumptions associated with ideal-dilute solutions, are widely used, and so, once the biochemical standard state has been formally defined, the remainder of this book will express variables and equations using these terms.

In addition to being real systems, biochemical reactions are characterised by a number of distinctive features: in particular, they almost universally take place

- in aqueous solution
- at a constant pressure at, or close to, 1 bar
- at a constant temperature, as appropriate to the organism, this being about 37 °C = 310.15 K for humans
- at a constant, buffered, pH ...
- ... which is usually at, or close to, pH 7.

Most biochemical processes take place under these conditions (noting, however, that a few processes require a pH different from pH 7 – for example, acid conditions around pH 2 are optimal for the digestive enzyme pepsin), and most laboratory experiments are conducted under these conditions so as to mimic the living context as closely as possible. The conventional standard states for components in an aqueous solution, as defined in Table 6.4, correspond reasonably well as regards pressure and temperature, but pH 7 presents a particular problem.

The nature of this problem was discussed on pages 562 to 566, and concerns the calculation of thermodynamic data for any reaction of the form

$$aA + bB \rightleftharpoons cC + hH^+ \text{ (aq)}$$

taking place in solution, and which involves hydrogen ions, either as a product (as shown, for which h is positive), or as a reactant (for which h is negative).

As we saw on pages 562 to 566, if we determine the concentration equilibrium constant K_r from equation (16.13d)

$$\Delta_r G^\ominus = -RT \ln K_r \tag{16.13d}$$

from the value of $\Delta_r G^\ominus$ computed as

$$\Delta_r G^\ominus = (c\, \Delta_f G_C^\ominus + d\, \Delta_f G_D^\ominus) - (a\, \Delta_f G_A^\ominus + b\, \Delta_f G_B^\ominus) \tag{16.14d}$$

using standard reference data for each of the standard molar Gibbs free energies of formation $\Delta_f G_i^\circ$, then the result represents the concentration equilibrium constant K_r for the reaction taking place, and buffered, at pH 0. Furthermore, all the inferences concerning the reaction directionality and spontaneity that can be drawn from the sign and magnitude of $\Delta_r G^\circ$ also apply to the reaction buffered at pH 0 – which is of little relevance to a reaction buffered at pH 7. For biochemical applications, it would be far easier if there were tabulated, standard, values of thermodynamic data from which we could directly obtain insight into the behaviour of reactions buffered at pH 7, rather than pH 0.

The central role of pH 0 within the conventional standards is fundamentally attributable to the definition of the standard state of the aqueous hydrogen ion $H^+(aq)$. As a solute in a solution, reference to Table 6.4 will show that the standard state for $H^+(aq)$ ions, if in an ideal solution, is a molar concentration $[H^+] = 1$ mol dm^{-3}, or, for a real solution, a molality $b_{H+} = 1$ mol kg^{-1}, both of which correspond to pH 0. The international biochemical community has therefore chosen to change this particular standard by defining the **biochemical standard** (sometimes known as the **biological standard**) as

> **The biochemical standard defines the biochemical standard molal activity of the aqueous hydrogen ion** $H^+(aq)$ **as** 10^{-7}.

Since all biochemical systems are real, the standard is formally defined in terms of the molal activity $a_{b,H+}$, which, as we saw on page 664, is related to the IUPAC definition of pH as

$$\text{pH} = -\log_{10} a_{b,H+} = -\log_{10} \gamma_{b,H+}(b_{H+}/b_{H+}^\circ) \tag{22.11}$$

The standard $a_{b,H+} = 10^{-7}$ therefore implies pH 7. If we ignore the very small difference between the IUPAC definition of pH, and the definition of pH in terms of molar concentrations $[H^+]$, as discussed on page 665, then an alternative definition of the biochemical standard is

> **The biochemical standard defines the biochemical standard molar concentration of the aqueous hydrogen ion** $H^+(aq)$ **as** 10^{-7} **mol dm**$^{-3}$.

A key consequence of the biochemical standard is that, in a number of important contexts involving both hydrogen ions and also atomic and molecular hydrogen, there is a change in a relevant reference level. This implies that many measurements made according to the biochemical standard are different from the corresponding measurements made using the conventional standards – just as a measurement of temperature of a given system gives one number (say, 25 °C) using the reference level 'the melting point of pure ice' of the Celsius scale, and a different number (298.15 K) using the reference level 'the absolute zero of temperature' of the ideal gas scale.

To make it quite clear which thermodynamic standards are being used in any given instance, measurements made under the biological standard are known as **transformed**, and will be designated in this book using the superscript symbol ', for example, G' and K_c', and measurements relating directly to the biochemical standard state will use the superscript symbol $^{\circ\prime}$, for example $G^{\circ\prime}$. We note that other sources might use different symbols such as G'° or $G^{\oplus}$ for transformed standards.

Using this notation, the key change introduced by the biochemical standard can be represented, in terms of the molar concentrations [H$^+$], as

$$[H^+]^{\ominus\prime} = 10^{-7} \text{ mol dm}^{-3} \quad (\text{pH } 7) \tag{23.1a}$$

in contrast to the conventional standards for which

$$[H^+]^{\ominus} = 1 \text{ mol dm}^{-3} \quad (\text{pH } 0)$$

both at the standard pressure $P^{\ominus} = P^{\ominus\prime} = 1$ bar $= 10^5$ Pa. Similarly, in terms of molalities, for the biochemical standard state

$$b_{H^+}{}^{\ominus\prime} = 10^{-7} \text{ mol kg}^{-1} \quad (\text{pH } 7) \tag{23.1b}$$

in contrast the to the conventional standards for which

$$b_{H^+}{}^{\ominus} = 1 \text{ mol kg}^{-1} \quad (\text{pH } 0)$$

23.4 Why this is important

To clarify why the difference between the conventional standards and the biochemical standard is important, consider a generalised reaction

$$aA + bB \rightleftharpoons cC + hH^+ \text{ (aq)}$$

taking place in an ideal solution, associated with a standard reaction Gibbs free energy change $\Delta_r G^{\ominus}$

$$\Delta_r G^{\ominus} = (c\,\Delta_f \mathbf{G}_C{}^{\ominus} + d\,\Delta_f \mathbf{G}_D{}^{\ominus}) - (a\,\Delta_f \mathbf{G}_A{}^{\ominus} + b\,\Delta_f \mathbf{G}_B{}^{\ominus}) \tag{16.14d}$$

and an associated concentration equilibrium constant K_r

$$\Delta_r G^{\ominus} = -RT \ln K_r \tag{16.13d}$$

in which all items are measured using the conventional standards.

Both the sign, and the magnitude, of $\Delta_r G^{\ominus}$ provide important information as to the nature of the reaction equilibrium. As regards the sign, if $\Delta_r G^{\ominus} < 0$, the reaction equilibrium constant K_r will be greater than 1, and so the equilibrium mixture will favour the products over the reactants; if $\Delta_r G^{\ominus} > 0$, the equilibrium mixture will favour the reactants over the products. Furthermore, the magnitude of $\Delta_r G^{\ominus}$ tells us how far the equilibrium lies to the left or the right: if $\Delta_r G^{\ominus}$ is more negative than about -40 kJ, then, for practical purposes the reaction proceeds to completion to the right as written; if $\Delta_r G^{\ominus}$ is more positive than about $+40$ kJ, the reactants are, in essence, stable, and no reaction takes place. Knowledge of $\Delta_r G^{\ominus}$ therefore enables us to infer much about the directionality and spontaneity of the reaction, and the relative stability of the reactants and products. And, for any reaction, knowledge of $\Delta_r G^{\ominus}$ is not difficult to acquire: given that $\Delta_r G^{\ominus}$ is defined as

$$\Delta_r G^{\ominus} = (c\,\Delta_f \mathbf{G}_C{}^{\ominus} + d\,\Delta_f \mathbf{G}_D{}^{\ominus}) - (a\,\Delta_f \mathbf{G}_A{}^{\ominus} + b\,\Delta_f \mathbf{G}_B{}^{\ominus}) \tag{16.14d}$$

then the value of $\Delta_r G^{\ominus}$ can easily be computed from the appropriate standard molar Gibbs free energies of formation $\Delta_f G_i{}^{\ominus}$ as tabulated in reference works.

For a reaction involving hydrogen ions, however, as we saw in Chapter 17 (see page 564), all the inferences drawn from the sign and magnitude of $\Delta_r G^\ominus$, and the corresponding value of K_r, apply only to a reaction buffered at pH 0. If, however, the reaction is buffered at a pH other than pH 0, the situation can be very different, as we are about to see.

So suppose that, for a particular reaction, $\Delta_r G^\ominus = 40$ kJ as measured using the conventional standards. This positive value of $\Delta_r G^\ominus$ implies that the reaction favours the reactants over the products, suggesting that the reactants are relatively stable. The extent of this stability may be quantified from equation (16.13d): at $T = 298.15$ K, $K_r = 9.82 \times 10^{-8}$, implying that the equilibrium mixture, in a solution buffered at pH 0, contains very few products, and is almost entirely unreacted reactants.

As this chapter will demonstrate, if any of the molecules, A, B, or C contain hydrogen atoms, then the values of the transformed standard molar Gibbs free energies of formation $\Delta_f \mathbf{G}_A^{\ominus\prime}$, $\Delta_f \mathbf{G}_B^{\ominus\prime}$ and $\Delta_f \mathbf{G}_C^{\ominus\prime}$ will be different from their conventional counterparts $\Delta_f \mathbf{G}_A^\ominus$, $\Delta_f \mathbf{G}_B^\ominus$ and $\Delta_f \mathbf{G}_C^\ominus$; furthermore, for molecular hydrogen $H_2(g)$, $\Delta_f \mathbf{G}_{H_2}^{\ominus\prime}$ is different from $\Delta_f \mathbf{G}_{H_2}^\ominus$. Since, as we shall see,

$$\Delta_r G^{\ominus\prime} = (c\,\Delta_f \mathbf{G}_C^{\ominus\prime} + d\,\Delta_f \mathbf{G}_D^{\ominus\prime}) - (a\,\Delta_f \mathbf{G}_A^{\ominus\prime} + b\,\Delta_f \mathbf{G}_B^{\ominus\prime}) \tag{23.2}$$

then $\Delta_r G^{\ominus\prime}$ will be different from $\Delta_r G^\ominus$, perhaps substantially so. For the reaction

$$a\text{A} + b\text{B} \rightleftharpoons c\text{C} + h\text{H}^+(\text{aq})$$

we have assumed that $\Delta_r G^\ominus = 40$ kJ, from which $K_r = 9.82 \times 10^{-8}$, implying that at pH 0, the reaction is not spontaneous, with the equilibrium lying far to the left. If it so happens that $\Delta_r G^{\ominus\prime}$ computes to, say, -40 kJ, then the change in sign tells us that the reaction is now spontaneous, with a value of $K_r' = 1.02 \times 10^7$. For the reaction buffered at pH 7 – which is the context to which K_r' applies – the equilibrium mixture is rich in products, not reactants, this being in complete contrast to the equilibrium mixture for the same reaction, buffered at pH 0, which is rich in reactants, not products.

To a biochemist, the relevant information is what happens at pH 7, and so it is far more satisfactory to have reference tables of values of $\Delta_f \mathbf{G}_i^{\ominus\prime}$ that enable us to determine $\Delta_r G^{\ominus\prime}$ directly, rather than to be obliged to use only reference tables that provide values for $\Delta_r G^\ominus$, from which direct inferences could be quite misleading. It is, of course, always possible to transform data defined using the conventional standards to biochemical data whenever we might need to – but it is far, far, easier to adopt the biochemical standard for biochemistry, to gather and tabulate values for $\Delta_f \mathbf{G}_i^{\ominus\prime}$ in the reference literature, and so be able to interpret all the corresponding results relating to $\Delta_r G^{\ominus\prime}$ and K_r' in the context of the real situation.

23.5 The implications of the biochemical standard state

23.5.1 What's different

At first sight, it might appear that changing the definition of the standard state of the aqueous hydrogen ion from pH 0 to pH 7 is quite trivial, and would have only limited impact. In fact, the consequences of this seemingly very slight change are extensive, for, although the equations of thermodynamics remain the same, many of the data items manipulated by those equations – the numbers that we measure and that we find in reference tables – are substantially different. In particular, as we shall demonstrate throughout this chapter, changing the standard state for the aqueous hydrogen ion from pH 0 to pH 7 has very wide implications, including

- changing the value of the standard molar Gibbs free energy of formation of every molecule that contains hydrogen
- changing the value of the thermodynamic equilibrium constant of every reaction that involves hydrogen ions
- changing the values of all standard reversible redox electrode potentials.

Given the number of molecules that contain hydrogen atoms, and the number of reactions that involve hydrogen ions or are redox reactions – especially as regards biochemistry – changing from the conventional standards to the biochemical standard has a very significant effect.

In particular, there are four changes that relate specifically to hydrogen:

- Firstly, to repeat the fundamental definition of the biochemical standard, the standard state of the aqueous hydrogen ion $H^+(aq)$ is defined as an ideal solution such that $[H^+]^{\ominus\prime} = 10^{-7}$ mol dm^{-3} (pH 7), in contrast to the conventional standard $[H^+]^{\ominus} = 1$ mol dm^{-3} (pH 0), with both standards using the same standard pressure $P^{\ominus} = P^{\ominus\prime} = 1$ bar.
- Secondly, under the conventional standards, the reference level $G^{\ominus}(H^+) = 0$ J mol^{-1} for the aqueous hydrogen ion $H^+(aq)$ is defined in relation to a solution of $H^+(aq)$ ions in the conventional standard state of $[H^+]^{\ominus} = 1$ mol dm^{-3}, corresponding to pH 0. Under the biochemical standard, the reference level $G^{\ominus\prime}(H^+) = 0$ J mol^{-1} is defined as corresponding to a solution of $H^+(aq)$ ions in the biological standard state $[H^+]^{\ominus\prime} = 10^{-7}$ mol dm^{-3}, pH 7. As a consequence, $G^{\ominus\prime}(H^+) = 0$ J mol^{-1} maps onto a value of $G^{\ominus}(H^+) = 7 RT \ln 10 = 39.956$ kJ mol^{-1}, and $G^{\ominus}(H^+) = 0$ J mol^{-1} maps onto a value of $G^{\ominus\prime}(H^+) = -7 RT \ln 10 = -39.956$ kJ mol^{-1}.
- Thirdly, as shown in Table 20.2, under the conventional standards, the standard reversible redox potential for the hydrogen ion, $\mathbb{E}^{\ominus}(H^+, H_2) = 0$ V, whilst under the biochemical standard, $\mathbb{E}^{\ominus\prime}(H^+, H_2) = -0.414$ V.
- Fourthly, as shown in Table 13.1, under the conventional standards, the molar Gibbs free energy of formation of molecular hydrogen $\Delta_f G^{\ominus}(H_2) = 0$ J mol^{-1}; under the biochemical standard, $\Delta_f G^{\ominus\prime}(H_2) = 79.9$ kJ mol^{-1}.

These last three statements, and the numbers 39.956 kJ, – 0.414 V, and 79.9 kJ mol^{-1}, appear to have jumped 'out-of-the-blue': the statements will all be verified, and the numbers proven, in due course.

23.5.2 What's the same

Although the impact of the change to the biochemical standard is significant, many quantities are unchanged, and have the same values in both the conventional and biochemical standards. In particular:

- All units of measure are the same: so, for example, a Joule is still a Joule.
- All fundamental constants, such as R, N_A, k_B, F…are unchanged.

- All actual measurements of mass m, mole number n, volume V, pressure P, and temperature T remain the same.
- The three measures of the actual quantity of a species i in solution – the mole fraction x_i, the molality b_i, and the molar concentration $[I]$ – are unchanged: $x_i = x_i'$, $b_i = b_i'$, and $[I] = [I]'$ for all species i, including the hydrogen ion $H^+(aq)$. This is consistent with the previous point: for example, the molar concentration $[I]$ is the ratio n_i/V, and since both n_i and V are unchanged, the ratio n_i/V must also be unchanged.
- The standard pressure, $P^\ominus = 1$ bar $= 10^5$ Pa, remains unchanged.
- The standard states for all pure solids, liquids and gases, including molecular hydrogen, as defined in Table 6.4, are unchanged.
- For all solutes i in solution, with the sole exception of the hydrogen ion $H^+(aq)$, the standard state $[I]^\ominus$, as expressed in terms of the molar concentration, is unchanged: $[I]^{\ominus\prime} = [I]^\ominus = 1$ mol dm^{-3} (or 1 mol m^{-3}) for all solutes i except $H^+(aq)$.
- The value of the standard molar Gibbs free energy of formation of every molecule that does not contain any hydrogen remains the same.
- The value of the thermodynamic equilibrium constant for all reactions that do not involve hydrogen ions $H^+(aq)$ remains unchanged (which includes reactions that involve molecules that contain hydrogen).
- The measurement of the standard emf of any electrochemical cell is unchanged, and so $\mathbb{E}_{cell}{}^\ominus = \mathbb{E}_{cell}{}^{\ominus\prime}$.

23.6 Transformed equations

All the equations of thermodynamics can be transformed from the conventional standard to the biochemical standard by applying a 'rule' that may be expressed as "add the symbol $'$ to each variable". It's then important to make sure that the data being used in any equation is consistent – if the equation is expressed in conventional format, all the data must be measured using the conventional standards; if the equation is expressed in the transformed format, all the data must be measured using the biochemical standard.

So, as examples, two important equations that refer to variables measured according to the conventional standards, are

$$\Delta_r G^\ominus = -RT \ln K_r \tag{16.13d}$$

and

$$\Delta_r G^\ominus = (c\, \Delta_f G_C{}^\ominus + d\, \Delta_f G_D{}^\ominus) - (a\, \Delta_f G_A{}^\ominus + b\, \Delta_f G_B{}^\ominus) \tag{16.14d}$$

which transform to become

$$\Delta_r G^{\ominus\prime} = -RT \ln K_r' \tag{23.3}$$

and

$$\Delta_r G^{\ominus\prime} = (c\, \Delta_f G_C{}^{\ominus\prime} + d\, \Delta_f G_D{}^{\ominus\prime}) - (a\, \Delta_f G_A{}^{\ominus\prime} + b\, \Delta_f G_B{}^{\ominus\prime}) \tag{23.2}$$

both of which refer to variables measured according to the biochemical standard.

THE BIOCHEMICAL STANDARD STATE

Transforming equations between the two standards therefore seems, and indeed is, very simple. But there is one, very important, subtlety. Equation (16.10d)

$$G_{c,D} = G_{c,D}^{\ominus} + RT \ln[D] \tag{16.10d}$$

defines the molar Gibbs free energy $G_{c,D}$ of a solute D in an ideal solution, in which the quantity of D is measured as the molar concentration [D] mol dm^{-3} (or mol m^{-3}), and $G_{c,D}^{\ominus}$ is the standard molar Gibbs free energy of D in solution, as defined by reference to the standard state $[D]^{\ominus} = 1$ mol dm^{-3} (or mol m^{-3}), at the standard pressure of $P^{\ominus} = 1$ bar. If we apply the 'rule' defined in the previous paragraph – "to transform an equation from the conventional standards to the biochemical standard, simply add the symbol ′ to each variable" – we obtain

$$G_{c,D}' = G_{c,D}^{\ominus\prime} + RT \ln[D]'$$

Unfortunately, however, this equation is *incorrect*. Not because the 'rule' is wrong; rather, because the equation to which the 'rule' is being applied – equation (16.10d) – is itself, strictly speaking, incorrect.

Why so? Because, as we saw on page 516, the term [D] in equation (16.10d) is 'shorthand' for the ratio $[D]/[D]^{\ominus}$, which just happens to equal [D] because, according to the conventional standards, $[D]^{\ominus} = 1$ mol dm^{-3} (or mol m^{-3}) for all solutes D. The fully correct form of equation (16.10d), however, is

$$G_{c,D} = G_{c,D}^{\ominus} + RT \ln([D]/[D]^{\ominus}) \tag{16.10b}$$

which transforms, correctly, to

$$G_{c,D}' = G_{c,D}^{\ominus\prime} + RT \ln([D]'/[D]^{\ominus\prime}) \tag{23.4}$$

in which not only has [D] transformed into [D]′, but $[D]^{\ominus}$ has also become $[D]^{\ominus\prime}$.

Since $[D]^{\ominus} = 1$ mol dm^{-3} (or mol m^{-3}) for all solutes when using the conventional standards, the ratio $[D]/[D]^{\ominus}$ is always equal to [D]. As a consequence, many of the conventional equations that refer to molar concentrations are written using [D] rather than $[D]/[D]^{\ominus}$ - for example, the concentration equilibrium constant K_c expressed as

$$K_c = \frac{[C]_{eq}^{c} [D]_{eq}^{d}}{[A]_{eq}^{a} [B]_{eq}^{b}} \tag{16.12c}$$

It is very easy to forget that each of these terms is in fact a ratio, and, as a result, to transform equation (16.12c) incorrectly as

$$K_c' = \frac{[C]_{eq}'^{c} [D]_{eq}'^{d}}{[A]_{eq}'^{a} [B]_{eq}'^{b}}$$

In fact, the correct transformation is

$$K_c' = \frac{([C]_{eq}'/[C]^{\ominus\prime})^c \, ([D]_{eq}'/[D]^{\ominus\prime})^d}{([A]_{eq}'/[A]^{\ominus\prime})^a \, ([B]_{eq}'/[B]^{\ominus\prime})^b} \tag{23.5}$$

Likewise, the Nernst equation

$$\mathbb{E}(\text{ox, red}) = \mathbb{E}^{\circ}(\text{ox, red}) + \frac{RT}{n_e F} \ln \frac{[\text{ox}]}{[\text{red}]} \tag{20.19a}$$

transforms as

$$\mathbb{E}'(\text{ox, red}) = \mathbb{E}^{\circ\prime}(\text{ox, red}) + \frac{RT}{n_e F} \ln \frac{[\text{ox}]'/[\text{ox}]^{\circ\prime}}{[\text{red}]'/[\text{red}]^{\circ\prime}} \tag{23.6}$$

For all solutes D, except the aqueous hydrogen ion H$^+$(aq), $[D]^{\circ} = [D]^{\circ\prime} = 1$ mol dm^{-3}, and so the ratio $[D]/[D]^{\circ} = [D]$ transforms into $[D]'/[D]^{\circ\prime} = [D]'$. Furthermore, for all actual concentrations, $[D]' = [D]$, giving the overall result that both $[D]/[D]^{\circ} = [D]$ and $[D]'/[D]^{\circ\prime} = [D]$. For the hydrogen ion, however, $[H^+]^{\circ} = 1$ mol dm^{-3}, but $[H^+]^{\circ\prime} = 10^{-7}$ mol dm^{-3}, and so $[H^+]/[H^+]^{\circ} = [H^+]$ transforms into $[H^+]'/[H^+]^{\circ\prime} = [H^+]'/10^{-7} = 10^7[H^+]' = 10^7[H^+]$. The transformation of equations that contain the ratio $[H^+]/[H^+]^{\circ}$ therefore results in factors of 10^7, 10^{-7} and $\ln 10^7 = 7 \ln 10 = 16.118$ appearing in some rather unexpected places. For example ...

23.7 $G(H^+)$ and $G'(H^+)$, and $\Delta_f G^{\circ}(H^+)$ and $\Delta_f G^{\circ\prime}(H^+)$

As we have seen, for an ideal solution of a solute D, for which the quantity dissolved is measured as the molar concentration [D] mol dm^{-3}, then, using the conventional standards

$$G_{c,D} = G_{c,D}^{\circ} + RT \ln \frac{[D]}{[D]^{\circ}} \tag{16.10b}$$

Dropping the subscript $_c$ to avoid clutter, and on the understanding that all reactions are taking place in solution with quantities measured as molar concentrations, then, for an ideal solution of hydrogen ions H$^+$(aq)

$$G(H^+) = G^{\circ}(H^+) + RT \ln \frac{[H^+]}{[H^+]^{\circ}}$$

Under the conventional standards, $[H^+]^{\circ} = 1$ mol dm^{-3}, and so

$$\boldsymbol{G}(H^+) = \boldsymbol{G}^{\circ}(H^+) + RT \ln[H^+] \tag{23.7}$$

In this expression, $\boldsymbol{G}(H^+)$ which $\boldsymbol{G}^{\circ}(H^+)$ represent absolute values, which, as we saw on pages 397 and 398, are not independently measureable. We therefore need to define a reference level which specifies the zero of the measurement scale. To do this, we firstly define $\boldsymbol{G}^{\circ}(H^+)$ as the value of the system's molar Gibbs free energy $\boldsymbol{G}(H^+)$ when the system is in its reference state, which is identical to the standard state, for which $[H^+]^{\circ} = 1$ mol dm^{-3}, and $P^{\circ} = 1$ bar, and so

$$\boldsymbol{G}(H^+, [H^+] = 1) = \boldsymbol{G}^{\circ}(H^+)$$

We then designate the reference level for measurement as $\boldsymbol{G}^{\circ}(H^+) = 0$ J mol^{-1}, and so equation (23.7) simplifies to

$$\boldsymbol{G}(H^+) = RT \ln[H^+] \tag{23.8}$$

THE BIOCHEMICAL STANDARD STATE

and the resulting value of $G(H^+)$ becomes 'the' value of G for the solution at molar concentration $[H^+]$ mol dm^{-3}.

On transforming equation (16.10b) to the biochemical standard, we derive

$$G_D' = G_D^{\ominus\prime} + RT \ln \frac{[D]'}{[D]^{\ominus\prime}} \tag{23.4}$$

For any solute D, with the sole, and important, exception of the aqueous hydrogen ion $H^+(aq)$, the standard molar concentration is unchanged. Accordingly, $[D]^{\ominus\prime} = [D]^{\ominus} = 1$ mol dm^{-3} (or mol m^{-3}), and so equation (23.4) may be written as

$$G_D' = G_D^{\ominus\prime} + RT \ln [D]'$$

for all solutes D other than the hydrogen ion. But for $H^+(aq)$ ions, $[H^+]^{\ominus} = 1$ mol dm^{-3}, corresponding to pH 0, but $[H^+]^{\ominus\prime} = 10^{-7}$ mol dm^{-3}, corresponding to pH 7. Accordingly, for $H^+(aq)$ ions, equation (23.4) becomes

$$G'(H^+) = G^{\ominus\prime}(H^+) + RT \ln \frac{[H^+]'}{[H^+]^{\ominus\prime}}$$

$$= G^{\ominus\prime}(H^+) + RT \ln \frac{[H^+]'}{10^{-7}}$$

$$= G^{\ominus\prime}(H^+) + RT \ln 10^7 \, [H^+]'$$

and so

$$G'(H^+) = G^{\ominus\prime}(H^+) + RT \ln [H^+]' + 7 \, RT \ln 10$$

In this expression, $[H^+]'$ represents the actual molar concentration of $H^+(aq)$ ions in a given ideal solution, which must be the same as the actual measurement $[H^+]$ of the same solution under the conventional standards. Accordingly, $[H^+]' = [H^+]$, and so

$$G'(H^+) = G^{\ominus\prime}(H^+) + RT \ln [H^+] + 7 \, RT \ln 10$$

Once again, the absolute values $G'(H^+)$ and $G^{\ominus\prime}(H^+)$ cannot be measured, and so we need to define a new reference level for the biochemical standard. For the conventional standards, as we have just seen, $G^{\ominus}(H^+)$ was defined as the value of $G(H^+)$ corresponding to the standard state $[H^+]^{\ominus} = 1$ mol dm^{-3}, and $P^{\ominus} = 1$ bar, and so, for the biochemical standard, we define $G^{\ominus\prime}(H^+)$ as the value of $G'(H^+)$ corresponding to a system in the reference state defined as identical to the biochemical standard state, for which the actual molar concentration $[H^+] = [H^+]^{\ominus\prime} = 10^{-7}$ mol dm^{-3}, also at $P^{\ominus} = 1$ bar, giving

$$G'(H^+, [H^+] = 10^{-7}) = G^{\ominus\prime}(H^+) + RT \ln [H^+] + 7 \, RT \ln 10 = G^{\ominus\prime}(H^+)$$

We then define the reference level as $G^{\ominus\prime}(H^+) = 0$ J mol^{-1}, and so, in general, for any actual molar concentration $[H^+]$

$$G'(H^+) = RT \ln [H^+] + 7 \, RT \ln 10 \tag{23.9}$$

Equation (23.9) determines the value of $G'(H^+)$, as measured under the biochemical standard, for a solution of any actual molar concentration $[H^+]$ mol dm^{-3}, whilst equation (23.8)

$$G(H^+) = RT\ln[H^+] \tag{23.8}$$

defines the value of $G(H^+)$, as measured under the conventional standard, also for a solution of any actual molar concentration $[H^+]$ mol dm^{-3}. If measurements of $G(H^+)$ and $G'(H^+)$ are taken of the same solution, then $[H^+]$ is the same, and so the measurements will differ by $7 RT \ln 10$. At any temperature T, the quantity $7 RT \ln 10$ kJ mol^{-1} is a constant, which evaluates to 39.956 kJ mol^{-1} at 298.15 K, this being the reference level shift between any measurement $G(H^+)$ using the conventional standards, and the corresponding measurement $G'(H^+)$ of the same system using the biochemical standard, just as the measurement of the temperature of a system using the Celsius scale differs by 273.15 °C from a measurement of the same system using the ideal gas scale. The relationship between $G(H^+)$ and $G'(H^+)$ is illustrated in Figure 23.1.

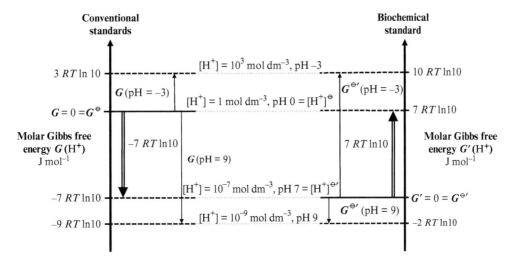

Figure 23.1 $G(H^+)$ and $G'(H^+)$. The left-hand axis shows values of the molar Gibbs free energy $G(H^+)$ for an ideal solution of H$^+$ (aq) ions as measured according to the conventional standards, and the right-hand axis $G'(H^+)$ as measured according to the biochemical standard. For the conventional standards, the reference level $G(H^+) = 0$ J mol^{-1} = $G^{\ominus}(H^+)$ corresponds to the reference state for which the molar concentration $[H^+] = [H^+]^{\ominus} = 1$ mol dm^{-3}, corresponding to pH 0; for the biochemical standard, the reference level $G'(H^+) = 0$ J mol^{-1} = $G^{\ominus\prime}(H^+)$ corresponds to the reference state $[H^+] = [H^+]^{\ominus\prime} = 10^{-7}$ mol dm^{-3}, pH 7.

Figure 23.1 identifies a number specific cases:

- For a solution for which $[H^+] = 1$ mol dm^{-3}, pH 0 = $[H^+]^{\ominus}$, then, according to equation (23.8), $G(H^+) = 0$ J mol^{-1} = $G^{\ominus}(H^+)$, corresponding to the standard state under the conventional standards. According to equation (23.9), for the same solution, $G'(H^+) = 7 RT \ln 10$ J mol^{-1}.
- For a solution for which $[H^+] = 10^{-7}$ mol dm^{-3}, pH 7 = $[H^+]^{\ominus\prime}$, equation (23.8) implies that $G(H^+) = -7 RT \ln 10$ J mol^{-1}; for the same solution, equation (23.9) gives $G'(H^+) = 0$ J mol^{-1} = $G^{\ominus\prime}(H^+)$, corresponding to the biochemical standard state.

- For a solution for which $[H^+] = 10^{-9}$ mol dm^{-3}, pH 9, $G(H^+) = -9\,RT \ln 10$ J mol^{-1}; for the same solution, $G'(H^+) = -2\,RT \ln 10$ J mol^{-1}.
- For a solution for which $[H^+] = 10^3$ mol dm^{-3}, pH -3, $G(H^+) = 3\,RT \ln 10$ J mol^{-1}; for the same solution, $G'(H^+) = 10\,RT \ln 10$ J mol^{-1}.

The general relationship between measurements $G(H^+)$ and $G'(H^+)$ on the same system can be derived by combining equations (23.8) and (23.9) as

$$G'(H^+) = G(H^+) + 7\,RT \ln 10 \tag{23.10}$$

Reference to page 399 will show that, under the conventional standards, the standard Gibbs free energy of formation $\Delta_f G^\ominus(H^+)$ of the aqueous hydrogen ion $H^+(aq)$, corresponding to a solution of hydrogen ions at the standard molar concentration of 1 mol dm^{-3}, is designated the reference level for the measurement of standard Gibbs free energies of formation of aqueous ions in general, such that $\Delta_f G^\ominus(H^+) = 0$ J mol^{-1} – as is consistent with the reference level $G^\ominus(H^+) = 0$ J mol^{-1}. Similarly, under the biochemical standard, given the reference level $G^{\ominus\prime}(H^+) = 0$ J mol^{-1}, the reference level $\Delta_f G^{\ominus\prime}(H^+)$ is designated as

$$\Delta_f G^{\ominus\prime}(H^+) = 0 \text{ J mol}^{-1} \tag{23.11}$$

corresponding, as usual, to a solution of hydrogen ions at molar concentration $[H^+] = 10^{-7}$ mol dm$^{-3} = [H^+]^{\ominus\prime}$.

23.8 $\mathbb{E}^\ominus(H^+, H_2)$ and $\mathbb{E}^{\ominus\prime}(H^+, H_2)$

As we saw in Chapter 20, for any redox half-reaction of the form

oxidised state $+ n_e e^- \rightarrow$ reduced state

in which charge-balance is assumed, there is an associated reversible redox, or electrode, potential $\mathbb{E}$, which may be expressed in terms of the Nernst equation as

$$\mathbb{E} = \mathbb{E}^\ominus + \frac{RT}{n_e F} \ln \frac{[\text{ox}]}{[\text{red}]} \tag{20.19a}$$

In this equation, [ox] and [red] are 'shorthand' for the quantity ratios [ox]/[ox]$^\ominus$ and [red]/[red]$^\ominus$, each raised to the appropriate stoichiometric power (see page 626). Also, $\mathbb{E}^\ominus$, the standard reversible redox, or electrode, potential, is the value of $\mathbb{E}$ when all reagents are in their standard states, so that [ox] = [ox]$^\ominus$ and [red] = [red]$^\ominus$, implying that $\ln$ [ox]/[red] = $\ln 1 = 0$.

Figure 20.5 (see page 607) shows a diagrammatic representation of the hydrogen electrode, which equilibrates gaseous hydrogen $H_2(g)$ with a solution of hydrogen ions $H^+(aq)$ as

$2\,H^+(aq) + 2\,e^- \rightarrow H_2\,(g)$

For this reaction, the oxidised state is $H^+(eq)$, and so

$$[\text{ox}] = \frac{[H^+]}{[H^+]^\ominus}$$

where, as we saw on page 627, there is no term for the electrons. The reduced state is molecular hydrogen $H_2(g)$, for which the appropriate measure of quantity is the partial pressure p bar, implying that

$$[\text{red}] = \frac{p\,H_2}{P^{\ominus}H_2}$$

The Nernst equation therefore becomes

$$\mathbb{E}(H^+, H_2) = \mathbb{E}^{\ominus}(H^+, H_2) + \frac{RT}{2F} \ln \frac{([H^+]/[H^+]^{\ominus})^2}{p\,H_2/P^{\ominus}H_2}$$

in which $\mathbb{E}(H^+, H_2)$ is the reversible electrode potential associated with a hydrogen electrode in which an ideal solution of $H^+(aq)$ ions, of molar concentration $[H^+]$ mol dm^{-3}, is in equilibrium with gaseous hydrogen at a partial pressure of p bar. Since the reagents are in different physical states, the ln term is a 'hybrid', as discussed on page 524.

Under the conventional standards, for molecular hydrogen $H_2(g)$, $P^{\ominus}H_2(g) = 1$ bar, and for aqueous hydrogen ions $H^+(aq)$, $[H^+]^{\ominus} = 1$ mol dm^{-3}; also, as shown in Table 20.1, the reference level for the measurement of standard electrode potentials is defined such that $\mathbb{E}^{\ominus}(H^+, H_2) = 0$ V. Accordingly, for an electrode operating at an arbitrary molar concentration $[H^+]$ mol dm^{-3} and at any partial pressure p bar, 'the' reversible electrode potential $\mathbb{E}(H^+, H_2)$ is given by

$$\mathbb{E}(H^+, H_2) = \frac{RT}{2F} \ln \frac{[H^+]^2}{p\,H_2}$$

Suppose that the electrode is operating under conditions such that $p = 1$ bar, and $[H^+] = 10^{-7}$ mol dm^{-3}, corresponding to pH 7. The corresponding reversible electrode potential $\mathbb{E}(H^+ \text{ pH } 7, H_2)$ is therefore

$$\mathbb{E}(H^+ \text{ pH } 7, H_2) = \frac{RT}{2F} \ln 10^{-14} = -\frac{7RT}{F} \ln 10$$

A solution of hydrogen ions at pH 7 and a pressure of 1 bar, however, is a standard solution under the biochemical standard, and so this electrode potential is the biochemical standard reversible electrode potential $\mathbb{E}^{\ominus\prime}(H^+, H_2)$

$$\mathbb{E}^{\ominus\prime}(H^+, H_2) = \mathbb{E}(H^+ \text{ pH } 7, H_2) = -\frac{7RT}{F} \ln 10 \tag{23.12}$$

At any temperature T, the quantity $-(7\,RT/F)\ln 10$ is a constant, which, at 298.15 K, has the value -0.41419 V, this being the number stated on page 680, but without explanation.

A consequence of equation (23.12) concerns the measurement of the standard reversible electrode potential $\mathbb{E}^{\ominus\prime}(\text{ox, red})$ of any redox electrode, using the biochemical standard. As we know, under the conventional standard, $\mathbb{E}^{\ominus}(\text{ox, red})$ is determined by measuring the emf $\mathbb{E}_{\text{cell}}^{\ominus}$ of an electrochemical cell in which the test redox electrode is the cathode and a (conventional) standard hydrogen electrode is the anode, as expressed by equation (20.10)

$$\mathbb{E}_{\text{cell}}^{\ominus} = \mathbb{E}_{\text{cat}}^{\ominus} - \mathbb{E}_{\text{an}}^{\ominus} = \mathbb{E}^{\ominus}(\text{ox, red}) - \mathbb{E}^{\ominus}(H^+, H_2) \tag{20.10}$$

If we transform equation (20.10), we derive

$$\mathbb{E}_{\text{cell}}^{\ominus\prime} = \mathbb{E}_{\text{cat}}^{\ominus\prime} - \mathbb{E}_{\text{an}}^{\ominus\prime} = \mathbb{E}^{\ominus\prime}(\text{ox, red}) - \mathbb{E}^{\ominus\prime}(H^+, H_2)$$

THE BIOCHEMICAL STANDARD STATE

The emf of the electrochemical cell, however, is an actual physical property of that cell, as determined by the materials comprising the cathode, and the hydrogen comprising the anode. As long as we use the same unit-of-measure, the volt, the numerical value of the emf must be the same under both standards. $E_{cell}^{\ominus}$ therefore equals $E_{cell}^{\ominus\prime}$, implying that

$$E^{\ominus}(ox, red) - E^{\ominus}(H^+, H_2) = E^{\ominus\prime}(ox, red) - E^{\ominus\prime}(H^+, H_2)$$

However, according to the conventional standard, $E^{\ominus}(H^+, H_2) = 0$; also, from equation (23.12), according to the biochemical standard, $E^{\ominus\prime}(H^+, H_2) = -(7\,RT/F)\ln 10$ V. Accordingly

$$E^{\ominus}(ox, red) = E^{\ominus\prime}(ox, red) + \frac{7RT}{F}\ln 10 \tag{23.13}$$

Any standard redox potential $E^{\ominus}(ox, red)$, as measured using the conventional standard, is therefore greater than the standard redox potential $E^{\ominus\prime}(ox, red)$ of the same electrode, as measured using the biochemical standard, by an amount $(7\,RT/F)\ln 10$ V, which equals 0.41419 V at 298.15 K. The quantity $(7\,RT/F)\ln 10$ V is therefore a reference level shift between the two standards, just like the use of the 'silver standard' discussed on pages 617 to 619.

Some values of the standard transformed reversible redox potential $E^{\ominus\prime}$ for some selected molecules of biochemical interest are shown in Table 23.1.

Table 23.1 Some standard transformed reversible redox potentials $E^{\ominus\prime}$ for selected biochemical redox reactions in aqueous solution, in accordance with the biochemical standard

Half-reaction	Standard transformed reversible redox potential $E^{\ominus\prime}$ V
$O_2\,(g) + 4\,H^+ + 4\,e^- \rightleftharpoons 2\,H_2O$	+ 0.816
Cytochrome a $(Fe^{3+}) + e^- \rightleftharpoons$ cytochrome a (Fe^{2+})	+ 0.290
Cytochrome c $(Fe^{3+}) + e^- \rightleftharpoons$ cytochrome c (Fe^{2+})	+ 0.254
Cytochrome b $(Fe^{3+}) + e^- \rightleftharpoons$ cytochrome b (Fe^{2+})	+ 0.077
Fumarate^{2-} + 2 H$^+$ + 2 e$^- \rightleftharpoons$ Succinate^{2-}	+ 0.031
Oxaloacetate^{2-} + 2 H$^+$ + 2 e$^- \rightleftharpoons$ Malate^{2-}	− 0.166
Pyruvate$^-$ + 2 H$^+$ + 2 e$^- \rightleftharpoons$ Lactate$^-$	− 0.185
FAD + 2 e$^-$ + 2 H$^+ \rightleftharpoons$ FADH$_2$	− 0.219
NAD$^+$ + H$^+$ + 2 e$^- \rightleftharpoons$ NADH	− 0.320
NADP$^+$ + H$^+$ + 2 e$^- \rightleftharpoons$ NADPH	− 0.324
2 H$^+$ + 2 e$^- \rightleftharpoons$ H$_2$(g)	− 0.414

Source: *CRC Handbook of Biochemistry and Molecular Biology*, 4th Edition, 2010, Section 6, pages 557 to 563.

Note that, for the half-reaction $O_2\,(g) + 4\,H^+ + 4\,e^- \rightleftharpoons 2\,H_2O$, $E^{\ominus\prime}$ is shown in Table 23.1 as + 0.816 V, and $E^{\ominus}$ is shown in Table 20.1 as + 1.229 V, sensibly in accordance with the general equation $E^{\ominus} = E^{\ominus\prime} + 0.414$ V.

23.9 $\Delta_f G^{\ominus\prime}(H_2)$

As we have just shown, according to equation (23.12)

$$\mathbb{E}^{\ominus\prime}(H^+, H_2) = -\frac{7RT}{F} \ln 10 \tag{23.12}$$

the standard reversible electrode potential $\mathbb{E}^{\ominus\prime}(H^+, H_2)$ of the hydrogen electrode, as measured using the biochemical standard, is $-(7\,RT/F)\ln 10$ V. Physically, this electrode potential is attributable to the redox half-reaction

$$2\,H^+\,(aq) + 2\,e^- \rightarrow H_2\,(g)$$

which is itself associated with a standard transformed reaction Gibbs free energy change $\Delta_r G^{\ominus\prime}(H^+, H_2)$.

As we saw on page 623, any standard reversible electrode potential $\mathbb{E}^{\ominus}$ V is related to the corresponding standard reaction Gibbs free energy change $\Delta_r G^{\ominus}$ according to equation (20.13)

$$-\Delta_r G^{\ominus} = n_e\,F\,\mathbb{E}^{\ominus} \tag{20.13}$$

which transforms as

$$-\Delta_r G^{\ominus\prime} = n_e\,F\,\mathbb{E}^{\ominus\prime} \tag{23.14}$$

For the standard hydrogen electrode, $n_e = 2$, and so equations (23.12) and (23.14) may be combined as

$$-\Delta_r G^{\ominus\prime}(H^+, H_2) = 2\,F\,\mathbb{E}^{\ominus\prime}(H^+, H_2) = -14\,RT \ln 10$$

from which

$$\Delta_r G^{\ominus\prime}(H^+, H_2) = 14\,RT \ln 10 \tag{23.15}$$

Now, for any general reaction

$$aA + bB \rightleftharpoons cC + dD$$

the transformed reaction standard Gibbs free energy change $\Delta_r G^{\ominus\prime}$ is defined as

$$\Delta_r G^{\ominus\prime} = (c\,\Delta_f \mathbf{G}_C^{\ominus\prime} + d\,\Delta_f \mathbf{G}_D^{\ominus\prime}) - (a\,\Delta_f \mathbf{G}_A^{\ominus\prime} + b\,\Delta_f \mathbf{G}_B^{\ominus\prime}) \tag{23.2}$$

and so, in the particular case of the half-cell reaction

$$2\,H^+\,(aq) + 2\,e^- \rightarrow H_2\,(g)$$

we have

$$\Delta_r G^{\ominus\prime}(H^+, H_2) = \Delta_f \mathbf{G}^{\ominus\prime}(H_2) - 2\,\Delta_f \mathbf{G}^{\ominus\prime}(H^+) - 2\,\Delta_f \mathbf{G}^{\ominus\prime}(e^-)$$

In this expression, the term $\Delta_f \mathbf{G}^{\ominus\prime}(H^+)$ is determined by equation (23.11)

$$\Delta_f \mathbf{G}^{\ominus\prime}(H^+) = 0 \quad \text{J mol}^{-1} \tag{23.11}$$

Furthermore, as we saw on page 627, since electrons do not accumulate, $\Delta_f \mathbf{G}^{\ominus}(e^-) = \Delta_f \mathbf{G}^{\ominus}(e^-) = 0$ J mol^{-1}, and so

$$\Delta_r G^{\ominus\prime}(H^+, H_2) = \Delta_f \mathbf{G}^{\ominus\prime}(H_2) \tag{23.16}$$

Equation (23.16) may now be combined with equation (23.15) to give

$$\Delta_f G^{\ominus\prime}(H_2) = 14\, RT \ln 10 \qquad \text{J per mol } H_2 \qquad (23.17)$$

At 298.15 K, $RT \ln 10 = 5.708$ kJ mol^{-1}, and so, at this temperature, $\Delta_f G^{\ominus\prime}(H_2) = 79.912$ kJ mol^{-1}, a number very different from its conventional counterpart $\Delta_f G^{\ominus}(H_2) = 0$ J mol^{-1}. As we shall show in the next section, the difference between $\Delta_f G^{\ominus}(H_2)$ and $\Delta_f G^{\ominus\prime}(H_2)$ is important, for it gives rise to a difference between $\Delta_f G^{\ominus}$ and $\Delta_f G^{\ominus\prime}$ for all molecules that contain hydrogen.

Firstly, however, we shall explain why there is a difference between $\Delta_f G^{\ominus}(H_2)$ and $\Delta_f G^{\ominus\prime}(H_2)$. To do this, we return to the fundamental definition of standard molar Gibbs free energies of formation, equation (13.44a)

$$\Delta_f G^{\ominus}(\text{molecule}) = G^{\ominus}(\text{molecule}) - \sum_{\text{elements}} G^{\ominus}(\text{element}) \qquad (13.44a)$$

in which, for any molecule, $\Delta_f G^{\ominus}$ (molecule) is expressed as the difference between the absolute value $G^{\ominus}$(molecule) and another absolute value, representing the sum of the appropriate values of $G^{\ominus}$(element) for each constituent element. Since neither $G^{\ominus}$(molecule) nor any $G^{\ominus}$(element) are directly measureable, we need a reference level, which, according to the conventional standards, is defined, as we saw on page 397, by designating $G^{\ominus}$(element) $= 0$ J mol^{-1} for all elements in their standard states, including both monatomic elements such as carbon, and diatomic molecular elements such as hydrogen.

If we represent the summation term in equation (13.44a) as the reference level $G_{\text{ref}}^{\ominus}$(molecule), then an alternative form of equation (13.44a) is

$$\Delta_f G^{\ominus}(\text{molecule}) = G^{\ominus}(\text{molecule}) - G_{\text{ref}}^{\ominus}(\text{molecule})$$

which transforms as

$$\Delta_f G^{\ominus\prime}(\text{molecule}) = G^{\ominus\prime}(\text{molecule}) - G_{\text{ref}}^{\ominus\prime}(\text{molecule})$$

If the molecule in question is the hydrogen molecule $H_2(g)$, and bearing in mind equation (23.17), we then have

$$\Delta_f G^{\ominus\prime}(H_2) = G^{\ominus\prime}(H_2) - G_{\text{ref}}^{\ominus\prime}(H_2) = 14\, RT \ln 10 \qquad (23.18)$$

in which $G^{\ominus\prime}(H_2)$ represents the absolute value of the molar Gibbs free energy of molecular hydrogen $H_2(g)$ under the biological standard, and $G_{\text{ref}}^{\ominus\prime}(H_2)$ the corresponding reference level. As we know, neither of these absolute values can be measured – but we do have a comparable measure for $G^{\ominus\prime}(H_2)$: under the conventional standards, $G^{\ominus}(H_2)$ is assigned the value of 0 J mol^{-1}. Even though absolute values cannot be measured, it makes sense to be consistent, and so, if $G^{\ominus}(H_2)$ is designated 0 J mol^{-1}, then $G^{\ominus\prime}(H_2)$ can be designated to be 0 J mol^{-1} too. Accordingly, equation (23.18) implies that

$$G_{\text{ref}}^{\ominus\prime}(H_2) = -14\, RT \ln 10 \qquad \text{J per mol } H_2 \qquad (23.19)$$

Under the conventional standards, the reference level for measurements of $\Delta_f G^{\ominus}$ for any molecule is defined such that $G_{\text{ref}}^{\ominus}$(element) $= 0$ J mol^{-1} for all elements, including hydrogen; for the biochemical standard, $G_{\text{ref}}^{\ominus\prime}$(element) $= 0$ J mol^{-1} for all elements, except hydrogen, whilst, for hydrogen, equation (23.19) tells us that $G_{\text{ref}}^{\ominus\prime}(H_2) = -14\, RT \ln 10$ J per mol H_2. This is a reference level shift, which applies to hydrogen only, such that the reference level under the biochemical standard is $14\,RT\ln 10$ J per mol H_2 more negative than the corresponding value under the conventional standards. This is very similar to the situation we discussed in relation to the hydrogen ion $H^+(aq)$, as illustrated in Figure 23.1, which shows that the reference level under the biochemical standard is $7\,RT\ln 10$ J per mol H^+ more negative than the reference level under the conventional standards. Given that molecular hydrogen is diatomic, $7\,RT\ln 10$ J per mol H^+ is the equivalent of $14\,RT\ln 10$ J per mol H_2, so these two downwards reference level shifts are consistent.

23.10 Transformed standard molar Gibbs free energies of formation $\Delta_f G^{\ominus\prime}$

As we have just shown, under the biochemical standard

$$\Delta_f G^{\ominus\prime}(H_2) = 14\,RT\ln 10 \qquad \text{J per mol } H_2 \qquad (23.17)$$

and

$$G_{ref}^{\ominus\prime}(H_2) = -14\,RT\ln 10 \qquad \text{J per mol } H_2 \qquad (23.19)$$

These relationships have an important impact on the transformed standard molar Gibbs free energies of formation $\Delta_f G^{\ominus\prime}$ of all molecules containing hydrogen, as we shall now show.

So, consider the generalised formation reaction of a molecule $X_x Y_y H_h$ from its constituent elements X, Y and hydrogen as

$$xX + yY + \frac{h}{2} H_2(g) \rightarrow X_x Y_y H_h$$

in which X and Y are any elements other than hydrogen.

If we firstly use the conventional standards, then $\Delta_f G^{\ominus}(X_x Y_y H_h)$ may be determined using equation (13.44a) as

$$\Delta_f G^{\ominus}(X_x Y_y H_h) = G^{\ominus}(X_x Y_y H_h) - \left[x\, G_{ref}^{\ominus}(X) + y\, G_{ref}^{\ominus}(Y) + \frac{h}{2} G_{ref}^{\ominus}(H_2) \right]$$

in which $G^{\ominus}(X_x Y_y H_h)$ is an absolute value, and the summation terms have been explicitly written in terms of reference values for each constituent element. As we know, under the conventional standards, the reference levels are defined such that $G_{ref}^{\ominus}(X) = G_{ref}^{\ominus}(Y) = G_{ref}^{\ominus}(H_2) = 0$ J mol^{-1}, and so

$$\Delta_f G^{\ominus}(X_x Y_y H_h) = G^{\ominus}(X_x Y_y H_h) \qquad (23.20)$$

Under the biochemical standard, we have

$$\Delta_f G^{\ominus\prime}(X_x Y_y H_h) = G^{\ominus\prime}(X_x Y_y H_h) - \left[x\, G_{ref}^{\ominus\prime}(X) + y\, G_{ref}^{\ominus\prime}(Y) + \frac{h}{2} G_{ref}^{\ominus\prime}(H_2) \right]$$

in which the reference levels are such that $G_{ref}^{\ominus\prime}(X) = G_{ref}^{\ominus\prime}(Y) = 0$ J mol^{-1}, but for hydrogen, according to equation (23.19)

$$G_{ref}^{\ominus\prime}(H_2) = -14\,RT\ln 10 \qquad \text{J per mol } H_2 \qquad (23.19)$$

and so

$$\Delta_f G^{\ominus\prime}(X_x Y_y H_h) = G^{\ominus\prime}(X_x Y_y H_h) + 7h\, RT \ln 10 \qquad (23.21)$$

However, for any molecule $X_x Y_y H_h$, the absolute value of the molecule's Gibbs free energy must be the same under both standards, implying that

$$G^{\ominus\prime}(X_x Y_y H_h) = G^{\ominus}(X_x Y_y H_h)$$

Furthermore, according to equation (23.20), under the conventional standards

$$\Delta_f G^{\ominus}(X_x Y_y H_h) = G^{\ominus}(X_x Y_y H_h) \qquad (23.20)$$

and so

$$G^{\ominus\prime}(X_x Y_y H_h) = \Delta_f G^{\ominus}(X_x Y_y H_h)$$

thereby enabling equation (23.21) to be written as

$$\Delta_f G^{\ominus\prime}(X_x Y_y H_h) = \Delta_f G^{\ominus}(X_x Y_y H_h) + 7h\, RT \ln 10 \qquad (23.22)$$

Equation (23.22) is another important result, and shows that, for any molecule $X_x Y_y H_h$, $\Delta_f G^{\ominus\prime}$ and $\Delta_f G^{\ominus}$ differ by an amount $7h\, RT \ln 10$ kJ mol^{-1}, as illustrated in Figure 23.2.

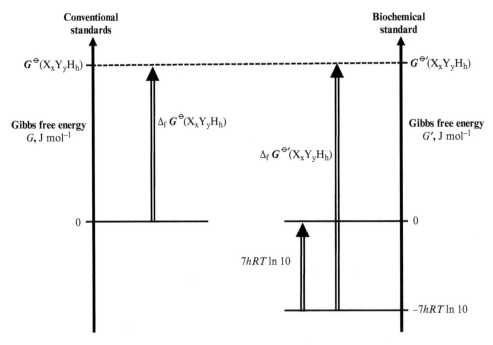

Figure 23.2 $\Delta_f G^{\ominus}$ and $\Delta_f G^{\ominus\prime}$. For any molecule $X_x Y_y H_h$, the absolute standard molar Gibbs free energy must be the same under both the conventional and the biochemical standards, as represented by the upper dashed line. The measurement assigned to the standard molar Gibbs free energy of formation, however, does depend on the standard used, as determined by the reference levels. Under the conventional standards, the reference level is defined such that $G_{ref}^{\ominus} = 0$ J mol^{-1} for all elements, as shown on the left. Under the biochemical standard $G_{ref}^{\ominus\prime} = 0$ J mol^{-1} for all elements except hydrogen, but for hydrogen, $G_{ref}^{\ominus\prime}(H_2) = -14RT \ln 10$ J per mol of H_2. For the molecule $X_x Y_y H_h$, the reference level for hydrogen is therefore displaced by $7hRT \ln 10$ downwards, as shown by the lower solid line. Accordingly, $\Delta_f G^{\ominus\prime}(X_x Y_y H_h) = \Delta_f G^{\ominus}(X_x Y_y H_h) + 7hRT \ln 10$.

Two special cases of equation (23.22) are

- If the molecule contains no hydrogen atoms, $h = 0$ and so $\Delta_f G^\circ(X_xY_y) = \Delta_f G^{\circ\prime}(X_xY_y)$. The standard molar Gibbs free energy of formation is therefore the same under both standards.
- For the hydrogen molecule $H_2(g)$ itself, $h = 2$, $\Delta_f G^\circ = 0$ J mol^{-1}, and so $\Delta_f G^{\circ\prime}(H_2) = 14RT \ln 10$ kJ mol^{-1}, as is consistent with equation (23.17).

Some values of $\Delta_f G^{\circ\prime}$ and $\Delta_f G^\circ$ for selected molecules are shown in Table 23.2.

Table 23.2 Some values of standard molar Gibbs free energy of formation as measured in accordance with conventional standards, $\Delta_f G^\circ$, and the biochemical standard, $\Delta_f G^{\circ\prime}$, for selected molecules at 298.15 K, and an ionic strength of zero

Molecule	$\Delta_f G^\circ$ kJ mol^{-1}	$\Delta_f G^{\circ\prime}$ kJ mol^{-1}
Acetone, C_3H_6O	−161.2	80.0
L-Cysteine, $C_3H_7O_2NS$	−339.8	−59.2
CO_2 (g)	−394.4	−394.4
Creatine, $C_4H_9O_2N_3$	−264.3	100.4
Ethanol, C_2H_6O	−181.5	58.1
D-Glucose, $C_6H_{12}O_6$	−917.2	−436.4
H_2 (g)	0.0	79.9
H_2O (l)	−237.2	−157.3
L-Leucine, $C_6H_{13}O_2N$	−356.3	167.2
O_2 (g)	0.0	0.0
Sucrose, $C_{12}H_{22}O_{11}$	−1551.8	−685.7
L-Tyrosine, $C_9H_{11}O_3N$	−387.2	68.8
Urea, CH_4ON_2	−203.8	−43.0

Source: *CRC Handbook of Chemistry and Physics*, 96th edn, 2015, Section 7, pages 7-19 to 7-21, and *Biochemistry: The Chemical Reactions of Living Cells*, Volume 1, DE Metzler and CM Metzler, Volume 1, Elsevier Academic Press, 2nd edn, 2003, pages 290–291.

As can be seen from Table 23.2, for molecules containing hydrogen, values of $\Delta_f G^{\circ\prime}$ are indeed different from the corresponding values of $\Delta_f G^\circ$, but the difference $\Delta_f G^{\circ\prime}(X_xY_yH_h) - \Delta_f G^\circ(X_xY_yH_h)$ is very specific: it equals $7h\,RT \ln 10$ J mol^{-1}, where h is the number of hydrogen atoms in the molecule of interest. At 298.15 K, this amounts to 39.956 kJ per mol of H atoms, or 39.956 h kJ per mol of molecules, each of which contains h hydrogen atoms. Also, we note that the values shown in Table 23.2 correspond to an ionic strength $I = 0$. As discussed in Chapter 22, the molal chemical potential μ_b of a component in a real solution depends on that component's molal activity $a_{b,D}$, which in turn is associated with a mean molal activity coefficient $\gamma_{b,\pm}$. According to equations (22.19) and (22.21), $\gamma_{b,\pm}$ is a function of the ionic strength, I, as defined by equation (22.20). For real solutions, values of $G^\circ = \mu^\circ$ are therefore different for solutions of different ionic strengths, and so the values shown in Table 23.2 correspond to $I = 0$.

23.11 $\Delta_r G^\ominus$ and $\Delta_r G^{\ominus\prime}$

We will now use the key result

$$\Delta_f G^{\ominus\prime}(X_x Y_y H_h) = \Delta_f G^\ominus(X_x Y_y H_h) + 7h\, RT \ln 10 \qquad (23.22)$$

from the previous section to explore the behaviour of a generalised chemical reaction, taking place in an ideal aqueous solution, of the form

$$a AH_p + b BH_q \rightleftharpoons c CH_r + d DH_s + h H^+ \qquad (23.23)$$

in which AH_p, BH_q, CH_r and DH_s are representations of water-soluble molecules that contain hydrogen atoms, and $H^+(aq)$ is the aqueous hydrogen ion. For the moment, we assume that none of AH_p, BH_q, CH_r and DH_s represent molecular water, H_2O, and so the role of water is to act as the solvent, rather than as a reactant or a product. Note that any one, and any combination, of the parameters p, q, r, s and h may be zero; furthermore, h may be negative, if hydrogen ions are consumed, rather than produced.

To ensure that reaction (23.23) balances chemically, all the hydrogen atoms must be accounted for, and so

$$ap + bq = cr + ds + h \qquad (23.24)$$

Reaction (23.23), as written, is not charged-balanced: in reality, of course, all reactions must be charged-balanced, so let us assume that the entities AH_p, BH_q, CH_r and DH_s are appropriately charged so that charge balance is indeed maintained.

As we saw in Chapter 16 (see page 423), reaction (23.23) will be associated with a value of $\Delta_r G^\ominus$, which, in accordance with the conventional standards, is given in general by

$$\Delta_r G^\ominus = (c\, \Delta_f G_C^\ominus + d\, \Delta_f G_D^\ominus) - (a\, \Delta_f G_A^\ominus + b\, \Delta_f G_B^\ominus) \qquad (16.14d)$$

and in this particular case becomes

$$\Delta_r G^\ominus = [c\, \Delta_f G^\ominus(CH_r) + d\, \Delta_f G^\ominus(DH_s) + h\, \Delta_f G^\ominus(H^+)]$$
$$- [a\, \Delta_f G^\ominus(AH_p) + b\, \Delta_f G^\ominus(BH_q)] \qquad (23.25)$$

In accordance with conventional standards, the reference level for the standard molar Gibbs free energies of ions is defined such that $\Delta_f G^\ominus(H^+) = 0$ and so equation (23.25) may be written as

$$\Delta_r G^\ominus = [c\, \Delta_f G^\ominus(CH_r) + d\, \Delta_f G^\ominus(DH_s)] - [a\, \Delta_f G^\ominus(AH_p) + b\, \Delta_f G^\ominus(BH_q)] \qquad (23.26)$$

As we saw on page 681, equation (16.14d) transforms as

$$\Delta_r G^{\ominus\prime} = (c\, \Delta_f G_C^{\ominus\prime} + d\, \Delta_f G_D^{\ominus\prime}) - (a\, \Delta_f G_A^{\ominus\prime} + b\, \Delta_f G_B^{\ominus\prime}) \qquad (23.2)$$

which, for the reaction $a AH_p + b BH_q \rightleftharpoons c CH_r + d DH_s + h H^+$, becomes

$$\Delta_r G^{\ominus\prime} = [c\, \Delta_f G^{\ominus\prime}(CH_r) + d\, \Delta_f G^{\ominus\prime}(DH_s) + h\, \Delta_f G^{\ominus\prime}(H^+)]$$
$$- [a\, \Delta_f G^{\ominus\prime}(AH_p) + b\, \Delta_f G^{\ominus\prime}(BH_q)] \qquad (23.27)$$

For four of the terms in equation (23.27), we now use equation (23.22)

$$\Delta_f G^{\ominus\prime}(X_x Y_y H_h) = \Delta_f G^\ominus(X_x Y_y H_h) + 7h\, RT \ln 10 \qquad (23.22)$$

As

$$\Delta_f G^{\ominus\prime}(AH_p) = \Delta_f G^\ominus(AH_p) + 7p\, RT \ln 10$$
$$\Delta_f G^{\ominus\prime}(BH_q) = \Delta_f G^\ominus(BH_q) + 7q\, RT \ln 10$$
$$\Delta_f G^{\ominus\prime}(CH_r) = \Delta_f G^\ominus(CH_r) + 7r\, RT \ln 10$$
$$\Delta_f G^{\ominus\prime}(DH_s) = \Delta_f G^\ominus(DH_s) + 7s\, RT \ln 10$$

However, as we saw on page 686, for the term relating to the hydrogen ion $H^+(aq)$

$$\Delta_f G^{\ominus\prime}(H^+) = 0$$

and there is no associated term of the form $7h\, RT \ln 10$.

Equation (23.27) therefore becomes

$$\Delta_r G^{\ominus\prime} = [c\, \Delta_f G^\ominus(CH_r) + d\, \Delta_f G^\ominus(DH_s)] - [a\, \Delta_f G^\ominus(AH_p) + b\, \Delta_f G^\ominus(BH_q)]$$
$$+ 7\,[(cr + ds) - (ap + bq)]\, RT \ln 10 \qquad (23.28)$$

For the reaction $aAH_p + bBH_q \rightleftharpoons cCH_r + dDH_s + hH^+$, the number of hydrogen atoms must balance, and so, as we have seen

$$ap + bq = cr + ds + h \qquad (23.24)$$

implying that

$$(cr + ds) - (ap + bq) = -h$$

Furthermore, according to equation (23.26)

$$\Delta_r G^\ominus = [c\, \Delta_f G_c^\ominus(CH_r) + d\, \Delta_f G_c^\ominus(DH_s)] - [a\, \Delta_f G_c^\ominus(AH_p) + b\, \Delta_f G_c^\ominus(BH_q)] \qquad (23.26)$$

Equation (23.28) therefore simplifies to the important result

$$\Delta_r G^{\ominus\prime} = \Delta_r G^\ominus - 7h\, RT \ln 10 \qquad (23.29)$$

Equation (23.29) is another key equation: for any reaction $aAH_p + bBH_q \rightleftharpoons cCH_r + dDH_s + hH^+$ taking place in an ideal solution, and which produces h mol of $H^+(aq)$ ions, $\Delta_r G^{\ominus\prime}$, as measured using the biochemical standard, is $7h\, RT \ln 10$ kJ *more negative* than $\Delta_r G^\ominus$, as measured using the conventional standards, for the same reaction. This implies that the equilibrium composition of the reaction buffered at pH 7 (as is relevant to the biochemical standard) is in general further to right (for the reaction as written) as compared to the equilibrium composition of the reaction buffered at pH 0 (as is relevant to the conventional standards). Furthermore, since, at 298.15 K, $7\, RT \ln 10 = 39.956$ kJ, if $h = 2$ – which is a reasonable value for a chemical reaction – then a reaction for which $\Delta_r G^\ominus$ is around + 40 kJ has a corresponding value of $\Delta_r G^{\ominus\prime}$ around – 40 kJ, as we saw in the example on page 678.

If the reaction does not produce any hydrogen ions, and so is of the form

$$aAH_p + bBH_q \rightleftharpoons cCH_r + dDH_s$$

then $h = 0$, and equation (23.29) implies that $\Delta_r G^{\ominus\prime} = \Delta_r G^\ominus$, showing that the standard Gibbs free energy of the reaction is the same under both the conventional and the biochemical standards, even if the reagents contain hydrogen atoms. Furthermore, if the reaction consumes hydrogen ions, and so is of the form

$$a\mathrm{AH_p} + b\mathrm{BH_q} + h\mathrm{H^+} \rightleftharpoons c\mathrm{CH_r} + d\mathrm{DH_s}$$

then h is a negative number, $-|h|$, and so equation (23.29) becomes

$$\Delta_r G^{\ominus\prime} = \Delta_r G^{\ominus} + 7|h|\, RT \ln 10$$

Equation (23.33), written as

$$\Delta_r G^{\ominus\prime} = \Delta_r G^{\ominus} - 7h\, RT \ln 10 \qquad (23.29)$$

therefore covers all reactions, with h being positive for reactions producing hydrogen ions, negative for reactions consuming hydrogen ions, and zero if no hydrogen ions are involved.

23.12 K_r and K_r'

As shown on page 523, for any generalised reaction, the change $\Delta_r G^{\ominus}$ in the reaction standard Gibbs free energy is related to the equilibrium constant K_r as

$$\Delta_r G^{\ominus} = -RT \ln K_r \qquad (16.13\mathrm{d})$$

which transforms into

$$\Delta_r G^{\ominus\prime} = -RT \ln K_r' \qquad (23.2)$$

Subtracting, we have

$$\Delta_r G^{\ominus} - \Delta_r G^{\ominus\prime} = -RT \ln K_r + RT \ln K_r' = RT \ln \frac{K_r'}{K_r} \qquad (23.30)$$

For the generalised reaction represented as

$$a\mathrm{AH_p} + b\mathrm{BH_q} \rightleftharpoons c\mathrm{CH_r} + d\mathrm{DH_s} + h\mathrm{H^+} \qquad (23.23)$$

then, according to equation (23.29)

$$\Delta_r G^{\ominus\prime} = \Delta_r G^{\ominus} - 7h\, RT \ln 10 \qquad (23.29)$$

and so

$$\Delta_r G^{\ominus} - \Delta_r G^{\ominus\prime} = 7h\, RT \ln 10$$

which implies, from equation (23.30), that

$$RT \ln \frac{K_r'}{K_r} = 7h\, RT \ln 10 = RT \ln 10^{7h}$$

giving

$$K_r' = 10^{7h} K_r \qquad (23.31\mathrm{a})$$

For a reaction taking place in an ideal solution, and for which the appropriate equilibrium constant is K_c, we therefore have

$$K_c' = 10^{7h} K_c \qquad (23.31\mathrm{b})$$

If such a reaction is of the form

$$aAH_p + bBH_q \rightleftharpoons cCH_r + dDH_s + hH^+ \qquad (23.23)$$

in which hydrogen ions are produced, then equation (23.31b) shows that the equilibrium constant K_c' is a factor of 10^{7h} greater than the equilibrium constant K_c, indicating that when the reaction is buffered at pH 7, the equilibrium lies far further to the right (for the reaction as written) as compared to the reaction buffered at pH 0 – just as we discussed on page 695. Likewise, if hydrogen ions are consumed, the parameter h is negative, $-|h|$, implying that $K_c' = 10^{-7|h|}K_c$. In this case, K_c' is a factor of $10^{7|h|}$ smaller than the equilibrium constant K_c, indicating that when the reaction is buffered at pH 7, the equilibrium lies far further to the left as compared to the reaction buffered at pH 0.

This is all totally consistent with the discussion on pages 561 to 566 in Chapter 17, where we explored the effect of buffering on the equilibrium of a reaction involving hydrogen ions, and taking place in solution, in terms of the pH-dependent parameter Λ, where, in the context of reaction (23.23), Λ is given by

$$\Lambda = \frac{[C]_{eq}^c \, [D]_{eq}^d}{[A]_{eq}^a \, [B]_{eq}^b} \qquad (23.47)$$

As we showed on page 565, Λ is related to K_c as

$$\Lambda \, (\text{pH } 7) = 10^{7h} \, K_c \qquad (17.56b)$$

for all values, both positive and negative, of the parameter h, implying that

$$\Lambda \, (\text{pH } 7) = K_c'$$

23.13 G_{sys} and G_{sys}' for an unbuffered reaction

In Chapter 14, we explored (see pages 417 to 428) how the total Gibbs free energy G_{sys} associated with the general gas phase reaction

$$aA + bB \rightleftharpoons cC + dD$$

varies with the extent-of-reaction parameter ξ, and we extended this analysis to a general reaction taking place in an ideal solution in Chapter 16 (see pages 517 to 520). In this section, we will explore the general ideal solution reaction

$$aAH_p + bBH_q \rightleftharpoons cCH_r + dDH_s + hH^+ \qquad (23.23)$$

to develop expressions for both G_{sys}, in accordance with the conventional standards, and also the transformed quantity G_{sys}', in accordance with the biochemical standards. This will demonstrate what happens when the same reaction is measured according to each standard.

An important assumption we make in this section is that reaction (23.23) is unconstrained, so that all components may adjust their composition as the reaction takes place, as determined by the reaction stoichiometry. In particular, the reaction is not buffered, and so the molar concentration $[H^+]$ mol dm^{-3} of hydrogen ions can vary freely, as will the pH. This section therefore examines the effect of using the two different standards of measurement on the same general reaction; the next section will explore the very different context in which the solution is buffered, implying that the molar concentration [H⁺], and the pH, are constrained.

As discussed in Chapters 14 and 16, and using the conventional standards, if the solution is ideal, at any instant during the reaction corresponding to an instantaneous value of ξ, the total Gibbs free energy $G_{sys}(\xi)$ of the system is the linear sum

$$G_{sys}(\xi) = a(1-\xi)\,G_c(AH_p) + b(1-\xi)\,G_c(BH_q) + c\xi\,G_c(CH_r) + d\xi\,G_c(DH_s) + h\xi\,G_c(H^+) \tag{23.32}$$

in which each $G_{c,i}$ is the instantaneous molar Gibbs free energy $G_{c,i}(\xi)$ of component i, with reference to the molar concentration as the standard state. As we noted at the start of this section, this reaction is not constrained at any specific pH, and so the quantity of hydrogen ions present in the reaction mixture can vary, as represented by the term $h\xi\,G_c(H^+)$, which increases with ξ as the reaction progresses.

For any component i in the ideal solution, at any instant

$$G_{c,i}(\xi) = G_{c,i}^{\ominus} + RT \ln \frac{[I(\xi)]}{[I]^{\ominus}} \tag{16.10b}$$

in which $[I(\xi)]$ is the instantaneous molar concentration of component i, and $[I]^{\ominus}$ is the corresponding standard molar concentration; as we saw in Chapter 22, for a real solution, the ratio $[I(\xi)]/[I]^{\ominus}$ is replaced by the molal activity $a_{b,i}(\xi)$, and $G_{c,i}^{\ominus}$ by the standard molal chemical potential $\mu_{i,b}^{\ominus}$.

Equation (16.10b) applies to all the reagents, noting that, for the hydrogen ion, $G_{c,i}^{\ominus}(H^+) = 0$ J mol⁻¹, and that for all components, including the hydrogen ion, $[I]^{\ominus} = 1$ mol dm⁻³. Accordingly, for the component AH_p

$$G_c(AH_p(\xi)) = G_c^{\ominus}(AH_p) + RT \ln[AH_p(\xi)]$$

At any instant during the reaction, as measured by the extent-of-reaction parameter ξ, the reaction mixture contains $a(1-\xi)$ mol of AH_p, and so, if the total volume of the solution is V dm³, the molar concentration $[AH_p(\xi)]$ mol dm⁻³ is

$$[AH_p(\xi)] = \frac{a(1-\xi)}{V}$$

giving

$$G_c(AH_p(\xi)) = G_c^{\ominus}(AH_p) + RT \ln \frac{a(1-\xi)}{V} \tag{23.33a}$$

Similarly

$$G_c(BH_q(\xi)) = G_c^{\ominus}(BH_q) + RT \ln \frac{b(1-\xi)}{V} \tag{23.33b}$$

$$G_c(CH_r(\xi)) = G_c^{\ominus}(CH_r) + RT \ln \frac{c\xi}{V} \tag{23.33c}$$

$$G_c(DH_s(\xi)) = G_c^{\ominus}(DH_s) + RT \ln \frac{d\xi}{V} \tag{23.33d}$$

$$G_c(H^+(\xi)) = G_c^{\ominus}(H^+) + RT \ln \frac{h\xi}{V} = RT \ln \frac{h\xi}{V} \tag{23.33e}$$

In equations (23.33), the various standard molar Gibbs free energies $G_{c,i}^{\ominus}$ are each constants, and so the only variable is the extent-of-reaction parameter ξ. Accordingly, equations (23.33) can therefore be used in equation (23.32) to determine $G_{sys}(\xi)$ explicitly as a function of ξ. On transforming to the biochemical standard, $G_{sys}'(\xi)$ for the same reaction may be written as

$$G_{sys}'(\xi) = a(1-\xi)\,G_c'(AH_p) + b(1-\xi)\,G_c'(BH_q)$$
$$+ c\xi\,G_c'(CH_r) + d\xi\,G_c'(DH_s) + h\xi\,G_c'(H^+) \tag{23.34}$$

in which, for each component i

$$G_{c,D}'(\xi) = G_{c,D}^{\ominus\prime} + RT \ln \frac{[I(\xi)]'}{[I]^{\ominus\prime}} \tag{23.5}$$

For the reactant AH_p, which contains p hydrogen atoms, we may draw on equation (23.22) to write

$$G_c^{\ominus\prime}(AH_p) = G_c^{\ominus}(AH_p) + 7p\,RT \ln 10$$

Also, the actual instantaneous molar concentration $[AH_p(\xi)]' = [AH_p(\xi)]$, and the transformed standard molar concentration $[AH_p]^{\ominus\prime} = 1$ mol dm^{-3}, implying that

$$G_c'(AH_p(\xi)) = G_c^{\ominus}(AH_p) + 7pRT \ln 10 + RT \ln[AH_p(\xi)]$$

As we have seen $[AH_p(\xi)] = a(1-\xi)/V$, and so

$$G_c'(AH_p(\xi)) = G_c^{\ominus}(AH_p) + 7pRT \ln 10 + RT \ln \frac{a(1-\xi)}{V} \tag{23.35a}$$

Similarly, for the reagents BH_q, CH_r and DH_s

$$G_c'(BH_q(\xi)) = G_c^{\ominus}(BH_q) + 7qRT \ln 10 + RT \ln \frac{b(1-\xi)}{V} \tag{23.35b}$$

$$G_c'(CH_r(\xi)) = G_c^{\ominus}(CH_r) + 7rRT \ln 10 + RT \ln \frac{c\xi}{V} \tag{23.35c}$$

$$G_c'(DH_s(\xi)) = G_c^{\ominus}(DH_s) + 7sRT \ln 10 + RT \ln \frac{d\xi}{V} \tag{23.35d}$$

For the hydrogen ions, however, we have

$$G_c'(H^+(\xi)) = G_c^{\ominus}(H^+) + RT \ln \frac{h\xi/V}{[H^+]^{\ominus\prime}}$$

in which $[H^+]^{\ominus\prime} = 10^{-7}$ mol dm^{-3}; also, as we saw on page 684, the reference level $G_c^{\ominus\prime}(H^+)$ is designated 0 J mol^{-1}, giving, in accordance with equation (23.9),

$$G_c'(H^+(\xi)) = 7\,RT \ln 10 + RT \ln(h\xi/V) \tag{23.35e}$$

Equations (23.35) may then be used in equation (23.34) to derive $G_{sys}'(\xi)$ as a function of ξ. Graphs showing the typical behaviour of both $G_{sys}'(\xi)$ and $G_{sys}(\xi)$ are shown in Figure 23.3: as will be seen, the curves shown in Figure 23.3 are very similar to those shown in Figures 14.1, 14.2, 14.3 and 16.6.

The distinctive feature of Figure 23.3 is that the shapes of the $G_{sys}(\xi)$ and $G_{sys}'(\xi)$ curves are the same, with the $G_{sys}'(\xi)$ curve directly above the $G_{sys}(\xi)$ curve. To determine the separation

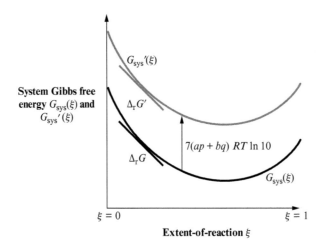

Figure 23.3 $G_{sys}(\xi)$ and $G_{sys}'(\xi)$ **for an unbuffered reaction**. For the unbuffered generalised reaction $aAH_p + bBH_q \rightleftharpoons cCH_r + dDH_s + hH^+$, taking place in an ideal aqueous solution, the lower, black, line shows how $G_{sys}(\xi)$, measured according to the conventional standards, varies with the extent-of-reaction parameter ξ. The upper, grey, line shows the corresponding values of $G_{sys}'(\xi)$ for the same reaction, measured according to the biochemical standard. As can be seen, the shapes of $G_{sys}'(\xi)$ and $G_{sys}(\xi)$ are identical, with $G_{sys}'(\xi)$ displaced above $G_{sys}(\xi)$ by a constant quantity $7(ap + bq) RT \ln 10$ J. Furthermore, for any value of ξ, the slope $dG_{sys}'(\xi)/d\xi$ of the $G_{sys}'(\xi)$ curve is equal to the slope $dG_{sys}(\xi)/d\xi$ of the $G_{sys}(\xi)$ curve, implying that $\Delta_r G' = \Delta_r G$.

between the curves, we may subtract equation (23.32) from equation (23.34) as

$$G_{sys}'(\xi) - G_{sys}(\xi) = a(1-\xi)[G_c'(AH_p) - G_c(AH_p)] + b(1-\xi)[G_c'(BH_q) - G_c(BH_q)]$$
$$+ c\xi[G_c'(CH_r) - G_c(CH_r)] + d\xi[G_c'(DH_s) - G_c(DH_s)] + h\xi[G_c'(H^+) - G_c(H^+)] \quad (23.36)$$

To simplify equation (23.36), we need to determine the difference $G_i' - G_i$ for each component i, and so, for the reagent AH_p, we may subtract equation (23.35a) from equation (23.33a) giving

$$G_c'(AH_p(\xi)) - G_c(AH_p(\xi)) = 7p\, RT \ln 10 \quad (23.37a)$$

Likewise, for the components BH_q, CH_r, DH_s and H^+

$$G_c'(BH_q(\xi)) - G_c(BH_q(\xi)) = 7q\, RT \ln 10 \quad (23.37b)$$

$$G_c'(CH_r(\xi)) - G_c(CH_r(\xi)) = 7r\, RT \ln 10 \quad (23.37c)$$

$$G_c'(DH_s(\xi)) - G_c(DH_s(\xi)) = 7s\, RT \ln 10 \quad (23.37d)$$

$$G_c'(H^+(\xi)) - G_c(H^+(\xi)) = 7\, RT \ln 10 \quad (23.37e)$$

We may now use the five equations (23.37) to substitute for each $G_{c,i}' - G_{c,i}$ in equation (23.36), giving

$$G_{sys}'(\xi) - G_{sys}(\xi) = a(1-\xi)\, 7p\, RT \ln 10 + b(1-\xi)\, 7q\, RT \ln 10 + c\xi\, 7r\, RT \ln 10$$
$$+ d\xi\, 7s\, RT \ln 10 + h\xi\, 7\, RT \ln 10$$

which simplifies to

$$G_{sys}'(\xi) - G_{sys}(\xi) = 7(ap + bq)RT\ln 10 + 7\xi[(cr + ds + h) - (ap + bq)]RT\ln 10 \quad (23.38)$$

But for the reaction

$$aAH_p + bBH_q \rightleftharpoons cCH_r + dDH_s + hH^+ \quad (23.23)$$

the balance of hydrogen atoms requires that

$$ap + bq = cr + ds + h$$

and so

$$[(cr + ds + h) - (ap + bq)] = 0$$

implying that equation (23.38) simplifies to

$$G_{sys}'(\xi) - G_{sys}(\xi) = 7(ap + bq)RT\ln 10 \quad (23.39)$$

At any temperature T, the term $7(ap + bq)RT\ln 10$ J for any given reaction is not only a constant, but also a constant determined solely by the reactants AH_p and BH_q. In particular, this quantity is independent of the products CH_r and DH_s, and is also independent of hydrogen H^+ (aq) ions, whether products resulting from the reaction, or reagents consumed by the reaction. Equation (23.39) therefore implies that a graph of $G_{sys}'(\xi)$ is displaced by a constant amount $7(ap + bq)RT\ln 10$ J above the corresponding graph of $G_{sys}(\xi)$, as illustrated in Figure 23.3.

A further implication of equation (23.39), and a consequence of the fact that, at any temperature T, the term $7(ap + bq)RT\ln 10$ is a constant, is that

$$\frac{dG_{sys}'}{d\xi} - \frac{dG_{sys}}{d\xi} = 0$$

implying that the slopes $dG_{sys}'(\xi)/d\xi$ and $dG_{sys}(\xi)/d\xi$ are equal for all values of ξ. As we saw on page 434, the slope $dG_{sys}(\xi)/d\xi$ of the $G_{sys}(\xi)$ curve is equal to the reaction Gibbs free energy change $\Delta_r G$ and so

$$\Delta_r G = \Delta_r G' \quad (23.40)$$

Furthermore, for a reaction taking place in an ideal solution, $\Delta_r G$ is given by equation (16.7c) and (16.18c)

$$\Delta_r G = -RT\ln K_c + RT\ln \Gamma_c = -RT\ln \frac{\Gamma_c}{K_c} \quad (16.17c)$$

where Γ_c is the concentration mass action ratio, which for the reaction $aAH_p + bBH_q \rightleftharpoons cCH_r + dDH_s + hH^+$, is given by

$$\Gamma_c = \frac{[CH_r]^c [DH_s]^d [H^+]^h}{[AH_p]^a [BH_q]^b} \quad (16.15c)$$

Equation (16.17c) transforms as

$$\Delta_r G' = -RT \ln K_c' + RT \ln \Gamma_c' = -RT \ln \frac{\Gamma_c'}{K_c'} \tag{23.41}$$

and since $\Delta_r G = \Delta_r G'$, equations (16.17c) and (23.41) therefore give

$$-RT \ln \frac{\Gamma_c}{K_c} = -RT \ln \frac{\Gamma_c'}{K_c'}$$

from which

$$\frac{\Gamma_c}{K_c} = \frac{\Gamma_c'}{K_c'}$$

and

$$\frac{K_c'}{K_c} = \frac{\Gamma_c'}{\Gamma_c} \tag{23.42}$$

Now, according to equation (23.31b)

$$K_c' = 10^{7h} K_c \tag{23.31b}$$

and so equation (23.42) implies that

$$\Gamma_c' = 10^{7h} \Gamma_c \tag{23.43}$$

Equation (23.43) is the relationship between the mass action ratio Γ_c' as measured using the biochemical standard, and Γ_c as measured using the conventional standards. In fact, this relationship can be demonstrated from the definition of Γ_c

$$\Gamma_c = \frac{[CH_r]^c \, [DH_s]^d \, [H^+]^h}{[AH_p]^a \, [BH_q]^b} \tag{16.15c}$$

which, more explicitly, is written as

$$\Gamma_c = \frac{([CH_r]/[CH_r]^{\ominus})^c \, ([DH_s]/[DH_s]^{\ominus})^d \, ([H^+]/[H^+]^{\ominus})^h}{([AH_p]/[AH_p]^{\ominus})^a \, ([BH_q]/[BH_q]^{\ominus})^b}$$

and transforms as

$$\Gamma_c' = \frac{([CH_r]'/[CH_r]^{\ominus})^c \, ([DH_s]'/[DH_s]^{\ominus})^d \, ([H^+]'/[H^+]^{\ominus})^h}{([AH_p]'/[AH_p]^{\ominus})^a \, ([BH_q]'/[BH_q]^{\ominus})^b} \tag{23.44}$$

As we know, for any component i other than the hydrogen ion H^+(aq), $[I] = [I]'$ and $[I]^{\ominus} = [I]^{\ominus\prime} = 1$ mol dm^{-3}; whilst for the hydrogen ion, $[H^+] = [H^+]'$, $[H^+]^{\ominus} = 1$ mol dm^{-3}, and $[H^+]^{\ominus\prime} = 10^{-7}$ mol dm^{-3}. Accordingly

$$\Gamma_c' = \frac{([CH_r]/1)^c \, ([DH_s]/1)^d \, ([H^+]/10^{-7})^h}{([AH_p]/1)^a \, ([BH_q]/1)^b} = \frac{[CH_r]^c \, [DH_s]^d \, [H^+]^h \, 10^{7h}}{[AH_p]^a \, [BH_q]^b}$$

and so

$$\Gamma_c' = 10^{7h} \Gamma_c \tag{23.43}$$

as expected.

One final point concerning Figure 23.3. As can be seen, both the $G_{sys}(\xi)$ curve, and the $G_{sys}'(\xi)$ curve, have minimum values at the same value of the extent-of-reaction parameter ξ. Since the minimum represents the reaction equilibrium state, the result that $\xi_{eq} = \xi_{eq}'$ implies that the equilibrium composition is the same for both $G_{sys}(\xi)$ and $G_{sys}'(\xi)$. Equation (23.31b)

$$K_c' = 10^{7h} K_c \tag{23.31b}$$

however, tells is that K_c' is a factor of 10^{7h} times greater than K_c, which implies that the equilibrium composition of a reaction under the biochemical standard is profoundly different from the equilibrium composition under the conventional standard. The equality of ξ_{eq} and ξ_{eq}', as shown by the position of the minima in Figure 23.3, and the implied equality of the corresponding equilibrium constants, seems to contradict equation (23.31b), and this can create significant confusion.

In fact, both Figure 23.3, and equation (23.31b) are correct: the possible confusion arises from the fact that Figure 23.3 and equation (23.31b) are describing different, but easily muddled, concepts.

Figure 23.3 represents the behaviour, as a function of the extent-of-reaction parameter ξ, of $G_{sys}(\xi)$ and $G_{sys}'(\xi)$, these being measures of the Gibbs free energy of the same system, as defined by the general chemical reaction

$$aAH_p + bBH_q \rightleftharpoons cCH_r + dDH_s + hH^+ \tag{23.23}$$

Because the reference levels associated with the measures of $G_{sys}(\xi)$ and $G_{sys}'(\xi)$ are different, the instantaneous values of $G_{sys}(\xi)$ and $G_{sys}'(\xi)$ are different, as shown by the fact that the curves shown in Figure 23.3 are at different positions on the vertical axis. The displacement between the two curves, however, is constant, this being attributable to the fact that the reference level shift between the two standards is constant, and independent of ξ.

Most importantly, however, the equilibrium composition of reaction (23.23) is a property of the reaction itself, and not of the measurement scales, and so the extent-of-reaction parameter ξ_{eq} that defines the equilibrium composition must have the same value, regardless of the standard, and the corresponding reference levels, used to measure the system's Gibbs free energy. As a consequence, although the values of $G_{sys}(\xi)$ and $G_{sys}'(\xi)$ are different, the minima of $G_{sys}(\xi)$ and $G_{sys}'(\xi)$ correspond to the same value of $\xi = \xi_{eq} = \xi_{eq}'$.

Whereas the graphs of $G_{sys}(\xi)$ and $G_{sys}'(\xi)$, as represented in Figure 23.3, refer to the *same* chemical reaction but measured according to two different scales, K_c and K_c' define the concentration equilibrium constants of *two different reactions* – or rather, the same reaction, but *taking place under two, very different, conditions*. That may not be obvious, but we have referred, on several occasions, to the fact that K_c represents the concentration equilibrium constant of reaction (23.23) buffered at pH 0, but K_c' refers to the same reaction buffered at pH 7. As we stated at the start of this section, the analysis represented by the $G_{sys}(\xi)$ and $G_{sys}'(\xi)$ curves shown in Figure 23.3 relates to an unconstrained, and therefore unbuffered, reaction, during which the pH changes as the reaction proceeds. K_c and K_c', however, describe the equilibrium compositions of a reaction in which the pH is constrained to be constant throughout the reaction, at pH 0 (for K_c), and pH 7 (for K_c'). In the context of Figure 23.3, the reaction pH is unconstrained, and so K_c and K_c' have no particular significance. Figure 23.3 and equation (23.31b) are therefore not in conflict. Given, however, the importance of buffered reactions, and of K_c', to biochemistry, the next section will explore the behaviours of $G_{sys}(\xi)$ and $G_{sys}'(\xi)$ for a buffered reaction.

23.14 G_{sys} and G_{sys}' for a buffered reaction

As we have just discussed, the previous section examined the behaviour of $G_{sys}(\xi)$ and $G_{sys}'(\xi)$ for the generalised reaction

$$a\,AH_p + b\,BH_q \rightleftharpoons c\,CH_r + d\,DH_s + h\,H^+ \tag{23.23}$$

taking place in the absence of any constraints, so that each of the components i in the reaction mixture, including the hydrogen ion $H^+(aq)$, is able to change its molar concentration [I] mol dm^{-3} as the reaction progresses, in accordance with the reaction stoichiometry. If, however, the reaction is buffered at a particular pH, the molar concentration [H$^+$] mol dm^{-3} is constrained to remain at [H$^+$] = 10^{-pH} = [H$^+$]$_{eq}$ mol dm^{-3} throughout the reaction, and, as we saw on pages 561 to 566, this influences the molar concentrations of all the other components, and also the composition of the equilibrium mixture.

Under the conventional standards, When the reaction is unconstrained, the concentration equilibrium constant, K_c is given by

$$K_c = \frac{([C]_{eq}/[C]^{\ominus})^c \, ([D]_{eq}/[D]^{\ominus})^d \, ([H^+]_{eq}/[H^+]^{\ominus})^h}{([A]_{eq}/[A]^{\ominus})^a \, ([B]_{eq}/[B]^{\ominus})^b} \tag{16.12c}$$

As we saw on page 563, if we define the parameter Λ as

$$\Lambda = \frac{([C]_{eq}/[C]^{\ominus})^c \, ([D]_{eq}/[D]^{\ominus})^d}{([A]_{eq}/[A]^{\ominus})^a \, ([B]_{eq}/[B]^{\ominus})^b} \tag{23.45}$$

then, under the conventional standards

$$K_c = \Lambda \left(\frac{[H^+]_{eq}}{[H^+]^{\ominus}} \right)^h \tag{23.46}$$

which transforms to the biochemical standard as

$$K_c' = \Lambda' \left(\frac{[H^+]_{eq}'}{[H^+]^{\ominus\prime}} \right)^h$$

Since, according to equation (23.47), Λ is defined in terms of components other than the hydrogen ion, $\Lambda' = \Lambda$; furthermore, [H$^+$]$_{eq}$ is an actual measurement, and so must equal [H$^+$]$_{eq}'$, implying that

$$K_c' = \Lambda \left(\frac{[H^+]_{eq}}{[H^+]^{\ominus\prime}} \right)^h \tag{23.47}$$

Under the conventional standards, [H$^+$]$^{\ominus}$ = 1 mol dm^{-3}, and so, from equation (23.46)

$$K_c = \Lambda(pH) \, [H^+]_{eq}^h \tag{23.48}$$

In this expression, since K_c is a constant, the parameter Λ must vary with [H$^+$]$_{eq}$, and therefore with the corresponding pH at which the reaction is buffered. The notation $\Lambda(pH)$ therefore recognises this, and represents the value of Λ at the pH corresponding to a given value of $[H^+]_{eq}$. As a particular case, if $[H^+]_{eq} = 1$ mol dm^{-3}, then $\Lambda(\text{pH } 0) = K_c$, so verifying that the value of K_c under the conventional standards corresponds to the reaction buffered at pH 0.

Similarly, under the biochemical standard, $[H^+]^{\ominus\prime} = 10^{-7}$ mol dm^{-3}, and so equation (23.47) becomes

$$K_c' = \Lambda(\text{pH}) \, 10^{7h} \, [H^+]_{eq}^{h} \tag{23.49}$$

If $[H^+]_{eq} = 10^{-7}$ mol dm^{-3}, $K_c' = \Lambda(\text{pH } 7)$, verifying that the transformed concentration equilibrium constant K_c' represents the equilibrium constant of the reaction buffered at pH 7; furthermore, since, according to equation (23.48), at any pH

$$K_c = \Lambda(\text{pH}) \, [H^+]_{eq}^{h} \tag{23.48}$$

implying, from equation (23.49), that

$$K_c' = 10^{7h} \, K_c \tag{23.31b}$$

as is consistent with equation (23.31b), as discussed on page 696.

Whilst the reaction

$$a\text{AH}_p + b\text{BH}_q \rightleftharpoons c\text{CH}_r + d\text{DH}_s + h\text{H}^+ \tag{23.23}$$

is in progress in a buffered solution, H$^+$(aq) ions are being produced in accordance with the stoichiometry of reaction (23.23), but rather than accumulating – as happens in an unbuffered solution – these H$^+$(aq) ions are consumed by the buffer, so 'pulling' the reaction as written to the right. This therefore predicts that, when hydrogen ions are a product, $K_c' > K_c$, as verified by equation (23.31b). Similarly, for a reaction of the general form

$$a\text{AH}_p + b\text{BH}_q + h\text{H}^+ \rightleftharpoons c\text{CH}_r + d\text{DH}_s$$

in which hydrogen ions are consumed, the parameter h in equation (23.31b) is negative, and $K_c' < K_c$ – as explained by the action of the buffer in 'pulling' the reaction as written to the left.

We will now derive an expression for the total Gibbs free energy $G_{sys}(\xi)$ associated with the general reaction

$$a\text{AH}_p + b\text{BH}_q \rightleftharpoons c\text{CH}_r + d\text{DH}_s + h\text{H}^+ \tag{23.23}$$

buffered at pH 0, as expressed by equation (23.32)

$$G_{sys}(\xi) = a(1-\xi)\,\mathbf{G}_c(\text{AH}_p) + b(1-\xi)\,\mathbf{G}_c(\text{BH}_q) + c\xi\,\mathbf{G}_c(\text{CH}_r) + d\xi\,\mathbf{G}_c(\text{DH}_s) + h\xi\,\mathbf{G}_c(\text{H}^+) \tag{23.32}$$

For the hydrogen ion, the last term may be written as

$$h\xi\,\mathbf{G}_c(\text{H}^+) = h\xi\,(\mathbf{G}_c^{\ominus}(\text{H}^+) + hRT\ln([H^+]/[H^+]^{\ominus}))$$

but since $\mathbf{G}_c^{\ominus}(\text{H}^+) = 0$ J mol^{-1}, and the reaction is buffered so that $[H^+] = 1$ mol dm^{-3} $= [H^+]^{\ominus}$ throughout the reaction, $h\xi\,\mathbf{G}_c(\text{H}^+) = 0$ J mol^{-1} for all values of ξ. Writing $G_{sys}(\xi)$ as $G_{sys}(\xi, \text{pH } 0)$, equation (23.32) therefore becomes

$$G_{sys}(\xi, \text{pH } 0) = a(1-\xi)\,\mathbf{G}_c(\text{AH}_p) + b(1-\xi)\,\mathbf{G}_c(\text{BH}_q) + c\xi\,\mathbf{G}_c(\text{CH}_r) + d\xi\,\mathbf{G}_c(\text{DH}_s) \tag{23.50}$$

THE BIOCHEMICAL STANDARD STATE

Similarly, in equation (23.34)

$$G_{sys}'(\xi) = a(1-\xi)\,G_c'(AH_p) + b(1-\xi)\,G_c'(BH_q) + c\xi\,G_c'(CH_r) + d\xi\,G_c'(DH_s) + h\xi\,G_c'(H^+) \quad (23.34)$$

for a reaction buffered at pH 7, $[H^+] = [H^+]^{\ominus\prime} = 10^{-7}$ mol dm^{-3}, and since $G_c^{\ominus\prime}[H^+] = 0$ (see page 684), the last term simplifies to

$$h\xi\,G_c'(H^+) = h\xi\,(G_c^{\ominus\prime}(H^+) + hRT\ln([H^+]/[H^+]^{\ominus\prime}) = 0$$

Writing $G_{sys}'(\xi)$ as $G_{sys}(\xi, pH\,7)$, equation (23.34) therefore reduces to four terms as

$$G_{sys}(\xi, pH\,7) = a(1-\xi)\,G_c'(AH_p) + b(1-\xi)\,G_c'(BH_q) + c\xi\,G_c'(CH_r) + d\xi\,G_c'(DH_s) \quad (23.51)$$

We may now use equations (23.33) and (23.35) in equations (23.50) and (23.51), giving the results shown in Figure 24.4.

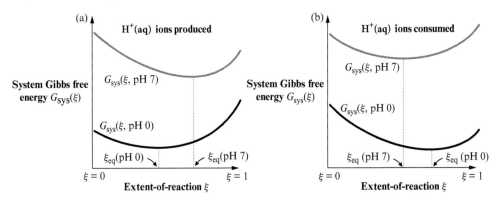

Figure 23.4 $G_{sys}(\xi, pH\,0)$ and $G_{sys}(\xi, pH\,7)$. For a reaction in which hydrogen ions are produced, (a) shows a schematic representation of the behaviour of the system Gibbs free energy $G_{sys}(\xi, pH\,0)$ when the reaction is buffered at pH 0, and of $G_{sys}(\xi, pH\,7)$ when the reaction is buffered at pH 7. The equilibrium extent-of-reaction $\xi_{eq}(pH\,7)$ is to the right of the value $\xi_{eq}(pH\,0)$, implying that $K_c' > K_c$. For a reaction in which hydrogen ions are consumed, $\xi_{eq}(pH\,7) < \xi_{eq}(pH\,0)$ and $K_c' < K_c$, as shown in (b). These diagrams are schematic in that, in reality, the difference in the shapes of $G_{sys}(\xi, pH\,0)$ and $G_{sys}(\xi, pH\,7)$ is much more pronounced: if H$^+$(aq) ions are produced as in (a), the real behaviour of $G_{sys}(\xi, pH\,7)$ is a line sloping downward to the right (as shown in Figure (14.1a)), with the reaction going to completion such that $\xi_{eq}(pH\,7) \approx 1$; if H$^+$(aq) ions are consumed as in (b), the real behaviour of $G_{sys}(\xi, pH\,7)$ is a line sloping upward to the right (as shown in Figure (14.1b)), with the equilibrium position to the far left, such that $\xi_{eq}(pH\,7) \approx 0$.

Figure 24.4(a) corresponds to a reaction in which hydrogen ions are produced, and Figure 24.4(b) to a reaction in which hydrogen ions are consumed; as can be seen, the graphs shown in Figure 23.4 are quite different from those shown in Figure 23.3.

In general, as we saw on page 700, for an unbuffered reaction, the $G_{sys}'(\xi)$ curve is positioned, in its entirety, directly above the $G_{sys}(\xi)$ curve, such that, at any temperature, the gap between the two curves is the constant quantity $7\,(ap + bq)\,RT \ln 10$ J. For a buffered reaction, however, as shown in Figure 23.4 and as verified by exercises 6 and 8, the $G_{sys}'(\xi) = G_{sys}(\xi, pH\,7)$ curve is both above the $G_{sys}(\xi) = G_{sys}(\xi, pH\,0)$ curve and also displaced to the right (for a reaction in which hydrogen ions are produced), or to the left (for a reaction in which hydrogen ions are consumed). As can be seen in Figure 23.4(a), a rightwards displacement of the $G_{sys}'(\xi) = G_{sys}(\xi, pH\,7)$ curve relative to the $G_{sys}(\xi) = G_{sys}(\xi, pH\,0)$ curve implies that the

WATER AS A BIOCHEMICAL REAGENT

transformed equilibrium extent-of-reaction parameter $\xi_{eq}' = \xi_{eq}(pH\ 7)$ is greater than the conventional parameter $\xi_{eq} = \xi_{eq}(pH\ 0)$; accordingly, the equilibrium composition of the reaction mixture at pH 7 is richer in products, rather than reactants, as compared to the equilibrium composition of the reaction mixture at pH 0. At pH 0, however, the equilibrium composition is also defined by K_c, whereas at pH 7, the equilibrium composition is defined by K_c'. Figure 23.4(a) therefore verifies that $K_c' > K_c$, as is consistent with equation (23.31b). Similarly, for a reaction in which hydrogen ions are consumed, as illustrated in Figure 23.4(b), $\xi_{eq}' = \xi_{eq}(pH\ 7) < \xi_{eq} = \xi_{eq}(pH\ 0)$, implying that the equilibrium mixture at pH 7 is richer in reactants, rather than products, as compared to the equilibrium mixture at pH 0. As a consequence, $K_c' < K_c$ – once again as is consistent with equation (23.31b) when h is negative.

23.15 Water as a biochemical reagent

The discussion so far in this chapter has assumed that all the components of the generalised reaction

$$aA + bB \rightleftharpoons cC + hH^+(aq)$$

are solutes in aqueous solution. However, for a reaction such as the hydrolysis of ATP represented as

$$ATP^{4-}(aq) + H_2O(l) \rightleftharpoons ADP^{3-}(aq) + HPO_4^{2-}(aq) + H^+(aq)$$

molecular water is a reactant, as well as the solvent.

So, for a generalised reaction of the form

$$aA + wH_2O \rightleftharpoons cC + hH^+(aq)$$

in which hydrogen ions are a product, and molecular water is a reactant as well as being the solvent, as we saw on page 524, we may express the equilibrium constant as a 'hybrid', K_r, as

$$K_r = \frac{([C]_{eq}/[C]^\ominus)^c\ ([H^+]_{eq}/[H^+]^\ominus)^h}{([A]_{eq}/[A]^\ominus)^a\ (x_{eq,H_2O}/x_{H_2O}^\ominus)^w} \quad (23.52)$$

in which the measure of quantity for water is the mole fraction. Since water, as the solvent, is present in vast excess, then, to very good approximation, $x_{eq,H_2O} = 1$; also, from the definition of the mole fraction standard state, $x_{H_2O}^\ominus = 1$, and so

$$K_r = \frac{([C]_{eq}/[C]^\ominus)^c\ ([H^+]_{eq}/[H^+]^\ominus)^h}{([A]_{eq}/[A]^\ominus)^a} = \frac{[C]_{eq}^c\ [H^+]_{eq}^h}{[A]_{eq}^a} \quad (23.53)$$

in which there is no explicit term for molecular water, even though water is a reagent.

On transforming equation (23.52) to the biochemical standard,

$$K_r' = \frac{([C]_{eq}'/[C]^{\ominus\prime})^c\ ([H^+]_{eq}'/[H^+]^{\ominus\prime})^h}{([A]_{eq}'/[A]^{\ominus\prime})^a\ (x_{eq,H_2O}'/x_{H_2O}^{\ominus\prime})^w} \quad (23.54)$$

All actual measurements are the same in both standards, and so $[A]_{eq}' = [A]_{eq}$, $[B]_{eq}' = [B]_{eq}$, $[H^+]_{eq}' = [H^+]_{eq}$, and $x_{eq,H_2O}' = x_{eq,H_2O} = 1$; also $[A]^{\ominus\prime} = [B]^{\ominus\prime} = 1$ mol dm^{-3}, $x_{H_2O}^\ominus = 1$, and

$[H^+]^{\ominus\prime} = 10^{-7}$ mol dm^{-3}. Equation (23.54) therefore becomes

$$K_r' = \frac{[C]_{eq}^c \, ([H^+]_{eq}/10^{-7})^h}{[A]_{eq}^a} = 10^{7h}\frac{[C]_{eq}^c \, [H^+]_{eq}^h}{[A]_{eq}^a} = 10^{7h} K_r \qquad (23.55)$$

In this expression for K_r', there is no term for the water; also equation (23.55) is in agreement with equation (23.31a).

23.16 The rules for converting into the biochemical standard

In summary, here are the 'rules' for converting a thermodynamic equation from the conventional standard to the biochemical standard:

- To compute a value for $\Delta_f G^{\ominus\prime}$ at 298.15 K for a molecule containing h hydrogen atoms, use the formula

$$\Delta_f G^{\ominus\prime} = \Delta_f G^{\ominus} + h \times 39.956 \qquad \text{kJ mol}^{-1} \qquad (23.22)$$

- To convert a thermodynamic equation, write it down in terms of the conventional standard.
- Ensure that all molar concentrations are expressed as ratios $[I]/[I]^{\ominus}$, and not as independent items $[I]$, and, similarly, that all molalities are expressed as ratios $b_i/b_i^{\ominus}$.
- Transform the equation, so that every variable X, as expressed in the conventional standard, becomes X', as expressed in the biochemical standard. Note that no changes are required for the temperature T, and constants, such as the universal gas constant R, and the mole number n.
- For all variables X, except the standard molar concentration $[H^+]^{\ominus\prime}$, and the standard molality $b_{H+}^{\ominus\prime}$, of the hydrogen ion $H^+(aq)$, set $X' = X$.
- For hydrogen ions only, set the standard molar concentration $[H^+]^{\ominus\prime} = 10^{-7}$ mol dm^{-3}, so that the ratio $[H^+]'/[H^+]^{\ominus\prime} = [H^+]/10^{-7} = 10^7 [H^+]$. Likewise, set the standard molality $b_{H+}^{\ominus\prime} = 10^{-7}$ mol kg^{-1}, so that $b_{H+}/b_{H+}^{\ominus\prime} = 10^7 b_{H+}$.
- If molecular water is a reactant, as either a reagent or a product, then $x_{H_2O}/x_{H_2O}^{\ominus} = x_{H_2O}'/x_{H_2O}^{\ominus\prime} = 1$.
- For real solutions, follow these rules, replacing the molar Gibbs free energy G by the chemical potential μ, and the molar concentration ratio $[I]/[I]^{\ominus}$, or molality ratio $b_i/b_i^{\ominus}$, by the molal activity a_b.

EXERCISES

1. Define the biochemical standard state.
2. Why is the biochemical standard state a useful concept?

3. With reference to Table 23.2, some values of $\Delta_f G^\circ$ and $\Delta_f G^{\circ\prime}$ are the same, and some are different. Why? For those molecules for which $\Delta_f G^\circ$ and $\Delta_f G^{\circ\prime}$ are different, interpret the difference $\Delta_f G^\circ - \Delta_f G^{\circ\prime}$ in terms of the corresponding molecular structure.

4. Question 3 of the exercises to Chapter 20 states
   Two biochemical redox half-reactions, which are important in respiration, are

   $NAD^+ + 2\,e^- + H^+ \rightleftharpoons NADH$ $\qquad E^\circ(NAD^+, NADH) = 0.094$ V

   $FAD + 2\,e^- + 2\,H^+ \rightleftharpoons FADH_2$ $\qquad E^\circ(FAD, FADH_2) = 0.195$ V

   Why are these numbers different from those in Table 23.2? Which numbers are correct? Justify your answers.

5. Suppose that a new branch of chemistry emerges focused on pH 3, and that the international community agree that it would be useful to define the 'acid standard' accordingly. For the generalized reversible reaction

   $aA + bB \rightleftharpoons cC + hH^+(aq)$

   and starting from the conventional definition of the concentration equilibrium constant K_c, show that

   $K_c'' = 10^{-3h}\, K_c$

   and that

   $\Delta_r G^{\circ\prime\prime} = \Delta_r G^\circ - 3h\, RT \ln 10$

   where the superscript " designates measurements according to the 'acid standard'. Show further that the reference level shift for the measurement of reversible electrode potentials is -0.177 V. Accordingly, what are the values of $E^{\circ\prime\prime}$ (NAD$^+$, NADH) and $E^{\circ\prime\prime}$ (FAD, FADH$_2$)?

6. This is a spreadsheet exercise to reproduce the graphs shown in Figures 23.3 and 23.4 from equations (23.32) and (23.34). As can be inferred from question 5, all of factors of 10^{-7} and $7\,RT \ln 10$ that appear in the equations relating to the biochemical state originate from the setting of the biochemical standard in terms of $[H^+] = 10^{-7}$ mol dm^{-3}, pH 7. These factors are therefore more generally written as $10^{-\text{pH}}$ and pH $RT \ln 10$. Accordingly, for a general reaction of the form

   $aAH_p + bBH_q \rightleftharpoons cCH_r + hH^+$

   (in which $d = 0$, and so there is no component DH$_s$), express equations (23.32) and (23.34) in terms of a general pH, and then write full equations for each $G_{c,i}$ and $G_{c,i}'$, resulting in equations for $G_{\text{sys}}(\xi)$ and $G_{\text{sys}}'(\xi)$ as functions of ξ.

   Towards the top left of the spreadsheet, define cells for R and T, and also for the generalised pH parameter, which equals 7 for the biochemical standard state. Then define a column of 101 rows for the extent-of-reaction parameter ξ, corresponding to values from $\xi = 0$ to $\xi = 1$ in increments of 0.01, with the adjacent column for values of $1 - \xi$. You now need to define the stoichiometry of reaction (23.23), as for example

   $4\,AH + 2\,BH_3 \rightleftharpoons CH_5 + 5\,H^+$

Charge balance does not matter, but balancing the number of hydrogen atoms does. You also need to define the values of $G_{c,i}^\circledast$ for each component, for example $G_c^\circledast(AH) = -8{,}000$ kJ mol^{-1}, $G_c^\circledast(BH_3) = -500$ kJ mol^{-1}, $G_c^\circledast(CH_5) = -10{,}000$ kJ mol^{-1}, and $G_c^\circledast(H^+) = 0$ kJ mol^{-1}. Note that the reaction is not intended to represent any specific reaction, nor do the values of $G_{c,i}^\circledast$ have any particular meaning, except that $G_c^\circledast(H^+) = 0$ kJ mol^{-1} complies with the conventional standards.

You now need to calculate $G_{c,i}(\xi)$ for each component, for each value of ξ from 0 to 1. To do this, you may wish to have a different column for each of the terms in the full expressions for each $G_{c,i}(\xi)$, with a column for $G_{c,i}(\xi)$ as the sum each of these terms. Once $G_{c,i}(\xi)$ has been determined for each of the four components, $G_{sys}(\xi)$ can then be computed as the sum.

Similarly, columns can be used to compute each $G_{c,i}'(\xi)$, and the corresponding sum $G_{sys}'(\xi)$. The resulting graphs of $G_{sys}(\xi)$ and $G_{sys}'(\xi)$ should have the properties shown in Figure 23.3, with $G_{sys}'(\xi)$ positioned a distance pH $(ap + bq) RT \ln 10$ directly above $G_{sys}(\xi)$ – the graphs actually shown in Figure 23.3 correspond to pH $= 0.30$.

To reproduce Figure 23.4, use equations (23.50) and (23.51), all the terms of which are already computed within the spreadsheet.

With this spreadsheet, explore what happens when
> you vary the parameter pH
> you vary the stoichiometry of the reaction, including reactions in which H$^+$(aq) ions are consumed, rather than produced
> you vary the standard molar Gibbs free energies $G_c^\circledast(AH)$, $G_c^\circledast(BH_3)$ and $G_c^\circledast(CH_5)$ – but not $G_c^\circledast(H^+)$.

As these numbers change, the graphs will move around: make sure you understand why the graphs are moving, and how the movements can be interpreted in terms of variables such as ξ_{eq}, ξ_{eq}', K_c, K_c', $\Delta_r G^\circledast$ and $\Delta_r G^{\circledast\prime}$.

7. For the unbuffered reaction

$$cCH_r + dDH_s + hH^+ \rightleftharpoons aAH_p + bBH_q$$

which is the reverse of reaction (23.23), and using a method similar to that shown on pages 697 to 701, derive expressions for $G_{sys}(\xi)$ and $G_{sys}'(\xi)$, and for the difference $G_{sys}(\xi) - G_{sys}'(\xi)$. Is the result the same as equation (23.39)? If it is, why? And if it isn't, why not? Explain why, for any unbuffered reaction, $G_{sys}(\xi) - G_{sys}'(\xi)$ is a constant, independent of ξ, and determined only by the reactants, or the products, alone.

8. With reference to equations 23.50 and 23.51, show that the difference between the curves shown in Figure 23.4 may be expressed as

$$G_{sys}(\xi, \text{pH 7}) - G_{sys}(\xi, \text{pH 0}) = 7 RT \ln 10 \, (ap + bq - h\xi)$$

How does this vary with ξ? What happens when $h = 0$? How would you explain this result? What is the corresponding expression for the generalised difference $G(\xi, \text{pH } x) - G(\xi, \text{pH 0})$? Verify these results by plotting graphs of $G_{sys}(\xi) - G_{sys}'(\xi)$ using the data generated in the last part of exercise 6. Also, show that the difference $dG_{sys}(\xi, \text{pH 7})/d\xi - dG_{sys}(\xi, \text{pH 0})/d\xi = -7hRT \ln 10$, and therefore that, for a buffered reaction that produces hydrogen ions, $\xi_{eq}(\text{pH 7}) > \xi_{eq}(\text{pH 0})$. What is the relationship between $\xi_{eq}(\text{pH 7})$ and $\xi_{eq}(\text{pH 0})$ for a buffered reaction that consumes hydrogen ions? Under what conditions is $\xi_{eq}(\text{pH 7})$ equal to $\xi_{eq}(\text{pH 0})$? Interpret these results in physical terms.

24 The bioenergetics of living cells

Summary

All biological systems create, and maintain, high degrees of order: proteins are more ordered than the corresponding amino acids, and organisms are more ordered than their constituent components. For this to happen, an endergonic reaction ($\Delta G_{end} > 0$) of synthesis must be coupled to an appropriately exergonic reaction ($\Delta G_{ex} < 0$) such that $\Delta G_{ex} + \Delta G_{end} < 0$. The study of coupled energy flows in biological systems is known as **bioenergetics**.

The free energy released by biological exergonic reactions can in general be captured either by the synthesis of molecules such as ATP and NADH (which are known as **energy currencies**), or by the creation of a **chemi-osmotic potential gradient** across a **biological membrane**. The ATP and NADH produced can then be used as substrates in the endergonic reactions of synthesis, whilst the chemi-osmotic potential gradient can be used to synthesise ATP from ADP and inorganic phosphate.

One of the most important sources of free energy – especially for animals – is glucose, which, if burned freely in oxygen releases 2930 kJ of free energy per mol. In biological systems, much of this free energy is captured as ATP in a series of metabolic pathways known as **glycolysis**, the **TCA cycle** and **oxidative phosphorylation**. Glucose itself is synthesised from atmospheric CO_2 in green plants in the process known as **photosynthesis**, in which energy from light is captured to synthesise ATP, which is then used to form glucose.

24.1 Creating order without breaking the Second Law: open systems

Living systems, such as cells, can appear to create order out of chaos. They build themselves into well-organised structures at all levels from the cell organelle to the organism, they carry out many complex functions, they replicate. Living cells, of course, do not break the Second Law of Thermodynamics – no system, living or non-living, can. Rather, living cells are thermodynamically open systems, allowing the transfer of materials and energy into and out of the cellular system. As a result, living cells can 'harvest' free energy from outside the system, free energy that is available as chemical, redox, electromagnetic, or thermal potentials, as well as free energy derived from various sources including geothermal heat, sunlight, organic molecules such as sugars made by plant cells, and even inorganic molecules found within rocks. This energy is then stored within cells in various forms, and subsequently utilised in a controlled manner within interlinked sequences of enzyme-catalysed reactions known as

Modern Thermodynamics for Chemists and Biochemists. Dennis Sherwood and Paul Dalby.
© Oxford University Press 2018. Published 2018 by Oxford University Press.

metabolic pathways. As we saw on pages 526 to 530, if an exergonic ($\Delta G_{ex} < 0$) spontaneous reaction that releases free energy can be coupled to a non-spontaneous endergonic ($\Delta G_{end} > 0$) reaction that requires free energy, then, if $\Delta G_{ex} + \Delta G_{end} < 0$, these two reactions, taking place simultaneously, can create local order without breaking the Second Law. Living cells do this in highly complex, and enormously successful, ways, as evidenced by the richness of the life we see all around us. Overall, the physical and chemical processes that take place within living things are all about the *organisation* of matter and energy, driven by exploiting the free energy and matter available in the environment. This chapter will therefore explore **bioenergetics** – the fundamental mechanisms for the capture, storage, transfer, release and use of free energy within biological systems.

24.2 Metabolic pathways, mass action and pseudo-equilibria

Chemical reactions within a cell are not isolated. Instead, the product of one reaction is typically the reactant of at least one other, to form a series of sequential steps called metabolic pathways. Cellular reactions are almost all enzyme-catalysed, with the rates for key steps under dynamic control. These control points define the starts and ends of distinct metabolic pathways, each of which directs and controls the breakdown of a particular nutrient molecule into common chemical building blocks, or the synthesis of a particular cellular product from those building blocks. Metabolic pathways are interlinked, and connected to the constantly changing availability of nutrients external to the cell. As a consequence, within the constantly changing open system of a living cell, very rarely are all chemical reactions actually at equilibrium, for the dynamics of cellular metabolism and the environment ensures that reactions are under continual flux. As we saw on page 8, however, the concept of 'equilibrium' is linked to the time horizon over which the system is being studied – and the shorter the period of observation, the greater the likelihood that an equilibrium analysis is applicable. In living systems, many enzyme-catalysed reactions are fast, and so can often be considered to be at '**pseudo-equilibrium**', with mass action (see pages 566 and 567) driving the net flux of reagents in one direction through the various sequential steps in any given pathway. The thermodynamics of a particular reaction within a cell can never be studied in isolation, but must always be taken in context of the associated metabolic pathways. Indeed, a reaction can be driven in a thermodynamically unfavourable direction when it is coupled by mass action to more favourable reactions.

24.3 Life's primary 'energy currency' – ATP

Adenosine triphosphate (ATP) is one of the main '**energy currencies**' within cells– an energy currency being an intermediate molecule created by a reaction for which ΔG is negative, and used to drive a reaction for which ΔG is positive. ATP is continuously synthesised 'here' and consumed 'there', and so acts as a means of transferring energy within a biochemical system, just as money is the means of transferring value within an economic system – hence the term 'energy currency'.

ATP is synthesised from adenosine diphosphate, ADP, and what is known as 'inorganic phosphate', represented as P_i; ATP, once formed, can be hydrolysed back to ADP and P_i. Since the hydrolysis of ATP, represented as

$$\text{ATP} + H_2O \rightleftharpoons \text{ADP} + P_i \qquad \Delta_r G_c^{\ominus} = -31 \text{ kJ} \qquad \Delta_r G_c^{\ominus\prime} = -57 \text{ kJ}$$

is exergonic ($\Delta G < 0$), the hydrolysis reaction is thermodynamically favourable to the right as written here, and it is the harnessing of this energy by other reactions that drives, for example, the synthesis of complex molecules such as proteins and nucleic acids. Conversely, the formation of ATP, in which the reaction proceeds from right to left as written, is not spontaneous of its own accord, and so must be driven by another source of energy, most fundamentally, the electromagnetic energy contained in light, as happens in green plants in the process of **photosynthesis**. ADP is not the only hydrolysis product derived from ATP – another is AMP (adenosine monophosphate), formed either by the loss of two P_i moieties from ATP, or the loss of one P_i from ADP.

As can be seen, $\Delta_r G_c^{\ominus}$ for the hydrolysis of ATP, as measured using the conventional standards, is different from $\Delta_r G_c^{\ominus\prime}$, as measured using the biochemical standards. In Chapter 23, we saw that $\Delta_r G_c^{\ominus}$ and $\Delta_r G_c^{\ominus\prime}$ are different only when a reaction either produces or consumes hydrogen ions. Since the reaction written as

$$\text{ATP} + H_2O \rightleftharpoons \text{ADP} + P_i$$

does not include any hydrogen ions, this should imply that $\Delta G^{\ominus}$ and $\Delta G^{\ominus\prime}$ should have the same values – which they don't. This apparent paradox is resolved by recognising that the reaction written as $\text{ATP} + H_2O \rightleftharpoons \text{ADP} + P_i$ is, in fact, a considerable simplification of a much more complex chemical system, which, in a living cell, comprises a mixture of, for example, ATP^{4-}, HATP^{3-}, $H_2\text{ATP}^{2-}$, $H_2PO_4^-$, HPO_4^{2-}, and also H^+ ions, as well as complexes with Mg^{2+} and Ca^{2+} ions too. Reported values of 'the standard Gibbs free energy for the hydrolysis of ATP' can therefore vary quite widely, depending not only on whether the data complies with conventional or biochemical standards, but also on the conditions in which a given measurement is made. Care must therefore be taken in identifying which value is being cited, and on which value is being used for any particular purpose.

The ATP hydrolysis reaction is associated with a concentration equilibrium constant K_c' as

$$K_c' = \frac{[\text{ADP}]_{eq}[P_i]_{eq}}{[\text{ATP}]_{eq}} = \exp\left(-\frac{\Delta_r G_c^{\ominus\prime}}{RT}\right) \qquad (24.1)$$

in which, as we saw on pages 567 and 568, there is no explicit term for molecular water, since water is both a reagent and the solvent. At $T = 298.15$ K, $K_c' \approx 10^{10}$, and so the equilibrium lies quite far to the right, implying that the hydrolysis of ATP is essentially complete.

Figure 24.1 shows how the free energy of the chemical system

$$\text{ATP} + H_2O \rightleftharpoons \text{ADP} + P_i$$

changes with the extent-of-reaction ξ. The extreme left of the horizontal axis, $\xi = 0$, represents a mixture of ATP and H_2O, and the extreme right, $\xi = 1$, a mixture of ADP and P_i; intermediate points correspond to mixtures of all four components, with the equilibrium mixture $\xi = \xi_{eq}$ corresponding to the minimum G_{eq} in the system's free energy. Figure 24.1 will be recognised

THE BIOENERGETICS OF LIVING CELLS

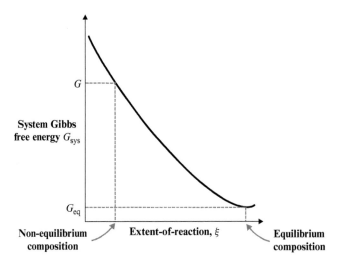

Figure 24.1 Energy is stored in a non-equilibrium system. At equilibrium, the free energy of the system is at a minimum, G_{eq}. If the system is not at equilibrium, the free energy will be some value $G > G_{eq}$. When the system moves from a non-equilibrium state to the equilibrium state, free energy $G - G_{eq}$ is released; alternatively, if the system can remain in the non-equilibrium state for some period of time, the non-equilibrium state can be thought of as storing energy $G - G_{eq}$.

as being similar to Figure 16.6, subject to the equilibrium composition in Figure 24.1 being further to the right.

As the diagram makes clear, when the system is in equilibrium, the free energy G_{eq} is a minimum; at any other composition, $G > G_{eq}$. It therefore follows that, when the system moves from any non-equilibrium state to equilibrium, a quantity of free energy $G - G_{eq}$ is released, this being 'available work' in the sense discussed on pages 386 to 390. Looked at in a rather different way, if the system can be 'held' in a non-equilibrium state for a certain time, then that state can be thought of as 'storing' energy of amount $G - G_{eq}$.

For the hydrolysis of ATP to ADP, the equilibrium lies far to the right, and so in practice, all non-equilibrium states correspond to an extent-of-reaction to the left of the equilibrium composition. This implies that, for any non-equilibrium state, $[ATP] > [ATP]_{eq}$ and $[ADP] < [ADP]_{eq}$. Furthermore, the ratio $[ATP]/[ADP]$, which is a characteristic of any (higher energy) non-equilibrium state, is greater than the ratio $[ATP]_{eq}/[ADP]_{eq}$, which is a characteristic of the (lower energy) equilibrium state. The ratio $[ATP]/[ADP]$ for any state, and – more significantly – the difference $[ATP]/[ADP] - [ATP]_{eq}/[ADP]_{eq}$, is therefore a measure of the amount of energy stored in that state, and of the amount of energy that could potentially be released.

In biochemical systems, the ratio $[ATP]/[ADP]$ is controlled at a value far from the (very small) equilibrium value, so ensuring that energy is stored, and potentially available for use by coupling the hydrolysis reaction to any of the very many cellular reactions that would otherwise be thermodynamically unfavourable, such as both DNA and protein synthesis. Physiologically, the ratio $[ATP]/[ADP]$ is typically in the range 10 to 1000 – a ratio that implies a plentiful store of chemical energy within ATP molecules, while also placing sufficient ADP into circulation to enable the cell to utilise it as a metabolite elsewhere.

ATP is sometimes referred to as a 'high-energy molecule', and the chemical bond that is hydrolysed when ADP is formed as a 'high-energy bond', ~P. These descriptions are both wrong and misleading. For the hydrolysis reaction ATP + H_2O ⇌ ADP + P_i, the magnitude of $\Delta_r G_c^{\ominus}$, using the conventional standards, is just 31 kJ for each mol of ATP hydrolysed – this being quite a small value, as can be seen by comparison to the figures shown in Table 6.8 and 6.9, which show bond dissociation energies and average bond energies: a C-H bond, for example, has a bond energy of some 400 kJ mol^{-1}. Indeed, it is *because* the magnitude of $\Delta_r G_c^{\ominus}$ is relatively small (and that of $\Delta_r G_c^{\ominus\prime}$ too) that ATP is so useful in acting as an energy currency – the small magnitudes of $\Delta_r G_c^{\ominus}$, and $\Delta_r G_c^{\ominus\prime}$ imply that the reaction ATP + H_2O ⇌ ADP + P_i is quite finely balanced, and can take place either to the right, hydrolysing ATP and releasing free energy, or to the left, synthesising ATP and consuming free energy, under relatively small changes in conditions, exactly as an 'ideal' energy currency should behave.

24.4 NADH is an energy currency too

ATP is not the only energy currency within cells – others are guanine triphosphate GTP, which is similar to ATP; two nicotinamide molecules, NADH (nicotinamide adenine dinucleotide) and NADPH (nicotinamide adenine dinucleotide phosphate); and a flavin derivative $FADH_2$, which is similar to NADH. Of these, NADH is probably the most widely used, and exists within cells as a redox pair with its oxidised counterpart NAD^+ according to the reversible reaction

$$NAD^+ + 2\,e^- + H^+ \rightleftharpoons NADH$$

which is endergonic as written, and exergonic in reverse.

For the hydrolysis of ATP, we can determine $\Delta_r G_c^{\ominus\prime}$ from an equilibrium constant defined in terms of molar concentrations according to equation (24.1), but for a reaction in which electrons are involved, using an equation of this form is not so intuitive, for it is not at all clear what a term such as $[e^-]$ might mean. This difficulty, however, is resolved if we invoke the principles of electrochemistry discussed in Chapter 20, where we saw (on page 619) that all redox reactions are associated with a standard electrode potential $\mathbb{E}^{\ominus}$, which is related to the value of $\Delta_r G_c^{\ominus}$ for the corresponding cell reaction by equation (20.8e)

$$\Delta_r G_c^{\ominus} = -n_e\,F\,\mathbb{E}^{\ominus} \tag{20.8e}$$

in which n_e is the number of electrons consumed, and F is Faraday's constant, 9.6485×10^4 C mol^{-1}, this being the magnitude of the electric charge carried by 1 mol of electrons. For a biochemical system, equation (20.8e) transforms as

$$\Delta_r G_c^{\ominus\prime} = -n_e\,F\,\mathbb{E}^{\ominus\prime} \tag{23.14}$$

and reference to Table 23.2 will show that, for the reduction of NAD^+ to NADH, $\mathbb{E}^{\ominus\prime} = -0.320$ V. Since two electrons are consumed

$$\Delta_r G_c^{\ominus\prime} = -2 \times 9.6485 \times 10^4 \times (-0.320)\,\text{J} = 61.8\,\text{kJ}$$

The formation of NADH from NAD^+ therefore requires 61.8 kJ per mol of NADH, implying that the oxidation of NADH to NAD^+ releases 61.8 kJ per mol of NADH, an amount comparable to the 57 kJ released by the hydrolysis of ATP.

To understand the thermodynamics of cellular metabolism more thoroughly we need to explore the ways in which ATP and NADH can be generated. There are two general ways in which this occurs: one directly as a result of enzyme-mediated catalysis, in a process known as **substrate-level phosphorylation**; and the second, indirectly, through the generation of a **chemi-osmotic potential gradient** across a **biological membrane** in a process known as **oxidative phosphorylation**. The bioenergetics of both of these mechanisms will be explored thermodynamically here.

24.5 Glycolysis – substrate-level phosphorylation of ADP

Cells derive their energy from the free energy released by oxidation of organic molecules, including sugars such as glucose. If glucose is simply burned – as can be demonstrated by rolling a sugar cube in cigarette ash, and then lighting the cube with a match – the reaction is:

$$\text{Glucose} + 6O_2 \rightarrow 6CO_2 + 6H_2O \qquad \Delta_c G^{\ominus} = -2930 \text{ kJ}$$

Since this is a combustion reaction, we use the symbol $\Delta_c G^{\ominus}$ rather than $\Delta_r G^{\ominus}$. Note that, in accordance with our discussion on page 693, since no hydrogen ions are involved in this reaction, $\Delta_c G^{\ominus\prime} = \Delta_c G^{\ominus}$

Burning sugars in the presence of oxygen leads to a large and rapid release of free energy in the form of heat. In contrast, cells control the release of this free energy using enzyme-catalysed reactions, harvesting the energy efficiently by coupling these reactions to the formation of ATP from ADP, and NADH from NAD^+.

The cellular oxidation of glucose follows two sequential metabolic pathways. The first pathway, known as **glycolysis**, takes place anaerobically – in the absence of molecular oxygen – and converts each molecule of glucose (which contains 6 carbon atoms) into two molecules of a 3-carbon intermediate, pyruvate. The second, subsequent, pathway, the **TCA cycle** (an acronym for the **tricarboxylic acid cycle**, also known as the **citric acid cycle** or the **Krebs cycle**), converts the pyruvate into carbon dioxide, and does require oxygen.

Figure 24.2 illustrates the glycolysis pathway, showing the ten steps by which each molecule of glucose, $C_6H_{12}O_6$, is split into two molecules of pyruvate, CH_3COCOO^-, coupled with the formation of 2 molecules of ATP and 2 molecules of NADH, so capturing some of the glucose's original chemical energy.

To examine the thermodynamics of the glycolysis pathway, we firstly consider the (hypothetical) pathway of the same reactions, but uncoupled from the consumption or generation of ATP and NADH. If the free energy of the original glucose system is represented as $G'(\text{glu})$, and the free energy of the system at the end of step X as $G'(X)$, then the change in free energy relative to glucose, $\Delta G'(X) = G'(X) - G'(\text{glu})$, is a measure of the extent to which the state of glycolysis as at the end of step X is thermodynamically favourable ($\Delta G'(X) < 0$), or unfavourable ($\Delta G'(X) > 0$), relative to the initial state of glucose. The behaviour of $\Delta G'(X)$ for each successive step is shown in Figure 24.3.

As shown in Figure 24.3, for the oxidation of glucose to pyruvate in a typical cellular cytoplasm, but ignoring ATP and NADH, the overall value of $\Delta G'$ would be about −320 kJ per mol of glucose. This is clearly favourable thermodynamically overall, but, in reality, glucose does not spontaneously decompose into pyruvate in the absence of glycolytic enzymes. The reason

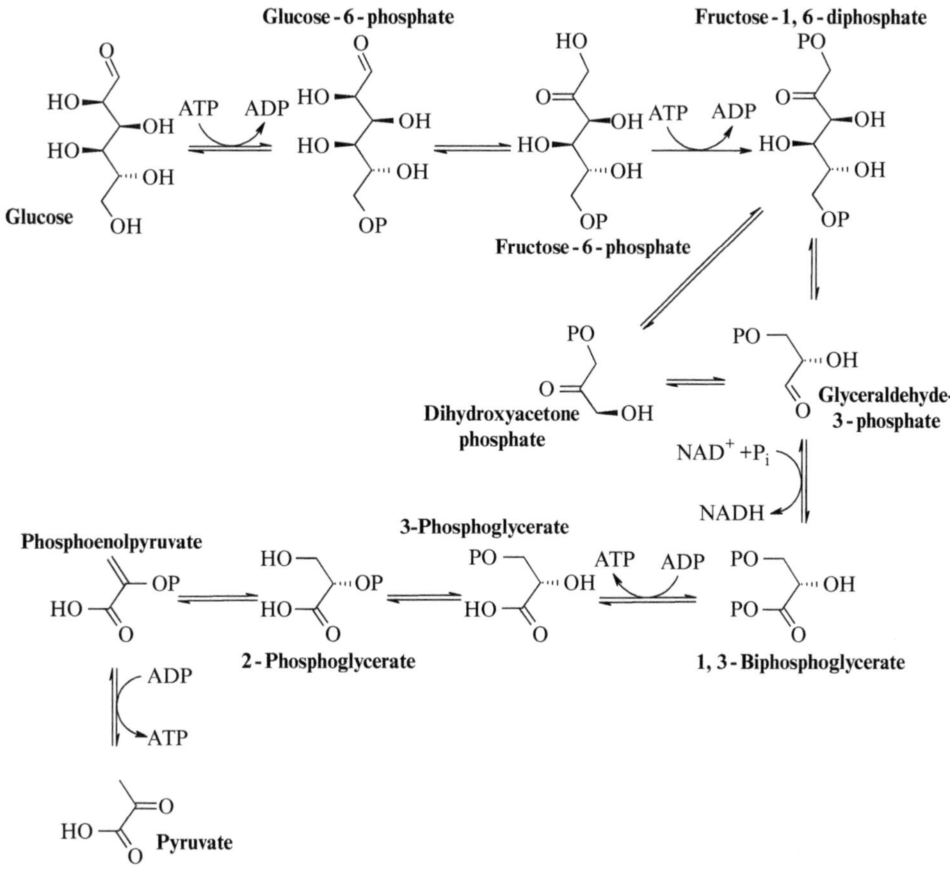

GLYCOLYSIS – SUBSTRATE-LEVEL PHOSPHORYLATION OF ADP

Figure 24.2 The glycolysis pathway of cellular metabolism. This pathway forms two pyruvate molecules from each glucose molecule, and uses the free energy released to generate 2ATP and 2 NADH molecules.

for this is also evident from Figure 24.3: as can be seen, in the absence of ATP and NAD^+, the first five states are all thermodynamically unfavourable, and these reactions are referred to as the 'preparatory phase' of glycolysis.

Figure 24.4 shows the impact of the enzyme-catalysed reactions which produce, or consume, ATP or NADH.

For the overall glycolysis pathway, one molecule of glucose, $C_6H_{12}O_6$, is split into two molecules of pyruvate, coupled with the transformation of two molecules of ADP into two of ATP, and also two NAD^+ into two NADH. As shown in Figure 24.4, the net production of two ATP molecules is, in fact, the overall effect of two earlier steps which each consume one ATP, followed by two steps which each produce two ATPs.

The very first step is the phosphorylation of glucose to glucose-6-phosphate, which ensures that the glucose does not diffuse back across the cell membrane. By itself, as shown in Figure 24.3, this reaction is thermodynamically unfavourable with $\Delta G' = +23$ kJ per mol of glucose, but by coupling this reaction to the hydrolysis of one mol of ATP, the $\Delta G'$ of -57 kJ per mol of ATP hydrolysed more than offsets the $\Delta G'$ of $+23$ kJ for the conversion of

THE BIOENERGETICS OF LIVING CELLS

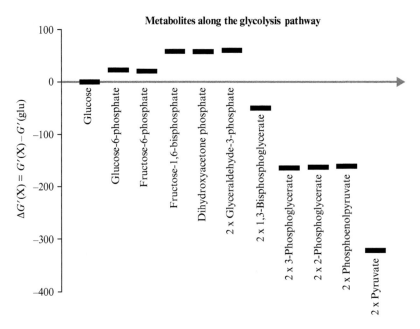

Figure 24.3 The bioenergetics of glycolysis (1). The diagram shows the free energy of each successive state along the glycolysis pathway from glucose to pyruvate, in the absence of any coupled reactions involving ATP or NADH, relative to the free energy of glucose as the baseline. Note that the step from dihydroxyacetone phosphate to glyceraldehyde-3-phosphate splits a 6-carbon molecule into two 3-carbon molecules.

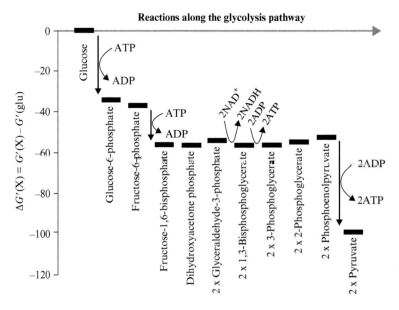

Figure 24.4 The bioenergetics of glycolysis (2). The diagram shows the free energy of each successive state along the glycolysis pathway from glucose to pyruvate, including the coupled reactions, relative to the free energy of glucose as the baseline. Note that the step that produces NAD^+ from NADH, and the subsequent step that produces ATP from ADP, are almost reversible, and so almost maximally efficient.

1 mol of glucose to glucose-6-phosphate to give a net $\Delta G'$ of –34 kJ, which is thermodynamically favourable. By coupling this reaction to the hydrolysis of ATP, the upward, endergonic, 'step' corresponding to this reaction in Figure 24.3 becomes the downward, exergonic, step in Figure 24.4.

Similar remarks apply to the second upward step in Figure 24.3, from fructose-6-phosphate to fructose-1,6-biphosphate, which becomes a downward step by coupling with ATP hydrolysis, as shown in Figure 24.4. Furthermore, the exergonic downward steps in Figure 24.3 remain as downward steps in Figure 24.4, but the size of the step is reduced as a result of coupling with the formation of NADH or ATP, thereby capturing some of the free energy released (in comparing the sizes of the downward steps between these two figures, note that the scales on the left-hand side of each figure are different).

When accounting for the consumption or generation of ATP and NADH in cellular glycolysis, we obtain the actual change in free energy for each reaction, relative to glucose. As illustrated in Figure 24.4, the overall $\Delta G'$ for the complete energy-coupled pathway is –98 kJ per mol of glucose, as compared to the value of –320 kJ for the uncoupled pathway, as illustrated in Figure 24.2. For the coupled pathway, the negative value –98 kJ for $\Delta G'$ is sufficiently favourable to drive the overall pathway forwards, whilst extracting 320 – 98 = 222 kJ of free energy in the net formation of 2 ATP and 2 NADH – in fact 222/320 = 69.4 % of the energy that is potentially available is captured and stored.

A notable feature of Figure 24.4 concerns the step from 2-glyceraldehyde-3-phosphate to 1-3 biphosphoglycerate, which captures energy as NADH, and the subsequent step from 1-3 biphosphoglycerate to 3-phosphoglycerate, capturing energy as ATP. In Figure 24.3, these two steps are both noticeably downward, but in Figure 24.4, the initial and final states of each step are almost at the same level, implying that, for each of these two steps, $\Delta G' \approx 0$. These reactions are therefore excellent examples of pseudo-equilibrium – reactions which are close to thermodynamic equilibrium, and which are associated with molar concentrations which are quite stable over time, despite the reality that the reactions are an integral part of a complex metabolic pathway through which much material is flowing. Furthermore, because these reactions are close to equilibrium, they are approaching thermodynamic reversibility, which, as we saw on page 143, implies that the process of energy capture as NADH and ATP is almost maximally efficient.

24.6 The metabolism of pyruvate

The end-products of glycolysis are ATP, NADH, and pyruvate. The primary use of the pyruvate is as the substrate in the aerobic TCA cycle, in which pyruvate is further oxidised to carbon dioxide, capturing more chemical energy, as will be described shortly. There are, however, many other cellular pathways that metabolise pyruvate, some being common to all organisms, whilst others are specialised to certain types of organism.

One especially well-known pathway, taking place within yeast, comprises two enzyme-catalysed steps, transforming pyruvate into ethanol and carbon dioxide according to the net reaction

$$CH_3COCOOH + NADH + H^+ \rightarrow C_2H_5OH + NAD^+ + CO_2$$

This is the key reaction harnessed by the brewing industry, and it is the released carbon dioxide that gives beer its fizz. The pathway also converts NADH, produced during glycolysis (see Figure 24.4), back into NAD^+, so replenishing the total quantity of NAD^+ within the cell, whilst simultaneously depleting the quantity of NADH. Since NADH is a reducing agent, an accumulation of NADH creates a progressively stronger reducing environment, which, if too strong, can disrupt the normal metabolism of the cell, and ultimately cause cell death. Maintaining the ratio $[NAD^+]/[NADH]$ within a narrow range is therefore essential for the maintenance of normal cellular activity, and this is one of the mechanisms by which this is achieved.

A further pathway for pyruvate metabolism is evident when humans run very fast, with relatively little breathing, as 100m sprinters often do. To produce the energy burst required by the sprinter's muscles, glucose is metabolised very fast, causing a rapid build-up of pyruvate. Simultaneously, the oxygen with the sprinter's body becomes depleted. As a consequence, the pyruvate cannot be metabolised by the aerobic TCA cycle, but follows an anaerobic pathway within muscle cells, mediated by the enzyme lactic dehydrogenase, to form lactic acid, as represented by the overall reaction

$$CH_3COCOOH + NADH + H^+ \rightarrow CH_3CHOHCOOH + NAD^+$$

It is this lactic acid that causes the very familiar muscle burn, and ultimately cramp.

24.7 The TCA cycle

Some micro-organisms oxidise glucose anaerobically, obtaining their ATP from glucose only through substrate-level phosphorylation as already described, but other organisms – including ourselves – can further oxidise the pyruvate product using molecular oxygen, in a process called **oxidative phosphorylation**.

This process begins with a metabolic pathway called the tricarboxylic acid cycle, also known as the TCA cycle, the citric acid cycle, or the Krebs cycle, as shown in Figure 24.5.

Pyruvate from glycolysis is first linked by an enzyme, pyruvate dehydrogenase, onto coenzyme A (CoASH) with the release of CO_2, to form acetyl CoA, capturing the free energy released in the production of one NADH. The acetyl CoA combines with oxaloacetate to form citric acid, and releases CoASH for recycling. The TCA cycle then operates through a sequence of seven enzymatic steps that together generate three NADH, one GTP (from which ATP can subsequently be synthesised), one $FADH_2$ (similar to NADH), and two further CO_2 molecules, as well as one oxaloacetate that replaces the oxaloacetate originally consumed. The cycle then starts again. For every turn of the cycle, all the components – oxaloacetate and the rest – are both consumed and replenished, and so their molar concentrations are conserved. Pyruvate, however, is consumed, and CO_2 produced, and so all the enzymes and associated molecules of the cycle might be thought of as acting collectively as a 'super catalyst', mediating the overall cycle reaction

$$CH_3COCOOH + 4NAD^+ + FAD + GDP \rightarrow 3\ CO_2 + 4NADH + FADH_2 + GTP$$

Despite its cyclic appearance, this is not a 'perpetual motion machine' that generates an unlimited supply of energy, stored as NADH, $FADH_2$ and GTP; rather, it is more like a water-wheel

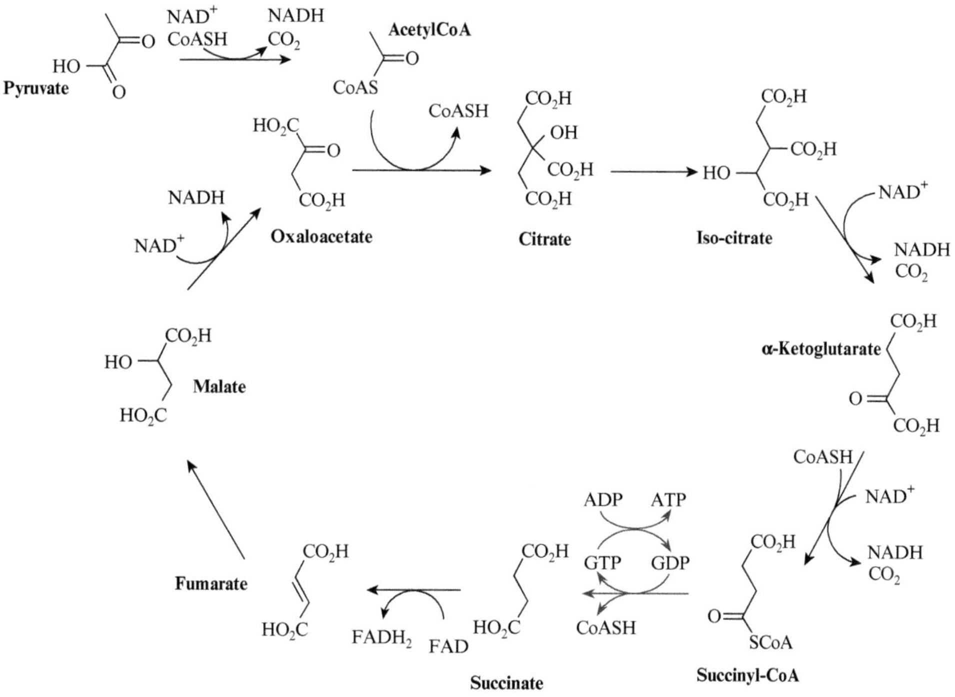

Figure 24.5 The TCA cycle. Pyruvate, produced from glycolysis, is shown at the top left, and combines with CoASH to form Acetyl CoA, which in turn reacts with oxaloacetate to form citrate. The cycle then proceeds as shown, resulting in the production of NADH, GTP, FADH$_2$ and CO$_2$, and also oxaloacetate, so replacing the molecule originally consumed.

that captures the energy of falling water to power a mill or an electricity turbine, as illustrated in Figure 24.6.

When water falls downhill, a state of higher potential gravitational energy transforms into a state of lower gravitational potential energy. For a natural waterfall, this energy difference is released in the form of the kinetic energy of the water, as heat, and as sound. If, however, the falling water drives a water-wheel, some of this energy is captured as kinetic energy of rotation – energy that can be harnessed to drive a mill-wheel or a turbine. By analogy, the TCA cycle captures the Gibbs free energy difference between high chemical free energy acetyl CoA and low chemical free energy oxaloacetate, as shown for each step in Figure 24.7.

Four features of Figure 24.7 are of particular interest. Firstly, the initial reaction of acetyl CoA and oxaloacetate to form citrate and CoASH is exergonic, and therefore spontaneous, so 'kick-starting' the cycle. Secondly, the cycle as a whole, starting with acetyl CoA and ending with oxaloacetate on each turn, is also exergonic, so the cycle will continue indefinitely as long as there is a sufficient supply of acetyl CoA, and the other intermediates such as NAD$^+$.

The third notable feature is the sequence from succinyl-CoA to succinate and fumarate, which recycles the CoASH required to form succinyl-CoA, as well as capturing energy in the form of GTP and FADH$_2$. As can be seen in Figure 24.6, the 'bars' associated with succinyl-CoA, succinate and fumarate are almost at the same 'level', implying that, for the reactions from succinyl-CoA to succinate, and from succinate to fumarate, the corresponding values of $\Delta G^{\ominus\prime}$

THE BIOENERGETICS OF LIVING CELLS

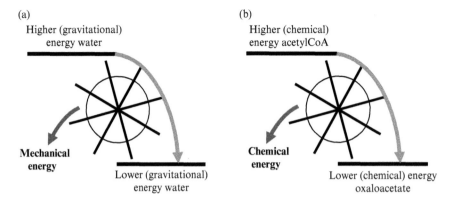

Figure 24.6 The TCA as compared to a water-wheel. In (a), water, at a higher level, with higher gravitational potential energy, falls over the water-wheel to a lower level, at a lower gravitational potential energy. The difference in the gravitational energy of the water is captured as the kinetic energy of rotation of the water-wheel, which can be harnessed as mechanical energy to drive a mill or power an electricity turbine. In (b), higher chemical energy acetyl CoA is transformed into lower energy oxaloacetate, and the difference in chemical energy can be harnessed by the biochemical TCA cycle in the form of the chemical energy stored in molecules such as NADH. Both systems operate as cycles, and as long as there is a supply of water, or a supply of acetyl CoA, the cycles will continue indefinitely. The operation of both cycles is spontaneous, and so complies with the Second Law; the First Law implies that, when the cycles are operating with 100% efficiency, the mechanical (or chemical) energy produced is equal to the difference between the gravitational (or chemical) energies of the upper and lower levels.

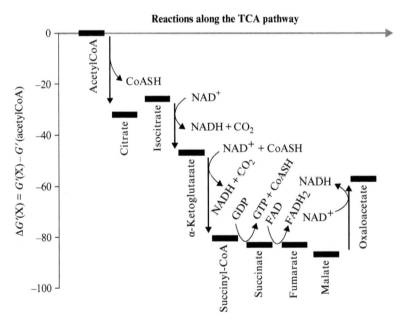

Figure 24.7 The bioenergetics of the TCA cycle. The diagram shows the free energy of each successive state along the TCA pathway from acetyl CoA to oxaloacetate, including the coupled reactions, relative to the free energy of acetyl CoA as the baseline.

are approximately zero. Like the glycolysis reactions discussed on page 719, these reactions are also approaching thermodynamic reversibility, implying that the energy capture by GTP and $FADH_2$ is almost maximally efficient.

As can be seen, however, the free energy levels associated with the sequence from succinyl-CoA, through succinate and fumarate to malate, are lower than the free energy of oxaloacetate, demonstrating the fourth feature: the implication that more energy would be released if the pathway were to stop at malate, rather than at oxaloacetate. Thermodynamically, however, this would not complete the cycle, which requires the final production of oxaloacetate to replace the molecule originally consumed; also, the final step from malate to oxaloacetate consumes NAD^+ to produce NADH, so helping to control the ratio $[NAD^+]/[NADH]$ within the cell, which as we saw on page 720, is metabolically important. So, although some energy appears to be 'wasted' in the upwards step from malate to oxaloacetate, there are good reasons why this step has evolved.

For the TCA cycle overall, each molecule of pyruvate metabolised results in the formation of 3 molecules of CO_2, 4 NADH, 1 $FADH_2$ and 1 GTP, which is the precursor of 1 ATP. Since each molecule of glucose results in two of pyruvate as a result of the glycolysis pathway, we may express the net outcome of the TCA, for each original molecule of glucose, as

2 Pyruvate + 8 NAD^+ + 2 FAD + 2 ADP → 6CO_2 + 8 NADH + 2 $FADH_2$ + 2 ATP

Similarly, for glycolysis, we have

Glucose + 2 NAD^+ + 2 ADP → 2 Pyruvate + 2 NADH + 2 ATP

and so, for the glycolysis and TCA pathways together

Glucose + 10 NAD^+ + 2 FAD + 4 ADP → 6CO_2 + 10 NADH + 2 $FADH_2$ + 4 ATP

24.8 Oxidative phosphorylation, and the chemi-osmotic potential

24.8.1 The origin of the chemi-osmotic potential

As we have just seen, as a result of the linking of the glycolysis and TCA pathways, each (six-carbon) glucose molecule is transformed into six (one-carbon) molecules of CO_2, coupled with 4 ATP, 10 NADH and 2 $FADH_2$. NADH and $FADH_2$, however, are reducing agents, and so a progressive build-up of these molecules, as driven by these two pathways, creates an increasingly strong reducing environment, which, as noted on page 720, can be very damaging. As we saw in Chapter 20, the ratio of the molar concentration (more correctly, activity – see page 663) of an oxidised state to that of the corresponding reduced state is associated with a redox potential, as given by the (transformed) Nernst equation

$$\mathbb{E}'(\text{ox, red}) = \mathbb{E}^{\ominus\prime}(\text{ox, red}) + \frac{RT}{n_e F} \ln \frac{[\text{ox}]'}{[\text{red}]'} \tag{23.16}$$

Within a cell, the ratios $[NAD^+]/[NADH]$ and $[FAD]/[FADH_2]$ are therefore associated with the corresponding redox potentials, redox potentials which become progressively more negative as glycolysis and the TCA cycle generate more and more NADH and $FADH_2$ on each cycle. As we are about to see, cells have an ingenious mechanism, called **oxidative phosphorylation** (alternatively, **cellular respiration** or just **respiration**), which transforms these redox potentials into chemical energy in the form of ATP.

A feature of both glycolysis and the TCA cycle is that they take place in solution – glycolysis within the cell cytoplasm, and the TCA cycle within the central matrix of the cell organelle, the **mitochondrion**. Oxidative phosphorylation is different: it is a process that takes place across a **biological membrane** known as the **respiratory membrane**. Biological membranes are important structures that perform a variety of functions, but one is especially important: the **lipid bilayer** of the membrane is a barrier to the free movement of dissolved molecules, and so can allow the molar concentrations of a given species to be different on each side. Furthermore, if the species in question is an electrically charged ion, a difference in the molar concentrations of the ions is also a difference in electric charge density, so creating an electrical potential difference across the membrane. The membrane can therefore maintain a gradient in both molar concentration and electrical potential, and it is the combination of these two types of gradient across a membrane that drives oxidative phosphorylation. This combination is known as the **chemi-osmotic potential**, a term introduced in the early 1960s by the English biochemist, Peter Mitchell, who was awarded the Nobel Prize in Chemistry in 1978 "for his contribution to the understanding of biological energy transfer through the formulation of the chemi-osmotic theory".

24.8.2 The key components of the respiratory membrane

In bacteria, the respiratory membrane is the bacterial plasma membrane, which is the inner of two membranes separating the cell from the outside world. For eukaryotes, such as ourselves, oxidative phosphorylation is located across the inner of the two membranes that surround the mitochondrion. A (very much!) simplified representation of the arrangement in the membrane is illustrated in Figure 24.8.

Four key components are required to establish a chemi-osmotic potential that can drive the formation of ATP from ADP:

- the membrane itself
- a cluster of proteins and associated bio-molecules, embedded within, and spanning, the membrane, known as the **electron transport chain**
- the **proton-motive force**, this being determined by the combination of the H^+ ion concentration gradient, and the associated electrical potential gradient
- a further cluster of bio-molecules, **ATP-synthase**.

24.8.3 The role of the membrane

Cellular membranes comprise a thin (~5 nm) closed surface surrounding, for example, a cell, or a cell organelle such as a mitochondrion. Structurally, membranes are composed primarily of a double layer (known as a **bilayer**) of lipids – molecules which in general have a polar moiety (such as a phosphate derivative of glycerol) at the 'head', linked to a 'tail' of two non-polar long-chain fatty acids (such as palmitic acid or oleic acid, whose chains have 16 and 18 carbon atoms respectively). Since the 'head' of each lipid is polar, and the 'tail' non-polar, lipids

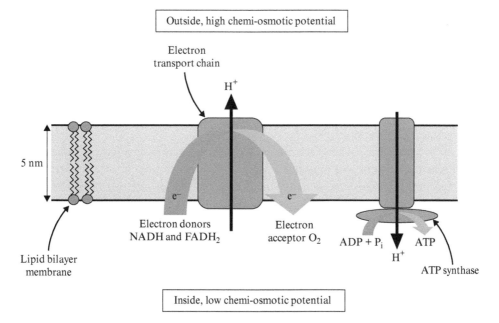

Figure 24.8 A much-simplified representation of the key components within a respiratory membrane. The lipid bilayer layer blocks the diffusion of water-soluble molecules and ions, but translocation across the membrane can happen as a result of the action of the embedded membrane proteins. The central feature of the diagram is a cluster of proteins and associated biomolecules, the electron transport chain, which collectively catalyse a pathway of coupled reactions oxidising the NADH and $FADH_2$ produced by glycolysis and the TCA cycle, and reducing dissolved molecular oxygen O_2 to H_2O, whilst simultaneously harvesting the free energy released to pump H^+ ions from the inside to the outside. Overall, this creates the chemi-osmotic potential attributable to both a concentration gradient ΔpH (high $[H^+]$ outside, low $[H^+]$ inside), and also an electrical potential gradient $\Delta\Psi$ (positive outside, negative inside). On the right is another protein cluster, ATP synthase, which couples the translocation of H^+ ions back into the cell to the synthesis of ATP, a process driven by the chemi-osmotic potential.

can self-assemble as tail-to-tail bilayer sheets, with the non-polar tails forming the interior, and the polar 'heads' on each external surface, interfacing with the polar aqueous cytoplasm on either side of the membrane.

Since the interior fatty acid chains are non-polar, molecules and ions that are soluble in water are insoluble within the membrane, which therefore forms an impermeable barrier, separating the interior volume contained within the closed surface from the outside. The impermeability of the membrane therefore creates the conditions which allow the inside and the outside to have different chemical compositions, with different molar concentrations of the same species, and different densities of electric charge if those species are charged ions.

Lipids, however, are not the only component of biological membranes: embedded within any membrane are a number of specialised proteins, co-enzymes and other biomolecules which can perform enzymic functions, and, importantly, influence – and so control – the concentrations of ions and molecules on either side of the membrane.

One way in which this can happen is as a result of a membrane protein that can change its molecular configuration so as to open, or close, a channel across the membrane, through which small molecules and ions might flow, driven by either a concentration or a charge gradient. When the channel is closed, gradients can be built up across the membrane; when the channel is open, the concentrations can equilibrate. A variation of this arises when there is a concentration gradient attributable to a molecule that is too large to pass through the channel, in which case water molecules can flow from the lower concentration to the higher concentration, so diluting the higher concentration and moving towards equilibrium. As we saw in Chapter 19, this flow of water is known as osmosis, and the membrane in this case is acting as a semi-permeable membrane.

In these two cases, the action of the membrane protein is one of opening a channel to allow the diffusive flow of ions and molecules, resulting in an equilibration of concentrations or charge densities. An alternative, and very important, function of some membrane proteins is to *create disequilibrium* across the membrane by increasing a concentration on one side, whilst simultaneously depleting a concentration on the other. So, suppose that a membrane protein catalyses a reaction on one side of the membrane that consumes, say, one H^+ ion. Suppose further that this reaction is coupled to another which releases one H^+ ion on the other side of the membrane. These two coupled reactions have the overall effect of depleting H^+ ions on one side of the membrane whilst simultaneously accumulating H^+ ions on the other. The end-result is the same as would have happened if an H^+ ion had diffused across the membrane, or through an open channel, against a gradient of concentration or charge (or indeed both). This process, known as **active transport**, is endergonic, and so is coupled to, for example, the hydrolysis of ATP.

By controlling diffusion, osmosis and especially active transport, a cell membrane can control the composition of the fluids on either side of the membrane, creating or dissipating concentration and electric potential gradients in accordance with the overall metabolism of the cell. The membrane is therefore a cell organelle in its own right – and one of great importance, for the membrane is instrumental in processes such oxidative phosphorylation, photosynthesis and nervous conduction, to name just three.

24.8.4 The electron transport chain

The **electron transport chain** is the name for a cluster of membrane-spanning proteins and co-enzymes which collectively catalyse a redox reaction in which NADH and $FADH_2$ reduce molecular oxygen to form water. This reaction which may be summarised as

$$10\ NADH + 2\ FADH_2 + 6\ O_2 + 10H^+ \rightarrow 10\ NAD^+ + 2\ FAD + 12\ H_2O$$

in which the stoichiometric coefficients have been chosen to consume six molecules of molecular oxygen, as required for the oxidation of a single molecule of glucose. This reaction is exergonic, and, as we shall see, the free energy released is coupled to the active transport of H^+ ions across the membrane, from the inside to the outside.

Redox reactions were explored in some detail in Chapter 20 (see in particular pages 628 to 631), where we saw that all redox reactions are characterised by a transfer of electrons, and a corresponding redox electrical potential. In the transformation from NADH to NAD^+, each NADH molecule releases two electrons as represented by the half-reaction

$$NAD^+ + 2\ e^- + H^+ \rightleftharpoons NADH \qquad\qquad \mathbb{E}^{\ominus\prime}(NAD^+, NADH) = -0.320\ V$$

which is written in the conventional direction for redox reactions (see pages 626 and Table 23.2), even though, in the case of the electron transport chain, the reaction takes place in the reverse direction.

Likewise, each $FADH_2$ molecule also releases two electrons

$$FAD + 2\,e^- + 2\,H^+ \rightleftharpoons FADH_2 \qquad \qquad \mathbb{E}^{\ominus\prime}(FAD, FADH_2) = -0.219 \text{ V}$$

These electrons are captured by the molecules in the electron transport chain, within which a series of reactions take place as the electrons are passed from one protein to the next (hence the term 'electron transport chain'), ultimately resulting in the acceptance of the electrons by molecular oxygen as represented by the half-reaction

$$O_2 + 4\,e^- + 4\,H^+ \rightleftharpoons 2\,H_2O \qquad \qquad \mathbb{E}^{\ominus\prime}(O_2, H_2O) = +0.816 \text{ V}$$

As we know, each glucose molecule produces 10 NADH and 2 $FADH_2$ molecules, which together release of a total of 24 electrons, so allowing 6 molecules of molecular oxygen to be reduced to water.

The formation of water from molecular oxygen, however, is not the only outcome of the electron transport chain. As electrons pass along the electron transport chain, successive reactions capture some of the free energy released at each step, and use this energy to achieve the net effect of translocating hydrogen ions across the membrane from the inside to the outside. The details of the processes involved will be found in the biochemistry texts; in (very brief) summary, this active transport occurs as the result of three separate, sequential, steps as illustrated schematically in Figure 24.9.

For the oxidation of NADH, the first step is the transformation of NADH into NAD^+, and the transfer of the electrons released to a molecule known as coenzyme Q. This exergonic process is coupled to the active transport of H^+ ions across the membrane from the inside to the outside: stoichiometrically, for each NADH molecule oxidised (and so for each pair of electrons released), four H^+ ions are actively transported.

In the second step, electrons from coenzyme Q are passed to another biomolecule, cytochrome c. This reaction too is exergonic, and for each pair of electrons passed from coenzyme Q to cytochrome c, two H^+ ions are actively transported from inside to out (the number two is somewhat uncertain – some sources suggest four).

In the third, final, step, the electrons captured by cytochrome c are used to reduce molecular oxygen to water, such that the flow of each pair of electrons is coupled to the active transport of four H+ ions from inside to out (once again, this number too is uncertain, with some sources reporting 'up to four').

The electron transport chain for $FADH_2$ differs in one respect from that for NADH: the first step oxidises $FADH_2$ to FAD, releasing two electrons, but, as shown in Figure 24.9(b), there is no associated active transport of H^+ ions across the membrane. Thereafter, the pathway followed is identical to that for NADH: the electrons released from $FADH_2$ are captured by coenzyme Q, and transferred to cytochrome c with the active transport of two (or perhaps four) H^+ ions for each pair of electrons, and the subsequent reduction of molecular oxygen is accompanied by the active transport of (up to) four H^+ ions per pair of electrons.

Stoichiometrically, each NADH molecule oxidised releases one pair of electrons, so driving the active transport of $4 + \sim 2 + \sim 4 = \sim 10$ H^+ ions as the electrons pass down the electron

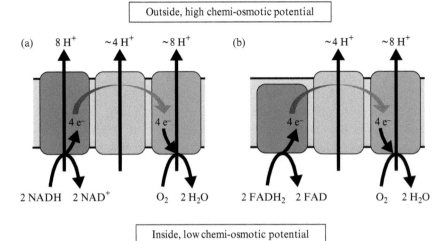

Figure 24.9 A schematic representation of the electron transport chain. When two molecules of NADH are oxidised to NAD$^+$, four electrons are released, and 8 H$^+$ ions are actively transported across the membrane, from the inside to the outside, as shown to the left of (a). The four electrons then pass through another component of the electron transport chain, actively transporting $\sim$4 H$^+$ ions. In the last component, the four electrons reduce one molecule of oxygen to water, coupled to the active transport of a further $\sim$8 H$^+$ ions. For the oxidation of FADH$_2$ to FAD, as shown in (b), the first component of the electron transport chain does not actively transport H$^+$ ions, whilst the subsequent two stages are identical to those for NADH. The net flow of H$^+$ ions from the inside to the outside creates both a concentration gradient, and a gradient in electrical potential, across the membrane, so that the inside of the membrane has a low, and the outside of the membrane a high, chemi-osmotic potential.

transport chain until ultimately reducing molecular oxygen to water. For each molecule of FADH$_2$, one pair of electrons also passes down the electron transport chain, but only $2 + \sim 4 = \sim 6$ H$^+$ ions are actively transported.

24.8.5 The bioenergetics of electron transport

As we have seen, each NADH molecule releases two electrons according to the half-reaction

$$\text{NAD}^+ + 2e^- + \text{H}^+ \rightleftharpoons \text{NADH}$$

This reaction has been expressed in the conventional direction, from an oxidised state to a reduced state, for which, as shown in Table 23.2, $\mathbb{E}^{\ominus\prime} = -0.320$ V.

Since

$$\Delta_r G_c^{\ominus\prime} = -n_e F \mathbb{E}^{\ominus\prime} \tag{24.2}$$

then, as we saw on page 715, $n_e = 2$ implying that $\Delta_r G_c^{\ominus\prime} = 61.8$ kJ. Accordingly for the reverse reaction written as

$$\text{NADH} \rightleftharpoons \text{NAD}^+ + 2\,e^- + \text{H}^+ \qquad\qquad \Delta_r G_c^{\ominus\prime} = -61.8\text{ kJ}$$

the sign of $\Delta_r G_c$ is reversed, and it is the reaction in this direction that takes place during electron transport.

OXIDATIVE PHOSPHORYLATION, AND THE CHEMI-OSMOTIC POTENTIAL

The electrons originating in NADH ultimately reduce molecular oxygen as

$$O_2 + 4e^- + 4H^+ \rightleftharpoons 2 H_2O \qquad \mathbb{E}^{\ominus\prime}(O_2, H_2O) = +0.816 \text{ V}$$
$$\Delta_r G_c^{\ominus\prime} = -315 \text{ kJ}$$

and so, for the overall reaction

$$O_2 + 2NADH + 2H^+ \rightleftharpoons 2H_2O + 2NAD^+ \qquad \Delta_r G_c^{\ominus\prime} = -438.6 \text{ kJ}$$

where $\Delta_r G_c^{\ominus\prime}$ is computed as $-315 + (2 \times -61.8) = -438.6$ kJ. The free energy change $\Delta_r G_c^{\ominus\prime}$ associated with the reduction of one mol of NADH is therefore $-438.6/2 = -219.3$ kJ.

We may carry out a similar calculation for FADH$_2$, for which

$$FAD + 2 e^- + 2 H^+ \rightleftharpoons FADH_2 \qquad \mathbb{E}^{\ominus\prime}(FAD, FADH_2) = -0.219 \text{ V}$$
$$\Delta_r G_c^{\ominus\prime} = 42.3 \text{ kJ}$$

$$O_2 + 4 e^- + 4 H^+ \rightleftharpoons 2 H_2O \qquad \mathbb{E}^{\ominus\prime}(O_2, H_2O) = +0.816 \text{ V}$$
$$\Delta_r G_c^{\ominus\prime} = -315 \text{ kJ}$$

and so, for the overall reaction

$$O_2 + 2 FADH_2 + 2H^+ \rightleftharpoons 2H_2O + 2FAD^+ \qquad \Delta_r G_c^{\ominus\prime} = -400 \text{ kJ}$$

where $\Delta_r G_c^{\ominus\prime}$ is computed as $-315 + (2 \times -42.3) = -400$ kJ, implying that the oxidation of each mol of FADH$_2$ releases 200 kJ of free energy.

The two reactions associated with the electron transport chain – the oxidation of NADH and of FADH$_2$ – are therefore both strongly exergonic, and it is this released free energy that drives the active transport of the H$^+$ ions across the membrane.

24.8.6 The chemi-osmotic potential and the proton motive force

The active transport of H$^+$ ions across the membrane from the inside to the outside has two effects.

Firstly, the molar concentrations [H$^+$] on either side of the membrane are such that $[H^+]_{out} > [H^+]_{in}$. Now, equation (16.10d)

$$G_{c,i} = G_{c,i}^{\ominus} + RT \ln[I] \qquad (16.10d)$$

tells us that the molar Gibbs free energy $G_{c,i}$ of any dissolved component I in an ideal solution increases with the molar concentration [I], where we note (see page 516) that [I] is 'shorthand' for the ratio [I]/[I]$^{\ominus}$. As we saw in Chapter 23, equation (16.10d) may be transformed to the biochemical standard

$$G_{c,i}' = G_{c,i}^{\ominus\prime} + RT \ln[I]' \qquad (24.3)$$

where [I]$'$ now represents the ratio [I]$'$/[I]$^{\ominus\prime}$ = [I]/[I]$^{\ominus\prime}$.

For a non-ideal solution, as we saw on page 663, instead of referring to the free energy G_i' and the molar concentration [I]$'$ we should refer to the chemical potential μ_i', and the activity a_i' as

$$\mu_i' = \mu_i^{\ominus\prime} + RT \ln a_i' \qquad (22.8b)$$

but the overall conclusion is the same: μ_i' increases with a_i', which itself increases with [I]'. In the current context, this implies that, for the H$^+$ ions on either side of the mitochondrial membrane, $G'(H^+, \text{out}) > G'(H^+, \text{in})$, and similarly $\mu'(H^+, \text{out}) > \mu'(H^+, \text{in})$.

The second effect concerns the positive charge associated with each H$^+$ ion: as the molar concentration [H$^+$] increases (or decreases), the charge density (the quantity of charge within any given volume) also increases (or decreases) accordingly. A principle from physics is that the electrical potential Ψ is proportional to the charge density, and so a further consequence of the active transport of H$^+$ across the membrane is to cause Ψ_{out} to be greater than Ψ_{in}.

We now define the **chemi-osmotic potential**, represented by the symbol G_{co}' (the subscript $_{co}$ representing 'chemi-osmotic'), as

$$G_{co}' = G'(H^+) + F\Psi \qquad (24.4a)$$

or, more strictly,

$$\mu_{co}' = \mu'(H^+) + F\Psi \qquad (24.4b)$$

F is the Faraday constant, as required to ensure that the units in equation (24.4a) are J mol^{-1}: the units of Ψ are V, and, as shown by equation (24.2)

$$\Delta_r G_c^{\ominus\prime} = -n_e F \mathbb{E}^{\ominus\prime} \qquad (24.2)$$

since the units of $\mathbb{E}^{\ominus\prime}$ are also V, the units of the product $F\mathbb{E}^{\ominus\prime}$, and so of the product $F\Psi$, are J mol^{-1}. For a dynamically stable respiratory membrane, since both $G'(H^+, \text{out}) > G'(H^+, \text{in})$ and $\Psi(\text{out}) > \Psi(\text{in})$, then $G_{co}'(\text{out}) > G_{co}'(\text{in})$: the chemi-osmotic potential is higher on the outside of the membrane than on the inside.

If we consider a change of state from an initial state in which 1 mol of H$^+$ ions are on the *outside* of the mitochondrial membrane to a final state in which that 1 mol of H$^+$ ions are on the *inside*, then there is a corresponding change in free energy, which we represent as $\Delta G_{co}'$, the **chemi-osmotic potential gradient**, given by

$$\Delta G_{co}' = [G'(H^+, \text{in}) + F\Psi(\text{in})] - [G'(H^+, \text{out}) + F\Psi(\text{out})] \qquad (24.5)$$

and so

$$\Delta G_{co}' = [G'(H^+, \text{in}) - G'(H^+, \text{out})] + F[\Psi(\text{in}) - \Psi(\text{out})] \qquad (24.6)$$

Since $G_{co}'(\text{out}) > G_{co}'(\text{in})$, the change in state represented by equation (24.6) corresponds to a flow of H$^+$ ions *down* the chemi-osmotic potential gradient; also, $G'(H^+, \text{out}) > G'(H^+, \text{in})$ and $\Psi(\text{out}) > \Psi(\text{in})$, implying that $\Delta G_{co}'$ is negative, and a measure of the free energy released.

As can be seen from equation (24.6), $\Delta G_{co}'$ is the sum of two components. The first term, $\Delta G'(H^+)$,

$$\Delta G'(H^+) = G'(H^+, \text{in}) - G'(H^+, \text{out}) \qquad (24.7)$$

is chemical, and, according to equation (24.3)

$$G_{c,i}' = G_{c,i}^{\ominus\prime} + RT \ln[I] \qquad (24.3)$$

is attributable to a difference in the molar concentrations [H$^+$] of H$^+$ ions on either side of the membrane. Since, for a dynamically stable mitochondrial membrane, [H]$_{\text{out}}$ > [H]$_{\text{in}}$, then

$G'(\mathrm{H}^+, \mathrm{out}) > G'(\mathrm{H}^+, \mathrm{in})$ and so $\Delta G'(\mathrm{H}^+)$ is necessarily negative: translocating H$^+$ ions from the outside to the inside of the mitochondrial membrane is exergonic.

The second term in equation (24.6), $F[\Psi(\mathrm{in}) - \Psi(\mathrm{out})]$, is electrochemical, and is attributable to the difference in charge density across the membrane. If we define $\Delta\Psi$ as

$$\Delta\Psi = \Psi(\mathrm{in}) - \Psi(\mathrm{out}) \tag{24.8}$$

then

$$F[\Psi(\mathrm{in}) - \Psi(\mathrm{out})] = F\Delta\Psi \tag{24.9}$$

$\Delta\Psi$, as defined according to equation (24.8), is known as the **membrane potential** – although the term 'membrane potential difference' is technically more correct, the term 'membrane potential' is widely used, and applies not just to respiratory membranes, but to any membrane across which there is an electrical potential difference; also, note that the definition of $\Delta\Psi$ is 'in – out', rather than the other way around. For the respiratory membrane in particular, H$^+$ ions, and hence positive charges, accumulate on the outside, and so $\Psi(\mathrm{out}) > \Psi(\mathrm{in})$. $\Delta\Psi$ is therefore always negative – in fact, for the mitochondrial membrane in its dynamic steady state, $\Delta\Psi$ is about -190 mV.

Expressing equation (24.6) in accordance with equations (24.7) and (24.8), we may write

$$\Delta G_{co}' = \Delta G'(\mathrm{H}^+) + F\Delta\Psi \tag{24.10}$$

Since both $\Delta G'(\mathrm{H}^+)$ and $F\Delta\Psi$ are always negative, $\Delta G_{co}'$ is also always negative, verifying that the flow of H$^+$ ions down the chemi-osmotic gradient is always exergonic. Conversely, any flow of H$^+$ ions against the chemi-osmotic gradient, from the inside of the mitochondrion to the outside, is endergonic, and can take place only when appropriately coupled to an exergonic reaction, such as the flow of electrons down the mitochondrial electron transport chain.

To quantify $\Delta G_{co}'$ for the mitochondrial membrane, we need a value for $\Delta\Psi$ and an expression for $\Delta G'(\mathrm{H}^+)$. As we have just seen, experiment shows that $\Delta\Psi$ is approximately -190 mV, and to determine $\Delta G'(\mathrm{H}^+)$, we may use equation (24.3)

$$G_{c,i}' = G_{c,i}^{\ominus\prime} + RT \ln[\mathrm{I}] \tag{24.3}$$

to give

$$\Delta G'(\mathrm{H}^+) = G_{\mathrm{in}}' - G_{\mathrm{out}}' = (G_{\mathrm{in}}^{\ominus\prime} + RT \ln [\mathrm{H}^+]_{\mathrm{in}}) - (G_{\mathrm{out}}^{\ominus\prime} + RT \ln [\mathrm{H}^+]_{\mathrm{out}})$$

where, as we saw on page 729, [H$^+$] stands for the ratio $[\mathrm{H}^+]'/[\mathrm{H}^+]^{\ominus\prime} = [\mathrm{H}^+]/[\mathrm{H}^+]^{\ominus\prime}$.

The standards $\Delta G^{\ominus\prime}$ are the same on both sides of the membrane, and so

$$\Delta G'(\mathrm{H}^+) = G_{\mathrm{in}}' - G_{\mathrm{out}}' = RT \ln [\mathrm{H}^+]_{\mathrm{in}} - RT \ln [\mathrm{H}^+]_{\mathrm{out}}$$

$$= 2.303 \, RT \, (\log_{10}[\mathrm{H}^+]_{\mathrm{in}} - \log_{10}[\mathrm{H}^+]_{\mathrm{out}})$$

$$= 2.303 \, RT \log_{10} \frac{[\mathrm{H}^+]_{\mathrm{in}}}{[\mathrm{H}^+]_{\mathrm{out}}}$$

In the ratio $[H^+]_{in}/[H^+]_{out}$, the implied quantities $[H^+]^{\ominus\prime}$ cancel out. Now

$$\log_{10} \frac{[H^+]_{in}}{[H^+]_{out}} = \log_{10}[H^+]_{in} - \log_{10}[H^+]_{out} = -pH_{in} + pH_{out}$$

$$= -(pH_{in} - pH_{out})$$

If we define

$$\Delta pH = pH_{in} - pH_{out} \tag{24.11}$$

then,

$$\Delta G'(H^+) = -2.303\, RT\, \Delta pH \tag{24.12}$$

Note that in equation (24.11), the pH gradient ΔpH is defined as 'in − out', exactly as the definition of $\Delta\Psi$ in equation (24.8). Since the result of the active transport of H^+ ions is to increase $[H^+]_{out}$ whilst simultaneously depleting $[H^+]_{in}$, for the stable dynamic state of the mitochondrial membrane, $[H^+]_{out}$ is always greater than $[H^+]_{in}$, implying that $pH_{out} < pH_{in}$. The mitochondrial ΔpH, as defined by equation (24.11), is therefore always positive. This in turn implies that, for the mitochondrial membrane, $\Delta G'(H^+)$ in equation (24.12) is, as expected, always negative: when an H^+ ion moves down the concentration gradient from the outside of the membrane to the inside, energy is released.

Equation (24.12) may now be used to substitute for $\Delta G'(H^+)$ in equation (24.10), allowing us to express the chemi-osmotic potential gradient $\Delta G_{co}'$ as

$$\Delta G_{co}' = -2.303\, RT\, \Delta pH + F\Delta\Psi \tag{24.13}$$

or in electrochemical terms as

$$\frac{\Delta G_{co}'}{F} = -2.303\, \frac{RT}{F}\, \Delta pH + \Delta\Psi = \Delta p \tag{24.14}$$

The ratio $\Delta G_{co}'/F$ is commonly known as the **proton motive force**, or **pmf**, Δp, the units-of-measure for which are volts, V.

At 298.15 K, $2.303\, RT/F = 59.2$ mV, and so an alternative version of equation (24.14) is

$$\Delta p = \Delta\Psi - 59.2\, \Delta pH \qquad\qquad \text{mV at 298.15 K} \tag{24.15}$$

For mitochondrial electron transport, the facts that ΔpH is always positive, and that $\Delta\Psi$ is always negative, implies that the pmf is always a negative number.

In fully respiring, dynamically stable, mitochondria, ΔpH is approximately $+0.5$ mV, and the membrane potential $\Delta\Psi$ is approximately -190 mV. Equation (24.15) therefore gives

$$\Delta p = -190 - 59.2 \times 0.5 = -219.6\, mV$$

and so $\Delta G_{co}' = -0.2196 \times 96.5 = -21.2$ kJ per mol of H^+ ions.

Now, on page 729, we estimated the value of $\Delta G'$ for the transfer of one electron pair from NADH to 0.5 O_2 as -219.3 kJ per mol of NADH oxidised, and this is the value of the maximum energy that can be captured from this reaction. We also know that the transfer from NADH of one electron pair, all the way along the electron transport chain, results in the active transport of 10 H^+ ions, and so a value of $\Delta G' = -219.3$ kJ per mol of NADH oxidised corresponds to a maximum free energy availability of -21.9 kJ per mol of H^+ ions actively transported. The value of the pmf, $\Delta G_{co}' = -21.2$ kJ per mol of H^+, measures the actual free energy captured in the form of the chemi-osmotic potential, from which we see that of the 21.9 kJ potentially available, 21.2 kJ are actually captured, whilst 21.9 − 21.2 = 0.7 kJ of free energy is dissipated. The efficiency of energy capture by the mitochondrial electron transport chain is therefore remarkably high at 21.2/21.9 = 97 %.

For mitochondria, the electrical potential difference $\Delta\Psi$ drives the translocation of ATP back out into the cytoplasm where it is needed, with a simultaneous 1:1 exchange of ADP in the opposite direction ready for ATP synthesis. In addition, and very importantly, the pmf is essential for the last stage in oxidative phosphorylation – the synthesis of ATP from ADP and P_i by an enzyme complex known as ATP synthase.

24.8.7 ATP synthase

ATP synthase is a complex of proteins that collectively span the respiratory membrane as illustrated in Figure 24.10.

The biological function of ATP synthase is to catalyse the synthesis of ATP from ADP, driving this endergonic reaction by harnessing some of the free energy released by the flow of H^+ ions from the outside to the inside, down the chemi-osmotic potential gradient. The mechanism by which this happens is remarkable, for it involves the physical rotation of one part of ATP synthase within another.

As illustrated in Figure 24.10, the membrane protein complex F_0 is attached to the 'stalk' of a second protein complex, F_1. The outside region of F_0 can bind H^+ ions, as a result of which the F_0 complex undergoes a conformational change causing some of the F_0 subunits, and the 'shaft' of the F_1 complex, to rotate; at the same time, H^+ ions are released on inside. The rotation of the 'shaft' then causes a series of cyclic conformational changes within the six (fixed) F_1 subunits, which, alternately, bind ADP and P_i and then release ATP. The shaft rotates in steps of 120°, and a complete three-step rotation produces three molecules of ATP.

The overall result is that the free energy released from an inwards flow of H^+ ions through the membrane, driven by the chemi-osmotic potential gradient, is coupled to the endergonic synthesis of ATP from ADP and P_i, as catalysed by conformational changes in the enzyme brought about by the rotation of the subunits within F_0 and F_1. ATP synthase is therefore a biological nanomachine – a biological system, just like a 'biological water-mill', that harnesses the energy stored in the chemi-osmotic potential into the mechanical energy required to cause the rotation of the 'shaft' and also the chemical energy required for the synthesis of ATP.

Note, however, that the 'water-mill' analogy cannot be taken literally, for there is no actual flow of H^+ ions through ATP synthase. Rather, one H^+ ion is bound at the outer surface of F_0, and another is released on the inside – a process which has the same overall result as the flow of an ion through a channel, but which is, in fact, the result of two separate processes. However, although there is no direct physical flow through a membrane channel, thermodynamically, the energy released by the 'notional' flow down the chemi-osmotic potential gradient is the energy source for the rotation, the corresponding configurational changes, and the resulting synthesis of ATP.

Stoichiometrially, the number of translocated H^+ ions associated with the synthesis of each ATP molecule depends on the specific physiological conditions, and is an active subject of research: current estimates are around 3 or 4 H^+ ions per ATP, this being 9 to 12 H^+ ions

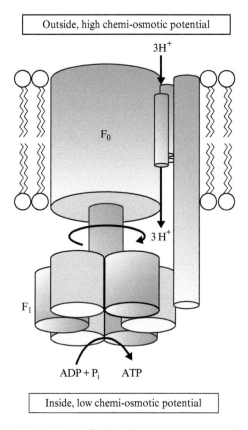

Figure 24.10 A schematic representation of ATP synthase. F_0 is a protein complex within, and spanning, the mitochondrial respiratory membrane. F_1 is another protein complex, within the interior of the mitochondrion, comprising a 'shaft' and a number of other components, including the six shown here that catalyse the conversion of ADP to ATP. The interior of the F_0 complex is attached to the 'shaft' of the F_1 complex, which can rotate within the six lower sub-units of F_1. A flow of H^+ ions into the F_0 complex from the outside causes a conformational change, resulting in the release of H^+ ions to the inside, and the rotation of the F_1 'shaft'. This rotation drives a cyclic series of conformational changes within the enzymic structures of F_1, thereby bringing about the production of ATP. Stoichiometrically, the translocation of ~ 3 H^+ ions is required for the synthesis of 1 ATP.

per rotation of the ATP synthase system. These numbers are sensible: as we saw on page 732, $\Delta G_{co}' = -21.2$ kJ per mol of H^+ ions, and so the flow of 2 mol of H^+ ions would in principle release 41.4 kJ, whilst a flow of 3 mol of H^+ ions releases 61.6 kJ. Given that the synthesis of one mol of ATP as

$$ADP + P_i \rightleftharpoons ATP + H_2O \qquad\qquad \Delta_r G_c^{\ominus\prime} = 57 \text{ kJ}$$

requires 57 kJ, a flow of two H^+ ions is insufficient, but a flow of three would work.

The upper limit can be estimated by considering how many H^+ ions are actively translocated across the membrane by the electron transport chain: as illustrated in Figure 24.11, the metabolism of one molecule of glucose results in the translocation of an estimated number of 112 H^+ ions, and the production of some 33 molecules of ATP, 2 of which are the result of glycolysis, 2 from the transformation of GTP, and 29 from oxidative phosphorylation. This suggests the maximum number of H^+ ions per molecule of ATP synthesised by ATP synthase is 112/29 = 3.9.

THE EFFICIENCY OF GLUCOSE METABOLISM

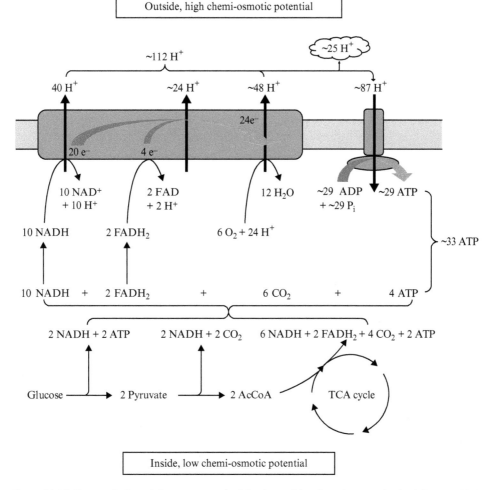

Figure 24.11 The metabolism of glucose. As a result of glycolysis and the TCA cycle, one molecule of glucose yields 10 NADH, 2 FADH$_2$, 6 CO$_2$ and 4 ATP. NADH and FADH$_2$ are the substrates for oxidative phosphorylation, the combination of the electron transport chain and ATP synthase. For each molecule of glucose, a total of 24 electrons pass down the electron transport chain, resulting in the active transport of approximately 112 H$^+$ ions across the membrane. The resulting chemi-osmotic potential drives the synthesis of ATP. Assuming that the chemi-osmotic potential attributable to 3 H$^+$ ions corresponds to the synthesis of 1 ATP molecule, and that one molecule of glucose results in the synthesis by ATP synthase of 29 molecules of ATP, then a total of 87 H$^+$ ions are utilised by ATP synthase, with the free energy of the other 25 H$^+$ ions being used for other purposes, or dissipated as heat.

24.9 The efficiency of glucose metabolism

24.9.1 The efficiency of oxidative phophorylation

As we saw on pages 733 to 735, the metabolism of one molecule of glucose through glycolysis and the TCA cycle, as summarised in Figure 24.11, results in the pumping of a maximum of 112 H$^+$ ions up the chemi-osmotic potential gradient, from the inside of the membrane to the outside – where we use the phrase 'a maximum of 112 H$^+$ ions' as a reminder that the number of H$^+$ ions actively transported across the membrane by each component of the electron transport chain is still somewhat uncertain.

If the synthesis of one molecule of ATP is coupled to the flow of 3 H$^+$ ions back down the gradient, then the maximum number of ATP molecules synthesised by oxidative phosphorylation for each molecule of glucose metabolised is 112/3 = 37 (rounding down to the nearest whole number). The actual number of ATP molecules synthesised depends on the specific physiological conditions, but there is now a wide consensus that a reliable estimate of the actual number is about 29. The efficiency of oxidative phosphorylation is therefore about 29/37 ≈ 78 %. Looked at in a different way, if the chain were 100 % efficient, the number of H$^+$ ions that need to be pumped is given by 29 × 3 = 87, suggesting that some 25 of the 112 H$^+$ ions actually pumped are 'wasted'. In fact, some of the free energy is required for other purposes such as the translocation of other ions and molecules (such as ATP, ADP and P$_i$) across the membrane, implying that fewer than 112 H$^+$ ions are available in principle for ATP synthase, whilst some H$^+$ ions leak back through the membrane, dissipating their energy as heat. If an allowance is made for the H$^+$ ions used for other purposes, then the number of H$^+$ ions available for ATP synthase is a smaller number than 112 – say, hypothetically, 102 – implying that the estimate of 112/3 = 37 molecules of ATP produced by the oxidative phosphorylation, for each molecule of glucose, is likely to be an over-statement of the theoretical maximum, and that the efficiency estimate of 78% is quite possibly an under-, rather than an over-, estimate.

24.9.2 The overall efficiency of the metabolism of glucose

To determine the overall efficiency of the metabolism of glucose – comprising glycolysis, the TCA cycle and also oxidative phosphorylation – we recall that, when one molecule of glucose is burned in air, 2930 kJ is dissipated as heat

$$\text{Glucose} + 6O_2 \rightarrow 6CO_2 + 6H_2O \qquad \Delta_c G^{\ominus\prime} = -2930 \text{ kJ}$$

If all this free energy captured as ATP, then since $\Delta_r G_c^{\ominus\prime}$ for the synthesis of ATP is 57 kJ mol^{-1}, the maximum number of ATP molecules that could be synthesised is 2930/57 = 51; the actual number is 2 directly from glycolysis, plus 2 from the TCA cycle (synthesised from GTP), plus about 29 from oxidative phosphorylation, giving a total of about 33 molecules of ATP per molecule of glucose. The total free energy captured by the synthesis of 33 mol of ATP is therefore 33 × 57 = 1881 kJ, giving an estimate of the overall efficiency of glucose metabolism as 1881/2930 = 64%. To put that figure into context, we note that the overall process of glucose metabolism – summarised in Table 24.1 – comprises (at least) 23 individual steps, implying that the 36% total energy loss, when spread over these 23 steps, represents an average loss of at most 1.6% at each step. The 64% efficiency of the biological process of glucose metabolism compares favourably to the efficiency of man-made processes: a typical petrol engine in a car achieves an efficiency of about 20%, and a wind turbine, about 40%.

One particular feature of glucose metabolism, as compared to burning glucose in air, is that burning in air is a single step, whereas glucose metabolism takes place over 23 steps – 11 in glycolysis, 9 in the TCA cycle and, at the very least, 3 in oxidative phosphorylation (in fact, there are many more steps, if we were to count all the electronic transitions within the electron transport chain). One of the very earliest principles we explored in this book was reversibility,

Table 24.1 Overall free energy changes for the complete metabolism of glucose in mitochondria, including the capture of free energy by ATP. The grand total of 1881 kJ for the free energy captured by ATP can be compared to the value of 2930 kJ for the free energy released by the combustion of glucose in oxygen, implying that the efficiency of free energy capture is 1881/2930 = 64 %. Note that the first column is expressed per mole of the first-named reactant in each row

	$\Delta_r G_c^{\ominus\prime}$ kJ mol^{-1}	Total $\Delta_r G_c^{\ominus\prime}$ kJ	Total $\Delta_r G_c^{\ominus\prime}$ as ATP kJ	Moles ATP per mole glucose
Glycolysis				
Glucose → 2 pyruvate	−320	−320		
2 ADP + 2 P$_i$ → 2 ATP	57	114	114	2
2 NAD + 4 e$^-$ + 2 H$^+$ → 2 NADH	54	108		
Subtotal		−98		
Pyruvate dehydrogenase				
2 pyruvate + 2 CoASH → Acetyl CoA + 2 CO$_2$	−87.5	−175		
2 NAD$^+$ + 4 e$^-$ + 2 H$^+$ → 2 NADH	54	108		
Subtotal		−67		
TCA cycle				
2 AcetylCoA → 4 CO$_2$	−319	−638		
6 NAD$^+$ + 12 e$^-$ + 6 H$^+$ → 6 NADH	54	324		
2 FAD → 2 FADH$_2$	42.5	85		
2 GDP + 2 Pi → 2 GTP	57	114	114	2
Subtotal		−115		
Oxidative phosphorylation				
10 NADH + 5 O$_2$ + 10 H$^+$ → 10 NAD$^+$ + 10 H$_2$O	−219.3	−2193		
2 FADH$_2$ + O$_2$ + 2 H$^+$ → 2 FAD + 2 H$_2$O	−200	−400		
29 ADP + 29 P$_i$ → 29 ATP	57	1653	1653	∼29
Subtotal		−940		
Grand total		−1220	1881	∼33

where we saw that a reversible change is necessarily quasistatic (see pages 30 and 123), taking place in an infinite number of infinitesimally small steps, and, as a result, reversible changes perform the maximum possible work (see page 143). The 23 steps of glucose metabolism is nowhere near an infinite sequence of infinitesimal steps; nonetheless, glucose metabolism is a real instance of the approach to reversibility, and hence efficiency, especially in comparison to the irreversible single step process of burning glucose in air.

24.10 Photosynthesis

24.10.1 The energy of light

So far, this chapter has looked in detail at the metabolism of glucose, and how the free energy stored in each glucose molecule can be harvested, in the form of ATP, for use elsewhere. This is just one example of bioenergetics, and there are many more. The details are of course different, but the principles are the same, in particular the principles associated with

- coupled reactions, in which the free energy released by an exergonic reaction can be coupled to the free energy required by an endergonic reaction so that an otherwise non-spontaneous reaction can take place
- the use of electrode potentials as a way of quantifying the free energy changes associated with redox reactions
- the use of the biochemical standards
- the interplay between biological structures and biochemical functions, as exemplified by the role of the membrane as a means of controlling the molar concentrations and electrical potentials on either side.

We will not examine any further pathways here, save for mentioning one pathway which is rather different – **photosynthesis**. The special feature of photosynthesis is the role of light: in photosynthetic organisms, light energy is used to drive the endergonic transformation of carbon dioxide and water into sugars – in essence, reversing glycolysis and the TCA cycle. Photosynthesis is of fundamental importance, for it is the primary means by which the energy of the sun is captured and made available for almost all living things.

The fundamental event in photosynthesis is the absorption of the electromagnetic energy contained in light by molecules, such as chlorophyll, present in green plants, algae and certain bacteria. All electromagnetic waves are characterised by a specific wavelength λ m and frequency ν Hz, such that $c = \lambda \nu = 3 \times 10^8$ m sec^{-1}, where c is the speed of light. Also, according to quantum theory, an electromagnetic wave of frequency ν Hz can be interpreted as a stream of **photons**, each associated with a **quantum** of energy given by $h\nu$ J, where h is Planck's constant, 6.626×10^{-34} J sec.

Many molecules naturally absorb electromagnetic radiation: it is this absorption, for example, that determines colour – the light we see is usually reflected from what we are looking at, and is composed of the wavelengths of light that have not been absorbed by the object in question. So, if all the light is absorbed, nothing is reflected, and so the object looks black (which is why windows in houses look black in daylight, when viewed from the outside); if all the light is reflected, the object appears as white, if some wavelengths are absorbed (say, the longer wavelength red waves), then we see wavelengths corresponding to the greens and blues. The molecules that absorb light in photosynthesis are therefore coloured – chlorophylls, for example, absorb blue and red light and so appear green; carotenes absorb blue and green, and appear red and orange; xanthophylls also absorb (a rather different mix of) blue and green, and appear yellow. Indeed, it is the presence of these **photosynthetic pigments** that explains why leaves in spring and high summer are green (dominated by the chlorophyll), and why we see such splendid colours in the autumn – during the autumn, the chlorophyll decomposes quite quickly, so that the colours of the carotenes and xanthophylls are revealed.

When a molecule absorbs light, the energy $h\nu$ carried by a photon 'promotes' an electron from its 'base' state to an 'excited' state of higher energy. The higher energy state is unstable, and so one of three things then happens. One possibility is that the electron returns to its base state, dissipating the energy as heat; a second is that the electron can fall to a lower energy state, releasing the energy as a quantum of light of rather lower energy (and hence longer wavelength and lower frequency), a phenomenon we refer to as **fluorescence**. A third possibility is a process in which the excited molecule transfers the energy (or much of it) to a neighbouring molecule, without radiating any light, in a process known as **resonance energy transfer** or **exciton transfer**. As we shall shortly see, resonance energy transfer plays a very important role in photosynthesis.

24.10.2 Chloroplasts and thylakoids

Very briefly, green plants contain cell organelles called **chloroplasts**, which, like mitochondria, are surrounded by a double membrane. The interior of the chloroplast – known as the **stroma** – hosts a variety of biochemical functions, including DNA replication and transcription, ribosomal protein synthesis, and, very importantly the transformation of molecular carbon dioxide into glyceraldehyde-3-phosphate, which then serves as a substrate for a number of pathways, including the formation of glucose. As the fundamental natural process for 'fixing' atmospheric CO_2 within the biomolecules of living systems, this pathway is of the utmost significance, as will be discussed shortly.

The stroma also contains structures known as **grana**, these being stacks of closed compartments, the **thylakoids**. Each thylakoid is surrounded by a single membrane, which contains the pigments that capture light energy in the **light-dependent reactions of photosynthesis**. The volume enclosed by, and so within, the thylakoid membrane is known as the **lumen**; the outer surface of the thylakoid membrane is in contact with the chloroplast stroma.

24.10.3 Photosystems I and II

The thylakoid membrane, like the mitochondrial inner membrane, contains a number of embedded membrane protein clusters, including an ATP synthase, and complexes named **photosystem I** and **photosystem II**. Photosystem I includes an array of membrane-bound proteins, called the **light-harvesting** (or **antenna**) **complex**, and also a **reaction centre** known as **P700**. The various components of the light-harvesting complex contain pigments which can absorb a variety of wavelengths of light, transferring the absorbed energy by resonance energy transfer to two chlorophyll molecules within P700. Similarly, photosystem II contains another reaction centre, **P680**, whose chlorophyll molecules accept energy captured by photosystem II's light-harvesting complex.

When P680 or P700 absorb sufficient harvested light energy, an electron is promoted to an excited state. The metabolic pathways collectively known as photosynthesis then capture the energy of that excited state, ultimately in the form of chemical free energy stored in ATP and NADPH (which is very similar to NADH). According to Planck's law, the energy carried by a photon of light of frequency ν is given by $h\nu$ J. If we take light of wavelength $\lambda = 600$ nm (and therefore of frequency $\nu = c/\lambda = 5 \times 10^{14}$ Hz) as representative of the light collected by the light-harvesting complexes associated with P680 and P700, then each photon carries an energy $h\nu = 6.626 \times 10^{-34} \times 5 \times 10^{14} \approx 3.3 \times 10^{-19}$ J, and so 1 mol of photons carry $6.022 \times 10^{23} \times 3.3 \times 10^{-19}$ J ≈ 200 kJ, where 6.022×10^{23} mol^{-1} is the Avogadro constant, N_A. Given that the free energy required to synthesise 1 mol of ATP from ADP + P_i is 57 kJ, we see that there is more than enough energy to synthesise ATP from light, provided that there is a mechanism to achieve this – which is precisely what photosynthesis does, according to two metabolic pathways: **non-cyclic photophosphorylation**, and **cyclic photophosphorylation**.

24.10.4 Non-cyclic photophosphorylation

Non-cyclic photophosphorylation is represented schematically in Figure 24.12.

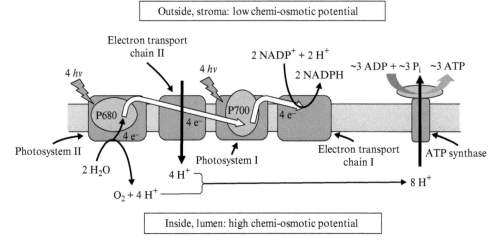

Figure 24.12 A schematic representation of non-cyclic photophosphorylation. Embedded within the thylakoid membrane are the protein complexes photosystems I and II, with their associated electron transport chains, and also ATP synthase. 'Inside' refers to the lumen within the thylakoid; 'outside' refers to the chloroplast stroma. Light, with energy $h\nu$, is absorbed by a pigment in photosystem II's light-harvesting complex, and the energy is transferred by resonance energy transfer to the reaction centre P680. The energy collected within P680 causes the excitation of electrons, which are captured by an electron transport chain, and replenished by splitting water. The electrons pass along the chain from photosystem II to photosystem I, where a second excitement takes place within photosystem I's reaction centre, P700, ultimately resulting in the formation, within the stroma, of NADPH. As the electrons pass from photosystem II to photosystem I, H$^+$ ions are translocated across the membrane from the stroma outside to the lumen inside, creating a chemi-osmotic potential which drives the formation of ATP outside the membrane, within the chloroplast. The stoichiometry shown corresponds to the flow of 4 electrons, and the resulting formation of O$_2$, 2 NADPH and 3 ATP.

The first event is the transfer of light from the light-harvesting complex of photosystem II, and the collection of this light within P680, causing excitation to the excited state P680*,

$$P680 + h\nu \rightarrow P680^*$$

The excited state P680* then releases an electron to the first component of an electron transport chain (similar to the chain we met in connection with mitochondrial oxidative phosphorylation), resulting in an electron deficient molecule P680$^+$

$$P680^* \rightarrow P680^+ + e^- \text{ (captured by the electron transport chain)}$$

The electron lost from the P680 molecule is re-instated by coupling the transfer of an electron to P680$^+$

$$P680^+ + e^- \rightarrow P680$$

with an enzymic reaction which splits water into molecular oxygen, H⁺ ions, and electrons as

$$2H_2O \rightarrow O_2 + 4H^+ + 4\,e^-$$

This reaction takes place on the inner surface of the thylakoid membrane, and so the H⁺ ions accumulate within the thylakoid lumen. The molecular oxygen ultimately escapes to the atmosphere, and it is this reaction that creates the oxygen that we, and all other animals, breathe. Overall, the originally excited P680 molecule is replenished, allowing the next cycle to take place, and each molecule of O_2 produced corresponds to the entry of 4 electrons into photosystem II's electron transport chain. As illustrated in Figure 24.12, as the electrons pass along the electron transport chain, their free energy is used to translocate H⁺ ions across the thylakoid membrane, depleting the H⁺ ion concentration in the stroma outside, whilst further increasing the H⁺ ion concentration within the lumen inside.

On reaching the end of the photosystem II electron transport chain, the electrons are captured by photosystem I, where a second excitation event takes place within P700, which also collects energy from its associated light-harvesting complex. These re-excited electrons then pass down a second electron transport chain, ultimately synthesising NADPH (similar to NADH) from NADP⁺ (similar to NAD⁺) on the outer surface of the thylakoid, within the chloroplast stroma

$$NADP^+ + H^+ + 2\,e^- \rightarrow NADPH$$

Stoichiometrically, 4 photons will excite 4 electrons within P680, which are replaced by the splitting of 2 H_2O and the formation of one molecule of O_2, and 4 H⁺, within the thylakoid lumen. The flow of 4 electrons along electron transport chain II drives the translocation of 4 H⁺ ions across the thylakoid membrane from the outside stroma to the inside lumen. A further 4 photons excite the electrons within P700, so enabling the formation of 2 NADPH, and consuming 2 H⁺, on the outside of the thylakoid. The overall reaction can therefore be summarised as

$$2\,H_2O + 8\,h\nu + 2NADP^+ + 2\,H^+_{out} \rightarrow O_2 + 2\,NADPH + 4H^+_{in} + 4H^+_{out \rightarrow in}$$

where 2 H^+_{out} represents the H⁺ ions consumed in the formation of NADPH in the chloroplast stroma outside the thylakoid; 4 H^+_{in}, the H⁺ ions produced by splitting water inside the thylakoid lumen; and 4 $H^+_{out \rightarrow in}$, the H⁺ ions translocated across the membrane, from outside to inside, by electron transport chain II.

Within the chloroplast stroma, outside the thylakoid, a total of 6 H⁺ ions are consumed, two in the synthesis of NADPH from NADP⁺, and four translocated across the thylakoid membrane. The stroma, however, is quite active biochemically, and so these H⁺ ions are replenished from other reactions, so that the stroma maintains a mildly alkaline pH around pH 8. But within the thylakoid lumen – which has a very small volume and a very limited biochemistry – the 8 H⁺ ions accumulate and make the lumen quite acidic, perhaps around pH 4, implying that there is a pH gradient $\Delta pH = pH_{in} - pH_{out} \approx -4$ across the thylakoid membrane.

As we saw on page 732, the chemi-osmotic potential gradient $\Delta G_{co}'$ may be expressed as

$$\Delta G_{co}' = -2.303\,RT\,\Delta pH + F\Delta\Psi \tag{24.13}$$

It so happens that the thylakoid membrane is quite permeable to a number of ions, including Mg^{2+} and Cl^-, and so the electrical effect of the accumulation of H^+ ions within the thylakoid lumen is largely cancelled out. This implies that, for the thylakoid membrane, $\Delta\Psi \approx 0$, and so

$$\Delta G_{co}' = -2.303\, RT\, \Delta pH \approx -2.303\, RT\, (-4) \approx 22.8\, kJ$$

This value, 22.8 kJ, for the chemi-osmotic potential gradient associated with the thylakoid membrane, can be compared to the corresponding value -21.2 kJ, for the mitochondrial membrane. The difference in the signs is attributable to the definition of $\Delta G_{co}'$

$$\Delta G_{co}' = [G'(H^+, in) - G'(H^+, out)] + F[\Psi(in) - \Psi(out)] \tag{24.6}$$

in which the chemi-osmotic potential $G' + F\Psi$ outside the membrane is subtracted from that for the inside of the membrane. For the mitochondrion, H^+ ions are pumped from the inside to the outside, so that the outside has the higher chemi-osmotic potential: $\Delta G_{co}'$ is therefore negative. For the thylakoid, H^+ ions are pumped in the other direction, from the outside to the inside, implying that the higher chemi-osmotic potential is on the inside, and that $\Delta G_{co}'$ is correspondingly positive.

What is particularly striking is the similarity in the magnitudes of $\Delta G_{co}'$, even though the values of ΔpH and $\Delta\Psi$ are quite different: for the mitochondrion $\Delta pH \approx 0.5$ and $\Delta pH \approx -190$ mV; for the thylakoid, $\Delta pH \approx -4$ and $\Delta\Psi \approx 0$.

In both cases, as shown in Figures 24.11 and 24.12, the chemi-osmotic potential drives the synthesis of ATP, and for photosynthesis, the overall stoichiometry has been determined experimentally as

$$2H_2O + 8h\nu + 2NADP^+ + 3\, ADP + 3\, P_i \rightarrow O_2 + 2\, NADPH + 2\, H^+ + 3\, ATP$$

Chemically, this may be decomposed into a reaction for the synthesis of ATP

$$3\, ADP + 3\, P_i \rightarrow 3\, ATP \qquad \Delta_r G_c^{\ominus\prime} = 3 \times 57 = 171\, kJ$$

and the two half-reactions

$$2\, H_2O \rightarrow O_2 + 4H^+ + 4e^- \qquad \mathbb{E}^{\ominus\prime} = -0.816\, V$$
$$\Delta_r G_c^{\ominus\prime} = 315\, kJ$$

and

$$2\, NADP^+ + 2H^+ + 4e^- \rightarrow 2NADPH \qquad \mathbb{E}^{\ominus\prime} = -0.324\, V$$
$$\Delta_r G_c^{\ominus\prime} = 125\, kJ$$

where $\Delta_r G_c^{\ominus\prime}$ for the redox half-reactions has been determined as

$$\Delta_r G_c^{\ominus\prime} = -n_e\, F\, \mathbb{E}^{\ominus\prime} \tag{24.2}$$

recalling (see page 614) that $\mathbb{E}^{\ominus\prime}$ is an intensive state function, and so takes the same value for all stoichiometries, and that the values of $\Delta_r G_c^{\ominus\prime}$ refer to the reactions as written.

We may therefore compute $\Delta G^{\ominus\prime}$ for the chemical reaction

$$2\, H_2O + 2\, NADP^+ + 2\, H^+ + 3\, ADP + 3\, P_i \rightarrow O_2 + 2\, NADPH + 3\, ATP$$

as $\Delta_r G_c^{\ominus\prime} = 315 + 125 + 171 = 611$ kJ, showing that the chemical reaction, by itself, is strongly endergonic, with the equilibrium position far to the left.

We now take account of the energy from the light, of a representative wavelength $\lambda = 600$ nm, $\nu = 5 \times 10^{14}$ Hz. For each mole of O_2 produced, 8 mol of photons are consumed (4 mol in photosystem II, 4 mol in photosystem I), and since $8N_A h\nu \approx 1600$ kJ, more than enough light energy is available to provide the 611 kJ required for the splitting of two molecules of water, and synthesis of 2NADPH and 3 ATP.

There is however, one point which we have not fully addressed. As can be seen from Figure 24.12, the formation of one molecule of O_2 is associated the splitting of two molecules of water, producing 4 H^+ ions within the lumen, and releasing 4 electrons to photosystem II's electron transport chain. As the electrons pass down the chain, four more H^+ ions are pumped across the thylakoid membrane from outside to in, giving a total accumulation of 8 H^+ ions within the lumen to drive the synthesis of three molecules of ATP. The stoichiometry of ATP formation, however, requires 3 H^+ ions for each molecule of ATP synthesised, implying that 9 H^+ ions are needed within the lumen. One H^+ ion is therefore 'missing'.

Where, then, is the missing ion? There are several possible explanations: one, for example, is the uncertainty in the stoichiometry of the various processes, but another lies in the existence of second photosynthetic process, acting independently of non-cyclic photophosphorylation, which uses the energy of light solely to effect the translocation of H^+ ions across the thylakoid membrane - a possibility we now discuss...

24.10.5 Cyclic photophosphorylation

As can be seen from the overall reaction

$$2 H_2O + 8h\nu + 2 NADP^+ + 3 ADP + 3 P_i \rightarrow O_2 + 2 NADPH + 2 H^+ + 3 ATP$$

the pathway for non-cyclic phosphorylation results in the formation of both NADPH and ATP, in a defined stoichiometric ratio, 2 NADPH : 3 ATP. Both these molecules share the property of being energy currencies, but there is one important difference: NADPH is a reducing agent, whereas ATP is neither oxidising nor reducing. Non-cyclic photophosphorylation therefore does two things simultaneously: it stores light energy as chemical free energy in the form of ATP and NADPH, whilst also making the stroma of the chloroplast progressively more reducing by virtue of the synthesis of NADPH. As we have noted (see, for example, page 720), too strong a reducing environment can be very damaging, and so cells have various pathways to control the ratio $[NADP^+]/[NADPH]$, just as the ratio $[NAD^+]/[NADH]$ is controlled. Furthermore, the thylakoid membrane has a second photosynthesis pathway, known as **cyclic photophosphorylation**, which can create a chemi-osmotic potential, and so synthesise ATP without producing any NADPH, as illustrated in Figure 24.13.

In non-cyclic photophosphorylation, as we saw in Figure 24.12, the excitation of an electron within P700 results in the activation of electron transport chain I, and the synthesis of NADPH. In cyclic photo-phosphorylation, the electron excited within P700 is not used to synthesise NADPH, but is captured by the electron transport chain of photosystem II, and then returns to P700, whilst simultaneously driving the translocation of one H^+ ion into the interior of the thylakoid, as shown on Figure 24.13. Since the electron lost from P700 after excitation is

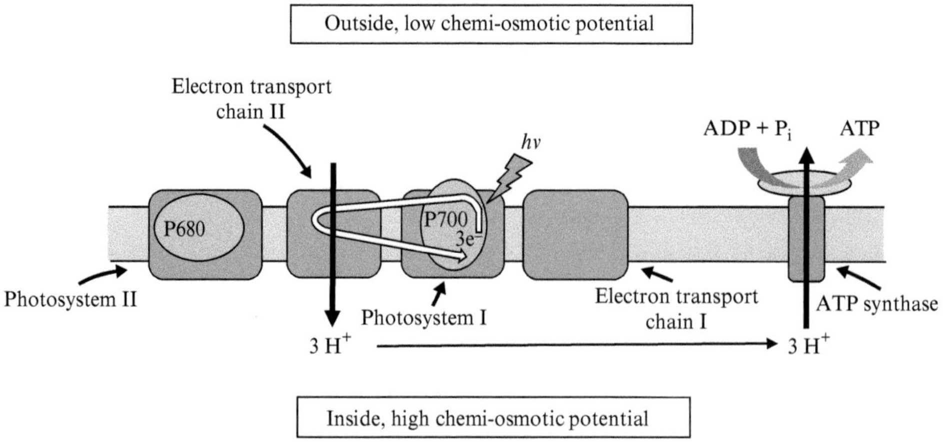

THE BIOENERGETICS OF LIVING CELLS

Figure 24.13 A schematic representation of cyclic photophosphorylation. Light excites an electron within P700, which is captured by electron transport chain II, and then returned to P700. As electrons pass down the electron transport chain, H^+ ions are translocated from outside the thylakoid to the inside, so creating the chemi-osmotic potential required for the synthesis of ATP.

replenished by the return from electron transport chain II, the process is cyclic; thermodynamically, the energy captured from the light is transformed into the chemi-osmotic potential attributable to the translocation of the H^+ ions, so creating the proton motive force required to produce ATP.

Cyclic photophosphorylation can therefore produce ATP, but without forming NADPH, and also without splitting water. Non-cyclic and cyclic photophosphorylation can operate independently, and so the cell can create whatever mixture of NADPH and ATP it needs at any time. Also, it is possible that a single photon could result in the translocation of a single H^+ ion, so providing the 'missing ion' noted on page 743.

24.10.6 The 'light-independent' reactions of photosynthesis

Overall, photosynthesis transforms light energy into ATP and NADPH, so building a store of chemical free energy that can subsequently be used in any metabolic process within the cell. Of especial importance within photosynthesising cells is a pathway comprising the so-called **'light-independent'** (or **'dark'**) **reactions of photosynthesis** (because they do not require light) in which molecular carbon dioxide and water are transformed into glyceraldehyde-3-phosphate by the **Calvin-Benson-Bassham** (alternatively, **CBB, Calvin** or **carbon-fixation**) **cycle**. The full details of this cycle will be found in a biochemistry text – we note here that the key step is the reaction catalysed by the enzyme known as **RuBisCO** in which molecular CO_2 is incorporated into the 5-carbon sugar ribulose-1,5-biphosphate (RuBP) to form two molecules of the 3-carbon molecule 3-phosphoglycerate. RuBisCO is a very important protein, for it is the only known enzyme that can capture molecular CO_2 from the earth's atmosphere and incorporate it into a biomolecule, a process known as CO_2 **fixation**.

Stoichiometrically, as catalysed by RuBisCO, 6 molecules of ribulose-1,5-biphosphate combine with 6 molecules of CO_2 to form 12 molecules of 3-carbon glyceraldehyde-3-phosphate, of which 2 can then be synthesised into (amongst other things) 1 molecule of 6-carbon glucose. The remaining 10 molecules of the 3-carbon molecule glyceraldehyde-3-phosphate (which collectively contain 30 carbon atoms) are then metabolised into 6 molecules of the 5-carbon molecule ribulose-1,5-biphosphate (which also collectively contain 30 carbon atoms), so replenishing those originally used, and completing the cycle.

Overall, the light-independent reactions of photosynthesis, followed by the synthesis of glucose from the 3-phosphoglycerate produced by the CBB cycle, can be summarised as

$$6\,CO_2 + 12\,NADPH + 18\,ATP + 12\,H_2O \rightarrow Glucose + 12\,NADPH^+ + 6\,H^+ + 18\,ADP + 18\,P_i$$

As we saw on page 741, for the absorption of 8 photons, non-cyclic photophosphorylation yields 2 NADPH + 3 ATP, and produces one molecule of O_2, implying that the absorption of 48 photons will result in the formation of 12 NADPH + 18 ATP, and 6 O_2. That provides enough NADPH and ATP for the formation of one molecule of glucose, so that, fundamentally, 1 mol of glucose is synthesised using the energy of 48 mol of photons. If the photons are typically of wavelength $\lambda = 600$ nm and $\nu = 5 \times 10^{14}$ Hz, then 48 mol of photons carry an energy of $48\,N_A h\nu = 48 \times 6.022 \times 10^{23} \times 6.626 \times 10^{-34} \times 5 \times 10^{14} \approx 9580$ kJ - more than enough for the 2930 kJ required for the synthesis of glucose

$$6\,CO_2 + 6\,H_2O \rightarrow Glucose + 6\,O_2 \qquad \Delta_c G^{\ominus \prime} = 2930\,\text{kJ}$$

and representing an efficiency of about $2930/9580 \approx 31\,\%$.

Finally, we mention one other form of light-related biochemical reaction – **bioluminescence**. This is the phenomenon in which living systems produce light, as may be familiar with the display shown by fireflies, and perhaps less familiar in relation to the often-spectacular behaviour of some marine species. Bioluminescence is attributable to the emission of light by a molecule in a suitably excited state, where the excitation is the result of coupling to an exergonic reaction such as the hydrolysis of ATP. So, in fireflies, for example, the enzyme luciferase catalyses the oxidation by molecular oxygen of the pigment luciferin as

$$\text{Luciferin} + O_2 + ATP \rightarrow \text{Oxy-luciferin} + CO_2 + AMP + \text{light}$$

in which ATP is hydrolysed to AMP (rather than ADP), and the reaction emits light.

EXERCISES

1. Define 'energy currency', and explain both the biochemical and the thermodynamic roles of molecules such as ATP. Why is the term 'high-energy bond' misleading?
2. Define
    - concentration gradient
    - pH gradient
    - electrical potential gradient
    - membrane potential
    - chemi-osmotic potential
    - proton motive force.

3. Explain the role of the mitochondrial membrane in creating, and maintaining, each of the phenomena mentioned in question 2. Imagine that you are carrying out some research, and that you need to design an experimental system that can be used to explore these phenomena, but without the use of any biological materials. To do this, you will need to select a suitable material for the artificial membrane, and to control, and measure, both the hydrogen ion concentration, and the electrical potential, on each side of the membrane. What are the necessary properties of the artificial membrane? And how might you design a suitable apparatus?
4. The 1997, two of the winners of the Nobel Prize in Chemistry were Paul Boyer, of UCLA, and John Walker, of Cambridge, for their work in relation to ATP synthase. Research what they did, and as a result, draw a much better diagram of ATP synthase than Figure 24.10!
5. You have been invited to give a one-hour presentation, entitled "How light is captured by green plants", to a class at a pre-university summer school. Prepare that presentation, including the text, drafts of any slides you would use, and drafts of any handout materials.
6. Grass is green. Discuss.

25 Macromolecular conformations and interactions

Summary

Under normal physiological conditions, the three-dimensional structure of any given **protein** is very specific and well-defined, in which the polypeptide chain is folded into a unique, stable, configuration, the **native state** N. Given that, in principle, the polypeptide could fold into many other states, the native state must represent the Gibbs free energy minimum of all possible states.

In the laboratory, it is possible to control experimental conditions such that a protein in its native state can reversibly change its configuration into an **unfolded state** U; furthermore, there are a variety of methods that can be used to monitor the unfolding and so derive the standard free energy of unfolding $\Delta_{unf} G^{\ominus}$, and the corresponding equilibrium constant K_{unf}. $\Delta_{unf} G^{\ominus}$ is a measure of the stability of the native state, as compared to the unfolded state, such that the more positive the value of $\Delta_{unf} G^{\ominus}$, the more stable the native state.

When in their native states, many proteins have the ability to bind with specific **ligands** – indeed all enzymes rely on this property to enable them to perform their correct functions. The equilibrium between a native protein, the free ligand, and the **protein-ligand complex** can be described in terms of a **dissociation constant** K_d, which is an inverse measure of the ligand-protein **binding affinity**: the smaller the value of K_d, the lower the concentration of the free ligand in the equilibrium solution, and so the more tightly bound the ligand (and *vice versa*). Measurements of K_d enrich our understanding of how proteins and ligands interact, which is of especial importance in the design of drugs.

This chapter also introduces some methods for study of the kinetics of protein folding, including **phi-value** analysis, which is based on comparisons between the structures of a given **wild-type protein**, and one or more **mutant proteins**.

25.1 Protein structure

Proteins are large, complex, three-dimensional molecules that are capable of very specific interactions with many other molecules, such as other proteins, DNA, RNA, and a wide variety of small molecules. Some proteins, called **enzymes**, are also capable of catalysing chemical reactions with very high specificity. These functions place proteins at the centre of almost all biological processes, including the catalysis of cellular reactions in metabolism, cellular dynamics, growth, organisation and structure, and also in mediating the interactions between cells to form organs. Understanding the structure and stability of proteins, and their functional interactions, is therefore of major importance in biology.

A detailed description of protein structure can be found in many biochemistry textbooks and will not be dealt with here beyond a basic introduction; the purpose of this chapter is to

Modern Thermodynamics for Chemists and Biochemists. Dennis Sherwood and Paul Dalby.
© Oxford University Press 2018. Published 2018 by Oxford University Press.

explore the forces that drive the folding of proteins into their final, stable, 'native' structure, as well as the thermodynamics that control these processes – processes which ultimately allow any given protein to perform its appropriate function.

Proteins are synthesised by cellular 'machines', called **ribosomes**, that read and decode a messenger-RNA (m-RNA) sequence, derived directly from the associated DNA sequence, to identify the appropriate encoded amino acid (from 20 possible), and then catalyse a polymerisation that condenses the next amino acid onto the previous one. A **polypeptide chain** is therefore formed like 'beads on a string', where each 'bead' is an amino acid, with the sequence of the final polypeptide chain corresponding directly to that encoded in the RNA sequence. This linear sequence is known as the **primary structure**.

Functional proteins are formed when one or more of these polypeptide chains folds up and assembles into a complex three-dimensional structure. The final structure – known as the **native** structure - is not random, and is in fact formed with remarkable precision into a unique structure that is almost entirely pre-defined by the polypeptide's amino acid sequence. There are two thermodynamic considerations that determine that only a single, unique, final structure is formed, even though there are an infinite number of possible structures. The first is the fundamental principle – as encapsulated in the 'Fourth Law'(see page 408) – that we have used many times in earlier chapters: the fact that the final native structure must be the global Gibbs free energy minimum on the 'landscape' of all possible conformations. This not only ensures that the final population of structures is homogeneous, with all molecules of the protein having the same structure, but also influences for how long a protein will exist inside a cell to perform its function, before it is no longer required and can be recycled back into its amino acid components.

The second concerns the process of protein folding, which is not a 'random walk' through all possible conformations, but rather a highly orchestrated sequence of events that begins with the formation of small islands of structure at specific points along the whole polypeptide chain. These induce the formation of stable elements called **secondary structures**, leading to more complex **tertiary structures**, which might then aggregate as **quaternary structures**. The thermodynamic stabilisation of structural intermediate states is therefore very important in directing the folding of a protein towards the final native structure.

In this chapter, we firstly explore the G-minimising principle, and show how the Gibbs free energy of a protein structure can be measured, so providing the evidence that the Gibbs free energy of the native structure is less than that of other, unfolded, structures. We then examine protein-ligand binding, and finally the kinetics of protein folding, so gaining insight into the process by which an unfolded polypeptide chain moves towards the final stable native state.

25.2 The thermodynamics of protein folding

25.2.1 How do proteins fold?

As we have just discussed, although all proteins are formed from long polypeptide chains of amino acids, under normal physiological conditions, many proteins adopt a compact, globular shape in which the polypeptide chain is **folded** into a very specific three-dimensional native structure. If the conditions are disturbed – for example, if some urea is added to a solution of an enzyme, or if the enzyme is gently heated – the native structure unfolds and the protein's enzymic function is lost. But if the urea is subsequently removed, or the solution gently cooled, the enzymic function can be restored, implying that the previously unfolded structure has returned to its original native state. This implies that the **native** N and **unfolded** U states are mutually interchangeable, according to a reversible reaction of the form

$$N \rightleftharpoons U$$

Under normal physiological conditions, the equilibrium is towards the left, with the protein in its native state; on the addition of urea, or under gentle heating, the equilibrium shifts to the right, favouring the unfolded state. When a protein is unfolded, it is necessarily not in its native state, and so is '**denatured**'. Some denatured states, however, are so denatured that they can never return to the native state, as, for example, happens when a protein is vigorously heated. In this chapter, we are concerned with reversible unfolding, rather than irreversible denaturation, and so will refer to the 'native' and 'unfolded' states, with the implicit assumption that these two states are mutually reversible.

Given the complexity of the polypeptide chain that is folding and unfolding, an important question concerns the nature of the folding-unfolding reaction. Does this take place through a sequence of different unfolded equilibrium states, each of which is thermodynamically stable? Or does one state 'flip' into the other in a single step, rather like a phase change in which a solid melts into a liquid? To distinguish between these two possibilities, suppose we can measure the degree of unfolding of a protein for progressively increasing values of an agent that causes unfolding, such as the concentration of urea present in the protein solution, or the temperature of the solution. If the folding-unfolding process were to pass through a sequence of thermodynamic equilibrium states, each corresponding to a different degree of unfolding which is stable for a specific range of urea concentration or temperature, then a series of such measurements would result in a 'stair-case' diagram, such as that illustrated in Figure 25.1(a); but if the process is more like a phase change between two states (see Figure 9.5 on page 296), the result would be a single sigmoid 'step', such as Figure 25.1(b).

As we shall see, most experimental results are predominantly of the form shown in Figure 25.1(b), implying that the folding-unfolding reaction is a co-operative, phase-like, change of state between a single thermodynamically stable native state N and a second, single, thermodynamically stable unfolded state U.

25.2.2 $\Delta_{unf} G^\ominus$ – a measure of protein stability

For the reversible reaction

$$N \rightleftharpoons U$$

between the native N state of a protein and its unfolded state U, under any conditions of temperature, pressure, pH, and so on, the native state N will be associated with a Gibbs free energy G_N, and the unfolded state U with a Gibbs free energy G_U. For the *unfolding* change

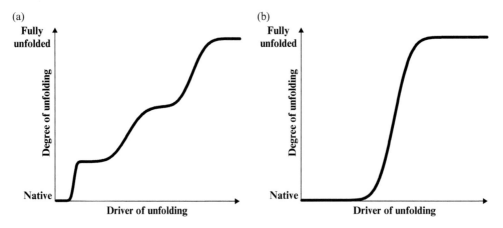

Figure 25.1 Two models of protein unfolding. The vertical axis represents the degree of unfolding of a protein, from the native state at the bottom, to the fully unfolded state at the top. The horizontal axis represents the value of an agent that causes unfolding, such as the concentration of urea in the protein solution, or the temperature of the solution. If the folding-unfolding process were to take place through a sequence of different unfolded equilibrium states, each of which is thermodynamically stable for the corresponding range of the unfolding agent, then we would expect a result such as (a); if the process is more like a phase change between two states, then the result would be (b).

from the native state N to the unfolded state U, we may define the corresponding Gibbs free energy of unfolding $\Delta_{unf} G$ as

$$\Delta_{unf} G = G_U - G_N$$

We note that the difference $G_U - G_N$ is sometimes represented as ΔG_{U-N}, where the subscript U-N represents 'U minus N' (importantly, *not* 'from U to N'). Our preference, however, and the notation we shall use here, is $\Delta_{unf} G$, for this makes explicit that the Gibbs free energy change refers to the process of unfolding, the change from the native state N to the unfolded state U, as well as being consistent with the standards shown in Table 6.5 (see page 177). Accordingly, for the *folding* change, from the unfolded state U to the native state N, the corresponding Gibbs free energy change is

$$\Delta_{fol} G = G_N - G_U = -\Delta_{unf} G$$

If, during the course of the unfolding reaction, the molar concentration of the native protein is represented as [N], and the molar concentration of the unfolded protein as [U], then the unfolding reaction may be associated with a mass action ratio Γ_{unf} as

$$\Gamma_{unf} = \frac{[U]}{[N]}$$

As we saw on page 552, for any reaction taking place in solution

$$\Delta_r G_c = \Delta_r G_c^{\ominus} + RT \ln \Gamma_c \tag{16.16c}$$

and so, for the protein unfolding reaction, we may write

$$\Delta_{unf} G = \Delta_{unf} G^{\ominus} + RT \ln \Gamma_{unf}$$

where $\Delta_{unf} G^{\circ}$ is the corresponding change in the standard Gibbs free energy for the folding-unfolding reaction.

When the native and unfolded states are in equilibrium, $\Delta_{unf} G = 0$, and the mass action ratio Γ_{unf} becomes the equilibrium constant K_{unf}, implying that

$$\Delta_{unf} G^{\circ} = - RT \ln K_{unf} \tag{25.1}$$

where

$$K_{unf} = \frac{[U]_{eq}}{[N]_{eq}} \tag{25.2}$$

in which $[N]_{eq}$ and $[U]_{eq}$ are the molar concentrations of the native and unfolded states in any given equilibrium mixture. Accordingly, equations (25.1) and (25.2) imply that

$$\Delta_{unf} G^{\circ} = - RT \ln \frac{[U]_{eq}}{[N]_{eq}} \tag{25.3a}$$

and

$$\frac{[U]_{eq}}{[N]_{eq}} = \exp\left(- \frac{\Delta_{unf} G^{\circ}}{RT}\right) \tag{25.3b}$$

For many proteins, the two-state description $N \rightleftharpoons U$ of native state stability is sufficient, as any intermediate states along the folding – unfolding trajectory are only transiently populated. For some proteins, however, additional terms are required in equations (25.1), (25.2) and (25.3) to include the population of intermediate states at equilibrium, whilst some other proteins do not fold reversibly, but form additional mis-folded and trapped states, or condense irreversibly into aggregates containing multiple proteins, as will be discussed in Chapter 25.

Given that proteins are important biomolecules, it might be thought that the standard Gibbs free energy change $\Delta_{unf} G^{\circ}$ and equilibrium constant K_{unf} should be expressed in terms of the biochemical standard states as $\Delta_{unf} G^{\circ\prime}$ and K_{unf}'. As discussed on page 681, however, measurements made using the biochemical standard state are different from those using the conventional standard state only when hydrogen ions are explicitly involved in the reaction of interest. For protein folding and unfolding, it is certainly true that the protein in solution interacts with the ions that might be present in the aqueous solvent, including hydrogen ions, and this interaction could well influence the relative stability of the native and unfolded states. Hydrogen ions, however, are not directly involved in the stoichiometry of the folding-unfolding reversible reaction $N \rightleftharpoons U$, and so conventional standards, as in equations (25.1), (25.2) and (25.3), can validly be used.

As it happens, the concept of the standard state as regards protein folding and unfolding is somewhat elusive, since there are no standard reference tables of, for example G°(native) and G°(unfolded) for any given protein, from which $\Delta_{unf} G^{\circ} = G^{\circ}$(unfolded) $- G^{\circ}$(native) can be computed, allowing the subsequent calculation of K_{unf} according to equation (25.1). This is quite different from the situation in conventional chemistry in which, as we saw on pages 180 and 398, tables of standards are an important aspect of every-day laboratory work. In the study of proteins, in practice, as we shall see shortly, values of K_{unf} are measured experimentally, and then equation (25.1) is used to compute the corresponding value of $\Delta_{unf} G^{\circ}$. Values of $\Delta_{unf} G^{\circ}$, and other related thermodynamic quantities such as $\Delta_{unf} H^{\ominus}$, are therefore determined by experiment, rather than from tables, and for this reason, some sources omit the $\ominus$ superscript. For clarity, however, and for consistency throughout this book, we will continue to use the notation $\Delta_{unf} G^{\ominus}$.

In general, $\Delta_{unf} G^{\ominus}$ will vary with temperature and pressure, but since proteins are studied in aqueous solution, $\Delta_{unf} G^{\ominus}$ will also vary with other system conditions, such as the pH, and also the molar concentrations of other components which might be dissolved in the solution. Indeed, as we shall shortly see, it is varying these conditions that we may gain experimental data on $\Delta_{unf} G^{\ominus}$ and related thermodynamic parameters.

A consequence of equation (25.1)

$$\Delta_{unf} G^{\ominus} = -RT \ln K_{unf} \tag{25.1}$$

is therefore that, under any given system conditions, if $\Delta_{unf} G^{\ominus} < 0$, then $K_{unf} = [U]_{eq}/[N]_{eq} > 1$, implying that, in any equilibrium mixture, the molar concentration $[N]_{eq}$ of the native state N will be smaller than the molar concentration $[U]_{eq}$ of the unfolded state U. Similarly, if $\Delta_{unf} G^{\ominus} > 0$, then $K_{unf} = [U]_{eq}/[N]_{eq} < 1$ and $[U]_{eq} < [N]_{eq}$, implying that the equilibrium mixture will favour the native state N over the unfolded state U; whilst $\Delta_{unf} G^{\ominus} = 0$ implies that $K_{unf} = 1$, and that $[U]_{eq} = [N]_{eq}$, and so the equilibrium mixture will contain equal molar concentrations of the two states. Accordingly, the more positive the value of $\Delta_{unf} G^{\ominus}$, the more stable the native state as compared to the unfolded state.

25.2.3 Measuring protein stability experimentally

Under normal biological conditions, the native state of a protein is stable, implying that $\Delta_{unf} G^{\ominus} > 0$. In the laboratory, however, for many proteins, it is possible to change the conditions – for example, by controlling the concentration of other components, such as urea – so that the protein can alter its structure, changing reversibly between the native and unfolded states, so enabling the thermodynamics of the reversible folding-unfolding reaction to be studied.

For example, a protein can be reversibly folded and unfolded by

- controlling the pH (most proteins are usually in their native state around pH 7, and become unfolded at extreme pH values)
- introducing chaotropes such as urea, $CO(NH_2)_2$; guadinium chloride, GdmCl, $C(NH_2)_3Cl$; or detergents (increasing the concentration of a chaotrope causes progressive unfolding)
- gentle heating, or sometimes cooling
- increasing the pressure.

If, for example, a solution of urea is added to a native protein, as the concentration of urea increases, the protein reversibly unfolds. As this happens, there are several experimental techniques that can be used to determine the degree of unfolding, and that allow measurement of the corresponding equilibrium concentrations $[U]_{eq}$ and $[N]_{eq}$ of the unfolded and native states. Two commonly-used spectroscopic methods that can detect the proportion of native protein present in a native-unfolded mixture are **circular dichroism** and **intrinsic fluorescence**. Circular dichroism detects the type of folded secondary structure present, and – importantly – can also measure the overall amount of secondary structure content relative to that in the fully native state. Complementary to this, intrinsic fluorescence intensity reports mainly on changes in the amount of tertiary structure present, again relative to that in the fully native protein. Both of these methods are very useful, for in the native state, the protein has a defined tertiary structure, and the maximum extent of secondary structure, whereas in the unfolded state, both tertiary and secondary structures are denatured and randomised.

Many spectroscopic instruments are now conveniently provided with thermal control, as well as auto-injectors (to enable controlled quantities of, for example, urea, to be introduced), so allowing for highly accurate step-wise perturbations between successive measurements.

25.2.4 Measuring protein unfolding using chaotropes

To measure the thermodynamic properties associated with protein unfolding, a typical experiment starts with a solution of the protein under study in its native state. Progressive quantities of a chaotrope such as urea are then introduced, increasing the molar concentration [Ur], so causing the protein to unfold and driving the reversible reaction

$$N \rightleftharpoons U$$

to the right. For any value of [Ur], spectroscopic measurements can be used to estimate the corresponding instantaneous equilibrium molar concentrations $[N]_{eq}$ and $[U]_{eq}$ of the native, and unfolded, protein. The ratio $[U]_{eq}/[N]_{eq}$ then determines the equilibrium constant K_{unf}, from which the $\Delta_{unf} G^{\ominus}$ can be computed using equation (25.1).

A typical profile for the unfolding of a protein by progressive addition of urea is shown in Figure 25.2, which has the general shape of Figure 25.1(b).

In this example, the protein structure was monitored using circular dichroism, which uses circularly polarised light to determine an optical property known as the **mean residue ellipticity**, a measure of the presence of secondary protein structures such as α-helices and β-pleated sheets. When a protein is in its native state, the mean residue ellipticity has a negative value; when the protein has become fully unfolded, the mean residue ellipticity becomes zero. The transition from the native state N to the unfolded state U is clearly shown as taking place between concentrations of urea of about 2 mol dm^{-3} and 4 mol dm^{-3}, and, as we saw on page 749, sigmoid shapes of this type are associated with phase-like transitions, the transition in this case being that from the protein's native state to the unfolded state. While other states exist along the pathway between these two states, they are only transiently-formed intermediates, which are never significantly populated at equilibrium. Unfolding transition profiles such as that shown in Figure 25.2 can be used to determine the relative populations of the native and unfolded proteins at any concentration of urea, so making it possible to determine the stability of the native state relative to the unfolded, as defined by the value of $\Delta_{unf} G^{\ominus}$.

Experimentally, we measure the intensity I of the circular dichroism signal at different concentrations [Ur] of urea, and we define I_N, I_U and I_{obs} as

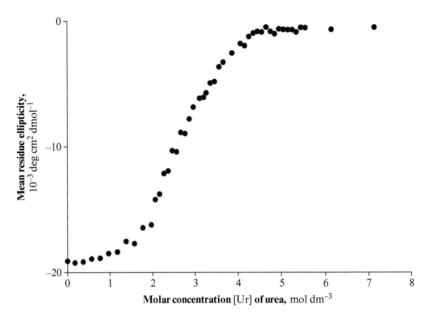

Figure 25.2 Measuring the unfolding of a protein. At low concentrations of urea, the protein is in its native state (as characterised by a negative value for the mean residue ellipticity), whereas at high concentrations of urea, the protein becomes unfolded (as characterised by a value of zero for the mean residue ellipticity). The sigmoid nature of this curve implies that the change from the native state N to the unfolded state U is a relatively sharp co-operative transition, rather like a phase change, and does not take place through a sequence of thermodynamically stable intermediate states.

- I_N: the circular dichroism signal attributable to a solution of the protein in its native state N, corresponding to a concentration [Ur] of urea of 0 mol dm^{-3}, as at the left-hand side of Figure 25.2.
- I_U: the circular dichroism signal attributable to a solution of the protein in its totally unfolded state U, corresponding to a high concentration [Ur] of urea, say, 8 mol dm^{-3}, as at the right-hand side of Figure 25.2.
- I_{obs}: the circular dichroism signal attributable to a solution of the protein at any intermediate concentration [Ur] of urea.

Measurements of I_N, I_U, and I_{obs} for any particular protein therefore enable us to **normalise** the data by computing the ratio $(I_{obs} - I_U)/(I_N - I_U)$ for all concentrations [Ur] of urea, so enabling the graph shown in Figure 25.2 to be transformed into the graph shown in Figure 25.3.

As can be seen, for low concentrations of urea, the ratio $(I_{obs} - I_U)/(I_N - I_U) = 1$, and for high concentrations of urea, the ratio $(I_{obs} - I_U)/(I_N - I_U) = 0$ – in fact, for any value of I_{obs}, this ratio is equal to the corresponding mole fraction x_N (some sources use the symbol f_N) of the native protein in the reaction mixture

$$\frac{I_{obs} - I_U}{I_N - I_U} = x_N$$

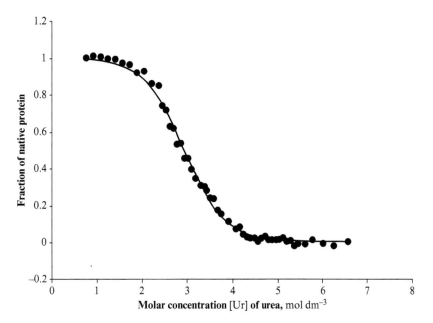

Figure 25.3 The mole fraction x_N of native protein for different concentrations [Ur] of urea. The vertical axis represents the ratio $(I_{obs} - I_U)/(I_N - I_U)$, this being a measure of the mole fraction x_N of the native protein in the reaction mixture.

If, at any point during the unfolding process, the equilibrium reaction mixture contains n_N mol of the native protein, and n_U mol of the unfolded protein, then, by definition, the equilibrium mole fraction x_N of the native protein is

$$x_N = \frac{n_N}{n_N + n_U}$$

If the volume of the solution is V dm^{-3}, then

$$x_N = \frac{n_N/V}{n_N/V + n_U/V} = \frac{[N]_{eq}}{[N]_{eq} + [U]_{eq}}$$

where $[N]_{eq}$ and $[U]_{eq}$ are the equilibrium molar concentrations of the native and unfolded protein. The experimentally observed ratio $(I_{obs} - I_U)/(I_N - I_U)$ is therefore a measure of both the equilibrium mole fraction x_N and also the concentration ratio $[N]_{eq}/([N]_{eq} + [U]_{eq})$. Furthermore, the equilibrium mole fraction x_U of the unfolded protein is given by

$$x_U = 1 - x_N = 1 - \frac{I_{obs} - I_U}{I_N - I_U} = \frac{I_N - I_{obs}}{I_N - I_U}$$

implying that x_U can also be determined directly from experimental data for I_{obs}, I_U and I_N.

Now, by definition,

$$x_U = \frac{n_U/V}{n_N/V + n_U/V} = \frac{[U]_{eq}}{[N]_{eq} + [U]_{eq}} = \frac{[U]_{eq}/[N]_{eq}}{1 + [U]_{eq}/[N]_{eq}}$$

and according to equation (25.3b)

$$\frac{[U]_{eq}}{[N]_{eq}} = \exp\left(-\frac{\Delta_{unf} G^\ominus}{RT}\right) \tag{25.3b}$$

implying that, at any intermediate concentration of urea, the instantaneous equilibrium mole fraction x_U of the unfolded state can therefore be expressed in terms of $\Delta_{unf} G^{\ominus}$ as

$$x_U = \frac{\exp(-\Delta_{unf} G^{\ominus}/RT)}{1 + \exp(-\Delta_{unf} G^{\ominus}/RT)} \quad (25.4)$$

For the reversible reaction $N \rightleftharpoons U$, the equilibrium ratio $[U]_{eq}/[N]_{eq}$ can vary from 0, corresponding to all the protein molecules being in the native state, to infinity, at which point all protein molecules are fully unfolded. The actual equilibrium can be controlled by the molar concentration $[Ur]$ of urea, from $[Ur] = 0$ mol dm^{-3}, for the totally native state $[x_N = 1, x_U = 0]$, to $[Ur] > 5$ mol dm^{-3}, for the totally unfolded state $[x_N = 0, x_U = 1]$.

As we have seen, x_U can be determined directly from experimental data as the normalised ratio $(I_N - I_{obs})/(I_N - I_U)$, and for any given value of x_U, we can use equation (25.4) to compute the corresponding value of $\Delta_{unf} G^{\ominus}$.

Figures 25.2 and 25.3 show that x_U, and hence $\Delta_{unf} G^{\ominus}$, vary with the molar concentration $[Ur]$ of urea. If $\Delta_{unf} G^{\ominus}$ for any given protein is determined experimentally for different values of $[Ur]$, it is found that $\Delta_{unf} G^{\ominus}$ decreases approximately linearly with $[Ur]$, as may be expressed in the form

$$\Delta_{unf} G^{\ominus} = \Delta_{unf} G^{\ominus}_{H_2O} - m_{unf}[Ur] \quad (25.5)$$

The (negative) slope of the line is the constant represented as m_{unf}, and the parameter designated as $\Delta_{unf} G^{\ominus}_{H_2O}$ is the value of $\Delta_{unf} G^{\ominus}$ in an aqueous solution, in the absence of urea, such that $[Ur] = 0$. This corresponds to the state in which the reversible reaction $N \rightleftharpoons U$ is as far to the left as is physiologically natural.

Reference to equation (25.4) will verify that when $\Delta_{unf} G^{\ominus} = 0$, $x_U = \frac{1}{2}$, implying that 50% of the protein is unfolded, and that the equilibrium mixture comprises the native and unfolded states in equal concentrations. At this point, $[Ur]$ is defined as $[Ur_{50\%}]$, and so equation (25.5) becomes

$$0 = \Delta_{unf} G^{\ominus}_{H_2O} - m_{unf}[Ur_{50\%}]$$

implying that

$$\Delta_{unf} G^{\ominus}_{H_2O} = m_{unf}[Ur_{50\%}] \quad (25.6)$$

Substituting $m_{unf}[Ur_{50\%}]$ for $\Delta_{unf} G^{\ominus}_{H_2O}$ in equation (25.5), we derive

$$\Delta_{unf} G^{\ominus} = m_{unf}([Ur_{50\%}] - [Ur]) = -m_{unf}([Ur] - [Ur_{50\%}])$$

and so equation (25.4) becomes

$$x_U = \frac{\exp\{m_{unf}([Ur] - [Ur_{50\%}])/RT\}}{1 + \exp\{m_{unf}([Ur] - [Ur_{50\%}])/RT\}} \quad (25.7)$$

By fitting equation (25.7) directly to the normalised experimental data, as shown in Figure 25.3, we may obtain values of $[Ur_{50\%}]$ and m_{unf}. These values can then be used into equation (25.6) to determine $\Delta_{unf} G^{\ominus}_{H_2O}$. So, in the example of Figure 25.3, the fit shown corresponds to values of $[Ur_{50\%}] = 2.69$ mol dm^{-3} and $m_{unf} = 4.53$ kJ mol^{-2} dm^3, implying, from equation (25.6), that

$$\begin{aligned}\Delta_{unf} G^{\ominus}_{H_2O} &= m_{unf}[Ur_{50\%}] \\ &= 4.53 \times 2.69 \\ &= 12.2 \text{ kJ mol}^{-1}\end{aligned}$$

This particular value is actually at the low end for proteins: $\Delta_{unf} G^{\ominus}_{H_2O}$ tends to be in the range 10 – 60 kJ mol^{-1}.

25.2.5 Measuring protein unfolding thermally

$\Delta_{unf} G^{\ominus}$ for proteins can also be obtained as a function of temperature by using spectroscopic measurements, with the step-wise unfolding of the protein in solution being driven by a slow increase in the system temperature, rather than resulting from changes in the concentration of a chaotrope such as urea. Figure 25.4 shows representative results from such an experiment, using fluorescence as the method of monitoring the unfolding.

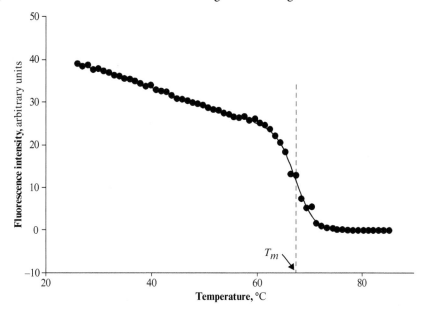

Figure 25.4 Thermal unfolding. The vertical axis is a fluorescence signal, which is relatively strong for a native protein, and weaker for an unfolded protein. As the temperature increases, the protein unfolds relatively slowly between 20 °C and 60 °C, and rapidly between 60 °C and 70 °C. The mid-point of the thermal transition is the melting temperature T_m, corresponding to the temperature at which the native and unfolded states are in equilibrium at equal molar concentrations.

To interpret the experimental data, we need to take into account the general variation of ΔG with temperature, and to do this, our starting point is the fundamental definition of ΔG as equation (13.13)

$$\Delta G = \Delta H - T\Delta S \tag{13.13}$$

Accordingly, for the unfolding of a protein according to the reversible reaction $N \rightleftharpoons U$ at a given temperature T_1, we may write

$$\Delta_{unf} G^{\ominus}_{T_1} = \Delta_{unf} H^{\ominus}_{T_1} - T \Delta_{unf} S^{\ominus}_{T_1} \tag{25.8}$$

and for the same reaction at a different temperature T

$$\Delta_{unf} G^{\ominus}_{T} = \Delta_{unf} H^{\ominus}_{T} - T \Delta_{unf} S^{\ominus}_{T} \tag{25.9}$$

Reference to page 193 will verify that the enthalpy terms $\Delta_{unf} H^{\ominus}_{T}$ and $\Delta_{unf} H^{\ominus}_{T_1}$ are related according to equation (6.29) as

$$\Delta_{unf} H^{\ominus}_{T} = \Delta_{unf} H^{\ominus}_{T_1} + \Delta C_P (T - T_1) \tag{25.10}$$

where $\Delta C_P = C_{P,U} - C_{P,N}$ is the difference between the heat capacity at constant pressure $C_{P,U}$ of the unfolded protein, and that of the native protein, $C_{P,N}$.

Similarly, a consequence of equation (9.42) on page 291 allows us to write

$$\Delta_{unf} S^{\ominus}_{T} = \Delta_{unf} S^{\ominus}_{T_1} + \Delta C_P \ln \frac{T}{T_1} \tag{25.11}$$

We may now use equations (25.10) and (25.11), in combination with equations (25.8) and (25.9), to express $\Delta_{unf} G^{\ominus}_{T}$ as a function of temperature as

$$\Delta_{unf} G^{\ominus}_{T} = [\Delta_{unf} H^{\ominus}_{T_1} + \Delta C_P (T - T_1)] - T \left[\Delta_{unf} S^{\ominus}_{T_1} + \Delta C_P \ln \frac{T}{T_1} \right] \tag{25.12}$$

To determine $\Delta_{unf} G^{\ominus}_{T}$ at any temperature T, we can be substitute equation (25.12) into equation (25.4)

$$x_U = \frac{\exp(-\Delta_{unf} G^{\ominus}/RT)}{1 + \exp(-\Delta_{unf} G^{\ominus}/RT)} \tag{25.4}$$

giving

$$x_U = \frac{\exp\left(-\frac{1}{RT} \left\{ [\Delta_{unf} H^{\ominus}_{T_1} + \Delta C_P (T - T_1)] - T \left[\Delta_{unf} S^{\ominus}_{T_1} + \Delta C_P \ln \frac{T}{T_1} \right] \right\} \right)}{1 + \exp\left(-\frac{1}{RT} \left\{ [\Delta_{unf} H^{\ominus}_{T_1} + \Delta C_P (T - T_1)] - T \left[\Delta_{unf} S^{\ominus}_{T_1} + \Delta C_P \ln \frac{T}{T_1} \right] \right\} \right)} \tag{25.13}$$

Just as we saw on page 756 in connection with the mid-point [Ur$_{50\%}$] of the urea induced unfolding, at the mid-point - the **melting temperature** T_m - of the thermally-induced unfolding, the native and unfolded states are present in equal concentrations, and so at $T = T_m$, $x_U = 0.5$, $\Delta_{unf} G^{\ominus}_{T} = 0$, and, from equation (25.9)

$$\Delta_{unf} G^{\ominus}_{T_m} = \Delta_{unf} H^{\ominus}_{T_m} - T_m \Delta_{unf} S^{\ominus}_{T_m} = 0$$

implying that

$$\Delta_{unf} S^{\ominus}_{T_m} = \frac{\Delta_{unf} H^{\ominus}_{T_m}}{T_m}$$

Setting $T_1 = T_m$ in equation (25.13), we now have

$$x_U = \frac{\exp\left(-\frac{1}{RT} \left\{ [\Delta_{unf} H^{\ominus}_{T_m} + \Delta C_P (T - T_m)] - T [\Delta_{unf} H^{\ominus}_{T_m}/T_m + \Delta C_P \ln(T/T_m)] \right\} \right)}{1 + \exp\left(-\frac{1}{RT} \left\{ [\Delta_{unf} H^{\ominus}_{T_m} + \Delta C_P (T - T_m)] - T [\Delta_{unf} H^{\ominus}_{T_m}/T_m + \Delta C_P \ln(T/T_m)] \right\} \right)}$$

which rearranges to

$$x_U = \frac{\exp\left(-\frac{1}{R}\left\{\Delta_{\text{unf}}H^\ominus_{T_m}(1/T - 1/T_m) + \Delta C_P(1 - T_m/T - \ln(T/T_m))\right\}\right)}{1 + \exp\left(-\frac{1}{R}\left\{\Delta_{\text{unf}}H^\ominus_{T_m}(1/T - 1/T_m) + \Delta C_P(1 - T_m/T - \ln(T/T_m))\right\}\right)} \quad (25.14)$$

As can be seen in Figure 25.4, the linear portions at the lower and higher temperatures of the sigmoidal plots obtained in protein thermal unfolding often have non-zero slopes. Equation (25.14) can therefore be modified with baseline terms $a_N + b_N T$ and $a_U + b_U T$ for the native and unfolded states, respectively, giving

$$x_U = \frac{(a_N + b_N T) + (a_U + b_U T)\exp\left(-\frac{1}{R}\left\{\Delta_{\text{unf}}H^\ominus_{T_m}(1/T - 1/T_m) + \Delta C_P(1 - T_m/T - \ln(T/T_m))\right\}\right)}{1 + \exp\left(-\frac{1}{R}\left\{\Delta_{\text{unf}}H^\ominus_{T_m}(1/T - 1/T_m) + \Delta C_P(1 - T_m/T - \ln(T/T_m))\right\}\right)}$$

(25.15)

This (somewhat clumsy!) expression can be fitted directly to the plot as shown in Figure 25.4. However, while $\Delta_{\text{unf}}H^\ominus_{T_m}$, and hence $\Delta_{\text{unf}}S^\ominus_{T_m}$, at $T = T_m$ can be determined as variables in the fit, it is important to use a pre-determined value of ΔC_P, as obtaining it as a variable in the fit will lead to high errors. Values of $\Delta_{\text{unf}}H^\ominus_{T_m}$ derived by fitting equations (25.14) or (25.15) to experimental data, rather than as obtained by other methods such as calorimetry, are, as we saw on page 446, **van't Hoff enthalpies** $\Delta_{\text{unf}}H^\ominus_{\text{vH},T_m}$ at $T = T_m$.

25.2.6 Evaluating ΔC_P for protein unfolding

As we have just noted, the practical use of equations (25.14) and (25.15) requires us to have a reliable value for ΔC_P in advance. A quick way to determine an approximate value of ΔC_P for protein unfolding is to use an average value of about 50 J deg^{-1} mol^{-1} per amino acid residue in the protein. This is fine as a rough estimate, but ΔC_P actually varies between proteins, in which case it can be measured experimentally. This can be achieved using **differential scanning calorimetry** (DSC) to obtain simultaneous values of $\Delta_{\text{unf}}H^\ominus$ and T_m across a range of pH values.

As ΔC_P is a function of temperature, but not ostensibly of pH, ΔC_P can be obtained from the slope of a plot of $\Delta_{\text{unf}}H^\ominus$ versus T_m. In principle, ΔC_P could be obtained directly from the difference between the pre- and post-unfolding baselines in a single DSC **thermogram** (see Figure 25.5, and also Figure 9.5), eliminating any potential effect of pH on ΔC_P. However, this is found to lead to greater errors due to extrapolation of poorly defined baselines.

The preceding discussion has assumed that ΔC_P is constant over the temperature range of interest. As shown in Chapter 6 (see pages 195 to 197), however, C_P in general does in fact vary with temperature, and so ΔC_P for protein unfolding will vary with temperature too, implying that it is often necessary to add an additional term to ΔC_P to account for this temperature dependence.

Also, we note that, theoretically, the value of $\Delta_{\text{unf}}H^\ominus$ as determined by DSC should be identical to the value of the van't Hoff enthalpy $\Delta_{\text{unf}}H^\ominus_{\text{vH},T_m}$ as determined by fitting equations (25.14) or (25.15) to experimental data of the form shown in Figure 25.4. In practice, there are sometimes discrepancies, which may be indicative of real phenomena that have not been

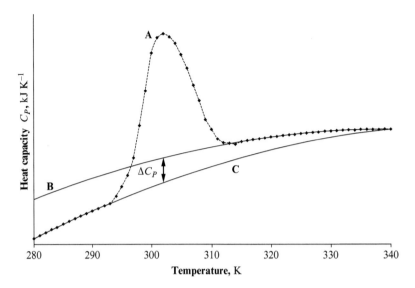

Figure 25.5 The measurement of ΔC_P for protein unfolding. Curve A shows a differential scanning calorimetry (DSC) thermogram for a protein as it changes from the native state (around 285 K) to an unfolded state (above about 310 K). Curve B fits the high temperature baseline, and curve C fits the low temperature baseline.

explicitly recognised in the theoretical models underpinning, for example, equations (25.14) and (25.15), such as the presence of thermodynamically stable intermediates between the native and fully unfolded states, or the possibility that the folding-unfolding process is not truly reversible. Careful experimental analysis, and the comparison of measurements derived by different methods, is therefore very valuable in enriching our understanding.

25.2.7 Calculating $\Delta_{unf} H^\ominus$, $\Delta_{unf} S^\ominus$ and $\Delta_{unf} G^\ominus$ for protein unfolding

For the example shown in Figure 25.4, the values obtained from fitting the data to equation 25.15, and assuming that ΔC_P = 26.6 KJ deg^{-1} mol^{-1}, are

$$\Delta_{unf} H^\ominus_{vH,T_m} = 685 \text{ kJ mol}^{-1}$$
$$T_m = 340.1 \text{ K}$$

$\Delta_{unf} H^\ominus_{vH,T_m}$ is the van't Hoff enthalpy at $T = T_m$, and so the corresponding value of the entropy $\Delta_{unf} S^\ominus_{T_m} = \Delta_{unf} H^\ominus_{vH,T_m}/T_m$ is 2 kJ mol^{-1} K^{-1}.

The temperature dependences of $\Delta_{unf} H^\ominus$ and $\Delta_{unf} S^\ominus$ can be obtained from the values of $\Delta_{unf} H^\ominus_{vH,T_m}$ and $\Delta_{unf} S^\ominus_{T_m}$ using equations 25.10 and 25.11 respectively, and the appropriate value of ΔC_P. At $T = T_m$, $\Delta_{unf} G^\ominus = 0$, and so $\Delta_{unf} G^\ominus$ can also be obtained at any temperature using equation 25.12, using the above values of $\Delta_{unf} H^\ominus_{vH,T_m}$, $\Delta_{unf} S^\ominus_{T_m}$, ΔC_P, and $T = T_m$. Accordingly, at 273 K for this protein, thermal measurements imply a value of $\Delta_{unf} G^\ominus_{273}$ = 12.4 kJ mol^{-1}, in close agreement with the value of 12.2 kJ mol^{-1} obtained on page 757 from data using urea to cause the unfolding.

25.2.8 The origins of $\Delta_{unf} H^\ominus$, $\Delta_{unf} S^\ominus$ and $\Delta_{unf} G^\ominus$ in protein unfolding

Figure 25.6 shows the same fluorescence data as in Figure 25.4 (the dots), and, in addition, the corresponding temperature-dependent values of $\Delta_{unf}H^\ominus$, $T\Delta_{unf}S^\ominus$ and $\Delta_{unf} G^\ominus$ for the experimental protein, plotted relative to the energy scale being to the right.

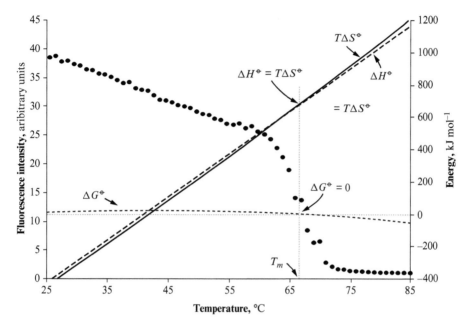

Figure 25.6 The temperature dependencies of $\Delta_{unf}H^\ominus$, $\Delta_{unf}S^\ominus$ and $\Delta_{unf} G^\ominus$ for protein unfolding. Both $\Delta_{unf}H^\ominus$ and $T\Delta_{unf}S^\ominus$ increase with temperature, and, over the range shown, are nearly equal in value. The difference $\Delta_{unf}H^\ominus - T\Delta_{unf}S^\ominus = \Delta_{unf} G^\ominus$ is therefore always a small number. At temperature $T = T_m$, $\Delta_{unf} G^\ominus = 0$ and $\Delta_{unf}H^\ominus = T\Delta_{unf}S^\ominus$, implying that the folded and unfolded states are in equilibrium in equal concentrations. At temperatures $T < T_m$, $\Delta_{unf} G^\ominus > 0$, $\Delta_{unf}H^\ominus > T\Delta_{unf}S^\ominus$ and the native protein is stable; at temperatures $T > T_m$, $\Delta_{unf}G^\ominus < 0$, $\Delta_{unf}H^\ominus < T\Delta_{unf}S^\ominus$ and the native protein spontaneously unfolds.

The $\Delta_{unf} G^\ominus$ plot is slightly curved, such that $\Delta_{unf} G^\ominus$ is positive at temperatures $T < T_m$, implying that the native form is more stable. At temperatures $T > T_m$, $\Delta_{unf} G^\ominus$ is negative, and so the unfolded state is more stable, with the equilibrium mixture becoming progressively richer in the unfolded state as the temperature increases beyond T_m. $\Delta_{unf} G^\ominus = 0$ at $T = T_m$, and the equilibrium mixture at that temperature has equal concentrations of both the native and the unfolded states.

The plots of $\Delta_{unf}H^\ominus$ and $T\Delta_{unf}S^\ominus$ are linear in Figure 25.6 as ΔC_P was assumed to be independent of temperature. A very striking feature of Figure 25.6 is that $\Delta_{unf}H^\ominus$ and $T\Delta_{unf}S^\ominus$ are near-identical throughout, and therefore both of very large value compared to $\Delta_{unf} G^\ominus$, which is of course the (small) difference $\Delta_{unf}H^\ominus - T\Delta_{unf}S^\ominus$. This numerically small value of $\Delta_{unf} G^\ominus$ implies that proteins are only marginally stable at temperatures less than T_m, poised on a knife-edge of large enthalpic and entropic contributions.

The enthalpy term $\Delta_{unf}H^\ominus$ for proteins originates mainly from an accumulation of the many molecular bonding forces that form upon protein folding. These are mostly from the non-covalent contributions of hydrogen bonds, van der Waals interactions, and electrostatic interactions between protein atoms. Hydrogen bonding interactions with water molecules in the unfolded state are lost when a protein folds, and might be expected to cancel out the gains from protein-protein hydrogen bonds. However, hydrogen bonds between a protein and neighbouring water molecules are slightly weaker than those between protein atoms. Some proteins contain additionally stabilising covalent bonds between sulphur atoms when folded. Folding of a protein can also constrain bond angles, and also torsion angles, in non-optimal positions, contributing to the enthalpy term and so slightly destabilising the native state.

The entropy term $T\Delta_{unf}S^{\circ}$ originates from two main sources. One is the decreased conformational freedom of the native folded protein compared to the unfolded state. The higher degree of order associated with the folded native state, as compared to the relative disorder of the unfolded state, decreases the system's entropy, favouring the unfolded state. Water, however, plays a much larger role that favours the protein native state. In the unfolded state, many hydrophobic moieties are exposed to the solvent, causing water to self-assemble around them into highly ordered ice-like structures with low entropy, and also with low enthalpy due to maximised hydrogen bonding. Protein folding buries the hydrophobic moieties internally, and releases the water molecules into a higher entropy state within the bulk solvent. Other hydrophilic moieties can hydrogen bond to water molecules in the unfolded state, and these are also broken upon protein refolding.

That $\Delta_{unf}H^{\circ}$ and $T\Delta_{unf}S^{\circ}$ are so finely balanced is a result of evolution. For any given protein, $\Delta_{unf}G^{\circ}$ only needs to be sufficiently negative for the protein to fold and function in physiological conditions, whilst being of a small enough magnitude to allow the protein to be readily degraded and recycled whenever appropriate.

25.2.9 Why understanding protein folding is important

We now understand the thermodynamics of protein folding sufficiently well that we can use computational methods to model protein dynamics with considerable accuracy. These methods rely upon so-called **potential functions** V that often comprise a summation of Gibbs free energy terms for each type of molecular force. For example, we may express ΔG_{unf} as the summation of a number of terms, each associated with a particular type of interaction

$$\Delta_{unf} G = \Delta G_{HB} + \Delta G_{VdW} + \Delta G_{Elec} + \Delta G_{Solv} + \Delta G_{Tor} + \Delta G_{BA} + \Delta G_{BL}$$

in which the subscripts refer to

HB: hydrogen bonds
VdW: van der Waals forces
Elec: electrostatic forces
Solv: solvation (entropy term)
Tor: torsion angle
BA: covalent bond angle
BL: covalent bond length

Each of these terms is typically based on an appropriate semi-empirical equation that models measurable physical inter-atomic potentials, these being mathematical functions that, in general, are such that the change in the potential is a measure of the corresponding value of ΔG –

which explains why the chemical potential μ, as introduced on page 404, is known as the 'chemical potential'.

So, for example, electrostatic potentials are of the form

$$V_{elec} = \frac{q}{4\pi \varepsilon_0 r}$$

where V_{elec} is the electrical potential at distance r from an electrical charge q, and ε_0 is a fundamental physical constant known as the 'permittivity of free space'. Another example is the **Lennard-Jones potential**, which closely models the observed van der Waals interactions between atoms in non-ideal (real) gases

$$V_{vdW} = 4\varepsilon \left[\left(\frac{\sigma}{r}\right)^{12} - \left(\frac{\sigma}{r}\right)^{6} \right]$$

in which ε and σ are constants for any given system, and r is the distance between two adjacent molecules. In this equation, the first term represents the repulsion between two adjacent molecules that arises when they are extremely close to each other, and the second term represents the somewhat longer-range inter-molecular attractive force, the presence of which – as we saw on pages 485 to 489 – explains why real gases can liquefy.

With current computational techniques, we can now extend the use of these potential functions to solve some very important problems, such as the calculation of the energy associated with protein-ligand interactions, and the prediction not only of possible binding sites for known **ligands**, but also of the types of chemical entities can interact with a given patch on a protein surface.

These methods form the basis of 'fragment-based' computational design of new drug molecules, where the aim is to design a chemical structure that will tightly bind to a specific protein surface. The protein surface is divided into small patches, and then screened computationally with a library of chemical fragments that are commonly found in organic molecules, with the objective of predicting those that bind most tightly to each patch. The individual fragments are then chemically stitched together firstly computationally, and subsequently experimentally, keeping them in the orientations and positions at which they bind to the protein surface. The resulting final drug candidate will therefore complement the topology and chemistry of the protein surface, and bind tightly.

Understanding the parameters that determine the nature and strength of the binding between a protein and a ligand is therefore central to the design of drugs. The following section therefore examines protein-ligand binding in some detail, and introduces some methods by which the interaction energies between proteins and ligands can be measured experimentally.

25.3 Protein-ligand interactions

25.3.1 Measuring protein-ligand interactions – the binding isotherm and K_d

Many proteins interact tightly with other proteins or small molecules, and with high specificity, such that only the right molecules will bind. While the nature of protein-ligand binding, and the specificity of binding, are notionally two separate things, the strength of protein-ligand binding, as measured by the **binding affinity**, is a result of the degree of complementarity of molecular shape, in combination with the strength of the inter-atomic bonding potentials, accumulated between the protein binding surface and the ligand. Inevitably, a greater summation of such interactions is only possible if the right molecule is bound. Consequently, the measurement of binding affinities is of particular value in enriching our understanding of many important biochemical processes. Additionally, a thermodynamic understanding of protein-ligand binding has had considerable impact on the ability to model these interactions, and therefore to predict potential molecules that could act as pharmaceutical agents due to their ability to bind to a specific protein.

Protein-ligand interactions are typically assumed to be two-state reactions at equilibrium

$$PL \rightleftharpoons P_{free} + L_{free}$$

where P_{free} is the free protein, not bound to the ligand, L_{free} is the free ligand, and PL is the **protein-ligand complex**. If under any particular conditions, the equilibrium molar concentrations of the free protein, free ligand, and protein-ligand complex are $[P_{free}]$, $[L_{free}]$ and $[PL]$, respectively (in which, to avoid clutter, we are dropping the subscript $_{eq}$), then this reversible reaction may be associated with an equilibrium **dissociation constant** K_d given by

$$K_d = \frac{[P_{free}][L_{free}]}{[PL]} = \frac{k_{off}}{k_{on}} \qquad (25.16)$$

where, as discussed on page 454, k_{off} and k_{on} are the rate constants for dissociation (the 'off-rate'), and association (the 'on-rate'), respectively. K_d is an inverse measure of the binding affinity between the protein and the ligand: the smaller the value of K_d, the lower the concentration $[L_{free}]$ of the free ligand in the equilibrium mixture, and so the more tightly bound the ligand; conversely, a higher value of K_d implies that the ligand is more loosely bound.

Many analytical techniques are available for monitoring the interactions between proteins and small molecules, so allowing the measurement of a signal which is proportional to the concentration $[PL]$ of the protein-ligand complex. Four particularly useful techniques are **surface plasmon resonance (SPR)**, **fluorescence**, **nuclear magnetic resonance (NMR)**, and **isothermal titration calorimetry (ITC)**. Each has its own relative merits and challenges, but overall, a typical experiment would involve titration of the ligand into the protein solution at constant temperature, resulting in the measurement of data which can be represented as a **binding isotherm**, an illustrative example of which is shown in Figure 25.7. Many proteins achieve remarkably high binding affinities as a result of a simultaneous combination of a very slow off-rate, and a relatively slow on-rate, implying that, with each titration of ligand, it is important that measurements are made only after sufficient time has elapsed to ensure that the system is truly at equilibrium.

25.3.2 How to determine K_d – case1: high ligand concentrations

For any protein-ligand binding reaction

$$PL \rightleftharpoons P_{free} + L_{free}$$

PROTEIN-LIGAND INTERACTIONS

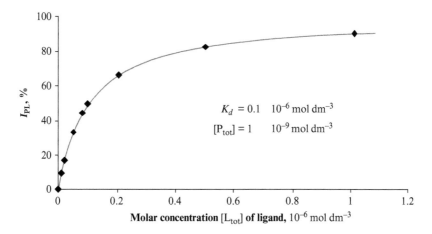

Figure 25.7 A representative binding isotherm. The horizontal axis represents the total molar concentration [L_{tot}] of a ligand L, which can bind to a protein P. The vertical axis represents the measurement of a signal, the intensity I of which depends on, and increases with, the concentration [PL] of the protein-ligand complex PL.

all the ligand is necessarily either in the free state or bound in the protein-ligand complex. The total ligand concentration [L_{tot}] may therefore be expressed as

$$[L_{tot}] = [L_{free}] + [PL]$$

Similarly, the total protein concentration [P_{tot}] is given by

$$[P_{tot}] = [P_{free}] + [PL]$$

and so, in equation (25.16)

$$K_d = \frac{[P_{free}][L_{free}]}{[PL]} = \frac{k_{off}}{k_{on}} \tag{25.16}$$

we may substitute [P_{tot}] − [PL] for [P_{free}] to derive

$$K_d = \frac{([P_{tot}] - [PL])[L_{free}]}{[PL]} = \frac{[P_{tot}][L_{free}]}{[PL]} - [L_{free}]$$

from which

$$[PL] = \frac{[P_{tot}][L_{free}]}{K_d + [L_{free}]} \tag{25.17}$$

If the ligand is present in considerable excess - for example such that the total concentration [L_{tot}] of ligand is at least a factor of 10 greater than the total concentration [P_{tot}] of protein – then [L_{tot}] ≫ [P_{tot}] = [P_{free}] + [PL], implying that [L_{tot}] ≫ [PL]. Furthermore, given that [L_{tot}] = [L_{free}] + [PL] and that [L_{tot}] ≫ [PL], it therefore follows that [L_{tot}] ≈ [L_{free}], as makes intuitive sense: when the ligand is in excess, very little of the ligand is bound, and most of the ligand is free. Accordingly, when [L_{tot}] ≫ [P_{tot}], equation (25.17) becomes

$$[PL] = \frac{[P_{tot}][L_{tot}]}{K_d + [L_{tot}]} \tag{25.18}$$

For some experimental methods, the observed signal I is attributable solely to the protein-ligand complex PL, rather than to the free components. Any observation I is therefore directly proportional to the corresponding concentration [PL]. If several measurements are made for increasing concentrations $[L_{tot}]$ of the ligand, whilst holding the total concentration $[P_{tot}]$ of protein constant, then, since K_d is also a constant, equation (25.18) predicts that [PL], and hence I, will increase hyperbolically from zero to a limiting value proportional to $[P_{tot}]$, as illustrated by the binding isotherm shown in Figure 25.7.

Furthermore, if the ligand concentration is in the range from at least ten-fold below to at least ten-fold above the expected K_d, then this enables a good fit of equation (25.18) to the majority of the measurements, from around 9% protein occupancy up until the protein becomes 99% saturated with ligand. K_d can then be obtained by fitting equation (25.18) to the experimental binding isotherm, such as that illustrated in Figure 25.7. With other measurement techniques, the observed signal I is a mixture (technically, the 'convolution') of two signals – one signal I_P from the free protein P_{free}, and the second a signal I_{PL} from the protein-ligand complex PL. In this case, equation (25.18) is modified to

$$[PL] = I_P + (I_{PL} - I_P) \frac{[L_{tot}]}{K_d + [L_{tot}]}$$

25.3.3 How to determine K_d – case 2: low ligand concentrations

In some circumstances, measurements can be made only when the total concentration $[P_{tot}]$ of protein is comparable to the total concentration $[L_{tot}]$ of ligand – as may happen, for example, if the ligand is less soluble (so limiting $[L_{tot}]$), or if relatively high concentrations of protein are required to enable an accurate measurement of the experimental signal. Accordingly, it is no longer the case that $[L_{tot}] \gg [P_{tot}]$, that $[L_{tot}] \gg [PL]$, and that $[L_{tot}] \approx [L_{free}]$. Equation (25.17)

$$[PL] = \frac{[P_{tot}][L_{free}]}{K_d + [L_{free}]} \tag{25.17}$$

however, remains true, and we know, for the ligand, that

$$[L_{tot}] = [L_{free}] + [PL]$$

and so

$$[L_{free}] = [L_{tot}] - [PL] \tag{25.19}$$

implying that

$$[PL] = \frac{[P_{tot}]([L_{tot}] - [PL])}{K_d + [L_{tot}] - [PL]} \tag{25.20}$$

Equation (25.20) is generally true, and for the special case in which the ligand is in excess, implying that $[L_{tot}] \gg [P_{tot}]$ and $[L_{tot}] \gg [PL]$, equation (25.20) reduces to (25.18), so confirming the discussion on page 765. For the general case, however, we must use equation (25.20), which can be re-arranged to give a quadratic equation in [PL] as

$$[PL]^2 - [PL]([P_{tot}] + [L_{tot}] + K_d) + [P_{tot}][L_{tot}] = 0 \tag{25.21}$$

one solution of which is

$$[PL] = \frac{1}{2}\left[([P_{tot}] + [L_{tot}] + K_d) - \sqrt{([P_{tot}] + [L_{tot}] + K_d)^2 - 4[P_{tot}][L_{tot}]}\right] \quad (25.22a)$$

Equation (25.22a) can be fitted to the experimental binding isotherm, and to determine the actual concentration $[L_{free}]$ of free ligand, equation (25.22a) may be combined with equation (25.19) to give

$$[L_{free}] = [L_{tot}] - \frac{1}{2}\left[([P_{tot}] + [L_{tot}] + K_d) - \sqrt{([P_{tot}] + [L_{tot}] + K_d)^2 - 4[P_{tot}][L_{tot}]}\right] \quad (25.23)$$

We note that, mathematically, the quadratic equation (25.21) has a second solution, with a positive square root as

$$[PL] = \frac{1}{2}\left[([P_{tot}] + [L_{tot}] + K_d) + \sqrt{([P_{tot}] + [L_{tot}] + K_d)^2 - 4[P_{tot}][L_{tot}]}\right] \quad (25.22b)$$

Expression (25.22b), however, is not physically feasible, as can be verified by considering how this expression behaves when the ligand is in excess so that $[L_{tot}] \gg [P_{tot}]$ and $[L_{tot}] \gg [PL]$. If we assume that $[L_{tot}] \gg [P_{tot}]$, then $([L_{tot}] + [P_{tot}] + K_d) \approx ([L_{tot}] + K_d)$; furthermore, since $[L_{tot}]^2 \gg [P_{tot}][L_{tot}]$, then it's very likely that $[L_{tot}]^2 \gg 4[P_{tot}][L_{tot}]$, and even more likely that $([L_{tot}] + [P_{tot}] + K_d)^2 \gg 4[P_{tot}][L_{tot}]$. Accordingly, equation (25.22b) reduces to $[PL] \approx [L_{tot}] + K_d$. By assumption, $[L_{tot}] \gg [PL]$, but since K_d is necessarily positive, the approximation $[PL] \approx [L_{tot}] + K_d$ is impossible. Expression (25.22b) therefore gives a non-feasible solution to the quadratic equation (25.21), and so equation (25.22a) is the only valid solution.

25.3.4 How to determine K_d – case 3: measuring a shift in protein stability

Consider a solution of a protein, such that the native state N and the unfolded state U are in equilibrium as

$$N \rightleftharpoons U$$

with equilibrium constant $K_{unf} = [U]/[N]$ (where [U] and [N] represent the equilibrium molar concentrations), and standard Gibbs free energy change $\Delta_{unf} G^\ominus = -RT \ln K_{unf}$.

Suppose that a ligand L is now added, a ligand that binds only with the native protein N according to the reversible reaction

$$NL \rightleftharpoons N_{free} + L_{free}$$

associated with the equilibrium dissociation constant $K_d = [N_{free}][L_{free}]/[NL]$ and standard Gibbs free energy change of dissociation $\Delta_d G^\ominus = -RT \ln K_d$.

The protein-ligand complex NL can also unfold, and if we assume that the unfolded form of the native protein N and the unfolded form of the protein-ligand complex NL cannot be distinguished, and have the same Gibbs free energy G_U, then, in the presence of the ligand L, the reversible unfolding reaction may be written as

$$N_{tot} \rightleftharpoons U$$

where N_{tot} represents the total quantity of the native protein present, in both the unbound form N_{free} and the bound form NL. This reaction is associated with a new, different, equilibrium constant $K_{unf,N+NL} = [U]/[N_{tot}]$, and a corresponding standard Gibbs free energy change

$$\Delta_{unf} G^{\ominus}_{N+NL} = -RT \ln K_{unf,N+NL}$$

where the subscript $_{N+NL}$ refers to the equilibrium between the mixture of three components - the free native protein N_{free}, the protein-ligand complex NL, and the common unfolded protein U.

The energy relationships for the unfolded protein U, the free native protein N_{free}, and the protein-ligand complex NL are illustrated in Figure 25.8.

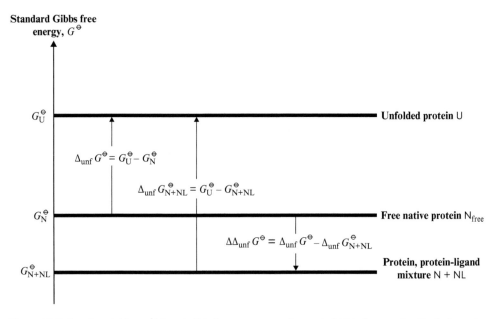

Figure 25.8 Protein unfolding and ligands. This diagram represents the standard Gibbs free energies $G^{\ominus}_N$ of a free native protein N_{free}, the corresponding unfolded protein U ($G^{\ominus}_U$), and a mixture of the free native protein and a protein-ligand complex ($G^{\ominus}_{N+NL}$). In this representation, the free native protein is more stable than the unfolded protein, and so $G^{\ominus}_N < G^{\ominus}_U$; also, the protein-ligand complex is more stable than the native protein, and so $G^{\ominus}_{N+NL} < G^{\ominus}_N$. The direction of each arrow indicates the direction of the corresponding ΔG, from an initial state to a final state.

For the reversible reaction $N_{tot} \rightleftharpoons U$,

$$K_{unf,N+NL} = \frac{[U]}{[N_{tot}]} = \frac{[U]}{[N_{free}] + [NL]} = \frac{[U]/[N_{free}]}{1 + [NL]/[N_{free}]} = \frac{K_{unf}}{1 + [L_{free}]/K_d}$$

and so

$$\frac{K_{unf}}{K_{unf,N+NL}} = 1 + \frac{[L_{free}]}{K_d}$$

implying that

$$\ln K_{unf} - \ln K_{unf,N+NL} = \ln\left(1 + \frac{[L_{free}]}{K_d}\right)$$

and that

$$-RT \ln K_{unf} - (-RT \ln K_{unf,N+NL}) = -RT \ln \left(1 + \frac{[L_{free}]}{K_d}\right) \qquad (25.24)$$

As we saw on page 676, $-RT \ln K_{unf}$ is equal to the standard Gibbs free energy $\Delta_{unf} G^\ominus$ of unfolding of the native protein as

$$\Delta_{unf} G^\ominus = -RT \ln K_{unf} \qquad (25.2)$$

Likewise, we may equate $-RT \ln K_{unf,N+NL}$ to the standard Gibbs free energy $\Delta_{unf} G^\ominus_{N+NL}$ of unfolding in the presence of a ligand as

$$\Delta_{unf} G^\ominus_{N+NL} = -RT \ln K_{unf,N+NL}$$

As shown in Figure 25.8, the difference between $\Delta_{unf} G^\ominus$ and $\Delta_{unf} G^\ominus_{N+NL}$ may be represented by the 'change in the change' variable $\Delta\Delta_{unf} G^\ominus$

$$\Delta\Delta_{unf} G^\ominus = \Delta_{unf} G^\ominus - \Delta_{unf} G^\ominus_{N+NL} = -RT \ln K_{unf} - (-RT \ln K_{unf,N+NL})$$

$$= -RT \ln \left(1 + \frac{[L_{free}]}{K_d}\right) \qquad (25.25a)$$

Equation (25.25a) draws on equation (25.24), and takes an alternative form as

$$K_d = \frac{[L_{free}]}{\exp(-\Delta\Delta_{unf} G^\ominus/RT) - 1} \qquad (25.25b)$$

Experimentally, it is possible to measure the standard Gibbs free energy change $\Delta_{unf} G^\ominus$ of protein unfolding in the absence of the ligand, and the corresponding value $\Delta_{unf} G^\ominus_{N+NL}$ in the presence of a ligand. This then enables us to compute $\Delta\Delta_{unf} G^\ominus = \Delta_{unf} G^\ominus - \Delta_{unf} G^\ominus_{N+NL}$, the value of which can then be used in equation (25.5b) to determine the dissociation constant K_d, and hence the binding affinity $1/K_d$, for the protein-ligand binding reaction. This last step also requires a value for $[L_{free}]$: as we saw on page 765, if $[L_{tot}] \gg [P_{tot}]$, then $[L_{free}] \approx [L_{tot}]$, which is known; if not, then we need to determine $[L_{free}]$ using equation (25.23).

25.4 Protein folding kinetics

25.4.1 Protein folding is determined by the amino acids

In 1961, the American biochemist Christian Anfinsen made a very important observation. He firstly added urea to a solution of the enzyme ribonuclease, causing it to unfold, and lose its enzymic catalytic properties. He then removed the urea, and discovered that the ribonuclease regained its function. He therefore inferred that the enzyme had re-formed to its original native state. Anfinsen was the first scientist to demonstrate that the folding and unfolding of an entire protein can be reversible, from which he deduced – very importantly – that it is the protein's amino acid sequence (rather than, for example, the way in which the amino acids are sequentially assembled) that is sufficient to determine the protein's final native structure. And the 1972 Nobel Prize for Chemistry was the result.

At the beginning of this chapter (see page 748), we stated that the native state represents the global free energy minimum on the landscape of all possible conformations. It is easy to see why this is the case when the native state and the unfolded state are at equilibrium, but how does the protein reach this state of equilibrium? What intermediate, transitory, reactions take place, and how fast are they?

25.4.2 Levinthal's paradox

Another major contribution to our understanding of protein folding was made by the American molecular biologist Cyrus Levinthal, who, in 1969, calculated that a protein of only 100 amino acids could form as many as 3^{198} potential stable conformations. He further established that if the protein were to sample these configurations at random, simply through thermal fluctuations, until the native state was obtained, then the process would take longer than the age of the universe! Molecular fluctuations are indeed fast (faster than picoseconds, so some 10^{12} configurations could in principle be sampled each second), but even that is not fast enough. And in reality, some proteins can fold in microseconds. This conundrum was dubbed 'Levinthal's Paradox'.

Folding is of course not paradoxical – it is very real. Accordingly, Levinthal inferred that protein folding is not random, but rather directed by the fast formation of stabilised local structures or nuclei along the polypeptide chain. The idea of protein folding pathways was born.

25.4.3 Protein folding kinetics and transition state theory

The simplest folding reaction may be represented as the formation of the native state N from the unfolded state U

$$U \rightleftharpoons N$$

We note that the folding reaction written in the direction $U \rightleftharpoons N$ is the other-way-around, as compared to the unfolding reaction $N \rightleftharpoons U$ that we have studied so far in this chapter. We trust this is not confusing: hitherto, we have been interested in unfolding, but in this section, our focus of attention is folding.

If the protein folding reaction were a classical chemical reaction, as we saw on page 457, as the folding reaction progresses from the unfolded state U to the native state N, we would expect there to be a transition state ‡ at some point along the reaction coordinate between U and N, as illustrated by the solid line in Figure 25.9.

The transition state ‡ is the high-energy state at the apex of the energy barrier that must be crossed for a reaction to progress from reactant to product. The higher the energy barrier – as measured by the activation energy – the slower the reaction, even if the folding reaction is thermodynamically favourable as determined by a negative value of $\Delta_{fol}G$. Kinetic measurements of protein folding and unfolding can therefore be used to determine whether such transition states exist, and also, potentially, their Gibbs free energies $G_{\ddagger}$. To do this, we will

PROTEIN FOLDING KINETICS

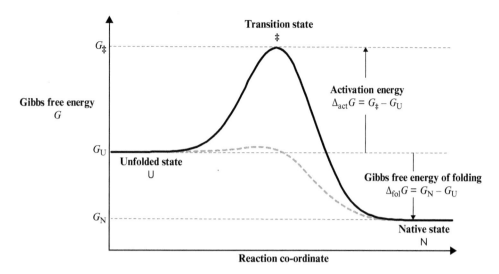

Figure 25.9 A kinetic representation of the folding reaction. This is a kinetic representation of the folding reaction, in which the folded state N is thermodynamically more stable than the unfolded state U, as shown by the fact that $G_N < G_U$, implying that $\Delta_{fol}G = G_N - G_U < 0$. Kinetically, however, the reaction has to cross the activation energy 'barrier', through the transition state ‡, and the greater the activation energy $\Delta_{act}G$, the slower the rate at which the folding reaction actually takes place. The solid line represents a folding reaction with a relatively high activation energy; the dashed line represents a folding reaction with a very low activation energy, a 'downhill slide'.

invoke **transition state theory**, a development of the concepts underpinning the Arrhenius equation, equation (14.38)

$$k(T) = Ae^{-\frac{E_a}{RT}} \qquad (14.38)$$

which, as we saw on pages 455 to 459, relates the rate $k(T)$ of a chemical reaction taking place at temperature T to the reaction's activation energy E_a.

Transition state theory seeks to gain a deeper understanding of the factors that determine the reaction rate $k(T)$, and was developed by considering how a chemical bond between two atoms might break. At any instant, the two atoms are vibrating along the bond, and the frequency of vibration is associated with a corresponding vibrational energy. If this vibrational energy is relatively low, the two atoms continue to vibrate; but as the vibrational energy increases, a threshold energy is reached – the transition state energy $E_‡$ – at which the bond will break. For the bond to break, corresponding to the formation of two separated atoms from the transition state, transition state theory uses Planck's law (energy = Planck's constant h × frequency) to express the rate constant $k^‡$ for the decomposition reaction in terms of the threshold vibrational frequency $= E_‡/h$ as (25.26)

$$k^‡ = \kappa \frac{E_‡}{h} \qquad (25.26)$$

In equation (25.26), κ is a constant of proportionality, known as the **transmission coefficient**. In general, κ can take any value between 0 and 1; physically, κ allows for the possibility that not necessarily all vibrations of energy $E_‡$ result in a breaking of the bond.

The mean vibrational $E_{\ddagger}$ of the excited state of a vibrating bond at temperature T can be shown to be $k_B T$, where k_B is the Boltzmann constant, and so

$$k^{\ddagger} = \kappa \frac{k_B T}{h} \tag{25.27}$$

If we now assume that the transition state ‡ of a reaction, and the starting (ground) state G, are in equilibrium as

$$G \rightleftharpoons \ddagger$$

then we may define an equilibrium constant K_{act} such that

$$K_{act} = \exp\left(-\frac{\Delta_{act} G^{\ominus}}{RT}\right) \tag{25.28}$$

where $\Delta_{act} G^{\ominus} = G_{\ddagger}^{\ominus} - G_G^{\ominus}$ represents the standard Gibbs free energy change associated with the activation from the ground state G to the transition state ‡.

One of the key results of transition state theory is to express the rate constant k_{act} for the forward activation reaction $G \to \ddagger$ as

$$k_{act} = k^{\ddagger} K_{act} = k^{\ddagger} \exp\left(-\frac{\Delta_{act} G^{\ominus}}{RT}\right) \tag{25.29}$$

For protein folding, the ground state G is the unfolded state U, and the rate constant k_{act} for the forward reaction is the rate of folding k_{fold}. If we use equation (25.27) to substitute for $k^{\ddagger}$, then the general equation (25.29) can be applied to protein folding in the form

$$k_{fold} = \frac{k_B T}{h} \exp\left(-\frac{\Delta_{act} G^{\ominus}}{RT}\right) \tag{25.30}$$

where, for protein folding, the parameter κ is assumed to be 1.

Typical values of k_{fold} under physiological conditions can range from <0.0001 sec^{-1} (implying that the time taken for folding, $1/k_{fold}$, is of the order of hours or days) up to at least 10,000 sec^{-1} (implying that the time taken for folding, $1/k_{fold}$, is of the order of 100 μsec). Using equation (25.30) at 298K, the standard Gibbs free energy of activation $\Delta_{act} G^{\ominus}$ can therefore range from about 100 kJ mol^{-1} to less than 50 kJ mol^{-1}. Even lower values of $\Delta_{act} G^{\ominus}$ are thought to be possible, implying that, for some proteins, there is only a very low 'energy hill' to climb, and that, kinetically, protein folding is in essence a 'downhill slide' from the unfolded state to the native state, as illustrated by the dashed line in Figure 25.9.

It is worth noting that protein folding is not the same as a vibrating bond, and so perhaps that transition state theory does not really apply. This may be true, but the equations we have discussed are still useful if the folding kinetics of a protein are compared between two different conditions – designated, say, as A and B - or as a result of mutation of the protein sequence. Accordingly, it is interesting and valid to determine the change in free energy of the transition state as

$$\frac{k_{fold,A}}{k_{fold,B}} = \frac{\exp\left(-\frac{\Delta_{act} G_A^{\ominus}}{RT}\right)}{\exp\left(-\frac{\Delta_{act} G_B^{\ominus}}{RT}\right)} = \exp\left(-\frac{(\Delta_{act} G_A^{\ominus} - \Delta_{act} G_B^{\ominus})}{RT}\right)$$

Conventionally, the numerator in the exponential term is represented as $\Delta\Delta_{act} G^{\ominus}$

$$\Delta\Delta_{act} G^{\ominus} = \Delta_{act} G_A^{\ominus} - \Delta_{act} G_B^{\ominus}$$

and so

$$\frac{k_{fold, A}}{k_{fold, B}} = \exp\left(-\frac{\Delta\Delta_{act} G^{\ominus}}{RT}\right) \quad (25.29)$$

25.4.4 Measurement of protein folding kinetics

Protein folding is typically unimolecular because each molecule can fold independently. The observed rate of folding is therefore not dependent on the concentration of the protein. If the protein folding reaction $U \rightleftharpoons N$ proceeds via a single transition state, and if the instantaneous molar concentration of the unfolded protein is represented as $[U]$, then the reaction will display kinetics that can be fitted to the differential equation

$$\frac{d[U]}{dt} = -k_{fold}[U]$$

which can be integrated as

$$\int_{[U]_0}^{[U]_t} \frac{d[U]}{[U]} = -k_{fold} \int_0^t dt$$

giving

$$\ln \frac{[U]_t}{[U]_0} = -k_{fold}\, t$$

or

$$[U]_t = [U]_0 \exp(-k_{fold}\, t) \quad (25.30)$$

In practice, the concentration $[U]$ of the unfolded state is not observed directly, but rather indirectly from a change in a circular dichroism or fluorescence signal, which may have contributions from both the unfolded and native protein. The observed signal I_t is therefore given by

$$I_t = I_{U,t} + I_{N,t}$$

where $I_{U,t}$ and $I_{N,t}$ are the contributions to the total signal attributable to the time-dependent concentrations of the unfolded and native species. However, provided that the contributions to the signal per mole of U and N do not change over time, or with varying compositions, then a single exponential change, as defined by equation (25.30), will still be observed.

Protein folding reactions can be initiated by diluting a chemically unfolded sample (this being a solution of a mixture of the protein and an appropriate chaotrope, such as urea, which causes the unfolding) with a solution that does not contain the chemical chaotrope (the 'folding buffer'). This can be done manually for slow-folding proteins, but for rapid-folding proteins, this is typically achieved using a stopped-flow instrument, as illustrated in Figure 25.10. This equipment pushes the two solutions rapidly through a mixer, in which the folding buffer dilutes the effect of the chaotrope, so causing the unfolded protein to re-fold. The solution

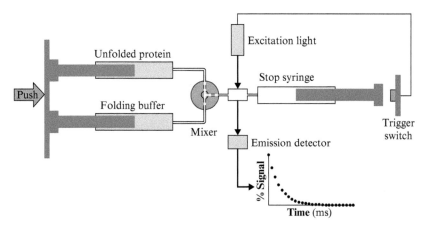

Figure 25.10 A diagrammatic representation of a stopped-flow instrument used to determine protein-folding kinetics.

containing the re-folded protein then passes through an observation cell, into a stop syringe, which pushes out onto a switch, triggering the measurements to be made. The solution of the re-folded protein can be monitored over time, typically using measurements of intrinsic fluorescence or circular dichroism.

The measurements can be repeated with different re-folding buffers, varying the final concentration of the chemical chaotrope. The unfolding reaction can also be measured at higher molar concentrations of the chaotrope, and together these lead to what is known as a **chevron plot**, as shown in Figure 25.11.

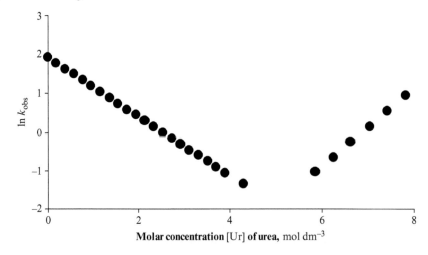

Figure 25.11 A chevron plot, showing observed folding (to the left) and unfolding (to the right) rate constants for a typical protein.

It can be seen that the folding and unfolding rate constants are at least approximately linearly dependent on the molar concentration [Ur] of the chaotrope. The two rate constants, k_{fold} and k_{unfold} can be combined into a single observed rate constant k_{obs}

$$k_{obs} = k_{fold} + k_{unfold} \tag{25.31}$$

This is necessary because experimentally, in the transition region where the protein is close to 50% unfolded, k_{fold} and k_{unfold} become similar, and so both folding and unfolding contribute to the observed kinetic rate constant.

Armed with the methods for measuring protein folding and unfolding, the folding pathways of many proteins have been analysed in detail, at a range of temperatures, pressures, pH values, and chemical chaotrope concentrations. Deviations from the simple chevron plot of Figure 25.11 have suggested stable intermediate states along the folding pathways of some proteins. The Gibbs free energies of these intermediates relative to unfolded, native and transition states, have also been determined experimentally in great detail.

25.4.5 Phi-value analysis

Intermediates and transition states are necessarily fleeting and unstable, and so the structure of these states cannot be determined by techniques, such as X-ray crystallography and nuclear magnetic resonance (NMR), that have been so successful in determining the structures of many native proteins, and some unfolded ones too. Information about the structures of intermediates and transition states therefore needs to be obtained by other means, one of which is the study of the effect of known mutations on the kinetics of protein folding and unfolding.

As we have seen, the native structure of a protein is determined by its amino acid sequence. If the amino acid sequence is changed by mutation, then the mutant protein is likely to adopt a different native structure. Sometimes, the mutation can be catastrophic, substantially changing the native structure and destroying, for example, the enzymic properties of the un-mutated protein (the wild-type). But some mutations – such as the replacement of a particular isoleucine by valine – cause only an incremental change, with the structure of the mutant being very close to the structure of the wild-type. If these incremental changes can be studied, then this can throw light on the wild-type's structure, particularly around the region influenced by the mutation.

Specific mutations can now be created quite easily using modern methods of molecular biology, allowing us to study the kinetics, and Gibbs free energy, of the folding and unfolding of the wild-type and the mutant, as represented in Figure 25.12.

As can be seen, the Gibbs free energies associated with the folding reaction path from the unfolded state, through the transition state, to the folded native state is different for the mutant, as compared to the wild-type, protein. For both the wild-type and the mutant, the standard Gibbs free energy $G_N^{\ominus}$ of the native state is less than the standard Gibbs free energy $G_U^{\ominus}$ of the unfolded state, implying that the native state is the stable form; the standard Gibbs free energies $G_{\ddagger}^{\ominus}$ of the transition states are higher, representing the 'energy hill'.

Furthermore, the changes in the standard Gibbs free energies of each state along the folding trajectory, as represented by the parameters α, β and γ shown in Figure 15.12, might be different, and the values of α, β and γ can therefore be used to probe whether specific interactions that are present in the native structure are present – or not – in the transition state or folding intermediates.

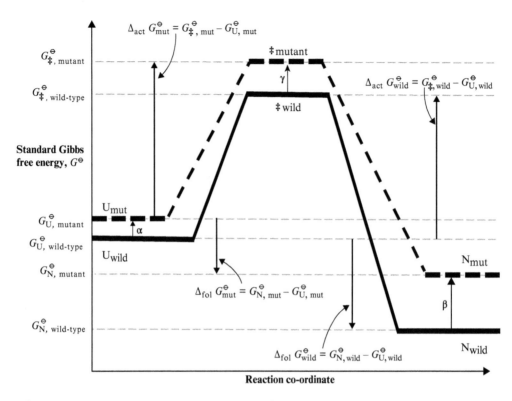

Figure 25.12 A representation of the Gibbs free energies of the unfolded (U), transition (‡), and native (N) states for a wild-type protein, and a mutant. The quantities shown as α, β and γ represent the shift in Gibbs free energy for the unfolded, native and transition states for the mutant, as compared to the wild-type, protein. The direction of each arrow indicates the direction of the corresponding ΔG, from an initial state to a final state.

With reference to Figure 25.12, for the wild-type, we may define the standard Gibbs free energy of folding $\Delta_{fol}G^{\ominus}_{wild}$

$$\Delta_{fol}G^{\ominus}_{wild} = G^{\ominus}_{N,wild} - G^{\ominus}_{U,wild}$$

and the standard Gibbs free energy of activation $\Delta_{act}G^{\ominus}_{wild}$, as

$$\Delta_{act}G^{\ominus}_{wild} = G^{\ominus}_{\ddagger,wild} - G^{\ominus}_{U,wild}$$

Similarly, for the mutant

$$\Delta_{fol}G^{\ominus}_{mut} = G^{\ominus}_{N,mut} - G^{\ominus}_{U,mut}$$

and

$$\Delta_{act}G^{\ominus}_{mut} = G^{\ominus}_{\ddagger,mut} - G^{\ominus}_{U,mut}$$

Furthermore, Figure 25.12 shows that the quantities α, β and γ are defined such that

$$G^{\ominus}_{U,mut} = G^{\ominus}_{U,wild} + \alpha$$
$$G^{\ominus}_{N,mut} = G^{\ominus}_{N,wild} + \beta$$
$$G^{\ominus}_{\ddagger,mut} = G^{\ominus}_{\ddagger,wild} + \gamma$$

The difference between the Gibbs free energies of folding for the wild-type and the mutant, $\Delta\Delta_{fol}G^{\ominus}$, is given by

$$\Delta\Delta_{fol}G^{\ominus} = \Delta_{fol}G^{\ominus}_{wild} - \Delta_{fol}G^{\ominus}_{mut} = (G^{\ominus}_{N,wild} - G^{\ominus}_{U,wild}) - (G^{\ominus}_{N,mut} - G^{\ominus}_{U,mut})$$

$$= (G^{\ominus}_{U,mut} - G^{\ominus}_{U,wild}) - (G^{\ominus}_{N,mut} - G^{\ominus}_{N,wild}) = \alpha - \beta$$

$\Delta\Delta_{fol}G^{\ominus}$ is negative when $\Delta_{fol}G^{\ominus}_{mut} > \Delta_{fol}G^{\ominus}_{wild}$, implying that the action of the mutation is to destabilise the native state of the wild-type protein, as might happen as a result of the removal of a favourable, stabilising, interaction.

Similarly, the difference between the Gibbs free energies of activation for the wild-type and the mutant, $\Delta\Delta_{act}G^{\ominus}$, is

$$\Delta\Delta_{act}G^{\ominus} = \Delta_{act}G^{\ominus}_{wild} - \Delta_{act}G^{\ominus}_{mut} = (G^{\ominus}_{\ddagger,wild} - G^{\ominus}_{U,wild}) - (G^{\ominus}_{\ddagger,mut} - G^{\ominus}_{U,mut})$$

$$= (G^{\ominus}_{U,mut} - G^{\ominus}_{U,wild}) - (G^{\ominus}_{\ddagger,mut} - G^{\ominus}_{\ddagger,wild}) = \alpha - \gamma$$

$\Delta\Delta_{act}G^{\ominus}$ is negative when $\Delta_{act}G^{\ominus}_{mut} > \Delta_{act}G^{\ominus}_{wild}$, implying that the action of the mutation is to increase the height of the 'energy hill' which needs to be 'climbed' when the unfolded state folds into the native state, so slowing down the rate of the folding reaction.

The **phi-value** ϕ_F (alternatively, Φ) is defined as the ratio

$$\phi_F = \frac{\Delta\Delta_{act}G^{\ominus}}{\Delta\Delta_{fol}G^{\ominus}} = \frac{\alpha - \gamma}{\alpha - \beta} = \frac{\gamma - \alpha}{\beta - \alpha} \tag{25.32}$$

ϕ_F can be measured experimentally, and the resulting value throws light on the structure of the transition state, as compared to the unfolded and native states. So, for example, if $\phi_F = 0$, then $\alpha = \gamma$, and quite often, these two values are both zero, as illustrated in Figure 25.13(a). A non-zero value of β indicates that the structural change attributable to the mutation affects the native state, but not the transition state, implying that the mutation impacts the 'downhill' path during which the protein folds.

By contrast, if $\phi_F = 1$, then $\beta = \gamma$, as illustrated in Figure 25.13(b). The implication here is that the structural change caused by the mutation affects the transition state, and so impacts the 'uphill', unfolding, path.

For most wild-type proteins and their corresponding mutants, $0 < \phi_F < 1$, although other values are sometimes observed. Studies of the value of ϕ_F for a number of different mutants of the same wild-type protein can enable the structures of intermediates and transition states to be determined, so complementing the knowledge derived from the structural studies of the more stable native states.

25.4.6 Protein folding pathways

In the last few decades, several competing models for protein folding pathways have been postulated and tested. These have been exotically named as 'framework', 'diffusion-collision', 'hydrophobic collapse', 'nucleation-condensation' and even the 'jigsaw' models. The diffusion-collision and nucleation condensation models now appear to be predominant. In the diffusion collision model, nuclei of structure are formed, and secondary structure elements form around

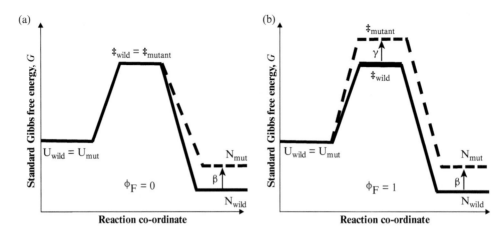

Figure 25.13 Two special cases of the phi-value ϕ_F. By definition, and using the parameters shown in Figure 25.12, $\phi_F = (\alpha - \gamma)/(\alpha - \beta)$. This figure shows two special cases, in both of which, for clarity, $\alpha = 0$: (a) illustrates the case in which $\alpha = \gamma$, and $\phi_F = 0$; (b) illustrates the case in which $\beta = \gamma$, and $\phi_F = 1$.

these. The secondary structure elements then diffuse around until they form a stable tertiary structure. In the nucleation-condensation model, a nucleus forms which contains elements of secondary and tertiary structure. This nucleus persists while the remaining structure condenses around it into the final folded state. Experimentally, different proteins appear to have characteristics mid-way along a continuum between these two models.

Protein folding models are also split into so-called 'classical-view' and 'new-view' models. The former is based on classical chemical kinetics (as described in this chapter) and invokes obligatory transition states and intermediates. Some of these states have indeed been characterised experimentally. However, the 'new view' concept was led by computational modelling of protein folding, and depicts folding from a large number of initial conformations (high entropy), down an energy 'funnel' to fewer native conformations (low entropy), via multiple routes, encountering kinetic traps, and deep energy wells as intermediates, and a narrowed funnel of multiple conformations that represent the transition state. These two views, along with the multiple proposed models, have been challenging to reconcile, but are finally beginning to converge in the concept of **foldons** (Englander and Mayne 2014).

Foldons are units of structure that are independently formed in two-state transitions. Proteins have been experimentally observed to fold by the formation of a sequential series of foldons, and can in some proteins form stably populated intermediates. Furthermore, the initially-formed foldons can represent the nuclei of both diffusion-collision and nucleation-condensation pathways. Finally, the apparent random search through multiple pathways invoked in the downhill energy funnels is relevant in the initial formation of foldons, but from that point onwards the folding pathway becomes more directed by intermediates and transition states best described by the classical view.

Further reading

Englander, S.W. and Mayne, L. (2014), The nature of protein folding pathways, *Proceedings of the National Academy of Sciences (USA)* **111**, 15873–15880.

EXERCISES

1. Define
   - native state
   - unfolded state
   - denatured state
   - primary structure
   - secondary structure
   - tertiary structure
   - chaotrope.

2. A material is steadily heated, at a constant pressure of 1 bar $\approx$ 1 atm, from 250 K to 350 K, and measurements are made of the material's molar enthalpy H, molar Gibbs free energy G, and molar heat capacity at constant pressure C_P as the temperature increases. Draw sketch graphs of the expected behaviour of each of H, G, and C_P, as a function of temperature for each of these materials:
   - copper
   - water
   - butter.

   How would you use these graphs to identify a phase change? What is the evidence that protein unfolding may be thought of as a phase change?

3. Write an essay entitled "Experimental methods for measuring the unfolding of proteins".

4. Describe the individual force-field terms that are generally used to generate potential functions for protein stability. Include a description of the molecular origins of each term included.

5. Write an essay entitled "Experimental methods for measuring protein-ligand interactions".

6. When you take your mobile phone charger out of your bag, the cable is always tangled. When a protein is naturally synthesised, the protein always assumes the 'right' three-dimensional structure. Discuss.

7. Write an essay entitled "Experimental methods for measuring the kinetics of protein folding".

8. Write an essay entitled "How the study of bacterial mutants has enriched our understanding of the kinetics and thermodynamics of protein folding, and the structure of proteins".

9. Write an essay entitled "Current thinking on the pathways of protein folding".

26 Thermodynamics today – and tomorrow

 Summary

A very important process is that of **self-assembly**, in which individual molecules spontaneously coalesce to form larger **aggregates**, two (very different) examples being the condensation of gaseous molecules into a liquid, and the formation of virus particles from the appropriate proteins and nucleic acids.

Some models that can be used to gain a deeper understanding of the mechanisms of aggregation are

- the **van der Waals equation**, by analogy to the formation of liquids from gases
- the two-step **Finke-Watzky model** of **nucleation** and **elongation**
- a development of this model in terms of three, rather than two, steps
- models that are based on the **contact energy** between adjacent sub-units in an aggregate, such as the proteins that **self-assemble** to form a virus **capsid**.

26.1 Self-assembly of large complexes

Our final chapter explores how individual molecules form larger complexes – a process that happens naturally and spontaneously, and so is known as self-assembly. This occurs in many systems, and through several different processes, including

- the formation of liquid droplets from gases
- the polymerisation of large molecules from smaller monomers
- the precipitation of solids, and crystal formation, from saturated liquids
- the more complex ordered assembly of carbon atoms in the C_{60} 'buckyball' and carbon nanotubes...
- ...and of biological molecules into nanomachines or viruses to name just a few of the more important.

In biology, many aggregates form to deliver essential functions – but sometimes, proteins self-assemble into highly-ordered complexes, called amyloids, which can be very dangerous, as associated with many medical disorders, particularly dementias such as "Mad-cow" or Creutzfeldt-Jakob disease, as well as Alzheimer's, Parkinson's and Huntington's diseases. Understanding the thermodynamics that underpin the assembly processes of large aggregates

Modern Thermodynamics for Chemists and Biochemists. Dennis Sherwood and Paul Dalby.
© Oxford University Press 2018. Published 2018 by Oxford University Press.

is therefore very important for the design of complex products with novel properties, for controlling the process to form a specific product, and also for the design of new drugs that can prevent or cure diseases associated with amyloids.

26.2 Non-ideal gases and the formation of liquids

The simplest self-assembly process is the formation of liquid droplets from a gas, a process in which the attractive interactions between gaseous molecules overcomes thermal motion, so causing previously independent molecules to bind together in aggregates of increasing size. One possible model for **protein aggregation** is therefore the process by which a gas condenses into a liquid.

As we saw on page 485, an ideal gas, by definition, is always in the gaseous phase, and can never become a liquid. For a liquid to be formed from a gas, there must be some form of attractive force between neighbouring molecules, which can hold the self-assembled molecules together, and resist the tendency for the molecules to separate as a result of thermal motion. One of the simplest, and most insightful, models for this process is expressed by the van der Waals equation

$$P = \frac{nRT}{V - nb} - \frac{an^2}{V^2} \tag{15.11}$$

in which the parameter b represents the molar volume of the molecules themselves, and the parameter a recognises the intermolecular attractive component of the van der Waals force.

Figure 26.1, which is based on the upper part of Figure 15.10, shows the general behaviour of the van der Waals equation below the critical temperature T_{crit}.

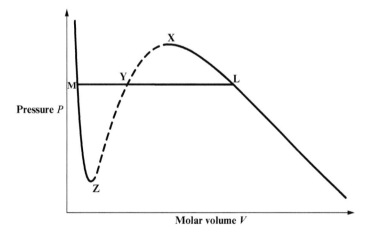

Figure 26.1 The self-aggregation of the molecules in a van der Waals gas to become a liquid. As a van der Waals gas, at a temperature T less than the critical temperature T_{crit}, is compressed isothermally, it will follow the P, V curve to state **L**. Under normal conditions, the gas will then condense into a liquid, corresponding to state **M**. States along the line segment **LX** correspond to supercooling, as might arise in the absence of suitable nucleation.

As discussed in detail on pages 493 and 494, a van der Waals gas at state **L** condenses spontaneously and reversibly into a liquid at state **M**, but only once **nucleation** has taken place, as aided by, for example, the presence of an impurity or a defect in the container surface. In the absence of nucleation, the gas can achieve a supercooled state, corresponding to any state along the line segment **LX**.

While any of the non-ideal gas equations could be used as the basis of modelling the equilibria of other nucleation and polymerisation reactions – for example, those of protein aggregation or viral assembly – such an approach would require significantly unrealistic oversimplifications: in practice, many mechanisms additional to van der Waals interactions are involved. Several alternative thermodynamic and kinetic schemes have therefore been derived to model biological aggregation, the most important of which we now describe.

26.3 Kinetics of nucleated molecular polymerisation

Many polymerisation processes, such as protein aggregation, transition metal nanocluster formation, and the self-assembly of viral particles, are driven by nucleation events. When these processes are monitored over time, they typically start with a **lag phase** associated with the slow formation of nuclei from a small number of initial **monomers**. This is then followed by the rapid addition of monomers to these nuclei, in a growth process also called **elongation**, until the **polymers** reach their stable length, this being **saturation** – as illustrated schematically in Figure 26.2.

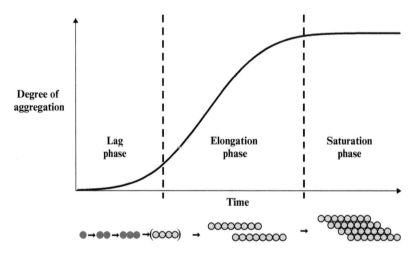

Figure 26.2 A representation of the time course for the evolution of aggregates from monomers. Nucleation takes place during the initial lag phase, leading to elongation and subsequently saturation.

The thermodynamics and kinetics of nucleation within an aggregation process are particularly interesting, and have been described by many different schemes. In the simplest, the initial nucleation step is the relatively slow approach towards equilibrium between monomers M and nuclei of a critical size M_{n^*}, where each **nucleus** comprises n^* monomers on average. The process by which n^* monomers M coalesce to form a nucleus M_{n^*} can be represented by the reversible reaction

$$n^*M \rightleftharpoons M_{n^*}$$

associated with a nucleation equilibrium constant K_{nuc} given by

$$K_{\text{nuc}} = \frac{[M_{n^*}]}{[M]^n}$$

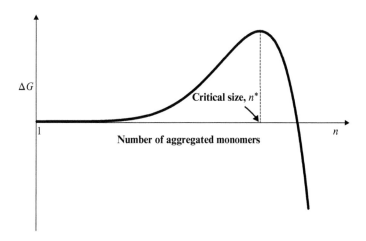

Figure 26.3 A representation of the Gibbs free energy change for nucleation-elongation to polymer length n. The horizontal axis represents the number of aggregated monomers, and the vertical axis represents the corresponding Gibbs free energy change $\Delta G = G_{\text{n-mer}} - G_{\text{monomer}}$. The curve shows a maximum at the critical size n^*. This implies that the formation of nuclei of size n^* is thermodynamically unfavourable, but once a nucleus has been formed, n-mers of length $n > n^*$ will form spontaneously.

As represented in Figure 26.3, the formation of nuclei is unfavourable, with each monomer added during the initial nucleation step having to climb a Gibbs free energy 'hill'. Just as we saw in connection with a supercooled non-ideal gas (see page 474), the Gibbs free energy maximum corresponds to a metastable state that is then readily perturbed by a nucleation event.

Only when the nuclei reach the critical size of n^* monomers does the addition of further monomers become energetically favourable such that the Gibbs free energy decreases; possible mechanistic reasons for this Gibbs free energy change are discussed later. The formation of this critically-sized nucleus marks the end of the lag phase, and the start of the elongation phase. The critical nucleus is formed at the highest point in the Gibbs free energy profile, which acts as a barrier to more extensive aggregation as the clusters of n^* monomers are thermodynamically the least stable species in the process. As the formation of nuclei is energetically unfavourable, they are only present at vanishingly low concentrations – concentrations that are often too difficult to observe directly. However, once even a very few nuclei have been formed, they are sufficient to nucleate further rapid growth.

Nucleated aggregation processes, as illustrated in Figures 26.2 and 26.3, are described well kinetically by the two-step **Finke-Watzky model**. This simplifies the complex set of aggregation reactions down into just two steps. The first step is an initial, slow, rate-limiting, nucleation step in which n^* monomers M coalesce into a critically-sized nucleus M_{n^*}, as represented by the reaction

$$n^*M \xrightarrow{k_{\text{nuc}}} M_{n^*} \qquad \text{nucleation step}$$

as associated with a reaction rate constant k_{nuc}.

The second step is the elongation step, which we write as

$$M + M_i \xrightarrow{k_{\text{el}}} 2M_i \qquad \text{elongation step}$$

as associated with a reaction rate constant k_{el}, and in which the symbol M_i is being used to represent not a specific nucleus of exactly I monomers, but rather any molecular nucleus comprised of I monomers, where i is any number greater than n^*. As can be seen from the way the elongation reaction is written, the product M_i of the reaction is also a reactant. Such a reaction is known as **autocatalytic**, in that the presence of the product M_i catalyses its own production.

The assumption that the elongation step is autocatalytic implies that any current nucleus M_i, which is formed from at least n^* monomers, can catalyse the conversion of a monomer into another nucleus. This somewhat ignores the need for the newly-formed nucleus actually to have the critical size n^*, but nonetheless it assumes that two kinetically-equivalent nuclei are formed from M and M_i. Despite this being an oversimplification, this model does fit much experimental data rather well.

The kinetics of aggregation are described in the two-step Finke-Watzky model by the equation

$$\frac{d[M]}{dt} = -k_{\text{nuc}}[M]^{n^*} - k_{\text{el}}[M][M_i] \qquad (26.4)$$

In this equation, the two terms on the right-hand side relate to the nucleation step $n^* M \to M_{n^*}$, and the elongation step $M + M_i \to 2 M_i$, respectively. The first term therefore captures the kinetics of the lag phase as the nucleus slowly forms to reach a critical concentration $[M_{n^*}]$, whilst the second term describes an exponential growth phase in which existing nuclei are elongated, while also catalysing the formation of new nuclei.

While equation (26.4) fits many different types of nucleated-aggregation process, including protein aggregation, the molecular basis of autocatalysed nuclei formation is not always intuitively described by equation (26.4). For protein aggregation, the two-step Finke-Watzky model can be enhanced by splitting the single elongation step into two steps – a simpler elongation in which a single monomer M is added to the aggregate M_i to form the longer aggregate M_{i+1}, plus a fragmentation step in which elongated nuclei break randomly at any one position to form two independent nuclei. This three-step process can therefore be described by a system of three kinetic equations as

$$n^*M \xrightarrow{k_{nuc}} M_{n^*} \qquad \text{nucleation step}$$

$$M + M_i \xrightarrow{k_{el}} M_{i+1} \quad \text{where } i \geq n^* \qquad \text{elongation step}$$

$$M_i \xrightarrow{k_{fr}} 2M_i \qquad \text{fragmentation step}$$

The third, fragmentation, step removes the need to assume an undefined autocatalytic mechanism, as the fragmentation is now increasing the number of nuclei.

In the fragmentation step, the average number $\langle n \rangle$ of monomers in the nuclei M_i is given by

$$\langle n \rangle = \frac{[M]_0 - [M]}{[M_i]}$$

where $[M]_0$ and $[M]$ are the initial and remaining free monomer concentrations.

The number of positions at which the break can occur in a given nucleus is one less than the number of monomers from which the nucleus is formed, namely, $\langle n \rangle - 1$, and so the molar concentration of breakable points, $[B]$, is given by

$$[B] = [M_i](\langle n \rangle - 1) = [M]_0 - [M] - [M_i]$$

The kinetics of this three-step model of protein aggregation are therefore described by the pair of equations

$$\frac{d[M]}{dt} = -k_{nuc}[M]^{n^*} - k_{el}[M][M_i] \tag{26.4}$$

$$\frac{d[M_i]}{dt} = k_{nuc}[M]^{n^*} + k_{fr}([M]_0 - [M] - [M_i]) \tag{26.5}$$

As for the two-step Finke-Watzky model, these equations describe a lag phase as the nucleus slowly forms to reach a critical concentration $[M_{n^*}]$, followed by an exponential growth phase – but in the three-step model, the growth phase is modelled as the elongation of existing nuclei, and also their random fragmentation into two new nuclei.

While both the two-step and three-step kinetic treatments fit much observed data well, they are both based on several simplifying assumptions. Indeed, many other more detailed schemes for protein aggregation kinetics have been described that include additional mechanistic steps, such as monomer pre-activation, branching, or heterogeneous nucleation onto the surface of the polymers. This variety of plausible models highlights the fact that even if the equations derived from a kinetic model fit the data well, this does not prove that the model actually represents the steps involved in the molecular mechanism. In circumstances such as these, where there are multiple theories, and where it's always possible to devise increasingly complex models, the best approach is to follow the advice of the fourteenth century philosopher William of Ockham: according to '**Ockham's razor**', the wise man uses the simplest possible model, based on the simplest set of assumptions and hypotheses, with this most simple model being made more complex only when warranted by experimental observation!

26.4 Molecular mechanisms in protein aggregation

You may ask "why is nucleation energetically unfavourable, but the elongation step energetically favourable?". This actually depends on what system is being observed, but for protein aggregation, a number of mechanisms have been proposed that explain this behaviour.

One general mechanism considers the number of contacts made by each monomer within the final aggregate. For example, when one monomer interacts with a second monomer to form a **dimer**, only one contact is made, and this has two effects. Firstly, the formation of the attractive bond between the two monomers is exothermic, and so the corresponding change in enthalpy $\Delta H_{1\to 2}$ (where the subscript $_{1\to 2}$ symbolises the formation of a 2-mer from two 1-mers) is negative; secondly, since the dimer is a more ordered state than two free monomers, there is a reduction in the system entropy, and so $\Delta S_{1\to 2}$ is negative. It is therefore possible that, at a given temperature T, $\Delta G_{1\to 2} = \Delta H_{1\to 2} - T\Delta S_{1\to 2}$ is positive, making the formation of the dimer from two monomers thermodynamically unfavourable. A third monomer could then form an interaction with both of the first two monomers, as represented by Scheme 1 in Figure 26.4, thereby increasing the enthalpy released ($\Delta H_{2\to 3}$ is now about double $\Delta H_{1\to 2}$, and so more negative), while less significantly affecting the entropy lost ($\Delta S_{2\to 3}$ for this step is still positive, but less than $\Delta S_{1\to 2}$). Even though, for the formation of the dimer from two monomers, $\Delta G_{1\to 2} = \Delta H_{1\to 2} - T\Delta S_{1\to 2} > 0$, it is now quite possible that $\Delta G_{2\to 3} = \Delta H_{2\to 3} - T\Delta S_{2\to 3} < 0$, implying that the formation of the trimer from the dimer is now thermodynamically favourable. According to Scheme 1, the dimer could be the thermodynamically unfavourable nucleus, with subsequent elongation being more favourable. Variations of this concept with the critical nuclei larger than dimers are also potentially possible.

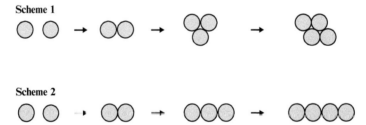

Figure 26.4 Two different schemes of molecular aggregation. In Scheme 1, monomers can add to existing aggregates making multiple contacts; in Scheme 2, each successive monomer makes only a single contact with the aggregate to which it binds. The greater the number of contacts associated with the addition of each successive monomer, the greater the enthalpy change on adding that monomer; the corresponding entropy change, however, is largely independent of the number of contacts.

An underlying problem with the largely enthalpy-driven mechanism described for Scheme 1, when applied to protein aggregation at least, is the possibility that the entropy change observed for the interaction of two or more proteins is actually positive. This is because of a rather significant contribution of water molecules that solvate the surface of proteins. These surface water molecules are typically in highly ordered, low entropy, structural states compared to bulk water, and so the association of two proteins leads to the de-solvation of the interacting surfaces, and an increase in entropy for the water molecules. This increase in solvent entropy is much larger overall than the loss of entropy associated with the protein molecules themselves.

For proteins, an additional possible mechanism that makes elongation both thermodynamically favourable, and also kinetically faster, is the occurrence of a structural rearrangement in one or more monomers that is thermodynamically promoted only once it is associated with one or more other monomers. This structural rearrangement could then create a more favourable binding site for the addition of new monomers during elongation. In the simplest case of linear polymerisation, as represented by Scheme 2 in Figure 26.4, in which aggregates contain only one site for addition of a new monomer, only the structural rearrangement mechanism can lead to nucleation, and this could conceivably occur in the dimer, trimer or even longer aggregates.

26.5 The end-point of protein aggregation

As shown in Figure 26.2, the rate of aggregation slows down in what is often termed the saturation phase. Essentially, the aggregation process continues until the free monomer and aggregated species reach a thermodynamic equilibrium, or until the free monomer is depleted. There is potential for confusion here, as protein aggregation is sometimes found to be reversible, and reaches a thermodynamic equilibrium, whereas in other cases, such as the protein aggregation that happens when you fry an egg, it is best described as an irreversible kinetic process. This is a little misleading as in theory there is no such thing as an irreversible reaction (in that an apparently 'irreversible' chemical reaction can always be described in terms of a reversible reaction associated with an extremely large, or extremely small, equilibrium constant), but in practice such reactions are rendered effectively irreversible by virtue of the reverse reaction being too slow, or improbable, compared to the forward reaction, under any achievable conditions.

26.6 The thermodynamics of self-assembly for systems with defined final structures

26.6.1 Virus structures

Biological systems are the masters of the self-assembly of a number of component parts into highly ordered super-structures – from protein complexes, DNA-associated proteins, and translational machinery, through to organelles, whole cells, organs, organisms and ultimately ecosystems. Two key processes are at work here. One involves *organisation* and requires a complex and dynamic network of reactions, such as metabolism, to convert energy sources (such as food and light) into an ordered, but ever-changing, system. The second process is *spontaneous self-assembly* in which multi-unit complexes are built from a relatively small number of different sub-unit types, which could be proteins, DNA, RNA or even cells.

Many biological systems self-assemble, reversibly, and spontaneously, into highly-ordered polymeric structures which have the important property that the end-points of the self-assembly process are well-defined in terms of the final number of subunits and the resulting geometric form. Not only does the process start with spontaneous self-assembly, the process also 'knows' when to stop. Virus particles are a classic example of this: a virus particle must self-assemble to form a macro-molecular complex that is sufficiently stable, so as to protect the genome packed inside, whilst also being somewhat dynamic, so that internal parts can transiently interact with the environment.

Figure 26.5 An electron micrograph of two icosahedral adenoviruses, alongside a cartoon representation of an icosahedron. https://commons.wikimedia.org/wiki/File:Icosahedral_Adenoviruses.jpg

Many spherical viruses are based on icosahedra with triangular facets, as seen in the electron micrographs in Figure 26.5 and as represented schematically in Figure 26.6.

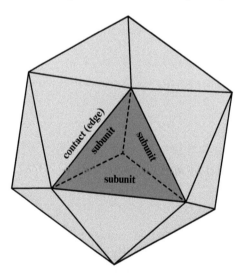

Figure 26.6 A regular icosahedron formed from 20 equilateral triangle facets. Viral capsids form icosahedra from the assembly of protein monomers or subunits. In the example shown, three proteins assemble to form one facet of the icosahedron.

In the simplest icosahedral virus, each triangular facet is comprised of three identical proteins (the 'subunits' as represented in Figure 26.6), and so the full icosahedron of 20 facets contains 60 proteins. The formation of relatively strong and specific interactions drives the overall geometry, which for viruses has often evolved to give a closed form – known as a capsid – such as a sphere. This is slightly different from the previous example of protein aggregation for which the interactions can be weak, or less specific, and not evolved to give closed higher-order structures such as spheres, but rather tend to give indefinitely extended fibrils, or otherwise disordered aggregates. In many viruses, the timing of self-assembly is controlled by environmental cues that induce conformational switches in the pre-activated monomer, giving activated forms that can then self-associate to form the nuclei for capsid assembly.

The kinetics of viral assembly, once activated, have characteristics similar to those of protein aggregation. The initial lag phase involves the accumulation of assembly intermediates up to a critical nucleus size, which is then followed by an elongation phase in which intermediates may combine, or be added to by individual monomers. Eventually, a population of full capsids is formed. The capsids are closed structures and have no vacant positions for further monomers to be added. The reaction therefore ends with a thermodynamic equilibrium that favours the most stable configuration, complete capsids.

Viral assembly, much like protein aggregation, is thermodynamically driven by the 'burial' of hydrophobic surfaces. The loss in entropy resulting from the association of subunits is therefore offset by the entropy increase obtained by the de-solvation of hydrophobic surfaces, in which many water molecules become dissociated. There is, of course, also a loss in enthalpy due to the newly formed van der Waals contacts between the hydrophobic surfaces of interacting proteins.

The kinetics and thermodynamics of viral assembly require some modification of the previous models for protein aggregation to account for the observations that

- virus particles are not open-ended, but form discrete and closed final structures, the capsids
- each capsid is formed from a single nucleus, whereas many protein aggregate particles can in theory form from a single nucleus.

26.6.2 Capsid stability

The stability of a capsid is more easily defined than that of a protein aggregate, as there is only one, well-defined, final state comprising N protein subunit monomers. The fundamental physical mechanisms underlying the assembly of capsids from subunits are similar to those driving the formation of critical nuclei from monomers in protein aggregation, suggesting that viral capsid formation is analogous to critical nuclei formation in aggregation, except that no further elongation can occur. If we represent the assembly of the capsid as the reversible reaction

N subunits $\rightleftharpoons$ capsid

then we may define an association constant K_{capsid} of assembly as

$$K_{capsid} = \frac{[capsid]}{[subunit]^N}$$

and the corresponding Gibbs free energy ΔG_{capsid} is therefore

$$\Delta G_{capsid} = -RT \ln K_{capsid} \tag{26.6}$$

26.6.3 Contact energy

A useful energy to measure is that associated with the formation of each protein interface, or contact with a neighbouring protein, within a completed capsid – the **contact energy**. To determine this, we need to recognise that each subunit can make only a finite number of contacts, leading to a well-defined number of contacts within a completed capsid, as can be inferred from Figure 26.6.

If c is the number of contacts that each protein subunit can make, and N is the number of protein subunits per capsid, then the total number of contacts in a capsid is $cN/2$, where the division by 2 is necessary as each contact is shared by two interacting protein subunits. The average Gibbs free energy per contact $\Delta G_{contact}$ is therefore

$$\Delta G_{contact} = \frac{\Delta G_{capsid}}{cN/2} = -RT \ln K_{contact} \tag{26.7}$$

where $K_{contact}$ is the association constant between individual protein subunits within a complete capsid. Combining equations (26.6) and (26.7), we have

$$-RT \ln K_{contact} = -\frac{RT \ln K_{capsid}}{cN/2}$$

implying that

$$K_{contact}^{cN/2} = K_{capsid} \tag{26.8a}$$

or

$$K_{contact} = K_{capsid}^{2/cN} \tag{26.8b}$$

Equations (26.8a) and (26.8b) can therefore be used to determine $K_{contact}$ from the experimentally measured capsid stability, as defined by K_{capsid}, at equilibrium.

We may also wish to consider the possibility that the subunits in each capsid can be rotated to give the same overall capsid structure, but with each individual subunit in a different relative orientation. This is known as **degeneracy**, and each distinct combination of subunit orientations has its own associated Gibbs free energy contribution to the final capsid population. For example, a capsid that has two different possible assembly configurations is two-times more likely to form, and so the equilibrium constant would be biased two-fold towards its formation. The correct determination of the Gibbs free energy contribution of an individual contact between subunits would therefore require that the $K_{contact}$ term, as determined as just described, is divided by a factor of two.

For a polyhedron in which each of N subunits can rotate independently to i positions, the total number of distinct conformations is given by i^{N-1}. Note that the result is not i^N, as this would lead to multiple counting of the same conformation – as may be verified by considering a theoretical two-faced polyhedron, in which each face can be oriented either to the left or to the right, as illustrated in Figure 26.7.

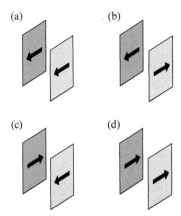

Figure 26.7 The number of different rotational conformations. A hypothetical structure is comprised of two subunits (as represented by the light and dark rectangles), each of which can exist in two conformations (as represented by the forward and backward arrows). There are a total of four possible configurations, but if configuration (a) is rotated through 180° about an axis perpendicular to the 'plane' of the subunits, the result is configuration (d) implying that, structurally, configurations (a) and (d) are identical; likewise, configuration (b) is identical to configuration (c).

As shown in Figure 26.7, our hypothetical two-faced polyhedron has a total of four possible overall conformations, but if conformation (a) is rotated by 180°, it becomes identical to conformation (d); likewise conformation (b) is identical to conformation (c). The number of distinct conformations in this case is therefore only two: (a) ≡ (d) and (b) ≡ (c). Extrapolating to the general case, the number of distinct conformations is therefore i^{N-1} for the entire polyhedron, or i^{N-1}/N conformations per subunit.

When the subunits can rotate, equation (26.8b) therefore becomes

$$K_{contact} = \frac{K_{capsid}^{2/cN}}{i^{N-1}/N} \tag{26.9a}$$

or

$$K_{contact}^{cN/2} \left(\frac{i^{N-1}}{N} \right) = K_{capsid} \tag{26.9b}$$

26.7 Towards the design of self-assembling systems

So far in this chapter, we have explored three characteristically different types of self-assembling systems. The first, the liquefaction of a non-ideal gas, results in a homogeneous final state, a liquid. The second, protein aggregation, leads in some cases to an ordered amyloid fibrillary (string-like) state, and in other cases to an amorphous precipitate, neither of which is generally welcome in biology or in healthcare applications. The third, viral capsid assembly, is a highly deterministic process that forms a distinct and finite structure that has a specific biological function. Whilst there exist many other examples similar to the first two cases, it is the third example which has a particular appeal as regards the design of materials and structures that can self-assemble in a controlled and well-defined manner. Understanding the kinetic and thermodynamic processes that control their assembly rates, specificity and final topology is crucial for exploiting these properties, as well as for designing new self-assembling systems for novel – and useful –purposes. This challenge forms the basis of considerable current research in the fields of nanotechnology and synthetic biology.

The RNA phage Qβ virus is yet one more example that can be readily reconstituted *in vitro* into its final icosahedral capsids. However, through simple mutations of the protein sequence, it was also found, under certain conditions, to form hexagonal sheets and rod-like tubes (Cielens et al. 2000). This was one of the first hints that the self-assembly of biological systems could be exploited to produce new complex geometric forms. The self-assembling nature of viral capsids, as well as their affinity to receptors on specific cell types, is now being used to our advantage in the design and manufacture of novel vaccines and gene therapy agents.

Other systems have recently been found that form self-assembled protein shells, called **bacterial microcompartments**, that are even larger than viral capsids. These are typically 'containers' for enzymatic reactions that need to be isolated from the rest of the cellular machinery, for example, because they produce toxic metabolic intermediates such as aldehydes, or in order to maintain local concentrations of metabolites. An important example of a bacterial microcompartment is the **carboxysome**, illustrated in Figure 26.8, which contains the CO_2-capturing enzyme RuBisCO (see page 744). Without this enzyme, plants, and therefore humans, would not exist. Only now are the contributions of different components of these systems being understood. By selectively removing one or more components, their impact on microcompartment assembly can be observed, while new structural forms, such as tubes and sheets, can also be created.

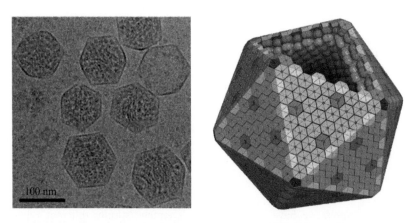

Figure 26.8 Self-assembly of carboxysomes - a bacterial microcompartment.
Image courtesy of Dryden et al., *Protein Science*, 2009.

The rules by which self-assembly is directed and controlled are now enabling synthetic biologists to design novel biologically-inspired structures for future applications in medicine. For example, protein sequences can be designed from scratch that then fold and self-assemble into pre-defined higher-order structures, such as spherical cages that mirror viral capsids.

A particularly interesting line of research is to exploit the encodable affinity of single strands of DNA to a reverse-complement partner strand to design novel three-dimensional structures of DNA, such as 'cages' with 'lids' that can be triggered to open and close. The design rules for artificial DNA super-structures are now so well understood that computer programmes have been written that tell us what DNA sequences to use in order to build a particular shape or structure of our choosing – as playfully shown in Figure 26.9!

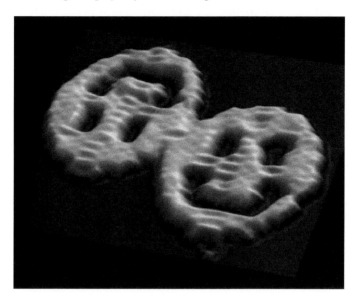

Figure 26.9 Self-assembly of computationally designed DNA molecules to form smiley faces.
Image courtesy of Paul W. K. Rothemund and Nick Papadakis.

Further Reading

Cielens, I., Ose, V., Petorvskis, I., et al. (2000) Mutilation of RNA phage Qβ virus-like particles: from icosahedrons to rods. *FEBS Letters* **482**, 261–264.

Dryden, K.A., Crowley, C.S., Tanaka, S., et al. (2009) Two-dimensional crystals of carboxysome shell proteins recapitulate the hexagonal packing of three-dimensional crystals. *Protein Science* **18**, 2629–2635.

Endres, D. and Zlotnick, A. (2002) Model-based analysis of assembly kinetics for virus capsids or other spherical polymers. *Biophysical Journal* **83**, 1217–1230.

EXERCISES

1. What is 'self-assembly'? Give at least five examples, from as widely different contexts as possible. Thermodynamically, how can self-assembly be interpreted and explained?
2. What is 'nucleation'? How does nucleation take place when a gas liquefies? What is the evidence that a process analogous to liquid nucleation plays an important role in the aggregation of proteins?

3. Describe the two-step Finke-Watzky model of protein aggregation. What experimental evidence is there that supports this model?
4. Do some research on protein aggregation to find some data which is fitted better by the three-step enhanced model, or by a different model. As a result of your research, write an essay entitled "The strengths and weaknesses of the Finke-Watzky model of protein aggregation".
5. Write an essay entitled "Current thinking on models of protein aggregation".
6. Write an essay entitled "Mechanisms for the self-assembly of virus particles".
7. Write an essay entitled "Current thinking on the self-assembly of bacterial microcompartments".
8. Do some research to find out, and understand, how the structures shown in Figure 26.9 – and other similar structures – were formed. Write an essay entitled "How to make DNA smile".

Glossary

Numbers in parentheses (...) indicate the main relevant page number within the text. Cross-references within the glossary are shown in *italics*.

α-**helix** (753): A *secondary structure* conformation of a *protein* in which the *polypeptide chain* is coiled into a right-handed helix, stabilized by the formation of *hydrogen bonds* between the N-H group of one *amino acid* with the C=O group of the *amino acid* (usually) four residues earlier in the chain.

Absolute entropy: see *Third law entropy*.

Absolute zero (44): The *temperature* equal to –273.15° C, regarded as the lowest conceivable *temperature*.

Acid, monoprotic: see *monoprotic acid*.

Acid dissociation constant K_a (538): The *equilibrium constant*, usually expressed as a *concentration thermodynamic equilibrium constant*, for the reversible dissociation of an acid in solution, so, for the *monoprotic acid* dissociation reaction AH (aq) $\rightleftharpoons$ H$^+$ (aq)+A$^-$ (aq)

$$K_a = \frac{[H^+]_{eq}[A^-]_{eq}}{[HA]_{eq}} \tag{17.12}$$

Activated complex (457): The high-energy complex formed when two species collide, corresponding to the intermediate state within a reaction when chemical bonds are broken and reformed.

Activated state: see *transition state*.

Activation energy E_a J mol^{-1} (456): The energy required to raise a molecule from its ground state to the *activated state*. See also *Arrhenius equation*.

Active transport (726): The *protein*-mediated translocation of an *ion* or molecule across a *biological membrane* against a concentration or charge gradient. The translocation is necessarily *endergonic*, and so can happen only if *coupled* to a corresponding *exergonic* process, such as the flow of electrons down an *electron transport chain*.

Activity $a_{b,i}$ (523): Alternatively, chemical activity. For a component *i* in an *ideal solution*, the *molar Gibbs free energy* G_i is related to the *molality* b_i as

$$G_i \text{ (in solution)} = G_i^{\ominus} \text{ (in solution)} + RT \ln (b_i/b_i^{\ominus}) \tag{16.9a}$$

where $G_i^{\ominus}$ is the *standard molar Gibbs free energy*, corresponding to the *standard state* of $b_i^{\ominus}$ = unit *molality*. Equation (16.9b), and similar equations for other physical *states*, can be generalized as

$$G_i = G_i^{\ominus} + RT \ln a_i \tag{16.19a}$$

and

$$\mu_i = \mu_i^{\ominus} + RT \ln a_i \tag{16.19b}$$

where a_i is the *activity* (sometimes 'chemical activity'), and μ_i is the *chemical potential*. See also *molal activity, mole fraction activity*, and *fugacity coefficient*.

Activity, molal: see *molal activity*.

Activity, mole fraction: see *mole fraction activity*.

Activity coefficient, mean molal: see *mean molal activity coefficient*.

Activity coefficient, molal: see *molal activity coefficient*.

Activity coefficient, mole fraction: see *mole fraction activity coefficient*.

Adiabat (57): A curve on a *P, V diagram* that connects corresponding values of *P* and *V* through which a *system* passes during a *reversible adiabatic change in state*.

Adiabatic (57): Associated with zero *heat* flow. An adiabatic system *boundary* is one which prevents the passage of *heat*, and an adiabatic *change in state* is one for which $đq = 0$. See also *adiabat*.

Adiabatic cooling (99): The cooling resulting from the *adiabatic* expansion of a *gas*.

Adiabatic demagnetization (361): A process in which a magnetized *paramagnetic* salt is allowed, reversibly, to demagnetize under *adiabatic* conditions, resulting in cooling, this being a key step in obtaining very low *temperatures*.

Affinity, binding: see *binding affinity*.

Affinity, electron: see *electron affinity*.

Agent, oxidising: see *oxidising agent*.

Agent, reducing: see *reducing agent*.

Aggregation: see *protein aggregation*.

Allotrope (173): At a given temperature and pressure, some materials can exist in the *solid phase* in different forms, for example, carbon in the form of graphite or diamond; phosphorus in the form of red phosphorus or white phosphorus. These forms are known as allotropes. In general, although two or more allotropes can exist, one allotrope is indefinitely *stable*, and represents the *thermodynamic equilibrium* state; the others are 'frozen' *metastable* states.

Amino acid (395): A small organic molecule with both a carboxyl -COOH group and also an amino -NH$_2$ group. Two amino acids can react to form a peptide bond -CO-NH- so creating a *dimer* from the two *monomers*, in which each of the original amino acids is now known as an amino acid residue, This process can continue indefinitely, resulting in a *polymer*, known as a *polypeptide chain*, which can comprise one hundred or more amino acid residues. See also *protein*.

Amphoteric (547): A molecule which can act as both an acid and a base, depending on the circumstances.

Amyloid (780): An aggregate of *protein* molecules which have been *folded* into the wrong shape, and so clump together in biologically 'unnatural' states, often with pathological consequences.

Anion (504): A chemical species, often present in aqueous *solution*, which carries a negative electric charge as a result of having a surplus of electrons, such as SO_4^{2-} (aq).

Anode (600): The *electrode* of an *electrochemical cell* at which an *oxidation half-reaction* takes place. In a *Galvanic electrochemical cell*, the anode carries a negative charge; in an *electrolytic electrochemical cell*, the anode carries a positive charge. The anode is conventionally shown on the left. See also *cathode*.

Anode potential E_{an} V (603): The *electrical potential* conferred upon an *anode* as a consequence of the anode *oxidation half-reaction*. See also *reversible electrode potential*.

Antenna complex: see *light-harvesting complex*.
Approximation, Henderson-Hasselbalch: see *Henderson-Hasselbalch approximation*.
Arrhenius equation (456):

$$k(T) = A\, e^{-\frac{E_a}{RT}} \qquad (14.38)$$

Atmospheric pressure: see *standard atmospheric pressure*.
ATP (173): Adenosine triphosphate, a ubiquitous biological *energy currency*.
ATP synthase (733): A *protein* straddling the *membranes* of *mitochondria* and *thylakoids* that uses the *Gibbs free energy* of the *proton motive force* to synthesise *ATP*.
Autocatalytic (784): A reaction in which a product of the reaction is also a reactant, so catalysing its own production, for example, the *elongation* phase in the *aggregation* of *proteins*.
Available work: see *useful work* and *maximum available work*.
Average bond energy: see *mean bond energy* and *bond dissociation energy*.
Avogadro constant, N_A (7): The number of particles in 1 mol = 6.022141×10^{23} mol^{-1}.
Avogadro('s) number: see *Avogadro constant*.
β-pleated sheet (753): A *secondary structure* conformation of a *protein* in which the *polypeptide chain* forms extended twisted sheets, stabilized by the formation of *hydrogen bonds* between the N-H group of one *amino acid* with the C=O group of another.
Bacterial microcompartment (792): A structure, within a bacterium, comprised of a *protein* shell, enclosing *enzymes* and other metabolites; examples are the *carboxysome* or the *thylakoid*.
Balance, charge: see *charge balance*.
Bar (15): Unit-of-measure of *pressure* equal to 10^5 Pa.
Base, monohydroxic: see *monohydroxic base*.
Base dissociation constant K_b (545): The *equilibrium constant* for the reversible dissociation of a base in solution, so, for the *monohydroxic base* dissociation reaction
BOH (aq) $\rightleftharpoons$ B$^+$(aq) + OH$^-$(aq)

$$K_b = \frac{[B^+]_{eq}[OH^-]_{eq}}{[BOH]_{eq}} \qquad (17.22)$$

Berthelot gas (50): A *gas* for which an appropriate *equation-of-state* is Berthelot's equation

$$\left(P + \frac{an^2}{TV^2}\right)(V - nb) = n\,RT \qquad (3.7)$$

Bilayer, lipid: see *lipid bilayer*.
Binding affinity (764): A measure of the strength of binding between a *ligand* and a *protein*. See also *dissociation constant*.
Binding isotherm (764): A graphical representation of experimental data measuring the *molar concentration* of a *protein-ligand complex* as the *molar concentration* of the free *ligand* increases.
Biochemical standard (677): Referring to measurements made in accordance with the *biochemical standard state*.

Biochemical standard state (674): Formally, the biochemical standard defines the biochemical *standard molal activity* of the aqueous hydrogen *ion* H⁺ (aq) as 10^{-7}, corresponding to pH 7; in practice, the biochemical standard defines the biochemical standard molar concentration [H⁺] of the aqueous hydrogen *ion* H⁺ (aq) as 10^{-7} mol dm⁻³. See also *conventional standard state*.

Bioenergetics (712): The study of energy flows within biological systems, and of the changes in *Gibbs free energy* associated with *metabolic pathways*, reactions and processes.

Biological membrane (724): A biological structure that surrounds, and therefore defines, an enclosed space: for example, the membranes that surround biological cells, mitochondria and chloroplasts. Biological membranes are usually formed from a *lipid bilayer* containing embedded *proteins*.

Biological standard state: see *biochemical standard state*.

Bioluminescence (745): The production of light as a result of a biological process, as exemplified by fireflies.

Boiling point, elevation of: see *elevation of the boiling point*.

Boltzmann constant, k_B J K⁻¹ (70): A universal constant defined as $k_B = R/N_A = 1.381 \times 10^{-23}$ J K⁻¹. See also *Avogadro constant* N_A and *gas constant R*.

Boltzmann distribution (68):

$$n(\mathcal{E}_i) = \frac{N}{k_B T} e^{-\varepsilon_i/k_B T} \tag{3.20}$$

Boltzmann's equation (342):

$$S = k_B \ln W \tag{11.2}$$

Bond, hydrogen: see *hydrogen bond*.

Bond dissociation energy D° J mol⁻¹ (187): The enthalpy change, usually measured at a temperature of 298 K, associated with the breaking of 1 mol of a given bond of some specific molecular entity, so creating two free radicals

$$X - Y (g) \rightarrow X^\bullet(g) + Y^\bullet(g) \qquad \Delta H^\ominus = D^\circ$$

See also *mean bond energy*.

Bond energy: see *bond dissociation energy* and *mean bond energy*.

Boundary (5): That which separates a *system* from its *surroundings*. A boundary may be physical or conceptual; it may be rigid or flexible. A given boundary has properties which determine what may, or may not, cross the boundary, therefore defining the characteristics of the corresponding *system*. An *adiabatic* boundary prevents the flow of *heat*; a *closed system* is defined by a boundary that permits the flow of *heat* and *work*, but prevents the flow of matter; an *isolated system* is defined by a boundary that blocks the flow of *heat*, *work* and matter; and an *open system* is defined by a boundary that permits the flow or *heat*, *work*, and matter.

Boyle's law (16): For a fixed mass of an *ideal gas* at constant *temperature*

$$PV = \text{a constant.} \tag{1.3}$$

Bridge, salt: see *salt bridge*.

Brønsted-Lowry theory (546): A theory offering a generalised definition of an acid as a donor of hydrogen *ions*, and a base as an acceptor, with the implication that all acid-base reactions can be expressed in terms of two *conjugate pairs* of acids and bases. So, for the forward acid-base reaction

$$HA\ (aq) + H_2O \rightleftharpoons H_3O^+(aq) + A^-(aq)$$

HA (aq) is the acid and H_2O the base; for the reverse reaction H_3O^+ (aq) is the acid and A^- (aq) the base. One conjugate pair is HA(aq), the *conjugate acid* to A^- (aq), the corresponding *conjugate base*; the other is H_2O, the *conjugate base*, to H_3O^+, the corresponding *conjugate acid*.

Buffer capacity β m^{-3} (555): A measure of a *buffer solution's* ability to resist changes in *pH*

$$\beta = -\frac{1}{V}\frac{dn_H}{dpH} \tag{17.33a}$$

Buffer solution (549): A *solution*, usually aqueous, of (often equal *molar concentrations* of) either a weak acid HA and a corresponding salt XA, or a weak base BOH and a corresponding salt BX. The *solution* has the property of being resistant to changes in *pH*.

Calorimeter (164): An instrument used for *calorimetry*.

Calorimetry (164): An experimental technique which measures *heat* flows, and the corresponding changes in *temperature*, so allowing for the determination of *internal energy* changes (at constant volume), *enthalpy* changes (at constant pressure), and *heat capacities*.

Calorimetry, differential scanning: see *differential scanning calorimetry*.

Calvin cycle: see *CBB cycle*.

Capsid (789): The *protein* shell within which the nucleic acids of a virus are contained.

Carbon fixation: see CO_2 *fixation*.

Carbon-fixation cycle: see *CBB cycle*.

Carathéodory statement of the Second Law of Thermodynamics (322): Two *adiabats* cannot cross. See also *Second Law of Thermodynamics, Clausius statement of the Second Law of Thermodynamics*, and *Kelvin–Planck statement of the Second Law of Thermodynamics*

Carboxysome (792): A biochemical *microcompartment* containing the enzyme *RuBisCO*, which is essential for the capture of carbon dioxide.

Carnot cycle (230): A *thermodynamic cycle* comprising four steps: a *reversible isothermal* expansion; a *reversible adiabatic* expansion; a *reversible isothermal* compression; and finally a *reversible adiabatic* compression.

Carnot heat engine (323): A device that operates according to a *Carnot cycle*, and so derives mechanical *work* from the flow of *heat* down a *temperature gradient*.

Carnot heat pump (324): A device that operates according to a reverse *Carnot cycle*, using mechanical *work* to cause *heat* to flow up a *temperature gradient*.

Carnot's theorem (325): The *efficiency* of a *reversible Carnot heat engine* is a function only of the *temperatures* of the *heat reservoirs* between which it operates, and is greater than the *efficiency* of any *irreversible* engine operating between the same two *reservoirs*.

Cathode (600): The *electrode* of an *electrochemical cell* at which a *reduction half-reaction* takes place. In a *Galvanic electrochemical cell*, the cathode carries a positive charge; in an *electrolytic electrochemical cell*, the cathode carries a negative charge. The cathode is conventionally shown on the right. See also *anode*.

GLOSSARY

Cathode potential E_{cat} V (604): The *electrical potential* conferred upon a *cathode* as a consequence of the cathode *reduction half-reaction*. See also *reversible electrode potential*.

Cation (504): A chemical species, often present in aqueous *solution*, which carries a positive electric charge as a result of having a deficit of elections, such as Cu^{2+} (aq).

CBB cycle (744): Alternatively, Calvin cycle, Calvin-Benson-Bassham cycle, carbon-fixation cycle. A cyclic *metabolic pathway* taking place within the *chloroplast stroma* in which the *Gibbs free energy* stored in *ATP* and *NADPH*, synthesized as a result of the *light-dependent reactions of photosynthesis*, is used to convert atmospheric CO_2 into the important biochemical intermediary glyceraldehyde-3-phosphate. A key feature of this cyclic *pathway* is the *fixation* of CO_2 by the enzyme *RuBisCO*.

Cell reaction (602): The *redox* chemical reaction taking place within an *electrochemical cell*, when the *reduction half-reaction* taking place at the *cathode* is suitably coupled with the *oxidation half-reaction* taking place at the *anode*.

Cellular respiration: see *oxidative phosphorylation*.

Celsius scale of temperature °C (44): Alternatively, Celsius scale of temperature. A scale of *temperature* such that, at *standard atmospheric pressure*, 0°C corresponds to the melting point of pure ice, and 100°C to the boiling point of pure water.

Centigrade scale of temperature: see *Celsius scale of temperature*.

Chain, electron transport: see *electron transport chain*.

Chain, polypeptide: see *polypeptide chain*.

Chain rule (641):

$$\left(\frac{\partial P}{\partial V}\right)_T \left(\frac{\partial V}{\partial T}\right)_P \left(\frac{\partial T}{\partial P}\right)_V = -1 \qquad (21.11)$$

Change, phase: see *phase change*.

Change in state (8): If two observations are made, at two successive times, on a *system* at *thermodynamic equilibrium*, and if the value of any *state function* has changed, then that *system* has undergone a change in state.

Chaotrope (752): An agent, such as urea, that causes a macromolecule, such as a *protein*, to unfold.

Charge balance (665): The requirement that a *solution* at *equilibrium* must be electrically neutral, and that the total positive charge, attributable to all positive *ions*, must equal the total negative charge, attributable to all negative *ions*.

Chemical activity: see *activity*.

Chemical kinetics (450): The study of how quickly, or slowly, chemical reactions take place, and their corresponding *rates of reaction*. See also *activation energy* and *Arrhenius equation*.

Chemical potential μ_i J mol^{-1} (404): The quantity

$$\left(\frac{\partial G}{\partial n_i}\right)_{P,T,n_{j \neq i}} = \mu_i = G_i \qquad (13.54)$$

Chemi-osmotic potential G_{co}' J mol^{-1} (730): The quantity derived from the *chemical potential* $G'(H^+)$ attributable to a local *molar concentration* of H^+ (aq) *ions*, and the local *electrical potential* Ψ, of a biological environment as

$$G_{co}' = G'(H^+) + F\Psi \qquad (24.4a)$$

See also *chemi-osmotic potential gradient* and *proton motive force*.

Chemi-osmotic potential gradient $\Delta G_{co}'$ J mol^{-1} (730): The difference between two *chemi-osmotic potentials*, for example, across a *biological membrane*

$$\Delta G_{co}' = [G'(\text{H}^+, \text{in}) + F\Psi(\text{in})] - [G'(\text{H}^+, \text{out}) + F\Psi(\text{out})] \quad (24.5)$$

See also *proton motive force*.

Chevron plot (774): A graph, with a chevron-like shape, derived from experimental data relating to the *kinetics* of *protein* folding and unfolding.

Chloroplast (739): An organelle within green plants, which is primarily responsible for *photosynthesis*.

Circular dichroism (753): The interaction between circularly polarised light and *proteins* in *solution*, measurement of which can give information on *protein folding* and unfolding. See also *mean residue ellipticity*.

Citric acid cycle: see *tricarboxylic acid cycle*.

Clapeyron equation (482): For any phase change at an external pressure P_{ex}

$$\frac{dP_{ex}}{dT_{ph}} = \frac{\Delta_{ph}H}{T_{ph}\Delta_{ph}V} \quad (15.7)$$

Clausius–Clapeyron equation (482): For liquid or solid $\rightleftharpoons$ vapour phase changes

$$\frac{d\ln p}{dT_{ph}} = \frac{\Delta_{ph}H}{RT_{ph}^2} \quad (15.8)$$

Clausius inequality (273):

$$0 \geqslant \oint \frac{dq}{T} \quad (9.11)$$

The equality holds for a *reversible* change, and leads to the definition of *entropy* as $dS = đq_{rev}/T$; the inequality holds for all *irreversible* – and hence real – changes. The use of an inequality introduces the asymmetry of *unidirectionality* that underpins *spontaneity*, and is the essence of the *Second Law of Thermodynamics*.

Clausius statement of the Second Law of Thermodynamics (304): It is impossible to create a device that operates in a *cycle* and has no effect other than the transfer of *heat* from a cooler to a warmer body. See also *Second Law of Thermodynamics*, *Kelvin–Planck statement of the Second Law of Thermodynamics*, and *Carathéodory statement of the Second Law of Thermodynamics*.

Closed system (57): A thermodynamic *system*, the *boundary* of which permits the exchange of *heat* and *work* between the *system* and the *surroundings*, but not the flow of matter.

CO_2 fixation (744): The process by which molecular CO_2 is incorporated into the biomolecule ribulose-1,5-biphosphate, as catalysed by the enzyme *RuBisCO* in the *carboxysome*.

Coefficient, Joule–Thomson: see *Joule–Thomson coefficient*.

Coefficient, mean molal activity: see *mean molal activity coefficient*.

Coefficient, molal activity: see *molal activity coefficient*.

Coefficient, mole fraction activity: see *mole fraction activity coefficient*.

Coefficient, transmission: see *transmission coefficient*.

Coefficient, virial: see *virial coefficient*.

Coefficient, volumetric expansion: see *volumetric expansion coefficient*.

Colligative properties (579): Thermodynamic properties of *solutions* that are associated with the number of dissolved particles, notably the *elevation of the boiling point*, the *depression of the freezing point*, and *osmosis*.

Complex, antenna: see *light-harvesting complex*.

Complex, light-harvesting: see *light-harvesting complex*.

Complex, protein-ligand: see *protein-ligand complex*.

Compressibility, isothermal: see *isothermal compressibility*.

Compressibility factor (652):

$$Z = \frac{PV}{RT} \tag{21.38}$$

Concentration, molar: see *molar concentration*

Concentration mass action ratio Γ_c (439): Alternatively, concentration reaction quotient Q_c. The *non-equilibrium* ratio

$$\Gamma_c = \frac{([C]/[C]^\ominus)^c \, ([D]/[D]^\ominus)^d}{([A]/[A]^\ominus)^a \, ([B]/[B]^\ominus)^b}$$

Concentration reaction quotient Q_c: see *concentration mass action ratio*.

Concentration thermodynamic equilibrium constant K_c (438): A *thermodynamic equilibrium constant* expressed in terms of the *molar concentrations* [I] mol m^{-3} of each reagent. Accordingly, for the generalized reaction $aA + bB \rightleftharpoons cC + dD$

$$K_c = \frac{([C]_{eq}/[C]^\ominus)^c \, ([D]_{eq}/[D]^\ominus)^d}{([A]_{eq}/[A]^\ominus)^a \, ([B]_{eq}/[B]^\ominus)^b} \tag{14.21a}$$

Conductor (592): A material, such as a copper wire, through which an electrical current can flow.

Conjugate acids and bases (439): According to the *Brønsted-Lowry theory* of acids and bases, for the generalised acid-base reaction

HA (aq) + H$_2$O $\rightleftharpoons$ H$_3$O$^+$ + A$^-$(aq)

HA (aq) is the conjugate acid to the conjugate base A$^-$ (aq); and H$_2$O is the conjugate base to the conjugate acid H$_3$O$^+$. See also *conjugate pair*.

Conjugate pair (439): According to the *Brønsted-Lowry theory* of acids and bases, for the generalised acid-base reaction

HA (aq) + H$_2$O $\rightleftharpoons$ H$_3$O$^+$(aq) + A$^-$(aq)

HA (aq) and A$^-$ (aq) are a conjugate pair, as are H$_2$O and H$_3$O$^+$ (aq). See also *conjugate acids and bases*.

Conjugate variables (299): Two *state functions* X and Y, where X is *intensive* and Y is *extensive*, such that the product $X\,dY$ represents a form of *work* đw or *heat* đq. So, for example, the *pressure* P and the *volume* V are conjugate variables since *reversible* P,V *work of expansion* đ$w_{rev} = P\,dV$; likewise, the *temperature* T and *entropy* S are conjugate variables since the *reversible heat* đ$q_{rev} = T\,dS$.

Constant, Avogadro: see *Avogadro constant*.

Constant, Boltzmann: see *Boltzmann constant*.

Constant, dissociation: see *dissociation constant*.

Constant, equilibrium: see *concentration thermodynamic equilibrium constant, molality thermodynamic equilibrium constant, mole fraction thermodynamic equilibrium constant* and *pressure thermodynamic equilibrium constant*.

Constant, Faraday: see *Faraday's constant*.

Constant, ideal gas: see *ideal gas constant*.

Construction, Maxwell: see *Maxwell construction*.

Contact energy (790): The *energy* associated with the formation of an interface between two neighbouring *proteins*, as is relevant, for example, to the formation of a virus *capsid*.

Conventional standard state (176). A *standard state* compliant with the IUPAC recommendations as shown in Table 6.4. See also *biochemical standard state*.

Conventional standards (169): Referring to measurements made in accordance with the IUPAC standards. See also *standard state*.

Cooling, adiabatic: see *adiabatic cooling*.

Coupled system (391): The equilibrium of an *endergonic* reaction A → B, for which $\Delta G_{A \to B}^{\ominus} = +x$ J, favours the reactant A. The formation of B from A is therefore not *spontaneous*, and any production of B appears to break the *Second Law of Thermodynamics*. But if, under suitable conditions, this reaction can take place simultaneously with an *exergonic* reaction C → D, for which $\Delta G_{C \to D}^{\ominus} = -y$ J, such that $-y + x < 0$, then the equilibrium of the coupled reaction A + C → B + D, $\Delta G_{A,C \to B,D}^{\ominus} = -y + x < 0$ lies to the right. As a result, B is produced, and the *Second Law of Thermodynamics* is upheld. By coupling reactions together, a reaction A → B, which would not happen *spontaneously* by itself, can be made to happen. This is of fundamental importance in biochemistry, where very many reactions, such as the synthesis of *proteins*, are *endergonic*, but can happen by coupling to *exergonic* reactions, such as the hydrolysis of the *energy currency* ATP.

Critical point (480): For any pure *gas*, the unique combination of *pressure* and *temperature* at which the P,V diagram shows a *point of inflexion* at which $(\partial P/\partial V)_T = (\partial^2 P/\partial V^2)_T = 0$. See also *critical pressure, critical temperature* and *critical volume*.

Critical pressure P_{crit} Pa (480): The *pressure* of a gas at the *critical point*. See also *critical temperature* and *critical volume*.

Critical temperature T_{crit} K (480): The highest temperature at which a *gas* can be liquefied solely by increasing the *pressure*. See also *critical point, critical pressure* and *critical volume*.

Critical volume V_{crit} m³ (491): The volume of a *gas* at the *critical point*. See also *critical pressure* and *critical temperature*.

Cryoscopic constant E_f K kg mol⁻¹ (578):

$$E_f = \frac{R(T_{\text{fus}}^*)^2 M_S}{\Delta_{\text{fus}} H} \tag{18.14}$$

See also *depression of the freezing point, ebullioscopic constant*, and *elevation of the boiling point*.

Crystallography, X-ray: see *X-ray crystallography*.

Currency, energy: see *energy currency*.

Cycle (36): Alternatively, *thermodynamic cycle*. A sequence of *changes in state* such that the system returns to its original state. If all the steps within the cycle are *reversible*, then the cycle as a whole is *reversible*. In any step within the cycle, however small, is *irreversible*, then the cycle as a whole is *irreversible*.

Cycle, Calvin: see *CBB cycle*.
Cycle, Calvin-Benson-Bassham: see *CBB cycle*.
Cycle, carbon-fixation: see *CBB cycle*.
Cycle, Carnot: see *Carnot cycle*.
Cycle, citric acid: see *tricarboxylic acid cycle*.
Cycle, Krebs: see *tricarboxylic acid cycle*.
Cycle, TCA: see *tricarboxylic acid cycle*.
Cycle, thermodynamic: see *cycle*.
Cyclic photophosphorylation (743): The *photosynthetic metabolic pathway* that captures light *energy* within the *reaction centre P700* of *photosystem I*, transferring the energy to an electron that then passes down the *electron transport chain* of *photosystem I*, before returning to *photosystem I*. During this cyclic process, H^+ ions are translocated across the *thylakoid* membrane, so creating a *proton motive force* which can drive the synthesis of *ATP* by *ATP synthase*. See also *non-cyclic photophosphorylation*.
Dalton's law of partial pressures (49): In a *system* comprising a mixture of *ideal gases*, each component i exerts a *partial pressure* p_i such that the total pressure P of the system is the linear sum of the *partial pressures* p_i

$$P = \sum_i p_i \tag{3.5}$$

Daniell cell (599): A *Galvanic electrochemical cell* in which the *cathode* is metallic copper in contact with an aqueous solution of Cu^{2+} (aq) *ions* (usually a solution of $CuSO_4$), and the *anode* is metallic zinc in contact with an aqueous solution of Zn^{2+} (aq) *ions* (usually a solution of $ZnSO_4$). If the two *electrodes* are connected by a *salt bridge*, a Daniell cell is represented as

$$Zn\,(s)\,|\,ZnSO_4(aq)\,||\,CuSO_4(aq)\,|\,Cu\,(s)$$

Dark reactions, of photosynthesis: see *light-independent reactions of photosynthesis*.
Data, normalised: see *normalised*.
Debye-Hückel law, extended: see *extended Debye-Hückel law*.
Debye-Hückel limiting law (667):

$$\log_{10} \gamma_{b,\pm} = -A|z_+ z_-|\sqrt{I} \tag{22.20}$$

Debye temperature (355): Experimentally, the *temperature* at which a graph of the *molar heat capacity at constant pressure* C_P of a *solid*, as a function of *temperature*, flattens out.
Degeneracy (790): In the context of the aggregation of *proteins*, the situation in which the *proteins* that form the subunits of a virus *capsid* can be positioned in different orientations.
Demagnetization, adiabatic: see *adiabatic demagnetization*.
Demon, Maxwell's: see *Maxwell's demon*.
Denatured state (749): A state in which the *polypeptide chain* of a *protein* is *unfolded*, in contrast to the *native state*. Some *unfolded* proteins may, under appropriate conditions, return to the *native state*, this being an example of *reversible* unfolding, but some unfolding is *irreversible*.

Depression of the freezing point (577): At any given pressure, the freezing point of a pure *liquid* is higher than the freezing point of a *solution* in which a *solute* is dissolved in that *liquid* as the *solvent*. This phenomenon is known as the depression of the freezing point. See also *colligative properties* and *cryoscopic constant*.

Derivative: see *partial derivative* and *total derivative*.

Derivative, partial: see *partial derivative*.

Derivative, total: see *total derivative*.

Desalination (588): The process whereby salt is removed from sea water, so as to produce drinking water, as achieved, for example, by *reverse osmosis*.

Diameter, effective: see *effective diameter*.

Dieterici gas (50): A *gas* for which an appropriate *equation-of-state* is Dieterici's equation

$$P(V - nb) = nRT \exp\left(-\frac{na}{RTV}\right) \tag{3.8}$$

Differential (77): Mathematically, an infinitesimal change dX. See also *exact differential*, *inexact differential*, *partial derivative*, *total derivative* and *total differential*.

Differential, exact: see *exact differential*.

Differential, inexact: see *inexact differential*.

Differential, total: see *total differential*.

Differential scanning calorimetry (DSC) (759): An experimental *calorimetry* technique which measures the difference between the rates of flow of *heat* into a sample, and into a reference material, as the *temperature* of both increases, under conditions in which, at any instant, the *temperatures* of the sample and the reference are the same. If the sample and the reference material differ on only one respect – for example, the presence of a *protein* in the sample – then the measurement is a direct measurement of the thermal properties, such as the *heat capacity*, of (in this example) the protein. The experimental results of DSC are usually represented as a *thermogram*.

Dilution, infinite: see *infinite dilution*.

Dilution law, Ostwald's: see *Ostwald's dilution law*.

Dimer (786): A molecule formed from two *monomers*, which might be the same (for example, the sugar maltose is a dimer of two glucose *monomers*), or different (the sugar sucrose is a dimer of glucose and fructose *monomers*).

Dissipative effect (139): A phenomenon which confers *irreversibility* on any given *path* between any two *equilibrium states*. Dissipative effects are associated with a *spontaneous* increase in a *system's entropy*, and arise from two different causes. Firstly, *irreversible* processes that transform *work* into *heat*, such as the effects of *dynamic friction* and *electrical resistance*; secondly, *irreversible* processes such as *molecular mixing*.

Dissociation constant K_d (535): The equilibrium constant for the *reversible* dissociation reaction $XY \rightleftharpoons X + Y$

$$K_d = \frac{[X]_{eq} [Y]_{eq}}{[XY]_{eq}} \tag{17.1}$$

Two special cases are the dissociation of an acid in *solution*, $HA \rightleftharpoons H^+(aq) + Y^-(aq)$, and the dissociation of a *protein-ligand* complex $PL \rightleftharpoons P_{free} + L_{free}$.

Dissociation constant, acid: see *acid dissociation constant*.

Dissociation constant, base: see *base dissociation constant*.

Dissociation energy: See *bond dissociation energy*.

Distribution, Boltzmann: see *Boltzmann distribution*.

DSC: see *differential scanning calorimetry*.

Dynamic equilibrium (451): Alternatively, microscopic dynamic equilibrium. A term used in *chemical kinetics* describing the *state* in which the *rate of reaction* v_f of the forward reaction is equal to the rate of reaction v_r of the reverse reaction. *Macroscopically*, this corresponds to a *state* of chemical *equilibrium*.

Dynamic friction (28): Alternatively, kinetic friction. A *force*, present at the contact area between two mutually moving adjacent surfaces, which resists the movement of either surface. For the movement to continue, *work* must be done to overcome the dynamic friction *force*. This *work* is inevitably *dissipated* immediately as *heat*. Dynamic friction is always present, and although its effects can be reduced by lubrication, it can never be removed totally. Dynamic friction is a principal cause of *irreversibility*. See also *static friction*.

Ebullioscopic constant E_b K kg mol^{-1} (575):

$$E_b = \frac{R(T_{vap}^*)^2 M_S}{\Delta_{vap} H} \tag{18.7}$$

See also *elevation of the boiling point, cryoscopic constant*, and *depression of the freezing point*

Effect, dissipative: see *dissipative effect*.

Effect, Seebeck: see *Seebeck effect*.

Effective diameter $å$ m, or nm (668): An estimate of the diameter of the sphere representing the size of an *ion* in solution.

Efficiency η (234): The efficiency η or (η^E) of a *heat engine* is defined as

$$\eta = \frac{\text{net } \textit{work} \text{ done by } \textit{heat engine} \text{ on one complete } \textit{cycle} \text{ of operation}}{\text{net } \textit{heat} \text{ transferred into the } \textit{heat engine} \text{ from the } \textit{hot reservoir}}$$

and the efficiency η or (η^P) of a *heat pump* is defined as

$$\eta = \frac{\text{net } \textit{work} \text{ done on } \textit{heat pump} \text{ during one complete } \textit{cycle} \text{ of operation}}{\text{net } \textit{heat} \text{ transferred from the } \textit{heat engine} \text{ into the } \textit{hot reservoir}}$$

Electrical current I A (592): A flow of electricity, through a *conductor*, driven by an *electrical potential difference*.

Electrical potential V (603): Just as *heat* flows *unidirectionally* and *irreversibly* from a high *temperature* to a low *temperature*, electrons flow *unidirectionally* and *irreversibly* from a low electrical potential to a high electrical potential (from low to high because electrons are negatively charged). Fundamentally, electrical potentials result from the local accumulation of positive or negative electrical charges, as happens, for example, at the *cathode* of a *Galvanic electrochemical cell* (which accumulates positively-charged *ions*, resulting in a positive electrical potential), and the *anode* (which accumulates negatively-charged electrons, resulting in a negative electrical potential). See also *electrical potential difference* and *electrode*.

Electrical potential difference E V (599): An accumulation of electric charge at a particular location, for example, on an *electrode*, confers an *electrical potential*. Two locations, at different *electrical potentials*–for example, the *cathode* and the *anode* of an *electrochemical cell*–are therefore associated with a corresponding electrical potential difference. If those locations are connected by a suitable electrical *conductor*, then the electrical potential difference will drive an *electrical current* through the *conductor*.

Electrical resistance R Ω (592): The ratio of the *electrical potential difference V* (V) across an electrical *conductor*, and the corresponding *electrical current I* (A), according to *Ohm's Law* $V = I R$. The lower the electrical resistance, the higher the *current* for any given *potential difference*.

Electrochemical cell (599): A device comprising two *electrodes*, of different materials, and their associated *electrolytes*. A *Galvanic* electrochemical cell is one that spontaneously creates an *electrical potential difference* between the two *electrodes*, so enabling an *electrical current* to be driven through an external *conductor*. This *electrical potential difference* is the result of a spontaneous chemical reaction, the *cell reaction*, which creates electrons at the *anode* and consumes electrons at the *cathode*. An *electrolytic* electrochemical cell is one connected to an external source of electrical energy, and in which the *cell reaction* is driven in reverse.

Electrode (596): If a rod of metallic zinc is in contact with a *solution* of Zn^{2+} (aq) *ions*, then a spontaneous *oxidation half-reaction*

$$Zn\ (s) \rightarrow Zn^{2+}(aq) + 2\ e^-$$

takes place in which metallic zinc passes into *solution*, leaving the corresponding electrons on the zinc rod. As a result of this accumulation of charge, the rod of metallic zinc acquires an *electrical potential*. The zinc rod is an example of an electrode, and the combination of the zinc rod, and the associated *solution*, known as the *electrolyte*, is an example of a *half-cell*. Electrodes can take many forms, and can be made of many materials, but all have the common property that a *half-reaction*, which may be an *oxidation* or a *reduction*, creates an accumulation of electric charge, so conferring an *electrical potential* on the electrode. See also *inert electrode, platinum electrode, redox electrode, reversible gas electrode, reversible hydrogen electrode, reversible metal electrode, reversible standard reference electrode* and *reversible redox electrode*.

Electrode, graphite: see *graphite electrode*.
Electrode, inert: see *inert electrode*.
Electrode, platinum: see *platinum electrode*.
Electrode, redox: see *redox electrode*.
Electrode, reversible gas: see *reversible gas electrode*.
Electrode, reversible hydrogen: see *reversible hydrogen electrode*.
Electrode, reversible redox: see *reversible redox electrode*.
Electrode potential E_{el} or E V (603): The *electrical potential* conferred on an *electrode* as a consequence of the accumulation of the electrical charge created by the appropriate *half-reaction*.

Electrode potential, standard: see *standard electrode potential*.

Electrolysis (609): The process in which an external electrical source forces an *electric current* through a suitable *electrolyte*, driving a chemical reaction that, under normal conditions, is not *spontaneous*, such as the splitting of water into molecular hydrogen and oxygen.

Electrolyte (596): A medium, for example, an aqueous *solution*, containing *ions*, and which allows the flow of *ions*. When an electrolyte is in contact with the surface of a suitable *electrode*, a *half-reaction* takes place which confers an electric charge, and the corresponding *electrode potential*, on the *electrode*. See also *salt bridge*.

Electrolytic electrochemical cell (609): An *electrochemical cell* which is connected to an external source of electrical *energy*, and in which the *cell reaction* is driven in reverse.

Electromotive force, $\mathbb{E}$ V (613): The maximum *electrical potential difference* that can be delivered by a given *electrochemical cell*, as achieved when the *electrochemical cell* is operating *reversibly*.

Electron transport chain (726): A cluster of *proteins*, and related bio-molecules, embedded within particular *biological membranes*, such as those of *mitochondria* and *thylakoids*, which transfer electrons from one component to the next in a sequence of *exergonic redox reactions*. The *Gibbs free energy* released can be captured by *coupling* to another reaction, such as the *active transport* of H^+ ions across the *membrane*, so creating a *chemi-osmotic potential*.

Electroplating (609): The deposition of a thin layer of, for example, an expensive metal (such as silver) onto the surface of a cheaper metal (such as copper), as achieved by constructing an *electrochemical cell* with the cheaper metal as the *cathode*, and the more expensive metal as the *anode*.

Elevation of the boiling point (572): At any given *pressure*, the boiling point of a pure *liquid* is lower than the boiling point of a *solution* in which a *solute* is dissolved in that *liquid* as the *solvent*. This phenomenon is known as the elevation of the boiling point. See also *colligative properties* and *ebullioscopic constant*.

Ellipticity: See *mean residue ellipticity* and *circular dichroism*.

Elongation phase (782): The phase in the *aggregation* of *proteins* during which *monomers* elongate *polymers* longer than the *nucleus*. See also *lag phase* and *saturation phase*.

Endergonic (391): A *change in state* for which the *Gibbs free energy* change ΔG is positive, $\Delta G > 0$.

Endothermic (154): A *change in state* for which the *enthalpy* change ΔH is positive, $\Delta H > 0$.

Energy (40): Expressed most simply, the capacity to perform *work*. But as this book testifies, the concept is rather broader and much, much deeper ...

Energy, activation: See *activation energy*.

Energy, bond: See *bond dissociation energy* and *mean bond energy*.

Energy, dissociation: See *bond dissociation energy*.

Energy, internal: See *internal energy*.

Energy currency (712): The hydrolysis of *ATP* $\rightarrow$ ADP + P_i is exergonic, with $\Delta G^{\ominus\prime} = -57$ kJ; its synthesis ADP + P_i $\rightarrow$ *ATP* is endergonic, with $\Delta G^{\ominus\prime} = +57$ kJ. These values of $\Delta G^{\ominus\prime}$ are large enough for *ATP*, or ADP+P_i to be stable, yet small enough for *ATP* to be synthesized, or hydrolysed, under appropriate conditions. By *coupling* a suitably *exergonic* reaction A $\rightarrow$ B to the synthesis of *ATP* as ADP+P_i $\rightarrow$ *ATP*, and then *coupling ATP* hydrolysis *ATP* $\rightarrow$ ADP+P_i to a second, *endergonic* reaction C $\rightarrow$ D, then the overall effect is the harnessing of the *Gibbs free energy* released by the decomposition of A into the formation of D, with the regeneration of ADP + P_i. *ATP* therefore acts as a transient 'carrier' of energy between the two reactions A $\rightarrow$ B and C $\rightarrow$ D. Hence, the metaphor of describing *ATP* as an 'energy currency'. Other energy currencies include $NADH \rightleftharpoons NAD^+ + H^+$, $NADPH \rightleftharpoons NADP^+ + H^+$ and $GTP \rightleftharpoons GDP + P_i$.

Enthalpy H J (149): An *extensive state function* defined as

$$H = U + PV \tag{6.4}$$

Enthalpy, van't Hoff: see *van't Hoff enthalpy*.

Enthalpy of combustion $\Delta_c H$ J (177): The *enthalpy* change for the reaction

  compound → total oxidation products

Enthalpy of dilution $\Delta_{dil} H$ J (177): The *enthalpy* change for the reaction

  solution at concentration [1] + *solvent* → *solution* at concentration[2]

Enthalpy of formation $\Delta_f H$ J (178): The *enthalpy* change for the reaction

  constituent elements → compound

Enthalpy of fusion $\Delta_{fus} H$ J (161): The *enthalpy* change for the reaction

  solid → liquid

Enthalpy of reaction, standard: see *standard enthalpy of reaction*

Enthalpy of sublimation $\Delta_{sub} H$ J (161): The enthalpy change for the reaction

  solid → gas

Enthalpy of vaporization $\Delta_{vap} H$ J (161): The enthalpy change for the reaction

  liquid → gas

Entropy S J K^{-1} (269): An *extensive state* function identified by the *Second Law of Thermodynamics*, and defined *macroscopically* as

$$dS = \frac{dq_{rev}}{T} \tag{9.1}$$

and *microscopically* by Boltzmann's equation

$$S = k_B \ln W \tag{11.2}$$

A system's entropy is a measure of the degree of disorder: the more highly ordered a system, the lower the system's entropy. For an *isolated system*, $\Delta S < 0$ is a criterion of *spontaneity*. See also *Boltzmann's constant k*, *multiplicity W*, and *spontaneous*.

Entropy, Third Law: see *Third Law entropy*.
Enzyme (747): A *protein* that catalyses a biochemical reaction.
Equation, Arrhenius: see *Arrhenius equation*
Equation, Berthelot's: see *Berthelot gas*.
Equation, Boltzmann's: see *Boltzmann's equation*.
Equation, Clapeyron: see *Clapeyron equation*.
Equation, Clausius–Clapeyron: see *Clausius–Clapeyron equation*.
Equation, Dieterici's: see *Dieterici gas*.
Equation, Henderson–Hasselbalch: see *Henderson–Hasselbalch equation*.
Equation, Kirchhoff's first: see *Kirchhoff's first equation*.
Equation, Kirchhoff's second: see *Kirchhoff's second equation*.
Equations, Maxwell: see *Maxwell relations*.
Equation, Nernst: see *Nernst equation*.
Equation, Shomate: see *Shomate equation*.
Equation, thermodynamic of state: see *thermodynamic equation-of-state*.

Equation, van der Waals: see *van der Waals gas*.

Equation, van't Hoff: see *van't Hoff equation*.

Equation, virial: see *virial equation*.

Equation-of-state (16): An equation describing the complete behaviour of any system in terms of that system's state functions, an example being the *ideal gas law*.

Equation-of-state, thermodynamic: see *thermodynamic equation-of-state*.

Equilibrium: see *equilibrium state*.

Equilibrium, metastable: see *metastable equilibrium*.

Equilibrium, pseudo-: see *pseudo-equilibrium*.

Equilibrium, stable: *see stable equilibrium*.

Equilibrium, thermal: see *thermal equilibrium*.

Equilibrium, thermodynamic: see *thermodynamic equilibrium*.

Equilibrium, unstable: *see unstable equilibrium*.

Equilibrium constant: An expression defining the quantities of reagents present in a chemical mixture at *thermodynamic equilibrium*. See also *concentration thermodynamic equilibrium constant, molality thermodynamic equilibrium constant, mole fraction thermodynamic equilibrium constant, pressure thermodynamic equilibrium constant,* and *thermodynamic equilibrium constant*.

Equilibrium constant, concentration: see *concentration thermodynamic equilibrium constant*.

Equilibrium constant, molality: see *molality thermodynamic equilibrium constant*.

Equilibrium constant, mole fraction: see *mole fraction thermodynamic equilibrium constant*.

Equilibrium constant, pressure: see *pressure thermodynamic equilibrium constant*.

Equilibrium extent-of-reaction ξ_{eq} (419): The value of the *extent-of-reaction* parameter corresponding to the *equilibrium* composition of a chemical reaction.

Equilibrium saturated vapour pressure p Pa (469): The *vapour pressure* exerted by a vapour in *equilibrium* with the corresponding *liquid* or *solid*.

Equilibrium state (8): A *state* of a *system* in which all *state functions* are constant over time, and for which all *intensive* state functions have the same values at all locations within the *system*.

Equilibrium thermodynamics (10): The study of the *thermodynamics* of *equilibrium states*.

Evaporation (468): The process by which molecules from a *solid* or a *liquid* 'escape' into the *gas phase*. See also *equilibrium saturated vapour pressure*.

Exact differential (11): For any change from an initial state [1] to a final state [2], the change ΔX in any *state function* X depends only on the value X_1 of X in the initial state, and the value X_2 of X in the final state: $\Delta X = X_2 - X_1$. Mathematically, this is represented by expressing an infinitesimal change in X as dX, where dX is an exact differential, designated by the symbol d. Importantly, the total change in X for a finite change in state, $\int_1^2 dX$, can be determined directly by integration as $X_2 - X_1$. See also *inexact differential*.

Exciton transfer: see *resonant energy transfer*.

Exergonic (390): A *change in state* for which the *Gibbs free energy* change ΔG is negative, $\Delta G < 0$.

Exothermic (154): A *change in state* for which the *enthalpy* change ΔH is negative, $\Delta H < 0$.

Expansion coefficient, volumetric: see *volumetric expansion coefficient*.

Extended Debye-Hückel law (668):

$$\log_{10} \gamma_{b,\pm} = -\frac{A|z_+ z_-|\sqrt{I}}{1 + B\text{å}\sqrt{I}} \qquad (22.21)$$

Extensive state function (6): A *state function* whose value varies with the quantity of material within the *system* (such as the system's *mole number* n) or extent of the system (such as the system's volume V). See also *intensive state function* and *state function*.

Extent-of-reaction ξ (419): A parameter that measures how far a chemical reaction has progressed, ad which can take values from $\xi = 0$ (the reaction hasn't yet started) to $\xi = 1$ (the reaction has gone to completion as written).

Extent-of-reaction, equilibrium: see *equilibrium extent-of-reaction*.

Factor, compressibility: see *compressibility factor*.

Fahrenheit scale of temperature °F (44): A scale of *temperature* such that, at *standard atmospheric pressure*, 32 °F corresponds to the melting point of pure ice, and 212 °F to the boiling point of pure water.

Faraday constant, F C mol^{-1} (595): The magnitude of the electric charge associated with 1 mol of electrons = 9.6485×10^4 C mol^{-1}). The Faraday constant is a positive number, even though the charge on an electron is negative.

Ferromagnetic: A material which becomes magnetised when placed in a magnetic field, and maintains its magnetic properties after the external magnetic field has been removed. See also *paramagnetic*.

Finke–Watzky model (783): A two-step *kinetic* model of *protein* polymer *nucleation* and *elongation*.

First law, Joule's: see *Joule's first law*.

First Law of Thermodynamics (93): *Energy* can be neither created nor destroyed. For any *closed system*, we may define an *extensive state function* U, the *internal energy* of the *system*, such that the change dU in the *system's internal energy* for any *change in state* is given by

$$dU = đq - đw \qquad (5.1a)$$

where $đq$ represents the total *heat* exchanged between the *system* and its *surroundings*, and $đw$ represents the total *work* done by the *system* on the *surroundings*, or by the *surroundings* on the *system*, during that *change in state*.

First thermodynamic equation-of-state (642):

$$\left(\frac{\partial U}{\partial V}\right)_T = T\left(\frac{\partial P}{\partial T}\right)_V - P \qquad (21.12)$$

Fixation: see *CO$_2$ fixation*.

Fluorescence (739): The phenomenon in which the absorption of ultra-violet radiation by a molecule results in the emission of visible light. See also *intrinsic fluorescence*.

Folded state (748): The *state* of a *protein* in which the *polypeptide chain* assumes a stable, ordered, three-dimensional configuration, likely to be associated with biological functionality. See also *native state* and *unfolded state*.

Foldon (778): A structural unit within a *protein* that is independently formed by a two-state transition, of importance in understanding the pathways by which proteins fold.

Force F N (238): The product mass x acceleration, in according with Newton's second law of motion.

Force, proton motive: see *proton motive force*.

'Fourth Law' of Thermodynamics (408): Under any given conditions, out of all the *states* that are accessible to a *system*, the *system* will occupy the *state* of lowest *Gibbs free energy*.

Free energy: see *Gibbs free energy* and *Helmholtz free energy*.

Free expansion: see *Joule expansion*.

Frequency factor: see *pre-exponential factor*.

Friction (128): A *force*, present at the contact area between two surfaces, which either prevents the mutual movement of the surfaces (*static friction*), or resists the movement of either surface (*dynamic friction*). See also *dissipative effect*.

Frictional heat (129): The *heat* inevitably generated by the performance of *frictional work*.

Frictional work (28): *Work* performed by a *system* against *friction*.

Fugacity f Pa (657): At any *temperature* T, the *molar Gibbs free energy* **G** of an *ideal gas* at *pressure* P is related to the *standard molar Gibbs free energy* **G**⦵ of the gas at the *standard pressure* P⦵ as

$$\mathbf{G} = \mathbf{G}^{\ominus} + RT \ln(P/P^{\ominus}) \tag{13.47d}$$

For a *real gas*, in order to preserve the form of equation (13.47d), the *pressure* P is replaced by the fugacity f giving

$$\mathbf{G} = \mathbf{G}^{\ominus} + RT \ln\left(f/P^{\ominus}\right) \tag{22.1a}$$

such that

$$f = \phi P \tag{22.3a}$$

in which ϕ is an empirically determined parameter, the *fugacity coefficient*.

Fugacity coefficient ϕ (658): An empirically determined, dimensionless, quantity relating the *fugacity* f of a *real gas* to the *gas's* actual *pressure* P as

$$f = \phi P \tag{22.3a}$$

Function (77): If the value of a *state function* X is observed to change when the value of another *state function* Y changes, and under no other circumstances, then, mathematically, X is a function X(Y) of Y. If the value of X changes when the values of either of two other state functions Y and Z change independently, then X is a function X(Y, Z) of the two variables Y and Z.

Function, conjugate: see *conjugate variables*.
Function, extensive: see *extensive function*.
Function, intensive: see *intensive function*.
Function, path: see *path function*.
Function, potential: see *potential function*.
Function, state: see *state function*.
Fundamental equation of thermodynamics (406):

$$dG = V\,dP - S\,dT + \sum_i \mu_i\,dn_i \tag{13.60}$$

Galvanic electrochemical cell (599): An *electrochemical cell* that *spontaneously* creates an *electrical potential difference* between the two *electrodes*, so enabling an *electrical current* to be driven through an external *conductor*.

Gas (5): A state of matter, or *phase*, identified by the material's ability to occupy all the available volume. See also *liquid* and *solid*.

Gas, Berthelot: see *Berthelot gas*.

Gas, Dieterici: see *Dieterici gas*.

Gas, ideal: see *ideal gas*.

Gas, perfect: see *ideal gas*.

Gas, van der Waals: see *van der Waals gas*.

Gas, virial: see *virial gas*.

Gas electrode (606): An *electrode* for which the corresponding *half-cell reaction* is that of a *gaseous* element, such as hydrogen $H_2(g)$, exchanging electrons with the corresponding aqueous *ions*, such as H^+ (aq).

Gas electrode, reversible: see *reversible gas electrode*.

Gibbs-Duhem equation (405):

$$\sum_i n_i \, d\mu_i = 0 \tag{13.59}$$

Gibbs free energy G J (374): An *extensive state function* defined by reference to the system's *enthalpy H*, *temperature T*, and *entropy S* as

$$G = H - TS \tag{13.10a}$$

For a *closed system* undergoing a *change in state* at constant *pressure* and *temperature*, $\Delta G < 0$ is a criterion of spontaneity. Gibbs free energy is also often referred to as just 'free energy'. See also *Helmholtz free energy*, and *spontaneous*.

Gibbs free energy, reaction: see *reaction Gibbs free energy*.

Gibbs-Helmholtz equation (382):

$$\left(\frac{\partial (G/T)}{\partial T} \right)_P = -\frac{H}{T^2} \tag{13.28}$$

Glycolysis (716): A *metabolic pathway* in which glucose, $C_6H_{12}O_6$, is transformed into two molecules of pyruvate, *coupled* to the synthesis of 2 *ATP* and 2 NADH.

Gradient, chemi-osmotic potential: see *chemi-osmotic potential gradient*.

Gradient, temperature: see *temperature gradient*.

Grana (739): A 'stack' of *thylakoids* within the *choloroplast*.

Graphite electrode (606): An *inert electrode* made of carbon, a good conductor of electricity, and a catalyst of the *electrode half-cell reaction*.

Half-cell (596): The combination of a single *electrode*, and an appropriate *electrolyte*. The corresponding *half-reaction* takes place on the surface of the *electrode*, at the *boundary* with the *electrolyte*. When two half-cells are connected together, they form an *electrochemical cell*.

Half-cell reaction: see *half-reaction*.

Half-reaction (596): Alternatively, half-cell reaction. A *redox reaction*, taking place on the surface of an *electrode* in contact with a suitable *electrolyte*, in which electrons are either generated (an *oxidation reaction* such as $Zn(s) \rightarrow Zn^{2+}(aq) + 2\,e^-$) or consumed (a *reduction reaction*, such as $Cu^{2+}(aq) + 2e^- \rightarrow Cu(s)$), so conferring an electric charge, and hence an *electrical potential*, on the electrode.

Hasselbalch: see Henderson–Hasselbalch equation and Henderson–Hasselbalch approximation.

Heat đq J (43): *Energy* that passes across the *boundary* of a *system* when the *system* and its *surroundings* are at different *temperatures*. Heat is a *path function*.

Heat, frictional: see *frictional heat*.

Heat, irreversible: see *irreversible heat*.

Heat, latent: see *latent heat*.

Heat, reversible: see *reversible heat*.

Heat capacity C_X J K^{-1} mol^{-1} (636): An *extensive state function* defined by a *partial derivative* of the form $(\partial U/\partial T)_X$ or $(\partial H/\partial T)_X$, representing the rate of change of a *system's internal energy U*, or *enthalpy H*, with respect to *temperature T*, while the *state function X* is held constant.

Heat capacity at constant pressure C_P J K^{-1} mol^{-1} (151): An *extensive state function* defined by the *partial derivative* $(\partial H/\partial T)_P$, representing the rate of change of a *system's enthalpy H* with respect to *temperature T*, while holding the system *pressure P* constant.

Heat capacity at constant volume C_V J K^{-1} mol^{-1} (107): An *extensive state function* defined by the *partial derivative* $(\partial U/\partial T)_V$, representing the rate of change of a *system's internal energy U* with respect to *temperature T*, while holding the system volume V constant.

Heat engine (313): A device that operates in a *thermodynamic cycle*, and so derives mechanical *work* from the flow of *heat* down a *temperature gradient*.

Heat of ...: see *enthalpy of* ...

Heat pump (314): A device that operates in a *thermodynamic cycle*, using mechanical *work* to cause *heat* to flow up a *temperature gradient*.

Heat reservoir (64): A body within the *surroundings* of a *system* of, in principle, infinite extent, which can, as required, provide *heat* to the system (so acting as a *heat* source), or accept *heat* from the system (so acting as a *heat* sink), while maintaining a constant *temperature*.

Helmholtz free energy A J (385): An *extensive state function* defined by reference to the system's *internal energy U*, *temperature T* and *entropy S* as

$$A = U - TS \qquad (13.34)$$

For a *closed system* undergoing a *change in state* at constant volume and *temperature*, $\Delta A < 0$ is a criterion of *spontaneity*. The Helmholtz free energy is different from the *Gibbs free energy* $G = H - TS$, and so the abbreviation 'free energy' is potentially ambiguous. Conventionally, the term 'free energy' is used only as an abbreviation for the *Gibbs free energy*, and should never be used as abbreviation for the Helmholtz free energy.

Henderson–Hasselbalch approximation (549): A form of the *Henderson–Hasselbalch equation* as

$$pH \approx pK_a + \log_{10} \frac{[A^-]_{eq}}{[HA]_i} \qquad (17.26)$$

applicable to the dissociation of a weak *monoprotic acid* HA, for which the initial molar concentration $[HA]_i$ is a reasonable approximation to the molar concentration $[HA]_{eq}$ of the undissociated acid in the equilibrium mixture.

Henderson–Hasselbalch equation (541): For the dissociation of a *monoprotic acid* HA

$$pH = pK_a + \log_{10} \frac{[A^-]_{eq}}{[HA]_{eq}} \quad (17.16)$$

See also *Henderson–Hasselbalch approximation*.

Henry's law (510): A variant of *Raoult's law*

$$p_i = x_i p_i^* \quad (16.1)$$

in which the proportionality between p_i and x_i is not the *vapour pressure* p_i^* of the pure component i, but a constant, the *Henry's law constant* $k_{H,i}$

$$p_i = x_i k_{H,i} \quad (16.2)$$

See also *ideal-dilute solution*.

Henry's law constant (510): The constant $k_{H,i}$ for component i in *Henry's law*

$$p_i = x_i k_{H,i} \quad (16.2)$$

Hess's law of constant heat formation (166): The *heat* released or absorbed in any chemical reaction depends solely on the initial reactants and final products, and is independent of the *path* taken during the reaction. This law is a consequence of the fact that the heat released or absorbed by a chemical reaction taking place at *constant pressure* is equal to the change in the system's *enthalpy*, and since *enthalpy* is a *state function*, the overall change is independent of the *path*.

Hydrogen bond (357): A weak electrostatic bond between a hydrogen atom, which carries a positive charge by virtue of being strongly bonded to an electronegative element such as oxygen or nitrogen, and a nearby electronegative atom, which carries a negative charge.

Hydrogen electrode, reversible: see *reversible hydorgen electrode*.

Hydronium ion (535): Alternatively, hydroxonium ion or oxonium ion. The positively charged *ion* H_3O^+ (aq) formed from the protonation of a water molecule, often abbreviated as H^+ (aq).

Hydroxonium ion: see *hydronium ion*.

Ideal (8): A *system* in which there are no inter-atomic or inter-molecular forces or interactions under any conditions. As a result, the mathematics describing the behaviour of the *system* is as simple as possible. The study of ideal systems provides very useful models for, and insights into, the behaviour of real systems. See also *ideal gas, ideal solution, real gas* and *real system*.

Ideal-dilute solution (510): A solution of a solute D in a solvent S, which is necessarily dilute (the *mole fraction* $x_D \ll$ the *mole fraction* x_S), and also complies with Henry's law

$$p_D = x_D k_{H,D} \quad (16.2)$$

in which p_D is the *equilibrium vapour pressure* of the solute D above the solution, and $k_{H,D}$ is the *Henry's law constant* for the solute D. See also *ideal solution* and *Raoult's law*.

Ideal gas (16): Alternatively, perfect gas. A *gas* that obeys the *equation-of-state*

$$PV = nRT \quad (3.3)$$

Molecules of an ideal gas occupy zero volume, are spherical, undergo elastic collisions, and have no mutual interactions, however far apart or close together. Collectively, these conditions imply that the entire volume of the system is available to the molecules, and no space is excluded; all the *energy* of the molecules is represented by the kinetic energy of translational motion, and no other modes of motion, such as rotational and vibrational motion, are present; and all molecules can move totally freely, and are not inhibited by any inter-molecular attractions or repulsions.

Ideal gas scale of temperature K (44): A scale of *temperature*, for which the *reference level*, 0 K, is the *absolute zero of temperature*, and the size of 1 K equals 1 °C.

Ideal solution (505): A solution such that all components *i* comply with Raoult's law

$$p_i = x_i p_i^* \tag{16.1}$$

in which p_i is the *equilibrium vapour pressure* of *i* above the solution, p_i^* is the *equilibrium vapour pressure* associated with the pure component *i* and x_i is the *mole fraction* of *i* within the solution, at all values of the mole fraction x_i from 0 to 1. The fundamental molecular assumption concerning an ideal solution is that the interactions between all pairs of adjacent molecules are equal. See also *non-ideal solution, ideal-dilute solution, Henry's law, negative deviation from Raoult's law* and *positive deviation from Raoult's law*.

Inequality, Clausius: see *Clausius inequality*.

Inert electrode (606): An *electrode* which, as a consequence of the appropriate *half-reaction*, does not dissolve into *solution*, nor are atoms deposited. During the half-cell reaction, the surface of the *electrode* therefore acts as a catalyst, mediating the half-cell reaction, rather than actively participating in the reaction. Inert electrodes are made from materials such as *graphite* and *platinised platinum*.

Inexact differential (26): For any change from an initial state [1] to a final state [2], the value of any *path function*, such as *heat q* or *work w*, depends on the *path* taken. Mathematically, this is represented by expressing an infinitesimal exchange of heat đq, or an infinitesimal quantity of work đw, as an inexact differential designated by the symbol đ. Importantly, the total heat exchanged for a finite change in state, $\int_1^2 đq$ cannot be expressed as $q_2 - q_1$; nor can the total work $\int_1^2 đw$ be expressed as $w_2 - w_1$; rather, these integrals need to be evaluated on a case-by-case basis, given knowledge of the nature of the *path*, for example, an *isothermal* path or an *adiabatic* path. See also *exact differential*.

Infinite dilution (173): A hypothetical *state* of a *solution* in which the *molality* or *molar concentration* of a *solute* is zero, and so the *solution* is infinitely dilute. Solutions at infinite dilution show *ideal* behaviour; also, as shown in Table 6.4, the *standard state* for a *solute* in a *real solution* is defined with respect to a *solution* at infinite dilution.

Inflexion, point of: see *point of inflexion*.

Inorganic phosphate (713): The entity represented by P_i in the hydrolysis of *ATP* as ATP + $H_2O \rightleftharpoons$ ADP + P_i. Whereas the molecules *ATP*, ADP and H_2O are well defined, in real biological systems, P_i is in practice a complex mixture of components including $H_2PO_4^-$ and HPO_4^{2-}.

Intensive state function (6): A *state function* whose value is independent of the quantity of material within the *system*, or the extent of the system, two examples being the *system's temperature T* and *pressure P*. The ratio of any two *extensive state function* – such as the molar volume $V = V/n$ – is always an intensive state function. See also *extensive state function*.

Internal energy U J (94): The *extensive state function* central to the *First Law of Thermodynamics*. Physically, the internal energy U, which is necessarily a *macroscopic* property of a thermodynamic *system*, represents the aggregate translational, rotational, vibrational, ... energies of the *microscopic* components from which the system is formed. Although it is impossible to determine the absolute value of any system's internal energy U, this is not a problem: what is important, and what can be measured using the *First Law of Thermodynamics*, is the change ΔU in a system's internal energy resulting from a *change in state*.

Intrinsic fluorescence (753): When certain molecules are excited by ultra-violet light, visible light is emitted. This is known as *fluorescence*. Within a *protein*, the *amino acids* tryptophan, and to a lesser extent phenylalanine and tyrosine, are naturally fluorescent, so conferring 'intrinsic' fluorescence on the protein. The characteristics of the fluorescence depend on, for example, the polarity of the local environment, and on how close a given tryptophan is to certain other *amino acids*, and so changes in intrinsic fluorescence can be interpreted in terms of the folding and unfolding of the *protein*.

Inversion curve (652): The curve on a T, P *diagram* that connects all the points at which the *Joule-Thomson coefficient* $\mu_{JT} = (\partial T/\partial P)_H = 0$, these being the maxima of a set of *isenthalps*. The inversion curve defines the zones on the T, P diagram corresponding to combinations of *temperature* and *pressure* in which a *gas* undergoing a *throttling process* either becomes cooler (in general, at lower *pressures*), or warmer (in general, at higher *pressures*).

Inversion temperature (652): The *temperature* T corresponding to the point on an *isenthalp* at which the *Joule-Thomson coefficient* $\mu_{JT} = (\partial T/\partial P)_H = 0$.

Ion (173): A chemical species, often present in aqueous *solution*, which carries an electric charge as a result of having either a surplus of electrons (a negative ion, or *anion*, such as SO_4^{2-} (aq)), or a deficit (a positive ion, or *cation*, such as Cu^{++} (aq)).

Ion-product of water: see *ionic product of water*.

Ionic enthalpy (184): The *enthalpy* of an *ion* as calculated assuming that the *enthalpy* of H^+ (aq) in its *standard state* is zero at all *temperatures*.

Ionic product of water K_w (536): Alternatively, ion product of water, or water ionisation constant.

$$K_w = [H^+]_{eq}[OH^-]_{eq} = 10^{-14} \tag{17.8}$$

Ionic strength (667):

$$I = \frac{1}{2}\sum_i z_i^2 \left(\frac{b_i}{b_i^\ominus}\right) \tag{22.20}$$

Irreversible (123): A *change in state* that takes place along a *path* that cannot be reversed such that both the *system* and the *surroundings* can be returned to their exact original states, as caused by the presence of *dissipative effects* such as *dynamic friction*. All real changes are irreversible. See also *quasistatic* and *reversible*.

Irreversible heat $đq_{irrev}$ J (143): A flow of *heat* associated with an *irreversible path*. For any *change in state*, the irreversible heat $đq_{irrev}$ is always less that the *reversible heat* $đq_{rev}$, $đq_{irrev} < đq_{rev}$, for some *internal energy* is inevitably lost in overcoming *dissipative effects* such as *friction*. See also *reversible heat* and *reversible path*.

Irreversible path: see *irreversible*.

Irreversible work đw_{irrev} J (142): A quantity of *work* associated with an *irreversible path*. For any *change in state*, the irreversible work đw_{irrev} is always less that the *reversible work* đw_{rev}, đw_{irrev} < đw_{rev}, for some *internal energy* is inevitably lost in overcoming *dissipative effects* such as *friction*. See also *reversible heat* and *reversible path*.

Isenthalp (651): A curve on a *T, P diagram* connecting points of constant *enthalpy H*.

Isobar (58): A curve on a *P, V diagram* connecting points of constant *pressure P*. See also *isobaric*.

Isobaric (58): A *change in state* taking place at constant *pressure P*. See also *isobar*.

Isochore (58): A curve on a *P, V diagram* connecting points of constant volume *V*. See also *isochoric*.

Isochore, van't Hoff: see *van't Hoff isochore*.

Isochoric (58): A *change in state* taking place at constant volume V. See also *isochore*.

Isolated system (57): A thermodynamic *system*, the *boundary* of which blocks the flow of *heat*, *work*, and matter.

Isotherm (57): A curve on a *T, P diagram, T, V diagram* or *T, S diagram* connecting points of constant *temperature T*. See also *isothermal*.

Isotherm, binding: see *binding isotherm*.

Isothermal (57): A *change in state* taking place at constant *temperature T*. See also *isotherm*.

Isothermal compressibility κ_T Pa^{-1} (89):

$$\kappa_T = -\frac{1}{V}\left(\frac{\partial V}{\partial P}\right)_T \tag{4.18}$$

Isothermal titration calorimetry (ITC) (764): An experimental *calorimetry* technique of particular importance in the measurement of the thermodynamic properties of *protein-ligand binding*. Like *differential scanning calorimetry*, isothermal titration calorimetry compares a sample with a reference, measuring the *heat* flows as a *ligand* is titrated into the sample, under conditions in which the *temperatures* of the sample and the reference are maintained constant.

Joule expansion (121): Alternatively, free expansion. The expansion of a *gas* into a vacuum.

Joule heating (592): The generation of *heat* resulting from the flow of electricity through an electrical *conductor* according to Joule's first law

$$\text{Heat} = I^2 R\tau \text{ J} \tag{20.1}$$

Joule heating is *dissipative* and *irreversible*, and is the electrical analogue of *dynamic friction*.

Joule's first law (592): An equation for the *heat* generated by the flow of electricity as

$$\text{Heat} = I^2 R\tau \text{ J} \tag{20.1}$$

Joule–Thomson coefficient μ_{JT} K Pa^{-1} (649): For a *gas* undergoing an *isenthalpic throttling process*, $\mu_{JT} = (\partial T/\partial P)_H$. Physically, the Joule-Thomson coefficient represents the change ∂T_H in *temperature* associated with the change ∂P_H in *pressure* resulting from the *throttling process*. See also *inversion curve* and *inversion temperature*.

Kelvin–Planck statement of the Second Law of Thermodynamics (311): It is impossible to construct a device that operates in a *cycle* and has no effect other than the performance of *work* and the exchange of *heat* with a single *reservoir*.

Kinetic friction: see *dynamic friction*.

Kinetics, chemical: see *chemical kinetics*.

Kirchhoff's first equation (197):

$$\left(\frac{\partial(\Delta_r H)}{\partial T}\right)_P = \Delta C_P \tag{6.33}$$

Kirchhoff's second equation (198):

$$\left(\frac{\partial(\Delta U)}{\partial T}\right)_V = \Delta C_V. \tag{6.36}$$

Krebs cycle: see *tricarboxylic acid cycle*.

Lag phase (782): The first stage in *protein aggregation*, during which *monomers* coalesce to form a *nucleus*. See also *nucleation*, *elongation phase* and *saturation phase*.

Latent heat (161): A term no longer used for the *heat* associated with *phase changes*, and now replaced by *molar enthalpy of fusion, vaporisation* or *sublimation*.

Law, [X]'s: see *[X]'s law*.

Le Chatelier's principle (448): If an external constraint is applied to a *system*, the system adjusts itself so as to oppose that constraint.

Lennard–Jones potential (763): A *potential function* describing inter-molecular *van der Waals forces* as

$$V_{vdW} = 4\varepsilon\left[\left(\frac{\sigma}{r}\right)^{12} - \left(\frac{\sigma}{r}\right)^{6}\right]$$

in which ε and σ are constants for any given system, and r is the distance between two adjacent molecules. The first term, which is positive and proportional to r^{-12}, represents the very short-distance repulsion between atomic nuclei; the second term, which is negative and proportional to r^{-6}, represents the attractive *force* that enables, for example, *liquids* to coalesce from a compressed *gas*.

Level, reference: see *reference level*.

Levinthal's paradox (770): The 'puzzle' articulated by Cyrus Levinthal, who calculated that a *protein* consisting of, say, 100 *amino acids* could, in principle, adopt any one of an astronomic number of alternative configurations. How does each *protein* 'discover' the 'right' configuration, and do so *spontaneously*, naturally, and also very quickly?

Ligand (763): A small molecule that binds to a larger molecule. See also *protein-ligand complex*.

Light-dependent reactions of photosynthesis (739): A *metabolic pathway*, taking place within the *membrane* of *chloroplast thylakoids*, in which light *energy* is captured by *photosystems I and II*, and transformed into the *energy currencies ATP and NADPH*. The *Gibbs free energy* stored in the *ATP* and *NADPH* is subsequently used in the *light-independent reactions of photosynthesis* to transform atmospheric CO_2 into the important biochemical intermediary glyceraldehyde-3-phosphate.

Light-harvesting complex (739): A cluster of *photosynthetic pigments*, associated with either *photosystem I* or *photosystem II* within the *thylakoid membrane*, which captures light, over a wide range of wavelengths, and transfers the *energy* by *resonance energy transfer* to the corresponding *reaction centre* — *P700* within *photosystem I*, or *P680* within *photosystem II*.

Light-independent reactions of photosynthesis (744): A *metabolic pathway*, taking place within the *chloroplast stroma*, in which the *Gibbs free energy* stored in *ATP* and *NADPH* synthesized as a result of the *light reactions of photosynthesis* is used to transform atmospheric CO_2 into the important biochemical intermediary glyceraldehyde-3-phosphate. See also *light-dependent reactions of photosynthesis*.

Lipid bilayer (724): The primary structural component of *biological membranes*, comprising two layers of phospholipids, arranged such that the polar 'heads' are oriented towards the aqueous environments on either side, with the non-polar 'tails' towards the interior. Lipid bilayers are therefore impervious to water, and water-soluble molecules and *ions*. Continuous lipid bilayers surround biological cells, cell organelles (such as *mitochondria* and *chloroplasts*) and smaller cellular structures (such as *thylakoids* and *carboxysomes*), so defining an 'inside' and an 'outside'. *Proteins* embedded with the lipid bilayer confer biological functions on the bilayer, such as the ability to transport molecules and *ions* against concentration and *electrical potential* gradients—as is central to *mitochondrial respiration*, and the *light dependent reactions of photosynthesis*.

Liquid (16): A state of matter, or *phase*, such that a given mass of material occupies essentially the same volume at a variety of *pressures*, but the material has very low resistance to shear stresses, and so, unlike a *solid*, can flow easily. See also *gas*.

Liquid junction potential (604): An electrical *potential difference* that arises at the junction between two different *electrolytes*. Liquid junction potentials reduce the *electromotive force* theoretically available from an *electrochemical cell*, and can be reduced by using, for example, a *salt bridge*.

Lumen (739): The interior of the *chloroplast thylakoid*, into which H^+ *ions* accumulate, as a result of the *light dependent reactions of photosynthesis*, which take place within and across the *thylakoid* membrane.

Macroscopic (4): Relating to readily observable properties, and which do not make any reference to, or relay on any assumptions relating to, any theories of atomic or molecular structure. See also *microscopic*.

Macrostate (334): A macroscopically measureable *state* of a *system*, as defined by the values of is *state functions*. See also *microstate* and *multiplicity*.

Mass action effect (567): The effect on a reaction's *equilibrium constant* attributable to, for example, the presence of one reagent in vast excess (as happens when water is both a reagent and also the *solvent*), or constraining one reagent's concentration to a particular value (as happens when a reaction involving hydrogen *ions* as a reagent takes place in a *buffer solution*).

Mass action ratio, concentration: see *concentration mass action ratio*.

Mass action ratio, molality: see *molality mass action ratio*.

Mass action ratio, mole fraction: see *mole fraction mass action ratio*.

Mass action ratio, pressure: see *pressure mass action ratio*.

Mathematical probability (339): In general, the ratio $\mathbb{P}_i$ of the number N_i of times a particular event i takes place to the total number of all possible events

$$\mathbb{P}_i = \frac{N_i}{\sum_i N_i}$$

In the context of the probabilistic interpretation of the *Second Law of Thermodynamics*, $\mathbb{P}_i$ is the ratio of the number W_i of *microstates* that can form a specific *macrostate* to the total number of all possible microstates

$$P_i = \frac{W_i}{\sum_i W_i} \tag{11.1}$$

Maximum available work (389): The maximum *useful work* that a *system* can perform. This *work* is necessarily *reversible*, and, for a *system* undergoing a *change in state* at constant *temperature* and *pressure* is equal to $-\Delta G$; for a change at constant *temperature* and constant volume, $-\Delta A$.

Maxwell construction (492): In relation to the liquefaction of a *van der Waals gas*, the rule that the areas in the two 'loops' of a graph of the *van der Waals equation* must be equal, as illustrated in Figure 15.10.

Maxwell relations (639): A set of four mathematical relationships, such as

$$\left(\frac{\partial T}{\partial V}\right)_S = -\left(\frac{\partial P}{\partial S}\right)_V \tag{21.6a}$$

in which the pairs of *conjugate variables* P and V, and S and T, are diagonally opposite each other.

Maxwell's demon (345): A "being with faculties so sharpened" that it can distinguish between molecules moving with different speeds. Might such a demon be able to break the *Second Law of Thermodynamics*?

Mean bond energy ϵ or $\Delta H^{\ominus}(X-Y)$ J mol^{-1} (187): The average value of the *gas phase bond dissociation energies*, usually at a *temperature* of 298 K, for all X–Y bonds of the same type within the same chemical species.

Mean molal activity coefficient $\gamma_{b,\pm}$ (666): Since *ions* can never be present by themselves, it is impossible to measure the *molal activity coefficients* γ_{b,X^+} of the *ion* X$^+$, and γ_{b,Y^-} of the *ion* Y$^-$. It is possible, however, to measure the product $\gamma_{b,X^+}\gamma_{b,Y^-}$, and so

$$\gamma_{b,\pm}^2 = \gamma_{b,X^+}\gamma_{b,Y^-} \tag{22.15}$$

Mean residue ellipticity (753): A measurement from data collected using *circular dichroism*, and which can be used to determine the degree of *protein* unfolding in an *equilibrium* mixture of the *native* and *unfolded states*: when the *protein* is totally *unfolded*, the mean residue ellipticity is zero; when the protein is in its *native state*, the mean residue ellipticity is a non-zero negative number.

Mechanics, statistical: see *statistical mechanics*.

Melting temperature T_m K (758): The temperature at which an *equilibrium* mixture of a *protein* in its *native* and *unfolded states* comprises equal *mole fractions* $x_N = x_U = 0.5$, implying that $\Delta_{unf}G^{\ominus}_{T_m} = 0$ and $K_{unf} = 1$.

Membrane: see *biological membrane* and *semi-permeable membrane*.

Membrane, biological: see *biological membrane*.

Membrane, respiratory: see *respiratory membrane*.

Membrane, semi-permeable: see *semi-permeable membrane*.

Membrane potential $\Delta\Psi$ V (731): The electrical potential difference across a *biological membrane*

$$\Delta\Psi = \Psi(\text{in}) - \Psi(\text{out}) \tag{24.8}$$

attributable to differences in charge density on either side of the membrane. See also *chemi-osmotic potential* and *proton motive force*.

Metabolic pathway (674): A sequence of biochemical reactions that collectively achieve an important overall result, such as the *glycolysis* pathway that transforms glucose into pyruvate, *ATP* and NADH, and the *light reactions of photosynthesis* that capture light energy as ATP and NADPH.

Metastable equilibrium (410): The description of the *thermodynamic equilibrium* associated with a *metastable state*. See also *stable state* and *unstable state*.

Metastable state (410): An *equilibrium state*, but one which is a local Gibbs free energy minimum, rather than the *state* with the lowest possible Gibbs free energy. In the following diagram, state A is a *stable state*, state B is a metastable state, and state C is an *unstable state*.

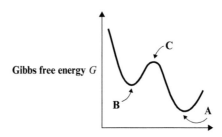

If the 'energy hill', represented by state C, is relatively 'low', then state B can change spontaneously to state A if the system is suitably disturbed (as happens, for example, to a *supercooled gas*); if, however, the 'energy hill' is relatively 'large', state B may be stable for long periods of time, and is known as a 'trapped' metastable state (for example, diamond, relative to graphite).

Microcompartment, bacterial: see *bacterial microcompartment*.

Microscopic (4): Relating to the behaviour of individual atoms or molecules. See also *macroscopic*.

Microscopic dynamic equilibrium: see *dynamic equilibrium*.

Microstate (334): The definition of a *system* as expressed in terms of the instantaneous positions, and momenta, of each individual particle within the *system*. In general, any given *macrostate* will correspond to very many different microstates, as specified by the *macrostate's multiplicity*.

Mitochondrion (724): A cell organelle found within eukaryotes, the *membrane* of which is the location of *respiration* - the *metabolic pathway* that captures the *Gibbs free energy* stored in NADH (produced as a result of *glycolysis* and the *TCA cycle*) to create a *proton motive force* that drives the synthesis of *ATP* by *ATP synthase*.

Mixing, molecular: see *molecular mixing*.

Model, Finke–Watzky: see *Finke–Watzky model*.

Molal activity $a_{b,i}$ (663): For a component i in an *ideal solution*, the *molar Gibbs free energy* $\mathbf{G}_i$ is related to the *molality* b_i as

$$\mathbf{G}_i \text{ (in solution)} = \mathbf{G}_i^{\ominus} \text{(in solution, } b) + RT \ln (b_i/b_i^{\ominus}) \qquad (16.9a)$$

where $G_i^\ominus$ is the *standard molar Gibbs Free energy*, corresponding to a the *solution* of unit *molality*. To preserve the simplicity of equation (16.9a) for *real solutions*, the molal activity $a_{b,i}$ replaces the *molality* b_i, so that *real solutions* can be described using the equation

$$\mu_i \text{ (in solution)} = \mu_i^\ominus \text{(in solution)} + RT \ln a_{b,i} \tag{22.8b}$$

such that

$$a_{b,i} = \gamma_{b,i} (b_i / b_i^\ominus) \tag{22.9}$$

where μ_i is the *chemical potential*, and $\gamma_{b,i}$ is the *molal activity coefficient*.

Molal activity coefficient $\gamma_{b,i}$ (663): For a component i in a *real solution*, an empirically-determined parameter relating the *molal activity* $a_{b,i}$ to the *molality* b_i according to

$$a_{b,i} = \gamma_{b,i} (b_i / b_i^\ominus) \tag{22.9}$$

Molal activity coefficient, mean: see *mean molal activity coefficient*.

Molality b_i mol kg^{-1} (171): A measure of the quantity of n_i mol of a *solute* i dissolved in a total mass M_S kg of a *solvent* S as $b_i = n_i/M_S$.

Molality mass action ratio Γ_b (521): Alternatively, molality reaction quotient Q_b. The non-equilibrium ratio

$$\Gamma_b = \frac{(b_C/b_C^\ominus)^c\ (b_D/b_D^\ominus)^d}{(b_A/b_A^\ominus)^a\ (b_B/b_B^\ominus)^b} \tag{16.15b}$$

Molality reaction quotient Q_b: see *molality mass action ratio*.

Molality thermodynamic equilibrium constant K_b (521): A *thermodynamic equilibrium constant* expressed in terms of the *molalities* b_i mol kg^{-1} of each reagent. Accordingly, for the generalized reaction $aA + bB \rightleftharpoons cC + dD$

$$K_b = \frac{(b_{C,eq}/b_C^\ominus)^c\ (b_{D,eq}/b_D^\ominus)^d}{(b_{A,eq}/b_A^\ominus)^a\ (b_{B,eq}/b_B^\ominus)^b} \tag{16.12b}$$

Molar mol^{-1} (7): A measurement associated with a single *mole* of material, such as the molar volume V m^3 mol^{-1}, the volume occupied by 1 mole of a material at a given *temperature* and *pressure*.

Molar concentration [I] mol m^{-3} or mol dm^{-3} (171): A measure of the quantity of n_i mol of a *solute* i dissolved in a total volume V m^3 (or dm^3) of *solution* as $[I] = n_i/V$. See also *molarity*.

Molarity (171): An alternative for the *molar concentration* [I], when measured in units of mol dm^{-3}.

Mole (7): 6.022141×10^{23} particles, such as atoms, molecules, or ions. See also *Avogadro constant*.

Mole fraction x_i (170): For a mixture containing n_i mol of component i, the mole fraction x_i is defined as

$$x_i = \frac{n_i}{\sum_i n_i} \tag{6.18}$$

Mole fraction activity $a_{x,i}$ (661): For a component i in an *ideal solution*, the *molar Gibbs free energy* G_i is related to the *mole fraction* x_i as

$$G_i \text{ (in solution)} = G_i^\ominus \text{(pure)} + RT \ln (x_i/x_i^\ominus) \tag{16.7a}$$

where G_i° is the *standard molar Gibbs Free energy*, corresponding to a the pure material *i*. To preserve the simplicity of equation (16.7a) for *real solutions*, the mole-fraction activity $a_{x,i}$ replaces the *mole fraction* x_i, so that *real solutions* can be described using the equation

$$\mu_i \text{ (in solution)} = \mu_i^\circ\text{(pure)} + RT \ln a_{x,i} \tag{22.6b}$$

such that

$$a_{x,i} = \gamma_{x,i}(x_i/x_i^\circ) \tag{22.5}$$

where μ_i is the *chemical potential*, and $\gamma_{x,i}$ is the *mole fraction activity coefficient*.

Mole fraction activity coefficient $\gamma_{x,i}$ (661): For a component *i* in a *real solution*, an empirically-determined parameter relating the *mole fraction activity* $a_{x,i}$ to the *mole fraction* x_i according to

$$a_{x,i} = \gamma_{x,i}(x_i/x_i^\circ) \tag{22.5}$$

Mole fraction mass action ratio Γ_x (437): Alternatively, mole fraction reaction quotient Q_x. The *non-equilibrium* ratio

$$\Gamma_x = \frac{(x_C/x_C^\circ)^c \, (x_D/x_D^\circ)^d}{(x_A/x_A^\circ)^a \, (x_B/x_B^\circ)^b} \tag{14.19}$$

Mole fraction reaction quotient Q_x: see *mole fraction mass action ratio*.

Mole fraction thermodynamic equilibrium constant K_x (437): A *thermodynamic equilibrium constant* expressed in terms of the *mole fractions* x_i of each reagent. Accordingly, for the generalized reaction $aA + bB \rightleftharpoons cC + dD$

$$K_x = \frac{(x_{C,eq}/x_C^\circ)^c \, (x_{D,eq}/x_D^\circ)^d}{(x_{A,eq}/x_A^\circ)^a \, (x_{B,eq}/x_B^\circ)^b} \tag{14.17a}$$

Mole number n_i (7): The number of *moles* of component *i* within a given *system*.

Molecular irreversibility (128): A *macroscopic* process which is *irreversible* as a result of a *microscopic* molecular phenomena such as *molecular mixing*.

Molecular mixing (93): A *dissipative effect* attributable to the *irreversible microscopic* mixing of atoms, molecules or *ions*, a process which *spontaneously* increases the *entropy* of the corresponding *system*.

Monohydroxic base (545): a base BOH that dissociates in aqueous *solution* to yield a single hydroxyl *ion* as

$$BOH\,(aq) \rightleftharpoons B^+(aq) + OH^-(aq)$$

Monomer (780): A (usually small) molecule that can react with a similarly small molecule, resulting in often long chains, called *polymers* – for example, *amino acids* are the monomers in the *polypeptide* chains of *proteins*, and nucleotides are the monomers in nucleic acids.

Monoprotic acid (538): an acid HA that dissociates in aqueous *solution* to yield a single hydrogen *ion* as

$$AH\,(aq) \rightleftharpoons H^+(aq) + A^-(aq)$$

(or, more correctly, a single *hydronium ion* $AH\,(aq) \rightleftharpoons H_3O^+(aq) + A^-(aq)$).

Motive force, proton: see *proton motive force*.

Multiplicity (337): The number of different *microstates* that correspond to a single *macrostate*. See also *thermodynamic probability*.

Mutant protein (775): A *protein* in which at least one *amino acid* residue is different from the population norm, usually resulting from a mutation on the corresponding DNA. See also *wild-type protein* and *phi-factor*.

Nanomachine (733): A structure, operating at a molecular scale, that directly transforms chemical *energy*, in the form of *Gibbs free energy*, into mechanical *energy*, in the form of physical movement. Muscular movement is a *macroscopic* manifestation of the co-ordinated action of countless nanomachines; another example is the action of the enzyme *ATP synthase*.

Native state (748): The naturally occurring *state*, under normal physiological conditions, of a *protein*, this usually being a very specific, *folded* structure. See also *unfolded state*.

Negative deviation from Raoult's law (507): A *system* of a *solute* D, dissolved in a *solute* S, or a *liquid* mixture of D and S, such that, at a given *temperature*, the *vapour pressure* p_i of either component is related to the *mole fraction* x_i, and the *vapour pressure* p_i^* of the pure component as

$$p_i < x_i p_i^*$$

See also *positive deviation from Raoult's law, Raoult's law, ideal solution*, and *non-ideal solution*.

Nernst equation (626):

$$\mathbb{E}(\text{ox,red}) = \mathbb{E}^{\ominus}(\text{ox,red}) + \frac{RT}{n_e F} \ln \frac{[\text{ox}]}{[\text{red}]} \qquad (20.19\text{a})$$

Non-cyclic photophosphorylation (740): The *metabolic pathway*, taking place within the *thylakoid* membrane, in which light *energy* is captured within *photosystem II*, causing an electron to be excited, and then transferred along *photosystem II's electron transport chain* to *photosystem I*, where a second excitation event occurs, ultimately resulting in the formation of NADPH. As the electron is transferred along *photosystem II's electron transport chain*, H^+ *ions* are translocated across the *thylakoid* membrane, so creating a *chemi-osmotic potential*, and a corresponding *proton motive force*, which ultimately drives the formation of *ATP* by *ATP synthase*. See also *cyclic photophosphorylation, light reactions of photosynthesis*, and *light-independent reactions of photosynthesis*.

Non-equilibrium state (10): A *state* of a *system* that is not in *equilibrium*, implying that the value of at least one *extensive state function* is changing over time, or that the values of at least one *intensive state function* are different at different points within the *system*.

Non-equilibrium thermodynamics (10): The study of the *thermodynamics* of *systems* that are not in *equilibrium*.

Non-ideal gas: see *real gas*.

Non-ideal solution: see *real solution*.

Normalised (754): In general, an experiment will measure the value x of a given parameter under a variety of conditions, resulting in a set of values x in a range between a minimum value x_{min} and a maximum value x_{max}. The normalised value x_{norm} corresponding to any experimental value x is defined as a ratio of the form

$$x_{norm} = \frac{x - x_{min}}{x_{max} - x_{min}}$$

such that all values of x_{norm} are between a minimum value of 0 and a maximum value of 1.

Nuclear magnetic resonance (NMR) (764): An experimental technique in which the interaction between radio frequency electromagnetic waves and atomic nuclei can be used to give information on molecular structures, and of especial use as regards *proteins* in less ordered states, for which other techniques, such as *X-ray crystallography* are less well-suited.

Nucleation (410): The process in which individual molecules coalesce into an *aggregate* of a critical size, such that further *aggregation* continues *spontaneously*. Two examples are the nucleation of *liquid* droplets as a *gas* condenses, and the nucleation of *protein* aggregates from *monomers*.

Nucleus (782): In the context of *polymer* chemistry, *protein aggregation* and virus self-assembly, an *aggregate* of a critical size, such that further *aggregation* continues *spontaneously*. See also *nucleation*.

Occam's razor: see *Ockham's razor*.

Ockham's razor (785): The principle that any theory used to explain a physical phenomenon should be as simple as possible, using the fewest possible number of explanatory variables, and making the fewest, simplest, assumptions, named after the medieval English philosopher William of Ockham (or Occam).

Open system (57): A thermodynamic *system*, the *boundary* of which permits the flow of *heat*, *work*, and matter.

Organodynamics (350): The application of the principles of *thermodynamics* to the behaviours of organisations, and of high-performing teams.

Osmosis (583): The flow of molecules of a *solvent*, through a *semi-permeable membrane*, from a *solution* containing a lower concentration of a *solute* into a *solution* containing a higher concentration of a *solute*. See also *van't Hoff equation (osmosis)*.

Osmotic potential: see *chemi-osmotic potential*.

Osmotic pressure Π Pa (587): The *pressure* that needs to be exerted on a *solution* of a given concentration of a *solute* within a *solvent* to prevent *osmosis* from a pure *solvent*. See also *van't Hoff equation (osmosis)*.

Ostwald's dilution law (542):

$$K_a = [HA]_i \frac{\xi_{eq}^2}{(1 - \xi_{eq})} \tag{17.18a}$$

Oxidant: see *oxidising agent*.

Oxidation (597): A chemical reaction in which a reagent loses one or more electrons, for example the oxidation of the stannous *ion* Sn^{2+} (aq) to the stannic *ion* Sn^{4+} (aq) as

$$Sn^{2+}(aq) + 2\,Fe^{3+}(aq) \rightleftharpoons 2\,Fe^{2+}(aq) + Sn^{4+}(aq)$$

The *anode* of an *electrochemical cell* is defined as the *electrode* at which an *oxidation half-reaction* takes place. See also *redox reaction* and *reduction*.

Oxidation half-reaction (601): A *half-reaction* in which the reactant loses one or more electrons, such as the oxidation of metallic zinc Zn (S) to a zinc *ion* Zn^{2+} (aq) as

$$Zn\,(s) \rightarrow Zn^{2+}(aq) + 2\,e^-$$

The *anode* of an *electrochemical cell* is defined as the *electrode* at which an *oxidation half-reaction* takes place. See also *reduction half-reaction*.

Oxidative phosphorylation (720): Alternatively, respiration or cellular respiration. The *metabolic pathway*, taking place in and across the *mitochondrial membrane*, in which electrons, originating in NADH and FADH$_2$ formed as a result of the *glycolysis* and *TCA cycle*, ultimately reduce molecular O$_2$ to H$_2$O. Electrons flow from NADH and FADH$_2$ to O$_2$ along the *mitochondrial electron transport chain*, causing the translocation of H$^+$ *ions* across the *mitochondrial membrane*, so creating a *chemi-osmotic potential*, and a corresponding *proton motive force*, thereby driving the formation of *ATP* by *ATP synthase*.

Oxidised state (597): A *state*, such as the *ion* Zn^{2+} (aq), which is electron-poor as compared to a corresponding electron-rich *reduced state*, such as metallic Zn (s).

Oxidising agent (620): Alternatively, oxidant. A molecule or *ion* that acts as the acceptor of electrons in a *redox reaction*. In accepting one or more electrons, the oxidising agent itself becomes reduced. The relative strength of an oxidising agent can be inferred from the corresponding *standard redox electrode potential*: the more positive that *potential*, the stronger the oxidising agent.

Oxonium ion: see *hydronium ion*.

P680 (739): The chlorophyll-based molecule that forms the *reaction centre* of *photosystem II*.

P700 (739): The chlorophyll-based molecule that forms the *reaction centre* of *photosystem I*.

Paradox, Levinthal's: see *Levinthal's paradox*.

Paramagnetic (357): A paramagnetic material is drawn into a magnetic field, and becomes weakly magnified, but loses its magnetism when the magnetic field is switched off. Paramagnetic salts are used in *adiabatic demagnetization*.

Partial derivative (79): If X is a *function* $X(Y, Z, \ldots)$ of two (or more) other variables $Y, Z, \ldots$, then the partial derivative of X with respect to Z, written as $(\partial X/\partial Z)_{Y,\ldots}$, represents rate of change of the variable X with respect to the variable Z, holding all the other variables $Y, \ldots$ constant. Mathematically, $(\partial X/\partial Z)_{Y,\ldots}$ is determined by differentiating $X(Y, Z, \ldots)$ with respect to Z assuming that all other variables are constant. See also, *total derivative*, and *total differential*.

Partial pressure (49): The *pressure* exerted by an in individual component *gas* within a gaseous mixture. See also *Dalton's law of partial pressures*.

Path (12): The specification of how a *change in state*, from an initial *equilibrium state* [1] to a final *equilibrium state* [2], takes place: for example, an *isothermal* path, an *adiabatic* path, a *quasistatic* path, an *irreversible* path, a *reversible* path. Importantly, the change $\Delta X = X_2 - X_1$ in any *state function* depends only on the initial and final equilibrium states and is independent of the path; in contrast, the value of a *path function* – such as the *heat* exchanged, or the *work* done, during the change in state – does depend on the path.

Path function (24): A phenomenon that can be measured only while a *change in state* is taking place, and whose value depends on the *path* taken during that *change in state*. Unlike a *state function*, a path function is not a property of a *state*; it is a property of the *path* between two *states*. Both *heat* and *work* are path functions.

Pathway, metabolic: see *metabolic pathway*.

Perfect gas: see *ideal gas*.

pH (664): A measure of the quantity of H$^+$ (aq) *ions* in a *solution* as

$$\text{pH} = -\log_{10} a_{b,\text{H}^+} = -\log_{10} \gamma_{b,\text{H}^+} (b_{\text{H}^+}/b_{\text{H}^+}^{\ominus}) \tag{22.11}$$

in which a_{b,H^+} is the *molal activity* of the H⁺ (aq) *ions*. To a very good approximation, equation (22.11) is equal to the more widely-used definition of pH in terms of the molar concentration [H⁺] as

$$pH = -\log_{10}[H^+] \tag{17.9}$$

Phase (154): A *state* of matter, such as *solid*, *liquid* or *gas*.
Phase, lag: see *lag phase*.
Phase change (154): Alternatively, phase transition. The *change in state* associated with a change in *phase*, such as the boiling of a *liquid* to form a *gas*. A phase change is associated with a significant change in the system's *enthalpy* over a small change in *temperature*, which, for an *ideal system*, is a discontinuity. This corresponds to a peak in the system's *heat capacity at constant pressure* C_P, when measured as a function of *temperature*, as represented on a *thermogram*. In many cases, the folding or unfolding of a *protein* may be interpreted as a phase change.
Phase transition: see *phase change*.
Phi-value ϕ_F (777): An experimentally determined parameter, defined as

$$\phi_F = \frac{\Delta\Delta_{act}G^\ominus}{\Delta\Delta_{fol}G^\ominus} = \frac{\alpha - \gamma}{\alpha - \beta} = \frac{\gamma - \alpha}{\beta - \alpha} \tag{25.32}$$

in which $\alpha = G_{U,mut} - G_{U,wild}$, this being the difference between the *standard Gibbs free energies* of the *unfolded state* of a *mutant* protein and the *wild-type*; β is the corresponding difference for the *native state*; and γ that for the *transition state*. Measurement of the phi-factor gives information on the structure of short-lived *transition states*.
Phosphate, inorganic: see *inorganic phosphate*.
Phosphorylation, oxidative: see *oxidative phosphorylation*.
Phosphorylation, substrate-level: see *substrate-level phosphorylation*.
Photon (738): According to quantum theory, electromagnetic radiation of wavelength λ m, and frequency ν Hz (such that $c = \lambda\nu$, where c is the velocity of light) is associated with a *quantum* of energy $h\nu$ J, where h is Planck's constant, 6.626×10^{-34} J sec. This *quantum* of energy suggests that light is behaving more like a particle than a wave, and so a photon is a 'particle' of light with energy $h\nu$ J.
Photophosphorylation (716): see *cyclic photophosphorylation*, *non-cyclic photophosphorylation*, *oxidative phosphorylation* and *substrate-level phosphorylation*.
Photosynthesis (738): The overall name for a number of *metabolic pathways* that collectively use energy from light to transform atmospheric CO_2 into biomolecules such as glyceraldehyde-3-phosphate and glucose, while splitting water into molecular O_2. See also *light reactions of photosynthesis* and *light-independent reactions of photosynthesis*.
Photosynthetic pigment (738): A light-absorbing molecule, such as a chlorophyll, carotene or xanthophyll. Clusters of these pigments are embedded in the *thylakoid membrane* as the *light harvesting complexes* that capture light for *photosystems I* and *II*.
Photosystem I (739): A cluster of *proteins*, and associated molecules, embedded in the *thylakoid membrane*. The *light-harvesting complex* transfers light energy by *resonance energy transfer* to a *reaction centre* containing the chlorophyll-based molecule *P700*. The captured light *energy* then promotes an electron, transferred to *P700* from the *electron transport chain* of *photosystem II*, to an excited state. The excited electron can then follow

one of two paths: either to be used to form a molecule of NADPH, or to pass back to the *electron transport chain* of *photosystem II*, a *pathway* known as *cyclic phosphorylation*.

Photosystem II (739): A cluster of *proteins*, and associated molecules, embedded in the *thylakoid membrane*. The *light-harvesting complex* transfers light *energy* by *resonance energy transfer* to a *reaction centre* containing the chlorophyll-based molecule *P680*. The captured light energy then promotes an electron to an excited state, which then passes down the *electron transport chain* of *photosystem II* to *photosystem I*. The electrons lost from *P680* are replenished by electrons released as a result of the splitting of water to molecular O_2 and H^+ *ions*, which accumulate within the *thylakoid lumen*; as the electrons from *P680* pass down the *electron transport chain*, H^+ ions are translocated across the *thylakoid membrane*, causing a further accumulation within the *lumen*. The resulting *chemi-osmotic potential* creates a corresponding *proton motive force*, which then drives the synthesis of *ATP* by *ATP synthase*.

Pigment, photosynthetic: see *photosynthetic pigment*.

pK_a (539):

$$pK_a = -\log_{10} K_a \qquad (17.13a)$$

where K_a is the *acid dissociation constant*.

pK_b (545):

$$pK_b = -\log_{10} K_b \qquad (17.20a)$$

where K_b is the *base dissociation constant*.

Platinised platinum (606): Platinum metal, the surface of which has been specially treated by a process known as 'platinisation', in which pure, clean, platinum metal is coated with fine particles of platinum metal, known as 'platinum black'. This creates a piece of pure platinum with a very much increased surface area, which can act as an *inert electrode*, mediating *redox half-reactions*.

Platinum, platinised: see platinised platinum.

Platinum electrode (606): An *inert electrode* made from *platinised platinum*.

Plot, chevron: see *chevron plot*.

pOH (537): A measure of the molar concentration [OH^-] of OH^- *ions* in a *solution* as

$$pOH = -\log_{10}[OH^-] \qquad (17.10)$$

See also pH.

Point, critical: see *critical point*.

Point, triple: see *triple point*.

Point of inflexion (489): A point on a curve $f(x, y, z)$ at which $(\partial f/\partial x)_{y,z} = (\partial^2 f/\partial x^2)_{y,z} = 0$. See also *critical point*.

Polymer (782): A molecule formed as a sequence of otherwise independent molecules—*monomers*—often in long chains, such as the *polypeptide* chain within a *protein*, which is a sequence of *amino acids*, or a nucleic acid, which is a sequence of nucleotides.

Polymerisation (780): The process by which a *polymer* is formed from the corresponding *monomers*.

Polypeptide chain (748): A *polymer*, usually a *protein*, formed as a chain of *amino acids*.

Polyprotic acid (540): An acid such as H_2A that dissociates in aqueous *solution* to yield more than one hydrogen *ion* as

$$H_2A\,(aq) \rightleftharpoons H^+\,(aq) + HA^-\,(aq)$$

followed by

$$HA^-\,(aq) \rightleftharpoons H^+\,(aq) + A^{2-}\,(aq)$$

Positive deviation from Raoult's law (507): A *system* of a *solute* D, dissolved in a *solute* S, or a *liquid* mixture of D and S, such that, at a given *temperature*, the *vapour pressure* p_i of either component is related to the *mole fraction* x_i, and the vapour pressure p_i^* of the pure component as

$$p_i > x_i p_i^*$$

See also *negative deviation from Raoult's law*, *Raoult's law*, *ideal solution*, and *non-ideal solution*.

Potential: see *potential function*.
Potential, chemical: see *chemical potential*.
Potential, chemi-osmotic: see *chemi-osmotic potential*.
Potential, electrical: see *electrical potential*.
Potential, Lennard-Jones: see *Lennard-Jones potential*.
Potential, liquid junction: see *liquid junction potential*.
Potential, membrane: see *membrane potential*.
Potential, standard electrode: see *standard electrode potential*.
Potential difference, electrical: see *electrical potential difference*.
Potential function (762): A mathematical function of the general form $V(x, y, z, \ldots)$ in which the variables x, y, z, …are relevant to the system in question, such as the co-ordinates of a point in space. In essence, the potential function $V(x, y, z, \ldots)$ defines the amount of *work* required to achieve the *system* configuration specified by given values of $x, y, z, \ldots$, and so is a measure of the *energy* associated with than configuration.
Potential gradient, chemi-osmotic: see *chemi-osmotic potential gradient*.
Pre-exponential factor A (456): Alternatively, frequency factor. A parameter in the *Arrhenius equation*

$$k(T) = A\, e^{-\frac{E_a}{RT}} \tag{14.38}$$

The dimensions of A are the same as those of the *reaction rate constant* $k(T)$ for the corresponding reaction.

Pressure P Pa (14): The *force* exerted per unit area. Pressure is an *intensive state function*.
Pressure, atmospheric: see *standard atmospheric pressure*.
Pressure, critical: see *critical pressure*.
Pressure, partial: see *partial pressure*.
Pressure, vapour: see *vapour pressure*.
Pressure mass action ratio Γ_p (432): Alternatively, pressure reaction quotient Q_p. The *non-equilibrium* ratio

$$\Gamma_p = \frac{(p_C/P_C^{\ominus})^c\,(p_D/P_D^{\ominus})^d}{(p_A/P_A^{\ominus})^a\,(p_B/P_B^{\ominus})^b} \tag{14.14}$$

Pressure reaction quotient Q_p: see *pressure mass action ratio*.

Pressure thermodynamic equilibrium constant K_p (428): A *thermodynamic equilibrium constant* expressed in terms of the *partial pressures* p_i Pa of each reagent. Accordingly, for the generalized reaction $aA + bB \rightleftharpoons cC + dD$

$$K_p = \frac{(p_{C,eq}/P_C^\ominus)^c \, (p_{D,eq}/P_D^\ominus)^d}{(p_{A,eq}/P_A^\ominus)^a \, (p_{B,eq}/P_B^\ominus)^b} \tag{14.12}$$

Primary structure (748): The specification of the sequence of *monomers* within a biological *polymer*, such as the sequence of *amino acids* along a *polypeptide* chain.

Principle of Le Chatelier: see *Le Chatelier's principle*.

Probability, mathematical: see *mathematical probability*.

Probability, thermodynamic: see *thermodynamic probability*.

Protein (395): A biological macromolecule formed from *amino acids*. Proteins perform a variety of biological functions, including catalysis—all *enzymes* are proteins. See also *primary, secondary, tertiary*, and *quaternary structure*.

Protein, mutant: see *mutant protein*.

Protein, wild-type: see *wild-type protein*.

Protein aggregation (781): The *spontaneous self-assembly* of individual *protein* molecules into larger structures, such as *amyloids* or virus *capsids*.

Protein-ligand complex (764): Many *proteins* bind, usually *reversibly* and transiently, with smaller molecules, *ligands*, so forming a protein-ligand complex.

Proton motive force Δp V (732): The ratio $\Delta G_{co}'/F$, where $\Delta G_{co}'$ is a *chemi-osmotic potential gradient*, and F is the *Faraday constant*.

$$\Delta p = \frac{\Delta G_{co}'}{F} = -2.303 \frac{RT}{F} \Delta pH + \Delta \Psi \tag{24.14}$$

Pseudo-equilibrium (712): Many *metabolic pathways* are complex sequences of reactions, all taking place simultaneously, often at different *reaction rates* at different parts of the *pathway*. Such systems are never in true *thermodynamic equilibrium*, and so the assumption of pseudo-equilibrium is a pragmatic simplification, allowing the equations of *equilibrium thermodynamics* to be applied to these systems, and to enable the interpretation of experimental results.

P, V diagram (31): A graph in which the vertical axis is the *system pressure* P, and the horizontal axis, the *system volume* V. The area between a given curve on the graph and the horizontal axis is the *reversible work* performed by the *system* on the *surroundings* (for an expansion), or on the *system* by the *surroundings* (for a compression) during a *change in state*.

P, V work of expansion (23): Alternatively, P, V *work*. *Work* done by the expansion of a *gas* against external pressure.

Quantum (738): According to quantum theory, the amount of energy $h\nu$ J associated with a *photon* of frequency ν Hz, where h is Planck's constant, 6.626×10^{-34} J sec.

Quasistatic (30): A *change in state*, which takes place along a *path* comprising an infinite sequence of infinitesimal steps between successive *equilibrium states*. All *reversible* paths are necessarily *quasistatic*, but it is (technically) possible for a quasistatic path to be *irreversible*.

Quaternary structure (748): Many biological structures, for example, *ribosomes*, are complexes of several components, such as *proteins*, each of which has its appropriate *primary*, *secondary* and *tertiary structure*. The quaternary structure defines the way in which the components are assembled in the final, multi-component, complex.

Raoult's law (505):

$$p_i = x_i p_i^* \tag{16.1}$$

See also *ideal solution, non-ideal solution, negative deviation from Raoult's law* and *positive deviation from Raoult's law*.

Raoult's law, negative or positive deviation from: see *negative deviation from Roault's law* and *positive deviation from Raoult's law*. See also *Raoult's law, ideal solution* and *non-ideal solution*.

Rate of reaction (450): The measure of how quickly, or slowly, a chemical reaction takes place. See also *activation energy* and *Arrhenius equation*.

Razor, Ockham's: see *Ockham's razor*.

Reaction, coupled: see *coupled reactions*.

Reaction, half-cell: see *half-cell reaction*.

Reaction, rate of: see *rate of reaction*.

Reaction centre (739): A cluster of molecules, central to *photosynthesis*, within which light energy, as originally captured by the *light-harvesting complex*, causes an electron to be excited to a higher *energy* level. See also *P680, P700, photosystem I* and *photosystem II*.

Reaction Gibbs free energy $\Delta_r G$ J (432):

$$\Delta_r G = \frac{dG_{sys}(\xi)}{d\xi} \tag{14.15}$$

Reaction quotient, concentration: see *concentration mass action ratio*.

Reaction quotient, molality: see *molality mass action ratio*.

Reaction quotient, mole fraction: see *mole fraction mass action ratio*.

Reaction quotient, pressure: see *pressure mass action ratio*.

Reaction rate constant k (452): A measure of the speed of a chemical reaction.

Real gas (50): A *gas* that does not comply with the *ideal gas equation-of-state*

$$PV = nRT \tag{3.3}$$

See also *Berthelot gas, ideal gas, Dieterici gas, van der Waals gas* and *virial equation of state*.

Real solution (173): A *solution* that does not comply with *Raoult's law*

$$p_i = x_i p_i^* \tag{16.1}$$

See also *ideal solution, ideal-dilute solution, Henry's law, negative deviation from Raoult's law*, and *positive deviation from Raoult's law*.

Redox electrode (606): An *electrode* in which an inert material, such as *platinised platinum*, mediates a *redox reaction* between two *ions* of the same element in different *oxidation* states such as

$$Fe^{+++} (aq) + e^- \rightarrow Fe^{++} (aq)$$

Redox reaction (603): A chemical reaction involving the transfer of electrons between two components, such that the component that loses one or more electrons (the electron donor, or *reducing agent*) is *oxidised*, and, simultaneously, the component that gains one or more electrons (the electron acceptor, or *oxidising agent*) is *reduced*.

Reduced state (597): A *state*, such as metallic Zn (s), which is electron-rich as compared to a corresponding electron-poor *oxidised state*, such as the *ion* Zn^{2+} (aq).

Reducing agent (620): Alternatively, reductant. A molecule or *ion* that acts as the donor of electrons in a *redox reaction*. In donating one or more electrons, the reducing agent itself becomes oxidised. The relative strength of a reducing agent can be inferred from the corresponding *standard redox electrode potential*: the more negative that *potential*, the stronger the reducing agent.

Reductant: see *reducing agent*.

Reduction (598): A chemical reaction in which a reagent gains one or more electrons, for example, the reduction of the ferric *ion* Fe^{3+} (aq) to the ferrous *ion* Fe^{2+} (aq) in the reaction

$$Sn^{2+} (aq) + 2\ Fe^{3+} (aq) \rightleftharpoons 2\ Fe^{2+} (aq) + Sn^{4+} (aq)$$

The *cathode* of an *electrochemical cell* is defined as the *electrode* at which a reduction *half-reaction* takes place. See also *redox reaction* and *oxidation*.

Reduction half-reaction (601): A *half-reaction* in which the reactant gains one or more electrons, such as the reduction of the copper *ion* Cu^{2+} (aq) to metallic copper Cu(s) as

$$Cu^{2+} (aq) + 2\ e^- \rightarrow Cu\ (s)$$

The *cathode* of an *electrochemical cell* is defined as the *electrode* at which a reduction *half-reaction* takes place.

Reference electrode, standard: see *standard reference electrode*.

Reference level (169): The definition of the baseline level of a given scale of measurement, against which other measurements can be compared. The measurement corresponding to the reference level is that of a suitably defined *reference state*. In most cases, the baseline level is the zero point on the scale: so, for the Celsius scale of *temperature* measurement, the original reference level, 0° C, corresponds to the temperature of an *equilibrium* mixture of pure water and pure ice at a *pressure* of 1.01325 bar, this being the *reference state*. In some circumstances, the reference level has a value other than zero – so, for example, for the *thermodynamic scale of temperature*, the reference level is the *triple point* of pure water, 273.16 K.

Reference level shift (175): The difference between two measurements, each made with respect to different *reference levels*.

Reference state (169): The *state* of a *system* for which a *reference level* of measurement is defined.

Relations, Maxwell: see *Maxwell relations*.

Reservoir: see *heat reservoir*.

Resonance energy transfer (739): Alternatively, exciton transfer. The transfer of *energy* from one molecule to another, without the radiation of light, as takes place within the *light-harvesting complexes* associated with *photosynthesis*.

Respiration: See *oxidative phosphorylation*.

Respiratory membrane (724): The *biological membrane* in, and across, which *oxidative phosphorylation* takes place.

GLOSSARY

Reverse osmosis (588): The process in which an external *pressure*, greater than a *solution's osmotic pressure*, is applied to that *solution*, so allowing the pure *solvent* to accumulate on the further side of a *semi-permeable membrane*. Reverse osmosis is important in the industrial *desalination* of sea water.

Reversible (123): A *change in state* which takes place along a *path* comprising an infinite sequence of infinitesimal steps between successive *equilibrium states*, and in the absence of all *dissipative effects* such as *dynamic friction*. A reversible change can be reversed so that both the *system* and the *surroundings* can be returned to their exact original *states*.

Reversible electrode potential E V (613): The *electrode potential* resulting from the corresponding *half-reaction* taking place *reversibly*.

Reversible gas electrode (606): An *electrode* which can bring into *equilibrium* a *gas* and a solution of its *ions*.

Reversible heat $đq_{rev}$ J (143): A flow of *heat* associated with a *reversible path*. For any *change in state*, $đq_{rev}$ is the maximum *heat* possible, and is inevitably greater than the *irreversible heat* $đq_{irrev}$: $đq_{rev} > đq_{irrev}$. See also *irreversible heat* and *irreversible path*.

Reversible hydrogen electrode (615): A *platinised platinum electrode* in which gaseous hydrogen H_2 (g) is in equilibrium with a *solution* of H^+ (aq) *ions*. If the *pressure* of the H_2 (g) is 1 *bar*, and the concentration of the H^+ (aq) is 1 mol dm^{-3}, then the *system* is in its *standard state* and the corresponding *electrode potential* is the *standard electrode potential*, this being the IUPAC *reference level* of 0 V.

Reversible path: see *reversible*.

Reversible redox electrode (617): An *electrode* which can bring into *equilibrium*, within *solution*, two ionic species of the same element, but of different oxidation states, for example, Fe^{2+} (aq) and Fe^{3+} (aq).

Reversible work $đw_{rev}$ J (142): A quantity of *work* associated with a *reversible path*. For any *change in state*, $đw_{rev}$ is the maximum *work* possible, and is inevitably greater than the *irreversible work* $đw_{irrev}$: $đw_{rev} > đw_{irrev}$. See also *irreversible heat* and *irreversible path*.

Ribosome (748): A cellular complex of *proteins* and RNA central to the synthesis of proteins – within the ribosome, the genetic code, as expressed on mRNA (which is derived from DNA), is 'read', so enabling *amino acids* to be assembled in the correct sequence.

RuBisCO (744): An *enzyme* within the *carboxysome* which catalyses the reaction between molecular CO_2 (1 carbon atom) and ribulose-1,5-biphosphate (5 carbon atoms) to form two molecules of 3-phosphoglycerate (3 carbon atoms each).

Salt bridge (604): An *electrolyte*, usually made from an agar gel containing an ionic salt such as KCl or NaCl, which provides electrical conductivity between two *half-cells*, while minimising *liquid junction potentials*.

Saturated vapour pressure: see *equilibrium saturated vapour pressure*.

Saturation phase (782): The final phase in the *aggregation* of a *polymer*, such as a *protein*, in which the *polymer* has achieved its stable length. See also *elongation phase* and *lag phase*.

Scanning calorimetry, differential: see *differential scanning calorimetry*.

Second Law of Thermodynamics (267): A spontaneous change proceeds, unidirectionally and irreversibly, from a non-equilibrium to an equilibrium state. For any system, we may define an extensive state function S, the entropy of the system, such that the change dS in the system's entropy for any change in state is given by

$$dS = \frac{đq_{rev}}{T} \tag{9.1}$$

where đq_{rev} is the quantity of heat exchanged, reversibly, at temperature T, during that change in state.

Second Law of Thermodynamics, Carathéodory statement: see *Carathéodory statement of the Second Law of Thermodynamics*.

Second Law of Thermodynamics. Clausius statement: see *Clausius statement of the Second Law of Thermodynamics*.

Second Law of Thermodynamics. Kelvin–Planck statement: see *Kelvin-Planck statement of the Second Law of Thermodynamics*.

Second thermodynamic equation-of-state (642):

$$\left(\frac{\partial U}{\partial P}\right)_T = -T\left(\frac{\partial V}{\partial T}\right)_P - P\left(\frac{\partial V}{\partial P}\right)_T \tag{21.13}$$

Secondary structure (748): Many biomolecules, such as *proteins* and nucleic acids, have an overall complex structure, known as the *tertiary structure*, within which are local domains of conformations known as α-*helices* or β-*pleated sheets*, which are stabilized by *hydrogen bonds*. These local domains are known as *secondary structures*. See also *primary structure, tertiary structure*, and *quaternary structure*.

Seebeck effect (594): When two electrical conductors, made of two different materials, are at two different *temperatures* and in mutual contact, then an *electrical potential difference* may be created at the junction between the two materials.

Self-assembly (780): The *spontaneous* assembly of a more complex structure from more simple components, as exemplified by the *spontaneous* formation of a virus *capsid* from its constituent subunit *proteins*.

Semi-permeable membrane (582): A membrane between two different *solutions* which allows molecules of the *solvent* to pass between the two *solutions*, but blocks the passage of *solute* molecules. See also *osmosis*.

Shomate equation (195):

$$C_P = A + BT + CT^2 + DT^3 + \frac{E}{T^2} \tag{6.31}$$

Solid (16): A *state* of matter, or *phase*, such that a given mass of material occupies essentially the same volume at a variety of *pressures*. A solid is also resistant to shear stresses, and so, unlike a *liquid*, does not flow easily, and maintains its original size and shape. See also *gas*.

Solute (170): A component within a *solution*, the *mole fraction* of which is less than the *mole fraction* of at least one other component.

Solution (170): A mixture such that the molecules of all components are distributed randomly, and can form a stable, equilibrium, homogeneous *phase*. The component of greatest *mole fraction* is the *solvent*; all other components are *solutes*.

Solution, buffer: see *buffer solution*.

Solution, ideal: see *ideal solution*.

Solution, non-ideal: see *non-ideal solution*.

Solvent (170): The component within a *solution* with the largest *mole fraction*.

Spontaneous (260): A *change in state* that takes place within a *system* of its own accord, and without being caused by an external stimulus: no phenomenon can be detected at the *boundary* of the system that can be attributed as a cause. A spontaneous change is *unidirectional*, and proceeds from a *non-equilibrium state* to an *equilibrium state*.

SPR: see *surface plasmon resonance*.

Stable equilibrium: see *stable state*.

Stable state (410): An *equilibrium state* corresponding to the lowest possible *Gibbs free energy*. In the following diagram, state A is a *stable state*, state B is a *metastable state*, and state C is an *unstable state*.

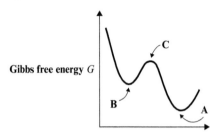

Standard (adjective) (168): As pertaining to the appropriate *standard state*.

Standard (noun) (168): A widely accepted convention, for example, as defined in the IUPAC 'Green Book'.

Standard, biochemical: see *biochemical standard*.

Standard, biological: see *biochemical standard*.

Standard atmospheric pressure $P^\ominus$ Pa (15): Alternatively atmospheric pressure or atmosphere. A *pressure* of 101,325 Pa, this being representative of the average *pressure* exerted by the earth's atmosphere at average sea level. See also *standard pressure*.

Standard electrode potential $\mathbb{E}^\ominus$ V (615): The *electrode potential* of an *electrode* operating such that all the components associated with the appropriate *half-cell reaction* are in their *standard states*. Under the *conventional standards*, the *standard electrode potential* $\mathbb{E}^\ominus$ (H^+, H_2) of the *reversible hydrogen electrode*, in which pure H_2 (g) at the *standard pressure* of 1 *bar* is in equilibrium with a solution of H^+ (aq) *ions* at a pressure of 1 bar and at the *standard molar concentration* of 1 mol dm^{-3}, corresponding to pH 0, is defined as 0 V at all temperatures, this being the IUPAC *reference level*.

Standard enthalpy of formation $\Delta_f H^\ominus$ J mol^{-1} (178): The *enthalpy change* associated with the formation of 1 mol of any molecule in its *standard state* from the corresponding constituent elements in their *standard states*. The *reference level* is defined such that the standard enthalpy of formation for all elements in their *standard states* is zero: $\Delta_f H^\ominus = 0$ J mol^{-1} for all elements.

Standard enthalpy of reaction $\Delta_r H^\ominus$ J (182): The *standard enthalpy* change for any chemical reaction expressed as the difference between the *standard enthalpies of formation* of the products and the *standard enthalpies of formation* of the reactants

$$\Delta_r H^\ominus = \sum_{\text{Products}} n_{\text{prod}} \Delta_f H_{\text{prod}}^\ominus - \sum_{\text{Reactants}} n_{\text{reac}} \Delta_f H_{\text{reac}}^\ominus \qquad (6.21)$$

Standard Gibbs free energy of formation $\Delta_f G^\ominus$ J mol^{-1} (397): The *Gibbs free energy change* associated with the formation of 1 mol of any molecule in its *standard state* from the corresponding constituent elements in their *standard states*. The *reference level* is defined such that the standard Gibbs free energy of formation for all elements in their *standard states* is zero: $\Delta_f G^\ominus = 0$ J mol^{-1} for all elements.

Standard Gibbs free energy of reaction $\Delta_r G^\ominus$ J (399): The *standard Gibbs free energy* change for any chemical reaction expressed as the difference between the *standard*

Gibbs free energies of formation of the products and the standard Gibbs free energies of formation of the reactants

$$\Delta_r G^\ominus = \sum_{\text{Products}} n_{\text{prod}} \Delta_f \mathbf{G}_{\text{prod}}^\ominus - \sum_{\text{Reactants}} n_{\text{reac}} \Delta_f \mathbf{G}_{\text{reac}}^\ominus \tag{13.45}$$

Standard pressure $P^\ominus$ Pa or bar (172): A *pressure* of 10^5 Pa = 1 bar, in accordance with the *standard states* define by IUPAC. See also *standard atmospheric pressure*.

Standard reversible electrode potential $\mathbb{E}^\ominus$ V (615): Alternatively standard reversible reduction potential. The *electrode potential* of an *electrode*, the materials of which are in their *standard states*, operating *reversibly*. The alternative term, standard reversible reduction potential, is attributable to the convention which represents *standard half-reactions* as *reductions*.

Standard state (172): A defined *state* for which reference measurements can be made, and published in reference tables. The IUPAC standard states, also known as the conventional standard states, are described in Table 6.4 (see page 172). See also *biochemical standard state*.

Standard state, biochemical: see *biochemical standard state*.

Standard state, conventional: see *standard state*.

Standards, conventional: see *conventional standards*.

State (6): The particular circumstances of a chosen *system*. If a system is in *thermodynamic equilibrium*, the state of the system is defined in terms of the values of the *system's state functions*.

State, change in: see *change in state*.

State, denatured: see *denatured state*.

State, equation of: see *equation-of-state*.

State, equilibrium: see *equilibrium state*.

State, folded: see *folded state*.

State, metastable: see *metastable state*.

State, native: see *native state*.

State, non-equilibrium: see *non-equilibrium state*.

State, reference: see *reference state*.

State, stable: see *stable state*.

State, standard: see *standard state*.

State, transition: see *transition state*.

State, unfolded: see *native unfolded*.

State, unstable: see *unstable state*.

State function (6): A function that describes the *state* of a *system* at *equilibrium*, irrespective of how the *system* arrived at that *state*. State functions are expressed in terms of *total differentials*. Importantly, the change $\Delta X_{1 \to 2}$ in any state function X for a change from an *equilibrium state* [1] to an *equilibrium state* [2] is independent of the *path* taken between those *states*. Examples are the *mole number n*, the *pressure P* and the *temperature T*. See also *extensive state function* and *intensive state function*.

Static friction (28): A *force*, present at the contact area between two surfaces, which resists the movement of either surface, for example, the *force* that prevents a ladder from slipping backwards. See also *dynamic friction*.

Statistical mechanics (5): The derivation of the *macroscopic* properties of *systems* from the *microscopic* statistical study of the dynamics of the *system's* individual atoms and molecules.

Stoichiometric coefficient (155): The numbers a, b, c, d... associated with the chemical reaction written as $aA + bB \rightleftharpoons cC + dD$. Stoichiometric coefficients are necessarily positive. See also *stoichiometric number*.

Stoichiometric number (160): For the reaction written as $aA + bB \rightleftharpoons cC + dD$, the stoichiometric numbers are the signed numbers $-a$, $-b$, $+c$, and $+d$, where a positive sign is associated with the *stoichiometric coefficient* of each product, and a negative sign with the *stoichiometric coefficient* of each reactant. The stoichiometric numbers therefore depend not only on the nature of the reaction, but also the direction in which the reaction equation is written.

Strength, ionic: see *ionic strength*.

Stroma (739): The interior space within the *chloroplast*. *Thylakoids* exist within, and are surrounded by, the stroma.

Structure, primary, secondary, tertiary, quaternary: See *primary structure, secondary structure, tertiary structure* or *quaternary structure*.

Substrate-level phosphorylation (716): The synthesis of *ATP* as a result of an *enzyme*-catalysed reaction taking place in the cytoplasm—for example, during *glycolysis*—and which is not driven by the presence of a *chemi-osmotic potential*, and the corresponding *proton motive force*, across a *membrane*, such as the *membrane* of the *mitochondrion* or *thylakoid*.

Supercooled gas (410): Alternatively, supersaturated gas. A *metastable state* that can arise when a *real gas* is cooled below the normal boiling point, but fails to liquefy as a result, for example, of the absence of suitable *nucleation*. See Fig. 15.2.

Superheated liquid (474): A *metastable state* that can arise when a *real liquid* is heated beyond the normal boiling point, but fails to vaporise. See Fig. 15.2.

Superheated solid (474): A *metastable state* that can in principle arise when a *solid* is heated beyond the normal melting point, but fails to melt. In practice, this phenomenon is rare.

Supersaturated gas: see *supercooled gas*.

Surface plasmon resonance (SPR) (764): An experimental technique that can detect events taking place on suitably prepared surfaces, as measured, for example, by changes in the refractive index of a surface layer. This technique is used to measure *protein-ligand* binding: the *ligand* is immobilized on the surface, and a solution of the *protein* of interest is allowed to flow over the surface. As the *protein* binds with the *ligand*, the refractive index of the surface layer changes.

Surroundings (5): That part of the universe outside of the chosen *system*, and separated from the *system* by the *system boundary*.

System (5): That part of the universe chosen for study, and separated from the surroundings by the system *boundary*.

System, closed: See *closed system*.

System, isolated: See *isolated system*.

System, open: See *open system*.

TCA cycle: see *tricarboxylic acid cycle*.

Temperature (43): An *intensive state function* defining the average *energy* of the molecules within a *system*, which in every-day terms, determines how hot the system is. If two adjacent *systems*, in *thermal contact*, exchange *heat* so that one *system* becomes hotter and the other cooler, then the *systems* were originally at different temperatures; if no heat is exchanged, then the *systems* are at the same temperature.

Temperature, critical: see *critical temperature*.

Temperature, Debye: see *Debye temperature*.

Temperature, melting: see *melting temperature*.

Temperature gradient (54): Any *temperature* difference $\Delta T = T_2 - T_1$. An 'uphill' temperature gradient is such that $T_2 > T_1$; a 'downhill' gradient is such that $T_2 < T_1$.

Tertiary structure (748): The complete description of the three-dimensional structure of a biological macromolecule, such as a *protein*. This is different from the *primary structure* (which is the linear sequence), the *secondary structure* (which is a partial three dimensional description, identifying only local features such as α-*helices* and β-*pleated sheets*), and the *quaternary structure* (which describes higher order structures, for example, multi-macromolecular assemblies).

Thermal contact (54): Two systems are in thermal contact when they are able to exchange *heat*.

Thermal equilibrium (9): Two *systems* are in thermal equilibrium when they exchange no *heat*, implying that the two systems are at the same *temperature*.

Thermochemistry (148): That branch of physical chemistry that deals with the *heat* exchanges accompanying chemical reactions.

Thermodynamic cycle (36): A sequence of *changes in state* such that the *system* returns to is original *state*.

Thermodynamic equation-of-state, first: see *first thermodynamic equation-of-state*.

Thermodynamic equation-of-state, second: see *second thermodynamic equation-of-state*.

Thermodynamic equilibrium (9): A *state* of a *system* such that all *state functions* have constant values over time, and all *intensive state functions* have the same value at all points within the system.

Thermodynamic equilibrium constant (428): An expression defining the quantities of reagents present in a chemical mixture at *thermodynamic equilibrium*, as determined by identifying the s*tate* of minimum *Gibbs free energy G*, under any given conditions of *temperature* and *pressure*. See also *concentration thermodynamic equilibrium constant*, *molality thermodynamic equilibrium constant*, *mole fraction thermodynamic equilibrium constant*, and *pressure thermodynamic equilibrium constant*.

Thermodynamic probability W (324): Or *multiplicity*: the number of *microstates* that correspond to a specific *macrostate*, as used in *Boltzmann's equation*.

Thermodynamic scale of temperature K (44): A scale of *temperature*, the *reference level* of which is the *triple point* of water, 273.16 K, and such that the size of 1 K equals 1 °C.

Thermodynamics (4): The branch of science concerned with *heat*, *work* and *energy*.

Thermodynamics, First Law of: see *First Law of Thermodynamics*.

Thermodynamics, 'Fourth Law' of: see *'Fourth Law' of Thermodynamics*.

Thermodynamics, Second Law of: see *Second Law of Thermodynamics*.

Thermodynamics, Second Law of, Carathéodory statement: see *Carathéodory statement of the Second Law of Thermodynamics*.

Thermodynamics, Second Law of, Clausius statement: see *Clausius statement of the Second Law of Thermodynamics*.

Thermodynamics, Second Law of, Kelvin-Planck statement: see *Kelvin-Planck statement of the Second Law of Thermodynamics*.

Thermodynamics, Third Law of: see *Third Law of Thermodynamics*.

Thermodynamics, Zeroth Law of: see *Zeroth Law of Thermodynamics*.

Thermoeconomics (348): The application of the principles of *thermodynamics* to the behaviours of economic *systems*.

Thermogram (759): A depiction of the experimental results of *differential scanning calorimetry*, in which the sample's *heat capacity at constant pressure*, C_P, is plotted as a function of *temperature*. Peaks on a thermogram represent *phase changes*, such as those associated with *protein* folding or unfolding. See Figure 25.5.

Third law entropy (355): Alternatively, absolute entropy. The value of the *entropy* of a substance as determined using the *Third Law of Thermodynamics* that the *entropy* of a pure crystalline substance is zero at the *absolute zero* of *temperature*.

Third Law of Thermodynamics (352): At the *absolute zero* of *temperature*, the *entropy* of all pure perfectly crystalline *solids* is zero; alternatively, the statement that at the *absolute zero* of *temperature*, the *entropy* change associated with any *isothermal* change is zero; alternatively the statement that it is impossible to reach the *absolute zero* of *temperature* in a finite number of operations.

Throttling process (648): A process in which a *gas* at a given *pressure* is forced through a small orifice, valve or porous plug into a region of reduced *pressure*. A reversible throttling process is a *change of state* at constant *enthalpy*. See also *Joule–Thomson coefficient*, *inversion temperature*, and *inversion curve*.

Thylakoid (739): A *membrane*-bound compartment within the *chloroplast*, often formed in 'stacks' called *grana*. The interior of the thylakoid is called the *lumen*, and the thylakoid systems is the location of the *light dependent reactions of photosynthesis*.

Total derivative (77): If X is a function $X(Y, Z, \ldots)$ of two (or more) other variables Y, $Z, \ldots$, then the total derivative of X, with respect to, say, Z, written as dX/dZ, is given by

$$\frac{dX}{dZ} = \left(\frac{\partial X}{\partial Y}\right)_{Z,\ldots} \frac{dY}{dZ} + \left(\frac{\partial X}{\partial Z}\right)_{Y,\ldots} + \ldots$$

in which terms of the form $(\partial X/\partial Y)_Z$ are *partial derivatives*. See also *total differential*.

Total differential (77): If X is a function $X(Y, Z, \ldots)$ of two (or more) other variables Y, $Z, \ldots$, then the total differential of X, written as dX, represents the change dX in the variable X attributable to simultaneous changes dY, dZ, … in all the other variables. Mathematically, dX is given by

$$dX = \left(\frac{\partial X}{\partial Y}\right)_{Z,\ldots} dY + \left(\frac{\partial X}{\partial Z}\right)_{Y,\ldots} dZ + \ldots$$

in which terms of the form $(\partial X/\partial Y)_Z$ are *partial derivatives*. See also *total derivative*.

T, P diagram (651): A graph in which the vertical axis is the *system temperature* T, and the horizontal axis, the *system pressure* P, of particular relevance to *throttling processes*.

T, S diagram (297): A graph in which the vertical axis is the *system temperature* T, and the horizontal axis, the *system entropy* S. The area between a given curve on the graph and the horizontal axis is the *reversible heat* gained by the *system* from the *surroundings* (corresponding to an increase in *entropy*), or lost from the system to the surroundings (corresponding to an decrease in entropy), during a *change in state*.

Transformed (677): A measurement made under, or an equation referring to, the *biochemical standards* is said to be 'transformed' as compared to the equivalent measurement made under, or equation referring to, the *conventional standards*.

Transition, phase: see *phase change*.

Transition state (457): Alternatively, activated state. A transient, intermediate, thermodynamically *unstable state* of higher energy, through which a *system* passes during a change from one *equilibrium state* to another. See Fig. 24.9.

Transition state theory (771): A theory of *chemical kinetics* in which the *reaction rate* is described in terms of the vibrational *energy* of the chemical bond between two atoms.

Transmission coefficient κ (771): A constant of proportionality used within *transition state theory*.

Transport, active: see *active transport*.

Transport chain: see *electron transport chain*.

'Trapped' metastable state: see *metastable state*.

Tricarboxylic cycle (716): Alternatively, TCA cycle or Krebs cycle. A cyclic *metabolic pathway* in which pyruvate, produced as a result of *glycolysis*, is transformed into molecular CO_2, while much of the *Gibbs free energy* released is captured by the *energy currencies* NADH, $FADH_2$ and GTP. The NADH and $FADH_2$ are then the substrates for mitochondrial *oxidative phosphorylation*, resulting in the formation of *ATP*.

Triple point (480): For any pure material, the single combination of *pressure* and *temperature* at which the three phases of *gas*, *liquid*, and *solid* are mutually in *thermodynamic equilibrium*.

Trouton's rule (294): The empirical observation that for a number of non-polar *liquids*, $\Delta_{vap}S = \Delta_{vap}H/T_{vap} \approx 85 - 95 \, J \, K^{-1} \, mol^{-1}$.

Unfolded state (749): The *state* of a *protein* in which the *polypeptide* chain assumes a random configuration.

Unidirectional (123): Describing a *change in state* that proceeds in one direction, but not the reverse direction. *Spontaneous* changes are necessarily unidirectional, proceeding from a *non-equilibrium state* to an *equilibrium state*.

Universe (5): The totality of the chosen *system* and the entire *surroundings*.

Unstable equilibrium: See *unstable state*.

Unstable state (411): An *equilibrium state* corresponding to the local maximum in Gibbs free energy, such as state C in the accompanying diagram. In principle, state C can be maintained, but, if the system is subject to even an infinitesimal disturbance, state C changes spontaneously to a neighbouring state, such as the *metastable state* B, or the *stable state* A.

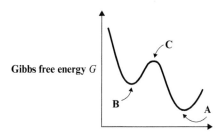

Useful work (27): Alternatively, available work. The *work* performed by a *system* other than any PV *work of expansion* done against the *surroundings*. See also *maximum available work*.

van der Waals force (488): A general term for the short-range attractive and repulsive forces between two neighbouring molecules or atoms, which explain, for example, why *real gases* liquefy when compressed: under compression, the average distance between any pair of gas molecules decreases, until a point is reached at which the attractive forces are greater than the average *thermal energy*, so causing the molecules to coalesce as a liquid. See also *Lennard-Jones potential*.

van der Waals gas (50): A gas that obeys van der Waals's equation

$$\left(P + \frac{an^2}{V^2}\right)(V - nb) = nRT \tag{3.6}$$

Van Slyke equation (560):

$$\beta = 2.303 \frac{K_a [\text{Buf}]_{\text{tot}} [H^+]}{(K_a + [H^+])^2} \tag{17.47}$$

van't Hoff enthalpy (446): The enthalpy change ΔH°, sometimes represented as $\Delta_r H^\circ_{\text{vH}}$, as determined by fitting experimental data to the *van't Hoff equation*, in contrast to being determined from other methods such as *calorimetry*.

van't Hoff equation (445): Alternatively, van't Hoff isochore.

$$\frac{d \ln K_p}{dT} = \frac{\Delta_r H^\circ}{RT^2} \tag{14.31a}$$

van't Hoff equation (osmosis) (588):

$$\Pi = RT \frac{n_D}{V} = RT\,[D] \tag{19.6}$$

van't Hoff isochore (445): An alternative name for the *van't Hoff equation*.

Vapour pressure p Pa (468): The *pressure* exerted by a *vapour*.

Vapour pressure, equilibrium saturated: see *equilibrium saturated vapour pressure*.

Virial coefficient (50): The (temperature-dependent) coefficients B, C ... in the *virial equation-of-state*

$$PV = nRT\left(1 + B\frac{n}{V} + C\frac{n^2}{V^2} + D\frac{n^3}{V^3} + \ldots\right) \tag{3.9}$$

Virial equation-of-state (50):

$$PV = nRT\left(1 + B\frac{n}{V} + C\frac{n^2}{V^2} + D\frac{n^3}{V^3} + \ldots\right) \tag{3.9}$$

Volatility (470): The tendency of a *solid* or *liquid* to evaporate. See also *evaporation*.

Volume, critical: see *critical volume*.

Volumetric expansion coefficient α_V K^{-1} (650):

$$\alpha_V = \frac{1}{V}\left(\frac{\partial V}{\partial T}\right)_P \tag{4.17}$$

Water ionisation constant: see *ionic product of water*.

Wild-type protein (775): A naturally occurring *protein*. See also *mutant protein* and *phi-factor*.

Work (21): A phenomenon observed at the *boundary* of a *system* when that *system* changes *state*, if and only if something occurs at that *boundary* (such as the motion of the *boundary*) which may be interpreted as motion against an external *force*. Work is a *path function*.

Work, available: see *useful work* and *maximum available work*.

Work, frictional: *see frictional work*.

Work, irreversible: see *irreversible work*.

Work, maximum available: see *maximum available work*.

Work, P,V of expansion: see *P,V work of expansion. useful work*.

Work, reversible: see *reversible work*.

Work, useful: see *useful work*.

Working substance (317): The material that is expanded and compressed during the operation of a *heat engine* or *heat pump* – so, for example, steam is the working substance in a steam engine.

X-ray crystallography (775): An experimental technique in which X-rays are diffracted by a specimen, usually in the form of a crystal. Analysis of the resulting diffraction pattern gives information concerning the structure of the specimen.

Zeroth Law of Thermodynamics (55): If two *systems* are each in *thermal equilibrium* with a third *system*, then they will be in *thermal equilibrium* with each other. Each of the three *systems* may be associated with an *intensive state function, temperature*, which, in this case, will take the same value for each *system*.

Bibliography

There are very many texts on thermodynamics in general, and on the various associated fields. Here are just a few ...

Reference works

CRC Handbook of Biochemistry and Molecular Biology. Roger Lundblad and Fiona Macdonald, CRC Press, 4th edition, 2010.

CRC Handbook of Chemistry and Physics. William Haynes, CRC Press, 97th edition, 2016.

IUPAC 'Gold Book'. Release 2.3.3b, 2017. Download available from https://goldbook.iupac.org/.

IUPAC 'Green Book'. RSC Publishing, 3rd edition, 2007.

Lange's Handbook of Chemistry. James Speight, McGraw-Hill Education, 17th edition, 2016.

Perry's Chemical Engineers' Handbook. Don Green and Robert Perry, McGraw-Hill Education, 8th edition, 2007.

The Properties of Gases and Liquids, Bruce E Poling, John M Prausnitz and John P O'Connell, McGraw-Hill, 5th edition, 2001.

Chemical thermodynamics

Four Laws That Drive the Universe. Peter Atkins, Oxford University Press, 2007.

The Laws of Thermodynamics: A Very Short Introduction. Peter Atkins, Oxford University Press, 2010.

Thermodynamics of Chemical Processes. Gareth Price, Oxford University Press, 1998.

Entropy

A Student's Guide to Entropy. Don Lemons, Cambridge University Press, 2013.

Entropy and the Magic Flute. Harold J Morowitz, Oxford University Press, 1997.

The Entropy Principle: Thermodynamics for the Unsatisfied. André Thess, Springer, 2011.

The Second Law. Peter Atkins, Scientific American Library, 1984.

Physical chemistry

Biophysical Chemistry. Alan Cooper, Royal Society of Chemistry, 2nd edition 2011.

Physical Chemistry Volume 1: Thermodynamics and Kinetics. Peter Atkins and Julio de Paula, 9th edition, Oxford University Press, 2010.

Chemical kinetics

An Introduction to Chemical Kinetics. Margaret Robson obson Wright, Wiley-Interscience, 2004.

Non-equilibrium thermodynamics

Modern Thermodynamics: From Heat Engines to Dissipative Structures. Dilip Kondepudi and Ilya Prigogine, Wiley-Blackwell, 2nd edition 2014.

The End of Certainty: Time, Chaos and the New Laws of Nature. Ilya Prigogine, Free Press, New York, 1997.

Biochemistry

Biochemistry. Jeremy M L Berg, John Tymoczko, Gregory Gatto Jr and Lubert Stryer, Palgrave Macmillan, 8th edition, 2015.

Biochemistry. Christopher Mathew, Kensal van Holde, Dean Appling and Spencer Anthony-Cahill. Prentice Hall Canada, 4th edition, 2012.

Molecular Biology of the Cell. Bruce Alberts, Alexander Johnson, Julian Lewis, Martin Raff, Keith Roberts, and Peter Walter. Garland Science, 6th edition 2014.

Principles of Biochemistry. Donald Voet and Charlotte Pratt, John Wiley & Sons, 4th edition, 2012.

Bioenergetics

Bioenergetics. David Nicholls and Stuart Ferguson, Academic Press, 4th edition 2013.

Introducing Biological Energetics. Norman Cheetham, Oxford University Press, 2010.

Protein folding

Structure and Mechanism in Protein Science: A Guide to Enzyme Catalysis and Protein Folding. Alan Fersht, WH Freeman & Co Ltd, 3rd edition, 1999.

Membranes

Membrane Structural Biology: With Biochemical and Biophysical Foundations. Mary Luckey, Cambridge University Press, 2nd edition, 2014.

Climate change

A Farewell to Ice: A Report from the Arctic. Peter Wadhams, Allen Lane, 2016.

Applied Thermodynamics for Meteorologists. Sam Miller, Cambridge University Press, 2015.

Climate Change: The Science of Global Warming and our Energy Future. Edmond A. Mathez and Jason E. Smerdon, Columbia University Press, 2009.

Gaia - The Practical Science of Planetary Medicine. James Lovelock, Gaia Books, 2nd edition, 2000.

The Planet Remade: How Geoengineering Could Change the World. Oliver Morton, Granta Books, 2015.

And ...

Catching Fire: How Cooking made us Human, Richard Wrangham, Basic Books, New York, 2009.

Information: A Very Short Introduction. Luciano Floridi, Oxford University Press, 2010.

Pharmaceutics: The Science of Medicine Design. Edited by Philip Denton and Chris Rostron, Oxford University Press Oxford 2013.

The Oxford Handbook of Philosophy of Time. Craig Callendar, Oxford University Press, 2013.

What is Life? Erwin Schrödinger, Cambridge University Press, Cambridge, 2012.

Websites

https://chem.libretexts.org/Reference/Reference_Tables

http://webbook.nist.gov/chemistry/

Index

p represents an entry on page p.
G(p) represents an entry in the glossary on page p
S*n*(p) represents an entry in the summary of chapter *n*, on page p.
T(p) represents an entry in a table on page p.

A

Absolute entropy G(795), S12(351), 352–357
  and computation of Gibbs free energy 498–500
  values T(356)
Absolute zero G(795), 44, 47
  approach towards 357–362
  and entropy S9(268)
  helium not solid at 355
  impossibility of reaching 44, 361–362
  temperatures less than 73–74
  and Third Law of Thermodynamics S9(268), 297
Acid 538, 538–544
  as hydrogen ion donor 545
  strong 539
  weak 539, 542–544, 550
Acid, conjugate. *See* Conjugate acid; Conjugate base
Acid, monoprotic. *See* Monoprotic acid; Acid
Acid dissociation constant K_a G(795), S17(534), 538–540
  values T(540)
Activated complex G(795), 457
Activated state. *See* Transition state
Activation energy E_a J mol⁻¹ G(795), S14(414), 455–459
  graphical interpretation of 456
  and half reactions 598
  See also Arrhenius equation
Active transport G(795), 726
  and reduction of molecular oxygen in oxidative phosphorylation 726
  See also Electron transport chain
Activity a_i G(795), 523, 657
  and chemical potential 523
  and real solutions 661, 663
  = fugacity for real gas 657
  = 1 for immiscible liquid 525–526
  = 1 for a solid 525–526
  = 1 for water 525
  See also Molal activity; Mole fraction activity; Fugacity coefficient
Activity, molal. *See* Molal activity
Activity, mole fraction. *See* Mole fraction activity
Activity coefficient, mean molal. *See* Mean molal activity coefficient
Activity coefficient, molal. *See* Molal activity coefficient
Activity coefficient, mole fraction. *See* Mole fraction activity coefficient
Activity mass action ratio Γ_r 524, 623, 624, 626–627, 628–631
Activity reaction quotient Q_r. *See* Activity mass action ratio
Activity thermodynamic equilibrium constant K_r 524–526, 619, 621, 624
  transformed 696–697, 707–708
Adiabat G(796), 57
  two adiabats cannot cross S10(304), 321–323
Adiabatic change or path G(796), 57
  and Carnot cycle 230, 231
  equation 224, 226
  expansion of an ideal gas into vacuum entropy change 278–280, 281
  and First Law of Thermodynamics 106
  and ideal gas 221–229
  key ideal gas equations 299
Adiabatic cooling G(796), 99
Adiabatic demagnetization G(796), S12(351), 358–361
Adiabatic work 220–223, 226–228
  and cooling 222
  and warming 223
ADP 395, 527, 567, 713
  and ATP synthase 733–734
  non-equilibrium [ATP]/[ADP] ratio as measure of energy stored 714
Aeolipile 58–59
Affinity, binding. *See* Binding affinity
Agent, oxidising. *See* Oxidising agent
Agent, reducing. *See* Reducing agent
Aggregation
  of proteins 674, S26(780), 780–787
  of viral capsid proteins 787–791
  See also Protein aggregation and polymerisation
Alexandria, Hero, or Heron, of 58
Allotrope G(796), 173
Alloy 503
α-helix G(795), 753
Alzheimer's disease 780
Amino acid G(796), 395, 527, 624, S24(711), 748, 775
  and ΔC_P for protein folding and unfolding 759
  and mutations 775
  sequence determines protein structure 769
Ammonia
  as a base T(546)
  and Brønsted-Lowry theory 546
  as a buffer 550
  liquefaction of 652
Ammonia, synthesis of
  catalysis of 459
  commercial process 449, 459
  composition of equilibrium mixture 441–443
  computation of $\Delta_r G_p^\circ$, K_c and K_p 440–441
  ΔC_p 447–448
  Haber-Bosch process 440
  and Le Chatelier's principle 448–449
  variation of K_p with temperature 443–444, 446, 448, 460
Amontons, Guillaume 46
Amphoteric G(796), 547
Amyloid G(796), 780, 781
Anfinsen, Christian 769
Anion G(796), 504, 666, 667
Anode G(796), S20(591), 600–601
  always on left 602
  defined relative to standard hydrogen electrode 615
  definition independent of material 601, 617
  electrode potential 603
  more positive standard electrode potential 616–617
  oxidation half-reaction at 600
  reversible electrode potential 613–614
  See also Cathode
Anode potential E_{an} V G(796)
  See also Reversible electrode potential
Antenna complex. *See* Light-harvesting complex
Approximation, Henderson-Hasselbalch. *See* Henderson-Hasselbalch approximation

INDEX

Arrhenius, Svante 456
Arrhenius equation G(797), S14(414), 455–459, 771
Assembly, macromolecular. See Macromolecular assembly
Asymmetry 132, 304
 and time 139, 304
Atmosphere
 and chemical reactions 34
 as 'thermal blanket' 396
 work done against 27
Atmosphere Pa (unit-of-measure) 15, 170
Atmospheric pressure Pa 15, 467
Atmospheric pressure, standard Pa. See Standard atmospheric pressure
ATP G(797), 148, 395, 527, 674, S24(711), 712–715
 fine balance of 715
 and glycolysis 570
 'high-energy molecule' misnomer 715
 hydrolysis 567, 568, 707, 713–715
 neither oxidising nor reducing 743
 non-equilibrium [ATP]/[ADP] ratio as measure of energy stored 714
 stoichiometry of ATP synthesis 734–735
 synthesis of, by ATP synthase 733–734
ATP synthase G(797), 724, 733–734
 as a biological nanomachine 734
 and efficiency of oxidative phosphorylation 736
 and photosynthesis 739
 schematic representation of 734, 740
 stoichiometry of ATP synthesis 734–735
 within thylakoid membrane 739
Autocatalytic G(797), 784
Available work 27
 See also Maximum available work; Useful work; Work
Average bond energy. See Mean bond energy; Bond dissociation energy
Average molecular energy, S3(43)
Avogadro constant N_A G(797), 7
 numerical value of 7, 170

Avogadro('s) number. See Avogadro constant

B

Bacterial microcompartment G(797), 792
Balance, charge. See Charge balance
Bar (unit-of-measure) G(797), 15
Barometer, Torricelli. See Torricelli barometer
Base 538, 545
 as hydrogen ion acceptor 546
 strong 545
 weak 545, 550
Base, conjugate. See Conjugate acid; Conjugate base
Base, monohydroxic. See Monohydroxic base; Base
Base dissociation constant K_b G(797), 545
Bassham, James 36
Battery, electrical. See Electrical battery
Beer 720
Belousov-Zhabotinsky reaction 262
Benson, Andrew 36
Berthelot gas G(797), 50
Berthelot's equation 50, 80
β-pleated sheet G(797), 753
Bilayer, lipid. See Lipid bilayer
Binding affinity G(797), S25(747), 764
 See also Dissociation constant
Binding isotherm G(797), 764, 765, 766, 767
Biochemical standard G(797), 615, 618, S23(673), 677–678
 and composition of equilibrium mixture of buffered reaction 707
 and composition of equilibrium mixture of unbuffered reaction 703
 and $\Delta_f G^{\ominus\prime}$ of any molecule containing hydrogen 680, 691–693
 and $\Delta_f G^{\ominus\prime}$ of molecular hydrogen 680, 689–691
 and $\Delta_r G^{\ominus\prime}$ of any reaction involving hydrogen 694–696
 and $\Delta_r G^{\ominus\prime}$ of any reaction not involving hydrogen 695

 and folding and unfolding of proteins 751
 G_{sys} and G_{sys}' for buffered reaction 704–707
 G_{sys} and G_{sys}' for unbuffered reaction 697–704
 importance of 78
 key implications as compared to conventional standards S23(673)
 and K_r and K_r' for any reaction 696–697
 and K_r' of any reaction involving hydrogen 696–697
 and Λ (pH 7) 697
 and reference levels of $\mathbb{E}^{\ominus}(H^+, H_2)$ and $\mathbb{E}^{\ominus\prime}(H^+, H_2)$ 680
 and reference levels of $G^{\ominus}$ and $G^{\ominus\prime}$ of H^+ (aq) ions 680
 rules for conversion from conventional (IUPAC) standards 708
 what's different 670–680
 what's the same 680–681
Biochemical standard state G(798), 176, 618, S23(673), 674, 675
 consequences of 679–681
 definition of 677–678
 See also Biochemical standard; Standard state
Biochemistry 10
 and buffers 562
 key features of biochemical reactions 676
 and pH 2 562, 676
 and pH 7 176, 562, 564, 615, 618, 676, 677
 and reactions that 'shouldn't happen' 395
 See also CBB cycle; Glycolysis; Light-dependent reactions of photosynthesis; Lactic acid; Light-independent reactions of photosynthesis; Oxidative phosphorylation; Pyruvic acid; TCA cycle
Bioenergetics G(798), 527, 674, S24(711), 712
 of electron transport 728–733

Biological membrane G(798), 724
 and active transport 726
 barrier to diffusion 724, 725
 and chemi-osmotic potential 724
 and chloroplast 739
 and control of charges and concentrations 725–726
 enabler and controller of concentration gradient 724, 725–726
 enabler and controller of electrical potential gradient 724, 725–726
 and lipid bilayer 724
 location of oxidative phosphorylation 724, 726
 location of photosynthesis 726, 739, 740, 744
 and mitochondrion 724
 and nervous conduction 726
 and osmosis 726
 and proteins 725
 respiratory 724, 725–726
 schematic representation of respiratory 725
 as semi-permeable membrane 583, 726
 and thylakoid 739
Biological standard state. See Biochemical standard state
Biology, synthetic. See Synthetic biology
Bioluminence G(798), 745
Boiling 67, 472–480
 criterion for 477–478, 572, 575
 difference from freezing 575
 See also Phase change
Boiling point, elevation of. See Elevation of the boiling point
Boltzmann, Ludwig 67, 342
Boltzmann constant k_B J K^{-1} G(798), S3(43), 70, S11(332), 34, 667
 See also Avogadro constant N_A; Ideal gas constant R
Boltzmann distribution G(798), 68
 and negative temperatures 73–74
 and vapour pressure 470
 visualisation 71–72
Boltzmann's equation G(798), S11(332), 341–344
Bomb calorimeter. See Calorimeter
'Bond, high energy'. See 'High-energy bond'

INDEX

Bond, hydrogen. *See* Hydrogen bond
Bond breaking
  endothermic general nature of 186, 188, 457
  enthalpy change of 186–188
  and transition state energy 771
Bond dissociation energy D^o J G(798), 187–188
  values T(188)
  See also Mean bond energy
Bond energy 185–190
  values T(188), T(189)
  See also Bond dissociation energy; Mean bond energy
Bond making
  enthalpy change of 186–188
  exothermic general nature of 186, 188, 457
Bond strength and weakness 189–190
Born-Haber cycle
Boundary G(798), S1(3), 5
  and causality 264, 266
  and closed, isolated and open systems S3(43), 57
  and heat, S3(43), 57, 265
  and information 265, 266
  and matter, S3(43), 57, 265
  and work S2(20), S3(43), 23, 57, 265
Boyle, Robert 16
Boyle's law G(798), 15–16
  breakdown of 16
  and work 31
Branca, Giovanni 58
Brewing 719
Bridge, salt. *See* Salt bridge
Brønsted, Johannes 546
Brønsted-Lowry theory G(798), 546
'Buckyball' 780
Buffer capacity β m^{-3} G(799), 554–561
Buffer solution G(799), S17(534), 549–553
  biological significance 549
  data 554
  'golden rules' 560
  influence on the equilibrium composition of a reaction 562, 565–566, 706–707
  Λ(pH$_{buf}$) factor 563–566, 697, 704–705
  for measurement of kinetics of protein folding and unfolding 773–774
  why a buffer solution maintains constant pH 552–553
BZ reaction. *See* Belousov-Zhabotinsky reaction

C

Calorie Cal or cal (unit-of-measure) 54
Calorimeter G(799), 164
Calorimeter, bomb. *See* Calorimeter
Calorimetry G(799), 163–164, 446, 484
Calorimetry, differential scanning. *See* differential scanning calorimetry
Calvin, Melvin 36
Calvin-Benson-Bassham cycle. *See* CBB cycle
Calvin cycle. *See* CBB cycle
Capsid G(799), S26(780), 789
  self-assembly of 789–791
  stability 789–791
Carathéodory, Constantin 322
Carathéodory statement of the Second Law of Thermodynamics G(799), S10(303), 321–323
  See also Second Law of Thermodynamics; Clausius statement of the Second Law of Thermodynamics; Kelvin-Planck statement of the Second Law of Thermodynamics
Carbon fixation. *See* CO_2 fixation
Carbon-fixation cycle. *See* CBB cycle
Carbon nanotube 780
Carboxysome G(799), 792
Carnot, Sadi 323
Carnot cycle G(799), 209, 230–236, 323
  efficiency of 234–235, 314, 324–325
  efficiency depends only on temperatures of hot and cold reservoirs 325
  and entropy 236
  heat flows 231–232
  net work 232–234
  P,V diagram 231
  T,S diagram 297–298
  work flows 231
Carnot heat engine G(799), S10(304), 323–328
  efficiency of 325–328
  optimum design 325
  and real engine 328–330
  See also Heat engine
Carnot heat pump G(799), 323–328, 357
  efficiency of 325–328
  and Maxwell's demon 345
  optimum design 325
  See also Heat pump
Carnot's theorem G(799), 325
Carotene 738
Catalyst 410, 457, 459
  enzymes as 459, 552, 556, 562, 674, 711, 712, 716, 717, 719, 747
  graphical interpretation of 457
  and graphite electrode 606
  industrial use of 459
  and platinum electrode 606
  and synthesis of ammonia 459
Cathode G(799), S20(591), 600–601
  always on right 602
  defined relative to standard hydrogen electrode 615
  definition independent of material 601, 617
  electrode potential 603
  less positive standard electrode potential 616–617
  reduction half-reaction at 600
  reversible electrode potential 613–614
  See also Anode
Cathode potential E_{cat} V G(800)
Cation G(800), 504, 667
Causality
  and spontaneity 263–266
  and system boundary 264, 266
Cawley, John 58
CBB cycle G(800), 36, 744
Cell potential S20(591)
Cell reaction G(800), S20(591), 602–603
  See also Half-reaction; Redox reaction
Celsius, Anders 44
Celsius scale of temperature t °C G(800), 44, 169
Centigrade scale of temperature 44
Cerium magnesium nitrate paramagnetic salt and adiabatic cooling 357
Chain, electron transport. *See* Electron transport chain
Chain, polypeptide. *See* Polypeptide chain
Chain rule G(800), S21(633), 640–641
  use of 647
Change in state G(800), S1(3), 8, 10–13
  adiabatic 57
  finite 11
  identification of 11
  infinitesimal 11
  isobaric 58
  isochoric 58
  isothermal 57
  measurement of 11
  path independence 12
  quasistatic 30
  representation of 31–34
  and work 23, 24
Chaotrope G(800), 752, 753–757, 773–775
Charge, electrical. *See* Electrical charge
Charge balance G(800), 537, 665
Charge density 730, 731
Charge number z_+, z_- S22(656), 667
Charles, Jacques 46
Chemical activity G(800). *See* Activity
Chemical equilibrium S14(413), 413–461, 421, 423–428, 439–430, S16(502)
  composition of equilibrium mixture S14(413), 423, S16(502)
  effect of pressure on 448–449
  effect of temperature on 448, 459–460
  extent-of-reaction parameter 422
  forced 565, 566, 567, 568
  graphical representation of 422
  kinetic interpretation of S14(414), 454
  and Le Chatelier's principle 448–449
  mathematical determination of 424–427
  multi-state 522–526
  of reaction in solution 517–522
  variation with temperature 444–448
  See also Non-equilibrium system
Chemical kinetics G(800), S14(414), 449, 450
  and chemical equilibrium S14(414)
  and thermodynamics S14(414), 449–455

INDEX

Chemical potential μ_i
J mol^{-1} G(800),
S13(366), 403–406
and (chemical) activity
523
and First Law of
Thermodynamics
406–408
and Gibbs free energy
404
and phase equilibrium
472
Chemical reaction 262–263,
S14(413), 415–417
away from equilibrium
432–436
buffered 549, 552–553
effect of pressure on
448–449
effect of temperature on
448, 458, 459–460
and extent-of-reaction
parameter 417–423
general equilibrium
condition 461
Gibbs free energy of
reaction mixture
416–417, 417–423,
424
graphical representation
of 420, 422, 435, 457
involving hydrogen
or hydroxyl ions
554, 561–566, 676,
697–707
key features of bio-
chemical reactions
676
kinetic and thermo-
dynamic influences
on 457–459
microscopic interpret-
ation of 450, 451, 456,
458, 460
response to external
constraint S14(414)
time dependence of
Gibbs free energy
417–423
See also Chemical
equilibrium,
Non-equilibrium
system; Reversible
chemical reaction
Chemical reaction, forward.
See Forward chemical
reaction
Chemical reaction, reverse.
See Reverse chemical
reaction
Chemical reactions,
coupled. *See* Coupled
reactions
Chemi-osmotic potential
G_{co}' J mol^{-1} G(800),
723, 730–733
Chemi-osmotic potential
gradient $\Delta G_{co}'$
J mol^{-1} G(801),

S24(711), 716,
730–732
comparison of
photosynthetic and
mitochondrial 742
and cyclic photo-
phosphorylation
743–744
harnessing of by ATP
synthase 733
and non-cyclic photo-
phosphorylation 740,
741–742
and photosynthesis 740,
741–744
values for mitochondrial
and photosynthetic
732, 742
Chemistry
as a process in which
the total Gibbs free
energy of a mixture
changes over time
as the reaction
progresses 402
Chevron plot G(801),
774–775
Chlorophyll 738, 739
colour of 738
Chloroplast G(801) 739
and biological
membrane 739
Circular dichroism G(801),
753–754, 773
Citric acid cycle. *See*
Tricarboxylic acid
cycle
Clapeyron, Émile 47
Clapeyron equation G(801),
S15(467), 481–485
Clausius, Rudolf 273, 304,
323
Clausius-Clapeyron
equation G(801),
S15(467), 482–485
and depression of the
freezing point 576,
577–578
and elevation of the
boiling point 572–574
graphical representation
of 484–485
Clausius inequality G(801),
S9(267), 272–273, 304
Clausius statement of
the Second Law of
Thermodynamics
G(801), S10(303),
304–309
comparison to Kelvin-
Planck statement
311
graphical interpretation
of 320–321
and irreversibility 316
Climate change 395–396

Closed system G(801),
S3(43), 57, 85–86
criterion for spontan-
eous change 275, 369,
370, 374–375
and entropy changes
367–374
and First Law of
Thermodynamics 95,
104–105, 106–107
and Gibbs free energy
374–375
and spontaneous
changes 369–374
CO_2 fixation G(801), 739,
744, 792
Coefficient, Joule-
Thomson. *See*
Joule-Thomson
coefficient
Coefficient, mean molal
activity. *See* Mean
molal activity
coefficient
Coefficient, molal activity.
See Molal activity
coefficient
Coefficient, mole fraction
activity. *See* Mole
fraction activity
coefficient
Coefficient, transmission.
See Transmission
coefficient
Coefficient, virial. *See* Virial
coefficient
Coefficient, volumetric
expansion. *See* Volu-
metric expansion
coefficient
Coenzyme A 720, 721
Coenzyme Q 727
Cold reservoir 64, 65,
S10(303), 311, 317
Colligative properties
G(802), 579
Colour 738
of photosynthetic
pigments 738
Combustion 95, 154–157,
160, 164, 165,
167–168, 177, 185,
201–204, 439–440
activation energy of 458
enthalpy of 177
Gibbs free energy of 396
and glycolysis 716, 737
self-sustaining 459
Complex, antenna. *See*
Light-harvesting
complex
Complex, light-harvesting.
See Light-harvesting
complex
Complex, protein-ligand.
See Protein-ligand
complex
Compressibility, isother-
mal. *See* Isothermal
compressibility

Compressibility factor Z
G(802), S21(634),
652–654
ideal gas 652
of real gas 653–654, 656
of van der Waals gas
652–653, 654
Concentration. *See* Molar
concentration
Concentration, molar. *See*
Molar concentration
Concentration mass action
ratio Γ_c G(802), 439,
522
and mass action effect
567
transformed 702–703
Concentration reaction
quotient Q_c. *See*
Concentration mass
action ratio
Concentration thermo-
dynamic equilibrium
constant K_c G(802),
438, S14(413), 521
and biochemical
standard 696–697
in buffered solution
S17(534), 564–566,
697
constrained S17(534),
565
constrained at pH 0 564,
565, 677
constrained at pH 7 565,
677
forced 565, 566, 567, 568
and K_c' 703, 705–707
and K_x 439
and K_p 438–439
and $\Lambda(pH_{buf})$ factor
564–566, 697
and mass action effect
566–567
and reaction rate
constants 455
of redox reaction
619–620
significance of pH 0 564,
677, 678, 679
transformed S23(673),
681–682
Conduction 54
Conduction, nervous. *See*
Nervous conduction
Conductor G(802), 592
Conjugate acid G(802),
545–548
Conjugate base G(802),
545–548
Conjugate pairs G(802), 547
Conjugate variables G(802),
299, 639
Conservation, law of. *See*
Law of conservation
Conservation of energy
53–54, 94, 129, 134,
304
Constant, Avogadro. *See*
Avogadro constant

Constant, Boltzmann. *See* Boltzmann constant
Constant, dissociation. *See* Dissociation constant
Constant, equilibrium. *See* Concentration thermodynamic equilibrium constant; Molality thermodynamic equilibrium constant; Mole fraction thermodynamic equilibrium constant; Pressure thermodynamic equilibrium constant
Constant, Faraday. *See* Faraday constant
Constant, ideal gas. *See* Ideal gas constant
Constrained equilibrium S17(534), 561–568
and Λ(pH$_{buf}$) factor S17(534), 565
and mass action effect 566–568
Construction, Maxwell. *See* Maxwell construction
Contact, thermal. *See* Thermal contact
Contact energy G(803), S26(780), 786, 790–791
Convection 54
Convention, sign. *See* Sign convention
Conventional standard state G(803), 176, S23(673)
Conventional standards G(803), 176, 675
key implications as compared to biochemical standard S23(673)
rules for conversion to biochemical standard 708
Cooling, adiabatic. *See* Adiabatic cooling
Copper electrode 596
See also Daniell cell
Coupled reactions 526–530
importance in biochemistry and biology 526, 527, S24(711), 712
See also Coupled system
Coupled system G(803), S13(366), 391–393, 527–530
graphical representation of 530
and life 674, S24(711)
See also Coupled reactions
Cramp. *See* Muscle cramp
Creutzfeldt-Jakob disease 780

Critical point G(803), 480, 490
Critical pressure P_{crit} Pa G(803), 480
of van der Waals gas 490–491
Critical temperature T_{crit} K G(803), 480, 489
of van der Waals gas 489–491, 492, 652, 658, 781
Critical volume V_{crit} m^3 G(803)
of van der Waals gas 490–491
Cryogenics 357
Cryoscopic constant E_f K kg mol^{-1} G(803), S18(571), 578
values T(578)
Crystalline solid
entropy of S12(351), 353, 362
formation of as example of self-assembly 780
Crystallography, X-ray. *See* X-ray crystallography
Currency, energy. *See* Energy currency
Current, electrical. *See* Electrical current
Cycle G(803), 36
See also Thermodynamic cycle
Cycle, Calvin. *See* CBB cycle
Cycle, Calvin-Benson-Bassham. *See* CBB cycle
Cycle, carbon-fixation. *See* CBB cycle
Cycle, Carnot. *See* Carnot cycle
Cycle, citric acid. *See* Tricarboxylic acid cycle
Cycle, irreversible. *See* Thermodynamic cycle
Cycle, Krebs. *See* Tricarboxylic acid cycle
Cycle, reversible. *See* Thermodynamic cycle
Cycle, TCA. *See* Tricarboxylic acid cycle
Cycle, thermodynamic. *See* Thermodynamic cycle
Cyclic integral 37
and path functions 39, 230
and state functions, 37–38, 230, 236

Cyclic photophosphorylation G(804), 740, 743–744
overall bioenergetics of 744
schematic representation of 744
Cytochrome c 727

D

Dalton's law of partial pressures G(804), 48–49, 159–160, 418, 451
Damped harmonic motion 250
Daniell, John 593
Daniell cell G(804), 599–600
chemistry of 602–603
and Nernst equation 624–625
operating as electrolytic cell 608–609
operating as Galvanic cell 607
reversal of electrode polarity 609
See also Electrochemical cell; Electrode; Half-reaction
Dark reactions, of photosynthesis. *See* Light-independent reactions of photosynthesis
Data, normalised. *See* Normalised
Debye, Peter 296, 355, 361
Debye-Hückel law, extended. *See* Extended Debye-Hückel law
Debye-Hückel limiting law G(804), 667
Debye temperature T_D or Θ_D K G(804), 355
Degeneracy G(804), 790
Degree 44
Demagnetization, adiabatic. *See* Adiabatic demagnetization
Denatured state G(804), 562, 749
Depression of the freezing point G(805), 479–480, S18(571), 575–579
and Clausius-Clapeyron equation 576, 577–578
and Raoult's law 578
Desalination G(805), 588
Detergent 752
Diameter, effective. *See* Effective diameter
Diamond
order of 355–356

'trapped' state 410
Diatomic gas
heat capacity at constant volume 108–109, 111–112
Dichroism. *See* Circular dichroism
Dieterici gas G(805), 50
Dieterici's equation 50
Differential G(805), 635
Differential, exact. *See* Exact differential
Differential, inexact. *See* Inexact differential
Differential, total. *See* Total differential
Differential scanning colorimetry (DSC) G(805), 759–760
Diffusion 581
Dilute solution 504, 514–517, 520–522
and mole fraction of water 171, 514
Dilution, infinite. *See* Infinite dilution
Dilution law, Ostwald's. *See* Ostwald's dilution law
Dimer G(805)
Dipole, magnetic. *See* Magnetic dipole
Disease, Alzheimer's. *See* Alzheimer's disease
Disease, Creutzfeldt-Jakob. *See* Creutzfeldt-Jakob disease
Disease, Huntingdon's. *See* Huntingdon's disease
Disease, "mad-cow". *See* "Mad-cow" disease
Disease, Parkinson's. *See* Parkinson's disease
Disorder S11(332), 333–334
and adiabatic demagnetisation 357–361
and Boltzmann's equation 341–344
and entropy S11(332), 357
measurement of 337–340
and multiplicity 337
overcome by energy flow 394
and phase changes 343–344
and time 347
See also Order
Dissipative effect G(805), S5(93), 138, 139
and biochemical processes 735, 736, 738
examples of 139, 249, 252–254
and irreversible paths 125, 141–143, 150
and Joule heating 594

INDEX

Dissipative heat S5(93)
Dissociation
  of acid 535–536
  of protein-ligand complex 764, 765–769
Dissociation constant K_d G(805), 535
  and binding affinity S25(747), 764
  of protein-ligand binding S25(747), 764
Dissociation constant, acid. *See* Acid dissociation contant
Dissociation constant, base. *See* Base dissociation constant
Dissociation energy. *See* Bond dissociation energy
Distribution, Boltzmann. *See* Boltzmann distribution
DNA 748
Drug design 674, 763, 764, 781
DSC. *See* Differential scanning calorimetry
Duhem, Pierre 405
Dynamic equilibrium G(806), 451
Dynamic friction G(806), 28, S5(93), 128, 248
  always acts against motion 128, 248
  characteristics of 128–129

E

Ebullioscopic constant E_b K kg mol^{-1} G(806), S18(571), 575
  values T(575)
Economics 10, S11(332), 348
Effect, dissipative. *See* Dissipative effect
Effect, Seebeck. *See* Seebeck effect
Effective diameter d m, or nm G(806), 668
Efficiency η G(806), 63
  of car patrol engine 737
  of Carnot cycle 234–235, 314
  Carnot engine is the most efficient engine or pump 328
  of electron transport 733
  of glucose metabolism 736–737
  of glycolysis 719, 737
  of heat engine S(10)304, 314
  of heat pump 314–315

$\eta_{rev} > \eta_{irrev}$ 330
  must be < 1 or < 100% 314
  of oxidative phosphorylation 735–736, 737
  real engine less efficient than Carnot engine 330
  of wind turbine 737
  and work 143
Einstein, Albert 355
Electrical battery 592, 593, 599
  chemical reaction within 594, 596
  useful life of 612
  See also Electrochemical cell
Electrical charge Q C 595
Electrical current I A G(806), 592
  and flow of electrical charge 595, 601
  flow of electrons in opposite direction 601
  and heat 592, 593, 594–595
Electrical heat 592–593
  See also Joule heating
Electrical potential E V G(806), S20(591)
Electrical potential difference V V G(806), 599, 603
  across biological membrane 724
Electrical resistance R Ω G(807), 592, 594–595
Electrical work 23, 387, 592–594
Electrochemical cell G(807), S20(591), 599–600
  representation of 602, 604
  thermodynamics of 621–623
  See also Electrolytic electrochemical cell; Galvanic electrochemical cell
Electrochemical cell, electrolytic. *See* Electrolytic electrochemical cell
Electrochemical cell, Galvanic. *See* Galvanic electrochemical cell
Electrode G(807), S20(591), 596
  copper 596
  all electrodes are redox electrodes 607
  gas 606

graphite 606
hydrogen 606–607
inert 606
metal 596, 605–606
platinum 606
redox 606–607
spontaneous accumulation of electrical charge 596, 598
standard hydrogen 614–615
zinc 596
See also Anode; Cathode
Electrode potential E, E_{el}, $\mathbb{E}$, $\mathbb{E}^\ominus$ V G(807), S20(591), 603
  and associated half-reaction 603
  intensive state function 604, 614
  and Nernst equation 614, 615–627
  and oxidising agents 620, 630
  and reducing agents 620, 630
  representation as $\mathbb{E}$ (ox, red) and $\mathbb{E}^\ominus$ (ox, red) 619, 626
  reversible 613
  standard reference level, biochemical standard 687
  standard reference level, conventional standards 615, 617–619
  standard reversible 614–619
  standard reversible hydrogen 615
  standard reversible redox 617, 625–626, 630
  standard reversible reduction 619
  standard reversible values T(616), T(617)
Electrolysis G(807), 609–611
Electrolyte G(807), S20(591), 596, 601–602
Electrolytic electrochemical cell G(808), 609–612
Electromagnetic energy 738
  See also Light
Electromotive force, $\mathbb{E}$ V G(808), S20(591), 613
  and maximum work 613
Electron
  accumulation on anode 596, 597, 600, 602
  consumption at cathode 598, 600, 602
  consumption by reduction half-reaction 600
  excitation of on absorption of a

quantum of light energy 739
molar concentration of 627
production by oxidation half-reaction 600
representation in Nernst equation 627, 687
Electron flow
  and dξ 622
  in opposite direction of electrical current 601
Electron transport chain of mitochondrial respiratory membrane G(808)
  bioenergetics of 728–733
  efficiency of 733
  within mitochondrial respiratory membrane 724, 726–728
  schematic representation of 728
Electron transport chains of thylakoid photosynthetic membrane
  of photosystem I 741
  of photosystem II 740
  schematic representation of 740
  and synthesis of NADPH 739–742, 743
Electroplating G(808), 609, 611–612
Electrostatic potential 762, 763
Elevation of the boiling point G(808), 169, 479, S18(571), 572–575
  and Raoult's law 574
Ellipticity. *See* Mean residue ellipticity; Circular dichroism
Elongation G(808), S26(780)
  energetically favourable 783, 786
  phase of polymerisation 782, 783, 784
  phase of virus capsid assembly 789
  step in Finke-Watzky model 784
emf. *See* Electromotive force
Endergonic G(808), S13(365), 391, 527, S24(711)
  See also Coupled reactions
Endothermic G(808), S6(146), 154
  effect of temperature 460
  nature of bond breaking 186, 188, 457
  and phase change 156

INDEX

Energy G(808), S1(3), S2(20), S3(43), 40
average and total 72–73
average molecular S3(43), 68–69
conservation of 53–54, 94
and First Law of Thermodynamics S5(93)
flow with respect to closed system 57
flow with respect to creating and maintaining order 350, 394
flow with respect to isolated system 57
flow with respect to making system hotter 394
flow with respect to open system 57
flow with respect to system boundary 57, 265, 266
as 'heat-in-waiting' 53, 54
of light 738
molecular 68
molecular distribution 67, 70–71
neither created nor destroyed S5(93), 94
of the sun 738
and temperature 70–71
as 'work-in-waiting' 40, 54
See also Gibbs free energy; Helmholtz free energy; Internal energy
Energy, activation. *See* Activation energy
Energy, bond. *See* Bond dissociation energy; Mean bond energy
Energy, dissociation. *See* Bond dissociation energy
Energy, electromagnetic. *See* Electromagnetic energy
Energy, internal. *See* Internal energy
Energy currency G(808), 674, S24(711), 712, 715, 743
and appropriate magnitudes of $\Delta_r G_c^{\circ}$ and $\Delta_r G_c^{\circ\prime}$ 715
Energy 'hill' 409–410, 456–457
graphical interpretation of 457
and polymerisation 783
and protein folding and unfolding 770–772, 775–776

See also Activation energy
Engine, Carnot. *See* Carnot heat engine
Engine, heat. *See* Heat engine
Engine, petrol. *See* Petrol engine
Engine, steam. *See* Steam engine
Enthalpy H J G(808), 81, S6(146), 148–150
absolute value 176, 179
constant for throttling process 649
definition 149
function of temperature only (ideal gas) 153, 209–210, 214, 644–645
as a function of two variables 81
and heat capacity at constant pressure 214–216
impossibility of measuring absolute value 176
molar 149
standard molar of an element 179
Enthalpy, van't Hoff. *See* van't Hoff enthalpy
Enthalpy change ΔH J S6(146)
of bond breaking and making 186–188, 457
and ΔS 370–373, 761–762
and ΔU 158–160, 164
and directionality 157
effects of hydrogen bonding in water 509–510, 762
and endothermic reaction S6(146), 154
and exothermic reaction S6(146), 154
and heat flow at constant pressure S6(146), 149, 150, 216
independence of reference level 176
key ideal gas equations 299
and magnitude of $T\Delta S$ 378
and maximum heat 150, 216
measurement of 163–164
and open system 408
of phase change 161–163, S9(268), 291–296, 373–374
and protein aggregation and polymerisation 786–787
and reaction stoichiometry 155

and spontaneity 150, 377–378
temperature variation 190–194, 197–198
unit-of-measure 155
Enthalpy-driven reaction S13(365), 377–378
and human activity 396
and protein aggregation and polymerisation 786
Enthalpy of combustion $\Delta_c H$ J G(809), 177
Enthalpy of dilution $\Delta_{dil} H$ J G(809), 177
Enthalpy of formation $\Delta_f H$ J G(809), S6(146), 177
Enthalpy of formation, standard. *See* Standard enthalpy of formation
Enthalpy of fusion $\Delta_{fus} H$ J G(809), 161, 177, 193–194
estimation of 484
values T(162), T(293)
Enthalpy of reaction $\Delta_r H$ J 177
relative magnitudes of $\Delta_r H$ and ΔC_P 193, 200
temperature variation 190–194, 197
See also Standard enthalpy of reaction
Enthalpy of reaction, standard. *See* Standard enthalpy of reaction
Enthalpy of sublimation $\Delta_{sub} H$ J G(809), 161, 163, 177, 193–194
estimation of using Clausius-Clapeyron equation 484–485
Enthalpy of vaporization $\Delta_{vap} H$ J G(809), 161, 177, 193–194
estimation of using Clausius-Clapeyron equation 484–485
temperature variation 193–194
values T(162),T(293)
Entropy S J K^{-1} G(809), 81
absolute S12(351), 355
at absolute zero S9(268), 297, 353
and adiabatic de-magnetisation 358–361
and Boltzmann's equation 342
and Carnot cycle 236
of closed system 275
of crystalline solid 353

$\mathrm{d}q_{\text{rev}}/T$ is a state function 226, 269–272
and economics 348
of gases, liquids and solids 295, 352
and heat capacity at constant pressure S9(268), 283
and heat capacity at constant volume S9(268), 283
and high performing teams 348–350
increase with temperature 283, 284, 297
and information 346
of isolated system, and criterion of spontaneous change S9(267), 275
of mixing 421
molar 269, 356
of open system 275
and order and disorder S11(332), 333, 352, 357, 360, 762
and phase change S9(268), 291–296
of proteins 762
and real change 273–275
and role of leadership 350
and Second Law of Thermodynamics S9(267), 268, 269
standard molar 355
standard molar values T(356)
T, S diagram 297–299
temperature dependence S9(268), 290–291, 294–296, 297, 354
theoretical calculation of 355
Third Law S12(351), 355
and Third Law of Thermodynamics S9(268)
and time S11(332), 347
of universe increases S9(267), 274, 304
Entropy, absolute. *See* Absolute entropy
Entropy, Hartley. *See* Hartley entropy
Entropy, Third Law. *See* Absolute entropy
Entropy change ΔS J K^{-1}
adiabatic expansion of an ideal gas into a vacuum 278–280, 281
and closed systems 367–374

INDEX

Entropy change ΔS J K^{-1} (Contd.)
  and ΔH 370–373, 761–762
  heat flow down temperature gradient 275–278, 281–282
  and human activity 396
  ideal gas equations 285–290
  and phase change S9(268), 291–296, 373–374
  and relative magnitudes of ΔH and $T\Delta S$ 377–378
  standard molar 357
  in surroundings 367–368
  values at phase change T(293)
  and water 509–510, 762, 786–787, 789
Entropy-driven reaction S13(365), 377–378
Enzyme G(809), 459, 461, 552, 556, 561, 562, S24(711), 711, 712, 716, 717, 719, 720, 725, 741, 747, 749
  ATP synthase 724, 733–735, 739–740
  lactic dehydrogenase 720
  and ligand S25(747)
  luciferase 745
  pepsin 676
  pH required 562, 674
  pyruvate dehydrogenase 720
  reversible folding and unfolding of ribonuclease 769
  ribonuclease 769
  RuBisCO 744, 792
Equation, Arrhenius. See Arrhenius equation
Equation, Berthelot's. See Berthelot gas
Equation, Boltzmann's. See Boltzmann's equation
Equation, Clapeyron. See Clapeyron equation
Equation, Clausius-Clapeyron. See Clausius-Clapeyron equation
Equation, Dieterici's. See Dieterici gas
Equation, Henderson-Hasselbalch. See Henderson-Hasselbalch equation
Equation, Kirchhoff's first. See Kirchhoff's first equation
Equation, Kirchhoff's second. See Kirchhoff's second equation
Equation, Nernst. See Nernst equation

Equation, Shomate. See Shomate equation
Equation, thermodynamic of state. See Thermodynamic equation-of-state
Equation, van der Waals. See van der Waals equation
Equation, van't Hoff. See van't Hoff equation
Equation, virial. See Virial equation
Equation-of-state G(810), 16, 50, 107
  Berthelot's equation 50
  Boyle's law 16
  and computation of work 31
  Dieterici's equation 50
  of ideal gas S3(43), 47, 50, 209
  of liquid 50
  and partial derivatives 79
  of real gas 20
  of solid 50
  van der Waals's equation 16, 50
  virial equation 50
Equation-of-state, thermodynamic. See Thermodynamic equation-of-state
Equilibrium 8–10
  and change ΔG in a closed system S13(365), 375
  criterion for in closed system S13(365), 375
  criterion for in isolated system S9(267)
  general condition for 461
  and measurement 9
  movement of non-equilibrium states towards 9, 45, 260, 261–263
  of phases S15(467), 472
  and time 8–9
  See also Chemical equilibrium; Equilibrium state; Thermal equilibrium; Thermodynamic equilibrium
Equilibrium, chemical. See Chemical equilibrium
Equilibrium, constrained. See Constrained equilibrium
Equilibrium, dynamic. See Dynamic equilibrium
Equilibrium, forced. See Constrained equilibrium
Equilibrium, metastable. See Metastable equilibrium
Equilibrium, phase. See Phase equilibrium

Equilibrium, pseudo-. See Pseudo-equilibrium
Equilibrium, stable. See Stable equilibrium
Equilibrium, thermal. See Thermal equilibrium
Equilibrium, thermodynamic. See Thermodynamic equilibrium
Equilibrium, unstable. See Unstable equilibrium
Equilibrium constant K G(810), S14(413), S16(502), S20(592)
  and $\Delta G^{\ominus}$ 461
  and $\Delta_r G_p^{\ominus}$ 523–526
  generalised K_r 523–526, 568, 619, 621, 624, 628
  molality K_b 437–439
  molar concentration K_c 437–439, 521–522
  mole fraction K_x 436–437, 519–520
  pressure K_p 428–430
  and reaction rate constants 455
  temperature variation 444–448
  transformed $K_c{'}$ and $K_r{'}$ 682, 696–697, 703–707, 707–708
Equilibrium constant, activity. See Activity thermodynamic equilibrium constant
Equilibrium constant, hybrid. See Hybrid equilibrium constant
Equilibrium constant, concentration. See Concentration thermodynamic equilibrium constant
Equilibrium constant, molality. See Molality thermodynamic equilibrium constant
Equilibrium constant, mole fraction. See Mole fraction thermodynamic equilibrium constant
Equilibrium constant, pressure. See Pressure thermodynamic equilibrium constant
Equilibrium extent-of-reaction parameter ξ_{eq} G(810), 422, 518, 528–530, 703, 706–707
  and acids 539, 542, 542–543
  and buffer solutions 550–551

and coupled reactions 529
difference between ξ_{eq} and $\xi_{eq}{'}$ for buffered reaction 707
and dissociation reactions 535–536
equality of ξ_{eq} and $\xi_{eq}{'}$ for unbuffered reaction 703
and Γ_p 432
graphical representation of 420, 422, 435, 518, 530
inequality of ξ_{eq}(pH 7) and $\xi_{eq}{'} = \xi_{eq}$(pH 0) for buffered reaction 707
and K_p 431, 432, 442–443
See also Extent-of-reaction parameter ξ
Equilibrium saturated vapour pressure p Pa G(810), S15(467), 469, 475
  and Roault's law 505
Equilibrium state G(810), 8–10
  and non-equilibrium state 254
  and Second Law of Thermodynamics S9(267), 268, 269
Equilibrium thermodynamics G(810), 10
Equivalent of heat, mechanical. See Mechanical equivalent of heat
Evaporation G(810), 467–468, 470
  See also Vapour pressure; Equilibrium saturated vapour pressure
Exact differential G(810), S1(3), and state functions 12
Excited state of an electron 738–739
  and dissipation of energy as heat 738
  and fluorescence 739
  and resonance energy transfer 739
Exciton transfer. See Resonance energy transfer
Exergonic G(810), S13(365), 390, 527, S24(711)
  See also Coupled reactions
Exothermic G(811), S6(146), 154
  effect of temperature 460
  and exergonic change 391

nature of bond making 186, 188, 457
self-sustaining combustion 459
Expansion, free. See Free expansion
Expansion, Joule. See Free expansion
Expansion coefficient, volumetric. See Volumetric expansion coefficient
Explosion 200, 203–206
Extended Debye-Hückel law G(811), 668
Extensive state function G(811), S1(3), 6
additive nature of 6–7
constant over time at equilibrium 9
equilibrium conditions 9
and the ideal concept 8
intensive molar equivalent 7
list of 634
non-additive examples 8
Extent-of-reaction parameter ξ G(811), 417–426, 518, 530, 697, 705
away from equilibrium 432, 518
$d\xi$ and flow of electrons in electrochemical cell 622
at equilibrium 422, 518
and Γ_p 433–434
graphical representation of 420, 422, 435, 518, 530, 700, 706, 714
and hydrolysis of ATP 713–714
See also Equilibrium extent-of-reaction parameter ξ_{eq}

F

Factor, compressibility. See Compressibility factor
$FADH_2$ 715, 720–723
and bioenergetics of electron transport 723–728
as a reducing agent 723
and reduction of molecular oxygen in oxidative phosphorylation 726–728
redox potential of [FAD]/[$FADH_2$] ratio 723–724
Fahrenheit, Daniel 44
Fahrenheit scale of temperature °F G(811), 44, 174
Faraday, Michael 593

Faraday constant, F C mol^{-1} G(811), 595
Ferromagnetic material G(811), 358
See also Paramagnetic material
Field, magnetic. See Magnetic field
Finke-Watzky model G(811), S26(780), 783–785
elongation step 784
nucleation step 783–784
Fire 147–148
First law, Joule's. See Joule's first law
First Law of Thermodynamics G(811), 55, S5(93), 93–144, 94
and adiabatic change 106
and chemical potential 409–408
and closed system 95, 104–105, 106–107
combined with Second Law of Thermodynamics S9(268), 280–282, 407
and conservation of energy 94, 129
and directionality 134
and friction 129–132, 134
and heat 94
and ideal gas 208
and isobaric change 114–116
and isochoric change 106
and isolated system 105
and isothermal change 114
and open system 95, 104, 406–408
questions addressed by 95
and reversible and irreversible changes 143–144, 260
sign conventions 96
and spontaneity 157
structure of 94
and thermodynamic cycles 101–102
and work 94
First thermodynamic equation-of-state G(811), S21(633), 642
Fixation. See CO_2 fixation
Flame 4, 200–203
Fluorescence G(811), 739, 764, 773
See also Intrinsic fluorescence
Folded state G(811), S25(747)
Foldon G(812), 778
Force F N G(812), 14
frictional 28
and pressure 21–22
and work 20–21

Force, proton motive. See Proton motive force
Forced equilibrium. See Constrained equilibrium
Forward chemical reaction S14(414), 416, 450
equilibrium condition 454–455
graphical representation of 453, 454
reaction rate of 450–454
Forward rate of reaction v_f S14(414), 450–454
graphical representation of 453, 454
'Fourth Law' of Thermodynamics G(812), S13(367), 408–411, 674
application to protein configuration 674, 748
Fragmentation step in three-step kinetic model of polymerisation 785
Free energy. See Gibbs free energy; Helmholtz free energy
Free expansion. See Joule expansion
Freezing 575
difference from boiling 575
'Freezing' of time 24
Freezing point, depression of. See Depression of the freezing point
Frequency factor. See Pre-exponential factor
Frequency of light v sec^{-1} 738
Friction G(812), 23, 28, 128–140
always acts against motion 128, 248
analogy with electricity 593–594
asymmetry of 129–132
and irreversible paths 125, 134–140, 142–144
mathematics of 132–134, 251
and real paths 140–144
and the thermodynamic pendulum 248–254
Friction, internal. See Internal friction
Frictional heat G(812), 51–52, 129
attributable to frictional work 129, 130–132, 134, 139, 142–144, 254, 394
and unidirectionality 129

Frictional work G(812), 23, 28, 51–53, 128
always positive 133, 134, 139, 252, 253
and direction of time 139–140
dissipation as heat 129, 130–132, 134, 139, 142–144, 254, 394
increases with time 134
mathematics of 132–134
and unidirectionality 129
Fugacity f Pa G(812), S22(655), 657–659, 668
Fugacity coefficient ϕ G(812), S22(655), 657–659
Function G(812), 77, 79
of one variable 77–78
of two (or more) variables 79–80
Function, conjugate. See Conjugate variables
Function, extensive. See Extensive function
Function, intensive. See Intensive function
Function, path. See Path function
Function, potential. See Potential function
Function, state. See State function
Fundamental equation of thermodynamics S(812), G13(367), 406

G

Gadolinium sulphate paramagnetic salt and adiabatic cooling 357, 361
Galvani, Luigi 608
Galvanic electrochemical cell G(813), 599, 607
and electromotive force 613
reversible operation 613
Gas G(813), 17
compressibility of 491, 647
degree of disorder 344, 356–357
entropy of 291–296, 352, 356
existence independent of associated liquid or solid 475
Gibbs free energy of 472–473, 475–478, 497–499
liquefaction of 474
molecular motions within 17–18, 95, 352
See also Ideal gas; Phase change; Real gas

INDEX

Gas, Berthelot. *See* Berthelot gas
Gas, Dieterici. *See* Dieterici gas
Gas, greenhouse. *See* Greenhouse gas
Gas, ideal. *See* Ideal gas
Gas, monatomic. *See* Monatomic gas
Gas, perfect. *See* Ideal gas
Gas, real. *See* Non-ideal gas
Gas, supercooled. *See* Supercoooled gas
Gas, van der Waals. *See* van der Waals gas
Gas, virial. *See* Virial equation-of-state
Gas constant. *See* Ideal gas constant
Gas constant, universal. *See* Ideal gas constant
Gas electrode G(813), 606
GdmCl. *See* Guanidinium chloride
Giauque, William 361
Gibbs, Josiah Willard 375
Gibbs-Duhem equation G(813), S13(366), 405
Gibbs free energy G J G(813), 81, S13(365)
 change of as a reaction progresses 402, 417–423
 and chemical potential 404
 of component in an ideal solution G16(502), 510–517
 criterion for phase equilibrium S15(467), 471–472, 494, 511, 581
 and 'Fourth Law' of Thermodynamics S13(365), 408
 G_{sys} and G_{sys}' for a buffered reaction 704 707
 G_{sys} and G_{sys}' for an unbuffered reaction 697–704
 graphical representation of chemical equilibrium 422
 graphical representation of chemical reaction 420, 422, 435
 and hydrostatic pressure 588–600
 of ideal solution 517–520
 impossibility of absolute measurement 397
 minimisation of S13(367), 408–411, S14(413), 422, G25(747), 748, 770

 of mixtures 401–408, S14(413), 416–417, 580–582
 and osmosis 582–583, 583–587
 and phase change 471–472
 pressure dependence of 584–586
 of solids, liquids and gases 472–478, 497–500
 time dependence of 417–423
 of van der Waals gas 384, 493–494
Gibbs free energy, reaction $\Delta_r G$ J. *See* Reaction Gibbs free energy
Gibbs free energy change ΔG J S13(365)
 and contact between proteins 790–791
 and coupled systems 391–393
 and equilibrium in closed systems S13(365), 375
 and Joule heating 594
 and maximum available work 389, 594
 and open system 406
 and polymerisation 783
 and spontaneity in closed systems S13(365), 375
 why reaction doesn't happen even when $\Delta G < 0$ S14(413-414)
Gibbs-Helmholtz equation G(813), S13(366), 382
Global warming. *See* Climate change
Glucose 57, S24(711)
 biological oxidation of 148, 716
 combustion in air 716, 736
 and glycolysis 570
 and maximum work 737
 as near quasistatic process 737
 overall efficiency of metabolism 736–737
 photosynthesis of 532
 and reversibility 736–737
 schematic representation of overall metabolism 735
Glycolysis G(813), 570, S24(711), 716–719
 efficiency of 719, 737
 location within cell cytoplasm 724
 metabolic pathway 717
 reversed in photosynthesis 738
 schematic representation of combined

metabolic pathway with TCA 735
 and TCA cycle 716, 720, 723
 thermodynamics of metabolic pathway 718–719
Gradient, chemi-osmotic potential. *See* Chemi-osmotic potential gradient
Gradient, temperature. *See* Temperature gradient
Grana G(813), 739
Graphite 173, 410
Graphite electrode G(813), 606
Gravitational work 23, S10(303)
Greenhouse gas 396
GTP 715, 720–723
Guanidinium chloride 752

H

Haemaglobin 674
Half-cell G(813), S20(591), 596–598
Half-cell reaction. *See* Half-reaction
Half-reaction G(813), S20(591), 596
 and activation energy 598
 and associated electrode potential 603
Harmonic motion. *See* Simple harmonic motion; Damped harmonic motion
Hartley entropy H 346
Hasselbalch. *See* Henderson-Hasselbalch equation; Henderson-Hasselbalch approximation
Heat dq J G(814), S1(3), S2(20), S3(43), 50–56
 and adiabatic change 57
 and Carnot cycle 231–232, 234–235
 conduction of 54
 at constant pressure 150
 at constant volume 106, 211–212, 214
 convection of 54
 $dq_{rev} > dq_{irrev}$ 144, 150, 217, S9(267), 272
 dq_{rev}/T is a state function 236, 269–272
 dissipation of S5(93), 139, 141–143, 249,

252–254, 594, 735, 736, 738
 and efficiency 234–235, 314
 and electricity 592
 and energy transfer 53, 54, 71
 and enthalpy 150, 216
 and First Law of Thermodynamics S5(93), 94, 97–99, 101–102
 flow across system boundary 57
 flow down temperature gradient 54, 55, 258, S10(303), 304–305, 316
 flow down temperature gradient entropy change 275–278, 281–282
 and folding and unfolding of proteins 752
 and friction 28, 51–53, 134
 impossibility of total conversion into work 311, 314, 316
 and internal energy 106, 211–212
 and isothermal change 57
 key ideal gas equations 299
 net heat exchange over a thermodynamic cycle 306–307
 as a path function 55–56
 as a process 54, 55
 radiation of 54
 reversible and irreversible S5(94)
 and reversible and irreversible changes 143–144
 reversible and Second Law of Thermodynamics 267
 sign convention 56, S5(93), 96, 307
 as a 'substance' 51–52
 and T, S diagram 299
 theoretical maximum 150, 216
 and thermodynamic cycle 56, 101–102
 unidirectional flow of 55, 258
 units of measurement 54
 and work 4, 51–53, 58–66, 234–235, 311, 592
Heat, dissipative. *See* Dissipative heat
Heat, frictional. *See* frictional heat
Heat, irreversible. *See* irreversible heat

Heat, mechanical equivalent of. See mechanical equivalent of heat
Heat, reversible. See reversible heat
Heat, specific. See heat capacity
Heat capacity C_X J K^{-1} mol^{-1} G(814), 636–637
Heat capacity at constant pressure C_P J K^{-1} mol^{-1} G(814), S6(146), 151, 210–218, 636–637
ΔC_P for protein folding and unfolding 759–760
difference C_P-C_V for ideal gas 153, 216
difference C_P-C_V for real gas 217, 645–648
discontinuity at phase change 291–296
and enthalpy 151–153, 214–216
and entropy S9(268), 283
and heat capacity at constant volume 153, 216
of ideal gas 153, 214–216
and ideal gas constant 153
polynomial expression 195, 354
ratio $\gamma = C_P/C_V$ 224, 225
of real gas 216–218
relative magnitudes of ΔC_P and $\Delta_r H$ 193, 200
Shomate equation 195–196
of solid 354–355
state function 152
temperature variation 195–197, 291
values T(152), T(217)
Heat capacity at constant volume C_V J K^{-1} mol^{-1} G(814), S5(93), 107–113, 151, 210–218, 636–637
as a 'capacity' 108
definition 107
of diatomic gas 108–109, 111–112
difference C_P-C_V for ideal gas 153, 216
difference C_P-C_V for real gas 217, 645–648
and entropy S9(268), 283
and heat capacity at constant pressure 153, 216

of ideal gas 110–112, 210–214, 216
and ideal gas constant 111
and internal energy 107–109, 210–214
molecular interpretation 111–112
of monatomic gas 108–109, 111–112
of polyatomic gas 109, 111–112
ratio $\gamma = C_P/C_V$ 224, 225
of real gas 216–218
state function or path function 112–113
temperature variation 195, 291
values 108–109, T(109), T(217)
Heat engine G(814), 65, 313–315, 317–321
Carnot engine has maximum efficiency 328
cold and hot reservoirs 317
efficiency of 314
efficiency of Carnot engine 324–328
efficiency of real engine 328–330
how a heat engine works 317–321
and non-equilibrium states 66
power stroke 317
real engine must be less efficient than Carnot engine 330
return stroke 317
working substance 317
See also Carnot heat engine
Heat of [...]. See Enthalpy of [...]
Heat pump G(814), 313–315
efficiency of 314–315
Heat reservoir G(814), 64, 65, 235, S10(303), 311, 317
and Carnot cycle efficiency 234
Heisenberg's uncertainty principle. See Uncertainty principle, Heisenberg's
Helium
and adiabatic de-magnetisation 358–360
not solid at absolute zero 355
Helmholtz, Hermann von 385
Helmholtz free energy A J G(814), 81, S13(366), 385–386

Henderson-Haselbalch approximation G(814), S17(534), 548–549
accuracy of 553–554
difference from Henderson-Hasselbalch equation 549
Henderson-Hasselbalch equation G(815), S17(534), 541–542
base equivalent 545
difference from Henderson-Hasselbalch approximation 549
and $\Lambda(\text{pH}_{buf})$ factor 563
Henry's law G(815), S16(502), 510–511, 662
Henry's law constant G(815), G16(502), 510, 662
Hero, or Heron, of Alexandria 58
Hess's law of constant heat formation G(815), S6(146), 165–168, 181–182, 191, 397
'High-energy bond' 715
'High-energy molecule'. See ATP
High-performing team. See Teams and teamwork
'Hill', energy. See Energy 'hill'; Activation energy
Hot reservoir 64, 65, S10(303), 311, 317
Huntingdon's disease 780
Hurricane 256–257
Hybrid equilibrium constant 524, 621
and water as a reagent 568, 707
Hydrogen
$\Delta_f G^{\ominus\prime}$ of any molecule containing hydrogen 680, 691–693
$\Delta_f G^{\ominus\prime}$ of molecular hydrogen 680, 689–691
Hydrogen bond G(815), 357, 510
and potential function 762
and protein folding 761–762
Hydrogen electrode 606–607
Hydrogen ion G(815)
and biochemical standard state S23(673), 680, 683–686, 686–689
and conventional standard state S23(673), 680, 683–686

and definition of pH 537, 664–665
molality, molar concentration and pH 664–665
standard ionic enthalpy 184, 185
standard reversible redox potential 613–617, 617–618, 680, 686–689
Hydronium ion G(815), 535
Hydrostatic pressure 585
and Gibbs free energy 588–600
Hydroxonium ion. See Hydronium ion

I

Ideal G(815), 8
See also Ideal-dilute solution; Ideal gas; Ideal mixture; Ideal solution; Ideal system
Ideal-dilute solution G(815), G16(502), 510
distinction from dilute ideal solution 511
Ideal gas G(815), S1(3), 16–17, S3(43), 208–256
and adiabatic path 221–229
assumptions 17, 48, 209
compressibility factor 652, 656
and Dalton's law of partial pressures 48–49
difference C_P-C_V 153, 216
enthalpy, function of temperature only 153, 209–210, 644–645 (proof)
entropy change equations 285–290
entropy change of adiabatic expansion into a vacuum 278–280, 281
equation-of-state S4(43),47, 50, 209
heat capacity at constant volume 110–112
impossibility of liquefaction 485–486
internal energy, function of temperature only 97, 110, 153, 209, 210, 643 (proof)
and isochoric path 218–219

INDEX

Ideal gas (Contd.)
  and isothermal path 220–221
  Joule-Thomson coefficient 651, 652, 656
  key equations 229
  only two state functions required for description 48
  and partial derivatives 79–80, 89
  properties of S1(3), 17, 48, 209, 485
  standard state 172
Ideal gas constant $R = 8.314$ J K^{-1} mol^{-1} 47
  and heat capacity at constant volume 111
Ideal gas law S3(43), 46–47, 209
Ideal gas scale of temperature T K G(816), 44
  equivalence to thermodynamic scale 47
  reference point 44, 47
Ideal mixture 85, 402
  no volume change on formation 8, 508
  See also Ideal solution; Real mixture ; Real solution
Ideal solution G(816), 174, G16(502), 503–511
  distinction from ideal-dilute solution 511
  Gibbs free energy G16(502), 511–517, 518–519
  no heat of mixing 509
  no volume change on formation 8, 508
  properties of 508–510
  standard state 172, 174, 664
Ideal system S1(3)
  concept 8
  linear addition of state functions 8, 49, 402
  and mixtures 8, 402
  and real systems 656
Inequality 144
Inequality, Clausius. See Clausius inequality
Inert electrode G(816), 606
Inexact differential G(816), S2(20)
  and integration 26
  and path functions 26
Infinite dilution G(816), 173, 664
Inflexion, point of. See Point of inflexion

Information
  flow across system boundary and causality 265, 266
  and Maxwell's demon 346
Information theory S11(332), 346–347
Inorganic phosphate G(816), 395, S24(711), 713
Integral, cyclic. See Cyclic integral
Intensive state function G(816), S1(3), 6
  constant at all locations at equilibrium 9
  constant over time at equilibrium 9
  electrode potential 604
  equilibrium conditions 9
  list of 634–635
  molar intensive state functions 7
  non-additive nature of 6–7
  relationship to extensive state function 7
  variation with gravity at equilibrium 589
Internal energy U J G(817), 81, S5(93), 94
  absolute value 176, 179
  and First Law of Thermodynamics S5(93), 94
  and friction 130–132
  function of temperature only (ideal gas) 97, 110, 209, 643 (proof)
  and heat 94, 97–99
  and heat capacity at constant volume 107–109, 210–214
  of ideal monatomic gas 96–97
  impossibility of measuring absolute value 176
  key ideal gas equations 299
  macroscopic interpretation of 97–99
  microscopic interpretation of 95, 96–97, 111–112
  molar 95, 97
  of real gas 97
  and reversible and irreversible paths 143–144
  and temperature 96–99
  and work 94, 97–99

Internal energy change ΔU J
  and ΔH 158–160, 164
  and heat at constant volume 106–107, 108, 211–212, 214
  independence of reference level 176
  and open system 408
  temperature variation 198
Internal friction
  and Joule heating 593–594, 603
International Union of Pure and Applied Chemistry. See IUPAC
Intrinsic fluorescence G(817), 753
Inversion, population. See Population inversion
Inversion curve G(817), S21(634), 651–652
Inversion temperature G(817), S21(634), 652
Ion G(817), 173, 184, 504
  See also Anion; Cation; Hydrogen ion; Redox reaction
Ion-product of water 536
Ionic enthalpy G(817), 184–185
  reference level 184
  values T(185)
Ionic product of water G(817), 536
Ionic strength G(817), S22(656), 665–668
Irreversibility G(817), S5(93), 123–126, S8(258), 315–317
  and asymmetry 132
  and Clausius inequality 273
  and equilibrium and non-equilibrium states 254
  examples 258, 261–263
  and friction 134–140
  key characteristics 260
  mixing 126–128
  property of paths not states 125
  and real paths S5(93) 123, 125–126, 138, 140, 254
  and reversible paths 123
  and Second Law of Thermodynamics S9(267), 268, 269
  and time S11(332), 347
Irreversibility, molecular. See Molecular irreversibility
Irreversible change or path. See Irreversibility

Irreversible heat dq_{irrev} J G(817), S5(94), S9(267)
  and reversible heat S5(94), 144, 150, 217, S9(267), 272
Irreversible work dw_{irrev} J G(818), S5(93)
  and reversible work S5(94), 142–144
Isenthalp G(818), 651–652, 656
Isobar G(818), 58
Isobaric change or path G(818), 58, 219–220
  and First Law of Thermodynamics 114–116
  and ideal gas 219–220
  key ideal gas equations 229
Isochore G(818), 58
Isochore, van't Hoff. See van't Hoff isochore
Isochoric change or path G(818), 58, 218–219
  and chemical reaction 108–109
  and First Law of Thermodynamics 106
  and ideal gas 218–219
  key ideal gas equations 299
Isolated system G(818), S3(43), 57
  and causality 266
  criterion for equilibrium S9(267)
  criterion for spontaneous change S9(267)
  and First Law of Thermodynamics 105
  and spontaneity 266, 274
  universe as example S9(267)
Isotherm G(818), 57
  two isotherms cannot cross 321
Isotherm, binding. See Binding isotherm
Isothermal change or path G(818), 57
  and Carnot cycle 230, 231
  and First Law of Thermodynamics 114
  and ideal gas 220–221
  key ideal gas equations 299
Isothermal compressibility κ_T Pa^{-1} G(818), 89, 491
  necessarily negative 491, 495, 647
Isothermal magnetisation 360
  See also Adiabatic demagnetisation

INDEX

Isothermal titration calorimetry (ITC) G(818), 764
IUPAC 169, 172, 174, 187
 definition of pH 665, 677
 standard state of real solution 664
 standard states 172, 174, 675

J

Joule J (unit-of-measure) 21, 54
Joule, James 52, 53, 110, 128, 132, 134, 139, 308, 316, 592, 593
 electrical appratus 592, 593
 mechanical apparatus 52, 53, 316
Joule expansion G(818), 121
Joule heating G(818), 592
 and ΔG 594–595
 dissipative 594
 and 'internal friction' 593, 603
 microscopic interpretation of 593, 594
 necessarily positive 592–593, 595
 and Seebeck effect 594
Joule-Thomson coefficient μ_{JT} K Pa^{-1} G(818), S21(633), 648–652
 of ideal gas 651, 652, 656
 as indicator of ideality 652
 and inversion curve 651
 of real gas 651, 652
Joule's first law G(818), 592

K

Kelvin K (unit-of-measure) 44, 47
Kelvin, Lord (William Thomson) 44, 311, 323, 345
Kelvin-Planck statement of the Second Law of Thermodynamics G(818), S10(303), 309–313
 comparison to Clausius statement 311
 graphical interpretation of 320
 and irreversibility 316
Kilocalorie kcal or Cal (unit-of-measure) 54
Kinetic friction. See Dynamic friction
Kinetics S14(414), 449, 450
 of polymerisation 782–785
 of protein folding S25(747) 769–775
 of virus capsid self-assembly 789
 See also Chemical kinetics
Kinetics, chemical. See Chemical kinetics
Kirchhoff's first equation G(819), S6(147), 197, 198
Kirchhoff's second equation G(819), 198
Krebs cycle. See Tricarboxylic acid cycle

L

Lactic acid 570, 720
 and muscle cramp 720
Lactic dehydrogenase 720
Lag phase G(819), 782
 of polymerisation 782, 783, 785
 of virus capsid self-assembly 789
Latent heat G(819), 161
Law, [X's]. See [X's] law
Law of conservation 51, 54
Le Chatelier's principle G(819), S14(414), 448–449, 459, 460, 469, 482
Lennard-Jones potential G(819), 763
Level, reference. See Reference level
Levinthal, Cyrus 770
Levinthal's paradox G(819), 770
Life
 and complexity 66
 and coupled systems 674, S24(711), 712
 and creation and maintenance of order 350, 395, 395–396, 674, S24(711), 711–712
 and non-equilibrium states 66
Ligand. See Protein-ligand binding; Protein-ligand complex
Light
 absorption by photosynthetic pigment 738–739
 energy associated with 738–739
 frequency of 738
 quantum of energy 738
 speed of 738
 wavelength of 738
Light-dependent reactions of photosynthesis G(819), 739–744
Light-harvesting complex G(819), 739, 740, 741
Light-independent reactions of photosynthesis G(820), 744–745
Lipid bilayer G(820), 724, 725
Liquefaction
 as example of self-assembly 780
 impossibility of for ideal gas 485–486
 molecular interpretation of 485–486, 489
 and nucleation 410, 474, 782
 of real gas 410, 474
 of van der Waals gas 489, 492–497
 of van der Waals gas as a model for protein self-assembly 781–782
Liquid G(820)
 always in equilibrium with vapour 475
 boiling 477–478
 (chemical) activity of immiscible 525–526
 compressibility of 491, 647
 degree of disorder 344, 356–357
 entropy of 291–296, 352, 356
 equation-of-state 50
 formation from gas as example of self-assembly 780
 formation of solid from as example of self-assembly 780
 Gibbs free energy of 472–473, 475, 497–500
 molecular motions within 95, 352, 468
 standard state 172
 and vapour pressure 468–470
 See also Phase change
Liquid junction potential G(820), 604–605
Lowry, Thomas 546
Luciferase 745
Lumen G(820), 739
 accumulation of H$^+$ ions within 741
 schematic representation of 740

M

Macromolecular assembly 674
Macromolecule 674
 configuration 674
 three dimensional shape 674
Macroscopic G(820)
 contrast with microscopic 4
 and macrostate S11(332), 334
 viewpoint 4
Macrostate G(820), S11(332), 334–337
 in economics 348
 and high-performing teams 349
 information required to describe 335
 mathematical probability of 339, 340, 341
 and multiplicity 337
 number of associated microstates 335, 336, 337, 339
 as observed 337, 340, 341
"Mad-cow" disease 780
Magnet, permanent 358
Magnetic dipole 358
Magnetic field 357–358
Magnetic work 23, 360
Magnetisation, isothermal. See Isothermal magnetisation
Mass action effect G(820), 566–567, 712
 and metabolic pathways 712
 and pseudo-equilibrium 712
 and water as a reagent 567–568
Mass action ratio, activity. See Activity mass action ratio
Mass action ratio, concentration. See Concentration mass action ratio
Mass action ratio, molality. See Molality mass action ratio
Mass action ratio, mole fraction. See Mole fraction mass action ratio
Mass action ratio, pressure. See Pressure mass action ratio
Mathematical probability $\mathbb{P}$ G(820), 339
Matter
 flow across system boundary S3(43), 57, 265, 266, 711
Maximum available work G(821), S13(366), 386–390
 and ΔG 389
 and electromotive force 613
 and Joule heating 594
 and metabolism of glucose 737
Maxwell, James Clerk 344, 345

INDEX

Maxwell construction G(821), 492–497
Maxwell relations G(821), S21(633), 637–639
  rules for 639
  use of 642, 644, 646, 650
Maxwell's demon G(821), 344–347
Mean bond energy ϵ or ΔH^* (X-Y) J mol^{-1} G(821), 187–190
  values T(189)
Mean molal activity coefficient $\gamma_{b,\pm}$ G(821), 666–667
Mean residue ellipticity G(821), 753
Mechanical equivalent of heat 52, 110, 132
Mechanical work 387
Mechanics, statistical. See Statistical mechanics
Melting 472–480
  explanation of 474
  of non-crystalline solid 296
  See also Phase change
Melting temperature T_m K G(821), 758–761
Membrane. See Biological membrane; Respiratory membrane; Semi-permeable membrane
Membrane, biological. See Biological membrane
Membrane, semi-permeable. See Semi-permeable membrane
Membrane potential Ψ or $\Delta\Psi$ V G(821), 730–731
  of mitochondrial respiratory membrane 731
  of thylakoid membrane 742
Messenger RNA (m-RNA) 748
Metabolic pathway G(822), 674, S24(711), 712
Metabolism
  of sugars 36
Metal electrode 596, 605–606
Metastable equilibrium G(822), 409–410
Metastable state G(822), 409–410, 497, 783
  and nucleation of a polymer 782
Methane
  combustion of 439–440
Microcompartment, bacterial. See Bacterial microcompartment

Microscopic G(822), 4
  contrast with macroscopic 4
  interpretation of adiabatic demagnetisation 359–361
  interpretation of chemical reaction 450–451, 456, 458, 460
  interpretation of dynamic equilibrium 451
  interpretation of failure of water and olive oil to mix 509–510
  interpretation of heat capacity at constant volume 111–112
  interpretation of ideal gas 17, 48, 209, 485
  interpretation of ideal mixture 507–510, 660
  interpretation of ideal solution 505, 507–510, 660
  interpretation of internal energy 95
  interpretation of Joule heating 593, 594
  interpretation of liquefaction 485–486, 489
  interpretation of mixing 127–128, 580–582
  interpretation of osmosis 583, 584, 587
  interpretation of paramagnetism 358
  interpretation of pressure 16–17
  interpretation of Raoult's Law 505, 507–510, 660
  interpretation of real gas 485–489, 658
  interpretation of real mixture 508–510
  interpretation of real solution 173, 508, 661, 662
  interpretation of temperature 67–74
  interpretation of van der Waals gas 486–489
  interpretation of vapour pressure 468–470
  and microstates S11(332)
  viewpoint 4
  viewpoint and statistical mechanics 4
Microstate G(822), S11(332), 334–337
  association with macrostate 335, 336, 337

  in economics 348
  and high-performing teams 349
  information required to describe 335
  and mathematical probability of macrostate 339
  and multiplicity 337
  and observed macrostate 337, 340, 341
Minimisation of Gibbs free energy S13(367), 408–411, S14(413)
Mitchell, Peter 724
Mitochondrion G(822), 724, 725, 739
  oxidative phosphorylation within membrane 724
  TCA cycle within matrix 724
Mixing 126–128, 261–262
  change of volume 8, 508
  enthalpy and entropy effects 509–510
  entropy of 421
  failure of water and olive oil to mix 509–510
  and Gibbs free energy 580–582
  heat of 509
  microscopic interpretation 127–128, 580–582
  unidirectionality of 123, 261–262
Mixing, molecular. See Molecular mixing
Mixture 85
  Gibbs free energy of 401–408, S14(413), 416–417, 417–423
  ideal assumption 402
  measures of 170–171
  microscopic interpretation of 507–510
  and solution 503
  tendency to ideal behaviour on dilution 509
  time dependence of Gibbs free energy 417–423
  See also Ideal mixture; Ideal solution; Real mixture; Real solution
Mixture, equilibrium. See Chemical equilibrium
Model, Finke-Watzky. See Finke-Watzky model
Molal activity $a_{b,i}$ G(822), S22(656), 663–664, 668
  approach to ideal behaviour 662

Molal activity coefficient $\gamma_{b,i}$ G(823), S22(656), 663–664
  approach to ideal behaviour 662
  and mean molal activity coefficient 666
Molal activity coefficient, mean. See Mean molal activity coefficient
Molality b_i mol kg^{-1} G(823), 171, 514, 668
  and IUPAC standard definition of pH 665
  of pure water 664–665
  preference over molar concentration for real solutions 633
  and real solutions 172, 173, 663–664
  and standard state of a real solution 172, 664
Molality mass action ratio Γ_b G(823), 521
Molality reaction quotient Q_b. See Molality mass action ratio
Molality thermodynamic equilibrium constant K_b G(823), 521
Molar mol^{-1} G(823), 7, 85
  enthalpy H J mol^{-1} 149
  entropy S J K^{-1} mol^{-1} 269
  Gibbs free energy G J mol^{-1} 375
  heat capacity at constant pressure C_P J K^{-1} mol^{-1} 151
  heat capacity at constant volume C_V J K^{-1} mol^{-1} 108
  Helmholtz free energy A J mol^{-1} 385
  internal energy U J mol^{-1} 94
  mass M kg mol^{-1} 7
  volume V m^3 mol^{-1} 48, 81
Molar concentration [I] mol m^{-3} or mol dm^{-3} G(823), 171, S14(414), 437, 514
  and biochemical standard 682
  of H$^+$(aq) ion, biochemical standard S23(673), 677–678
  of H$^+$ (aq) ion, conventional standard S23(673), 172, 173, 614–615
  of pure water 537
  preference for molality for real solutions 663
  and real solutions 173, 663
  transformation to biochemical standard 682

Molarity G(823), 171
Mole G(823), 7
　See also Avogadro constant
Mole fraction x_i G(823), 170, 418, 436, 503, 668
　and mole fraction activity 661
　of water in dilute solution 171, 514
Mole fraction activity $a_{x,i}$ G(823), S22(655), 668
　approach to ideal behaviour 662
　magnitude of and Raoult's law 661
　and real solutions 661
Mole fraction activity coefficient $\gamma_{x,i}$ G(824), S22(655), 661
　approach to ideal behaviour 662
　magnitude of and Raoult's law 661
Mole fraction mass action ratio Γ_x G(824), 437, 439, 519
Mole fraction reaction quotient Q_x. See Mole fraction mass action ratio
Mole fraction thermodynamic equilibrium constant K_x G(824), 437, 439, G16(502), S16(502), 519
　and K_c 439
　and K_p 437, 439
Mole number n G(824), 7
Molecular energy 68
　average S3(43), 68–69, 111
　average and total 72–73
　distribution 67, 70–71
　and temperature 70–71
Molecular interpretation of thermodynamic phenomena. See Microscopic
Molecular irreversibility G(824), 128, 140
Molecular mixing G(824), S5(93), 127–128, 261–262
Molecular velocity 17–18, 67
Molecular weight
　measurement of 579
Monatomic gas
　average molecular energy of S3(43), 68–70, 96
　heat capacity at constant volume 108–109, 111–112
Monohydroxic base G(824), 545
　See also Base

Monomer G(824), 780, 782–785, 786–787, 788–789
　polymerisation of as an example of self-assembly 780
Monoprotic acid G(824), 539
　See also Acid
Motion, harmonic. See Simple harmonic motion; Damped harmonic motion
Motive force, proton. See Proton motive force
Multiplicity W G(825), S11(332), 337–340
　and Boltzmann's equation 342
　and mathematical probability 339
　as a measure of order and disorder 337
　and observed macrostate 337
　and thermodynamic probability 342
Muscle cramp 720
Mutant protein G(825), S25(747), 775–777

N

NADH 674, S24(711), 715–716
　and bioenergetics of electron transport 723–728
　cellular control of [NAD$^+$]/[NADH] ratio 720, 723
　as a reducing agent 720, 723
　and reduction of molecular oxygen in oxidative phosphorylation 726–728
　redox potential $E^{\ominus}$ (NAD$^+$, NADH) 632, 709
　redox potential $E^{\ominus}$ (NAD$^+$, NADH) 668
　redox potential of [NAD$^+$]/[NADH] ratio 723–724
NADPH 715
　cellular control of [NADP$^+$]/[NADPH] ratio 743
　as a reducing agent 743
　synthesis during photosynthesis 740, 741
Nanomachine G(825), 345, 733
　as example of self-assembly 780
Nanotube, carbon. See Carbon nanotube

Native state G(825), S25(747), 748, 749
　measurement of 753–759
　and mutation 775
　stability 752–753
　structure determined by amino acid sequence 769
　structure not random 748
Negative deviation from Raoult's law G(825), 506–507, 660–661, 668
Nernst, Walter 624
Nernst equation G(825), S20(592) 623, 624–627, 723
　and ratios of [NAD$^+$]/[NADH] and [FAD]/[FADH$_2$] 723
　representation of electrons 627, 687
　transformation of S23(673), 683
Nervous conduction 726
Net heat
　equivalence with net work over thermodynamic cycle 307
　over thermodynamic cycle 306–307
　and T,S diagram 299
Net work 61
　and Carnot cycle 231
　equivalence with net heat over thermodynamic cycle 307
　over thermodynamic cycle 306–307
　and T,S diagram 299
Newcomen, Thomas 58, 63, 66
Newton's First Law of Motion 17
Newton's Second Law of Motion 237–238, 242
Nobel prize 10, 296, 355, 445, 456, 588, 624, 724, 746, 769
Non-conservative function 376–377
Non-cyclic photophosphorylation G(825), 740–743
　'missing' H$^+$ ion 743, 744
　overall bioenergetics of 742–743
　overall stoichiometry of 742–743
　schematic representation of 740
Non-equilibrium state G(825), 9, 10, 33
　and heat engine 66

　and life 66
　maintenance over time 66
　movement towards equilibrium 9, 47, 254, 260, 261–263
　and Second Law of Thermodynamics S9(267), 268, 269
Non-equilibrium system 432–436
Non-equilibrium thermodynamics G(825), 10
Non-ideal gas. See Real gas
Non-ideal solution. See Real solution
Non-quasistatic path 33–34, 128
　See also Irreversibility; Quasistatic; Reversibility
Non-spontaneous change and coupled systems 391–393
　how to make them happen 391–393, 527–530
　why they happen 391
Non-volatile solute 571–572
　measurement of molecular weight 579
Normalised G(825), 754
Nuclear magnetic resonance (NMR) G(826), 764, 775
Nucleation G(826), 410, 474, S26(780), 782
　energetically unfavourable 783, 786
　phase in polymerisation 782–784
　step in Finke-Watzky model 783–784
Nucleus G(826)
　critical size of 782–783, 784, 789
　for liquefaction of a gas 474
　and polymerisation 782–787
　for self-assembly of virus capsid 789

O

Occam's razor. See Ockham's razor
Ockham's razor G(826), 785
Ohm, Georg 593
Ohm's law 593, 595
Oil, olive. See Olive oil
Olive oil
　failure to mix with water 509–510

INDEX

Open system G(826), S3(43), 57
  criterion for spontaneous change 275
  and ΔG 406
  and ΔH 408
  and ΔU 408
  and First Law of Thermodynamics 95, 104, 406–408
  and First and Second Laws of Thermodynamics 407
  and living things 57, 711
Order S11(332), 333–334
  and adiabatic demagnetisation 357–361
  and Boltzmann's equation 341–344
  of diamond 355–356
  and energy flow 350, 394
  and entropy S11(332), 352
  of gases, liquids and solids 355–357
  and high-performing teams 349
  and human activity 395–396
  and life 395, 674, S24(711), 711–712
  maintained or increased by energy flow 394
  measurement of 337–340
  and multiplicity 337
  and phase changes 343–344
  of proteins 395, 762
  and time 347
  of water 509–510
  See also Disorder
Organodynamics G(826), 348–350
Ørsted, Hans Christian 593
Osmosis G(826), G19(580), 582–583
  across biological membrane 726
  biological significance 583
  and Gibbs free energy 582–583, 584–587
  microscopic interpretation of 583, 584, 587
Osmosis, reverse. See Reverse osmosis
Osmotic potential. See Chemi-osmotic potential
Osmotic pressure Π Pa G(826), G19(580), 583–588
  of sea water 588
Ostwald's dilution law G(826), 542, 543
Oxidant. See Oxidising agent

Oxidation G(826), 147, 148, S20(591), 597
Oxidation half-reaction G(826), 597, 600
  at anode of electrochemical cell 600
Oxidative phosphorylation G(827), S24(711), 716, 720, 723–735
  efficiency of 735–736, 737
  location within and across respiratory membrane 724
  and redox potentials associated with [NAD$^+$]/[NADH] and [FAD]/[FADH$_2$] ratios 724
Oxidised state 597, 598
Oxidising agent G(827), 620, 630
Oxonium ion. See Hydronium ion

P

P680 G(827), 739
  excitation of 740
  schematic representation of 740
  and splitting of water 740–741
P700 G(827), 739
  excitation of 741, 743
  schematic representation of 740
Papin, Denis 58, 63
Paradox, Levinthal's. See Levinthal's paradox
Paramagnetic material G(827), 357–358
Parkinson's disease 780
Partial derivative G(827), S4(76), 79
  and ideal gas 79–80, 89
  importance of 86–88
  meaning of 87–88
  and specific path 103
  and state functions 82–84
Partial pressure G(827), 49, S14(413), 416, 668
Pascal Pa (unit-of-measure) 14
Pascal, Blaise 14
Path G(827), S2(20)
  and state functions 12
Path, non-quasistatic. See Non-quasistatic path
Path, quasistatic. See Quasistatic path

Path function G(827), S2(20), 24
  and cyclic integral 39, 230
  difference from state function 24
  and heat 55–56
  and heat capacity at constant volume 112–113
  and inexact differential 26
  link to state function 39–40, 56, 269, 299
  list of 634
  and partial derivatives 103
  path dependence 25–27
  and state function 101–102
  and thermodynamic cycles S2(20), 38–39
  and work 23–27
Pathway, metabolic. See Metabolic pathway
Pathway, of protein folding. See Protein folding and unfolding
Pendulum, thermodynamic. See Thermodynamic pendulum
Pepsin 676
Perfect gas. See Ideal gas
Permanent magnet 358
Permittivity of free space ε_0 C V^{-1} m^{-1} 763
Petrol engine
  efficiency of 736
pH G(827), 537, 664–665
  biological significance 549, 564
  difference between values based on molality and molar concentration 663
  and enzyme activity 549, 562
  and folding and unfolding of proteins 752
  gradient across mitochondrial respiratory membrane 732
  IUPAC definition 664, 677
  and standard states 176
Phase G(828), 467–500
Phase, lag. See Lag phase
Phase change G(828), 154, 161–163, 177
  and chemical potential 472
  not chemical reaction 163

  and Clapeyron equation 481–482, 483
  and Clausius-Clapeyron equation 482–483
  discontinuity in values of heat capacity at constant volume 291, 294
  effect of pressure 478–480
  endothermic 154, 156
  enthalpy change of 161–163, S9(268), 291–296, 373–374
  entropy change of S9(268), 291–296, 343–344, 373–374
  exothermic 154
  and folding and unfolding of protein 749, 750, 753–754
  and Gibbs free energy 471–472
  melting of non-crystalline solid 296
  molar entropies of 293
  and order 343–344
  spontaneity of 373–374
  thermodynamic data for 293
  and unidirectionality 262
Phase equilibrium 467–500
  criterion for S15(467), 471–472, 494, 511, 581
Phi-value ϕ_F G(828), G25(747), 777–778
Phosphate, inorganic. See Inorganic phosphate
Phosphorylation, oxidative. See Oxidative phosphorylation
Phosphorylation, substrate-level. See Substrate-level phosphorylation
Photon G(828), 738
  energy of 738
Photophosphorylation G(828)
Photosynthesis G(828), 36, 57, 154, 532, S24(711), 713, 738–745
  chemi-osmotic potential of 742
  'missing' H$^+$ ion 743
  overall bioenergetics of 745
  overall stoichiometry of 742–743
  as primary means of capturing energy of the sun 738
  as reversal of glycolysis and TCA cycle 738

INDEX

Photosynthetic pigment G(828), 738
  location within thylakoid membrane 739
  and photosystems I and II 739
Photosystem I G(828), 739, 741
  schematic representation of 740
Photosystem II G(829), 739, 740
  schematic representation of 740
Pigment, photosynthetic. *See* Photosynthetic pigment
pK_a G(829), 539
  values T(540)
pK_b G(829), 545
  values T(546)
Planck, Max 311, 342
  Kelvin-Planck statement of the Second Law of Thermodynamics S10(303), 311
Planck's constant h J sec 738, 771
Planck's law 739, 771
Platinised electrode. *See* Electrode
Platinised platinum G(829), 606
Platinum, platinised. *See* Platinised platinum
Platinum black 606
Platinum electrode G(829), 606
Plot, chevron. *See* Chevron plot
pmf. *See* Proton motive force
pOH G(829), 537
Point, critical. *See* Critical point
Point, triple. *See* Triple point
Point of inflexion G(829), 489
Polyatomic gas
  heat capacity at constant volume 109, 111–112
Polymer G(829)
  melting of 296
  See also Polymerisation
Polymerisation G(829), 780, 782
  See also Protein aggregation and polymerisation
Polypeptide chain G(829), S25(747), 748
Polyprotic acid G(829), 540
  See also Acid
Population inversion 73–74
Positive deviation from Raoult's law G(830), 506–507, 660–661
Potential. *See* Potential function

Potential, chemical. *See* Chemical potential
Potential, chemi-osmotic. *See* Chemi-osmotic potential
Potential, electrical. *See* Electrical potential
Potential, electrode. *See* Electrode potential; Electromotive force
Potential, electrostatic. *See* Electrostatic potential
Potential, Lennard-Jones. *See* Lennard-Jones potential
Potential, liquid junction. *See* Liquid junction potential
Potential, membrane. *See* Membrane potential
Potential difference, electrical. *See* Electrical potential difference
Potential function G(830), 762
Potential gradient, chemi-osmotic. *See* Chemi-osmotic potential gradient
Power stroke 317, 319
  must be at higher temperature than return strike 319
Pre-exponential factor G(830), S14(414), 456
Pressure P Pa G(830), 13–16
  definition 14
  effect on chemical reaction 448–449
  effect on phase change 478–480
  and folding and unfolding of proteins 752
  and fugacity 657
  molecular interpretation 17–18
  variation with gravity at equilibrium 589
  See also Partial pressure
Pressure, atmospheric. *See* Atmospheric pressure
Pressure, critical. *See* Critical pressure
Pressure, osmotic. *See* Osmotic pressure
Pressure, partial. *See* Partial pressure
Pressure, vapour. *See* Vapour pressure; Equilibrium saturated vapour pressure
Pressure mass action ratio Γ_p G(830), 432–436, 439

  graphical representation of 435
  and K_p 432, 434–435
  and reaction Gibbs free energy $\Delta_r G$ 432, 434
  and ξ 434–435
  and ξ_{eq} 432
Pressure reaction quotient Q_p. *See* Pressure mass action ratio
Pressure thermodynamic equilibrium constant K_p G(831), S14(413), 427–428, 428–430
  and $\Delta_r G^{\circ}$ 428, 430–432
  and Γ_p 432, 434Ű–435
  interpretation of 429–430
  and K_c 439
  and K_x 437, 439
  and reaction rate constants 455
  and reaction stoichiometry 429
  'shorthand' version 429
  and Γ_p 432, 434–435
  variation with temperature 444–448, 460
  and ξ_{eq} 431, 432, 442–443
Prigogine, Ilya 10
Primary structure G(831), 748
Principle of Le Chatelier. *See* Le Chatelier's principle
Prize, Nobel. *See* Nobel prize
Probability, mathematical. *See* Mathematical probability
Probability, thermodynamic. *See* Thermodynamic probability
Protein G(831), 148, 527, 583, S24(711), S25(747), 747
  aggregation of 674, 780–793
  α-helix 753
  β-pleated sheet 753
  binding with ligands S25(747), 763
  within biological membrane 724–725
  denatured 749
  and drug design 763, 764
  folding of 748, 748–763
  foldon 778
  and 'Fourth Law' of Thermodynamics 748
  and hydrogen bonds 761–762
  kinetics of folding S25(747)

  and minimisation of Gibbs free energy S25(747), 748, 770
  misfolding 751
  mutations 775
  native structure of 748
  nucleation of structure 777–778
  order of 395, 762
  primary structure of 748
  quaternary structure of 748
  secondary structure of 748, 753, 777–778
  stability 752
  structure determined by amino acid sequence 769, 775
  synthesis of 748, 748–763
  tertiary structure of 748, 753, 778
  three-dimensional structure of 674, 747, 748
  unfolding of 296, S25(747)
  uniqueness of structure 674, S25(747)
  See also Native state; Protein aggregation and polymerisation; Protein folding and unfolding; Unfolded state
Protein, mutant. *See* Mutant protein
Protein, folded state. *See* Folded state
Protein, native state. *See* Native state
Protein, wild-type. *See* Wild-type protein
Protein aggregation and polymerisation G(831), 780, 782
  elongation phase 782, 783, 784
  elongation step in Finke-Watzky kinetic model 784
  enthalpy-driven model 786
  entropy effects 786–787
  Finke-Watzky two-step model 782–785
  fragmentation step of three-step model 785
  Gibbs free energy change during 783
  kinetics of 782–785
  lag phase 782, 783
  nucleation phase 782, 783
  nucleation step in Finke-Watzky model 784
  nuclei of critical size 782–785

Protein aggregation and polymerisation (*Contd.*)
  saturation phase 782, 783, 787
  of self-assembling systems 791–794
  three step kinetic model 784–785
  of virus capsids 787–791
Protein folding and unfolding 748–763
  action of chaotropes 749, 753–757, 773–775
  action of detergents 749
  action of guanidine chloride 749
  action of heat 749
  action of pH 752
  action of pressure 749
  action of urea 749, 753–757, 773
  analysis of kinetics using transition state theory 770–773
  and biochemical standard 751
  'classical view' model for pathway of folding 778
  ΔC_P 759–760
  'diffusion collision' theory for pathway of folding 777
  effect of mutations on kinetics of folding and unfolding 775
  enthalpy and entropy effects 761–762
  experimental measurement of 752–760
  foldons 778
  'framework' theory for pathway of folding 777
  and hydrogen bonds 761–762
  'hydrophobic collapse' theory for pathway of folding 777
  'jigsaw' theory for pathway of folding 777
  kinetics of 769–778
  and Levinthal's paradox 770
  as measured using circular dichroism 753–755
  as measured using differential scanning calorimetry 759–760
  and measurement of binding affinity and dissociation constant 767–769
  misfolding 751
  'new view' model for pathway of folding 778
  'nucleation condensation' theory for pathway of folding 777, 778
  order and disorder 762
  pathways of folding 770, 775, 777–778
  'phase change' or 'staircase'? 749, 750, 753–754
  and phi-value analysis 775–778
  not a random process 770
  rate of folding 772–773, 774–775
  rate of unfolding 774–775
  reversibility of 748, 761–762, 769
  theories of 777–778
  thermodynamics of 748–762
  transition state 770–772, 775–776
  unimolecular assumption for kinetics 773
Protein-ligand binding
  binding affinity 764
  binding isotherm 764, 765, 766, 767
  dissociation constant 764, 765–769
  and drug design 764
  energy of 763
  high ligand concentrations 764–766
  low ligand concentrations 766–767
  as measured using fluorescence 674
  as measured using isothermal titration calorimetry 674
  as measured using nuclear magnetic resonance 674
  as measured from protein folding and unfolding 767–769
  as measured using surface plasmon resonance 674
  measurement of 763–769
  'off-rate' 764, 765
  'on-rate' 764, 765
  thermodynamics of 764–769
  two-state model 764
Protein-ligand complex G(831), S25(747), 764
Protein self-assembly S26(780), 780–793
  enthalpy and entropy effects 789
  importance of understanding 780–781
  liquefaction of van der Waals gas as model for 781–782
  and polymerisation 782–787
  and self-assembling systems 791–792
  and virus structures 787–791
Proton motive force Δp V G(831), 724, 732
  value for mitochondrial respiratory membrane 732
Pseudo-equilibrium G(831), 712, 719
Puzzle, thermochemical. *See* Thermochemical puzzle
P,V diagram G(831), S2(20), 31–34
  and isobaric change 58
  and isochoric change 58
  and T,S diagram 299
  of steam engine 62
P,V work. *See* P,V work of expansion
P,V work of expansion G(831), S2(20), 23
  against the atmosphere 27
  that cannot be used 27
Pyruvate dehydrogenase 720
Pyruvic acid 570
  anaerobic pathway 720
  metabolism of 719–720

Q

Quantum of energy $h\nu$ J G(831), 738
  absorption by a molecule 739–740
Quasistatic G(831), S2(20), 30
  chemical reaction 31
  flow of heat 30
  and reversible and irreversible paths 125–126
  work 30
Quasistatic change or path 28–31, 125–126
  characteristics of S2(20), 30
  and glucose metabolism 737
  graphical representation 31–34
Quaternary structure G(831), 748

R

Radiation 54
Raoult, François-Marie 505
Raoult's law G(832), G16(502), 504–510, 571, 660
  and depression of the freezing point 578
  and elevation of the boiling point 574
  and Henry's law 510
  microscopic interpretation of 505, 507–510, 660
  and mole fraction activity 661
  negative deviation from 506–507, 660–661, 668
  negative deviation shown by solutions of ions 661, 668
  positive deviation from 506–507, 660–661
Raoult's law, negative or positive deviation from:
Rate constant, reaction. *See* Reaction rate constant
Rate of reaction G(832), S14(414), 450–454
  and Arrhenius equation 455–456
  equilibrium condition 454–455
  graphical representation of 453, 454
Rate of reaction, forward. *See* Forward rate of reaction
Rate of reaction, reverse. *See* Reverse rate of reaction
Razor, Ockham's. *See* Ockham's razor
Reaction, cell. *See* Cell reaction
Reaction, chemical. *See* Chemical reaction
Reaction, coupled. *See* Coupled reactions
Reaction, half-cell. *See* Half-cell reaction
Reaction, rate of. *See* Rate of reaction
Reaction, redox. *See* Redox reaction
Reaction centre G(832), 739
  schematic representation of 740

INDEX

Reaction Gibbs free energy $\Delta_r G$ J G(832), 432–436

Reaction quotient, activity. *See* Activity mass action ratio

Reaction quotient, concentration. *See* Concentration mass action ratio

Reaction quotient, molality. *See* Molality mass action ratio

Reaction quotient, mole fraction. *See* Mole fraction mass action ratio

Reaction quotient, pressure. *See* Pressure mass action ratio

Reaction rate constant k G(832), S14(414), 452
 and Arrhenius equation 455–456
 equilibrium condition 454–455
 and K_c 455
 and K_p 455

Reactivity, of a bond 189–190

Reactivity, of a molecule 179–181

Real change or path S5(93), 140–144, 273–275
 and entropy 274
 and irreversibility S5(93), 138

Real gas G(832) 16, 50, 173, 485–500, S22(655), 656–659
 approximation to ideal behaviour 209
 compressibility factor 652–654, 656
 effect of attractive forces 485–486, 489, 658
 fugacity 657–659
 Gibbs free energy of 658–659
 indicator of ideality 652, 656
 internal energy of 97
 Joule-Thomson coefficient 651, 652
 and Lennard-Jones potential 763
 liquefaction of 485–497
 microscopic interpretation of 485–489, 658
 standard state 172, 658–659

Real mixture
 heat of mixing 509
 volume change 8, 508

See also Ideal mixture, Ideal solution; Real solution

Real solution G(832) 173, S22(656), 659–668
 heat of mixing 509
 microscopic interpretation of 173, 507–510, 661, 662
 and molality 173
 preference of molality over molar concentration 663
 standard state 172, 173, 664
 tendency towards ideal behaviour on dilution 509, 664
 volume change 8, 508

Redox electrode G(832), 606–607

Redox potential 617
 associated with [NAD$^+$]/[NADH] ratio 723–724
 associated with [FAD]/[FADH$_2$] ratio 723–724
 conventional values T(617)
 and oxidative phosphorylation 724
 and oxidising and reducing agents 620
 transformed values T(688)
 See also Nernst equation

Redox reaction G(832), 603, 619–620, 628–631
 and concentration thermodynamic equilibrium constant 619–620
 and oxidative phosphorylation 726–729

Redox system 603
 See also Redox potential; Redox reaction

Reduced state G(833), 597, 598

Reducing agent G(833), 620, 630
 and FADH$_2$ 723
 and NADH 720, 723
 and NADPH 743

Reductant. *See* Reducing agent

Reduction G(833), S20(591), 598

Reduction half-reaction G(833), 598, 600
 at cathode of electrochemical cell 600

Reference electrode, standard. *See* Standard reference electrode

Reference level G(833), 169, 174–176

arbitrary nature of 169, 174–176, 617, 675
and biochemical standard 618, 685, 692, 703
and conventional (IUPAC) standards 675, 703
of standard enthalpy of formation 179
of standard Gibbs free energy of formation 397, 399
of standard ionic enthalpy 184
of standard reversible electrode potentials 617–619
of temperature scales 44, 47

Reference level shift G(833), 175–176, 618
and biochemical standard 685, 688, 691, 703
measurements of difference independent of 175–176, 618

Reference state G(833), 169

Reflections on the Motive Power of Fire, by Sadi Carnot 323

Refrigerant 357

Refrigerator 260, 306, 357

Relations, Maxwell. *See* Maxwell relations

Reservoir. *See* Heat reservoir

Reservoir, cold. *See* Hot reservoir

Reservoir, hot. *See* Cold reservoir

Resistance, electrical. *See* Electrical resistance

Resonance energy transfer G(833), 739
and photosystems I and II 739

Respiration G(833), 148, 154, 350
See also Oxidative phosphorylation

Respiratory membrane G(833), 724–728
and ATP synthase 733–734
key components 724
location of oxidative phosphorylation 724
schematic representation 725, 734

Return stroke 317, 319
must be at lower temperature than power stroke 319

Reverse osmosis G(834), 588

Reverse rate of reaction v_r S14(413), 450–454
graphical representation of 454

Reversibility G(834), 13, S5(93), 123–126
biological example 719, 723
criteria for reversible change S5(93), 125–126
and metabolism of glucose 736–737
property of paths not states 125

Reversible change or path. *See* Reversibility

Reversible chemical reaction S14(413), 416, 450
equilibrium condition 454–455
graphical representation of 454
reaction rate of 450–454

Reversible electrode potential E V G(834), 592, 613–614
See also Standard reversible electrode potential; Standard reversible redox potential; Standard reversible reduction potential

Reversible gas electrode G(834), 606, 614–615

Reversible heat dq_{rev} J G(834), S5(94)
and entropy 269
and irreversible heat S5(94), 144, 150, 217, S9(267), 272

Reversible hydrogen electrode G(834), 615

Reversible path. *See* Reversibility

Reversible redox electrode G(834), 617

Reversible redox potential. *See* Standard reversible redox potential

Reversible work dw_{rev} J G(834), S5(93)
and irreversible work S5(94), 142–144

Ribonuclease 769

Ribosome G(834), 674, 748

RNA 748
messenger, m-RNA 748

RNA phage Qβ 792

RuBisCO G(834), 744, 792

Rumford, Count (Benjamin Thompson) 51–52, 58, 110, 128, 129, 132, 134, 139, 315, 316, 593–594

S

Salt bridge G(834), 604–605

Saturated vapour pressure. *See* Equilibrium saturated vapour pressure

INDEX

Saturation phase of polymerisation G(834), 782, 787
Savery, Thomas 58, 63, 66
Scale of temperature 44, 47
Scanning calorimetry, differential. *See* Differential scanning calorimetry
Schrödinger, Erwin 674
Sea water
  desalination 588
  ionic composition 588
  osmotic pressure of 588
Second Law of Thermodynamics G(834), 55, 94, S9(267), 267–299, 268
  apparent contravention by Maxwell's demon 345, 347
  combined with First Law of Thermodynamics S9(268), 280–282, 407
  and creation and maintenance of order 350
  and direction of time 139, 347
  and directionality 134, 254
  and energy flow 350
  and increase in entropy of universe 274
  and life 711–712
  non-equilibrium state to equilibrium state 254
  and open system 407, 711
  and reversible and irreversible paths 138, 254
  and spontaneity 157
Second Law of Thermodynamics, Carathéodory statement. *See* Carathéodory statement of the Second Law of Thermodynamics
Second Law of Thermodynamics, Clausius statement. *See* Clausius statement of the Second Law of Thermodynamics
Second Law of Thermodynamics, Kelvin-Planck statement. *See* Kelvin-Planck statement of the Second Law of Thermodynamics
Second thermodynamic equation-of-state G(835), S21(633), 642
Secondary structure G(835), 748
  α-helix 753

β-pleated sheet 753
Seebeck effect G(835), 594
Self-assembly G(835), S26(780), 780–781, 787–788
  See also Protein self-assembly
Semi-permeable membrane G(835), G19(580), 582
Shift, reference level. *See* Reference level shift
Shomate equation G(835), 195
  values of Shomate coefficients T(196)
Sign convention for heat 56, S5(93), 96
  for work 26–27, S5(93), 96
Simple harmonic motion 236–239, 242, 248
Solid G(835), 656
  always in equilibrium with vapour 475
  behaviour of C_P 354–355
  (chemical) activity of 525–526
  compressibility of 647
  crystalline 297, 344, 362
  degree of disorder 344, 355–356
  entropy of 291–296, 297, S12(351), 352, 353, 357
  equation-of-state 50
  formation of from liquid as example of self-assembly 780
  Gibbs free energy of 472–473, 475, 497–500
  molecular motions within 95, 352
  standard state 172
  and vapour pressure 469–470
  See also Phase change
Solute G(835), 170, 503
  measures of 170–171
  and molality 173
  non-volatile 571–572
  standard state 172, 173
  See also Ideal solution; Ideal-dilute solution; Real solution
Solution G(835), 170, 503
  dilute 504
  measures of 170–171
  and molality 173
  standard state 172, 173
  See also Ideal solution; Ideal-dilute solution; Real solution
Solution, buffer. *See* Buffer solution
Solution, ideal. *See* Ideal solution
Solution, non-ideal. *See* Non-ideal solution
Solution, real. *See* Non-ideal solution

Solvent G(835), 170, 503
  measures of 170–171
  standard state 172
  See also Ideal solution; Ideal-dilute solution; Real solution
Spectroscopic measurement 752, 753
Speed of light c m sec^{-1} 738
Spontaneity G(835), S8(258), 258–266, S10(303)
  and asymmetry 132
  and causality 262–266
  and Clausius inequality 273
  and closed systems 369–374
  and coupled systems 391–393
  criteria for S6(146), 157
  criterion for in closed system S9(267), 275, 369, 370, 374–375, 376–377
  criterion for in isolated system S9(267), 274
  criterion for in open system S9(267), 275
  and enthalpy 157
  examples 258, 261–263
  flow of heat and performance of work S10(303)
  and Gibbs free energy 374–375, 422
  graphical representation of chemical reaction 420, 422, 435
  and heat flow down a temperature gradient 258, S10(303), 304–305
  and isolated system 266, 274
  key characteristics 260
  and minimisation of Gibbs free energy 422
  and mixing 582
  and non-conservative function 376–377
  and non-spontaneity 391
  of phase changes 373–374
  and relative magnitudes of ΔH and $T\Delta S$ 378
  and Second Law of Thermodynamics S9(267), 268, 269
  and signs of ΔH and ΔS 371–373
  and supercooled and superheated states 497
  and time S11(332), 347
  unidirectionality of 132

why change doesn't happen even when $\Delta G < 0$ S14(413-414)
SPR. *See* Surface plasmon resonance
Stability, of a molecule 179–181
Stable equilibrium 409–410, 422
Stable state G(836), S6(146)
Standard (adjective) G(836)
  See also Specific types
Standard (noun) G(836), 168–178
  arbitrary nature of 168, 174, 617–618, 675
  biochemical 675
  conventional (IUPAC) 675
  temperature variation 177–178
Standard, biochemical. *See* Biochemical standard
Standard atmospheric pressure Pa or bar G(836), 15
Standard electrode potential E V G(836), S20(592), 614–619
  values T(616), T(617)
Standard electromotive force E_{cell} V S20(592), 615, 616, 618, 630
Standard enthalpy of formation $\Delta_f H^\ominus$ J mol^{-1} G(836), 178–183
  of an element 179, 180
  function of temperature only 179
  and Hess's law of constant heat formation 181–182
  and reactivity of a molecule 179–180
  reference level 179
  and stability of a molecule 179–180
  and stoichiometry of reaction 183
  use of 181–182
  values T(180)
Standard enthalpy of reaction $\Delta_r H^\ominus$ J G(836), S6(146)
  computation of 181–183
  key equation 182
  relative magnitudes of $\Delta_r H^\ominus$ and ΔC_P 200
  temperature variation 198–200
  unit-of-measure 183
Standard Gibbs free energy of formation $\Delta_f G^\ominus$ J mol^{-1} G(836), S13(366), 397, S23(673)
  of molecular hydrogen under biochemical standard 680, 689–691

of molecular hydrogen
under conventional
(IUPAC) standards
397, 680
of molecule containing
hydrogen under
biochemical standard
680, 691–693
of a molecule containing
hydrogen under
conventional
(IUPAC) standards
680
values T(398)
values under bio-
chemical standard
T(693)
Standard Gibbs free energy
of reaction $\Delta_r G^\ominus$ J
G(836), 399, 426–427
and biochemical
standard 678–679,
694–696
and equilibrium
constant 428,
430–432
interpretation of
430–432, 678–679
transformed S23(673),
681
Standard hydrogen
electrode 614–615
Standard pressure $P^\ominus$ Pa or
bar G(837), 172
Standard redox potential
$E^\ominus$ V S20(592), 617
and biochemical
standard 680,
686–688
conventional values
T(617)
transformed values
T(688)
Standard reversible
electrode potential
$E^\ominus$ V G(837),
615–617, T(616),
T(617)
Standard reversible redox
potential $E^\ominus$ V 617,
T(617)
Standard reversible
reduction potential
$E^\ominus$ V T(616), T(617),
619
Standard state G(837),
171–174, 176
biochemical 675
conventional (IUPAC)
definitions 172, 675
Standard state, biochemical.
See Biochemical
standard state
Standard state, conven-
tional. See Standard
state
Standards, conventional.
See Conventional
standards
State G(837), S1(3), 6

State, change in. See Change
in state
State, denatured. See
Denatured state
State, equation of. See
Equation-of-state
State, equilibrium. See
Equilibrium state
State, excited of an electron.
See Excited state of an
electron
State, folded. See
Equilibrium state
State, metastable. See
Metastable state
State, native. See Native
state
State, non-equilibrium.
See Non-equilibrium
state
State, reference. See
Reference state
State, stable. See Stable state
State, standard. See
Standard state
State, transition. See
Transition state
State, 'trapped'. See
'Trapped' state
State, unfolded. See
Unfolded state
State, unstable. See Unstable
state
State function G(837),
S1(3), 6
and change of state 11
conditions for thermo-
dynamic equilibrium
S1(3), 8
constant at different
locations S1(3), 9, 589
constant over time S1(3),
8, 24, 589
and cyclic integral
37–38, 230, 236
definition and
description 5–7
difference from path
function 24
dq_{rev}/T is a state
function 236,
269–272
and energy 40
and entropy 236,
269–272
and equilibrium 8
and exact differential 12
extensive S1(3), 6–7
extensive per unit mass
S1(3), 7
and heat capacity at
constant pressure 150
and heat capacity at
constant volume
112–113

and ideal concept 8, 49
intensive S1(3), 6–7
linear addition of for
ideal mixtures 8, 49,
85, 508
link to path function
39–40, 56, 269, 299
and mixtures 85
no 'memory' 12
non-linear addition for
real mixtures 8, 508
and partial derivatives
82–84
path independence 12,
25
and path function
101–102
and reversible paths 13
and surroundings 13
and thermodynamic
cycles S2(20), 37–38
two state function
principle 48, 81, 85,
86, 97, 102, 209, 282,
401, 635
variation with gravity at
equilibrium 589
State function concept
12–13
Static friction G(837), 28,
128
Statistical mechanics
G(838), 5
Steam engine 4, 58–66
P,V diagram of 62
Stoichiometric coefficient
G(838), 155, 160–161,
182–183
Stoichiometric number
G(838), 160–161, 566
Strength, ionic. See Ionic
strength
Stroke, power. See Power
stroke
Stroke, return. See Return
stroke
Stroma G(838), 739
depletion of H^+ ions
within 741
mildly alkaline
conditions within 741
schematic representation
of 740
synthesis of NADPH
within 741
Structure, primary,
secondary, tertiary,
quaternary. See
Primary structure;
Secondary structure;
Tertiary structure;
Quaternary structure
Substance, working. See
Working substance
Substrate-level phosphor-
ylation G(838), 716,
716–719
Sugar
metabolism of 36
See also Glucose;
Glycolysis

Sun
as fundamental energy
source 395, 738
Supercooled gas G(838),
410, 474, 497, 783
Supercooled liquid 474
Superheated liquid G(838),
474, 497
Superheated solid G(838),
474
Supersaturated gas. See
Supercooled gas
Surface plasmon resonance
(SPR) G(838), 764
Surface work 23
Surroundings G(838),
S1(3), 5, 13
entropy changes in
367–368
Synthetic biology 674,
792–793
System G(838), S1(3), 5
of constant mass 80–81,
85–86
System, closed. See Closed
system
System, ideal. See Ideal
system
System, isolated. See
Isolated system
System, non-equilibrium.
See Non-equilibrium
system
System, open. See Open
system
System, redox. See Redox
system
Systems, coupled. See
Coupled systems
System boundary. See
Boundary

T

TCA cycle 36, 570,
S24(711), 716, 719,
720–723
efficiency of 737
and glycolysis 716, 720,
723
location within
central matrix of
mitochondrion 724
metabolic pathway 721
reversed in photo-
synthesis 738
schematic representation
of combined
metabolic pathway
with glycolysis 735
thermodynamics of
metabolic pathway
722
Teams and teamwork
S11(332), 348–350
and manufacturing 349
and an orchestra 349
and role of leader 350
and sport 349

INDEX

Temperature G(839), S3(43), 4, 43–46, 70
  and average molecular energy, S3(43), 70–71, 96–97
  Celsius scale 44
  centigrade scale 44
  effect on chemical reaction 448, 458, 459–460
  equivalence of ideal gas and thermodynamic scales 47
  Fahrenheit scale 44
  ideal gas scale 44
  and internal energy 96–99
  and internal energy of ideal gas 97, 110
  negative 73–74
  scales 44
  and Second Law of Thermodynamics S9(267)
  thermodynamic scale 44
  variation of standards 177–178
  and 'Zeroth' Law of Thermodynamics 55
Temperature gradient G(839), 54, 275
  heat flow down 54, 55
  unidirectionality of heat flow down 55
Temperature, critical. See Critical temperature
Temperature, Debye. See Debye temperature
Temperature, melting. See Melting temperature
Tertiary structure G(839), 748
Thermal contact G(839), 54
Thermal energy 67–73
Thermal equilibrium G(839), 9, 45
  and thermodynamic equilibrium 9
Thermochemical puzzle 166–168
Thermochemistry G(839), 148, 156, 166–167
  Hess's law of constant heat formation 165–168
Thermodynamic cycle G(839), S2(20), 35–38, 230
  Carnot cycle 230–236
  equivalence of heat and work over cycle 307
  and First Law of Thermodynamics 101–102
  and heat 56, 306–307
  irreversible 125
  and path functions S2(20), 38–39
  reversible 121, 123–125, 298
  and state functions S2(20), 37–38
  and work 39, 306–307
Thermodynamic equation-of-state, first. See First thermodynamic equation-of-state
Thermodynamic equation-of-state, second. See Second thermodynamic equation-of-state
Thermodynamic equilibrium G(839), S1(3), 9
  conditions for S1(3), 9
  and thermal equilibrium 9
  and time 8–9
Thermodynamic equilibrium constant G(839)
  See also Activity thermodynamic equilibrium constant; Concentration thermodynamic equilibrium constant; Molality thermodynamic equilibrium constant; Mole fraction thermodynamic equilibrium constant; Pressure thermodynamic equilibrium constant
Thermodynamic pendulum 209, 236–254, 257
Thermodynamic probability W G(839), 342
  See also Multiplicity
Thermodynamic scale of temperature T K G(839), 44
  equivalence to ideal gas scale 47
  reference point 44, 47
Thermodynamics G(839), S1(3), 3
  and chemical kinetics S14(414), 449–455, 457–459
  definition of 3
  does not take account of time 458
  intellectual framework of 5
  limitations of 449
  macroscopic viewpoint of 4–5
  scope of 4
Thermodynamics, First Law of. See First Law of Thermodynamics
Thermodynamics, 'Fourth Law' of.
  See 'Fourth Law' of Thermodynamics
Thermodynamics, fundamental equation of thermodynamics.
  See Fundamental equation of Thermodynamics
Thermodynamics, Second Law of, Carathéodory statement. See Carathéodory statement of the Second Law of Thermodynamics
Thermodynamics, Second Law of, Clausius statement. See Clausius statement of the Second Law of Thermodynamics
Thermodynamics, Second Law of, Kelvin-Planck statement. See Kelvin-Planck statement of the Second Law of Thermodynamics
Thermodynamics, Third Law of. See Third Law of Thermodynamics
Thermodynamics, Zeroth Law of.
  See Zeroth Law of Thermodynamics
Thermoeconomics G(840), 347–348
Thermogram G(840), 759–750
Thermometer 45
Third law entropy. See Absolute entropy
Third Law of Thermodynamics G(840), S9(268), 296–297, S10(351), 351–362
Thompson, Benjamin (Count Rumford) 51–52, 58
Thomson, William (Lord Kelvin) 44, 311
Throttling process G(840), S21(633), 648
  at constant enthalpy 649
Thylakoid G(840), 739
  and ATP synthase 739
  chemi-osmotic potential gradient across 741–742
  comparison with mitochondrion 742
  membrane components 739
  membrane potential of 742
  permeability of membrane to ions 742
  pH gradient across 741
  and photosystem I 739
  and photosystem II 739
  schematic representation of membrane 740, 744
  splitting of water at inner surface 741
  synthesis of NADPH on outer surface 741
  translocation of H^+ ions across membrane during photosynthesis 741, 743
Time S11(322)
  direction of 139–140, S11(332)
  and entropy 347
  'freezing' of 24
  and friction 134
  and path functions 24
  and state functions 8, 24
  and thermodynamic equilibrium 8–9
  unidirectionality of 132
Titration 561, 629
  and measurement of protein-ligand binding 764
Toricelli barometer 467
Toricellian vacuum 467, 469
Total derivative G(840), 77, 86
Total differential G(840), 77
T, P diagram G(840), 651
T, S diagram G(840), 297–299
  of Carnot cycle 297–298
  interpretation of in terms of heat and work 299
  and P, V diagram 299
Transformed data and equations G(841), S23(673), 681–683
  and biochemical standard 677
  common error associated with 682
  and molar concentrations 682–683
  notation 677
  reversible redox potentials 686–688
  standard reversible redox potentials, values T(688)
Transition, phase. See Phase transition
Transition state G(841), 457, 770–772, 775–776
Transition state theory G(841), 771
'Trapped' state 409–410
Transmission coefficient κ G(841), 771
Transport chain. See Electron transport chain

INDEX

Transport, active. *See* Active transport
'Trapped' metastable state. *See* Metastable state
Tricarboxylic cycle. *See* TCA cycle
Triple point G(841), 44, 480, 485
   of water 44
Trouton's rule G(841), 294

U

Uncertainty principle, Heisenberg's 144
Unfolded state G(841), 749
   difference from denatured state 749
   measurement of 753–759
Unfolding of proteins 296
Unidirectionality G(841), 123, 254, S8(258)
   and asymmetry 132
   of chemical reactions 123, 157, 262–263
   and Clausius inequality 273
   and enthalpy 157
   of expansion of gas into vacuum 123, 261
   of flow of heat 55, 258–259
   of friction 128, 132
   and irreversibility 123, 254
   key characteristics 260
   of mixing 123, 126–128, 261–262
   non-equilibrium state to equilibrium state 254
   of phase changes 262
   and Second Law of Thermodynamics S9(267), 268, 269
   of spontaneous changes 132
   of time 132, S11(332)
Unit-of-measure 168–169, 170
   of enthalpy change 155
   of standard enthalpy of reaction 183
   and standards 675
Universal gas constant. *See* Ideal gas constant
Universe G(841), S1(3), 5
   and direction of time 347
   entropy increase S9(267), 274, 304
   as isolated system S9(267), 274
Unstable equilibrium 409, 411
Unstable state G(841), 409, 411

Urea 749
   agent of protein unfolding 749, 753–757
   and measurement of kinetics of protein folding and unfolding 773
Useful work G(842), 27, 58–59, 62, 64–66, S10(303)
   See also Maximum available work; Work

V

Vacuum
   adiabatic expansion into 121–125, 261, 343
   adiabatic expansion into, entropy change 278–280
   expansion into 13–14, 340–341
   no work done expanding against 22, 25
   and steam engine 63–65
Vacuum, Toricellian. *See* Toricellian vacuum
van der Waals force G(842), 468, 488, 761, 762, 763
van der Waals gas G(842), 50, 486–497
   compressibility factor 652–652, 654
   critical parameters 490–491
   Gibbs free energy of 384, 493–494, 658–659
   liquefaction of 489, 492–497
   liquefaction of as model for protein self-assembly 781–782
   microscopic interpretation of 486–489
   parameter values T(488)
van der Waals's equation 16, 50, 90, S26(780)
   See also van der Waals gas
Van Slyke, Donald 560
Van Slyke equation G(842), 560–561
van't Hoff, Jacobus 445, 588
van't Hoff enthalpy G(842), 446
van't Hoff equation G(842), S14(414), S14(416), 445, 460
van't Hoff equation (osmosis) G(842), S19(580), 588
van't Hoff isochore. *See* van't Hoff equation

Vaporisation
   enthalpy of 161, 162, 177, 193–194, 468, 293
   microscopic interpretation of 468
Vapour
   always in equilibrium with liquid or solid 475
   See also Equilibrium vapour pressure; Vapour pressure
Vapour pressure p Pa G(842), S15(467)
   above solution 504–508
   and external pressure 470
   and Gibbs free energy 475–478
   increase with temperature 469
   microscopic interpretation of 468–470
   and Raoult's law 504–505
   and solids 469–470
   See also Equilibrium saturated vapour pressure
Vapour pressure, equilibrium saturated. *See* Equilibrium saturated vapour pressure
Variable, conjugate. *See* Conjugate variables
Velocity, molecular
   distribution 67
   and pressure 17–18
Virial coefficient G(842), 50
Virial equation-of-state G(842), 50
Virus 674, S26(780), 780, 787–791
   kinetics of self-assembly 789
   See also Capsid
Volatility G(842), 470
Voltage E V
   of electrical battery 595
   and Joule heating 595
Volume V m
   additivity for ideal mixture 8, 508
   non-additivity for real mixture 8, 508
Volume, critical. *See* Critical volume
Volumetric expansion coefficient α_V K^{-1} G(842), 89, 650

W

Wasted work 28, S10(303), 311, 314, 390

Water
   amphoteric property 547
   as a biochemical reagent 707–708
   and biology 583
   boiling point 44
   (chemical) activity of 525
   as a conjugate acid-conjugate base pair 548–549
   density at 298.15 K 665
   electrolysis of 610
   enthalpy effects 762
   entropy effects 509–510, 762, 786–787, 789
   failure to mix with olive oil 509–510
   freezing point 44
   heat capacity at constant pressure 152
   influence on protein folding 762
   ion-product 536
   ionic product 536
   ionisation 536–538
   ionisation constant 536
   and the mass action effect 566–567
   maximum density of 64
   molar concentration of pure 568
   mole fraction of in dilute solution 171, 514
   as a monoprotic acid 539–540
   and protein folding and unfolding 762
   as a reagent 566–567
   splitting of in photosynthesis 740–741
   triple point 44
Water, sea. *See* Sea water
Watt, James 58, 59, 64, 66
Wavelength of light λ m 738
What is Life? 674
Wild-type protein G(843), S25(747), 775–777
   See also Mutant protein; phi-value
Wind turbine
   efficiency of 736
Worcester, Marquis of 58
Work G(843), S1(3), S2(20), 20–40
   adiabatic 221–223, 226–228
   against 'internal friction' 594, 603
   against the atmosphere 27
   and Boyle's Law 31
   'by' and 'on' 22, 26–27
   that cannot be used 27
   and Carnot cycle 231, 232–234

and change in state 23, 24
computation of 31
đw_{rev} > đw_{irrev} 143–144, 299, 330
definition 21, 23
done by an expanding gas 21–22
and efficiency 143, S10(304), 314
electrical 23
and electricity 592–594, 603
and expansion into a vacuum 22, 25
and external force 22
and First Law of Thermodynamics S5(93), 94, 97–99, 101–102
frictional 23, 28, 51–53
gravitational 23
and heat 4, 51–53, 54, 592
key ideal gas equations 299
magnetic 23
maximum 4, 143
net 61, 231
net work done over a thermodynamic cycle 306–307
and path functions 23–27
and quasistatic paths 28–31
reversible and irreversible S5(94), 142–144
sign convention for 23–27, S5(93), 96, 307
surface 23
and system boundary 23, 57
and T,S diagram 299
and thermodynamic cycle S2(20), 38–39, 101–102
useful 27, 58–59, 62, 64–66, S10(303)
wasted 28, S10(303), 311, 314
See also. P,V work of expansion
Work of expansion. See P,V work of expansion
Work of expansion, P,V. See P,V work of expansion
Work, irreversible. See Irreversible work
Work, frictional. See Frictional work
Work, reversible. See Reversible work
Work, useful. See Useful work
Working substance G(843), 317

X

Xanthophyll 738
X-ray crystallography G(843), 775

Y

Yeast 719–720

Z

Zero point, of measurement. See Reference level
Zeroth Law of Thermodynamics G(843), S3(43), 55
Zinc electrode 596
See also Daniell cell